WITHDRAWN

JOHN F KENNEDY SCHOOL

THE SHOCKING HISTORY OF ELECTRIC FISHES

The Shocking History of Electric Fishes

From Ancient Epochs to the Birth of Modern Neurophysiology

Stanley Finger
Marco Piccolino

OXFORD
UNIVERSITY PRESS

OXFORD
UNIVERSITY PRESS

Oxford University Press, Inc., publishes works that further Oxford University's
objective of excellence in research, scholarship, and education.

Oxford New York
Auckland Cape Town Dar es Salaam Hong Kong Karachi
Kuala Lumpur Madrid Melbourne Mexico City Nairobi
New Delhi Shanghai Taipei Toronto

With offices in
Argentina Austria Brazil Chile Czech Republic France Greece
Guatemala Hungary Italy Japan Poland Portugal Singapore
South Korea Switzerland Thailand Turkey Ukraine Vietnam

Copyright © 2011 by Oxford University Press, Inc.

Published by Oxford University Press, Inc.
198 Madison Avenue, New York, New York 10016
www.oup.com

Oxford is a registered trademark of Oxford University Press
All rights reserved. No part of this publication may be reproduced,
stored in a retrieval system, or transmitted, in any form or by any means,
electronic, mechanical, photocopying, recording, or otherwise,
without the prior permission of Oxford University Press.

Library of Congress Cataloging-in-Publication Data

Finger, Stanley.
A shocking history of electric fish : how they changed science and medicine /
Stanley Finger, Marco Piccolino.
p. ; cm.
Includes bibliographical references and index.
ISBN 978-0-19-536672-3 1. Neurosciences—History. 2. Electric fishes—History.
I. Piccolino, Marco. II. Title.
[DNLM: 1. Electric Fish—physiology. 2. Neurophysiology—history.
3. Electric Organ—physiology. WL 11.1]
RC338.F56 2011
573.9′7—dc22
2010032840

---

9 8 7 6 5 4 3 2
Printed in USA
on acid-free paper

To our wonderful and understanding families,

We dedicate this book with love.

Acknowledgments

The authors would like to thank the many people who helped them with this project. Deserving special mention for providing materials and/or translations are: Edda Bresciani, Douglas Brewer, Sergio Donadoni, Bill Needle, Wolfhart Westendorf (Egypt); Alessandro Bausi (Ethiopia), Asad Ahmed, Gerrit Bos, Lutfallah Gari, Ivan Garofalo, and Gül Russell (Middle Eastern Cultures); Giacomo Magrini (Greek and Latin); Jenny Smith (Latin); and Gabriele De Angelis, Maryska Suda, and Nick Wade (German). Thanks also go to Armelle Debru for her suggestions and helpful criticisms with the chapters dealing with the classic Greek and Roman world, to Giuliano Ranucci for philological and linguistic advice, to Lucia Faedo for help with the Roman art, and to Frank Egerton and Paolo Tongiorgi for historico-zoological expertise. We would be remiss if we did not also mention Peter Koehler, who helped with the Dutch science, Germana Pareti for fruitful discussions on the history of German electrophysiology, and Lucia Tongiorgi-Tomasi for stimulating advice on the relationships between art and science in the Renaissance, and for the many books she directed to our attention.

Marco Piccolino is especially indebted to Marco Bresadola for his many years of collaborative studies on the history of electrophysiology, on which the chapters on Luigi Galvani and Alessandro Volta are based. He further acknowledges the generous support of the Bakken Library in Minneapolis, where he had the chance to spend a very fruitful period of bibliographic research.

This book also would not have been possible without the intelligent and kind assistance of many librarians, too many to acknowledge. In particular, we remember the generous help given by Rupert Baker, Sabina Fiorenzi, Livia Iannucci, Elizabeth Ihrig, Alessandra Lenzi, and Lilla Vekerdy, as well as others working in the libraries of the University of Pisa, Washington University in St. Louis, the Scuola Normale Superiore of Pisa, and the Bibliothèque Nationale of Paris.

We also wish to single out Joan Bossert, our always encouraging and very knowledgeable editor at Oxford University Press, for all that she has done from start to finish to help us in a myriad of ways. Joan, we fill and raise our glasses and sing praises to thee, both as an editor and as a friend!

We recognize that we might have inadvertently left out the names of other people who were extremely important to us while we were conducting our research, collecting illustrations, and agonizing over how best to organize and express our thoughts. To these unnamed helpful individuals, please realize that your help was also greatly appreciated. And to everyone who has helped us with this project, let us say that, if errors of omission and commission had been made, we take full and complete responsibility for what we wrote.

Lastly, we have to express our thanks to those institutions around the world that have digitized thousands and thousands of books and generously conceded them to open access through the latest technologies. Old books and new technologies are not in opposition. They have now come together to allow a better diffusion of cultural heritages to people everywhere, filling in a temporal space of thousands of years, as Galileo Galilei had predicted in closing the *Giornata Prima* of his monumental *Dialog on the Two Chief Systems of the World*.

Preface

The theme of the story we shall narrate about the amazing powers of three strange fishes is unique and begs to be told. These fishes, which can produce powerful electric shocks capable of numbing the hands, arms, and in some cases even the torsos of those who dare to touch them, are the flat saltwater torpedo rays, of which there are many types, various species of electric catfishes found in murky African rivers, and the so-called "electric eel," a favorite of children and adults who flock to see them in the large aquariums of some cities. The latter are eels in name only; they belong to a different order of fishes and are found exclusively in the equatorial rivers and ponds of the warmer parts of South America (e.g., Surinam, Venezuela, Brazil).

Unlike the electric eel, which Europeans first encountered in the 1500s, African catfishes and torpedoes have pictorial and written histories that take us back to Egyptian and Greek antiquity, respectively. In Roman times, Claudian even composed a poem about torpedoes, in which he put forward the emotions produced by the strange powers of these rays in particularly expressive way. "Who has not heard of the untamed art of the wonderful torpedo and of the powers that win it its name?" he asked. In his poem, the disk-shaped ray is presented as a small but dreadful monster of the sea, ready to devour other fishes and also to terrify unwary fishermen who happen to capture it—which might account for why "dreadful" was substituted for "wonderful" in some later versions of Claudian's text (*diræ* instead of *miræ* in the Latin original).

That torpedoes could be thought of as both wonderful and dreadful is meaningful. For centuries, these "narcotizing" (from the Greek word νάρκη transliterated as *nárkē*) first used to describe these fish) or "torporific" (the Roman equivalent) fishes occupied the realm of the fantastic, characterized simultaneously as wonderful and feared, and, as might be expected, truly fantastic. They captured the imaginations of the laity and scholars from classic antiquity, through both the Middle Ages and the Renaissance, into the modern era.

With their benumbing powers, especially the ability to stun at a distance in water, these fishes were for a long time well beyond the science of the day. Poorly understood and thought of possessing magical or occult properties, they were included in the ancient "bestiaries" alongside fiery dragons, dangerous griffins, and basilisks, those serpent-like, mythological creatures that scorched the earth, left a trail of deadly venom, and instantly killed with their breath or even with only a glance.

With the emergence of modern scientific thinking during the 1600s, most of the fabled wonders of the past began to be exposed as figments of the imagination and were exiled from the new natural history books. But torporific fishes, with their wonderful and dreadful "art," did more than just survive the scrutiny of the natural philosophers of the epoch; they now drew even more attention. How, asked the new natural philosophers, who believed in the revealing powers of experiments, could a torpedo, a Nile catfish, or a South American eel exert its actions at a distance on the smaller fishes they eat, or even on a human holding a wet line or a net with the offending fish still beneath the water?

The search for clues that could lead to an understanding of the nature of the powerful bolts of these fishes is basic to the story we are going to narrate. Electricity, as we know, is one of the fundamental forces of nature, and it enters into the constitution of matter in an essential way. But in contrast to other forces or energies (e.g., mechanical, gravitational, chemical, heat, light), we cannot perceive it in a direct way, because we, unlike some fishes and a few other animals, lack a specific "electric sense." There is indeed only one condition in nature in which electricity expresses itself in a form that humans can be aware of in an immediate way, and this is precisely in the shocks of a few kinds of electric fishes. When we see lightning or hear a thunder, in contrast, we do not perceive electricity in an immediate way, we only perceive the flash of light or the sound that electricity produces.

Despite the fact that the numbing and painful effects of the discharges have been known since antiquity, it took a great scientific endeavor, with open minds and many experiments, to recognize that the fishes' shocks are, in fact, electrical. Difficulties seemed to preclude this conclusion even when electricity began to be studied intensively during the middle of the eighteenth century, which is when the first suggestions were made that these fishes, in a Zeus-like way, seemed to be throwing bolts of electricity. After all, how could moist bodies generate, store, and release electricity? How could living organisms do such things without harming themselves? And then there was the bothersome fact that these animals lived in an electrically conductive medium, something that seemed to defy the known laws of physics, not only of physiology!

Several important "electricians" of the Enlightenment expressed these and other difficulties surrounding the electrical hypothesis. In a letter dated 1775, which we shall examine in Chapter 16, we read, for example, that "when a Gentleman can so give up his reason as to believe in the possibility of an accumulation of electricity among conductors sufficient to produce the effects ascribed to the Torpedo, he need not hesitate a moment to embrace as truths the greatest contradictions that can be laid before him." This letter was written by and sent to a Fellow of the famed Royal Society of London, one of the most important scientific organizations at the time.

The people, the experiments, and the historical events that overcame common sense and this negative *Zeitgeist* to enable three strongly electrical fishes to become electrical makes for a storyline that is the equal of any that might appear in an historical novel. How this happened forms the largest part of the story we shall tell. Our story, however, will not end with the recognition that a few strange fishes are electrical. This is because thinking and changing perceptions about these creatures had major ramifications in many fields.

With regard to physiology, research on these strongly electric fishes represented the departure point on the path that has led researchers to discover that electricity is the mysterious "fluid" of even our nerves, leaving older ideas about "animal spirits" by the wayside. When we look at a beautiful landscape or listen to pleasant music, when we express emotion, and when we ponder new information, such as that we hope to provide in this book, we now know that these processes are based on minute electric signals flowing within the circuits of our nervous systems. If we consider, moreover, that the even the pulsations of our heart are controlled electrically, we cannot but acknowledge the fundamental importance of "animal electricity," or what we now call physiological or bio-electricity, in our own bodies and those of other animals that also do not shock.

A broader role for animal electricity stems from late in the nineteenth century and is closely associated with Luigi Galvani, who was well aware of what had been discovered with electric fishes before beginning his own experiments on frogs, and later on torpedoes caught off the Italian coast. Galvani was trained as a physician, and he felt his new conception of how the nerves and muscles might work also had highly relevant medical applications. He viewed epilepsy as an electrical storm and thought that certain other nervous disorders are also due to electrical blockages. Today, many clinical tests are based on electrical signals that can be detected from the body's surface. One need only consider the electroencephalogram (EEG), the electrocardiogram (EKG), and those clinical tests used for tracing the flow of sensory stimuli (e.g., visual images, sound) from the periphery into and within the central nervous system.

Of course, electricity also has a therapeutic side to it. By Galvani's time, electricity from machines had been used in attempts to make paralyzed muscles contract, for painful conditions (e.g., gout), and to try to restore sight to the blind and hearing to the deaf. Today, we can think of the life-saving actions of cardiac pacemakers and defibrillators, as well as how electricity is used for stimulating muscles in bedridden patients so they will not atrophy, and for enhancing new bone growth.

In this book, we shall connect fish electricity to other ideas and developments of this sort. Specifically, we shall show how live electric fishes were used to treat medical disorders in ancient times, long before anyone was even thinking about electricity. We shall also look at how fish therapeutics had been presented as evidence for these fishes being electrical during the eighteenth century. And on the physics side, which is yet another dimension of the story we shall tell, we shall show how Alessandro Volta's revolutionary pile or electrical battery was modeled after the "electric organs" of torpedoes and electric eels, a fact the Italian physicist emphasized when introducing his "artificial electric organ" in 1800. It is not an exaggeration to state that, whenever we turn on a light or a computer, look at a TV, or use any other electrical device, there is a fragment of electric fish history behind these activities, not to mention the fundamentals of modern chemistry.

As more became known about these fishes during and following the Enlightenment, they gained a different sort of attractiveness. Naturalists and natural philosophers with fairly easy access to them ventured forth to study them, whereas inquisitive minds far from their natural habitats either made trips to where they could be studied, asked others on location to conduct specific experiments on them, or paid to have them shipped to places where they could be properly studied. In a very real way, almost everyone who had heard that these fishes seemed to be electrical now wished to study them and probe their deepest secrets.

One such person was John Walsh, an Englishman who, with the help of Benjamin Franklin, provided some key experimental evidence for the electric nature of shocks during the 1770s. In the scientific diary of his trip to the French coast, where torpedoes are plentiful, Walsh penned: *"Je l'ai donté"* in his shaky French. He had just spent some intensive time studying the local torpedoes, and many (but by no means all) of his tests for fish electricity were successful. Basically, Walsh was playing with Claudian's words, and he was ecstatic, feeling that he had now "tamed" the "indomitable art of the wonderful torpedo," meaning that he had obtained good evidence of its electrical nature. In the Baconian tradition of the Royal Society, of which Walsh was a Fellow, to acquire this knowledge also meant conquering a force of nature. Within a few years, Walsh would have a live electric eel in London, and he would draw a visible spark from it—the most convincing evidence of fish electricity to date.

Contrary to some common portrayals, in which scientists are presented as cold, withdrawn, and isolated from the culture around them, we hope to show that scientific endeavors to understand electric fishes were often rich in *humano aroma* (human flavor). Our story involves real people with a myriad of motives and personalities, and a constellation of exciting events, some fruitful and others leading to blind alleys. Some of this can be presented in narrative form, with the paths to the discoveries described by the players themselves.

We shall try to shed light on these discoveries, on the personalities involved, on their individual motives, and on their thought processes, in this colorful saga. We shall also do our best to convey the frustrations encountered and, as with John Walsh, the jubilation that can accompany a thrilling discovery that might support an exciting new idea.

Yet at the same time we also realize that, as extensive as this book is, there may be parts of this story and alternative interpretations of the material that we have unknowingly omitted, as well some notable references, new and old, that might have escaped our notice. With such things in mind, we shall offer an *apologia* here for what we might have failed to do in this book, despite the long and hard work underlying it. But at the same time, and on a more positive note, this would also seem to be the place to express the hope that these errors of omission and possible issues of interpretation will have the positive effect of stimulating more historical writings on these incredible fishes.

As you now join us on this voyage of discovery, it is our wish that by reading what we have written, you, the reader, will feel some of the joys and pleasures that we found when researching this subject, especially when we came across things we had not anticipated or known. In addition, we also hope that you will be able to experience some of the same intellectual excitement that we felt when trying to understand the thinking behind what we were encountering, and when we were assembling the various pieces to complete this long-overdue book on a topic that traverses so many disciplines and has so many significant ramifications. With these thoughts in mind, let us step back in time and begin our amazing journey.

Contents

Part I: Introduction
 1. The Allure of Electric Fishes: Humboldt's Obsession   3

Part II: Ancient Cultures
 2. The Shocking Catfish of the Nile   19
 3. Torpedoes in the Greco-Roman World: Pt. 1. Wonders of Nature Between Science and Myth   29
 4. Torpedoes in the Greco-Roman World: Pt. 2. From Therapeutic Shocks to Theories of the Discharge   45
 5. Byzantine and Islamic Writings   64

Part III: Late Middle Ages to the Early Modern Period
 6. Torpedoes: From the Scholastics through the Renaissance   87
 7. Rediscovering the Torporific Catfish   112
 8. The "Eels" of South America   127
 9. From the Occult to Mechanical Theories of the Discharge   137

Part IV: The Emergence of Fish Electricity
 10. The Electrical World of Benjamin Franklin   163
 11. Animal Spirits and Physiology   179
 12. First Steps Toward Fish Electricity   191
 13. The Dutch, the Eel, and Electricity   201

Part V: The Royal Society And The Coveted Spark
 14. Edward Bancroft's Guiana Eels and London Connections   217
 15. John Walsh's Scientific Journey   230
 16. The Royal Society and Interdisciplinary Science   243
 17. Out of the Guianas: The American Philosophical Society and the Eel   258
 18. Alexander Garden: A Linnaean in South Carolina and Captain Baker's Eels   269
 19. Sparks in Darkness and the Eel's Electrical Sense   281
 20. Public Knowledge: Newspapers, Magazines, and "Shocking" Poetry   292

Part VI: From Fish to Nerve Physiology and Back
 21. Galvani's Animal Electricity   307
 22. Electric Fishes in Volta's Path to the Battery   326
 23. Galvanism *Contra* "Voltaism": Electric Fishes and the "Unsolvable" Dilemma   339
 24. Electric Fishes in the Nineteenth Century   351
 25. The Changing Neurophysiological Setting   375
 26. Understanding the Shock Mechanisms: A Twentieth-Century Odyssey   394

Epilogue   416

APPENDIX: Names with Birth and Death Dates   421

References   429

Index   459

PART I
Introduction

Chapter 1
The Allure of Electric Fishes: Humboldt's Obsession

> The eels, stunned by the noise, defend themselves by the repeated discharge of their electric batteries.
>
> Alexander von Humboldt and Aimé Bonpland
> 1811 (1852/1971 trans., Vol. 2, p. 115)

The first person to dismount from his horse that sweltering day was a short, slender young man with blond hair and an aristocratic countenance. A taller and stouter European with rather coarse features followed his lead. Their weary entourage followed the example, happy to be able to stretch and rest. The group included guides and Indians, who took care of the pack animals and served as porters and laborers.

The foreign travelers must have presented a most unusual sight on March 19, 1800, when they were welcomed by officials in the small town of Rastro de Abaxo, in what was then Spanish Guyana and is now Venezuela. Don Miguel Cousin, the administrator of the *Real Hacienda*, had received advance notice and was pleased to have the distinguished visitors, who had made their way from Calabozo, a cattle-trading outpost of about 5,000 people, and before that, Caracas (Fig. 1.1). The new arrivals had crossed mountains and the flat river plains and swamps of the *Llanos* to reach his village. The nearby terrain had been almost featureless, the midday sun so blistering that the men had to travel at night, and the stinging and biting insects, insufferable. With alligators and snakes by the waterholes, even washing was dangerous.

Don Miguel must have been astonished to discover what the pack animals were carrying, above and beyond the necessary food, clothes, and medicines. The two Europeans had packed some 42 state-of-the-art scientific instruments, each in a velvet-lined box. There were barometers, hygrometers, telescopes, sextants, thermometers, chronometers, compasses, scales, magnetometers, electrometers, microscopes, and much, much more.

These instruments might have made Don Miguel somewhat suspicious, especially since the two men leading the caravan did not talk with each other in Spanish. As might be befitting of spies, they had all of the tools needed for reconnaissance and conducting a survey that could result in detailed maps of these normally restricted lands. But if he had some initial doubts, they abated when he was shown passports and papers coming from the *Primera Secretaria de Estado* to his Majesty, the King of Spain. The official documents included instructions to the administrators of the Spanish colonies, informing these officials that they were not to interfere with the scientific studies of the foreign guests. Further, they were granted the rights to collect mineral, plant, and animal specimens of all types.

The two men, of course, had introduced themselves before showing Don Miguel their documents. The thin blond man with impeccable manners and a good understanding of languages was Alexander von Humboldt, a young German baron with training in the sciences (Fig. 1.2).[1,2] Aimé Alexandre Goujaud, a military surgeon who had been given the name "Bonpland" by his father because of his love of botany, was his plant-collecting companion (Fig. 1.3).

Pleasant conversations with the two men followed, and their intentions within the sweltering inland region close to the Equator were made clearer. Officially, the Spanish government wanted them to do several things. One was to ascertain the source of the Orinoco River, and another was to determine whether the Rio Orinoco and the Rio Negro, a tributary of the Amazon River, join together. These were matters of considerable importance for commerce with the region. Some Indians and missionaries had mentioned an Orinoco and Rio Negro connection, and so had Sir Walter Raleigh, among the early explorers. But did these rivers really join in a way that could be exploited commercially, or was the natural Casiquiare Canal just another fanciful tale in a long string of unsubstantiated stories from the New World?

The two men showed great enthusiasm when they talked about nature, and they expressed the hope that their studies would help unravel some of the deeper mysteries of the natural world. As the thinner man had explained to a friend before setting sail, he wished "to find out how the forces of nature interact upon one another and how the geographical environment influences plant and animal life." "In other words," he explained, "I must find out about the unity of nature" (trans. in Botting, 1973, p. 65).

---

[1] Birth and death years of the people mentioned in this book are presented in the Appendix.

[2] For biographies of Humboldt, see Taylor, 1859; Bruhns, 1873; Löwenberg, Avé-Lallemant and Dove, 1873; Terra, 1955; Bitterling, 1959; Beck, 1959–61; Botting, 1973; Kellner, 1963; Helferich, 2004.

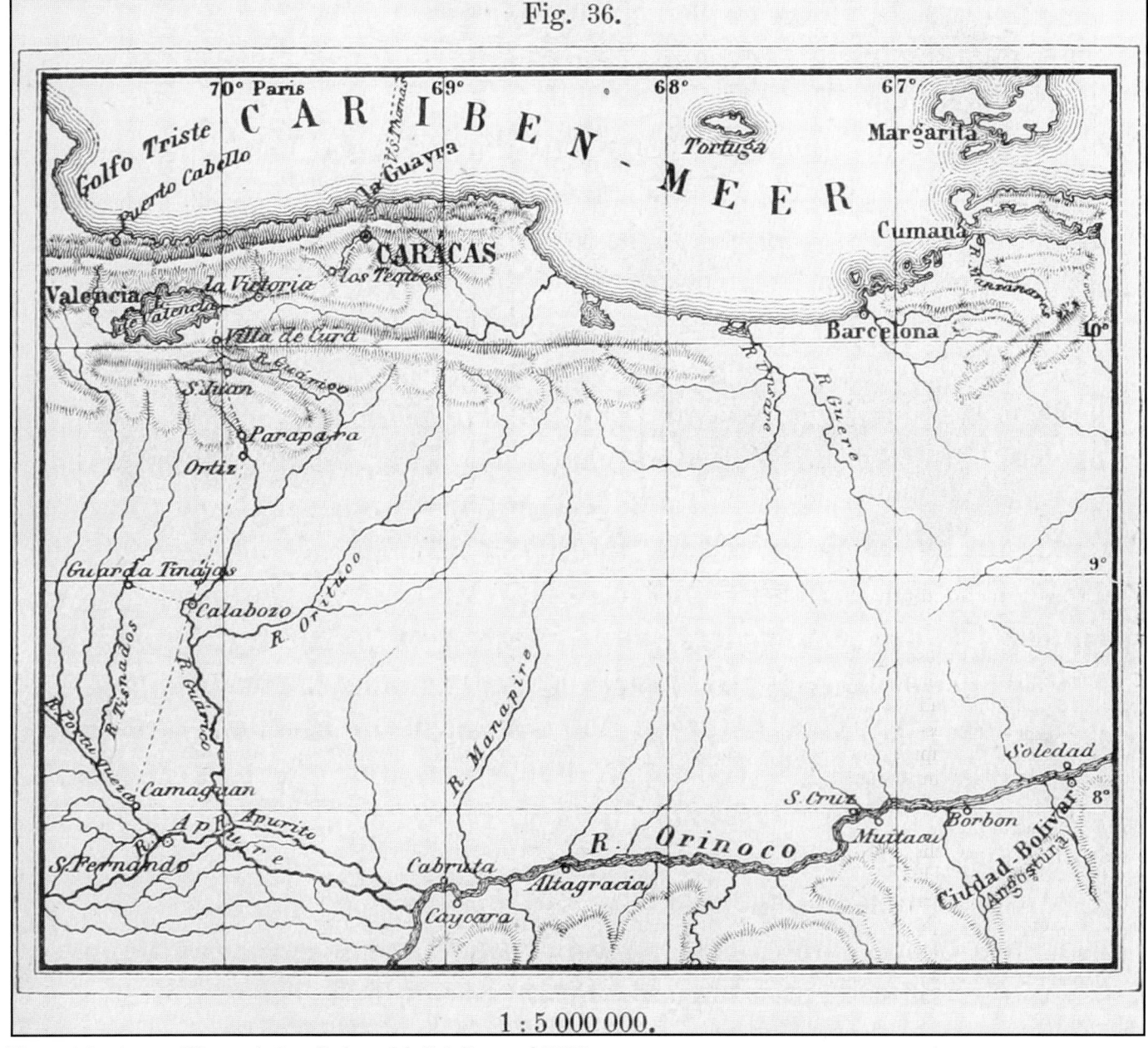

Figure 1.1. A map of Venezuela from Sachs and du Bois-Reymond (1881).

The travelers probably also discussed a most unusual person they had just met in Calabozo. He also showed a love of science, even though he lived in intellectual isolation. He was particularly interested in electricity, sharing this trait with the thin German aristocrat who headed the caravan. The latter would write:

> We found at Calabozo, in the midst of *Llanos*, an electrical machine with large plates, electrophores, batteries, electrometers; an apparatus nearly as complete as our first scientific men in Europe possesses. . . . All these articles had not been purchased in the United States; they were the work of a man who had never seen any instrument, who had no person to consult, and who was acquainted with the phenomena of electricity only by reading the *Traité* of Sigaud de La Fond, and of the *Memoirs* of Francklin [*sic*]. Mister Carlos del Pozo, the name of this enlightened and ingenious man, had begun to make electrical machines of the cylindrical type, employing large glass jars, after having cut off their necks. . . . I had with me some straw electrometers, pithballs, and gold leaf; also a small Leyden jar which could be charged by friction according to the method of Ingenhousz, and which served for my physiological experiments. Señor del Pozo could not contain his joy on seeing for the first time instruments, which he had not made, and which appeared to be copied from his own.[3]

---

[3] Humboldt and Bonpland 1820, vol. VI, p. 103–4. Unless otherwise specified, additional quotations from Humboldt will be drawn from this source, which is in French (and is only one of the numerous texts in which Humboldt reported on his experiments on the electric eels). For the English translation we have referred to the edition in three volumes published in 1852, by checking it with reference to the original and revising it when needed.

Figure 1.2. Alexander von Humboldt (1769–1859), the famous German explorer and scientist, when he was 26 years old.

But just why did these foreigners set their sights on the tiny out-of-the-way village of Rastro de Abaxo? The baron explained that this had much to do with one of nature's most unusual creatures—one feared by the natives, yet one that had drawn the rapt attention of naturalists and even the lay public. Their immediate scientific objective was to study the "electric eels" that were supposed to be so plentiful in this vicinity.

As men of science, the visitors knew that these eels were not really eels, but had been called eels only because of their slimy bodies and elongated shapes. Unlike true eels, they breathe mainly through their gills, although they also gulp air, and cannot survive long out of water. Nevertheless, they were typically referred to as the eels of Guyana or Guiana (or of Surinam or of Cayenne), terms that specified where they were found.

Among European taxonomists and natural philosophers,[4] this fish was formally called *Gymnotus* and classified under *Gymnoti,* terms derived from two Greek words: *gymnos* meaning "naked" and *notos* meaning "back." This name drew from the fact that these fish lack dorsal fins. Carl Linnaeus (Linné), the Swedish father of taxonomy, had more specifically classified them as *Gymnotus electricus*. This was because they could emit violent shocks, much like those that could be released from some of the electrical devices that the European travelers carried with them. In 1864, because there was some confusion about several fishes sharing the name *Gymnotus,* a new name for the genus was

---

[4] This phrase was frequently used up to the beginning of the nineteenth century to designate individuals we would now call scientists. Hence, we shall use this term in all but the final chapters of this book, although there will be some instances where we shall use the more modern term "scientist" to clarify or to make a specific point.

Figure 1.3. Aimé Alexandre Goujaud, "Bonpland" (1733–1858), the botanist who accompanied Humboldt to the New World.

proposed, one with only one member: *Electrophorus* (Gill, 1864). Today, the "eel" is known as *Electrophorus electricus*.

Indeed, the unique feature of this fish is its extremely powerful discharge. The shocks from large, healthy eels can measure in the range of 600 to 700 volts at 1 ampere (Keynes, 1956: Grundfest, 1957). These feared members of the knifefish family were also given some early names reflective of this power. They were sometimes called "torporific eels" and "tremble fish" (*trembladores* in Spanish, which becomes *tembladores* in Humboldt's narrative, *anguilles tremblantes* in French, *Siddervis* in Dutch) by those familiar with the numbness, pain, and muscle tremors their shocks could cause. Of course, there were many colorful local names for them as well. The Tamarac Indians of South America, for example, called them *arimna,* which the early explorers were told means "something that deprives of motion." To the extent possible, we shall use names common to the epoch and the culture for these fish and the other electric fishes in this book.

CAPTURING THE EELS

The earlier stop in Calabozo was disappointing when it came to obtaining these fish. They were present in the sluggish rivers and pools, but the Chayma Indians were terrified of handling live ones, and the Guyaqueries, the best Indian fishermen in the region, were not particularly eager to do this either. Everyone knew that, when threatened, livid eels would come together to shock *en masse*. They also knew that each would continue to release its painful discharges for a surprisingly long time, before eventually tiring and swimming away to rest. With their powerful shocks, as

Humboldt and Bonpland were informed, these three-foot eels could kill a person and even a horse.

As a result, they had been presented with only a single, "much enfeebled" eel on the previous stop. Even offering the Indians a generous two piastres for every vigorous eel they could provide did not produce the specimens they wanted. Hence, Humboldt and Bonpland moved on to Rastro de Abaxo, hoping for better luck. Now, after introducing themselves, eating, and a brief rest, they again mounted their horses. This time, however, they followed some local Indian fishermen to the Caño de Bera, a nearby marsh surrounded by Indian figs, fragrant mimosas, and various other exotic flowers and plants that the stockier man described in his notebooks. They had been assured that some eels could be captured in this swamp, and they optimistically set their equipment near a promising pool, so as to avoid transporting the eels back to town, which could damage and weaken them, and thereby affect the results of their planned physiological experiments.

But given their previous experiences, they were still perplexed about how they would get healthy specimens for these studies. Humboldt, who had waited for years for this golden opportunity, would later write:

> To catch the *Gymnotes* with nets is very difficult, on account of the extreme agility of these fish, which bury themselves in the mud like snakes. We would not employ the *Barbasco,* that is to say, the roots of the *Piscidea erithryna,* the *Jacquinia armillaris,* and some species of *Phyllantus,* which, when thrown into the pool, intoxicate or benumb the eels. These methods have the effects of enfeebling the *Gymnotes*. (pp. 108–109)

The natives, however, had a plan—one that was far from the mindsets of the visitors. The best way to catch the *tembladores,* they explained, would be *embarbascar con cavallos*—literally meaning they would stupefy the fish with horses. With this they galloped away, leaving the two Europeans, who had learned some Spanish but were probably not sure quite what this meant, somewhat bewildered.

After some time had passed, Humboldt and Bonpland detected a rumbling noise that seemed to increase in intensity. Looking up, they saw a rapidly expanding cloud of dust. After a while, they were able to see horsemen driving a herd of about 30 wild horses and mules toward the marsh. The animals, prodded on by drivers waving sticks and shouting loudly, were driven into the warm, stagnant pool not far from where they were located.

For what happened next, we can do no better than read what Humboldt wrote in his narrative about that amazing day in March 1800:

> The extraordinary noise caused by the trampling of horses' hoofs, makes the fish issue from the mud, and excites them to attack. These yellowish and livid eels, resembling large aquatic serpents, swim on the surface of the water, and crowd under the bellies of the horses and mules. A contest between animals of so different an organization presents a very striking spectacle. The Indians, provided with harpoons and slender reeds, surround the pool closely; and some climb up the trees, the branches of which extend horizontally over the surface of the water. By their wild cries, and the length of their reeds, they prevent the horses from running away and reaching the bank of the pool. (pp. 109–110)

The result was a battle reminiscent of those described in the great epic poems of the distant past—clashes involving powerful giants and strange mythological creatures (Fig. 1.4):

> The eels, stunned by the noise, defend themselves by the repeated discharge of their electric batteries. For a long interval they seem likely to prove victorious. Several horses sink beneath the violence of the invisible strokes, which they receive from all sides, in organs the most essential to life; and stunned by the force and frequency of the shocks, they disappear under the water. Others, panting, with mane erect, and haggard eyes expressing anguish and dismay, raise themselves, and endeavour to flee from the storm by which they are overtaken. They are driven back by the Indians into the middle of the water; but a small number succeed in eluding the active vigilance of the fishermen. These regain the shore, stumbling at every step, and stretch themselves on the sand,

Figure 1.4. The epic battle between horses and eels in the *Llanos* of Venezuela (from a vignette illustration of du Bois-Reymond, 1848–84, vol. II, first part).

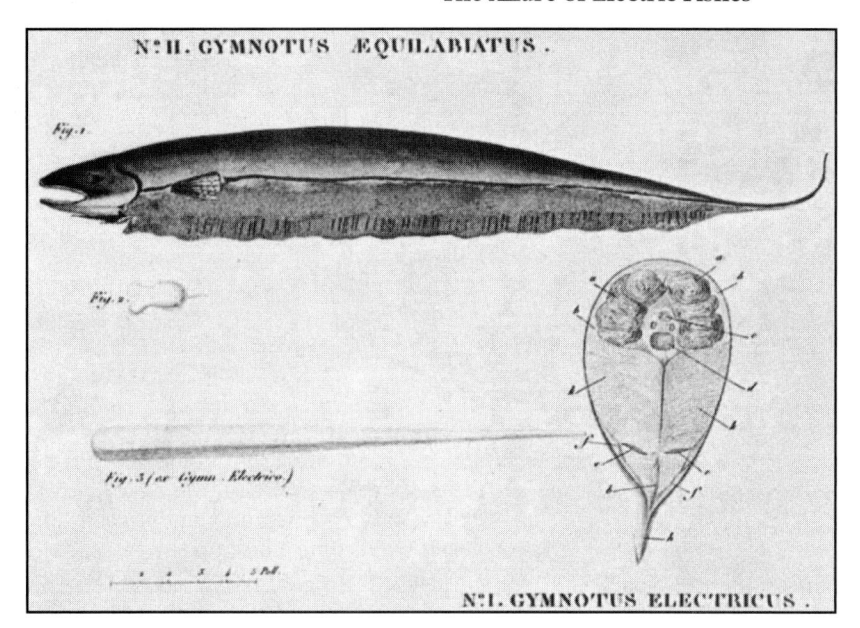

Figure 1.5. Humboldt's illustration of some of the fishes he studied in what is now Venezuela. In the lower part of the image is a cross-section of the *Gymnotus electricus* (the term for the electric eel at that time) showing the electric organs. The upper part shows a different type of knife-fish, which lacks the power of producing strong electric shocks (from Humboldt & Bonpland, 1811).

exhausted with fatigue, and with limbs benumbed by the electric shocks of the *gymnoti*. (pp. 110–11)[5]

Two horses experienced terrifying deaths in just the first five minutes. But Humboldt was unsure whether they had been killed by the electrical discharges or because they were unable to keep their heads above the water, with the other horses and mules pushing them down in the frenzy. Nevertheless, he was convinced that every horse and mule would have been killed, one by one, had the battle been allowed to continue with the horses and mules having no chance of escape. "But by degrees, the impetuosity of this unequal combat diminished, and the weary *gymnoti* dispersed . . . to repair the galvanic force they have lost" (pp. 111–2).[6]

Within minutes, the two Europeans had five healthy eels to study, with additional specimens being captured soon after this time (Fig. 1.5). The biggest eel measured a few inches over 5 feet (over 1.5 meters) in length, with the Indians telling them that even larger ones could be found in these waters. An eel of almost 4 feet (about 1.2 meters) in length, Humboldt noted, weighs about 12 pounds (about 5.5 kilograms).

ONE OF SEVERAL ELECTRIC FISHES

Electric fishes had been an obsession of naturalists, physicists, physicians, and philosophers during the second half of the eighteenth century. But as Humboldt, with his broad education and love of history knew, this fascination was not new and the South American river eels are not the only

---

[5] The original communication dealing with the eels experiments was sent by Humboldt to the French *Académie des Sciences* on September 27th 1806 as documented by a letter still extant in the Humbolts' file of the *Académie* archives.

[6] There has been some discussion about the truthful character of the scene depicted by Humboldt in his narrative (see Sachs 1879, 1871). The method used by these locals to capture these fish was certainly not widespread, but was, in fact, sometimes used in the Oronoco region. It was probably derived from the relatively common practice of letting unmanned horses or mules wade into the marshes or rivers to see if these feared fish are present. People would follow them into the water once assured the eels were not there.

fishes that can numb, although they are the most powerful. Two other types of fishes were also on his mind when he did his experiments.

One was a silurus, more commonly known as a catfish, which inhabits some of Africa's warm rivers and streams (Fig. 1.6). Humboldt briefly alluded to this fish as "the electric fish of the Nile" in the narrative of his trip to South America. A few years earlier, German naturalist Johann Friedrich Gmelin (1789) had named it *Silurus electricus* (Fig. 1.7). Before Humboldt would return to Europe, however, Bernard Germain de Lacépède, a leading French naturalist, would rename it *Malapterurus electricus*: *Malapterurus* because of its soft adipose fins (from the Greek: *malakon* = soft, *pteron* = fin, *oura* = tail), and the second word, as with the eels, its ability to discharge what seemed to be electricity (Lacépède, 1803).

In Humboldt's day, only one species of electric catfish was recognized. Since his time, there has been an increasing number of species identified, with 19 known at the time of this writing, 14 of which were described within the past few years (Norris, 2002; Nelson, 2006). There is even a dwarf

Figure 1.6. *Malapterurus*, the electric silurus or catfish, from a nineteenth-century popular zoological treatise (from Figuier, 1869).

Figure 1.7. Johann Friedrich Gmelin (1748–1804), who gave the catfish that Humboldt called "the electric fish of the Nile" the name *Silurus electricus* (in 1789).

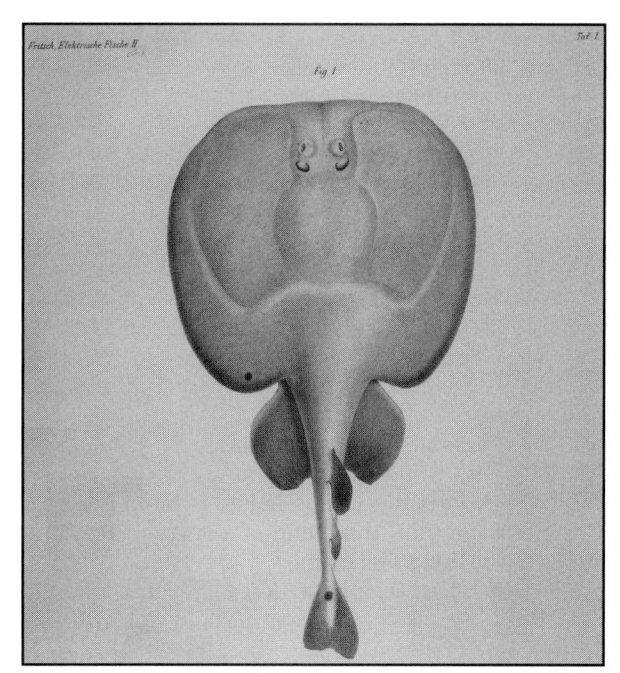

Figure 1.8. A top view of a torpedo, or electric ray (from Fritsch, 1890, vol. II).

variety, *Malapterurus microstoma*. All of the electric catfishes, the *Malapteruridae*, come from Africa's warm rivers (e.g., the Nile, the Niger) and lakes (e.g., Lake Chad, Lake Kainji, Lake Tana). The species of catfish that Humboldt referred to as "the electric fish of the Nile" can deliver painful jolts of up to 350 volts.

The ancient Egyptians living along the Nile depicted this catfish in a very realistic way on the walls of several tombs dating from the Age of Pyramids, about 4,400 years before Humboldt arrived in the New World (see Chapter 2). During the Age of Exploration, the Portuguese "rediscovered" torporific catfishes in Ethiopia (the source of the Blue Nile) and other parts of Africa (see Chapter 7). In the 1600s, before anything was known about fish electric power, they described their stunning powers in letters and books, and what they wrote clearly astonished others reading these accounts back home in Europe and elsewhere in the world.

In addition, Europeans had long been familiar with several kinds of saltwater torpedoes (Bini, 1967) (Fig. 1.8). These disk-shaped rays with relatively small tails were given this name from the Latin because they could induce "torpor," meaning inertia or languor. The earlier Greek word had been νάρκη (*nárkē*), from which we get the modern words "narcotics," "narcotizing," and "narcolepsy," the latter being a sleep disorder in which brief episodes of sleep unexpectedly intrude, even during daytime activities.

The shocks of the smaller torpedoes common to the Mediterranean Sea, however, are usually only about 50 volts (Grundfest, 1957; Fessard, 1958). Some much larger and considerably more powerful varieties, however, have been caught in the warm coastal waters of southern Europe, although they are relatively more common in the Atlantic Ocean (see Chapter 16). The ancient Greeks and Romans, who fished the southern seas, wrote a great deal about the local torpedoes (see Chapters 3 and 4), and Humboldt, ever the scholar, mentioned several of their names and alluded to their writings when describing his coveted eels.

Humboldt also knew that some kinds of torpedoes could reside far from the Mediterranean. During the 1770s, some large ones made headline news when they had been netted in the cold waters off the British Isles, and some less weighty cousins had been caught off both the Atlantic and Pacific coasts of the Americas. Yet Humboldt now learned something quite new about these rays: they could, in fact, swim into the sluggish freshwater rivers of South America. This surprising realization took place after he had asked some Indians in Calabozo to bring him some live *tembladores*. The first fish they brought back was actually a torpedo that had made its way up a small river. This torpedo was small and feeble when captured, so much so that the daring explorer found its shocks almost imperceptible.

A PHILOSOPHER WITH A SCIENTIFIC AGENDA

Just why was Humboldt willing to endure so much, and even risk his life, to make his way to these eels? What were the issues of his day that brought him to the interior of South America? And what specific questions did he hope to answer now that he had access to his coveted fish? After all, they and their shocks had been described many times before 1800, first in the wild and then under better conditions after several were shipped alive to North America and then to London during the 1770s. To begin to answer these questions we have to backtrack to understand Humboldt's interests, philosophy, and willingness to take risks, as well as his mindset and the scientific *Zeitgeist* in which he thrived.

Alexander von Humboldt had been born into a prominent Berlin family. He had been provided with the best tutors in his youth. He then studied the sciences (physics, chemistry, mineralogy, etc.) and medicine at the University of Frankfurt/Oder (*Viadrina*) and Göttingen. He also attended the Academy of Commerce in Hamburg and the Academy of Mining in Freiberg (Saxony), and spent varying periods of time at the universities in Berlin, Hamburg, and Jena.

By the time he finished his formal studies in 1792, Humboldt had mastered several foreign languages and had a good understanding of geology, astronomy, botany, anatomy, and physiology. He was also unusually talented when it came to instrumentation, and showed a penchant for collecting, measuring, and organizing large amounts of material.[7]

Stimulated by what he was reading about scientific explorations, he was driven by a need to make great discoveries of his own. Initially, he traveled through Prussia as an inspector of mines, and afterwards into Holland, England, Austria, Poland, Italy, and Switzerland. These wanderings led to significant publications in various fields, including a treatise about minerals by the Rhine and one on the vegetation in the mines of southern Germany.

Like many other scientists in the second half of the eighteenth century, Humboldt was fascinated by electricity, a force that was just beginning to be understood. The Latin term *electricitas*, which would become electricity in English, had appeared in England only 200 years before he had his eels. William Gilbert (1600), the physician of Queen Elizabeth I, made recourse to this word in order to designate the capability of several materials to attract light bodies when appropriately rubbed. In old times it was assumed that this property belonged uniquely to amber (*ēlektron* in Greek), a fossil sap largely used for jewelry.

Electricity attracted more attention when machines began to be constructed that could produce it at will, using friction. The first such devices utilized globes of sulfur, but during the eighteenth century glass balls, tubes, and disks became more popular. A second technological breakthrough took place in the 1740s. It was the discovery that electricity could be stored in foil-coated glass bottles called "Leyden jars" (see Chapter 10).

With these technological advances, both highly trained scientists and inquisitive amateurs began to study the nature of electricity and its pervasiveness in the natural world (Ritterbush, 1964). Benjamin Franklin, although better remembered as an eighteenth-century American diplomat, philosopher, and writer, was extremely important in this regard. He invented new tools for studying electricity (e.g., "magic squares" or "Franklin squares" that were flat, metal-coated panes of glass that worked like Leyden jars). He also showed how lightning could be captured in Leyden jars with tall metal poles or special kites, and that this feared and terribly destructive force was qualitatively the same as frictional electricity (see Chapter 10).

Franklin's pointed lightning rod protected houses from nature's wrath, his theory of electricity seemed to explain phenomena better than other theories, and the *Zeitgeist* he helped create in the mid-1700s led him and others to wonder whether there are additional sources and manifestations of electricity in nature. Some experimental natural philosophers and naturalists began to look at torpifying fishes very differently in this new setting. The powers of these fishes had never been satisfactorily explained, and the sensations they produced really felt like shocks from the new electrical devices. In addition, these men now discovered that the same substances that could transmit man-made electricity, such as metal rods and human bodies, could also convey their discharges. In contrast, glass and other non-conductors of electricity seemed to block their various effects. Thus, the possibility that at least a few fishes might be electrical began to be entertained in the middle of the eighteenth century.

The ramifications of this new way of thinking about some of nature's living creatures would be enormous. The fish research soon opened the door to the possibility that electricity might be the mysterious "fluid" that courses through the nerves and causes muscles to contract, even in people. It also bore on diseases. One idea was that blockages obstructing the normal flow of electricity might underlie paralyses, as contrasted with electrical "storms," which could account for seizure disorders. Not surprisingly, therapeutic electricity using frictional machines and Leyden jars also emerged at mid-century (see Chapter 10). Related to this fascinating development, yet somewhat forgotten, the ancient Romans had found that live torpedoes could be used to treat painful conditions, such as headache and gout, although they could only loosely speculate about the nature of the fish discharges, which they generally thought acted somewhat like a cold poison.

But is electricity really the mysterious nerve force, or might this be going too far? Indeed, do these rays, catfishes, and eels really emit electricity, or might it only be something that superficially resembles true electricity? To say the least, fish electricity was a "highly charged" subject during the second half of the eighteenth century. After all, everyone knew that electricity would quickly spread across moist tissues. This being the case, how could there *not* be the physiological equivalent of chaos in a fish's body, with the electricity running helter-skelter and the nerves being short-circuited? This recurrent thought seemed to be at odds with the kinds of fine motor acts they could perform, such as snapping up a small stunned fish. Hence, what was then thought about the physics of electricity precluded many natural philosophers from even thinking about these eels, torpedoes, and catfishes being electrical with an open mind.

Further, no one had ever reported that any of these fishes could affect pith balls on strings, even though these balls should move about or "dance" when they approach an electrified object. This sort of experiment had been repeatedly tried, including with more sophisticated electrometers than those using pith balls, but always without success. There was also the matter of seeing a spark or hearing a crackling sound. The absence of a biological thunder-and-lightning show was a major concern, because these sensory phenomena were thought to be basic features of charged or electrified bodies.

Nevertheless, some scientific barriers had, in fact, crumbled before Humboldt set off for the New World. A particularly significant one was crossed in 1776, when John Walsh showed fellow members of the Royal Society of London that electric eels do, in fact, generate sparks that could be seen across a minute gap in a wire in a darkened room (see Chapter 19; Piccolino, 2003). Indeed, a good case had been

---

[7] With his obsession for measuring and prospecting anything and anywhere, Humboldt is one of the protagonists of a recent novel of Daniel Kehlmann, the title of which is, in the English translation, "Measuring the World" (Kehlmann, 2005).

made for fish electricity before June 5, 1799, the date on which Humboldt and Bonpland departed on the Spanish frigate *Pizarro* for the New World.

But in this era well before there were electrical recordings, some questions remained, one of which had drawn Humboldt's rapt attention. In particular, he wanted to know if the fish discharges are truly electrical, meaning not qualitatively different from frictional or atmospheric electricity, or whether they are somewhat different, showing only some of the features of true electricity but not others. This question of a single electricity or a family of related fluids had been bothering Humboldt for years, and he recognized that only by doing experiments with electric fishes would he be able to come forth with a good answer.

VOLTA'S ARTIFICIAL ELECTRICAL ORGAN

On the very same day in which the eels were being captured for Humboldt in Rastro de Abaxo in such dramatic circumstances, thousands of miles away fish electricity was on the mind of another great researcher of the Enlightenment, Alessandro Volta (Fig. 1.9). Surely, what was happening in his aristocratic mansion near the shores of a beautiful Alpine lake on March 19, 1800, was not as epically dramatic as the battle between eels and horses that Humboldt would feature so intensively in his narratives. This notwithstanding, the anxiety and feverish ardor with which Volta (a physicist of Como and professor at the nearby University of Pavia) was writing his long scientific letter was comparable to Humboldt's intellectual excitation in the New World. For both men, these singular fishes would reveal intimate secrets of nature.

In Como, where he had taken refuge because of the war between French and Austrians for the conquest of Lombardy, Volta was attempting to complete a letter announcing to Joseph Banks, the President of the Royal Society, the discovery of a most amazing device he had just made—his battery (Fig. 1.10). As we shall see in Chapter 22, beyond any most wishful hope of its inventor, this device would have extraordinary consequences for the progress of science and technology for centuries to come.

In the letter that Volta would send the next day, the "great physicist of Pavia" (as he was called, because of the importance of his previous research and inventions) attempted to explain the electricity of fishes and also the much less powerful electricity that a doctor of Bologna, Luigi Galvani, believed he had discovered in the muscles and nerves of frogs (and other ordinary animals), and described in 1791. Galvani had dubbed this electricity "animal electricity" and he assumed it to be similar to the animal electricity discovered by John Walsh in torpedoes and electric eels many years before. Volta was convinced that both the electricity discovered by Galvani and that of electric fishes as demonstrated by Walsh were simply generated by the contacts of dissimilar electrically conductive substances present in animal bodies. In particular, animal electricity corresponded to the electricity that he had produced by putting together two dissimilar metals, and had been able to measure in a most convincing way in 1796 (see Chapters 22 and 23; see Polvani, 1942; Piccolino & Bresadola, 2003).

Without question, Volta's new force resembled common electricity in many ways: in its physical properties, in its

Figure 1.9. Alessandro Volta (1745–1827) in an engraving of 1814 reproduced in the first volume of the *Edizione Nazionale* of his works (Volta, 1918).

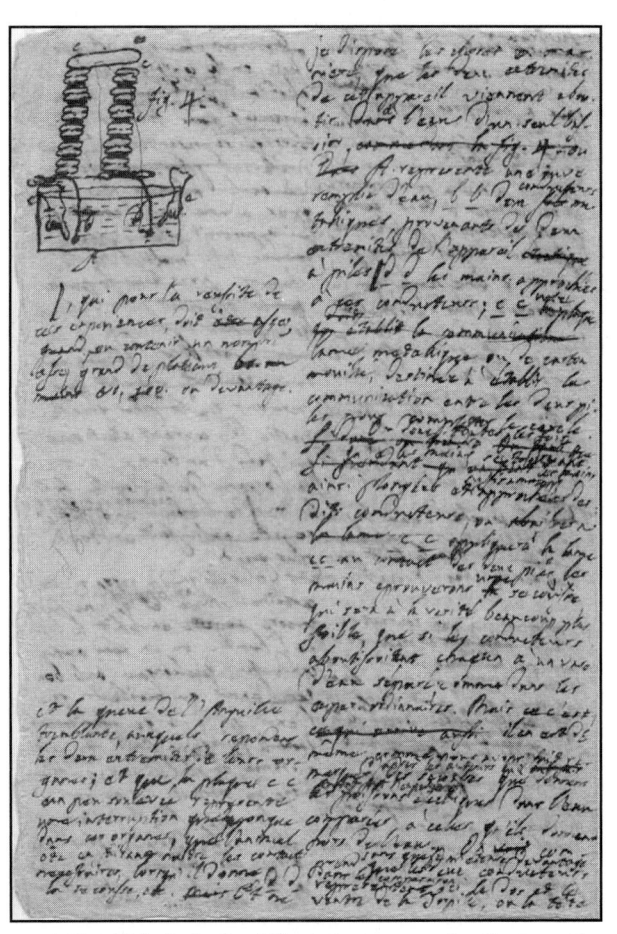

Figure 1.10. Volta's drafts of the letter announcing the invention of the pile or battery, which he modeled on electric fish organs (from Volta's manuscripts at the Istituto Lombardo, Milan ©).

ability to stimulate frog muscles to contract, and in the tingling sensations it could produce. Volta was so deeply convinced of the identity of his metallic electricity and animal electricity (and particularly the electricity of fishes) that, in his letter to Banks, he called his new instrument the *organe électrique artificiel*—that is, the artificial version of the natural organ of fishes.

To Volta's chagrin, "galvanism" was the name others began to apply to the force produced by different metals, and this bothered him for two reasons. First, it implied that he had not discovered a new source of electricity, but rather something different. And second, "galvanism" suggested that another man, Luigi Galvani, had made the discovery.

Galvani's (1791) memoir on animal electricity, which appeared in a periodical published by Bologna's *Accademia delle Scienze,* had caused quite a stir. In the seventh volume of its *Commentarii* dated 1791 (though actually published at the beginning of 1792), he theorized that electricity is not just found in a few torporific fish. Rather, it exists in subtler amounts in frogs, farmyard animals, and even humans—essentially that it is the long-misunderstood agent that flows through the nerves and is stored in the muscles, enabling them to contract when stimulated.

Volta agreed with Galvani that electricity from a frictional machine could stimulate nerves and cause muscle contractions in animals. But he could not accept his countryman's theory that more than a few fishes possess intrinsic electricity, much less the Leyden jar and wiring model Galvani seemed to be so naïvely proposing. Even today, Volta's position is not always properly understood. He never argued against the idea that some fishes could be electrical. When it came to these fishes, he was greatly influenced by John Walsh, whom he had met in London; by John Hunter, who dissected the electric organs of some of Walsh's torpedoes and electric eels; and by fellow physicist Henry Cavendish, who worked on the physics of the discharge, and was, like Walsh and Hunter, a member of the famed Royal Society of London (see Chapter 16).

But whereas a few fishes seemed to have specialized electrical organs, nothing like these structures could be found in any other known animal, including humans. This observation led Volta to question Galvani's conclusions, and he became increasingly critical of Galvani when he realized that his countryman's experiments also had methodological flaws. The biggest problem, Volta contended, was that Galvani and his associates used different metals, thus providing an external source of electricity. In other words, it was his belief that the electricity in Galvani's experiments came from outside the animal, not from inside its body. And with this insight, Volta rejected Galvani's theory about electricity being intrinsic to more than just a few very odd fish.

HUMBOLDT'S INITIAL FORAYS WITH ELECTRICITY

In 1792, 23-year-old Alexander von Humboldt set out to replicate some of the experiments Galvani and Volta had performed, and to conduct some new electrical experiments of his own. Several issues drove his efforts, the most important of which was whether all animals, including people, might function by "animal electricity," as Galvani claimed. A related issue was whether animal electricity is identical in all respects to electricity from frictional machines and from the atmosphere. And a third issue pertained to the electrical force Volta claimed could be produced when two or more metals are combined: Is this so-called electricity the same as the electricity coming from known sources, as Volta believed, or is it different? There were other questions as well, but these were foremost on Humboldt's mind when he started his experiments.

Humboldt had little idea of how many electrical experiments he would end up conducting. His initial thought was perhaps hundreds, but this was a gross underestimate. In all, he performed some 4,000 experiments over a 5-year period!

In some of his experiments, he did not use living matter, but in others he used plants and animals in his circuits. The latter included a wide variety of invertebrates: snails, worms, crabs, crayfish, leeches, and many common insects. Volta had indicated that invertebrate muscles and the involuntary muscles of vertebrates might not be sensitive to electricity from metals, so this was on Humboldt's mind when doing some of these experiments (see Piccolino & Bresadola, 2003). But Humboldt also believed in a fundamental similarity of the "animal substance" (*thierische Stoff*) throughout the animal kingdom, from insects up to humans, giving these studies additional meaning (Humboldt, 1797, Vol. 1, p. 283).

He also experimented on the usual laboratory animals: birds, rabbits, dogs, and, of course, a huge number of common pond and river frogs. In one of his experiments, he employed frogs together with a touch-sensitive plant, the *Mimosa pudica,* thought by some to have electrical properties because its leaves would contract when touched (Humboldt, 1797, Vol. 1, p. 250; also see Turner, 1746; Chapter 12). When he could, he also assessed the effects of electricity on different fishes. In 1794, for example, he studied the effects of galvanic stimulation on heartbeat of some fishes caught in Poland's Vistula River. Nevertheless, he did not have access to electric fishes at this time in his life.

Humboldt's most spectacular and memorable experiments were those performed on his own body. Many of these experiments were benign, as exemplified by one in which he connected a frog to a zinc armature in contact with his tongue, in order to determine if he would perceived a taste sensation when the frog moved, which he could then attribute to circulating animal electricity. Yet other electrical experiments were extremely painful and some even damaged his body. In this domain, he performed several experiments on his aching jaw after a tooth had been extracted to ascertain whether over-stimulating the nerve would help diminish the pain (it had the opposite effect). He also conducted electrical experiments on his wounded hand and carefully documented others on his shoulders, which earlier had been made raw by blistering plasters. Describing one of these gruesome experiments, he wrote: "The blisters [on my back] were cut and contact was made with the zinc and silver electrodes, . . . and Frogs placed upon my back were observed to hop" (Humboldt, 1797, vol. I, p. 331).

Humboldt even replicated an almost unimaginable experiment conducted by Franz Karl Achard, placing a zinc armature into his mouth and one of silver into his rectum. When connected with a wire, this arrangement produced abdominal pains, stomach contractions, involuntary evacuations, and even visual sensations. "What struck me more in the Achard experiment, which I have confirmed several times on myself," he explained, "is that, by inserting the silver more deeply into the rectum, a bright light appears before both eyes" (Humboldt, 1797, vol. I. p. 334).

Humboldt's self-experimentation might be hard for us to imagine today. But using one's body for the advancement of science had been common in earlier electrical studies (Schaffer, 1994) and was actually encouraged in Germany during the Romantic Era—although not to the point where one's life would be put at risk. The prevailing thought among many German *Naturphilosophen*[8] was that, much as a scientist can learn more about himself by studying nature, he could also learn the secrets of nature by studying himself. With this guiding philosophy, the human body had become a laboratory instrument—a sensitive "electroscope" for detecting electricity and estimating its quantity and qualities. But unlike an electroscope, the human body was also a metaphor for nature itself (Strickland, 1998).

What Humboldt did to himself was equaled if not exceeded by what Johann Wilhelm Ritter did to his own body (Strickland, 1998). Humboldt had been so impressed when he met Ritter that he took him under his wings and asked him to scrutinize every aspect of his huge 1797 manuscript on electricity before it went to press. True to the cause, Ritter did confirm Humboldt's findings in important ways. But he was so weakened by what he did to himself in these and later experiments that he died a martyr on the altar of science at just 34 years of age.

With Ritter's help, Humboldt prepared his notes for publication, hoping his book would establish his name in the field and become a landmark in the sciences. His two-volume book appeared in German in 1797 and was titled *Versuche über die gereizte Muskel- und Nervenfaser nebst Vermuthungen über den chemischen Process des Lebens in der Their- und Pflanzenwelt* ("Research into the Irritated Muscle- and Nerve Fiber with some Assumptions about the Chemical Process of Life in the Realm of Animals and Plants"). An abridged French edition was published in 1799, while Humboldt was planning his trip to the Americas.

Humboldt's conclusions were somewhat different from Galvani's and Volta's. He supported some of what both Galvani and Volta were proposing, but not everything they were claiming. Although several different metals (e.g., silver and zinc) could clearly serve as a powerful stimulant for contraction (Galvani's and Volta's position), metals were, in Humboldt's opinion, not obligatory for this physiological response (Galvani's position). Under conditions of heightened reactivity (e.g., particularly lively preparations or preparations treated with stimulants), he could obtain contractions simply by connecting a nerve to a muscle. Movements could also be elicited by connecting a nerve to a muscle through animal and even plant tissue.

These experiments went against Volta's stance that Galvani's experiments could be explained by metals alone. They were in accord with some observations Galvani published in 1794 (albeit anonymously), and suggested that there really is animal electricity. Nevertheless, Humboldt was not fully supportive of Galvani. Specifically, he was not convinced that the intrinsic animal fluid is identical to true electricity in all of its features. Notably, the animal force, like the metallic fluid, would not pass through animal bone and certain other substances, whereas the electricity produced with frictional machines seemed to possess this attribute.

Thus, Humboldt looked upon this intrinsic animal fluid as something other than true electricity. It was, to his broad mind, a particular expression of a vital life force (*Lebenskraft*) that is common to all animals and to a certain extent plants. This was a seminal feature of his worldview. This internal life force, he believed, would build up in the muscles and could be triggered by an internal or an external stimulus, such as a spark from an electrical machine, causing a muscle movement.

Although he disagreed with Volta about how the physicist interpreted some of Galvani's experiments and his dismissal of the idea that all animals possess an internal fluid with electrical features, Humboldt recognized the great importance of what Volta had discovered—namely, that a fluid superficially resembling electricity could be produced by putting dissimilar metals in contact with one another. But, he noted, Volta's metallic force, like animal electricity, could not be transmitted through dry animal bones, and it too failed to cross a candle flame.

Less than certain about whether these differences would hold up, Humboldt referred to Volta's force with a noncommittal term. He called it *Metallreiz*, meaning "metallic irritation." Still, he lavished praise on Volta, the genius behind this new discovery, which he prophesized would be of great utility for further investigating animal and human physiology, and perhaps might also be important in therapies. In some of his experiments, he assembled several preparations of animal tissues and metals one on top of the other in an ordered way, hoping to increase the irritating force he was studying. They resembled the assemblies of metallic and humid disks that would lead Volta to the invention of his electric battery a few years later. Humboldt would afterwards regret that although he came so near to this epochal invention himself, he has missed it (Kettenmann, 1997). Interestingly, the term he used to designate his assemblies was *Kette* (literally "chain"), the same word that would be used in German to designate the elements of the electric battery (see Chapter 24).

THE NEED FOR ELECTRIC FISH

Humboldt was still uncertain about "animal electricity" and whether there is just one electricity or several fluids closely resembling true electricity as the long eighteenth century drew to a close. If anything, his findings had been pointing to the possibility of a family of electrical fluids rather than a single fluid. Since the electric fishes seemed to have more of this universal fluid than other animals, he felt that some well-conducted experiments on these unusual fishes could shed more light on the situation.

---

[8] In the context of German science of the time, this expression, literally meaning "philosophers of nature," sometimes had specific connotations. It often referred to German natural philosophers and also philosophers *tout court* (e.g., Johann Gottlieb Fichte and Friedrich Wilhelm Schelling), who endeavored to understand the basis of life. In some cases, it was specifically used to designate those philosophers who assumed that the phenomena of life could not be explained solely by the laws of physics and chemistry. This was true of Humboldt, who brought a special vital force into the picture he painted of God's glorious kingdom.

Humboldt's logic was straightforward. If the discharges could pass through all of the substances that could transmit true electricity, he would have stronger evidence to support the idea of a single, universal electricity. But if their discharges could be blocked by charcoal, bone, flame, and perhaps certain other substances that could convey frictional or atmospheric electricity, such findings would further suggest that what he and others had been calling animal electricity or galvanism should not be thought of as (true) electricity.

The problem Humboldt now faced was how to obtain some of these fishes for these needed experiments. The idea of trying to bring catfishes from Africa, eels from the rivers in Surinam, or torpedoes from the Mediterranean Sea to Germany did not make sense for several reasons, including the low probability that they could survive such a trip and the cold climate that would await them. These thoughts left Humboldt with only one alternative—namely traveling to places where these fishes lived and doing the needed experiments on site.

Humboldt understood that the easiest path to follow was to head south into Italy. In his 1797 book on electricity, he wrote about the importance of a trip south, specifically mentioning the need for experiments using animal bone:

> Possibly, on the occasion of a future visit in some southern region of the world, I might have the chance of investigating with the use of animal bone, whether the particular matter circulating in the *Raia torpedo* is intercepted, like the Galvanic [metallic] one, or conducted, as is [true] electricity. (Humboldt, 1797, Vol. 1, p. 452)

Notably, Humboldt had planned to go to Naples with his brother Wilhelm, a linguist, two years before his two volumes on electricity even appeared (Fig. 1.11). This would have given him access to torpedoes, volcanoes, and the region's culture and archaeology, which also held great interest for him. His excitement was enhanced following discussions with the great Romantic poet and scientist, Johann Wolfgang von Goethe, who had visited Italy and was then very interested in optics and color theory (Goethe, 1791-92, 1810). Wilhelm had introduced his philosophical brother to Goethe in 1794, and although Goethe had a reputation for being cold and aloof, the two men met several times and enjoyed each other's company (Eckermann, 1836). Alexander, not one to miss an opportunity, even conducted some electrical experiments on frog preparations with Goethe and his brother Wilhelm (Humboldt, 1797, vol. 1, pp. 76–77; Botting, 1973).

Unfortunately, the Naples trip did not come off as expected, because of the fighting between the French and the Austrians in politically fragmented Italy. Yet Humboldt did manage to visit Volta at his home in the northern Italian town of Como, where they discussed electricity and also performed some electrical experiments together—although again not on an electric fish.

Naples was not the only destination to elude Humboldt in the 1790s. He also tried to go to Egypt, this time with Aimé Bonpland. This excursion might have given him access to the electric catfish of the Nile. Yet his efforts to join the large army of savants following Napoleon Bonaparte's expeditionary forces twice failed to materialize, owing to transportation problems and, of course, the fighting.

When upon further inquiry he was offered the opportunity to go to some parts of the New World controlled by Spain, he could not have been happier. In addition to having a chance to study a powerful electric fish, he would have the opportunity to observe features of nature that had been hidden from civilized men and to discover a myriad of new forms of life. In South America, he would be able to see the intricacies and universals of the great divine design, and to appreciate the pervasiveness of the life force he felt was basic to God's great plan.

THE SOUTH AMERICAN EXPERIMENTS

On March 19, 1800, Humboldt finally had his eels for study, and he began his gross measurements, physiological studies, and dissections, all with the aim of better understanding whether their discharges are truly electrical. After experiencing the shocks, he would write: "I do not remember having ever received from the discharge of a large Leyden jar a more dreadful shock than that which I experienced by imprudently placing my feet on a *gymnotus* just taken out of the water." He added that he "was affected during the rest of the day with a violent pain in the knees, and in almost every joint" (p. 118).

Humboldt thought that the eels were releasing their shocks intentionally, a contention made by others before him. He reasoned that, if the discharges are intentional, they should cease upon severing the nerves from the brain to the electrical organs. A cut from his knife confirmed this prediction.

He also found that shocks could be transmitted through most of the usual conductors of electricity, including metal rods and people holding hands, and not through the standard array of non-conductors. This set up his long-awaited experiments

Figure 1.11. Wilhelm von Humboldt (1767–1835), Alexander's scholarly brother.

with substances that had revealed possible differences between animal, metallic, and true electricity.

As had been true in Europe, his new experiments again revealed differences between animal electricity and that designated as true electricity. Specifically, even though he was now using very powerful eels, the animal force again failed to pass through pieces of charcoal or a flame. Using his own body as an electrometer, he felt nothing when he included himself in the circuit, and he was unable to see or record anything to the contrary. Thus, although the force of the mighty *Gymnotus* seemed to feel like electricity, and appeared to obey some of the same laws as electricity, the resemblance was less than total.

Nevertheless, this conclusion, which was supportive of something like electricity but not supportive of true electricity, would not be the one Humboldt would convey to his readers after returning to Europe from the Americas (Humboldt, 1806, 1807, 1819, 1820; Humboldt & Bonpland, 1811). As noted, a landmark development was taking place in Italy on the very day the German baron set foot in the small Guyana town of Rastro de Abaxo. Volta had come forth with his "pile" or battery, and with it he was able to multiply the weak electricity usually obtained by joining different metals together, producing significantly stronger effects.

For Humboldt, who met with Volta soon after his return to Europe in 1804, this new tool was of paramount importance, because it suggested that the differences he found between true electricity, metals, and the animal force did not necessarily mean that these forces had to be qualitatively different. Instead, the physical and perceptual effects obtained by multiplying subtle electricity, combined with his respect for Volta's acumen, led him to write that electric fish are probably electrical, even though their discharges might be too weak to pass through some substances. Thus, Humboldt told the world that there is probably true electricity in all life forms, as well as when certain metals are joined together, even if he still had some lingering doubts.

Near the end of this book, we shall return to Volta to see in detail how his new invention was based on what he had learned from Walsh, Hunter, and Cavendish about the electric organs of some fishes. We shall also examine how important Volta's invention turned out to be for naturalists trying to understand these fish and for natural philosophers willing to take another look at nerve physiology.

AN ODYSSEY INTO THE PAST

Humboldt became an international celebrity in the opening decades of the nineteenth century as a result of his exciting and dangerous travels through the New World and his magnificent illustrated volumes about his scientific explorations. Containing innumerable observations and descriptions, and thousands of newly discovered flora and fauna, what he described traversed a great number of fields: botany, geology, mineralogy, zoology, oceanography, ethnography, astronomy, meteorology, art, paleontology, geography and geographical economics, ecology, and more.

Humboldt was a master of scientific prose, and it is not surprising that his stories both enchanted and excited wide audiences. And of all his writings, the material that more than any other captured his readers' imaginations was his encounter with the South American eels. The colorful scene he painted of horses fighting them in the turgid waters of a wild and distant land was reproduced many times verbatim or summarized in numerous books and periodicals of the epoch.[9]

Schoolchildren were excited by the epic battle between horses and eels, and many fantasized about going on their own great voyages of scientific exploration upon growing up. Charles Darwin, who would show a penchant for voyages, collecting, and thinking about unifying themes in nature, would praise Humboldt as "the greatest traveling scientist who ever lived." As a young naturalist, Darwin took an English edition of Humboldt's narrative with him when he set sail in 1831 on what would be his landmark 5-year voyage on the *Beagle*.

The emotional impact of the great horse-and-eel battle was raised to an even higher level by the images used to illustrate Humboldt's writings. Artists produced magnificent, action-packed graphics showing frantic, terrified horses being attacked by writhing eels releasing their powerful electrical charges. It is not by chance that one of these images was used to illustrate the frontispiece of the second volume of Emil du Bois-Reymond's *Untersuchungen über thierische Elektricität*. Appearing in the period 1848–1884, this two-volume text has been hailed as the most important nineteenth-century book in electrophysiology. Fully appreciating the importance of electric fish in the development of his chosen field, du Bois-Reymond not only studied them himself (see Chapter 24), but also selected an image of an electric eel and a torpedo for the frontispiece of his first volume. As he explained, "Every scientist is a descendent of Humboldt."

In retrospect, Humboldt played important roles in the sciences, and deservingly, mountains, animals, and waters have been named after him. With regard to electric fishes, he wrestled with some of the most important issues about them and animal electricity in his day, and he did so in an exceptionally colorful, vibrant, and publicly visible way. He was also exceptionally well read in many fields, and was aware of the efforts of some of the talented individuals who studied electric fish in antiquity and just before him. These men, he knew, provided the broad shoulders on which he stood, while attempting to contribute to the growing body of knowledge about some of nature's most unusual creatures. In brief, these are among the reasons why we have chosen Humboldt and the *Zeitgeist* around 1800 to introduce this volume.

From this starting point, we shall now look in detail at what was known about electric fishes prior to this very exciting time, and what was learned about them during and after this time. In the chapters that follow, we shall proceed

---

[9] The phrase "Humboldtian science" has become current by modern science historiographers to indicate the particular character of Humbold's scientific approach (see for instance Nicolson 1987; and Jardine, Secord, & Spary, 1996). Although it is difficult to provide a short definition of this expression, it is important to remark how, in Humboldt's science, the quantitative approach to the study of natural phenomena was combined with the descriptive and taxonomic tradition of the eighteenth century naturalism. This occurred within a framework of harmony between the forces of the physical and living world. Humboldt's compulsive attitude for scientific measurement is the subject of a recent piece of historical fiction *Die Vermessung der Welt* (translated as "Measuring the World": Kehlmann, 2005).

chronologically, beginning with what was known about electric fishes in murky antiquity. Our initial emphases will be on how these unusual fishes were described, on how they were incorporated into ancient medicine, and on how people in the distant past tried to explain their seemingly inexplicable powers.

After working our way through Medieval times and the Renaissance, during which time these fishes were often associated with occult or magical powers, our focus will become more physiological. Here we shall deal with a number of natural philosophers and examine what they thought about the shocks emitted by these fishes, especially when it became more fashionable to dissect and do experiments on them. This road will take us into the eighteenth century, which is when the idea that these fishes might emit electricity came to the fore, literally sparking the physiological revolution in which Humboldt participated.

The story of how electric fishes came to be viewed as electrical has a great number of twists and turns, and, as suggested above, many significant ramifications. As we shall see, it is deeply embedded in cultural history and has interweaving plots and subplots, recurrent mistakes, and more than a few statements that can be baffling to modern readers. It is also a history marked by clever experiments and remarkable insights. It bears on practical and theoretical medicine, on the scientific disciplines now recognized as biology, physiology, and physics, and even on subjects as diverse as magical thinking and technological innovations.

Surprisingly, the curious, wondrous, and long history of electric fishes has not been explored in depth in the past. And although there have been some eye-catching journal articles and chapters written about the early history of electric fishes, they have tended to be few in number, with significant omissions and with minimal social or scientific context (e.g., Brazier, 1984; Kellaway, 1946; Lanza, 1881; Wu, 1984; Moller, 1995; Musitelli, 2002; Whittaker, 1992).

By more slowly weaving our way from the distant past through the twentieth century, when scientists would finally understand how these fishes manage to do what they do to give off shocks in their different habitats, many questions will be raised and many answered. Some of the most interesting ones are: How old are the earliest known records, pictorial and written, of an electric fish? How did ancient writers describe their shocks, and to what were they ascribed? When and how did these fishes find their ways into the healing arts? What diseases were their shocks thought to cure, and what is the evidence for such claims? How did natural philosophers first measure their powers, and what instruments did scientists use in later periods? And of particular importance: Just when did natural philosophers begin to think that the powers of these strongly discharging fishes are electrical, what cultural changes led to this new way of thinking, and what kinds of findings were presented as proofs?

With these questions and others in mind, and with stories involving a cast of hundreds, some famous and others long forgotten, let us now examine what can happen when science and culture come together to deal with a phenomenon that seemed wondrous if not impossible not all that long ago—the ability of living creatures to discharge electricity and even shock their victims, and not just by direct contact, but at a distance. Hence, without further ado, let us now see how some strongly electric fishes slowly became electrical and triggered new ways of thinking about animal physiology, causing a scientific revolution which would be the rival of the great political upheavals that were occurring at about the same time.

PART II
Ancient Cultures

Chapter 2
The Shocking Catfish of the Nile

> To expel the *setet*:
> Head of the *djedeb* fish.
> This will be boiled in oil/fat
> and applied to the man on his flesh.
>
> *(Ebers Papyrus,* No. 304; see Westendorf, 1999)

The earliest records showing or describing any of the electric fishes come from ancient Egypt. They depict what Humboldt called "the electric catfish of the Nile"—a smooth-skinned fish with a head as long as it is high, tiny eyes, an adipose dorsal fin without rays, and a round snout with three pairs of whiskers (Fig. 2.1). This fish, the *Malapterurus electricus* of the scientific literature (also indicated with a variety of other names, such as the electric silurus), is usually gray or brown with some bluish tinting and black blotches on its back, and with red, orange, and yellow markings on its ventral fins. Still encountered along the Nile by modern fishermen, it can grow over 2 feet (about 60 centimeters) in length.

Malapteruridae, the growing family of electric catfishes, are the only catfishes lacking dorsal fins. They range over the entire Nile and can be found in its slower water near reed beds, on bottoms with mud and loam, and at times in deeper water. Their habitats also include other sluggish rivers, tributaries, and lakes in tropical Africa, such as the warm Niger and Volta Rivers, and Lakes Chad and Tanganyika.

The prevailing Arabic name for these fishes is *ra'ad,* which literally means "thunder-fish," although Egyptian fishermen are more likely to associate the *ra'ad* with the shaking it can elicit (Brewer & Friedman, 1989, p. 70).[1] In this regard, there is another Arabic word for these fishes—*ra'ash,* a term generally meaning "trembler." Nevertheless, just what these fishes might have been called in pre-dynastic or in early dynastic Egypt is uncertain (Thompson, 1928).

As with the other electric fishes, the power of these catfishes to produce their "thunder" comes from a system of flat cells of muscular origin, arranged in an orderly fashion and controlled by the nervous system (see Chapter 25). In these catfishes, unlike torpedoes and the electric eel, the specialized organ is situated just below the skin, and extends from the head to the tail. The shocks of a healthy large specimen can exceed 300 volts. Its discharges can be used for defense and to stun smaller fishes, and they can cause another fish to regurgitate its meal, which could then be pilfered (Boulenger, 1907, p. 400).

Electric catfishes are not the only catfishes in the Nile. It has been estimated that over 30 types of catfishes may exist over the full extent of this river, which had been home to well over a hundred species of fish in ancient Egyptian times (Boulenger, 1907; Greenwood, 1976; Brewer & Friedman, 1989). Today we know that there are at least 18 species of catfishes in Lake Nasser alone, which was formed in 1970 when the Aswan High Dam was constructed to control the Nile and harness its power.

Although these other catfishes differ from the *ra'ad* in size, shape, habitat, and behavior (e.g., some species swim upside-down and are surface feeders, others are much larger or smaller), the fact that there are many other catfishes in the Nile can make it difficult to identify the specific type of catfish depicted in many archaeological findings. This is particularly true where the image is presented in a highly stylized way, as with just a few lines. In some of the more realistic depictions, the cylindrical shape, fins, and somewhat bloated appearance of *Malapterurus* make it somewhat easier to distinguish them from the others, and it is to these illustrations, associated with tombs of powerful early Egyptians, that we now turn.

REALISTIC DEPICTIONS

Malapterurus-like images can be found on bas-reliefs depicting fishermen and hunters from the Old Kingdom, dating back to Dynasty V tombs, circa 2465–2323 B.C. (for illustrations, see Gaillard, 1923). Several can be found at Saqqara (Sakkara), near the ancient city of Ineb-Hedj ("The White Wall"), the ancient metropolis better known today as Memphis, the name given to it by the Greeks.

Located at the root of the Nile Delta just south of present-day Cairo, Memphis was founded by the legendary figure Menes (c. 3050 B.C.; now identified with King Aha), a king or

---

[1] As a matter of fact the Arabic terms *ra'ad* and *ra'ash* derive from a same root indicating movement. In English something similar happen with the word "bolt" which can mean both thunder and violent motion.

Figure 2.1. A realistic depiction of an electric catfish, of which many types are now recognized (from Schinz and Brodtmann, 1836). Electric catfish common to the Nile appear on ancient Egyptian tomb drawings.

perhaps even a composite of several pre-dynastic kings, associated with the unification of Upper and Lower Egypt (Malek, 1986; Hawass, 2008; also see below). It served as the seat of government and a religious center during the Old Kingdom. Saqqara was the largest necropolis for the city, composed of burial grounds that spread for miles along the Nile's west bank. The name *Saqqara* derives from the god Sokaris, who was venerated as the protector of the cemeteries.

Although little remains of ancient Memphis, several important landmarks can still be seen at Saqqara (Lauer, 1976). One is the famous step-pyramid designed by the grand vizier (a high minister of state), "patron of scribes," and priest-magician-healer, Imhotep, for his pharaoh Nesu Neterikhet or Djoser (see Forty, 1998, for more on the pharaohs) (Fig. 2.2). Imhotep lived during Dynasty III, about 2650–2575 B.C. This was just before the more famous stone pyramids of Giza were erected during Dynasty IV (ca. 2575–2467 B.C.), often thought of as "the Age of Pyramids."

The *mastaba* (table-like) Tomb of Ti is one of the major sites at Saqqara. French archaeologist Auguste Mariette discovered it in 1865 (Lauer, 1976, pp. 31–55). Ti had been a wealthy nobleman, and he had been responsible for overseeing the sun temples of Neferikara and Nyussera (Niuserre) during Dynasty V, about 2400 B.C. On one of the many reliefs in this tomb, we see Ti in a boat not far from the shore, with papyrus plants by the water's edge (for the panorama, see Lauer, 1976, p. 20) (Fig. 2.3a). Ti is standing and is depicted

Figure 2.3. (a) The Ruler Ti participating in a hippopotamus hunt. This scene is one of many realistic scenes that can be found on the walls of his tomb in Saqqara (c. 2400 B.C.). *Malapterurus,* the electric catfish, can be seen right under the back of smaller boat on the left. (b) A closer look at the depiction of *Malapterurus* (photos by Wendy Finger).

much larger than any of the other figures, as would be befitting the most important person in the scene. He is watching some men in a second boat thrusting spears into the water. Along with some other river fish, a *ra'ad* is realistically depicted under one of the boating poles (Fig. 2.3b).

Spear or bident (harpoon) fishing from a skiff made of reeds was a popular method of fishing at this time. The spear was typically 9 to 12 feet (about 2.5 to 3.5 meters) long and a line could be attached to it for easier retrieval. Most historians believe, however, that this scene depicts not a fishing trip but a hippopotamus hunt, since a hippopotamus is visible in the water along with the *ra'ad* and various other fish (Lauer, 1976, pp. 50–51). Hippopotamuses can be destructive of crops and dangerous to river dwellers, which are among the many reasons they were feared and hunted in ancient Egypt. To his contemporaries, Ti could well have been displaying his power and ability to conquer chaos and evil.

The *ra'ad*, with its distinctive shape and lack of dorsal fins, can also be seen in other scenes in this same tomb, which show everyday fishermen on the Nile. Again, a variety of fishes can be made out in these realistic scenes, the *ra'ad* never being presented alone (see Gaillard, 1923, pp. 2–4 and Plates I–III). These tomb paintings show us that the

Figure 2.2. The Step Pyramid (c. 2600 B.C.), the most famous site at the burial complex of Saqqara.

Egyptian fishermen had many ways to catch fishes. In addition to spearing, they could cast small nets, drag larger seine nets, employ wicker basket traps, and tend lines with hooks. Although the fishing rod did not come into its own until later, Old Kingdom fishermen also had simple devices for winding their lines, and they were known to use copper hooks (Radcliffe, 1974; Brewer & Friedman, 1989; Van Neer, 1989).

A fish that looks very much like a *ra'ad* can also be observed on the walls of a double tomb from about 2430 B.C. in Saqqara (the dating on this tomb is currently imprecise; it is in some places listed as Dynasty IV and in other as Dynasty V). Because its two occupants are shown on the walls standing face to face, it previously had been postulated that they were brothers or twins. Hence, the tomb of Niankh-Khnum and Khnum-Hotep, discovered in 1964, was at first called the "Tomb of the Two Brothers." Today it is believed that this tomb was built for two unrelated priests of the Sun Temple, who also served as royal servants. Their titles were "Prophets of Ra" and "Heads of the Manicurists of the Great House." This site is among the most spectacular ones at Saqqara, and once again we find a *ra'ad* depicted in a very realistic way along with other river fishes.

The mastaba burial complex of Mereruka (Meri) is a third site with bas-reliefs that seem to show this type of catfish (Duell, 1938). French archeologist and geologist Jacques de Morgan excavated this spectacular complex in the 1890s. It too is located in the necropolis of Saqqara. Mereruka served Teti, the first pharaoh of Dynasty VI, and was also his son-in-law. Mereruka had some 84 civil and religious titles, ranging from Chief Lector-Priest and Grand Vizier to Chief Judge and Scribe of the Divine Books.

This important burial place contains over 30 rooms, two thirds of which are for Mereruka, with the remainder for his wife (Her-watet-khet or Watet Hathor) and son (Meriteti). The walls have many hunting and fishing scenes, and they show numerous animals and fishes that can be identified with confidence. Even within the catfish grouping, several different species can be made out by such features as how they are swimming, their fin structures, body and head shapes, whiskers, and the like (Fig. 2.4).

Malapterurus is depicted in several places, including in the very first room, Chamber A-1. Here Mereruka and his wife are shown in a light papyrus boat spearing fishes in a marsh. Nearby, some men in two small boats are engaged in a hippopotamus hunt, while a variety of captured fishes are carried off in baskets elsewhere on the panorama (Duell, 1938, Plates 10 and 11). There are wonderful images in Chamber A-4 of fishermen using many different techniques to catch their quarries. Some are shown in a boat hauling in a large seine, others fish with hooks and line, and still others use hand nets and traps. Here too we can find good images of the *ra'ad* (Duell, 1938, Plates 42, 43, 45).

With reference to another site and time period, a catfish that seems to be *Malapterurus* can also be seen at the east end of the north wall of the chapel of the Tomb of Senbi. This tomb is located in the Necropolis of Meir, on the west bank of the Nile near the modern town of el-Qusiya. Senbi held the important positions of Nomarch (provincial ruler of the region) and "Overseer of Priests" for Amenemhat I, who lived from 1991 to 1962 B.C. and was a pharaoh associated with Dynasty XII of the Middle Kingdom (Blackman, 1914, Plate IV). This tomb drew attention in 1914, when the Egyptian Exploration Fund published an illustrated book about it. Containing scenes showing vase making, agricultural activities, Senbi hunting with his dogs in the desert, and other activities, it was excavated a few years later. What looks very much like a *ra'ad*, based on its shape and whiskers, can be seen swimming under a reed boat in one of these lifelike scenes.

In summary, realistic images of what is almost certainly *Malapterurus* can be found on the walls of some of the most famous ancient Egyptian tombs. They typically show this fish with other aquatic creatures in fishing, hunting, or

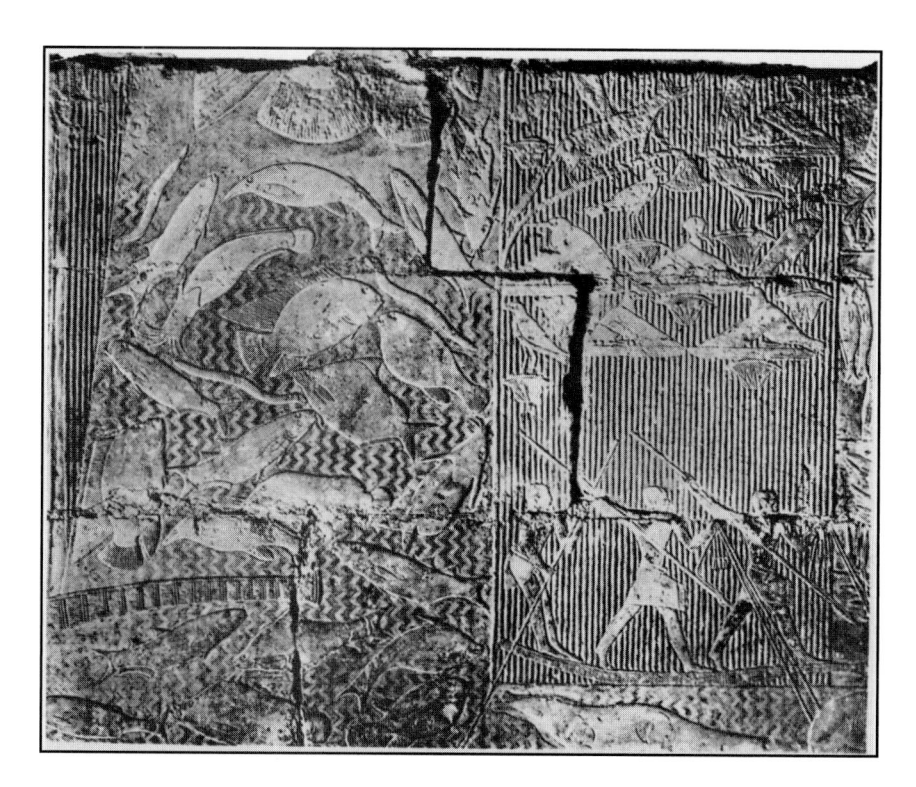

Figure 2.4. Various fish images and a hunting scene from the Tomb of Mereruka at Saqqara. *Malapterurus* can be identified on some of the panels on this tomb.

river scenes. Nevertheless, there are no accompanying explanations to explain why this type of catfish, or any other fish, might have been chosen for these scenes. As a result, scholars can do little more than make educated guesses, knowing the limited amount they do about the people living along the Nile thousands of years ago, including their diets and beliefs.

MAGIC, RELIGION, AND TOMB ART

It is often stated that these tomb illustrations depict everyday life along the Nile—scenes that would have been witnessed by Ti and others. It is also known that fishes formed an important part of the diet for the Ancient Egyptians, with the possible exceptions of the priesthood in certain places, and perhaps some royalty (Breasted, 1906-07; Brewer & Friedman, 1989; Darby et al., 1977; Nunn, 1996).

Without question, religion, magic, and mythology were important to the Egyptians. James Breasted, an American pioneer in the field of Egyptology, once wrote: "It is difficult for the modern mind to understand how completely the belief in magic penetrated the whole substance of [Egyptian] life.... It constituted the very atmosphere in which the men of the early oriental world lived" (Breasted, 1912, pp. 290–291). The Egyptians, Breasted asserted, believed that their gods could change the forces of nature and repel demon-caused diseases. They could also grant the worthy dead a future life akin to the one each had been living.

Today, it is widely accepted that the scenes they painted on the tomb walls would have reflected their religious beliefs. The thought is that these images served the deceased in various ways in the next life. From this religious or magical perspective, the fishing scenes in the Old and Middle Kingdom tombs could have represented a steady source of much-needed food to sustain the deceased and/or his or her servants in the next life. In this context, royal tombs and those of important others often contain tables upon which real offerings could be placed for the occupants' eternal life. In addition, the fish could have been drawn on the walls as a source of income for the deceased—as a way of providing them with the means to care for their tombs or for other goods and services into eternity (Duell, 1938).

These broad theories about how such tomb paintings might have served the deceased in the afterlife do not preclude the possibility that certain images signified very specific things to the superstitious Egyptians. As noted, the hippopotamus symbolized evil and chaos. In this context, we know that the Ancient Egyptians equated the yearly retreat of the Nile River with the creation of the world. As might be expected, when the river began to recede, some fish were left exposed in the shallow pools. Thus, with the yearly flooding and the greening of the earth that followed each year, some fishes became symbols of creation, renewal, and eternal life—important components of Egyptian life, mythology, and religious thinking (Hornung, 1982; Van Neer, 1986, 1989; Brewer & Friedman, 1989).

The Egyptians further believed different fishes symbolized different things, just as did land animals, and this finer distinction also entered into their art and artifacts (Caton-Thompson & Gardner, 1934; Wendorf & Schild, 1976; Wenke et al., 1983, 1988; Brewer, 1986; Hassan, 1986; Brewer & Friedman, 1989). Mullets, for instance, were the messengers of Hapy, the Egyptian flood god. As for mouth-breeding tilapia, they were associated with the creator god Atum, who took his seed into his mouth and then spat it out, thereby creating the world. With their vivid colors during the breeding season, these fish were closely associated with the sun, which the ancient Egyptians worshipped.

Catfishes had a place in this mythology. Since they were often encountered in muddy waters, it was believed that they could guide the solar disk (sun) and boats through the darkness of the underworld. Some New Kingdom tombs even contain depictions of catfish-headed human figures. These figures are thought to assist the god Aker, who oversees the nightly course of the great solar disk. Nevertheless, these particular images more closely resemble the catfish *Heterobranchus* than they do *Malapterurus* (Gamer-Wallert, 1970; also see below).

Within their broader mythology, catfishes also tended to be closely associated with Bastet (Bast, Ubasti), the sun goddess, who became an important national symbol at about 950 B.C., with her festival being among the most popular in ancient Egypt (Mercante, 1978). She is usually depicted as a woman with a cat's head. Bastet was Re's daughter, Re being yet another deity associated with the sun.

The myth of Osiris and how his evil brother Set (the god of the desert) dismembered his body to assume the throne of Egypt is among most popular of the ancient Egyptian myths. We mention it here because some variations of the story also have a catfish link. One version of this myth has Osiris' body cut into 14 pieces, which were then scattered or thrown into the Nile by Set, so he could not be made whole again. One of these parts was his penis, which the Egyptians believed was eaten by a fish (although Plutarch says a crocodile and another version says Set). The Nubians in particular, but not uniquely, specify a catfish, making it even more important in their mythology. Isis, Osiris' wife, goes on to find 13 of his parts and resurrects her husband *sans* his penis. In one version of the story the demilitarized deity is given a penis made of clay, while in another it is made of gold. In all cases, Osiris chooses to rule the underworld, leaving his earthly kingdom to his son Horus, the powerful falcon-headed god, whom Isis now somehow magically or immaculately conceives. Horus later goes on to defeat Set in an epic battle in which he loses an eye.

Whether *ra'ad* was associated with the same deities as the other catfishes, with one specific deity, or with any deities at all, is far from clear. Egyptologists are still looking for good evidence that would link the *ra'ad*'s unique powers to the supernatural—firm evidence that has been anything but forthcoming. Consequently, even though ancient fisherman must have known that a *ra'ad* could stun smaller fish at a distance, and numb the hands of those fishermen who mistakenly or otherwise dare to handle them, we still know very little from Egyptian art and artifacts about how these magical powers might have entered into their elaborate belief systems.

THE EBERS PAPYRUS

Few papyri of scientific or medical significance have survived from ancient Egypt. But among the most important are the

Edwin Smith Papyrus, which deals with the treatment of wounds, and the *Ebers Papyrus,* which provides over 800 simple or compound prescriptions from at least 40 different sources for various diseases and disorders.

Edwin Smith acquired both papyri in Luxor, once a part of the ancient city of Thebes, in 1862. He then sold the larger papyrus to Georg Ebers in 1872. Although both date from the period 1500–1600 B.C. and use a form of script called hieratic, these papyri are believed to be based on considerably older lost writings, as well as information passed on from generation to generation through oral tradition.

There is nothing in the *Edwin Smith Papyrus* about the medical use of the *ra'ad,* dead or alive (Breasted, 1930). The 68-foot *Ebers Papyrus* is, however, much more intriguing, although some translations can be very misleading. Most notably, in the often-consulted Bryan (1930/1974) translation into English, which is based on a widely criticized German translation, the reader will come across remedies calling for "electric eels," even though they are native to South America and nothing was known about electricity at the time of the papyrus! The call for an eel can be found under "Organic Remedies" (p. 35) and under "To Prevent Burn Wounds," where the reader is instructed to warm its head in oil and then apply it to a burned body part (p. 68).

A much better German translation of the *Ebers Papyrus* has recently been completed (Westendorf, 1999). In several organic remedies, the head, skull, or perhaps brain of a fish, noted in Egyptian hieratic as $n^c r$, is mixed with honey, oil, or fat to make a salve. Here the Egyptian word $n^c r$ is presented in German as *Wels,* meaning catfish in English.

This catfish salve is recommended for treating a sickness in half the head in Ebers No. 250. This might be a reference to *hemicrania continua,* a condition associated with daily, mild to moderate, unilateral headaches, which can exacerbate more painful migraines on the affected side. Whether "half the head" might (also) have signified front versus back headaches in the *Ebers Papyrus* is, however, uncertain.

The catfish salve is also recommended for leg or shin pain (Ebers No. 128), and for removing thorns embedded in the flesh (Ebers No. 730). But again, this salve does not call for a living catfish, only a dead one, and there is no wording calling specifically for a *ra'ad* to treat these painful conditions.

Prescription No. 304 is the most interesting in this regard. Here the remedy for treating what seems to be inflammations (*setet*) calls for a fish called *ddb* or *djedeb* in Egyptian. According to the translator, this word derives from a word meaning "sting." Thus, this recipe might have called for a *Malapterurus* (or perhaps a sea torpedo). The ambiguous *djedeb*'s head is again burned to ash, mixed with oil or fat, and applied to the flesh.

Hence, the prescriptions in the *Ebers Papyrus* do not call for the shocks of a living catfish. But what can we make of the fact that there are several remedies that require catfish heads? From what we know about the magical world of the ancient Egyptians, we cannot exclude the possibility that a part of a dead animal might have the same significance or properties as a living animal. In this context, we can only wonder whether the Egyptians were calling for the head of the unidentified stinging fish (*ddb*), because this fish could use its astonishing magical powers to repel dangers when alive, could cause numbness, or for some other reason related to its shocks.

NARMER'S CATFISH HIEROGLYPH

Hieroglyphic writing grew out of pictorial art. It functioned to convey messages that would outlast and extend beyond the spoken word. The term itself originated with early Greek travelers, who recognized that these pictorial images were employed for sacred (Greek = *hiera*) inscriptions that were sculpted (Greek = *gluphe* engraving, carving) onto Egyptian temples and religious monuments.

Egyptologists divide hieroglyphs into various types. Ideograms or sense-signs are pictures that can stand for whole words or concepts. In contrast, phonograms convey certain sounds. Silent determinatives, a third type of hieroglyph, are signs used to clarify the meanings of words that would otherwise be ambiguous. Anticipating Benjamin Franklin, who wrote in *Poor Richard's Almanack* for the year 1736 that "Fish and Visitors stink in 3 days" (Finger, 2006a, p. 29), the ancient Egyptians sometimes used a hieroglyph of a fish as a determinative—to signify a bad smell!

Stylized images of catfishes can be found among early Egyptian hieroglyphs. They date back to Narmer, who is thought by some to have preceded Menes (King Aha), living approximately 3150 B.C.

The Narmer plate or palette, which dates from 3200–3100 B.C., is the most famous archaeological finding showing catfish hieroglyphic images (Fig. 2.5). In 1898, British archaeologist James E. Quibell discovered this piece in Hierakonpolis (Greek for the ancient city of Nekhen), then an important settlement and possibly the capital of the Upper Kingdom. Some Egyptologists have interpreted the various images on the two sides of this greenish slate plate as showing how Narmer won a battle that helped bring the two kingdoms together (he wears two different crowns: a white one from Upper Egypt on the obverse and the red crown of Lower Egypt on the reverse). Others, however, think the images on the plate are a symbolic display of the king's power, or perhaps relate to a festival or to the year in which this object was made and presented to the temple.

There is more agreement about the *serekh* (a precursor of the cartouche for presenting a ruler's name) at the top of both the front and back of the plate. It depicts a catfish above a chisel, and these two images are believed to be phonetic, not ideographic, hieroglyphs. The catfish is read as $n'r$ or $n^c r$ and the chisel is read as $m^c r$ or mhr. Together, they represent the two-syllable name by which this king is now known: $n'r$-$m^c r$ or Narmer.

The catfish *serekh* also appears on a mace head and on potsherds found at Hierakonpolis. In addition, it can be observed on seal impressions from the Dynasty I tombs of the Kings Den and Ka at Abydos, where Narmer is presented as a predecessor, and where he might have also built a tomb (his own grave is believed to be at Umm el-Qaab, Abydos, where another of his *serekhs* has been found). There have also been Narmer *serekh* findings from other Egyptian sites, including Tel Ibrahmin Awad and Minshat Abu Ormar.

It is thought that, with the passage of time, a catfish alone began to signify Narmer's name—the chisel no longer being critical for the full phonetic connection—and that its image was eventually reduced to just a few lines. This sort of thinking has been applied to a shard presumably imported into the Land of Canaan (Israel) from Egypt some 5,000 years ago. It shows a stylized catfish in a *serekh* without

Figure 2.5. The front and back of the Narmer palette (c. 3200–3100 B.C.) with a drawing showing the catfish image (from the top center, both front and back sides) more clearly. (The palette is in the Egyptian Museum in Cairo; Dr. Bill Needle kindly provided the drawing.)

a chisel, which is nonetheless thought to signify Narmer (Levy et al., 1995).

We know that early Egyptian kings took the names of powerful animals, and Narmer might have been viewed as some sort of a "Catfish King," much like there was a predynastic Scorpion King, whose dreaded venomous sting could slay or repel enemies, both big and small. But whether a catfish was chosen because of its benumbing powers, because of its mythological association with the sun and its deities, or because some Nile catfish can be large and aggressive, is uncertain (Gamer-Wallert, 1970; Altenmüller, 1973).

A closely related problem is whether Narmer's name should be associated with a specific catfish or catfish in general. If in some way it were related to a *ra'ad*, the catfish symbol could have stemmed from the impressive ability of this fish to stop both the hunter and the hunted at a distance, stunning them into submission and, in the case of people, forcing them to drop their weapons and surrender.

Early in the twentieth century, some scholars thought that *Malapterurus* was, in fact, the specific type of catfish associated with Narmer's *serekh*. In one authoritative book, we read that this species of catfish was "Well-known to the ancient Egyptians, who have depicted it on their mural paintings and elsewhere." There is even an accompanying footnote specifying that it is "Represented on the great slate palette of Narmer" (see Boulenger, 1907, p. 398). In accord with these thoughts, there are Egyptologists who now contend that the chisel or *m'r* part of Narmer's name could mean "painful" or "stinging" (Helck & Westendorf, 1982; Westendorf, personal communication; see Gardiner, 1978, for more on hieroglyphs). If so, the king's symbolic name could be interpreted as "Stinging Catfish"—terms that would accurately describe *Malapterurus*.

Nevertheless, many Egyptologists are not willing to make this association. Further, even those who brought it up have their doubts, and emphasize that they are only raising a possibility without being in the least dogmatic (Helck & Westendorf, 1982; Westendorf, personal communication).

Also, many Egyptologists have opined that the image on the Narmer plate is a member of the common *Claridae* family, not the dreaded *Malapterurus* (Lortet & Gaillard, 1908, pp. 134–135; Gaillard, 1923, p. 78; Gamer-Wallert, 1970, p. 10; Brewer & Friedman, 1989, p. 63). Among the *Claridae*, attention has been drawn to the previously mentioned genus *Heterobranchus*, a large non-shocking catfish known for its dual dorsal fins, lack of a prominent dorsal fin spine, rounded caudal fin, and distinctive facial whiskers.

Hence, the idea that a stinging catfish is on Narmer's *serekh* is just an intriguing possibility. It is based on indirect evidence and reasoning by analogy, and cannot be regarded as a fact, although it makes for a good story. In reality, the images on the Narmer artifacts are so generalized that it is difficult to say anything more than that they depict a catfish. And because some images amount to nothing more than a few cut lines in a piece of clay, they might not even bring a catfish to mind, were it not for certain other findings against which these items have been compared (see Levy et al., 1995).

THE HIEROGLYPHICA OF HORAPOLLO

Over time, the Egyptians lost the ability to read their ancient hieroglyphs, and this led to speculation about what the various images meant. Early in the Christian era, a prevailing thought was that the ancient Egyptians had developed a writing system in which each image conveys a thought or a concept. As an example, a bee would symbolize a king, because it can make sweet honey yet also give a very painful sting when threatened.

The two *Hieroglyphica of Horapollo*, dating from the early Christian era, embrace this orientation and have drawn considerable scholarly attention. Horapollo (from *Horus*, the falcon-headed Egyptian god, and *Apollo*, the Greco-Latin god of oracles) Niliacus was a fifth-century A.D. priest who lived in Egypt, converted to Christianity, and wrote in Greek. The *Hieroglyphica* associated with his name consists of two parts or books, and contains "explanations" for some 189 Egyptian hieroglyphs. It is believed, however, that the more allegorical second book of the two was written or edited by another person, most likely another "Greek or

Alexandrian, whose thought and language were Greek" (Boaz, 1993, pp. 17–18).

Both volumes of the *Hieroglyphica* provide an amalgam of speculations about earlier Egyptian thinking based in part on the newer Greco-Roman philosophy that flourished particularly in the declining phase of the ancient culture. It included a religious-philosophical movement known as *Gnosis*, and other cultural trends that emerged in pagan culture and also in the Christian religion at its beginning (notably Hermetism, see Chapter 9). A variety of texts were produced within the framework, including lapidaries, bestiaries, and magic and medical books (Boaz, 1993; Grafton, 1993).

One of the assumptions of this cultural and religious orientation was that people could learn important lessons from the beasts, plants, and stones. This was because it was maintained that everything in the human world conveyed deep messages ultimately derived from God's wisdom, although they could be difficult to interpret. These themes were amply diffused. An echo of them can be found also in the writings of Aelian, who wrote about torpedoes and other animals in the third century A.D. (see Chapter 4). Within this perspective, understanding animal behavior was one of the keys for unlocking the secrets of the hieroglyphic code.

After the *Hieroglyphica* were reportedly found on the Greek island of Andros in 1422, Renaissance translators emphasized this moralistic orientation, adding to and altering the work so it would have greater appeal. The *Hieroglyphica* also created a fad, inspiring artists and scholars to produce hieroglyphic-like symbols to replace their own signatures. When John Pringle (1775a,b) of the Royal Society handed John Walsh the 1774 Copley Medal for his recent work on fish electricity, he even mentioned how "philosophers of the early ages" used these torpifying fishes as "an emblem, or an hieroglyphic." No one wanted to appear like a man with the head of an ass—the hieroglyph now signifying an ignorant person.

The great appeal of the *Hieroglyphica* during the Renaissance was intimately connected with some of the most characteristic aspects of the culture of the times. There was a strong interest in magic and in the idea of a superior form of learning and wisdom that is reserved for privileged individuals, not common people. Because of its depth and secretiveness, this wisdom had to be represented with ambiguous forms that only wise men or magicians could interpret. Due to its vague symbolism and multiplicity of possible significations, hieroglyphs seemed to be an ideal way to convey this magical and secretive wisdom.

Moreover, with explicit reference to Egypt, hieroglyphs exerted another strong attraction on the great minds of the Renaissance. It was then believed that a few great ancient civilizations had achieved superior wisdom and knowledge. In particular, the Egyptians were believed to have this primeval knowledge (*prisca sapientia*). This was the domain of the Egyptian priests and magicians, including the mythical and divine Hermes Trismegistus ("Thrice Great") (see Rossi, 1960, 2006; Yates, 1964: see Chapter 9).

Among the people captivated by Horapollo's *Hieroglyphica*, first printed in 1505 along with Aesop's *Fables*, was Holy Roman Emperor Maximilian. In 1500, he received a gift copy of the *Hieroglyphica* illustrated by the great German painter and printmaker Albrecht Dürer. All told,

more than 30 editions and printings of the *Hieroglyphica* came forth in just the sixteenth century.

In 1950, George Boaz translated a 1727 Greek and Latin edition of the *Hieroglyphica* into English. His translation with notes has been reprinted multiple times (e.g., Boaz, 1993) and historians still frequently cite it. In his translation of the second book, we are told that the author "felt the need of adding what has also been written before by others, though there is no explaining it." We then encounter 28 animals that bear on human nature, including several aquatic creatures. For instance: (a) the eel is said to represent a man hostile to everyone, because it does not live in the company of other fish, but in isolation; (b) the octopus signifies a person who has squandered necessities and superfluities badly, for it is wasteful with its food; and (c) oysters symbolize a man yoked to a woman, because they are born clinging to one another.

It is in this context that we find:

104. A MAN SAVING OTHERS FROM DROWNING. When they wish to indicate a man saving many others from drowning, they draw an electric ray. For when the ray sees many fish unable to swim, it attracts them to itself and saves them. (pp. 94–95)

There are obvious difficulties with the Boaz translation, one being the anachronistic adjective "electric." A second major problem pertains to the word "ray." The *Hieroglyphica* that had been discovered in the fifteenth century used the Greek word *nárkē*. In the ancient world, this word was used generically for any fish that could narcotize or numb. Thus, although *nárkē* could have signified the flat torpedo rays caught off the Egyptian coast, it could also have signified the narcotizing catfish, *Malapterurus*, found in the Nile. As we have seen, there are hieroglyphs of Nile catfishes, such as those on the Narmer palette. Yet there appear to be none of a torpedo ray, raising questions about the fish referred to by the author of the original *Hieroglyphica*.

In the original *Hieroglyphica* manuscripts that arrived in Florence in the fifteenth century there were no images, but they soon appeared in the editions of the work that were printed at a great pace after Gutenberg invented his press for mass printing. The images tend not to be realistic, being symbolic "emblems" of what the animal or the object signifies in a moral or religious sense, rather than attempts to reproduce it in a faithful way. Typically, the torpedo is represented in a way that suggests a different fish (Fig. 2.6).

But what about the underlying idea that this hieroglyph is about a "protector of fishes"—that it is symbolic of a kindly person who might save others from a watery death? To the best of our knowledge, this idea has yet to be found in ancient Egyptian writings. This suggests that the author might have used what he had garnered elsewhere about saltwater rays, the subject of many stories that circulated in Rome and throughout the Empire closer to his own time (see Chapters 3 and 4).

In particular, it had long been held that surprised fishermen would drop what they were holding when benumbed by a torpedo through a wet net, allowing all of their captured fishes to swim to freedom. Of course, a similar if not stronger effect could occur with *Malapterurus*, with discharges three or four times as powerful as the small

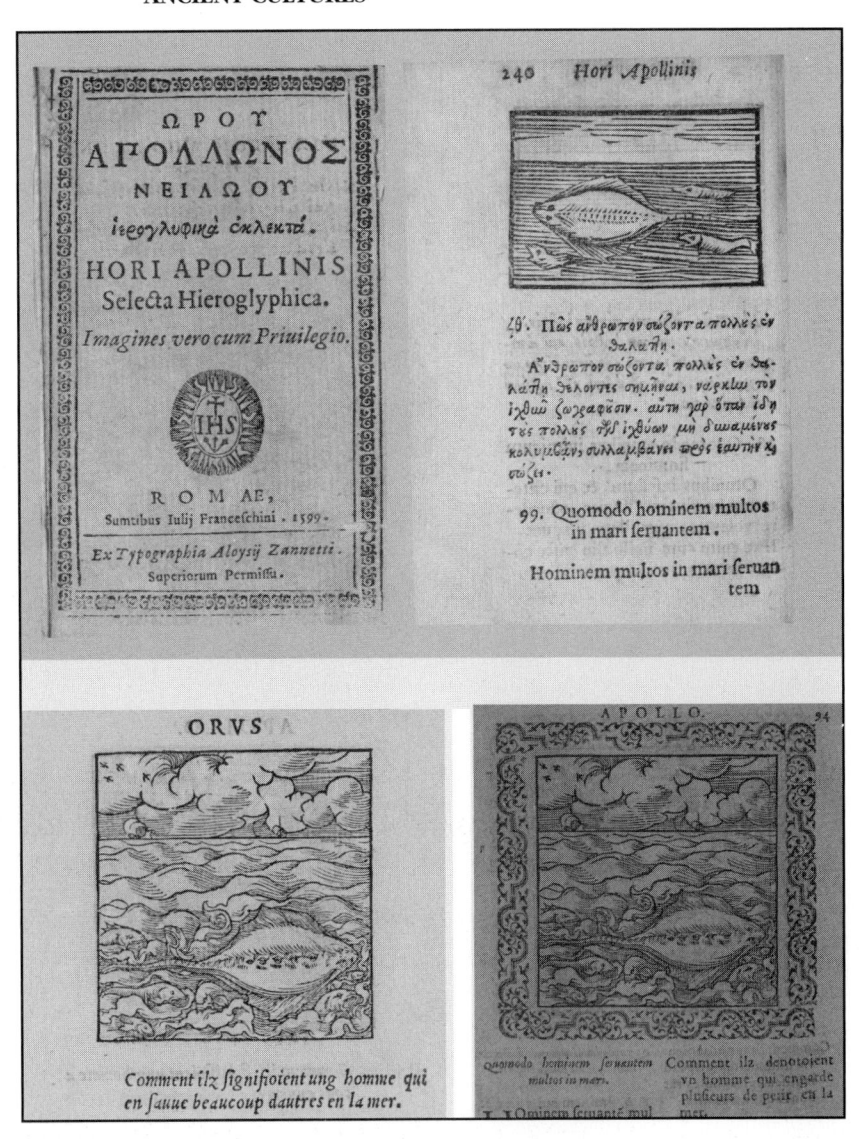

Figure 2.6. Top: The title page of a 1599 abridged Greek and Latin edition of Horapollo's *Hieroglyphica*, with a non-realistic depiction of the torpedo. Bottom: Similar non-realistic images of the torpedo from two French editions of the work, printed in 1543 (on the left) and in 1574 (on the right).

Mediterranean torpedo rays. In fact, Pietro Valeriano (1555, pp. 212–213) was among the Renaissance scholars to write that this is how the torpedo got its reputation among Egyptian priests for being a protector of smaller fish. He did this in a book also titled *Hieroglyphica*, a major work of Renaissance culture that circulated widely across Europe (Figs. 2.7 and 2.8).

Yet whether the catfish legend started with a catfish or with a ray bearing the same ambiguous name is unclear. Just where and when the legend originated are also unknown. Hence, the legend might not have originated in ancient Egypt and might have been constructed by later writers, as was likely the depicted hieroglyph itself. An alternative explanation for the legend is that ancient Greeks and Romans gazing into the shallow coastal waters might have seen torpedoes swimming among stunned smaller fish. Again, the same thing could apply to fisherman seeing a *Malapterurus* swimming among smaller immobilized fishes in the Nile River. If the observers missed seeing the larger fish subsequently eating the stunned smaller ones, this would have conveyed the image of a big fish saving smaller ones unable to swim from harm.

Yet another possibility has to do with an observation reported by Aristotle or perhaps his follower Theophrastos, and repeated by others after them, including Pliny the Elder (see Chapters 3 and 4). In the Aristotelian zoological treatise *Historia animalium,* it is stated that a large torpedo had been observed with about 80 embryos in it (see Chapter 3). The sea torpedo is ovoviviparous, and finding the young in its body might also have conveyed the mistaken idea that it was protecting smaller fish.

All told, it is hard to conclude very much from the *Hieroglyphica of Horapollo,* with the copy discovered not providing an image of the hieroglyph or a physical description of the fish. What is written about the *nárkē* that saves other fish could pertain to the sea torpedo, which could be caught off Egypt's Mediterranean coast and could swim into the mouth of the Nile, but does not appear on known hieroglyphs, or to *Malapterurus,* which does appear on tomb drawings and perhaps is represented in a stylized way on Narmer's seal. It is also entirely possible that the original author of this cryptic passage melded information about two different fishes into one, focused on providing a moral lesson to us all.

Figure 2.7. The frontispiece and the initial dedicatory page of the *Hieroglyphica* of Valeriano from a Latin edition printed in Basel in 1556. The work was dedicated to the Lord of Tuscany, Cosimo I de' Medici, who was greatly interested in the Hermetic aspects of Renaissance culture.

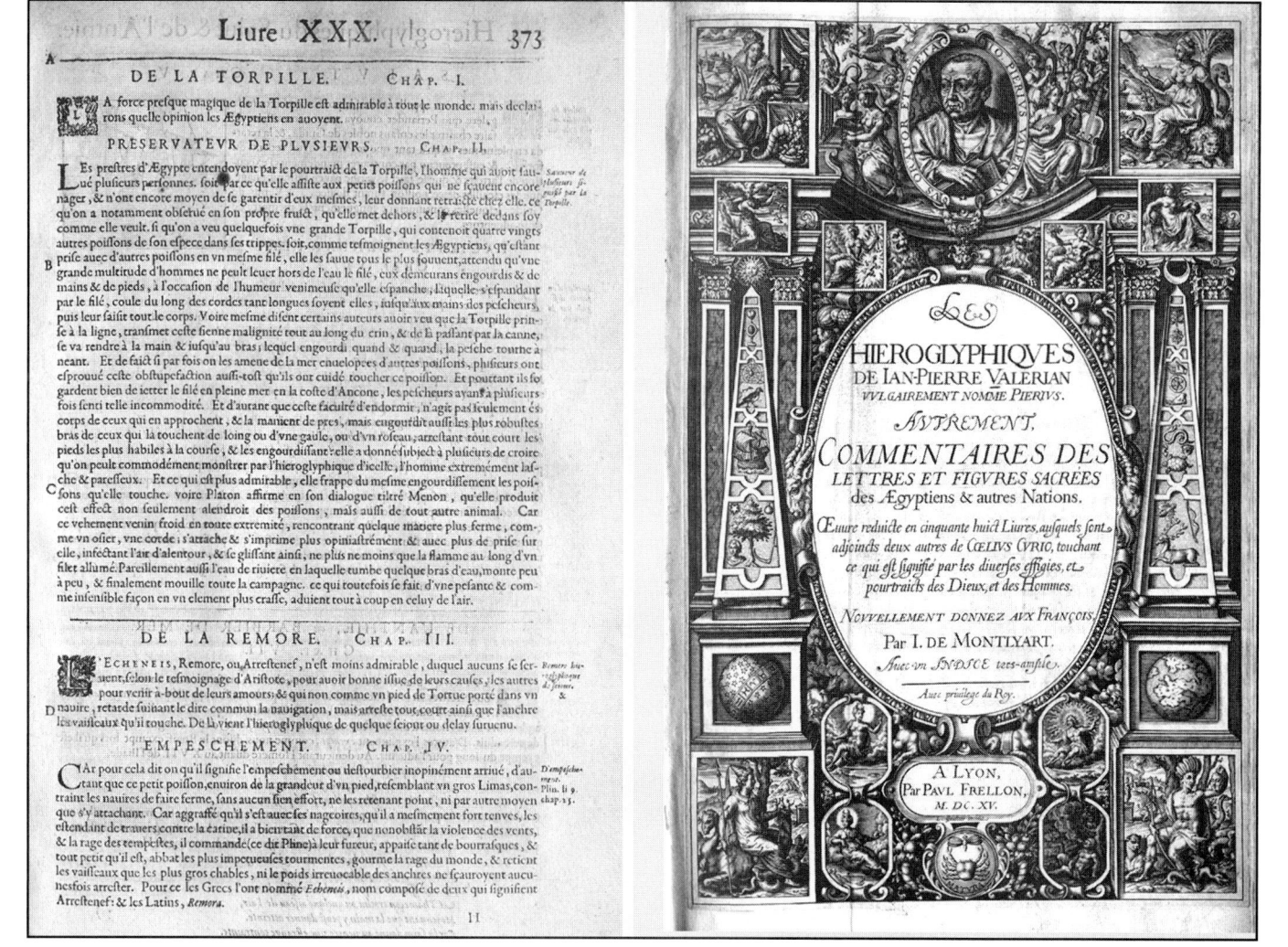

Figure 2.8. The frontispiece of a 1615 French edition of Valeriano's *Hieroglyphica* (right) with, on the left, his passage about the torpedo legend (from Valeriano, 1615).

MUMMIES AND SACRED ANIMALS

The Egyptians had many sacred animals (e.g., falcons, cats), and they even had some sacred fish. Mullets, for example, were venerated at Elephantine (Radcliffe, 1974). These sacred animals were sometimes mummified and placed in tombs of the powerful and wealthy.

Although some large and impressive catfish mummies have been unearthed, they are of another species (*Heterobranchus bidorsalis*). Some smaller catfish mummies have also been discovered, including an exceptionally well-preserved one at the Rosicrucian Museum in San Jose, California, but its species has yet to be identified. Thus, there is no hard evidence to suggest that *Malapterurus* was as sacred as, or more sacred than, other catfishes to the ancient Egyptians.

Similarly, catfishes have now been found listed on some late period temple inscriptions as being among a select few taboo fishes (Brewer & Friedman, 1989, p. 17). But here too the *ra'ad* is not singled out within the grouping.

Extensive excavations at Tell el-Amara, between Memphis and Thebes, have now revealed the remains of thousands of fishes eaten over the ages, including tilapia and several types of catfishes (Luff & Bailey, 2000). These remains are interesting because they suggest that *Malapterurus* might not have been a part of the normal diet here, at least for some time. This finding, however, could relate to several factors above and beyond religious taboos, including fishermen not wanting to handle it and its relative scarcity compared to other fishes and even certain catfishes.

In fact, how little we know about *Malapterurus* stands in contrast to how much we now know about another Nile catfish, *Synodontis*. This more common catfish was caught and eaten in great numbers, serving as a reliable source of high-quality food for the masses (Luff & Bailey, 2000). It also seemed to have had a place in medicine, at least as seen by some scholars (see Gammer-Wallert, 1970, p. 69; Darby, Ghalioungui, & Grivetti, 1977, pp. 379–380).

In the next two chapters, we shall travel across the Mediterranean Sea to southern Europe, where torpedo rays were described in several important treatises, some dating back to the fifth century B.C. With the Greeks and Romans, we can and shall go well beyond what can be gleaned about electric fish from the Egyptians, whose art and artifacts are a challenge to understand, and whose telegraphic writings and individual hieroglyphs can also be very difficult to comprehend.

Beginning with the ancient Greeks, we shall find good descriptions of how sea torpedoes behave, their numbing powers, and how they were used as easily digested food for the sick. We shall also see that their shocks were used therapeutically during the Roman era, and that ancient scholars tried to explain their baffling effects, including their astonishing ability to torpify even without direct contact.

Chapter 3
Torpedoes in the Greco-Roman World: Pt. 1. Wonders of Nature Between Science and Myth

> Would it not have been quite sufficient only to cite the instance of the torpedo, another habitant also of the sea, as a manifestation of the mighty powers of Nature? From a distance, from a considerable distance even, and if only touched with the end of a spear or staff, this fish has the property of benumbing even the most vigorous arm, and of riveting the feet of the runner, however swift he may be in the race.
>
> Pliny the Elder, 1st century A.D.
> (*Naturalis historia*, Book XXXII, 2;
> Trans. in Pliny the Elder, 1857, Vol. 6, p. 4)

The ancient Greeks and other people living along the coast of the warm Mediterranean Sea and its branches were very familiar with torpedoes, "a mean and groveling animal armed with lightning, that awful and celestial fire, revered by the ancients as the peculiar attribute of the father of their gods" (Fig. 3.1). This was the way they were described in 1774 by Sir John Pringle, then President of the Royal Society of London, when he awarded John Walsh the Copley Medal for his research on the electrical properties of the shocks of the torpedoes he had studied on a trip to France (Walsh, 1773; see the last line in Pringle, 1775a,b). In his discourse, Pringle (and later Humboldt, in the narrative of his voyage to South America) mentioned this fact and also alluded to the wide distribution of these flat fishes.

In this chapter, we shall present and discuss how torpedoes are presented in the texts of some of the most important authors from classical antiquity, in particular Hippocrates, Plato, Aristotle-Theophrastus, and Plutarch from the Greek epoch, and Pliny the Elder from the Roman era. Specifically, we shall look at how torpedoes were first used as a food for the sick, how they began to be associated with immobility in literature, what early natural philosophers wrote about them, and how they became the material of myths and legends. In addition to Galen, about whom more will be said in the next chapter, the authors to be mentioned here had very large followings, and they contributed in significant ways to how these fishes would be perceived later on, notably during the Middle Ages and Renaissance. Their enduring and colorful legacies would be substantially modified only with the emergence of modern science, particularly as a consequence of Walsh's experimental endeavors, together with those of his colleagues at the Royal Society of London, during the 1770s.

Some matters must be clarified at the outset. The first is that designating authorships can be confusing for some of the texts discussed in the present chapter, this being a common thread running through the literature prior to the advent of the printing press (Johannes Guttenberg's press dates from the 1430s). This is particularly true when we examine the collected material loosely attributed to Hippocrates and Aristotle. Without question, these collections include pieces not written by these better-known and highly revered authors, whose names alone would sell more texts, as even vendors in Greco-Roman antiquity knew. Rather, they reflect the contributions of various members of their philosophical schools and, with names omitted, at times it can be hard to distinguish between masters and followers. This is why we linked Aristotle and Theophrastus together in the preceding paragraph, and why we shall frequently use generic terms, such as "Hippocratic" and "Aristotelian," throughout this book.

A second issue to consider is that all of the ancient texts that have come down to us in manuscript form can reflect the actions of generations of copyists, commentators, and translators. These actions and the general transmission process could have produced modifications and alterations of the original texts, some minor but others significant. Not being able to identify these changes represents another difficulty and more uncertainty in the interpretation of these documents.

Our third important and related point is that relatively few texts from classical antiquity have survived the ravages of time, with significantly fewer surviving in their entirety. We can do no more than guess what might have been stored along with the known Hippocratic writings in the lost library of Alexandria, or what might have been lost when a large number of Galen's manuscripts perished during a fire in Rome in his own lifetime.

Hence, it is important to remember that in our treatments of torpedoes (or *Malapterurus*) in antiquity, our sampling must, by default alone, represent only a percentage of what

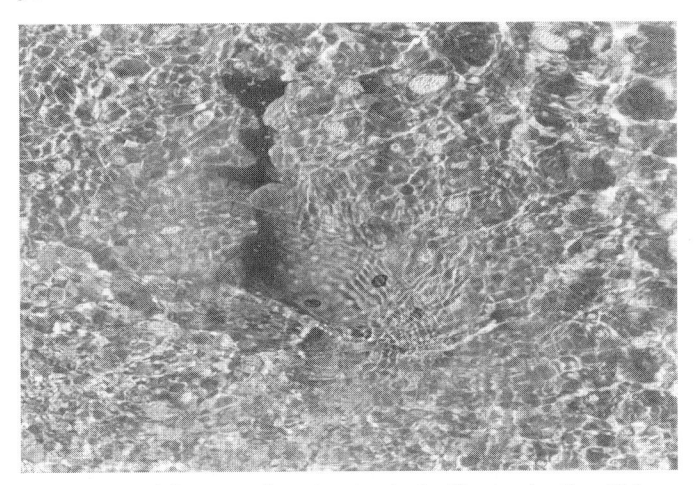

Figure 3.1. A live torpedo swimming in the Tyrrhenian Sea. This fish is the spotted type, *Torpedo oculata* (or *Torpedo torpedo* according to current zoological nomenclature), characterized by the presence on its back of prominent blue spots (normally five) surrounded by black rings simulating the eyes.

had actually been written about these fishes, and that much is based on fragments or parts of projects, and attributions by others. With the original documents long lost, with full context denied, with the editing and personal agendas of those who followed, not to mention faulty translations, this literature has significant challenges, to say the least. Tying the origins of some ideas to specific authors with any suggestion of certainty can, indeed, be dangerous.

This said, we shall concentrate on two themes concerning torpedoes in this and the next chapter. One is the use of torpedoes for treating various medical disorders in Greco-Roman times. The other involves various hypotheses put forward to account for the shocks and their conduction through various media, including water and rods. These themes are related, and they represent the two main aspects of classic culture in which some writers from antiquity, within their own frameworks, came near to anticipating some of the achievements associated with modern medicine and science.

TORPEDOES AS SLEEPERS AND TREMBLERS

The Greeks must have known about several types of electric rays, although extant texts treat them as if they constituted a single kind of fish (today ichthyologists recognize about 60 species of electric rays, belonging to 10 genera, with numbers still growing; see Carvalho & Randall, 2003). They were usually described as νάρκη, which can be transliterated into *nárkē* (or less commonly *narce*), from which several modern words have been derived—three of the more common ones are "narcotic," "narcosis," and "narcotize." The Greek word was based on the fish's ability to cause numbness.[1]

Torpedo is the Latin term for these rays, *torpille* is the French, and *torpedine* is the Italian (and Spanish) term. All convey an image of numbness or torpor. One of the popular English equivalents for torpedo, namely numb-fish (or numb-ray), has this same meaning, as is also true for the French words *dourmillouse* or *dormilieuse* (or the similar *durmigliousa*, used in Genoa), which relate to the word sleep. The same meaning is present even the distant Chinese ideographic designation, 痹仔, meaning "little thing that causes sleep (or numbness)," and the similar Japanese designation 痺鱏 (シビレエイ in the syllabic or *kana* version, to be read as *shibire-ei*), which means sleeper (or numbing) ray.[2]

Not only are these rays able to cause numbness and loss of sensation, they can also cause involuntary muscle contractions, and these effects are also reflected in the nomenclature surrounding them. In English, cramp-ray and cramp-fish (crampfish) are common names for members of the torpedo family, and they correspond with the Italian *crampo* (used in Bologna) and *sgranfo* (used in Venice). More than just cramps, they can also cause muscular tremors or "vibrations," and this is reflected in another lengthy list of descriptive terms for these fishes. This list includes the French words *tremble* or *trembleur*, the Spanish *tremblador, tremulosa, tremielga* (with their numerous variations), and the Portuguese *tremedeira* and *tremelga*. For those speaking or writing in German, we have *Zitterrochen*, among the Dutch, *Sidderrog*, and in Italian, *tremola* and its numerous varieties (see Piccolino, 2005a).

The early literature on the torpedo can be confusing because, as already mentioned, the Greek and Roman writers, and those who followed them, generally used the same term for both the torpedo and the torporific African catfish (for more on this subject, see Thomson, 1928, 1947, pp. 169–172). After all, both fishes have numbing powers—that is, both can narcotize. And although some ancient writers included

---

[1] On the basis of an analysis of ancient Greek texts, Armelle Debru (2006, 2008) has provided evidence that the fish name derives from a term used to indicate the sensation (and not the reverse). The word νάρκη, with the meaning of numbness, can be found in older authors (it appears for instance in Homer's *Iliad*, dating from around the eighth century B.C.), whereas the first available occurrences of νάρκη as a fish dates from the fifth to fourth century B.C.

[2] There are popular names, especially for the torpedo, that do not allude to either the numbing or to the trembling actions consequent to the fish shock. One of the most curious, frequently listed in the zoological compilations of the Renaissance as being used in Rome, is *battipotta* (sometimes spelled *battipota* or *battiporta*; see Chapter 6). By a curious series of circumstances we have been able to establish the very likely origin of the term (Piccolino, 2010). The first part of the word derives from the verb *battere*, meaning "to hit," "to strike," and so forth. The second part of the word, *potta*, is a popular (and vulgar) name used to indicate the female sex, especially in the central regions of Italy (and particularly in the region of Rome, the Latium and in Tuscany). At least in Tuscany, the expression *battipotta* is still used in some places (for instance in Leghorn) to indicate the hidden pocket that working-class women used to have on the inner face of their waist apron, in order to keep money and wallets safe. It was particularly used during World War II to hide black-market items. Due to its position, the pocket may, at least in principle, hit the genital region of the woman bearing the apron. This accounts for the term *battipotta*. Because of its flat and round shape, the torpedo resembles to the *battipotta* pocket, and this can explain why it was called by this name in Rome. To support our interpretation there is also another expression, *fotterigia* ot, also listed among the names used for the torpedo in Rome (see for instance Giovio, 1524; Pier Andrea Mattioli, 1565). This term relates to the verb *fottere*, a vulgar expression to indicate sexual intercourse (equivalent to "to fuck" in English). We are indebted to Mario Cardinali, director of the *Vernacoliere* of Leghorn for linguistic assistance in looking for the derivation of *battipotta*. *Battipotta* is no longer used to denote the fish in Rome, the common term being now *tremola*. An interesting Italian popular name used in a very limited zone of the sea coast at the south of Rome, from Terracina to Formia, is *treppina* or *trippina*. As we shall discuss in Chapter 6, this is a linguistic archeological relic of the term used by low-class people in the Latin world. A rather wide list of popular names of the torpedo can be found in Rolland, 1919, vol. XI, pp. 164–167.

the word for "sea" in the fish's name, θαλαττία *(thalattía)* or θαλασσία *(thalassía)*, in order to distinguish the saltwater torpedo from the river catfish νάρκη ποταμία *(nárkē potamía*, meaning river torpedo), this was not always the case.

As might be expected, we encounter the phrase *nárkē thalattia* (and its variations) much more frequently than *nárkē potamía* in these texts, because the ancient Greco-Roman authors were more familiar with sea torpedoes than with torporific African catfishes. It should be noted, moreover, that because of the uniqueness of the term employed to describe both the numbing fish and numbness as a symptom, some authors felt obliged to use additional or other specifications in order to avoid ambiguities. They wrote, for instance, something like "*nárkē* the animal" or "*nárkē* the fish" to designate the shocking fish. As we shall see in the next chapter, this is the case in Galen's writings, among others.

Thus, the historian must look carefully at the wording, the description, and its full context before making a judgment about the species of fish in the ancient writings. Faced with ambiguities of this sort, this will how we shall proceed, beginning with the writings of the Hippocratic physicians, Plato, and the Aristotelians, who very clearly had the flat sea torpedo in mind.

THE *CORPUS HIPPOCRATICUM* AND TORPEDOES AS NOURISHMENT AND MEDICINE

Hippocrates was born in 460 B.C. on Cos, a small island off the Doric coast of modern Turkey, just as ancient Greece was entering the celebrated Golden Age (Fig. 3.2). He began to study medicine under the tutelage of his father, but then moved to Athens, the cultural center of the Greek city-states. After a plague broke out in Athens and the Peloponnesian War between Athens and Sparta began, he set forth to teach, practice, and write about medicine in other locations. Details about his life are sketchy, although there is agreement that he died near Larissa, a city in Thessaly in what is now Turkey. The exact date of his death is uncertain, but it is thought to be about 370 B.C.

The type of medicine associated with Hippocrates is secular, not religious (Sigerist, 1961; Temkin, 1991; Jouanna, 1999; Finger, 2000, pp. 21–37). He taught that the best physicians are skilled craftsman guided by nature, not religiously inspired priests. By his time, some Greek philosophers, following in the footsteps of Thales of Miletos, Pythagoras, and Empedocles, had begun to study natural phenomena as understandable events. As would be true of other investigators for a very long time, these first investigators of nature were not scientists in the modern sense of the word, but philosophers in a broad sense.

The view that dominated the Greek and Roman culture and persisted through Middle Ages into the Renaissance was that everything is composed of four basic elements: air, earth, fire, and water. It was also believed that the body is composed of four "humours," meaning fluids associated with these elements: blood, black bile, yellow bile, and phlegm. These four humors, in turn, had different qualities: they could be wet or dry and hot or cold. The job of the physician was to keep the humors in balance (*eucrasia*) and to

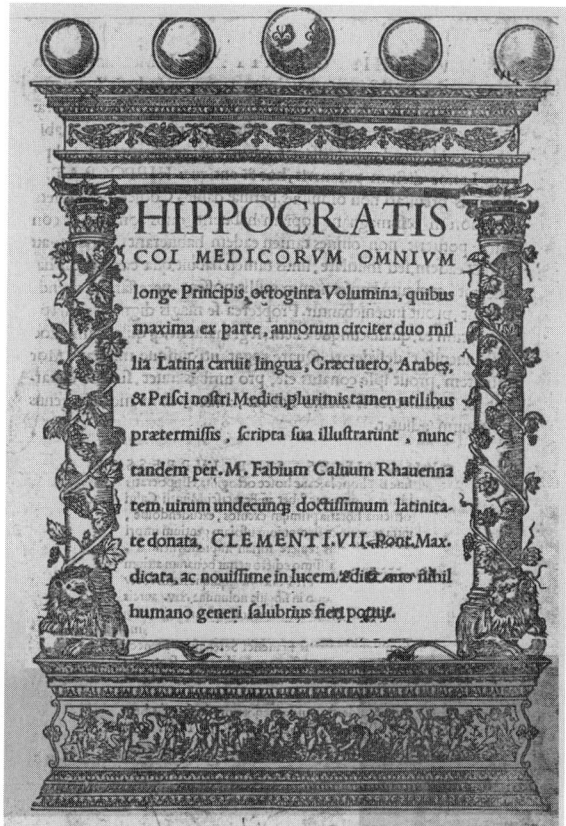

Figure 3.2. A first-century A.D. coin from the island of Cos depicting the head of Hippocrates (c. 460–377 B.C.) and the title page of a sixteenth-century edition of the *Corpus Hippocraticum* (Hippocrates, 1525).

restore the balance when it was upset (*dyscrasia*). By following their directives, good health could be maintained and, if necessary, restored.[3]

For some two millennia, the restoration of a perceived imbalance largely involved treating with opposites, which made intuitive sense. Hence, a person with chills would be warmed, while one with a raging fever would be cooled. Medicine also involved decreasing or expelling a humour that was believed to have accumulated to too high a level. Thus, for a person with a fever and a flushed appearance, bloodletting was a treatment of choice. In contrast, for deficiencies, such as a pale appearance signifying inadequate blood, a food rich in this needed humour, as exemplified by blood-rich liver, would have been a recommended, perfectly logical remedy.

The Hippocratic physicians primarily prescribed rest, proper exercise, and environmental changes to maintain health and to help sick patients. As noted, they also had great faith in dietary changes. These remedies characterize the *Corpus Hippocraticum*, a collection of medical tracts, manuals, speeches, notes, and fragments dating from approximately 450 to 350 B.C.

Deciding exactly which items came from Hippocrates as opposed to his followers has generated considerable debate. Indeed, neither Hippocrates nor his followers signed any of these works, which were collected around 300 B.C. for the new library in Alexandria, Egypt, intended to be the most complete storehouse for all forms of knowledge anywhere the world (Canfora, 1990).

The *nárkē* has a place in the *Corpus Hippocraticum*. It appears in the second of three treatises called *Regimen* (Περὶ διαίτης or in Latin, *De victus ratione*). These treatises deal with diet and exercise, favorite themes of the Hippocratic healers, who believed that the physician should begin his treatments with those remedies least likely to cause harm. Thus, the first step was to try dietary changes and to balance rest with exercise. Medications, these healers maintained, should be tried only if diet, rest, and exercise fail. Physical interventions for diseases were very much last resorts.

How to regulate the diet so there would be neither deficiencies nor excesses (called *plethoras*) was the main issue addressed in these treatises. A secondary but related issue was the need for personal control (Jones, 1979, pp. xxxviii–lv.). This interventionism stands in marked contrast to an earlier, more fatalistic orientation, in which it was assumed that diseases would naturally run their courses.

In contrast to *Regimen I,* which is more concerned with the elemental nature of man, *Regimen II* looks at the characteristics of various foods, both plant and animal (Jones, 1979, pp. 297–365). Galen, the second-century physician who found praising others difficult, questioned whether *Regimen I* was even written by Hippocrates (Galen, 1969, p. 103). But he considered the second book worthy of the "Father of Medicine" himself, contending: "My opinion—and I swear to it by God—is that Hippocrates has made this account of his so clear and obvious that not even a child, much less anyone else would find it obscure" (Galen, 1969, pp. 87).

Of the 53 species of animals mentioned in this second tract, more than one third are fishes. One, mentioned in section XLVIII, bears the name νάρκη and almost assuredly refers to the sea torpedo, because it is inserted in a discussion about sea fishes. This may be the earliest, unambiguous reference that we currently have in writing to the torpedo, even though the discussion is only about its nutritive value and not its narcotizing shocks.

In translation, we read:

> The fishes that frequent stony places are almost all light, as are the thrush fish, the hake, the gudgeon and elephitis. These are lighter than those that move from place to place, for those remaining quiet have a rare and light flesh, but those that wander and are wave-tossed have a more solid and deeper flesh, being much battered by the toil. The torpedoes, skates, turbots and such-like them are light. All those fish that feed in muddy and marshy places, such as the mullet, cestreus, eels, and the like are heavier, because they feed upon muddy water and other things which grow therein. (Jones, 1979, p. 321; here revised)

This short passage leaves no doubt that the flesh of the *nárkē,* being soft and easily digested, was eaten in ancient Greece. Moreover, the Hippocratic physicians, who abhorred heavy regimens, recommended them for preserving health and treating certain disorders. For example, they called for the tender flesh of a boiled torpedo to be served at breakfast for the disorder known as φθίσις (*phthísis*), meaning consumption, a progressive wasting away. It was also used for other disorders associated with poor digestion and declining fitness. It was even recommended for gynecological diseases, along with other sea fishes (e.g., skates, scorpion fish, congers, eels, turbots, gobies).

This is not to say that the torpedo was the favorite food of the Hippocratic healers. They usually recommended barley gruel, because it is extremely nutritious and easily digested. Galen pointed this out several hundred years later in his essay, *On Regimen in Acute Diseases in Accordance with the Theories of Hippocrates* (Galen, 1969). He questioned whether the Hippocratic physicians might have overdone the gruel diet at the expense of more nourishing but still easily digested foods, praising torpedoes in this context.

The Hippocratic physicians do not mention the shocking powers of the torpedo in their surviving writings. This omission, barring the appearance of new evidence, would strongly suggest that torpedo discharges were not yet a part of mainstream medicine. The call for live torpedoes in treatment regimens would go out during the Roman era, as we shall see in Chapter 4.

As noted by Armelle Debru (2006, 2008), despite the absence of any mention of shock of the νάρκη fish in the *Corpus Hippocraticum*, there are ample discussions of νάρκη as a symptom, meaning torpor and numbness, reduction of movement, and diminished sensitivity. In particular, the

---

[3] Although ancient medicine was dominated by the theory of humours and qualities, the idea of the four humors, four qualities, and four elements being associated with many aspects of nature and life crystallized particularly during the transmission of ancient medicine through the Middle Ages. The theories present in the Hippocratic and Galenic corpus may not strictly correspond to the four-humour conception.

Hippocratic physicians associated the numbness with the physiological and pathological effects of cold, an idea that Galen would elaborate on and relate to nerve function (see Chapter 4).

PLATO'S SOCRATES–TORPEDO ANALOGY

Plato was born to an aristocratic family in Athens around 428 B.C., just before Hippocrates died, and he was originally named Aristocles. He received the nickname *Platon* ("broad") while at school because he had wide shoulders. It is likely that he became familiar with the great philosopher Socrates while he was young. His intellectual mentor's execution in 399 B.C. (for corrupting the youth of Athens) had a profound effect on him, leading him to travel, to a life of contemplation, and to dedicate himself to philosophy.

Plato's school of learning has been called the Academy because it was founded at Akedemia (in about 385 B.C.), a sanctuary of Athena, the goddess of wisdom. Here, north of Athens, Plato prepared people for the ruling class, enabling them to answer questions posed by others. As was true of Socrates, Plato was strongly interested in moral aspects of philosophy, which he taught using the dialectic method.

Plato died in 348 B.C., and his *Dialogues* remain among the most widely read works in Western philosophy. But because he viewed natural philosophy as an inferior sort of knowledge, what he wrote about the torpedo added little to the corpus of scientific knowledge. Still, it does have importance because it reveals that the torpedo's powers were known in his day, and likely very well known, being very different from those of the other fishes that were being sold in the markets for food.

Thus we find Plato bringing up the torpedo in his *Meno*, a dialogue on the meaning of virtue and whether this admirable trait could be taught. In terms of the larger picture, the *Meno* addresses the origin of ideas and reflects Plato's belief that reason will lead to the highest intellectual insights. The main characters in this dialogue, date unknown, are Socrates and Meno.

In one of their discussions, Meno, who seems ignorant of Socrates' didactic style of teaching, complains that he personally has no trouble discussing virtue with others, but that his thoughts desert him and he gets perplexed in Socrates' presence. Socrates, who is now compared to a torpedo ray, has benumbed his mind! Specifically, after being asked yet again by Socrates how he defines virtue, Meno answers:

> O Socrates, I used to be told, before I knew you, that you were always doubting yourself and making others doubt; and now you are casting over me your spells and magic potions and words, and I am simply getting full of doubts. And if I may venture to make a jest upon you, you seem to me both in your appearance and in your power over others to be like flat sea torpedo (νάρκη τῇ θαλαττίᾳ), who torpifies those who come near him and touch him, as you have now torpified me, I think. For I am benumbed in my soul and my mouth, and I do not know how to answer you; and though I have been delivered of an infinite variety of speeches about virtue before now, and to many persons—and very good ones they were, as I thought—at this moment I cannot even say what virtue is. And I think that you are very wise in not voyaging and going away from home, for if you did in other places as you do in Athens, you would be cast into prison as a magician.[4] (Translation in Plato, 1892, p. 39, sect. 80; here revised)

Plato's passage is notable for two reasons. First, it is the earliest text we came across alluding to the effects of the torpedo's shocks. And second, it is notable because magic is mentioned in this context. Specifically, it brings up spells, bewitchments, and enchantments in its description of Socrates' powers, here presented as akin to those of a torpedo. This wording reminds us that magic was still a very important part of ancient Greek life, despite the trend toward rationalism shown by many intellectuals. Magic in the Greek world (and in the Roman) included a myriad of gods with fantastic powers, fortune-tellers, diviners, the use of charms, witchcraft, necromancy, lucky and unlucky days, and the like (Thorndike, 1923). In religion and mythology, both animals and herbs were thought to possess specific magical properties, and occult "science" can be found in a number of Plato's other writings, most notably in his *Timaeus*, an influential text that even contributed to the renewed interest in magic during the Renaissance.

Plato's description of Socrates' torpedo-like powers was written when torpedoes were increasingly gracing beautiful pieces of Greek pottery, providing further evidence that torpedoes were well known throughout mainland Greece and the Greek colonies at this time. In the next chapter, we shall look at some of these pieces of pottery and discuss their uses.

ARISTOTELIAN NATURAL PHILOSOPHY

In contrast to Plato, the Aristotelians focused more on natural philosophy, and especially on what we would today call biology and the life sciences. Aristotle was, in fact, the greatest natural philosopher of antiquity. Born in Stagira (Thrace) in 374 B.C., he went to study at Plato's Academy while still in his teens, and he remained in Athens for nearly 20 years. He then served Philip of Macedonia as his agent at Assos (in Asia Minor), where he might have set up a school for scholars. He met Theophrastus of Eresos either at this school or on Lesbos, where he fled when the Persians captured Assos. He then spent about 7 years in Pella doing diplomatic work for Phillip, where Theophrastus might well have joined him.

Alexander the Great ascended to the throne when Phillip was assassinated in 336 B.C., and he set forth to consolidate his father's growing but rebellious empire. The era starting with the defeat of the Persians in 334 B.C., and ending when Rome conquered Egypt in 30 B.C., is called the Hellenistic period, from the ancient Greek word *Hellenes*, the term by

---

[4] The similarity between Socrates and the flat torpedo, which Meno alluded to in a fairly humorous way, also could have had a physical connection. This is because the philosopher had a snubbed nose, something pointed out in many descriptions from his time and in Plato's *Dialogues*.

which these people described themselves. It was with the ascent of Alexander that Aristotle returned to Athens, where he started his Lyceum, a school for administrators to serve their new ruler. Here he philosophized, wrote, and taught for about 12 years. After Alexander died, and afraid of being put on trial as a Macedonian spy, he fled Athens yet again, 322 B.C. being the year of his death.

Intrigued by all facets of natural philosophy, Aristotle studied zoology, anatomy, physiology, embryology, and psychology, in addition to physics, geography, geology, astronomy, and meteorology (Lanza & Vegetti, 1971; Gotthelf, 1985). In quantitative terms, his biological texts represent about one third of his extant works. Most of them, and particularly those indicated by him as Περὶ ζώων, today better known by their Latin name, *Historia animalium*, were among his first writings, dating from soon after he left Plato's Academy.

These writings reflect an orientation very different from that of his revered master. In his *Timaeus*, Plato attempted to find the *clavis universalis* ("universal key") for interpreting the world, based on deductive, logico-mathematical discourses (see Rossi, 2006, for more on the theme of the "universal key" across the centuries). Within this framework, he tended to limit and simplify the complexity of reality. In contrast, Aristotle focused on investigating the variety and complexity of the natural world, relying on personal observations and experimentation, as well as collecting information from other sources.

Despite their different approaches, Aristotle was nevertheless guided by some of the same principles that entered into Plato's conception of reality. Among them were two ideas expounded in Plato's *Timaeus* within a metaphysico-religious framework: one was that the world is admirably organized on the basis of beauty, order, and finality; and the other was the idea of "plenitude" and the concept of hierarchy in its constitution. With Aristotle, the ideas of plenitude and hierarchy would lead to the assumption that there is a continuum in nature, from inorganic matter to plants, animals, and ultimately humans. In the form of a *scala naturae* (literally "ladder of nature"), the notion of a "great chain of being" would permeate Western culture through the Renaissance. This concept would draw renewed and unprecedented attention in the mid-nineteenth century, albeit in a different framework, with Charles Darwin's (1859) theories about evolution (see Lovejoy, 1936).

To accomplish his vast research program, Aristotle commissioned fishermen, hunters, apothecaries, butchers, farmers, and others all over the known world to provide him with specimens and observations, while also critically considering material passed on by oral tradition. These sources were essential to the fact-finding process, and the people he relied on were sometimes the only ones with good access to the most remarkable secrets of nature. Aristotle's texts, in fact, are the first important works that give voice to these overlooked possessors of practical and technical information.[5]

According to Pliny the Elder, the great Roman collector of information from the first century A.D. (about whom much more will be said below), some specimens studied by Aristotle were sent to him by Alexander the Great during his campaigns in the far-off and unexplored Eastern countries. Contemporary scholars, however, have questioned this "fact" (Jaeger, 1948; Romm, 1989).[6] Notably, the original texts of the *Historia animalium* were accompanied by a series of images, collectively referred to as Ἀνατομαί (anatomies), thought to be the oldest or among the oldest naturalistic works containing illustrative materials.

In Aristotelian terminology, *historia* means detailed observations of facts considered in their particularities. These texts were indeed viewed as collections of particular subject materials, preliminary to more theoretical and scientific discussions aimed at investigating the causes and principles underlying such phenomena. In this regard, these books were like volumes of an open encyclopedia, to which new observations and new information could be added (somewhat like what is now happening with some Internet encyclopedias). Aristotle later entered into more elaborate discussions on living beings in his more philosophically and physiologically oriented writings, especially his *De partibus*, *De generatione*, and *Parva naturalia*, in which he explored general interpretative principles in greater depth.

Even more than with other texts that have been ascribed to Aristotle, there have been serious debates about the botanical and zoological pieces that have long been associated with Aristotle's name (Grayeff, 1956; Egerton, 2001). The problem stems from the fact that the works of Aristotle, Theophrastus, and others associated with them (collectively known as the Peripatetic philosophers, because they walked while thinking and discussing) were moved and hidden to prevent them from being seized some 70 years after Aristotle died. The remains of these early Hellenistic works resurfaced about 100 B.C. and were taken to Rome after the fall of Athens in 86 B.C. There they were packaged and circulated a few decades later, some 275 years after Aristotle's death.

The collected pieces appeared under Aristotle's name, which helped to give them authority and unquestionably raised sales. Yet they were in reality a collection of texts by various authors, probably assembled by members of the Peripatetic school. With identities not divulged, scholars can only surmise which pieces came directly from Aristotle, which belonged to Theophrastus, and which had been written by other disciples. In addition, historians can only

---

[5] Although the emergence of modern science seems to be based largely on a reaction to Aristotelian culture, it is with the scientific revolutions of the seventeenth century, and particularly with Galileo, that the technical culture of engineers, craftsmen, and other possessors of practical knowledge became fully integrated into mainstream science (see Rossi, 2002).

[6] Pliny refers particularly to the case of an elephant accurately described in the *Historia animalium*. According to ancient sources (for instance, Athenaeus of Naucrathis) Alexander also funded Aristotle's research with an enormous amount of money. Among the specimens supposedly sent by Alexander there were various exotic species unknown to Western scholars and appearing to them as bizarre and monstrous. This may be one of the reasons why, starting from about the third century A.D., the figure of Alexander became associated with certain myths and legends that stemmed from the idea that far-off lands are the territories of the wondrous and the monstrous. The "Romance of Alexander" became one of the most popular epics during the Middle Ages, particularly when the great emperor was "transformed" into a Christian hero fighting for the faith, and his myth became contaminated with other popular legends. Alexander's myth had been revived in the Middle Ages by the circulation of some apocryphal letters, written in various languages, addressed to Aristotle, on the model of one circulating in the third century. As we shall see in Chapter 7, apocryphal texts and letters would play a similar role in another epic at this time. For an account of the various legendary letters of Alexander, see Gunderson (1980).

speculate about how they might have been altered in one way or another by later Peripatetic philosophers, who continued to work within the general Aristotelian theoretical framework. All that can be said with certainty is that the *Corpus Aristotelicum* (really better termed the *Corpus Peripateticum*) clearly represented a group effort, with sources rooted in Athens, in islands off the mainland, and even in distant Alexandria.

It is important to note, moreover, that most (or even the totality) of the surviving texts attributed to Aristotle come from writings conceived as drafts for verbally teaching those within his circle (i.e., his *acroamatic* or *esoteric* texts). Apparently nothing has remained of those texts prepared for diffusion among the general public (his *exoteric* texts). This accounts for the generally low literary quality of most of the writings of the *Corpus* (when compared, for instance, to Plato's), and in places for its lack of clarity. The view that the *Historia animalium* had been an "open text" further helps us to understand the particularly varied nature of the observations present in it.

Some scholars examining the marine biology in the *Historia* have focused considerable attention on the possible role played in this domain by Theophrastus, who headed the Lyceum for some three dozen years after Aristotle left (Fortenbaugh & Gutas, 1992; Fortenbaugh, Huby, & Long, 1985; Fortenbaugh, Huby, Sharpless, & Gutas, 1992; Fortenbaugh & Sharpless, 1988). Having grown up on the island of Lesbos, Theophrastus was especially knowledgeable about fishes, and much of the information about marine species in the *Historia* seems to have come from that island, where Aristotle spent only 1 year. There is indeed a wide consensus now that Books IX and X are not from Aristotle's hand but were probably written by Theophrastus.

Unfortunately, important texts of the Peripatetic school dealing with nature have been lost, and we know of their existence only by indirect citations of other authors (e.g., Athenaeus of Naucratis, see next chapter). For us, one particularly notable loss is a treatise on the torpedo (Περί νάρκης) written by one of Aristotle's close followers, Clearchus of Soli. Other texts of Theophrastus also seem to have been lost, and what remains of the most ancient commentaries on the work of Aristotle's and his immediate disciples is, regrettably, incomplete.

With the *Historia* and even more in subsequent biological texts, Aristotle and his followers initiated their searching for basic principles underlying the apparent variety and complexity of living nature. They assumed that everything in the world has a purpose, that there is never anything (neither a species nor even a body part) that is superfluous or without utility. Also, there is an order to things. As already mentioned, living things appear above inanimate things, with animals higher than plants, with fish above insects, and mammals above fish and reptiles (which feel cold to the touch), on the hierarchy.

These authors do not, however, attempt to create a rigid classification of the biological items described. Even the language used to identify their subjects is of the sort generally employed by laymen. This approach is very different from that which led Carl Linnaeus to create his "modern" classification system in the 1700s, with double names in a fixed, non-vernacular, language, such as *Gymnotus electricus,* for the electric eel (see Chapter 18).

With these important considerations in mind, let us now examine the passages about the sea torpedo in the Aristotelian zoological treatise, *Historia animalium*. As with other marine creatures, these passages cover material that would now fall under the general heading of biology. Absent from this treatise are any lines associating the torpedo with medicine. We do not even find the torpedo prescribed as an easily digested food here. This reflects the important fact that, unlike the Hippocratic physicians, Aristotle did not focus on medicine. Indeed, the *Historia* is the first extant Greek work having a distinctly biological character. After Aristotle, some of the knowledge contained in this great work would become a part of medicine. Only during the Renaissance, and more particularly during the Enlightenment, would some new works in biology return to this earlier autonomy from medicine.

In Book II, we come across some of the structural characteristics of the torpedo ray, such as its gill structure and the position of its gall bladder. Book V, in contrast, covers how torpedoes breed. Breeding behavior represents an important aspect of Aristotelian biology, fully developed in the tract *De generatione animalium*. The Aristotelians were first to make a clear distinction between oviparous, viviparous, and ovoviviparous animals, classifying torpedoes (and other cartilaginous fishes) in the last category. Here we read that a large torpedo (probably a *Torpedo nobiliana*) had been observed with about 80 embryos in it.[7] As noted in the previous chapter, this ability to carry live young might have contributed to the belief that the torpedo may protect smaller fish.[8]

A passage included in Book IX (or Book VIII in other editions) of the Peripatetic *Historia animalium* is particularly interesting given the focus of our book, since it bears on the torpedo's benumbing actions. The torpedo's power is here described as a "specialized function" (τεχνικά), a central theme of Aristotelian biology. This power is a facet of its intelligence. It is a stratagem (σόφισμα) comparable to that used by other fishes and marine animals, particularly the sea-frog or frog-fish (*Rana pescatrix*, literally "frog that catches fish," also called the fishing-frog or sea-devil, now classified as *Lophius piscatorius*), with which the torpedo is associated in this text.[9] Both fishes are very conscious of their respective capabilities, which allow them to capture fast-moving prey despite the slowness of their own movements (and notwithstanding the fact that they spend most of their time hidden in the sands at the bottom of the sea).

---

[7] This statement is not legendary. Charles Lucien Bonaparte, the nineteenth-century naturalist who described and named the *Torpedo nobiliana,* wrote that he had personally observed one of these big fish bearing more than 60 fetuses in its belly (Bonaparte, 1841, unnumbered pages of Chapters 16 and 24).

[8] Of interest here is a notation by Roman scholar Claudius Aelianus (or Aelian) in Book IX in his ΠερίΖώων Ἰδιότητος ("On the Nature of Animals"), a treatise written in Greek in the third century A.D. He writes that if a torpedo is caught when it is pregnant, parturition can be induced by pouring seawater into the vessel where it is contained (see next chapter).

[9] Curiously, Aristotle classifies the frog-fish among the cartilaginous fish, probably on the basis of its external similarity to the torpedo and to other "flat" rays. The association between torpedo and frog-fish, due to their unusual and crafty ways of catching smaller fish, would continue for centuries. Over time, the torpedo would become even more closely associated with the remora, a small fish that supposedly could stop big ships. But whereas the torpedo's power to stun is real, the remora's is imaginary. More will be said about the remora later in this chapter, when we look more closely at Pliny's writings. For more on the history of the association of remora and torpedo across the centuries, see Copenhaver, 1991.

As a proof that these slow-moving fishes actually do catch faster prey, it is remarked, "they are often caught with mullets in their interior, the swiftest of fishes" (Section 620b).

The frog-fish does this by attracting other fishes through two special processes, the text says, issued from its eyes. This is how the torpedo captures its prey, according to one of the many English translations of the *Historia animalium*:

> And the torpedo by causing numbness in whatever small fishes it intends to overcome, catching them by the means which it possesses in its body, feeds on them; it hides itself in the sand and mud, and catches all the fish that swim towards it and become numbed as they are carried near; this has been actually observed by some people....
> And the torpedo is seen to cause numbness in humans too. (Aristotle, 1965, pp. 311–312)

This brief but remarkable passage is the source of most subsequent descriptions of how torpedoes feed and defend themselves, appearing again and again in texts and compilations prior to the beginning of the modern era. Of great significance is how purposeful a torpedo can be when it comes to exercising its astonishing power. As noted, this is its intelligence—a useful gift of nature granted to the fish, so it can survive and flourish in its underwater world, which has unique inhabitants and features. Allusions to the torpedo's intelligence would resurface not only in subsequent naturalistic treatises, but also in moralistic writings dealing with animal behavior, from antiquity to the start of the Darwinian era.[10]

Even though the general description of the torpedo's power seems fairly clear here, there are problems relating to the wording. Starting with the earliest direct translations from the Greek, a torpedo's power is generally said to reside in its body (σώματι). This is the case, for instance, in a translation by Theodorus Gaza, a Greek scholar living in Italy, that dates from 1476 (only about 20 years after the Gutenberg Bible was first printed) (Fig. 3.3).[11] The extant manuscripts, however, bear the term στόματι, meaning *mouth*. Moreover, when designating the fish's shocking capability, a term is encountered that is vague, namely τῷ τρόπῳ, meaning attitude, manner, or disposition (this is translated in the aforementioned quotation as "means").

Some editors have emendated τῷ τρόπῳ to τῷ τρόμῳ, meaning "tremor." This corresponds to the trembling effects of the torpedo's shock, reflected in many popular names for these fish.[12] Among the other emendations proposed for τρόπῳ, we find ῥόπτρῳ, meaning trap (Aristotle, 1969). Present in a relatively recent French edition of this text, this emendation seems less justified linguistically (see Renna, 1995).

Thus, the power of the torpedo alluded to in the *Historia animalium* has been interpreted in a variety of ways. They include "the means which it possesses in its body," "the tremor residing in its mouth," "the benumbing faculty it has in its body," "the power of the shock residing in its body," "the proper manner" by which it catches other fish with its mouth, and "the trap it has in its body," among others.

In subsequent Greco-Roman scientific literature, different and generally more characterizing names will be used to describe the special feature of these rays. The terminology will include δύναμις, meaning power; ἀλκή, meaning force; πάθος, signifying affection (strong feeling), pain; φάρμακον, a term for a drug, poison, or a magical potion; and ἀπορροάς, which can be translated as effluvia. The variations in terminology reflect various interpretations of the phenomenon, as seen through the eyes of different authors in different places and at different times (Renna, 1995).

A TRANSMISSIBLE POWER

There is a characteristic of the torpedo's shock that is widely represented in the scientific and literary productions of the Greco-Roman era, and one that would continue to draw attention into more modern times, often in a lively way. This is the ability of torporific fishes to transmit their effects to animals and humans without direct physical contact. Whether through water, metal rods, wet nets, or other intervening bodies, or even sometimes through a vacuum, their discharges were still found to immobilize, numb, and deter other fishes—and even humans.

Most likely, fishermen were the first to notice these distance effects, since they used tridents to spear or harpoon torpedoes, nets to capture them, and ropes, rods, and metal hooks in their endeavors. In short, fishermen in ancient Greece and Rome, much like fishermen today, fished using a variety of conductive materials, from brass and iron hooks, to silk, flax, and even horsehair lines and nets moistened with seawater, which were likely held by damp hands.

Theophrastus was perhaps the first of many authors to refer to the transmissibility of the shocks through intervening bodies. He alluded to them in a book, unfortunately lost, dealing with "animals that bite or envenom." Athenaeus of Naucratis, an Egyptian-born Greek writer of the second to third century A.D., quotes the relevant passage concerning the transmissibility "through wooden rods (ξύλων) and tridents" in Book VII of his *Deipnosophistae* ("The Learned at the Banquet"), a text about which more will be said in the next chapter.

---

[10] Cicero, the great Roman orator and philosopher who lived in the first century B.C., wrote about the torpedo's purposeful character in his *De natura deorum* (Cicero, 1958, II). Through Cicero's eyes, the world is magnificently adorned by "the Providence of Gods," with all sorts of plants and animals that are endowed with powers capable of ensuring their existence and protection (Egerton, 1973). The torpedo ray is one of his examples, and although he does not mention the fishing-frog along with the torpedo, he associates torpedoes with cuttlefish (*sepia*), writing, "tutantur atramenti effusione sepiae, torpore torpedines" ("Cuttlefish defend themselves by ink effusion, torpedoes by torpor") (Cicero, 1958, II, 47, 50). Cuttlefish are also mentioned in the Aristotelian *Historia animalium*, although not for their ink-emitting protective actions. Here the point is that they capture prey with their tentacles, in a way somewhat similar to the fishing-frog. In contrast to the Aristotelians, Cicero's interest in torpedoes, cuttlefish, and other animals provided with useful survival mechanisms is decisively moral and religious. For more on his life and thinking, see Douglas (1968), Bailey (1971), and Everitt (2001).

[11] As with other early printed books, the first edition of Gaza's translation of the *Historia animalium* (included with the other Aristotelian treatises of zoological character in a volume titled *De animalibus*) has no page numbers, chapter numbers, or chapter titles. The reference to the torpedo's power is present in Book IX and reads "Torpedo pisces quos appetit afficit, ea ipsa, quam suo in suo corpore continet, facultate torpendi" ("The torpedo affects the fishes she covets, with that proper benumbing faculty she contains in her body") (Aristotle, 1476; for more on Gaza's translation and its numerous editions, see Beullens & Gotthelf, 2007).

[12] This is the case, for instance, of Nikolaos Savva Pikkolos (or Piccolos), a leading figure in Greek and Bulgarian culture, who published a Greek edition (Περί ζῴων) of the *Historia* in 1863 (Aristotle, 1863), and D'Arcy Thompson, who pioneered a mathematic approach to biology, author of the Oxford English translation (Aristotle, 1910).

Figure 3.3. The initial page of the books on the animals of the *Corpus Aristotelicum* in the Latin edition of Theodor Gaza published in Venice in 1476, with, on the right, a portrait of the Byzantine scholar (see Aristotle, 1476).

Some 400 years after Aristotle and Theophrastus were together, the transmission of the torpedo's shock through water resurfaced in a new, extremely influential treatise on nature. This was Pliny the Elder's *Naturalis historia,* translated into English as "Natural History," a first-century A.D. text about which considerably more will soon be stated. Despite the profound differences existing between the zoological writings of Pliny and those of Aristotle and Theophrastus from the earlier Hellenistic period, this is not surprising, because Pliny drew heavily from the *Historia animalium,* although largely through later Hellenistic compilations (Lanza & Vegetti, 1971).[13]

As with the author(s) of the *Historia animalium,* Pliny looked upon the torpedo's power as a type of intelligent behavior. To describe this trait (in Book XXXII, 1, of his *Naturalis historia*), he used the Latin word *sensus* (in the accusative form of *sensum*), which means both sensory capability and intelligence. This is a power that the animal is well aware of and can control.

In addition to admitting the transmission of the shock, through water, without a direct contact (*procul*), and even from a distance (*e longiquo*), Pliny, most likely on the basis of Theophrastus, also pointed to the transmissibility of the shock through some solid bodies, specifically a spear or rod (*hasta virgaque*). With this remarkable ability, as we have seen in the quotation presented at the front of this chapter, "this fish has the property of benumbing even the most vigorous arm, and of riveting the feet of the runner, however swift he may be in the race."

Shortly after Pliny the Elder, and probably also relying on Theophrastus, Plutarch of Chaeronea also pointed to the transmission of the torpedo's shock through water. Plutarch was a well-traveled scholar, who visited Rome, although he spent most of his life in Greece, where he was born in 46 A.D. (Russell, 2001). Best remembered for his biographies comparing famous Greeks with Romans (*Parallel Lives*), Plutarch wrote on a variety of subjects with great clarity and appeal, even for non-philosophers. It is in his *Moralia,* a collection of some 78 essays and speeches written in Greek on customs and mores, that we encounter the torpedo.

---

[13] It is interesting to note that some later authors, such as the Byzantine scholar Theophylact Simocatta (who died in 640 A.D.) and Michael Glycas (who flourished in the twelfth century) also included "air" among the conductive media (Simocatta, 1835 pp. 11, 95; Glycas, 1836, p. 70; see Renna, 1995). This was generally connected to the idea that the torpedo's shocks could be due to the emission of some effluvia, a hypothesis present in Plutarch (see below).

The fish appears in a dialogue, generally referred to with the Latin title *De sollertia animalium*, in which the discussion is on whether water or land animals are cleverer or craftier. This framework, and the fact that he cites the torpedo and the fishing-frog as being among the craftiest animals, reflect how the earlier Peripatetic philosophers also influenced his thinking.[14] In translation:

> You know, of course, the property of the torpedo: not only does it paralyse all those who touch it, but even through the net it creates a heavy numbness in the hands of trawlers. And some who have experimented further with it report that, if it is washed ashore alive and you pour water upon it from above, you may perceive a numbness mounting to the hand and dulling your sense of touch by way of the water which, so it seems, suffers a change and is first infected. Having, therefore, an innate sense of this power, it never makes a frontal attack or endangers itself; rather, it swims in a circle around its prey and discharges its effluvia as if they were darts, and thus poisons ($\varphi\alpha\rho\mu\acute{\alpha}\tau\tau o \upsilon \sigma \alpha$) first the water, then through the water the creature which can neither defend itself nor escape, being held fast as if by chains and solidified ($\pi\eta\gamma\nu\acute{\upsilon}\mu\epsilon\nu o\nu$). (Plutarch, 1957, XII, 978, p. 435; translation revised)

Plutarch's description of the torpedo's remarkable powers is considered among the best from the classical era. Besides pointing to the transmissibility of the torpedo's shock in its sea habitat, his passage is noteworthy for describing the conduction through water under somewhat artificial conditions, and for the clarity with which he presents the idea that the shocks might be conveyed by some sort of poisonous emanation from the fish's body.

Although Plutarch's passage would be quoted or restated using modified wording in many later texts, his place in the history of science and medicine pales next to that of Pliny the Elder. Hence, it is to Pliny the Elder that we must now return, to understand his objectives, to examine his *Naturalis historia* in more detail, and to begin to show how this text helped to create the intellectual environment that made torpedoes and their powers so wondrous and even mythical in his day and through the Renaissance.

PLINY AND HIS *NATURALIS HISTORIA*

In contrast to Hippocrates, Plato, Aristotle, Theophrastus, and Plutarch, Pliny the Elder was born on Italian soil in 23 A.D. Further, he wrote in Latin, not Greek, although Greek tended to be the favored (more erudite) language for most highly educated individuals who wrote treatises about science and medicine during the Roman era (e.g., Galen).

On the political front, Greece was now a province of the Roman Empire, having been annexed in 146 B.C. Yet, as the Latin poet Horace put it, "Graecia capta ferum victorem cepit," meaning, "Having been conquered, Greece conquered the ferocious conqueror" (Horace, Epist. II, 1; see Horace, 1825, p. 222). To this famous statement Horace added, "et artes intulit agresti Latio," meaning, "and it brought arts [e.g., fine arts, literature, science, etc.] into rural Latium." With these words, he properly signified the importance that Greek culture had for future developments after Rome became the dominant power.

Pliny's scientific and literary activity took place when the Empire was close to the acme of its power, and he is generally considered the greatest Roman naturalist. He and Aristotle certainly exerted the most influence on the study of natural world from the Classical period through the Renaissance. Pliny's *Naturalis historia,* a truly monumental and encyclopedic treatise of 37 books, was written between 77 and 79 A.D. It was shown to the world after his death by his nephew and literary executor, Pliny the Younger, and is the largest single work from the Roman Empire that has survived into the present.

Pliny the Elder (from now on, Pliny *tout court*) was a compulsive compiler of information and, even more, a voracious reader and tireless worker (Beagon, 1992; French & Greenaway, 1986; Healy, 1999). "Perire omne tempus arbitrabatur, quod studiis non impenderetur" was what his nephew wrote of him, meaning that he considered every moment of his life not devoted to intellectual activity a lost moment (Ep. 3,5; see Pliny the Younger, 1832, p. 218).

Unlike Aristotle and his followers, however, Pliny based his work mainly on literary sources, and very few of the "facts" listed in his writings are the outcomes of his own observations (Thorndike, 1923, Ch. 2; Conte, 1984; French & Greenaway, 1986; Healy, 1999). It is somewhat paradoxical that, having collected most of his observations indirectly, he died in 79 A.D., as is well known, while personally observing the eruption of Mount Vesuvius[15] (Fig. 3.4). It must be added, however, that Pliny was in the area commanding the Roman fleet, which was then attempting to rescue a populace endangered by the erupting volcano. He never did complete a final revision of his massive *Naturalis historia*, his last and only work that we have.

In the preface to his *Naturalis historia,* Pliny claims that he had gathered 20,000 facts from 2,000 books, and he cited almost 500 Greek and Roman authorities. He opined: "True glory consists of doing what deserves to be written, and writing what deserves to be read." Many centuries afterwards, another gifted writer and tireless worker, also an important figure in our story, "pilfer'd" (his term) from Pliny when assembling the 1738 edition of his popular *Poor Richard's Almanack*. There we find Benjamin Franklin, just before he began his electrical experiments, adopting same thought, but now presenting it in a rhyme:

> If you wou'd not be forgotten
> As soon as you are dead and rotten,
> Either write things worth reading,
> Or do things worth the writing.

(Labaree, 1960, p. 194)

---

[14] Plutarch's sources are not given, but Aristotle, Theophrastus, and Clearchus of Soli, a fourth-century Aristotelian, are good possibilities. As already mentioned in the text, Clearchus of Soli had written a treatise on the torpedo, which unfortunately has been lost. We know about this text only through a citation of Athenaeus of Naucratis, who, however, says that he does not remember the explanation of the shock provided by Clearchus, because "it was too complicated" (see next chapter).

[15] As we shall see in Chapter 24, unlike Pliny, Humboldt would watch Vesuvius erupt in 1805, and he would live to write about it.

term "electric ray"[16]—we read: "It is indeed reported that the electric ray, the plaice and the sole hide through the winters in the ground, that is, in a hole scraped out at the bottom of the sea" (p. 24); "There is a second class of flatfish that has gristle instead of a backbone, for instance, rays, prickly rays, angel-sharks, the electric ray" (p. 40); and "The torpedo knows her power and does not herself possess the torpor she inflicts; she hides by plunging into mud, and snaps up any fish that have received a shock while swimming carelessly above her" (p. 56). In his next sentence, Pliny adds: "No tender morsel; is preferred to the liver of this fish," before bringing up the frog-fish and other slow-moving fishes that, because of their cunning, "are found with mullet, the swiftest of all fish, in their belly." Other passages deal with when torpedoes breed (Part 84), their surprisingly small eggs, and the fact that females can be "found having broods numbering eighty" (p. 75).

Nevertheless, he mixes facts with conjectures, particularly when he alludes to the transmissibility of the torpedo's shock through a spear or a rod (see the previous section). Here he is certainly factual when he mentions that the fish's discharge can be felt through certain intermediaries (especially metals and moist objects). But this same passage also contains a remark that goes beyond empirical science and is more theoretical, although it is presented without qualification or anything else to suggest that it is not a fact. He writes that the torpedo's effect is due to "just an odor or a kind of emanation (*aura*) of its body." Seemingly based on the theory of corpuscular emission (see Plutarch's clearer passage above), this thought is clearly conjectural, not an observable fact.

The possible "scientific" meaning (in the modern sense) of what Pliny presents in his passage about stopping a swift runner in his tracks, the passage of the discharge through intermediaries, and "the breath from the creature's body" is also severely undermined by its context. This passage appears in the introductory part of Book XXXII, which is the second book devoted to remedies and therapies derived from animals living in aquatic habitats (*aquatilia*). It is preceded by a long passage illustrating the absurdly imaginary power of "a small fish of half a foot size," this being his *mora* (the Latin word for "delay") (Fig. 3.5). This small fish is indicated in the Aristotelian texts as ἐχενηΐς (*echeneis*), Greek from the words "to hold" and "a ship"), and it is now called a *remora*.[17]

We are first told that the remora can slow down and even block the motion of big ships. The myth behind this statement was that, by adhering to Marc Antony's praetorian warship, this tiny fish allowed Octavian's forces under Marcus Vipsanius Agrippa to emerge victorious in 31 B.C., in the Roman civil war battle of Actium! This battle took place in the Ionic Sea near the Greek coast.[18]

Figure 3.4. The death of Pliny the Elder during the eruption of the Vesuvius (from the frontispiece of a 1779 French edition of Pliny's *Historia naturalis*).

To the modern reader, Pliny's *opus* comes across as an indiscriminate mixture of natural science, fables, and just wishful thinking, into which he adds morals, values, and his own personal beliefs. The upside of Pliny's amalgam is that it provides a glimpse into what was commonly believed about the natural world in the second half of the first century A.D., both the verifiable facts noted by the *experti* and the myths, to which he was quite susceptible. The downside is that a lot of what he writes is misleading and even absurd (at least to us), even though he repeatedly assails others for their *vanitas* or nonsensical beliefs. Even more disheartening, the real facts are often tightly associated or put next to unverified thoughts, which can at times undermine the scientific value and relevance of what he managed to get right.

In his volume on aquatic animals (IX), Pliny is accurate in the parts in which he mentions the physical features and habits of the torpedo, clearly drawing largely from the observant Aristotelians. For example, in the popular Jones (1963) and Rackham (1983) translations—where we find *torpedo* in Latin across from pages in English using the anachronistic

---

[16] Fortunately this does not occur in all scholarly translations. For example, this is not the case in Bostock and Riley's translation (1855–57).

[17] Pliny writes about this fish, capable of stopping ships and used, by analogy, in magic and medicinal practices (also in Book IX, 41).

[18] Marc Antony's defeat was due to the fact that Agrippa had smaller and more maneuverable vessels under his command, better-trained and better-equipped fighting men, and superior timing, but this is hardly the stuff of myths and legends. The legend about the remora (present among classic authors only in Pliny) was also nourished by another episode, also mentioned by Pliny, concerning the Emperor Caligula, whose ship (and only it among the ships of the fleet) would have been stopped by one of these fish between Astura and Antium, near the coast

Figure 3.5. A mid-seventeenth-century image of a remora (*Echeneis naucrates*) painted by an unknown painter at the Grand Duke's court in Florence (© "Gabinetto Disegni e Stampe" of the Uffizi Gallery of Florence).

Pliny then goes on to say that because of the remora's slowing power, it is used in Greece as an amulet to prevent premature deliveries and, even more curiously, sometimes to facilitate childbirth. Given the remora's extraordinary ability to stop even large ships in motion—and, as he adds soon after, the power of the torpedo to act at a distance—he writes that we should not limit our hopes about what could be achieved with remedies. In Pliny's vision of the world, this is particularly true of natural remedies, due to the extraordinary power of nature (*potentia naturae*). His mention of the remora and the torpedo in this specific context is obviously intended to make the most incredible therapeutic effects involving the use of aquatic animals or their parts that much more plausible.

Pliny's passage about the torpedo occurs next, and it is followed by some myths about the sea-hare (today's *aplysia*, a simple but important animal in modern neuroscience), which serves as his third example of how spectacular natural phenomena can be—but which also detracts from what he had written about torpedoes. If a pregnant woman looks upon a sea-hare, he assures his readers, an abortion will result (3). Moreover, a protection against this effect is to bear a salted male "fish" of the same species within a bracelet as an amulet. But this is not all. He goes on to state that people poisoned by a sea-hare will emanate a fishy odor and die within a few days, although events transpire differently in the Indies. In those lands, it is the men who are terribly venomous to the fish, which die upon simple contact with a human finger!

Besides its diffusion by exhalation, the transmission of torpedo's shock through a spear or a rod might also have appeared less trustworthy to some knowledgeable readers if they had read what Pliny had also written about the terrible power of the basilisk in an earlier book (VIII, 33) (Fig. 3.6).[19]

Figure 3.6. The basilisk as depicted in two different ways in the ancient bestiaries, and also included, with minor variations, in various zoological treatises of the Renaissance. Top: Shown as a snake, roughly corresponding to the animal described by Pliny. Bottom: Depicted as a composite dragon-bird, an image that dominated medieval imagery (from Aldrovandi, 1640).

There he states that this monstrous creature is a serpent "produced by the province of Cirene" (in Africa) and that it could cause great harm at a distance or (like the torpedo) through intermediaries. "It destroys," he continues, "all shrubs, not only by its contact, but those even that it has breathed upon; it burns up all the grass too, and breaks the stones, so tremendous is its noxious influence." It was once believed, Pliny writes, "that a man on horseback killed one of these animals with a spear, and the poison, running up the weapon, killed, not only the rider, but the horse as well."

Thus, Pliny's uncritical acceptance of myths, legends, fantasies, and theories, and his habit of mixing them with verifiable facts, stands in striking contrast with the material in the Aristotelian texts making up the *Historia animalium*. Pliny's approach is all-inclusive and his objective is to provide an encyclopedia of knowledge that would also have entertainment value. In contrast, Aristotle, Theophrastus, and the other Peripatetic philosophers personally studied and, with the help of others, collected facts about nature, in order to derive unifying principles. As put by Paula Findlen,

---

of Rome. When the remora was brought onboard, Caligula expressed his indignation that such a small and apparently powerless fish could counteract the endeavour of some 400 oarsmen. Pliny considered the fish as a bad omen to the Emperor, because Caligula was assassinated in Rome, soon after the event, in 41 A.D. Although the episode of the remora is reported only by Pliny, several bad omens announcing Caligula's death are present in the texts of other Roman authors (and particularly in Chapter LVII of his biography included by Suetonius in *De Vita duodecim Caesarum*, known in English as "The Lives of the Twelve Caesars") (Suetonius, 1543).

[19] Despite its small size, the basilisk (from a Greek name meaning "little king," because of a spot on its head resembling a crown) was considered the most dangerous of the snakes, capable of killing with a single glance. This legend undoubtedly relates to the belief that snakes can freeze their prey with their hypnotic glances, and it might also have a basis in the reports on snakes that we now know can project their venom through the air (e.g., spitting cobras). The imagery of this feared creature changed during the centuries, taking on bird-like wings, a beak, and clawed legs, thus enriching the bestiaries of the Middle Ages. It appears in a great number of myths and on innumerable crests and seals, where we often find it glaring and releasing its last deadly breath while dying at the end of a brave knight's spear.

a leading historian of natural history, these approaches clashed, although both orientations contributed to the sciences in monumental ways:

> Pliny's appeal lay in his expansive rather than synoptic approach to knowledge; in its most essential sense, his philosophy of nature completely undermined the premise of Aristotelian philosophy. While Aristotle helped naturalists reduce nature to First Principles, creating a formal philosophy of nature, Pliny allowed them to revel in the particularity and infinity of the world. . . . The format of *Natural History* reminded collectors (and readers in general) that no detail of nature was so insignificant that it deserved neglect. (Findlen, 1994, p. 62)

This conclusion can be shared by anyone who goes through the many facts, anecdotes, and *mirabilia* scattered throughout the 37 books of Pliny's monumental *Naturalis historia*. Its organization, as well as the number and the diversity of what is described or narrated, shows the unsystematic character of Pliny's approach to natural phenomena, his unrestrained compulsion to include everything he had read or heard, his low standards for inclusion, and his inability or unwillingness to distinguish between fact and fantasy. The criteria by which facts and descriptions are assembled change continuously, and his digressions are frequent. We have seen some of this in Book XXXII, which contains the remora and sea-hare myths, and the passage about the torpedo's abilities. This book dealing with nature's remedies starts off by listing the remedies derived from single aquatic animals, before the organization suddenly shifts, with many prescriptions thereafter listed under specific diseases and/or the bodily parts affected.

Pliny was a compulsive bibliophile, a man who felt that there is no book so bad that it could not have some usefulness. He also had an obsessive tendency to summarize all sorts of information in a multitude of classification files (Conte, 1984). At times, in fact, his expositions seem to follow the ordering of the classification cards in a particular file more than some clear logic or systematic criteria. More than a book of science or a treatise on nature in the Aristotelian or modern sense, his *Naturalis historia* comes across as a depository. It is a massive inventory of pieces of information and stories of variable validity, with a particular proclivity to the wonderful and legendary, somewhat like the endless stories told by Scheherazade in *One Thousand and One Nights*.

It is with this mindset that we now turn to Pliny's medical prescriptions using torpedoes, some of which have an occult flavor. Since he also brings up the silurus or catfish, we shall rhetorically ask whether he was specifically pointing to *Malapterurus*, the electric catfish, in his prescriptions, and whether he even knew about these creatures, which dwell in the Nile River and some other warm rivers and lakes of Africa.

PLINY'S TORPEDO PRESCRIPTIONS

Although some of Pliny's prescriptions using torpedoes or torpedo parts may seem strange to us, others are easily understood. For example, following in the footsteps of the Hippocratic physicians, he informs readers that a diet containing torpedoes could help with digestive problems: "The bowels are relaxed by the silurus, taken with its broth, by the torpedo, taken in food, by the sea cabbage" (XXXII, 31).

But then he further writes: "Splenic trouble is treated by the application of the fish sole, of the torpedo, or of the turbot, but the fish then is put back living into the sea" (32). And, "An application of the torpedo to the intestinal region reduces morbid procidence there" (33). These two prescriptions are extremely vague, and given that there is no mention of a shock or discharge of any sort here, it would be speculative to conclude that he is calling for the shocks from a healthy living fish, even in the first case, in which he specifies a living fish.

With some of his other torpedo remedies and treatments, the stars quite literally have to be aligned, and there is no question about the fish being dead. Consider what he writes about the torpedo's brain as a depilatory agent, a prescription for which he specifies the exact day to begin treatment:

> Superfluous hair is removed by blood, gall, and liver of the tunny [tuna], whether fresh or preserved, by the liver too when beaten up, mixed with cedar oil and stored in a leaden box. In this way, slave boys were prepared for market by Salpe the midwife. . . . There is also used the ash of the crab or the sea scolopendra with oil, the sea anemone beaten up in squill vinegar, or the brain of the torpedo applied with alum on the sixteenth day of the moon. (47)

In yet another remedy, the timing is based on the heavens and no internal or external application of the fish is called for, giving it even more of an occult or magical flavor. Here, in what is reminiscent of one of his fanciful (and seemingly contradictory) statements about the remora (see above), we read: "What I find about the torpedo is also wonderful: that, if it is caught while the moon is in Libra and kept for three days in the open, it makes parturition easy every time afterwards that it is brought into the room" (36).

Also intriguing is his statement that the torpedo's gall can stem excessive sexual craving. Here the torpedo is directly linked with the remora, with its slowing powers, as well as with slow-moving hippopotamuses! Venery is inhibited, he writes, by "the echeneis [remora], and by the skin taken from the left side of the forehead of a hippopotamus attached as an amulet in lamb skin, or the gall of the torpedo, while it is still alive, applied to the genitals" (50). This remedy is particularly interesting, because the ancients more often prescribed fishes as sexual stimulants, not to slow or diminish sexual lusting.

During the seventeenth century, Francesco Redi decided to test what Pliny had written above, which many people still might have believed. With what had to have been a wry smile, he wrote that he experimented with torpedo gall and found that it was without effect "on that horn used by men (as our Boccaccio says) in their butting" (Redi, 1671, p. 51).

We shall have much more to say about Redi when we discuss his mechanical theory of the nature of the discharge in Chapter 9. And in our next chapter, we shall look at treatments from Roman times that clearly called for shocks from torpedoes. For now, let us address the question of whether *Malapterurus* appears in Pliny's medicine, or anywhere else in his writings for that matter.

MALAPTERURUS IN PLINY?

As noted above, Pliny had written: "The bowels are relaxed by the silurus [catfish], taken with its broth, by the torpedo, taken in food, by the sea cabbage" (XXXII, 31). Although it may be tempting to assume that he is alluding to *Malapterurus* in this context, because this type of catfish has torporific powers like torpedo rays, it should not be forgotten that he also mentions the completely unrelated, safe-when-touched, and non-torpifying sea cabbage (sea kale) in this sentence, in which, moreover, the fishes (and the plant) are no longer alive when ready to be ingested.

The thought that he might be associating a specific silurus with the torpedo ray also fails to garner support from other sections of his *Historia naturalis*. In his medical passages, Pliny tells us that a dead silurus, or a part of one, can be used for treating sciatica, voice problems, chilblains (a condition associated with skin irritations), burns, and drawing sharp weapons from the flesh (XXXII, 28, 30, 33, 36, 40, 43). These remedies are taken as foods, burned or pulverized and applied as pastes, and even injected in the rectum. But nowhere here does he specifically mention *Malapterurus* or even bring up torpedo rays with their related powers in life. Indeed, he even writes about catfishes in the plural, noting, "Siluri also, especially the African, are said to make easier the birth of children" (XXXII, 46; compare to what he had written about torpedoes above; in Chapter 5 we shall have more to say about to this legendary catfish attribute).

Further, Pliny writes nothing about the shocks or unusual powers of a catfish found in Egypt, or anywhere else where *Malapterurus* is known to reside. He does, however, bring up an unspecified catfish in the context of what others had to say about the sources of the Nile, a river that contains many different catfishes in addition to the electric variety:

> The sources of the Nile are unascertained, and, traveling as it does for an immense distance through deserts and burning sands, it is only known to us by common report. . . . It rises, so far indeed as King Juba[20] was enabled to ascertain, in a mountain of Lower Mauritania, not far from the ocean; immediately after which it forms a lake of standing water, which bears the name Nilides. In this lake are found the several kinds of fish known by the names of alabeta, coracinus, *and silurus;* a crocodile also was brought thence as proof that this really is the Nile, and was consecrated by Juba himself in the temple of Isis at Caesarea [modern Cherchell, Algeria], where it may be seen at the present day. (V, 10; Italics ours)

Even though Pliny provides no more information about this silurus, and was incorrect about a body of water flowing east from the Atlas Mountains into the Nile,[21] he was keenly aware of the fact that catfishes could be caught in various places. In addition to what he wrote about Africa and Egypt (then not considered a part of Africa), he knew that catfishes thrived in southern European rivers and lakes, and he even mentioned a catfish from the Canary Islands (VI, 37). He further pointed out that they are among the few types of fishes common to both fresh and salt water (XXXII, 53).

He also recognized that there are considerable physical differences among the catfishes, although he does not provide names. He notes, for example, an unusually large variety, comparable (presumably in size) to a tuna, which can be caught in the Nile (IX, 17). This catfish could well have been *Heterobranchus bidorsalis,* which can grow 5 feet in length and, as was pointed out in the last chapter, has been found mummified in Egyptian tombs.

Yet with all this information, there is not a single word in Pliny's *Historia naturalis* about any silurus being torporific. His numerous medical, biological, and geographical statements, as well as what he did not write, are not supportive of the idea that he was specifically pointing to *Malapterurus* with its special powers in life in multiple places or even anywhere in his text. Further, there is little to suggest that he was including *Malapterurus* under the generic term "torpedo." He is, or appears to be, denoting common sea torpedoes in places where he brings up torpedoes, although a few of his passages might border on the ambiguous.

These points should not be construed as meaning that Pliny knew nothing about *Malapterurus*. While this is a distinct possibility, especially given its conspicuous absence in his *Historia naturalis* and his unrestrained propensity to put everything he had read or heard on paper, the point we wish to make is only that he did not single out this catfish from other catfishes, or from sea torpedoes, in his massive encyclopedia.

PLINY'S LEGACY

Pliny's *Naturalis historia* is a bibliographic treasure chest. As noted, he cited almost 500 ancient authors, and he even provides a long list of them at the end. He also gives numerous references to extant works and, perhaps of even greater importance to scholars, to texts that have since disappeared. Using these resources, he conveys the beliefs, customs, and traditions of his time, which are poorly represented in more scholarly writings. This is particularly the case for the literature on magic, which he presents with many details, sometimes extremely macabre.

---

[20] Juba II, King of Numidia late in the first century B.C., who moved to Mauritania.

[21] Many Romans erroneously believed in a branch of the Nile that received the melting snows from the Atlas Mountains and flowed from west to east. This branch supposedly cut across and under Mauritania's deserts into Egypt, causing the springtime floods. Most likely, King Juba II was alluding to a lake in northwest Africa that would not flow into the Nile, and which would not have been home to *Malapterurus*—not to Ethiopia, which the Romans also recognized as a source of the Nile, and where this catfish can be found (see Chapter 7). In this context, it should be noted that the Roman province of Mauritania, named for the Moors who inhabited it, is most closely associated with the modern countries of Libya, Algeria, and Morocco.

Despite Pliny's unreliability and the generally low level of the scientific information he provides, the books making up his *Naturalis historia* would be widely read and some of his passages would be repeatedly referenced in Roman times and for centuries thereafter. Notably, his lines about the torpedo's shocks and some of his torpedo prescriptions would be passed down through the centuries into the modern period.

Compared to the scholarly Aristotelian *Historia animalium*, Pliny's *Naturalis historia* appealed to general readers because it was filled with colorful myths, fables, and legends, and because Pliny was a very good storyteller. In fact, interest in it during the Middle Ages and Renaissance was even greater than it was in his own lifetime. This had much to do with the moral character of the work, largely inspired by Stoic philosophy and centered on the idea of a rather animated and anthropomorphic nature. It grips people with its descriptions of natural forces encumbering over them, and, of course, its myriad of wonders and prodigies, among which is the torpedo. Pliny's nature, a somewhat pantheistic deity, may be providential but, as seen with his torpedo and even more so with his feared basilisk, it is also fraught with great dangers.[22]

Pliny's vision, and particularly the way in which it would be received in the Middle Ages through Christian scholars, would have an enormous influence on Western culture. In particular, the legends and myths that flourished during the Middle Ages drew heavily from Pliny's *mirabilia*, including minerals with strange but fascinating names (e.g., Chrysolampsis, Chrysopis, Cetionides, Daphnean, Diadochos, Diphyes); his amulets, miraculous fountains, and trees of life; and his monstrously deformed men (e.g., his dog-headed Cynamolgi, meaning dog-milkers). To this list we can add numerous other inhabitants of far lands with exotic customs (e.g., his Cisori, Logonpori, Usibalchi, Cispi, Isbeli, Perusii, Ballii or Orciani, Commori, Berdrigae, Pharmacotrophi, Chomarae, Choamani, Murrasiarae, Mandruani, Chorasmi, Gandari, Pariani, Zarangae, Arasmi, Marotiani).

Among Pliny's *mirabilia*, there also are many unforgettable animals, some with strange names (e.g., ichneumon, catoblepas, crocotae, corocottae, leucrocotae, leontophon, mantichorae) and some endowed with the most extraordinary powers. In fact, it is through Pliny, even more than through Aristotle, that scholars of the Middle Ages would be told about torpedo rays and their wondrous powers.

The *Historia animalium* was one of the first books to be disseminated with the advent of the printing press, and it had astonishing popularity (Fig. 3.7). In the words of one historian, it long reigned as "the most popular natural history ever published," with a remarkable 500 whole or partial editions from 1469, shortly after the invention of the

Figure 3.7. The title page of a 1543 Italian edition of Pliny's *Naturalis historia*.

printing press, to 1800 (Gudger, 1924). Pliny's influence remained profound through the Age of Exploration, when explorers set forth for distant lands with their mercantile and religious agendas (see Chapter 7, on the torporific African catfish) through the early modern period, as can be surmised from the dating just given.

During the seventeenth century, although some writers would still be citing Pliny's authority rather uncritically, others would set forth to falsify many things he had written (see Chapter 10). John Pringle (1775a,b), one of the shining stars of the eighteenth-century Enlightenment, would provide a brief history of electric fish and specifically cite what Pliny wrote about the torpedo when handing John Walsh the Copley Medal for his work on fish electricity. After Pringle, George Leclerc de Buffon and George Cuvier, two great French naturalists, would praise Pliny for collecting the knowledge of his epoch (e.g., astronomy, physics, geography, agriculture, commerce, medicine, magic, fine arts, history of nations, and much more) and for putting it together into one great encyclopedia. As Georges Cuvier (Vol. 35, p. 70) wrote in 1823, in his article "Pline" for the *Biographie Universelle*, Pliny's *Naturalis historia* was "l'encyclopédie de son temps" ("the encyclopedia of his age").

By 1800, torpedoes had left Pliny's fabulous world of prodigies and wonders, thanks in part to the research efforts of John Walsh and his colleagues at the Royal Society of London, although the association with electricity now made

---

[22] It can be particularly threatening to humans when they try to modify and adulterate natural productions with their arts, especially for luxury and lust. This contempt for luxury and lust was a significant discussion in Roman society at its political and economic acme, and it is an important moral theme in the *Naturalis historia*. In Book XXXII, the one dealing with the remora and torpedo, Pliny develops this theme, especially with regard to the vast use of *purpura* (purple), the expensive pigment extracted from the mucus of the murex sea snail. It is in the context that Pliny (IX, 57) mentions how Cleopatra, in one of her banquets with her lover Mark Antony, squandered an enormous fortune by swallowing two extremely precious pearls after having partially dissolved them in vinegar.

them exciting in a decidedly new way. But to appreciate how this transpired and the ramifications of this development, we must first look more closely at how learned men in the Greco-Roman era tried to explain their remarkable actions, while also promoting the utility of their discharges in the practice of medicine. We shall also have to examine other aspects of the torpedo's shocks in order to provide a more complete picture of the science and imagery associated with these rays and to provide a better foundation for appreciating exactly what was bequeathed to subsequent ages.

Chapter 4
Torpedoes in the Greco-Roman World: Pt. 2. From Therapeutic Shocks to Theories of the Discharge

Who has not heard of the untamed art of the terrible torpedo and of the powers that win it its name?

(From Claudian's poem, *Torpedo*, c. 400 A.D.)

As announced at the end of the previous chapter, we shall now deal with two fascinating subjects. The first is how torpedo shocks began to be used in medicine as a form of "electrotherapy" *ante litteram*. The second relates to what the ancients speculated about the nature of the discharge and its transmission in the absence of any idea at all about the force we now call electricity. In addition, we shall also comment on how these numbing fishes appeared in the arts during Greco-Roman times. With regard to the fine arts, we shall examine some of the pottery and mosaics depicting torpedoes, and relate what is illustrated to dietary ideas and significant events. Also interesting, given our major objective in writing this book, is how the torpedo is featured in the poetry of the epoch. Here we shall see how verses describing the "chilling" effects of their discharges mirrored the science and theories of the day. Thus, this chapter will continue to show how these unusual fishes entered into multiple aspects of the culture of the epoch (e.g., medicine, physiology, physics, food, and poetry). Yet it will also show how the ancients could only speculate about their powers, based largely on analogies and the physics of the day. As we shall see, the benumbing powers of live torpedoes were paradigmatic of a class of physiological and physical phenomena that were both extremely challenging and of immense importance to philosophers and physicians interested in science.

THE BIRTH OF TORPEDINAL THERAPY

The medical use of a torpedo as a dietetic food, as recommended by Hippocrates and his followers, and the various medical preparations derived from dead torpedoes, amply described in Pliny, appear to be exclusively of historical or documentary interest. The therapeutic uses of the shocks of live torpedoes, about which Pliny gives only vague indications, are far more interesting in this regard. This important development in the history of medicine is associated with Scribonius Largus, who was born in Sicily around 3 B.C. and probably died in 54 A.D. Few facts are known about his life, and they can be briefly summarized.

Scribonius accompanied the Emperor Claudius on his expedition across the English Channel in 43 A.D. to make Britannia a province of the Roman Empire. He served as a physician and showed a strong interest in pharmacological agents, especially botanicals (Baldwin, 1992; Nutton, 1995; Pellegrino & Pellegrino, 1988). But although he treated members of the royal household, he was probably neither Claudius' own physician nor the physician of his wife Messalina, as has sometimes been claimed. Stertinius Xenophon, who also went on the British expedition and was drawn to drugs and magic (and who helped to murder Claudius), held the exalted position of *archiatros* (first physician). This was the highest rank a physician could hope for at this time—one that involved demonstrating astonishing cures, but also one associated with unabashed self-promotion and even humiliating the competition when proclaiming authority (Mattern, 1999).

Scribonius Largus was known for his high ethics and for promoting genuine humanism in his medical philosophy, traits possibly acquired from reading Cicero. The imperial ministers held him in high esteem at a time when various medical sects fought with each other about the best medicines and how they should be used (Hamilton, 1986). To some extent, this was because of his *misericordia* (compassion, mercy), but it was also because he favored only those interventions firmly associated with expectations of successful outcomes. Among the Hippocratic maxims, those that seem to have meant the most to him were, "First do no harm" (*Primum non nocere*) and "Medicine is the science of healing, not of harming" (Hamilton, 1986; Pellegrino & Pellegrino, 1988).

Scribonius Largus' medical philosophy and fluency in Greek caught the attention of Gaius Julius Callistus, who became Claudius' legal secretary. Callistus helped him rise professionally, and in return Scribonius dedicated his *Compositiones medicae* to him (Fig. 4.1). This treatise was probably written in 47 or 48 A.D. (the standard edition is

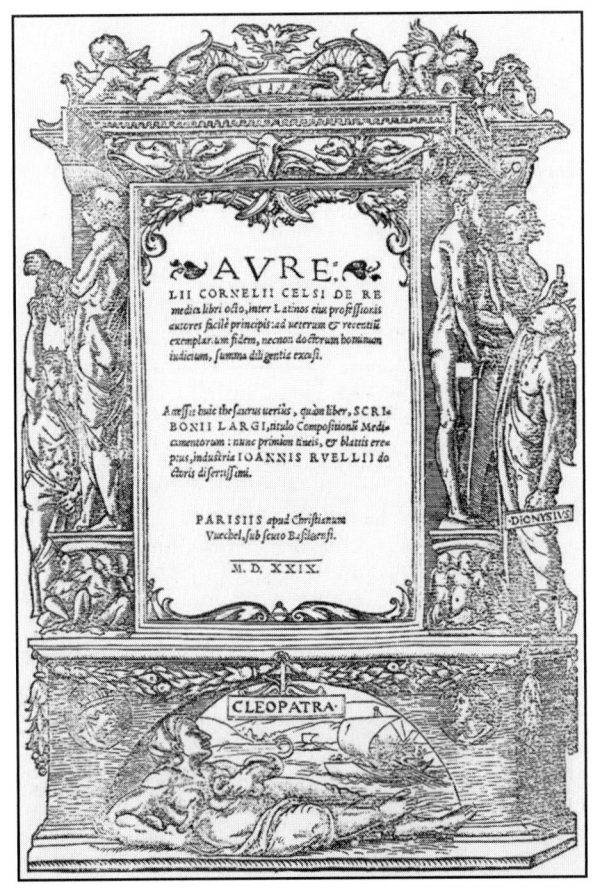

Figure 4.1. A sixteenth-century edition of Aurelius Celsus' medical treatise, which also includes Scribonius Largus' *Compositiones medicae* (from Celsus et al., 1529).

that edited by Sconocchia in 1983; also see Sconocchia, 2001). It is a collection of remedies and recipes composed while Scribonius was away from home with few books to consult. It followed other Roman medical compendiums, which literate citizens consulted to treat themselves, their families, and their slaves.

Many prescriptions in the *Compositiones medicae* were taken from Scribonius' teachers (e.g., Tryphon), other practitioners that he knew, and connections at court. He also consulted slaves and women who practiced the healing arts. His writings include a poison antidote for Augustus, a drug for Tiberius' colic, and various ointments and dressings for others. In all, he described 271 compounds using 306 substances. Most remedies were from plants; 36 used minerals, and just 27 came from animals—one being the torpedo.

The treatments are presented in a logical sequence, beginning with problems of the head and ending with those of the foot and toes (a similar organization can be found in the Edwin Smith Papyrus from ancient Egypt). Accordingly, he starts with headache and ends with gout, which can cause excruciating pain in the toes. Interestingly, the torpedo's shocks are mentioned only for these two disorders, the common denominator being chronic pain.

The passage on treating headaches reads as follows:

> Headache even if it is chronic and unbearable is taken away and remedied forever by a live torpedo placed under the spot that is in pain, until the pain ceases. As soon as the numbness has been felt the remedy should be removed lest the ability to feel be taken from the part. Moreover, several torpedo's of the same kind should be prepared because the cure, that is, the torpor which is a sign of betterment, is sometimes effective only after two or three. (trans. from Scribonius Largus, 1983, XI, p. 18)

The final statement about the torpor being a sign of betterment is in accord with parts of the *Corpus Hippocraticum* dealing with the relationship between numbness and pain, and could well have been inspired by these earlier writings (Debru, 2006, 2008). This relationship, which Galen would discuss in detail (see below), is particularly well expressed by the Hippocratic aphorism "a moderate torpor relieves pain" (νάρκη δὲ μετρίη ὀδύνης λυτική: Aphorisms, 5, 25: see Littré, 1839–61, vol. 4, p. 540).

As for gout, this painful disorder was endemic among Roman aristocrats and royalty at this time (Copeman, 1964; Rodnan & Benedek, 1963). Celsus, who was a contemporary of Scribonius as well as other Roman writers, relates it to tendencies to indulge in rich foods (we now know meats are a source of uric acid, which builds up in the joints of gout patients) and to drink too much wine (alcohol competes with uric acid for excretion by the kidneys). Additionally, but unknowingly, wealthy Romans also consumed dangerous amounts of kidney-damaging lead from additives used to sweeten their wines, the use of lead pipes, and storage vessels and other household implements made with this element, which they associated with Saturn, the gloomiest of their gods (Gilfillan, 1965, 1990; Nriagu, 1983a,b). As a result, the ancient literature contains many excellent descriptions of *podagra* (ποδάγρα, literally meaning "foot grabber"), a disorder mostly relating to gout.[1]

Anteros, a court official of Emperor Tiberius, charged with the administration of the inheritance and legacies left to the Emperor (*procurator hereditatium*), was born in 42 B.C. and died in 37 A.D. While at the seashore, he accidentally stepped on a live torpedo, which was probably partially buried in the wet sand. To his amazement, he discovered that its shocks numbed the pain in his foot caused by his *podagra*. In his *Compositiones Medicae*, Scribonius advices the use of the torpedo's shock for both types of *podagra*: the hot one (*podagra calda*), characterized by hotness and red swelling, and the cold one (*podagra frigida*), where the pain is not associated with these signs of inflammation:

> For both types of *podagra* a live black torpedo should, when the pain begins, be placed under the feet. The patient must stand on a moist shore washed by the sea, and he should stay like this until his whole foot and leg up to the knee is numb. This takes away present pain and prevents pain from coming on, if it has not already arisen. In this way Anteros, a freedman of Tiberius, Procurator of the inheritances (*supra hereditates*) was cured. (Scribonius Largus, *Compositiones*, CXLII, see Sconocchia, p. 79; modified)

---

[1] Although *podagra* certainly includes the condition that we now know as gout, in ancient medicine the term might also have indicated arthritic conditions involving foot articulations, particularly in the case of the so-called *frigida podagra* (i.e., the type of foot disorder not associated with signs of inflammation, such as redness and tumefaction).

Figure 4.2. The title pages of three of the many editions of Dioscorides' medical works published in the first half of the sixteenth century. These works were among the most influential medical texts during the Middle Ages and the Renaissance (from Dioscorides, 1516, 1527, and 1529, respectively).

Today, we know that peripheral stimuli transmitted by rapidly conducting nerve pathways can block pain sensations, which utilize mainly slowly conducting nerve fibers (e.g., the "gate control theory" of Melzack and Wall, 1965, 1996; Wall & Sweet, 1967). We also know that painful stimuli can cause the release of endogenous opioids in the brain, which can diminish subsequent pain sensations. In addition, there is a large literature showing that electrical shocks can subdue the dull, throbbing aches and pains caused by chronic diseases, including various types of headaches, sometimes for lengthy periods of time (White & Sweet, 1969; Slavin, 2008). From these facts of modern medicine, the claim that applying living torpedoes helped Romans suffering from headaches and chronic gout pain does not seem far-fetched.[2]

Thus, it would be a mistake to dismiss Scribonius Largus' new cure as just another worthless remedy, of which there were many in the ancient world. At the same time, however, we must remember that it has been repeatedly shown that strong beliefs could affect clinical outcomes. Thus, the electric fish therapeutics recommended by Scribonius Largus could well have been effective for physiological and biochemical reasons, but also because of expectations and beliefs, or perhaps because both physical and mental factors were brought together in his practice.

In later chapters, we shall see that the first routine applications of electricity to bedside medicine would generate debates on the efficacy of the various treatments proposed. Whether the electricity came from machines or from electric fishes, some practitioners, their followers, and their patients promoted the "cures" as miraculous. In some circles, they were even hailed as a panacea for treating virtually any disease or disorder. Yet to others, such as Benjamin Franklin (see Chapter 10), they were viewed as cures for some conditions (hysteria) and not others (palsies). Some individuals even branded all electrical medicine a worthless remedy. Besides the possibility of charlatanism, there is little doubt that, in some circumstances, the extraordinary efficacy of the electric treatments noted in some patients was due at least in part to placebo effects, whereas in others such cures seemed to have good scientific backing.

Soon after Scribonius Largus brought forth the story of Anteros, other writers began to look upon the torpedo as a living therapeutic marvel. One such scholar was Pedanius Dioscorides of Anazarbos (a town of the Ancient Cilicia, now Turkey), who was born in the first century of the current era and traveled widely. Dioscorides is best remembered as the author of an important pharmacopoeia, which was very popular and had prescriptions that were widely copied into the Renaissance and even later (Riddle, 1985) (Fig. 4.2). In his Περί ὕλης ἰατρικῆς (literally, "On Medical Matter," generally indicated with the Latin title, *De materia medica*), written in Greek near the middle of the first century, he emphasized the therapeutic uses of various plants (some 537 types!) (Fig. 4.3). But like Scribonius, he also devoted space to mineral and animal cures.

Following Scribonius, Dioscorides mentions how the torpedo could be used to treat chronic headache pain. But in addition, he extends the torpedo's use to a condition that

---

[2] In addition to the application of the shock of a live torpedo, Scribonius recommends the torpedo as a therapeutic food. It is to be mixed with bread and administered for a condition associated with acutely reduced vision and accompanied by vertigo and headache (a particular type of migraine according to modern medical terminology; see *Compositiones*, XCIX).

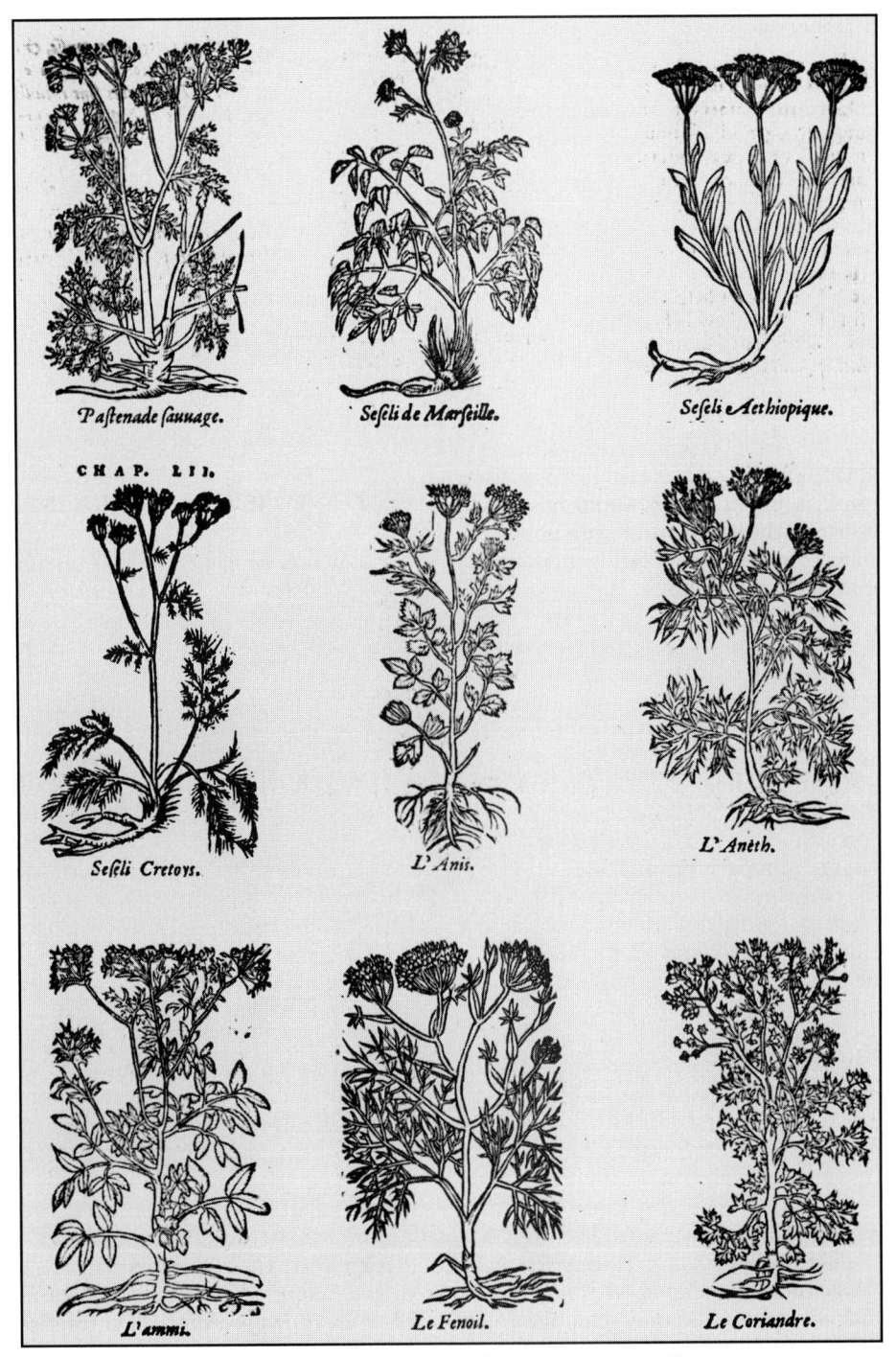

Figure 4.3. Images of therapeutic herbs from a 1559 French edition of Dioscorides' *Materia medica*.

affects "the seat." His compound statement about these two conditions is just a sentence long. As presented in an English translation from 1655 (but appearing in print only in the twentieth century), it reads: "The Sea Torpedo being applyed in griefs of long continuance about ye head, doth assuage the fiercenesse of the grief: the same too applyed doth stay up the seate, being overturned, or else falled down" (from Dioscorides, Book II, 17; see Gunther, 1959, p. 97). Using newer wording, we provide this alternative translation: "The sea torpedo, when applied for chronic pain about the head, lightens the severity of the pain; and being applied to the prolapsed seat, gathers it up [or draws it in]."

But what exactly did Dioscorides mean when he wrote about a fallen or turned-out (ἐκτρεπομένην) seat, which his followers, using Latin rather than Greek terms, have called *prolapsus ani*? Prolapsus is a condition characterized by slippage or displacement of an organ or part of an organ downwards or outwards. There is little doubt that with the term seat (ἕδραν), Dioscorides was referring to the anus, and that he was thus suggesting the live torpedo as a remedy for a disorder involving tissue at least partially protruding through the anus.

Some modern proctologists contend that Dioscorides was probably referring to patients with prolapsed or external hemorrhoids, two relatively common conditions associated with dilated (varicose) veins in the anal region. Hemorrhoid formation is associated with straining during bowel movements, pregnancy, childbirth, constipation, diet, and aging, to name but a few variables. If the hemorrhoids form internally and prolapse, or form externally, sufferers typically experience discomfort when sitting, pain during difficult bowel movements, and bleeding.

Given how common hemorrhoids were and still are, this interpretation of what was meant by a fallen seat or *prolapsus ani* makes a certain amount of sense. But could Dioscorides have meant something else with his wording, or might he have been thinking about multiple conditions that could affect the seat, some of which would have been difficult to distinguish from others then and even now?

The symptoms of advanced rectal prolapse are similar to those of external or prolapsed hemorrhoids. Rectal prolapse, however, originates higher in the system, is not due to dilated veins, and occurs far less frequently. In it, the ligaments and rectal and sphincter muscles weaken, allowing varying amounts of the lower end of the colon (i.e., the rectum) to fall from place and to protrude through the anus in advanced cases.

Thus, when Dioscorides wrote about prolapse of the seat, he could have been referring to hemorrhoids and/or to rectal prolapse, knowing what we do about diagnoses and the embryonic status of pathology in his time. Although less likely, he might even have included lumps in the anal region associated with scar tissue, a consequence of poorly treated but common anal fissures. But could torpedoes have helped with any of these conditions?

As with Scribonius' recommendations, Dioscorides' statement about treating the symptoms of *prolapsus ani* with a living torpedo has some scientific support. First, we now know that electricity can cause the anal sphincter muscles to relax (the opposite of what would be expected with most muscles), making it easier to have bowel movements, and thereby reducing a major cause of both hemorrhoids and rectal prolapse (e.g., Hopkinson, 1972, 1975; Keighley & Matheson, 1981). Second, mild electricity might exercise and strengthen other muscles. Third, electric shocks could reduce the pain and discomfort associated with any of the aforementioned conditions, much as they can diminish headache and gout pain. And fourth, medical researchers have shown that even mild electrical currents, when properly applied, can cause hemorrhoids to shrink by constricting blood vessels (Izadpanah & Hosseini, 2005).

Unfortunately, Dioscorides' torpedo prescription is so brief that it tells us next to nothing about his recommended methods, such as how frequently and for how long the torpedo should be applied to the anal region, or anything about the size, vigor, or kind of torpedo to be used. He does even not tell us whether the torpedo's beneficial effects are long-lasting. In fact, although he writes that the torpedo should be of the marine type (νάρκη θαλαττία), he does not even specify that the fish should be alive. For us, this clearly appears to be the case from the context, but it might not have been so evident to some readers in the distant past, who might not have known much about torpedoes and would not have had the physiological understanding that emerged in modern times.

Despite this paucity of information, Dioscorides was extremely influential, and at least in this regard he can be compared to Pliny. Many centuries later, after the invention of the printing press, many editions of his *De Materia medica* were published, and his recommendations were repeated in herbals and books of medicine well into the early modern period (see, for instance, Lovell, 1661, p. 191).[3] Indeed, even in Roman times, his writings were an important source for other writers, one being Galen, who was by far the most highly educated and important physician-scientist of the Roman era (Renna, 1995).

GALEN'S TORPEDINAL MEDICINE

Galen was born in 129 A.D. in Pergamon, one of the greatest centers of Hellenistic culture on the beautiful northwestern coast of what is now Turkey. The exact date of his death is uncertain, but it is believed to be about the year 200. He studied philosophy, science, and medicine at home and at centers of learning in Greece, Italy, and coastal Egypt, all of which were controlled by Rome. At this time, Christianity was rapidly gaining in importance, and Galen's vast scientific and literary production has been viewed as an attempt to counteract the progressive spiritual and cultural hegemony of the new religion (see Garofalo & Vegetti, 1978).

Galen moved to Rome and served as the personal physician to four successive emperors. He also wrote hundreds of books, although many were lost when the library in which they were stored burned in 191. One of his ambitions was to unify Roman medicine, which was fragmented by many sects with different orientations. He firmly believed that medicine had to be based on dissection and experimentation, although he did not shy from theory (Sarton, 1954, Finger, 2000; Rocca, 2003).

We can find many things about the torpedo in Galen's writings. We mentioned one in the previous chapter when we discussed a Hippocratic treatise on diet, in which the torpedo was recommended as a light, easily digested food for people with a myriad of internal disorders. Galen provided commentary on what Hippocrates wrote in a treatise that translates as *On Regimen in Acute Diseases in Accordance with the Theories of Hippocrates*. This treatise, originally written in Attic Greek, has been lost, but an Arabic translation believed to be authentic has survived (Galen, 1969). Here he tells us that torpedo flesh is more nourishing than the barley gruel recommended by Hippocrates for patients suffering from acute diseases, yet not the most nourishing of foods, being so soft. He recommended slowly adding it to a sick person's diet, which should, in fact, begin with gruel (pp. 91–92).

This is not Galen's only text dealing with torpedoes as a medicinal food. In the third book of his *De alimentorum*

---

[3] It is perhaps worth reminding here that in Voltaire's *Candide*, the protagonist, after having been severely injured, "was cured in three weeks by a worthy *surgeon* with ointments originally prescribed by Dioscorides" (Voltaire, 1759/1947, p.24).

facultatibus ("On the Properties of Foods"), Galen again agrees with the earlier Hippocratic physicians that the flesh of the torpedo is "not difficult to digest" (πεπτομενὴν οὐ χαλεπῶς) (Kühn VI, p. 737).[4] Here he points out that most cartilaginous fishes have their fleshier parts near the tail, a fact he feels is particularly true for torpedoes. This statement is certainly correct, because the rest of the torpedo's body is filled with a gelatinous and insipid matter, which we now know corresponds to its electric organs. Galen advised torpedo's flesh particularly for cold and dry temperaments.

In yet another treatise, *De puero epileptico,* meaning "On the Epileptic Boy," he singled out the torpedo as the only suitable cartilaginous fish for the diet of young epileptics (Fig. 4.4). In his words: "Among fishes, those living near the rocks are indeed very suitable; also those living in open sea, indicated as pelagic, can be taken; moreover, among the cartilaginous the torpedo is about the only one suitable and convenient" (Kühn XII, p. 373; see also Galen, 1586, Classis VII, folium 180).

Of even greater importance for us, however, is what Galen wrote about the torpedo's shocks in therapeutic settings. Here it is important to know that Galen was always highly skeptical about what others had written, looking upon most physicians, other than Hippocrates, as poorly educated and basically ignorant. As a result, he liked to try remedies on his own body before endorsing them.

His comments on the therapeutic use of the torpedo for headache and *prolapsus ani* reveal much about his hands-on approach. Specifically, he experimented first with dead torpedoes and found that they did not work for either disorder, whereas further experimentation convinced him that live fish could ameliorate headaches:

> The whole torpedo, I mean the sea animal, is said by some to cure headache and reduce the prolapsed seat when applied. I indeed tried both these things and found neither to be true. Having thought that the torpedo should be applied alive to the person who has the headache, and that it could possibly an anodyne [i.e., analgesic] remedy and could free the patient from pain, as happens with remedies which that numb (ναρκοῖ) the sensibility, I found it to be so. And I think that he who first tried this did so with a similar idea in mind. (Kühn, 1821–33, XII, p. 365[5])

This passage would continue to be cited for many centuries to come. An analysis of Galen's words, and a comparison with Scribonius and Dioscorides, is of interest here. First, such an exercise reveals that Galen probably had not read Scribonius (or at least not the original *Compositiones medicae* as we now have them). This is because Scribonius specifies in an unambiguous way that it is necessary to use a live torpedo in order to alleviate pain (*torpedinem nigram vivam,* meaning a live black torpedo). Scribonius also clearly states that the effect depends on the numbness produced by the shock.[6] Moreover, Scribonius does not include *prolapsus ani* in this torpedinal context, although Galen does, and Scribonius discusses treating *podagra*, which Galen does not.

These facts suggest that Galen probably derived much of his information from Dioscorides rather than from Scribonius (Nutton, 1995; Renna, 1995; see below). Not only do Galen and Dioscorides not mention treating *podagra,* while bringing up *prolapsus ani,* Dioscorides does not explicitly mention the need for a live fish in the treatment. His open-ended wording could account for Galen's trials with dead fish, which proved unsuccessful. Further, the possibility of Galen drawing his information from Dioscorides is substantiated on a linguistic basis. The two authors use similar wording in referring to the "seat" (ἕδραν), and also to its prolapsed situation (ἐκτρεπομένην in Dioscorides and ἐξεστραμμένην in Galen; see Renna, 1995).

Figure 4.4. The title page of one of the tomes of a Venetian edition of the works of Galen. This particular tome, the *VII Classis* ("Seventh Class"), is devoted to therapeutics. It includes *De puero epileptico*, in which the Greek scholar prescribes the flesh of torpedo in the regimen for young epileptic patients (from Galen, 1576).

GALEN'S "COLD VENOM" THEORY

The final part of the passage from Galen (above) is of special interest, because it sheds light on the thinking that led him (and possibly others before him) to see if the benumbing

---

[4] Following the scholarly convention generally used in quoting Galen's texts, we shall normally refer to the edition of his complete works edited in the period 1821–33 by Karl Gottlob Kühn in 20 volumes. These volumes have been reproduced several times in facsimile. Interestingly, in addition to being a classicist, Kühn was a physiologist and pathologist, and he was also interested in animal electricity. As a young man, he translated and edited a work on this theme by the eighteenth-century French physicist Pierre Bertholon (see Bertholon & Kühn, 1788). We shall look more closely at Bertholon, a passionate promoter of the medical application of electricity, later in our book.

[5] The English translations of Galen's works from Kühn are ours.

[6] To describe the effect of fish application, Scribonius uses the verb *torpere* (to numb) in the case of *podagra* and *obstupescere* (to become stupefied) in the case of headache.

effects of the torpedo could actually be an analgesic remedy. To appreciate his thinking, we have to consider how numbness, sensibility, and pain (and also coldness) were related to each other in Greco-Roman times. This subject was already present in the *Corpus Hippocraticum*, and Galen developed it even further in his writings, as has been pointed out by Armelle Debru (2006, 2008).

Specifically, Galen denied that numbness should be considered a form of pain. This is because, as the Hippocratic writers had already noted when dealing with the effects of moderate coldness, numbness actually reduces pain by interfering with nerve function. Galen also felt that numbness should not be considered a sensation, because it reduces sensibility. Again referring to Hippocrates, one of his few heroes (Aristotle being another), Galen wrote that numbness (νάρκη) could be considered nothing other than "a non-negligible coldness" (Ψύξις οὐκ ἀγεννής). We find this thought in one of Galen's most important treatises, his *De locis affectis*, meaning "On the Affected Places" (Book II; Kühn, VIII, p. 71).

Having now defined numbness (νάρκη) as an affection combining pain and coldness, Galen notices that such a condition is also produced by nerve compression and, moreover, "by the torpedo [living] animal of the sea" (ἔτι τῆς νάρκης τοῦ θαλαττίου ζώου: Kühn, Tome VII, p. 109). A similar correspondence between numbness, the physiological action of cold, and the effect of a torpedo's shocks can be found in the Book I of another of his major treatises, *De symptomatum caussis* ("The Causes of Symptoms"; Kühn, VII, pp. 108–109). In this way Galen established an association between *nárkē* (meaning both numbness and the numbing fish) and coldness and its physiological consequences (i.e., reduced sensation and movement).

This association forms the basis of the idea that the torpedo's effects could be due to a refrigerating force or some sort of cold venom. This thinking fit well within the framework of Greco-Roman medicine, in which the qualities of "coldness" and "hotness" were tightly associated with the four elements and the four humours (see Chapter 3). As we shall see later in this chapter, these ideas about frost (*frigus*) and cold venom (*gelido veneno*) would make their way into a poem by Claudian dealing with torpedoes.

The materialization of a refrigerating action or force (δύναμις) into a venom might be in part derived from a passage of Book VI of *De locis affectis*, in which Galen mentions the power of the torpedo in association with the venom of the spider and scorpion. As pointed out by Armelle Debru, in these cases the torpedo is invoked (together with the two animals) in order to show that a great physiological and pathological effect—one possibly involving the entire organism—could be produced by an apparently minimal cause.

Galen first considers the spider and the scorpion, which are capable of producing generalized and violent effects through a minimal quantity of venom injected into an extremely restricted region of the body. Next he turns to the torpedo with these words:

> Some [physicians] even believe that, through the action of their power (δυνάμει), some matters could alter nearby bodies by simple contact. Such a nature is encountered in the sea torpedoes. They have a power (δύναμιν) so strong that, through the trident of the fishermen, the alteration is transmitted to the hand, which soon gets numbed (ναρκώδη). (Kühn, VIII, pp. 421–422)

After mentioning the transmission of torpedo's power through a fishing trident, Galen turns to the lodestone ("Heraclea stone") as an analogy for how the trident can transmit its power. In addition to attracting nearby iron bodies, he writes, a lodestone can communicate its power to those iron bodies, which in turn can affect other bodies further down the line in the sequence. As we shall see, the analogy with a magnet would continue to be of great importance as the ancients pondered the nature of the torpedo's shock, its transmission, and its broad effects.

In *De locis affectis*, Galen considered the similarity between the animate (torpedo) and inanimate (lodestone) forces exclusively within a pharmacological/toxicological context. Here is evidence that a body can acquire a power (or a quality: δύναμις) of an adjacent body through contact. As a consequence, one body can become capable of communicating the power to other bodies. In Galen's schema, however, this transmission of δύναμις does not necessarily imply any transmission of substance.[7]

Due to its transmissible numbness, the torpedo provided Galen with a model to account for generalized diseases, even in the absence of any clear conception of the mechanisms underlying their production. In fact, in another of his treatises (*De usu respirationis*), Galen considers the torpedo's power as a model for properties that can be demonstrated and experienced, but for which it is difficult to provide a causal explanation. Within the context of an imaginary discussion, in which he replies to adversaries who wish to promote an alternative view on respiration, he explains:

> I shall tell you it, if you will tell me first the cause whereby sea fishermen are numbed when they capture a live torpedo. If you do not have anything to say on that, let me invoke the numbing power (δύναμιν) of the animal, so strong as to bring its painful action, even through the transfixing trident, up to the hands of the fishermen. You should therefore admit that this property of the fish belongs to those kinds of qualities and powers (ποιότητας καὶ δυνάμεις), of which one is numbness (νάρκην), another sopor [really *sopore* in the Italian sense, meaning stupefaction, dizziness], others coldness and putrefaction, others some other kinds of harm. Would thus you deny some similar force to air? (Kühn, IV, p. 497)

"UN-NAMEABLE QUALITIES"

The passage of Galen just quoted reflects an important discussion present in classical science that would continue until the emergence of modern science (and also nowadays, although within a different framework). This discussion concerns a group of "qualities" or "forces" about which the causes (and the mechanisms) appear difficult to identify.

---

[7] As A. Debru remarked, the possibility that a body can acquire the quality of an adjacent body allowed Galen to account for several generalized diseases. Among them there is a form of female hysteria, due to stagnation of "female sperm" in the uterus in the absence of sexual intercourse. As a matter of fact, because knowledge about blood circulation was lacking (as was the case in classical physiology before the Harvey's discovery in the seventeenth century), it was difficult to explain how a small quantity of poison (or a disease arising in a localized region of the body) could rapidly diffuse to affect the entire organism.

These qualities were dealt with in the first book of the *Problemata* ("Problems"), a pseudo-Aristotelian treatise now attributed to Alexander of Aphrodisias, a famous commentator of Aristotle, who was active in the third century A.D. (Merlan, 1970). They were designated as ἄρρητοι, a term literally meaning "un-nameable" or "not to be said" (but also including the connotation of "mysterious" or "occult"). Typical of them were the powers of the magnet and the torpedo. In the *Problemata* we read:

> Why does the magnet-stone attract only iron? Why does the substance called "amber" draw up to it only chaff and dry straws, making them stick together? . . . No one is ignorant of the marine torpedo. How does it numb the body through a string? . . . I might prepare you a list of many such things, which are known only by experience and are called un-nameable properties by the physicians. For the peculiar un-nameable thing asserted of each of them suffices for an explanation of the causes. (Trans. in Copenhaver, 1991, p. 380)

As can be gleaned from the end of this passage, when the true cause of a phenomenon is impossible to ascertain, the simple assertion of one of the un-nameable qualities with reference to the phenomenon studied was considered a sufficient explanation of it.

Galen took these qualities into account mainly to explain a series of physiological and medical phenomena produced by drugs, herbs, or animals, including the torpedo. He considered them as properties "which act through the entire substance." As noted by Brian Copenhaver (1991), the author of a perceptive study on the history of the remora and torpedo over the centuries, Galen wished to signify with this expression that these qualities belong to the entire body which produce them and cannot be assigned to one specific part of it.

An important feature of the discussions concerned the ways some of these forces or qualities could act at a distance—how they could be transmitted through media of various kinds. As mentioned, most of the qualities or forces Galen considered concerned the organic world. In ancient science, however, the discussions ranged wider than this. They also concerned purely physical phenomena, some being the transmission of light and of heat, the magnet's attractive actions, and various kinds of astronomical and astrological influences. And they also had, as we might expect, implications that bore on magic and related fields.

The problem of the existence of a vacuum loomed large within these discussions. Some philosophers and physicists (e.g., Democritus, Epicure, and the atomists in general) argued that there is a vacuum in nature, whereas others strongly challenged this idea (e.g., Aristotle, most of his Peripatetic followers, and the Stoics).

At the extremes, there were two opposite conceptions. One was based on the idea that any transmission must necessarily imply the propagation of some form of corpuscular matter; the other was founded on the notion that qualities or forces could be transmitted without any actual passage of matter. In the latter conception, the qualities or forces were considered attributes (or forms) of the matter, whether inorganic or organic. By themselves, the qualities or forces were devoid of any spatial dimension or configuration, and thus they lacked any physical substantiality. To use Galen's phrase, they were located in the "entire substance" of the body possessing them.

The transmission of the power of the torpedo, together with magnetic attraction—two phenomena relatively easy to verify but very difficult to interpret—are recurring issues in Galen's writings and in the *Problemata* within the discussion about "un-nameable" qualities and, more generally, about transmission at distance.

THE TORPEDO, THE COSMOS, AND VISION

One of the oldest texts in which the transmission of the torpedo shock was invoked within a physical framework concerned the propagation of light and heat. Of this text, we know the likely title (Περί κενοῦ: "On the Vacuum") and the author, Strato (or Straton) of Lampsacus, but we do not possess the original. We only have Hero of Alexandria's compendium of it, which he inserted in the *proemium* of his *Pneumatica* (for a modern English version, see Woodcroft, 1971).[8]

Strato, who lived between the fourth and third century B.C., was a follower of Aristotle and succeeded Theophrastus as director of the Lyceum. In opposition to Aristotle, who denied the existence of the vacuum, he assumed that a vacuum could exist in the form of microscopic empty spaces interspersed within the body of essentially continuous matter. In this regard, Strato's concept of a "discontinuous vacuum" is profoundly different from the continuous vacuums of the atomists (e.g., Leucippus, Democritus, and Epicurus), who envisioned an empty space where the atoms could freely move.

Hero accepted Strato's theory. In translation, Strato's passage, in which he brings up the torpedo, reads in Hero as follows:

> Besides, the presence of the vacuum in the water appears evident from the fact that, when poured into water, wine spreads and diffuses everywhere though the water. If there were no vacuum, in water, that would be impossible. Even a light passes across another light. If, indeed, one lights many lamps, all objects become more luminous, because the rays cross each other through all sides. But, on the other hand, [the heat] also penetrates through bronze, and all other bodies, exactly as it occurs for the torpedo of the sea (τῆς νάρκης τῆς θαλασσίας). (See Nix & Schmidt, 1899, p. 26)

After Strato and Hero, the aforementioned Alexander of Aphrodisias brought up the propagation of the torpedo's shock in two texts (besides the one already quoted and of somewhat uncertain authorship). One of these texts has come down to us in the original form (Alexander's commentary on Aristotle's *Meteorologica*), whereas we can appreciate the other only indirectly. It appears within the commentaries to

---

[8] Hero of Alexandria, who lived in the first century A.D., was a mathematician and engineer in what had been Greek and was now Roman Alexandria on the Mediterranean coast of Egypt. At the time, Alexandria was one of the most important centers of Hellenistic culture (particularly in medicine and the physical sciences). It was also the site of the most important library of the ancient world. In addition to his mathematical and physical formulations, Hero contrived various technical devices, and even anticipated modern engines, some of which had immediate utility.

De caelo, written by Simplicius of Cilicia, who lived in the sixth century A.D.[9]

In the first of these works, the transmission of the torpedo's shock is brought up within the framework of a cosmological problem typical of the Aristotelian conception of the world. In classical cosmology, the earth is at the center of the universe, which comprises a series of concentric rotating spheres (or orbs or skies) bearing the planets (which includes the sun) and the stars. The earthly world (which comprises all the space and matter situated under the first sphere [i.e., the sphere of the moon]) is formed from the four elements: earth, water, air, and fire. It is the domain of transformation and mutation (and also of life and death). In contrast, the sphere of the moon and all the other spheres with their planets and stars (the supra-lunar world) are constituted by a fifth element (*quinta essentia*) or ether, a purely crystalline and perfect matter, completely immutable and impassible (ἀπαθές).

Because of its immutability and impassibility, the sphere of the moon cannot acquire any external qualities. One of the consequences of this is that it cannot be heated when the sun's rays pass across it. This is the difficulty Alexander wrestled with in his commentary on the *Meteorologica*. Thus being such things, Alexander reasoned, how is it possible for the heat of the sun to reach the earth and increase its temperature? In other words, could a body transmit a quality or "affection" without being affected itself?

Alexander's answer to this question is yes, it can, and it was based on two analogies. The first was derived from the observation that, by letting the sun's rays pass through globes of glass of suitable form filled with water, it is possible to ignite a pile of straw, even in the absence of a substantial increase in the water's temperature. The second analogy is more interesting from our point of view, because it has to do with the torpedo:

> Even fishermen say that they know when they have a torpedo in their fishing nets, because they feel a sense of torpor in their hands when they draw in the nets. This happens notwithstanding [the fact] that the nets do not receive the same affection, before the hands. (Alexander of Aphrodisias; modified, from Hayduck, 1888, p. 18)[10]

The torpedo analogy used to solve this cosmological problem occurs twice in Alexander's text, as reported by Simplicius in his commentaries on Aristotle's *De coelo*. In the first case, it involves a consideration of physiological relevance. To account for the transmission of the sun's heat to the earth, and before invoking the example of the torpedo, Alexander points out the need for adequate "receptivity" in the body that receives the heat. To such purpose he says (according to Simplicius) that it is because of a lack of a proper receptivity that the "hearing faculty" does not perceive chromatic sensations when struck by light. Whereas the receptive property is needed in the final, affected body, for Alexander it is not, however, necessary in the intermediary body that acts exclusively as a transmission device. Alexander now considers the torpedo and, in addition to the numbness transmitted to men, in this case he also mentions the transmission through water to fishes swimming over them (ὑπερνηχομένους), which, he writes, "are said to sink down certainly without the intermediate water getting numbed" (οὐ δήπου τοῦ μεταξὺ ναρκῶντος ὕδατος: see Heiberg, 1894, p. 373).

The first of Alexander's torpedo analogies reported by Simplicius does not bring up the possible igniting actions of the sun's rays on ordinary earthly objects (present in Alexander's commentary to the *Meteorologica*). This action is, however, alluded to in the second passage in Simplicius, in which the torpedo is again mentioned within the framework of the cosmological problem of the sun's heating actions on earth. Here Simplicius writes:

> And nevertheless, to solve such a problem, Alexander states, both above and now, that many affections are produced through vehicles which, however, do not suffer the same affection they transmit to the affected matters; as for instance—he notes—the dry wood, which, takes fire by the sun's action, without any change in the medium. In the previous cases he had produced as evidence the affection produced by the sea torpedo (θαλασσίας νάρκης) on man, in the absence of any numbness in the fishing nets. (p. 440)

As noted by Rescigno (2000), Alexander of Aphrodisias probably derived his information on the transmission of the torpedo's shocks from the writings of Theophrastus and of other members of the Peripatetic school; the same writings were also the sources for Plutarch and Pliny the Elder. Alexander, in turn, was the probable source for Plotinus, a Greek philosopher, who was born in Egypt and lived in the third century A.D. (partially overlapping Alexander). Plotinus studied in Alexandria, traveled throughout the Mediterranean world, and eventually settled in Rome. He was a spiritual and moral philosopher, who distrusted the sensory or material world. He was largely inspired by Plato and strongly accentuated the mystical aspects of Plato's philosophy. For these reasons, his writings would play an important role in the resurgence of a Platonism with strong esoteric connotations in the Italian Renaissance (Neoplatonism), particularly after Marsilius Ficinus translated his main work, the *Enneades* (also spelled *Enneads* in English translation), in Florence during the first half of the fifteenth century.

Although likely inspired by Alexander's texts (and similarly dealing with the nature of intermediate bodies in the transmission process), Plotinus does not bring up the torpedo in a cosmological context.[11] Instead, he is concerned

---

[9] Readers of Galileo Galilei know well that Galileo has derived the name of one of the three discussants in his two main works, the *Dialogo sopra i Massimi Sistemi del Mondo* (1632) and the *Discorsi e Dimostrazioni matematiche sopra due nuove Scienze* (1638), from Simplicius. As with the historical Simplicius, Galileo's Simplicio is also a commentator and a supporter of Aristotle.

[10] As discussed by Andrea Rescigno (2000) in an article on the sea torpedo in Alexander of Aphrodisias and Plotinus, Alexander's conclusions were strongly criticized by Johannes Philoponus, another of the commentators of Aristotle, who lived in the sixth century A.D. (at the same time as Simplicius). Among the arguments invoked by Philoponus was the fact that, if we assume that the fishing nets are not numbed because they cannot be conscious of it, we should also conclude that, under the action of heating or cooling agents, other inanimate bodies (e.g., stones, metals, woods) do not warm or cool, because they cannot perceive what is happening in them.

[11] See Rescigno (2000) for more on the exchanges among Alexander, Plotinus, and other scholars of these times, who were dealing with the transmission of torpedo shock.

with a problem of sensory physics and physiology, namely the role played by the medium by which the visual stimulus (χρώματος [i.e., color]) is transmitted.[12] In this regard, it bears a particular relationship with the passage in which Simplicius reports Alexander's words about the absence of chromatic sensations when light strikes the "hearing faculty."

Plotinus shares Alexander's view that it is not necessary for the intermediate medium to feel or participate in the same affection that it transmits. With regard to vision, he even goes beyond Alexander, claiming that affections could be transmitted in the absence of any medium whatsoever. As a matter of fact, he writes in the fourth book of the *Enneades* that, the more media are dense, the more they interfere with vision, whereas with tenuous media we can see more clearly. As with Alexander, Plotinus holds that what is necessary for an affection to be transmitted is a natural correspondence—a "sympathy" (συμπάθεια) between the body producing the affection and the one receiving it.

This is the long passage from the *Enneades* where Plotinus invokes the example of the torpedo in the context of vision: evidence that an affection (in his words, πάθος, meaning pathos) can be transmitted even through an "indifferent" body:

> But it is not perhaps necessary that also the medium suffers the affection, because the eye—which is by nature capable of suffering it—actually suffers it; or maybe if the medium suffers it, it does it in a different way. As a matter of fact, even the rod that is between the torpedo-fish and the hand is not affected in the same way as the hand; and certainly there too, if a rod or a hair [i.e., a line] were not in between, the hand would not be affected. Or even this might be disputed: for if the torpedo-fish gets into a net, the fisherman is said to receive a numbness (ναρκᾶν). The reasoning brings us to what we call sympathy (συμπαθείας). Because of a natural correspondence, or because it possesses something in common [with another], a substance suffers an affection from another; if the medium does not share such common nature, it does not suffer the affection or does suffer it in a minimal way. This being [the nature of] things, a substance which is by nature capable of receiving an affection, would suffer it much more if there were no medium, even if the medium were of a nature capable of somehow suffering that affection. (Plotinus, *Enneades IV,* Book V, 1; from Volkmann, 1884, pp. 102–103; also see Plotinus, 1995, p. 283)

The association of the transmission of the torpedo's shock within the context of sensory physiology is also present in other Greco-Roman philosophical texts. Among these authors we find Porphyrius, a pupil of Plotinus who lived in the third century A.D., and Calcidius, who lived a century later (see Renna, me1995).[13]

In a more physical context, the discussions on the transmission of the torpedo's shocks at this time can be considered the closest approximations we know of to the problem of electric conduction in the classical epoch. As we have previously stated, there was no notion of electricity at this time. The term ἤλεκτρον ("electron") only designated amber, which—when adequately rubbed—was known to attract chaff and other light bodies (by some "un-nameable" quality proper to it). The similarly mysterious power of the torpedo was then the only available "electricity" for which the transmission could be observed.

AELIAN'S MYTHS ABOUT TORPEDOES AND WATER

As we have seen, the transmission of the torpedo's shocks was of great interest to ancient philosophers. Nevertheless, the available texts seem to refer only to spontaneous observations of the phenomenon. What is mentioned is generally based on the reports of fishermen, and in many cases the absence of the author's own direct experiences is evident from the wording. Thus, we often find the phrases "it is said," "people say," and the like.

In this context, and in the absence of a real notion of electricity, the ancients did not seem to attempt to distinguish between the media capable of transmitting the power of these fishes and those agents that we would today label as "non-conductive" bodies. Indeed, as we pointed out in the previous chapter, later scholars, including Theophylact Simocatta and Michael Glycas, have assumed—most likely on the basis of the texts of earlier authors—that the shocks could even be transmitted through air.

Aelian is among the ancient scholars who based most of what they wrote on what they read or heard. We mentioned him in the previous chapter (Footnote 8) as a source of an anecdote about the pregnant torpedo, specifically that it can be made to deliver its offspring by pouring water on it. Although born in Praeneste, not far from Rome, near the end of the third century A.D., Aelian wrote his major works in Greek. The mention of the pregnant torpedo appears in one of these texts, his Περί ζῴων ἰδιότητος, Latinized as his *De animalium natura* (In English, "On the Nature of Animals" or "On the Characteristics of Animals"; see Aelian, 1959) (Fig. 4.5). This work is a collection of stories and anecdotes, and sometimes myths and fables, about animals, often with moral intentions, directed to a relatively wide readership.

---

[12] As shown by Rescigno (2000), it is very likely that Alexander developed the analogical argument by putting together both the cosmological argument (i.e., the heating action of the sun on the earth) and the physiological one (i.e., the problem of the medium in vision), albeit in a lost text. To support his interpretation, Rescigno invokes a text by Themistius, a Greek scholar born in what is now Turkey, who lived in the fourth century A.D. Unfortunately, Themistius' text (again a commentary on Aristotle's *De coelo*), originally written in Greek, has arrived to us in an incomplete form: that is, in a Latin translation derived from an Arabic one, which in turn was based on a Hebraic translation of the original.

[13] The problem of the role of the medium in the process of vision would be one of the most debated issues stemming from the epoch of Aristotle, and the debate would continue until relatively recent times (it would be considered, for instance, by Goethe in his controversial but important studies on vision; see Goethe, 1810). Moreover, although in a very different (and decidedly scientific) form, the problem of the receptivity (and of the sympathy) of the sensory organ with relation to a specific affection (or energy) would resurface in the famous doctrine of "specific nervous energies" most closely associated with the nineteenth-century German physiologist Johannes Müller. According to this doctrine, which has laid the groundwork for modern sensory physiology, any sensory receptor or organ is specifically accommodated to respond to a specific type of sensory stimulus and will generate a specific sensation (see Meulders, 2001; Finger & Wade, 2002a,b; Wade, 2003; also Chapter 22).

Figure 4.5. The title page of a 1562 Latin edition of Aelian's zoological poems.

The first mention of the torpedo in this work occurs in a short passage where the power of the torpedo is linked to that of the remora. This occurs in a chapter devoted to the wondrous properties of more or less fantastic animals (which might suggest a derivation from Pliny). For both fishes, Aelian plays with the similarity that existed between their names and the effects they could produce:

> The fish known as torpedo (νάρκη) produces the effects expressed by its name and induces numbness (ναρκᾶν) in everybody who touches it. In a similar way, the remora is so called because it stops the ships (ἐχενηΐς ἐπέχει τὰς ναῦς) and delays them. (*De animalium natura*, I, 36: from Hercher, 1858, p. 12)

A more ample account of the torpedo appears in Chapter 14 of Book IX. This short chapter is entirely devoted to the torpedo. Despite the relative richness of the details provided, however, Aelian introduced some surely misleading information when he wrote:

> When I was only a boy, I heard from my mother that if somebody touches the torpedo, his hands undergo the affection expressed by its name (ἐκ τοῦ ὀνόματος πάθος). Afterwards, I was informed from expert men that, if one touches the net in which it was caught, he is totally torpified (ναρκᾷ πάντως). If, moreover, somebody puts it alive in a vessel, and pours seawater on it, if the fish is pregnant and near the parturition time, it will deliver. Furthermore, if this water were poured from the vessel on the hand or foot of a man, the hand or the foot would be unavoidably torpified. (From Hercher, 1858, p. 153)

The last statement on the torpifying action of the water coming from a container in which a torpedo is kept suggests that Aelian was reporting an uncontrolled observation. It was probably based on a hypothesis of the torpedo's power derived from the texts of previous authors, such as Plutarch, Pliny, and even perhaps Galen.[14]

DIPHILUS AND EXPERIMENTALLY LOCALIZING THE TORPEDO'S POWER

Diphilus of Laodicea, a Greek scholar who seems to have lived in the first century A.D., provides one of the few passages in ancient science in which a property of the torpedo shock seems to have been derived from genuine experimentation. We know of what he did from Athenaeus of Naucratis, about whom much more will be said later in this chapter. In Book VII of Athenaeus' *Deipnosophistae*, we find these lines:

> In his commentary on Nicander's *Theriaca*, Diphilus of Laodicea says that not the entire body of the animal produces the numbness (νάρκην), but only a part of it. He alleges that he has arrived at this conclusion on the basis of many experiments (πείρας πολλῆς) (Book VII, 314d).[15]

It is likely that there were more details on the body part(s) responsible for the shocks in Diphilus' lost text. Also, that text might have been the source for subsequent authors, including the two poets, Oppian and Claudian, to whom we shall next turn. Both assigned the torpedo's power to the "sides" of the fish: λαγόνες in the Greek original of Oppian, and *latus* in the Latin of Claudian.

Oppian, moreover, used the term κερκίδες, meaning the shuttles (or spools) of the loom to designate the part of the fish responsible for the shock. The electric organs actually have shapes that roughly correspond to loom shuttles. This observation strongly suggests that Oppian's wording was actually derived from reported experimental investigations. Whether the investigator was Diphilus himself, or whether Oppian derived his information from others involved in these experiments, is unclear, the extant evidence being both indirect and fragmentary.

---

[14] It is indeed extremely improbable (although not totally impossible) that a torpedo can produce a perceived shock at the very moment when the vessel in which it is contained is emptied, while the flow of the water establishes a conductive path between it and the hand and the foot on which the water is poured.

[15] The best available edition of the *Dehipnosophistae* is the four-volume set edited by Luciano Canfora, Christian Jacob, and others, published in the original Greek with an Italian translation (see Athenaeus, 2001). The reference we give is to the original Athenaeus book and its section, which permits locating the passage in other scholarly editions.

THE NIGHTMARISH AND WONDERFUL TORPEDO OF POETS

The poets Oppian and Claudian bring us to the more literary side of electric fishes. Yet their works also tell us quite a bit about popular beliefs and the degree to which some of the "science" discussed above permeated the culture. Needless to say, their lines also add more color to the picture we have been painting, since poets can have unique ways of looking at nature's oddities.

Oppian (Oppianos, Oppianus) of Corycus in Cilicia wrote his verses during the second century A.D. It is commonly held that when he went to Rome and presented his poems to Marcus Aurelius, the emperor was so pleased that he gave him a piece of gold for each line and took him into his favor. Unfortunately, Oppian died during a plague shortly thereafter at the age of 30.

Oppian's huge Ἁλιευτικά or *Halieutika* comprises five books with some 3,500 lines (Figs. 4.6 and 4.7). After being translated into Latin during the Renaissance, it was translated into English and many other languages, continuously attracting scholarly interest (Oppian, 1722; Gow, 1968; Bodson, 1981). It deals with the habits and characteristics of fishes, and gives instructions for fishing. As for his sources, some of Oppian's material was gathered from personal observations, although he also borrowed from others (e.g., Xenophon of Athens), hoping to provide advice to fishermen and to amuse those wanting to learn more about fishes and fishing.

Oppian's work was dedicated to "Antoninus," presumably Marcus Aurelius, and his son Commodus, and it is often cited as the most complete treatise on fishes and fishing prior to the Renaissance. What he wrote about torpedoes provides one of the best synopses of what was known about them from antiquity. In an intelligent way, he incorporated most of the available information derived from some of the better sources of the epoch, and expanded the account of the torpedo's effects with poetic charm toward the territories of the wondrous and troubling.

Oppian calls the torpedo the fish "with the appropriate name" (ἐτήτυμον ὄνομα νάρκη; *Halieutika*, I, 105). He dedicates a long passage of Book II of the *Halieutika* to its behavior. Here he emphasizes the contrast between the fish's sluggish behavior and apparently weak body, and its dreadful force (ἀλκή). As noted, the "spools" on the sides of its body underlie this force. The torpedo, Oppian remarks, is conscious of this "gift received from God," and it uses it with great astuteness and boldness to catch and pitilessly devour its prey. As Galen and other authors have pointed out, Oppian also tells his readers that the torpor produced by the fish leads both to deprivation of movement and reductions of sensitivity.

The immobilizing actions of a torpedo's shocks come across with intense force in Oppian's poetry, being compared to the imagery of a terrible nightmare. The torpedo's prey is unable to swim away in its attempt to escape its demise, much as a sleeper might feel an imaginary paralysis that would prevent him or her from fleeing from some imminent danger. By this time, the image of the dreadful nightmare with a person trying to escape imminent danger and unable to do so was well established in both Greek and Latin poetry, although it was not associated with the torpedo.[16]

To experience this nightmarish association, let us read from Book II of *Halieutika*. Our source for an English translation will come from the eighteenth century. Although the poetry differs in places from the Greek original, it still provides its haunting flavor:

The Pow'r of latent Charms the Cramp-Fish know,
Tho' soft their Bodies, and Motion slow.
Unseen, foreboding Chance and future Prey,
The crafty Sluggards take their silent Way.
Stretcht from each Side they point their magick Wands,
Whose icy Touch the strongest Fin commands;
Quick thro' the whole it shoots the rushing Pain,
Freezes the Blood, and thrills in ev'ry Vein;
Strikes all that dare approach with strange Surprize,
Stiffens the Fin, and dims the mazed Eyes.
Conscious of secret Pow'r, a Gift divine,
On Sands, as dead, the Cramp-Fish lies supine,
Thus careless stretcht a wide Destruction makes,
And wand'ring Shoals without her Labour takes.
Fixt sudden they the numbing Torpor feel;
The Parts contract, the Fluids all congeal.
No more the busy Messengers of Sense
Motion around, and conscious Life dispense;
Nor flowing Streams the circling Heart diffuse,
But the chill'd Parts forget their former Use.
While urg'd by pleasing Hopes, to fresh Repast
The willy Cramp-Fish moves with awkward Hast.
Oft, as the nimble Swimmers heedless pride
In active Course, and curling Streams divide,
They lifeless stretch by sudden Pains confin'd,
And secret Chains the fetter'd Captives bind,
No more they wanton dive, or giddy roam,
Vault on the Seas, and vex the rishing Foam;
Dull Rest they now, and fatal Slumbers love,
Nor backward can retreat, nor forward move.

As when in Dreams imagin'd Forms appear,
When dreaded Sounds we distant seem to hear,
Or shady Ghosts with silent Horror rise,
And Spectres glare before sleeping Eyes,
Fearful of coming Ills we sweating lie,
And willing would from fancy'd Dangers fly:
Rooted we stand, the Heart incessant beats,

---

[16] We find it in the second book of Homer's *Iliad* (verses 199 following) where, during the duel between Hector and Achilles, the image of the nightmare serves to signify the impossibility of Hector fleeing and of Achilles reaching him. In the concluding part of the *Aeneid* (Book XII, verses 908 and following), Virgil makes recourse to a somewhat similar image, during the duel between Aeneas and Turnus, to signify the impossibility of Turnus, who was immobilized in action, fighting effectively against the Greek hero.

In the context of nightmarish images, it is perhaps interesting to mention here what was said about the torpedo by Artemidorus Daldianus, a professional dream interpreter, who lived c.140 A.D. in Greece. Artemidorus wrote a work in Greek that he titled *Oneirokritika* (commonly translated as *Interpretation of the Dreams*), one of the most important texts of this type from the Greco-Roman world. In Chapter XIV of Book II, he wrote that, as for other flat cartilaginous fish, dreaming of a torpedo is a bad omen because it indicates imminent danger (κινδύνουσ) and scheming or machinations (ἐπιβουλάς) (see Reiff, 1805, vol. I, p. 107). It is possible that the idea that the torpedo is a sign of schemes and plots derives from the paralyzing action of its shocks, which we find so well expressed in Oppian.

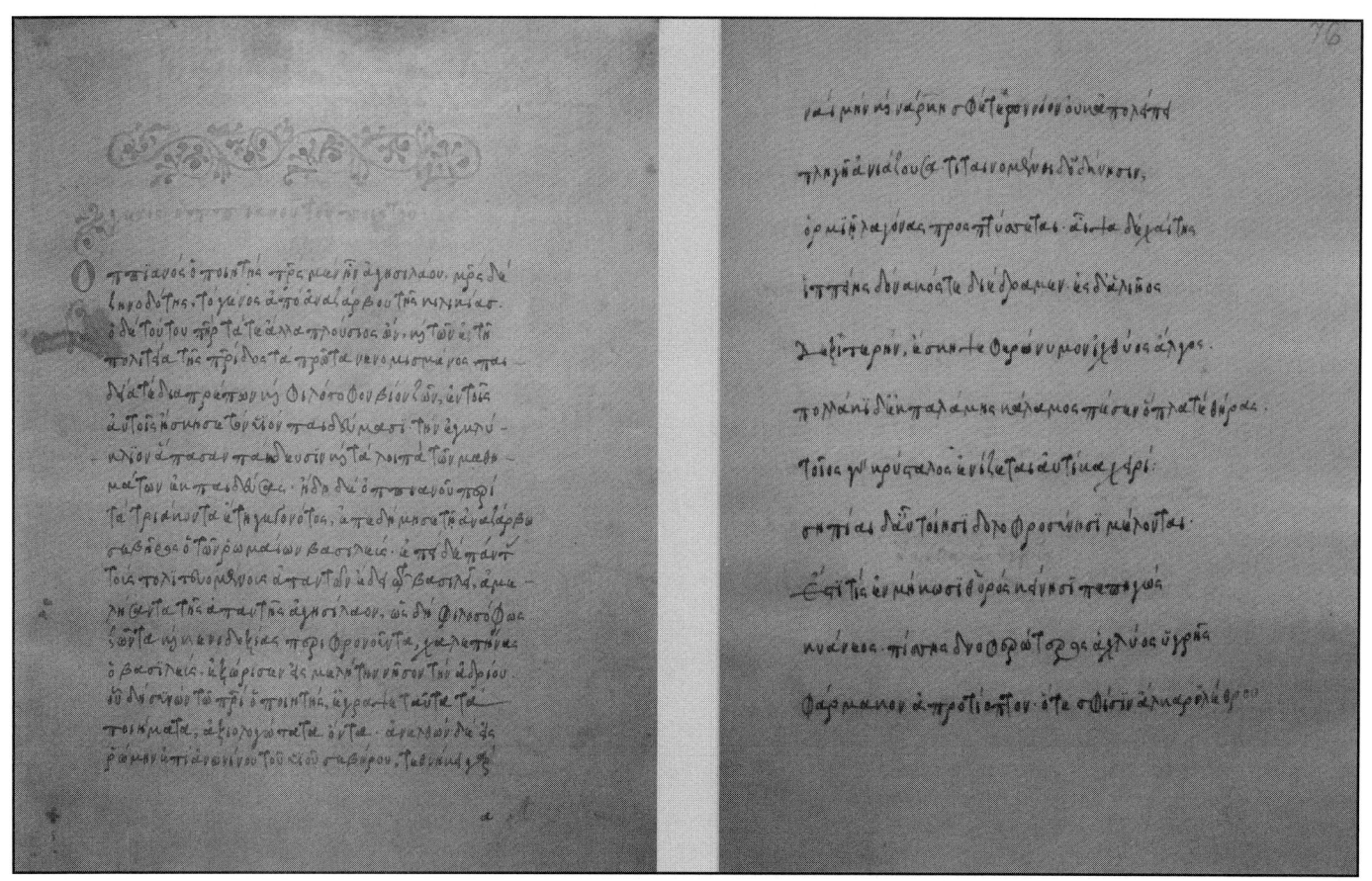

Figure 4.6. Two pages of a sixteenth-century Greek manuscript of Oppian's *Halieuticks*. The torpedo is mentioned as *nárkē* in the first line of the page at the right. (Courtesy of The Bakken, Minneapolis; © The Bakken).

And hasty Strokes the quicker Pulse repeats.
Lab'ring to move we seem to strive in vain,
Wile pond'rous Clogs the struggling Feet retain.

With such a binding Force the Cramp-Fish stays
The swiftest Fish, and strikes with dizzy Maze.
One Touch of her's dams up the vital Flood,
Contracts the Nerves, and clots the stagnate Blood.

(Oppian, 1722, pp. 64–65: *Halieutika*, Book II, verses 56–85)

In Book III of *Halieutika*, the torpedo is listed among those fishes that can make an angler remorseful when captured. Here Oppian gives an account of the transmission of the shock through the line (χαίτης ἱππείης, horse hairs in the original) and the fishing rod (δόνακος, meaning reed):

The Cramp-Fish, when the pungent pain alarms,
Exerts his magick Pow'rs and poison'd Charms.
Clings around the Line, and bids th' Embrace infuse
From fertile Cells comprest his subtil Juice
Th' aspiring Tide its restless Volumes rears,
Rolls up the steep Ascent of slipp'ry Hairs,
Then down the Rod with easy Motion slides
And entering in the Fisher's Hand subsides.
Of every Joint an icy Stiffness steals,
The flowing Spirit binds, and Blood congeals
Down drops the Rod dismist, and floating lies,
Drawn captive in the Turn, the Fish's Prize.

The idea of coldness (here "an icy Stiffness") as a vehicle and the mechanism of the fish's shock is present in the original Greek, where we find the term κρύσταλλος, meaning ice, icy water. In contrast, the phrase a "subtil Juice" is an invention of the translator based on eighteenth-century physiological concepts or is due to a corrupted literary transmission. In the original, what runs up to the hand of the angler is simply and interestingly pain (ἄλγος).

Our second poet, Claudius Claudianus, or Claudian, followed Oppian by several centuries, living from about 370 to 410. Unlike Oppian, he was brought up on the Nile Delta, and specifically in Alexandria, the great cultural center for Hellenism in Egypt. He visited Rome in 394, became court poet to Emperor Honorius, and was also associated with Stilicho (Cameron, 1970; Platnauer, 1922). A 5-year stay in Milan interrupted his stay in Rome.

Claudian also differs from Oppian in the nature of his output. He is best remembered for his idylls, or short pastoral poems, which were written in Latin. But like Oppian, torpedoes fascinated him, which is why he devoted one of his poems about nature to the torpedo (Fig. 4.8).

Titled *De torpedine* ("The Torpedo"), this short poem begins with a rhetorical question (see Claudian 1922, pp. 276–279, for the entire poem in both Latin and English; this poem also appears in other editions [e.g., Claudian, 1985]). In Latin:

Quis non indomitam dirae torpedinis artem
audiit et merito signatas nomine vires?

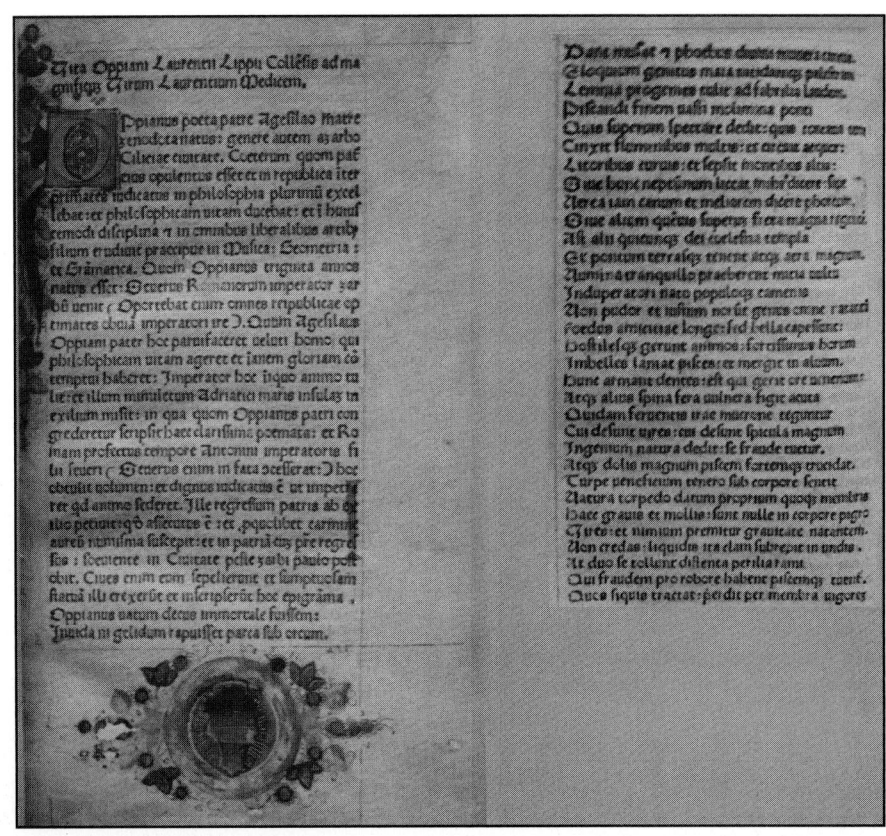

Figure 4.7. Two pages of a Florentine codex dated 1478 of a Latin edition of Oppian's *Halieuticks*. The torpedo is mentioned in the lower part of the page on the right. (Courtesy of The Bakken, Minneapolis; © The Bakken).

And in English:

Who has not heard of the untamed art of the terrible torpedo and of the powers that win it its name? (Fig. 4.9)

In our Foreword, we have already alluded to these words, which are of special historical importance to us because they will later be associated with John Walsh and his experiments on the electrical nature of the torpedo's shocks. Walsh's torpedo experiments took place in France in 1772, and his words, which show his familiarity with Claudian's poem, will be put into context in Chapter 14. For now, the important point is that we again encounter many of the themes concerning the torpedo that we have considered in this and the previous chapter in this poem. Indeed, we can also visualize Oppian as one of Claudian's sources.

Claudian followed Galen in stating that the torpedo makes the body feel numb in the same way as exposure to cold. He adopted the theory that the fish produces these effects by some sort of a *gelido veneno* (cold venom), and he is even more explicit about this than Oppian. But like Oppian, he suggests a localization of the structures responsible for the fish shock, this being in the "sides" of the fish.

There are other similarities with Oppian that tie Claudian to him, or at least to common sources, all of which show that some of the notions that we have been discussing about the torpedo had wide diffusion when these men lived. These even include a reference to the blood, which may coagulate following a shock, and the presentation of the fish as a small but astute monster, conscious of its power and ready to immobilize and devour other fishes that come near it.

Particularly well developed in the last part of the poem is the reference to the wretched angler, who happened to catch one of these rays and now regrets it.

In translation, select parts of Claudian's piece on the torpedo read:

Its body is soft and its motions slow. Scarcely does it mark the sand o'er which it crawls so sluggishly. But nature has armed its flanks with a numbing poison and mingl'd with its marrow chill to freeze all living creatures, hiding as it were its own winter in its heart. The fish seconds nature's efforts with its own guilefulness, knowing its own capabilities, it employs cunning and trusting to its powers of touch lies stretched full length among the sea-weed and so attacks its prey. It stays motionless; all that have touched it lie benumbed. Then, when success has crowned its efforts, it springs up and greedily devours without fear the living limbs of its victims.

Should it carelessly swallow a piece of bait that hides a hook of bronze and feels the pull of the jagged barbs, it does not swim away nor seek to free itself by vainly biting at the line; but artfully approaches the dark line and, though a prisoner, forgets not its skill, emitting from its poisonous veins an effluence which spreads far and wide through the water. The poison's bane leaves the sea and creeps up the line; it will soon prove too much for the distant fisherman. The dead paralysing force rises above the water's level and climbing up the drooping line, passes down the jointed rod, and congeals, e'er he is even aware of it, the blood of the fisherman's victorious hand.

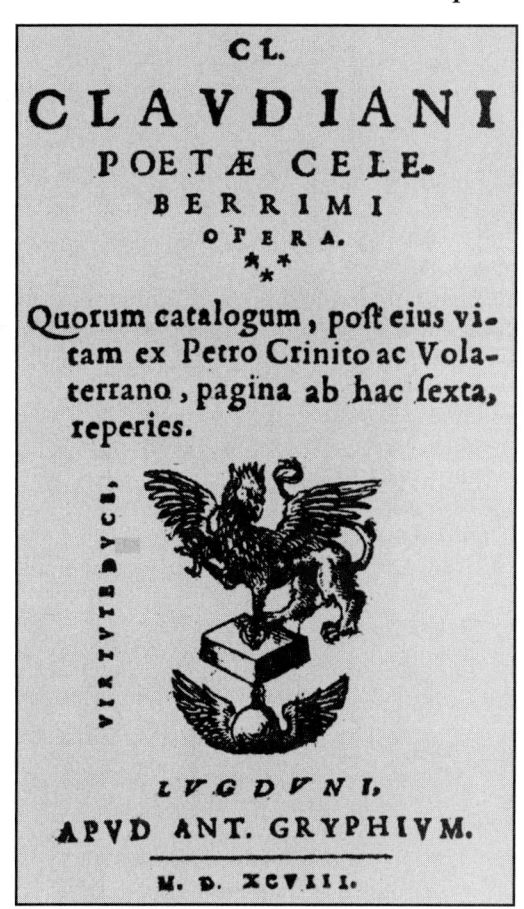

Figure 4.8. The title page of a 1598 edition of Claudian's poems.

He casts away his dangerous burden and lets go his rebel prey, returning home disarmed without his rod. (Claudianus, 1922, Part XLIX)

ATHENAEUS AND THE BIBLIOMANIAC GOURMANDS

Let us now return to Athenaeus of Naucratis and look more carefully at his *Deipnosophistae* (his "Banquet of the Learned," or better, "The Learned in the Banquet"; for an English translation, see Athenaeus, 1929) (Fig. 4.10). As with Oppian, the exact dates of Athenaeus' life are not known, but from secondary sources and an analysis of his material we can place him in the late second and into the third century A.D.

We have already noted that Oppian, a Roman author, wrote in Greek, whereas Claudian, a Greek from Hellenistic Egypt, wrote in Latin. With his life and work, Athenaeus represents an even more notable case of this mutual interchange of cultures, and of the difficulty of defining definite boundaries within the Roman *œcumene*. Born in Naucratis, a flourishing Greek colony near the mouth of the Nile, certainly educated in Alexandria, a great center of Hellenistic culture, and bearing his Greek origin within his own name, Athenaeus spent most his life in Rome, where he dedicated his main work to his mentor and sponsor, the Roman Laurentius, a high officer who might have been in charge of supervising religious ceremonies for the imperial administration (Braund & Wilkins, 2000).

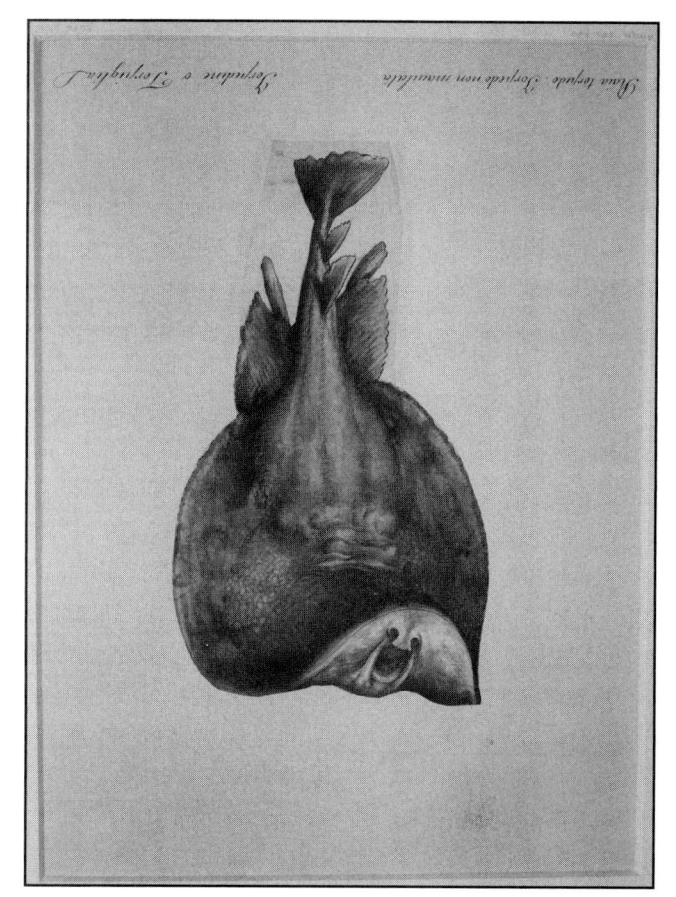

Figure 4.9. A torpedo depicted by an anonymous painter of the seventeenth century as a dreadful sea creature which might have been inspired by Claudian's poem. The image, based on a pen drawing colored with watercolors, represents the fish in the typical attitude it can take when, being very lively, it is held by the tail. The torpedo which is of the non-spotted variety, can correspond either to the *Torpedo marmorata* or to the much rarer giant *Torpedo nobiliana* (i.e. the "black torpedo" used by Scribonius in his medical application). The original image, belonging to the collection of the Medici Grand Duke's family, has been mounted in the nineteenth century upside-down on a paperboard with the addition of some Latin and Italian or Tuscan vernacular names: *Raja torpedo. Torpedo non maculata. Torpedine* or *Torpiglia*. (© Museo degli Uffizi, Florence, Gabinetto Disegni e Stampe).

The *Deipnosophistae* was written in Greek, even though it deals with the sumptuous banquets that were held periodically in Laurentius' house in Rome. The idea of writing a text with reference to an important social ceremony (as were such banquets during the Roman Empire) was clearly inspired by Plato's *Symposium* and related texts, including those of Xenophon and Plutarch. In the *Deipnosophistae*, Athenaeus reports the conversations occurring among Laurentius' guests to a friend named Timocrates. These dialogues take place both at the banquet (*deîpnon*) and in the drinking phase that follows it (*sympósion*). Those participating in the event are *literati* and the upper crust of the Greek community in Rome. One such person is Galen.

Although the theme of food is the thread uniting the many disparate subjects covered during the banquet, the deep theme of the *Deipnosophistae* is principally the culture of the great ages of Greek civilization, already in its decline by Athenaeus' time. The *Deipnosophistae* can and should be considered a precious repertory of this culture. Thus, although extremely rich, sumptuous, and socially important, the

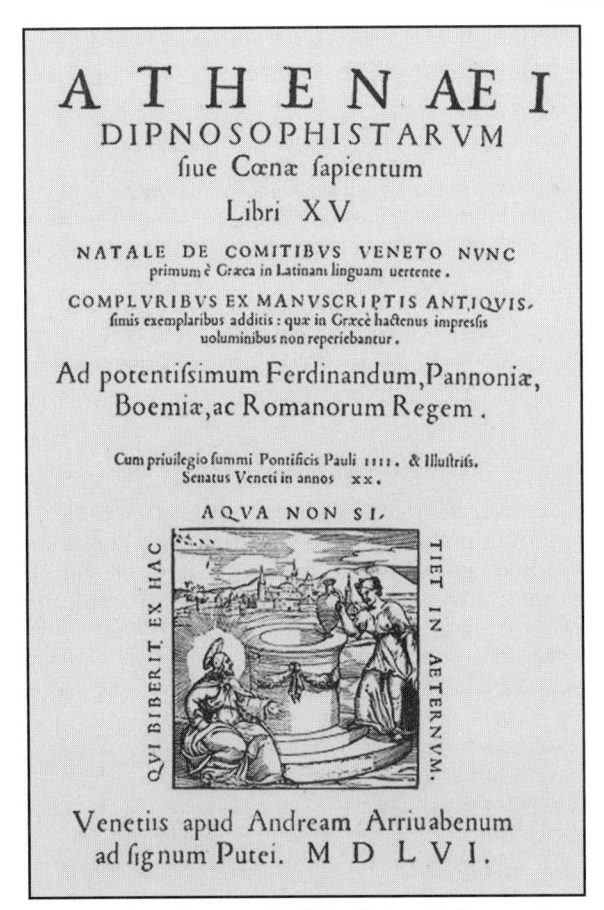

Figure 4.10. The title page of a 1556 edition of Athenaeus' *Deipnosophistae*.

banquets of the *Deipnosophistae* were fundamentally cultural events held in a mansion holding one of the most important libraries in Rome. The ambiance reminds us of the Library and the Museum of Alexandria, where scholars used to have learned conversations while taking their meals together. The host of the *Deipnosophistae* was himself a scholar and bibliophile, and every guest was expected to bring several books (generally in the form of scrolls) to be used during the philosophical and literary conversations.

Athenaeus cites thousands of names and hundreds of books and, for some of them, what we find in the *Deipnosophistae* is the only source we have for what has since been lost.[17] As we mentioned, Athenaeus is the only source of the work of Clearchus on the torpedo. He is also the source for Diphilus of Laodicaea, who localized the torpedo's power in only a part of its body. The scheme of the narration, with continuous digressions, where a guest refers to information that some other person has told him after reading a book in which the author speaks of the work of a second or possibly a third author, has led some modern scholars to look upon the *Deipnosophistae* as an ancient hypertext, in which the reader is invited to navigate by following the various erudite conversations (Jacob, 2001).

---

[17] Even the *Deipnosophistae* have not survived in their entirety. There is an ongoing debate among scholars about how incomplete is the extant text is relative to the original (Jacob, 2001).

Some of these conversations involve the torpedo. Athenaeus refers to the sea torpedo with the term *nárkē*, but he does not normally specify that it is the sea-ray, except in the literary quotation from Plato's *Meno*. In fact, he is most interested in the torpedo as a food. We find, for example, the torpedo served with many others delicacies in a sumptuous banquet described in the poem *The Banquet*, which Athenaeus attributes to Philoxenos of Citera (Book IV, 147b).

The most ample reference to the torpedo, including some words about its numbing power, can be found in Book VII. Here it is mentioned with other rays and cartilaginous fishes prepared for the banquet in "Hebe's marriage" of Epicharmus (of Megara), the founder of the comedy (in Section 286c). In Section 304a, Athenaeus alludes to a work of Antiphanes, *The Hairstylist*, where a man from the country is told not to like seafood and to eat only those fishes living close to shores, one being the torpedo. Also in the same book (313f), there is mention of a book attributed to "Plato [Comicus] or Cantharus" where we find "a boiled torpedo is a really delicious dish." Soon after this, Athenaeus mentions a verse from the *Phanus* of Menander, noting the different spelling of the fish name used by the author (*nárka* instead of *nárkē*). In translation, we find the following wording here: "a certain torpor is creeping over my skin." He also refers to Icesius, who remarks that this fish is not particularly nutritious.

It is after citing Icesius that Athenaeus provides more information of immediate interest for us, since it concerns the behavior of the torpedo and its shocking power. As a matter of fact, besides a brief allusion in Book III to the pricking effect of the *nárkē* on the delicate fingers of a woman (III, 107c), it is only in Book VII that we find a serious mention of its ability to transmit its benumbing property. First Athenaeus cites a lost treatise of Theophrastus on *Animals that Live in Holes*, where it is said that the torpedo likes to stay underground because of the cold. Next he cites another lost work of the same author, *On Poisonous Animals and on Animals that Sting*, where we find mention of the transmission of the torpedo's shock through wood and harpoons. It is at this point that he refers to the work on the torpedo of Clearchus of Solis and, moreover, to the commentary of Diphilus of Laodicaea on Nicander's *Theriaca* (which we have discussed).

The zoological/physiological reference to the torpedo is, however, just an intermezzo, because he again comes back to its gastronomical aspects with short quotations of Archestratus, Alexis the comic, and Demetrius. From these we gather that, in contrast to how we feel about eating torpedoes now, in those days the (sea) torpedo was generally considered a delicious food.

Notably, Athenaeus uses the word *nárkē* to denote the first fish on a rather long list of fishes found in the Nile River. Although it usually has been assumed that he was referring to an electric catfish here, this is not totally certain, because torpedo rays can swim into the mouths of rivers, as was discovered by the astonished Alexander von Humboldt when he visited South America in 1800. Thus, Athenaeus might have been alluding to the sea torpedo, not *Malapterurus*. In support of this possibility, he writes that the Nile *nárkē* is "the most delicious" of all the listed fishes. His wording is consistent with the fact that, as mentioned in his *Deipnosophistae*, the sea torpedo was so highly estimated that it was served in sumptuous banquets, whereas he tells us that a silurus or catfish is generally considered a low-quality fish. It should also be noted that although Athenaeus also includes the *silourous* on his list of Nile fishes, he does not allude to any particular type of catfish or mention anything about benumbing catfish discharges. Indeed, the reference to the torpedo's benumbing power is rather scanty in the *Deipnosophistae*, as we have noted.

The torpedo passages in Athenaeus' *Deipnosophistae* show the encyclopedic and fragmentary character of the information contained here and why it is open to speculation. Despite the documentary importance of the text, the author was not motivated by a scientific interest when he described it, or for that matter other fishes, animals, and plants. He was most interested in them as food, and possibly also as medicines. His passages about the *nárkē* are ambiguous at best, and although some modern authors have inferred that he was in places referencing an electric catfish with this term, our own analyses suggest that this may not necessarily be the case. Further, he writes only in a very general way about catfishes, giving us no reason to think that he was singling out *Malapterurus* when discussing catfishes.

TORPEDOES AND THE FINE ARTS

In closing this chapter, and so as to leave no stone unturned, let us now examine the torpedo on ancient objects of art. In addition to the writings of philosophers, physicians, naturalists, "physicists," poets, and bibliomaniac gourmands, the artistic production of the epoch can tell us something about how the torpedo was perceived and how well it was known in the ancient Mediterranean world. In this regard, the fine arts quite literally can add more color to the canvas that we are trying to paint.

Torpedoes were already gracing beautiful pieces of Greek pottery during the first half of the fourth century B.C., when Plato was comparing Socrates' logic and words to a torpedo's riveting effects in his *Meno*. Utilitarian items, such as plates, bowls, and vessels, with torpedoes (typically accompanied by other fishes, porpoises, octopuses, squid, and the like) were made in mainland Greece and in factories elsewhere in the Greek world, and they were a prized export. Many complete pieces and considerably more fragments have survived the ravages of time, and they can be found in various museums and private collections around the world.

Among the most studied of these items are the "red-figured fish plates" produced in Athens and other parts of what is now Greece, and in a number of colonies in *Magna Graecia* (see McPhee & Trendall, 1987). In fact, estimates are that over 80 percent of these plates were produced in southern Italy. It is estimated that there are at least 1,000 in museums and private collections at this time, with numbers still growing.

Attic or Athenian red-figured fish plates, which are circular and have red fishes on black backgrounds, date back to about 400 B.C. They were in vogue for only about half a century (black-glazed fishes plates began to be produced in Athens at about the same time, and even in the same factories, and they remained in production into the second century B.C.). The red-figured plates, some associated with myths, such as the abduction of Europa, and others with just marine creatures ("plain" or "decorated"), have depressed centers and are approximately 7 to 9 inches (19 to 23 centimeters) in diameter.

Largely produced for export, especially to the Greek city-states on the Black Sea, they depict various marine animals so realistically that most are readily identifiable. These creatures include various fishes (e.g., bream, mullet, perch), crustaceans (e.g., cuttlefish, squid, octopuses), and mollusks, all of which were considered good eating. Moreover, the fact that these plates (ancient name unknown) were often found in cemeteries strongly suggests that they were primarily used for grave ceremonies or offerings.

Torpedoes seem to be absent from the Attic plates, but they do appear on the plates that began to be manufactured in southern Italy. These red-figured fish plates were strongly influenced by what the Attic artisans produced, and might even have begun with migrants (who would be called *Italiotes*) from mainland Greece at the start of the fourth century. Red-figured plates from this part of the Greek Empire were extremely popular during the fourth century, although the black-glazed fish plates from this same region would also remain in production far longer.

These plates have now been classified geographically (Sicilian, Campanian, Apulian), by styles (spots, use of white and other colors, waves on brims), by the materials used, and even by workshops or artists. Today, one artist of significance, who might have worked both at Campania (the region of *Neápolis*—Naples, deriving its name from Greek words meaning "new city") and at Paestum, has even been dubbed "the torpedo painter" (McPhee & Trendall, 1987, pp. 98–101).

In general, the plates from southern Italy are more decorated, detailed, and colorful than the Attic plates. The shapes of these plates are, however, similar to those from the Greek mainland, and the smaller ones, such as those made by "the torpedo painter," again seem to have been used at burial sites. Some of the larger plates might also have served in this capacity, although they could also have been used for serving fishes under happier circumstances.

Torpedoes are the most frequently depicted non-bony (cartilaginous) fishes on these plates, as well as on other pottery from this part of the world (Fig. 4.11). They are often shown with dark dots on their bodies, suggesting that the model could well have been the torpedo species now designated *Torpedo torpedo*. This fish was previously called *Torpedo oculata* because it bears a variable number of spots (normally five) simulating multiple eyes (*oculi* in Latin) on its back. It is the most common torpedo caught along the southern Italian shores, and this could account for its likeness on these dishes, although the number and arrangement of the spots on the pottery do not always correspond to the patterns on real fish. In this context, we should note that the disk-like shape of this torpedo and the intriguing presence of spots simulating eyes on its back made its image easy to stylize, which could have enhanced the appeal of these plates to buyers. Spotted torpedo images appear with particular frequency on Campanian ware and are always accompanied by other marine creatures on these fish plates.

Early in the nineteenth century, famed German anthropologist and naturalist Johann Friedrich Blumenbach, who was very interested in animal electricity, illustrated

Figure 4.11. A red-figured fish plate of the Roman period showing a torpedo along with other fishes. This plate comes from a factory in Apulia, a southern region of Italy. (© Ministero dei Beni Culturali, Museo Archeologico di Napoli).

one of these red-figured fish plates in a book of naturalistic images (Blumenbach, 1810). His chosen plate depicts a torpedo along with two perch and is typical of the plates manufactured in the Campanian region during the fourth century A.D.

These plates inform us that people living in the Golden Age of ancient Greece were very familiar with torpedoes, although unlike Plato's comparison of Socrates to a torpedo, these artifacts seem to have had little to do with the unique properties of these disk-like saltwater creatures or, for that matter, any special symbolism. Judging by the marine vertebrate and invertebrate company they keep on these plates, what seems likely is that they must have been considered gastronomically desirable, above and beyond what the Hippocratic physicians and others might have been proclaiming about them being easily digested and therefore medically useful. This would be in accord with the passages of the *Deipnosophistae* mentioned above.

Turning to the Roman era, it is clear that these artisans adopted much from the Greeks. Notably, they also produced dishes, pitchers, and bowls depicting torpedoes. In addition, archaeologists have discovered some beautiful torpedo images on Roman mosaics. Wishing to be selective, and knowing that this subject alone could make for a book, we shall focus on just two important examples.

The best known of these magnificent Roman mosaics was discovered in the ruins of Pompeii (House VIII, 2, 16; see Pugliese Carratelli et al., 1998, fig. 29; also shown in Feder, 1978; Kellaway, 1946; Moller, 1995; Wu, 1984) (Fig. 4.12a). It now resides in the Museo Archeologico Nazionale of Naples (Cat. No. 120177). Its size is 90 centimeters (35 inches), it is square and it beautifully depicts some 20 fishes found in the Mediterranean Sea and its branches, along with a squid and a large octopus. The torpedo, with its distinctive circular body, is positioned near the top center of this mosaic. The central scene depicts the capture of a lobster by an octopus, which might allude to a fishing technique based on the use of an octopus, still used by some fishermen today.

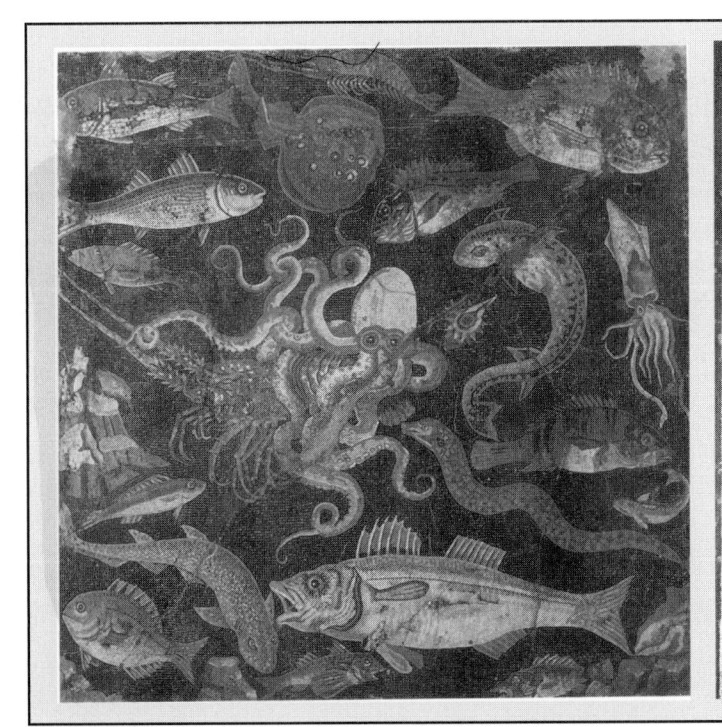

Figure 4.12. Two similar mosaics from Pompeii (now at the National Archeological Museum of Naples) depicting a sea scene, based on the capture of a lobster by an octopus. The mosaic on the left (a), which is better preserved, comes from "House VIII," whereas that on the right (b) comes from the "House of the Faunus." As mentioned in the text, the central scene alludes to a fisherman's technique for capturing lobsters that is still practiced in the region of Naples (© National Archeological Museum of Naples, Italy).

There are other mosaics based on similar representations of fish and other sea animals in Pompeii. In particular, one found in the House of the Faun is like that described above, differing mainly in the presence of its rich, polychromatic frame (see *Pompei, Pitture e Mosaici*, vol. VI, fig. 30) (Fig. 4.12b).

Although the two Pompeii mosaics are the most beautiful and best-preserved mosaics of this type, many others exist in other places with basically the same scene. They can be seen in Palestrina, Rome, Ampurias, and Aquileia in Italy, in Sousse in Tunisia, and in other sites of the Mediterranean world as well (see Meyboom, 1977 and 1978; Donati & Pasini, 1997). This suggests that there might have been some painted model(s) that circulated through the Mediterranean world.

More generally, mosaics and other artistic handicrafts with aquatic scenes (including fishes and other animals) were very common in the Greek and Roman worlds, and they are collectively referred as to Nilotic (from the word Nile) art, although typical Nilotic scenes also incorporate plants and sometimes even human images. Artisans radiating out from Egypt and other places in North Africa traveled to southern Europe and brought their art forms with them, this being yet another facet of the cultural interchanges that were occurring in the Mediterranean world. In this regard, it is interesting to see some features in the Nilotic art that are reminiscent of those that we had described when discussing ancient Egyptian tomb paintings in Chapter 2.

Roman fishermen, with their lines and nets, and fishmongers would have been very familiar with the sea dwellers on these art forms. The fact that the fins of some fishes are shown collapsed would suggest that artisans might have been using at least some freshly caught fishes when working on these mosaics and/or the model(s) for them (Thompson, 1947). It is also entirely possible that they decorated seafood shops. It is not hard to envision how these saltwater creatures could have graced the tables of those living along the coasts of southern Europe during Roman times, much as they did during the preceding Greek era.

Finally, mention must be made of the mosaics at the Villa Romana de Casale, a few miles from the town of Piazza Armerina in Sicily. As with the pottery and mosaics just mentioned, the torpedo is found along with many other fishes here. These mosaics date from about 320 A.D. and depict an incident in the legend of Arion, a Greek poet who won a singing contest in Sicily but was taken prisoner while sailing back to Corinth. Facing execution, he threw himself into the sea with its torpedoes rather than submitting passively to his captors. He was saved from various sea monsters and virtually certain death by a dolphin that carried him safely to land (see Whittaker, 1992, p. 486, for an illustration).

As we have mentioned in this and the previous chapter, myths and legends about animals with strange powers would circulate from ancient times through the Renaissance. Pliny the Elder, whose all-inclusive writings long appealed to wide audiences, would be especially important in this regard, but he would not be the sole source of fanciful information. Mixed with these myths, there would also be solid naturalistic observations and intriguing theories, such as Galen's ideas about a cold agent. It is by no means an exaggeration to say that we could not begin to appreciate what would follow into the early modern period without having a good understanding of Greek and Roman philosophy, science, and medicine, and an appreciation of the various ways in which the torpedo entered into their cultures, including in various art forms.

Chapter 5
Byzantine and Islamic Writings

> Your books have brought us back, as if to our home, when we were roving and wandering in our own city like strangers, so that we might sometimes be able to understand who and where we are. You have laid open the lifetime of our country, the description of the ages, the laws of sacred matters and of priests, learning both domestic and public, the names, kinds, functions and causes of settlements, regions, places, and all matters both human and divine.
> (Cicero's praise of Marcus Terentius Varro. From *Academica posteriora*, I, 3, trans. in Brunet, 2006, p. 8)

> The ray is a sea fish; Romans call it *tourpaína*; if boiled when fresh, its broth has a laxative action when taken alone or with wine; if eaten it frequently is beneficial to the stomach and stimulates love desires.
> (*Kyranides*, Fourth Book, Entry B; see Kaimakis, 1976, p. 246)

In this chapter, we shall deal with information about electric fish through most of the Middle Ages, a period loosely defined as starting with the end of classical antiquity, more or less in the fifth century A.D., and ending around the fifteenth century. The exact dates or cornerstones of the Middle Ages vary according to different historians, in part because its key developments were far from uniform across Europe. For now, we shall confine ourselves to how information about electric fish was received and transmitted up to a period referred to as the "Twelfth-Century Renaissance," a period of change that would lay the groundwork for the later Italian (and European) Renaissance (see Haskins, 1939).

In addition to works written in Greek and Latin, we shall examine works written in Arabic and related Middle Eastern languages in this chapter. This is because Middle Eastern writers contributed to science and medicine in important ways. Also, it was mainly through these writers that Greco-Roman ideas, some intact and others now somewhat modified, were retransmitted back to the West.

To situate the culture of the Middle Ages in proper context, and thus to appreciate how knowledge about electric fishes would be handed down to scholars in later periods, we must begin with some information about the final phases of the Roman Empire and the emergence of the Middle Ages. This is a complex history, but one fundamental to understanding what facets of the old culture were transmitted to the new, and what new developments would appear during this so-called intermediate period in the West.

THE END OF THE ROMAN EMPIRE AND THE START OF THE "DARK AGES"

Starting from about the third century A.D., the Roman Empire began a slow decline that, through a complex series of events, would culminate with its dissolution. Its demise would initiate an epoch of political fragmentation and great instability, the Early Middle Ages, which in turn would lead to the formation of new political and linguistic realities. Importantly, Latin would slowly cease to be the universally spoken language and, amidst the conquests and alliances that would take place, the foundations of modern European nations would begin to emerge with their specific languages.

One of the most significant events of this epoch would be the progressive integration of a multiplicity of new populations coming from the territories situated in the east and north of Europe into Western and especially Mediterranean civilizations. Another would be the emergence of Christianity as a religious and political institution of great importance. Also significant would be the separation of the eastern division of the Roman Empire from the western part. This was the ultimate consequence of the attempts by several emperors to reduce the political importance of the Roman senatorial aristocracy by moving the center of decisional power away from Rome.

The western empire collapsed somewhat officially with the deposing of the last Roman emperor, Romulus Augustus, in 476 A.D. In contrast to what happened in the fragmented west, the eastern division of the empire remained relatively cohesive until the beginning of the early modern period. Through a complex series of events, the eastern empire had become a separate political and administrative entity with its center in the Greek city of Byzantium. Byzantium was renamed Constantinople after the Emperor Constantine greatly expanded and embellished it, and it is today's Istanbul.

The Byzantine Empire differed culturally from the Roman Empire to the west. Especially at the beginning, it displayed more continuity with Greek culture. One of the greatest achievements of the Eastern Empire occurred during the sixth century, when the Eastern Roman Emperor Justinian established the *Corpus juris civilis*, a body of civil

Figure 5.1. An ancient coin showing the Roman Emperor Justinian I (Justinian the Great; c. 482–565).

laws that would form the foundation for future legislation in almost all European nations (Fig. 5.1).

Although the Eastern Empire exerted its main political and administrative dominion on the eastern regions of the Empire (i.e., eastern Europe, neighboring regions of Asia, Egypt), for a long time its influence also extended to a much wider area. This included the Adriatic territories of Italy[1] and its southern regions (from Naples to Sicily), all of North Africa, and vast parts of southern Spain. This influence was especially wide during Justinian's attempts to reunify the old Roman Empire. Nevertheless, there was a progressive reduction in the cultural exchanges between the Byzantium Empire and Western Europe.[2]

During the declining phase of Rome and the western part of Europe, the Catholic Church developed and grew stronger, and eventually most of the West fell under its influence, which increased its authority, not only as a world religion but also politically. It was under the authority of the Roman Catholic Church that Pope Leo III crowned Charlemagne, previously the King of Franks, Emperor of the Holy Roman Empire, in the year 800. With the western empire having been in disarray for some four centuries, this was a momentous event in the history of the Middle Ages.

Because the eastern empire was tied to Greek language and culture, it played an important role in preserving original works from classical Greece. Moreover, under Constantinople's influence (and particularly with regard to the oriental diffusion of Christianity), various other important cultures emerged on the basis of differences in language, geography, and religion (e.g., Armenian, Syriac, Egyptian, Ethiopic), which interacted in various ways in a cultural pool that the great Italian orientalist Enrico Cerulli (1968) referred to as the *Oriente Cristiano* (Christian Orient).

We owe some of the first translations of Aristotle's works (initially in Syriac and afterwards in Arabic) to this cultural pool.[3] Some classical texts would be eventually translated from Arabic into Latin, sometimes even through Hebrew intermediaries. Indeed, Jewish scholars also played an important role in the development and diffusion of culture during the Middle Ages, particularly in the fields of philosophy and medicine. This occurred not only in the Middle East, but also in Europe itself. In Spain, Jews worked together with Arabs and Christians translating Greek texts from the classical period and also important Arabic texts, making Spain an intercultural center for European scholars, particularly after the twelfth century.[4]

The interesting point for us is that, in the course of these sometimes-complex translation (or transmission) routes, ancient texts were modified and occasionally enriched with comments and annotations that incorporated the translator's or editor's knowledge. This was an integral part of the cultural heritage that would be transmitted forward. Interestingly, when faced with the difficulty of translating Arabic words that they could not easily interpret, the translators sometimes transliterated the Arabic wording, so that many Arabic words became a part of the European vocabulary, particularly in the sciences.

Texts were much less likely to be translated directly from Greek into Latin at this time, but this sometimes occurred, notably in the southern regions of Italy where, because of the political influence of Constantinople, the Greek language was somewhat better known, even after the separation of Western Europe from the Eastern Empire was clear-cut. Dioscorides' *De materia medica* was among the first medical works translated from Greek into Latin. These translations appeared in manuscripts produced in Beneventum (a town

---

[1] In opposition to *Longobardia*, the territories occupied by the *Longobardi* (the Lombards), this region, which had its chief town in Ravenna, was called *Romània*, from which derives the present name *Romagna*.

[2] This separation was accentuated by religious differences, which led to a complete separation of the Eastern (Orthodox) and Western (Catholic) Church with the Great Schism of 1054. Besides this main contrast, theological problems, and particularly the understanding of the Trinity dogma, led to the formation of various sects (e.g., Nestorians, Jacobites) and national churches, particularly in the oriental countries. Further, the conflicting relations between the imperial powers and the barbarian populations had religious connotations, particularly because many barbarians adhered to Arianism, which differed from orthodoxy about the interpretation of the Trinity.

[3] Some old Greek texts of a religious and magical nature have come down to us only through these intermediate "oriental" translations, including some versions made in the monasteries of far-off Ethiopia.

[4] To translators operating in Spain we owe, for instance, translations of most of the main medical works of Hippocrates and Galen, and also important Arabic medical texts, including the *Canon* of Avicenna (ibn Sīnā). Among the most important translators were Gerard of Cremona, who translated a number of originally Greek and Arabic texts into Latin, including works of Aristotle and Ptolemy (notably his main astronomical text, the *Almagest*); Michael Scot, who translated the zoological treatises of Aristotle (and also astronomical texts) from Arabic into Latin; and Hermann of Carinthia, known mainly for his translations of Aristotle's philosophical works.

in today's Campania region of Italy), and they date from the ninth to the eleventh century (see Haskins, 1928, 1939; Crombie, 1952).

The rapid rise of Islam, which spread from the Arabian peninsula, particularly during the eighth century, was another of the landmark events of the Middle Ages, and one that would greatly affect relations between the Occident and the Orient. In the West, its influence was particularly strong in southern Spain, which became one of the main centers of Islamic culture (*al-Andalus*). Spain would eventually play a key role in the intellectual exchange between the Islamic and Christian worlds and, as just mentioned, it would become a center for translating ancient texts.

In addition, Sicily, which had long been under Arabic domination, was an important center for cultural exchanges between Christianity and Islam.[5] Sicily played a particularly important role in intercultural transmission when Palermo became the seat of the magnificent *Magna curia* (literally "Big Court") of Emperor Frederick II of Hohenstaufen, grandchild of Roger II (and also of Frederick I Barbarossa). Besides favoring translations and producing original works in various languages,[6] Frederick II was himself a scholar, and he wrote an important piece on falconry, in which he started to dispute Aristotle's authority in zoology, stressing the importance of personal experiences in this field of science (see next chapter). The Sicily of this enlightened sovereign would contribute in an important way to a more advanced phase of the Middle Ages, which would pave the way for subsequent cultural progress in Europe (see Chapter 6).

Returning to the Islamic world, in addition to developing a specific literature, mainly in Arabic, it contributed in significant ways both to the transmission of classical information and to developments in medicine, mathematics, and the other sciences, including astronomy and geography. The conflict that developed between the Islamic and Byzantine Empires would eventually lead various Western European countries to launch a series of military campaigns known as the Crusades to bring the Holy Land, which it had lost in the year 972, back into the Christian fold. The Crusades started at the end of the eleventh century and were military disasters for the West. Nevertheless, they led to cultural exchanges between the warring sides, and Greek and Arabic books, sometimes the spoils of war, were brought back to Western Europe with profound consequences. This was also a period in which many technical advances and other cultural and material objects made their way from the East to the West.[7]

The Middle Ages were never uniform. Even the appropriateness of the phrases used to indicate them, somewhat suggesting the idea of a dark, stagnant, or intermediate period between the glorious ancient ages and the glittering Renaissance, can be contested. Within the millennium of its traditional duration, there are certainly conspicuous differences and specific periods, sometimes indicated as "Renaissances," to denote the political and cultural "re-births" characterizing them.

Among these are the so-called Carolingian and Ottonian Renaissances, spanning the ninth and tenth centuries, which were mainly localized phenomena. In our introduction to this chapter, we also alluded to a "Twelfth-Century Renaissance," which had a broader impact on Europe. In addition to the building of the magnificent Gothic cathedrals to proclaim the glory of God, this period was characterized by the founding of important teaching institutions, including the first universities, and by the flourishing of Christian philosophy (with the golden phase of the Scholastics). This "Renaissance" will be the subject of the next chapter.

In the initial transition phase immediately following the collapse of the Roman Empire, the Middle Ages were undoubtedly a "Dark Age" in imminent danger of losing its Greco-Roman traditions and culture. This very real threat was perceived by members of the surviving cultural aristocracy, who were in part the successors of the learned Roman classes, integrated with Romanized barbarian scholars. This happened particularly after the sack of Rome by the Visigoths in 410, which was seen by many as the true end of a millenary world that was supposed to be perpetual (Rome being the "eternal city" of the Latin authors).[8]

The danger that an entire culture could disappear forever with the collapse of the Roman Empire was felt so strongly that later scholars (e.g., Isidore of Seville in the sixth century, see below) dedicated themselves in an indefatigable way to preserving all knowledge of the ancient world. This attempt to preserve and transmit the ancient culture manifested itself in several ways, one being the compilation of encyclopedias. These texts tried to arrange all possible knowledge in a condensed and easily transmissible way. In this endeavor, scholars of the so-called Dark Ages were continuing a legacy that had already begun in Roman times, as we shall now see.

---

[5] Under Norman King Roger II, Muhammad al-Idrisi wrote his *Geographia* (*Kitab nuzhat al-mushtaq*) in Palermo. This was one of the most important geographical works of Arabic culture, illustrated by a famous map called *Tabula Rogeriana* after Roger. In the Norman court of Sicily, around 1154, Eugene of Palermo translated the treatise on optics that the great Alexandrian astronomer Ptolemy had written in Greek in the second century A.D. (his translation was from the Arabic into Latin). Under the immediate successors of Roger II, William I and William II, this activity continued at an intense pace, with many of the works of Euclid, Ptolemy, and Proclus translated. It was during this period that the *Pneumatica* of Hero of Alexandria was translated. In it, as mentioned in the previous chapter, there is the reference to Strato of Lampsacus about the transmission of "affections" and of the shock of the torpedo.

[6] Michael Scot, a Master Theodore of Antioch, and a Master John of Palermo were among the translators and scholars active at Frederick's court. The importance of the cultural activity of Frederick, his semi-oriental style of life, and his love of Islamic culture justify a statement expressed by the Italian Arabist Michele Amari. He wrote that Frederick II was "one of the two baptized sultans of Sicily, to whom Italy owes a non-negligible part of her civilization," the other being Roger I, Frederick's great-grandfather (see Amari, 1868, p. 365).

[7] One of the dramatic events of the Crusades was the savage sacking of Constantinople by the Europeans in 1204. Among the many goods that the Crusaders pillaged were books that were taken back to Europe, leading to a renewed interest in Greek culture. The sack of Constantinople was the beginning of an irreversible decline for that city and the Eastern Empire. The Europeans lost the city to the Ottoman Turks after a siege in 1453.

[8] In the emergent Christian culture (and particularly in the writings of Saint Augustine, the prolific fourth-century scholar who exerted the greatest influence in the oldest part of the Middle Ages), the crisis of the Roman world led to rethinking all history from a more religious perspective. This new thinking helped to preserve hope in human destiny, notwithstanding the visible destruction and misery that was occurring at the time.

CLASSIC ENCYCLOPEDIAS AND CULTURAL TRANSMISSION TO THE MIDDLE AGES

As noted in the two previous chapters, the spirit of scientific investigation that had marked Greek civilization from the time of Aristotle and his successors was already in decline when the Roman Empire was at its peak. This was particularly true for the study of the animal kingdom. Although curiosity about the world and its wonders was certainly an important aspect of Roman culture, it did not lead to any important new investigations of a genuine zoological character.

Undoubtedly, a growing interest in anatomy and physiology characterized some (but certainly not all) approaches to medicine during the Roman epoch. This led to a continuation of the study of animal anatomy and physiology, which had flourished together with many other fields of science and culture during the Hellenistic period, especially in Alexandria. Galen, who studied in Alexandria (among other places), stands out for his experimental observations, such as showing how ligations and severing nerves could affect certain functions in live animals, and more generally for his firsthand approach to medical problems (as we discussed in the context of his torpedoes prescriptions in Chapter 4). There was, however, nothing in Roman culture corresponding to the inquisitive attitude that Aristotle had for living nature—the spirit of inquiry that had led the Aristotelians to their important texts on animals. Those texts, as we have seen, did not arise from another interest, such as medicine. Rather, they stemmed from zoological and ecological interests in living nature *per se*, without another overriding agenda.

As remarked by some perceptive exponents of Roman culture, including Cicero, what characterized Roman society with respect to the Greek world was not really science but *religio*. This term denotes a complex of cultural and social practices—the rites, laws, customs, social organization, philosophical attitudes, and religious beliefs—all of which tended to create cohesion within the Roman world ("cohesion" being one of the original meanings of the word *religio*). This characteristic of the Roman world also contributed to Roman philosophies, and particularly to the way in which Stoicism was elaborated in Rome, with moral and human aspects of intellectual endeavors greatly accentuated.[9]

Besides their possible relevance for medicine, and particularly as a source of remedies, animals were, in fact, now mostly of interest for their philosophical and moral implications. That is, their behaviors and natures had symbolic overtones of relevance to people—a symbolism that could be exaggerated and even considered fantastic (as in fantasy). This tradition dates back to Aesop, the sixth-century B.C. author of popular fables, and to Seneca's *Quaestiones naturales*. It was greatly reinforced with the circulation of the *Physiologus*, which began in about the third century A.D. The latter, highly influential work was derived from Greek literature, and it was translated into Latin, Syriac, Ethiopic, Armenian, Georgian, and afterwards into most of the Romance languages

Figure 5.2. A page of the so-called *Physiologus* of Bern, a ninth-century illuminated codex of the text that was one of the main sources for the medieval bestiaries. (From the Burgerbibliothek of Bern, Switzerland)

of Europe. It became one of the main sources for the fanciful bestiaries of the Middle Ages (Fig. 5.2).

Along with the most important works of Roman culture, such as Pliny's *Naturalis historia*, there was no clear-cut distinction between verifiable sound information and the legends and fantasies concerning nature's various components, from humans to animals, and down to plants and minerals. As with Pliny and Athenaeus, moreover, and apart from genuine medical texts, the typical literary works in which animals appeared were compilations and like encyclopedias.

The encyclopedia-like works attempted to include, in a more or less condensed form, all the knowledge of the epoch, from history to cosmology, from rhetoric to geography, and, of course, poetry, religion, agriculture, medicine, magic, zoology, botany, mineralogy, and more. Unfortunately, because few have survived intact, we have only a fragmentary and indirect idea about the production and circulation of these works, some of which have come down to us in the form of "etymologies"—that is, works directed to ascertaining the deep and true meaning (*etúmon*) of words, generally based on an alphabetical listing of subjects and words examined.[10]

---

[9] As pointed out by Giovanni Di Pasquale (2004), the scientific interests of Romans, other than in medicine, concerned mainly technology and mechanics. This led to noteworthy practical and theoretical progress in the field of construction and new and improved machines.

[10] Ἐτυμολογίαι had indeed flourished in Greek culture since the fourth century B.C. with the writings of Heraclides, Ponticus, and the later works of Apollodorus of Athens and Demetrius of Ixion (second century B.C.). But they had a particularly wide diffusion in Latin culture, where the first important writer of this literary production seems to have been Aelius Stilo Praeconimus, who lived from about 154 to 74 B.C. Marcus Terentius Varro, who lived just before Cicero and received great praise from him for his literary production (see the opening citation of this chapter) was one of the most important Roman encyclopedia writers. He probably wrote more than 70 works, comprising more than 600 books. Origenes, one of the first important Christian writers (the so-called Church

One of the reasons for this enhanced book production, and especially for the diffusion of compilations, was the utilitarian importance of a proper cultural background for members of the upper social class, especially those engaged in public life.[11] The epitomizing attitude underlying the production of Roman encyclopedias became particularly intense in the later phases of Roman Empire, when economic and social instabilities threatened this culture.

The compilers of these encyclopedias frequently did not have direct access to the original works from which to derive their information, and so they relied on other compilations. Small wonder that—with the simplifying and shortening attitudes underlying the encyclopedic texts, the huge number of passages, and of the impossibility of mastering all fields of inquiry—these texts were filled with errors of omission and commission, errors destined to be transmitted to other cultures.

Although medicine was one of the more cultivated sciences, especially in Galen's hands, it shared the destiny of other aspects of Roman science and culture, becoming the subject of compilations that might even be consulted by laymen for practical and educational purposes. Aurelius Celsus' *De medicina* is one of the most important medical tracts from the Roman world. Celsus was a scholar of Gallic origin who lived in the first century A.D., and his *De medicina* was only one part of a far greater encyclopedic work dealing with various aspects of classical culture. Celsius was not, however, trained in medicine or a practicing physician.[12]

The compilations and encyclopedic works that flourished during the classical era, and especially those from Roman times, epitomized not simply what was known, but also what "ought to be known," by any person of culture. But the predominance of such works over the production of original writings, especially in the natural sciences, still reflected the decline of basic research in Roman society.[13]

Another part of the literary production that also reflected the decline of Aristotelian science is the diffusion of texts known as commentaries, which were devoted to interpreting and discussing texts of particular importance. In previous chapters we cited some such works, including those of Alexander of Aphrodisias and Simplicius of Cilicia, where we found information about the debate concerning the transmission of forces or "affections" at distance. Writing commentaries instead of original works would represent one of the main intellectual endeavors through the Middle Ages, and this trend would continue almost to the beginning of the early modern era (that is, to the seventeenth century).

With this information providing the background we need, let us now turn to electric fishes, starting with the writing of Marcellus of Bordeaux, an author who lived some of his life when the two divisions of the Roman Empire still had a single ruler. We shall then turn to Aëtius of Amida, Alexander of Tralles, and Paulus of Aegina, physicians more connected with the Eastern Empire, and then to (Saint) Isidore of Seville, a bishop and writer from Visigoth Spain. All of these individuals were compilers, and all of their compilations found large audiences. We shall also consider a magico-medical text, the *Kyranides*, which enjoyed great popularity for about a millennium, starting from its compilation in around the third century A.D. After doing these things, we shall look at some writings from the Middle East to see how this earlier knowledge was preserved and transformed before it was transmitted back to the West through complex and sometimes tortuous transmission paths (Fig. 5.3).

Figure 5.3. A 1567 edition of the medical treatises written by the most important physicians ("the Princes of medical art") after Hippocrates and Galen, which includes the works of most of the Roman and Byzantine scholars dealing with the torpedo.

Fathers) from the third century, wrote more than 6,000 books, according to Saint Hieronymus (Jerome). Most of the Roman encyclopedias have been partially or totally lost, but we know the authors of many of them. Among the Latin writers who wrote encyclopedias or compilation works, or were important sources for the encyclopedias of the Middle Ages, there are (besides Varro and, of course, Pliny) Cornelius Celsus, Suetonius, Julius Solinus, Nonius Marcellus, a Servius author of many commentaries, Martianus Capella, Macrobius, Cassiodorus, Boëthius, and many others.

[11] Compilation works were particularly needed as reference texts for the numerous public speeches that Roman citizens (*cives*) would give in the course of their civil careers. In a shortened and easily digested form (*compendiosa brevitas*), these works aimed at providing citations, references, and other important information for these and other occasions. They also provided useful material for didactic purposes in the numerous schools that flourished throughout the Roman Empire.

[12] Celsus does not include the torpedo among the therapeutic treatments in his work.

[13] Even poetical works, such as those of Oppian, Aelian, and Claudian, represent an aspect—albeit a very peculiar one—of the tendency to collect and present interesting or useful information in a form that would be accessible and easy to remember. Poetry, with its rhythms and cadences, favors mnemonic learning, which was of great importance in times when few books were produced (because of the long time and difficult work needed to produce and copy manuscripts), and when only a small part of society could afford them and knew how to read them. In this context, it is worth remembering that Lucretius' *De rerum nature*, perhaps the most important "scientific" text of Roman culture, was written using a poetic form.

Figure 5.4. Emperors of the Late Roman period, from an illuminated fourteenth-century codex of a chronicle written by Byzantine scholar Constantine Manasse (c. 1130–c. 1187). The emperors are represented in chronological order from right to left: Theodosius I (Flavius Theodosius or Theodosius the Great), Valentinian I, Gratian, Valens Jovian, and Julian (the Apostate).

MARCELLUS OF BORDEAUX

Marcellus (Empiricus) of Bordeaux's writings served as an important source for prescriptions and therapeutics for centuries (Must, 1960; Stannard, 1973). His *Liber de medicamentis*, written in Latin sometime between 395 and 410, was dedicated to Flavius Theodosius (Theodosius the Great), the reigning emperor late in the fourth century, whom he dutifully served. After reigning in the East (from 378 to 392) and sharing power in the West (with Gratian and Valentinian), Theodosius was the last emperor of both the Eastern and Western Roman Empires, and the ruler who proclaimed Christianity the official state religion (Fig. 5.4). The two parts of his combined empire would permanently split apart after he died in Milan in 395 (Williams & Friell, 1994).

Marcellus tells us that his 36-chapter book of *materia medica* was written to provide medical advice for the emperor's sons, should they not have access to a well-trained physician. As for Marcellus's own background, few facts are available other than that he was from Gall and was considered very knowledgeable in the healing arts. There is no conclusive evidence that he trained as a physician or was even a Christian.

Most of Marcellus's prescriptions were taken from earlier sources, including Scribonius Largus and his copyists. Yet he melded the past with his own cultural values and drew from the local Celtic healing tradition, putting magical recipes and remedies in his texts. By giving plant names rather than just descriptions, with his greater reliance on the polypharmacy (most of his recipes have over 20 ingredients), and with his use of magical formulas and charms, Marcellus took Greco-Roman medical treatments along a somewhat different path than most previous authors. Yet at the same time many of his changes amounted to little more than rearranging the order of standard ingredients, adding or subtracting some agents, or varying quantities.

Marcellus was not one for giving physiological or philosophical rationales for his treatments. His concern was whether something worked or not, and he was not bashful about advertising the cures associated with some of his more successful treatments. His non-theoretical orientation, which is laid out in the preface of his book, is why the descriptor *Empiricus* was occasionally added to his name.

Marcellus did not make any earth-shattering changes in the use of live torpedoes for headaches. In fact, in the first chapter of *De medicamentis*, dedicated to the treatment of head pains (*Ad capitis dolorem*), he copied the torpedo recipe virtually word for word from Scribonius Largus, inserting it in a long list of remedies, most of which included, as was often the case for Marcellus, various herbs and animal preparations, dietetic advice, and other treatments (see Marcellus, I, 11; e.g., in Helmreich, 1889, p. 28). Following Scribonius on headache, Marcellus clearly states that the treatment should involve a live black torpedo (*torpedo nigra viva*), unlike Dioscorides (see Chapter 4).

Marcellus also called for a live torpedo when treating *podagra*. This remedy is presented in Chapter XXXVI under *Ad podagram calidam et frigidam* ("For Hot and Cold Podragra"), which actually deals with both foot (*podagra*) and hand (*chiragra*) pains. In Section 46 of this chapter, which starts with a preparation for *podagra* that calls for *oesypum* (a grease from unwashed wool), woman's milk, and *cerussa* (white lead), we encounter wording that is very close to what Scribonius had written about foot pain—again specifying the need for a live black torpedo.[14]

Specifically, Marcellus recommends a live black torpedo for what he refers to as both types of foot pain (i.e., the hot and cold varieties). Gout, which can make a toe or toes hot to the touch and red in color, can readily be identified as the hot variety. The cold *podagra* more likely relates to other conditions, such as arthritis, rheumatism, or arthrosis (an articular illness that produces fibrosis or degeneration of the cartilage), although physicians at this time lacked the knowledge to make the finer pathological distinctions we

---

[14] There is a slight difference here between Scribonius, who calls for a live black torpedo (*torpedinem nigram vivam*), and Marcellus, who changes the order and writes *torpedinem vivam nigram* (a black live torpedo). However, as pointed by Sconocchia (2001, p. 268) the expression can also change in Scribonius according to the manuscripts and recensions.

Figure 5.5. The title page of 1542 abridged edition of Aëtius' medical treatise and the passage (on the right) in Latin and Greek dealing with the "sea torpedo" prescriptions for patients with headache and *prolapsus ani*.

have today. Indeed, the generic term for all of these conditions, gout included, was ἀρθρῖτις ("arthritis") (Copeman, 1964).[15]

In summary, Marcellus is one of the transmitters of the earlier culture who, in the case of the torpedo, faithfully reproduced the information found in the original source (probably having direct access to a copy of the original work). He did not transform, comment, or integrate what he received—his objective was simply the "copying and pasting" of the earlier information. As we shall see, others would not necessarily follow his *modus operandi*.

AËTIUS OF AMIDA

Aëtius of Amida lived between 530 and 560, during the reign of Eastern Roman Emperor Justinian I (Justinian the Great), who tried to reunite the two empires but failed in this endeavor (Bury, 1958; Evans, 2005). It is believed that Aëtius was a relatively high-ranking court physician and a learned man who had studied in Alexandria. In his medical treatises, he tells us that he visited various places, including Cyprus, Solis (the town of Clearchus), Jericho, and the Dead Sea (for more on his life and writings, see Kudlien, 1970).

His large medical encyclopedia, now referred to as his *Libri medicinales*, was traditionally called *Tetrabibloi* or, in English, "Sixteen Medical Books," because it consisted of four volumes containing four books (or λόγοι "discourses") each, although this division was made by later editors (Fig. 5.5) This medical encyclopedia contains excerpts from Galen, Dioscorides, Soranus, Oribasius, and others, although in some cases these came via indirect transmission. As such, Aëtius provides relatively little that is new, although he did make some changes, filtering out what he regarded as misguided or unessential information and occasionally adding material from his practice as an Alexandrian physician.

Aëtius' medical treatises are of great historical importance because they provide information on the healing practices of the time (particularly in the fields of surgery, gynecology, and ophthalmology), including the use of drugs.

---

[15] The distinction between gout and rheumatism goes back to the time of Hippocrates, although the term "gout" would make its appearance at a later date, originating as *gutta* (meaning "drop") with the "barbarians" of the Roman Empire, who thought an excess of a humour dropped or flowed into a joint weakened by an excess of food, wine, venery, or some combinations of these factors (Copeman, 1964). In Alexander of Tralles, in fact, mention is made of four types of *podagra*: (a) sanguineous, with the joints feeling hot due to an overheating of the blood; (b) bilious, where the joints become red due to too much gall (yellow bile) in the blood; (c) phlegmatic, where the joints are painful but not hot, red, or swollen, due to phlegm in blood; and (d) melancholic, which is associated with cold joints, due to too much black gall (bile). Here the first two are hot varieties, whereas the last two are not and can be classified as cold in a simple two-category way. As for treatments, the Hippocratic physicians typically used cold remedies for hot disorders. Thus, we find applying cold water being among their recommendations for gout (see Schnitker, 1936), which is in accord with the philosophy of treating with opposites to restore a state of balance (*eucrasia*).

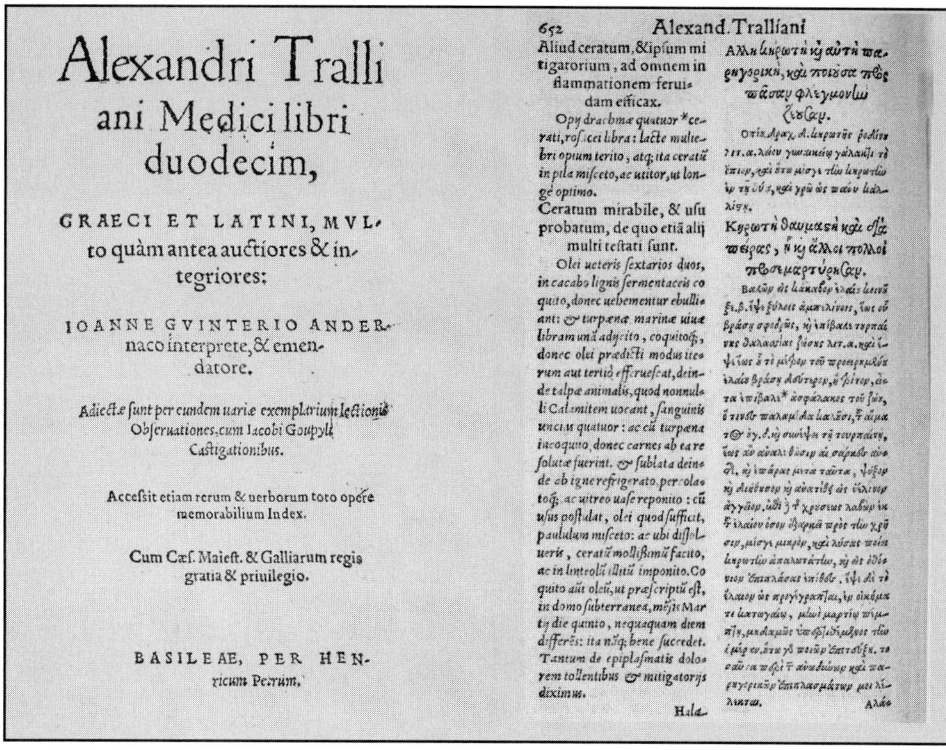

Figure 5.6. A 1556 edition of Alexander of Tralles' medical treatise with the passage, in both Latin and Greek, dealing with the analgesic use of a plastering based on an ointment prepared from the sea torpedo. The torpedo is indicated with the Greek term τούρπαινα (*tourpæna* in Latin), clearly inspired from the Latin *torpedo*.

They are also noteworthy for being among the Byzantine medical texts that would be consulted for generations to come—in some countries up until the nineteenth century. Subsequent editions of his text, however, show modifications made not by him, but by copyists (early on), editors, and translators.

Aëtius recommends the torpedo as a dietetic food and finds it useful for various conditions (following the tradition of Hippocrates and Galen, likely through the mediation of Oribasius, physician to the Emperor Julian "the Apostate") (see Fig. 5.4). Moreover, in an alphabetical listing of fish and other ingredients for medical remedies, he succinctly mentions the use of a live torpedo (Νάρκα ζῶσα) for persistent headache (πόνον κεφαλῆς, literally "harassment of the head") and *prolapsus ani* (προσπίπτοντα δακτύλια). Importantly, he states that a dead torpedo is either totally ineffective or at best only poorly effective (ἢ οὐδ᾽ ὅλως ἢ ἐπ᾽ ὀλίγον ταῦτα ποιεῖ; Aëtius, *Libri Medicinales*, II, 185; see Olivieri, 1935, vol. I, p. 221).

The call for a live torpedo, together with no mention of one as a treatment for *podagra*, suggests that Aëtius might have derived his information from Galen, possibly through an intermediate source. It is less likely that his main sources were Scribonius or Dioscorides: the former because he did not write about *prolapsus ani* and instead mentioned the use of a live fish for *podagra;* the latter because he did not explicitly declare the need for a live fish.[16]

ALEXANDER OF TRALLES AND THE MAGIC OF THE TORPEDO

More is known about Aëtius' contemporary, Alexander, who was born in Tralles, Lydia, in 525 (now Aydin in Turkey), and practiced medicine and taught in Rome, dying at about the age of 80 (Brunet, 1933). He was a member of a prominent family, with his father and two brothers being celebrated physicians and other brothers being important scholars in different fields. One of them, Anthemius, was among the architects who built the magnificent Saint Sophia Cathedral (later the Hagia Sophia or Ayasofya mosque) in Byzantium.

From Alexander's main treatise, *Therapeutica* ("Twelve Books on Medicine"), and particularly from the section dealing with epilepsy in Book I, we know that he learned his medical remedies in several Mediterranean and Near East locations, including Armenia, Thrace, Corfu, Cyrenaic, Gallia, Tuscia (Tuscany), and, of course, Rome. His books, including his *Therapeutica*, were originally written in Greek and were afterwards translated into various languages, including Arabic, Latin, and French (not to mention more modern translations into various European languages) (Fig. 5.6). The older Latin editions are among the most informative, because of the corruptions that have taken place over the ages with editing, translating, and copying.

Alexander's writings served as a basis for instruction for centuries. Of particular importance are his texts dealing with surgery, since he had direct experiences, owing to his service as a military surgeon, and his ophthalmologic entries. His books are considered the main source for Roger of Salerno's *Practica chirurgica,* a twelfth-century text that would be important for medicine's revival in the Western world.

---

[16] The word used to indicate the "seat" here is different from that used by Dioscorides and Galen (ἕδραν). This would suggest that the derivation was indirect.

Alexander's therapeutic principles are largely inspired from Hippocrates and Galen, with primary importance given to the doctrine of humours, temperaments, and qualities. This notwithstanding, he does not hesitate to criticize medical authorities based on his own experiences. He refers to a famous statement attributed to Aristotle: "Plato is my friend, but truth is a better friend." In this context, he frequently declares that he has personally practiced or "witnessed" the remedy with which he is dealing, and he often mentions the place where he has learned or seen the particular treatment.

The torpedo is present in Alexander's extant medical books, but in a somewhat unexpected way. First at all, there is no mention of any treatment based on the fish in the (lengthy) first book of *Therapeutica*, in which he deals with a multiplicity of head diseases (including epilepsy, delirium, lethargy, and melancholia) or other conditions involving head (e.g., hair loss, a common entry for treatment in ancient medical texts). For headache (cephalalgy or hemicrania, or other situations, including sun stroke and hangover) and for other head diseases, he lists, however, a great number of other remedies. Some call for a woman's milk, the blood of killed gladiators, pigeon or dog feces, minerals, and (as might be expected) various plant preparations. But nowhere here does he mention the torpedo. It is not even among the foods he recommends for treating epileptic children, even though he discusses diets appropriate for epilepsy in great detail, with some fishes recommended and others that must be avoided.

In contrast to the absence of the torpedo in his prescriptions for headache, and also for *prolapsus ani*, we find the torpedo in two passages of Book XII (XI in older editions), a book entirely devoted to *podagra* and arthritis. There is nothing in these passages, however, concerning the use of a live torpedo and specifically its shock. A live fish is, in fact, mentioned in both passages, but only as an ingredient to be put into a receptacle, where it is then boiled with oil and water. His objective is to make a therapeutic ointment.

We quote the second and somewhat shorter passage, which, as often happens with Alexander, might seem like a witch's brew mixed with some astrology to modern eyes, though he and his contemporaries certainly would not have viewed it this way. Here the preparation is intended as a cure and a means to prevent *podagra* and conditions involving joint pain:

Cast a live torpedo (Νάρκην ζῶσαν) into a bronze receptacle, add oil, and place the whole into a pot with water, so that the oil will not burn and is not consumed. Add a narcissus plant, picked during the last quarter of the moon. Once mixed with the creature, cook the whole animal until it is completely dissolved and the bones are bare. Then simply carefully extract the remaining oil from the water in the marmite. All these operations should take place during the waning of the moon. Then use the oil to anoint the sick three times a day. If the patient suffers joint pains, the treatment will cure him; if he feels no pain, it will guard him against all future discomfort. These anointments should be practiced for three days, while the moon is in complete decline, for the measures will not succeed at any other time. (see Puschmann, 1963; vol. II, p. 581; translated from Burnet, 1937, p. 261, revised)

This recipe is somewhat typical for this author from Asia Minor. His cures often start with what the ancients had written. He then adds magical elements, a healthy dose of astrology, and occasional occult practices, such as boiling live animals in oil to extract juices thought to have special properties. To modern readers, the weird and "witchy" impression of this treatment or preventative for foot pain is accentuated by the next one listed by Alexander. It bears the annotation, "*Another* [remedy] *through experiment*" [i.e., verified]: "Apply on the affected part a bandage spotted with the first menstrual blood of a virgin and one would not suffer podagra any longer."

In his *Practice of Medicine,* Alexander even recommended using a ring with a figure of Hercules strangling a lion as a cure for colic. This, of course, is not to deny that some of his recipes contain ingredients that could reduce inflammation or lessen pain—for some surely did, as for instance those based on opium.

Alexander's recipe for boiling a torpedo to make a salve for treating joint pain reflects how complex such cures had become by the sixth century, and how a number of factors converged to make certain formulations more attractive than others to Byzantine healers. His association of the *nárkē* fish with the narcissus plant is particularly intriguing. This association probably stems from the inebriating properties of the narcissus flower, but it might also be connected to the linguistic similarity of the terms and probably to their same etymology.

Regarding the linguistic problems connected to Alexander's torpedo prescriptions, there is also an interesting one in the second passage in which he deals with this fish for treating foot pain. The relevant section in Book XII is titled (in translation), "*A cerate* [plastering] *wonderful and proved by experience and also recommended by others.*" The preparation, which is intended as an analgesic treatment and not a prophylactic remedy, is described in more detail than the other one, which it still resembles. Among the notable differences is the need to add the blood of a mole (παλαμίδα) to the receptacle in which a (previously live) torpedo is being cooked. The mole blood here substitutes for the narcissus. In this case, his astrological flavoring comes forth in the remark that the preparation (to be applied in the form of a plastering) should be done on the "fifth day of the month of March."

What is particularly unusual is the fact that, despite the similarity of the two preparations, the fish is not indicated as *thálassia nárkē* but with a different term clearly inspired from the Latin term *torpedo,* namely *tourpaínēs thalassías* (τουρπαίνης θαλασσίας).[17] The fact that Alexander could use

---

[17] This is generally indicated as the first occurrence of this rare word in the Greek scientific literature, as noted by the fact that specialized Greek lexicons list it only in Alexander and Paulus Aegineta. As we shall see, however, the term occurs in the *Kyranides*, a text that is surely earlier than those of Alexander and Paulus, and in its first redaction might even have been known to Galen. The term caused a series of problems to the first translators. Having never encountered the word, and being uncertain that it was the same as *nárkē,* they simply transliterated the Greek, thus creating new Latin expressions, such as *turpaena* or *turpana marina,* or French phrases, such as *turpena de mer.* As to funny stories with translations, the editor of the current recension of Alexander's text, in translating the expression into German, correctly writes *Zitterochen* (the ordinary word for torpedo in German). Wrongly, however, he suggests that it might correspond to "Gymnotus electricus" [i.e., the electric eel]; see Puschmann, 1963, vol. II, p. 574). The eel is found only in South America and there is no possibility that Alexander might have known about it. Moreover, in the original text there is a clear indication of

this expression is not intriguing *per se*, because he might have encountered it in Italy, France (Gallia), or the Dalmatian regions during his travels. The intrigue stems from the fact that this wording is used just a few pages before the usual Greek term, *nárkē*, yet in a similar context. Among the possible reasons for the switching, we might think about an intervention of a copyist, or an integration of the part containing the new term from an available Latin version. The original text (or possibly an oral source) was likely of popular origin, because *tourpaína* is more like the regional names used late in the Latin world (and still used nowadays in some places) than the proper Latin term, *torpedo*.[18] Whatever the explanation, the significant point is that *tourpaína* might have entered into Alexander's text very early on—a possibility suggested by examining the writings of Paulus Aegineta, another Byzantine physician, and consulting an older text, the *Kyranides* or *Cyranides*.

But before leaving Alexander, some additional comments seem necessary, starting with the witchery and magic in his prescriptions. As we shall see, particularly when dealing with the *Kyranides*, these are characteristics of classical medicine that intensified with the decline of the Roman Empire and during the Byzantine era, especially at the level of popular medicine. Before discounting Alexander's attitude on the basis of a form of antiquated medicine of the Dark Ages, however, we must raise the possibility that he was aware that suggestion and placebo effects could be significant, particularly with some patients. This is suggested in his writings and it could partially justify the magical part of his art.

As to the absence of the power of a live torpedo among his prescriptions for headache and *podagra*, it is possible that he had dealt with them in texts since lost. Alternatively, he might have experimented on them without obtaining the desired effects, therefore choosing not to mention them. In any case, the great diffusion of Alexander's medical texts through the Middle Ages seemed to contribute to the idea that the efficacy of the torpedo as a remedy against pain might depend more on complex preparations from a dead fish than on the shocks from a live one. In other words, his ointments probably played a role in the shift away from live torpedoes as recommended by Scribonius and Galen. Interestingly, his ointments are, in fact, somewhat reminiscent of those we encountered in the Ebers Papyrus, although the early Egyptian prescriptions were not this complex (see Chapter 2).

PAULUS AEGINETA: BETWEEN LIVE AND DEAD TORPEDOES

We shall now examine the texts of Paulus Aegineta (or Paul of Aegina), who was partly influenced by Alexander of Tralles and partly preserved more faithfully the older tradition on electric fishes. Paulus was also a physician and his life overlapped that of Alexander, being born on the Greek island of Aegina and living from about 625 to 690 A.D.

This author is best known for a compendium in seven books, with no specific title in the ancient manuscripts, but generally indicated as *Epitomes iatrikes biblio hepta*, usually translated as his "Medical Compendium in Seven Books."[19] Paulus practiced medicine in Alexandria, where he remained after its conquest by the Arab armies. He was revered during his lifetime as a great authority on medicine and surgery (particularly war or injury surgery), and his writings, which drew heavily on ancient Greek and Roman texts, remained very influential in the Arabic world, and more generally in the Middle East, for some eight centuries.

In addition to a first known Arabic translation made in the ninth century, his influence on Islamic medicine was due to the exceptionally wide circulation of the medical works of al-Zahrāwī, better known in the West as Albucasis, a physician who lived from 936 to 1013. This is because Albucasis' works (which were translated into Latin by Gerard of Cremona in the twelfth century) drew heavily from Paulus, particularly in the surgical sections. Later, Paulus' writings were among the first Greek-language medical treatises to be published using the original language, together with the work of Hippocrates and Galen, in 1528 in Venice. At the beginning of the seventeenth century, they were still among the medical books used for teaching at the Sorbonne in Paris.

Like Aëtius, and in accord with the Hippocratic writers and Galen, he recommends the torpedo to his sick patients as an easily digested food (Fig. 5.7). In the last book of his treatise, in which drugs and treatments ("simples") are listed alphabetically, he also prescribes the sea torpedo (Νάρκη θαλασσία) in the case of headache, stating it should be used when still alive (ἔτι ζῶσα). He further writes that the fish "is said to remedy (προστέλλειν) the prolapsed seat" (Book VII, 3, 10; see Heiberg, vol. II, p. 245).[20] Moreover, he remarks that "the oil in which the living animal has been boiled, when rubbed in, allays the most violent of the arthritic pains."

The third book of his treatise contains a lengthy section (78) devoted to "Gout and Arthritis" (Περὶ ποδάγρας καί ἀρθρίτιδος). Like many others before him, including Galen, Paulus looked upon "unwholesome food and drinking much wine," as well as "unseasonable use of venery," as being the most significant causes of gout and arthritic complaints. The Greeks had imagined Aphrodite being seduced by Dionysus, whereas the Romans changed the names of gout's parents to Venus and Bacchus.[21]

---

the sea habitat of the fish, whereas, as we well know, the electric eel is a freshwater fish. When Puschmann was first editing Alexander's works (in the second half of the nineteenth century), electric fishes were well known in Europe and particularly in Germany, thanks especially to the work of du Bois-Reymond (see Chapter 24).

[18] As a matter of fact, *tourpaína* (or *toupaína*) bears a relatively minor similarity to *torpedo* compared to the similarity it has with the regional expressions still used in Italy and Dalmatia (and also in France). This is the case for instance of *treppina*, used in the south of today's Latium. Such terms might have derived from *terpina* through the common phenomenon of metathesis characterizing the passage from Latin to Romance languages. Also similar to *tourpaína* is the term *trpigna* used in Split, on the Adriatic coast of Dalmatia (today Croatia). Split (i.e., *Spalatum*) was the place where Diocletian built a magnificent palace. *Spalatum* remained under a strong linguistic influence of Latin even while Greek prevailed as the language of the Byzantine Empire.

[19] The current edition of the original text is that edited by Heiberg and published in two volumes in 1921–24.

[20] Notably, to indicate the "seat," Paulus uses the same word (ἔδραν) present in Dioscorides and Galen. It is possible that he derives his information directly or indirectly from Galen, because he also suggests that the effect may be a consequence of the property of the fish to produce torpor.

[21] Both the Greeks and Romans, and now the Byzantines, agreed that the joint swelling and pain could be kept at bay by bland diets, limited wine intake, and even more regrettably, less lovemaking (Schnitker, 1936). These were among the

Figure 5.7. The title page of a 1532 Latin edition of the medical treatises of Paulus of Aegina with, on the right, the page containing the prescription of an analgesic plastering based on a sea torpedo's ointment. The torpedo is mentioned as *turpena*.

Among the various remedies suggested by Paulus to treat this condition we find the torpedo. In contrast to Scribonius and Marcellus, but like Alexander of Tralles, Paulus did not, however, prescribe a live fish. Instead, he too recommends an ointment. This corresponds to his aforementioned remark on the analgesic properties of the "torpedo oil" against "arthritic pains." For the treatment of the paroxysmal phase of bilious gout, he advises various "refreshing" treatments, including herbs of a cold nature, such as nightshade, purslain, henbane, houseleek, marsh-lentil, endive, poppy, knotgrass, rose-leaves, and the like. Then he writes:

When the pain is great and difficult to endure, we must have recourse to preparations of saffron, of horned poppy, and of opium, triturating them with oxycrate, and then using them in the form of liniments, or of cataplasms [plasters] with crumbs of bread. Galen makes use of this one: Of poppy-juice [*opium*], dr[achmae] 4; of saffron, dr[achma] 1; triturate with milk of a cow or of a goat; or add the inner part of bread, and having softened it with a little rose-oil, use for a cataplasm. Or triturate the opium and saffron with the milk, and add to the rose cerate [a thick, externally applied ointment]. *And the cerate made from the torpedo [tourpaínēs] is active in the paroxysms and the oil of the same has similar properties.* (Paulus Aegineta, Book III, Sect. LXXVIII. Translation in Adams, 1844, vol. I, p. 659; revised and italics ours;[22] vol. 1, p. 301)

plethoras that could predispose certain body parts to serious problems. Gouty conditions occurred when such excesses allow a humour to become thick and turgid, and to build up in the joints. In Paulus' words: "When, therefore, the humour is seated in the joints of the feet, the complaint is called *podagra* [gout]; but when the cause is diffused over all the joints of the body, we commonly call it arthritis" (Book III, Sect. LXXVIII, p. 657; Teubner Edition I, p. 299). The "noxious" humour most often associated with gout, Paulus further tells us, is yellow bile. When it accumulates, the toes become red and it "occasions a sensation of heat and acute pain." This calls for dietary and behavioral changes, and in severe attacks there "should be the evacuation of the offending humour by purgative medicines."

[22] Paulus' recipe has many interesting components that are worthy of further comment. Two are noted painkillers, albeit under other conditions. One is the torpedo, which can give shocks that can affect pain perception, but it is here employed after it has been killed. The other, of course, is opium, a source of morphine. Notably, Paulus did not suggest smoking the opium or administering it orally. Rather, it was put into his salve, perhaps because he believed it could also have topical effects. Another ingredient deserving comment is saffron, which

This passage is very likely derived from Alexander of Tralles, and not only because of the use of the torpedo as an ingredient for a cerate to be used to alleviate *podagra*. It is especially so because both authors use the word *tourpaínēs*. It is likely that Paulus, with his epitomizing attitude and having not personally investigated the treatment, preferred to keep the original word, uncertain about the nature of the fish indicated by Alexander. As a matter of fact, he might not have even been aware that it was a fish or an animal, because he (unlike Alexander) does not even specify that the preparation must be obtained by cooking a live *tourpaína*. Paulus might have derived his information through an intermediate text, likely also a compilation.

KYRANIDES: MAGIC, NATURE, MEDICINE, AND SEXUALITY IN A HERMETIC TEXT

As noted earlier, the doubling of the unique shocking sea-ray into a conventional Greek *nárkē* and into an unusual *tourpaína* (actually spelled in a variety of ways) of clear Roman ascent occurs in a text written well before both Alexander of Tralles and Paulus Aegineta. It is found in the *Kyranides* or *Cyranides*, a composite work in the magico-hermetic tradition (see Chapter 9). This work, consisting of four books (actually four *Kyranides*, but with different numbers in some editions), was progressively compiled, with additions, transformations, and corruptions of various sorts, mainly in Greek but also in Latin (and afterwards in Arabic and other languages), starting back in imperial Rome.

Its primitive nucleus actually dates from the Hellenistic period (probably dating to around 200 A.D.), but its compilation and influence on other writings continued until much later phases of the Middle Ages.[23] It remained a reference text for magical and medical practices up to the modern epoch, existing in a variety of manuscripts, despite the fact that many have been lost, even in relatively recent times. It is largely based on previous or contemporary works or compilations that are more or less scientific in nature, such as those of Theophrastus, Strabo, Dioscorides, and Marcellus of Side, but it progressively incorporates additional magical elements.[24]

This is a text in which practices and beliefs of oriental origin merge with classical Greco-Roman knowledge and rites, revealing inspirations from pagan, neo-Platonic, Gnostic, and even Christian sources. Its composite character and particularly the oriental *coté* are declared in a rather explicit way in the introduction to the first *Kyranide*, with a reference to three more or less legendary authors: Kyranus, King of Persia, of rather uncertain identification; Hermes Trismegistus, the mythical god-scholar of Egyptian-Greek tradition, who supposedly received his mystical knowledge from the angels, composing "a book of natural virtues, formed by two books of sympathies and antipathies" (Ruelle, *Lapidaires,* vol. III, p. 34); and Harpocration of Alexandria, probably a real scholar, possibly imbued of Christian culture, who is said to have contributed to the first *Kyranide* on the basis of a medical text engraved on a Syriac stele, which he had found during a voyage. The Harpocration reference clearly points to the Alexandrian culture on which the work is based.

We shall deal in some detail with this work, which had an enormous diffusion during the Byzantine era and later Middle Ages, because it will allow us to penetrate the uncertain boundaries existing at these times between medicine, especially popular medicine, magic, and even religion.

The first *Kyranide* has a definite magico-hermetic character and a symbolic structure. As with most hermetic texts, it opens with the caution that the type of knowledge it contains should not be divulged to laymen. It is divided in 24 sections, each of which is defined by one of the letters of the Greek alphabet, and puts together four things (a plant, a bird, a stone, and a fish) united by their names, which begin with the letter. This procedure is somewhat like the kabala and is known as "litteromantic." The magical, divinatory, and medical powers of the four components are described, and particularly the way of preparing remedies of varying complexity from them, namely amulets with magical formulas and imprecations. These amulets are generally prepared by engraving the figure of one or more of the other components on the specific stone.

The other three books are somewhat more conventional medical texts, based on the use of animals in remedies (quadrupeds, birds, and fishes in the second, third, and

---

today refers to threads obtained from the crocus flower, but in the past might have referred to the plant itself. Crocuses were listed among the plants used to treat various disorders in antiquity, including gout. In particular, one species was highly regarded in ancient medicine: the colchicum crocus. Long exported from the meadows of Colchis, a temperate region on the Black Sea from which its name is derived, the corm of the colchicum plant can be very effective in curing gout when taken internally (Hartung, 1954; Wallace, 1968). Neither the Greeks nor the Romans, however, suspected that colchicum had specific anti-gout properties, and it was not always taken internally—as is the case with Paulus. It was generally viewed it as a purgative for ridding the body of harmful matter.

The idea that colchicum could be a "specific" for fighting gout and other arthritic diseases might even have emerged with Alexander of Tralles, who used the corm of the crocus plant (hermodactyl) as a substitute for opium (Copeman, 1964; Schnitker, 1936). During modern times, colchicine, a very effective anti-gout drug, would be derived from it. Knowing that ingesting the corm of the plant could cause stomach problems, the Byzantines recommended taking milk with it, recognizing that it could reduce this unwanted side effect. Because his concoction was not swallowed, Paulus would not have had to deal with stomach issues, although it is notable that he still called for milk in his formulas. In retrospect, Paulus' salve would not have cured anybody's gout. Whether his paste or ointment might have temporarily reduced its symptoms is less certain. But if the physician reduced certain foods and wine at the same time, as was suggested by Galen, the underlying disorder could well have been diminished, providing the needed illusion that his torpedo prescription really did work, an illusion that could only be strengthened by Paulus' reputation as a great healer.

[23] The *Kyranides* are considered the largest extant texts of a magico-medical literature known as *Phūsika*. This literature was probably inaugurated by a Greek scholar of the third century B.C., known as Bolos of Mendes, the Democritic (or Bolos Democritus, not to be confused with the atomistic philosopher: Mendes is the Greek name for the ancient Egyptian city of Djedet). Bolos' texts had a great diffusion and were an important source for Pliny, Aelian, possibly Galen, and perhaps others. His medical and magical theory was based on the concept of occult properties, and on "sympathies" and "antipathies" between plants, animals, parts of the body, etc., an alternative title of his *Phūsika* being *On Sympathies and Antipathies*. Only a few fragments of his texts are extant, and they seem not to include the torpedo. The torpedo is, however, mentioned in the extant fragments of a poem on fishes written by Marcellus of Sida, one of the physicians belonging to this magico-medical tradition. The torpedo is listed in a long enumeration of fish that uses the ordinary Greek term *nárkē*, and there are no prescriptions based on it in the extant texts (see Morel, 1591, for a Latin version of the electronic edition of the *Thesaurus Linguae Grecae* at the following address: http://stephanus.tlg.uci.edu/inst/textsearch for the original Greek text of Marcellus' poetic fragments). For more on Bolos, the *Kyranides*, and magico-medical texts of antiquity, see Wellman, 1928, 1934; Cox, 1983; Festugière, 2006.

[24] The reference edition of the *Kyranides* is that of Kaimakis (1976). The Greek text together with a French translation can be found in Ruelle (1898). Old Latin and French texts can be found in Delatte (1942).

fourth *Kyranide*, respectively). Besides producing medical and other beneficial body effects (for instance, reversing hair loss or procuring a particularly attracting color of the skin), the preparations listed in the various *Kyranides* can ensure victory in war, protection against other dangers, invisibility, divine and illusionary capabilities, and much more.

Often the preparations serve as remedies to allay pain, with gout, arthritis, and headache among the recurrent painful conditions to be treated. Diseases of the eye are also particularly important, with preparations derived from the gall of various fishes being valued as especially effective. Talismans or other preparations derived from birds, and particularly from eagles with their exceptional sight, make vision more acute.

Even more than with pain and diseases, the *Kyranides* are about love and sexuality. They include methods to avoid or to facilitate conception, to assist in childbirth, to counteract prolapse or retroversion of the uterus, to prevent excessive menstrual bleeding, and to re-establish the normal female cycle. Several preparations are used to increase sexual desire in men and women (these sometimes based on the dried penis of an animal). There are also methods for attenuating or potentiating virility. Among the latter, the "Aphrodite belt," a complex amulet prepared by using an opsian stone and stones derived from the head of a particular fish (the correspondent of the opsian among the sea animals) is said to be particularly effective. The Aphrodite belt makes men effeminate and sexually impotent; eating the fish stone can even render a man homosexual. If a woman carries the belt on her, no man would be able to have sexual intercourses with her, because he would be incapable of having an erection.

Among the more effective remedies to enhance love and sexuality are those requiring a fox. In particular, the beverages derived from the right testicle of the animal are a love philter for women and those from the left one are effective on men. Various preparations derived from the fox's testicles and from its penis increase male virility. To make sexual intercourse safe, one has to introduce the fox penis into a bladder and write a specific magic formula on the bladder using Smyrna ink.

Preparations obtained from the right and left testicle of an animal can produce diametrically opposite effects, as occurs, for instance, in the case of the weasel. The right testicle of this predator, mashed and mixed with myrrh and applied to the woman as a pessary (vaginal suppository), favors conception. On the contrary, the left one, put inside the skin of a mule on which particularly magic words have been written, has anti-contraceptive effects. Interestingly, readers are invited to verify the preparation experimentally on birds before trying it personally.[25]

In view of the general character of the *Kyranides*, it is no wonder that information concerning the torpedo is contained in this work, mixing science and magic together in strange ways. The fish is mentioned in two passages of the fourth *Kyranide*, the one dealing with remedies derived from sea animals, which can be translated as "Book of Hermes Trismegistus on the Scientific Knowledge and Influence of Marine Animals, Sea Fish, Composed by Asklepios, his Disciple."

The first occurrence, found near the beginning of this book, is, however, not listed under *nárkē*, the usual Greek name of the fish, but as "On the Ray" (Περὶ βάτου). It seems evident that the compiler of this *Kyranide* is not aware that the fish he is dealing with is the sea torpedo. This is how the short passage "On the Ray" reads in English:

> The ray (βάτος or βατίς) is a sea fish; Romans call it tourpaína (τούρπαιναν);[26] if boiled when fresh, its broth has a laxative action, when taken alone or with wine; if eaten frequently is beneficial to the stomach and stimulates love desires (ἀφροδίσια).

In the second passage (corresponding to the Greek letter "n"), the fish is indicated with the proper Greek name *nárkē*. The various prescriptions listed correspond more closely to those we have covered in this and in previous chapters, although with evident transformations from the original sources.

> The torpedo [*nárkē* or *nárka* according to the version] is a sea fish that many call *marga*; it relieves pain if applied when still alive to those who suffer headache; boiled still alive in oil until dissolution, it allays the pains of arthritics if used as a liniment after having been filtered. When calcined and reduced to ashes, if used as a dry powder, it resets the fall of the seat [i.e., *prolapsus ani*]. Its fat, spread out on wool and placed on the seat, block ascensions (ἀναδρομάς) of the uterus.[27] If a woman smears her sexual areas with this fat, her husband would no longer have sexual relations with her.

A comparison of the two passages is of significant interest. This is particularly so because they show how information can be transformed as it passes through compilations, and how different elements, even of a contrasting character, can be fused together in a rather incoherent way. Such transformations, needless to say, can be due to errors by the copyists or editors, to difficulties in reading a manuscript, and to

---

[25] Among the most disconcerting prescriptions is one suggested for healing a condition indicated as strangling (or suffocation) of the uterus (πνῖγας). The penis of a bear should be inserted into the woman's vagina in such a way that its extremity would fit into the mouth of the uterus. Also unnerving for modern sensibilities are the remedies based on animal urine and feces. Here we find that the dung of the cat, used as a liniment, can cure mild fevers or block hair loss, and that camel or sheep excrements taken orally can be effective against dysentery or colic.

[26] Or *toupaína* or *toulpaína* according the editions. In an old Latin version, edited by Delatte in 1942, the term is *tulpena*.

[27] It is possible that the author of the fourth *Kyranide* is here alluding simply to the retroversion of the uterus, a relatively common pathological condition (as assumed by some translators). It is also possible, however, that he is referring to a supposedly real upward wandering movement of the uterus, considered to be the cause of hysteria by Egyptian physicians and also alluded to by Plato in the *Timaeus*. In its ascending movements, the uterus might enter the chest, causing "a choking deliquium" and possibly even compress the trachea, supposedly causing suffocation. The word hysteria comes from *hystera* (ὑστέρα), the Greek term used to indicate the uterus. Medical and medical history students are sometimes reminded the episode of the German-born Austrian surgeon Theodor Bilroth, who alleged the relation of hysteria with the uterus in order to undermine the possibility, proposed by Freud, that men could also become hysterics. The male hysteria idea was accepted by Jean-Martin Charcot, with whom Freud studied in the 1880s, and by others well before this time. (For more on the history of hysteria, see Foucault, 1965; Vieth, 1970; Porter, 1987; Trillat, 1995.)

insertions of heterogeneous texts or annotations by the owner of a particular manuscript, to name but a few factors. For this type of text, these processes can have major implications, mainly because of their wide diffusion among even ordinary people. These people used them as readily available home medical handbooks or formularies, and did not abstain from adding anything that might have seemed useful into them.

One of the most striking aspects of the comparison between the *tourpaína* and the *nárkē* passages is that concerning the influence of the torpedo on sexual desire. Frequently eating the *tourpaína*, we read, has excitatory actions, whereas a preparation based on the fat of the *nárkē* does the opposite—having a sexually repellent action on men. Pliny described a similar action, which he could have transmitted to the *Kyranide,* perhaps in an indirect or otherwise transformed way, or, more likely, both Pliny and the author of the *Kyranide* might have derived their information from an older source. As for the sexually stimulating action of frequently eating the *tourpaína*, this might be related to the many fish prescriptions in classical medicine for this purpose (we also find it in later texts, see below). The allusion to the laxative action of the *tourpaína* broth may also be related to Pliny (or to his source or sources).

The term *tourpaína* establishes a relation with Alexander of Tralles and Paulus Aegineta, although the prescriptions using the *tourpaína* in the *Kyranide* are totally unrelated to those for the fish with same name found in these two Byzantine authors. This suggests two possibilities: either Greek texts preceding the *Kyranides* were already using the Latin-derived word, or the author of this work took his inspiration directly from a Latin source (possibly an oral one, and very likely one belonging to popular practices; see Footnote 18). One or the other of these possibilities might also have occurred for Alexander and Paulus, possibly independently with regard to the *Kyranides*.

The second or *nárkē* passage in the *Kyranides* puts together almost all of the older prescriptions with the fish that we were able to find and adds some more uses. Since it also deals with headache, it is unlikely that Dioscorides might have been its only (or main) source. In the case of headache, the direction that the fish should be alive is clear-cut. As for it affording relief from arthritic pains, it is less elaborate than what we found in Alexander but reflects a similar source, particularly if one considers the indication that the fish should be cooked in oil "when still alive."

The prescription of the *nárkē* torpedo preparation for *prolapsus ani* contrasts with the absence of this usage in Alexander and Paulus. The use of a liniment obtained from dead torpedoes, indicated by the *Kyranides*, also contrasts with Aëtius' annotation that a dead fish is totally ineffective in this condition. In the *Kyranides* there is, moreover, the idea that a preparation from the fish can be effective for a misplaced uterus, something we have not encountered in previous texts. The anti-aphrodisiac effect of the fat of fish applied on the sexual zones of a woman is also particularly notable.

Due to the multiplicity of manuscripts circulating during the late Imperial era and Byzantine era, the possibility remains that the *Kyranides* (and the other texts that we have examined in this chapters) derived their information, in one form or another, from other intermediary compilations. What is certain is that the information they transferred to later ages did not contain any trustable new facts of significance. Instead, they made the torpedo more magical and suitable for occult practices. The fact that they preserved at least part of the legacy from the classical era is, however, of considerable importance.

ISIDORE OF SEVILLE: A CHRISTIAN PLINY AND THE PRESERVATION OF KNOWLEDGE

Let us now turn west to Isidore of Seville, who lived a little later than Alexander and Paulus. Isidore, even more than the others, seemed to devote his entire life to the preservation and transmission of classical and Christian cultures across the Dark Ages. As in the case of most of the intellectuals of his period, he too based his work mainly on previous compilations.

In contrast to most of the aforementioned writers, at least from Marcellus to Paulus, Isidore of Seville, considered the Pliny of his age, was not a physician. Further, unlike these authors, whose works were exclusively of a medical character, Isidore's compilations were general encyclopedic works aimed at preserving without limits all previous knowledge.

Born around 560, Isidore was a bishop in Spain at the time of Visigoths, and he played an important role in both the religious and political affairs of his epoch. He belonged to an important family of the Hispano-Roman aristocracy, possibly originating from Cartagena in southern Spain. Two of his elder brothers had been bishops, and the eldest, Leander, was instrumental in converting the Visigoths from Arianism to Christian orthodoxy.

Isidore wrote a great number of texts in Latin dealing with all branches of knowledge, such as grammar, history (e.g., of the Goths and other barbarian populations), philosophy, theology, and religion (heresies, lives of saints, rules for monastic orders). Some of his books were inspired, even in their titles, by classical works. This was the case, for instance, of his *De natura rerum* ("On the Nature of Things"), *De viris illustribus* ("On the Illustrious Men"), and *Quaestiones* ("Questions"). He was especially interested in words and numbers, and his main work, *Etymologiae*, a huge encyclopedic compilation in 20 books, is based on the origins and meanings of words (Fig. 5.8). It includes almost all fields of knowledge from ancient epochs to his Christian one.

The *Etymologies* start with grammar and rhetoric. They then go into mathematics, medicine, civil and religious laws, theology, history, and social organization. Later, they deal with the natural world in its various aspects, and with a variety of other matters, eventually ending with home and agriculture instruments.

Even more than in Pliny, the separation of trustable information from legends and fantasies is very poor in Isidore's books. In particular, some of his word derivations are totally without basis, and sometimes decidedly laughable. Consider the word "wine" (*vinum*), for instance. It is called by this name, Isidore tells us, because it replenishes the veins.

Despite the general unreliability of the information contained in the *Etymologiae*, Isidore's work still had an enormous importance, being probably the best-read treatise in the first part of the Middle Ages, particularly in teaching institutions. As such, it contributed in a very significant way to the imagery of the Middle Ages. His legends of exotic animals and

Figure 5.8. Left: Saint Isidore of Seville with his pupil and first editor of his texts, Bishop Braulio (from a medieval codex at the Abbey of Einsiedeln, Switzerland). Right: The page of a 1483 edition of Isidore's *Etymologiae* dealing with the torpedo.

strange human beings (partially derived from Pliny and other classic authors) are especially noteworthy in this regard.

The work's main cultural and historical importance derives from the author's conscious attempt to save both classical and Christian culture during a period in which there was a real risk of losing the knowledge accumulated over the centuries. It is not by chance that Braulio, the first editor of the *Etymologiae* (published posthumously), who was Isidore's friend and the Bishop of Saragossa, recalled Cicero's praise of Varro with reference to Isidore's work (see the opening citation of this chapter). In the celebrated prose of the great Latin orator, this praise illustrated the importance of efforts aimed at preserving culture and civilization—the life-long intellectual endeavor of the Bishop of Saragossa (later St. Braulio).

The ninth book of the *Etymologiae* is dedicated to animals and it is here that we find a short reference to the torpedo based largely on Pliny. The book starts with the idea, common to Christian scholars, that Adam had imposed the true names on all the animals, Hebrew being the common language of all human beings before the Great Deluge. As with the case of most other etymologies, the derivation of the common Latin denominations of most animals is completely fanciful. For instance, Isidore writes that *armenta* (i.e., herds of big animals, such as horses and cows) are so called because they are *apta armis* (suitable for weapons), being usually employed in war. In a similar way the *agnus* (lamb) is so named because, better of any other young animal, it easily *agnoscit* (recognizes) its mother!

In a long digression on horses, Isidore inserts the story that beautiful horses are placed in front of gravid mares so that they would discharge beautiful offspring, with the remark that similar things could also happen with women (a belief that would persist over many centuries, with Albertus Magnus mentioning it in the twelfth century and Ulisse Aldrovandi bringing it up three centuries later; see Chapter 9). Isidore further considers griffons to be real animals; these legendary beasts have the body of a lion but possess wings and faces like eagles. He also refers to castors, stating that hunters try to kill them because of the medical importance of their testicular extracts, and maintaining that they will castrate themselves in order to save their lives.

Although Isidore derives these and many other legends concerning real and imaginary animals from classic authors (e.g., Pliny), he is more original in his derivations of animal names, some of which are quite fanciful. In the case of the brief mention of the torpedo inserted in Book IX, however, he is correct about the derivation of the fish name. He writes: "The torpedo is so called because it has a numbing action (*torpescere faciat*) if somebody touches it when alive." As to the transmission of the shock, here he quotes Pliny, writing that the shock can be conveyed "through a spear or rod," and that it has the power "of riveting the feet of the runner, however swift he may be in the race." He also mentions that "even the aura of the fish body may weaken the limbs," again probably deriving his information from Pliny.

Although there is no mention of the medical properties of the torpedo's shocks in Isidore's text, the information he transmits to posterity about the torpedo is neither more nor less fanciful than what his predecessors, and specifically Pliny, had written. The underlying problem is similar to that mentioned in the case of Pliny's *Naturalis historia* (Chapter 3). Surrounded by mythological, legendary, and magical animals in his *Etymologiae*, the torpedo comes across as an almost mythical wonder of nature, as opposed to a very real fish with unusual capabilities.

EARLY MIDDLE EASTERN WRITINGS

The writers of the Middle East blended Byzantine and other values with their own Islamic ideals. They too were more interested in the practice of medicine and medical botany than in zoology (Egerton, 2002). Further, as was true of the Byzantines and other scholars of the first parts of the Middle Ages, their surviving scientific and medical writings remained heavily reliant on Galen, Dioscorides, and other authorities from classical antiquity, whom they cited and quoted extensively.

The so-called Golden Age of Islam is usually dated from about 800 to about 1200. By this time, scholars in the Middle East had effectively become the trustees of Greco-Roman scientific and medical material. For example, in Jundi-Shapur, in what is now Iran, ancient Greek texts were being translated by exiled Nestorian Christians into Syriac and then into Arabic. These and other translations, in turn, stimulated additional books on science and medicine, works still based on Greco-Roman material.

The *ra'ad*, *raâda*, or *raadah* (literally "trembling"), which could refer to the Nile catfish or torpedo, is mentioned in several important Golden Age and later Middle Eastern treatises. Because a detailed survey of the Islamic literature is exceedingly difficult, given how many texts have not been translated from the Arabic and are difficult to access, we shall only examine a sample of this material. It is important to keep in mind that this sample is not random; rather, it comes from books that, in translation or in their original languages, have been deemed important enough to be preserved, copied, and later reprinted, even in the West.

'Ali ibn 'Abbās, the tenth-century Persian physician Latinized as Haly Abbas, is one of the earliest of these important Middle Eastern authors. He was also sometimes called 'Ali ibn Abbās al-Majūsi by his foes, this added last name meaning "the Magician." Around 980, he completed his medical encyclopedia, his *Kāmil as-Sinā'a at-Tibbiyya* (literally "The Complete Book of the Medical Profession"), which after being dedicated to the emir was sometimes commonly called the *Kitab al-Maliki* ("The Royal Book") or *Liber Regalis* in Latin. This work is divided into discourses, the first grouping dealing with theory and the second with practical medicine.

Constantinus Africanus made a partial translation of this work into Latin in 1087, and it was a founding text of the *Schola Medica Salernitana* in Salerno, Italy. Stephen of Pisa (also called Stephen of Antioch) made a more complete and better translation in 1127, which was reprinted in Venice in 1492 and 1523. It was read by Geoffrey Chaucer, who cited it in his *Canterbury Tales,* which dates from the end of the fourteenth century.

This encyclopedia includes a rather long section on numbness, and it is here that we find a reference to the *ra'ad*. The passage is based on classic Galenic physiology, and in translation it reads:

> Numbness occurs as a result of injury to the sensory nerves. This is due to a plethora or a deficiency. Sometimes this injury occurs in the nerves of sense and of movement, so that movement becomes difficult because of the sensory deficiency. The cause of this damage is either something that destroys the sensory faculty, so it becomes weaker, as happens with bad fevers and when one is near death. Thus, in this situation, the faculty weakens and becomes incapable of sensing. It can also occur as a result of damage to the instrument, namely the nerves, and one of its causes can be intense cold. This follows the sting of an animal that has cold venom; for example, this happens in the case of the *ra'ad*, the fish called *barq* [lightning]. Drinking poisonous numbing medicines, such as opium and hemlock, can also cause it. (from 'Abbās, 1877 Arabic ed., Vol. 2, Ch. 25, Sect. 29, p. 274)[28]

Thus, we find the narcotizing or torpifying fish (or fishes) still associated with the quality of cold in the Middle Eastern literature of the tenth century, much as had been the case for Galen when he practiced medicine in Rome during the second century. Most likely, what 'Ali ibn 'Abbās wrote can be traced back to Galen, albeit not necessarily directly.

Ibn Sina, who is best known in the West by his Latinized name, Avicenna, is an even better known Middle Eastern author. He was a Muslim scholar born in 980 in Transoxiana, a part of Muslim Central Asia (later Caesarian Russia), although he spent much of his professional life in Persia (Goodman, 1992). He turned to medicine at the age of 17 and, having access to a royal library for his studies, published his first book at age 21. It is believed that he wrote about 450 tracts, with 150 surviving works dealing with philosophy and about 40 dealing with medicine. He died in 1037 (Fig. 5.9).

His two most famous medical works are *The Book of Healing*, a huge philosophical and scientific encyclopedia,

---

[28] It is possible that *barq* is the consequence of an incorrect copying of a term used in the Arabic transliteration of the Greek word *nárkē*. As a matter of fact, in Arabic the characters "b" and "n" differs only for the position of a diacritic dot. If this is the case, it is purely accidental that the word present in this edition of the 'Abbās' works means "lightning", a term that, as we know, would be of great significance for the fish shock (we thank Ivan Garofalo for pointing out this possibility).

Figure 5.9. Avicenna (on right) with Hippocrates (left) and Galen (center). Avicenna borrowed heavily from Galen, and his *Canon* figured prominently in both Eastern and Western medicine prior to the Renaissance. This illustration dates from the sixteenth century.

and his *Qanun* (a Persian and Arabic word meaning "The Law"). In English (and similarly in many other European languages), the latter is referred to as his "Canon of Medicine." It seems to have been completed around 1025, and the earliest known edition of it dates to 1030 (Fig. 5.10). This work is again based heavily on Galen, although it also incorporates Indian medicine (e.g., the teachings of Sushruta and Charaka), earlier Islamic medicine, and the author's own personal experiences. It served as a standard medical text at many Islamic universities through the eighteenth century, and its Latin editions were used in European medical schools for hundreds of years. (Paracelsus, demanding change, burned Avicenna's books during a student celebration in Basel, on St. John's Day in 1527, in one of the most colorful episodes of the tumultuous Renaissance.)

The Canon of Medicine is divided into five volumes. The first deals with general principles; the second with simple drugs; the third with diseases of particular organs; the fourth

Figure 5.10. A page of the second volume of the oldest Arabic codex of Avicenna's *Qanun* preserved at "The Institute of Manuscripts" of Azerbaijan National Academy of Sciences, with, on the right, a sixteenth-century woodcut representing Avicenna as "the Prince of Medicine."

with diseases that spread from one location to other parts of the body; and the fifth with compound medicines. *The Canon* was translated into Latin and Hebrew in the thirteenth century, and afterwards published in a number of languages, including Hebrew on a press in Naples in 1491–92 (see Leibowitz, 1954).

Avicenna mentions the *ra'ad* in Book 2, the one dealing with simple drugs, among which are the fishes (*samak* in Arabic). At the start of the passage, he mentions sea fishes that are easily digested. He then follows earlier writers from the Roman period, including Scribonius Largus, Dioscorides, and Galen, in recommending the torpedo for headaches. But unlike some other important Islamic scientific and medical writers, he does not mention the name or names of his sources in this context. His brief statement can be translated from the Arabic or Latin into English as follows: "When a *ra'ad*, a live one, approaches the head of someone suffering from headache, it numbs the sensation of head pain" (Avicenna, Book 2, Sect. 10; see Avicenna, 1564, p. 361).

Hence, we have a return to the fish's shocks, as opposed to ointments and broths, in therapy with Avicenna. That he is specifying headache or head pain is clear from the Arabic, where we have term *al-masdou*, meaning a person with headache. It is also consistent with the Latin translation cited above, where the word *soda* appears. *Soda* is a word that does not exist in classical Latin. It was invented by a translator by transliterating the Arab word *sudaa*, referring to head pain. This fact is important because some later writers (e.g., ad-Damari), modern historians (e.g., Kellaway, 1946), and electric fish researchers have maintained that Ibn Sina was contending that epileptic seizures could be treated with live torpedoes—a mistaken impression that unfortunately has become part of the electric fish literature (e.g., see below, and also Leibowitz, 1957).

'Abd al-Latif[29] al-Bagdadi, sometimes just referred to as Allatif in the West, also merits attention. He was born in Baghdad in 1162 and studied the humanities, natural philosophy, and medicine in what is now Iraq. He then visited Egypt, where he met with the Jewish physician and scholar Maimonides, became one of the circle of learned men who were close to Saladin, and taught medicine and philosophy.

The best known of his writings in the Western hemisphere is his "Account of Egypt" in two parts, this being the English title used for an Arabic manuscript discovered in 1665 by orientalist Edward Pococke, which is now in Oxford's Bodleian Library. This manuscript was translated into Latin in 1800 and from the Latin to French (with commentaries) by Antoine Isaac, Baron Silvestre de Sacy, in 1810, and now has an English translation (Zand, Videan, & Videan, 1965). What is noteworthy here is that this text is not just based on others' descriptions, but on personal observations and interviews, best exemplified by his vivid description of a famine caused when the Nile River failed to flood one year (Silvestre de Sacy, 1810; Boll, 1874).

With regard to the *ra'ad* or *raâda*,[30] what Abd al-Latif al-Baqhdadi wrote in his "Account of Egypt" came from a fisherman, and it translates as follows:

Among the animals peculiar to Egypt, we should not forget the fish known as *raâda*, because no one can touch it while it is alive without experiencing trembling, which is impossible to resist. This trembling sensation is accompanied by cold, excessive numbness, and a crawling feeling in the limbs, so that it is impossible for one to remain in an upright position or hold anything. This stupefaction is communicated through the arm, the shoulder, and the entire side, however light the contact with the fish might have been. A fisherman who had fished the *raâda* assured me that, when such a fish is caught in a net, the fisherman feels the same distinct effect without his hand touching the fish, and similarly at the distance of a span from it. When dead, the *raada* loses this power. This fish is one of those that have no scales; its flesh contains few bones and is very fatty; its skin is a finger thick and is easily removed. The *raâda* is not good to eat. They are large and small, weighing from one to twenty *roti*.[31] People accustomed to bathe in water where this fish is found say that the mere breath of the *raâda* is sufficient to produce such numbness of the body that the person affected can scarcely keep from going down. This fish is common in the lower regions of Egypt and Alexandria. (Based on Silvestre de Sacy's 1810 French translation and Zand, Videan, and Videan's 1965 English translation)

There is little doubt that this author is here describing the *Malapterurus* (i.e., the electric catfish) and not a torpedo ray. First, he specifies that it is a fish "peculiar to Egypt," whereas torpedoes are ubiquitous sea fish, found throughout the Mediterranean Sea and its tributaries. Second, even if common torpedoes can swim into the Nile Delta, they do not cause such severe numbing and trembling effects, because of their relatively low-intensity shocks. In the Mediterranean region, the capture of the large and very powerful *Torpedo nobiliana* is indeed extremely rare. Third, although 'Abd al-Latif al-Baqhdadi does not give a description of the fish's shape that would allow us differentiate it physically from a torpedo, the annotations concerning the fattiness of the skin and its thickness and the easiness of its removal strongly suggest that it is an electric catfish. These characteristics have to do with the fact that, unlike other electric fishes, the electric organ of the *Malapterurus* is situated in the subcutaneous tissue and is surrounded by richly developed fatty tissue. Lastly, he tells us that the *raâda* "is not good to eat." When discussing Athenaeus' *Deipnosophistae* in the previous chapter, we pointed out that he and his gourmands felt torpedoes were worthy of being served at sumptuous banquets. In contrast, *Malapterurus*, although not clearly singled out from other catfishes in the *Deipnosophistae*, appears to have been considered a rather undesirable, lower-quality fish.

---

[29] His first name means "the slave of the Gentle," the Gentle being one of the names of God in Islam.

[30] In the Arabic, we find *al-ma 'rufanti bi-r-ra'ad*, meaning "known as ra'ad." In one partial English translation, the *bi-r-ra'ad* was combined into *barada* (Zand, Videan, & Videan, 1965).

[31] The *roti* is a unit of weight used in Islamic countries, equal to about a pound (0.45 kilograms).

Mention must also be made of Abu-l-Walid Muhammad ibn Ahmad ibn Rushd, also better known in the West by his Latin name, Averroës. He was born in 1126 into a family of legal scholars in Cordoba, and worked there and in Seville. Although Spanish by birth, his heritage was Arabic and he died in North Africa, in today's Morocco, in 1198.

Ibn Rushd was a polymath, and he made contributions to many fields while also practicing medicine. His most-cited medical text is his *Kulliyat* (a title that literally translates as "Generalities" but actually means "Theories of Medicine" or "Theoretical Medicine"). This work has been published as his *Colliget* in Latin and English. He was particularly influenced by the writings of Aristotle and Plato, and wrote commentaries on them in Arabic. Some of these treatises were translated into Latin and Hebrew, and today he is considered a rationalist and a founding father of secular thought in Western Europe (Leaman, 1998; Fakhry, 2001).

Ibn Rushd mentions the *ra'ad* in one of his commentaries on Aristotle's writings, his treatise on Aristotle's *De caelo*. This is his *Jawāmi' al-samā' wa-l-'ālam*, better known as his summary of *De caelo et mundo*, or "Summary of the Sky and Cosmos" (Ibn Rushd, 1994). Here he discusses astronomy and how celestial objects can affect one another, a subject that had long interested natural philosophers (North, 1989). Indeed, Averroës mentions the torpedo within the same cosmological context that we described in the previous chapter when discussing Alexander of Aphrodisias, whom he specifically cites.

At issue in Chapter 2 in the section titled "Nature of the Planets" is why the sun heats the earth but seemingly not the moon. Here Averroës discusses the angle of the sun's rays and features of heated bodies, such as their heaviness. As had Alexander of Aphrodisias, Averroës concludes that some things can be affected by certain powers, whereas others cannot, even though they can still act as intermediaries. This property, he tells us, is widespread in nature.

> We say that it appears that the matter regarding this [heating at a distance] is, as Aristotle says, due to the large size and thickness of the sun. As for the specifics of the heating of the planets by the sun, this has to do with the emanation of the rays. For it appears that the rays, when they are reflected, heat the bodies that we have by a divine capacity. When the lines of the rays occur on the heated bodies at right angles, this would cause the body to be heated more than would be the case otherwise.... When the rays affect a heavy body, it will heat even more. The planets send out rays, not because of fire, as some people used to think, for the rays do not exist because of fire. For this reason, the fire that appears in the orbit of the moon is not caused by rays. The illumination of the rays occurs because fire attaches to other matter. Likewise, fire does not heat in so far as it is fire.... We only say that is among its specific properties to heat bodies when it is reflected, because it does not heat the moon, or anything else among the celestial bodies around it, because it is not in their nature to accept heat. However, that which is heated by the intermediaries of celestial bodies, its state is, as Alexander [of Aphrodisias] says, like the powers of the sea fish that can numb the hands of the fisherman by the intermediary of the net, without numbing the net itself, and like the state of many similar things. (Trans. from Ibn Rushd, 1994, p. 65, by Lutfallah Gari)

THE AMBIGUITIES IN AL-QAZWINI

During the thirteenth century, there was a decline in scientific and medical scholarship in the Middle East. This change coincided with the fighting that was then taking place with the invading European armies. The Christians had already begun to drive the Moors out of southern Spain (Toledo was conquered in 1085; Cordoba would fall in 1236, and Seville in 1248), and the Crusades (1095–1291) were underway to regain the Holy Land, creating turmoil. The resulting decline, however, was neither rapid nor uniform, and some notable works still came forth from the region, some more accurate than others.

Abu Yahya Zakariya' ibn Muhammad al-Qazwini, a mid-thirteenth-century physician who practiced in Persia and Baghdad, was one of the writers of this post-Golden Age, and he mentioned the torpedo. It appears in his popular cosmography, *Ajā'ib al-Makhlūqāt wa-Gharā'ib al-Mawjūdāt*, which can be translated as "Marvels of Things Created and Miraculous Aspects of Things Existing." This work was composed in Arabic in 1263, with an enlarged second edition that dates from 1275. As the title implies, it is filled with wonders, many mythological, including creatures that are partly animal and partly human, which would be incorporated into works by other Middle Eastern authors (see below).

Under the heading "Water Animals" (no part or section headings are given) we find: "If a woman carries any portion of this fish around on herself, her husband would not be able to bear being separated from her" (al-Qazwini, 1983). It is unclear whether al-Qazwini is specifying the saltwater torpedo or the African catfish in this passage, but it is possible and even probable that it is the catfish (see the quotation at the end of this section, where a different al-Qazwini associates it with a fish of the Nile).

As discussed, Pliny, who is cited elsewhere in al-Qazwini's text, had written that applying the gall from a live torpedo to the genitals could diminish excessive sexual craving. In the *Kyranides* there are contrasting indications of the aphrodisiac effect of the torpedo *tourpaína* and the anti-aphrodisiac effect of the *nárkē* fish. Despite the opposite conclusions about the venereal actions of the fish preparation between al-al-Qazwini's *ra'ad* and the *nárkē* of the *Kyranides*, the somewhat stylistic similarity of the two texts suggests that al-Qazwini might have derived some of his information, perhaps in an indirect and corrupted way, from the *Kyranides*. In addition to these Western sources, and even more importantly, he tapped many Middle Eastern sources, all of which use a word like *ra'ad*, which can mean the shocking catfish, the ray, or both narcotizing fishes.

The aphrodisiac effects of the electric catfish would be made clearer in subsequent writings, indirectly supporting the idea that al-Qazwini's statement might have been based on *Malapterurus*. Among later authors, the Dominican missionary João dos Santos traveled to Africa at the end of the sixteenth century and wrote that the natives use the skin of a tremble fish found in the Sofala River in fetishes and for sorceries (Santos, 1609, p. 39; Purchas, 1625/1905, Vol. IX, pp. 227–228; see Chapter 7). Also, Carsten Niebuhr, in his book on the fish observed by Petrus Forsskål in the Middle East, writes that, according to local tradition, "a salted Torpedo held in the hand is a very strong stimulant," very likely meaning a stimulant of sexual desire (Forsskål/Niebuhr,

1775, p. 16). Although this comment appears in Latin under the heading *Raja* (meaning ray), we are told that this torpedo comes from the Nile and has six barbells (*tentacula*) on its face. These facts make it clear that Forsskål was writing about *Malapterurus*.

In this context, it should be noted that a species of electric catfish is still associated with ideas of this sort in the country of Mali. There it is called *N'tigin* and is associated with the female (a deadly water snake is its male counterpart) and the vagina. Its dried skin is an ingredient in a beverage that is administered to assist in childbirth (Brett-Smith, 1994, pp. 127–128).

The following passage can also be found in al-Qazwini, and its last line is of particular significance:

> It is a small fish, but it is very numbing. If it falls in the fishing net of one of the fishermen while he is grabbing the rope of the fishing net, it makes him feel so cold that he will start shivering. Usually once the fishermen feel it in their fishing nets, they attach it to one of the trees until it dies. And when it dies it becomes no longer effective. Physicians in India use it to cure diseases that cause high fever, but in other countries no one knows how to use it.

The last line about torpedoes being used therapeutically in India goes well beyond what we have discovered in Galen or the other Greek, Roman, and Byzantine writers that we have cited. As we shall see in Chapter 7, the term "India" can have multiple meanings. If al-Qazwini had been referring the land that is considered India today, he could not have had *Malapterurus* in mind. This is because there are no electric catfishes in India, whereas India's coastal waters are home to several types of *Narcinidae* (numbfishes) and *Torpedinidae*, including *Torpedo panthera* (the panther or leopard torpedo) and *Torpedo sinuspersici* (a variety of marbled torpedo). But al-Qazwini does not tell us the sources behind his statement about treating fevers in India, and so we have to ask ourselves whether al-Qazwini's reference to India could have signified some other place.

In fact, three different Indias were recognized at this time, with not all writers agreeing on their boundaries. One was India Major, from Malabar through the East Indies (i.e., China); the second was India Minor, from Malabar to Sind (a province in today's India); and the third was India Tertia, which included the east coast of Africa. In some conceptions, India Tertia, the westernmost of these Indias, includes today's Ethiopia, the source of the Blue Nile and a habitat of the electric catfish. According to one writer (Hiob Ludolf), the Ethiopians were still using it to treat fevers in the seventeenth century, although we are compelled to add that this frequently cited "authority" never visited Ethiopia and does not divulge his source or sources for this statement (see Chapter 7).

There is also third statement about the *ra'ad* in al-Qazwini that demands another look, because it too is rather bewildering and has created considerable confusion. He writes, "Ibn Sina said: If you get the *ra'ad* close to the person who suffers from fits [seizures, epilepsy] it will numb his senses immediately" (al-Qazwini, 1983 edition, p. 187; trans. by Lutfallah Gari). The problem here is that Ibn Sina or Avicenna did not recommend the *ra'ad* for seizures; rather, he targeted patients with headaches, as we have seen.

Very likely this error in al-Qazwini stems from how similar the Arabic letters "d" and "r" are to each other, and this relates directly to how close the Arabic words for seizure disorder and headache are to each other.[32] The Arabic *al-masdou* refers to a person with headache, whereas *al-masrou*, the term encountered in al-Qazwini's text, refers to a person with seizures. Whether the mix-up stemmed from a faulty edition of Ibn Sina, from al-Qazwini misreading what his predecessor had written, from al-Qazwini's copyists, or perhaps some other source, is unknown. Without question, this promotion of a wonderful cure for "epilepsy" (as put in the later electric fish literature) provides the mistaken suggestion that a spectacular new medical finding related to the living fish emerged from the Middle East. Until shown otherwise, we think it most likely that this common historical misperception stemmed from an error.

Finally, and adding to the confusions that we have associated with al-Qazwini, we must point out that there was not one but two al-Qazwini's. Not only did they overlap each other, but they both wrote books that included passages about the *ra'ad*. The slightly later author is Hamdullah al-Mustafa al-Qazwini, who was born in about 1281. An industrious compiler, he composed his *Nuzhatu-l-Qulub* ("Hearts' Delight") in about 1340 (approximately 77 years after the text just discussed), and he used Abu Yahya Zakariya' ibn Muhammad al-Qazwini as one of his sources (Stephenson, 1928a, b)

The younger al-Qazwini's *Nuzhatu-l-Qulub* is again an encyclopedia, or more fittingly a "popular educator," with three *maqalas* (parts). The third *martaba* of first *maqala* covers the animal kingdom, and *Samak* (fishes) is a single entry on the alphabetical list (after *Sulafāt*, meaning turtle), although 37 varieties of fishes are mentioned. The *Ra"ādat* or *ra'ad* is the tenth and some of his wording is identical to that of the first al-Qazwini:

> Tenth, *Ra'ādat,* a small fish, and extremely cold; so that when they fish for it its coldness numbs the fisherman and it carries the line out of his hands; and hence the fishermen make fast the line to something before throwing it towards the fish; and this characteristic of the fish persists till it dies. It is useful in hot diseases; and if either of a married couple carry any portion of one about them, the other will not be able to bear being separated from him (or her) for a single instant. It is plentiful in the Egyptian Nile. (trans. in Stephenson, 1928a, p. 55)

Hamdullah al-Mustafa al-Qazwini does not mention anything about headaches or epilepsy in this passage. But he helps clear up some of the mysteries in the text of his predecessor of the same name. Specifically, he states that the fish he is describing is common to the Nile River. This fact strongly suggests that he is writing about electric catfish.

SOME LATER ISLAMIC WRITERS

Abu Muhammad Abdullah Ibn al-Bītār (or Ibn al-Baitar) was born in Malaga, a part of Muslim Spain, late in the twelfth century. He was a learned botanist, who traveled

---

[32] We thank Lutfallah Gari and Gül Russell of Texas A & M University, Asad Ahmed of Washington University, and Ivan Garofalo of the University of Siena for helping us here and with other Arabic translations.

along the northern coast of Africa and to Asia Minor, and he became chief herbalist to the Egyptian governor around 1225, whose region was later extended to Damascus. He is best remembered today for an encyclopedia about botanicals and one on medicine, and he died in Damascus in 1248.

Ibn al-Bītār specified his sources when bringing up the medical attributes of torpifying fishes in his text *Al-Jāmi' li mufradāt al-adwiya wa-l-aghdhiya*, which translates as "Compendium of Simple Drugs and Foods." In this part of his encyclopedia, which is arranged alphabetically, under the word *ra'ad* he first cites Galen, Dioscorides, and Paul of Aegina, all of whom focused on the sea torpedo:

> Galen said: It is a numbing sea animal. Some people mention, that if you get it close to the head of a person who suffers from a headache, it will calm the headache. Also, it cures *maqada* [anal prolapse]. If you get it close to the backside of the patient, it can return the protrusion to the normal position. But when I tried it in both cases, it did not work! So I thought of bringing it close to the head of a person with headache, while the animal was still alive, because I thought that in this state it would be a medicine that would calm the headache in the place of other medicines for headache. And when I did, I found it to be beneficial for as long as it [the fish] was alive!

> Dioscorides said: It is a numbing fish. If you put it on the head of person who suffers from headache, it can calm the intensity of his pain. And if it is carried to the backside, it can cure *prolapsus ani* as well.

> Paulus of Aegina said: The oil that you cook the *ra'ad* in can be used to relieve joint pain, by rubbing it on the painful areas.

We then read that this author had personally seen torpedoes off the coast of Spain and in the Nile. But his passage also raises questions, because he tells us that the Spanish fish he is describing could be poisonous when eaten, although the Nile fish (perhaps *Malapterurus*, but which, based on its coloring and weaker shocks, seems more likely to have been a ray) is not. In translation, Ibn al-Bītār writes:

> In Andalusia—on the shores of my city of Malaga, I saw a wide flat fish called "al-Arouna" that lives in the sea and comes to the shore. Its back has the same colour as the *ra'ad* I saw in Egypt in the Nile, and in the area underneath it has a white colour. It numbs anyone who catches it, as does the *ra'ad* of Egypt—or even more so, but the difference is: it is not edible! Once, some ignorant people didn't know about it; they grilled it and ate it. After just an hour all of them were dead. (Ibn al-Bītār, 1874 edition, Vol. 2, p. 141; trans. by Lutfallah Gari)

This brings us to Muhammad Ibn Musa Kamal ad-din ad-Damiri, who seems to have blindly accepted what his predecessors, especially al-Qazwini, had written. This author, who was born in 1351 and died in 1405, lectured at the al-Azar University in Cairo and wrote *Hayāt-al-Hayawān* ("The Book of the Lives of the Animals"; ad-Damiri, c. 1370). This book deals with the wonderful features and properties of animals, including their uses in medicine (for a translation, see ad-Damīrī, 1906–08). It is an alphabetically organized zoological lexicon or encyclopedia.

Under *Samak*, the word for fish, we find general and specific statements about fish and other water creatures, including *ar-ra'âdah* (*ra'ad*). Following earlier authors, ad-Damiri writes:

> Among the kinds of fish, there is the *ra'ad* and it is small in size. When it falls in the net and the fisherman touches its rope, his hand will tremble. Fishermen know about this property and, therefore, when they sense it [the fish in the net] they tie the rope of the net to a peg or a tree until the fish dies. When it dies it loses this property. (From the Arabic in ad-Damiri, 1975, Vol. 1, pp. 567–568; for a slightly different translation in English, see ad-Damīrī, 1908, Vol. 2, pp. 68–69)

After briefly presenting some poetry in which the poet snipes at a critic and compares him to a *ra'ad* fish, ad-Damiri presents what al-Qazwini wrote about physicians in India. To quote: "Physicians in India use it to cure diseases characterized by excessive heat [or high fever], but in other countries, it is not possible to use it [in them]." He then includes his version of the mistranslated or erroneously copied phrase from ibn Sina, in which the Arabic word for "fits" has replaced the word for "headache." "Ibn-Sina states that, if the *ra'ad* is placed near the head of a person in a fit [epileptic attack], when it is alive, it will have a beneficial effect on him."

Another thought found in al-Qazwini, and one that is obviously fanciful, appears immediately after the headache statement and concludes what ad-Damiri has to say about electric fish. He writes: "If a woman carries a portion of this fish on herself, her husband will not be able to bear being separated [to part] from her" (ad-Damīrī, 1908, Vol. 2, pp. 69).

Thus, there is really nothing new in ad-Damiri, who draws upon others and unknowingly mixes facts with fancy, as do other writers from this period. As put by one ad-Damiri scholar, "Ancient Greek medical traditions, the development of them by Arabic scientists, the medical views of the Prophet Mohammad, and popular quackery and practices subsisting from time immemorial among the Arabs—all these are interwoven in the medical chapters of ad-Damiri's *Kitah hayāt al-hayawān*" (Somogyi, 1957, p. 67).

Indeed, in a way that is reminiscent of some of our Byzantine authors, ad-Damiri shows no skepticism when he further writes that "deafness is removed with the blood of a wolf mixed with walnut oil"; "riding an ass is beneficial in scorpion-stings"; "removing a sleeping dog and urinating in its place cures colic"; "eating of the brain of the wild cow is beneficial in paralysis"; and "the foot of the frog, hung on the patient, relieves gout" (trans. in Somogyi, 1957, pp. 70, 74, 80).

Clearly, there had been a significant decline in empirical and verifiable medical and scientific facts since Galen's death at the end of the second century A.D. Yet as fanciful as natural philosophy had become in the hands of some Middle Eastern authors, the scientific landscape was in many ways not very different in the West until the late Middle Ages, the time period that we cover in our next chapter.

PART III
Late Middle Ages to the Early Modern Period

Chapter 6
Torpedoes: From the Scholastics through the Renaissance

> Torpedo, once called *Nárke* by the Greeks, ... because it brings torpor to the hands by its own coldness ... is known to the people of Bordeaux and among them it is called *Tremble*, as if to say tremulous, because with its slimy, smooth skin, which is soft and cold, it makes hands tremulous by contact with it.
> (Pierre Belon, 1553, p. 89)

This chapter will be devoted to knowledge about electric fishes in the period from the beginning of the twelfth century to the end of the sixteenth century—that is, from the Golden age of the Scholastics in the late Middle Ages through the Renaissance. As we shall see, there would be no striking advancements in electric fish science: no significant breakthroughs in understanding the mechanisms of the shocks, their transmission, or their medical applications. In this regard, the material we shall cover will have features in common with the Byzantine and Middle Eastern cultures discussed in the previous chapter. Nevertheless, there would be new mindsets, a new appreciation of living nature, and other developments, including the advent of beautifully illustrated fish books and a call for more accurate information, which will help set the stage for the scientific revolution of the seventeenth century.

Some of our readers might wonder why we are dedicating yet another chapter to such a scientifically "unproductive" time. Why, in brief, are we not skipping directly to the productive and "luminous" phase of electric fish research that started in the second half of the seventeenth century, when new facts and ideas began to emerge with the experimental science of Francesco Redi and Stefano Lorenzini? One reason we can give is the need for background information to help us see how ideas about these fishes were maintained and how they changed in broader scientific and cultural contexts. But more than this, thoughts that might appear entrenched, negligible, or even decidedly wrong can often be found on the paths leading to momentous scientific discoveries. Hence, it is important for historians, particularly when dealing with the evolution of scientific ideas, to discuss the influences of the past in as clear and detailed a way as possible. This means carefully and broadly examining a myriad of historical factors, including those that might appear sterile or misguided on first glance, when discussing the evolutionary and at times revolutionary development of an idea—in this case, fish (and more generally animal) electricity.

Thus, the material that we shall now examine is not just important for the history of Western civilization, but can be related in a direct way to the story we are telling. The social, cultural, and even artistic changes that we shall now describe have significance, and some of them will lead to changes in thinking about our strange fishes, which in turn will culminate in new ways of explaining their powerful and perplexing powers. For these reasons, let us now turn to developments in the history of electric fishes starting with the end of the Middle Ages and working through the Renaissance.

EARLIER CULTURAL LEGACIES

As noted, the Byzantines and their Middle Eastern and Islamic counterparts did not add a wealth of new scientific facts to the electric fish story, although some of these writers extended what had been written about torpedoes to another type of narcotizing or torporific fish, an African river catfish, which was given a similar name and was not sharply distinguished from the sea torpedoes. In other domains, however, these writers tended to accept information uncritically and even corrupt the "facts" passed down from classical sources. This is particularly true with regard to the properties of the shocks and the therapeutic uses of these fishes. Stated somewhat differently, the Aristotelian tradition of carefully describing living nature became increasingly contaminated with imaginary elements, legends, and magic, while therapies veered away from those prescriptions written during the Roman Period that specified living torpedoes.

Specifically, live torpedoes had been used to treat a variety of painful conditions, including headache, foot pain (including gout), and "prolapse of the seat" in the texts of Scribonius and Galen. Although probably offering no more than temporary pain relief for most patients, this kind of treatment represented a sort of electrotherapy *ante litteram*—a medical use for electricity before there was any knowledge whatsoever of electricity. In texts such as the *Kyranides* and in the writings of Alexander of Tralles and Paulus of Aegineta, torpedoes continued to find a place in medicine. But their medical applications no longer involved the shocks of a living fish.

Rather, they were based on popular medico-magical practices, such as the thought that fish parts or fish ointments would have similar attributes and wondrous effects. From a scientific viewpoint, this was a regression from earlier thinking, which had a firm empirical basis. We need only think about Anteros' personal experiences on the seashore and Galen's experimentalism.

As with the Byzantines, there were no significant advances in the Middle East when it came to the physiology of the electric fishes or their therapeutic uses, not even during the Golden Age of Islamic culture. The writings of scholars from this part of the world were based mainly on classical Greco-Roman texts (e.g., Dioscorides, Galen), on Byzantine compilations (e.g., Paulus), on magical traditions, and on their own cultural beliefs.

Thus, the once-strong interest in zoology characterizing Aristotelian science declined during the Late Classical Era. The representations of nature that increasingly dominated the Western and Eastern cultures now had more of the flavor of Pliny, with facts mixed with myths and the inclusion of folklore. The emphasis was mostly on the religious aspects of culture. In particular, following Saint Augustine and other Church Fathers, Christians of the Early Middle Ages were now viewing the world in a manner consistent with their new religious conceptions. Living nature was a wondrous part of God's glorious kingdom to be sure, but animals were now viewed as signs or symbols (*signacula*) rather than as objects worthy of human investigation in their own right.

All of this notwithstanding, the Middle Ages cannot be brushed aside as a time of total stagnation. Great cultural changes with wide ramifications, even in science and medicine, took place at this time. People would witness the creation of public hospitals, teaching institutions, new technologies, and significant changes in the arts. In the latter domain, there would be imposing new architectural structures (some of the most beautiful palaces, churches, and monuments of the age were created during this time), and splendid mosaics and fascinating frescoes came forth from artists' studios to glorify the new structures. Still, cultural interests were dominated more by religious philosophies and the "arts" of the so-called *trivium* (i.e., grammar, dialectic, and rhetoric) than by mathematics or experimental science.

The philosophy that flourished during the Byzantine Era was important for the transmission of the earlier classical heritage, for its transformation, and for its broad influence. As with Alexander of Aphrodisias and Simplicius of Cilicia (mentioned in Chapter 4, in our discussion of the mechanisms of shock transmission),[1] the Byzantine commentators were not passive with regard to the earlier knowledge. For instance, John Philoponus of Alexandria, the sixth-century Christian author of important commentaries on Aristotle, was critical of Peripatetic conceptions of movements and vacuums. Simplicius amply mentioned the problems Philoponus encountered and, after these problems came to the attention of Western scholars in the thirteenth century, they contributed to the scientific progress in the field of physics. During the sixteenth century, they stimulated Giovanni Battista Benedetti, who, while also questioning Aristotle, would demonstrate that the speed of falling bodies is not proportional to their weight, and that objects tend to fall at the same speed in a vacuum.[2]

The introduction of the system of numerals, derived from the Hindu world, was among the greatest contributions of Islamic scholars to scientific progress. Europeans slowly adopted this system, ultimately leading to the development of modern mathematics. After a period of stagnation, in which medieval mathematics had been based only on the notions contained in compilations (e.g., of Boethius, Isidore, and Bede), Western mathematics resumed its progress around the end of the tenth century through the efforts of Gerbert d'Aurillac (the future Pope Silvester II), who was influenced by developments from Islamic science. Gerbert reintroduced the abacus into Europe after it had disappeared from the Western world with the decline of Classical Era (he also reintroduced the armillary sphere, an important astronomical tool). Significant progress in mathematics really began, however, only during the twelfth century with Leonardo Fibonacci (Leonardo of Pisa), who was strongly influenced by Arabic contributions and used their system of numerals. His imprint would spread from Italy to other European countries, especially to England, France, and Germany.

During the thirteenth century, the translation of Arabic texts dealing with optics (particularly of those of al-Kindi, Alhazen, Avicenna, and Averroës) was also of great importance. It stimulated the study of optics in the West. The optical works of Erazmus Witelo (Vitelo), John Pecham, Robert Grosseteste and Roger Bacon collectively represent one of the most significant developments in European culture prior to the scientific revolution of the seventeenth century (see Haskins, 1939; Crombie, 1952; Lindberg, 1976).

The *Quaestiones naturales* by Adelard of Bath must also be mentioned in this context, because they further reflect how the European cultural revival derived from contacts with the Middle East. Adelard was a twelfth-century English scholar who journeyed to southern Europe and the Orient, which allowed him to become acquainted with important Arabic texts (Adelard, 1998, and see Crombie, 1952). Upon his return, he described an imaginary dialogue with his nephew, who typified a religious school student. The young man sticks to the old culture and to classical and Christian authorities, and he refuses to reason when dealing with questions concerning nature and world. The culture to which the nephew makes reference is almost uniquely that of the Latin authors taught in the schools, mainly through compilations and treatises of rhetoric, and the way he discusses the natural world is purely a logical one based on theological views without reference to physical problems and natural causes. Adelard opposes his nephew's attitude and supports the new science that is emerging in the most

---

[1] During the Byzantine period, various scholars dealt with the problem of transmission of the shock within the general framework of the transmission of forces and affections at distance. These scholars included Simocatta and Glycas (see Chapter 3).

[2] To counteract Aristotle, Benedetti conceived a very acute "thought experiment" capable of proving the absurdity of the logical consequences of Aristotle's claims that the speed of falling bodies is proportional to their weight. His argument was developed and perfected by Galileo in his *Dialogo sopra i Massimi Sistemi del Mondo*, playing a significant role in the new physics of motion elaborated by the great Pisan scientist. For more on Benedetti, see Maccagni (1985).

advanced institutions of his period. That science is largely influenced by developments from the Islamic world.

Even if genuine scientific interest and experimental observations never entirely disappeared during the Middle Ages, it is clear that Arabic science and philosophy—and particularly developments in optics, mathematics, and astronomy—still played significant roles in the revival of Western science and culture, starting during the twelfth century.

THE MIDDLE AGES REVIVAL

Complex and multiple historical factors account for the great blossoming of European culture that began in the twelfth century, the period in which Adelard wrote his *Quaestiones naturales*. Compared to the preceding Dark Ages, with its barbarian invasions, political instability, and extremely precarious conditions, Adelard's era was characterized by improved social and political institutions. This was a time marked by the consolidation of important centers of power (the resurgence of the Western Roman Empire, various flourishing kingdoms with active cities, and, of course, the Church with its central and peripheral institutions) and economic progress, the latter due to improvements in agriculture and various aspects of manufacturing and technology. It was also a time of great advances in navigation and cartography, as well as in literature, the fine arts, and architecture (with magnificent Gothic cathedrals first appearing in France and afterwards in England and other European countries).

This progress had its counterpart in the evolution of new cultural institutions. One was the development of religious schools associated with great abbeys, monastic centers, and cathedrals, such as those in Montecassino, Bec, Cluny, Canterbury, Fulda, St. Gall, Lyons, Orleans, Rheims and Chartres. The first universities also appeared at this time in various towns and cities, including Salerno, Bologna, Paris, Montpellier, and Oxford. Lay schools also began to appear, especially in Italy (see Haskins, 1928, 1939).

In these cultural centers, and particularly in the cathedral schools, an intense and lively intellectual life developed, which in turn led to a progressive accentuation of the role of reason as a source of knowledge, particularly for nature in its various forms. The idea of a world as a *liber involutus* ("convoluted book") that can be interpreted only with recourse to faith assisted by the authority of holy books and of the Church Fathers (essential for deciphering God's messages) slowly gave way to the conception of "the book of nature." Nature would be the book that people could read and interpret on the basis of their own intellectual faculties, without the immediate assistance of the Bible, the Church Fathers, or other religious writings.

One of the events contributing to this slow change of mindset was a renewed interest in Aristotle, more specifically in his comprehensive intellectual endeavors, spanning philosophy, cosmology, physics, biology, psychology, and the like. This new interest provided evidence of the great potential of the human intellect, even when unaided by the Christian religion.

Efforts began to be made by some of the greatest theologians of the epoch to incorporate the Aristotelian vision of the world within the framework of their own Christian beliefs. Thus, Albertus Magnus, Thomas Aquinas, and other great men of the cloth now strove to promote Nature as a wonderful creation of God. As such, animals again became a subject worthy of study—even during a period of deep spiritual values, and one with a greater focus on the salvation of the immortal soul than on an earthly existence.

Studying the Aristotelian texts stimulated intellectual discussions in various fields of knowledge. One of its consequences was the development of logic in the mainstream medieval philosophical movement known as Scholasticism (from *schola*, meaning school). In addition to Adelard of Bath, Albertus Magnus and Thomas Aquinas, the Scholastics included Peter Abelard, John Duns Scotus, Roger Bacon, William of Ockham, and many others. These thinkers would be influential in changing the culture of the West.

Without a doubt, Scholasticism sometimes resulted in sterile intellectual exercises, with debates about such things as whether angels have a gender. But it also promoted discussions of great significance, including those in the fields of physics and cosmology, where it led to new theories of motion in its attempts to solve the problems inherent in Aristotle's conceptions of both terrestrial and celestial physics. One of the most important outcomes of this discussion came in the fourteenth century, when John Buridan and Nicole of Oresme developed the theory of *impetus*, which paved the way for the concept of inertia, a concept of great importance for Galileo, Newton, and others.

The trustworthiness of knowledge of natural matters was among the more earthly issues that greatly interested the Scholastics. With the deeper study of Aristotle's writings and other Greek texts, it appeared as if a sure kind of knowledge could be obtained by deduction, starting with some generally accepted principles or axioms. The problem was whether it was possible to arrive to a trustable and certain knowledge through the "inductive method," which starts with observations based on the senses and then works up to conclusions thought to have broad applicability. The discussion on this point was central to Scholasticism, and it involved Robert Grosseteste, Albertus Magnus, Roger Bacon, Duns Scotus William of Ockham, Nicholas d'Autrecourt, and others. This matter also had medical implications, as can be seen in the works of Petrus Hispanus (later Pope John XXI), and afterwards in the texts of the "medical logicians," such as Pietro d'Abano and Jacopo da Forlì.

This focus on induction led some of great thinkers of the day to stress the importance of observation and experiments to understand nature, despite the predominance of logic and dialectics in mainstream Scholasticism. Among the consequences of this attitude was criticism of Aristotle, since it became clear that conclusions derived from direct observations did not always support the views expressed by the great philosopher of antiquity. Paradoxically, as noted by Alistair Crombie (1952), Aristotle occupied center stage in the intellectual discussions into the scientific revolution in two contrasting ways, somewhat like a "tragic hero." On the one hand, he was the classical scholar who was contributing the most to the revival of natural philosophy and the new interest in the natural sciences (particularly zoology) during the late Middle Ages and the Renaissance. But on the other, he would increasingly be considered the main hindrance to scientific progress, since many people were

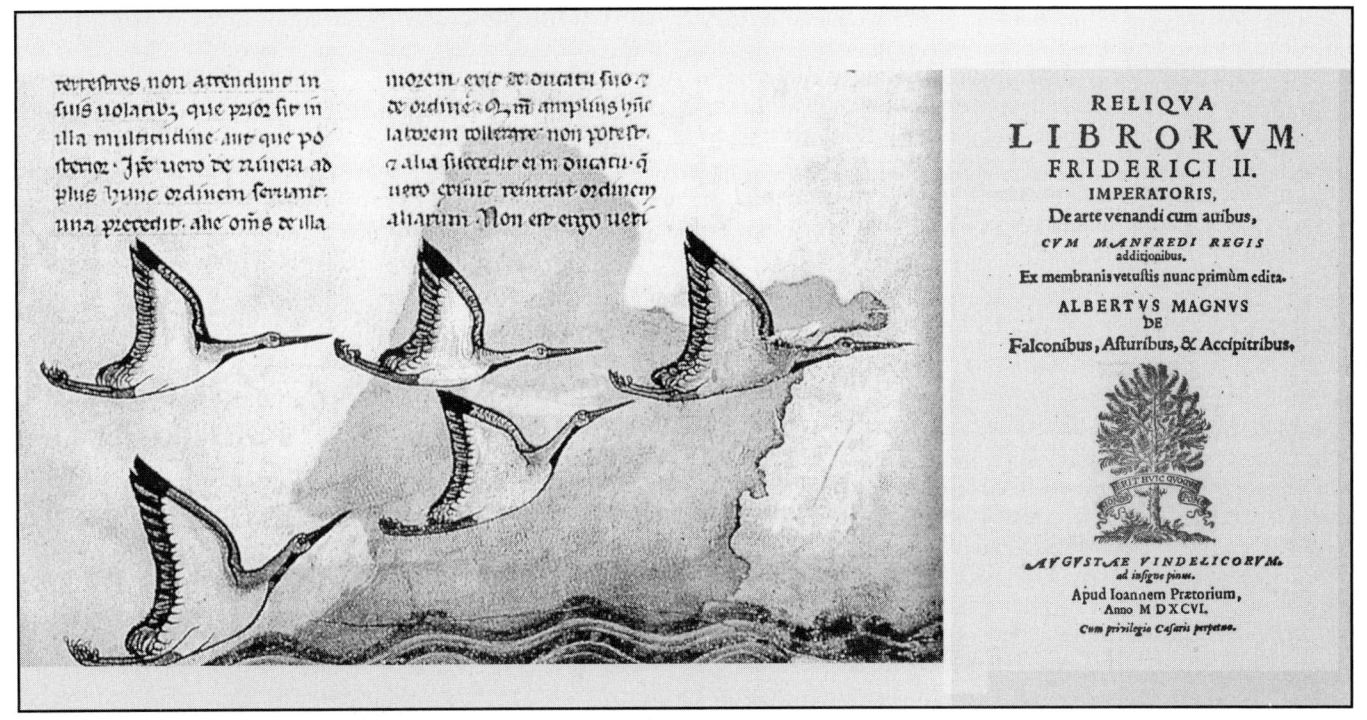

Figure 6.1. A page of the original manuscript of *De arte venandi cum avibus* by Emperor Frederick II (now at the Bibliotheca Vaticana in Rome), with the title page of a printed edition of the treatise published in 1596.

reluctant to abandon his conclusions and renounce to his great authority—even when seeing things with their own eyes and realizing the fallacies and inadequacies of some of his statements.

Curiously, one of the first medieval texts containing significant criticism of Aristotle, and one signifying the importance of direct observation, was written not by a Scholastic philosopher but by an amateur natural philosopher. This man was Frederick II, the Holy Roman Emperor and King of Sicily. Being interested in falconry, Frederick wrote a treatise called *De arte venandi cum avibus* ("The Art of Hunting with Birds"), partially inspired by Arabic sources. His treatise has a great number of beautiful images, and it contains important observations on bird zoology and ethology (Fig. 6.1). Breaking with the past, it even includes controlled experiments of their behaviors. Frederick II frequently remarks that Aristotle made mistakes when it came to the birds, saying he was "a man of books." By this he meant that Aristotle relied too much on indirect evidence, hearsay, and earlier literary sources than his own careful observations.

We can also find criticisms of Aristotle in Albertus Magnus. Despite his love of Aristotle and the fact that he was inspired by him, Albertus (and Thomas Aquinas after him) contended that, should there be a conflict between direct experience and Aristotle's word, it is experience that should be trusted. Both men also held that there could not be any real conflict between reason and faith in the matters of nature, since both come from God. Although they advised their followers to follow the Bible and religious authorities in matters of faith and morality, when it came to medicine, physics, and the other sciences, they believed that observant physicians and great scholars may, in fact, be more accurate than their beloved Church Fathers.

THE ZOOLOGICAL TEXTS OF THE LATE MIDDLE AGES

What we have stated in the previous section about the intellectual revival that characterized the late Middle Ages should not foster the expectation that something new was about to appear in the domain of electric fishes or even more generally in zoology. There were only a few forward-looking zoological texts in this era, such as the one written by Frederick II of Sicily. Also, there were very few medical treatises or even herbals with new knowledge concerning the human body and the curative properties of plants. Medieval zoology was, in brief, still largely dominated by the legendary and the wondrous, with strong moral and symbolic overtones. Hence, most new texts remained basically encyclopedic works in which correct information and sometimes even new personal achievements were presented side by side with fabulous narratives about the extraordinary powers of legendary animals, miraculous plants, and magical stones.

This was, for instance, the case with the widely circulated texts of Rabanus Maurus, Gervase of Tilbury, Thomas of Cantinpré, Vincent of Beauvais, Bartholomeus Anglicus ("Bartholomew the Englishman"), and others—some specifically mentioning the torpedo in their compilations[3] (Figs. 6.2 and 6.3). Indeed, wonders and the wondrous in their many

---

[3] Vincent de Beauvais uses the word *narco*, likely derived from the Greek *nárkē*, in his *Speculum naturale*. He quotes Aristotle and Pliny as his sources on the torpedo, but probably also has Claudian in mind when he writes that the "numbness and insensibility" is transmitted through the fishing line and rod up to the hand of the fisherman, who is obliged to abandon the hook in order not to have his entire body numbed (*Speculum naturale*, Book XVII, Chapter 75; see Beauvais, 1624, column 1289). Cantinpré, who instead uses the common Latin term *torpedo*, derives his information from Pliny and Isidore (*De natura rerum*, Book VII, Chapter 82; see Cantinpré, 1973, p. 273).

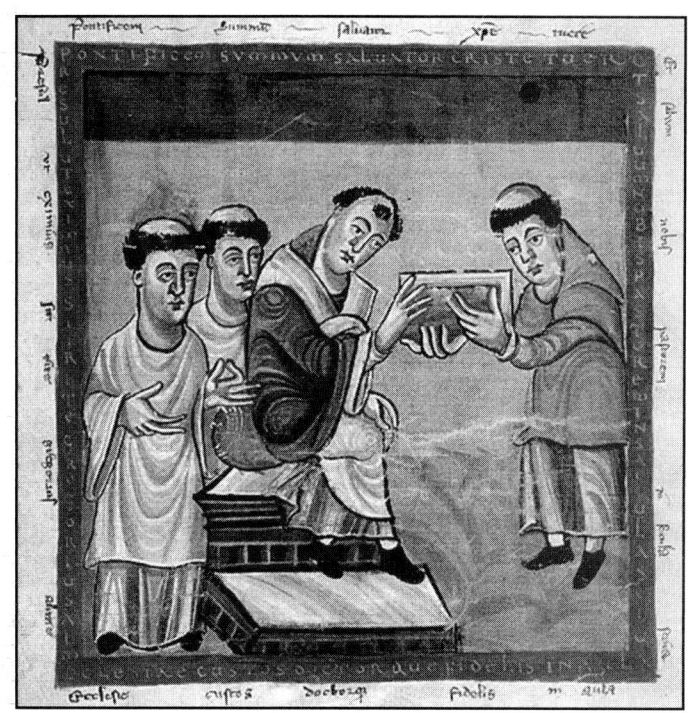

Figure 6.2. Benedictine monk Rabanus Maurus (c. 780–856), Abbot of Fulda, presenting one of his works to the Pope. Rabanus (or Hrabanus) was a prolific author of erudite compilations. Among his more popular works is his *De universo libri xxii, sive etymologiarum opus*, written in the style of Isidore's *Etymologiae* (from a codex at the Austrian National Library).

acceptances remained among the most appealing themes of late medieval culture.[4] Nevertheless, even these compilations were an advance over the earlier treatises, chronicles, and encyclopedias that had covered the marvels of nature. This is because their authors tried to obtain at least some of their information through personal experiences or from trustworthy persons—meaning qualified witnesses to at least some of the "facts" they now were choosing to narrate.

Accounts of "true" voyages to the Far East must be counted among the medieval texts in which verifiable trustworthy information can be found alongside the fabulous. William of Rubruck, Odoric of Pordenone, Giovanni da Pian de Carpine, and especially Marco Polo are among the notable early voyagers. Being based on personal experiences, their texts differ from false portrayals of trips that never took place, such as the celebrated voyage of John Mandeville. They also differ, of course, from the legendary voyages of the heroes of many medieval myths.

Fabulous portrayals of animals and the exotic natural world clearly flourished in the rich and colorful literature of the Middle Ages, and it can be safely said the extant encyclopedias and accounts of real trips provided some of the material for the famous medieval cycles and legends with their fantasy voyages to strange lands.[5] Indeed, in the medieval

---

[4] Works on agriculture and husbandry are among the few medieval texts in which animals are mentioned without much indulging in the fabulous and wondrous. Such works were written by Walter of Henley (an Englishman) in old French and by Pietro Crescenzi (an Italian) in Latin. For more on such works, see Crombie (1952).

[5] Historically and from a literary point of view, some these voyages were connected to a tradition of legendary voyages initiated late in the Classical Era. We here point to a romance of Alexander the Great originally written in Greek in the

Figure 6.3. A page with the torpedo mentioned as *narcos*, from a thirteenth-century codex of Vincent de Beauvais' *Speculum naturale*, originally at the Cistercian monastery of Cambron in Belgium (© The Bakken).

imagination, with its ideas of normality and perfection centered even geographically and intellectually on Rome, distantly situated exotic places, being remote and not Christian, were the lands of monsters, including strange and inferior types of beings, and other wondrous and often dangerous life forms.

We shall not deal here with these aspects of Middle Ages imagery, because they will be considered in some detail in Chapter 7. In that chapter, we shall encounter them in the context of the "discovery" of the African electric catfish by Europeans hoping to find the lost land of Prester John. Instead, we shall now limit ourselves to a few late medieval texts of a more "scientific" character that mention the saltwater torpedo rays, which brings us back to the great Scholastic, Albertus Magnus.

ALBERTUS AND THE STUPEFYING TORPEDO

The way in which the torpedo appears in Albertus' writings is interesting for our story because it typifies the attitude of

---

third century A.D. by an unidentified author (indicated as Pseudo-Callisthenes), and in ancient epochs translated and elaborated in a variety of other languages, including Latin, Armenian, Georgian, Ethiopian, and Syriac.

that phase of Western culture in which a new interest in natural things had just emerged, largely prompted by the rediscoveries of Aristotle and Pliny. At the same time, what Albertus writes about this shocking fish shows the limits of medieval science when compared to the later Renaissance. Between Albertus' time and the sixteenth century, when specialized ichthyologic treatises would appear (see below), many things would happen, leading to significant changes in thinking. Although still tied in some respects to medieval culture, and especially to the Aristotelian hierarchy (from inanimate objects up to plants, animals, and humans), the Renaissance, with its profound transformations, would be destined to provide even more fertile soil for critically dealing with the realities of the biological and physical world.

Albertus Magnus, who was born in Germany in about 1193, was originally called Albertus of Cologne (or Lauingen or of Bollstädt) (for more on his life and works, see Paszewski, 1968; Stannard, 1978; Weisheipl, 1980; Egerton, 1983; Kitchel & Resnick, 1999). He studied at the University of Padua, joined the Dominicans, returned to Germany to teach theology, and then went to the University of Paris, where he overlapped Roger Bacon (who dubbed him "Magnus" because of his immense erudition). Afterward, he made his way to Cologne, where he died in 1280 (Fig. 6.4). Along with his best-known student, fellow Dominican Thomas Aquinas, Albertus, as mentioned above, tried to incorporate Aristotelian natural philosophy and science within the framework of his own Christian beliefs.

During Albertus' lifetime, most of the works of the Aristotelian corpus, including those on philosophy and the natural world, had been (or were being) translated from Greek or Arabic into Latin. When his Dominican brothers asked him to explain Aristotle to them in Latin, he set forth to paraphrase just about everything written by or attributed to Aristotle. More than this, he added commentaries based on the knowledge of other scholars following Aristotle, his own experiences, and contemporary sources he deemed reliable. In addition to Aristotle, some of his sources were Pliny, Galen, and Avicenna, and he was also indebted to one of his own students, Thomas of Cantimpré, who had just completed his own encyclopedia of nature.

Albertus was the most prolific science writer of the thirteenth century. In fact, he is considered the first Western scholar to add significant amounts of new material to what had been handed down on nature from the past. Some of his treatises, and particularly an abridged version of his text on natural secrets, his so-called "Small Albertus," circulated widely, the latter being translated into multiple languages, especially after the invention of the printing press.

Like Aristotle, Albertus wrote about many different animals within the broader context of universal principles. As for nature's oddities, what he stated about them often resembled the descriptions found in the bestiaries and encyclopedias of his day, which presented individual animals, real and imaginary, with little in the way of thoughtful new philosophical analyses or reasoning. After all, the more unusual the species, the less likely its strange powers could be explained with the general laws in which he deeply believed.

In his *De animalibus* ("On Animals"), which would be mass printed in 1478, soon after the invention of the printing press, and also in his *Quaestiones super de animalibus*, Albertus covers such things as the distinctions between the sexes, the generation of perfect animals, and formal causes of an animal's body from the four elements of old (see Geyer, 1955). Following Aristotle, he also deals with degrees of perfection, with humans being "the most perfect animal," followed in descending order by quadrupeds, birds, aquatic animals, serpents (reptiles and amphibians), and vermin (mostly insects). In each of these categories, he gives brief descriptions of a large number of species—most real but some mythological—again much like the bestiaries of his day. He also contends with various other things, including whether parents could possibly determine the gender of a child to come, and equally interesting from a modern perspective, whether life expectancy is longer for men or women.

Albertus deals with the torpedo in various parts of his *De animalibus*. In a passage in the third treatise of Book VII, the torpedo is presented as one of the examples of the *solertia* (acumen, skillfulness) of some sea creatures.[6] Following another tradition dating from Aristotle's time, the torpedo is listed together with the fishing-frog—another slow-moving fish that has managed to survive because of its craftiness

Figure 6.4. Albertus Magnus teaching his pupils, as depicted in a Renaissance-era woodcut.

---

[6] This treatise is clearly inspired from the essay contained in Plutarch's *Moralia* and is generally referred to as *De sollertia animalium*, even if the derivation may be indirect (see Chapter 3 for a reference to the torpedo in Plutarch's text).

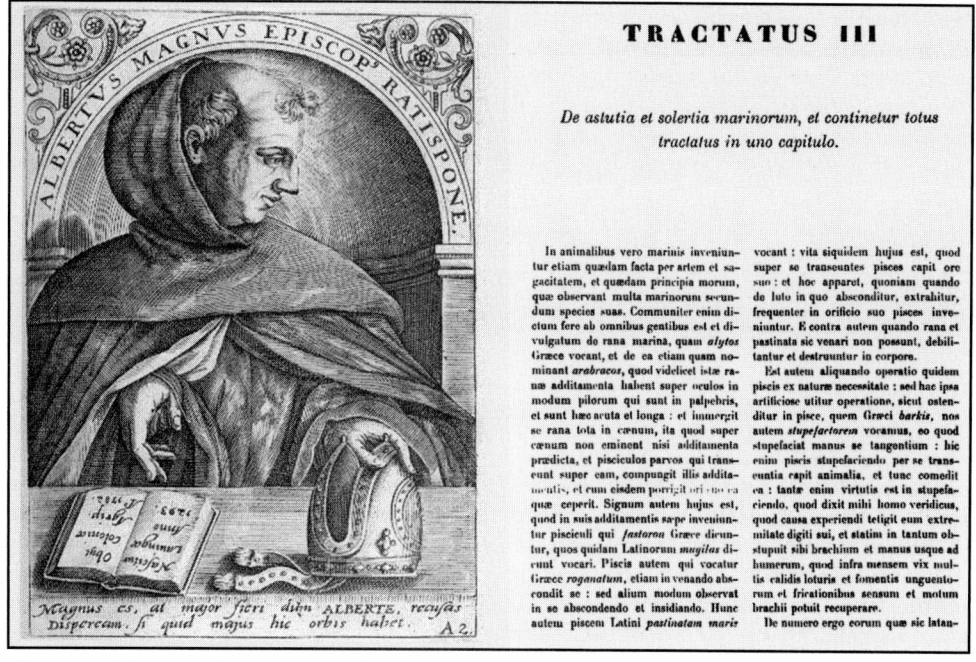

Figure 6.5. Albertus Magnus with a page from an 1891 edition of his treatise on animals dealing with the torpedo, where the fish is indicated with the distorted Greek name *barkis* and with the Latin name *stupefactor*.

(see Chapter 3). This is how this passage reads in translation from Latin:

> There is, moreover, sometimes a given operation of the fish, which derives from the necessity of nature. But this operation is also carried out in an artful way, as it appears in the case of the fish that Greeks call *barkis* and we call torporific [*stupefactorem*],[7] because it induces torpor in the hands of those who touch it. It is by inducing torpor that this fish captures animals that pass near it and instantly eats them. So great is its virtue in inducing torpor that a trustworthy man who touched the fish with the extremity of his finger for the sake of experience had his arm so numbed that he had to undergo treatments for more than a month with hot baths, applications of warm ointments, and with friction [rubbing], in order to recover his sensibility and movement. (Translated from Borgnet, 1891, vol. 11, p. 455) (Fig. 6.5)

Among the notable points of this passage is the strange way Albertus writes the Greek name *nárkē*. This is, however, typical of the linguistic and scientific inaccuracies characterizing most texts on nature from this time, and particularly the compilation works hoping to include all knowledge. In the previous book from the same work, in which he deals with the anatomy of the fish reproductive system, he had used the even more distorted form of *barachi* when transliterating the Greek term (*De animalibus*, VII, 2, 1; see Albertus Magnus, 1891, p. 343).[8]

A second point concerns the extraordinary effect of the torpedo he likely obtained from a brother of his own religious order, a man he considered a reliable source (see second quotation below). The shocks of the torpedoes commonly caught off the coast of Italy, France, and other parts of southern Europe are rather small and usually cause localized tingling effects that do not spread much beyond the hand and last only minutes. Although it is possible that Albertus' unfortunate acquaintance had an encounter with a much larger and more powerful species of torpedo (*Torpedo nobiliana*), a sensory and motor loss requiring treatments for a month would still have to be regarded as most unusual. Indeed, he will tell us in another place that this man took 6 months to heal (see below). Thus, we have to wonder whether his source might have been exaggerating to make his point, whether he might have been suffered a mental disorder (e.g., hysteria or what would now be called a conversion disorder) stemming from the event, or whether he might have endured a stroke or heart attack upon being shocked. As we shall see, the fanciful idea that the shock of the torpedo can have such astonishing consequences would appear again and again in the electric fish story.

A second passage in which Albertus mentions the torpedo can be found under "Aquatic Animals," one of the categories in his *De animalibus*. Following a long-established tradition, he includes more than fishes here. Notably, his list contains invertebrates (e.g., starfishes) and other marine creatures, ranging from turtles to incredible sea monsters (e.g., Syrenae; women with bird-like claws and wings, and scaly tails). Among the 139 aquatic animals in his Book 24, we find the

---

[7] It is possible that with *stupefactorem* Albertus is putting into Latin a vernacular name (the French word *dourmillieuse*?!), because he implicitly writes that is the term in use at his time. In a previous book, in addition to using a slightly different transliteration of the Greek term (see below), he writes that, in Latin, some people call the fish *torpedo*.

[8] In this passage, following the Aristotelian tradition, Albertus mentions the torpedo that was found with 80 fetuses inside its body. As to the curious transliteration of *nárkē* into *barkis* or *barachi*, it is to be noted that the transformation of the Greek consonant "n" into "b" was rather frequent in medieval manuscripts, and the "eta" (appearing at the end of *nárkē*) was pronounced in Greek as the Latin "i" starting during the first part of the Middle Ages. It is more likely, however, that Albertus' curious spelling of *nárkē* might reflect in part the Arabic word *barq* sometimes used to denote the torpedo, as found for instance in the *Kāmil as-Sinā'a at-Tibbiyya* of Ali ibn 'Abbās (see Chapter 5).

following brief entry (No. 127) on the torpedo, repeating some of the information in the entry just mentioned:

> TORPEDO: The torpedo is the fish that we have called the *stupefactor* in preceding books. Hidden in the mud, this fish seizes and devours fishes that approach it. It numbs [*stupefacit*] anyone touching it, no matter how fast he might withdraw his hand. It does so with such power that one of our brethren, just poking at it with the tip of his finger, touched it and just barely regained sensation in his arm after six months by means of warm baths and unguents. Pliny and Isidore say that it numbs if just touched with a spear, and that breezes blowing from this same fish numb those standing nearby. It has, however, a very tender liver. (See Albertus Magnus, 1999, Vol. 2, Book 24, p. 1704; trans. revised)

Albertus might have borrowed the information on the transmission through the spear and the allusion to the breeze blowing from the fish directly from Pliny or indirectly from Isidore, who borrowed from Pliny. As pointed out in Chapter 3, Pliny had written that "from a considerable distance, even and if only touched with a sphere or staff, this fish has the power of . . . riveting the feet of a runner, however swift he may be in the race." It is again to be noted, however, that the consequences of the shock on the hapless victim go well beyond anything that even Pliny (with his vivid imagination and low standards of inclusion) had written, since more than half a year was needed for the poor man to recover from its effects.

Although Albertus was a man of his time, alluding to fantastic stories about the animals, plants, and stones that were so much a part of medieval imagery, he did, in fact, try to distance himself from most fantastic tales, much as he shunned supernatural explanations for natural phenomena. In his section on Syrenae, for example, he begins by writing: "Poets' tales tell us that sirens are sea monsters." In other words, he is aware that the descriptions of these creatures did not come from known reliable sources.

Summing up, Albertus neither brings up the basis of the torpedo's powers nor alludes to its obscure (occult) properties anywhere in his writings. Nowhere does he enter into a philosophical discussion of cause and effect. Instead, he just mentions bits of what Aristotle and then Pliny had written about the torpedo, and then indirectly supports Pliny's statement about its powers with testimony from one of his contemporaries, whom he obviously viewed as reliable, and whose remarkable story was based on personal experience.

Being men of the cloth, Albertus Magnus and his followers Thomas Aquinas and Roger Bacon taught in ways that conveyed the certainty of their beliefs (Daston & Park, 1998). They played down the oddities of nature and focused on Aristotelian universal principles, not particulars that inspired wonder because they defied understanding or categorization. Their teachings, as can be imagined, had broad social and cultural ramifications, most notably among their devout brothers, who preached and taught in seminaries around Europe.

Nevertheless, the torpedo, along with other wondrous animals with hidden or unknowable properties, continued to hold a place in medieval culture. Medical formulas (*materia medica*) and treatments (*practica*) continued to call for strange creatures with occult properties, even though these animals were no longer a part of church-sanctioned natural philosophy. And without question, awe for these fable-like creatures continued to run deep among the common folk. Men, women, and children still responded emotionally to their own sensory experiences and to unsettling stories they heard from others, and they passed this information on to friends and family, even if their religious leaders and some of the greatest thinkers of the day were now viewing wonder as a sign of vulgarity or ignorance.

THE CULTURE OF THE RENAISSANCE

As mentioned at the beginning of this chapter, a new understanding of electric fishes would have to wait until the second half of the seventeenth century. This notwithstanding, and despite the fact that the interest in animals during the next centuries would revolve to a large extent around the study of classical sources, important changes did occur with the Renaissance (Durant, 1953; Zophy, 1997).

Interestingly for our story, one of the fields in which this was to happen in a particularly remarkable way is in the study of fishes, leading to the birth of ichthyology as a separate area of science in the sixteenth century. After the epoch of Albertus, Aquinas, Adelard, Bacon, and other great scholars of the Middle Ages revival, the Western World would witness many further transformations. Although sometimes loosely ascribed to the rebirth of learning or Renaissance in Italy (or Renaissance *tout-court*), as if there were no antecedents, what occurred really started with the changes that had been initiated by the Scholastics. Now, however, the thirst for new learning would intensify, it would cross a greater number of cultural and geographical boundaries, and it would have a more secular or earthly orientation to it. Many historically significant factors and events, so many and with a complexity that is impossible to detail here, contributed to these changes, but some demand special recognition.

Although it is always difficult to tell when a great transformation of human history starts (often making it unsafe to divide history into clearly distinct eras), the Renaissance is generally said to have begun during the 1300s in Florence. It had much to do with the commerce, wealth, and leadership in that part of Italy, and particularly with the great development of the arts, and especially of painting.

From a cultural point of view, two somewhat contrasting factors contributed to the new orientation. One was the need for a fresher and more direct view of nature. This demand would stimulate intense research based on direct personal observations. The other was an intense and accurate study of the classical texts, a movement initiated by humanists Francesco Petrarca and Lorenzo Valla with great philological understanding. The philological accuracy of the Renaissance humanists contrasts with the linguistic shabbiness of the Middle Ages compilers and encyclopedia writers, a shabbiness that is present even in Albertus.[9] By looking to the past with fresh eyes and with a new orientation demanding

---

[9] A particularly remarkable example of Albertus' philological and zoological sloppiness is the confusion that he makes between the sea urchin and the remora, on the basis of the similarity of the Greek names of the two sea dwellers (*echinus*, actually *escynus* in Albertus' text, for the urchin, and *echeneis* for the "shipholder" or remora fish; see Copenhaver, 1991).

greater precision, it is not far-fetched to say that the leaders of the Renaissance were laying the groundwork for the future.[10]

A renewed interest in the real world, combined with other historical motives, helped to contribute to the new accuracy in the representations that emerged in pictorial art, first in Italy and then in other countries, especially in the German-speaking and the Flemish parts of Europe. A breaking advancement in this domain was the invention of "scientific" perspective by Filippo Brunelleschi, Paolo Uccello, Masaccio, Piero della Francesca, Andrea Mantegna, and others—the graphical technique by which painters could overcome the limits of the bidimensionality of a canvas by forcing a third dimension to enter through the frame of the painting.[11] Perspective was, however, not simply a powerful artistic method. It also corresponded to a philosophical and moral attitude toward the world itself. Paintings became a sort of "window" for observing nature, and the wood frame delimiting the canvas (*quadro* in Italian) served as a metaphor for a real window, through which one could scan the external world.

Continuing a tradition that had emerged in the late Middle Ages, particularly with herbals, Renaissance artists studied objects of nature and represented them with such accuracy that many botanicals and animals, even uncommon and exotic ones, still can be recognized in their paintings. In this context, the earliest voyages, made for commercial purposes and with the desire to discover new lands to be exploited for the richness they might possess, served as an important stimulus for what would take place. The new descriptions from Asia, Africa, and the Americas helped to generate an even greater intellectual interest in nature, especially exotic wonders. The natural objects that were brought back by ships, and even more the plants and animals that remained in far-off countries, demanded faithful reproductions that could be studied by scholars in distant places.

Among the art that typified the new interest in the exotic is a famous woodcut of a rhinoceros made by Albrecht Dürer in 1515 (Fig. 6.6). Although rather inaccurate from a zoological point of view (due to the fact that the German artist likely based his work on a written description of an animal that had arrived in Lisbon a year earlier), his graphical representation of the wondrous animal portrays an intense pictorial realism. As pointed out by art historian Ernst Gombrich in an essay published in 1960, realism is not simply based on the direct observation, but needs some interpretative code on the part of the artist. In other words, realism is at least as much in the eye of the beholder as in the reality itself.

Importantly, scholars, physicians, and humanists fully recognized artists for their skills during the Renaissance, employing them and elevating their social status (previously, they had been viewed more as ordinary craftsmen). Leon Battista Alberti, Leonardo da Vinci, and Michelangelo

Figure 6.6. Albrecht Dürer's celebrated 1515 woodcut depicting the rhinoceros.

Buonarroti are among the great artists of this period who stand out for being in great demand and for their diverse contributions. As expressed by Benedetto Varchi (1962–63, I, p. 39) in his *Lezzione* on the *paragone*[12] (comparison) of painting with sculpture: painting was now "of a very great utility in the sciences, as we can see in the book of Anatomy of Vesalius, in the forty-eight images of the sky of Camillo della Golpaia, and in the Herbal book of Fuchs, and much better and more naturally in those painted for the most illustrious Duke of Florence by Francesco [Ubertini] Bachiacca."

Ulisse Aldrovandi, one of the sixteenth-century naturalists we shall meet later in this chapter, even considered the possible therapeutic uses of the figurative arts in his writings. He advised husbands to have the faces of very handsome men and women painted in their bedrooms, while suggesting images of fountains and water courses for people afflicted with dry disorders or dryness (*arsura*: Aldrovandi manuscripts, cited from Olmi, 1992 pp. 26–27).[13]

The higher, recently acquired, social status that some artists now achieved prompted more explicit reflections on the methods of representation, and they helped stimulate an interest in the past along with a fresher look at the natural world. Scholars interested in science not only had to have knowledge of art, but artists had to know something about the sciences, and particularly anatomy. As put by Vincenzo Danti, the interaction would promote "the proclivity to perfection, which the study of science can add to art" (Barocchi, 1962, p. 257).

The contribution of artists to the renewal of the scientific attitude that characterized the Renaissance (and helped to lay the grounds for the later scientific revolution) can be

---

[10] For a view of Renaissance culture, the reader is referred to the works of Eugenio Garin, Frances Yates, and Paolo Rossi, among others.

[11] Although some forms of perspective can be found in the artistic production of older periods (for instance, Hellenistic and Roman art, in the mosaics of Ravenna, and in medieval art, notably in Giotto), there is little doubt that the perspective based on a single centric vanishing point and on accurate mathematical calculations was an invention of the Italian Renaissance.

[12] The *paragone* between two arts in order to establish which one holds primacy over the other was represented in Renaissance poetry, music, painting, and sculpture, among the arts most often considered. Beautiful *paragoni* asserting the superiority of painting over all other arts were written by Leonardo in his *Trattato della pittura*. A fundamental *paragone* also asserting the primacy of painting (over sculpture) was written by Galileo on behalf of his friend, Tuscan painter Ludovico Cigoli (see Panofsky, 1956).

[13] This was in the line with the old tradition we encountered when dealing with Isidore of Seville (see the previous chapter).

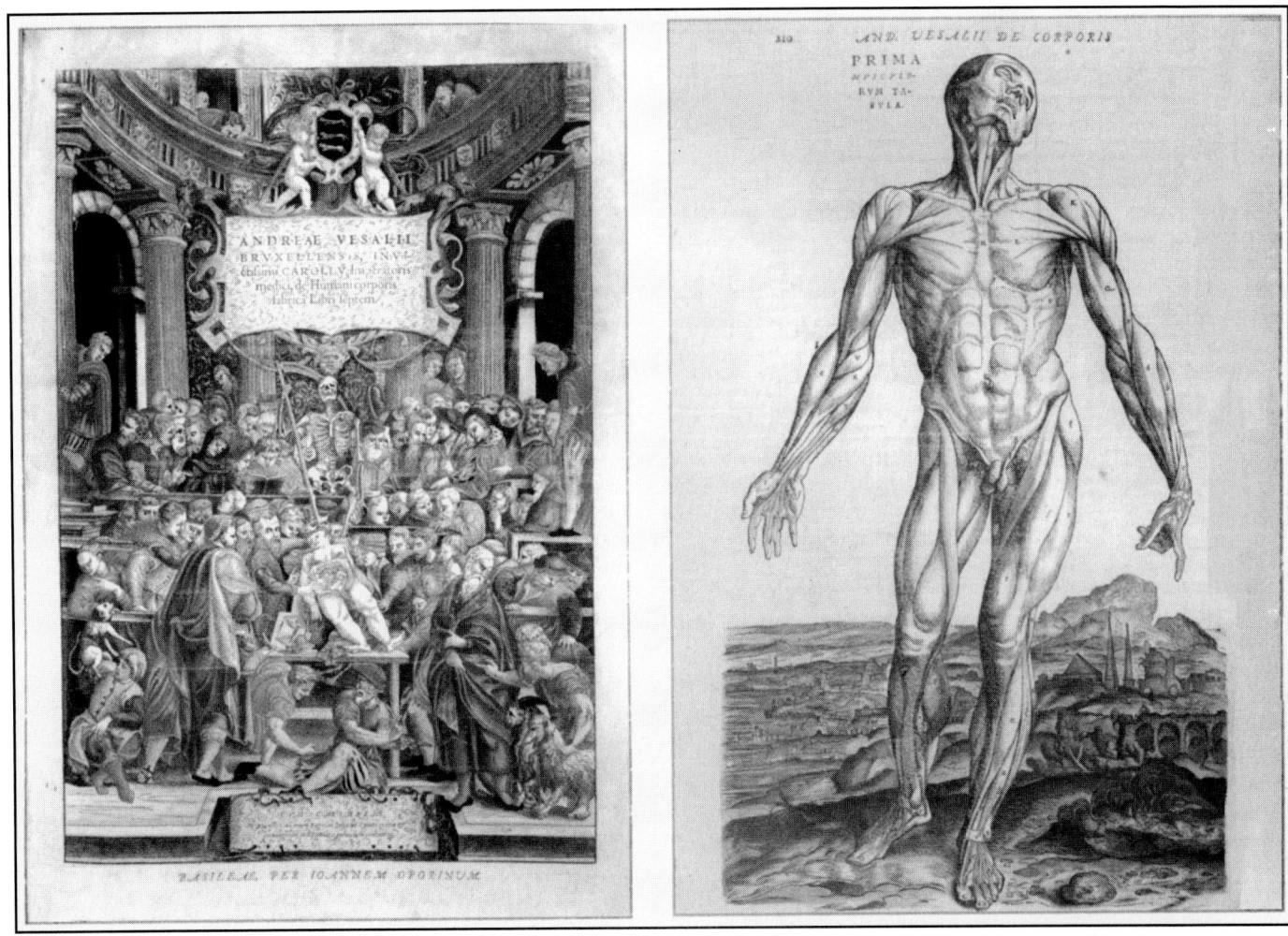

Figure 6.7. The title page and an image depicting the muscles from the original (1543) edition of Vesalius' *De fabrica corporis humani*.

perhaps best appreciated by considering the work of Andreas Vesalius, who published his famous *Fabrica corporis humanis* in 1543, the same year as Nicolaus Copernicus' astronomical treatise presenting his heliocentric conception of the universe (for more on Vesalius, see O'Malley, 1964). The figures that illustrate Vesalius' (1543) work are masterpieces of scientific illustration (Fig. 6.7). They required the work not only of an expert anatomist (as Vesalius undoubtedly was) but also of artists who knew anatomy and could solve the complex problems that a morphological representation of human body would present. There is no doubt that, in the absence of the extraordinary figures that came from Titian's (Tiziano Vecellio's) studio, Vesalius' *Fabrica corporis humanis* would not have contributed as much as it did to the development of modern anatomy, and the same can be said for some of his followers (Figs. 6.8 and 6.9).[14]

The same can be said about other fields, as for instance the study of plants. German botanist Leonhart Fuchs even inserted portraits (in doctoral robes) of the artists who collaborated with him in his *Historia stirpium*, this being his way of emphasizing how important they had been for his botanical studies (Fig. 6.10).

Besides the natural curiosity for anything new and surprising, there were other reasons why the discovery of new lands led to a change in the relationships between people and the natural world. In the providential vision of the world that had dominated the Middle Ages, humanity occupied a central position, and not only cosmologically. Hence, it seemed odd that inanimate objects, plants, and animals could have existed for men (meaning Christians of the Western world, the only ones reputed to posses the complete amount of humanity) without these men ever having an acquaintance with them. The new discoveries of the Renaissance would seed a change away from this particular attitude, one that would be fully achieved only with Copernicus, Galileo, and the scientific revolution they and others helped trigger.

Not to be overlooked in this evolving landscape, the means of communication were now changing dramatically. Previously, scribes hand-copied manuscripts, sometimes working for months or even years on a single book, which might then be read by very few individuals. But in 1454, less than two decades after the process of papermaking was perfected, Johannes Gutenberg began printing Bibles on his

---

[14] After Vesalius, anatomical images of very high scientific and artistic quality were produced for Hieronymus Fabricius' (ab Acquapendente) and Giulio Casseri's anatomical treatises. The painted versions of Fabrici's *tabulae pictae* are of unsurpassed beauty and clarity but gathered dust over the centuries in the Marciana Library of Venice. They were rediscovered at the beginning of the twentieth century and only recently have been exhibited to the public, after undergoing restoration (see Rippa Bonati & Pardo-Tomás, 2004).

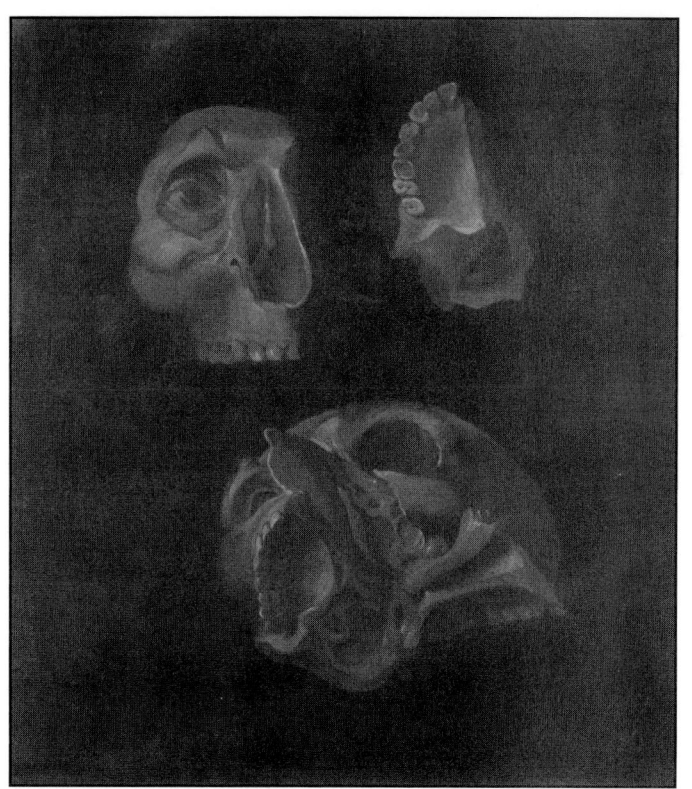

Figure 6.8. An anatomical image from Hieronimus Fabricius ab Aquapendente, Professor of Anatomy in Padua and one of the foremost anatomists of the Renaissance. This image is part of the collection of Fabricius' *Tabulare pictae* ("painted plates") preserved at the Marciana Library in Venice. It was based on the collaboration of Fabricius with important painters of the Venetian school. (© Ministero dei Beni Culturali, Biblioteca Nazionale Marciana, Venezia)

transformed winepress in Mainz, Germany. In less than a century, there were at least 1,000 printers in more than 200 locations using moveable type.

Although Bibles were by far the most frequently printed books, treatises on science and medicine also found rapidly growing audiences. In 1469, the first printed edition of Pliny's *Historia naturalis* came off a press in Venice, and there were at least 39 editions prior to 1500 (Gudger, 1924). Just a few years later, Celsus' encyclopedia came forth, followed by Latin translations of books by Aristotle, Theophrastus, and Dioscorides, all of which included passages about the wondrous torpedo (Sarton, 1955). Pliny's writings found an especially receptive audience at this time, and his marvelous tales and descriptions were especially influential in laying out the course that natural history would (and would not) follow.

Some of the first publishers (for example, Aldo Manuzio in Venice) were themselves humanistic scholars. Further, the wider circulation of their books, as well as the need to provide relatively accurate editions, stimulated intense philological treatises on the ancient texts. The printing industry and artists' studios now came together in new ways, with printers and naturalists needing suitable illustrations for their books, and with artists developing techniques for the mass reproduction of images.

Going hand in hand with the printing industry, glasses or spectacles now became available (Rosen, 1956; Rosenthal, 1996). They made the printed word clearer to people with visual problems, notably people who were eager to read but found the printed word blurry, and particularly older persons affected by presbyopia. Inexpensive eyeglasses that bridged the nose helped increase the demand for books, including natural histories, and fed into the thirst for renewed learning.

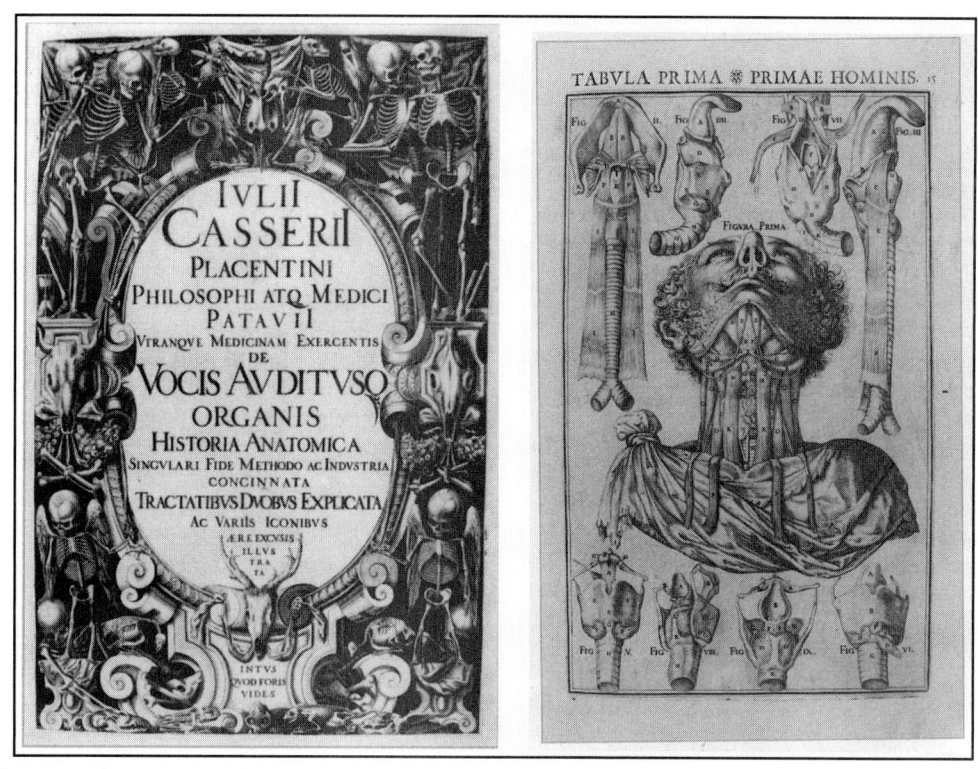

Figure 6.9. The title page and an anatomical plate of the organs of speech from a treatise on hearing and voice production published in 1600 by Fabricius' pupil and successor, Giulio Cesare Casseri. Casseri's images involved collaborating with important artists of the epoch, including possibly Jacopo Ligozzi, one of the greatest painters of anatomical and naturalistic images of the epoch.

Figure 6.10. Two images from the *De historia stirpium* by Leonhart Fuchs, first published in Basel in 1542. On the left, two artists drawing a plant from nature, and on the right, a realistic depiction of the marijuana plant, *Cannabis sativa*. Comparing this image with the plant figures in the Megenberg's herbal (see Fig. 6.2) shows the progress of botanical iconography between the fifteenth century and the seventeenth century.

It is in this cultural framework that a new interest in the science of zoology emerged, one in which the study of animals was no longer just an offshoot to medicine or tied to morality. Moreover, wondrous animals were no longer marginalized, and wonder was no longer looked upon as a sign of ignorance. In these regards, there was a partial return to the Aristotelian tradition of carefully studying both odd and common animals during the late Renaissance, and this included fishes.

A LITERARY DIGRESSION

Before we turn to the fish books that came forth in the sixteenth century, and especially in the second half of that century, a very important point must be made. It is that the torpedo was finding its way into more than science and medicine books during the Renaissance, reflecting just how much this fish had become interwoven into various parts of the culture. To appreciate this development, we can do no better than turn to the writings of a man whose genius was much more closely associated with moral philosophy and the art of writing. This individual is Michel Eyquem de Montaigne, the deep-thinking "father of the essay," which he personally viewed as test of his intellectual faculties and a vehicle for providing his philosophical thoughts (Frame, 1958).

Montaigne's thinking about torpedoes, or really man's inability to understand everything in nature, can be found in his *Apology for Raymond Sebond*, which was written about 1576 and published along with other essays in 1580 and then reprinted (Montaigne, 1652. Raymond Sebond, who had died almost 150 years earlier, had been a Spanish scholar and Regius Professor of Theology at Toulouse. He felt one could come closer to understanding God by studying His creatures and the physical world, an idea promoted by the Scholastics. In his essay, which really deals very little with Sebond, Montaigne holds that man, at least man without divine grace, is no wiser, better, or happier than the animals in the woods or those we would find at a farm. His essay calls for candid self-examination, and it displays his own great skepticism (*Qué sçay-je?*—What do I know?) as it deals with vanity and human presumptions.

Montaigne discusses several things that seem beyond man's capacity to imitate or even fully appreciate, and here he shows his familiarity with Pliny and the ancients, as well with some of the issues of his own day. Specifically, he mentions "the little fish that the Latins call *remora*," which could stop a ship, how a chameleon can quickly change its color, and the remarkable flight of birds. All of these creatures have certain abilities that are "different and keener than ours," he writes. And the same is true of torpedoes.

Montaigne is clearly perplexed and very much humbled by the torpedo's amazing powers, reflecting how little he and his contemporaries really knew about the discharges of these fishes. But being familiar with what had been written centuries earlier, he repeats the ancient phrase, "this chill of hers," although he does not go on, as did Galen and the Galenists, to tie the cold to some sort of poison. Specifically, he tells us that a torpedo:

> has the property, not only of putting to sleep the limbs that directly touch her, but of transmitting through nets and seines a sort of numbed heaviness to the hands of those who move and handle them. Indeed they say further that if you pour water upon her, you feel this sensation, which climbs uphill to the hand and numbs the touch through the water. The power is marvellous, but it is not useless to the torpedo; she senses it and uses it, in this way, that to catch the prey she seeks, you see her hide under the mud, so that the other fish, gliding past above, struck and benumbed by this chill of hers, fall into her power. (Montaigne, 1580; trans. in Frame, 1958, pp. 344–345)

As we shall now see as we turn to the ichthyology books that were coming off the press, not only was the torpedo generating intense philosophical (theoretical) interest at this time, but some natural philosophers were searching for, or at least open to, better explanations for its "chilling" powers. But, at the same time, as reflected in Montaigne's

Figure 6.11. Paolo Giovio with a page from his *De romanis piscibus* (1524) dealing with the torpedo.

essay, the real basis or cause of the torpedo's effects would remain far from understood. There would be many allusions to its hidden or occult powers in these books, a topic we shall return to in much greater detail in Chapter 9.

PAOLO GIOVIO'S BOOK ON FISHES

Interestingly, five men, all born within a few years of each other (between 1507 and 1522), wrote illustrated books dedicated exclusively or largely to fishes, their achievements leading to the creation of modern ichthyology. These pioneers are Pierre Belon, Guillaume Rondelet, Ippolito Salviani, Conrad Gessner, and Ulisse Aldrovandi. Representing France, Italy, and Switzerland, their works are now widely regarded as landmarks in the history of ichthyology (Gudger, 1934; Cuvier, 1995; Egerton, 2003).

Before them, however, Paolo Giovio (Giovius), an Italian polymath (famous mainly for his historical works and collection of portraits), had published a small book on the fishes of Rome, which although often overlooked also deserves comment. Dated 1524, this book seems to be the first printed work since ancient times devoted entirely to fishes. But Giovio's *De romanis piscibus* does not contain illustrations. Further, it had a limited and non-systematic arrangement, being to a large extent a book on fishes as food and a source of pleasant stories and anecdotes (in an Epicurean and even facetious style).

Still, Giovio's book is interesting for our electric fish story because it contains an entry on the torpedo. In Chapter 28, for the first time in a printed book to the best of our knowledge, he gives some vernacular names of his fishes, thus inaugurating modern ichthyonomy (fish naming) (Fig. 6.11). This is how Giovio's *De torpedine* reads in translation:

> Torpedo is wonderful for its shape [*effigie*] and power, because, this is what people say, when caught in the nets, it numbs fishermen's hands before being touched. For this reason, Venetians call it Cramp [*Sgramfum*], because it numbs the affected arm. Romans, moreover, call it alternatively Battipotam or Fotterisiam, but indeed frequently they say Oculatellam [from *occhio* meaning eye] because on its back there are five blackish [*subnigros*] ocelli. It is reddish [*subrufa*] in its ventral part [*prona parte*], white [*candida*] in its dorsal [supine] part. Averroës and other philosophers supposed that torpedo affects the hands through the same property [*qualitate*] by which iron is attracted by loadstone. On the other hand, Galen, in the third Book of *De alimentis* mentioned the torpedo among the fish on which we feed, although nowadays only very poor people eat it. (Giovio, 1524, pp. 100–101)

Despite of the paucity of his scientific information, Giovio displays an attitude that would be particularly well developed by later sixteenth-century ichthyologists. Specifically, he combines what he has found in classical sources, notably in Pliny and his followers, with new research from the field. In his case, he interrogated fishermen and visited fish markets, mainly to gain more knowledge about fish names.[15]

In Chapter 3 we discussed the meaning and possible origins of both the Venetian term *sgranfo* and the Roman words *battipotta* and *fotterigia*. Giovio provides these names in a slightly different (somewhat Latinized) form. Also Latinized is the term *oculatellam*, used in place of the popular Roman word *occhiatella*, which is still used, albeit rarely, in the region of Rome to indicate the *Torpedo ocellata*. A small point concerns the low estimation of the torpedo as food in

---

[15] The way that Giovio describes the effects of the torpedo's power on fishermen makes it rather clear that he likely did not personally experience the fish's shocks. Giovio's frequent visits to fishermen and fishmongers of Rome are fairly well documented, as is his collaboration with Rondelet and Salviani in Rome, whose work will soon be discussed (see Olmi & Tomasi, 1993, p. 47).

Rome at the time of Renaissance. According to Giovio, only very poor people now consumed this fish. This was a considerable change from what we had encountered during Athenaeus' epoch, when it was served to gourmands!

Giovio makes a notable error when he indicates that the ventral part of the fish is reddish, whereas the dorsal part is white. In reality, just the opposite is true: the back is darker and more reddish than the belly in the two species of torpedoes commonly caught off the coast near Rome (*ocellata* and *marmorata*). His error might have stemmed from a trivial cause (e.g., a misprint, a typographical error) or it might be that he had not personally seen torpedoes and confused what he heard, although this seems less likely. Giovio's error is more comprehensible when we realize that many Renaissance scholars interested in fishes were first and foremost classical scholars, more at ease in book-filled libraries than when doing experimental and observational work outdoors or even in research cabinets.

ILLUSTRATED FISH BOOKS

Giovio's book on fishes was in reality only a kind of literary amusement in the Renaissance style, aimed mainly at obtaining the favors of an important cardinal, the dedicatee of the work. Things were different for Pierre Belon, Guillaume Rondelet, Ippolito Salviani, Conrad Gessner, and Ulisse Aldrovandi, the later ichthyologists who were very serious about science and devoted great energy to their illustrated treatises on fishes.

Unlike Paolo Giovio, these ichthyologists not only exchanged considerable information with one another, but also interacted extensively with other naturalists both near and far away. Their interactions represented a further change in how science was now starting to be pursued. Such cooperation would, in fact, lead to the formation of more scientific societies.

> A feature of natural studies, which seems to become more prominent in the sixteenth than in preceding centuries, is cooperation between different individuals. They send one another specimens or at least drawings or written descriptions of strange animals and unfamiliar herbs, which they have run across. Gessner maintained a correspondence with Turner, whom he had met at Zurich as Turner returned from Italy with the doctoral degree, and many other persons on such matters. . . . This cooperation in science by men of diverse nationalities, professions, religions and even philosophies is indeed impressive. (Thorndike, 1941, Vol. 7, p. 267)

There are multiple reasons why these five late Renaissance scholars decided to devote themselves to studying and writing books on fishes at this unique moment in time. A relatively important one is linguistic or philological, and it has to do with the immense number and variety of fishes that were then being recognized. The large and growing number of fishes caused great difficulties for scholars attempting to produce philologically correct texts in accord with the zoological works of the Aristotelians and even Pliny.

That the number of known fish species was already astonishingly huge is reflected in a poem written by Luigi Pulci a century earlier. Called *Morgante Maggiore*, this poem contains these lines:

> Fishes are more than stars . . .
> and they are of so many manners,
> that they cannot be said with one hundred languages

There is long list of fishes in the *Morgante*, although Pulci, with poetic license, lumps sea and river species together. The torpedo is not, however, to be found on his list. It is also generally absent from the lists of fishes present in the cookbooks of the time, most likely because of the low esteem that the Renaissance gourmets had for its flesh, as we have just noted (Folena, 1963–64).

Faced with the need to translate Greek fish names into Latin and to identify the fishes mentioned by Pliny and other classical authors, sixteenth-century ichthyologists (who were humanistic scholars, zoologists, or both), felt the need to investigate the matter first hand, to the extent possible.[16] But in dealing with their fishes, they encountered problems similar to those experienced by Theodorus Gaza a century earlier, when he set forth to produce the first Latin version (from the Greek) of the Aristotelian treatise *Historia animalium*, which would be a reference for ichthyologists.

These men traveled around, mainly through Italy and Greece, asking fishermen, fishmongers, and any informed people on the subject for information, trying not to influence them with their own knowledge. Along with this mainly ichthyonomic investigation, they also pursued anatomical or zoological observations on a variety of sea species, and especially on those mentioned by classical authors, to resolve difficult issues and old controversies. These issues might concern the extraordinary powers traditionally attributed to some species or problems of identification. For some fishes, notably the remora or *echeneis*, they had to deal with both of these thorny issues.

Another important factor leading to these books relates to the fact that although kings and princes had monopolized collecting rare natural objects prior to the twelfth century, things were different by the fifteenth century. Now wealthy men indulged in collecting as a way of displaying their own wealth, erudition, and power. This was especially true in Italy, where collections were symbols of culture and status, and where the first science museums appeared (Bedini, 1965; Findlen, 1994). To quote one historian of the period: "Through the possession of objects, one physically acquired knowledge, and through their display, one symbolically

---

[16] In addition to Belon, Rondelet, Salviani, Gessner, and Aldrovandi, there were other scholars who contributed to fish studies in the sixteenth century. Among them are the Italian humanist Ermolao Barbaro, who wrote a series of *Corollarii* to Dioscorides, which appeared in an edition of the works of the Cilician physician published in 1529 in Strasbourg. One of Barbaro's *Corollari* concerned the torpedo. It is interesting because it contains what may be the first mention in a printed text of the Greek term τουρπαίνα (transliterated as *tourpaena* by Barbaro, who correctly attributes it to Paulus Aegineta). Pierre Gilles was another important contributor, and his Latin edition of Aelian's zoological treatise was first published in 1533 (with an appendix of French and Latin popular fish names). Gilles collaborated with Gessner, and a subsequent edition of Aelian's work (also containing the Greek text) was edited with him and was published posthumously in 1611. The work of Francesco Massari (or Maser) is also notable. In 1542, he published a book commenting on and correcting the ninth book of Pliny's *Naturalis historia*, where there is again a discussion of the torpedo with a reference to the *turpena* almost literally derived from Barbaro.

acquired the honor and reputation that all men of learning cultivated" (Findlen, 1994, p. 3).

The museums that now formed contained many ordinary fishes and animals, as well as rather common mineral and vegetable specimens, mixed with manmade artifacts (for more on collectors and the later evolution of museums, see Chapter 13). But they also had prized exotica, including the supposed remains of legendary animals, such as basilisks, unicorns, and satyrs, and also human monsters, such as conjoined twins. The guiding idea was to understand and present nature on an Aristotelian continuum, while managing the abundance of new material that came in from all around Europe and increasingly from voyages to strange new lands.

These factors, and others as well, led erudite Europeans to catalogue natural phenomena more systematically than before. These catalogues and books became increasingly specialized, and over time included more species. As can be imagined, these writers were particularly enamored with strange creatures: the stranger the better. And with this love of oddities, they "shifted the marvels of nature from the periphery to the center . . . and they reclaimed for natural philosophy not only wonderful phenomena but also the emotion of wonder itself" (Daston & Park, 1998, p. 160).

Iconography, or graphic representation, was of great importance to these men. The copper engravings used by Salviani for his plates were of exceptional quality and were also published apart from the main text. His torpedo picture is among the best of the epoch. The wood or copper engravings used by Gessner were in some cases the result of collaborating with Albrecht Dürer, who had already produced extraordinarily realistic images, one of the best being the stag beetle now displayed at the Getty Museum in Malibu, California.

The importance of collaborating with skilled artists, a landmark of Renaissance science, is also reflected in the works of other ichthyologists and zoologists of the epoch. Ulisse Aldrovandi, for example, tried to hire several artists, both fellow Italians and foreigners, in order to have faithful reproductions of the fishes and land animals for his various treatises. Jacopo Ligozzi, the court painter for the Grand Duke of Florence, Francesco I de' Medici, was one such artist (Tongiorgi-Tomasi, 2000) (Fig. 6.12). He had an extraordinary talent for naturalistic representation, the best of his epoch, in Aldrovandi's words, "in drawing and painting figures."

These Renaissance zoologists fully understood that the need for faithful artistic images went beyond just portraying various fishes and animals in ways that could be accomplished with words alone, although verbal descriptions were extremely important. Because many unusual specimens were now being found in far-off lands and could not be seen throughout Europe, they were asked to "counterfeit" nature (in a positive sense), so their realistic images could serve as substitutes for the rare objects that might be found only in some collector's cabinets, if even there. In this context, we find Gonzalo Fernandez de Oviedo y Valdez wishing he could have a talented artist, such as "Berreguete . . . or Leonardo da Vinci, or Mantegna" to faithfully paint a botanical he had come across in South America—an exotic tree that words alone could not adequately describe.

The need for faithful representations made pictorial realism a high priority, even though Renaissance naturalists soon became aware of the impossibility of producing images

Figure 6.12. A detail with a butterfly (*Saturnia pyri*) from the botanical image depicting the *Euphorbia* as painted in the second half of the sixteenth century by Jacopo Ligozzi for the Grand Duke of Florence. (© Gabinetto Disegni e Stampe of the Uffizi Gallery of Florence)

capable of capturing the biological mutability and complexity of the living world. The demand also led to requests for sharing, when good illustrators were found and exceptional pictures were produced. In 1564, for instance, Aldrovandi sent Gessner a letter asking him for help in hiring "two or three most excellent painters" from his Swiss town or elsewhere, because he too needed good artists for naturalistic illustrations. As we shall see, Gessner reproduced Rondelet's figures in his own work, while Aldrovandi copied Salviani's magnificent torpedo, albeit with more modest results.

BELON'S BOOKS

The first of the five illustrated fish book authors, naturalist and physician Pierre Belon (Petrus Bellonius Cenomanus, i.e., from Le Mans), was born near Le Mans and studied medicine in Paris. Between 1546 and 1550, he went on an extensive scientific journey to Italy, Greece, Asia Minor (Turkey), Egypt, Arabia, Judea, and other locales to study and document nature. One of his purposes was to compare what he could see with his own eyes with what earlier authorities had described. His patron, Cardinal François de Tournon, supported his trip, which included a stop in Rome, where he met with Rondelet and Salviani.

Belon's first book on ichthyology appeared in 1551. Titled *L'histoire Naturelle des Estranges Poissons Marins* . . . ("The Natural History of Strange Marine Fishes with the True Picture and Description of the Dolphin and of other Creatures of its Species"), it has been hailed as the first modern book on "fishes," broadly defined. This is because Belon (1551) was mainly interested in porpoises, and 38 of this work's 55 leaves were on this mammal, which he compared to terrestrial mammals. He also included sections on the hippopotamus and the nautilus in his illustrated volume.

In 1553, Belon published a second work in Latin, his *De aquatilibus libri duo* . . . ("Two Books on Water Creatures with Illustrations Depicting as far as Possible their Living Likeness"). Although he again included more than fishes, he described many more aquatic creatures than in his first book. Further, there were 110 realistic drawings accompanying his detailed descriptions, with 20 more aquatic animals not illustrated. Belon's realistic illustrations in these volumes were applauded and marked a turning point in the evolution of ichthyology books.

Two types of torpedoes are mentioned in his second publication. One is simply called *Torpedo,* and the other *Torpedo oculate*, corresponding to Giovio's *oculatella* (i.e., *Torpedo ocellata* or nowadays *Torpedo torpedo*), a term, as previously stated, stemming from the fact that it has spots resembling eyes on its back. Belon mentions the ancients but is most interested in detailing the external features of these two torpedoes, although he also briefly comments on how they taste. To appreciate what he covered and his penchant for details, we here present translations of his two passages, beginning with his *Torpedo* (without spots), which, still following in the footsteps of the Aristotelians, is associated with *Rana marina,* meaning the fishing-frog (see Chapter 3).

> Torpedo, once called *Nárkē* by the Greeks, both terms accounting in themselves for the nature of fish, because it brings torpor to the hands by its own coldness. The common people of Greece call it *Margotirem*.[17] The fish is almost unknown to the Venetians: if it goes there sometimes, it has the name of *Sgramphus;* in Rome it is known by no other name than *Occhiatella*. It frequents the muddy seashore and is rare in the Gallic Ocean; it is known to the people of Bordeaux and among them is called *Tremble,* as if to say tremulous, because with its slimy, smooth skin, which is soft and cold, it makes hands tremulous by contact with it. At first sight, it reminds one of the frog-fish (*Rana marina*) or the stingray (*Pastinaca*), except it is much rounder. The lower part (*prona pars*) is ash-like whitish (*ex cinereo albicat*), and its upper part gets whiter (*magis canescit*).[18] It has no snout, but in its place has a notch impressed between the two sides in the anterior part of its body; it has a shorter tail than the ray, pressed in at the front and arranged as with lampreys, for it cuts through the sea *in contrarium*. At its base, two little fins sprout out that greatly help in swimming. It has small eyes on the dorsal surface, situated as in the ray. It has a broad mouth in the shape of a crescent moon on the ventral side, the mouth being fortified only on the forward part with blunt teeth arranged in many rows. It has duplex gills, each with five orders, and a very big gall bladder for the size of its body. In other respects, it has such tough flesh that it has a very disagreeable taste unless it is fried. Even when fried, it does not become firm and is cooked with the greatest difficulty. People think that it drives out a fever when it is applied to the ends of limbs, just like the freshwater *Tinca* [Tench], which we shall describe later on. The following picture shows the shape of the torpedo, both in prone and supine positions. (p. 89)

Turning now to *Torpedo oculata*:

> People criticizing Pliny, because he called this kind of torpedo, which has six and sometimes seven spots or less on its back reminding one of eyes, *Oculate* [*Oculatam*], are astonishingly mistaken. They assert that this nickname should be applied rather to the *Melanurus* [a type of sea fish]. But if the *Melanurus* receives the name *Oculate* amongst the Latins, because of the beauty and size of its eyes, what prevents us from also calling this type of torpedo *Oculate*, because of the appearance of its many eyes, even though it differs a great deal from the former? Indeed, the very word signifies the number of eyes, rather than their size, although it can also have both meanings. And the common people of Rome show well enough how much these critics, who are a little too inflexible in my opinion, are mistaken. Even today they call the torpedo *Occhiatelle*, although no one has taught them this. But I shall write about this more fully in reference to the *Melanurus*. Now this Torpedo is like the other, not only in size and shape, but also in the softness of its flesh. It differs from it only in its five, six or sometimes seven spot-like eyes. It exhibits these so exactly in the middle of its back that no judge would think a painter could have drawn them more correctly or precisely. In other respects, it has small real eyes, which resemble those of the *Torpedo* described above. (p. 92)

Belon included three illustrations to accompany what he wrote about these two kinds of torpedoes. He shows both types of torpedoes: a dorsal and ventral view of the first variety, and a dorsal view of the *Torpedo oculata*, albeit with its five *ocelli* in rather unrealistic locations (Fig. 6.13).

Belon really limited himself to an external analysis of the fish shape, his sole interior anatomical annotation being about the great size of the gall bladder.[19] Even when describing the fish's appearance, however, he missed the presence of two penises, even though both of his illustrated torpedoes were males. The sexual organs of male torpedoes correspond to the bipartite uteruses of the females.[20]

---

[17] The term was evidently derived from the *nárkē*, through a linguistic transformation that had already partly occurred by the time the fourth book of the *Kyranides* was composed. As mentioned in the previous chapter, the author of the book wrote that the *nárkē* was called by many *marga*.

[18] The derivation from Giovio is evident with the expression as *prona pars* and *supina*. In our translation, Belon seems to repeat Giovio's error about its superior part being whiter than its inferior part. However, the wording is somewhat ambiguous with *albicans* and *canescit*, which are difficult to render properly in modern English. It is possible that Belon, who apparently had investigated the fish directly, purposely used ambiguous wording perhaps because he did not wish to contradict or offend Giovio in print.

[19] As already remarked, especially when dealing with the *Kyranides* in the previous chapter, torpedo bile was frequently used as a medication.

[20] The presence of two sexual appendices in the male torpedo had been already recognized in the Aristotelian corpus when dealing with the copulation among cartilaginous fishes (*Historia animalium*, V, 540^b).

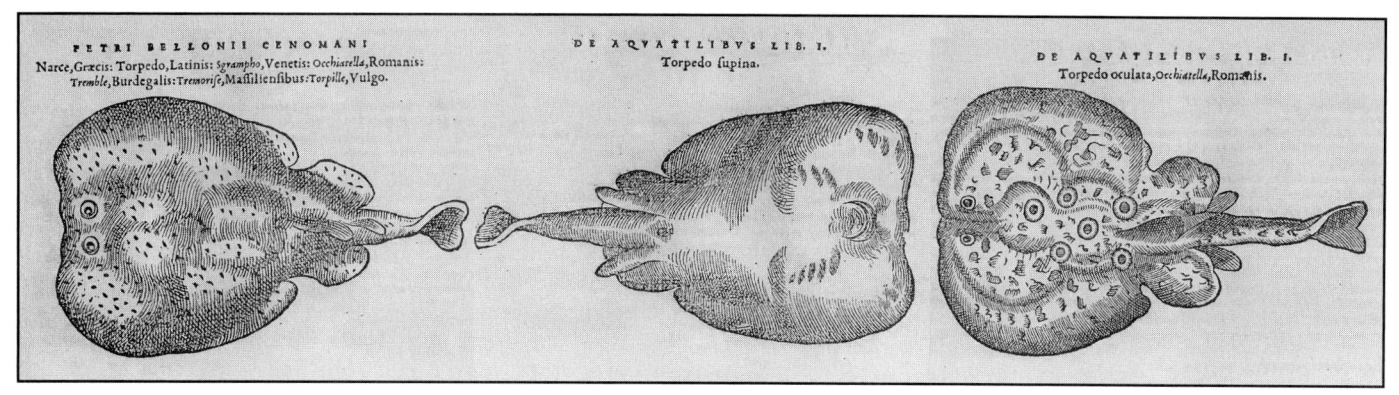

Figure 6.13. Images of torpedoes from the *De aquatilibus* of Pierre Belon (1553). These are likely the first images of the fishes to appear in a printed book. The two figures on the left depict Belon's "Torpedo" (dorsal and ventral views) and the one on the right shows his "Torpedo oculata," the variety with "eyes" on its back.

Further, there are no fresh new explanations for the torpedo's effects on these pages. With one foot mired in the past, Belon refers only to its cold or chilling actions, an ancient thought that is consistent with the idea that it could be used to treat a burning fever (i.e., the Greco-Roman theory of restoring balance by treating with opposites). It is nevertheless notable that he uses the phrase "people think" in his sentence about employing a live torpedo to treat a fever—showing that he, for one, was less than completely willing to endorse this often-repeated claim. Here, at least, we sense a call for more verification.

RONDELET'S TEXT

In 1554, a year after Belon's second book came out, Guillaume Rondelet, Regius Professor of Medicine at the University of Montpellier in the south of France, published his *De piscibus marinis*, one of his two texts on fishes (the other being his *Universae aquatilium historiae pars altera*, in 1555). Rondelet had trained in Paris and Montpellier, and he served as Cardinal de Tournon's personal physician (Oppenheimer, 1936; Cole, 1944). As was the case with Belon, the Cardinal assisted him with his travels, and he studied fishes in Italy, France, and elsewhere in Western Europe.

More skilled in anatomy than Belon, Rondelet described some 244 species of fishes and provided better illustrations. In his 583-page book, he tried to distinguish among his chosen fishes by differences in markings, morphology, habitat, parts, taste, and the like. Compared to Belon, Rondelet deals with the torpedo (and other fishes as well) with more history, philological erudition, and scientific information. He tries to separate four different kinds of torpedoes, three being of the *oculata* variety and one being without "eyes" (Fig. 6.14).

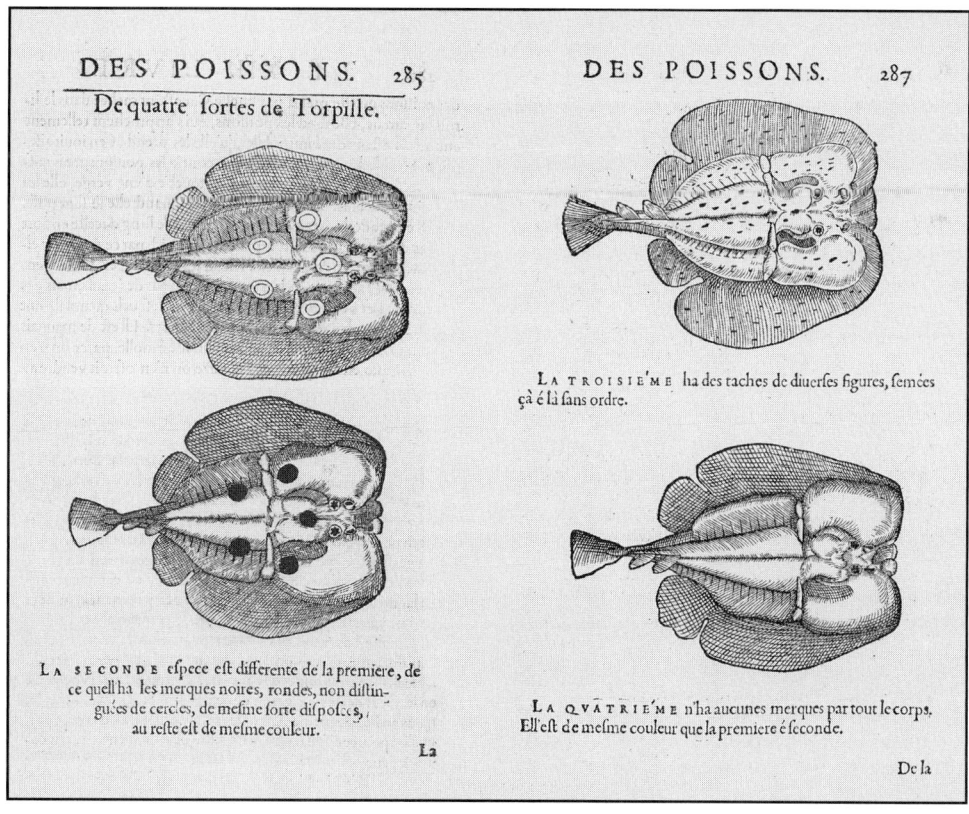

Figure 6.14. Images of the various types of torpedoes depicted by Rondelet in the 1558 French edition of his book on fishes.

His separation of the three types with these eyes on their backs was, however, based on zoologically negligible differences.

Rondelet deserves credit for describing both the external and internal anatomy of his torpedoes in far greater detail than his predecessor. He also brings forth more vernacular fish names. Among them we find *tremorize*, used in the Genoa Riviera, and *dourmilleuse*, a term used in Marseilles, as well as also the current French name, *torpille*. Continuing to go well beyond Belon, Rondelet also mentions most of the major ancient authors who had written about the torpedo, with ample quotations in both Greek and Latin from their texts. His coverage includes Theophrastus, Galen, Pliny, Paulus, and many others.

He cites Galen on the ability of the live torpedo to relieve headaches. But he does not mention Scribonius and he does not allude to the usefulness of the shocks for treating *podagra* or *prolapsus ani*. As to Paulus, Rondelet refers to his boiled torpedo prescription for treating joint pain. He also mentions Pliny on torpedo preparations for facilitating parturition and diminishing sexual drive in men, but here he at least starts off by declaring his astonishment.

With regard to the culinary value of the fish, he first refers to the views of the ancients, who favored it as a delicious food. But then he gives his own opinion, stating, "those who disapproved of the torpedo among fish judged correctly," basically confirming what the earlier Renaissance ichthyologists had written (p. 361).

In contrast to Belon, Rondelet shows a greater interest in the mechanism of the shock. What he writes is still tied to Galen, but it goes beyond just attributing it to a cold quality, and translates as follows:

> Although the summer's heat was at its peak when I tried to move my hand on it, I actually felt a sensation of cold from a torpedo long dead . . . and so I would judge that Galen was quite right to count contact with a living torpedo among the causes of numbness. . . . The cause of this numbness is cold, which is also true of opium, mandrake and henbane, yet it is not cold alone but also some unseen power naturally innate in the torpedo. For Galen also seems to ascribe this power of the torpedo not only to cold but also to its obscure faculty. (From Rondelet, 1554, pp. 360–361; Trans. in Copenhaver, 1991, p. 384)

Rondelet addressed the benumbing qualities of these creatures, not just citing what the ancients had written, but on the basis of his own observations in Venice and elsewhere. His wording is particularly interesting because it brings together Galen's well-known theory of coldness (a manifest quality) with the idea that the torpedo also possesses a *caeca quaedam vis* [a blind or invisible power] or *obscura facultas* [obscure faculty] that mainstream science cannot explain—an occult quality to which Galen had also alluded (see Chapter 4).

With his relatively accurate descriptions of the external and internal gross anatomy of these fishes, with his annotations on the mechanism of the shock, and especially with his broad and erudite treatment of the history and literature concerning the torpedo, Rondelet sets a high bar for his fellow ichthyologists. What is of relatively low quality is the illustrative material in his work. His four torpedo images deviate from what one would observe with real torpedoes and are disappointing given how well he covers the torpedo verbally in his book.[21]

SALVIANI: HIS FISH IMAGE AND TEXT

In the domain of illustration, the third writer (chronologically) in this group, Ippolito (Hyppolyto) Salviani, took the next needed step forward. Salviani was born in Città di Castello near Urbino and educated in Rome. He taught at the University of Rome and served as physician to three popes. His work on fishes, his *Aquatilium animalium historiae*, appeared in parts between 1554 and 1558 (with a separate publication of the illustrations in 1557) (Fig. 6.15).

He mentions 99 species of fishes (with Greek, Latin, and common names) and also includes the octopus and squid. But rather than illustrating his books with woodcuts like the other ichthyologists, he turned to brass engravings for 93 species. Salviani's (1554) book has been given high scores for its beautiful art, and his picture of the torpedo is one of the best in his book (see Fig. 6.15). But his text has been criticized for being less than perfectly accurate and for conveying images (some graphically) of unsubstantiated fantastic creatures (Castellani, 1975).

With regard to the torpedo's powers, Salviani accepted both *frigiditas*, the cold quality, along with the aforementioned idea of *qualitas caeca*, some sort of an occult force. In this cause-and-effect domain, he clearly followed Rondelet, with whom he consulted in Rome, and, of course, Galen. As a matter of fact, apart some small differences, almost all of Salviani's description and historical treatment of this fish are derived from his French colleague. Very likely, however, he also studied the fish personally, because he provides anatomical annotations (e.g., on the oblong and bicornuate uteruses) that are absent in Rondelet.

Salviani's logic for bringing up an occult quality is notable. He points out that when a torpedo dies, it seems colder to the touch, even though it loses its remarkable powers. Hence, its power must derive at least in part from another, unknown force. He associates this occult quality with the whole torpedo rather than a specific part, quoting Galen extensively. While sharing the opinion of the great physician from Pergamon, he also mentions the ideas of Diphilus (via Athenaeus) and Oppian in his text.

GESSNER'S MASSIVE ICHTHYOLOGY WORK

Konrad (Conrad) Gessner (Gesner, Gesnerus), the fourth writer in this cluster, was born in Switzerland. Possessing a doctorate from Basel (awarded in 1541), he was one of the most important scientific and medical writers of the 1500s, and was universally regarded as one of the most learned men of his day (Fischer, 1966; Serrai, 1990). He practiced medicine in Zurich and was lecturer in physics at the Carolinum (Fig. 6.16).

---

[21] Among the interesting discussions in Rondelet, there is one in which he deals with the remora, making clear the confusion made by the ancients on different sea species under the name *echeneis*. After dismissing the ancient stories of its powers as too exaggerated or fabulous to be believed, he attempts to explain the real mechanical effects that a remora can have by adhering to a ship's rudder with a suction mechanism.

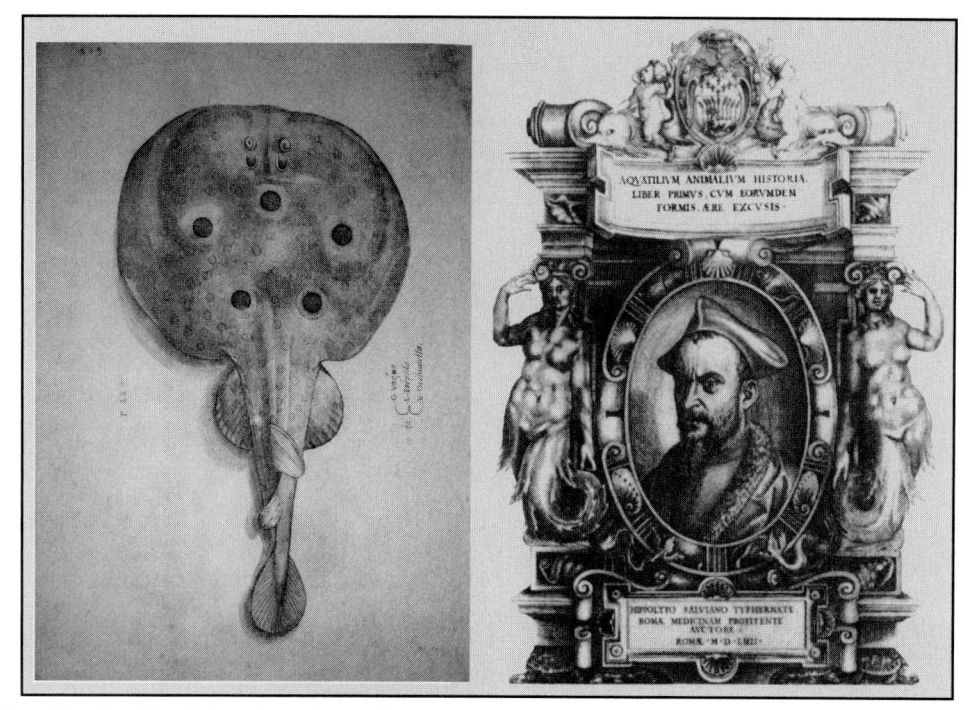

Figure 6.15. *Torpedo ocellata (occhiatella)* as shown by Ippolito Salviani in the 1557 edition of his *Aquatilium animalium formae*, along with his portrait from the title page of the 1554 edition of *Aquatilium animalium historiæ*. A similar image of the torpedo also appears in this second work.

His books are like encyclopedias, and his *Historiae animalium* comprises four volumes published between 1551 and 1558, with a fifth published posthumously in 1587. Fishes appear in Volume IV of the Latin version (the German translation has four volumes). Even by modern standards,

Figure 6.16. Gessner's portrait as depicted in an old engraving reprinted in Pizzetta (1893).

his *De piscium et aquatilium animantium natura* is enormous, comprising about 1,300 pages. Here he presents a huge number of fishes and other aquatic animals in alphabetical order, based on their Latin names. There are also 900 woodcuts, drawn both by Gessner and an assembly of artists working well into the night at his home. One notable artist, as already mentioned, was Albrecht Dürer. Gessner's zoological treatises were published in various forms and underwent numerous editions.

Like his predecessors, Gessner was well acquainted with the classical literature on the torpedo, and had even published an edition of the works of Claudius Aelianus in collaboration with Pierre Gilles, another sixteenth-century ichthyologist (see Footnote 14). He included what many authorities of the Greco-Roman past had written, along with newer material about the torpedo. He also knew Belon and Rondelet personally, having first met them in 1540 when he went to study medicine at Montpellier, and he borrowed from their newly published books. He also drew from Salviani, consulting some 250 books overall, in addition to tapping a massive network of foreign correspondents (including Aldrovandi in Bologna).

Gessner strove to get to his facts right, wishing to avoid the ridiculous in his books, although he was not always successful. For example, he included eight pages on the unicorn and provided a beautiful illustration of a horse with a narwhal-like tusk in his text on animals (Fig. 6.17). He did, however, break from the ancients when discussing the torpedo, telling his readers that its shock is not as strong as the ancients had made it out to be. In this domain, he also deviated from Albertus Magnus, who, as remarked above, mentioned a colleague whose full recovery upon touching a torpedo had taken 6 months.

Pierre Gilles, whose specialty was fishes in the French Mediterranean, and who had written a short treatise giving their Latin and French vernacular names (which he added to his Latin translation of Aelian), was Gessner's reliable

Figure 6.17. A page from the first volume of Gessner's *Historia animalium* (1551) showing the fanciful image of the unicorn (above that of a giraffe).

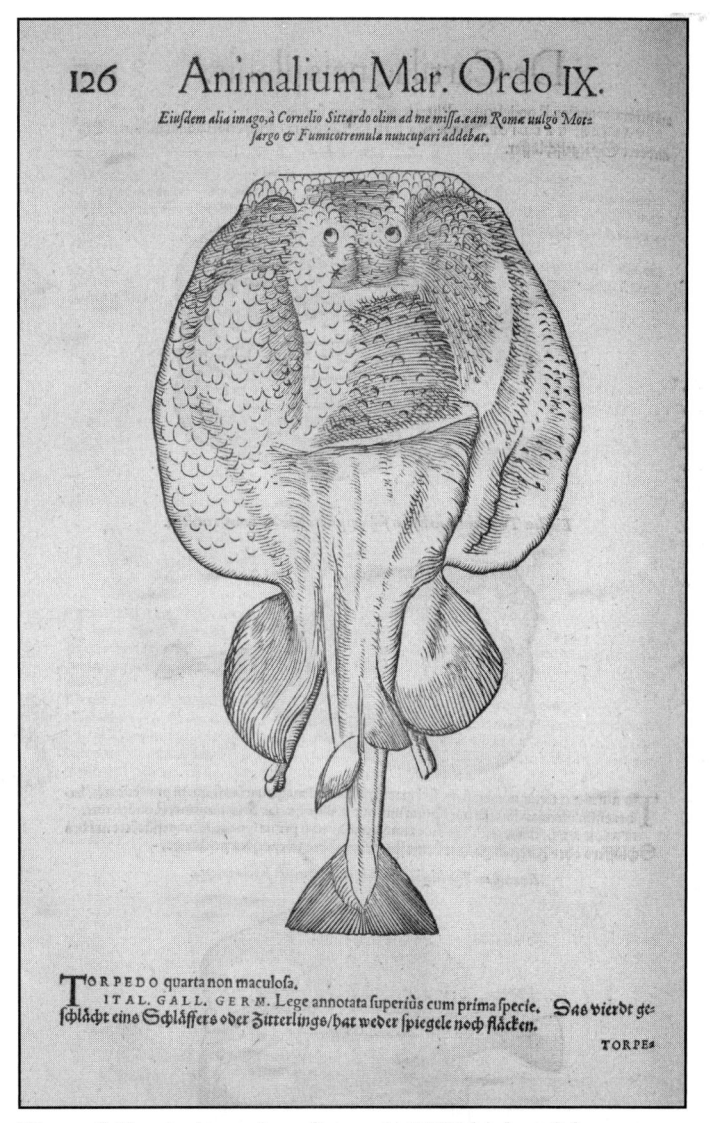

Figure 6.18. An image from Gessner's (1558) book on fishes showing the torpedo depicted from an image provided by Gessner's colleague Cornelius Sittardus.

main source for what the shocks really feel like. After noting that the consequences of the shocks are not nearly as intense as the ancients had declared, with their effects also disappearing much more rapidly, he mentions the torpedo's names in Marseilles and Genoa, Latinizing them as *turpiliam* and *tremorizam*.

In all, the torpedo section of Gessner's *De piscium* occupies a rather lengthy 12 pages (Gessner, 1558, pp. 1182–1194). A good part of it is simply a reproduction of the accounts previously given by Belon and Rondelet. He includes seven illustrations, four of which are the same as those provided by Rondelet, with the other three being original.

One, depicting a torpedo corresponding to one of the non-spotted varieties of his French associate, is also presented in a reproduction provided by Cornelius Sittardus, a physician probably from the Netherlands, who practiced in Rome.[22]

This illustration is comparable in quality and other characteristics to Rondelet's (Fig. 6.18) Nevertheless, Gessner had major problems with another illustration he ordered while he was in Venice. It was so dissimilar to those of Rondelet, and the matter bothered him so much, that he voiced his complaints. As a matter of fact, it is possible that the fish represented might not even be a torpedo[23] (Fig. 6.19). In any case, we have here an instance where the collaboration between an artist and a scientist was less than successful, not meeting the standards of the High Renaissance.

Even more than his colleagues, Gessner was an internationally recognized classical scholar and lexicographer, having published in 1545 his *Bibliotheca universalis*, a reasoned catalogue containing almost 15,000 entries. In his *De piscium*, he provides a deep historical account of the literature on

---

[22] Sittardus, one of the colleagues of Gessner mentioned as a provider of naturalistic illustrations, made many of Gessner's figures, which are still extant as watercolors. Sittardus was not, however, the artist behind the figures that Gessner received from Gijsbert van der Horst, a Dutch physician also practicing in Rome (see Holthuis, 1996).

[23] This fanciful Venetian torpedo in Gessner might be at the origins of the poorly realistic representations of the torpedo in the *Hieroglyphica*-type works that flourished in the sixteenth century (see Chapter 2).

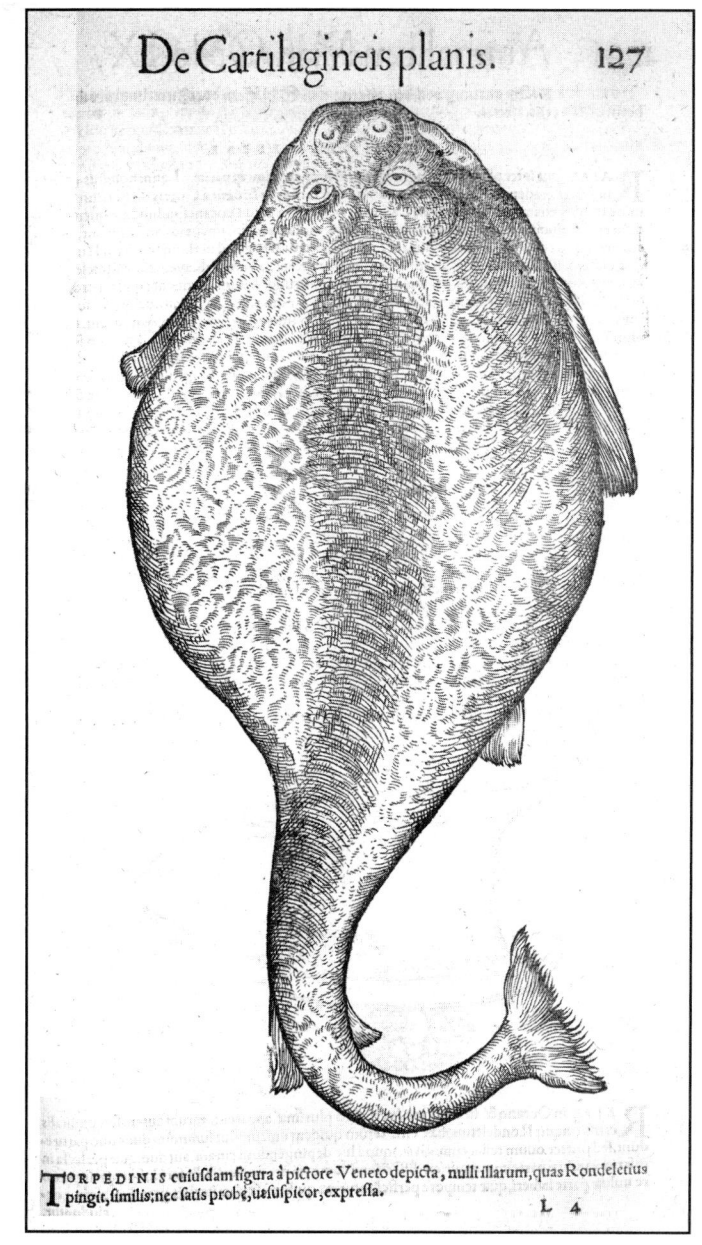

Figure 6.19. A less-than-realistic image of a torpedo that had been painted for Gessner by a Venetian artist. This image was printed in the 1558 edition of Gessner's book on fishes.

torpedoes, referring to an enormous variety of sources, both ancient and contemporary, many not even appearing in Rondelet. Being a polyglot and cognizant of various ancient languages besides Greek and Latin, he also gives the Arabic and Hebrew names for the torpedo, using the original lettering in the latter case.

Gessner might be the first of the ichthyologists of his epoch to quote Scribonius and also Marcellus Empiricus (of Bordeaux) on the use of the torpedo's shocks for treating headaches and *podagra*. He also seems to be the first author to present the Greek term that he spells τύρπαινα (*túrpaina*) in a modern ichthyologic treatise—this being a term that we encountered in various spellings in the *Kyranides* and in the writings of some Byzantine physicians (see Chapter 5), and one that he might have drawn from Ermolao Barbaro (see Footnote 15). Gessner states that this word was derived from the Latin, although he declares himself incapable of deciding whether it resulted from a corruption of the correct name *torpedo* or from the fact that the fish was used to alleviate the pain of *podagra* (the former being more likely; see Chapter 4).

In the context of the names for the torpedo in their various acceptances, Gessner brings up other terms, including *torpigo*, used by a thirteenth-century physician of Salerno, Matthaeus Sylvaticus (whose work *Opus pandectarum medicinae* was popular in the previous century). He also mentions that the Arabic term appeared in a medical treatise written by an Eastern scholar named Serapion (but of somewhat uncertain identity) during the Middle Ages, which went through many editors.[24] And he provides a long list of vernacular names from various sources.[25]

When it came to the nature of the torpedo's powers, Gessner, like his immediate predecessors, was inclined to admit that the shock depends not only on cold (or a similar physical agent), but also on a certain "blind" [*coeca*, in the sense of occult] quality. Here he mentions both Salviani and Galen. He does not abstain, however, from referring to those ancient authors (e.g., Diphilus and Oppian) who believed that the torpedo's power resides in specific regions of its body. This contrasts with Galen's idea that the fish's occult qualities belong to its entire body.

ULISSE ALDROVANDI: EXTENDING ARISTOTLE AND PLINY

With Ulisse Aldrovandi we come to the last of the great Renaissance scholars who wrote extensively on fishes. Aldrovandi taught logic and natural history for four decades at the University of Bologna, where he had received his medical degree in 1553, and at the University of Padua, which he had previously attended (his studies also took him to Pisa). Although he also wrote a notable text about fishes, what he did in the field of natural history is so extraordinary and important for understanding the setting in which he lived, and the changes that would follow, that it merits special attention (Fig. 6.20).

---

[24] Varying with the edition, Serapion's treatise might include, in addition to the Arabic term *tead* or *thead*, a variety of vernacular names, including *tremolo*, *tremoriza*, *batti potta*, *fotterigia*, and *turpille*, which were very likely introduced by the editors. The rich manipulation of old texts, which characterized some of the first printed editions of this classic of medicine, makes it sometimes difficult to distinguish what was in the original text and what was added later on to update the work. In a similar way, we can find many vernacular names of the torpedo in the *Commentarii* to Dioscorides of Pietro Andrea Mattioli, published in 1555. The same is true with the *Enarrationes* ("Narrations") to Dioscorides' works published in 1558 by Amatus Lusitanus (i.e., the Portuguese Jewish physician, João Rodrigues de Castelo Branco). The new ichthyologists not only interacted with each other but almost assuredly exchanged information with the editors and commentators of classical books of medicine and zoology.

[25] Among the vernacular names that he was probably the first to record, we find the Spanish terms *tremielga* and *hugia* (which he says he received from an unnamed correspondent). Among his more curious names, we find the term *gallina*, apparently used in the French region of Provence, and two strange terms, used, in Sittardus' words, in Rome: *mostargo* and *fumicotremula*—the first likely derived from a comparison of the fish's shocks with the irritating actions of mustard, and the second combining *tremula* (i.e., trembler) with a vernacular modification of *formica* (i.e., ant) and alluding to the *formicolio* (i.e., the tingling sensations produced by a weak fish shock; a word corresponding to the French *fourmillement*). Gessner also attempts to create German names for the torpedo "by similarity." Among those proposed, *Zitterfisch* or *Krampffisch* have enjoyed more fortune in popular usage than *Schlafferfisch* or *Zitterling*.

Figure 6.20. Aldrovandi's portrait from the 1599 edition of his *Ornithologiae* and the title page of his 1613 edition of his book on fishes and cetaceous.

Aldrovandi was a cousin of Pope Gregory XIII, who took the reins of the Roman Catholic Church in 1572. In that same year, his own remarkable career took off when he was given a "dragon" captured near Bologna. He illustrated it, dissected it, and wrote a book titled *Dracologia*, in which he presented his specimen as a natural phenomenon without giving it metaphysical overtones (e.g., such as being an omen of tragedies to come). Aldrovandi's *draco*, actually a forgery fabricated by using the skin of a ray, is still extant in his museum, now housed in the buildings of the University of Bologna.

With a huge inheritance to draw on, Aldrovandi never practiced medicine. Instead, he edited Greek and Roman accounts of natural history, directed a botanical garden and a museum in Bologna, and worked endlessly on his treatises.

Aldrovandi's *museo* (literally a place to venerate the Muses, the guardians of knowledge; also called his *teatro*, *microcosmo*, and *archivio*) was his passion. By 1595, he estimated that he had some 18,000 different objects in his collection, in addition to having paintings and woodcuts made of an additional 5,000 natural objects (Olmi, 1992; Olmi & Tongiorgi-Tomasi, 1993; Findlen, 1994; Tongiorgi-Tomasi, 2000). His collection was initially housed in several rooms of his family estate in Bologna, where it became a not-to-be-missed site for royalty, special guests, and tourists of stature, who were interested in the natural world, its diversity and fringes, and *materia medica*. Aldrovandi referred to his *museo* as the eighth wonder of the world, and with Gessner's death in 1565 ending what he viewed as a competitive undertaking in Switzerland, scholarly friends and "foreigners" who saw his collection hailed it as the most complete ever assembled.

Aldrovandi's contemporaries called him the "Bolognese Aristotle" and the Pliny of his time. With his reverence for the ancients and exceedingly high opinion of himself, he welcomed such titles. Never one for modesty, Aldrovandi inscribed these words (in Latin) under his portrait: "This is not you, Aristotle, but an image of Ulysses: though the faces are dissimilar, nonetheless the genius is the same." He backed up these pompous words by pointing to 13 volumes of his own well-illustrated writings (Tugnoli Pattaro, 1981). And although Gessner might have been one of his main sources, he referenced him and others only infrequently (Gessner's name is completely omitted from his volume on fishes), an action that further amplified his stature in many quarters but probably did not endear him in others (Petit & Théodoridès, 1962).

Aldrovandi's chosen mission was to complete the Aristotelian classification of nature by seeing things with his own eyes, touching them with his own hands, dissecting specimens, and putting them on display for others to confirm. Moreover, with his empirical orientation and by using his senses, Aldrovandi strove to verify and correct Aristotle, whom he criticized for not personally checking everything he wrote about living nature.

As for Pliny, if this Roman naturalist could surpass Aristotle by collecting "20,000 noteworthy facts," would it not be admirable to surpass the great Pliny with an even greater collection of facts? Notably, Aldrovandi also proclaimed himself the new Galen and, following the great voyages of the Age of Exploration, compared himself to Columbus, unabashedly stating that he would be the best man (read "philosopher") in all of Europe to understand the significance of the newly acquired material!

In all, Aldrovandi wrote some 400 treatises about his specimens. His collection would grow to approximately 20,000 items by the end of the century—a number that by itself matched the number of Pliny's "noteworthy facts." Aldrovandi had a great taste for art and artistic representation, and in his writings and letters he shows an acute awareness of the complexity of the task facing scientists and artists in their attempts to produce faithful images from nature.

His texts, like his museum, attempt to follow the Aristotelian hierarchy and progression of nature. Here we find animate objects given higher status than inanimate ones, plants situated below animals, and humans above the animals. But along with his real animals, and like Pliny, Albertus Magnus, and others, there are some notable imaginary marvels (e.g., unicorns), as well as monsters, two being a fabulous ass-headed man and a man with horns and wings.

One of the problems that he faced was that not all of his animals fit neatly into his Aristotelian classification scheme—that is, in the intellectual framework he chose to follow. In part, this was because some of the specimens were not real, being composites and forgeries made by unscrupulous suppliers—men who knew that they could make tidy profits by showing him their "rarities" and even threatening to sell them to competitors. But it was also because some of his real specimens were, simply put, really quite odd (see Ashworth, 1990).

His published works are filled with illustrations, some of which were made by exceptional artists and engravers, which was now expected in volumes on natural history, but which also made these books difficult to produce and expensive to purchase. In fact, only four of his volumes were actually published in his lifetime (three on birds and one on insects), and when he died in 1605 he left numerous manuscripts behind, in addition to his collections, bequeathing them to the Senate of Bologna. (The state dutifully continued to show his collection, which remained a centerpiece of culture in the city and, as noted above, was ultimately housed in the museum of the University of Bologna.) Further, eight more of his treatises were published after his death, following his instructions.

The images printed in Aldrovandi's volumes, as in most scientific books of the age except Salviani's, are based on xylography (i.e., woodcuts). This technique is useful for reproductions, but it leads to images that are not of the highest artistic quality. On the other hand, his original paintings, most of which were made with watercolors and tempera (according to a technique developed by Renaissance artists for naturalistic representations), are in many cases of exceptional quality, and among the extant ones there is a marbled torpedo with a list of its names (Olmi & Tomasi, 1993; see Fig. 6.21).

As intimated, Aldrovandi wrote a separate treatise on fishes, one in fact called *De piscibus* ("On Fish"), and it appeared posthumously in five volumes in 1613. This is the first printed book on fishes that includes only fishes (no porpoises, squid, octopuses, etc.). Moreover, and in contrast to Gessner, who chose to present his material alphabetically, Aldrovandi (1613) tried to bring those fishes together that seemed to belong in the same scientific groupings.

He tells us in his 1586 autobiography that his personal interest in natural history began with a visit to a Roman fish market in 1549–50. "In that same time that [Guillaume Rondelet] was in Rome, I began to be interested in the sensory knowledge of plants, and also of dried animals, particularly the fish that I saw in the fish markets" (translated in Findlen, 1994, p. 175). He also enjoyed fishing and was a beachcomber, and he later described his excursions to the shores of the Adriatic as a most enjoyable way to expand his collection. Again in Aldrovandi's words: "Thirty years ago, while fishing at Ravenna, three miles from where my brother was then Abbot, by the luck of the sea I found many bizarre marine creatures on the beach" (Aldrovandi, ms. 21, Vol. IV, c. 91r).

Aldrovandi also had a truly extensive network of contacts, and they provided him with fishes that were caught or washed up on the shores with notes about them. One of these men was physician Costanzo Felici of Rimini, who in

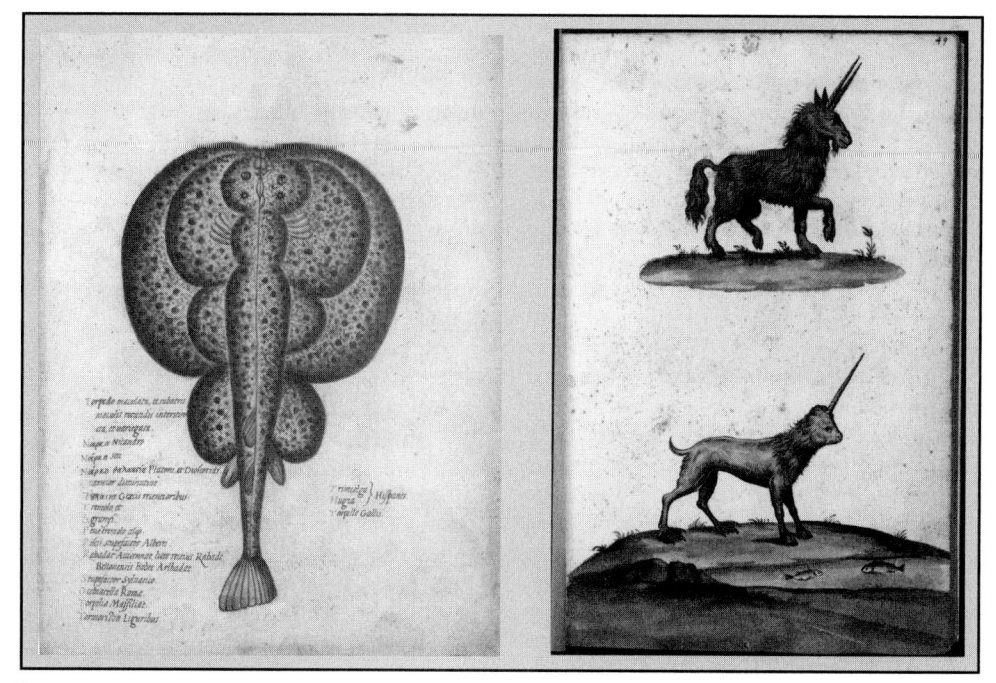

Figure 6.21. Left: The painted torpedo from the unpublished collection of Aldrovandi's colored animal and plant images now at the Library of Bologna University. Right: Two images of the unicorn from the same collection. Although nicely decorated, the torpedo image is not very realistic, and it is unlikely that the artist painted it from nature. The unicorn, of course, is a fictional animal. (From Olmi & Tongiorgi-Tomasi, 1993) (© Biblioteca Universitaria di Bologna)

1557 told him: "So that you will not marvel at my facts when I send you the most ordinary things, I warn you that my intention is to send you almost every sort of fish, rare or common, that one catches in our port" (Felici, 1982, p. 39). Another was Pietro Fumagalli, who in the fall of 1558 stated that he "will purchase everything that I find, having the fishermen inform me of the type of fish and the time at which it was caught to the extent that they can. Thus I will slowly acquire most, if not all, of those fishes you requested" (Aldrovandi, ms. 38, Vol. III, c. 2r). He even asked Ippolito Salviani to scour the markets in Rome for him, providing the author of the *Aquatilium animalium historiae* with a list of his 40 most wanted fishes. Whether he caught his own fishes, found them along the shores, received them from friends and correspondents, or bought them from local fishermen or in markets (where he found he always learned more about fishes), torpedoes were among the items he wanted in his cabinet and among the fishes that he most wanted to illustrate.

The torpedo is the first fish Aldrovandi deals with among the flat cartilaginous species in Chapter 45 of Book III of his ichthyologic treatise. He is comparable only to Gessner for the erudition shown in providing old names of the torpedo, in citing past and contemporary authors, and in critically reviewing the literature on the subject. In the latter domain, he corrects what others have written, including Salviani on the river torpedo mentioned by classical authors. Like Gessner, he provides the Greek term τυρπαινα (sic) and, in a strongly dubitative form, suggests that it could be the same as the *nárkē* or torpedo. Overall, however, he adds little to what previous authors had written and he does not leave us with the sense that he had personally experimented with a torpedo, touched one, or even observed a live specimen, even if he might have.

He also follows previous authors, from ancient to his own time, when he discusses the torpedo's powers. Basically, he was just as confused as Gessner had been on this matter. To quote from Aldrovandi: "Whatever the ray's numbness may be, it arises from a certain obscure and occult quality and not from a manifest one, as can be proved from reason and from the authority of the most respected authors." Clearly, he was not thinking that it simply releases a cold poison, or he would have said so. But its force was still mysterious, hidden from the eyes of the man who considered himself the greatest natural philosopher of the era, and for that matter, of all times.

A BRIEF NOTE ON MAGIC AND OCCULT QUALITIES

As we shall see in Chapter 9, what drew sixteenth-century ichthyologists to "occult qualities" as an "explanation" for the torpedo's shocking powers has much to do with one of the most important themes of the Renaissance. This is the revival of magic, a development strongly influenced by translations of the Hermetic corpus from Greek into Latin. Marsilio Ficino (Marsilius Ficinus) undertook this arduous task in Florence in 1463, and his efforts were printed under the title *Pimander* in 1471 (Hermes, 1471).

Even more than in the ichthyologic treatises, other zoological texts, and medical books, the torpedo would repeatedly be mentioned in Renaissance magic and philosophy books as evidence of the existence of occult qualities or forces. This was typically done to substantiate or discuss the efficacy of magical practices based on the manipulation of the forces of nature, according to the Renaissance concept of *magia naturale* or natural magic.

The modern science of Galileo would diametrically oppose magical views and allusions to occult forces. But in another way, another great protagonist of the scientific revolution, Sir Francis Bacon, who also strongly opposed magic, would invoke modern science as a way of realizing, in a pure and non-demoniac way, the aspirations of the Renaissance *magi*—namely their aforementioned ambition to control the forces of nature. In a somewhat ambiguous form, *magia naturalis* could be a synonym for science in the sixteenth and seventeenth centuries, as we shall see in Chapter 9, where we shall look in considerably more detail at magic and occult forces, and the reactions to these notions by Galileo, Redi, and other early modern scientists.

LOOKING BACK ON THE SIXTEENTH CENTURY

Reading the pages on the torpedo written by the sixteenth-century ichthyologists makes for an interesting exercise, and it leaves us thinking about many things. On one side, it is clear that these scholars were looking at the natural world and the texts of the ancients with fresher eyes and better intellectual tools. The dissemination of books made possible by the invention of the printing press and other factors now allowed scholars to become more familiar with the works of the ancients, and to have more advanced philological and linguistic methodologies for studying them. Some also had access to research cabinets of anatomy or "physics," networks of helpful assistants, and skilled artists at their call. They would personally study fish, and would visit fishermen and fish markets, to have information on these aquatic creatures and inquire about their regional names, which would help them interpret the older texts.

These things notwithstanding, and despite their repeated assertions of the importance of directly studying nature, it is clear that a great part of what they wrote was not derived from personal study, but from previous encyclopedias and collations. The texts were really presentations of the older material that were now garnished with some new observations, but not with bold new thoughts.

Especially in the texts of Gessner and Aldrovandi, erudition dominates over personal investigations in clear ways. In fact, the drive to go outdoors and investigate the open world with fresh eyes seems diminished under the excess of erudition. Using a metaphor recurrent in Galileo Galilei, "the book of the world" to be read directly with one's own eyes was still very much "a world of books," a world to be consulted within the walled confines of a library. In this regard, Aldrovandi's project to expand the vastness of the natural world with his writing, his *museo*, and his concept of a *teatro del mondo*, still had the somewhat stuffy atmosphere of a *Wunderkammer* or cabinet of wonders.

Galileo effectively conveys this impression in another context with these words:

> We enter in the study of some little man with a taste for curios, who has been pleased to fit it out with things that

have something strange about them, because of age or rarity or for some other reason but are, as a matter of fact, nothing but a bric-a-brac: a petrified crayfish, a dried-up chameleon, a fly and a spider embedded in a piece of amber; some of those lithe clay figures which are said to be found in the ancient tombs of Egypt. (Galileo's letter to the painter Lodovico Cigoli, written June 26, 1612; trans. in Panofsky, 1954, pp. 9–10)

The hope inherent in the Renaissance naturalists, namely that their studies and investigations could provide a vision of reality encompassing the entire living (and even inanimate) world, was overly ambitious. Besides not having the intellectual and research tools that would emerge with the scientific revolution of the seventeenth century, there was no way that a single scholar, or even a genius with a network of learned colleagues, could tackle the vastness of the natural world. This shortcoming led these ambitious men to trust indirect accounts and to substitute more or less faithful images sent from distant lands (and sometimes copied from other images) for real specimens. This was particularly the case for exotica coming not from Europe but from the distant Indies, the Western and the Eastern ones, which were now being explored.

Even the declared proposal to reject the fanciful and legendary animals of the ancient and medieval texts is not fulfilled. As was true of the reports from the first Middle Ages travelers to the East Indies and China (from Rubruck to Marco Polo), the accounts of the first navigators to the New World (including those of Columbus) continued to contain descriptions of extraordinary animals and of monstrous men that the Renaissance naturalists, in the absence of a personal and direct experience, could not quickly dismiss as misrepresentations and fantasies.

Still, the Renaissance texts that we have examined have to be regarded as a significant advancement over the bestiaries and other compilations of the Middle Ages. Even in the encyclopedic naturalistic works, such as those by Gessner and Aldrovandi, one nevertheless has the perception of a profound intellectual change (and not only because of the greater philological and also scientific precision) with respect to the texts of the Middle Ages. These new works were deeply different from the medieval compilations, as for instance the encyclopedic treatise of Vincent de Beauvais.

In the three volumes of the *Speculum historiale* and the fourth volume of the *Speculum naturale*, Vincent, following the tradition of Isidore and Bede, attempted to include and systematize the entire knowledge of the world into a limited number of volumes (see Fig. 6.3). The monks of the medieval abbeys and the students of the *scholae* of the age could possibly find in them all that was needed, from theology to the remedies for diseases, from the lives of saints to the techniques of agriculture. They were thus not pushed to go farther and investigate the world with their own eyes. The medieval universe could be contained in the space of a relatively few volumes also because, as already mentioned, it was a closed universe *per se*, with God providing men with all that they needed, and having finalized to their necessities all the objects of the physical, animate, and even extraterrestrial world.

The world of the Renaissance was certainly no longer such a closed universe as it had been in the Middle Ages. With the discovery of new lands, and with the new objects from these lands being brought into view, these Renaissance scholars realized that there are many more things to discover and investigate, including plants and animals that could not easily be inserted into the Aristotelian vision of the world that had dominated philosophical thinking for so long. The broader and somewhat unsuspected vastness of the human world was also revealed by discoveries and rediscoveries of old texts, stimulated at least in part by the demands of the already flourishing editorial industry (as stated, about a century after Gutenberg, Gessner could list more than 10,000 books in his monumental *Bibliotheca*).

Thus, despite their limits and their emphases on the past and erudition, Renaissance scholars broke through some of the boundaries of the medieval world and set the stage for future changes, including a better understanding of electric fishes. In particular, curiosity and wonder, which were considered signs of ignorance and even sinful to the Scholastics, had once again become fashionable and were now thought of as virtues characterizing learned men. As we shall see when we turn to the scientific revolution (Chapter 9), the new goal would be to learn enough about nature to take the wonder out of wonder, and to come forth with better explanations than those based on hidden or occult forces. In a curious way, this new role for wonder would actually serve to take us back to Aristotle, who had embraced wonder as a stimulus for the proper study of nature.

In the next two chapters, we shall follow the electric fish story in two directions that stemmed from the navigations that were now taking Europeans far from home. First, we shall visit Africa, and particularly Ethiopia, one of the lands of the *nárkē potamia* (i.e., the "river torpedo," actually the electric catfish almost unknown to western *literati*), which Athenaeus, some Islamic authors, and others had alluded to, albeit in fairly vague or ambiguous ways. We shall then move to South America, the land of a newly discovered eel-like fish, which would possess the mysterious force of a *nárkē* in an even more intense and frightening way. After our trips to Africa and South America, we shall return to the Renaissance notion of occult properties, this serving also as a basis for fully appreciating the advent of some new, decidedly mechanical, explanations for the shocks of these fishes.

Some of what we shall be narrating in these chapters will take place while the ichthyologists of the sixteenth century were writing their books. Moreover, some of these men will play a role in these stories too. One is Paolo Giovio, who about a decade after writing his *De piscibus*, would translate some letters allegedly sent from Ethiopia by the legendary "Prester John" to the Pope, in the framework of one of the historical circumstances that would help to make the electric catfish much better known in Europe, as we shall now see.

Chapter 7
Rediscovering the Torporific Catfish

> In these Rivers and Lakes is also found the Torpedo [African catfish] ... it so tormenteth the body of him which holds it, that his Arteries, Joints, Sinewes, all his Members feele exceeding the paine with a certaine numbnesse.
> Father Antonio Fernándes, 1610 (Translated in Purchas, 1625)

VOYAGES TO DISTANT LANDS

As shown in the previous chapter, the Renaissance in the West was a period in which the limits of the world were broadened in many ways. Geographically, this was a time marked by great navigations toward far and unexplored lands with new wonders to behold. In this chapter, we shall focus on Africa and particularly Ethiopia to look at how these voyages and this widening of the world during the Age of Discovery drew more attention to electric fishes.

The Age of Discovery or Age of Exploration began in the fifteenth century. Early in this century, Infante Henrique o Navegador (Prince Henry the Navigator or Seafarer), the father of European worldwide explorations, sent expeditions from Portugal down the western coast of Africa, which was largely unknown to Europeans (Beazley, 1894; Russell, 2000). His main goals were to find the sources of the African gold trade, to expand Portuguese trade in general, and to subdue North African pirate attacks on the Portuguese coast. With his patronage, his sea captains now circumvented the ancient Muslim trade routes across the Sahara Desert and sailed down the West African coast to the Gambia River (1456) and Cape Palmas (1459–60), discovering regions rich in gold and slaves.

This was also the century of Portuguese explorers Bartolomeu Dias, who sailed around the southernmost tip of Africa (Cape of Good Hope) in 1488; of Fernão Magalhães (Ferdinand Magellan), who circumnavigated the globe while searching for the Spice Islands; and of Christopher Columbus (Cristoforo Colombo), who made three voyages to the "Other World," soon to be called *Mundus Novus,* meaning "New World," in the 1490s (Boxer, 1991). It was also the century of Vasco de Gama, the first sailor to go from Portugal around the Cape of Good Hope to India, returning in 1499 with exotic spices from the East.

The roots for the many voyages that took place late in the fifteenth century and during the sixteenth century can be traced to earlier explorations, including Marco Polo's journeys to Asia, which took place from 1271 to 1295 (see Chapter 6 and below). The newer voyages were aided by improvements in shipbuilding, as exemplified by the carrack and then the caravel, multi-mast ships developed by the Portuguese, which could sail throughout the Mediterranean with greater safety and even into more perilous oceans.

With better ships, mapmakers, and instruments for guidance, the Portuguese dominated maritime exploration for some time. They established colonies in India (Goa, Calcutta, Bombay), the Far East (Macau, Timor), and South America, and also made inroads into various parts of Africa, including the Gold Coast, Mozambique, and Ethiopia.

The Age of Exploration had mercantilism and dreams of riches based on new trade roots and valuable commodities at its core. But it had other dimensions and ramifications as well, two being politics and religion. In the latter context, Henry the Navigator hoped to find the lost Christian empire of Prester John, about which much more will be said in this chapter, while the missionaries who followed also hoped to save souls with conversions. For King Manuel I of Portugal, the conquest of distant lands was tied to the search for a lost religious utopia, and it had the spirit of a crusade to expand the Kingdom of God.

In this chapter, we shall examine how the silurus or electric catfish (*Malapterurus electricus*), so beautifully depicted on Egyptian tomb paintings (Chapter 2), and alluded to by some scholars in the Middle East (Chapter 5), was "discovered" by Europeans who were in Africa in the Early Modern Era. Our emphasis shall be on the Portuguese missionaries in Ethiopia, and how their findings were described and disseminated. As we shall see, the early descriptions focused on the ability of this fish to torpify, while providing minimal morphological information. This will cause some people to confuse the elongated African river catfish with disk-like sea torpedoes, as well as with the electric eel that was just being discovered in South America (Chapter 8).

ETHIOPIA IN THE 1500s

In 1494, Pope Alexander VI assigned the eastern regions of the rapidly expanding world to Portugal and the western regions (including the Americas) to Spain. This division occurred shortly after the Portuguese were trying to establish diplomatic ties with the Ethiopians. Soon more Portuguese with economic, political, and religious agendas were setting sail for this important part of Africa, where the locals had been practicing Christianity for more than one thousand years (Conti Rossini, 1928; Doresse, 1957).

Largely because of Middle Age legends, Ethiopia appeared extremely rich and powerful to many Europeans, who were fascinated by what they had heard. Early in the sixteenth century, after a long period in which the region was almost closed to Westerners, the Christians in Ethiopia sought a military alliance with the Portuguese against the Muslims, who were threatening to invade. Delegations were now exchanged and the Portuguese sent an embassy to Ethiopia that included Francisco Alvares (or Alvarez), chaplain to King Manuel I. This delegation reached Ethiopia in 1520 and came back to Lisbon about 6 years later bringing letters written in 1524 by the Ethiopian emperor, Lebna Dengel ("Incense of the Virgin"), addressed to Pope Clemens VII and Portuguese King João III (Conti Rossini, 1894, 1941; Crummey, 2000).

The Ethiopian monk Zagazabo (Ṣägä Z ä' Äb, meaning "Grace of the Father") joined those members of the embassy returning to Portugal, serving as a delegate of the Ethiopian emperor. At the same time, João Bermudes, a barber-surgeon with the Portuguese Embassy, about whom more will be stated below, remained in Ethiopia as a hostage for Zagazabo. Upon his arrival in Lisbon, Zagazabo was interrogated by the Inquisition, which, upon considering the nuances of his Ethiopian faith and how Ethiopian religious practices differed from orthodox Catholicism, did not permit him to have religious sacraments.

This marked the beginning of a complex and long dispute concerning the orthodoxy of the Ethiopian religion, one destined to have disastrous consequences for the landlocked African country. Zagazabo eventually wrote a defense of the Ethiopian faith. It was translated into Latin and printed in 1541 by the Portuguese humanist Damiao de Goís, a pupil of Erasmus of Rotterdam, in pamphlet form. The Inquisition also opposed it. Gois' (1541) publication included a Latin translation of Lebna Dengel's letters to the Pope, which Paolo Giovio (mentioned for his book on fishes in the previous chapter) had prepared in 1533.

Father Alvares was the first European to provide a firsthand account of the country. His book was initially censored by the authorities and was finally published in 1541, one year after his death. As might be expected, Alvares focused his attention mainly on the local religion, having noticed that Ethiopian Christianity included many Jewish and Islamic rituals, such as circumcision and fasting rites. His book was intended to be a "true history," one that would contrast with the largely fabricated accounts of the African country. Nevertheless, his statements about abundant sources of gold, precious stones, and other natural resources had much more impact on many of his readers than his descriptions of the general poverty and difficult life conditions of the natives, raising expectations even higher.

The feared Muslim invasion led by Ahmad ibn Ibrahim al-Ghazi (nicknamed Gragn, "The Left-Handed") began in 1528, while Alvares was in Ethiopia (Conti Rossini, 1894; Wallis Budge, 1928; Whiteway, 1902). Many Ethiopians were killed, their villages pillaged, and their churches burned. Facing the overthrow of his kingdom, Emperor Lebna Dengel sent João Bermudes back to Europe to seek military help (Bermudes, 1565).

In 1541, Cristóvão da Gama (the son of Vasco da Gama, the discoverer of the sea route to India) sailed from Goa, India, to Ethiopia with his 400-man army equipped with modern firearms. The invasion ended 2 years later, when Ahmad was killed after a series of battles that had first gone poorly for the Portuguese and the Ethiopian Christians they had been trying to help. The account of the story of the Portuguese expedition given by one of its members, Miguel de Castanhoso (Castanhoso, 1564), tended to accentuate the epic character of the events and particularly the heroic behavior of Cristóvão da Gama, who was killed by Gragn. It also enhanced the achievement of the Portuguese soldier who, wishing to avenge his captain, killed Ahmad with a single firearm shot, reminiscent of how the champions of bygone eras might have fought.

The outcome of the war stirred up more interest in Ethiopia, which was now drawing even more attention because of the seemingly "heretical" nature of Ethiopian Christianity. Hence now, even more than before, the Church in Rome set its sights on converting the wayward Ethiopians to their own form of Catholicism—a task entrusted to the recently founded Society of Jesus (Jesuits). In 1554, at the request of the King of Portugal, Pope Julius III appointed João Nuñes Barreto, a Portuguese Jesuit, to become the first Catholic Patriarch of Ethiopia. He also designated two new bishops to aid him in his mission, one being Andrés de Oviedo (Caraman, 1985; Pennec, 2003; Romo, 2006).

As was true of his ancestors, Ethiopian Emperor Gelawdewos understood the stabilizing role long played by the Ethiopian Church in his vast empire, which included people of different races, languages, and customs. The apparent Judaism tinting their religious practices was an essential aspect of their heritage. After all, the Ethiopian kings traced their lineages back to the legendary King Menelik, supposedly the son of King Solomon of Israel, and Makeda, the Queen of Sheba. Moreover, the Ethiopians considered their country to be the "true Israel," where God's mission, abandoned by the Hebrews because of their sins, rightfully continued. Unwilling to forsake the ways of his ancestors or to view his religious beliefs as misguided, but thankful to the Portuguese for the military assistance, Claudius received Oviedo politely when he arrived in 1557, but made only minor concessions (Conti Rossini, 1941; Caraman, 1985).

The situation deteriorated when Emperor Gelawdewos died 2 years later. The new Ethiopian emperor first imposed severe restrictions on the missionaries, and then exiled the Jesuits to an isolated place they named Fremona, in memory of Saint Frumentius (d. 383), the Syrian monk who had introduced Christianity to Ethiopia in the fourth century. Oviedo died at Fremona in 1577, followed by all of the remaining Jesuit missionaries before the century ended.

Although these Jesuits had been living in a part of Ethiopia where the Nile originates and the electric catfish is found, their surviving writings reveal nothing about this freshwater fish. This would change with the second Jesuit mission, decreed in 1588 but not started until 1603.

THE FIRST PUBLISHED WESTERN DESCRIPTION OF THE AFRICAN "TORPEDO"

Other missionaries were also attempting to introduce Catholicism in Africa while the first Jesuit mission was in Ethiopia. Among them were some Dominicans. Having been excluded from Ethiopia in favor of the Jesuits, the Dominicans concentrated their actions in other places on the vast continent. In this context, Portuguese Dominican priest João Dos Santos was successful in converting many natives in Mozambique to Christianity between 1586 and 1597. During this period, he observed a torporific river fish, which he then described in a book he was writing—the Dominicans preceding the Jesuits in this descriptive achievement.

Dos Santos had been born in 1564 in Evora, the site of one of the most important universities in Portugal. He was ordained a priest in 1584 in his native town, but spent most of his active life in missions abroad, dying in 1622 in Goa. The book where the numbing fish is first mentioned was given the title *Ethiopia oriental e varia historia de cousas notaveis do Oriente* ("Oriental Ethiopia and Various History of Remarkable Things about the Orient") and was published in 1609 (Fig. 7.1).

The passage about the presence of this fish, called *peixe tremedor* (tremble fish) by the Portuguese on the Sofala River in Mozambique, translates as follows:

> A certain species of fish lives in the freshwater rivers of this region. It is called a tremble fish by the Portuguese dwellers and Thinta by the Africans [Cafres], having the property that nobody can hold it in his hand when it is alive. If someone touches the fish, it causes him so much pain in his hand and entire arm and so disrupts the joints that he immediately lets the fish go free. When the fish dies, however, it is like any other fish, and it is very tasty and esteemed as food. The natives say that enchantments are made from the skin of this fish, and also that it is a strong medicine against the colic, when it is roasted and ground and taken in a glass of wine. The largest fish of this species has the dimension of a cubit. This fish has a skin like that of a smooth hound fish [dogfish: *cação* in the original] almost black, which is very rough and thick. (Trans. from Dos Santos, 1609, p. 39)

We can be reasonable sure that the fish Dos Santos described was *Malapterurus* and not the sea torpedo for three reasons. One is that it was found in a freshwater river far from the ocean, where torpedo rays are not found. A second is the far greater strength of its discharge when compared to the common saltwater torpedoes. And a third is what he wrote about fish's skin being rough and thick like that of a hound fish; the skin of a sea torpedo is smoother and considerably slimier.

Even though Dos Santos' book has a title that starts with the word Ethiopia, we should emphasize that Ethiopia was not among the countries he visited during his missionary activity. The use of the word in the book's title stems from the fact that the geographical term "Ethiopia" was used differently in Dos Santos' day. It could refer to a far larger part of Africa, or to India, or to the extensive lands between them. Herodotus of Halicarnassus, the fifth-century B.C. historian and traveler, for example, designated Ethiopia the "the land of people with sunburned faces," and India as "Oriental Ethiopia." During the Middle Ages, the Ethiopia in Africa was considered the most western of the three Indias.

THE SECOND JESUIT MISSION TO ETHIOPIA

In 1588, Pedro Paez (Paéz, Pêro Pais), a cultured monk born near Madrid but educated in Portugal, was asked to head the second Jesuit mission to Ethiopia. Because of a series of

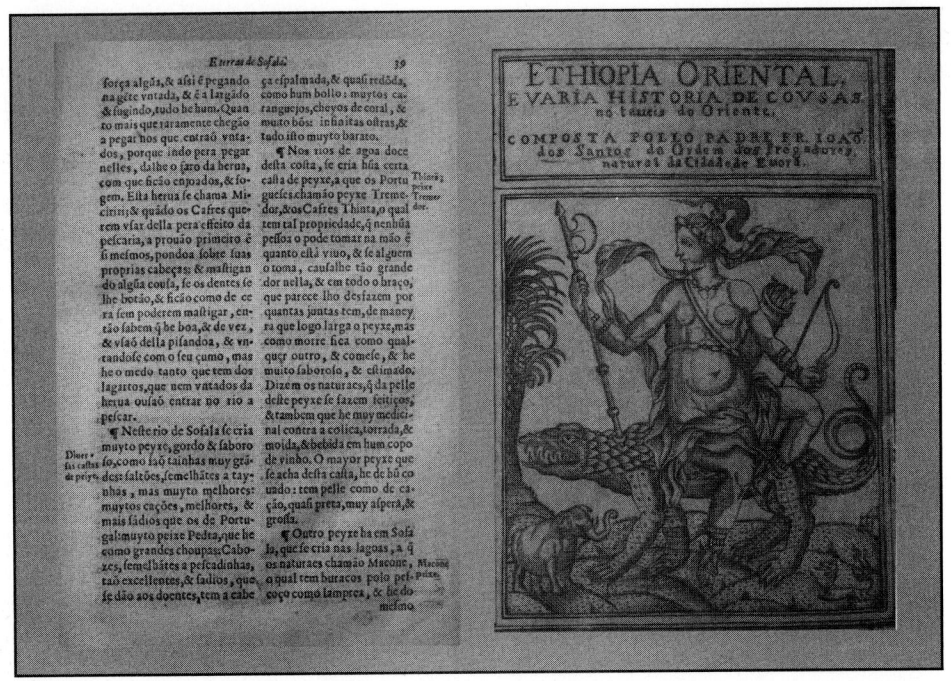

Figure 7.1. The title page of João Dos Santos' *Ethiopia Oriental*, published in 1609, and the page providing the first description of *Malapterurus*, the electric catfish, by a European. This fish was called *Thinta* by the natives and *peixe Tremedor* (tremble fish) by the Portuguese on the Sofala River (in today's Mozambique).

misadventures, including being held captive in Yemen for several years, he did not set foot in Ethiopia until the beginning of the seventeenth century, and only on his second attempt (Paez, vols. 2–3 in Beccari, 1903–17; for more on Paéz, see Camaran, 1985; Bishop, 1998; Reverte, 2001; Pennec, 2003; Alfonso Mola & Martínez Shaw, 2004). Paez arrived at the Jesuit camp in Fremona during the spring of 1603, assisted at the end of his trip by João Gabriel, the Ethiopian-born captain of the Portuguese soldiers, whose mother was Ethiopian and father Italian (see Conti Rossini, 1941).

Paez and his men worked diligently and intelligently to promote Catholicism. He drew on his exceptional linguistic abilities (e.g., Amharic, Ge'ez, Arabic, Persian, Hebrew) and understanding of Ethiopian society and history, and he made inroads with the Ethiopian leaders that resulted in important religious conversions (including of Emperor Susenyos) and a gift of land in northwest Ethiopia on the banks of Lake Tana, the source of the Blue Nile, where the emperor was building the new capital city, Gorgora. There, Paez eventually succeeded in erecting a monumental Catholic cathedral, the first church in Ethiopia built in stone following Western construction techniques.

While accompanying the emperor with his army in their military campaigns, Paez had the occasion to visit the sources of the Blue Nile, where *Malapterurus* can be found. He was, in fact, the first known Westerner to write a truly vivid account of the fish, this eventually appearing in his *Historia da Ethiopia* (see Paéz, 1905–06). This is perhaps the most important text of the early historiography of Ethiopia, with some of the first translations from Ge'ez and Amharic taken from fundamental writings of Ethiopian culture, such as the *Kebra Nagast* ("Glory of the Kings"), a text having an importance for Ethiopian society and religion comparable to the Bible.

Unfortunately, Paez's book (like many other Jesuit missionary writings in Ethiopia) was not published until the beginning of the twentieth century (see Beccari, 1903–17). Further, another Jesuit, Balthasar Telles, incorporated only some parts of Paez's manuscript in his *Historia geral da Ethiopia a alta ...* (Telles, 1660). In contrast to Paez (and to other Jesuits whose material was also included in his work), Telles never visited Ethiopia.

After Paez's death, Antonio Fernandes (or Fernández) led the mission, working with the emperor and receiving other Jesuits sent to Ethiopia. Two of the newcomers were Manoel de Almeida and Jerónimo Lobo, who would write accounts of Ethiopia. As with Paez, their narratives would not be published at the time of completion, but Telles would include them in his *História geral da Ethiópia* and they would be published at later times (Lobo, 1728; Beccari, 1903–17).

The religious conflict between the converted emperor and the Ethiopian traditionalists entered its most dramatic phase after Afonso Mendes, the new Catholic Patriarch, arrived in 1626. Appointed by Rome, Mendes lacked the skills and moderation of Paez. He tried to abolish all local religious and ritual practices that did not conform to Roman Catholicism, imposed his own church's religious celebrations, and strove for a complete conversion of the people. His actions and demands were so inflexible that they led to fratricidal wars, which in 1634, during the reign of Susenyos' son Fasilides, culminated in the persecution and expulsion of all Catholics. These tragic events ended all bonds that had been forged between Ethiopia and Europe in this era (Conti Rossini, 1940; Doresse, 1957; Pennec, 2003).

FERNANDES' LETTER

Antonio Fernandes, Paez's collaborator who acted as the chief of the Jesuit mission until the advent of Mendes, played an important role in the story of the African river fish, because he provided the material on which the first printed mention of the fish in the Jesuit writings from Ethiopia is based.

Following the steps of Paez, Fernandes, who was born in Lisbon, worked diligently with his superior on church and diplomatic matters. He learned the local language, won favor with the emperor, and traveled to various parts of the country. In 1610, just 1 year after Dos Santos' book was published, Fernandes composed a long annual letter on the state of the mission and on Ethiopia, which included the catfish. This letter, written in Portuguese, was sent from Dambia to the *Visitator Indiarum,* the Jesuit overseeing the province from his base in Goa.

This letter was not included in the collection of "annual letters" from India, which were published by Fernão Guerreiro beginning in 1603 in Lisbon (Guerreiro, 1611). Only a part of it remains in a handwritten transcription in the Jesuit archives in Rome, and it does not include the passage about the fish. The missing section did, however, appear in Latin in a book published 5 years after the letter had been written (Fig. 7.2). That book's author, the Jesuit Nicolao Godinho, stated he was providing a faithful translation of passages in Fernandes' letter.

In English translation, the "torpedo fish" section of the letter presented by Godinho and mistakenly dated 1620 (because of a misprint) reads as follows:

> The torpedo fish, well known for the strongly debated opinions and controversies of the philosophers, can also be found in these rivers and lakes. On the basis of the experiences of many people, it appears that this fish is endowed with such a contrivance that, if handled by somebody with his hand, it does not cause problems while it remains still. On the other hand, if the fish moves a little, it tortures the body of the person who holds him so much that the arteries, joints, nerves, connections of the body parts, in short all the components of his body, feel an intense affliction accompanied by stupor. As soon the fish is released from the hand, the pain and stupor fade away. Superstitious Ethiopians believe that this fish has the power of chasing demons from human bodies, as if these [evil] spirits were sensible to the same afflictions that torture corporeal bodies. People say (I have no direct experience of this) that, if a torpedo is placed among dead fish, and if it then moves, the fishes that have contact with it will be animated to a somewhat arcane motion, as if they were alive. Let those who investigate natural things try to ascertain the cause of this phenomenon, and to search for the force of the motion transmitted by the torpedo to the dead fish. These fish can be found in abundance in Nile River at the extreme limits of the Province

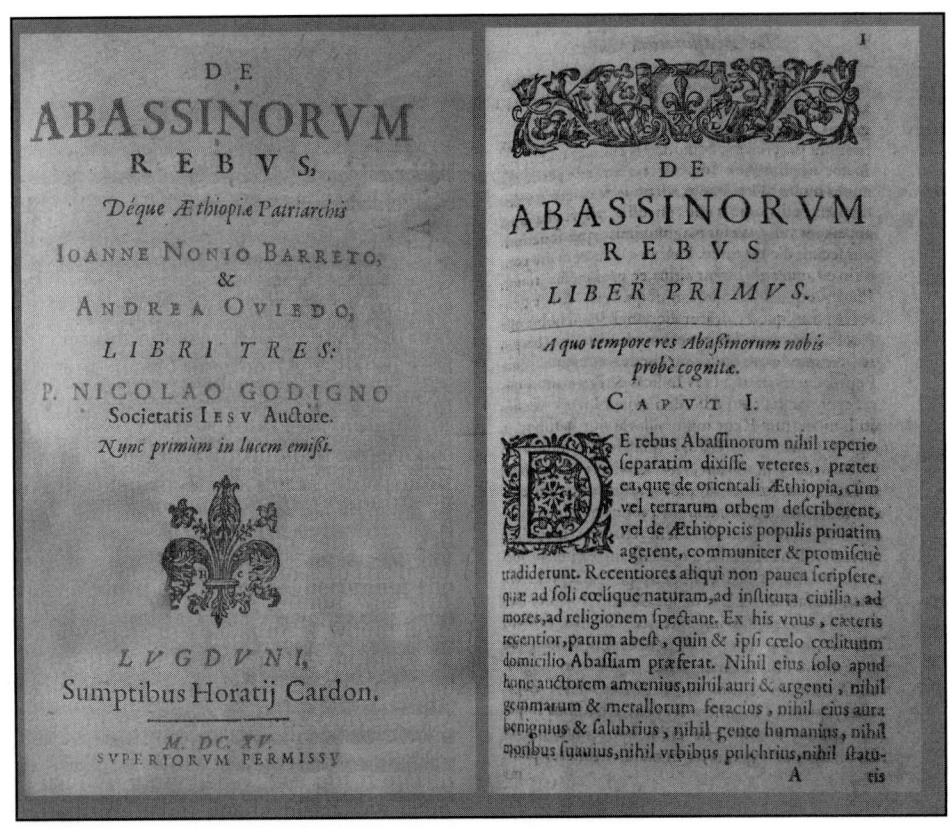

Figure 7.2. The title page and first page of Nicolao Godigno's *Abassinorum rebus* of 1615.

of Goyam, where there is a bottomless pool with perennial and marvellous sources of churning waters. Here the Nile River begins. (Godinho, 1615, Book I, Chapter XVII, p. 18) (Figs. 7.2 and 7.3)

Given the thousands of miles from the mouth of the Nile to where the fish was found, this fish could not be a ray that swam from its saltwater abode up the river. But again, there is no physical description of this African "torpedo," and some of what Fernandes wrote (e.g., "The torpedo fish, well known for the strongly debated opinions and controversies") could have left people envisioning the well-known saltwater ray. In brief, some readers might have not realized that Fernandes was describing a very different fish.

What Fernandes wrote about the locals using this fish for chasing demons is particularly interesting. It can be tied to the fact that his letter was largely concerned with Ethiopian superstitions and religious errors. As for its ability to induce motion in dead fishes (a possibility that the Jesuit considers with evident caution), we must take into account that an electric stimulus can induce motion in dead animals as long as their tissues remain excitable (cold-blooded animals such as fishes, amphibians, and reptiles can last for days or even weeks, provided the tissues are kept moist). Thus there likely is an element of truth in this seemingly incredible statement, even though it appears together with a statement about a superstitious attribute of the fish (the ability to chase demons). It is one of the frequent instances in which the "natural magic" intrinsic to genuine scientific observations seems to go beyond the black magic of wizards and sorcerers.

Fernandes tied the pain and numbness to movements of the fish, although he was more than willing to let others figure out the source and nature of the fish's remarkable powers. There was no broadly accepted explanation for the movements of any torpedo fish at this time. Nevertheless, he might have been thinking that the action is mechanical, because the fish moved when it released its effects, as a consequence of the tremor caused in its victim. Hence, the movement of the fish could have been perceived as the cause of the surprising effect (see Piccolino, 2005a).

Mechanical explanations for the torpedo's effects would soon emerge in Europe. They would be stimulated by mechanical conceptions of physical and physiological phenomena, prompted by Galileo Galilei and René Descartes with their new approaches to the sciences. In the closing decades of the seventeenth century and within this new philosophical framework, Italian experimental natural philosophers would propose a mechanical action as the most likely explanation for the torpedo's strange effects, a development we shall examine in detail in Chapter 9.

THE MONK URRETA AND PRESTER JOHN'S FABULOUS EMPIRE

The events leading to the publication of Godinho's book in 1615, where parts of Fernandes' letter can be found, are tightly connected to a Dominican publication that had just appeared. As previously mentioned with regard to Dos Santos, the Dominicans were also sending missionaries to Africa, although not to Ethiopia, and an intense rivalry existed between the Jesuits (mostly Portuguese in Ethiopia) and the Dominicans.

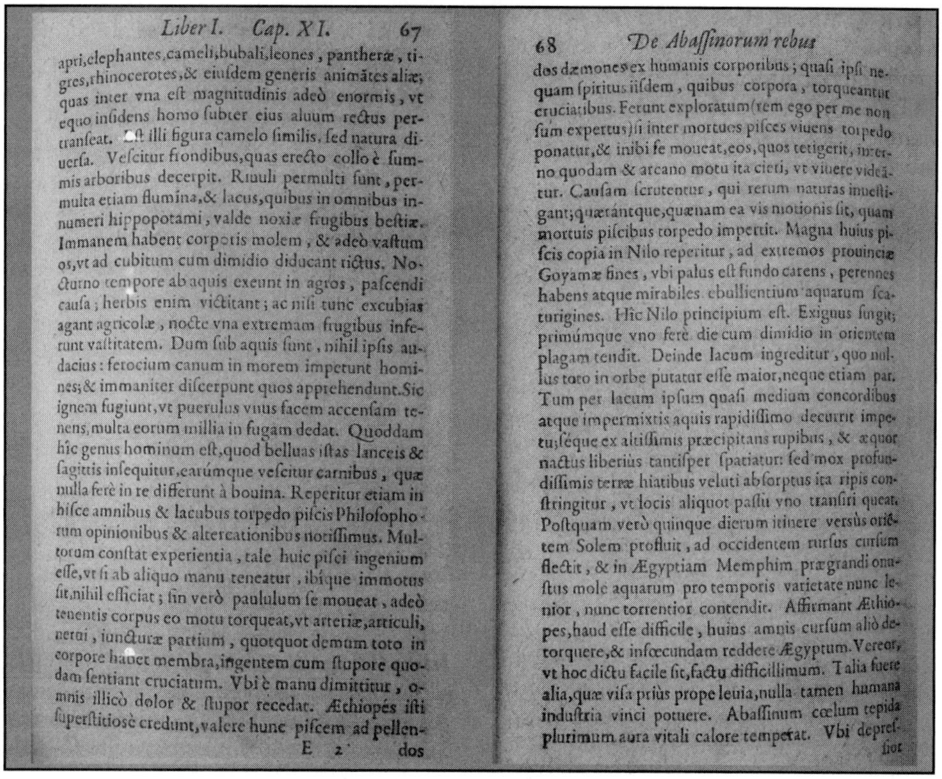

Liber I. Cap. XI. 67

apri,elephantes,cameli,bubali,leones, pantheræ, ti-
gres,rhinocerotes,& eiuſdem generis animãtes aliæ;
quas inter vna eſt magnitudinis adeò enormis, vt
equo inſidens homo ſubter eius aluum rectus per-
tranſeat. Eſt illi figura camelo ſimilis, ſed natura di-
uerſa. Veſcitur frondibus,quas erecto collo è ſum-
mis arboribus decerpit. Riuuli permulti ſunt, per-
multa etiam flumina,& lacus,quibus in omnibus in-
numeri hippopotami, valde noxiæ frugibus beſtiæ.
Immanem habent corporis molem, & adeò vaſtum
os,vt ad cubitum cum dimidio diducant rictus. No-
cturno tempore ab aquis exeunt in agros, paſcendi
cauſa; herbis enim victitant; ac niſi tunc excubias
agant agricolæ, nocte vna extremam frugibus infe-
runt vaſtitatem. Dum ſub aquis ſunt, nihil ipſis au-
dacius: ferocium canum in morem impetunt homi-
nes;& immaniter diſcerpunt quos apprehendunt.Sic
ignem fugiunt,vt puerulus vnus facem accenſam te-
nens,multa eorum millia in fugam dedat. Quoddam
hîc genus hominum eſt,quod belluas iſtas lanceis &
ſagittis inſequitur,earúmque veſcitur carnibus, quæ
nulla ferè in re differunt à bouina. Reperitur etiam in
hiſce amnibus & lacubus torpedo piſcis Philoſopho-
rum opinionibus & altercationibus notiſſimus. Mul-
torum conſtat experientia, tale huic piſci ingenium
eſſe,vt ſi ab aliquo manu teneatur, ibíque immotus
ſit,nihil efficiat; ſin verò paululum ſe moueat, adeò
tenentis corpus eo motu torqueat,vt arteriæ,articuli,
nerui, iuncturæ partium, quotquot demum toto in
corpore habet membra,ingentem cum ſtupore quodam ſentiant cruciatum. Vbi è manu dimittitur, o-
mnis illicò dolor & ſtupor recedat. Æthiopés iſti
ſuperſtitiosè credunt,valere hunc piſcem ad pellen-
E 2 dos

68 De Abaſſinorum rebus

dos dæmones ex humanis corporibus; quaſi ipſi ne-
quam ſpiritus iiſdem, quibus corpora, torqueantur
cruciatibus. Ferunt exploratum(rem ego per me non
ſum expertus)ſi inter mortuos piſces viuens torpedo
ponatur,& inibi ſe moueat,eos,quos tetigerit, interno quodam & arcano motu ita cieri, vt viuere videa-
tur. Cauſam ſcrutentur, qui rerum naturas inueſti-
gant;quærántque,quænam ea vis motionis ſit, quam
mortuis piſcibus torpedo impertit. Magna huius piſcis copia in Nilo reperitur, ad extremos prouinciæ
Goyamæ fines, vbi palus eſt fundo carens, perennes
habens atque mirabiles ebullientium aquarum ſca-
turigines. Hîc Nilo principium eſt. Exiguus ſurgit;
primúmque vno ferè die cum dimidio in orientem
plagam tendit. Deinde lacum ingreditur, quo nul-
lus toto in orbe putatur eſſe maior,neque etiam par.
Tum per lacum ipſum quaſi medium concordibus
atque impermixtis aquis rapidiſſimo decurrit impe-
tu;ſéque ex altiſſimis præcipitans rupibus, & æquor
nactus liberiùs tantiſper ſpatiatur: ſed mox profun-
diſſimis terræ hiatibus veluti abſorptus ita ripis con-
ſtringitur, vt locis aliquot paſſu vno tranſiri queat.
Poſtquam verò quinque dierum itinere versùs orie-
tem Solem profluit, ad occidentem rurſus curſum
flectit, & in Ægyptiam Memphim prægrandi onu-
ſtus mole aquarum pro temporis varietate nunc le-
nior, nunc torrentior contendit. Affirmant Æthio-
pes,haud eſſe difficile, huius amnis curſum aliò de-
torquere,& infœcundam reddere Ægyptum. Vereor,
vt hoc dictu facile ſit,factu difficillimum. Talia fuere
alia,quæ viſa priùs prope leuia,nulla tamen humana
induſtria vinci potuere. Abaſſinum cœlum tepida
plurimum aura vitali calore temperat. Vbi depreſ-
ſior

Figure 7.3. The pages of Nicolao Godigno's *Abassinorum rebus* showing his Latin translation of the part of Antonio Fernandes' letter of 1610 describing Ethiopia's freshwater "torpedo" (i.e., its electric catfish).

During the opening decades of the 1600s, the Jesuits, although the newer religious order (founded around 1550), were making significant progress in Persia, Ceylon, India, China, Tibet, Japan, and other targeted locations, including Ethiopia. Their Ethiopian successes stimulated Luis de Urreta, a Spanish Dominican monk, to write a polemic work that was published in two volumes, in 1610 and 1611.

Urreta's 1610 text was titled *Historia Eclesiastica y Politica, Natural y Moral, de los Grandes y Remotos Reynos de la Etiopia* (literally "History Ecclesiastical and Political, Natural and Moral, of the Great and Remote Kingdoms of Ethiopia") (Fig. 7.4). His objective was to glorify his own order's accomplishments in Ethiopia while diminishing those of the Jesuits. It was followed in 1611 by a second volume dealing specifically with the Dominican presence in Ethiopia and the history of Ethiopian saints.

Urreta (1610, 1611) made the case for an ancient Dominican presence in the country so as to bolster his contention that the Dominicans should be regarded as the rightful missionaries to Ethiopia. Moreover, to undermine the Jesuits' claim that they had introduced Catholicism in the country, he presented the Ethiopian religion as an ideal form of Christianity, very much in the spirit of the Apostles and the Holy Gospel, and thus not in need of sweeping changes.

Urreta was more at ease writing about stories and legends involving the saints than about verifiable historical facts, and had a great proclivity for verbose discourses. In the spirit of the Catholic Counter-Reformation, he aimed at stirring religious emotions by creating a fictitious, idealized representation of Ethiopia, drawing heavily on legends from the past without much regard for historical plausibility. Specifically, he adapted a legend that had dominated Western culture during the Middle Ages, one with diffusion and impact comparable to the legends of Parsifal, the Knights of Round Table, and Roland. This was the legend of Prester John. As with other medieval myths, this legend grew over time, had many variations, and had great historical and literary impact (Zarncke, 1879–83; Doresse, 1957; Slessarev, 1959; Wagner, 2000).

The term "Prester John," in its Latin form *Presbiter Iohannes*, first appeared in a chronicle written by Otto von Freising, an important German bishop and historian of the first half of the twelfth century. He recorded a story he had heard in 1145 from Hugh, a bishop of the French-Syrian town of Jabala (modern Gébal, near Laodicea) (Hofmeister, 1912). Hugh told him that Prester John was a rich descendent of the Kings Magi and a venerated Christian leader then living in far-off lands (beyond Persia and Armenia). After defeating the Persians and Medians in Ecbatana (Hameda, Iran), Prester John aspired to help the Christians in Jerusalem. But unable to cross the Tigris River, he was forced to return home.

More soon began to be heard about this Prester John. In 1177, for example, a papal physician coming back from the Holy Land reported about a Christian sovereign in India (most likely the African India) known as Prester John. Following this report, Pope Alexander III even wrote a letter in Latin addressed to "King John . . . the illustrious and mighty king of the Indias, the most holy of priests," and asked the physician to deliver it. Allusions to a powerful Christian king of the Far East lands also appeared in various old chronicles.

But more than anything, what transformed some vague allusions into one of the most important legends in this era

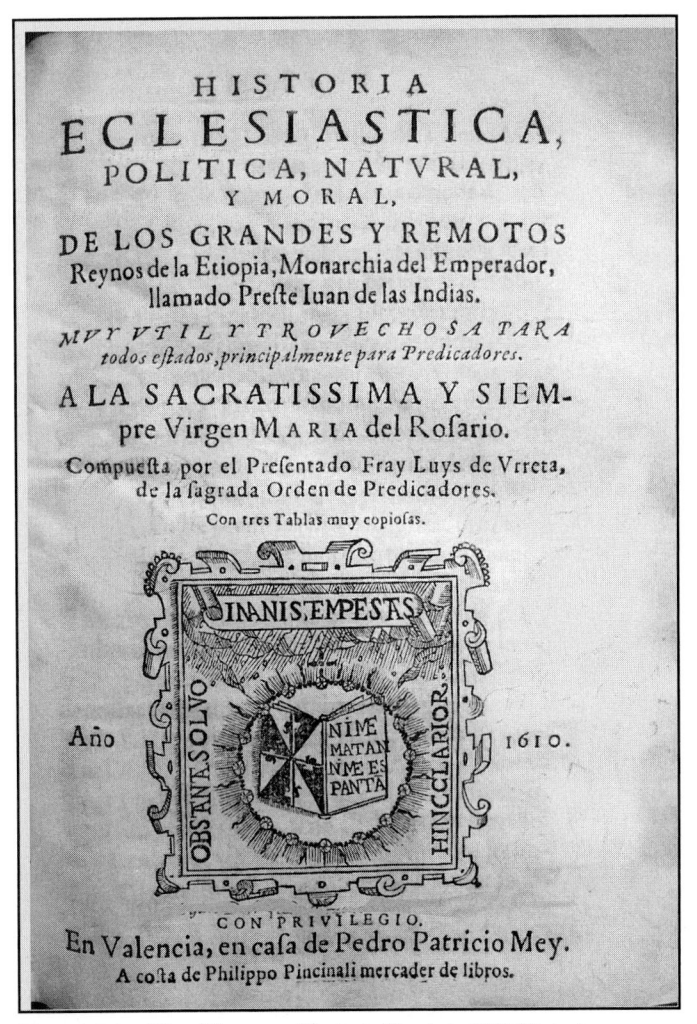

Figure 7.4. The title page of Luis de Urreta's 1610 *Historia eclesiastica*. The author exaggerates Dominican achievements while playing down those of the Jesuits in Ethiopia in this fanciful volume, which includes the legend of Prester John's lost kingdom.

was the arrival, starting in about 1165, of letters signed by a "Prester John of the Indias" and addressed to the Pope and Europe's and Byzantium's most Christian of sovereigns. In these letters, Prester John presented himself as the lord of the largest empire on earth (extending over the three Indias) and a true defender of the faith. He had 72 kings paying tribute to him and wonderful treasures, including magical stones, a water source that could keep one young and healthy, and palaces constructed from gold and precious stones. There were also exotic animals and monsters in his miraculous kingdom, all of which obeyed his every command, and people who lived without envy and subsisted on manna from heaven. With his powerful armies Prester John fought incessantly against the enemies of the Christian religion. Yet despite of his power and riches, he remained a pious and humble man, preferring the simple title "Priest" to any other.

Word of these letters spread around Europe, and they were translated into many languages and edited in various ways. Although they were apocryphal (most likely written by a German bishop follower of Frederick Barbarossa), they stimulated tremendous interest in the possibility that the legendary empire might exist. And with that thought, the search for Prester John began in earnest.

Signs of his presence were sought in the reports of the first European voyagers to the Orient, including those of the thirteenth-century Franciscan monks Giovanni da Pian del Carpine and William of Rubruck (sent by Pope Innocent IV and Saint Louis IX of France, respectively). They were also looked for in the writings of Marco Polo and those of many others thereafter. Some of these early travelers reported things they heard about various Prester Johns, but none encountered hard evidence of his mythical empire in the Near or the Far East (Olschki, 1937; Zaganelli, 1992).

Still, people never stopped believing in these legends, and the search gradually shifted from Asia to Africa, and especially to Ethiopia. This new location for the lost kingdom was bolstered by reports from pilgrims traveling to the Holy Land, by tales heard among sub-Saharan Africans, and by Western clergymen in contact with African priests (Conti Rossini, 1928; Cerulli, 1943–47; Lefevre, 1944, 1945, 1947).

With its new Ethiopian location, the lingering Prester John legend fit beautifully into Urreta's project, which portrayed not poverty but a wonderful Christian kingdom in this part of Africa. Moreover, Juan de Baltasar (Baltazar, Balthasar, etc.), a self-styled Ethiopian monk who was now lodging at Urreta's convent in Valencia, Spain, supported the Spanish Dominican in his scheme (Fig. 7.5). In the *Prologo* to his *Historia eclesiastica* of 1610, Urreta described Baltasar as "Commendador of the Order of St Antony Abbott, and member of the guard of the King of Ethiopia, called Prester John of the Indias." Baltasar claimed to be of noble blood and said he was in Europe on a diplomatic mission to convey important communications from Prester John to the Pope, the King of Portugal, and other important personages (Lefevre, 1944, 1945, 1947). In his book, Urreta stated repeatedly that the wonders he was describing were supported by Baltasar as a personal witness, or by authoritative documents provided by him.

It is hard to tell whether Urreta knew that Baltasar was just making up fanciful stories, which man was more gullible, or precisely what was going through Urreta's mind. But two things are certain. One is that Baltasar was not a fictitious person and the other is that Urreta's books, with their detailed and vivid descriptions, took the Jesuits associated with Ethiopia and Europe by storm.

Urreta's descriptions of the wonders of Mount Amarà are exemplary. This was presented as the site of a magnificent library housing books given by Solomon to the Queen of Sheba and other magnificent treasures, where the heirs of the imperial family were educated. There were also splendid palaces and beautiful churches on the mountain, and a myriad of living wonders with extraordinary qualities, such as a fruit that could ensure exceptionally long life.

As can be imagined, the idealized Ethiopia described by Urreta was not the Ethiopia known to the Portuguese Jesuit missionaries, who were actually living there. They knew nothing about this fictitious Mount Amarà, or many of the other things Urreta described. Nor was this the Ethiopia of their superiors, the men who received their letters and reports, such as the one sent by Fernandes to Goa. Consequently, the Jesuit leaders felt compelled to react swiftly.

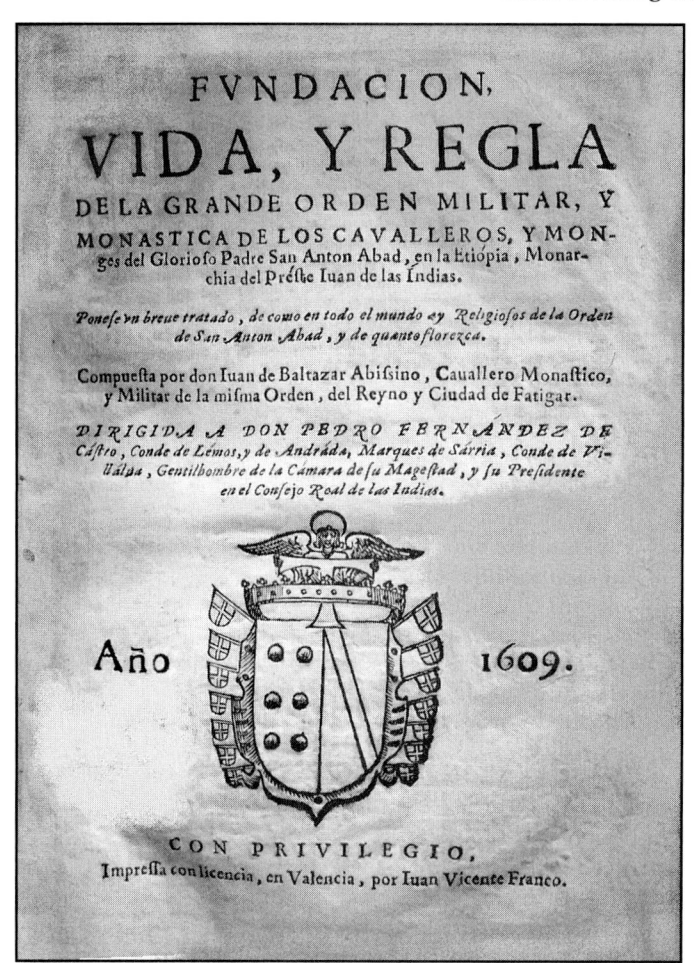

Figure 7.5. The cover page of a text by Juan de Baltasar dated 1609. Baltasar was a self-styled Ethiopian monk claiming to be of noble blood. He contended he had communications from his monarch for the Pope and the King of Portugal, and he provided many fables and stories for Urreta.

GODINHO'S BOOK

Nicolao Godinho's book from 1615 served as a somewhat official refutation by the Jesuits of Urreta's claims. Godinho was a Portuguese Jesuit priest living in Rome, and he served the Jesuits as the reviser of works to be published in Portuguese. With this assignment, he had access to documents from around the world and was in a location that allowed him to meet people who could provide additional information for his rebuttal.

Godinho's declared intention was to refute each of Urreta's fantasies one by one. Yet he never mentioned the Dominican by name in his book, which was written in Latin and bore the title *De Abassinorum rebus, déque Aethiopiae Patriarchis Ioanne Nonio Barrreto & Andrea Oviedo,* meaning "On Abyssinian Matters, and on the Patriarchs Johannes Nuñez Barreto and Andrés Oviedo." It was in this book, printed 4 years after the publication of Urreta's first volume on Ethiopia, that Godinho included the passage dealing with the "African torpedo," which was taken from Fernandes' letter of 1610, together with other material from the same letter dealing with real plants, animals, and fishes at the headwaters of the Blue Nile (Godinho, 1615, Book I, Chapter XVII). This book, far more than Dos Santos' tome, really introduced Europeans to the presence of a torpedo deep in Africa, far from the salty waters lapping the mysterious continent's coasts.

The main reason for this impact was the controversy between Urreta and Jesuits, which attracted the attention of the European *literati*. This factor combined with the growing interest within both political and religious circles (on both the Catholic and Protestant sides) in Ethiopia and its unique form of Christianity.

PAEZ'S UNPUBLISHED REACTION

Padre Pedro Paez's *Historia da Ethiopia* was specifically written at the request of his superiors in Goa to refute Urreta, although as already stated, it long remained in manuscript form. Paez, who headed the second Jesuit mission, saw the land through perceptive eyes, deeply understood the historical situation of Ethiopia, read Ethiopian sacred books and chronicles, and interrogated commoners and aristocratic Ethiopians (including the emperor). And without question, he found Urreta's writings "apocryphal" and "fable-like."

Paez's refutation was a much more accurate account of a misunderstood country, its "problematic" church, and its leadership. To cite but one example, he wrote that in contriving his ideal Mount Amarà, Urreta was actually taking liberties with the term *Amhara*, which designated a vast and historically important region of Ethiopia. He used it to designate a mountainous stronghold, in reality the Amba Guixên, which he then populated with plants, animals, and structures from his fertile imagination. On these rugged hills, Paez related, a visitor can find only mud buildings and churches with straw roofs, just a single bitter fruit among the few bits of greenery, and very few animals (Paez in Beccari, 1903–17, vol. II, pp. 121–131). He also explained that Urreta had greatly misinterpreted the theological and ritual aspects of the Ethiopian religion, pointing out that he could find no evidence of convents filled with Dominican monks in the country, and much, much more.

Importantly for the electric fish story, Paez included a passage about a river fish that could only be *Malapterurus* in his manuscript. This passage translates as follows:

> There is a great quantity of many kinds of fish here, and fat, because the fish find good food here. And among them there is one we call torpedo in Latin, and that the natives call *Adenguêz*, which means "fright." This is because, as they say, if one seizes it with his hand and the fish moves, it causes "fright," because it seems as if all of his bones would come apart. This happened to some Portuguese who told me about it, and notably to João Gabriel, who, while fishing with some fellows, caught a fish of more than a span of length, without scales, which was rather similar to a smooth hound fish [*caçam*], and motionless when he brought it in. He took the fish in his hand to release the hook, but as soon as the fish moved he quickly let it go, because his bones up to his teeth started shaking and seemed to be moving away from his body. Were he not seated he probably would have fallen down. He came back to himself soon after, and realized which kind of fish it was.

His passage continues:

> And in order to play a joke on one of his attendants, he asked him to remove the fish from the hook. As soon the attendant took it in his hand, the fish moved and he immediately fell over, without knowing what had happened to him. When he got up, he said: what could I possibly have done upon you, sir, for you to frighten me in this way? The captain and many of the others of the party laughed out loud, seeing how much the attendant was astonished, not knowing what had happened to him. They waited for the fish to die before taking it off the hook; the captain told me that he regarded it as certain that when the fish does not move it does not produce this effect, because he felt nothing until the fish moved. He added that another Portuguese fisherman caught one of these fish a cubit in length. (Paez, in Beccari, 1903–17, vol. I, pp. 285–286; vol. II, p. 260)

Thus, Paez repeats some of what had been related by Fernandes in 1610 and published by Godinho in 1615, such as the fish inducing its frightening effects only when it moved. But Paez also provides more details, one being that it had been touched and immediately dropped by none other than João Gabriel, the captain of the Portuguese troops in Ethiopia—a man he considered a reliable witness.

With regard to the similarity between the African fish and the hound fish as mentioned by Dos Santos, Paez makes it clear, albeit in an implicit way, that this similarity also refers to the shape of the fish, and not simply to the characteristic of its skin. A morphological similarity with an elongated fish dispels any possibility that the African river fish could be a flat torpedo ray. Still, there is no notation about the most distinguishing feature of a catfish, the presence of facial barbels.

Although Paez's assembled material remained in manuscript form for centuries, it was copied (at least two copies have survived) and circulated among a few specialists. It was of particular interest to scholars interested in the sources of the Blue Nile (e.g., Athanasius Kircher, who included Paez's mention of the Nile's sources in his *Oedipus Aegyptiacus*, published in 1652–54, and James Bruce, 1790).

ALMEIDA AND TELLES

The *Historia geral da Ethiopia a alta . . .*, a book published in 1660 by Balthasar Telles, contains another passage describing this fish. Telles was also a Jesuit and his book elaborated on previous works, the most important being an unpublished text by Manoel de Almeida, who collaborated with Paez and drew heavily on Paez's manuscript (Fig. 7.6). The fish is described among the animals living in the Tacazee River, which originates about 40 miles south of Addis Ababa and flows into Lake Zeway. After reporting the presence of crocodiles and hippopotamus (indicated as "marine horses"), based on Almeida's description, Telles writes (in translation):

> There is also a lot of fish in this river. Father Manoel d'Almeyda testifies that he was showed here, in a trough, the fish that, in consequence of its effect is called in Latin Torpedo; and that seizing it by his hand, he received soon

Figure 7.6. The frontispiece of Baltasar Telles' *Historia geral da Ethiópia a Alta*, a book published in 1660 that elaborated on earlier works, including Manoel de Almeida's important but unpublished text in which the river torpedo is described.

> in the same hand, and in the forearm, a remarkable shock, to such a point that in great hurry he released it, with no wish of continuing anymore such a troublesome experience. (Telles, 1660, p. 21)

Telles' text is a fairly faithful translation of the original passage of Almeida's manuscript (finally published at the beginning of the twentieth century by Beccari, using the manuscript conserved at the British Museum). In 1954, Beckingham and Huntingford published an English edition of Almeida's work based on a different manuscript. (now at the School of Oriental and African Studies in London). In this edition, the translation is as follows:

> There are many fish in this river. The first time I crossed it the men who accompanied us gave me one of the kind the Latins call torpedo. Many people, and I too, tested its power in a trough of water. I squeezed it in my hand under the water. I felt my hand so weak and powerless that I let it go very hurriedly. (Almeida c. 1645, translated and edited by Beckingham & Huntingford, 1954, p. 31)

MORE THAN JUST SOME PORTUGUESE REPORTS: JOBSON'S GAMBIA

During the 1600s, descriptions and first-hand accounts of the torporific African river fish came largely from the Portuguese. But they were not the only ones to notice this fish during this century. The fish also caught the attention of Richard Jobson, a writer about whom next to nothing is known. The few bits of information that we have about Jobson suggest he might have been an Irishman (he was clearly very familiar with Ireland) who sailed from Dartmouth, England, to the River Gambia in 1620; that he was a Protestant; that he believed that King Solomon's land of Ophir (noted for its wealth in the Old Testament) was in the Gambian region; that he despised the Afro-Portuguese traders on the river but was not about to involve himself in the slave trade; and that he returned to England in 1621 (see Gamble & Hair, 1999).

Jobson's role on ship is also shrouded in mystery. He described himself in one petition as a "poore seaman," was addressed as "Master Richard Jobson" and "Capt Jobson" by another writer (namely Purchas), and signed the Stationers' Register as "Captaine Jobson." Whether "Captaine" was a military title, a naval title, or a reflection of the fact that he did lead a small group of men farther up river, is unclear. But without question, he was by no means master and commander when his ship left England, and most likely not a merchant, given his disdain for the merchants aboard ship. More likely he was a skilled subordinate (possibly in surveying) or a schooled gentleman associated in some way with the Guinea Company, which financed the trip to Africa.

The 700-mile River Gambia begins in modern-day Guinea and runs through Gambia to the Atlantic Ocean on the West African coast. The Portuguese were the first Europeans to make their way to the River Gambia, this being in the mid-1400s. They quickly penetrated several hundred miles upriver, thinking the interior rich in gold and eager to be involved with the lucrative trade. From that time well into the seventeenth century, they protected their territory and monopolized trade in the region, killing or driving off unwanted intruders.

English merchants began to sail up the river in 1618, with the first two voyages proving disastrous. Jobson's voyage up the River Gambia was the third in the series and it involved two ships. This time the survival statistics were better, although the trip into Africa's interior still resulted in many deaths (about a third of the 80 to 100 men died, mostly from disease) and yielded only a tiny amount of gold.

Jobson's book on Gambia appeared in 1623, 2 years after he returned to England. Chronologically, *The Golden Trade: Or, A Discovery of River Gambia* stands out as the first book to describe the African catfish in English (Fig. 7.7).

Buckor Sano, an African trader in the interior village of Tinda, had told Jobson that deeper in the interior, some 2 months' travel from the West African coast, he would find Timbuktu (Mali). There, he would discover even the houses covered in gold—a story quite comparable to the one circulated by Urreta about Prester John's lost kingdom in Ethiopia.

Even though Jobson never found his rich city with its gold houses, or anything else that would have made him fabulously wealthy, he did make a name for himself by writing about his adventures on the river. And in these writings he provided a wonderful description of one of its river fishes, which the crew had netted and was called *tingoo* by the native Mandinka (unfortunately it would be called an "electric eel" in the introduction to a recent edition of his book, even though it is correctly identified as

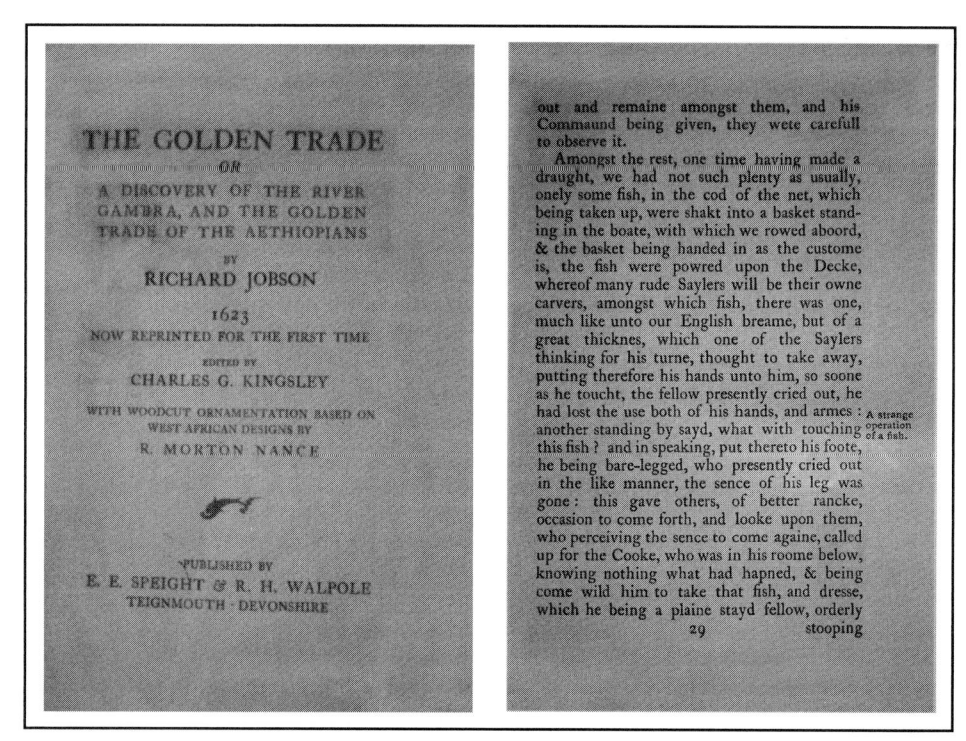

Figure 7.7. The cover page of a reprint of Richard Jobson's 1623 text, *The Golden Trade: Or, A Discovery of River Gambia*, with the first page of the section dealing with how a live torporific catfish could cause numbness.

Malopterurus [sic] in a footnote to the text; see Gamble & Hair, 1999, pp. 33, 94).

The episode took place at Kasang and the description, although challenging to follow because of the poor spelling and grammar (even for this era), reads as follows:

> Amongst the rest, one time having made a draught, we had not plenty as usually, onely some fish, in the cod [bag] of the net, which being taken up, were shackt into a basket standing in the boate, with which we rowed aboord, & the basket being handed in as the custome is, the fish were powred upon the Decke, whereof many rude Saylers will be their owne carvers amongts which fish, there was one, much like our English breame, but of a great thickness, which one of the Saylers thinking for his turne, thought to take away, putting therefore his hand unto him,

Without a period anywhere to be found, the same run-on sentence continues with a description of how this fish made the poor sailor and two others on the boat wish they had never touched it:

> so soone as he toucht, the fellow presently cried out, he had lost use both of his hands, and armes: another standing by sayd, what with touching this fish? and in speaking, put thereto his foote, he being bare-legged, who presently cried out in the like manner, the sense of his leg was gone: this gave others, of better rancke, occasion to come forth, and looke upon them, who perceiving the sense to come againe, called up for the Cooke, who was in the roome below, knowing nothing what had hapned, & being come wild him to take that fish, and dresse, which he being a plaine stayd fellow, orderly stooping to take up, as his hands were on him, suncke presently upon his hindere parts, and in the like manner, made grievous mone: he felt not his hands, which bred a wonderfull admiration amongst us:

The fish was well known to a native, who could only laugh at what he saw, as we read in the conclusion of this rather remarkable sentence:

> from the shore at the same time was comming a Canoe aboord us, in which a Blacke man called Sandie, who in regard he had some small knowledge of the Portingall tongue, had great recourse amongst us, we brought him to the fish, and shewed it unto him, upon sight whereof, he fell into laughter, and told us, was a fish they much feared in the water, for what he touched hee num'd, his nature being to stroke himselfe upon another fish, who presently he likewise num'd, and then pray'd upon him, but bid us cut of his head: and being dead, his vertue was gone, and he very good to eate. (Jobson, 1623, pp. 29–30)

Jobson's fish in the River Gambia was not an isolated rarity. Another description of it would follow in the next century (Moore, 1738; see Chapter 15).

In addition, there was a short notation in the final quartile of the century on the presence of "the *Reade* [i.e., the *ra'ad*], or the Cramp-Fish" in Egypt by a German, Johann Michael Wansleben. He was the author of *The Present State of Egypt,* a book published in English in 1678 (and in French under the name Vansleb, sometimes also catalogued as Vansleben; Wansleben, 1678; Vansleb, 1677). Wansleben had been in Egypt twice and seemingly also knew about the catfish first-hand.

BRINGING MALAPTERURUS TO A WIDER AUDIENCE

An important part of the story of the African torpedo deals with reports written by people who had first-hand experience with this fish or were in the area and spoke with those who had experienced it. Some of these individuals were Dos Santos, Fernandes, Paez, Almeida, and Jobson. But compilations of reports that were edited and packaged together were also important means of spreading the word. These increasingly popular collections of regional, continental, or world travels did not appeal just to the clergy and those interested in religion, although they were part of the target audience. Natural philosophers, physicians, government officials, and inquisitive outsiders avidly read them. Hence, compilations and commentaries also helped to disseminate information about nature's living wonders: the plants, animals, and fishes in distant places. If the subject matter were unusual enough to stir the imagination, even better.

The books of Samuel Purchas stand out in this regard. Purchas was born in 1577 and graduated from St. John's College, Cambridge, with an M.A. degree in 1600 (for more on Purchas' life, see the "Publisher's Note" in Vol. 1 of the 1905–7 reprint under Purchas, 1625). He then served as vicar of Eastwood from 1604 to 1614. Purchas' vicarage was close to Leigh on the Thames, a flourishing seaport. Its location allowed him to meet sailors as they disembarked from their ships, and others waiting to sail off again to foreign lands. Engrossed by what he heard from the seafarers, Purchas began to keep records of their narratives, and he never lost his desire for a good travel story.

In 1614, Purchas became chaplain to the Archbishop of Canterbury and rector of St. Martin's, Ludgate, and he received an Oxford University divinity degree the next year. But his real fame came from the travel books he was now starting to write. The first of these books was *Purchas His Pilgrimage*, a survey of the people and religions of the world (Purchas, 1613). It led to *Purchas his Pilgrim*, dealing with the degeneration and regeneration of man (Purchas, 1619). Purchas' most famous undertaking, however, was his *Hakluytus Posthumus, or Purchas his Pilgrimes*, which has the subtitle, *Contayning a History of the World in Sea Voyages and Lande Travells by Englishmen and Others*. This four-volume work appeared in 1625 and was based on the unpublished papers of Richard Hakluyt, an English writer and collector of travel manuscripts, whom he had been assisting up until his employer's death in 1616 (Purchas would die 10 years later).

In an introductory section to this book called "To the Reader," Purchas relates that he never traveled even 200 miles away from Thaxted, the Essex town in which he was born. As for the text itself, it is divided into two parts: the New World (the Far East, Australia, the Americas) and the Elder or Old World (Europe, Africa, parts of Asia). Each comprises 10 books, written to provide entertainment "at no great charge."

In the Old World volumes dealing with Africa, one can find long sections of Friar Ioão dos Sanctos' [sic] book, including his description of the fishes caught in Mozambique, with a side notation reading "Historie of Fishes. The Torpedo" (Purchas, 1625, Part II, p. 1546). Purchas did not know that author, but he seemed to know Jobson, and one can find Jobson's story about the cook who was tricked into touching this fish in the same volume (p. 1568; see Gamble & Hair, 1999, for information about Purchas–Jobson interactions). The side notation for this passage is "Torpedo, Tremedor, or Thinta."

The word "torpedo" also appears in a section called "Description of the Countries, and Severall Regions, Religions, and Abassine Opinions." It is in this section that Purchas presents a translation of Fernandes' letter from 1610, as drawn from Godinho 5 years later. It appears in abridged form and we include it here to show how Purchas edited it and the English he used in 1625:

> In these Rivers and Lakes is also found the Torpedo, which if any man hold in his hand, if it stirre not, it doth produce no effect; but if it move itself never so little, it so tormenteth the body of him which holds it, that his Arteries, Joints, Sinewes, all his Members feele exceeding the paine with a certain numbnesse: and as soone as it is let go out of the hand, all that paine and numbnesse is also gone. The Superstitious *Abassines* believe that it is good to expell Devils out of the humane bodies, as if it did torment Spirits no lesse than men. They say that if one of these alive bee laid amongst dead Fishes, if it there stirre itself, it makes those which it toucheth to stirre as if they were alive. There is great store of this kind in the *Nilus* [Nile], in the furthest parts of Goyama, where there is a Meere or Fenne without bottome, welling and admirably boyling forth waters continually, whence the *Nilus* springeth. (Purchas, 1625, Vol. VI, pp. 1182–1183)

The Fernandes' letter is also reported with slightly different wording in the fourth edition of *Purchas his Pilgrimage, or Relations of the World and the Religions* (Purchas, 1626) (Fig. 7.8). This work presents more about Urreta's fictitious account of Ethiopia and various references to his informant Baltasar, both of whom are strongly criticized. In fact, Purchas says that he was tempted to exclude Urreta's narration because it is so unreliable.

Purchas was not the only compiler to help spread the word about the fish. Another important contributor was Hiob Ludolf, a German scholar who is considered the father of Ethiopian studies. Ludolf had studied philology in Erfurt, Germany, and in the Dutch town of Leiden. He also traveled and interacted with others to increase his linguistic skills, becoming familiar with some 25 languages, including Ethiopian. He learned this language from Abba Gorgoreyos, an Ethiopian monk from Amhara, whom he had met in Rome while searching for some documents at the request of the Swedish court.

In contrast to Purchas, whose books date from the early-seventeenth century, Hiob Ludolf's *Historia Æthiopica sive brevis & succincta descriptio regni Habessinorum, quod vulgò malè Presbyteri Johannis vocatur* (literally "Ethiopian History or a Short and Succinct Description of the Kingdom of Abyssinians, Which is Incorrectly Said to be Prester John's") is more scholarly and came out later in the century in Latin and other languages (Ludolf, 1681, 1682) (Fig. 7.9). It too became well known and it underwent numerous

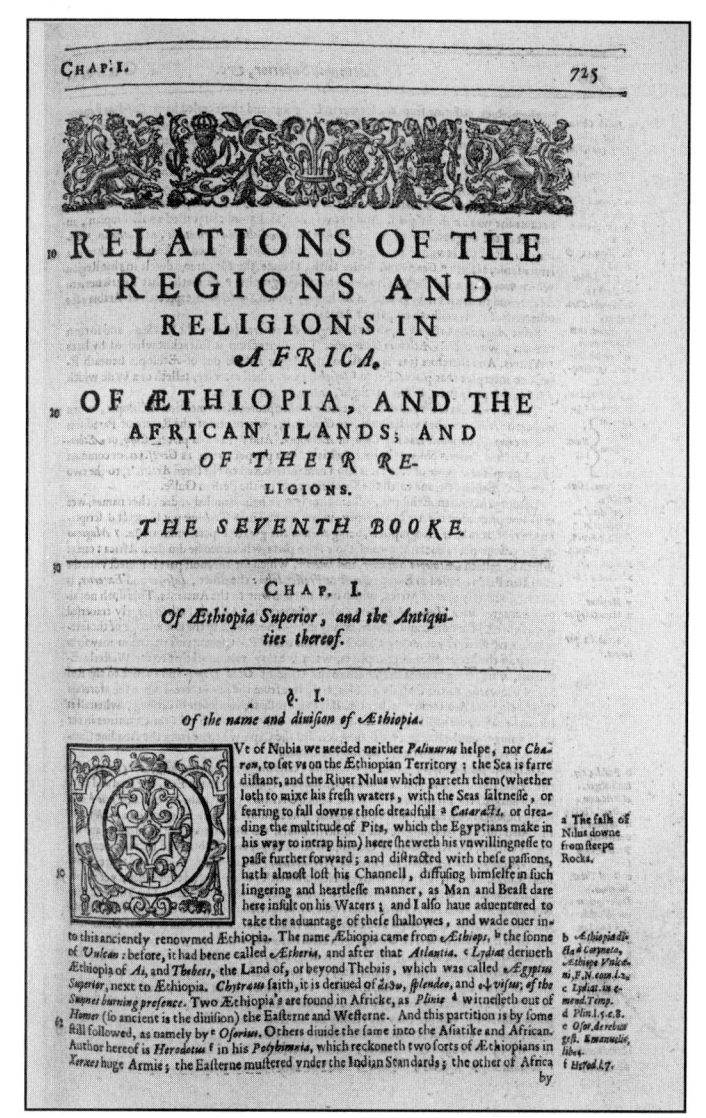

Figure 7.8. The first page of a chapter on Ethiopia from the 1626 edition of *Purchas his Pilgrimage, or Relations of the World and the Religions*. This work presents Fernandes' letter dealing with the strange river fish and Urreta's fictitious account of Ethiopia (which he criticizes). It also mentions Dos Santos' account of the African fish.

editions in various languages. Further, Ludolf later published an important companion piece in Latin, his *Commentarius*, which appeared in 1691.

As was true of Purchas, Ludolf never visited Ethiopia. He just read everything he could find about the region, writing his *Historia Æthiopica* and *Commentarius* in the German city of Frankfurt-am-Main, where he had "retired" in 1678.

Ludolf's description of the African torpedo in his 1681 volume is based mainly on Fernandes' letter in Godinho's book and on Almeida's account as published by Telles. He also mentions Dutch writer Olfert Dapper, a contemporary who wrote about Africa and called this fish a *Drillvisch* and a *Zitterfisch*. Dapper (1670) had been the author of a collection of travel books, first published in Dutch. His account of Ethiopia was largely based on Telles.

In the 1682 English edition of the *Historia Aethiopica*, it is written that the "*Torpedo* is very remarkable, frequent in Africa," and that "if it be touched with the hand, it strikes a most intolerable Trembling into the Members. This *Peter*

Figure 7.9. Hiob Ludolf and the accompanying title page of his 1691 *Commentarius*, which followed his *Historia Æthiopica* of 1681. Ludolf, an expert of oriental idioms, is considered the founder of the modern study of Ethiopian language and history, although he never visited Ethiopia himself. He provides a detailed account of the previous reports on the Ethiopian torpedo (i.e., the electric catfish). The most important European center for Ethiopian studies, established at Hamburg University, was named after Ludolf (*Hiob Ludolf-Zentrum für Äthiopistik*)

[*sic*, really "Father"] *Almeyda*, the Jesuit experimenting, paid for his knowledge." He also states:

> The *Habessines* cure *Quartan* and *Tertian* Agues with it. The manner thus, the Patient is first bound hard to a Table, after which the Fish being applied to his joynts, causeth a most cruel pain over all his Members, which being done, the fit never returns again. A severe Medicine, which perhaps would not be unprofitable to those that are troubled with the Gout, in regard some say that Disease is to be Cured by Torment. Those *Ethiopians* would certainly believe it, who affirm, that the Venue of this Fish will dispossess a man of the Devil himself. And yet if you touch this Fish with a Spear or a Wand, the sinews of it, though very strong, presently grow numb and the Feet of it, though otherwise a swift runner, lye as if they were bound, as Plinie reports. Which Modern Writers testifie to be no untruth. (Ludolf, 1682, Bk. I, p. 62)

By referring to Pliny the Elder and bringing up the ancient legend of how the (sea) torpedo's discharge could stop a swift runner (see Chapter 4), Ludolf appears to be confusing what he had read about two very different fishes, the African catfish and the torpedo ray. Also, there is also no mention of treating painful gout in the aforementioned Portuguese writings, although gout was one of the disorders that Scribonius Largus and other Roman physicians thought could be treated with live sea torpedoes, as some sixteenth-century ichthyologists had noted in their treatises (Chapters 5 and 6). The use of the fish to treat Quartan and Tertian Agues, [two forms of malaria], is new, although the idea that torpedoes could be used to treat fevers does have some history (Chapter 6).

Ludolf's confusions are even more apparent in his *Commentarius*, where he writes:

> It is also commonly found in Africa. About it *Caspar Barlaeus* in his very beautiful work *De rerum in Brasilia gestis* p. 134 writes, *It is assured that there is here the Torpedo, called Puraquam by the Barbars [i.e., the natives], because it induces torpor in the members, and even if you touch it with a stick: after being killed it loses its venom and can be eaten.* This experience confirms that were true the things the ancients wrote about this fish and among them Pliny Lib. 32. c.1. & 10. (Ludolf, 1691, pp. 160–161)

By now bringing Caspar Barlaeus (Van Baerle) into the picture, Ludolf effectively melded what he had read or heard about three different kinds of "torpedoes" into an amalgam. This is because Barlaeus was a Dutch scholar who wrote about South America and notably Brazil, where the rivers are devoid of electric catfish but contain torporific "eels," this being in the mid-seventeenth century (Van Baerle, 1660).

Ludolf's works thus had both positive and negative aspects. In retrospect, the most obvious negative is that he was confusing two and sometimes three different fishes with his commentaries. A similar confusion was, however, frequent. It can also be observed in Purchas, where, in an account of the electric eel in Brazil, the reader is referred to both the flat torpedo and to Jobson's description of the African catfish (see Purchas, 1625, IV Part, p. 1314).

This shortcoming reflected two unfortunate facts. One is the aforementioned use of the word "torpedo" to indicate any fish capable of producing a strong commotion. The other is that the texts consulted by both Purchas and Ludolf lacked a picture of the "African torpedo," or even a good physical description of its features (e.g., its barbels). On the positive side, even with his errors Ludolf drew a great attention to the strange fish that had been described by the Portuguese earlier in the century, because of the great popularity and status of his books.

ETHIOPIAN CHRISTIANITY AND RELIGIOUS INTOLERANCE

Neither Purchas nor Ludolf was a Catholic, but both showed interest in the traditional faith of the Ethiopians and in Catholic attempts to convert Ethiopians to their religious orthodoxy. Putting this in context, it should be noted that Protestants were very interested in the Ethiopian religion and in the Jesuit missions to Ethiopia during the seventeenth century, because they viewed what was happening as another instance of the unyielding attitude of the Roman Catholic Church toward other churches during the Counter-Reformation.

This Protestant interest had its parallel in the previous century with some Catholic liberals. In 1541, stimulated by the arrival of the Ethiopian ambassador in Lisbon, Damião de Góis, a Catholic who had been a student of Erasmus of Rotterdam, took the occasion to promote his views about religious tolerance. In his treatise, Góis (1541) included various relevant texts dealing with European–Ethiopian relations, most notably a defense of Ethiopian religion written by Ethiopian ambassador himself (Pennec, 2003; Marcocci, 2005).

Further, two publications of somewhat apocryphal authorship were among the most notable editorial events in the highly charged atmosphere of Ludolf's time. As the long title of the first one made clear, it concerned "The Rebellions and Bloudshed occasioned by the Anti-Christian practices of the Jesuits and other Popish Emissaries in the Empire of Ethiopia" and was based "on a manuscript written in Latin, by Jo. Michael Wansleben, a learned papist" (Wansleben, 1679). This was the same Wansleben who had mentioned "the *Reade* or the Cramp-Fish" in an account of his travels to Egypt (1678).

Wansleben was Ludolf's student, an expert of oriental languages, and a Lutheran. He had been sent to Ethiopia by the Duke of Saxony to obtain direct and trustworthy information about the Jesuit mission. Although he could not travel beyond Egypt, in 1663 he wrote a treatise on the subject, basing it on the information he had obtained during his voyage (Bausi, 1989). Although he now converted to Catholicism and became a Dominican, his treatise on Ethiopia circulated and was used in the anti-Jesuit pamphlet bearing his name, preceded by a foreword that clarified the reasons for the great Protestant interest in Ethiopia.

After discussing how the Catholic Church promotes "subversion by the force of the Arms" when she does not succeed in proselytism "by the force of the Arguments," the anonymous author of the foreword explains:

This is a Truth which more than one Age and Nation had sadly experienced, but none ever had more reason to abhor and deprecate than Ours. Yet the well-ordered Governments of our British *Church and State is not the sole Object of the* Roman *envy, not hath England been the only Scene of the* Popish *Cruelty. Not to mention the known and memorable Instances of* Paris, Piedmont, Ireland &c. Ethiopia, *a Country little known and less frequented by the* English, *has felt the smart of* Rome's *malice, and bears fresh Scars of the* Jesuits *Treachery.* [Italics in original]

The second notable editorial event occurred in 1670 with the publication of a book in London titled *The Late Travels of S. Giacomo Baratti, an Italian Gentleman, into the Remote Countries of the Abissins [sic], or of Ethiopia Interior* (Baratti, 1670) (Fig. 7.10). According to the English editor (signed G. D., probably for George Doddington), this was a translation of a book published twice in Italy with "universal applause." Yet both Baratti and his travels were born of fantasy, there being only an apocryphal English edition (Tedeschi, 1992). The contents of the book, and particularly the contrasts made between the Ethiopian and Roman Catholic Church, and the inclusion of documents from the previous century (e.g., a defense of Ethiopian religion taken from Góis' Latin version), shows clearly that a Protestant wrote it as an anti-Catholic polemic. In some ways, Baratti's

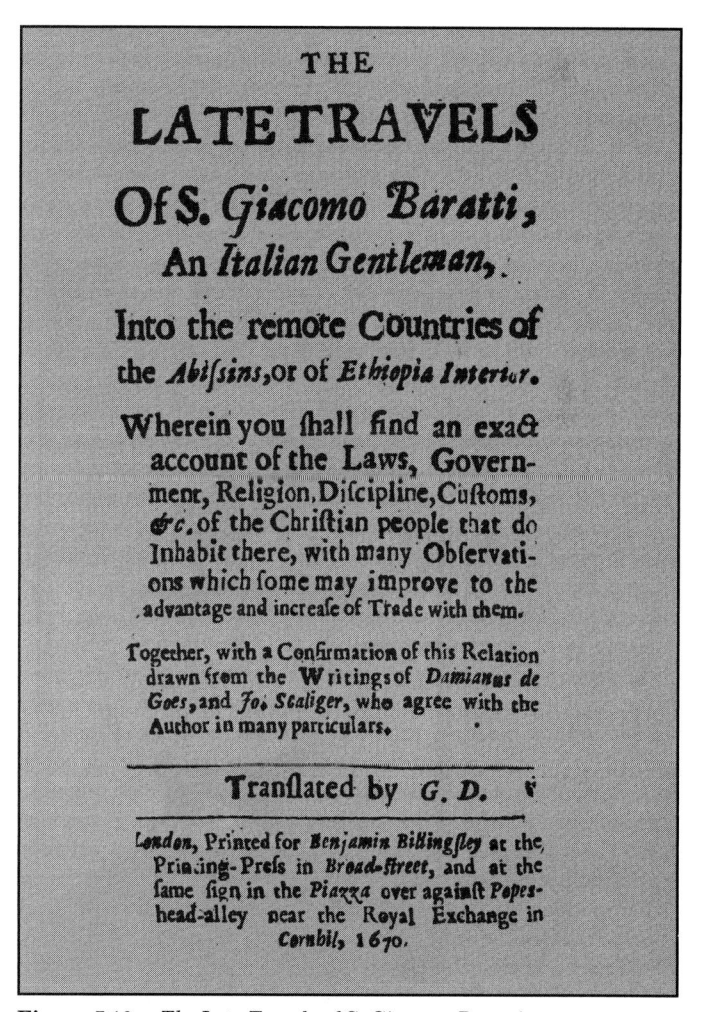

Figure 7.10. *The Late Travels of S. Giacomo Baratti*, an anti-Catholic polemic published in London in 1670. Presented as a translation of a book first published in Italy, both Baratti and his travels were made up.

book is the Protestant counterpart of Urreta's book, although it lacks the verbosity and rhetoric of the work by the Spanish monk.

The great interest in Ethiopia's religion, which had drawn so much attention in the sixteenth century, would continue into the seventeenth century. As a result, people who read monographs or collections about Ethiopia became increasingly more aware of a river fish functionally similar to the flat sea torpedo. This fish, however, had a different appearance and was capable of producing a stronger effect than the small sea torpedoes that were common in the Mediterranean.

MYTHS AND REALITY IN THE SEVENTEENTH CENTURY

In contrast to Urreta's writings, the Jesuits tried to separate myth from reality when it came to Ethiopian natural history. This more realistic attitude was applied to the African river torpedo. Without question, this fish seemed to have magical powers, much like some of the creatures that Urreta described in his books about Prester John's kingdom. Europeans (read "reliable witnesses") willing to touch a healthy specimen could confirm the pain and numbness this toothless river fish could cause. Hence, statements about its benumbing powers were never questioned; rather, they were always presented as factual, even if the fish's power might have seemed inexplicable.

In contrast, the European authors distanced themselves from the more questionable statements made by the locals or "naturals." The Africans were not considered trustworthy witnesses, as they tended to be superstitious, gullible, and guided by faulty belief systems. Thus, Dos Santos wrote, "The natives say that enchantments are made from the skin of this fish," without endorsing this claim. Similarly, Godinho emphasized that "Superstitious Ethiopians" are the ones who "believe that this fish has the power of chasing demons from human bodies." Even more telling, this author informs his readers: "I had no direct experience of this." Even Ludolf, for all of his confusions, distanced himself from the claim that this fish could cure quartan and tertian agues, terms signifying different forms of malaria.

Turning to the fish itself, despite the problems caused by the absence of detailed information about it, such as its distinctive shape and facial barbels, there can be no doubt that Dos Santos, Fernandes, Paez, Almeida, and the others involved, could only be describing *Malapterurus*, the electric silurus or catfish. This is because the fish they wrote about was found in rivers and lakes far from the saltwater habitat of torpedo rays. Also, the other fish to which it was compared physically (e.g., "hound fish," "English breame") do not look anything like rays, and nowhere is the river fish said to have the appearance of a ray. The strong intensity of the discharge, which even caused the captain of the Portuguese military contingent to release it, would also suggest *Malapterurus*. The shocks elicited from the small torpedo rays found throughout the Mediterranean region are considerably gentler and normally cause only a mild tingling of the hand and wrist.

To summarize, the African "torpedo" or *Malapterurus* seemed unknown to Europeans prior to the Age of Exploration, even though it had been depicted on Egyptian tomb paintings and seems to have been described in a few texts from the Middle East. Its "discovery" by Europeans took place against a background of trying to find trade routes to India, legends about a lost city with fabulous treasures, and interactions with Christians in Africa, who had belief systems that had to be brought into conformity with those of the Roman Catholic Church. Although the functional attributes of this fish were nicely described, the physical descriptions were surprisingly minimal.

Notably, there was no theorizing about the nature of the fish's powers, other than that its effects seemed to be associated with preceding movements. In a sense, this absence of theorizing is not surprising, since the early reports did not come from natural philosophers who studied animals and fishes, but primarily from men of faith who were far more interested in converting wayward souls.

Thus, the attraction that the African "torpedo" had for Europeans did not occur in a cultural or scientific vacuum. Further, there were other notable explorations taking place at this time. Some of these explorations took Europeans to South America, where a far more powerful tremble fish resembling an eel was encountered in the sluggish, warm rivers. It is to these newly discovered fish that we now turn—the frightening "eels" that Humboldt wrote could panic horses and mules, and possibly even cause their deaths with their strong and repeated discharges.

Chapter 8
The "Eels" of South America

> I was even more surprised to see a three to four foot fish, similar to an eel, as fat as a leg, which the fishermen call a conger, that by simple touch with a finger or the tip of a stick, so numbs the arm and that part of the body closest to it that one remains about 15 minutes without being able to move.
> (Jean Richer, 1693, p. 70)

VOYAGES TO SOUTH AMERICA

After our long journey to the mythical lands of Prester John and the Queen of Sheba, where we encountered the mysterious African torpedo, we must now set sail for an another distant land, where we shall find another fish endowed with the same mysterious force that had been puzzling naturalists from ancient times into the Renaissance. As we shall now see, the first bits of news about the existence of the new "tremble fish" began to reach Europe when zoology was being reshaped by exotica from newly discovered lands, and when the illustrated fish books were being published. None of the late-sixteenth-century ichthyologists, however, mention this new shocking fish, which would be dubbed the "electric eel" in the second half of the eighteenth century. Further, it would take a century before it would be included in zoological works.

In parallel to what was described in the previous chapter on how the African catfish was found in the rivers of Ethiopia and Gambia, the discovery of the eel stemmed from voyages to the New World that involved trade routes and the promise of riches, as well as from missionaries and early settlers venturing into the territories that were opened. The history of European travels in this part of the world started with Columbus and his search for New World riches. During his fourth voyage, which dated from 1498–1500, Columbus sighted the coast of South America near Trinidad, but he mistook it for yet another island. Pedro Álvares Cabral (Fig. 8.1), who intended to establish commercial relationships and introduce Christianity wherever he went, accidentally discovered the coast of central Brazil on April 22, 1500.

Also at about 1499–1500, Amerigo Vespucci, the Florentine after whom the whole continent was named, encountered the coast of Guiana. At the time, he was flying the Spanish flag and was hoping to sail around the southern African coast into the Indian Ocean. Vespucci then headed south, discovering the mouth of the Amazon River. On his next voyage in 1501, this time sailing under the Portuguese flag, he sailed down the Brazilian coast, claiming to go far as Patagonia at the tip of South America. The Spaniard Alonso de Ojeda was another early explorer of the Atlantic coast of South America, and he worked his way up from Guiana to Panama in 1500–1501.

The European conquests of South America began shortly after these voyages. Francisco Pizarro, a Spaniard, sailed along the western coast of South America in 1527, hoping to find Inca gold, and then returned to capture and execute Inca king Atahaulpa, whose empire was already decimated by deadly smallpox. At about the same time, others conquered Columbia (Nikolaus Federmann, Sebastián de Benálcazar, and Jiménez de Quesada), Peru (Diego de Almagro), and other South American territories, making it safer for more missionaries, settlers, administrators, and soldiers to sail for South America, where the Europeans would quite literally be shocked by the torporific eels that frightened the natives that bathed and fished in the slow-moving rivers and their tributaries and pools.

Interest in the newly discovered lands was a boon for the printing industry, leading to the publication of reports of voyages, maps, letters from America, histories of conquests, and missionary achievements (Fig. 8.2). These helped to spread the knowledge of the New World across Europe.

EARLY DESCRIPTIONS FROM VENEZUELA AND BRAZIL

The Europeans found the eels across a fairly wide area of northern and central South America, a region known for its hot, insect-infected climate and muddy waters. Specifically, they were encountered with some frequency in the rivers of the territories of the countries now known as Brazil, Venezuela (where Humboldt studied them in 1800), Guyana (or Guiana), and Surinam (also spelled Suriname: Figs. 8.3 and 8.4). Early records indicate that they might even have been found farther south in Argentina (Termeyer, 1781; Asúa, 2008).

Gonzalo Fernández de Oviedo y Valdez, who spent time in the Americas as an administrator and collector of information for the Spaniards, was one of the first Europeans to learn of

Figure 8.1. Pedro Alvares Cabral (1497–1520), the Portuguese nobleman-explorer who discovered the central coast of Brazil in 1500.

this fish and to write about it. He wrote a general and natural history of the regions he visited, starting his project in 1515. Summaries and parts of his book appeared in 1526 and 1535, and everything was essentially completed in 1548, after he returned to Spain and was appointed Historian of the Spanish Indies. But although more of his material was published in 1550, it was not until 1851, more than three centuries after he had finished his project, that Oviedo y Valdez's entire book was published.

This Spanish writer, who modeled himself after Pliny and cited the Roman naturalist repeatedly, mentioned the shock of a fish found in the Huyaparí (Orinoco) River in the Province of Paria, a part of present-day Venezuela (Oviedo y Valdez, 1959, pp. 193–194). He states that:

> Pliny, speaking about the water animals, said that the torpedo, when touched with a spear or rod, even from a distance, could affect any strong or brave arm, and feet running at full speed. But this author does not speak of the form of this animal. And our Spaniards, who are in the Indies and have run into it, do not know its name, but they speak of its looks and manner. And so . . . you will find, reader, that in the Huyaparí River they caught a fish like a moray eel, colored, as wide as the wrist of a man's arm and as long as four hands. They caught it with net and put it on land, and as long as it was still living, touching it with a spear, a sword, or a stick at any distance, instantly caused intense pain in the arm, which was so tormented with pain and became sluggish that it was smart to let go of it. Many Spaniards experienced this and

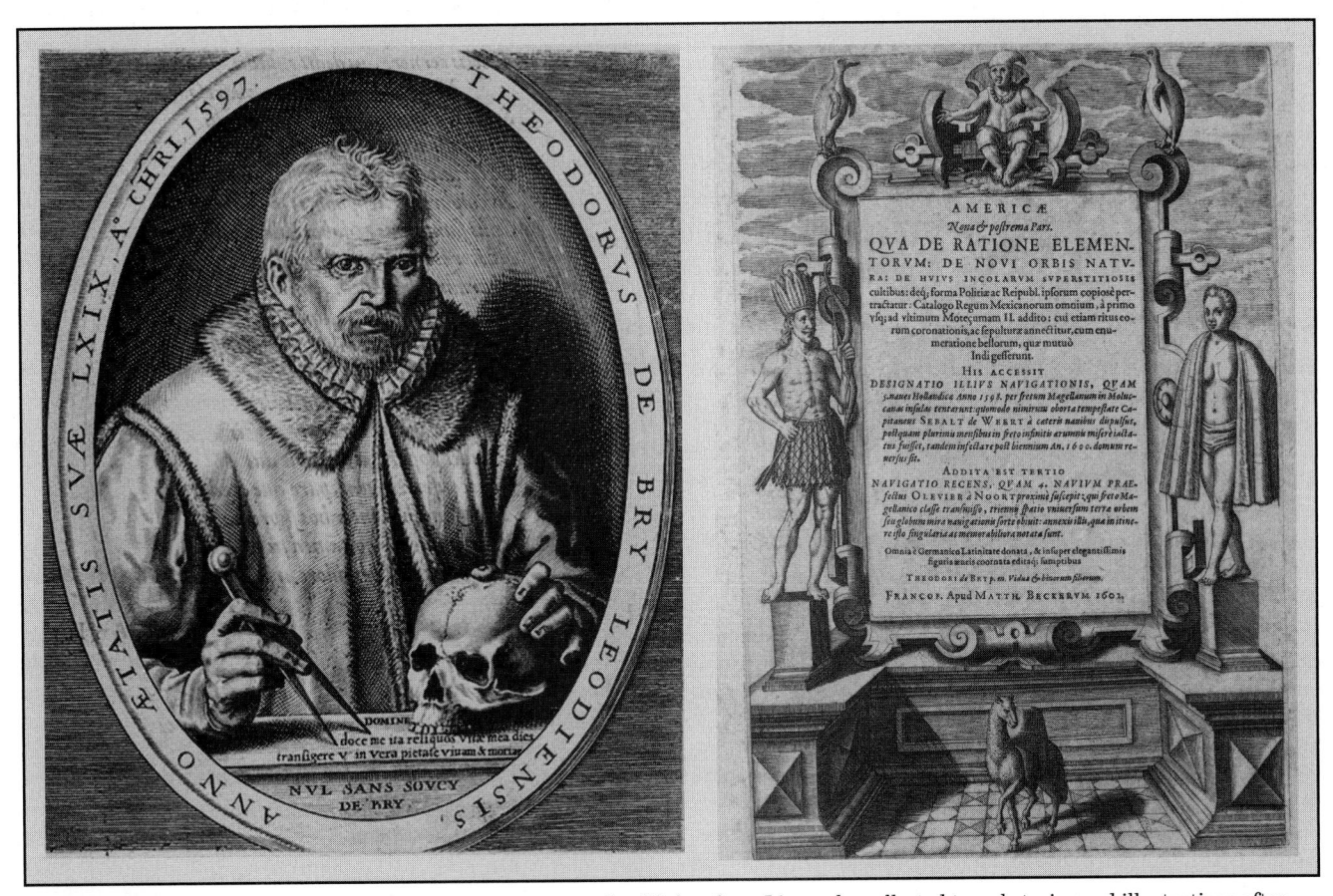

Figure 8.2. (Left) Theodore de Bry (1528–1598), an engraver and publisher from Liege who collected travel stories and illustrations after meeting Richard Hakluyt in London. His most famous set of books was *Les Grands Voyages*, a collection sometimes called "The Discovery of America." (Right) The cover page of a volume of the Latin edition of this collection shows American natives (see Bry et al., 1602).

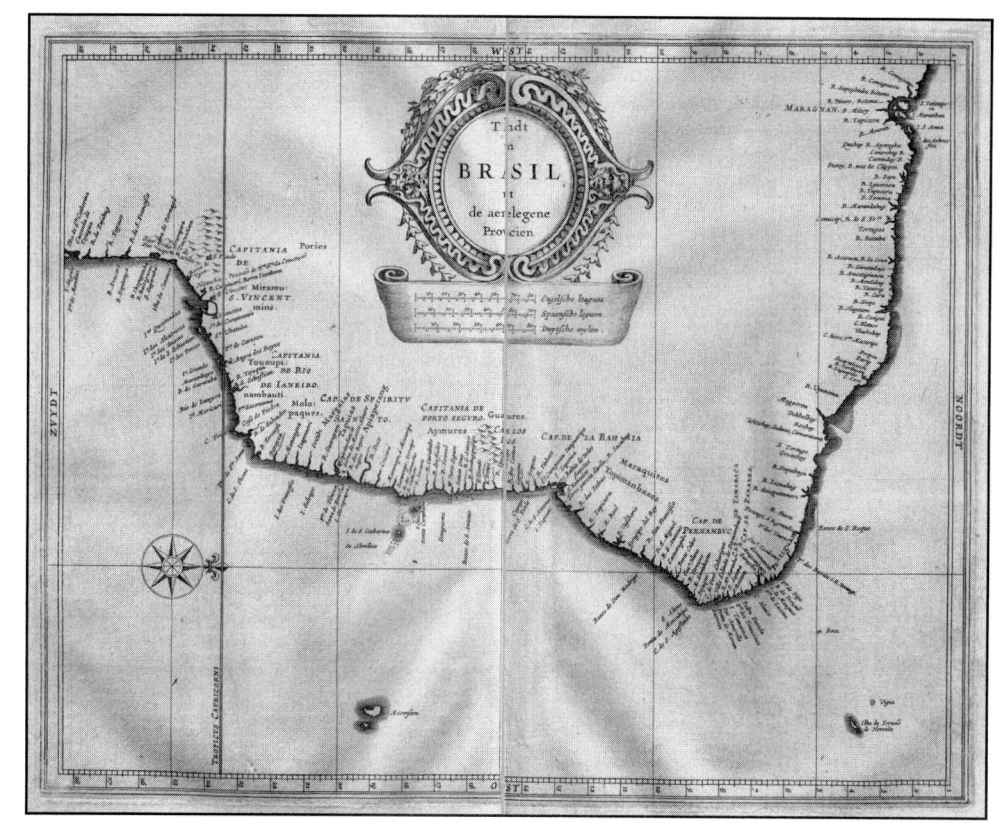

Figure 8.3. An early map of the coast of Brazil (from Laet, 1625).

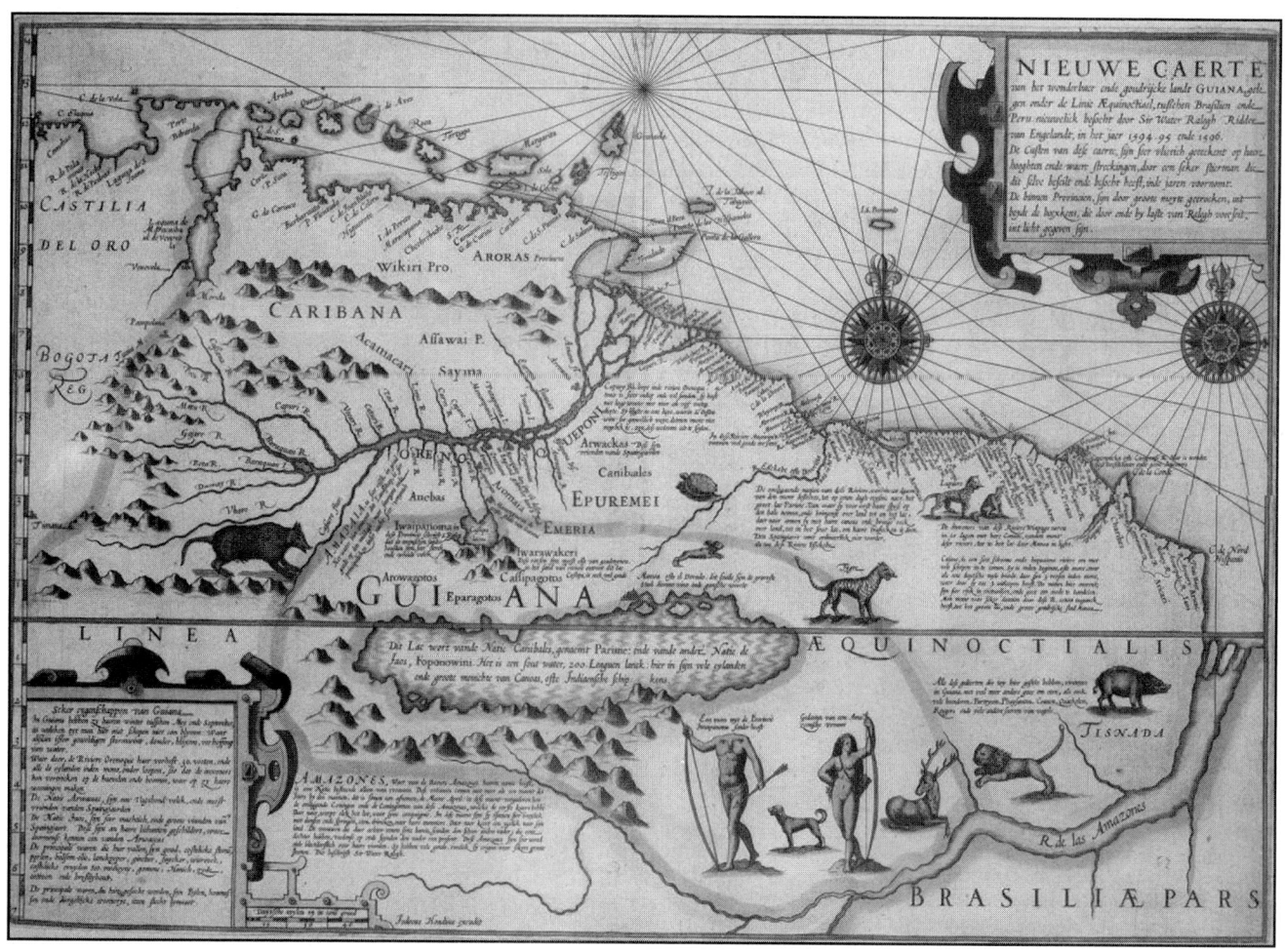

Figure 8.4. An early Dutch map showing Guiana (from Hondius, 1598).

so many wanted to know its secret, that while grabbing it to have the experience, they killed it. And after it was dead the property died with it, and it no longer gave any pain or upset to anyone who touched it. (Trans. from Oviedo y Valdez, 1959, pp. 193–194)

Fernão Cardim, a Jesuit priest, learned about the numbing effects of the South American eel later in the sixteenth century and wrote about them in 1585 (see Cardim, 1925, p. 68; Cardim, 1965, p. 489). Samuel Purchas included a translated version of his passage on the eel in a section of his *Hakluytus Posthumus, or Purchas his Pilgrims,* published in 1625 (see the previous chapter for more on Purchas and this book). It appeared as "A Treatise on Brazil, Written by a Portugall which had Long Lived There," with Purchas lamenting that he did not know the name of the author of the treatise he had read. Making some of the same points as Oviedo y Valdez had previously made, it reads:

Puraque is like the Scate, it hath such a vertue that if any touch it, he remaineth shaking as one that hath the Palsie, and touching it with a sticke, or other thing it benummeth presently him that toucheth it, and while he holdeth the stick over him, the arm that holdeth the sticke is benummed, and asleepe; it is taken with flue-nets, and with casting Nets it maketh all the bodie tremble, and benummes it with the paine, but being dead it is eaten, and it hath no poison. (Purchas, 1625, Part IV, Book VII, p. 1314; 1905 reprint, Vol. XVI, p. 488)

The Franciscan missionary Claude d'Abbeville was yet another early visitor to the New World who described the eel, and he did so in his 1614 chronicle dealing with the Isle of Maranhão in northeastern Brazil (Fig. 8.5). Written in French, his section on the *Pouraké,* here translated into English, reads:

The *Pouraké* is impressive among all the fishes found in rivers and other sweet waters: it is much bigger than a thigh and about four feet in length. In addition, it provides a very pleasant sight, because of the diversity of its colors, being variegated with red, blue, green and white. It is its custom not to care about being hit by your sword and will not move, no matter how you hit him, particularly because its flesh is so soft that it adapts to the hits without being torn. If it moves to a small or large extent while being hit, it would make your arm numb and painful, and you would be moved back four or five feet; you would fall on one side and your sword would fall on the other side. This was the experience, at his own expense, of one of our gentlemen friends. (Abbeville, 1614, p. 246)

Bernabé Cobo, who held a position at the Jesuit College of San Marcos in Lima, also described a "torpedo of the Indies." It appears in his 1653 text, *Historia del Nuevo Mundo* (Cobo, 1956 reprint, p. 307).

Thus, there are several good descriptions of the "eel" in Venezuela and in Brazil that date from the sixteenth to the middle of the seventeenth century. Nevertheless, as was true for the catfish in Ethiopia and Gambia, the early explorers and missionaries who encountered them, or heard about them from others, devoted only a few lines to these strange fish and did not discuss how they were able to affect those who touched them, much less indirectly with swords and poles.

The situation is more interesting and at the same time rather confusing with the report of a Dutch expedition to Brazil provided by Georg Marcgraf (or Margrave), a German

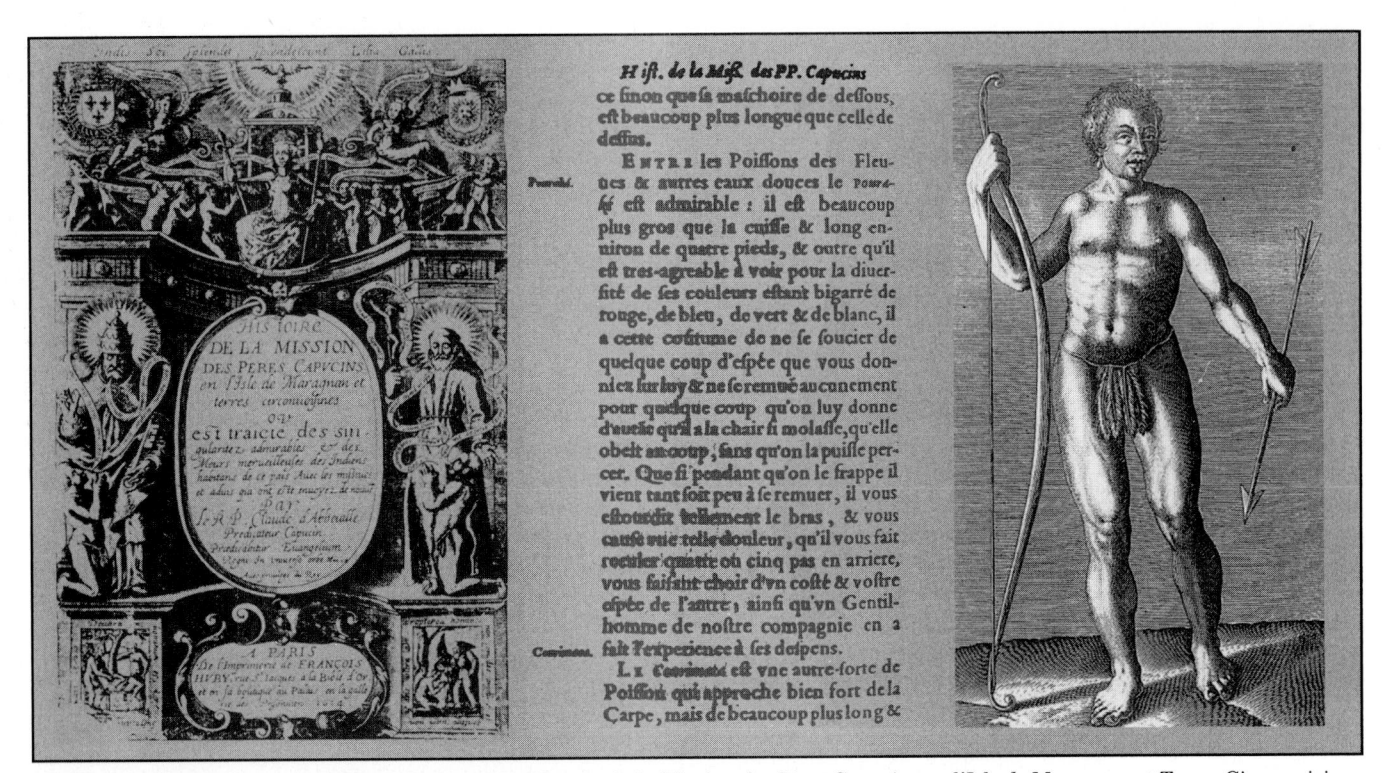

Figure 8.5. (Left) Cover page of Claude d'Abbeville's *Histoire de la Mission des Pères Capucins en l'Isle de Maragnan et Terres Circonvoisines,* first published in 1614. (Right) His description of a torporific fish (electric eel) from northeastern Brazil and an illustration of a native.

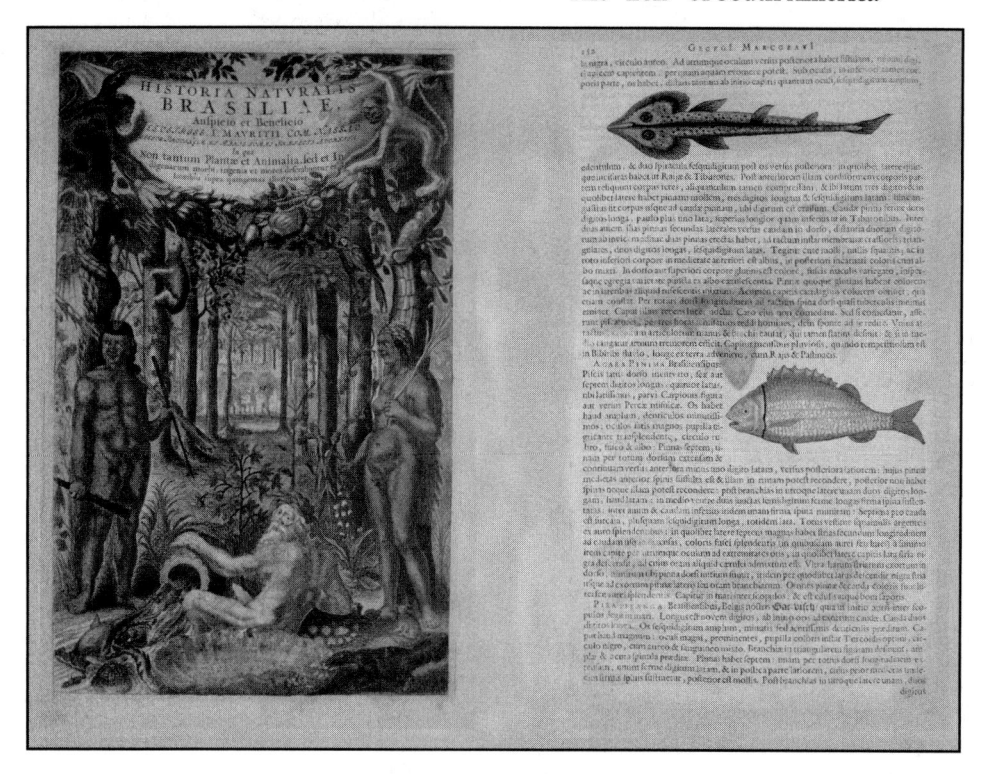

Figure 8.6. The frontispiece of Georg Marcgraf's (Margrave's) 1648 *Historia naturalis Brasiliae*, with an illustration of the torporific *puraquê* represented as a curious, elongated ray-like fish.

naturalist who eventually settled in Leyden, and Wilhelm Piso, a Dutch physician. Species that would later be known as electric fish are mentioned twice in their monumental *Historia Naturalis Brasiliae* (1648) (Fig. 8.6). The first mention, due to Piso, is less than three lines long, and it is inserted in a short section on poisons, mainly devoted to the sting-ray (*Pastinaca*). Piso simply writes that the harm (*malo*) made by the Brazilian torpedo is similar to that of the European fish and lasts one or two hours (p. 44).

The second mention, due to Marcgraf, is more intriguing (pp. 151–152). The fish is indicated with the local name *puraquê* (and also as *peixe viola*, the name used by the Portuguese because its shape is similar to a viola, the musical instrument). No doubt, the local names refer to the electric eel (Read, 1945). This notwithstanding, the description, apparently rather accurate, does not correspond to the eel, since the fish is considered to be of the cartilaginous type and its shape, illustrated by a curious figure, is roughly similar to that of a torpedo, even though its tail is much longer. This at least justifies Marcgraf's comparison with the popular musical instrument.

It is remarked that the fish is not suitable as food, and if eaten it cause faintness of three hours' duration. As to the fish's shocking power, Marcgraf describes it as causing a clattering (*crepitum*), if touched, and writes that it produces tremor if touched in the middle of its body. The fish, Marcgraf continues, is found in the river, but this does not appear to be an important criterion of distinction from the sea torpedo, because the naturalist also remarks that the same occurs with rays and sting-rays (*pastinacae*) (p. 152).

Despite the detailed description of the fish, it is possible that Marcgraf never saw the *puraquê* himself and instead drew his information from a *Description of the New World* first published in Dutch and afterward, in 1633, in Latin by Joannes de Laet, a Dutch geographer and naturalist, who served as a director of the West India Company (Fig. 8.7).

There is an important difference, however, because Laet writes that, once dead, the fish can be eaten without any problem. The figure that is given by Laet is somewhat similar to that of Marcgraf, but Laet assures his readers that it was made by observing a live fish (*ad vivum*).

The conclusion that can be derived from the reports of Marcgraf and Laet is that, even during the epoch of Galileo and Harvey, naturalistic and geographic treatises still depended heavily on indirect evidence. What is disappointing is that Laet's and Marcgraf's figures could have been the first published illustrations of the electric eel, but they are so ambiguous that about another century would have to pass before readers would have an accurate graphic representation of this fish. As we shall see in Chapter 13, the Dutch will still get the credit for this achievement.

GEORGE WARREN'S DESCRIPTION OF SURINAM

Explorations of the region now associated with the small South American countries of Surinam, Guiana, and Guyana began in the 1590s with the voyages of Sir Walter Raleigh, his close friend Lawrence Keymis, and others. Nevertheless, these explorers did not venture far inland. During the 1600s, however, French, Spanish, Portuguese, British, and Dutch settlers began building plantations in the fertile areas along the major rivers, which gave them easy access to the Atlantic Ocean for trade (Fermin, 1769, 1781). The colony of Surinam dates from 1651 and was founded by Francis, Lord Willoughby, governor of Barbados.

In 1667, a London publisher advertised a small book dealing with Surinam, which was growing in economic importance. It was by George Warren and it bore the title *An Impartial Description of Surinam upon the Continent of Guiana in America, with a History of Several Strange Beasts, Birds, Fishes, Serpents, Insects, and Customs of that Colony, &c.*

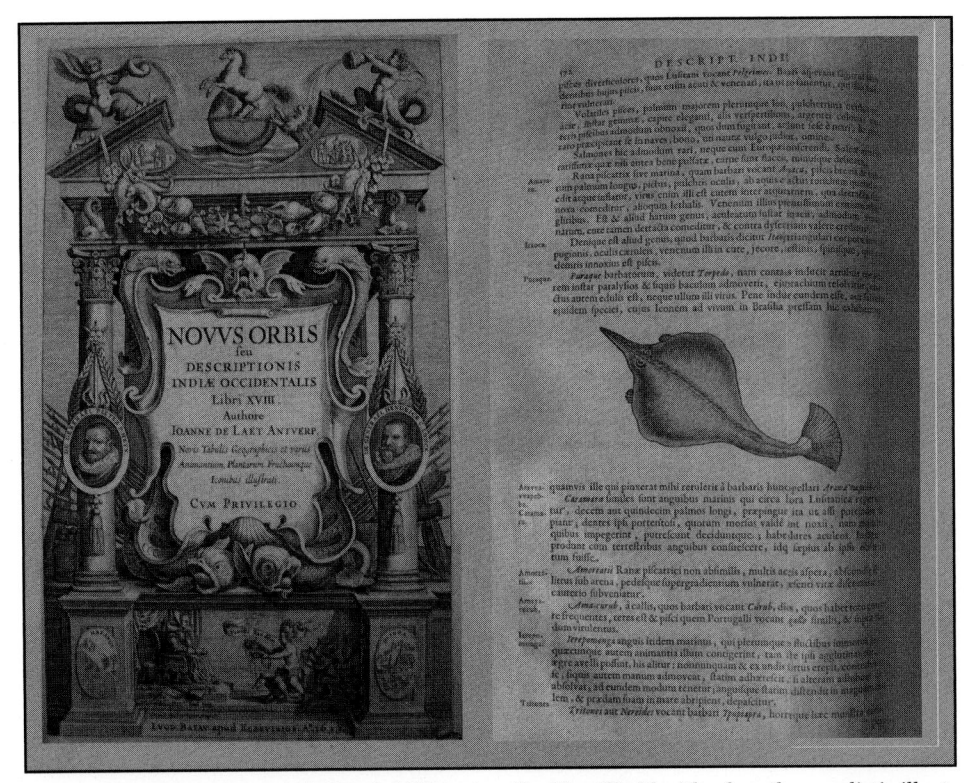

Figure 8.7. The frontispiece of Dutchman Joannes de Laet's 1633 text on the New World with a less-than-realistic illustration of the *puraque* fish (also spelled with an accent on the e in some early publications), which was probably imitated by the illustrator of Marcgraf.

Warren had been in Surinam for 3 years, "not without many hazards to my self," and his 28-page book described numerous wonders of this exotic land, including its eels.

Warren introduced the *Numb Eele* on the second page of his book, knowing that it would fascinate readers and draw them in because of "the strangeness of its nature." His account is more vivid and chilling than those that preceded it. He informs his readers that if a living numb eel, also referred to as a *Paling* and a "river torpedo," managed to touch any living creature, it:

> strikes such a deadness into all the parts, as for a while renders them wholly useless, and insensible, which is believ'd, has occasioned the Drowning of several persons who have been unhappily taken, as they were Swimming the River [Surinam]: It produces the like effect if it is touched with the end of a long Pole, or one man immediately laying hold of another so benumb'd: The Truth of this was experienced. One of them being taken and thrown upon the Bank, where a Dog spying it stir, catches it in his Mouth, and presently falls down, which the Master observing, and going to pull him off becomes motionless himself; another standing by, and endeavouring to remove him, follows the same fortune; the *Eele* getting loose, they Return quickly to themselves. (Warren, 1667, p. 2)

APHRA BEHN'S OROONOKO

It is difficult to know how many people read Warren's book, but one person who seemingly did was a British woman who was already a well-known novelist, poet, and playwright. Her name was Aphra Behn, and she was more than capable of handling herself in what had been a man's world. More than two centuries later, Virginia Woolf would call her "the first professional woman author in English" (Gallagher & Stern, 2000, p. 3; also see Duffy, 1977; Goreau, 1980). Notably, John Dryden was the only British playwright able to boast more productions on stage than Behn had during the 1670s.

Behn was born in 1640 and died in 1689, and only late in her career did she turn to dramatic prose. *Oroonoko; or, The Royal Slave* is by far the piece for which she is best remembered today. This stirring novel on the horrors of the African slave trade was published in 1688, a year before she died, and it quickly became a bestseller. It has also remained a very popular book, especially among students and academics.

Much of the novel takes place in Surinam, now a growing but still relatively small colony in the Guianas. At the time, the settlers from different countries battled among themselves for ports and prime land for planting in the region. The transplanted Europeans, along with some North American entrepreneurs, also battled with the native Indians. And all were endlessly dealing with bands of fugitive African slaves who had been brought in to work the sugar and coffee plantations but had managed to escape and sometimes made brutal, daring raids on the plantations and their European overseers.

Behn visited Surinam in the 1660s, at the onset of the Second Anglo-Dutch War (1665–67), in which England tried to end Dutch domination of foreign trade. This war would end with the Treaty of Breda, in which the Dutch would acquire Surinam and the British would acquire New Amsterdam, thereafter to become New York. She might have been in the region as a spy for King Charles II, who ruled Britain during the Restoration period. The novel takes place at the end of the period in which the British maintained control of the economically promising colony.

Oroonoko was a multilingual African, a *"Gallant Slave . . . who]* address'd himself, as if his Education had been in some *European* Court." A prince among his people, he was tricked into capture by a smooth-talking slave trader. Unlike large numbers of his fellow captives, he survived the voyage across the Atlantic Ocean to "a Colony in *America*, called *Surinam*, in the *West Indies.*"

Surinam's plantations were producing large quantities of valuable sugar cane, coffee, and cocoa for export during Behn's visit. Since the English could not subdue the local Indians, who also happened to prove helpful in other ways, the West African slave trade became vital for the economic survival of the lucrative sugar plantations. But with the slaves far outnumbering the Europeans, who often treated them very cruelly, there were inevitable rebellions. Behn's novel ends tragically when a bloody revolt led by the once-princely Oroonoko leads to his horrible execution (for more on slave revolts in Surinam, see Stedman, 1796/1988; Fermin, 1769, 1781).

Before Oroonoko turned on his masters, sealing his fate, he had been treated far better than the other slaves because of his eloquent speech and knowledge of philosophy and world events. Hence, he was allowed to spend considerable time conversing with his masters and with Behn, who put herself in the novel as a visitor to the plantation of the man who owned him. During one of their more carefree interludes, the discussion turned to how adept the natives seemed to be at swimming and fishing, and to Oroonoko's own fishing skills:

> At other times he would go a Fishing; and discoursing on that Diversion, he found we had in that Country a very strange Fish, call'd a *Numb Eel* (an Eel of which I have eaten) that while it is still alive, it has a quality so Cold, that those who are Angling, though with a Line of never so great a length, with a Rod at the end of it, it shall, in the same minute that the Bait is touched by this *Eel*, seize him or her that holds the Rod with benumb'dness, that shall deprive 'em of Sense, for a while; and some have fall'n into the Water, and other drop'd as dead on the Banks of the Rivers where they stood, as soon as this Fish touches the Bait. (Behn, 1688/2000, p. 80)

In her next sentence, Behn writes:

> Caesar [the slave name given to Oroonoko], us'd to laugh at this, and believ'd it impossible a Man cou'd loose his Force at the touch of a Fish; and cou'd not understand that Philosophy, that a cold Quantity should be of that Nature.

> However, he had a great Curiosity to try whether it wou'd have the same effect on him it had on others, and often try'd, but in vain; at last, the sought for Fish came to the Bait, as he stood Angling on the Bank; and instead of throwing away the Rod, or giving it a sudden twitch out of the Water, whereby he might have caught both the *Eel*, and have dismist the Rod, before it cou'd have too much Power over him; for Experiment sake, he grasp'd it but the harder, and fainting fell into the River, and being still possest of the Rod, the Tide carry'd him senseless as he was a great way, till an *Indian* Boat took him up; and perceiv'd, when they touch'd him, a Numbness seize them, and by that knew the Rod was in his Hand; which with a Paddle (that is, a short Oar) they struck away, and snach'd it into the Boat, *Eel* and all. (Behn, 1688/2000, p. 80)

This escapade in the otherwise tragic life of the royal slave would seem to have been based on Warren's account. And here too there is a happy ending, at least for the slave Oroonoko, who recovers completely:

> If *Caesar* were almost Dead, with the effect of this Fish, he was more so with that of the Water, where he had remain'd the space of going a League, and they found they had much a-do to bring him back to Life: But, at last, they did, and brought him home, where he was in a few Hours well Recover'd and Refresh'd; and not a little Asham'd to find he shou'd be overcome by an *Eel;* and that all the People, who heard his Defiance, wou'd laugh at him. But we cheered him up; and he, being convinc'd, we had the *Eel* at Supper; which was a quarter of an Ell about, and most delicate Meat; and was of the more Value, since it cost so Dear, as almost the life of so gallant a Man. (Behn, 1688/2000, p. 80)

In retrospect, Behn's repeated mentioning of a cold quality suggests that she might have familiar with Galen's theory (Chapter 2), most likely through the publications of others who alluded to a feeling of cold numbness in their writings. Fourteen hundred years earlier, Galen, and Claudian soon after him, attributed the torpedo's benumbing powers to some sort of venom, with effects like those associated with freezing. Behn, however, limited her pen to the cold sensations themselves, without bringing up a venom or toxin, agents well known to the settlers, who had to endure poisonous snakes and other dangerous animals and insects in the Guianas.

Behn never used language suggestive of the newer mechanical theories of the torpedo's shocks (see Chapter 9). The simplest explanation for this omission is that Behn, who was not a scientist, was probably not following the emerging scientific literature—a chore that would have been arduous for her, because the newer mechanical explanations for the torpedo's shocks had just emerged in Italy and the seminal writings of the pioneering Italians (e.g., Redi, Lorenzini) had not yet been translated for English readers.

As for electricity, nowhere does Behn call the discharge electrical or compare its effects to anything electrical, including to shocks from frictional machines or flashes of lightning. She does, however, accept as factual that the fish discharges can be transmitted through a series of intervening bodies, a feature that would have been very difficult to reconcile with the discharge of a poison, the release of small particles that could enter the pores of the skin, or rapid muscular movements.

Looking back, its is still fascinating to think that perhaps the first person to bring the South American eels into thousands of European homes and imaginations was not an explorer, a natural philosopher, or a physician, but a female novelist with a strong moral sense about slavery and no scientific agenda. Needless to say, what Behn did not include in her book is also notable. In addition to not addressing the nature of the eel's powers, she does not state that the fish discharges were being used in medicine, either by the native Indians or by the European settlers who, with their slaves, suffered injuries and were smitten with a wide range of disorders in the New World.

THE PLAGIARIST

In 1695, Adriaan van Berkel, a Dutchman who made three journeys between Fort Nassau and Fort Kyk-over-al, the capitals of Berbice and Essequibo in the Guianas, brought more attention to the eels, particularly in the Netherlands. In contrast to Aphra Behn, who seems to have used Warren as a source for her novel, Van Berkel (1695) literally lifted Warren's full-paragraph description of the eels and included it word for word, without attribution, in his own *Amerikaansche Voyagien*.

Just why Van Berkel resorted to such overt plagiarism is a bit hard to discern, since he was a "general collector" and would likely have encountered these eels himself, both while traveling through Guiana with the Arowak Indians and while living with the Dutch, who were building plantations and using the eel-infested Berbice and Essequibo Rivers for shipping sugar and other products out into the Atlantic Ocean and ultimately to European ports. Historians speculate, however, that Van Berkel might never have visited the colony of Surinam, yet another thriving colony about which he wanted to include some facts for his book. This colony was, in fact, the subject of the second part of his book, and it is here that he pilfered from Warren and used his predecessor's paragraph on the eels, albeit now presenting this material in Dutch.

A number of different river and pond fishes are mentioned in this book, as well as various ways of catching them. Van Berkel mentions using hooks and lines and baskets; shooting fishes with bows and arrows; and using natural toxins (most likely the roots of the haiari plant):

> When we wanted to dish up a meal of fish, orders were given to the Indians around the Fort to stop up a creek: which being done, we went there at low water, and stirred it down to the bottom with a certain kind of timber antagonistic to the temperament of the fish: as a results of which such a quantity came to the surface that we, keeping the most delicate, gave the remainder to the people. (Van Berkel, 1695, Roth trans. 1941, pp. 91–92)

Van Berkel is very specific about the methods used to catch certain freshwater fishes. But when presenting Warren's description of the dangerous eels, he added no more information of his own. Hence, it is left up to the reader to envision the method or methods the Indians used to catch the eels, which we read could kill a swimmer, paralyze a dog, and even exert its effects on others who touch a victim at the time of the action. It is also left up to the reader to imagine how this fish could do such things, as Van Berkel's source never delved into possible reasons for the eel's powers—and neither did he.

Van Berkel's book was not translated into English until the 1920s, when it came out serially in the *Daily Chronicle*, a British Guiana newspaper, and then as a single publication in book form (Roth, 1941). But his book in Dutch still helps to show how these strange creatures were becoming better known in some parts of Europe, where for a myriad of compelling reasons there was a tremendous interest in distant lands and exotica.

JEAN RICHER'S VOYAGE

Up to this point, the descriptions of the eel that were sent to Europe did not come from men associated with any of the great European scientific academies that were now forming. In closing this chapter, mention has to be made of an internationally known scientist with such an affiliation, who came across the eels in the 1670s. Indeed, the French astronomer Jean Richer was the first great European man of science to write about the eel and to have his work presented to a major scientific society. Richter visited South America during the 1670s, where he was sent by the Académie Royale des Sciences.

Richer had several objectives when sailing to equatorial America, one of which was to study solar and planetary motions and distances, while similar measurements were being made in Paris. The trip gave him the opportunity to determine longitudes at different places (Olmsted, 1942, 1960). Latitude, the distance to the Equator (0 degrees latitude), was easily determined by measuring the position of the North Star, located directly above the North Pole (90 degrees). Determining east–west coordinates was decidedly more difficult.

One promising strategy was based on setting one clock at the time at which the sun would be directly overhead at a known European location, and then finding how much time would elapse for it to be at its zenith at distant places. After Christian Huygens invented an accurate pendulum clock in 1657, this seemed possible. So the French ministry first sent Richer with two Dutch pendulum clocks, one of which was set to London time, to North America to measure time differences; the basic idea being that each hour of time difference would equal 15 degrees of longitude.

Richer sailed from La Rochelle, but the *St. Sebastien* encountered severe storms, which disabled his clocks by the time he arrived in New England. After returning to France, the clocks were fixed and he sailed off again, this time on a merchantman of the French West India Company bound for the French colony of Caïenne (Cayenne, now the capital of Guyana).

Arriving in 1672, he studied the fixed stars and planets, and the solstices and equinoxes. And he subsequently determined that his clocks with pendulums that were fixed to oscillate every second actually lost time near the Equator. This finding implied less gravitational attraction at the equator than in northern locations, suggesting that the earth is not a perfect sphere (i.e., that it bulges out more at the equator).

The shape of the earth was in question during the second half of the seventeenth century, and Richer's measurements clearly indicated that the earth is not a sphere. Citing gravitational and centrifugal forces, as well as measurements, Newton would contend in his *Principia mathematica* of 1687 that the earth is an oblate sphere or "rotating elipsoid" flattened by its poles. But not everyone would agree with this position, especially in France, where opinions were divided. Jean Dominique (Gian Domenico) Cassini and his son Jacques, for example, would contend that the polar axis is longer than the equatorial axis. The topic was one of tremendous interest, and with further measurements in the first half of the eighteenth century, especially by the French, the rotating elipsoid idea, with the earth bulging at the equator, would in fact be found to be correct.

Richer returned to Paris in 1673. He presented some findings and his description of the eels in a written piece dated 1679, but it did not appear in that year. Rather, it became part of a larger volume that was not published until 1693 (Richer, 1693). His lines were repeated verbatim in his *Observations Astronomiques et Physiques faites en l'Isle de*

Caïenne, published 33 years after his death by the Académie des Sciences. In this 1729 publication, under Article VII, *Remarques sur quelques Animaux et Poissons* ("Remarks on some Animals and Fish"; Richer, 1729, pp. 325–326), as well as on page 70 of his 1693 publication, we find the following lines, here translated into English:

> I was even more surprised to see a three to four foot fish, similar to an eel, as fat as a leg, which the fishermen call a conger, that by simple touch with a finger or the tip of a stick, so numbs the arm and that part of the body closest to it that one remains about 15 minutes without being able to move. It even causes dizziness, which would make you fall to the ground, unless you prevented the fall by lying down on the ground first. Afterward, you come back to your original state. I was witness to this effect and I felt it, having touched the fish with my finger one day when I met up with some savages, who still held one alive after having wounded it with an arrow and having pulled it out of the water with the same arrow. I could not find out from them what the fish was called: they said that when this fish hits the other fish with its tail, it puts them to sleep and they are then eaten, which is easy to believe once you witness the effect it produces when touched. (Fig. 8.8)

A THEORETICAL VOID

One of the most important similarities between the early reports described in this chapter and those from about the same time on the African catfish is the absence of a theoretical discussion. There is really nothing in these first reports to explain their mechanisms and remarkable powers, especially with intervening bodies. This void is understandable, however, from several perspectives.

One is that most of the writers, with Richer being the most notable exception, were not men with strong backgrounds in science. Rather, many of the descriptions came from missionaries, who were amazed by God's creations but were not scientifically inquisitive or sophisticated enough in the sciences to be able to provide good explanations for their powers. The same can be said about some of the early South American administrators, such as Gonzalo Fernández de Oviedo y Valdez, and, of course, Aphra Behn, who was a woman of letters and not a physician or a natural philosopher.

In addition, and perhaps of equal importance knowing that some of these individuals might have had access to scientific writings, there really was no widely accepted theory that could account in an effective way for the actions of these fishes at this particular moment in time. Ancient hypothesis based on venoms, although still discussed, were no longer being embraced as good explanations for the powers of the torporific fishes. Alternatively, the idea of "occult qualities," which has a long history and was alluded to by some Renaissance writers, was deeply philosophical. As such, the early voyagers and missionaries, with their more practical agendas, would not have been drawn to it. Indeed, it was scarcely an explanation at all for an increasing number of seventeenth-century natural philosophers.

During the time period from the first accounts of the South America tremble fish to Richer's observations the sciences were, however, beginning to undergo dramatic changes that would have important consequences for our electric fish story. Indeed, the landscape was clearly changing when the French physicist was finishing the report of his voyage. In fact, the scientific revolution, which would be associated with Copernicus, Francis Bacon, and Galileo, would eventually bring about new explanations for the mysterious powers of all of the torporific fishes.

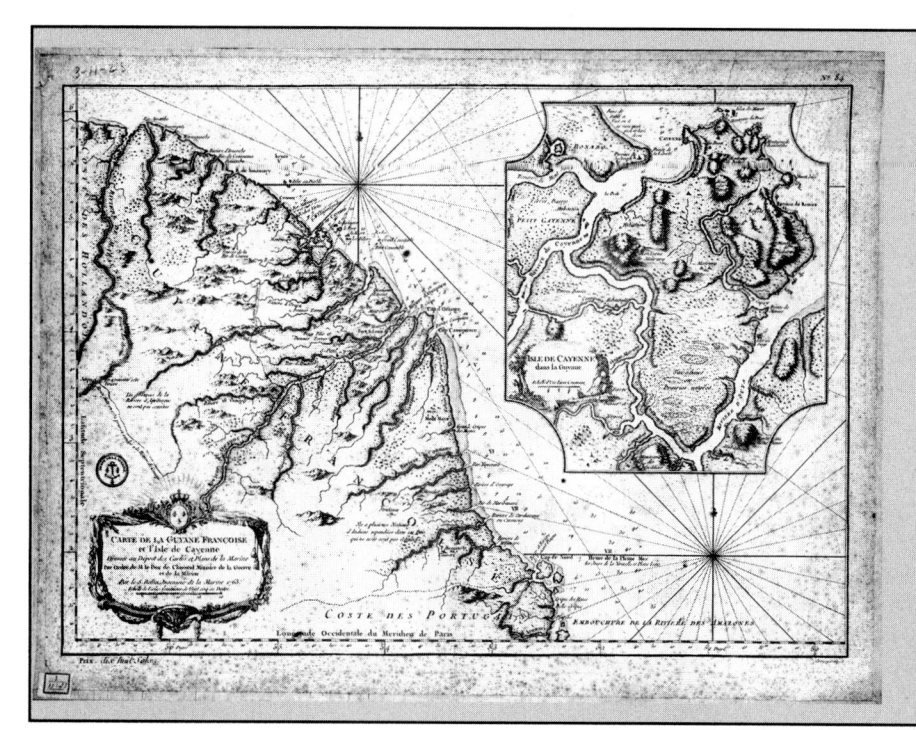

Figure 8.8 Left: A map of the French Guiana with the Island of Cayenne, made in 1663, a few years before Richer's voyage. Right: The initial page of Richer's report on the trembling eel published in the 1729 tome of *Mémoires* of the French Académie.

The new explanations, which would come forth in the second half of the seventeenth century, would be mechanical in nature. The powers of these fishes would be tied to rapid contractions of specialized muscles, and in some instances to the mechanical actions of their envisioned corpuscular emissions. Although faulty, these new ideas, which would dominate the scientific literature for about a century, would represent an important intermediary step between ancient ideas about cold venoms and various thoughts about occult qualities, on the one side, and the revolutionary conception that a few very unusual fishes, starting with the eels of South America, are, in fact, electrical, on the other.

In our next chapter, we shall examine how this new conception of nature would emerge in a world still haunted by notions of the occult and related magical forces, and we shall show some of the ways in which this new way of thinking would be applied to the shock mechanisms of these fishes. The interesting point is that, in a period in which criticism was emerging about occult forces as a scientific explanation for natural phenomena, the powers of the electric fishes were still viewed by some natural philosophers as possible *de facto* evidence for the existence of these forces. In this context we shall meet Athanasius Kircher, the famous Jesuit who attacked the new science of Galileo and his followers. And we shall see that it would not be just by chance that the first allusion to the torpedo's power as a mechanical phenomenon would be presented by a follower of Galileo, namely Francesco Redi, in a letter addressed to Kircher, one in which his "old" science will be strongly criticized.

Chapter 9
From the Occult to Mechanical Theories of the Discharge

> These are the experiments I have done. From them I conclude that the numbing power of the torpedo does not proceed from its entire body, but from a specific part, i.e., the falciform muscles that I have described above.
>
> Stefano Lorenzini (1678; p. 110)

This chapter will primarily examine the new theories about the mechanisms of the torpedo discharge that emerged as a consequence of the scientific revolution of the seventeenth century. These theories will be centered on mechanical actions, and specifically violent contractions of special muscles (*musculi falcati* or falciform muscles) in the fish's body. The basic idea is that rapid muscular contractions could, in some way, produce the torpedo's well-known benumbing effects.

From the viewpoint of modern science, these theories were largely flawed. As we now know, the torpedo's shocks are electrical and not mechanical in nature. But while this is clear now, the new mechanical explanations for the torpedo's powers were clearly advances over the idea of a cold poison, which can be traced back to Galen's era, and over the numerous vague references to occult or hidden qualities, which became popular during the Renaissance but really explained nothing (Chapter 6). In brief, they were based on a "new" and more demanding sort of science and on new ways of thinking.

Following the stimulating observations of Francesco Redi in Italy, two basic mechanical ideas about the torpedo's shock were being considered in the decades prior to the advent of the electrical hypothesis. One assumed that sudden contractions of the *musculi falcati* would lead to an immediate and explosive discharge of minute corpuscles capable of penetrating the nerves and blocking their functions. Italian scientist Stefano Lorenzini formulated this idea in 1678. The second basic idea did not include the intervention of corpuscular emissions. Instead, the assumption was the violent contractions of the *musculi falcati* could by themselves affect the nerves and muscles. During the eighteenth century, this last theory was closely associated with Frenchman René-Antoine Ferchault de Réaumur, who presented a landmark *mémoire* on the subject in 1714, although Giovanni Alfonso Borelli anticipated it in a work published posthumously in 1680–81.

The new science promoted by Galileo and other great protagonists of the scientific revolution was mainly based on experiments and sensory observations (*sensate esperienze*, meaning sensible and sound experiences). It may be surmised, therefore, that planned experiments and detailed observations of the fish when they produce their torporific effects were what led Lorenzini and Réaumur to their mechanical explanations. The reality is not, however, quite this simple. As several researchers would point out in the second half of the eighteenth century, torpedoes really do not contract in a significant way at the moment of their discharges.

Thus, the transition from occult qualities to mechanical theories was not due just to the collection of new, easily observed, experimental facts. Although this is part of the story, it was also the result of a change in thinking about the natural world. Not only were older, inadequate ideas being questioned, but philosophers and scientists were now more willing to entertain new theories that might help them to explain their findings better. To use the language of Thomas Kuhn (1962), author of the influential book *The Structure of Scientific Revolutions*, what was taking place was indeed a "paradigm shift."

The scientific revolution of the seventeenth century was heavily biased towards mechanics, and the leaders of the movement attempted to explain both physical and also living phenomena on its basis, with an emphasis on matter in motion. Galileo, for example, successfully applied his mechanical ideas, including the principle of local movements, both to the cosmos and the terrestrial world. It was from a mechanical perspective that post-Galilean natural philosophers attempted to explain the movement of animals, as well as other physiological phenomena. With this new orientation, the torpedo became more mechanical and its phenomena were interpreted in this way, although, as mentioned, this fish really does not move when delivering a shock.

Both Lorenzini's and Réaumur's erroneous conclusions are comprehensible only within the mechanical paradigms of the scientific revolution. It is nonetheless astonishing that both scholars declared that they had observed movements of the fish's body at the moment of the shock. Even more perplexing is that Réaumur (1714, p. 351) had written he had perceived a movement, which "was so quick (*prompt*), that even more attentive eyes cannot perceive it." Said in a different way, he wrote that he had *seen* what could only be described as an *invisible* movement! And it was on the basis of this invisible movement that he elaborated his mechanical theory, which was pretty much unchallenged during the first half of the eighteenth century. Clearly, theoretical assumptions still were important during these now more empirical times, sometimes trumping, although at other times succumbing to, what the eyes could see and the hands could feel.

Before looking at Lorenzini's and Réaumur's ideas in detail, let us backtrack to the concept of occult qualities. This return to an aforementioned Renaissance theme deeply associated with electric fishes will help us to understand the paradigm shift that resulted in the Early Modern Era, the age of mechanics. As we shall now show, there was a rather "weird," dark, and gloomy side to the Renaissance—an epoch simplistically thought of as a luminous age, during which great minds effectively and totally brushed away the dark and entangling cobwebs of medieval culture.

This voyage back in time will reveal the extent to which the new scientists (*novatores*) of the seventeenth century really did break with the philosophical and scientific thinking that characterized Renaissance cultures, particularly as concerns occult qualities. Yet, as we stressed in Chapter 5, there can be no doubt that the men of the Renaissance, with their inquisitiveness about nature, philological studies, specialized texts, and new vision of art, contributed in substantial ways to intellectual and scientific revolutions of the seventeenth century. In an essay published about 50 years ago, art historian Erwin Panofsky (1954) indeed showed in most a convincing way how rooted even Galileo Galilei was in Renaissance culture, even though he was one of the main critics of occult forces as viable explanations for natural phenomena.

HERMETIC THINKING

One of the most remarkable contrasts between the Renaissance and the early modern period has to do with the role of the magic and the occult (Thorndike, 1923–58; Yates, 1964; Rossi, 2006). The allusion to occult qualities as an explanation for the torpedo's shock, which can be found in the works of most of the sixteenth-century ichthyologists that we met in Chapter 6, has a long tradition rooted in classic antiquity. The Renaissance ichthyologists, however, did not endorse this view simply because of their admiration for the classical world or interest in classical science, although both should be considered contributing factors. Rather, they did so also and perhaps even more because there was a revival of interest in magic, which enhanced the belief in occult forces and in the possibility that these forces could be manipulated by superior minds, men who could succeed in acquiring secrets of the "art."

During the Renaissance, magic was no longer considered a questionable practice or a private matter scarcely (if at all) tolerated by the Church, or as a belief system embraced mostly by members of the uneducated lower class. It was now viewed as a powerful tool for dominating the secret forces of nature, and also as an explanation for a wide range of phenomena that seemed hard to understand. It was only in the twentieth century, starting with the seminal studies by the great German art historian Aby Warburg and his followers, that Renaissance magic has emerged in a new light, showing the complexity of the culture and emotional-intellectual drives of this fascinating historical period (Warburg, 1912/1922, 1999).

In its original acceptance, and particularly in the works of Galen, the idea of occult qualities did not always have the strange connotations that we now associate with the concept or with magic. For Galen, occult qualities merely designated the qualities of a body that are difficult to understand with experimental techniques and reason. In his view, these qualities depend on the actions of a body in its totality: they could not be ascribed to, or explained on, the basis of a specific part.

Discussions about occult qualities continued into late Roman culture and then into the Byzantine era. They permeated writings from the Middle East and the West during medieval times. One need only consult the texts of Alexander of Aphrodisias, Plotinus, Simplicius, or Averroës, to name some of the earlier writers covered in this book. They discussed occult forces when dealing with the transmission or influence of a force or agent (e.g., light, heat) at a distance—a force that could not be explained by the simple propagation of matter. The torpedo, as we have seen, was frequently invoked in the framework of these discussions.

Along with these "scientific" discussions, occult qualities were also considered in more magical contexts, again with the torpedo and the remora frequently mentioned together. This more mystical tradition also had important antecedents, as we have already noticed when dealing with Pliny (see Chapter 3). In many respects, we can even find an implicit admission of the existence of occult forces in this context in the works of Albertus Magnus and Thomas Aquinas, the two Scholastic authors who at the same time declared their aversion to such magic.

The revival of magic in Renaissance culture was not simply a continuation of this interest. As mentioned in Chapter 6, what transpired had to do with a well-defined historical event, the translation of the *Corpus Hermeticum* by Marsilio Ficino in the second half of the fifteenth century (Fig. 9.1). Ficino, one of the greatest scholars of Italian humanism, was fluent in Greek, having studied it in his younger years. He had been stimulated by the discourses of Byzantine Neo-Platonic philosopher Gemistus Pletho, which had arrived in Italy at the time of the Ferrara-Florence Council, summoned by the Pope to reconcile differences between Eastern and Western Christianity.[1] Among the texts subsequently arriving in Florence were some Greek manuscripts, supposedly written by a most ancient and learned scholar, partially Egyptian and partially Greek, Hermes Trismegistus.[2]

---

[1] This Council was a landmark in the history of relationships between Europe and Ethiopia because of the presence of a delegation of Ethiopian prelates, who were interested in their Catholic colleagues and the learned society of Florence (see Lefevre, 1944, 1945, 1947).

[2] Trismegistus, meaning thrice great, alluded to the fact that Hermes was great as priest, philosopher, and lawmaker.

Figure 9.1.   Marsilio Ficino (1433–1499) and a 1549 Italian edition of the *Pimander* (Mercurio Trimegisto, 1549).

Hermes had been considered a prophet and a god (identifiable with the Greek god Hermes and the Egyptian god Thoth). But prior to this time, only one text of the Hermetic corpus had been known in a Latin translation (attributed to Apuleius of Madaura),[3] and this was the *Asclepius*, a magical and esoteric work. Since Grand Duke Cosimo de' Medici was strongly interested in astrology, he ordered Ficino to abandon his translation of Plato's texts and instead to dedicate himself to translating the *Hermetic Corpus* (see Yates, 1964).

Ficino's translation was published in 1471 with the title of *Pimander* (this being the name of the *Nous* or *Mens*, the supreme faculty of intuitive knowledge, showing Hermes the path he needed to follow to achieve ultimate knowledge and the ecstasy of supreme illumination; Hermes, 1471). The *Pimander* circulated widely, and its basic message was that the human soul should try to purify and elevate itself (following an ascending path) in order to gain primeval unity with cosmic and astral forces.

The path of elevation, with its strongly religious connotations, is mystical and emotional, not rational, and through it one could achieve the creative and divine powers lost when corrupted by sin. Here was a means of acquiring magical powers for controlling the forces of nature.

One of the most important aspects of Hermetic thinking is that such knowledge should remain within a circle of wise and superior spirits. That is, it should not be divulged to laymen, because it could have dangerous consequences on simple, non-receptive minds (this accounting for today's meaning of the word "hermetic").

The philosophical and religious conceptions underlying Hermetic thinking were derived partially from the *Timaeus*, mainly through Plotinus' interpretation of Plato's thinking. Here, supreme light is a metaphor for perfect knowledge, from which comes the term "illumination" to describe the final stage to be reached. This, in effect, made it a solar cult, with Apocentyn, the "ideal town" of the initiates, being a "City of the Sun," built following a complex geometrical and astrological scheme to favor the sun's good influence.

Moreover, Hermetic philosophy is based on the idea of a correspondence between the microcosms and the macrocosm. It involves an intimate relationship or "sympathy" between stones and plants and the cosmic and animated forces presiding over the movements of planets and with other astral forces. Thus, the plants and stones sympathetic with Venus can have the same influences as the planet Venus, and thus keep Saturn's gloomy influence at bay. Ficino would point this out in a letter to the young Lorenzo di Pierfrancesco de' Medici (see Gombrich, 1945 p. 16). Additionally, music was considered a powerful tool for propitiating the beneficial cosmic forces.

Sympathies (and their opposite, antipathies) had places in Hermetic medicine. Preparations using stones, plants, or animals, sometimes on the basis of a superficial resemblance between the natural object and a given part of the human body, now achieved broader acceptance. Drugs based on certain nuts, for example, were called for to treat brain diseases, because the brain and the nut kernel resemble each other. Indeed, medicine was tightly linked to astrology and alchemy within this broader framework; it was a perfect field for the application of this sort of magic.

In *Pimander* and in his subsequent works (e.g., his *De vita coelitus comparanda*[4]), Ficino, who was a priest, attempted to

---

[3] Apuleius was the author of the *Golden Ass*, one of the classics of Latin esoteric literature, based on the myth of Cupid and Psyche.

[4] This title is difficult to translate but can be rendered as "On Arranging Life in a Heavenly Manner," or as "The Way of Procuring Celestial Favor."

Christianize Hermetism. He did this partially by transforming Hermes into a prophet who announces Jesus, and by assimilating him with Moses.[5] He also pointed to the extreme antiquity of Trismegistus, his teaching being the highest and oldest expression of primeval knowledge and religion, a *prisca sapientia* or *prisca theologia*, a theme central to the philosophical and religious debates of the Renaissance.

For the sake of brevity, we cannot go deeper into Hermetic thinking. But before closing this digression, a few more important points must be made. One concerns the extent to which Hermetism permeated various aspects of Renaissance life. Here, for example, we can point to how it influenced the fine arts. Suffice it to say that some of the greatest masterpieces of Renaissance painting (e.g., Botticelli's *Venus* and *Primavera*) are pictorial expressions of Hermetic ideas based on astrological magic (see Gombrich, 1945; Yates, 1964). Hermes, in a partially Christianized acceptance, is also represented in the famous mosaics on the pavement of Siena Cathedral, and even in the frescoes painted by Pinturicchio for the decoration of Borgia's rooms in Vatican palaces. As first shown by Warburg in his seminal study on the "Decans" of the Schifanoia Palace in Ferrara, it is practically impossible to interpret the symbolism of the many frescoes and paintings of this age without a reference to Hermetism (Warburg, 1912/1922 and 1999). Further, it is certainly impossible to understand the meaning of many Renaissance emblems, or even to decipher the messages conveyed on the frontispieces of many books published in this epoch, without this essential knowledge.

As mentioned, medicine too was affected by Hermetic revival, thus reinforcing the magical and astral aspects already present in ancient and medieval medical traditions. Ficino was, in fact, also a physician. He used Hermetic tools and rites in his medical practice with the Florentine aristocracy, and his *De vita* was conceived of as a medical text. The penetration and persistence of Hermetic medicine is even attested to in the propitiatory rites Tommaso Campanella performed in the Vatican Palace in 1628, hoping to protect Pope Urban VIII from the bad omens that seemed to plague him (see Campanella, 1629).[6]

Another point to be stressed is that, unlike what Ficino and other Renaissance scholars assumed, the Hermetic texts were neither extremely ancient nor the works of a single author. They were produced between the second and the fourth century A.D. and comprised multiple writings created within the framework of the philosophical and religious movement known as *Gnosis,* which was based on a largely emotional and intuitive apprehension of supreme knowledge.[7]

Hermetism was, moreover, not really linked to Egypt and its priests. Egypt and its religion were evoked only as literary and mystic fictions within the cultural framework of the crisis of the classical world; it was a kind of refuge or depository for the most ancient and purest cult of the gods. We have already mentioned one aspect of this fictitious "Egyptianism," namely the production of the *Hieroglyphica of Horapollo*, which was at its core an attempt by a person (or persons) familiar with Greek culture to fabricate and interpret certain hieroglyphs, and which was done in ways that would have astonished the ancient Egyptians (see Chapter 2). Not surprisingly, the *Hieroglyphica of Horapollo* was translated and eagerly read by Renaissance scholars, leading to a revival of interest in hieroglyphs, considered as a symbolic writing particularly appropriate to convey the *prisca sapientia*.

Piero Valeriano, as noted, wrote one of the main works dealing with hieroglyphs within the framework of Renaissance Hermetism. He attempted to provide an explanation for the fictitious hieroglyph representing the torpedo that is verbally described in the *Hieroglyphica of Horapollo*.

THE TORPEDO AND OCCULT QUALITIES DURING THE RENAISSANCE

It was a custom of almost all Renaissance scholars interested in magic to mention the torpedo together with the remora and other more or less fantastic animals. They did this to substantiate the view that there are some phenomena in nature that could be accounted for by mysterious forces, as opposed to known physical or material mechanisms. Although this had also occurred during the classical era and throughout the Middle Ages, starting with Ficino the mention of the torpedo together with prodigious animals (and *mirabilia*) becomes almost constant. In his *De vita*, the Florentine humanist set the stage in a particularly expressive way for transforming the torpedo into a philosophical tool for substantiating a magical vision of medicine and life:

> Who does not know that the occult virtues of things, which are called "specifics" by physicians, come not from an elementary nature but from celestial ones? Therefore rays [from the heavens] can impress on images wondrous and occult powers ... [that] can arise much more quickly than in various mixtures of elements and elemental qualities.... Is it not said that certain families among the Illyrians and the Triballi kill people by looking at them when they are angry, and that certain women in Scythia do the same? And catoblepas and regulus [i.e., basilisk] snakes destroy people by shooting rays from their eyes. The marine torpedo also suddenly benumbs the hand that touches it, even at distance through a rod, and by contact alone the echinus fish[8] is said to retard a great ship. The phalangia in Apulia transform and quickly benumb the spirit and the mind with a secret [*occulto*] bite of some sort. How does a rabid dog have its effect if no bite appears? What about the broom, and moreover what about the arbute? Is it not true that by a very slight contact they activate the venom and rabies? Will you then deny that celestial bodies can perform wonders with the rays of their eyes as soon they gaze on us? (Ficino, 1471,

---

[5] On his side, Moses, being assimilated with Hermes and with Zoroaster, was seen as a great magician during the Renaissance.

[6] Urban VIII was the pope during the Galileo conflict, and Campanella, a philosopher, profoundly imbued with astral and magical conceptions, is particularly known for his *Apologia pro Galileo*, published in 1622.

[7] It is within the same cultural context that the *Kyranides* is situated. The first text in particular is imbued with Hermetic thinking. Interestingly, contrary to what Ficino assumed, despite the superficial allusions to Jesus in the Hermetic texts, the Gnosis flourished mainly as a pagan movement against the emergence of Christianity and was thus, fundamentally, anti-Christian.

[8] Here Ficino repeats an error, already present in Albertus, calling the remora *echinus* (i.e., with the name of the sea-urchin). As mentioned, this error was due to the similarity between *echinus* and *echeneis*, the Greek name for the remora.

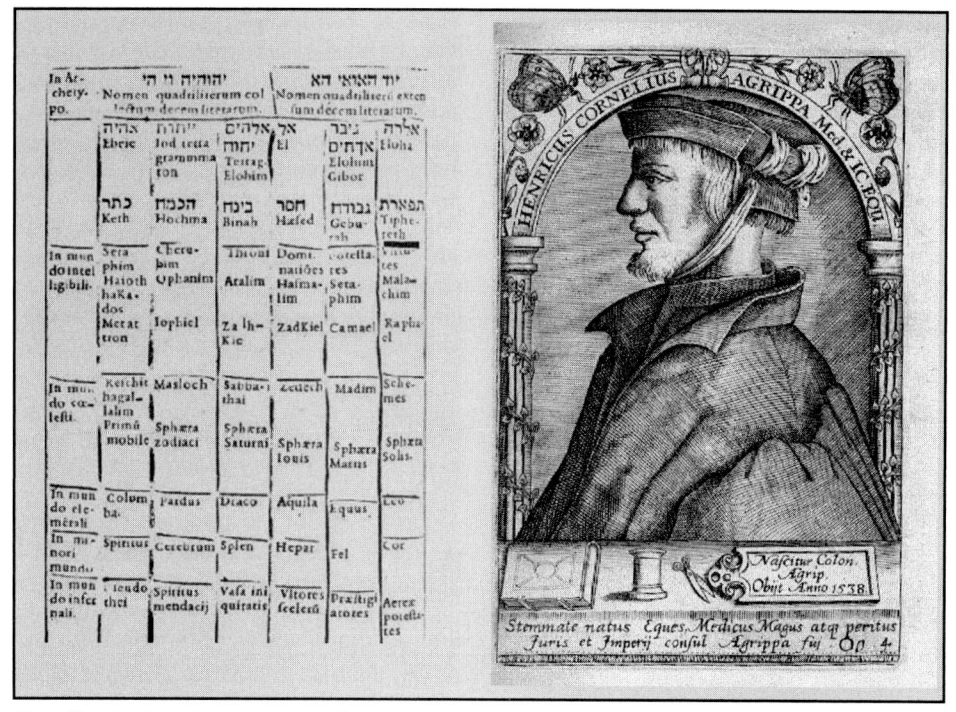

Figure 9.2. Henricus Cornelius Agrippa (1486–1535) and a page from his *De occulta philosophia*, a work devoted to natural magic that mentions the torpedo under "Lunary fishes" (Agrippa, 1533).

III book, cp. 16; translation partially based on Copenhaver, 1990, pp. 275–276).[9]

Cornelius Agrippa von Nettesheim (1533) was one of Ficino's followers (Fig. 9.2). The torpedo is mentioned in the first book of his popular *De occulta philosophia*, which is devoted to "natural magic" (i.e., that concerning the lower or "elemental world," which is under the influence of the "intellectual world," which is itself influenced by the superior or "celestial world," the subjects of his second and third books, respectively). Following a principle dominating Renaissance magic, which is linked to a resurgence of Plato's views, the natural world is animated by an *anima mundi* (soul of the world). Astrological factors are at play here, with the planets and stars linked to "occult virtues," which are hidden in natural objects.

Together with the remora, the fishing-frog, and other sea species, *torpedo marina*, or sea torpedo, is listed under "Lunary fishes" (i.e., sea dwellers that are influenced by the moon and can, in turn, convey its influences; Chapter XXIV). An even more interesting allusion to the torpedo, again with the remora and other linked sea species (e.g., starfish and the "sea-hare," a type of *Aplysia* or sea-slug), is present in Chapter XLVI, which deals with "Natural Alligations and Suspensions," meaning those natural objects that act though physical contact or when carried about, and which can be used either for sorceries or therapeutic purposes. Agrippa explains the rationale of this type of action at the beginning of the chapter in a passage that is worth quoting *in extenso* because it reflects the Renaissance belief in magic and connects enchantments with medicine:

> When the Soul of the World by its virtue doth make all things that are naturally generated or artificially made to be fruitful, by infusing into them celestial properties for the working of some wonderful effects, then things themselves not only when applied by suffumigations, or collyries, or ointments, or potions, or any other such like way, but also when they, being conveniently wrapped up, are bound to or hanged about the neck, or in any other way applied, although by never so easy a contact do impress their virtue upon us. By these alligations, therefore, suspensions, wrappings up, applications, and contacts, the accidents of the body and mind are changed into sickness, health, boldness, fear, sadness, and joy, and the like. They render them that carry them gracious or terrible, acceptable or rejected—honored and beloved or hateful and abominable. (*De occulta philosophia*, XLVI. Translation revised, from Agrippa, 1898s, vol. I, pp. 139–40)[10]

After mentioning that it is through an influence of the "alligation" type that, in the botanical world, the vital virtue will be transmitted when grafting a tree, or in a fecundation of a palm tree, Agrippa comes to the torpedo (and other sea species endowed with mysterious powers) with these words:

> In like manner we see that the cramp-fish, or torpedo, being touched afar off with a long pole, doth presently stupefy the hand of him that toucheth it. And if any shall

---

[9] As in the cases of Ficino, some of the works of the Renaissance (and of later periods as well) dealing with magic and occult uses are written in a particularly elaborate style with a particular abundance of rhetorical questions, useful to substantiate the thesis expounded in the lack of more solid arguments. In his harsh criticism of the old culture, Galileo also addressed these stylistic aspects.

[10] We have chosen this translation because of its old English style, which may convey the rhetorical aspects of Agrippa's convoluted prose. It is, however, full of errors that we have tried to emendate. Notably, Agrippa's term *torpedo marina* is translated as "tortoise" in the passage on "Lunary fish."

touch the sea-hare with his hand or stick, he will presently run out of his wits [*animu defectu incurrit*]. Also, if the fish called stella, or starfish, as they say, being fastened with the blood of a fox and a brass nail to a gate, evil medicines can do no hurt to any in such house. Also, it is said that, if a woman take a needle and soils it with dung, and then wraps it up in earth in which the corpse of a man was buried, and shall carry it about her in a cloth which was used at the funeral, no man shall be able to have sexual intercourse with her, so long as she hath it about her.[11]

Many other Renaissance magicians and philosophers also argued for the existence of occult forces in the natural world by pointing to the torpedo, and many also employed it their *magia naturalis*. They include Pietro d'Abano, Giulio Cesare Scaligero, Girolamo Fracastoro, Jean Fernel, Bernardino Telesio, Jacopo Zabarella, Girolamo Mercuriale, Giordano Bruno, Tommaso Campanella, Jean Riolan, Giulio Cesare Vanini, Gerolamo Cardano, and others.[12] Although we cannot deal with each of these individuals here, special mention must be given to Pietro Pomponazzi, because he was critical of some of these conceptions and was at the crossroads of the cultural tensions.

Pomponazzi was a professor at Padua, Ferrara, and Bologna, and one of the most important Aristotelian philosophers of the Renaissance. On the basis of his studies of the *Peripatetic corpus*, and influenced by the celebrated Aristotelian commentator, Alexander of Aphrodisias, he showed that Aristotle was not a kind of Christian philosopher *ante litteram*, although this view had been fostered by Albertus Magnus and Thomas Aquinas. When he remarked that Aristotle had contended that the human soul is mortal while the cosmos is eternal, he found himself in trouble with the Inquisition (this also occurred with other Aristotelian philosophers of the epoch who shared such views).

Pomponazzi deals with occult qualities in a philosophical and medical context in his *De naturalium effectuum causis sive de incantationibus* ("On Spells or Causes of Natural Effects"), written in 1520 and widely circulated in manuscript form before being mass printed in 1556 (with enormous influence in Italy and abroad). His intention was to provide natural, as opposed to supernatural, causes for various effects, including those usually ascribed to demonic (or angelic) interventions.[13] He does not completely dismiss the possibility of demonic interventions to modify the course of natural and human events, most likely because the Church taught that demons do exist.[14] But if they exist, he writes, they should produce their effects by operating through forces that are part of the natural world, and according to the laws governing these natural forces.[15] Yet it is unlikely that this could happen, continues Pomponazzi, because demons are not supposed to know the laws of nature like natural philosophers and physicians do. There is, moreover, no real need to bring up the intervention of demons, because the phenomena some have attributed to these spirits can be explained simply on the basis of natural forces, although some could be more difficult to account for than others.

To substantiate this view, Pomponazzi invokes the classical ensemble of more or less legendary animals, considered, at least since Pliny's time (along with lodestones, medicinal plants, etc.), as evidence for the extraordinary possibilities of nature. Here we find our well-known torpedo, together with the remora, basilisk, and other frightful serpents. Because,

> if a fish the size of a finger size can hold back a loaded ship of two hundred feet or more, driven by wind and oar; if the torpedo-fish can stun at so great a distance with no manifest action [*insensibiliter*]; or the regulus or basilisk can infect so great an amount of air; and many such kinds of serpents do unbelievable things frequently described by philosophers and physicians; ... then it seems to me that if demons do this by applying actives [forces] to passives, then humans can understand such things naturally as well, from which it also follows that many who were considered magicians and necromancers, like Peter of Abano and Cecco d'Ascoli, may have had no commerce with unclean spirits, and, with Aristotle, they may even believe that demons do not exist. (Sect. 20; Pomponazzi, 1556, pp. 42–43; translation based on Copenhaver 1992, p. 397).

In addition to Pliny, Pomponazzi invokes the authority of Albertus on the extraordinary powers of stones, plants, and animals. He does this to situate his views within the authoritative Christian tradition, since Albertus was a saint

---

[11] This passage may remind people familiar with Italian opera of the air of Verdi's *Il Ballo in Maschera*, in which Amelia goes to a cemetery to collect a herb in order prepare an enchantment capable of reducing the fateful love attraction Riccardo exerts on her: *Ecco l'orrido campo ove s'accopppia/al delitto la morte!* (Here's the ghastly field where delict is joined to death).

[12] Girolamo Cardano, a Jesuit, physician, natural philosopher, and collector, is particularly interesting. Born in Pavia in 1501, Cardano was a friend of Leonardo da Vinci and did some teaching at the universities of Pavia and Bologna. It is estimated that he wrote over 200 works, using wonders as a basis for his studies of the nature of the universe, which he viewed as filled with rare, occult, and chance effects. He boasted that he had solved some 240,000 problems in his lifetime, earning him the Latin accolade of *vir inventionum* (the man of inventions). Cardano made some regrettable mistakes, including confusing torpedoes with other fishes in his *De subtilitate rerum* ("The Subtlety of Things") of 1560. Rondelet (1554) would criticize him for these mistakes in his *De piscibus marinis*. Cardano later corrected some of his errors. To his credit, he drew attention to the possibility that special organs might be responsible for the torpedo's "craft" (Cardano, 1557, Book VII, Ch. 3).For more on Cardano, see Cardano (1560, 1557, 1562) and Marcus (1983). For the views of Mercurialis, see Mercurialis (1571, Book I, Ch. 23, p. 96). Of course, allusions to the occult would not end abruptly in the early 1600s, as can be seen in the writings of Villalpando Francisco Torreblanca (1623) and Lazare Meyssonnier (1639) (also see Monceaux, 1683).

[13] One of Pomponazzi's aims was to disprove the supernatural nature of the many supposed miracles that happened in his era. In this regard, Pomponazzi is an anticipator of the libertine thinking of the Enlightenment.

[14] Pomponazzi has been criticized for adhering to the doctrine of the "double-truth"—holding two opposing views in the same argument: one based on faith and the teachings of the Church and the other derived from experience and reasoning. But the double-truth was of great importance because it permitted a degree of intellectual freedom during a period of stringent religious controls (Pine, 1968). Pomponazzi's writings also display another attitude typical of the age, the so-called "honest dissimulation."

[15] For Pomponazzi, the same ideas hold for angelic interventions. After Galileo, we will find Pomponazzi's thoughts echoed in the idea that, if a superior agent is involved in the physiological operations of a body, it must operate according to the law of nature. This idea is particularly well expressed by Marcello Malpighi in a famous metaphor expounded in his *Opera posthuma*, where he writes about an angel moving a machine, who must operate according to the laws of mechanics. In the case of Malpighi (and of other post-Galilean scientists) this argument is mainly used to convey the thought that science should limit itself to immediate laws and not go into prime causes (Malpighi, 1698, p. 212; see Piccolino, 2005b, pp. 131–179).

and one of the most celebrated Catholic scholars. From this perspective, it was dangerous to refer only to Pietro d'Abano and Cecco d'Ascoli, two medieval physicians and philosophers brutally persecuted by the Church for "commerce" with demons. Interestingly, Pomponazzi also mentions two great authorities, namely Galen and Boethius, among those suspected of having practiced demonology. This occurred, he writes, because of the great achievements they attained with their science.

As with Agrippa, Pomponazzi's attitude toward the magical aspects of natural phenomena is strongly pervaded by Neo-Platonic influences, especially as derived from Ficino. One of the notions he takes from the Florentine humanist is that human powers can be even more extraordinary that those of other life forms or objects. In this context, he develops the concept that humans can influence other humans, in addition to acting on natural forces, through their psychic and mental powers. Here he brings up the possibility that imagination and suggestion can affect our own bodies, as well as the bodies and minds of others. To support his view, he cites both sound evidence (e.g., the possible physiological consequences of strong emotions and passions, the effects of attitude on one's health) and a plethora of questionable traditional beliefs. Although some of Pomponazzi's ideas are wishful thinking (e.g., that a mother's strong wishes during a pregnancy could have specific effects on her baby), we can find much that is basic to psychotherapy today in his Renaissance text.

Pomponazzi's writings also reflect another significant facet of Renaissance culture, one that will be further developed in the post-Galilean epoch and will directly bear on electric fishes. When accounting for the effects of suggestion, imagery, and natural objects, he hypothesizes the emission of minute material corpuscles. Through these exhalations, men with highly developed therapeutic powers could treat diseases and wounds without ordinary physical contact. In a similar way, he writes: "rhubarb has the virtue to purge gall; however, rhubarb does not purge gall if it not activated by the natural heat and transformed in vapors" (p. 44).[16]

Vapors could account for the captivating powers exerted by witches or snakes, the dangerous effects of an envious person or of a menstruating woman, and numerous other actions, including the effects of cerates and poultices (both beneficial and maleficent). If material bodies can produce these surprising effects, just think of what could be expected from the human soul! But even in the case of the soul's effects, the mechanism is a material one: "Indeed, in our opinion, the soul operates these things in no other way than through an alteration and by vapors transmitted from her," which are affected by a given virtue or malice (p. 50). Interestingly, we find leprosy and other contagious diseases among the affections that can be transmitted through vapors or exhalations. Pomponazzi is well aware of the possibility of transmission by contagion, and he invokes contagious diseases as evidence of the effects that vapors can have on the body.

The concepts of contagion and emissions are actually ancient notions. Galen warned people to stay clear of marshes, swamps, and sewers on sticky, sweltering days, and to avoid the stench of rotting corpses and decaying vegetation. He even theorized that "pestilential seeds" might be carried in the air from these sites, and even from person to person (Nutton, 1983). These seeds of contagion, he thought, could cause havoc, especially in poorly regimented bodies.

Indeed, the idea of virulent "seeds" even precedes Galen. Lucretius and Varro, two writers from the previous century, used comparable terminology (Nutton, 1983). The concept of pestilent seeds, however, had attracted very little attention in the Renaissance before Girolamo Fracastoro revived it. Fracastoro published his *Syphilis sive de morbo gallico* in poetic form in 1539, and his *De contagione* in 1546, and, as we have noted, Pomponazzi's *De incantationibus* appeared in print in 1556.

Fracastoro's and Pomponazzi's texts reflect a renewed interest in atomism, the ancient doctrine basic to the philosophies of Democritus and Leucippus. This revival had a lot to do with the publication of Lucretius' poem *De rerum natura*. Because of his materialistic views of nature, and because he refused to accept God's hand in the final organization of world, *De rerum natura* became one of the most prohibited books in the history of Western civilization. Nevertheless, after humanist Poggio Bracciolini rediscovered it in 1417, and it began to be printed (initially in 1473), it also became one of the most influential texts in the history of science (Beretta & Citti, 2008).

It was within this cultural climate that Fracastoro and Pomponazzi wrote about the diffusion of diseases and other things that could be influenced by material corpuscular emissions. Inherent to these views was the possibility that drugs and venoms could also act in this way—as could the torpedo. From this perspective, we can now understand why there are repeated references to torpedoes in Renaissance texts on venoms and various related matters (e.g., in those of Ferdinando Ponzetti, Jacques Grevin, Andrea Bacci, Eberhard Gockel, Matías García, Johan Lindestolpe, and others).[17] Atomism would remain a popular theme during the scientific revolution, and it can be found in the writings of Galileo, René Descartes, Sir Isaac Newton, Robert Boyle, physician Thomas Sydenham, and even Benjamin Franklin (for his corpuscular theory of colds and flu, see Finger, 2006, pp. 151–164).

From this viewpoint, the importance of atomism in general, and of Lucretius in particular, was both broad in scope and remarkably profound. Together with other significant intellectual and social factors, it helped shape a mental and practical attitude toward the world that differed dramatically from that dominating before the Renaissance. Not to be overlooked, it allowed natural philosophers to separate themselves from some principles of ancient philosophy that had long hindered the development of science and medicine.

---

[16] Pomponazzi's mention of rhubarb in the context of occult properties of natural objects is significant, because rhubarb was one of the medical plants Galen thought could possess these kinds of properties, identified as *idiotetes arretoi* (un-nameable or indescribable qualities) in Galen's terminology. Galen resorted to indescribable qualities because he found nothing in the manifest characteristics of the plant that could account, on the basis of the medical theories he endorsed, for the bile-purging actions of rhubarb.

[17] For more on Grévin and Ponzetti, as representative, see Grévin 1567–68, and Ponzetti, 1521.

The new theories of the torpedo shocks that would emerge in the post-Galilean period would resound with atomistic conceptions. As we shall see, Galileo himself favored an atomistic approach to science, and there is even an explicit reference to his atomism in Stefano Lorenzini's text of 1678, where he presents his corpuscular conception of the torpedo's shock. But before relating atomism to the emergence of modern science and to the torpedo in particular, we must briefly draw attention to some other factors during the Renaissance that helped the *novatores* to lay the groundwork on which they would construct their new science—developments that would allow them to go beyond some of the tenets of Aristotelian physics and philosophy.

Among the intellectual factors, we must point to the discovery (or rediscovery) of many scientific texts of the classical period, and especially the works of physicists and mathematicians from the Hellenistic or Alexandrian period. Among these texts we find those of Archimedes, Euclid, Apollonius of Perga, Apollonius Rhodes, and Hero of Alexandria, and they and others were intensively studied. In the works of some of these authors, there are fundamental problems of classical physics (such as movement, gravity, and vacuums) that are considered in radically different ways from the Aristotelians.

The Renaissance also witnessed the development of new technologies in many fields and the construction of more sophisticated machines. These devices revolutionized various fields, some being manufacturing, hydraulics, pneumatics, metallurgy, mining, and navigation. Importantly, by abandoning an attitude that ascended up to Plato and his student Aristotle, who considered technical and practical knowledge (*téknē*) inferior to science (*epistēmē*), and therefore not worthy of great intellects, Galileo, Newton, and other protagonists of the scientific revolution, like Leonardo da Vinci during the Renaissance, were extremely interested in technologies. Their interests went beyond just using scientific and mathematical elaborations to construct better mechanical devices. They also believed that by studying well-crafted machines, scientists could better understand physics and mechanics. Indeed, the study of mechanics could be applied to the machinery of the human body and to the parts of other animate and motile bodies.[18]

Coming to the atomism in Galileo's intellectual elaboration, it is interesting to note that he might have endorsed atomism early on in an anti-Aristotelian perspective (Camerota, 2004 and 2008). In the course of his work, Galileo considered atomism on various occasions and in several respects. One was the problem of gravity, which Aristotle had attributed to the tendency of bodies to go toward their natural sites (the center of the earth for the heaviest ones and the "sphere" of fire for the lightest ones). Galileo refuted this view and, within an atomistic framework, he considered the heavier bodies to be "those bodies that contain a greater number of particles within a narrower space" (Galileo, *De motu antiquiora*, manuscripts published in Galileo's *Opere*, 1890–1909, vol. I, p. 253). He refuted, moreover, the Aristotelian conception of heavy and light bodies *per se*. Instead, he assumed that the tendency to ascend or descend is due to the action of the medium in which a body is immersed, citing the atomists and Archimedes. A body goes up if it contains a smaller number of particles than the medium in a unit of space, and it falls down if the opposite is true.

These views were strongly anti-Aristotelian because they were based on the assumption that vacuums exist, a notion that the great Greek philosopher strongly opposed. Similar conceptions to those of Galileo can be found in the texts of the Greek atomists and in Lucretius' *De rerum natura*, a work that he might have known either directly or indirectly through the courses taught by Francesco Buonamici at the University of Pisa. Indeed, many of the atomistic passages of Galileo's discussion on the nature of gravity have linguistic and conceptual similarities with passages in Lucretius' poem that deal with the same subject.

Galileo provided an atomistic explanation in a different circumstance in 1611–1612 (after he first published his telescopic observations), during a debate with a group of Florentine Aristotelians on the reasons why some bodies float in water. Aristotle attributed buoyancy mainly to the shape of the bodies, whereas Galileo advocated Archimedes' conception, which was based on the concept of specific gravity. Within the framework of this discussion, particularly in order to account for the effects of changing the water temperature, Galileo referred to the actions of minute igneous particles, explicitly using atomistic terminology. The expressions recurrent in his texts for these minute igneous particles are *atomi ignei*, *atomi del fuoco*, *atomi calidi*, and *sottilissimi atomi* (meaning igneous atoms, atoms of fire, hot atoms, and very thin atoms, respectively. This terminology would recur in Lorenzini's new mechanical theory of torpedo shocks, which would be based on the intervention of corpuscular emissions (see below).

Galileo also deals with atoms from a mathematical point of view in other texts—for example, when discussing what we would now call infinitesimal calculus. For us, however, what is particularly interesting and important is how Galileo implied the intervention of atoms when discussing issues of sensation. For this, we turn to his *Saggiatore*, an important polemic work published in 1623, during the pontificate of Urban VIII, whom we mentioned above (on Galileo and the senses, see Piccolino, 2005b, and Piccolino & Wade, 2008a,b).

An analysis of the nature of heat and thermal sensations is the immediate departure point for Galileo's discussion of sensory mechanisms in Chapter 49 of the *Saggiatore*. He tackles the problem from general principles by first considering the properties of natural objects, which he separates into two distinct categories (later corresponding to primary and secondary qualities). The first category concerns the spatial aspects of material things, such as position, dimension, and movement, which he considers to be objective and intrinsic attributes of reality. For Galileo, the existence of these attributes cannot be doubted, and they are intimately associated with the objects themselves. His second category involves those properties that, in his opinion, have no real existence *per se*, and according to him exist only in individuals endowed with the sensory systems capable of detecting them.

---

[18] Galileo was one of the protagonists of the scientific revolution who pointed particularly to the importance of the knowledge acquired by expert craftsmen for scientific investigations of natural phenomena. One of his main works, the *Discorsi e dimostrazioni matematiche sopra due Nuove Scienze*, which lays the ground of modern statics and mechanics, begins with a famous *Incipit, Largo campo di filosofare*, which is an extraordinary eulogy of the engineers (*proti*) of the Venice arsenal.

In Galileo's words:

> I say that, as soon as I conceive a piece of matter, or a corporeal substance, I feel myself necessarily compelled to conceive along with it, that it is bounded, and has this or that shape, that in relation to some other body it is either small or large; that it is in this or that place, and in this or that time; that it is in motion or at rest; that it either touches or does not touch some other body; and that it is one, few, or many; nor can I separate it from these states by any act of the imagination. But I do not feel my mind forced to conceive it as necessarily accompanied by such states as being white or red, bitter or sweet, noisy or quiet, or having a nice or nasty smell. On the contrary, if we were not guided by our senses, thinking or imagining would probably never arrive at them by themselves. This is why I think that, as far as concerns the object in which these tastes, smells, colors, etc. appear to reside, they are nothing other than mere names, and they have their location only in the sentient body. Consequently, if the living being (*animale*) were removed, all these qualities would disappear and be annihilated. (Galileo, 1623, pp. 196–197)

To clarify the distinction between the two types of attributes, Galileo brings up the sensation of tickling:

> I move one of my hands, first over a marble statue, and then over a living man. As far as concerns the action which comes from the hand, it is one and the same for each subject, and it consists of those primary accidents, namely motion and touch; and these are the only names we have given them. But the animate body which receives these actions, feels different affections depending on which parts are touched. For example, when touched under the soles of the feet, on the knees, or under the armpits, in addition to the ordinary sensation of touch, there is another sensation to which we have given a special name, by calling it tickling. This affection belongs wholly to us, and not a whit of it belongs to the hand. And it seems to me that it would be a serious error if one wanted to say that, in addition to the motion and the touching, the hand had in itself this distinct capacity of tickling, as if tickling were an accident which inhered in it. (pp. 197–198)

With his mechanical views, Galileo had no difficulty in conceiving how tactile or auditory sensations could be produced on the basis of the spatial characteristics of the object or of movement (undulating vibrations in the case of sound sensations). To account for other sensations, such as thermal, gustatory, olfactory, and visual ones, which could not be explained on purely mechanical grounds, he envisioned "minimal corpuscles" or simply *minima* emanating from external objects. These minute bodies, he contended, could act as intermediaries for stimulating specific sensory systems.

In this respect, Galileo was clearly reviving, although in a more modern way, the doctrines of sensation that Democritus, Epicurus, and particularly Lucretius had promoted long before his time. Nevertheless, we must add that Galileo only uses terminology that is strongly atomistic (e.g., *atomi realmente indivisibili*, meaning "really indivisible atoms" and *minimi quanti*) when dealing with light. For the other senses, and particularly for thermal sensations, he uses the less specific terms, including *ignicoli* (fire corpuscles), *minimi ignei* (igneous minima), *minimi* (minima), *particelle minime* (minimal particles), and *corpicelli minimi* (minimal corpuscles).

In Galileo's view, these minimal corpuscles can vary in their subtlety and in the speed of movement, so that some might be lighter and move faster than others, this affecting the specific senses they can excite. The sensation can be pleasant or unpleasant, mainly "according to the multitude and speed of those minima." He tells us, moreover, that the thinnest and fastest are involved in luminous sensations. Specifically, when speaking of the "flying particles" responsible for several types of sensation, he writes that "when their ultimate and highest resolution into truly indivisible atoms is reached, light is created" (p. 201).

The important point in Galileo's philosophical elaboration on senses is that he stresses in a particularly clear way that there is nothing inherently sensorial in the minimal corpuscles responsible for most perceptions. Nor are there any differences between the corpuscles responsible for the different sensations, other than two: the degree of corpuscle dissolution and the speed of movement (if there is no movement, there is no sensation). In denying any additional differences between the corpuscles responsible for the different sensations, he goes beyond what the ancients had written, because the scholars of ancient times believed that atoms could have different shapes that could affect sensation (e.g., sharp-pointed atoms cause pain). Galileo's thoughts about corpuscles, movements, and sensations would find a receptive audience in Lorenzini, who, as we shall see, would incorporate these various notions into his theory of the torpedo's shocks.[19]

As with the case of gravity and the problem of buoyancy, Galileo's elaboration on senses is also strongly anti-Aristotelian. This is because Aristotle had assumed sensory qualities to be objective attributes of objects, this being particularly true for those types of sensations (e.g., colors, tastes, and smells) that Galileo considers to be devoid of any real existence ("pure names") in the absence of the sentient individual. For Aristotle, these inherent sensory attributes of reality ("proper sensibles") are specific qualities of objects aimed at interacting with specific sensory systems. They cooperate with other sensibles, such as those involved in spatial sensations (e.g., form, position of a shape) that Aristotle calls "common sensibles." In Aristotle's view, the world has a specific language finalized to interact with sensory systems, and particularly with the senses of humans, the highest level of living beings. Galileo's sensory theories, on the contrary, point to the non-existence of any attribute of reality specifically aimed at interacting with specific senses, since atoms are devoid of sensory connotations and produce

---

[19] Galileo's views, which in part extend the conceptions put forward by contemporary authors (e.g., Estevao Rodrigues de Castro, Professor at Pisa University), are extremely modern. They correspond to some of the basic principles underlying modern sensory physiology in that nothing in the external world is considered to have an inherent sensory specification, and sensory connotations are acquired only when external energies interact with a specific sensory system. In the words of a perceptive contemporary vision scientist, Richard Gregory: "We should realize quite clearly that without life there would be no brightness and no colour. Before life came, especially higher forms of life, all was invisible and silent though the sun shone and the mountains toppled" (Gregory, 2005, p. 85).

sensations only because of their movements and states of aggregation.[20]

As already mentioned, the atomistic theory of sensation developed by Galileo was of great importance for Lorenzini's hypothesis of electric fish discharge, based on the emission of corpuscles produced by the contraction of the *musculi falcati*. Lorenzini would situate the power of the fish well outside the domain of "occult qualities" for the first time with a modern flavor and with recourse to experiments, using this atomistic approach.

As with all historical transitions, the shift from Renaissance "occult qualities" to early modern mechanical and atomistic theories of nature, and specifically to corpuscular theories of the fish discharge, was less than abrupt. Prior to Galileo, we had Pomponazzi's exhalation conception of how the torpedo produces its effects, although elaborated in a different context. Further, corpuscular theories of the shock were put forward by Gassendi and Boyle prior to Lorenzini (see Boyle, 1738). However, although they were more atomistic, their thoughts were not based on direct experiments on torpedoes, and they did not involve definite mechanical actions. In these domains, Lorenzini clearly took the next important steps. As we shall see, the theory of torpedo shock elaborated by Athanasius Kircher just before Lorenzini's is also based on a material exhalation, although the general framework in which it is developed is still impregnated with old scientific conceptions.

As for the general importance of atomism in the emergence of the scientific revolution, a particularly significant aspect of what occurred has to do with the epistemological sterility of the ancient but lingering Aristotelian dualistic conception of "matter and form." This conception was used to account for many aspects of the physical (and metaphysical) world. The possibility of explaining the quality and actions of a natural object, from stones to animals (and also of accounting for the specific physiology of living organism), on the basis of specific or "substantial" forms, had precluded natural philosophers from investigating the mechanisms underlying the production of various phenomena.[21] That is, any phenomenon and any property could be accounted for simply by saying that a given object is endowed with its "substantial form" or some related but still hollow term, such as "virtue," "affection," or "quality."[22]

There is no such simple explanation for the properties of things within an atomistic context, and particularly within the Galilean conception, which assumes that corpuscles differ only in their density and movement. To account for the physical properties of matter (and also for a variety of physiological phenomena), it is important to know how atoms are assembled, and thus the structure of objects, the mechanics of their interactions at microscopic and macroscopic levels, and other aspects of their motions. Hence, with regard to living organisms, the atomistic paradigm served as an important stimulus for anatomical investigations, both with the naked eye and at a microscopic level.

In particular, microscopic investigations are especially important to account for those physiologic mechanisms that elude simple mechanical explanations based on the visible configurations of the parts. In this domain, nervous functions stand out, because the nerves, unlike the muscles, do not move in a visible way when they are active. This is one of the reasons for the great importance of microscopic investigation in the post-Galilean era, as reflected in the works of Giovanni Battista Hodierna, Marco Aurelio Severino, Marcello Malpighi, Robert Hooke, Anton van Leeuwenhoek, and many others.[23]

Finally, it is notable that, besides having developed the telescope for his astronomical investigations, Galileo also invented the microscope and used it starting in 1610 to observe the structure of minute insects. The term "microscope," however, did not exist at this time; Johannes Faber, a member of the *Accademia dei Lincei*, to which also Galileo belonged, would come forth with this term in 1624.[24] The impulse to study the internal structure of objects also came from another of the great fathers of the scientific revolution, Francis Bacon, who pointed to the *schematismus latens* or a hidden configuration of objects, which he opined should be investigated to account for their properties.

A CONSERVATIVE PRIEST, A FLORENTINE POLYMATH, AND THE TORPEDO

Old ideas have ways of continuing for surprisingly long times, even during periods of relatively rapid change. By way of example, we can mention that Aristotelian philosophy and science were still basic in many European universities until late in the eighteenth century. Moreover, despite the Copernican revolution in astronomy, the Ptolemaic cosmological system, with the earth at the center of the universe, was taught even later, particularly in religious institutions, in some cases until the beginning of the twentieth century. No wonder, therefore, that despite the work of Bacon, Galileo, Descartes, and other notables, some scholars of the late-seventeenth century still adhered to a Hermetic and magic vision of the world—one dominated by

---

[20] Galileo was promoting a revolutionary shift in sensory physiology as important of that produced by the Copernican revolution in cosmology. In both respects, humankind was deprived by its privileged position in the universe. We were no longer at the physical center of the cosmos, and not at the center of reception of a specific language of a nature specifically adapted to reach us by way of the senses.

[21] See Copenhaver (1990) for a good discussion of the concept of "substantial form" in the Aristotelian philosophy of the Middle Ages. This text analyzes the concept in the context of "occult qualities" in general, and specifically of the power of the remora and torpedo.

[22] No wonder this terminology and the underlying conception was the target of the harsh criticism of Galileo, particularly in his *Dialogo sopra i due Massimi sistemi del mondo* and *Il Saggiatore*. People familiar with Molière know how one satirizes this attitude in his *Malade imaginaire* ("Imaginary Invalid"), when one of his characters accounts for the sleepiness produced by opium by saying, "*Quia est in eo virtus dormitiva, Cujus est natura sensus assoupire*" ("Because in it there is a sleep-inducing virtue, the nature of which is to make senses sleepy").

[23] Interestingly, one of the first works of microscopic anatomy, Severino's *Zootomia deomocritea*, alludes in its title to one of the founders of atomism, the Greek philosopher Democritus of Abdera. Hooke's classic work is his *Micrographia: Or some Physiological Descriptions of Minute Bodies Made by Magnifying Glasses*, first published in London in 1665. For more on Leeuwenhoek and his "wonderful little animals," see Dobell, 1960.

[24] Galileo was strongly interested in studying living bodies, as he announced in a letter written in 1610, where he speaks about a work he had planned to title *De animalium motibus*. Galileo's program of animal investigations would be pursued by his direct and indirect followers, and particularly by Giovanni Alfonso Borelli and Marcello Malpighi. Borelli's important work published posthumously in 1680–81, his *De motu animalium*, recalls Galileo's title (although a similar title had also been used by classical authors).

astrological influences and by angelic and demonic forces—with an *anima mundi* permeating both the macrocosm and the microcosm, and with occult qualities still being raised to account for numerous terrestrial and celestial phenomena.

Athanasius Kircher, a German Jesuit, was one such person, and a very prominent scholar (Reilly, 1974; Godwin, 1979). Kircher spent much of his life in Rome, where had been summoned in 1632, the year Galileo's trial started, to teach in the famous *Collegio Romano* (the Jesuit school now called Gregorian University). He would serve four popes and befriend two emperors, and he would publish 38 treatises in his lifetime, with his Jesuit followers providing many more publications based on his teachings.

Kircher's intellectual activity developed along many lines, from linguistics to geology, optics, and theology, and it was strongly impregnated by the Hermetic and animistic ambience of the earlier Renaissance culture. Kircher lived during an epoch in which there was a great interest in oriental languages, with Rome as the main center of these studies (he was only about 20 years older than Hiob Ludolf, the linguist and Ethiopian scholar mentioned in Chapter 7).[25] This knowledge was one of the reasons behind his involvement in a translation of a Coptic-Arab text brought to Rome from Egypt, which led him to the interpretation of hieroglyphs (he was only partially successful) and more generally to Egyptology.

To Kircher, nature was like an ancient Egyptian hieroglyph waiting to be interpreted, but divinely inspired. His sacred mission was to try to decipher the hidden meanings within the great book of nature, meaning to understand God's universe in a Jesuit context. To the extent that he could travel or have specimens sent to him, he would, of course, use his own eyes. But when this was not possible, he would rely on expert on-site witnesses, with a preference for other Jesuits trained in Rome before departing on their sacred missions.

But although he had access to the latest methodologies and technologies, including new microscopes, Kircher never really appreciated what they were showing. Rather, he continued to promote and defend many ancient thoughts, including Hermetic ideas. His books are filled with the *mirabilia* of the Middle Ages and Renaissance, along with new items and new information about people inhabiting distant lands, the Jesuits being in a particularly privileged position to receive such information from the various Indias.

Although admired by some scholars of his period, Kircher became the perfect target for the *novatores*, who wished to found modern science on totally different grounds and wished to purge it of conceptions based on occult qualities. Some of these men were reunited in the *Accademia del Cimento* (Middleton, 1971). This society was founded in 1657 by Prince Leopoldo of Tuscany and by some of Galileo's students, including Evangelista Torricelli, the first person to demonstrate experimentally the existence of the vacuum and to establish the nature of atmospheric pressure (which he measured with the "Torricellian tube," the first modern barometer).

This accounts for why one of the greater followers of Galilean science in Florence, Francesco Redi, a court physician, chose to present his critical remarks about Kircher's sort of science in a book addressed to the Jesuit. This book was published in 1671 with the title *Esperienze intorno a diverse cose naturali e particolarmente a quelle che ci sono portate dalle Indie* ("Experiments on Various Natural Things and Particularly on those Brought to Us from the Indias").[26] Besides its general importance for the history of scientific thinking, Redi's work is of particular significance for the history of electric fish research, because it is the first text in which a new hypothesis for the torpedo's power appears, based on a personal study of the ray.

Since the torpedo's power was one of the natural objects most often invoked in Renaissance magic to support the existence of occult qualities, it is not surprising that Redi studied the fish and mentioned it in a work aimed to undermine the old science. The specific reason why the torpedo was mentioned in a book written as a letter to Kircher has to do with the fact that, as could be expected, the Jesuit, like so many others, had also dealt with the fish in some of his writings. Notably, Kircher dealt amply with the torpedo in his *Magneticum naturae regnum* ("The Magnetic Kingdom of Nature"), published in 1667. Although strongly laced with the occult, magic, factual errors, and fanciful thoughts, this work also provided many solid facts, and in it he even praised Galileo, the Lynceans,[27] and members of the *Accademia del Cimento* for some of their achievements (Freedberg, 2002).

Alexander Fabiano, Kircher's correspondent in Los Angeles (then Mexico), stimulated him to write what he did about the torpedo in his *Magneticum naturae regnum*. Fabiano wrote about "two sea animals with round-shaped

---

[25] The particular situation of Kircher as an influent member of Vatican intelligentsia, and as a Jesuit, allowed him to be acquainted with many books and manuscripts. Notably, he was the first to publish a written account of the discovery of the source of the Blue Nile, based on the manuscript works of Pedro Paes on Ethiopia (see Chapter 7).

[26] In the words of one historian, "The real importance of Kircher's biological work was that it provoked Francesco Redi to refute his uncritical conclusions" (Egerton, 2005, p. 135).

[27] In 1603, Federico Cesi, a Roman prince, founded the *Accademia dei Lincei* (Academy of the Lynx-Eyed) with several of his friends (Freedberg, 2002). His objective was to reform natural philosophy. Cesi demanded that members of his group not be slaves unto Aristotle, or to any other philosopher for that matter, but to be free and noble intellects when endeavoring to comprehend the natural world. The members of the new society aspired to see the world with the exceptionally keen vision of the lynx, and they agreed to use what they saw to create a fresh new natural philosophy. Galileo was one of the Lyncei, and he applied mathematics, physics, and reasoned analyses to his experimental natural philosophy. But the activities of the rest of the group were not particularly mathematical, and overall they still had much in common with Aldrovandi, whose dream was to have a massive museum and to come forth with a definitive encyclopedia of nature. There was, however, a major difference between the Lyncei and Aldrovandi, who was still alive when the group was founded. Whereas Aldrovandi had always worked within an Aristotelian framework, Cesi, Galileo, Johannes Schmidt, and Francesco Barberini, who was the organization's patron, wanted to use their collections and new tools (e.g., the microscope) to dismantle Aristotelian natural history. Cesi called his planned encyclopedia his "Theater of Nature," and it strongly emphasized what could be seen with the naked eye and under the microscope. In 1624, Galileo had given him an *occhialino* (i.e., a microscope, literally "little eyeglass"), so he could see the fine details of his specimens. The Lyncei disbanded in 1630 with Ceci's death. The latter event put a damper on promoting new ways of thinking for several decades. The Jesuits, who had a more religious agenda, would now play a more visible role in science, and Atanasio Kircher would emerge as the leading Jesuit natural philosopher of this era (Reilly, 1974; Godwin, 1979).

shells" (*dos conchas redondas d'un animal marino*) endowed with extraordinary and wondrous properties.

> Even not taking it, but simply touching it, suddenly it attracts, or stirs up, or disturb the humors, or the blood, in such a way that it seems that the arm gets numbed and swells, and gets senseless, and from there [the effect] diffuses to the entire body, so that it seems that all the body gets disjointed, and dies, being unclear that this happens because of an attractive, commoting, or venomous property. (Kircher, 1667, p. 121)

According to Kircher's correspondent, the wondrous sea animal operates like a "loadstone on iron." The "round-shaped shells" pointed to a shellfish endowed with the power of releasing some irritating liquid, and Kircher assumed that it might have swallowed a torpedo. In other words, its extraordinary properties derived from the torpedo, a fish that is a "magnetic animal" (a category that also includes the remora).

With this idea in mind, Kircher reported what he knew about the torpedo on the basis of classical authorities (with an ample quotation from Oppian) and newer information. In particular, he mentions the extraordinary power of Godinho's Abyssinian torpedoes, and especially their power to re-animate recently dead fish with their shocks. He also reported the results of his own experiments with torpedoes (*in nostris quidem Torpedinibus*), in which he confirmed the fairly strong intensity of the shocks, although he noted that the effects do not last long. On the basis of his observations, after omitting innumerable secondary details about the fish (*omissis innumeris circa hunc piscem nugamentis*), he starts to consider the nature of the shock, assuming it must involve a narcotic quality (*narcotica quadam facultate*) capable of inducing numbness (*torporem*), which does not correspond in a precise way to cold (*non prècise frigoris effectum esse dico*).

In Kircher's view, the fish is capable of blocking the senses, nerves, and muscles, possessing a quality capable of doing that (*sensibus imperviam qualitatem nervis & musculis contrariam*). After diffusing out from the fish body in a spherical way (*sphaerice diffusa*), if it encounters nerves or muscles, it penetrates and numbs them by acting on some spiritual substance (*spiritosam substantiam*) contained inside them. It is similar in its acts and diffuses like the quality of some venomous snakes (*venenorum quorundam serpentum qualitas*), which can diffuse not only with an actual bite, but also through "spears, sticks, swords" (he is here evidently alluding to the basilisk and other wondrous snakes), inducing convulsions, spasms, and comparable signs and symptoms.

At the conclusion of his "physiological" considerations Kircher writes:

> I say that, when irritated, the torpedo acts in no other way than by propagating with an insensible motion (*insensibili motu*), in a spherical way (*in orbem*), this narcotic quality; in this way, it freezes with a certain specific coldness placed in it (*specifica quadam frigiditate ipsi indita*) the blood and the spirit, which is present in a latent way in veins, muscles, and nerves; from this freezing effect (*congelationem*) numbness, spasms, and even convulsions may ensue. I do not see any difficulty in conceiving the cause of this effect. (p. 125)

Despite his context and reference to the traditional *frigiditas*, Kircher's hypothesis shares some features with those put forward by Redi and Lorenzini. Specifically, it assumes that some motion of the animal is the starting process for the shock and the emission of a quality that, although not defined physically, appears to be of a material type. We shall also find other aspects of Kircher's theory in Lorenzini's hypothesis, as for instance the blocking action of the emission on nerves. Further, Kircher does not seem to assign the torpedo's power to an occult quality, at least not in an explicit way. What is occult and invisible in his theory are the movements responsible for the emission of the torporific quality. But this, as already briefly mentioned, would also be the case with Réaumur, one of the great naturalists of the *Siècle des Lumières*.

FRANCESCO REDI: THE *CIMENTO*, VIPERS, AND INSECTS

The considerations developed in the first part of this chapter on the special status of the torpedo as a natural object invoked to support the existence of "occult qualities" account for the great interest in the mechanism of its power during the early phase of the scientific revolution. This interest was particularly strong because this fish appeared to be the only living object that had resisted the scrutiny of experience and the ravages of time: the remora, the basilisk, and other fabled animals had by this time lost their significance or were soon to die at the hands of the *novatores*.

Because of these factors, and in consideration of the fundamental role played by Galileo and his followers in promoting the new intellectual attitude characterizing modern science, there is little wonder that torpedo was the target of special attention by the scientists of the Tuscan court. This was particularly the case after the establishment of the *Accademia del Cimento* in 1657, this being the first scientific institution organized to pursue experimental investigations, as demanded by the new science. The importance of experimental methods in the activity of the academy is indicated by its name. *Cimento*, meaning experiment, assay, or test, is a word derived from the activity of goldsmiths, who used it to denote the assaying necessary to establish the quality of gold. Indeed, the motto of this short-lived society was *provando e riprovando*, meaning "testing and retesting" (or replicating).

Although the interest in the torpedo is understandable in the Tuscan science of this period, it is still amazing to discover how many scientists made anatomical or physiological experiments on the torpedo during the second half of the seventeenth century in Tuscany (see below). In a sense, the torpedo had become a "cult object" for the *novatores*, who wished to debunk the myths and overturn the explanations they had inherited from other scientists.

The work of Francesco Redi is particularly important in this context, in part because of his scientific stature, but also because his approach to the torpedo was part of his larger plan to topple the old science with its legends, unsupported evidence, *mirabilia*, and occult qualities. In contrast to the natural philosophers of the Renaissance, who normally wrote in Latin, Redi used Italian, displayed a brilliant style, and possessed a wry sense of humor, satirizing old beliefs with well-chosen literary or historical references. His inspiration came from Galileo, the creator of modern scientific prose in Italian literature, who chose the vernacular for his main works. This was because, contrary to the Renaissance scholars, Galileo aimed his scientific and cultural messages at all people endowed with investigative curiosity, irrespective of academic or other affiliations.

Redi found writing easy, having an immense erudition and being a poet and a linguist, among his many talents (Bernardi & Guerrini, 1999). He had feasted on William Harvey's (1628) treatise on the circulation of the blood, appreciated new instruments and methodologies, and believed good experiments would guide natural philosophers along the rocky road to ultimate truth. Redi wrote that he had become more and more convinced with every passing day that he should not accept what others have written, unless he could see a phenomenon with his own eyes and conduct experiments of his own.

In the dedicatory epistle of his *Osservazioni intorno alle Vipere* ("Observations on Vipers") he explains: "Every day I am assuring myself in my proposition to not put much faith in natural matters, except in what I see with my own eyes, and unless I can confirm by repeated and re-repeated experiment those matters that I did not clarify personally by experimenting" (Redi, 1664, p. 5). Further, a quotation in Arabic characters inserted along with its Italian translation on the initial page of his most important work, his *Esperienze Intorno alla Generazione degl'Insetti* ("Experiments on the Generation of Insects"), translates as: "The one who makes experiments enhances knowledge, the one who is credulous [gullible] increases the error" (Redi, 1668).

As with Galileo and the other protagonists of the scientific revolution, Redi was against the principle of authority, which he often addressed with irony, such as when he introduced his experiments on vipers. He wrote that the first part of the animal to be tested experimentally was the gall, because one of the attending persons strongly asserted that the venom was contained there, as witnessed by ancient scholars. Among the authors that man cited, Redi writes, were Galen, Pliny, Avicenna, Rasis, Alì the Abbot, Albucasis, Guglielmo di Piacenza, Santi Arduino, the Cardinal of Saint Pancras, Bertruccio Bolognese, Cesalpino, and Giulio Cesare Claudino. This long list was a satire aimed at traditional medical and pharmacological treatises, which were based on long lists of authorities, rather than on direct experimental evidence.

Redi's association with the Medici court is important, because the Medicis provided encouragement and financial backing for his studies. Ferdinando II de' Medici, who ruled from 1621 to his death in 1670, was the Grand Duke of Tuscany at the time. Extremely interested in the sciences, he gave his natural philosophers animals that died in captivity or were killed on hunts, and he provided them with opportunities to travel. His predecessors, Francesco I and Ferdinando I, had previously helped Aldrovandi in his endeavors, and in 1669 he had even traveled to London, where he discussed scientific developments with members of the Royal Society (Findlen, 1994).

Most of Redi's endeavors took place in an improvised laboratory at the vibrant Medici court. There he oversaw endless dissections on animals that were verified by other natural philosophers and men of high stature (see below). It has been written that "By the mid-sixteenth century, whenever learned patricians who professed an interest in nature converged, a dissection was sure to occur" (Findlen, 1994, p. 209). This statement is, of course, an over-generalization, since not every naturalist or natural philosopher was a dissector. But it does reflect how important dissections had become, especially in the more progressive parts of Italy.

Redi's first book of importance was his *Osservazioni Intorno alle Vipere* of 1664. He had decapitated some 250 vipers in order to find the source of the venom, and he used his microscope to examine the stingers of numerous scorpions, also noted for their poison. By so doing, he discovered that a viper's venom is stored in sacks associated with its fangs.

In his *Esperienze Intorno alla Generazione degl'Insetti*, which appeared 4 years later, Redi (1668) dispelled the notion of spontaneous generation, which was one of Father Kircher's most famous claims. Kircher had placed dead flies in honey water and watched as new flies emerged "spontaneously." He also claimed that ox dung spontaneously produces worms that become bees. When Redi covered his dishes containing dead flies or fresh dung, so that insects could not swarm and lay eggs in these breeding grounds, he failed to find any evidence for spontaneous generation. His experiments in this domain set the stage for another Italian, Abbé Lazzaro Spallanzani, and later Frenchman Louis Pasteur, to show even more convincingly that spontaneous generation was a flawed theory.

Redi's science covered far more than snake venoms and spontaneous generation—39 volumes of his manuscripts can be read in Florence alone. The impression given is that everything in nature was of interest to him, and this included the torpedo.

As mentioned, the results of some of his experiments were inserted in a text written in the form of a letter addressed to Kircher, first published in 1671 (Fig. 9.3). The occasion of the book was a series of letters sent by the Jesuit to Redi, in which Kircher mentioned some experiments made personally by him and a colleague proving the efficacy of some stones as antidotes against snake venoms. These extraordinary stones were said to come from the Indias, and to have been extracted from the heads of particular serpents.

Redi starts his book with a series of experiments on animals (cocks in particular) that disprove the presumed efficacy of the stones, not only against snake venoms, but also against other venoms or poisons, notably those discharged by scorpions. He then narrates an episode that occurred at the court of the Grand Duke, in which he exposed a charlatan who tried to sell a preparation based on some extraordinary herbs, supposedly capable of protecting against injuries produced by firearms.

Continuing on this theme, Redi scrutinizes the power of experimental science to expose the pretended therapeutic efficacies and properties of stones, herbs, animals, and really all the *mirabilia* of medieval and Renaissance imagery. It is in this context that he comes to the torpedo, the ubiquitous member of fantastic bestiaries since ancient times.

REDI'S TORPEDO AND MECHANICAL THEORY

It was the custom of the Tuscan court to spend large parts of the year in Pisa or in Leghorn, where the Grand Duke had palaces and villas, both in the town and in the countryside.

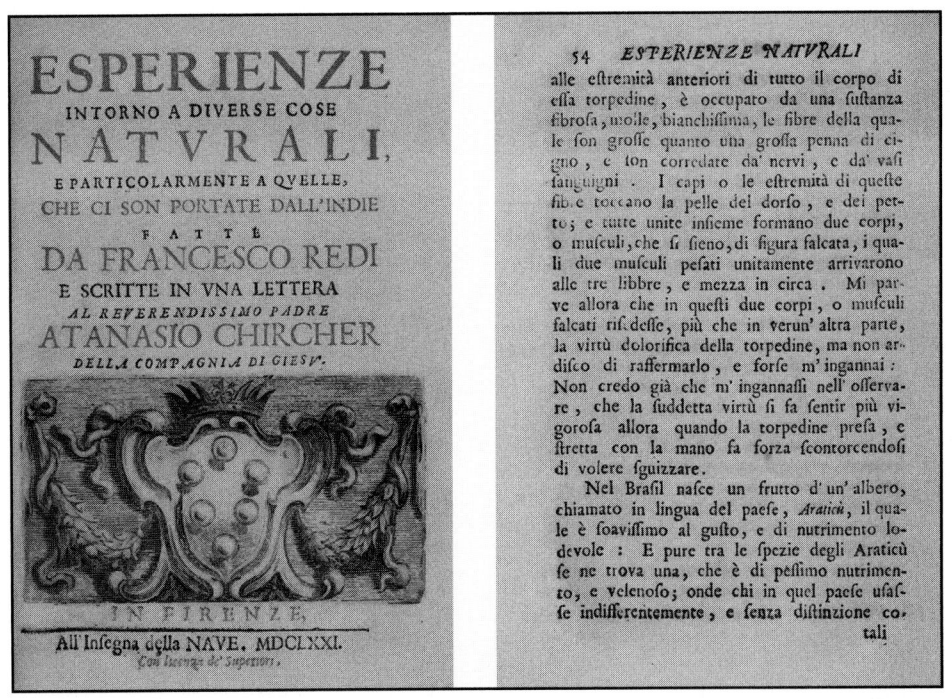

Figure 9.3. Francesco Redi's text of 1671, written in response to Father Kircher, in which he associates the torpedo's powers with its falciform muscles.

There, many animals and fishes were available for experimentation, and court scientists and foreigners traveled to Tuscany in great numbers to study them. Indeed, Tuscany was considered as a Mecca of biological research in the post-Galilean era.

Being the court physician, Redi usually accompanied the Grand Duke to the coast, and he profited greatly by being able to carry out experiments on animals that were not readily available in Florence. It was in this way that he was able to conduct a series of experiments that he recorded in his laboratory protocols, which are still extant at the Marucelliana Library in Florence (MMSS Redi 32, ff. 177–188). These manuscripts include the observations he carried out on several varieties of fishes caught between Pisa and Leghorn, including torpedoes.

Redi's extant observations on the torpedo were all made in the Harbor of Leghorn and correspond to two periods with a lengthy gap between them: 1666–67 and 1677. Redi's personal account of the torpedo in the dissertation letter he addressed to Father Kircher is based only on those observations made in the first period. As noted, this was published in 1671 (and thus well before the second series of experiments).

Most of the experiments recorded in these protocols consist of anatomical dissections, but from the annotations it appears that Redi was also able to work with live torpedoes. The experiments on live animals were intended to provide an interpretation of the torpedo's mysterious power. A reference *en passant* to the use of a live torpedo is given in the protocol dealing with the experiments carried out on March 24, 1666, where Redi writes: "The dissection on this torpedo started when it was alive at 5 p.m." (f. 184 verso).[28]

We know from both the manuscripts and other sources that the experiments Redi made in 1676 were carried out in collaboration with his pupil, Stefano Lorenzini (see below). Although the extant protocols from the first period do not allude to collaborations with other scholars, there is evidence that they were made with Nicolaus Steno (Niels Stensen), the Danish scholar (and eventually Catholic bishop) who spent a long period at the Medici court. Notably, there is a strong similarity between the anatomical description of the torpedo given in Redi's letter to Father Kircher, which was published in 1671, and the account of the fish given by Steno in a memoir he published in 1675 (Steno, 1675, pp. 223–225; Guerrini, 1999).[29]

Redi presents his anatomical and physiological observations with humility and almost apologetically. He writes that they were carried out on a single torpedo caught on March 14, 1666, and that he was forced to work "in a hurry, and, as people say, in a rough way," because of his "many occupations."[30] Although in the Marucelliana Library manuscripts there are no extant protocols of this day (i.e., March 14th), there is, as mentioned, a description of a torpedo studied on March 24, 1666. It is on the basis of this other torpedo that Redi provides one of the most famous passages in his published account of the fish—his description of the special organs he believed to be responsible for its power (see below).

There is no description of the shock caused by this second torpedo in these manuscripts, although the fish was described

---

[28] *"Questa tremola si cominciò ad aprire viva a ore 17"*.

[29] Guerrini assumes that Redi and Steno worked on just a single torpedo. This assumption, which is partially justified by the way Redi presents his description of the fish in the letter to Kircher, is, however, incorrect. A comparison of the Marucelliana manuscripts and the letter to Kircher reveals that Redi's description is based on at least two torpedoes. Steno's description includes the anatomical details of both torpedoes described by his colleague.

[30] *"ma per le molte occupazioni lo feci in fretta e, come si suol dire, alla grossolana"* (Redi, 1671, p. 50).

as alive when the dissection was started. Thus, Redi's report on the shocking effects of the torpedo could well have been based on the fish caught on March 14. In all, the physiological annotations represent only a minor fraction of Redi's account, which is mainly devoted to an accurate anatomical description of the external and internal structure of the fish.

With regard to the shock, Redi writes: "I had scarcely touched and pressed it with my hand than my hand started to tingle (*informicolare*) and the same also happened to my arm and to the entire shoulder. At the same time, I experienced such a disagreeable trembling with a painful and acute sensation in the elbow joint that I was obliged to withdraw my arm immediately" (Redi, 1671, p. 48). He adds that the numbness became less perceptible as the animal approached death, and that the torpedo became incapable of producing numbness only after it was definitely dead (i.e., after about 3 hours of experimental trials). He explains, moreover, that he could not prove that the shock could be transmitted by indirect contact through water or other intervening materials. He theorizes that this is probably because of the weakness of the fish, and does not refute transmission through intermediaries, something asserted by fishermen since ancient times.

Redi's description of the internal organs of the fish is much more accurate than those provided by the ichthyologists of the previous century, and it includes weighing all organs that he could extract from the fish's body. In particular, he points to the great size of the gall bladder, which had been noted by others, and verifies that, contrary to what Aldrovandi had asserted, gall is not responsible for the numbing property of the fish. In this context, he also relates that he considers Pliny's statements about the anti-aphrodisiac properties of the gall unsupportable.

For our story, the most salient outcome of Redi's dissection is his description of what we now know to be the electric organs. In translation, we read:

> All the space of the torpedo's body that is situated between the gills and the head, and between the places where the fins and the anterior extremities of the entire body are situated, is filled with a fibrous substance, which is soft and very white. The fibers of this substance are big, as much as a swan quill, and are provided with nerves and blood vessels. The heads and the extremities of these fibers come into contact with the skin of the back and of the belly; together they form two bodies, or possibly two muscles of falciform shape (*due corpi, o musculi, che si sieno, di figura falcata*). Having being weighed these muscle were found to be about three and half pounds. (pp. 53–54)

This passage is a rather faithful transcription of what Redi annotated during the dissection of the torpedo studied on March 24, 1666 (MM Redi, 32, Folium 183 Recto). The passage that follows in the published text is, on the other hand, not present in the manuscripts, and might be related to the observations made on the torpedo of March 14.

> It seems to me then that the pain generating virtue (*la virtù dolorifica*) of the torpedo was situated (*risiedesse*) in these two bodies or falciform muscles (*musculi falcati*) more than in any other place, but I do not venture to assert it, and perhaps I deceived myself. I do not believe though to deceive myself in saying that the abovementioned virtue makes itself more manifest when the torpedo, having been caught, and pressed with the hand, forcedly tries to wriggle out by twisting itself (*fa forza scontorcendosi di voler sguizzare*). (p. 54)

Although Redi appears to be very careful in asserting the dependency of the shock on the muscular effort of the fish, and particularly cautious in attributing the torpedo's power to the falciform bodies, this passage is considered the first printed proposal of the mechanical theory of the torpedo shock. It appears to be the first new scientific theory to emerge since antiquity, and it offered a viable scientific alternative to ancient notions and vague concepts and tautologies that really explained nothing, such as "occult qualities." Importantly, this theory could be tested (and falsified) with the methodology of modern science. It would take a little more than a century, though, for it to be disproved (see Chapter 15).

Besides the clarity of the anatomical description and the allusion to mechanics, it seems likely that the fame Redi achieved with his account is partly due to linguistic factors, and most specifically to his coining the phrase *musculi falcati*. This phrase or one like it with a different spelling (e.g., *muscoli* instead of *musculi*), would recur again and again in the subsequent scientific literature, both in Italian and in other languages, even Russian.

In alluding in his dubitative way to the possibility of a mechanical origin of the torpedo's shock, Redi was certainly influenced by the mechanical *Zeitgeist* of his age. In addition to Galileo, René Descartes had advocated mechanical views in his explanations of natural phenomena, including various physiological processes. Descartes had expressed so much confidence in mechanics that, in his *Principia philosophiae* of 1644, he contended that it could explain every fact of nature, including the oddities. Redi's torpedo was undoubtedly one of the most striking natural oddities that the natural philosophers of the scientific revolution wanted to understand.

It is also possible, however, that Redi, like most of the scientists in his Tuscan circle, did not trust Descartes' *aprioristic* approach to science. Further, he might have been influenced by earlier descriptions, such as the one by Godinho, who pointed out that the African torpedo moves at the instant of the shock. In fact, he might even have learned what Godinho had written by reading the texts of his correspondent and adversary, Father Kircher. As we shall see, the most important immediate reason underlying Redi's mechanical hypothesis for the torpedo was likely the influence of his colleague at the Cimento, Giovanni Alfonso Borelli.

STEFANO LORENZINI

Nothing can better illustrate how atomism acted as a "red thread," which allowed the torpedo to pass from the old culture of *mirabilia* and occult qualities to the new science of Galileo and his followers, than the initial pages of Lorenzini's *Osservazioni intorno alle Torpedini*, published in 1678 (Fig. 9.4). The incipit of the book is *La saviezza di Democrito* (meaning the wisdom or sapience of Democritus), and the first two pages praise the Greek atomistic philosopher.

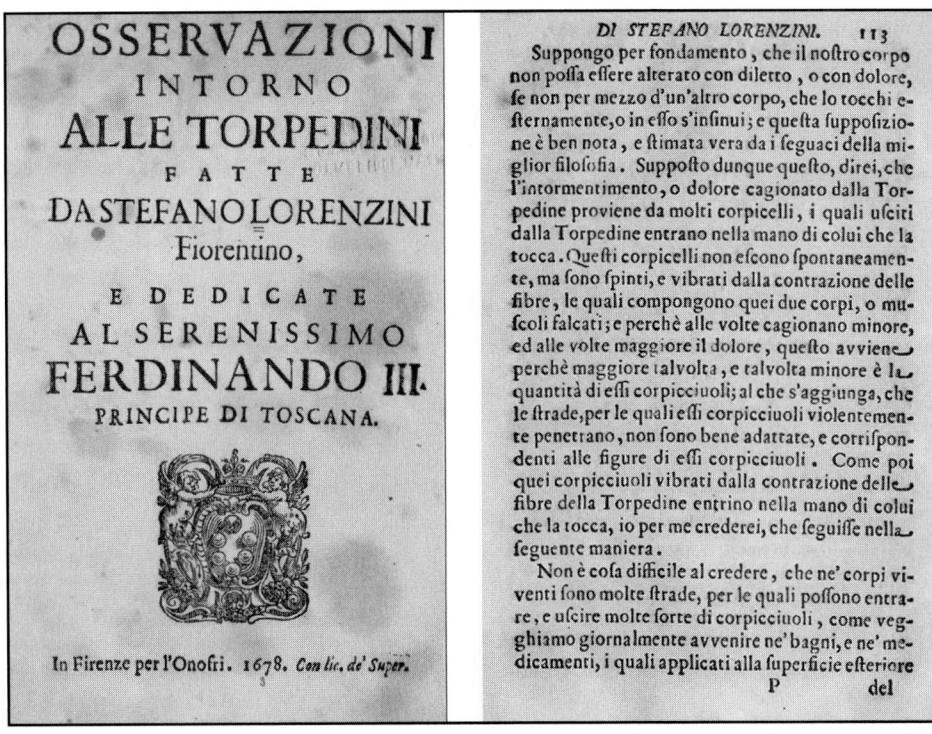

Figure 9.4. The title page of Stefano Lorenzini's 1678 text on the torpedo and a page with his corpuscular theory. Lorenzini envisioned the torpedo emitting minute corpuscles that could paralyze another fish and even enter a touching hand.

Lorenzini's book is the first text devoted entirely to the torpedo, and the first one published that really delineates a genuinely modern theory for its shock. Moreover, it is the first scientific book on the fish that richly illustrates the internal anatomy, and particularly the electric organs, based on microscopic observations (Fig. 9.5).

As a matter of fact, despite the great historical importance of Redi's previous passage on the torpedo, it is clear that Redi did not explicitly propose any theory for the torpedo's shock. If we disregard its context, Redi's passage really does not contain anything more on the mechanical hypothesis than the reports of the Jesuit missionaries in Ethiopia, who remarked that the African torpedo produces its effect only when it moves, whereas nothing happens when it remains still.

We say this not to praise Lorenzini at the expense of Redi. This would be unjustified, especially when we add two important considerations. The first is that Lorenzini was

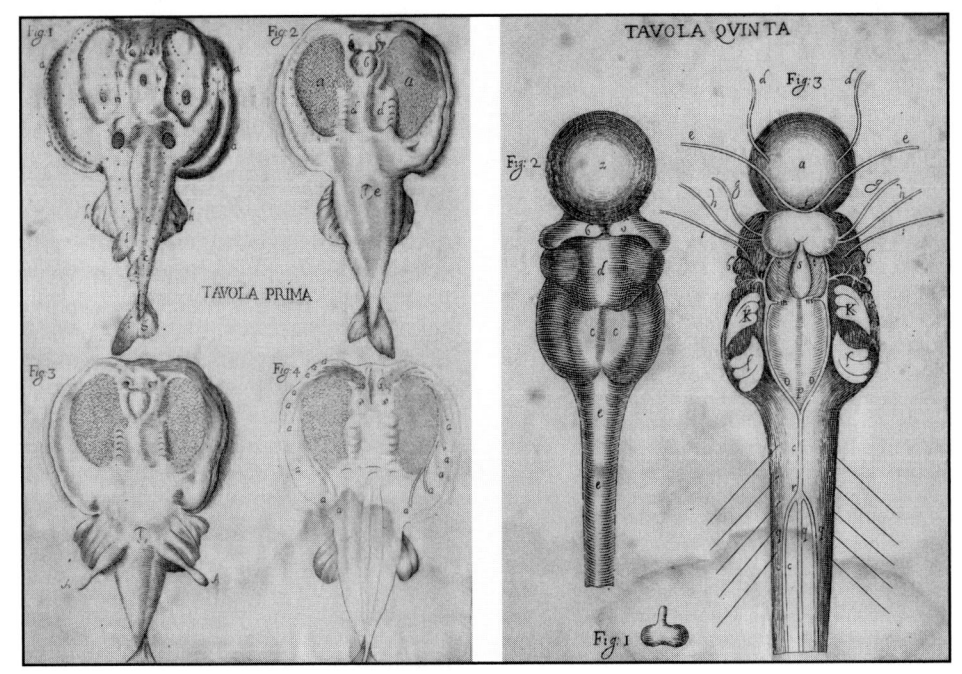

Figure 9.5. Two anatomical drawings from Lorenzini's *Osservazioni intorno alle torpedini* (1678). The figure on the left shows the shocking organs and that on the right the source of the nerves to those organs.

Redi's pupil, and the second is that Redi might have contributed in a substantial way to Lorenzini's book (see Guerrini, 1999).[31] It is indeed likely that the decision to publish the book only with Lorenzini's name was the consequence of a complex situation involving rivalries among scientists and courtesans at the Grand Duke's court (Guerrini, 1999; Bernardi, 2008; see also the correspondence of Marcello Malpighi published by Adelmann in 1975).[32] These rivalries led to heavy consequences for Lorenzini, who was involved in the disgrace of his brother Lorenzo, a mathematician. Both were long detained in a jail in Volterra.

Stefano Lorenzini was born in Tuscany some time after 1652 and died some time after 1700, his birth and death dates being uncertain (Guerrini, 1999). Trained in medicine, and although he was at one time Redi's pupil, he also studied under Marcello Malphigi and Nicolaus Steno. Among his many notable achievements is his description of the receptors in some fish for what would later be identified as electrical currents. Today, these electroreceptors bear an eponym; they are called the "ampullae of Lorenzini."[33]

Lorenzini had the opportunity to study torpedoes at Leghorn and Pisa with Redi and other talented naturalists (besides Redi, he mentions Holger Jacobsen and Christopher Bartholin) in the period 1676 to 1678—that is, about 10 years after Redi's first important research on the torpedo. Most of his book on the torpedo is devoted to his anatomical investigations of its internal organs, and to discussions about some aspects of anatomy within the context of the biological and zoological conceptions of the time. Only the last 14 pages (pp. 104–118), accounting for about one tenth of the book, deal with the shock mechanisms. Referring to earlier reports about the torpedo's powers, he writes that Godinho's account of how the torpedo could reanimate dead fish must be regarded as "fabulous" (in the sense of fable-like) (pp. 105–106).

The first personal experiments Lorenzini described date from 1676 and took place in Pisa. They involved six live torpedoes that the Grand Duke made available. The torpedoes, put together in a vessel filled of water, appeared to be "vigorous, very vigorous" (p. 106). Thinking in terms of a "great torporific and painful effluvium" (*grande effluvio stupefattivo e dolorifico*), he assumed that he should be numbed when immersing his hands in the water. Nothing, however, happened. Further, no effect was perceived by touching the torpedo with a rod. These results were confirmed by other attending persons and with other torpedoes over the following days. Only when a torpedo was touched directly with a hand, whether in the water or when removed from it, could a painful sensation be perceived. Lorenzini relates that the pain never went beyond his own elbow, whereas for Redi, it extended to the shoulder (p. 107). He explained that the painful and numbing sensation was like that produced by a blow to the elbow, a sensation that similarly lasts a short time and has no lasting consequences.[34]

On page 108, the experiment is clearly described within a mechanical framework. The passage is worth quoting *in extenso*, being of particular interest if we keep in mind that, as we have stated, the shock is really not accompanied by any contraction of the fish's body:

> It is to be remarked that in doing these trials I felt the pain and the numbness only when there was a contraction of the fibers composing the two bodies or falciform muscles [*muscoli falcati*] consequent to the touching and clenching [*toccando e strignendo*] of those bodies. I did not feel any alteration when these fibers did not contract and remained still. In is to be noted, moreover, that the numbness varies with the degree of the contraction. As a matter of fact, the torpor is very great when the fibers contract vigorously, and it propagates not only to the hand but up to the entire elbow. When the fibers do not contract so vehemently [*tanto veementemente*], but a little more slowly, then we feel something like a swarming [*informicolamento*] in the entire hand, with no diffusion outside the hand. And when they contract very slowly [*lentissimamente*], then we feel something like a convulsive motion limited to the fingers; this is reiterated any time the fibers contract. And when the fibers do not contract at all, there is no alteration at all in the touching hand.

The logical conclusion of these observations is expressed as follows:

> Therefore the cause of the torpor, or numbness, is the contraction of the fibers composing those falciform muscles. This would appear manifest to anybody wishing to do the trial. (pp. 108–109)

Lorenzini was so convinced that it is necessary to touch and clench the fish's body directly in order to feel the effect, that he obliged fishermen, who asserted the possibility of transmission through fishing nets and other intermediate bodies, to conduct experiments in his presence. The results, Lorenzini wrote, confirmed his conclusion and the opinion that some of the extraordinary effects attributed to these fish are due to the fantasy and gullibility of those who report them.

---

[31] It is possible that the praise of Democritus at the beginning of Lorenzini's books was due to Redi, because in a manuscript version of the text in the Marucelliana Library of Florence we can recognize Redi's handwriting (see Guerrini, 1999). Redi continued the experiments on the torpedo and other fishes after publishing his letter to Kircher and, as already noted, in these experiments he collaborated with Lorenzini. Parts of Redi's manuscripts on the torpedo are in Lorenzini's handwriting. Moreover, in some of his manuscripts Redi writes that a particular observation was made by "Signor Lorenzini," or that he leaves to Lorenzini a more complete description of the fish's anatomy.

[32] In the case of the torpedo studies, one of the possible reasons of the polemic among the scientists of the Grand Duke's court was the fact that, before Redi, a mechanical theory of the shock had been proposed by Borelli on the basis of work made in collaboration with Lorenzo Bellini. Borelli, Malpighi, and Bellini were indeed strongly critical of the anatomical and physiological conclusions of the torpedo studies of both Redi and Lorenzini. It is also possible that Lorenzini's books incorporated some material on the fish that Bellini communicated to him (see Guerrini, 1999, and Adelmann, 1975).

[33] The description of the structures that would be later recognized as electroreceptors is given by Lorenzini on pages 7–11 and 23–30 of his book.

[34] The passage in which Lorenzini remarks on the similarity between the numbness induced by the torpedo and that caused by a blow to the elbow is very similar to the mention of the torpedo's effect inserted by Lorenzo Bellini in his *Gustus organum* (1665, pp. 98–99). Because Lorenzini did not quote him in this context (and for other reasons as well), Bellini expressed his complaints to Malpighi in a letter dated March 7, 1679 (see Adelmann, 1975, vol. I, pp. 793–797). In this letter Bellini tells Malpighi that he had informed Redi about "the place and the structure which makes the tremor" of the torpedo on the basis of information received many years before from Borelli. He also criticizes the way in which Lorenzini describes "the muscle that makes the stupor" in his *Osservazioni* (p. 795).

He went even further in disproving these beliefs. He helped fishermen to draw in the net in which a torpedo had been caught, again not perceiving any effect. He also put his shoed foot on one without consequence. And he conducted experiments to show that, contrary to Godinho's assertion, the torpedo is not able to reanimate dead fishes with its powers (pp. 109–110).

Lorenzini's description of his torpedo experiments is remarkable, because almost nothing of what he writes corresponds to what we know today about the torpedo's shock! We now are well aware that the shock is transmitted through water and other bodies (including wet fishing nets and wet rods). We also know that recently deceased fishes can move when put in direct or indirect contact with a live torpedo thorough a conductive medium.

What is most astonishing, however, is Lorenzini's repeated assertions about the movements of the fibers of the falciform bodies at the moment of the shock. We now know from well-controlled experiments that there is no visible movement whatsoever of the electric organ, or for that matter, of any other body part. Most likely the movements perceived by Lorenzini can be accounted for by an action that does not start with the fish but with the experimenter, namely the mechanical transmission of the tremor back to the fish from the hand touching it (see Piccolino, 2005b). Lorenzini's report would appear to provide a very powerful illustration of how theories or general conceptions can affect the results of scientific experiments, or what some researchers today refer to as the "theory-ladenness" of experimental observations.

The last pages of Lorenzini's *Esperienze* are devoted to developing a hypothesis of the mechanism underlying the torpedo shock. As previously noted, he conceived of his hypothesis within an atomistic framework largely inspired by Galileo's theory of sensations. His elaboration was in accord with the medical conceptions of his age, which were also inspired by mechanical advances and have been called iatromechanics (*iatros* meaning physician in Greek, thus medical mechanics) or iatrophysics.

In the *Zeitgeist* of the scientific revolution, and from the use of microscopes to investigate animal (and plant) bodies, one of the consequence of atomism in the field of physiology was to assume the presence of extremely minute pores, canaliculi (tiny canals), and microscopic glands. These microscopic structures were believed to underlie a variety of functions, such as secretion, filtration of body fluids, and the diffusion and distribution of minute particulate matter in motion. The pores in the skin, for example, were assumed to function in the exchange of minute particles or effluvia with the surrounding air, even though this exchange might not be detectable by sensory mechanisms. This was the case, for instance, for what was called *perspiratio insensibilis* (insensible perspiration), a concept introduced by Santorio Santorio, a colleague of Galileo's in Padua and the first physician to apply rigorous quantitative methods to the study of physiological process.

Sensory processes and nervous conduction were among the other functions that iatromechanically oriented physicians associated with pores and canaliculi. As already alluded to by Galileo in his *Saggiatore*, sensations could arise from the penetration of minute corpuscles or atoms inside specialized sensory structures. Some scholars supposed that the canaliculi running through the nerves were filled with a tenuous fluid that would convey subtle corpuscular matter along the nerve pathways. Turning to the pathological side of this model, nervous function could be hindered by blockages of the nervous fluid in various ways, including intrusions of corpuscles of certain sizes and shapes. Disrupting the normal flow of the fluid could result in tremors, paralyses, insensibility, pain, and the like.

Lorenzini set forth within this medical and physiological *Zeitgeist* to construct his corpuscular theory for the torpedo's shock. His departing point is the principle (basic to the new science) that no action could be produced in a living body without a material agency. As he declares at the outset:

> I assume as a founding principle [*fondamento*] that our body can be modified, either in a pleasant or painful way [*con diletto, o con dolore*], only through another body that touches it or penetrates inside it. This supposition is well known, and valued as veridical, by the followers of the best philosophy. (p. 113)

On this basis, Lorenzini assumes that the numbing and painful actions of the torpedo are due to the release of many corpuscles [*molti corpuscoli*] that enter through the hand in contact with the fish's body. Notably: "These corpuscles do not come out in a spontaneous way, but are pushed, and vibrated by the contraction of the fibers, which compose these bodies or falciform muscles" (p. 113).

The pores and pathways through which the fish's corpuscles travel are not adapted to their shapes, causing numbing and painful sensations. These paths are nevertheless the same, Lorenzini remarks, as those involved in the penetration of various medicaments and the actions of some remedies [as for instance the application of cantharides [*canterelle*]. Under normal conditions, the entrance of corpuscles adapted to the pores and paths of the body will produce normal and pleasant sensations. But that is not the case here. Thus, the torpedo's effects can be accounted for by the corpuscles it emits and the features of the affected body, because the corpuscles might be of a shape and/or size unsuited to the pores and canals that they enter, and/or because they are emitted "in such a great quantity, and with such a vigorous impetus" [*in così gran quantità, e con impeto cotanto gagliardo*] that they trigger painful sensations (p. 116).

After mentioning (almost *verbatim*) Galileo's passage from the *Saggiatore*, in which he discusses his hypothesis of thermal sensations, Lorenzini writes something that directly connects his views to those of the great scientist of Pisa:

> I know very well that this doctrine would not appear novel to those who aspire only to the noble investigation of the truth, since in the past it was discovered by that great and luminous mind [*grande lume*] and restorer of the Philosophy, our *sig. Galileo Galilei*, and taught to us with unspeakable clearness [*indicibil chiarezza*] in his most exquisite *Saggiatore*. (p. 116)

Lorenzini's book enjoyed great success and was translated into English and French. It was one of the books that, as we shall see, John Walsh consulted when he was planning his experiments on the torpedo, and he even lent it to Franklin, who guided him in his work at La Rochelle. Walsh most probably owned the Italian edition of the book, because when he quoted Lorenzini in a paper on the larger British torpedoes, he referred specifically to an Italian edition (Walsh, 1774b, p. 465; see Chapter 15).

OTHER TORPEDO SCIENTISTS AT THE MEDICI COURT

Other scientists also spent time studying Italian torpedoes under the auspices of the Grand Duke of Tuscany (Guerrini, 1999). Some were themselves Italians, and this subgroup included Giovanni Alfonso Borelli, Marcello Malpighi, and Lorenzo Bellini.

Giovanni Alfonso Borelli, whose treatise on muscular motion (*De motu animalium*) is filled with mathematics, geometry, physics, and mechanics, deserves special mention. He was an indirect disciple of Galileo and a professor at Pisa University in the period 1656–67. His life research program was to apply the laws of the mechanics to the understanding of the physiological processes of humans and animals, as well as other things, along the lines indicated by Galileo.[35]

In his *De motu animalium*, this imaginative Neapolitan scientist elaborated his views on nerve conduction and muscle contraction, which were long considered and frequently cited as the bases for comprehending these physiological processes. On the subject of nerve conduction, he denied that it was based on a flow of animal spirits, as implied by Galenic physiology. With other scientists of his epoch, he hypothesized that the nerves are filled with a subtle fluid, and that this fluid (different from the nerve's nutritious juice) could produce contractions of the muscles through a process like fermentation and ebullition (Borelli, 1680–81, vol. II. pp. 26–29). Importantly, in Borelli's schema nervous conduction did not imply a material flow of the nerve fluid from one to the other end of the nerve, but was analogous to the progress of elastic undulation (pp. 317–331).

Further, Borelli is important in the history of life sciences because he also oriented the scientific endeavors of Marcello Malpighi in a modern, "Democritean" direction (i.e. based on the revival of the atomistic doctrine stemming from Democritus), as Malpighi himself acknowledged in his *Opera posthuma* (Malpighi, 1698; Belloni, 1967). The collaboration between Malpighi and Borelli, started during the period of Malpighi's professorship in Pisa (1656–59), was exceptionally fruitful. This stemmed in part from the fact that Borelli, a great theorist, was not a skilled experimentalist, whereas Malpighi, who did not have a mathematical and physics background, was one of the greatest anatomists of his epoch. Together, these two scientists could give one other those things that they lacked.

Borelli deals with the torpedo in Chapter 20 of the second volume of his *De motu animalium*, which is devoted to a discussion of the causes of tremors in animals. Curiously, the action of the fish is discussed in Proposition 219, together with the effects induced by the porcupine (*Istrix*), here viewed as two instances of "fabulous narrations." From the outset, he makes clear that what he writes about the two animals is based on personal experiences (*De utroque animal enarrabo ea, quae proprijs oculis vidi*).

His experiments on the torpedo were made in collaboration with Malpighi and Bellini at the time of Redi's first experiments, and therefore well before Lorenzini. But Borelli did not publish his results in his lifetime, inserting them in his *De motu animalium*, which was printed posthumously as two volumes in 1680 and 1681. As mentioned in an earlier footnote, only a short reference to these experiments appeared in Bellini's book from 1665. From the extant correspondence and from other documents, it is clear that Borelli, Malpighi, and Bellini belonged to a group that was in competition with the one headed by Redi (see Bernardi, 2008). It is indeed probable that the views expounded by Redi in his letter to Kircher were largely derived from conceptions of the fish shock that had been developed by Borelli first.

Having introduced the power of the torpedo as supposedly due to a venomous force (*Torpedinis Piscis vi venemata*), Borelli denied that it actually poisons, which is not surprising given his orientation and the changing times. He based his conclusion on the rapidity of the production and cessation of the tremor and stupefaction induced by the fish, which in his opinion appear only when the fish contracts. No harmful effect is produced, he emphasized, in the hand holding the torpedo when fish remains still (*eo tempore, quo quescit, manum prorsus non laedit;* p. 441).

In an implicit way, he connects the shock of the fish to the condition of the experimenter's nerves and muscles by stating that stronger effects are felt when the part of the hand that compresses the torpedo's chest is rich with nerves and muscles. He also comments on an English anatomist (identifiable as John Finch), who asserted in the presence of the Grand Duke that he felt the pain induced by the torpedo for 2 days. In his opinion, this was an exaggeration. He adds, however, that the intensity of the shock depends on the particular way the fish is held, and he suggests an involvement of the ligaments and tendons of the hand in the transmission of the torporific power of the fish, which he attributes to concussion-like contractions of its muscles.

When comparing Borelli's and Redi's opinions on the torpedo, we find that they are basically similar. Thus, it is possible, as already intimated, that Redi, who published his results in 1671, drew from Borelli (who by this time had left Pisa for Rome, where he served a member of the academy founded by Queen Christina of Sweden, who had left her country and embraced Catholicism). If this is the case, Borelli, more than Redi, should get the credit for first proposing a mechanical explanation for the torpedo's powers.

It is also possible that Lorenzini, when accentuating the importance of corpuscular emissions in the production of a torpedo's shock, wanted to propose a model different from that of the Neapolitan. This is, however, speculative. What is certain is that *De motu animalium* had an enormous diffusion (in 1685 a second edition appeared, modified by Jean Bernouilli, which was reprinted several times). In brief, many books, journals, and even lay magazines, in Italy and abroad, presented Borelli's mechanical opinions of the torpedo, as can best be appreciated by looking at what was written in the London periodical *British Apollo*, which received an inquiry in 1711 that read:

> Q. Gentlemen, *Their being a Fish (very uncommon in* Lisbon) *called by name a* Nummfish, *which if taken hold of while alive, immediately* Numms the Joints; so that

---

[35] Borelli envisioned, among other things, a submarine that could be rowed beneath the waves. He was also one of the most influential mathematicians of his age. He translated the conic section of the treatise by Apollonius Rhodes into Latin (this was one of the most important and difficult mathematical works of Greek science), furthering the advancement of mathematics in Europe. His theoretical work on planetary motion anticipated the theory of universal gravitation, as Newton (1728) himself recognized in his *De mundi systemate*.

the Part which is touch'd is void of any feeling, during the time one holds it. I'll desire you inform me in your next which will ever, lay an Obligation on

Your Admirer and Humble Servant, J.S.

The answer to the question was based on Borelli, whose name is given in Latin:

> A. This is better known to the English by the name of *Crampfish,* called in the Latin *Torpedo,* a description of which see in Mr. Ray's Edition of *Willoughby's Icthyographia.* The strange benumbing faculty it is endued with is not easily accounted for: only it is to be observed that one may touch it without any Danger, either when it is Dead or when it lyes still. Hence, *Alphonsus Borellus* concludes that the Stupefaction does not proceed from any poisonous Quality proceeding from the Body of the Fish, but merely from the violent Trembling and Convulsions the Fish is put into, especially when touched near the Throat, where there are such a vast Multitude of Nerves and Muscles. *(British Apollo,* January 3, 1711)

As already mentioned, there were also foreign guests who went to Tuscany and participated in the torpedo experiments. Lorenzini mentioned some of them in his *Osservazioni* (see above). They included Danes Nicolas Steno, who lived in Tuscany for many years, and Christopher Bartholin (Cristofano Bartolini) and Holger Jacobsen (Oligero Giacobeo), who spent shorter periods at the Grand Duke's court.[36] Some of them (e.g., Steno and Jacobsen) published the results of their torpedo studies, but they are mainly anatomical in content. Nevertheless, Lorenzini tells us that Bartholin and Jacobsen attended physiological demonstrations and personally experienced the shocks.

As for Steno, one of the most eminent anatomists and thinkers of the time, he worked closely with Redi. In fact, they performed so many dissections that one historian described what occurred as a "furious slaughter of every living creature within reach of the court to impress Grand Duke Ferdinando II and his courtiers with their mutual virtuosity" (Findlen, 1994, p. 219). Steno (1673) probably studied the same ray that Redi had dissected and then written about (Guerrini, 1999). In 1673, Steno published his findings on the anatomy of the torpedo. The nerve bundles to the paired muscles responsible for the torpedo's effects were particularly intriguing to him.

Steno was an empiricist and did not like to go beyond the boundaries of what he could see with his science. In this regard, he was very critical of René Descartes and Thomas Willis for loosely theorizing about the nerves and muscles, as we shall see in Chapter 11. Also interested in many other aspects of scientific enquiry (e.g., paleontology and geology), he was nevertheless drawn to Catholicism, and he became a bishop and extremely spiritual after his conversion.

RENÉ-ANTOINE FERCHAULT DE RÉAUMUR

With Réaumur we leave the seventeenth century and enter the century of the Enlightenment, the *Siècle des Lumières.* René-Antoine Ferchault de Réaumur was not one of the dissectors to visit the Tuscan court. He was born in 1683 and spent most of his life in Paris, dying in 1757. Although far from Italy, he found merit in the mechanical theory of muscle contractions that came forth from that country. He would, in fact, be the person most closely associated with this sort of thinking about the torpedo throughout the eighteenth century, and he had a large following.

A pillar of French science, Réaumur came from the old Huguenot town of La Rochelle on the Atlantic coast of France, where torpedoes abound (Wheeler, 1926;Birembaut *et al.* 1962). He went to Paris in 1703 and became a member of the *Académie Royale des Sciences* in 1708, when only 24 years old. Held in esteem by his colleagues, he served as director of the *Académie* 12 times and as its sub-director 9 times. With his dedication to science, he corresponded with like-minded people all over the world, not being shy about asking for specimens, descriptions of plants and animals, and answers to questions that were bothering him.

Réaumur had extremely broad interests, including practical metallurgy, instrumentation (he invented a thermometer in 1831), and natural history (Egerton, 2006). In the latter domain, he wrote on various topics, including shells, pearls, silk, and polyps. He remains best known for his *Mémoires pour Servir à l'Histoire des Insectes* of 1734–42, a set of six volumes in which he included amphibians, reptiles, and various marine invertebrates, in addition to insects. Darwinian biologist Thomas Huxley (1900, vol. 1, p. 480) once stated, "I know of no one who is to be placed in the same rank [with Darwin] except Réaumur."

Besides his own studies, Réaumur is also known to science historians because Dutch physicist Pieter van Musschenbroek addressed the letter to him in 1746 in which he first described the instrument that would soon be called Leyden jar. This was the first electrical capacitor in history, and what Musschenbroek wrote to his correspondent was dutifully presented to the French *Académie.* This device soon became widely known, and it would play a very important role in our electric fish story (see Chapter 13).

Réaumur studied torpedoes (he called them *trembles,* using a vernacular name that would also recur in Walsh's experimental protocols) caught off the coast of Poitou, close to La Rochelle. He presented his thoughts about them on November 4, 1714, following its convention of waiting several years. The published *mémoire* is a modified version of the manuscript in Réaumur's own hand still extant in the archives of the *Académie,* where he was "Perpetual Secretary" (Fig. 9.6).

Réaumur dismissed the ancient version of the idea that the torpedo releases some sort of spirit or poison, or what Johann Baptist van Helmont had viewed as a narcotic ejaculation, the thought that it somehow "throws the poison of its glance" (van Helmont, 1648, 1621, p. 541; also see Thorndike, 1923–58, pp. 227–230). And he also found it difficult to accept

---

[36] In 1674, Danish physician Thomas Bartholin wrote a treatise that translates as *On Medical Travel* for his sons Caspar and Christopher and his nephew Holger Jacobsen, who were about to travel south (Jacobaeus, 1675, 1686). Thomas had been to France and Italy in 1640–1645 and advised these young men that physicians should be able to see things outside the boundaries of local experience, making them perfect "visitors" to distant places. As translated from his Latin, "no one puts much faith in the authority of a physician who has not set foot outside of his native land" (Bartholin, 1674; 1961 trans., p. 50).

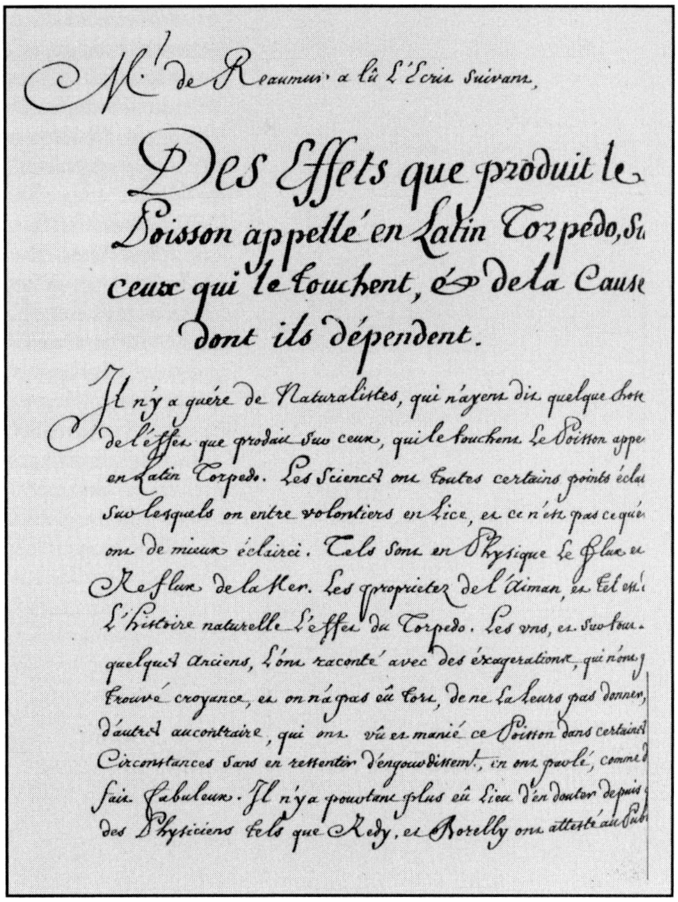

Figure 9.6. The first page of René-Antoine Ferchault de Réaumur's 1714 memoir on the torpedo in the handwritten records from the archives of the *Académie*. Réaumur's mechanical ideas about the torpedo's actions would be widely accepted during the first half of the eighteenth century.

Lorenzini's (1678) theory that contractions of special muscles somehow release minute *corpicciuoli* or noxious particles that could be transmitted to and into another body. He even devoted several pages of his *mémoire* to confute Lorenzini's view, based on corpuscles that Réaumur indicates as *torporifiques* (very likely drawing his inspiration from the *energia torporifica* of Borelli, who is constantly referenced in Réaumur's *mémoire*).

The problem Réaumur had with Lorenzini's theory stemmed, at least in part, from the fact that he found that the torpedo's effects could be felt, although in an attenuated way, through a fishing line and other intermediaries. In his words, "in touching a Torpedo with a walking stick of an ordinary length, it seemed to me that I have felt, although in a light degree, the action of this fish" (Réaumur, 1717, p. 357). Corpuscular transfer simply could not account for this phenomenon.

Thus, Réaumur concluded that the fish must jolt its prey or tormenters by direct and extremely rapid motions of the muscles that make up its specialized organs. This was basically what Borelli and Redi (often indicated as *Borelly* and *Redy* in Réaumur's paper) had said about a half century before. Also like Borelli and Lorenzini, Réaumur compared the effects produced by the fish to that of a sudden blow to the elbow.

In Réaumur's version of the theory, the shock is produced at the moment that the dorsal surface of the fish's body, which is usually flat or even concave in the preparatory phase, suddenly becomes convex. This is a consequence of a contraction, which is so rapid that it cannot be perceived with the eyes (*un mouvement si prompt, qu'il est impossible aux yeux les plus attentifs de l'apercevoir;* p. 351). Most likely, Réaumur was misinterpreting the concave position assumed instinctively by the fish in the preparatory phase preceding a shock, when troubled by the presence and actions of an experimenter.

Réaumur remarks that the effects of the torpedo's shocks are of short duration, and in this context he comments on the anecdote provided by Borelli about the "learned English Anatomist" who exaggerated the duration of the painful sensation caused by the fish. He notes that, besides the action of imagination, the effect might have been due to the fact that, as Borelli had mentioned, the anatomist was affected by a paralytic tremor (pp. 348–349).

The accurate microscopic study that Réaumur conducted on the *musculi falcati* is of particular importance for his theory. In addition to providing a figure in which these muscles are faithfully represented, he describes their components as cylinders composed of a multitude of fibers staked one above the other (Fig. 9.7). In his mechanical conception, these cylinders behave somewhat like elastic springs (*ressorts*). When the torpedo contracts its muscles, making its back concave, these springs become mechanically "charged," thus ready to release their energy in a sudden way when the fish relaxes its body and releases them. The explosive character of the torpedo's movement in passing from the concave to the convex positions is compared to a bullet. "The movement of a musket ball," he writes, "is possibly not as rapid [*prompt*] as that of the flesh returning to its original position, or at least it is more difficult to detect than the other" (p. 351).

Réaumur was aware of the existence of the other shocking fish. He refers to Richer's account of the South American eels in the *Académie*'s proceedings (and to the preliminary report of the *Académie*'s Secretary, Duhamel). He fails to realize, however, that the African torpedo mentioned "in the History of Abyssinia" is a very different fish. This fish, he writes, is capable "not only of killing the fishes that are alive," but also, a most amazing fantasy, "to give life back to dead fishes" (p. 359).

RÉAUMUR'S SPHERE OF INFLUENCE

Even more than the conceptions of Borelli, Redi, and Lorenzini, Réaumur's mechanical hypothesis of the torpedo's shock dominated the *Siècles des Lumières* before being questioned in the 1750s and 1760s. The death knell was sounded for Réaumur's hypothesis in the 1770s, when Walsh conducted his experiments, first on torpedoes and then on electric eels. But this is jumping ahead, and at this juncture it is important to give examples of some of the scientists and writers who cited Réaumur's theory prior to the second half of the eighteenth century to illustrate its broad acceptance.

One such person was his famous colleague at the *Académie*, Charles-Marie de la Condamine, a mathematician, physicist, explorer, and geographer. Born in the first year of the new century, the *Académie des Sciences* sent him across the Atlantic in 1735 to follow up on Jean Richer's work (Chapter 8), specifically asking him to gather more data on whether the earth is, in fact, wider at the equator

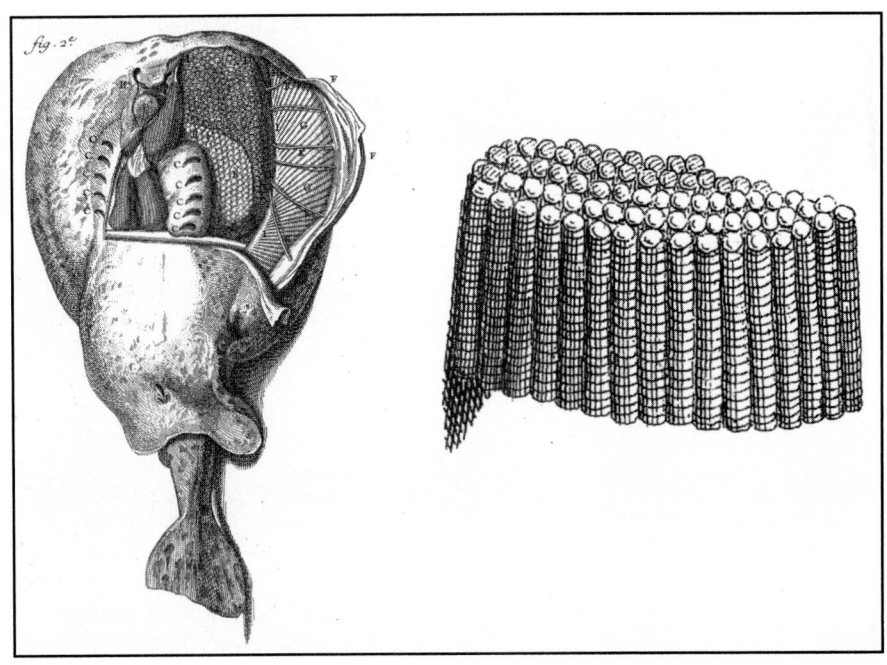

Figure 9.7. Réaumur's (1717) illustration of the torpedo's *organes électrique*s and their construction.

than at the poles. At the same time, they sent a second expedition to Lapland. Both groups would find distortions in the earth's shape, bulges, supporting Newton's theory.

Eight years after arriving in South America, and upon completing all he could do on the question of the earth's shape in the Province of Quito (now Ecuador), he decided to raft down "a route almost unknown"—the treacherous Amazon River. He was optimistic, hoping to map the entire navigable part of the river, "collecting observations of all sorts which I could make in such an unknown land" (Condamine, 1745; translated in McConnell, 1991, p. 3). This exceedingly dangerous voyage took place in 1743–44, after which he sailed back to Paris, where he presented his findings to the *Académie des Sciences* (McConnell, 1991).

Condamine's writings were so highly regarded that even Benjamin Franklin cited some of them in the 1750 edition of

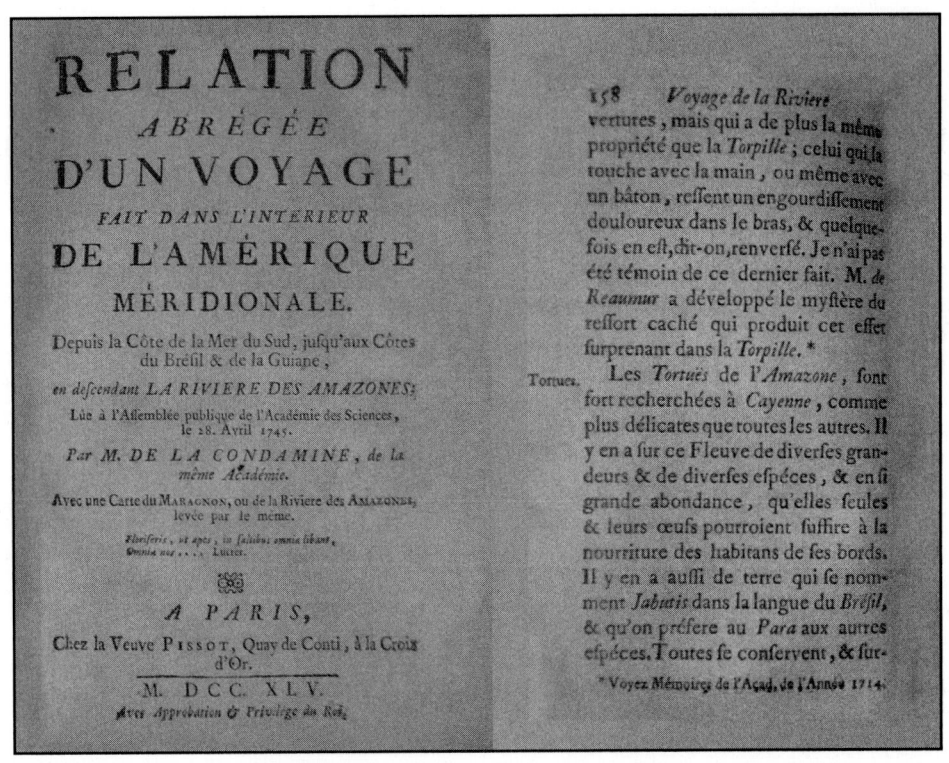

Figure 9.8. Charles-Marie de la Condamine's 1745 publication *Relation Abregée d'un Voyage...*, which includes some lines about a torporific river fish from South America that would soon be called the electric eel. Condamine endorsed Réaumur's mechanical explanation and did not link the powers of this fish with electricity.

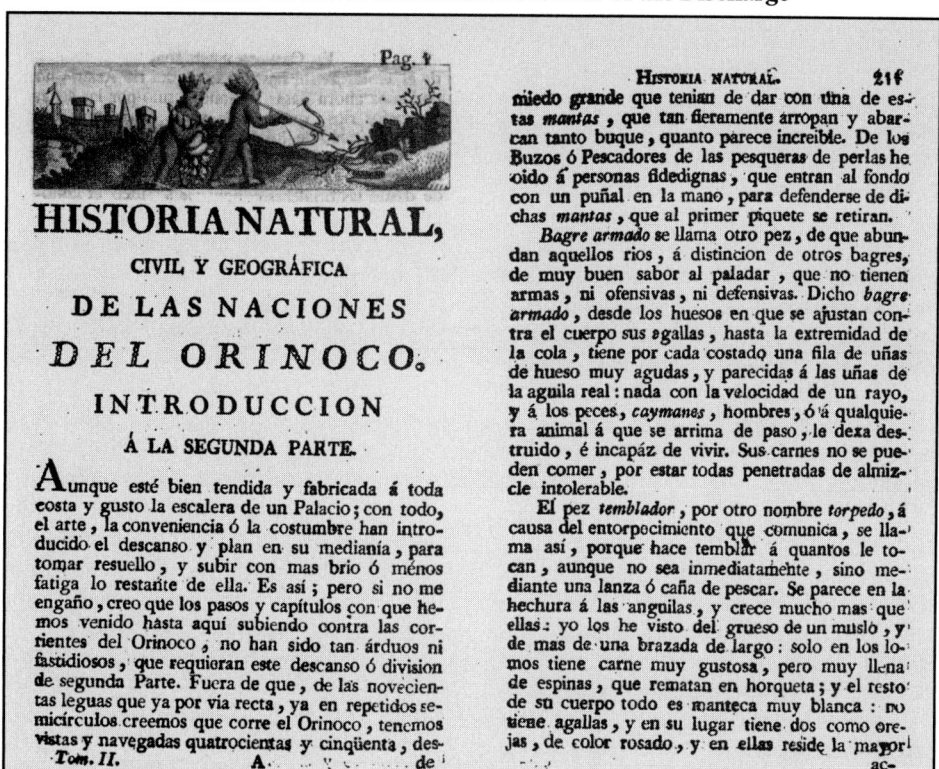

From the Occult to Mechanical Theories of the Discharge

Figure 9.9. The part of Gumilla's (1745) book on the Orinoco region of South America with the opening lines of his description of *el pez temblador*.

Poor Richard's Almanack (Labaree, 1961a, p. 445; Finger, 2006a, p. 60). Franklin, who was at the time promoting the implantation of scabs from smallpox sufferers into cut skin as a means of preventing the dreaded disease, was especially impressed by Condamine's (1754, 1759) observation that many lives had been saved along the Amazon River by missionaries who inoculated the Indians. Humboldt would later cite Condamine repeatedly in the narrative of his own travels through the Americas, which included his experiments on the South American "eels."

Condamine mentioned these eels in several publications. They can be found in a short report from 1745 and in a lengthier memoir submitted to the *Académie* in the same year, but published 4 years later (Condamine, 1745, 1749; both bear the title *Relation Abregée d'un Voyage...*) (Fig. 9.8), yet curiously the eels do not appear in the full account of his voyage, which appeared in 1751 (Condamine, 1751).

Translating from the *Histoire de l'Académie des Sciences pour l'Année 1745*, we read:

> I have seen a fish called the *Puraqué* in the region of Parà that has a body similar to the lamprey, with a great number of openings, and additionally has the same property as the *Torpedo*. Anyone who touches it with a hand or even with a stick feels a painful numbness in the arm, and sometimes, it is said, falls backwards. I have not been a witness to this last effect, but the examples are so common that there is no reason to doubt it. M. de Réaumur has solved the mystery of this hidden elastic force [*ressort cache*].[37] Without a doubt, similar mechanics operate in the Lamprey under question (La Condamine, 1749, p. 466).

Condamine's mention of Réaumur is significant, and not just because he was a fellow member of the *Académie*. It shows that Réaumur's mechanical theory was very much "the explanation" for the sea torpedo's effects in the 1740s, and that scientists believed that similar mechanics probably operated in other torporific fish, including the South American eels.

In this general context, what Joseph Gumilla published in 1745 is also worthy of attention (Fig. 9.9). Gumilla was a Jesuit who saved souls in the Orinoco River basin, but more than this, he studied natural history. His passage on *el pez temblador* (the tremble fish) appeared in a treatise published in 1745, and in it he even mentions a specific body part that could, he believed, cause its terrible effects.

Gumilla did not specifically mention muscle actions, but he did point to "two pink structures resembling ears" that have the features of muscles. Thus, in parallel to what European scientists were writing about the falciform muscles responsible for the torpedo's effects, Gumilla was stating that a comparable paired organ had also been identified in another location in the eel. Although his choice of organs would ultimately be found to be incorrect, his description of how he was drawn to this organ is interesting, as is the information he provides on the eel's hunting technique:

> The *temblador*[38] fish, also called *torpedo* because of the numbness it communicates, is given its name because it causes trembling in all who touch it, even when the

---

[37] The term *ressort*, meaning mechanic elastic spring, is the same as that used by Réaumur in his 1714 memoir on the torpedo.

[38] The most common spelling of the name is *tremblador* rather than *temblador*.

contact is not direct, but occurs through a spear or rod fishing. It looks structurally like eels, but it grows much more than they do; I have seen one the thickness of a thigh and more than an arm in length. It has meat only in the loin, which is very tasty, but it is very full of spines that end in a fork. The rest of the body is like very white butter. It has no gills, and instead it has two pink structures resembling ears, and these possess the greatest means for producing numbness—so much so that, after one is dead, the Indians cut them off before putting it in a recipient [*olla*] or to bake or roast without causing trembling; but if they touch these ears, they will tremble and have dulled senses. All the body is solid except for a distance of one palm below the mouth, where no intestine, except the craw, and directly next to it, the outlet for the faeces, will be found.

Neither caimans nor other large fish stay in the pools or backwaters of the river where the tremble fish stay, because they fear them. Their way of catching mid-sized fish is to get close to them as they pass, and stupefy them and swallow them at will. But it prefers smaller sardines, and the way in which it catches them is unusual. Upon recognizing them, it follows them until they are next to the bank and, at that point, it forms its body into a semicircle, with the head and the tip of the tail at the bank. All those sardines touched as the formation is made, and all attempting to leave the semicircle that are touched, are torpified, and the mouth has ample time to gulp all of them down—I say gulp because it has no teeth. (Gumilla, 1745, pp. 211–212)

Leaving the eels and returning to the torpedo, Chamber's *Cyclopaedia: Or, An Universal Dictionary of Arts and Sciences* of 1743 made it clear that Réaumur had settled the matter of the torpedo's actions. He had "cleared the point" as far as the editors were concerned (Chambers, 1743).

Similarly, in Volume XVI of the massive French *Encyclopédie* of 1765, there is an article on the *Torpille* by Louis de Jaucourt, a French scholar trained in science at Cambridge and medicine at Leiden. Jaucourt was the most prolific contributor to the *Encyclopédie,* and with regard to the torpedo's mechanism of action, Réaumur's explanation was again provided.

By 1765, the year of Jaucourt's entry in the *Encyclopédie,* an alternative explanation for the actions of this group of specialized fishes was, however, starting to be entertained. The new contention was that these fish are electrical. Indeed, many natural things, including lightning and even some minerals, were starting to be linked to electricity at this time. This had happened within a new *Zeitgeist,* with initially the Dutch in both the colonies and in the homeland, and afterwards Edward Bancroft, Benjamin Franklin, and the Royal Society of London playing central roles in it.

Réaumur's ghost and stature would haunt these investigators as they conducted their electrical experiments. As a matter of fact, Bancroft, who proposed an electric theory of the discharge of electric eels in 1769, felt it necessary to address Réaumur's hypothesis in an extremely critical way to make his point. Nevertheless, when presenting his electrical theory, Bancroft drew from Réaumur's language. The *corpicelli* invoked by Lorenzini to account for the torpedo's shock, the *corpuscles torporifiques* that Réaumur rejected, would become, in Bancroft's words, "torporific, or electric particles." How and why this will happen will be the subject of our next chapters.

PART IV
The Emergence of Fish Electricity

Chapter 10
The Electrical World of Benjamin Franklin

> Let the experiment be made.
> (Benjamin Franklin in a letter to John Lining, 1749; Labaree, 1962, pp. 523–524)

In the previous chapter we saw how theories involving rapid muscle contractions replaced ancient ideas about cold vapors and venoms, and vague notions of occult (hidden) forces, when accounting for the effects of torporific fishes. These mechanical theories emerged in the final decades of the seventeenth century, and with Réaumur in France, they continued to have substantial backing in the first half of the eighteenth century.

The first suggestions that these fishes just might function in another way, by electricity, would appear around 1750. This significant change in thinking would emerge in the context of the tremendous attraction that electricity held for academic scientists and even amateurs in the 1740s and 1750s. In this *Zeitgeist*, a wide variety of phenomena, in both the animate and inanimate world, would be increasingly perceived as electrical.

This chapter examines this phase of the history of electricity, and by so doing it sets the stage for the understanding how torpedoes, a family of African catfishes, and the South American eel literally became electrical. It also introduces the reader to a number of technological innovations and to Benjamin Franklin, who would not only play a major role in understanding the nature of electricity, but would also be influential in showing that torpedoes are, in fact, electrical. Franklin began his studies of the electrical world in Philadelphia but became an active member of the Royal Society of London when he moved to England in 1757, and it was there more than anywhere else that major developments in understanding fish electricity took place. Hence, it is also fitting that we present historical information about the Royal Society in this chapter, emphasizing its commitment to overthrowing falsehoods with experiments, and its belief that even the smallest facts might lead to useful knowledge.

THE EARLY HISTORY OF ELECTRICITY

The scientific history of electricity is relatively new. It is usually said to have started in 1600 with William Gilbert, a British natural philosopher and later physician to Queen Elizabeth I (Fig. 10.1). In his book dealing with the Earth as a magnet, his *De Magnete*, Gilbert used the term *electricam*, a Latin term that would serve as the basis for the English words "electricity," "electric," and "electrical" (Gilbert, 1600) (Fig. 10.2). He based his new word on the Greek term for amber (ἤλεκτρον), a fossil sap long known to attract feathers and straw when rubbed. In his text, Gilbert tried to separate electricity from magnetism and advocated "trustworthy experiments," as opposed to "guesses and opinions," reflecting the faith he had in experiments.

Figure 10.1. William Gilbert (1544–1603), who attempted to distinguish between electricity and magnetism.

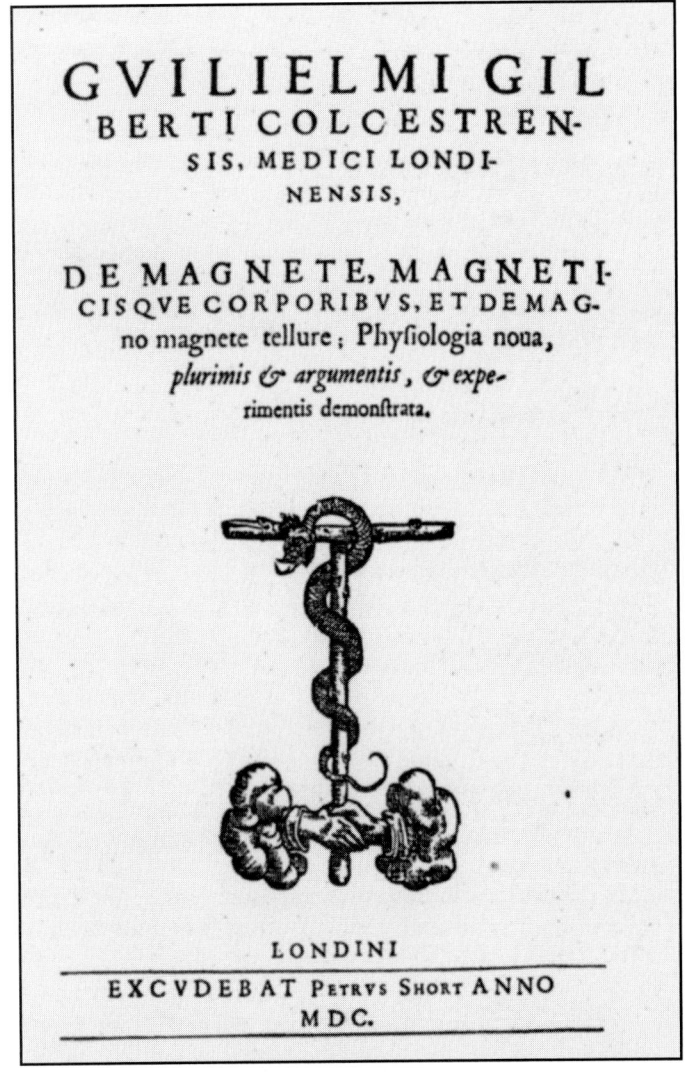

Figure 10.2. The frontispiece for Gilbert's *De Magnete* of 1600.

Figure 10.3. Otto von Guericke (1602–1686), the German inventor of a famous machine for producing static electricity.

The second half of the seventeenth century saw the advent of machines that could create electrical sparks by friction, and this technological development opened the door for more advanced scientific studies of electricity. The first static electricity machines came from Germany (Heilbron, 1979; Walker, 1936). After reading Gilbert on electricity, Otto von Guericke of Magdeburg (Fig. 10.3) began to experiment with a sulfur globe, about "the size of a child's head," that he pierced with an iron shaft, rotated, and rubbed with his hands, a leather pad, or a wool cloth. In 1672, he wrote that his charged globe would repel threads and move suspended feathers, sometimes with crackling sounds and even sparks (Guericke, 1672) (Fig. 10.4).

Francis Hawksbee took the next important steps, publishing his seminal observations in the opening decade of the next century (Hawksbee, 1709). He rotated an evacuated glass cylinder on a spindle and found that it glowed when his hands rubbed it (Fig. 10.5). His charged glass vessels were capable of generating sparks and could cause thin pieces of leaf brass to move, even when a foot away. They could even make a second glass vessel light up.

Hawksbee also experimented with suspended muslin and woolen threads, and noted that they could reveal the direction of the charge by how they reacted when brought close to an electrified body. With their divergent angular actions, suspended threads served as the first physical instruments used to detect and estimate the "degree of electricity" (Walker, 1936). A few decades later, elder pith balls suspended on string would replace them. John Canton would introduce such an electrometer in 1753 and illustrated it a year later (Canton, 1753, 1754).

In 1731, Stephen Gray, a silk dyer who moved to London from Canterbury, showed that electricity from machines using glass could be transmitted over threads and wires (Gray, 1731). In some of his experiments, the charge was conducted several hundred feet by hemp packthread that was suspended. In addition, he distinguished between substances that could conduct electricity and those that could not, the latter of which could serve as protective insulation.

Gray became well known to the public for his wondrous demonstrations, some of which showed that the human body could be safely electrified. His most famous crowd-pleaser involved a charity school boy, who was suspended horizontally from the ceiling with silk cords (Fig. 10.6). After being "electricised" with a glass cylinder held near his feet, the boy attracted feathers and brass strips to his body.

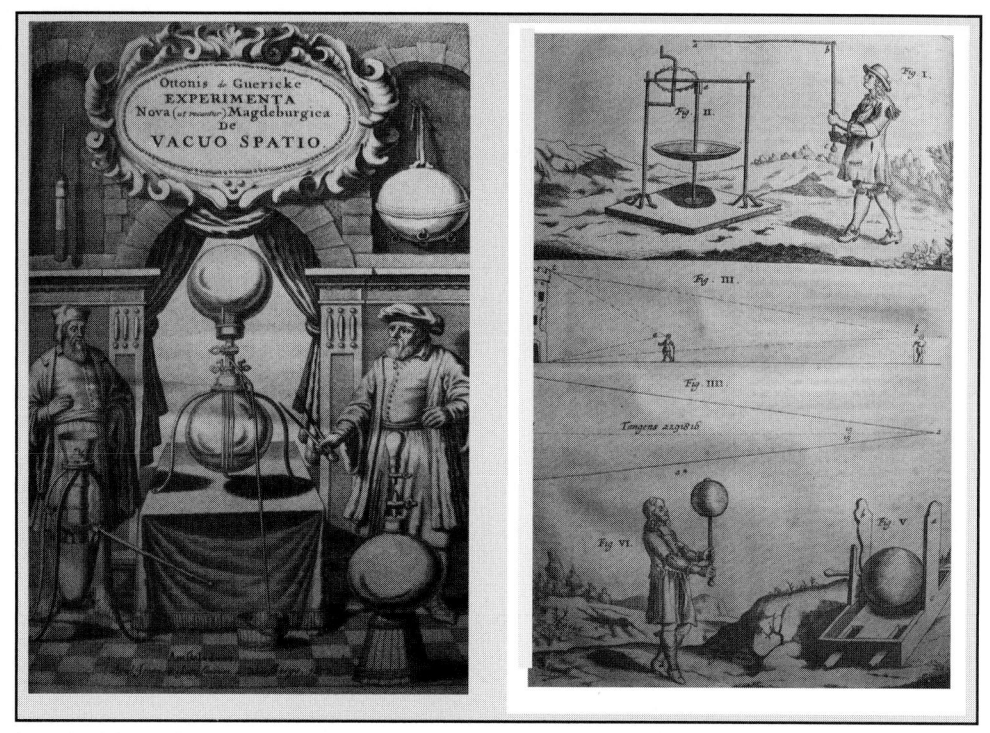

Figure 10.4. The title page of Guericke's 1672 treatise with a page illustrating his electric experiments.

To the amazement of those present, crackling sparks could safely be drawn from the boy's charged body without causing him any distress.

More powerful electrical machines now began to appear, most using glass plates and tubes. Still, it was an entirely new and unexpected invention that made studying electricity easier, leading to a better understanding of its physical properties, its now more predictable actions, and its uses in medicine. This small invention could store and release shocks on demand, and it was soon dubbed the "Leyden jar," a term embraced by natural philosophers.

THE LEYDEN JAR

The Leyden jar made its debut in the mid-1740s (Dorsman & Grommelin, 1957; Hackmann, 1978; Heilbron, 1979). Made of glass, covered with foil, and filled with water or lead shot, the jar comprised an inner and an outer conductor of electricity separated by a non-conductor (Fig. 10.7). Although primitive, it functioned as a capacitor or condenser. It could collect a charge from a metal rod called the "prime conductor," which was attached to an electrical generator, such as a rotating glass disk that would produce electricity by friction. The charge could be maintained and even transported before being released, which was accomplished by connecting the inside to the outside of the bottle through a conducting circuit, which could include wires, animals, or even people holding hands.

Pieter (Petrus) van Musschenbroek, a follower of Newton and a professor of mathematics and physics at the University of Leiden (founded in 1573 and in the past sometimes spelled Leyden), is often credited with the jar's invention, this being early in 1746. But Andreas Cunaeus, a visiting student, and Jean Nicolas Sébastien Allamand, another Leiden physicist,

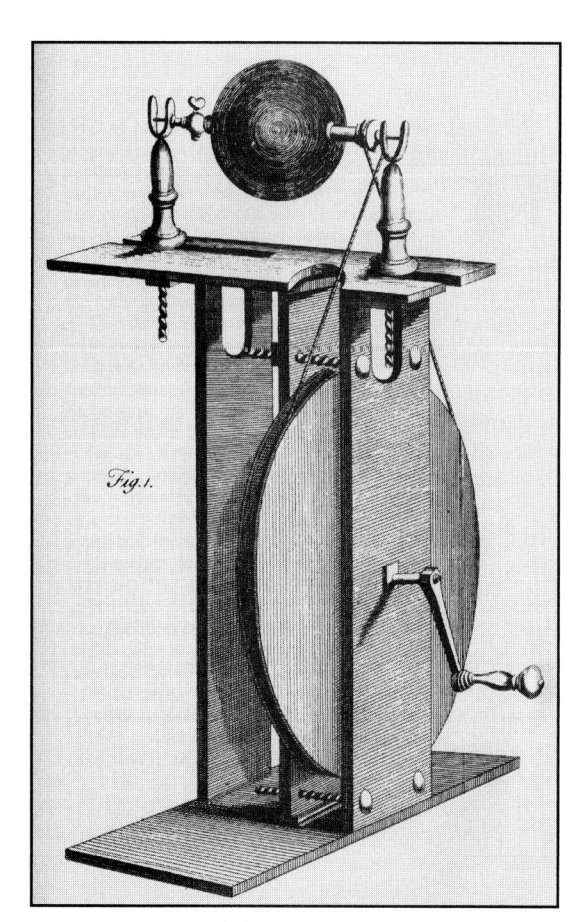

Figure 10.5. Francis Hawksbee's (1666–1713) frictional apparatus for producing electricity (Hawksbee, 1709).

Figure 10.6. Stephen Gray's (1666–1736) suspended boy experiment, showing how sparks could be drawn from his charged body (nose) by bringing a pointed finger close to it (from Nollet, 1746).

Figure 10.7. A Leyden jar with the metallic arc that connects the inner and outer plates resulting in discharge of its electricity and the production of sparks (from Nollet, 1746).

were also involved with the discovery in the Dutch city known for its science and medical school. (In Chapter 13, we shall return to Allamand and Van Musschenbroek, who were among the first scientists to contend that the eel and probably torpedoes are electrical.)

The jar also had a more distant parent. Ewald Georg von Kleist, a cathedral dean from Kammin (Kamien), a town in Pomerania (then Prussia, now Poland), came forth with a similar device in November 1745. He used a glass bottle filled with water that had a nail protruding through its cork stopper. He then hooked the nail to an electrostatic machine to charge the bottle. When he held the outside of the bottle and touched the nail, the bottle discharged through his body. Yet because Kleist was not as public or as well placed as van Musschenbroek, he is less well recognized for the discovery, which was also accidental and took place a few months earlier than the one in Leiden, where he had, in fact, been a student.

Being a foreign member of the *Académie des Sciences* in Paris, Van Musschenbroek used to send scientific reports to Réaumur, his contact there. The report addressed to Réaumur at the beginning of 1746 was mainly concerned with meteorological observations made in Leiden. At the end, however, Van Musschenbroek added an extremely interesting passage, just because, he said, "I see that there is empty space at the end of this paper." His communication was read and was recorded in the registers of *Académie* of the *séance* of January 12, 1746 (Fig. 10.8). In translation, Van Musschenbroek's annotation reads as follows:

I will communicate to you a new but terrible experiment, that I will advise you not to try personally nor will I do it again, not even for the entire Kingdom of France; I who have tried it and have survived by the mercy of God. I was doing some research in order to reveal the forces of electricity. To that purpose, an iron bar, *AB*, had been suspended with two threads of blue silk, and a glass globe that was spun rapidly had been placed near one of the extremities, *A*. It had been rubbed with the hands and communicated its electric force freely to the iron bar *AB*. On the other end, *B*, a brass wire, *CD*, hung freely. With my right hand, *F*, I was holding a round glass vase partly filled with water. With my left hand, *E*, I was trying to draw the crackling sparks that issued from the electrified iron bar toward my finger. Suddenly, my right hand, *F*,

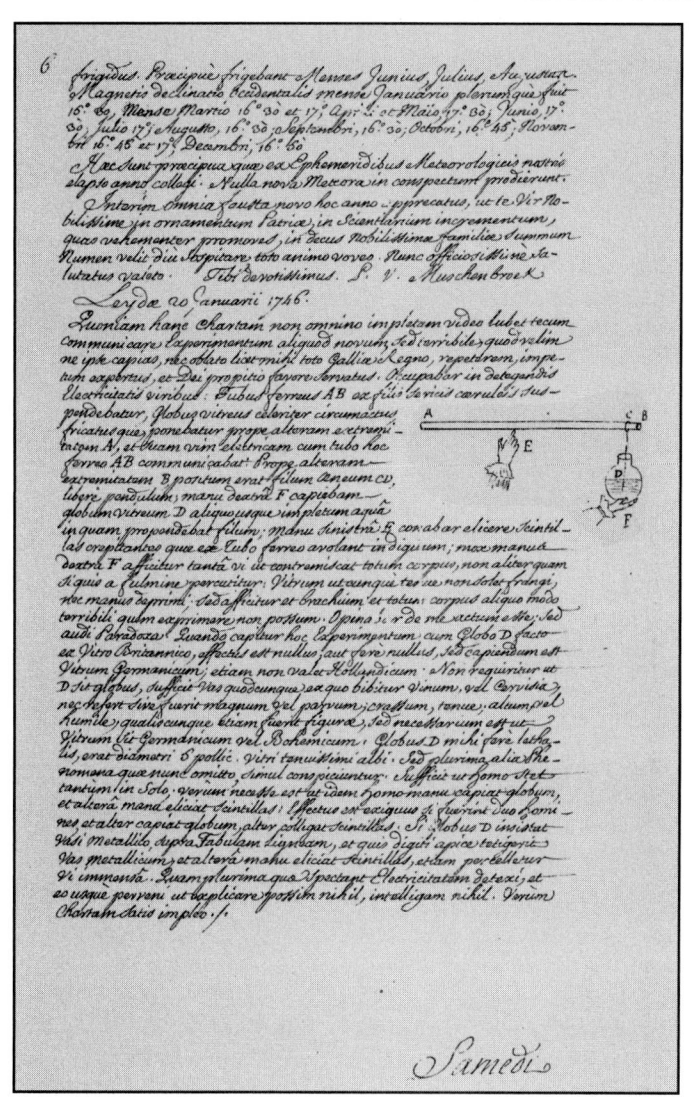

Figure 10.8. Pieter van Musschenbroek's 1746 communication informing Réaumur, his French correspondent, of his accidental invention of the device for storing electricity that would be called the Leyden jar (from the original handwritten recordings of the French *Académie*).

was hit with so much commotion that my body was rattled as if struck by a thunderbolt. My limb and entire body were terribly affected in a way that I cannot express. (Van Musschenbroek, 1746, p. 6)

WONDERS AND AMUSEMENTS

Abbé Jean-Antoine Nollet (Fig. 10.9), a non-ordained theology graduate from the University of Paris in 1724, was the official court electrician to Louis XV, an instrument maker, a lecturer and demonstrator, a royal tutor (for the Dauphin de France), and a powerful figure in the *Académie des Sciences* (Quegnon, 1925; Torlais, 1954; Bertucci, 2006). He was also the individual who translated Van Musschenbroek's letter from Latin into French, confirmed the Dutch physicist's claim, and coined the term "Leyden jar" for the terrifying, wondrous, but obviously important device.

Nollet consciously "made a business out of the upper classes' demand for entertainment and novelty" (Bertucci,

Figure 10.9. Abbé Jean-Antoine Nollet (1700–1770), the leading French electrical scientist at mid-century. Nollet was also known for his spectacular demonstrations.

2006, p. 197), and he demonstrated the Leyden jar in France in several marvelous performances that captured the public imagination. Before royal onlookers at Versailles, he constructed a circuit with 180 hand-holding grenadiers, who leaped in unison to its discharge. He used an entire convent of Carthusian monks for an even more spectacular demonstration before the king: over 600 held iron rods, forming a chain about 1.1 miles (about 2 kilometers) long, and they "gave a sudden spring" when the circuit was completed.

Although Nollet was a serious experimental natural philosopher, having come forth with a new theory of electricity and published a course on experimental physics that dealt with the known laws of electricity, and although electricity was the most fashionable branch of physics by the late 1740s, it was still more a source of amusement than a scientific enterprise for most philosophers at mid-century. In Germany, people flocked to see oxen toppled by Leyden jars, while a muscular Irish bishop buckled before onlookers in a London frolic. In addition, the *cognoscenti* were making "halos" appear by using static machines to raise hair, and performers were setting glasses of brandy aflame with sparks from the lips of amazed volunteers. For those of an inquisitive nature, the "electrical kiss," "electrical stars," and even "electrical rain" simply had to be personally experienced.

Electricity was virtually synonymous with spectacle at this time, with its renewed emphasis on wonder and curiosity (Bertucci, 2006; Bertucci & Pancaldi, 2001; Hackmann, 1978; Heilbron, 1979; Schaffer, 1983, 1993). And realizing that money could be made with these shows, some ambitious entrepreneurs took their shows to the British North American colonies, where Benjamin Franklin attended one that lacked an audience but changed his own life (Fig. 10.10).

Figure 10.10. Benjamin Franklin (1706–1790), whose Philadelphia experiments increased the understanding of the nature of electricity and helped make it a science.

FRANKLIN AND ELECTRICITY

The first time Franklin witnessed a demonstration with electricity was in 1743, just before the invention of the Leyden jar. At the time, he was a colonial postmaster whose work, coupled with a desire to see family members, took him from his adopted city, Philadelphia, north to his birthplace, Boston. While reading the *Boston Globe*, he came across a "Course of Experimental Philosophy" to be given by a Dr. Adam Spencer of Edinburgh (Cohen, 1990, Finger, 2006a; Franklin, 1996 reprint; Heathcote, 1955; Heilbron, 1977).

Franklin was disappointed when too few subscribers materialized for Spencer to give his lectures. But he was intrigued enough to pay for a private showing. He would write that Spencer's electric experiments "were imperfectly perform'd, as he was not very expert; but being on a subject quite new to me, they equally surpris'd and pleased me" (Franklin, 1996 reprint, pp. 120–121). Franklin so enjoyed what he saw that he encouraged Spencer to take his show to Philadelphia, where he advertised it and served as his ticket agent. Two men (William Black and John Smith), who were there in 1744, tell us that Spencer demonstrated "Sparks of Fire Emitted from the Face and Hands of a Boy Suspended Horizontally, by only rubbing a Glass Tube at his Feet" (Cohen, 1943, p. 6). He also showed how thin strips of brass and gold could be thrown in motion when brought close to a charged body.

Although Spencer piqued Franklin's interest in the wonders of electricity, it was a Londoner, Peter Collinson, who transformed Franklin into a serious electrical scientist. Collinson, a member of the Royal Society and a botanist and a merchant, became the organization's unofficial liaison with the British North American colonists during the 1730s. In addition, he was a Quaker who looked kindly on Pennsylvania with its large Quaker population, and he generously helped Franklin's Library Company acquire books and scientific instruments, which he shipped over from London.

In one of these shipments, Collinson sent the Library Company a three-foot glass tube as wide as a man's waist, with instructions on how to rub it so as "to obtain sparks that could set spirits or oil afire." His package included an anonymous article on electricity from *Gentleman's Magazine* that mentioned the breakthroughs of Guericke, Hawksbee, and Gray (Anon., 1745). This article from 1745 was a rather bad translation of the anonymous review on electricity that Swiss polymath Albrecht von Haller, the "most learned of Europeans," had written in the same year for the *Bibliothèque Raisonnée des Ouvrages des Savans de l'Europe*, a journal published in Amsterdam (Cohen, 1943; Heathcote, 1955; Heilbron, 1977).

"Your kind present of an electric tube, with directions for using it," Franklin wrote to Collinson in 1747, "has put several of us on making electrical experiments. . . . For my own part, I never was before engaged in any study that so totally engrossed my attention and my time as this has lately done" (Labaree, 1961a, pp. 118–119).

Franklin now had local glassblowers make additional tubes, accepted more gifts of electrical equipment, and purchased all of Spencer's apparatus (Fig. 10.11). He also invited other "amateurs" to work with him on experiments (Franklin was self-educated, having only a few years of primary schooling). These experiments increasingly took up so much of his time that, in 1748, he turned over the day-to-day operations of his lucrative printing business to one of his employees, David Hall, so he could pursue his science in a manner befitting a proper gentleman.

His group's first important discovery was that points are special in throwing off "the electrical fire." Franklin tells us that this discovery, made "by my ingenious friend Thomas Hopkinson," led to a series of experiments confirming that pointed bodies, such as needles, are far better than balls and "blunts," not just for releasing electricity, but also for attracting it.

The outstanding practical application to come forth from this insight was the pointed lightning rod, which Franklin described a letter to Collinson in 1750. Collinson shared Franklin's letter with members of the Royal Society and published it in *The Gentleman's Magazine* as "A Curious Remark about Electricity; from a Gentleman in America." As with his other inventions, Franklin gave it as a gift to humankind, rather than patenting it to obtain substantial profits.

In his letter to Collinson about points, Franklin also questioned prevailing theories of electricity, specifically those coming out of France. Charles François de Cisternay DuFay, who had become a member of the *Académie des Sciences* in 1723 after being sponsored by Réaumur, and who had been given the prestigious position of *Intendant de la Jardin du Roi* (King's Garden including collections of nature), had argued that there are two types or streams of electricity. DuFay (1735) called them vitreous and resinous,

object will be "plus," or "electricised positively," if it has an "over quantity" or excess of the electrical matter, and will be "minus," or "electricised negatively," if it has an "under quantity" or deficiency of this matter. There is a neutral condition, he further hypothesized, one not associated with attraction or repulsion, in which the particles are equally distributed. He further maintained that, although the electrical fluid might move from one location to another, its overall quantity remains constant.

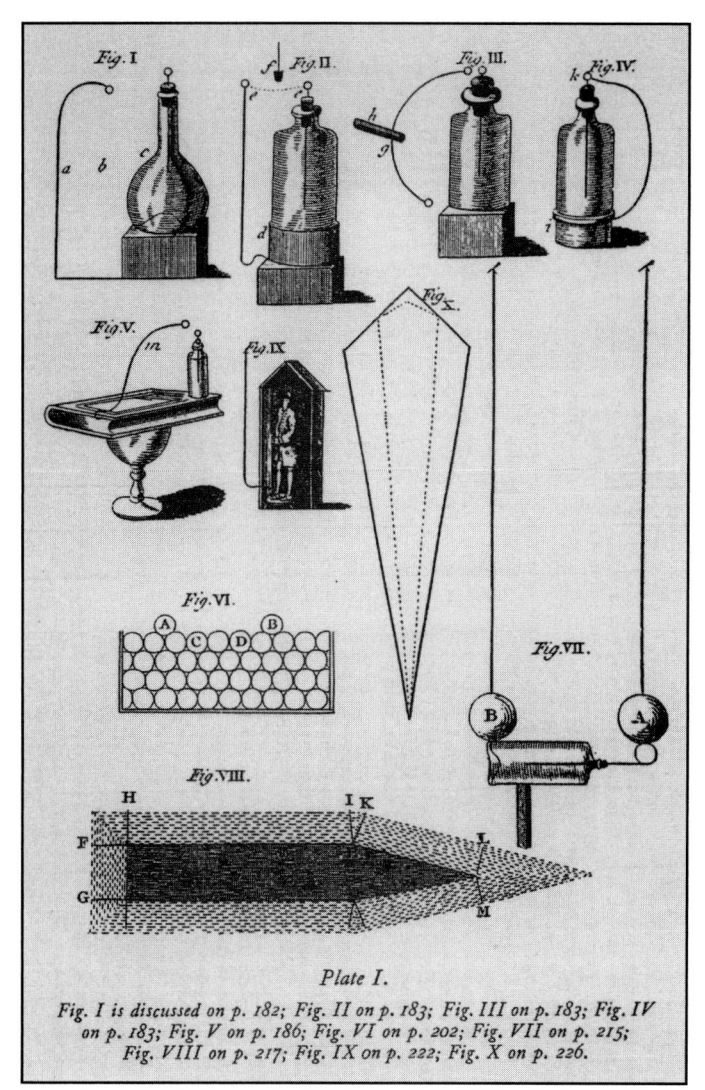

With his new theory of electricity, Franklin felt that he could now explain how "Mr. Musschenbroek's wonderful bottle" works (Labaree, 1961a, pp. 156–165). With a charged Leyden jar, there is an abundance of electrical fluid on the inside of the glass and a corresponding deficiency on the outside. Completing the circuit restores the equilibrium. Franklin went on to show that a bottle is not even necessary for this effect. He could collect electricity with a flat pane of glass having thin metal foil on its top and bottom. This flat condenser became known by several names, including *quadratum magicum* (magic square), "fulminating pane," "Franklin pane," and "Franklin square" (Fig. 10.12). Along with the Leyden jar, electric fish researchers would soon include it in their arsenals.

Franklin also felt he could also explain "Thunder Strokes," the eighteenth-century term for thunder and lightning storms, with his new theory. The dark wet clouds and land below must have different charges, he concluded, with the clouds usually being in a negative state. Thus, lightning would shoot up from the overcharged earth to the undercharged clouds, and an explosive sound would takes place as the neutral condition or equilibrium is restored. Researchers, as we shall see, would soon be assuming that fish should also flash and pop if electrical, albeit in a miniature way, this being fundamental to all highly charged discharging bodies.

In 1749, Franklin began to think about how he might prove that lightning and man-made electricity might be

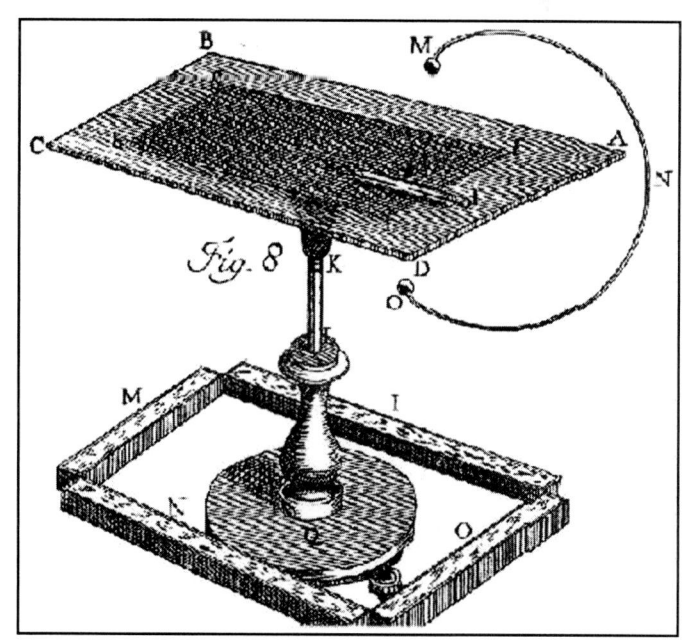

Figure 10.11. Franklin's (1751) illustration of some of his electrical equipment.

because stroking glass (along with certain minerals and fur) and stroking amber (along with paper and silk) were known to have opposite effects (attraction or repulsion) on certain objects, such as thin strips of gold leaf. He theorized that objects repel each other when they have the same type of electricity, but attract each other if they each have different types.

In 1733, DuFay accepted Nollet as one of his students. After some time with DuFay at the *Jardin du Roi*, Nollet gravitated to Réaumur. Nollet then became a private lecturer and worked his way up the ladder, acquiring various private, university, and governmental positions. Importantly, Nollet now rejected DuFay's electrical theory in favor of his own, which was based on two electrical streams traveling in opposite directions. Attraction, he argued, is due to an affluent stream carrying attracted matter toward a charged object, whereas an effluent stream flowing out from pores in the electrified object repulses or pushes objects away from the charged body. With this theory, Nollet felt he could explain all the established facts about electrical bodies.

Across the Atlantic, Franklin was also thinking in terms of extremely subtle particles of matter, but he reached a different conclusion about electricity. He theorized that an

Figure 10.12. Franklin's "magic square," the flat version of the first electric capacitor (from Beccaria, 1753).

qualitatively the same, as some people were already postulating. His thoughts on this subject were summarized in a letter to a South Carolina physician and natural philosopher, John Lining:

> Electricity agrees with lightning in these particulars: 1. Giving light. 2. Colour of the light. 3. Crooked direction. 4. Swift motion. 5. Being conducted by metals. 6. Crack or noise in exploding. 7. Subsisting in water or ice. 8. Rending bodies it passes through. 9. Destroying animals. 10. Melting metals. 11. Firing inflammable substances. 12. Sulphurous smell.
>
> The electrical fluid is attracted by points. We do not know whether this property is in lightning. But since they agree in all particulars wherein we can already compare them, is it not probable they agree likewise in this? Let the experiment be made. (Labaree, 1962, pp. 523–524)

Instructions on how the experiment might be made were published in 1750. Franklin's "sentry box" experiment, with a tall rod for capturing the lightning, was conducted near Paris by Thomas-François d'Alibard in the spring of 1752, but Franklin did not know about this successful achievement until later that summer (Cohen, 1990, pp. 66–158) (Fig. 10.13). Hence, in a variation of the crucial experiment, he proceeded to capture lighting using a specially constructed kite (Fig. 10.14). His personal success was mentioned that October in his own newspaper, *The Pennsylvania Gazette* (Labaree, 1961b, pp. 366–367), and in more detail by Joseph Priestley in *The History and Present State of Electricity* (1765, 1775), which Franklin oversaw and edited before it went to press. Priestley, who became very close to Franklin, called the kite experiment "the greatest, perhaps, that has been made in the whole compass of philosophy, since the time of Sir Isaac Newton" (Labaree, 1961b, p. 180) (Fig. 10.15).

Franklin showed that lightning could charge a Leyden jar, kindle spirits, and do everything that could be done with rubbed glass tubes (Labaree, 1962, p. 69). In terms of the bigger picture, he proved that electricity is not just a man-made force, but a natural one too, helping to lay the groundwork for another revolutionary idea and the theme of this book, that at least some living organisms might be electrical.

FOREMOST AMONG ELECTRICIANS

Peter Collinson and John Fothergill saw to it that Franklin's early communications on electricity were read before the Royal Society of London and published by Edward Cave, who quickly erected a pointed lightning rod above his London office. Cave edited *Gentleman's Magazine* but thought Franklin's communications were so important that he published them as a separate pamphlet, first in 1751 and then in various updated editions (Franklin, 1751; e.g., Franklin, 1774, this being the fifth edition) (Fig. 10.16). There were also editions in French and other languages.

Franklin's *Experiments and Observations on Electricity, Made at Philadelphia in America* contained many new electrical terms: plus and minus, positive and negative, electrical shock, electrified, charged, charging, uncharged, discharge, electrical battery, electrical fluid, Leyden bottle (and phial), armature, brush, conductor, condenser, non-conducting, and non-electric. Franklin even coined the word "electrician" (Yerkes, 1993).

His new theory of electricity, his lightning rod, his pamphlet, and his terminology helped Franklin to become the leading and best-recognized electrician in the world at this time. He was awarded Masters of Arts degrees from Harvard and Yale in 1753. This was followed by a doctorate from the University of St. Andrews in 1759 for his "successful Experiments with which he hath enriched the Science of Natural Philosophy and more especially of Electricity which heretofore was little known" (Labaree, 1965, p. 279). Thereafter, even though he never even attended college, he was known and addressed as Dr. Franklin (Finger, 2006a).

German philosopher Immanuel Kant called Franklin "the new Prometheus," whereas Priestley referred him as "the father of modern electricity" (Willcox, 1973, p. 259). One of his biographers, however, probably put it best when he wrote: "He found electricity a curiosity and left it a science," colorfully describing Franklin as being "an American Adam in his electrical garden" (Van Doren, 1938, pp. 171, 173).

This is not to say that there were no animosities, for there were. In the *Autobiography of Benjamin Franklin*, there is a telling passage about the reception of his papers in Paris. There he had some supportive and helpful friends (e.g., Buffon, DuFay's replacement at the *Jardin du Roi*), but now also an

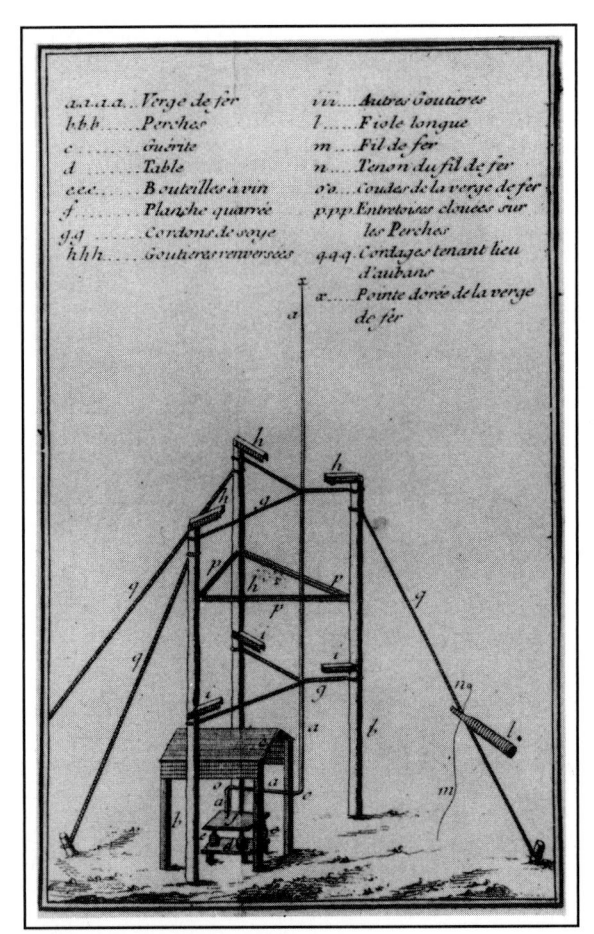

Figure 10.13. A "sentry box" with a tall pole for capturing electricity from the heavens (from Franklin, 1751).

Figure 10.14. Franklin's kite experiment as illustrated by Currier & Ives. Franklin is shown with his bastard son William, who was actually 20 at the time, serving as a helper and as a witness.

enemy in Nollet, whose ideas were seriously challenged and who felt Franklin's theory lacked "truly physical causes" (Riskin, 1998). Franklin, who found Nollet too opinionated and could not agree with him that everything must be explained mechanically, would later write:

> A copy of them [my papers] happening to fall into the hands of the Count de Buffon, a philosopher deservedly of great reputation in France, and, indeed, all over Europe, he prevailed with M. Dalibard to translate them into French, and they were printed at Paris. The publication offended the Abbé Nollet, preceptor in Natural Philosophy to the royal family, and an able experimenter, who had form'd and publish'd a theory of electricity, which then had the general vogue. He could not at first believe that such a work came from America, and said it must have been fabricated by his enemies at Paris, to decry his system. Afterwards, having been assur'd that there really existed such a person as Franklin at Philadelphia, which he had doubted, he wrote and published a volume of Letters, chiefly address'd to me, defending his theory, and denying the verity of my experiments, and of the positions deduc'd from them.
>
> I once propos'd answering the abbé, and actually began the answer; but . . . I concluded to let my papers shift for themselves, believing it was better to spend what time I had in making new experiments, than in disputing about those already made. I therefore never answered M. Nollet, and the event gave me no cause to repent my silence; for my friend M. le Roy, of the Royal Academy of Sciences, took up my cause and refuted him; my book was translated into the Italian, German, and Latin languages; and the doctrine it contain'd was by degrees universally adopted by the philosophers of Europe, in preference to that of the abbé. (Franklin, 1996 reprint, p. 122)

Thus, Franklin's theory had its opponents. But overall it achieved broad acceptance, especially by open-minded electrical scientists in France, in Italy, and in many other places, including in the mother country, England.

THE ROYAL SOCIETY OF LONDON

As indicated, Franklin's experiments in Philadelphia became well known to members of the Royal Society of London, although he rather unjustly complained, "It was, however, some time before those papers were much taken notice of in England" (Franklin, 1996 reprint, p. 122). Actually, Peter Collinson and John Fothergill presented his formal letters to the Royal Society in a timely way, as he had hoped they would do, while William Watson (Fig. 10.17) the prestigious organization's leading electrical scientist at the time, reviewed his pamphlet in the 1751 *Philosophical Transactions*, writing:

> Upon the whole, Mr. Franklin appears in the work before us to be a very able and ingenious man; that he has a head to conceive, and a hand to carry into execution, whatever he thinks may conduce to enlighten the subject-matter, of which he is treating: and although there are in this work some few opinions in which I cannot perfectly agree with him, I think scarce any body is better acquainted with the subject of electricity than himself. (Watson, 1751, pp. 210–211)

Franklin was honored with the Copley Medal from the Royal Society in 1753. He was nominated Fellow of the

Figure 10.15. Joseph Priestley (1733–1804), Franklin's protégé and author (with Franklin's help) of an early history of electricity.

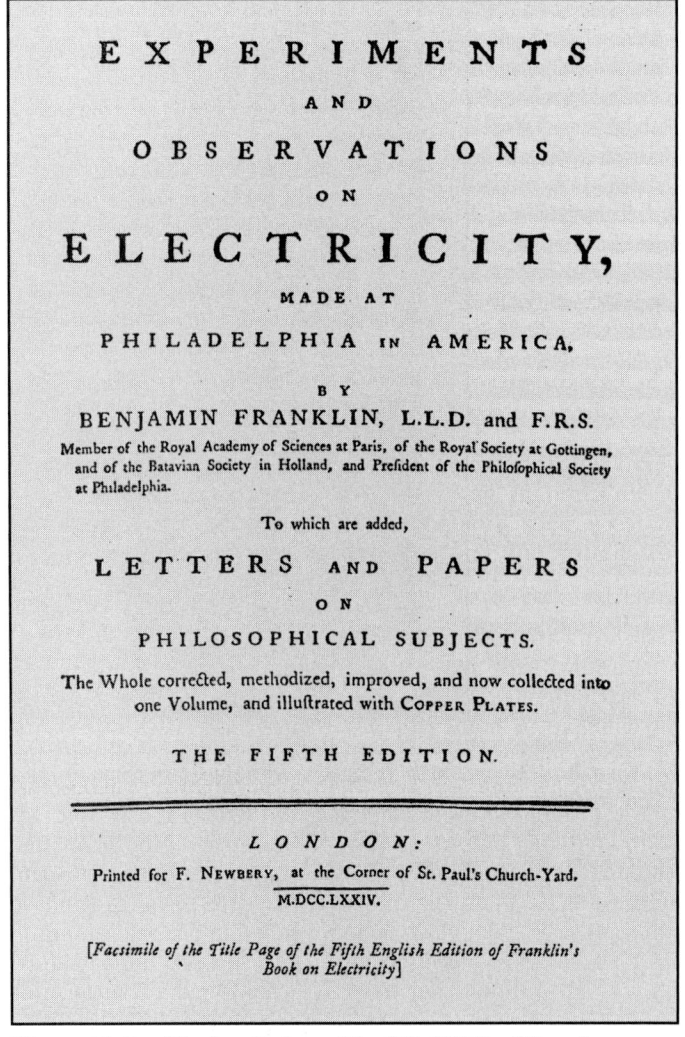

Figure 10.16. The frontispiece of the fifth (1774) edition of Franklin's pamphlet on electricity.

Royal Society in 1756 (being from the British North American Colonies, he was not considered a foreign associate), and when he arrived in London a year later he signed the register and became "official." Franklin always viewed this affiliation as one of the great honors of his life, in part because of the many friends he made through the organization and their support for his ideas, but also because of what the society represented since its inception.

The Royal Society was organized in 1660 and chartered by Charles II during the Restoration period to promote useful, natural knowledge that could benefit humankind. In the preamble to its *First Charter*, dated 1662, the founders of this organization emphasized the need to conduct "actual experiments" with good instruments to determine nature's truths (Hartley, 1962). The motto selected by the founders is telling: it is *Nullius in Verba*, meaning "on the word of no one."

Francis Bacon had sounded the clarion call to look at the world anew when the century began (Crowther, 1950; Eiseley, 1973; Pérez-Ramos, 1988; Rees, 2000a, b). As a young man, Bacon had gone to France to study French and Roman law. There he came upon the writings of Pierre de la Ramée, a sixteenth-century reformer who attacked Aristotelian teachings as misguided and faulty. When he returned to England, Bacon practiced law, took a seat in Parliament, befriended Queen Elizabeth I, and rose politically at her court. He was knighted after Elizabeth's death in 1603 by her successor, James I, King of England, Scotland, and Ireland (who reigned from 1601 until 1625). In 1618, he became Baron Verulam, Lord Chancellor, and 3 years later, Viscount St. Alban (Fig. 10.18).

Bacon's rank and political positions provided him with an ideal platform for presenting his new ideas about science and learning. Indeed, one of his aphorisms, dating from 1597, is "knowledge is power." He laid out his program to reform natural history, natural philosophy, and learning in general between 1605 and 1620 (Findlen, 1997).

Bacon's program involved starting over. His emphases were on careful observations and experimentation, with experiments being the most powerful scientific tool for settling debates. His method was inductive, starting with raw data collected bit by bit, as opposed to being guided by speculations and loose theorizing. With careful reasoning, theories and inferences might eventually emerge from the

Figure 10.17. William Watson (1715–1787), one of the electricians at the Royal Society of London.

Figure 10.18. Francis Bacon (1561–1626), one of the great fathers of the scientific revolution of the seventeenth century.

growing storehouse of facts, an understanding of which might also result in a useful discovery that could benefit humankind. Hence, Bacon was advocating more than a strict empiricism. He was calling for analyses of the new findings, however small, however odd, believing that broader truths would emerge if reasoning could be combined with good raw material.[1]

Bacon's endeavors centered on natural history (essential to natural philosophy and "the foundation of all"), and in his call to record the knowledge of the world, he emphasized the need for collecting, including what had been considered the marvels and wonders of nature (Findlen, 1994, 1997). He wanted these specimens and phenomena to be displayed and subjected to proper scientific study in order to debunk the "fables and popular errors" of the past and to establish lasting truths. As stated in his *Novum Organum:*

> For we are not to give up the investigation, until the properties and qualities found in such things as may be taken for miracles of nature be reduced and comprehended under some Form or fixed Law; so that all the irregularity or singularity shall be found to depend on some common Form, and the miracle shall turn out to be only in the exact species differences, and the degree, and the rare occurrence; not in the species itself. Whereas now the thoughts of men go no further than to pronounce such things the secrets and mighty works of nature, things as it were causeless, and exceptions to general rules. (Bacon, 1857–74, vol. 2, p. 28)

Thus, as Bacon saw it, the natural history of particulars would play an unprecedented role in the new undertaking, not only as the permanent foundation of natural philosophy, but also as a constant check upon abstraction and generalization. In this context, Bacon sometimes referred to natural history as a "warehouse," one that must be constantly replenished and drawn upon, if natural philosophers were ever to fathom the deepest secrets of nature (Daston & Parke, 1998, p. 224).

Bacon (1620/2000; 1857-74) severely criticized his predecessors for trying to complete the work of Aristotle. He also chastised them for their devotion to other ancient authorities, especially Pliny, who had numerous faulty "facts" that he never verified and no consistent criteria for inclusion in his often-cited works (see Chapter 3). He also took aim at the natural philosophers of the previous few centuries for not fully appreciating the unity of nature, which included natural oddities. He emphasized that all nature deserves further study, because even the strangest life forms could have utility in "art," a term that then included medicine and technology.

Bacon's ideas on how the sciences must advance beyond the "Pillars of Hercules" (a boundary of the ancient world) can be found in a number of his books, including *The Advancement of Learning*, published in 1604, and *Novum Organum*, dating from 1620. Bacon's books, however, contained few experiments, and this led William Harvey, the famed discoverer of the circulation of the blood, to quip that Bacon wrote natural

---

[1] On Francis Bacon in particular and, more in general, on the scientific revolution, the reader is referred especially to the important works on these themes of Paolo Rossi (Rossi, 1957; 1998; 2002, 2006).

philosophy "like a Lord Chancellor" rather than a scientist (Crowther, 1950, p. 11). Nevertheless, it was Bacon's words, rather than what he did, that drew followers and changed the face of science in Britain and around the world.

The change, however, was not immediate. His thoughts started to draw serious attention only in the 1640s, and they were not really embraced until the 1660s with the formation of the Royal Society of London. They then spread like a flame throughout Western Europe, as well as to the North American colonies.

With Bacon's noble program in mind, many seventeenth-century men set forth to gather facts about the weather, diseases, plants, minerals, and animals, with great attention being given to new and unusual species and material. "By the 1590s, when Bacon began to mention natural history in his writings, the increased circulation of printed texts, the growing presence of naturalists in London and the desire for cultural emulation all made natural history attractive to the English nobility" (Findlen, 1997, p. 246).

Nevertheless, only a select few individuals would go beyond the raw facts, as can be seen by perusing the records of the Royal Society of London and the *Académie Royale des Sciences* in Paris at this time. But some had the training or the intellect to do so, and these thinkers tried to explain what the facts meant in terms of the bigger picture, while also hoping to find utility with a better understanding of natural causes. In a very real way, the reasoning of some of the great men of the era, such as Benjamin Franklin, helped to take the wonder out of wonders (e.g., lightning), even though their chosen phenomena still garnered great curiosity.

As a consequence of various scientific, social and historical events occurring in the seventeenth and eighteenth century, naturalists and natural philosophers were thus given many reasons to study electricity and, although still not tied to electricity as mid-century approached, to look more closely at torporific fishes. After all, an understanding of electricity and the true powers of these unusual fishes could shed light on nature in general, and the insights gained might be of service to humanity. At the very least, some of the "vulgar errors" of the past could be exposed, such as ringing church bells to protect against lightning strikes.

VULGAR ERRORS SCRUTINIZED

Thomas Browne was one of Bacon's most important followers in England. Trained in Oxford and Leiden, he was a well-traveled physician who dedicated himself to debunking "vulgar errors"—including wonders, marvels, curiosities, and miracles—long passed off as truths. Like Bacon, one of his major targets was Pliny, although he also took aim at mindless, gullible people, who blindly and naïvely accepted what the revered ancients and other authorities of the past had written. He was particularly incensed by those who made their livings by marketing the false marvelous to the ignorant, as if they were truths.

Browne's *Pseudodoxia Epidemica, or, Enquiries into Very many Received Tenets, and commonly Presumed Truths* first appeared in 1646. This book was so popular that it underwent many editions, and it helped to set the stage for the formation of the Royal Society, which, as we have noted, would be dedicated to exposing falsehoods and determining truths.

Browne realized that solid, empirical science combined with reason is essential for getting to the truth. He dismissed many legends with varying degrees of wit and charm, starting with minerals and vegetables and moving on to humans and the cosmos. The labels of some of his sections are themselves revealing, a few being: "That a Diamond is made soft, or broke, by the blood of a Goate," "Of Fayrie stones," "That an Elephant hath no joints," and "Of the Unicornes horne."

Browne discusses torpedoes in Chapter XXVII, which deals with some 13 miscellaneous falsehoods. He does not believe for a minute that dead animals can still exert the powers they exhibit when alive. Much as the emerald-green light dies with the glowworm, "Thus also the torpedo which alive hath power to stupefy at a distance, hath none upon contraction being dead, as Galen and Rondeletius particularly experimented" (Brown, 1646, p. 178).

One of Browne's heroes was William Harvey (1628), who, with his discovery of blood circulation, had set in motion a revolution that overthrew the entire system of old physiology that dated from the times of Hippocrates and Galen. This, as we know, was based on the idea of the four humours and, notably, on the continuous production of blood necessary to replace that which was used for nourishing the organs of the body. It was Harvey's approach, based on replicable experiments, that Browne valued far more than the words of the so-called great authorities of the past.

SOME MYTHS CONTINUE

Bacon's call for facts, Browne's words about "vulgar errors," and Harvey's experimentalism play important parts in our story. So does Sir Isaac Newton's way of seeing the world. Newton became president of the fledgling Royal Society and promoted his philosophy with a strong fist. On the subject of occult qualities, he had this to say:

> To tell us that every Species of Things is endow'd with an occult specifick Quality by which it acts and produces manifest Effects, is to tell us nothing. But to derive two or three General Principles of Motion from Phaenomena, and afterward to tell us how the Properties and Actions of all corporeal Things follow from those manifest Principles, would be a very great step in Philosophy, though the Causes of those Principles were not yet discover'd. (Newton, 1704, Query 31)

Newton was bothered by the fact that some people were still alluding to occult properties early in the eighteenth century. He was also bothered by some of the unsubstantiated myths that were still circulating. Although he did not single out torporific fishes, he could have. One need only look back at Robert Lovell, a compiler of massive amounts of information, who studied botany, zoology, and mineralogy at Oxford and then moved to Coventry, where he practiced medicine until his death in 1690.

Lovell's *Panzooryktologia sive, Panzoologicomineralogia* came off the press in 1661, 15 years after Browne's book had appeared and shortly after the Royal Society was organized. Written while he was still in Oxford, it was an herbal or pharmacopoeia, and in it he cites the claims of some 250 authors, without verification. It includes a section on the unicorn with wonderful details, including that the "flesh is bitter and unfit to be eaten," and that it has "a white and smooth horne, serving to expel and dissolve all poyson, if put into the water" (Lovell, 1661, pp. 125–126).

The part on the "Cramp-fish Torpedo" appears later in the book (pp. 191–192). Interestingly, Lovell writes that these fishes come from "Nilus and muddy places of the Sea." Although these words would suggest that he will group the torporific catfishes and sea torpedoes together, all of his references are to flat saltwater rays, suggesting that he might not have recognized the differences between these fish in their river and sea locations.

Drawing from the ancients, most notably Pliny, and their Byzantine and Middle Eastern followers, Lovell writes that torpedoes can "loosen the belly, and help the hepatick disease when boiled." He then mentions uncritically that they can help epileptics and, when "boiled in oile, and used with a little wax and oile it helps the gout." Inflammations can also be treated with these creatures, he informs his readers, adding that its skin "helps the falling out of the matrix," and another ancient idea, that its gall applied to the genitals can reduce excessive venery!

To Newton and the men of the new Royal Society, and to someone like Browne, Lovell's compilation would have been viewed as a relic from the past—one reflecting blind adherence to the authorities of the past, and one laced with fiction and magic. Its pages are filled with classification errors, and even its underlying philosophy was in some circles questionable: that every animal, plant, and mineral must cure some disorder.

From these perspectives, it is easy to see why a Royal Society had to be formed and why the leadership in the mid-1700s wanted Franklin—the brilliant but skeptical electrical scientist who understood the importance of good instruments, careful experiments, and hard facts—to join its ranks. It is also easy to understand why Franklin, with his ideals about helping others and love of good fellowship, was so happy to become a member and then a leader of this esteemed organization, which was filled with like minds who shared his values and views, not always politically but at least about good science.

MEDICAL ELECTRICITY

Electricity, which was such an exciting topic in the middle of the eighteenth century, was at this time also affecting the practice of medicine. In fact, the topic of medical electricity was under discussion at the Royal Society in 1757, when Franklin arrived in London and began to attend the meetings of this organization.

The use of electrical machines for treating various disorders with mild shocks or by drawing sparks from a charged human body did not stem from the idea that torpedoes—used therapeutically in antiquity—might be electrical. Rather, medical electricity began shortly *before* natural philosophers began to speculate that some fishes might be electrical, and well before this association became an established fact.

The guiding idea had much to do with the eighteenth-century search for utility in just about everything natural philosophers investigated. Long tied to philosophy (e.g., the Greco-Roman idea that all parts and things must have some usefulness), finding uses for even seemingly mundane discoveries was a mission of the Royal Society and related organizations in the seventeenth century.

Franklin knew that controlling electricity with grounded lightning rods was not really a use for electricity. Nor were natural philosophers willing to take Franklin seriously when he wrote tongue-in-cheek to Peter Collinson in 1747:

In going on with these Experiments, how many pretty systems do we build, which we soon find ourselves oblig'd to destroy! If there is no Use discover'd of Electricity, this, however, is something considerable, it may *help to make a vain Man humble*. (Labaree, 1961a, p. 171)

Historians tend to agree that the use of electrical machines in medicine started with Johann Gottlob Krüger, a professor of philosophy and medicine in Halle (Licht, 1967; Piccolino & Bresadola, 2003). Krüger had been asked by his students to lecture on electricity and, in 1743, he told his students that:

all things must have a usefulness: that is certain. Since electricity must have a usefulness, and we have seen it cannot be looked for either in theology or in jurisprudence, there is obviously nothing left but medicine. . . . The best effect would be found in paralyzed limbs. (Krüger, 1744, p. 5)

When making this statement about paralyzed limbs, Krüger might have been thinking about how accidental electrical shocks, a common occurrence among people experimenting with electricity at this time, could make muscles twitch and contract involuntary. In 1744, the year his lectures were published (Fig. 10.19, left and center), Christian Gottlieb Kratzenstein, one of his students, tried using electricity on two patients with movement disorders, most likely due to arthritic conditions. Fifteen minutes of electrification allowed a woman to overcome a contraction in one finger, whereas a man with two lame fingers was able to play his harpsichord once again.

Kratzenstein (1745) described these successes in a monograph published the next year and, as word spread, the new cure quickly drew an assortment of medical practitioners (see Fig. 10.19, right). To a certain extent, this had a lot to do with how alluring electricity was at the time. But it also had much to do with the fact that traditional medical practices were still largely rooted in bleeding, purging, and other heroic remedies, and were largely ineffective when it came to treating paralyses, tremors, deafness, and other sensory disorders.

Just a few years later, Jean Jallabert, a professor of philosophy and mathematics in Geneva, reported that he applied shocks from a Leyden jar to a locksmith, a man who had lost movement on his right side after being hit with a hammer years earlier (Jallabert, 1748). Although the facts of this case would be questioned, his findings raised hopes about medical electricity even higher.

The theory of the day was that electrical treatments could unblock blockages that underlie impaired function, including turgid fluids affecting the nerves, and that it could even tighten flaccid nerves. In the 1745 *Gentleman's Magazine* article that Collinson sent to Franklin, we read: "It has already been discover'd, or believ'd to be so, that electricity accelerates the movement of water in a pipe." Readers were further informed that electricity could enhance the pulse and cause blood to flow more freely from a cut vein. "There are

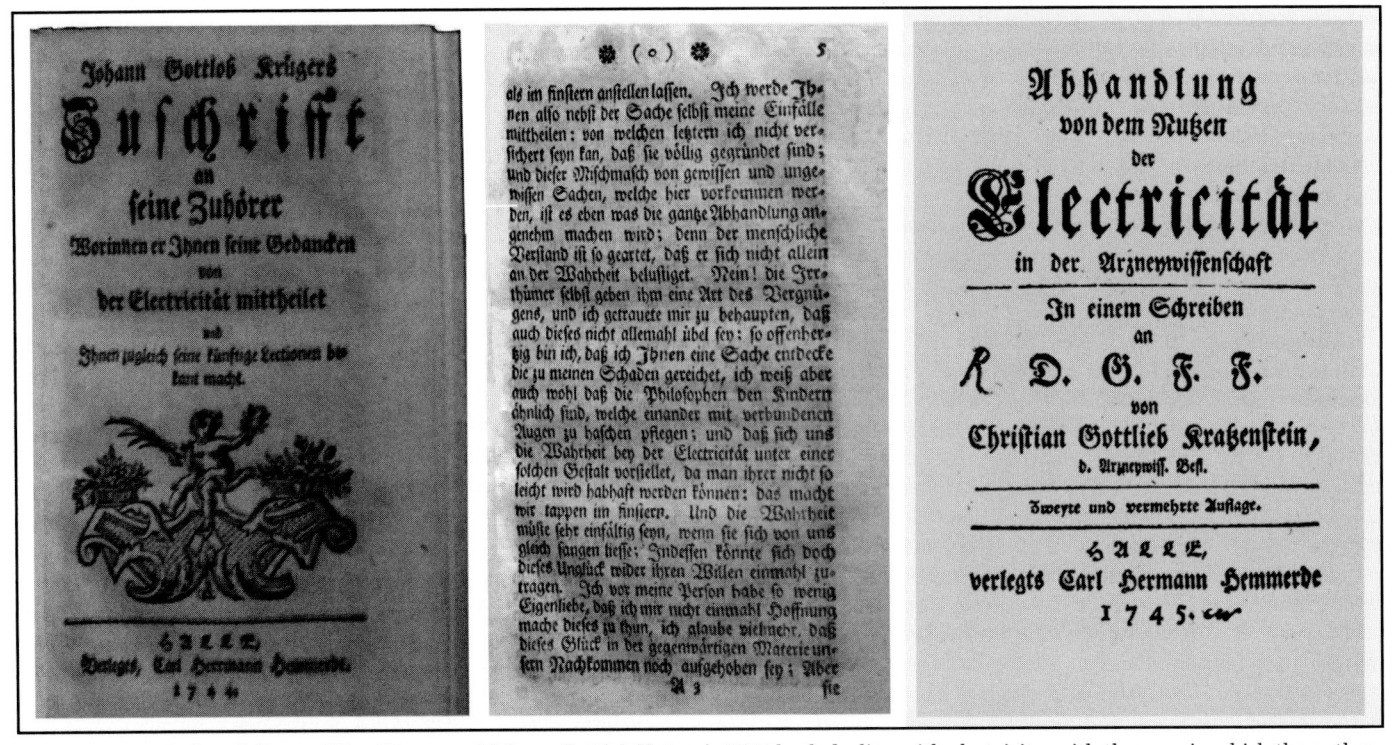

Figure 10.19. Left and Center: The title page of Johann Gottlob Krüger's 1744 book dealing with electricity, with the page in which the author expounds his reasoning why electricity should have utility in medicine. Right: The title page of the 1745 monograph of Krüger's student, Christian Gottlieb Kratzenstein, reporting some of the first successes of the medical application of electricity.

hopes," its author concluded, "for finding a remedy for the sciatica or palsy" (Anon., 1745, p. 197).

Clearly, little was known about nerve physiology in the mid-eighteenth century, a subject that we shall examine in more detail in the next chapter. But not everyone who tried medical electricity was driven by theory. Many electrotherapists, and certainly those who did not attend a medical school, focused only on the fact that it could be administered safely, and the associated claim that it often worked (Fig. 10.20). Franklin was one such person, and even before he had become a member of the Royal Society, he set out to see with his own two eyes whether medical electricity is effective or is just another quack remedy.

Although Franklin did not leave a casebook, his surviving letters show that he treated or oversaw the treatments of several very important colonials, beginning in the late 1740s (Finger, 2006a, pp. 80–101, 2006b, c, 2007). One was James Logan, who was William Penn's secretary, one of the most scholarly men in the colonies, a chief judge of Pennsylvania's Supreme Court, and an acting governor of that state.

Figure 10.20. Two images illustrating how electricity could be applied to sick or disabled patients. There were numerous publications proclaiming beneficial effects of electricity in the second half of the seventeenth century, although there were also notable failures (from Adams, 1785 and 1792).

Logan had had a history of strokes and, in 1749, suffered a serious stroke that left him paralyzed, speechless, and helpless. Another of his patients was Jonathan Belcher, who had governed Massachusetts and was governing New Jersey in 1750, when "stricken with the palsy." Neither Belcher nor Logan responded favorably to the electrical treatments, nor did numerous others with palsies of long duration whose names may never be known, since Franklin did not identify them.

In contrast, Franklin had a notable success with a young woman with hysteria (Finger, 2006a, pp. 102–108; Finger 2006b, c, 2007). Case C.B. had a lengthy history of "convulsions" and "almost the whole train [of] hysteric symptoms." Because she had not responded to medicines, Franklin and Cadwalader Evans, then a young Philadelphia physician, treated her with gentle shocks that resulted in a long-term cure. This was in 1752, and Evans published the case in 1757, the very year that Franklin began to attend the meetings of the Royal Society (Evans, 1757).

At those meetings, one of the first papers Franklin heard was about how a woman with hysterical palsy was cured with therapeutic electricity (Brydone, 1757). The claim was made in a formal letter to John Pringle, a leader and future president of the Royal Society who would become very interested in the history of electric fish research and would even write a small treatise on it. Patrick Brydone, a Scot who would become better known as a travel writer, and a future member of the Royal Society, wrote the letter, which was endorsed by renowned Edinburgh physician Robert Whytt, who had become a member of the organization in 1752 (Brydone, 1757).

After presenting the documents describing the successful case, Pringle asked Franklin to present his own findings. Franklin's formal letter to the Royal Society was written late in 1757 and published in the *Philosophical Transactions* the next year (Franklin, 1758). Mentioning just his own palsy cases, and not bringing up how he cured C.B.'s hysteria, he explained that he drew "a Number of large strong Sparks from all Parts of the affected Limb or Side" and also applied shocks to palsied patients who had sought his help in Philadelphia. There were some immediate beneficial effects, but they proved to be transient, he explained, "so that I never knew any Advantage from Electricity in Palsies that was permanent" (1758, p. 482). (Notably, almost the same wording would be applied to electric fishes by one of the natural philosophers in Franklin's circle, Edward Bancroft; see Chapter 14.)

Franklin was neither the only natural philosopher who was skeptical about electricity being a general cure-all by the middle of the century, nor the only one concerned about some of the claims being made about its utility for treating certain disorders. At about the same time, Abbé Nollet, who would become his enemy, was expressing similar concerns about the new cure, rightfully recognizing that it might be less than the panacea or cure-all everyone was hoping had finally been found.

In 1746, with the assistance of two other physicians, Nollet (1746) first used a Leyden jar to apply electricity to, and draw sparks from, three paralytics in a Paris hospital. His initial results showed promise. But if he had been optimistic at first, he became less so after 1748, as he continued his research on some paralyzed soldiers at the *Hôtel Royal des Invalides*. These findings suggested that the new cure held considerably less promise than he originally thought.

Nollet's suspicions increased in 1749, when he went to Italy to investigate the fantastic claims made by Giovanni Francesco Pivati, a Venetian who put medicines, such as Peruvian balsam, camphor, and opium, in sealed glass tubes or Leyden jars. He then either passed electricity though the tubes to release these virtues or had his patients hold the tubes while being electrified, convinced that the medicated particles inside the glass would now pass through its pores, and that all a patient had to do was to breathe or receive the electrical effluvia. These new ways of applying medicines, Pivati maintained, allowed him to cure many disorders (Bertucci, 2006, 2007). Here he pointed to an old bishop, who had lost the use of his hands and legs. After being treated, he was able to clap his hands energetically, and even jump down a flight of stairs like "a vigorous young man" (Pivati, 1747, p. 40).

Nollet returned to Paris from his 9-month trip to Italy late in 1749. He had discovered that the experiments performed by Pivati and his followers were flawed—that the cures were imagined and matters of the mind, not the result of an effective new way of delivering medicines (Nollet, 1749a, b). Franklin, who followed what was happening in Europe, agreed with Nollet's assessment, writing to an American physician friend that after "Reading in the Transactions the Accounts from Italy and Germany . . . I was persuaded they were not true" (Labaree, 1961a, p. 483).

After Nollet's (1749a, b) findings about the medicated tubes used in Italy and Franklin's (1758) letter on the common paralytic disorder were published, opinions about medical electricity varied tremendously. Some individuals still saw it as a panacea, others viewed it as quackery, and yet others, following Franklin, took a middle-of-the-road approach, feeling it might be useful for some conditions but not for others (Bertucci, 2001a).

John Wesley, the cleric who turned to electricity to heal bodies while he tried to save souls, now became more cautious in his statements. In *The Desideratum: Or Electricity made Plain and Useful*, which first appeared in 1760, Wesley, like Nollet, admitted that electricity was not quite the cure he at first thought it would be for paralyses. In this context, he wrote: "Again, in some Experiments, it helps at the very first, and promises a speedy Cure: But presently the good Effect ceases, and the Patient is as he was before" (Wesley, 1760, p. v). Later in his book, after presenting the case of a man who seemed to have been successfully treated by another electrotherapist for a paralytic stroke, Wesley (p. 63) sounded even more like Franklin, adding: "I have not yet known any Instance of this Kind. Many *Paralytics* have been helped: But, I think, scarce any *Palsy* of a Year standing has been thro'ly cured."

If the number of positive cases in the literature can be used as a barometer of what most people thought or at least hoped, the bias toward medical electricity was still far more positive than negative. *The Gentleman's Magazine* overflowed with communications touting the cure (Locke & Finger, 2007), while British hospitals (e.g., Middlesex Hospital in 1767) placed orders for the latest electrical machinery (Bertucci, 2001b). In France, more than 25 articles on curing paralyses with electricity appeared between 1750 and 1780 in just one journal, the *Journal de Médecine*.

And in the German-speaking countries, a bibliography from an encyclopedia published in 1777 included more than 100 pages of medical electricity titles (Hochadel, 2001).

THE NEW *ZEITGEIST*

Medical electricity and the various claims being made about it at mid-century are significant for our story for several reasons. Most importantly, coming at about the same time as Franklin's discovery of atmospheric electricity, they enhanced the perception that knowledge of electricity could explain just about everything in nature (Ritterbush, 1964). People were now beginning to think of electricity as more than just the cause of thunder and lightning storms. This seemingly all-pervasive force might also be responsible for the aurora borealis, earthquakes, meteors, the actions of certain minerals, the reactions of touch-sensitive plants, and much, much more. To the Reverend John Wesley (1760), electricity was the very "soul of the universe."

In addition to contributing to this new and exciting *Zeitgeist*, the various findings associated with electrical medicine indicated that much still remained to be learned about this force, both in terms of its physical properties and its potential utility. And not to be overlooked, the claims being made about medical electricity would help set the stage for parallel experiments using live fishes. Above and beyond the idea that the discharges from these fishes just might help the injured and afflicted, some scientifically minded individuals would view the results of these parallel therapeutic trials as evidence for or against animal electricity. We shall examine fish therapeutics in this new setting in Chapters 13 and 14, after we show how these fishes first began to be associated with electricity. But before even doing this, we must first discuss thoughts about nerve and muscle physiology up to the middle of the eighteenth century. For it is only by doing this that we can begin to appreciate why there was such an interest in fish therapeutics in the second half of the eighteenth century and, even more importantly, how the discovery of fish electricity would lead to a revolution in physiology.

Chapter 11
Animal Spirits and Physiology

> We are still more uncertain about what relates to the Animal Spirits. Are they Blood, or a particular Substance separated from the Chyle by the Glands of the Mesentery? Or may they not be derived from a serous sources? Some compare them to spirit of wine, & it may be speculated whether they are not the Matter of Light. The Dissections which we ordinarily perform cannot clear up any of these difficulties.
>
> (Steno, 1669/1772, pp. 207–208)

As we saw in the previous chapter, advances in the science and technology of electricity created a new scientific attitude by the middle of the eighteenth century, one in which natural philosophers began to see many phenomena in new ways. This changing perception applied to physiology, and particularly to how the nerves and muscles might work. In this chapter, we shall look at older theories of nerve and muscle physiology, at how the idea that these structures are electrical first emerged—and at some of the controversies this generated. In the next chapter, we shall show how increased knowledge about electric fishes made it more likely that at least some living animals might function electrically, ultimately leading to Galvani's landmark theory that all animals, including people, are electrical, even if they are not capable of generating shocks and sparks. From a broader perspective, the present chapter deals with the start of a fundamental shift away from classical science—one that will lead to a great revolution in the biological sciences, a paradigm shift akin to that of Harvey's discovery of the circulation of the blood.

ANCIENT THEORIES OF NERVE FUNCTION

Ideas about how the nerves might work began to emerge prior to the time of Hippocrates (Ochs, 2004). About 500 B.C., Alcmaeon of Croton philosophized that there must be special channels for bringing sensory information to the brain, some large, as exemplified by the optic nerve, and others too small to be seen (Stratton, 1917; Codellas, 1932). Croton, a town in the south of Italy, was at the time an important center of Greek culture.

It was not until the 4th century B.C., however, that the Alexandrian dissectors distinguished between the nerves, tendons, and blood vessels. There were still confusions, but the theory that the nerves are conduits for conveying information to and from the brain became much more accepted. The underlying idea was that the flow of *pneuma psychika*, or "animal spirits," through hollow nerve channels underlies sensation and movement.

Earlier Greek ideas about the brain and nerves were adopted, modified, and packaged in a very systematic way by Galen during the Roman Era (Manzoni, 2001; Rocca, 2003). Spirits, Galen taught, originate from digested food that is carried from the intestines to the liver by blood vessels. The liver then distributes these nourishing spirits throughout the body by its contractile movements. Some of these spirits are carried to the heart by the veins, where they are heated. When they mix with air from the lungs, they are transformed into "vital spirits." Those vital spirits that ascend by the arteries to the brain are transformed yet again, this time into animal spirits in a network of blood vessels at the base of the brain called the *rete mirabile* (Galen, who dissected only animals, did not recognize that humans do not possess this "wonderful net"). These animal spirits are housed in the ventricles and, when called upon, are sent coursing through the nerves to perform their various sensory and motor functions.

The doctrine of nerve function was a basic part of Galen's physiology, a physiology based heavily on humours and qualities, and one that would dominate medical culture until the modern era. Although broad acceptance of Galen's physiology over the centuries was one of the factors hindering the progress of medicine, his nerve doctrine had its merits, particularly because it pointed to the brain as the center for sensation, motion, and intellectual functions. This contrasted with the conception of Aristotle, who considered the heart as the origin of nerves and the center of sensations, movement, and emotions.

One kind of evidence provided by Galen in support of a pneumatic flow of spirits within the hollow nerve core was an experiment in which ligatures (or compressions) were applied to the *reversivi*, his term for the recurrent laryngeal nerves, of a living animal (usually a squealing pig, although he also studied these nerves in many other animals). This maneuver, demonstrated before the elders of Rome, resulted in an immediate and dramatic cessation of the squealing, which Galen interpreted to be a consequence of the blockage of the flow of the animal spirits from the brain to the muscles of the larynx. To the great astonishment of the onlookers, loosening the ligature or the experimenter's fingers usually permitted the sounds to resume, whereas cutting the nerves resulted in permanent deficits (Galen, 1586; for more on the history of the recurrent laryngeal nerves, see Kaplan et al., 2009).

Galen's physiology can be confusing because it has both pneumatic and mechanical components (e.g., sensory nerves are softer than motor nerves and thereby better able to receive impressions). Nevertheless, his version of the hollow nerve theory, with its emphasis on animal spirits, would dominate science and medicine for the next 1,500 years. It can be found intact in the writings of the Church Fathers (e.g., Nemesius, Augustine), in the works of Middle Eastern compilers and commentators (e.g., Hunain ibn Ishaq, Avicenna, Alhazen) and in the books of the Scholastics, including Albertus Magnus and Thomas Aquinas.

One of the reasons for the acceptance of Galen's nerve doctrine by various religious cultures (Christian, Jewish, and Islamic) was that a subtle pneumatic spirit appeared to fit well with the idea of an immaterial soul. In Galenic physiology, hollow parts of the brain play important roles, and the ventricles (supposed to be three in number and of a perfect spherical shape) were viewed as receptacles of the most sublime components of human nature. The use of the term "cortex" to designate the external part of the brain surrounding the hollow central cavities betrays the accessory protective function assigned to the solid part of the brain. Also, the cortex seemed less apt to be a material substratum of the soul because of its rich content of earth, the most heavy and least noble of the four elements of classical physics.

The only real modification of the schema made prior to the Renaissance may be the notion that different ventricles govern perception, thinking, and memory (Manzoni, 1998). Nevertheless, Galen might have suggested even this development, since early backers of this theory, who could have known about some of his writings that have been lost, typically cite him. Many of Galen's texts perished in 191, when a fire engulfed the library in Rome that housed his material, making the origins of the theory of ventricular localization difficult to discern.

THE RENAISSANCE AND EARLY MODERN PERIODS

Despite his inventive mind, accomplished dissections, and first realistic drawings of the ventricles (based on injecting molten wax into the decapitated heads of oxen), Leonardo da Vinci, one of the greatest minds of the Renaissance, adhered to this time-honored theory (O'Malley & Saunders, 1952). So did Andreas Vesalius, who in the sixteenth century promoted a new human anatomy based on actually dissecting human bodies—an anatomy that challenged Galenic texts that were based on animal dissections (Vesalius, 1543; Singer, 1952; O'Malley, 1964). Despite his intense criticism of Galenic anatomy, which he felt did not apply well or generalize to humans, Vesalius did not stray from the physiological tradition when he wrote that the ventricles are cavities where animal spirits originate, when inhaled air is added to vital spirits from the heart.

Not even William Harvey's (1628) revolutionary work on the circulation of the blood served as the death knell for this theory. Although Harvey rejected the ancient idea about heart function and blood movement, he continued to accept the idea of animal spirits traversing through hollow nerves.

Some writers who followed Harvey, however, began to modify the animal spirits schema to account for Harvey's discovery, and new anatomical and physiological variants of the theory began to appear. Nevertheless, the linking of animal spirits with the brain, nerves, and muscles remained intact. Much as the heart moved the blood, the brain pumped animal spirits through the nerves and, at least in some conceptions (e.g., that of Henricus Regius), this was done in a way that allowed them to be refreshed and used again (Clarke, 1968; Clower, 1998).

René Descartes, one of the leaders of the seventeenth-century scientific revolution, added the tiny pineal gland to the machinery (Descartes, 1649, 1662; Finger, 1995, 2000, pp. 69–83) (Fig. 11.1). He wrote that from its central location close to the ventricles, this gland could make new animal spirits from the tiniest of particles in the blood, receive existing spirits arriving from peripheral nerves, and send both newer and older spirits back out to the muscles to inflate them as if they were balloons (Fig. 11.2). According to the French philosopher, who was inspired by a general mechanical theory of the universe, the nerves contain hollow tubules too small to be seen with the naked eye, as well as

Figure 11.1. Etching of French philosopher René Descartes (1596–1650), after a painting by Frans Hals. Although Descartes wanted to deal with fundamental truths, his theory of animal spirits was entirely speculative.

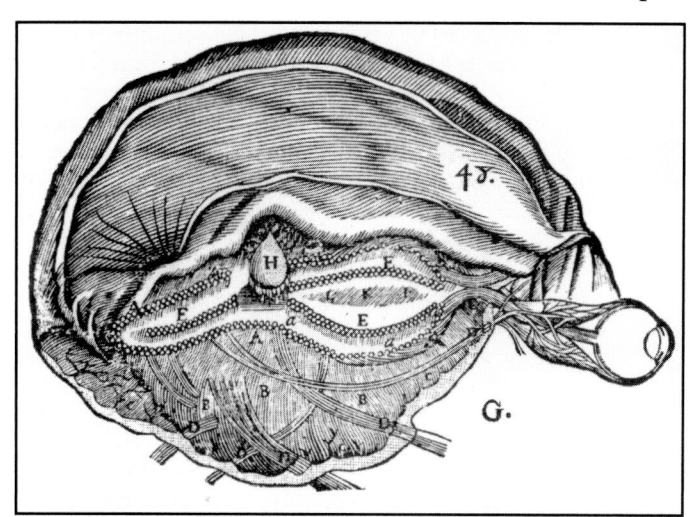

Figure 11.2. A figure from Descartes showing a cross-section of the brain. In his schema, the pineal gland, labeled H, releases animal spirits into the ventricles, from which they may pass through imagined valves (small circles) in the walls to enter the hollow nerves to the muscles.

fine cords for opening tiny valvules in the walls of the ventricles, where the animal spirits are stored.

Animals were automata to Descartes, who did some dissections of his own. But soul-possessing humans can think and thereby override some of the basic reflexive programs (e.g., we can move a leg voluntarily). Although Descartes' machine-like approach to the body and its parts had considerable impact (Rosenfeld, 1940), his solution to the mind–brain problem was widely criticized, even in this epoch, as wishful thinking, meaning not stemming from experiments. As a result, although he had many mechanically and mathematically oriented followers, the novel pineal gland part of his theory was never really incorporated into mainstream scientific or medical thinking.

In contrast, a different variation of the ancient theory arose shortly after the great French philosopher's death, and it caught on in Britain and elsewhere. It was based in part on the medical chemistry of sixteenth-century Swiss physician "Paracelsus" (Theophrastus Bombastus von Hohenheim) and his seventeenth-century follower, Jean Baptiste van Helmont (Figs. 11.3 and 11.4). These innovators and their followers rejected ancient theories and proposed alternatives based on chemical properties. The nature

Figure 11.3. Paracelsus (Theophrastus Bombastus von Hohenheim; 1493–1541), who attempted to overthrow Greco-Roman medical notions with his new chemistry, which was called iatrochemistry (literally "medical chemistry") (from an old engraving at the Galileo Museum in Florence).

Figure 11.4. Jean Baptiste Van Helmont (1579–1644), one of many seventeenth-century medical figures who questioned old ideas about animal spirits and tried to think in terms of a new chemistry (from Helmont & Helmont, 1652).

of the nervous agent, which had been considered akin to a subtle air, changed progressively to that of a more compact fluid endowed with properties that varied with the specific theory.

In England, Thomas Willis adopted some of Paracelsus' chemical ideas during the second half of the seventeenth century, although what he promoted is increasingly being recognized as a melding or a synthesis of chemical and mechanical approaches to physiology and medicine (Debus, 1966; Isler, 1968; Dewhurst, 1982; Finger, 2000; Wear, 2000) (Fig. 11.5). Willis (1664, 1672, 1676-80) hypothesized that vital spirits are extracted from the blood vessels at the base of the brain (soon to be called "the circle of Willis") and are chemically distilled into animal spirits, which are stored in the gray matter making up the brain's shell (Fig. 11.6). These spirits are then secreted through the white matter (nerve pathways) to perform their functions, which include two kinds of motor acts. The steady stream of spirits from the "Cerebel" (cerebellum) regulates automatic, unconscious actions, such as those essential for digestion, whereas the irregular stream from the "Brain" (cerebrum) governs volitional acts. With Willis, the emphasis shifted from subtle spirits in the hollow ventricles to the chemistry of more solid matter.

William Croone, the London physician after whom the Royal Society's Croonian Lectures on Muscular Motion were named, came forth with a fairly similar (deductive) theory at the same time as Willis did. In his Latin text *De ratione motus musculorum*, translated as "On the Reason of the Movements of the Muscles," Croone (1664, 2000) reasoned that animal spirits, again perceived as tiny particles, must make up the nerve fluid. This liquor, he contends, can "carry out all functions of sensation as well as of movement." "Driven out of the branches of the nerves by the violent movement of the fibers," it can trigger a violent chemical reaction when injected into muscles bathed with arterial blood (2000, p. 73).

Croone's theory of nerve and muscle action was, like many others at the time, an amalgam of chemical and mechanical concepts. For example, he postulated that the mind is able to thrust these subtle spirits into appropriate

Figure 11.5. Thomas Willis (1621–1675), the Oxford and later London physician. Willis was one of the most influential medical figures of the seventeenth century. His physiology mixed old ideas about animal spirits with new ideas based in chemistry and mechanics.

nerves and push arterial blood "copiously" into specific muscles, which in the revised edition of his text he viewed as made up of many tiny bladders. He also writes about hydraulics and the power of the mind, although he leaves much to the imagination. He never provides details about how the mind

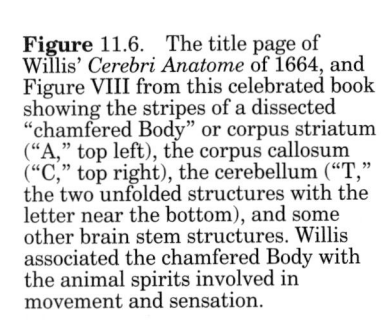

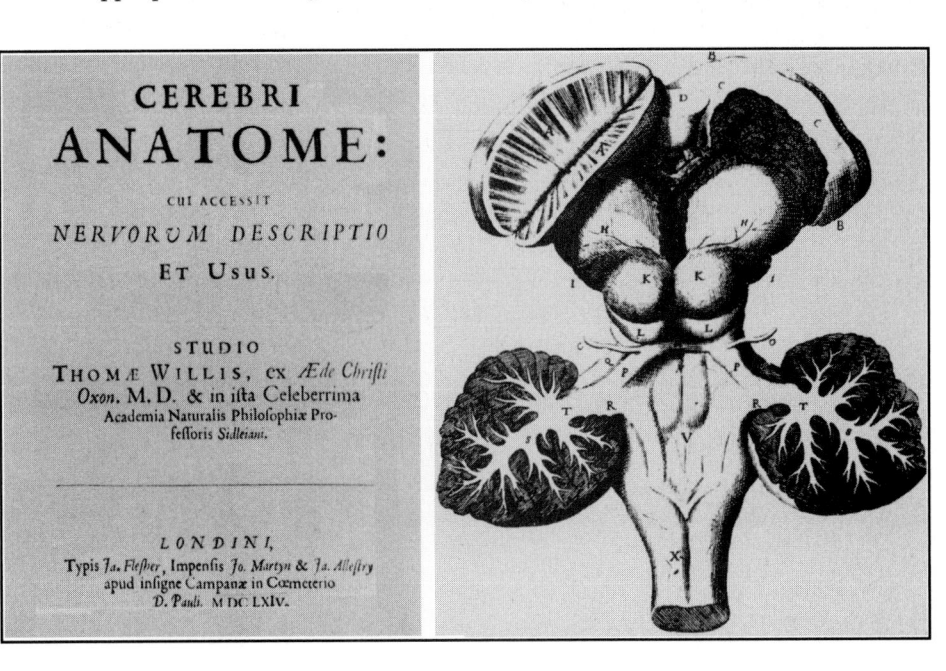

Figure 11.6. The title page of Willis' *Cerebri Anatome* of 1664, and Figure VIII from this celebrated book showing the stripes of a dissected "chamfered Body" or corpus striatum ("A," top left), the corpus callosum ("C," top right), the cerebellum ("T," the two unfolded structures with the letter near the bottom), and some other brain stem structures. Willis associated the chamfered Body with the animal spirits involved in movement and sensation.

drives the machinery of the body or, for that matter, about his envisioned components for fermentation, effervescence, or "ebullition." His point is only that a mixture of nerve and blood components causes muscle inflation or expansion, resulting in tendon pulling, which in turn can account for bodily movements.

Marcello Malpighi, the seventeenth-century Italian microscopist, also accepted the idea of a fluid that moves through the central white matter and the peripheral nerves (Fig. 11.7). Malpighi (1666) prepared his brains by boiling them, pealing off the meninges, and pouring ink on them. Within his general theory of the glandular structure of bodily organs, he reasoned that the brain is made of a multitude of seed-like "cortical glands." This idea of the brain as a gland would persist for a long time in science and medicine.

Similarly, John Mayow, with his "Nitro-aerial" particles (which he tells us "fit the character of Animal Spirits"), also supported a physical-chemical version of the theory. In 1674, he wrote that these particles combine with "salino-sulphureous" matter in the blood to effervesce and cause muscle contractions (Mayow, 1674).

Figure 11.7. Marcello Malpighi (1628–1694), the Italian microscopist who discovered the capillaries and provided the first clear demonstration of the alveolar structure of lungs (from Malpighi, 1698). Malpighi conceived of the brain as an aggregate of microscopic glands secreting a juice. In his view this nerve juice would perform the functions traditionally attributed to animal spirits.

But although these men moved the spirits from the air to a fluid environment, they saw themselves as merely clarifying the theory of animal spirits, which was now modified so it could fit better with the physics of the day. As put in 1733 by George Cheyne, a Scot with a very successful medical practice in London: "The doctrine of Spirits, to explain the animal Function and their Diseases, has been so readily and universally receiv'd from the Days of the Arabian Physicians (and higher) down to our present Times, that scarce one (except here and there a Heritick of late) has call'd this Catholick Doctrine into Question" (Cheyne, 1733, p. 74).

VIBRATIONS?

William Croone viewed the actions along a nerve as analogous to the vibrations of a tightly stringed musical instrument, and there were other analogies to vibrating bodies at the time (Wallace, 2003). Nevertheless, it was Sir Isaac Newton who did the most to popularize vibration theory. He gave the theory a more modern, mechanical flavor and applied it to the nerves. What he presented was unlike Croone's theory. For Newton, the vibrations were thought to occur among the particles found in the nerve core, in keeping with his belief in "a certain most subtle spirit," or ether, that can be found throughout nature.

These ideas can be found in the *General Scholium* of Book 3 of 1713 and later editions of his *Principia Mathematica*. Newton raised the possibility that nerve actions might be a consequence of vibratory or elastic movements of an ethereal medium penetrating the solid parts of nerves. This medium was thought to be similar to that involved in electric and magnetic phenomena. In concluding the *Scholium generale* of the *Principia*, however, he regretfully admits, "these are things that cannot be explained in few words, nor are there enough of experiments that are required for an accurate determination and demonstration of the laws by which this spirit operates" (Newton, 1713, p. 484). Interestingly, in an interleaved copy of the 1713 edition of the *Principia*, belonging to Newton himself and annotated by him, the "spirit" of this statement is qualified as "electric & elastic" (*spiritus electrici & elastici*; see Newton, 1972, vol. II, p. 765).

Newton also mentioned vibration theory at the end of his *Opticks*, a work that first appeared in 1704. In later editions, he added more on vibrations, particularly in the Queries of the last part of the work (Fig. 11.8). For example, in Query 24 of the 1730 edition we find:

> Is not Animal Motion perform'd by the Vibrations of this Medium, excited in the Brain by the power of the Will, and propagated from thence through the solid, pellucid and uniform Capillamenta of the Nerves into the Muscles, for contracting and dilating them? I suppose that the Capillamenta of the Nerves are each of them solid and uniform, that the vibrating Motion of the Ætherial Medium may be propagated along them from one end to the other uniformly, and without interruption: For Obstructions in the Nerves create Palsies. (Newton, 1730, p. 328)

Newton focused on how vibrations could account for visual color perception. To quote from Query 14 of the same edition: "May not the harmony and discord of Colours arise

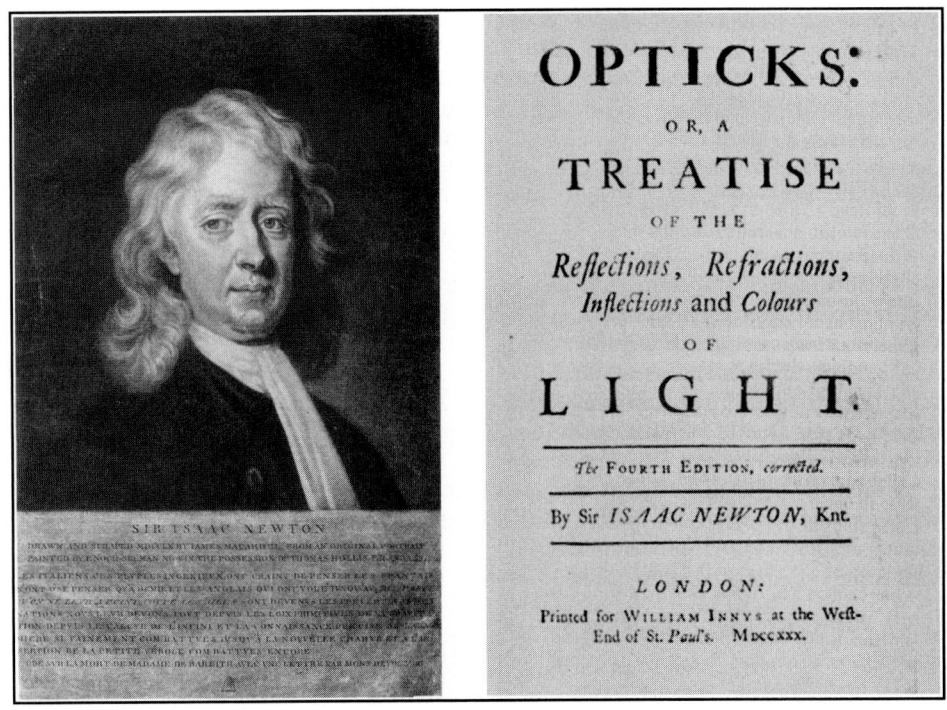

Figure 11.8. A portrait of Sir Isaac Newton (1643–1727) with the fourth edition of *Opticks* (1730), one of the books in which he presented his theory of nerve vibrations.

from the proportions of the Vibrations propagated through the Fibres of the optic Nerves into the brain as the harmony and discord of Sounds arise from the proportions of the Vibrations of the air?" (p. 320). Since the ether of Newtonian physics was also the medium of electrical and magnetic phenomena, this vibration idea would in its own way help to stimulate electrical theories of nerve conduction, as we shall soon see.

David Hartley, a practicing English physician without formal medical training in England, viewed nature from Newton's shoulders and made nerve vibrations central to his thinking, explaining that "My chief Design . . . is . . . to explain, establish and apply the *Doctrine of Vibrations and Associations*" (Hartley, 1749, p. 5). He combined Newton's theory of vibrations with John Locke's psychological theory of associations to explain such things as memory and perception (Aubert & Whitaker, 1996; Buckingham & Finger, 1997; Glassman & Buckingham, 2007; Smith, 1987; Wade 2005). He did this as an attachment to a 1746 treatise on treating kidney stones (an ailment from which he suffered, as did Franklin and many other notable men of epochs past), and in his far more famous essay, *Observations on Man: His Frame, his Duty, and his Expectations*, 3 years later (Hartley, 1746, 1749).

To quote from the latter source:

> We are to conceive, that the Vibrations thus excited in the Aether will agitate the small Particles of the medullary Substance of the sensory Nerves with synchronous Vibrations, in the same manner as the Vibrations of the Air in Sounds agitate many regular Bodies with corresponding Vibrations or Tremblings. (Hartley, 1749, pp. 21–22)
>
> Sensory Vibrations, by being often repeated, beget in the medullary Substance of the Brain, a Disposition to diminutive Vibrations, which may also be called Vibratiuncles and Miniatures, corresponding to themselves respectively. This Correspondence of the diminutive Vibrations to the original sensory ones, consists in this, that they agree in Kind, Place, and Line of Direction; and differ only in being more feeble, i.e. in Degree. . . . For since Sensations, by being often repeated, beget Ideas, it cannot but be that those Vibrations, which accompany Sensations, should beget something which may accompany Ideas in like manner. (p. 58)

Like Newton, Hartley never implied that the nerves had to move back and forth, like a harp or violin string. He viewed the minute vibrations as being like "the tremblings of the particles of sounding bodies," and they were confined to particles within the nerves themselves. Interestingly, he even mentions torpedoes in his *Observations*:

> From hence we may proceed to consider the Numbness occasioned by the Stroke of *the Torpedo*. For the Oscillations of this Fish's Back may neither be isochronous in themselves, nor suitable to those which existed previously in the Hand; and yet they may be so strong, as not only to check and overpower those in the Part which touches the Fish, but also to propagate themselves along the Skin, and up the Nerves, to the Brachial Ganglion, and even to the Spinal Marrow and Brain; whence the Person would first feel the Stupefaction ascend along the Arm to the Shoulder, and then fall into a Giddiness, and general Confusion, as is affirmed to happen sometimes. Some effects of Concussions of the Brain, and perhaps of the Spinal Marrow, also of being tossed in a Ship, . . . and other violent and unusual Agitations of the Body, seem to bear a Relation to the present Subject. (Hartley, 1749, p. 133)

Hartley admitted that unlike his doctrine of associations, of which he felt certain, his physiological theory based on Newton's loose thoughts about vibrations was speculative and less defensible. Among other things, very little was known about the substance of the nerves or the nerve core. Nevertheless, thinking about vibrating matter within the nerves can be viewed as a reaction to the problems inherent in the older theories of animal spirits. In this regard, and in terms of the physics of motion, it can be argued that Newtonian vibrations were taking philosophers closer to electrical explanations of nerve actions than they had been at any time in the past.

PROBLEMS WITH EARLIER THEORIES

By Hartley's time, each of these notions of nerve action—ethereal spirits, fluids, and vibrations—seemed to be even more at odds with careful observations. Even worse, they were being challenged by findings with microscopes and from clever experiments. These challenges to existing theories would also help set the stage for the new idea that the mysterious nerve agent is, in fact, electricity.

On the morphological front, microscopic studies were revealing that the nerves are not hollow. Anton van Leeuwenhoek (1674), who believed in nerve spirits, was distressed when he failed to observe a hollow canal in the optic nerve of a cow, and he also examined peripheral nerves (Fig. 11.9). After all, Galen had contended that such an opening could be seen with the naked eye, simply by holding a severed optic nerve up to the sun. Yet even with his newest lenses and microscopes, he could not detect the canal. Blinded by theory, he reasoned that the hollowness must be present in life, only to disappear as specimens dry out. Clearly, abandoning a time-honored theory was no easy matter, especially without a worthy alternative to fill the void.

Subsequent microscopists also searched for hollow openings, leading to more challenges against the hollow nerve concept. If anything, the nerves seemed to be filled with a viscous fluid, which severely weakened the idea that the nerves might inflate the muscles. But if the nerves are filled with a viscous fluid, how could spirits possibly be transmitted from brain to muscle with such rapidity?

The experimentalists by this time were enumerating bigger problems for the theory of animal spirits, ones that could not be easily explained away as due to inadequate microscopes or dried specimens. Jan Swammerdam, a Dutch naturalist, performed some of the most significant experiments in this context during the 1660s. Notably, he put a frog leg muscle with its attached nerve in an enclosed chamber. A thinner exit tube at the top contained a bubble, which could rise and fall with changes in the volume of the muscle below it. He then stimulated the distal end of the nerve by tugging at it with an attached wire. To his amazement, he found that there was no upward movement of the bubble, as expected with the older theory of injected air or the thought that fluid might somehow gorge the muscles.

Importantly, Swammerdam showed that contractions could also take place without the nerves having a direct link to the brain. Yet in theory, the nerves supposedly delivered the spirits. Interestingly, he obtained muscle contractions using metals (e.g., wires, brass hook), which might have been a source of electricity, in some of his of experiments (for more on metals as a source of electricity, see Chapters 21–23).

Swammerdam's findings made him skeptical of prevailing theories, but he was unable to come forth with a good alternative for how the nerves might work. Unfortunately, his work was not that well known during his lifetime, because his findings were printed in Dutch, and because he left science to devote himself to God. It was not until 1758 that Herman Boerhaave translated his writings into Latin (Swammerdam, 1758, pp. 122–125). In contrast, Swammerdam's use of frog preparations for physiological studies of the nerves and muscles did have a significant impact and became increasingly popular for future generations.

Others were soon performing experiments comparable to Swammerdam's, and some were on humans. One of the best known of these experimenters was Francis Glisson, Regius Professor of Physics at Cambridge and a founder of the Royal Society. In an experiment conducted shortly before he died in 1677, Glisson took a large, oblong glass tube that was closed at one end, had a "strong brawny man" insert his arm in it, and closed the arm opening with bandages. He then filled the tube with water through an attached upright funnel. The expectation was that water should be displaced into the funnel-like tube when the man's enclosed muscles flexed. Yet Glisson (1677) found that no more water was displaced when the muscle flexed than when the muscles were

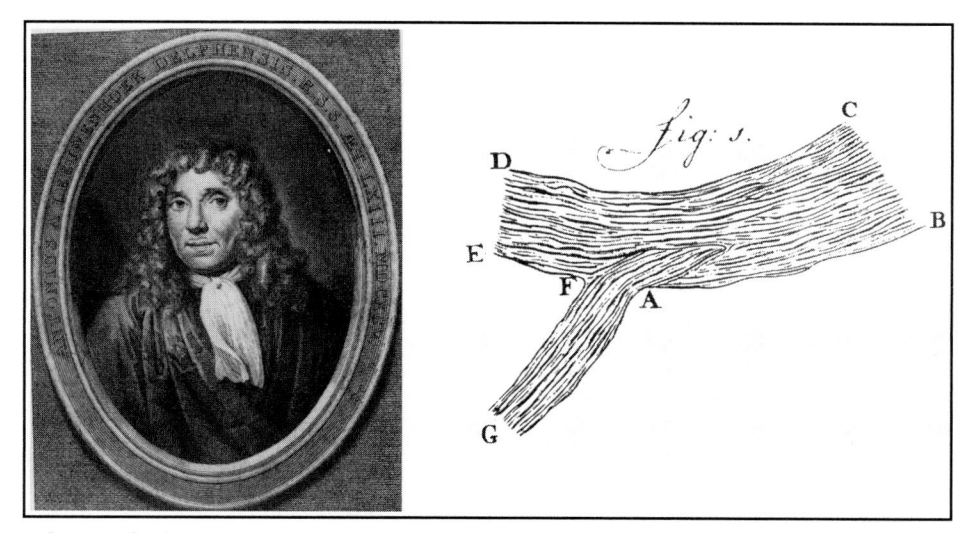

Figure 11.9. Anton van Leeuwenhoek (1632–1723) and an image from a peripheral nerve dissection (from Van Leeuwenhoek, 1719).

Figure 11.10. Nicolaus Steno (Stensen; 1638–1687), the Danish physician and later Catholic bishop, who felt that there was too much speculation about the nerves and brain, especially with regard to animal spirits. On the right side, the first page of the 1772 edition of his 1668 discourse on the brain, in which he expressed his ignorance about animal spirits.

relaxed—if anything, the water level decreased when it should have increased.

Based on his experiments, Glisson dismissed the idea of spirits swelling the muscles during contraction. Instead, he suggested that the nerves must stimulate the muscles by irritating them in some way, a theory that Albrecht von Haller would develop further in the mid-1700s (see below). Two years later, Richard Lower (1669), who was active in the nascent Royal Society, supported Glisson. He wrote that muscles just become smaller and harder, and do not visibly swell during contraction.

Nicolaus Steno, the Danish physician-priest who studied torpedoes in Italy and had befriended Swammerdam, and then led him deeper into Catholicism and away from science, also obtained negative findings (Fig. 11.10). He did some experiments to study muscle contraction and used geometry, concluding that the volume of the muscle remains the same whether relaxed or contracted, there being, therefore, no influx of air or juice.

Steno, along with Malpighi and a growing number of scientists in this epoch, opposed loose theorizing. He was especially critical of Thomas Willis for his unfounded speculations, and when it came to Descartes, he wrote: "Descartes' method is praiseworthy, but blameworthy is a philosophy where the author forgets his own method and takes that for granted which he has not yet proven by reason" (see Snorrason, 1968, p. 83). In 1665, before a group of intellectuals who were meeting in the Paris home of Melchisédech Thévenot, Steno lamented about how little was really known about the brain and animal spirits:

We are still more uncertain about what relates to the Animal Spirits. Are they Blood, or a particular Substance separated from the Chyle by the Glands of the Mesentery? Or may they not be derived from a serous source? Some compare them to spirit of wine, & it may be speculated whether they are not the Matter of Light. The Dissections which we ordinarily perform cannot clear up any of these difficulties. . . . It is true, on the other hand, that all that both the Ancients and Moderns have told us about the Brain is so full of disputes, that, as many books as there are on the anatomy of that part, there are just as many difficulties, disputes, doubts and controversies. (Steno, 1669/1772, pp. 207–208 and 229–230)

Giovanni Alfonso Borelli (1680–81) (Fig. 11.11) added fuel to the fire by performing a revealing experiment of a different sort. He immersed a live animal's limb in a vat of water and cut across some of its muscles. No bubbles came out, even though the muscles were violently agitated by the struggling animal. Based on what he failed to see, he rejected the idea of nerve spirits inflating the muscles, while assuming that waves of some sort must be propagated through the fluid in the nerve core, without the progression of matter.

Experiments of this sort left many scientists wondering just how the nerves might really work. In theory, a build-up of air filled with spirits or a blockage of the internal juices should cause the nerves to swell just behind the knot. The experimental findings were inconsistent with this expectation. Worse, when the nerves were cut, drops of fluid (and bubbles) did not materialize.

There were other problems with the theory of juices too. One was that anatomists were unable to find valves in the nerves or brain that might regulate the flow of these juices, much like blood, which flows in one direction. Another was that the nerves are so small that, even if they secrete some sort of a juice, it would not be able to fill a muscle. And a third was the growing realization that the nerve fluid seemed so viscous that it was hard to envision how it could be involved with rapid actions. The theory of animal spirits as *pneumata* or even a liquid was no longer as tenable as it first seemed to be.

Figure 11.11. A portrait of Giovanni Alfonso Borelli (1608–1679) with the second volume of his *De motu animalium* (1780–81), in which he presented and discussed his experiments on nerve conduction and muscle contraction.

But what about ideas based on nerve vibrations? This notion just seemed wrong to most scientists from the start. The problem envisioned was that the nerves are soft and pulpy, unlike objects that would vibrate when struck and might spring back when cut. For these reasons, Leyden physician Herman Boerhaave, in a lecture published in 1743, referred to vibration theory as just another "repugnant" idea. Quoting Boerhaave: "There is therefore no Face of Truth in that Opinion, which asserts the Nerves to perform all their Actions by Vibrations, like those which arise from striking a tense Chord" (Boerhaave, 1743, Vol. 2, p. 310).

Similarly, his student Albrecht von Haller (1769) was left scratching his head (Zimmermann, 1755; King, 1966) (Fig. 11.12). The nerve vibration idea did not make sense to him either. In Haller's words:

> If an irritated nerve flickers in the manner of an elastic cord, which trembles upon being plucked, this would happen if the nerve were made of hard fibers and tied at its extremities to solid bodies. The fibers should, moreover, be tense, because neither soft cords, nor those that are not tense, or those that are not well fastened, are ever observed to tremble. But all the nerves are at their origin medullar and very soft, and exceedingly far from any kind of tension.... Finally, that the nerves are destitute of all elasticity is demonstrated by the experiment in which a nerve cut in two neither shortens nor draws back its divided ends to the solid parts. (Haller, 1771, pp. 212–213)

Haller's scientific authority constituted an important reason why vibration theories did not attract many supporters (see Duchesneau, 1982, for more on Haller and Enlightenment physiology). As we shall note at the end of this chapter, Haller would conduct his own important experiments on nerve and muscle physiology and promote a different kind of theory, while still accepting the notion of animal spirits. His ideas would also help set the stage for the events that would soon take place in physiology, even though the new

Figure 11.12. Albrecht von Haller (1708–1777), the eighteenth-century authority on physiology, who was skeptical about nerve vibration theories and animal electricity.

paradigm would generate experiments that would challenge some of his basic views.

THE IMPACT OF ELECTRICITY

With the rapid development of electrical science in the first half of the eighteenth century, electricity would now become an increasingly attractive agent for nerve conduction. There were several reasons for this emerging development. After all, was there any other agent that could travel over distances with such rapid velocity? And was any other force known to have such an effect on the vital processes of the body? Moreover, this agent seemed to be present throughout the universe, and it seemed reasonable to assume that it must play a role in living nature.

Electricity also fit well with Newtonian physics, which emphasized particles in motion. The electrical fluid or fire, as assumed by Franklin and many others, comprised subtle particulate matter. And much like the movement of the electrical fluid in the atmosphere, if there is electrical fluid in the animal body, it too must be in continuous motion. This motion might be caused by any of a number of factors, such as the circulating blood rubbing on the walls of the blood vessels, which some individuals thought could also account for bodily heat (Hackmann, 1972).

Since these electrical particles can travel or vibrate through the ether and some solids, whether the nerve cores are hollow or filled with wet pulpy material became less important. That electricity seemed to be especially tied to, or associated with, the nerves was the more important thought.

Not to be overlooked, and as had been discussed in the previous chapter, electricity seemed to have therapeutic properties. It was not that it could cure every disease or disorder, although some practitioners might have promoted it as a panacea. But there were enough reports of its positive effects on the body to catch everyone's attention and to raise the idea that some disorders might be due to inadequate amounts of electricity or blockages affecting the flow of electricity, which the skilled therapist now supplied from an external source.

Stephen Hales, the Cambridge-educated theologian and natural philosopher, first hinted that electricity might be the mysterious fluid of the nerves in 1733 (Fig. 11.13; for more on Hales, see Clark-Kennedy, 1929; Allan & Schofield, 1970). Hales had been studying how pressure from the pulsating arteries to the brain might affect the nerves, and specifically whether the pulsation has any effect on the muscles. Using animals, he concluded that arterial pressure could not provide enough force in the nerves to produce muscular actions. Searching for an alternative explanation for muscle movements, Hales wrote about a "wonderful and hitherto inexplicable Mystery of Nature . . . a vibrating *electrical Virtue* [that] can be conveyed, and freely act with considerable Energy along the surface of animal Fibers, and therefore on the Nerves" (Hales, 1733, pp. 58–59; italics ours).

Hales' thoughts clearly stemmed from Newton, and specifically from his reading of the Queries in the second edition of Newton's *Opticks*. He brought up electricity, much as did Newton, not to proclaim the identity of the nerve force, but more as an analogy. In this context, he agreed with Newton, who had written in the *General scholium* of

Figure 11.13. Stephen Hales (1677–1761), an English natural philosopher who made important discoveries in various field of physiology and chemistry, was one of the first to propose the involvement of electricity in nervous conduction.

his *Principia*: "But these are things that cannot be explained in a few words, nor are we furnished with that sufficiency of experiments which is required to an accurate determination and demonstration of the laws by which this electric and elastic spirit operates."

Others followed with similar statements. For example, Browne Langrish, in his Croonian Lecture for the year 1747, opined that "The surprising discoveries which have been made of late years by a variety of experiments upon electricity, do in some measure give us an idea of the great subtlety and velocity of the nervous fluid" (Langrish, 1747, p. 31).

The statements became more direct in the next decade. Tommaso Laghi of the University of Bologna wrote that muscle motion is probably produced by the flow of an electrical fluid that originates in the brain (Laghi, 1757a; see Piccolino, 2003a, pp. 16–17).

Father Giambattista Beccaria, another Italian and a strong supporter of Franklin's thinking, was also open to this idea. In 1753, he wrote a treatise on electricity. He covered some of his own experiments (e.g., applying electricity to the leg muscles of a live bird) and advances in medical electricity, writing:

The speed with which the electrical vapor moves, changes direction, stops and races forth again seems consistent with the speed and changes in animal sensations and motions. The singular ease in which it passes across—in general, through electrical bodies by communication [i.e., electrically conductive bodies], and in particular, through the nervous and muscular parts of animals—is consistent with the ease with which the mutations induced in organs by various objects are conveyed to the seat of sensation; it is also consistent with the agility with which other motions correspondingly ensue in the body. And the contractions and dilatations caused in the muscles by an electrical spark or electrical shock are arguments, perhaps even decisive ones, for substantiating the above-mentioned conjecture. (Beccaria, 1753, pp. 126–127: Trans. in Pera, 1992, p. 58, revised)

Still, it was a long jump from the observation that muscles could respond rapidly to an electrical charge to the claim that electricity underlies nerve action. Most scientists understood that the evidence was indirect and shaky, and that there were still theoretical problems to overcome.

Albrecht von Haller, while recognizing some of the virtues of the theory of electrical conduction, laid out the physical and physiological difficulties with this line of thought. He did this in various works in clear and effective ways. One of the biggest issues that bothered him was the idea that electricity could not be confined to the nerves (Home, 1970). Knowing that moist living tissues are electrically conductive, Haller could not conceive of how there could be electrical imbalances inside an animal's body—imbalances needed for the conduction of electricity along the nerves.

Also, Haller believed that nothing would prevent electricity from leaking out of the nerves onto surrounding tissues with maladaptive consequences. Since electricity could not be restricted to individual muscles once released, it could not account for fine, coordinated movements. As Haller wrote in 1762 in the fourth volume of his influential *Elementa physiologiae*:

Suppose now that the ischiatic [sciatic] nerve is well full of electric matter; suppose, moreover, that you want to move only the big toe, and you wish to do it in the correct way. For that purpose it would be necessary that electricity were in the nerve of the big toe more than in the main trunk of the [ischiatic] nerve, and for this, the electric fluid should move from the main trunk into that particular branch. In no way can it happen, however, that the flow would be conveyed uniquely toward the big toe; likewise, it could not happen that all the other muscles of the foot would have less abundance of electric fluid, even though by action of the will we would keep quiet their nervous branches. If then they receive the same abundance of fluid as the big toe, they will be put in a similar motion. Even though they had before less abundance of it, nevertheless they could not remain in a resting state. (Haller, 1762, p. 380)

A third and important objection to the electrical hypothesis stemmed directly from some experimental results. One of the most significant experiments involved tying a thread around a nerve, which did not abolish the flow of electricity along the nerve. Haller would comment that "a ligature on the nerve takes away sense and motion, but cannot stop the motion of a torrent of electrical matter" (Haller, 1786/1966, p. 221). This "ligature objection" would be raised for years to come.

These were legitimate issues, given the moistness intrinsic to the body and findings showing that electricity could spread across the body. In fact, was any part of the body even designed to produce the electricity? Hence, many natural philosophers viewed the idea of animal electricity as a dead end. It was an area to be skillfully avoided by those scientists taking the time to think the issues through.

HALLER'S IRRITABILITY THEORY

Haller's own theory, which became very influential in the history of physiology, holds that muscle fibers are endowed with a specific property or force, one that he called "irritability" (for more on Haller's theory, see Miller, 1939; Duchesneau, 1982; Frixione, 2007). This term and idea stemmed from the writings of Francis Glisson (see above), but it has even earlier roots (Temkin, 1964). Basically, irritability is the capacity to react to a stimulus with movement. As Haller stated in a discourse held at the Royal Academy of Göttingen in 1752, "I call that part of the human body irritable, which becomes shorter upon being touched by some external agent; very irritable if it contracts upon a slight touch, and the contrary if only by a violent touch it is put in movement" (see Haller, 1753 p. 116). Thus, muscles contract when properly stimulated, because they are irritable.

In an experimental situation, these contractions could be elicited with electricity or with some other agent. But, Haller contended in 1752, it must be an agent other than electricity in normal physiological situations. He further stated that, in contrast to the muscles, the nerves are "sensible," meaning they respond to stimuli with various sensations. This feature is revealed in animal experiments using electricity and other painful stimuli by various signs of pain and discomfort.

Haller performed experiments on hundreds of animals between 1746 and 1752, before concluding that irritability is a specific property of muscles independent of any external or internal agency, and that sensibility is inherent to the nerves. As he saw it, this revelation holds the same importance for living organisms as Newton's gravity does for the physical world.

Like Newton, who hypothesized a universal attractive force but did not attempt to explain the mechanism of its actions, Haller did not attempt to explain irritability. Nor did he tie it to the soul. In this context, he wrote that he was "convinced that the source of both [irritability and sensibility] lies concealed and beyond the power the knife and microscope. Indeed, I do not hazard many conjectures beyond the knife and the microscope, as I abstain willingly from teaching what I am ignorant of myself" (Haller, 1753, p. 115). Nevertheless, he went on to attribute it to the "gluten" in the muscle core:

But with the muscular fiber being composed of gelatin or gluten and earthly particles, it may be asked whether the Irritability resides in the gluten or in the elemental part. It appears most probably to reside in the former, because, when it is pulled, the gluten shrinks and get shorter, whereas, on the contrary, dry earth in no way resumes its shape once mutated, and being extremely brittle, when its parts are separated, they remain so. (p. 152)

The idea that specific physiological processes could be studied in animal body parts separated from the entire body was a particularly important methodological consequence of Haller's thinking. As we shall see nearer the end of this book, this message and Haller's ideas would be heard in

Bologna, where Luigi Galvani would undertake his research program with "prepared" frogs. Also, Haller's idea that nervous action is only the stimulus, or "exciting cause," behind muscle motion, and not the "effective cause" of the movement, anticipates the idea that different parts of a complex organism (even an artificial one) might interact on the basis of control commands rather than of energies.

In the chapters that follow, we shall show how electric fishes became increasingly electrical in the eyes of many naturalists and experimental natural philosophers during the second half of the eighteenth century. And we shall see that, with this growing realization, attention would steadily shift to several closely related but extremely important issues related to the nervous system and its physiology.

Two, as might be expected, are where the electricity comes from and how it might be confined to specific body parts, so as not to run helter-skelter throughout the animal's body causing chaos. A third issue, presented when we introduced Humboldt in the first chapter, is whether fish electricity is really identical to frictional or atmospheric electricity, or just closely related to "true" electricity. And then there is the most important issue of all, which will stem from the discoveries that will be made with electric fishes and will relate directly to the groundwork we have just presented in this chapter about animal spirits. It is whether animals that do not torpify might also function by electricity. That is, is more subtle electricity the nerve force within a frog's or a sheep's body, and the basis for our own physiology? With these important questions in mind, let us start by examining the first hints that some living organisms, and more specifically some fishes, just might be electrical.

Chapter 12
First Steps Toward Fish Electricity

> Ce poisson ne seroit-il point electrique? ("Could this fish not be electrical?")
> (David Etienne Choffin, 1749–50, p. 81)

This chapter examines how some fishes first became associated with electricity. It focuses on the 1750s, when the scientific study of electricity was in vogue and new discoveries about electricity, as well as weakly substantiated claims about it, were capturing the public imagination. The thought that some fishes might be electrical offered an alternative to the various mechanical hypotheses described in Chapter 9, as well as to more speculative, occult, and magical conceptions.

But although electricity could account for more findings, such as the transmission of the effects through various intermediaries, it seemed to defy common sense and clearly had people also wondering whether such thinking went contrary to the laws of physics. Thus, following the first steps towards electricity that we shall report in this chapter, fish electricity was by no means broadly accepted. Rather, the associations made at mid-century put the idea into play and only stimulated scientists to come forth with more and better evidence for such claims.

Before turning right to the mid-eighteenth-century literature, we shall begin with some comments on what is sometimes regarded as the first allusion to electricity in the fish literature. It involves a dissertation written at the end of the seventeenth century and then a widely read book published in 1712 by the same author, who compared a torpedo's discharge to lightning.

KAEMPFER'S ANALOGY TO LIGHTNING

Englebert Kaempfer (Kämpfer) was a multifaceted individual: he was a physician, a naturalist, a linguist, a collector, and a daring explorer (Bowers, 1966; Carrubba & Bowers, 1982; Meier-Lemgo, 1968). The son of a Lutheran pastor in Lemgo, a town east of Bielefeld in Germany, Kaempfer was born in 1651 and educated in several cities, including Danzig and Cracow, receiving his doctorate in philosophy from Cracow in 1673. He also spent 4 years at Königsberg, studying natural history and academic (essentially Galenic) medicine.

Finding inadequate career opportunities in Germany, Kaempfer moved to Uppsala in 1680, where he impressed the Swedish faculty. While in Sweden, he was offered the position of court scholar to King Charles XI of that country. Instead, he chose to become secretary to a Swedish mission to the Persian court. In March 1683, he left Stockholm for Isfahan to promote commerce and hasten Persia's departure from the Ottoman Empire.

After completing his diplomatic and commercial assignments, Kaempfer served as a physician for the United East Indies Company of the Netherlands. Before returning to Western Europe (Holland) in October 1693, he visited Russia, Persia, India, Siam, the East Indies, and Japan. As expressed by two scholars who have examined his life and work in great detail, "He was probably the most learned and surely the most widely traveled scholar of his era" (Carrubba & Bowers, 1982, p. 264).

Kaempfer's observations on several torpedoes captured in the Persian Gulf occurred between March 1684 and June 1688, while he was with the Swedish embassy, and then during a lengthy sickness that nearly killed him and prevented him from leaving. Because the locals had little use for the torpedoes they inadvertently caught, he "was able to obtain them from fishermen for a small price" (Kaempfer, 1712, p. 509).

He called the variety of torpedo he had access to at the port of Gamron (Bandar Abbas) *Torpedo sinus persici*, meaning "Torpedo of the Persian Gulf" (today's *Torpedo panthera*), and states the Persians called it لهرس ماهي (*lehrs mahii*). Kaempfer knew that this mottled electric ray differs from the northern Mediterranean torpedoes described by various ancient authors, and he cites Aristotle, Pliny, Galen and others in this context. But like the torpedoes of the ancients, it too could give numbing and painful shocks.

Kaempfer did not publish the notes he made while he remained abroad. But in 1694, after returning, he included some of his findings in the doctoral thesis he submitted to the faculty of medicine at the University of Leiden. In translation, his thesis was titled *Ten Exotic Observations*, and observation IV was his *Torpedo sinus persici* (translated in Bowers & Carrubba, 1970, pp. 287–289).

He presented his material in more detail in the Fascicle III ("Curious Physico-Medical Observations") of a book published in 1712 with the full title *Amoenitatum exoticarum politico-physico-medicarum fasciculi V...* ("Five Fascicles of Exotic, Political, Physical and Medical Entertainments..."). This work is of great historical importance for the knowledge

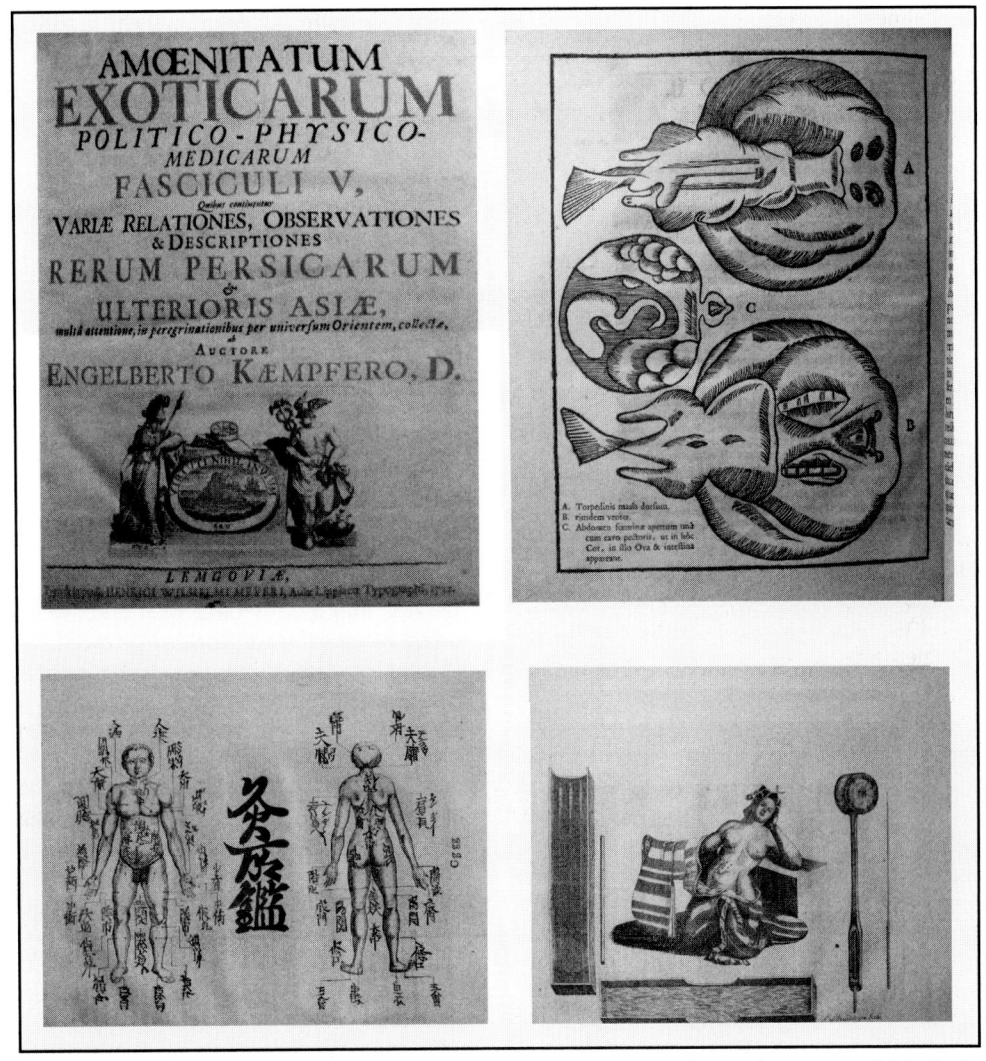

Figure 12.1. Top: The cover page to Englebert Kaempfer's 1712 book and his illustration of *Torpedo sinus persici* (today's *Torpedo panthera*). Kaempfer's report of his voyage to the East is of great historical importance because it provided important information on the culture and customs of various oriental countries, including China and Japan. Bottom: Among the many aspects he dealt with there was the Chinese practice of acupuncture and its application to various medical conditions.

of Persia and other Asiatic countries (and notably China and Japan), and it circulated widely and was frequently cited (Fig. 12.1).

The section on the torpedo (pp. 505–515) opens with a discussion of its external and internal anatomy, and includes an illustration. Kaempfer tells us that the rays possessed tiny eyes, lacked scales, never exceeded 18 inches in diameter, and were only about 2 inches thick in the middle. He then described the nervous system, other internal parts, and the "floating" eggs carried by the female of the species.

With the anatomy thus covered, he turned to "the miraculous faculties that the Persian torpedo exhibited and that I observed in the open air: faculties that have been occasionally and uncritically described by scientists or reported by other people" (p. 513; p. 271 in the English translation by Carrubba & Bowers, 1982, which we shall follow with minor modifications in the rest of the chapter). Kaempfer was more than intrigued by the benumbing shocks, writing:

> The torpedo emits its power with a sort of momentary belching or a certain convulsive motion of the viscera, whereby it dilates the spiracles of the abdomen and absorbs air; with the same effort it simultaneously thrusts out its dreadful virus into the air.... When grasped in water, the torpedo's strike is less powerful, either because the water intercepts the force or because in its proper element the torpedo is not thoroughly provoked.... To be sure, the livelier, the larger, and the more recently captured torpedo numbs more powerfully and frequently. Again, the female torpedo appeared to me to strike with more effect than the male. (pp. 513–514; p. 272)

As for the utility of this force:

> Fishermen believe that nature has endowed the torpedo with the ability to induce numbness as a weapon with which it can render aggressors and neighbors powerless and torpid. Aristotle asserts this, Pliny agrees, and swift anchovies, which I have on a number of occasions found with other small fish in the stomachs of torpedoes, are proof. (p. 515; pp. 273–274)

Kaempfer devotes several lines to describing what the shock feels like and what it can do to a person's limbs, some of which read:

> When handled, the torpedo strikes hardest at the arms and shoulders; similarly, when annoyed by a foot (even

one protected by a shoe), it directs its dreadful numbing force chiefly at the knees, shins, and thighs. Complaints about increased palpitation of the heart come more from those struck on the foot than from those struck on the hands.... The numbness induced is not the sort felt in a sleeping limb, but a sudden condition that instantly travels through the touching part and penetrates the citadel of life and breath. Then it overwhelms the whole body and mind, as it seizes the sinewy and bony parts, such as the hands, shins, and elbows. In a word, you would think that your major joints were broken and limp, especially of that member which first received the expelled vapors. (p. 514; pp. 272–273)

Kaempfer did not, however, believe that the effect could be mediated through nets or rods. He wrote: "Fisherman deny that a torpedo caught in a net can be felt by drawing on the ropes," and "certainly the numbness is not conducted to the hands of a person provoking the fish with the staff (*scipionem*), nor the hands of a person provoking the fish with a spear or rod (*hasta virgaque*), as Pliny writes and scholars believe" (p. 514; pp. 272–273). Further, while dismissing a truth (its actions through intermediaries), he also helped perpetrate a new myth, namely that a torpedo's actions would not be felt if a person takes a deep breath and holds it while touching the fish. He learned about this defense from an African man, and he then tested it on himself and wrote that he confirmed it with others:

A certain African from the group of spectators boldly lifted up the torpedo repeatedly and held it without any sense of horror. I asked him to explain how he did this. He said: "Take a deep breath filling the lungs and hold it, being careful not to exhale. For as long as you can manage this, you will feel no harm from the fish." I tested this method with success. Others, to whom I revealed this new discovery, also succeeded in avoiding harm by holding their breath. (pp. 514–515; p. 273)

Kaempfer theorized that holding one's breath "drove off the miasma emanating from the fish," envisioning a torpedo releasing "a dreadful venom into the air." This idea can be traced to Pliny, the first-century author who wrote in his *Naturalis historia* that a torpedo could benumb a vigorous arm and rivet the feet of a swift runner "by smell alone, and by what I may call the breath from the creature's body" (see Pliny, 1963, book XXXII, Part II; see Chapter 3). It was also consistent with the theory of propagation of diseases by miasmas or exhalations that was influencing the medicine of the epoch. Girolamo Fracastoro had proposed this theory in the 1500s for the transmission of a disease that would later be called syphilis (Fracastoro, 1530). The conceptual background of this theory was the atomism of classical science, which had a great revival during the Renaissance, due in good measure to the wide circulation of *De rerum natura* by the Roman poet Lucretius. As we noted in Chapter 9, atomism also underlies the corpuscular conception of a torpedo's discharge as put forward by Lorenzini. The theory of the Italian naturalist was, however, clearly different from Kaempfer's idea about the torpedo's poisonous breath.

Within his idea of the torpedo's action, Kaempfer could easily account for the protective effect of holding one's breath. After all, by not breathing in a torpedo's horrible exhalations and emanations, there should be some protection against its benumbing powers.

John Walsh, whose many electrical experiments on French torpedoes will be examined in Chapter 15, would prove the breath-holding idea false. Thus, Kaempfer's assertion might be discounted as just the psychological consequence of suggestion or imagination. It remains possible, however, that there is some truth in what Kaempfer wrote. This is because the skin's electrical resistance can change with alterations in the respiratory pattern (an effect that is exploited in some psychotherapeutic treatments).

In addition, the difference between Kaempfer's and Walsh's observations could be accounted for by differences in the intensities of the shocks produced by the torpedoes studied. Walsh's torpedoes might have had more vitality, allowing their shocks to be felt even with changes in skin resistance, whereas Kaempfer's torpedoes might have been weaker, although his aforementioned descriptions of their effects on the human body also give the impression that they might have been stronger. If they were, in fact, weaker, this explanation would also fit with the shock's lack of transmission through intervening bodies, as reported by Kaempfer. Even though this too might depend on various other conditions, in Kaempfer's case the spear or rod could have been made at least partially of wood (and presumably dry) and, moreover, the current might have been shunted because of improper electric insulation. During the period in which Kaempfer made his observations, there was no clear distinction between electrically conductive and insulating bodies. This might be why he did not specify in detailed terms the construction of the tools he used to test the transmissibility of the shock. Adding to an already complex situation, it is also possible that Kaempfer had a conceptual bias against the possibility of transmission through an intervening body, because it did not fit with the idea that a torpedo's discharge is of an aerial nature (a bias also found in Lorenzini).

There is nothing about any medical uses for torpedoes in Kaempfer's texts. His statement that the Persians had no use for them is significant in this regard, especially in the light of all of the other information he painstakingly acquired.

Kaempfer is sometimes cited as "the earliest writer to compare the effect with that of lightning and therefore the first to link the phenomenon with electricity" (Walker, 1937, p. 88). Examining his words in context is important here. In both his dissertation and his book of 1712 we read: "So powerful and so swift is the force of the horrifying exhalation that like a cold bolt of lightning it shoots through the handler" (p. 273).

So Kaempfer did use the word lightning (*fulgure*). But he associated it with a cold quality (*frigido*) in a manner reminiscent of Galen's cold venom theory (Chapter 4), in contrast to the fiery displays or igniting actions normally associated with it. Kaempfer's intention was, however, to do *no more* than to draw an analogy; that is, to point out that two forces, one from the heavens and the other from torpedoes, share some similarities—both are remarkably rapid and both can hurt. To think that he was claiming that a fish discharge and lightning also share a similar physical agency, electricity, goes well beyond what he actually wrote in his often-cited sentence. (In this context, it is also notable that the work "spark," and its French and Italian equivalents, *étincelle* and *scintilla*, were sometimes used in eighteenth-century electrical science to denote the sharp and rapid sensations associated with an electrical discharge, even in the absence of any luminous phenomenon.)

Kaempfer used the word "lightning" only this one time and, as stated, he nowhere suggested that the heavenly force is an electrical event. Although the idea that lightning might be electrical was beginning to be discussed during Kaempfer's lifetime (he died in 1716), this was well before Franklin proposed his sentry box experiment and captured lightning with his kite to show that electricity and lightning are one and the same force (Chapter 10). Yet perhaps after reading Kaempfer, especially closer to the middle of the eighteenth century, when electrical experiments were drawing so much attention and the electrical nature of lightning had been firmly established, others might have been more inclined to think that the shocks generated by these unusual fishes just might be electrical.

THE FLOUNDER ELECTRICAL?

In this context, Robert Turner, a British natural philosopher, is sometimes cited for being the first scientist to come out strongly and unequivocally in favor of the possibility of animal, and specifically fish, electricity (Turner, 1746; Ritterbush, 1964, pp. 35–36; Delbourgo, 2006, p. 329). Turner's thoughts appeared in his 1746 book *Electricology: Or a Discourse upon Electricity*. Yet what he claimed sometimes went so far beyond his supporting evidence that he left his more knowledgeable readers bewildered and even amused.

In contrast to Kaempfer's book, Turner's book was published when people were beginning to think of electricity as the cause of a wide variety of natural phenomena, from the aurora borealis and earthquakes to the growth ("motion") of vegetables and even human generation (Stukeley, 1750; Bertholon, 1786). His bold thesis was that living organisms can be more than just affected by electrical effluvia—they may function by electricity. He informs his readers that he will now "proceed to apply *Electricity* to the solution of some surprising *Phaenomena*, which have not as yet been accounted for, in any satisfactory manner" (p. 25).

He points to "the Sensitive Plant, whose Frame and Texture, is so very nice and Tender, that, at the least touch of the Finger, it will contract its Leaves, as if sensible of the Contact" (p. 25). The reference is to the *Mimosa pudica*, a plant with touch-sensitive leaves that had been the subject of some poetry by Matthew Prior just a few decades earlier (see Johnson, 1892, p. 90). In his 1719 poem, titled *Solomon on the Vanity of the World*, Prior did not associate electricity with the mimosa, but he wrote:

> Whence does it happen, that the plant which well
> We name the sensitive should move and feel?
> Whence know her leaves to answer her command,
> And with quick horror fly the neighbouring hand?

Turner answers Prior's question with a simple explanation: electricity! The plant builds up an electrical virtue (charge) that is discharged when touched, much like the completion of a circuit involving the newly invented Leyden jar. The leaves, he maintains, will remain distended until the "Quantity of the Virtue" again builds up.

Going beyond the botanical, Turner now turns to fauna, and specifically to torpedoes and their discharges. He boldly states that the most viable theory behind their actions is also one based on electricity:

> Some Authors have rais'd the Effects produced by this Fish to a Miracle; other attribute it to a Pair of Muscles in its Back, which upon contracting and suddenly dilating again, strike the Finger with great Force, and so cause that benumbing Sensation; but this (tho' the most prevailing Opinion) cannot be the Cause; for the *Torpedo* has been found to have this Property, after it has been dead for some Time. Some solve the Appearance one Way, and some another; but the most rational and satisfactory Account for the Cause of this benumbing Property of the Fish, is deducible from the Principles of *Electricity*, by supposing it naturally *endued* with this *Electric Quality*, which rushes out very forcibly upon the Finger of them who touch it. (pp. 27–28)

The problem Turner faced was how to support his statement experimentally, as demanded of scientists during the Enlightenment. Unable to obtain live torpedoes, and quite possibly never even feeling or even seeing one, Turner turned to experimenting on flounders, which are common enough and are flat like the rays. He then proceeded to electrify a hapless flounder, so as to create an "artificial Torpedo." He noted that, when touched with his finger, it gave him "a stroke, or something like a Numbness" that affected his hand, which quickly dissipated. It was directly comparable to what had been described with real torpedoes!

He then tried touching his charged flounder with different objects, writing: "As the *Torpedo* conveys its numbness to the Hand, through *hard* and *dense* Bodies; but not thro' *soft* ones: So when the *Flounder* is touch'd with a long piece of *Iron*, the benumbing Stroke is felt, but nothing is perceiv'd when touch'd with a *soft* Stick, tho' but six Inches long" (pp. 28–29). Further:

> If the *Torpedo* be touch'd thro' the Interposition of any thin Body, as Cloth etc. the Stroke is felt considerably: In the same manner it happens to the Person's Hand, who touches the Flounder with a Glove on. The Parallel might be carried further, but this I hope is sufficient to clear up this surprising *Phaenomenon*, and to set the Matter of the *Torpedo's* being an *electrified* Fish by Nature (which may arise from its Texture, etc.) beyond all further Dispute. (p. 29)

But how can a torpedo "be *electrified* by Nature"? He answers his rhetorical question by stating that an all-pervasive electrical ether must be accumulated through the pores:

> Nor is this Supposition at all ridiculous, for not only a few inanimate Bodies, as *Amber, Glass*, etc. but many animate ones, as Cats, Horses etc. and even some Men themselves have been found to be Electric. If so, why may not some Vegetables? Some Fish? Etc. (pp. 29–30)

What Turner proved, of course, was no more than that a fish, like a human, could hold an electrical charge and release it when touched directly or with certain intermediaries. When he argued that the shock of a torpedo is electrical, it was based on these facts and on the sensations he just experienced with his charged flounder. He did not provide direct evidence that a torpedo is electrical, much less that it gets its electricity from its watery environment. Thus, Turner's bold conclusion was hardly justified by his experiments and was no more than an unsubstantiated speculation. Still, in the bigger scheme of things Turner must be given some credit for bringing the hypothesis of animal (fish) electricity to the fore.

The idea of creating artificial devices capable of simulating the behavior of electric fishes would be recurrent in eighteenth century. It would appeal to physicists in North America, England, and Italy, who would try to make artificial torpedoes of different kinds (Chapters 16, 17, 22). As mentioned in our introductory chapter and will be discussed in more detail near the end of this book, Alessandro Volta was convinced that his newly invented electric pile (battery), more than any previous device, could imitate the physical mechanisms underlying torpedo or eel shocks. In a sense, Turner's artificial flounder can be thought of as an early artificial electric fish based on a living animal.

Turner's flounder also brings to the fore the need to consider the difficulty that existed in the electric science of the Enlightenment when it came to recognizing electricity of genuine physiological origin, as compared to physical electricity that is applied to an animal body (or arising from a physical process, such as rubbing the skin). This confusion is particularly notable in the works of a French electrician, Father Pierre Bertholon (1786), who popularized the phrase *électricité animale* ("animal electricity"), the term itself having been used earlier by Walsh in 1772 and by John Hunter in 1775 (see Chapters 15 and 18).

The difference between a genuine animal electricity and a purely physical electricity present (or stimulated) in an animal's body would be acutely recognized in 1782 by Volta, in an important letter addressed to Anne-Pauline Lenoir de Nanteuil, an aristocratic French lady interested in the study of electricity (see Volta, 1918, pp. 8–12; also Chapter 22). According to Volta, the expression "animal electricity" should denote uniquely "a kind of electricity essentially linked to life itself, and inherent in some function of animal economy." It was a term ill suited to those forms of electricity that could be produced "by rubbing the back of a cat, by currying a horse," or to the electricity "that has been observed arising spontaneously from the feathers of a living parrot."

Before leaving Turner, one other point must be made about his approach. As with many scientists during the Enlightenment, Franklin being a prime example, his goal was not just to understand things; it was to find uses for them, especially to help suffering humanity. In this regard, Turner concludes his book with some statements about medical electricity, which in his mind started with electric fishes and was now just beginning to be practiced with electrical machines:

> For several People positively declare, that they have received great Benefit by *Electricity,* when terribly afflicted with *Rheumatic Pains,* in the *Arms, Legs,* etc. Others with the most violent pains in the *Teeth,* have been instantly reliev'd. Several who have let *Blood,* have, when *Electrified,* bled more freely than before. But how, and in what manner this most wonderful and surprizing *Power of Nature,* may be apply'd to the Service of the *Human Body* with Success, must be left to the Judgment of the *Learned* and *Judicious* Physicians; who, as being well acquainted with the Constitution and Texture of the *Human* Body, are consequently best qualified to consider and determine it. (pp. 39–40)

INGRAM'S 1750 PUBLICATION

The South American eel was drawing increasing attention during the 1740s, although not in an electrical context (e.g., (Barrère, 1741, p. 169). Nevertheless, electricity began to enter the eel picture in 1750, in a short article that appeared in *The Student; or, The Oxford Monthly Miscellany*. This new English periodical was intended to favor correspondence and respect between *literati*. It was in no way a scientific journal. Its contents were of the most diverse character. In it one could find a speech delivered by the Bishop of Oxford, followed by a *Letter to Eugenio* contrasting *Wit and Good Nature,* and then by *The Speech of an Old OAK to an Extravagant Young HEIR as He was Going to be Cut Down*. Particularly frequent and significant were poetic compositions or exercises. Some were by celebrated authors, such as Alexander Pope, whereas others were by lesser personages, including Christopher Smart, the author of the *Jubilate agno,* an intriguing composition probably written while the author was in a madhouse (see Fitzgerald, 1968; Bertelsen, 1992).

In one of this periodical's first issues, interspersed among the speeches, philosophical tales, elegies, and odes, we find "New Experiments Concerning the TORPEDO." Dale Ingram, who was born in 1710 and died in 1793, was the author (Fig. 12.2). Ingram was a London surgeon and would afterwards publish several medical works. He tells us that he had lived in Surinam, "a colony once belonging to the English, but exchanged with the *Dutch* some years since for *New York*." Ingram had been in the service of Jan Jacob Mauricius, Governor-General of the Dutch colony from 1742 to 1751.

In 1745, while in Paramaribo, Ingram had the chance to observe "a live *Torpedo*." Like many others at the time, he used the term "torpedo" generically, actually describing an electric eel. He tells us it "was about an ell long, in shape not much unlike a large eel, but more flat, with a head considerably bigger, and a dark lift down his back" (Ingram, 1750a, p. 50):

> I then attempted to touch him with my fore finger, having stretched out my arm, and at the fame time I steadfastly kept my eyes fixed upon him to observe what motions he might make. Immediately to my great surprise and confusion and, as quick as lightning, my elbow received such a strong repelling force accompanied with such anguish that I thought my fore arm would have fallen off. But what is very wonderful, the fish never stirr'd, and my finger was scarce within an inch of touching him. (p. 50)

That Ingram knew about Réaumur's (1717) mechanical explanation for torpedo's shock is suggested by his interest in detecting possible movements of the fish at the time of discharge. It is also consistent with some experiments he performed to see if the discharge could be transmitted through certain intervening objects. With regard to the transmissibility of the effect through intermediaries, he writes:

> The next experiment I made with an iron hoop taken off from an old *Madeira* wine pipe. When streighten'd it was near fix feet long. With this I attempted slowly at arms

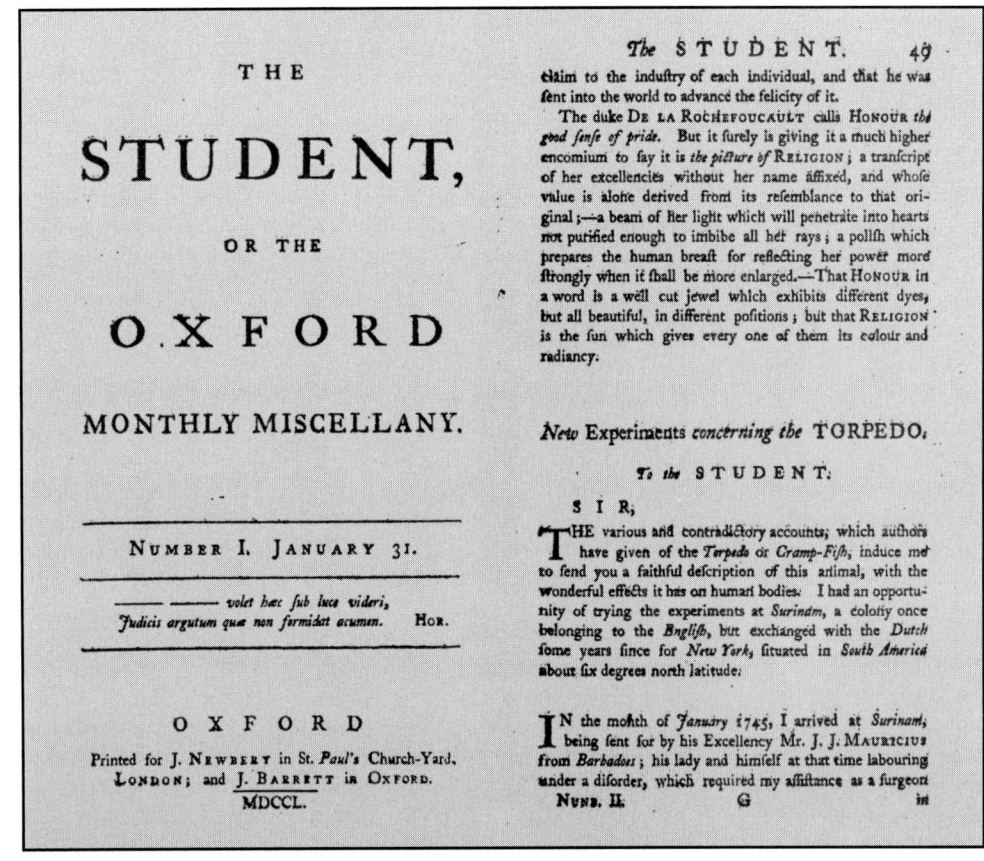

Figure 12.2. Images of the January 1750 issue of *The Student*, the Oxford monthly magazine where Dale Ingram published his report on the eel of the Surinam. Ingram's article is likely the first publication reporting experiments that suggest the possible electric nature of the fish.

length to touch the *Torpedo*; but before I could reach him, the iron twirled out of my hand with a resistless force, as when a learner is in fencing disarmed of his foil by a master of that science. (p. 50)

With this simple experiment, Ingram felt he could "overthrow the false notions of those who assert, that the *Torpedo* can have no effect on the human frame, where there is an intervening body" (p. 51). Such effects through intermediaries, he adds, are in agreement with reports by gentlemen-planters, who "assured me, that in angling [with an iron hook] the rod has frequently leaped out of their hands." This observation is reminiscent of what the poet Claudian wrote of the sea torpedo in about 400 A.D. (Chapter 4). Ingram added: "I tried the same experiment with a common stick, but my arm felt no pain, though I moved the fish about the tub with it" (p. 51).

Although Ingram correctly recognized that a metal rod, a good conductor of electricity, could transmit the eel's painful shock, whereas wood, a non-conductor, could not, he did not seem to possess a good understanding of the electrical transmission. What seemed important to him was only that there is a metallic component in the pathway, as evidenced by what he wrote about fishermen using iron hooks. His limited knowledge about what was involved also appears in a footnote, where he mentions what the Reverend Richard Walter had included about a sea torpedo encountered off Mexico in 1742.

Walter (1748), about whom more will be said below, had recently sailed with Commodore George Anson to the New World. Ingram had quoted him as stating: "I myself had a considerable degree of numbness convey'd to my right arm through a walking cane which I rested on the body of the fish." Ingram comments: "This account, tho' seemingly contradictory, may be reconciled with the above, if we can consider that Mr. Walter's walking cane had doubtless an iron spike or socket *ferrel* at the end of it" (p. 51). By invoking Walter's observation, Ingram seems not to have had a clear idea of the difference between the conductive role of metals and other possible effects (e.g., the capacitive action of metallic armatures) in electrical experimentation.

In addition, Ingram also seemed to believe some of the exaggerated tales he had heard about the South American eels, writing:

What is something more extraordinary and worth our notice is, that if a woman under her natural healthy evacuations should by accident touch this fish, they immediately cease, and the person falls into great anxiety succeeded by a jaundice or dropsy, and sometimes both, frequently terminating in a short time in death. Mr. MAURICIUS had an *Indian* woman, who languished some weeks and at last died by such an accident. (p. 51)

But what is perhaps most interesting about his article, and what makes his few pages of text especially worthy of our attention, is that he employed the word "electric" when referring to the discharge. His sentence using this word reads: "On enquiring what was the method used in taking these animals, I was told that the *Indians*, as soon as they discover where they are, immediately seize them by their back and grasp them with great force, which defeats all their electric energy or spring" (pp. 51–52).

Ingram's phraseology is nevertheless confusing. When using the words "electric energy," was he starting that the shock is, in fact, electrical? Alternatively, was he using the word "electric" simply to point out that the discharge, whatever it might be, is released with a surprising quick "spring"? Unfortunately, he does not mention electricity again, making it difficult to discern precisely what he really was thinking at this time. To some of Ingram's translators and followers, however, there was no ambiguity—they viewed his wording as a clear statement supporting fish electricity.

The Student was sold for 6 pence and advertisements for it appeared in some London newspapers. One was in the *Whitehall Evening Post or London Intelligencer,* and it provided readers with its Table of Contents in its February 27, 1750, issue. Item III, found right after "Religion the Basis of true Honour," was titled: "New Experiments concerning the Torpedo; by a Gentleman lately arrived from the Indies" (*Whitehall Evening Post,* 1750).

General culture magazines were quick to disseminate Ingram's article. The rapid spreading of cultural and scientific information beyond the limit of the specialized academic publications was, in fact, one of the hallmarks of the "material culture" of the Enlightenment (Daston, 1991; Darnton, 2003). His text was translated into German and published in 1750 in the *Hannoverische gelherte Anzeigen,* one of the general culture magazines printed in the city of Hamburg (Ingram, 1750b). This translation would later be reprinted in the *Neue physikalische Belustigungen,* a compendium of natural science out of Prague (Ingram, 1770). In these publications, the key phrase, "electric energy or spring," would become *elektrischer Kraft und Wirkung,* meaning "electric force and effect." Notably, this interpretation seems to be justified by Ingram's awareness of mechanical theories of the fish's shocks, and by his distrust of these theories.

Along more academic lines, Ingram's work was discussed in relation to electricity and electric medicine in an untitled letter signed "St. B. S." to Johann August Unzer, then editor of the medical journal *Der Artz,* also published in Hamburg (St. B. S., 1760). While the appetite for Ingram's material in this German city might be viewed as surprising, Hamburg was an important center of the German Enlightenment and, as some people of wit liked to think of it, "the most German of English cities"!

The 1770 German version of Ingram's paper was also cited in the German and French editions of Marcus Elieser Bloch's monumental works on fishes, which date from late in the eighteenth century (Bloch, 1786-87, 1785-97). The influential Bloch wrote "that this fish must have an electric atmosphere around it, because as he [Ingram] wished to touch it with a metallic hoop, even before the contact, his arm was so strongly shocked that he threw the hoop down" (Bloch, 1786-87, vol. I, p. 229). He also wrote that Ingram "received the shock even before touching the water," which went beyond what Ingram had actually written (p. 237).

WALTER AND ANSON'S TORPEDO REVISITED

As mentioned, Ingram's paper was stimulated by what Richard Walter had included in his book about a sea torpedo, even though the word "electricity" was never used in the description. The background behind what Walter did is as follows. From 1741 to 1744, he sailed with George Anson and the British Navy, taking aim at the Spanish and their prize possessions in the Pacific. Although the great majority of the British sailors did not survive the ordeal, keeping the Reverend Walter more than busy tending to the dead and dying, Anson's voyage was considered a military and strategic success. It was celebrated as an important epic by the English, even leading to an anonymous poem in *The Gentleman and Lady's Palladium and Chronologer for the Year of our Lord 1755* (Anon., 1755).

Walter saw to it that *Voyage Round the World in the Years 1741, 2, 3, 4 by George Anson, Esq.,* was published in 1748 (Walter, 1748). It immediately became a best-seller, with nine English editions by 1756. It was also translated into French, German, Dutch, and many other languages. References to it are widespread in the literature of the epoch and the passage on the torpedo was frequently included. For example, in 1757 this passage was inserted in the *Histoire Générale des Voyages*, a multi-volumes collection of travel narratives edited by Antoine François Prévost (Prévost et al., 1746-59, tome, 15, p. 329), and in the 5th Tome of the Section on Minerals of Georges-Louis Leclerc, Compte de Buffon's *Histoire Naturelle,* one of the most widely read scientific works of the Enlightenment (Buffon, 1788, p. 47). Undoubtedly Walter's book directed interest to torpedoes and helped stimulate the thought that their shocks might be electrical in nature, in an epoch in which fish electricity was rising to the fore.

In its entirety, the original passage on the torpedo reads as follows:

> And we here, and in other places, met with that extraordinary fish called the *Torpedo* or numbing fish, which is in shape very like the fiddle-fish, and is not to be known from it but by a brown circular spot, of about the bigness of a crown piece, near the centre of its back. Perhaps its figure will be better understood, when I say it is a flat fish, much resembling the thorn-back. This fish, the *Torpedo*, is indeed of a most singular nature, productive of the strangest effects on the human body: For whoever handles it, or happens even to set his foot upon it, is presently seized with a numbness all over him, but which is more distinguishable, in that limb which was in immediate contact with it. The same effect too will be, in some degree, produced by touching the fish with any thing held in the hand; since I myself had a considerable degree of numbness conveyed to my right arm, through a walking cane which I rested on the body of the fish for a short time only; and I make no doubt but I should have been much more sensibly affected, had not the fish been near expiring when I made the experiment: and it is observable, that this influence acts with most vigour upon the fish's being first taken out of the water, and entirely ceases as soon as it is dead, so that it may be handled or even eaten without any inconvenience. I shall only add, that the numbness of my arm upon this occasion did not go off on a sudden, as the account of some Naturalists gave me reason to expect, but diminished gradually, so that I had some sensation of it remaining till the next day. (Walter, 1756, pp. 264–265) (Fig. 12.3)

It is notable that the numbness described lasted into the next day, whereas Ingram, with a normally more powerful

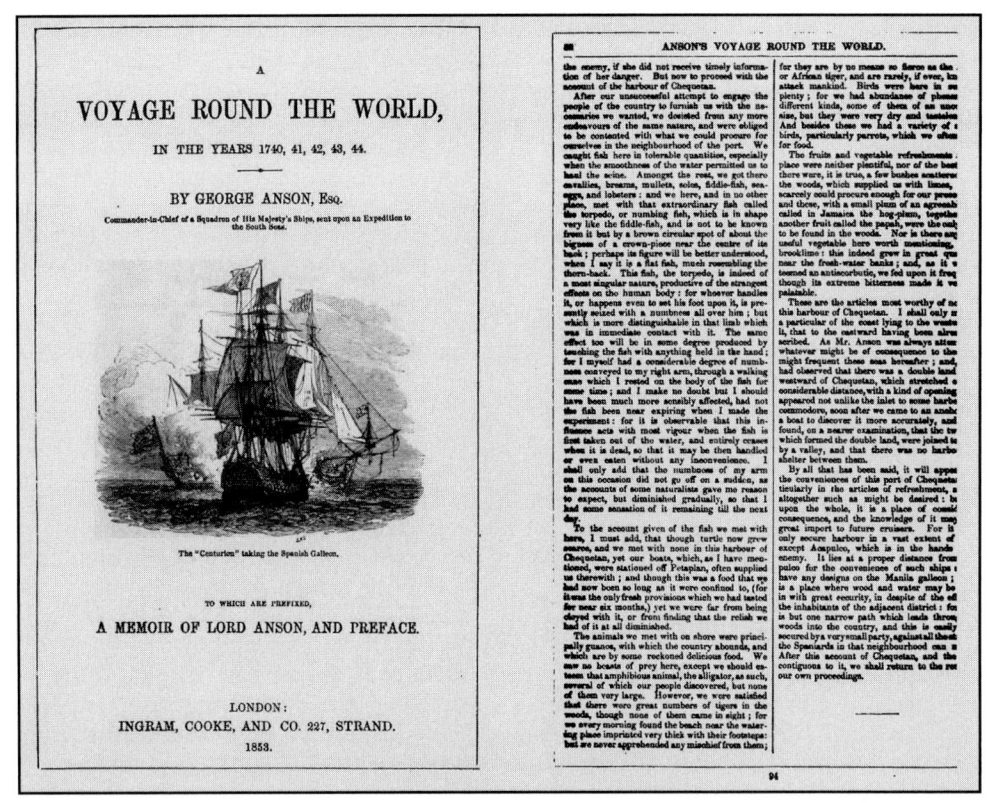

Figure 12.3. A late (1853) edition of the Walter's book, with the page dealing with experiments on the torpedo carried in Mexico during Anson's expedition.

eel, stated his own painful sensation lasted but a minute. Ingram did add, however, that others "have asserted that the anguish continues many hours." These diverse statements continue a long history of differences concerning the duration of the effects of the shocks of these various fishes on the human body.

Looking back, Peter Kolben, a German astronomer who lived in South Africa from 1704 to 1713, wrote in his account of a torpedo caught off the Cape of Good Hope: "I have not found that this numbness lasted as long as some persons pretend; never it has lasted more than a half a hour; after one or two minutes, it is at its maximum, after which it decrease slowly, step by step" (Kolben, 1741, vol. III, p. 127). Similarly, in his *Mémoire* on the sea torpedo, Réaumur stated that "the pain of this numbness is not of long duration; insensibly it diminishes; after some instants it is completely waned" (1717, p. 348). Yet in the second volume of his *De motu animalium*, Giovanni Alfonso Borelli mentions that an English anatomist "suffered for two days the effect of a torpedo's contact" (Borelli, 1680-81, vol. II, , pp. 441–442). The anatomist in this case was John Finch, who was in Tuscany late in the seventeenth century (see Chapter 9).

Needless to say, these divergent findings, much like those involving intermediaries, could be related to a myriad of factors. Some are the vitality of the fish, its size, the specific species, how it is touched, and even the health of the subject experiencing the shocks. In the case of Finch, Borelli supposed that his exaggerated sensitivity to the shocks might be related to the "paralytic tremor" affecting the English anatomist. Further, it is likely Ingram's eel was debilitated, since the full discharge of a healthy eel would ordinarily be much stronger, and consequently would be felt long after the pain from a small Italian torpedo would have dissipated.

Returning to the now-changing *Zeitgeist*, the account of the torpedo found in Walter's book was presented in a French periodical, the *Amusemens Philogiques*, intended to educate young aristocrats. After quoting from it extensively, and appropriately attributing it to Anson, David Etienne Choffin (1749-50, p. 81), Professor of French at the University of Halle in Germany, concluded with a question: *Ce poisson ne seroit-il point electrique?* ("Could this fish not be electrical?") This question would soon resonate with others (Fig. 12.4).

ADANSON'S AFRICAN CATFISH

In 1750, when Ingram published his article and Choffin asked his question about torpedoes being electrical, all of Europe knew about the newly invented Leyden jar. Moreover, many natural philosophers and physicians had personally experienced the stunning discharge of electricity that it could release. For those individuals also possessing first-hand knowledge about electric fishes, the two discharges would have seemed "shockingly" comparable.

One very observant and inquisitive person was Michel Adanson, a French traveler and naturalist, who lived in the African country of Senegal from 1749 to 1753. His description of a river fish found in Senegal dates from September 1751, so it is possible but unlikely that he would have known what Ingram and Choffin had written when he conducted his own experiments. It is more likely, however, that he learned about of Ingram's and Choffin's reports, and perhaps those of others, when he wrote up his experiments, because he did not publish

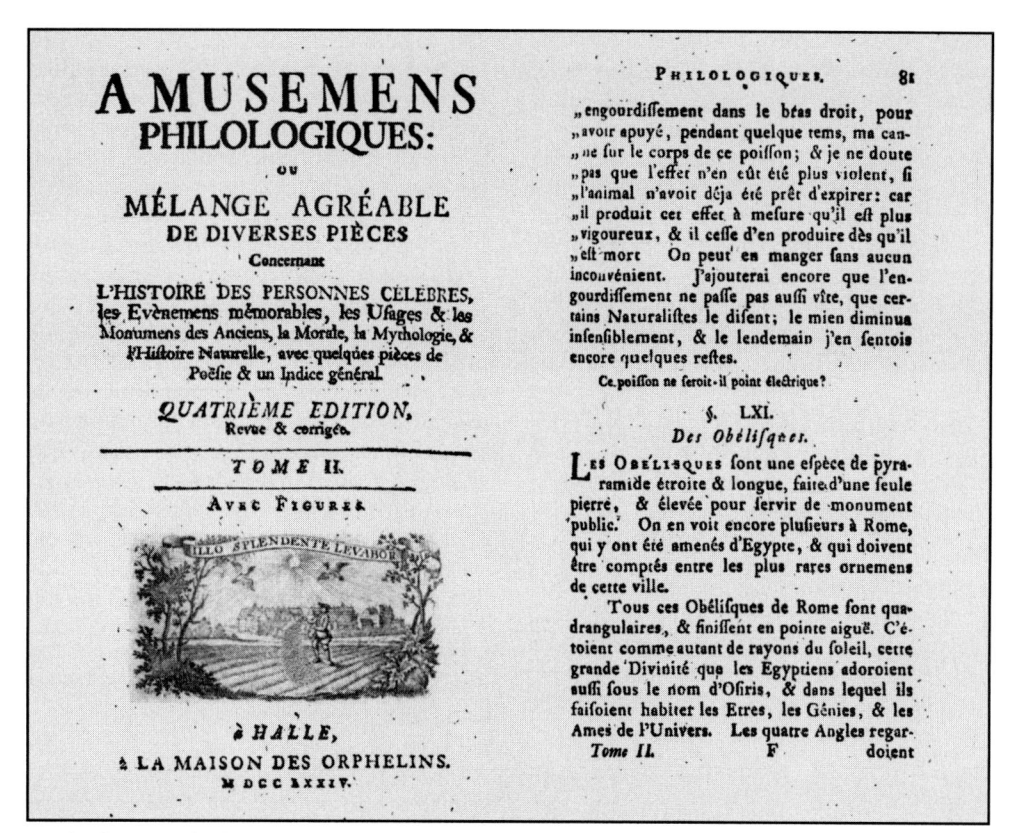

Figure 12.4. The general culture work of David Etienne Choffin where the author concludes his report dealing with Anson's observation on the torpedo with the query about its possible electric nature.

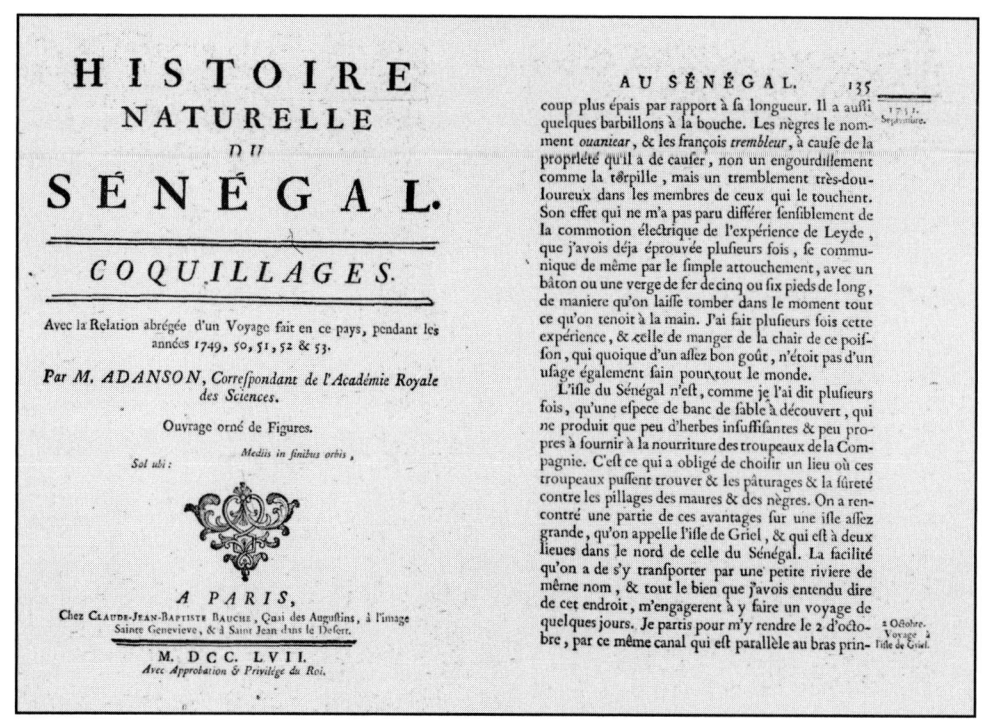

Figure 12.5. Adanson's book on Senegal with the description of the shocking freshwater fish and the report of experiments suggesting the electric nature of the shock. The observations were made in 1751 but the book was published in 1757.

the *Histoire Naturelle du Sénégal* until 1757, with an English edition printed 2 years later (Adanson, 1757, 1759). It is clear that Adanson had already experienced the shock of the Leyden jar when he encountered the river fish (Fig. 12.5).

His paragraph of interest is labeled *Poisson trembleur* ("Trembling fish" in English). In terms of physical characteristics, he tells his readers that the fish is "round without scales, and smooth as an eel, but much thicker in proportion to its length" (1757, p. 134). Quite importantly for the distinction of this fish from an eel (and particularly from an electric one), he adds: "it has some barbels at its mouth" (p. 135). This fundamental characteristic was unfortunately omitted in the 1759 English edition.

Adanson writes "the negroes call it *ouaniear*, and the French *trembleur* from the effect it produces, which is not a numbness like that arising from the cramp-fish, but a very painful trembling in the limbs of those who touch it." His next sentence is the most important of all. Here he writes:

> Its effect did not appear to me to differ sensibly from the electrical shock ["*commotion*"] of the Leyden experiment, which I had felt several times, and it is communicated in the same manner by simple contact, with a stock or iron rod of iron five or six feet long; so as to make you instantly drop whatever you hold in your hand (p. 135).

He goes on to state that he has "tried this experiment several times." And he concludes by telling his readers that he has also eaten this fish, which tasted good but "is not equally healthy for all constitutions."

Adanson's fish would not always be identified correctly. John Pringle (1775a; Chapter 16), when giving an address to the Royal Society in 1774, stated that Adanson had described an electric eel, like those found in South America. This could have been because he did not consult Adanson's book in its original French, but instead relied on the English edition. Thus, he would have read that it had some features like the eel, but not the fact that Adanson (p. 134) had also written "*Il a aussi quelques barbillons à la bouche*"—that his *poisson trembleur* has barbels by its mouth, this being a distinctive feature of catfishes. *Sans* this important sentence, an unsuspecting reader could easily be confused.

Bloch (1786-87, 1785-97, 1801), the aforementioned authority on fishes, also confused Adanson's fish with the South American eel, stating that the *Gymnotus* had been encountered in the rivers of Senegal. It remains possible that this error also occurred because Bloch too did not read the French edition of Adanson's book, and because he might even have used Pringle as a source. Another possibility is that he was obsessed with the eel far more than with these catfishes, because of what the Dutch in the New World and in the Netherlands had written about them in the 1750s and 1760s. Interestingly, his two-volume book appeared in 1786–87, shortly after French zoologist Pierre Marie Auguste Broussonet (1784) had provided naturalists the first really good description of the catfish.

With these developments as a prelude to what will follow, we now turn to the Dutch, with their collectors of exotic specimens in Amsterdam, talented physicists in Leiden, and brethren in South America. As we shall see, the Dutch knew more about these eels than anyone else at this time.

Chapter 13
The Dutch, the Eel, and Electricity

> The experiment was done with an eel called a tremble fish, and what I had written to you about it in my previous letter is true. It produces the same effect as the electricity that I felt with you, while holding in a hand a bottle [Leyden jar] that was connected to an electrified tube by an iron wire.
> Laurens Storm van 's Gravesande, 1754 (in Allamand, 1756a)

The Dutch were in a particularly privileged position to appreciate the parallel between the discharge of the most powerful electric fish known, the South American eel, and that of a Leyden jar.[1] The eels thrived in the rivers of their South American colonies; their talented natural philosophers had introduced the famous electrical bottle that had taken the world of physics by storm (Chapter 10); and they had wealthy and avid collectors desirous of obtaining exotica and showing off their specimens (Müsch, 2005). Moreover, that some of the most important steps towards fish electricity would involve Dutch scientists, both along the "Wild Coast" of the Guianas and in the Netherlands proper, should not be surprising when one considers the rich cultural heritage, means of communicating ideas, and flourishing nature of Dutch science in the mid-1700s.

With the Dutch of the 1750s and 1760s, the discharges of the eels discovered in the 1500s would become more electrical in many previously incredulous minds, and the idea that a few other fishes might also be electrical would gain needed credibility. But as is so often the case with the early history of a new idea, not every question that was being asked was about to be properly or fully answered, and over time some of the excitement associated with the Dutch and their South American eels seemed to fade into oblivion.

STORM VAN 'S GRAVESANDE AND JEAN ALLAMAND

Jean Nicolas Sébastien (Jan Nicolaas Sebastiaan) Allamand was one of the Dutch physicists associated with Pieter van Musschenbroek and the discovery of the Leyden jar (see Chapter 10; also Brazier, 1984; Dorsman & Grommelin, 1957; Hackmann, 1978). Born in Switzerland in 1731 and trained for the ministry, Allamand (Fig. 13.1) moved to the Netherlands in 1738, where he was first employed as a private teacher.

He became professor of philosophy in Franeker (Friesland) in 1747, and 2 years later became professor at Leiden University, where Pieter van Musschenbroek did his work.

Allamand was the first professor in the Netherlands to lecture on natural history. Also well known for his research on experimental physics and electricity, he was affiliated with many important European scientific societies and routinely communicated with scientists around the world. Knowing about Condamine's report, which mentioned Réaumur's (1717) mechanical theory favorably (Chapter 9), and probably aware of Ingram's (1750a,b) letter on the eels (Chapter 12), Allamand wanted to know more about the South American eels—and especially the nature of their discharges. He therefore tried to obtain a specimen for his cabinet and also asked Laurens Storm van 's Gravesande, a Dutchman living in Guiana, for additional information about live eels.

The Dutch began to establish settlements along the "Wild Coast," meaning in the Guianas, at the end of the 16th century for commercial and strategic reasons. Some were destroyed several times: first by the Spaniards, and later by the English and French. Only after the end of the War of the Spanish Succession (1702–13) did it become possible to expand the Dutch plantations in Essequibo, Demerara, and Berbice, where sugar cane and other exportable crops were grown. These Dutch settlements were located west of Surinam, which was controlled by the English at the time.

Laurens Storm van 's Gravesande governed the colony of Essequibo, founded in 1616 on the river of the same name, between 1742 and 1772 (Fig. 13.2). Prior to this time, he had served as secretary and bookkeeper for the Dutch West-Indische Compagnie in the colony. After the Demerara region was opened to free settlement, he was also made *Directeur-Gereraal* of that territory, while his son, Jonathan Samuel Storm van 's Gravesande, became its Commandeur. Under their direction, Demerara's plantations outnumbered the 140 plantations of Essequibo. By comparison, the number of plantations in Surinam, which would be conquered and exchanged for New-Netherlands (New York) by the peace

---

[1] Some of the material presented in this chapter appears in an earlier article written by the two authors and Peter Koehler (Koehler, Finger, & Piccolino, 2009).

Figure 13.1. Jean Nicolas Sébastien Allamand (1731–1787), the Dutch physicist who communicated with Laurens Storm Van 's Gravesande (1704–1775) in the Guiana colonies about eel electricity.

treaties of Breda (1667) and Westminster (1674), grew from about 400 in 1750 to about 500 in 1790 (see Den Heijer, 2002).

An energetic and enlightened man who enjoyed studying the wonders of nature, Laurens Storm van 's Gravesande was an ideal person to comply with Allamand's request. Moreover, there were family and personal ties. Allamand had been entrusted with the education of the two sons of Willem Jacob van 's Gravesande, Laurens' uncle, who had been a professor of astronomy and mathematics at Leiden, and one of the first teachers of Newtonian science in a continental European university. Allamand had also edited the collected works of Willem Jacob.

The governor referred to Allamand as his "good friend" in his informative return letter from the New World, which was composed in 1754. Upon receiving this letter in French, Allamand translated it into Dutch, added commentary, and published under his own name. It came out in 1756 in the journal of a scientific society that had been formed in the town of Haarlem four years earlier, the *Verhandelingen uitgegeeven door de Hollandsche Maatschappye der Weetenschappen te Haarlem*. Appearing 3 years after Allamand joined this society, 2 years after he published a note on medical electricity in the same journal (Allamand, 1754), and 1 year after he was elected into the Royal Society of London, the title of this article can be translated as "Effects that an American Fish Causes on those that Touch It" (Fig. 13.3).

In translation, it reads as follows:

Almost two years have past since I received a fish from Mr. 's Gravesande, general director of the Volksplanting [people plantation] of Issequebo [Essequibo]; a fish that the inhabitants of the place consider a kind a eel; although basically it is a fish, called Gymnoti by Artedi and Carapo by Marcgraf. The gentleman to whom I have shown it, Professor GAUBIUS, has had a figure of it made in copper and also provided a description in the new volume of the large book by SEBA, which will appear soon; thus I can be spared the trouble of describing it here. But what Mr. 's Gravesande has since written to me with respect to the effects of touching this fish is so peculiar that I do not doubt that people will receive this story with considerable satisfaction.

When the gentleman sent it to me, he wrote that it was a kind of tremble fish [the literal translation of *siddervis*, the Dutch term often used to denote the ordinary torpedo], the effect of which he had himself felt. This fish has, with respect to its form, nothing in common with the usual tremble fish, which is a kind of ray.

I asked Mr. 's Gravesande to be so kind as to examine the situation from the start with accuracy, and to inform me of his observations with more details. I could not ask anyone more capable of giving me satisfaction in this matter than this gentleman since, when he is not engaged in the welfare of the Volksplanting, of which he is the manager, he passes his time becoming fully acquainted with everything remarkable that is brought forth from the areas where he stays.

Examine how he answered me in a letter, dated Rio Issequebo the 22nd of the slaughtering season [November] 1754.

"The experiment was done with an eel called a tremble fish [or torpedo], and what I had written to you about it in my previous letter is true. It produces the same effect as the electricity that I felt with you, while holding in a hand a bottle [Leyden jar] that was connected to an electrified tube by an iron wire. But what will surprise you more is that Mr. CHARLES BOLTON of Barbados, who took two living fish upon his departure from here, found great interest to see and touch these fish at Barbados, and observed that several persons suffering from gouty pains in some limb were cured within two or three minutes after touching the tremble fish. This experiment has been repeated several times, but always with the same result.

"If our skippers would have the slightest inclination to oblige, I would send you a few living [eels]; it is very easy to feed them with small crabs, which is their usual food. Mr. Bolton has ordered the person who replaces [when away] him to send one with each ship that leaves from here. This seems to me to be a natural rarity, worthy of examination in more detail.

"If one touches the fish, it does not give off fire or sparks, similar to the apparatus for electricity. But for everything else it is the same; yes, even much stronger, because if the fish is big and lively, the shock produced by the animal will throw anyone who touches it to the

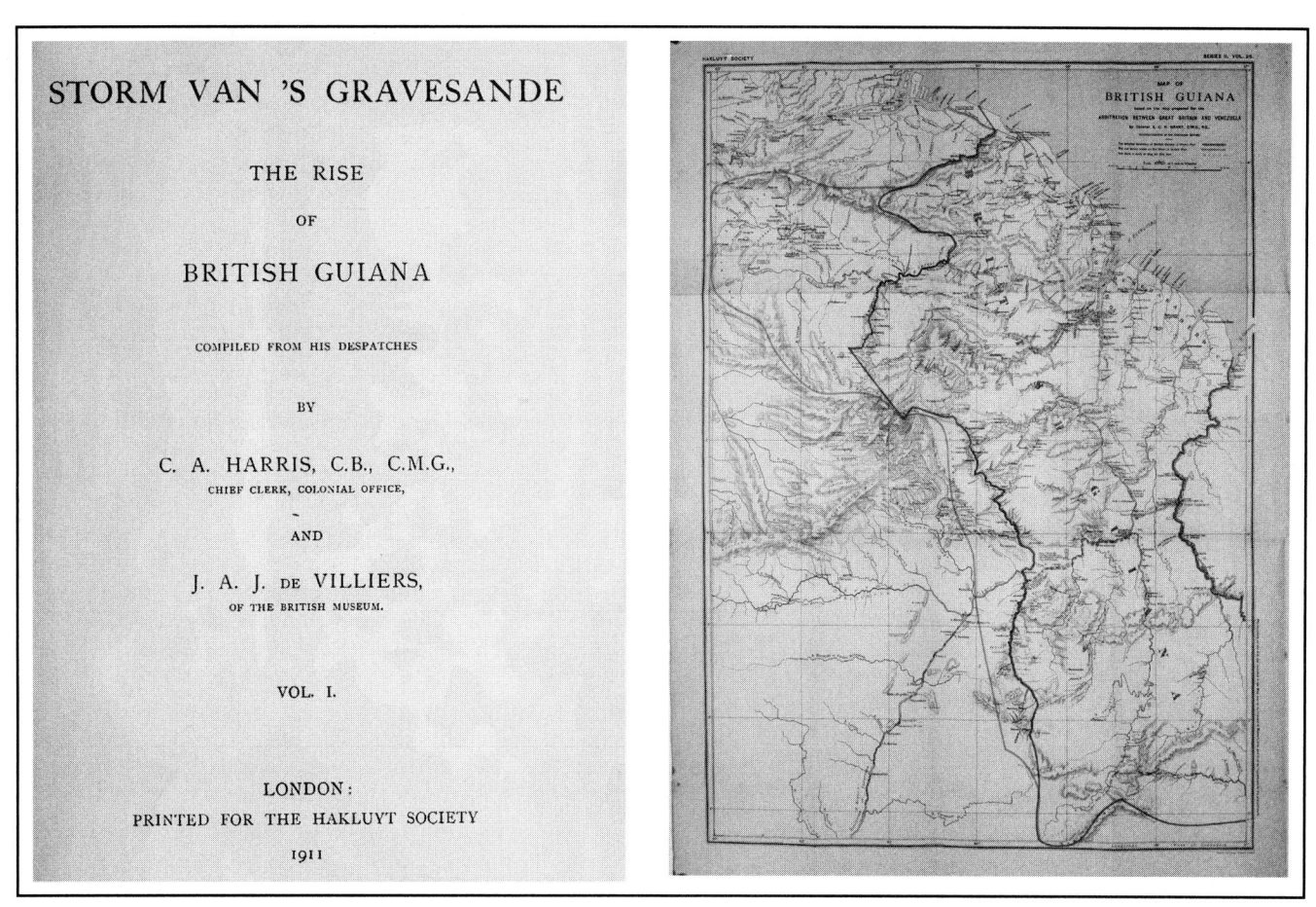

Figure 13.2. A book published on 1911 based on reports by Laurens Storm van 's Gravesande, Director of the Dutch South American colonies of Essequibo and Demerara, shown in the map on the right. Van 's Gravesande displayed a strong interest in natural philosophy, stayed in contact with leading naturalists and experimental natural philosophers back in the Netherlands, and conducted some of the first experiments on the electric eel. In this book there is, however, no report of these experiments.

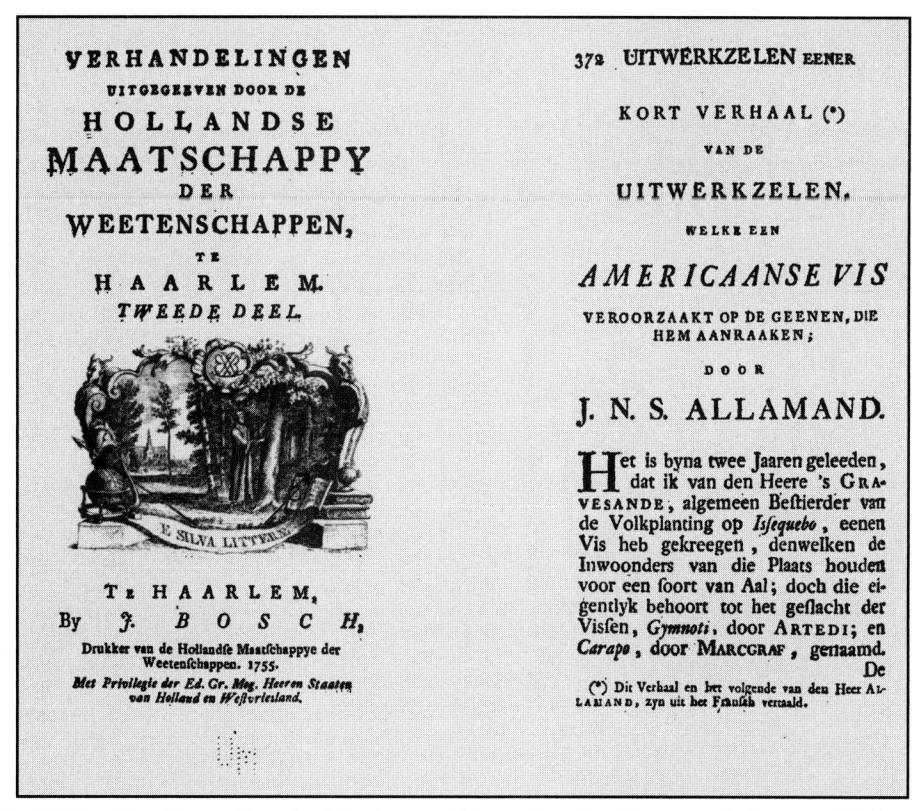

Figure 13.3. Allamand's (1756) publication on the eels of Surinam, which was based on Storm van 's Gravesande's letter from South America.

ground, without exception, and one feels it throughout the whole body; all joints crack without leaving the slightest harm. All this happens in an instant.

"However, here we have a certain Mr. VAN DER HEIDEN, who dares to hold the fish in his hand and does not feel anything.

"One sees very few of these eels in the river, but they are in all small creeks, and if one finds them, one may be certain that one will not see any fish within of eight or ten rods around it. They are good to eat: the Indians indulge others with it and they are considered a delicious snack; but if they are alive, they fear for it as for the devil."[2]

Allamand now provides this commentary:

It appears from this story that there is good reason to give this fish the name tremble fish, considering that the effect upon touching it does not differ from that of the common torpedo, other than it [the eel's shock] is much stronger.

The primary cause on which it [the shock] depends in the latter fish [torpedo] isn't a mystery anymore, since Mr. DE REAUMUR has provided a clear and neat explanation in the memoirs of the [French] Royal Academy of Sciences for the year 1714. It would be very useful to know whether this primary cause is the same in our Gymnotus. Dissection of the animal being the only means to teach this to us, I shall devote myself to it as soon as I have received another one, for which I have written, and which has been promised to me. I have not been able to dissect the one that has been sent to me, since it is a too valuable addition to the cabinet of our High Academy.

I take the opportunity to advise those who have properties in Issequebo or in the Berbices, where this same fish is found, to send for some living specimens, as they will very easily be granted the favor from the skippers of the vessels, who are dependent on them [shipping their products for income]. And the latter will be richly refunded for the care they will provide, since in Europe these fish retain the same feature that they have in America, notably of curing gouty pains; a quality that truly deserves to be examined, and which will probably be similar to that of the Electricity, which occasionally has been applied with good results for the cure of this disease, but mostly has not brought about any effect at all.

For the rest, this fish has been known for many years: Mr. Richer has seen them in Cayenne and even felt the numbness it may cause. DU HAMEL speaks about it in his History of the Academy for the year 1677 [actually for the year 1674]. There is, he says, another fish, three or four feet long, not different from the Conger-eel or Sea-eel, which, when touched with the finger, yes even with the end of a stick, brings a trembling to the arm, and brings about a dizziness which he [Richer] experienced himself.

Beyond all arguments, this fish, which was the subject of this story, is the same as our Gymnotus, the Figure of which indeed resembles that of the Conger-eel. It is true that its length is not more than of almost two feet, however, one has written me from America that some have been found that are twice as large. (Allamand, 1756a; Trans. from Koehler, Finger, & Piccolino, 2009)

Thus, we find the eel's discharge being directly compared to a Leyden jar in Laurens Storm van 's Gravesande's letter to Allamand. Chronologically, this comparison appeared 1 year before Adanson (1757) published his book in which an African catfish's discharge was compared to that of a Leyden jar (Chapter 12). It is of interest to read that Laurens Storm van 's Gravesande and Allamand had even experienced the explosive discharge of a charged Leyden jar together, while in the Netherlands (Fig. 13.4).

Knowing what a discharging Leyden jar and an electric eel can feel like was of great importance to the Dutch governor and his correspondent, because it had long been recognized that things that feel the same are likely to be of the same nature. Hence, sensation was an important tool in the physiology and physics of the time. Similarly, knowing that the impact of a fully charged large Leyden jar could knock a person to the ground, as happened on several occasions to Benjamin Franklin (Finger, 2006a, pp. 109–113; Finger & Zaromb, 2006), Allamand would also have been excited to read that "the shock produced by the animal will throw anyone who touches it to the ground."

When Laurens Storm van 's Gravesande mentioned that the eel's discharge could cure "gouty pains," the governor was again alluding to an important parallel between the discharges of these fishes and what was now being claimed for man-made electricity. Allamand would also have understood

Figure 13.4. A Leyden jar like those used to compare their shocks to the discharges of the South American eel. Starting in the 1750s, Dutch natural philosophers would write that the eel's shocks feel like Leyden jar discharges (e.g., Allamand, 1756a; Van der Lott, 1762). Adanson (1757) would make a similar statement about the African catfish at this time, and the assumption grew that torpedoes would be no different.

---

[2] In the Dutch journal, quotation marks appear to the left of every line.

this, having had, as we have already noted, just published a paper on medical electricity in the same journal as the one now publishing the eel article (Allamand, 1754). The fact that these fish could be eaten without ill effects also relates to electricity, at least by excluding another theoretical possibility: this observation made it highly unlikely that the eels might be using a stored poison with numbing effects. As mentioned before, during the second century, Galen had suggested that torpedoes might release a cold poison (Chapter 4), and his theory was still garnering some attention, even with the advent of newer, mechanical explanations.

In summary, Allamand learned that the tremble fish discharge is very much like one from an electrical device, including its remarkable speed ("All this happens in an instant"). But he was also told about one difference: The fish "does not give off fire or sparks similar to the apparatus for electricity." Given this difference, and very likely because of personal friendships (notably between Van Musschenbroek and Réaumur, both of whom were still alive), Allamand opted to be both cautious and extremely polite in his conclusions. Hence, when Allamand concluded that the eel and torpedo probably function in the same way, he did not come right out and say that both species must be electrical. Instead, he told his readers that Réaumur has "provided a clear and neat explanation" *for the torpedo*, and called for further studies, including a careful dissection of an eel, stating comparative anatomy should help to clarify the incomplete picture.

DUTCH COLLECTORS AND THEIR ZOOLOGICAL CABINETS

The attraction for exotic or rare fishes to be included in what were called *Kunstkammern* or *Wunderkammern*, meaning "cabinets of arts" or "cabinets of wonders," also played a significant part in the growing interest in electric eels in the mid-eighteenth century. These cabinets had already started to become popular in the sixteenth century (Olmi, 1992; Bredekamp, 1995; Impey & MacGregor, 2001). Over time, they would grow and evolve into modern museums (Fig. 13.5).

For Amsterdam collector of biological specimens Albert(us) Seba (Fig. 13.6), who was mentioned by Allamand, the interest in collecting was partly commercial. Seba, who had died 20 years earlier, had been a pharmacist, a natural philosopher, and a member of numerous scientific societies (including the Royal Society of London). He amassed what were considered to be some of the most important collections of *naturalia* of his time. Seba sold his first collection to Czar Peter the Great, who had visited Amsterdam in 1716 (Müsch, 2005). He then started a new collection, specimens of which made their way to the British Museum (Gronov, 1854).

In 1734, Seba began a monumental catalogue of his specimens, one that would be enriched with 446 magnificent copper engravings, making what he possessed even more valuable (see Seba, 1734–65, 2005). Swedish naturalist Peter Artedi, who was Linnaeus' teacher and is sometimes called the father of modern ichthyology, and Frederik Ruysch, who also sold his collection of anatomical specimens to Peter the Great, helped with the project. But Artedi's accidental death and Seba's own death in 1736 interrupted the project after the second volume was completed.

The third volume, the one dealing with fishes and other marine creatures, was eventually published in 1758, thanks to the efforts of Hieronymus David Gaubius (Fig. 13.7), who was also mentioned by Allamand. Gaubius served as Seba's curator, and he incorporated a large amount of Artedi's material and included an image of the *Gymnotus* mentioned by Allamand with some text in the third volume (Fig. 13.8).

Laurens Theodor Gronov (Gronow, Gronovius), who came from a very scholarly German family and lived in the Netherlands, also understood the thrill and the business of collecting, and he had an electric eel in his museum (Fig. 13.9). Like Seba, he had an interest in fishes, and he wanted to publish a book showing the magnificent collection begun by his father. Gronov's (1754–56) *Museum ichthyologicum sistens piscium indigenarum et quorundam exoticorum* contains a description of the eel he acquired (described as Item 169 in the collection) (Fig. 13.10). An updated description of his specimens, including the eel, appeared in his *Zoophylacium*, a magnificently illustrated volume published posthumously (Gronov, 1763-64). (Some of Gronov's specimens are still at the Natural History Museum in London; for more, see Gronov, 1854; Wheeler, 1958.)

In 1758, 2 years after Allamand's article, Gronov wrote an article for a Dutch magazine (*Uitgezogte Verhandelingen*) that had already published translations of three of Franklin's letters on electricity, two of which were addressed to Peter Collinson in London. Collinson, in turn, would sponsor Gronov's election to the Royal Society of London in 1763, and Gronov, in response, would dedicate his *Zoophylacium* to him.

Gronov's piece, *Beschryving van den Siddervis of Beef-Aal*, is a Dutch translation of the second part of a letter written in Latin that he had received from South America. He did not mention his contact's name, but 20 years later it became clear that he, like Allamand, was corresponding with Laurens Storm van 's Gravesande, although he might not have been his only source (Garn, 1778, p. 19).

This publication opens with a general description of the external shape of the fish. Gronov (1758, p. 472) draws attention to the numerous openings on the skin, which he associates with the slimy substance covering its body. He then raises the possibility that these openings could also be a way "by which possibly also the electric action of this fish comes out."

Gronov's choice of words stems from the answers he had received to a list of 25 questions about the eels and their very unpleasant effects. In fact, much of his publication covers the answers, as he did not dissect his own prized specimen to learn more. Among other things, he informs his readers that:

1) A person in water can be drowned by the shock of the eel.
2) The shock's effects typically last less than a minute.
3) There are individual differences in sensitivity, though everyone is affected.
4) Repetitive shocks decrease in intensity, but they strengthen again after the fish has had a rest.
5) The shocks can be felt by touching any body part, but it seem stronger from certain areas, such as the belly and tail.
6) The shocks can be felt just by touching the water, even if the fish is swimming a rather long distance away.

Figure 13.5. A view of the Museum (*gazophylacium*) of Levinus Vincent, a rich merchant of Haarlem who loved and collected a great variety of *naturalia* and *artificialia* and displayed them to the public (from Vincent, 1719).

7) It is not necessary to touch the fish with both hands in order to feel the shock.
8) The shocks will be felt even when a hand is covered by oil or wax.
9) The shocks can also be conveyed through wood, though they feel less intense than with direct contact.
10) They feel stronger if the wood tool has a metallic insert.
11) The pain is unbearable if one touches the eel with a gold ring.
12) The eel will kill other fish swimming with it.

Gronov concludes this communication by stating that the eel may be useful in medicine. He writes that "various accurate surgeons" informed him that the shocks could restore functions to nerves (theoretically) blocked by obstructions. He specifically mentions its use in Barbados by Mr. Bolton, directing interested readers to Allamand's (1756a) publication.

In summary, Gronov provided some new observations about the eels, and his work is also of historical importance because it drew more attention to this fish and its possible electrical nature. But being a collector, he did not show great awareness of the physical laws of electric conduction (much like Ingram before him; see Chapter 12). This is revealed by his "astonishment" to find that the shock is not transmitted through certain electric insulators (such as a lacquered wooden rod). Also, other than touching the fish with a gold ring, he did not consider different metals independently, but instead only pieces of metal inserted into the end of his wooden rod. These experiments would not have satisfied those physicists promoting a more rigorous electrical science.

Two years after this article and the third volume of Seba's *Thesaurus* appeared, Gronov's article appeared in its original Latin (Gronov, 1760). Interestingly, the phrase *Elektrikaale werktuigen* ("electric apparatus") used in the Dutch article in connection with the openings in the skin was presented as *motus tremuli* ("trembling motion") in Latin. This change raises a number of questions: Was Gronov still open to mechanical explanations? Did he think that fish electricity had to be the result of a mechanical action? Or did the decision to use *Elektrikaale werktuigen* in the Dutch version come from the editor's desk?

In 1763, the matter seemed settled. In the first fasciculus of his *Zoophylacium*, Gronov unequivocally wrote about an "electric force" (*vim electricam*). Further, he not only applied this explanation to the eel, but he extended it to the sea torpedo too.

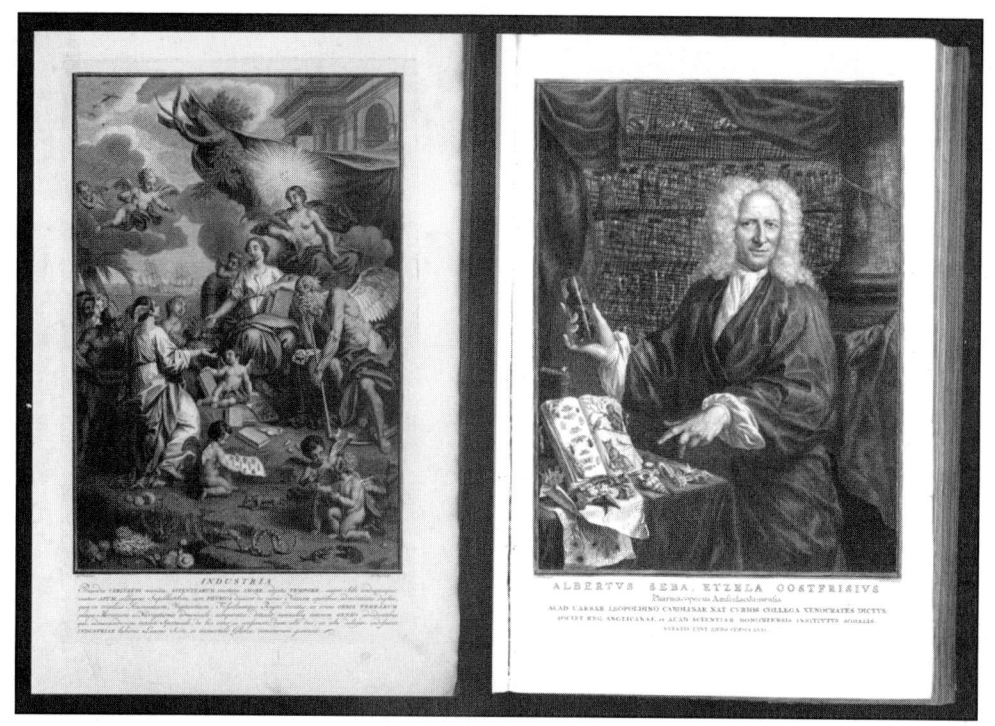

Figure 13.6. A portrait of Albert(us) Seba (1665–1736), the Dutch collector of *naturalia*, and the rich, symbol-filled frontispiece to one of the volumes of his *Locupletissimi rerum naturalium thesauri accurate descriptio*, which comprised four volumes published between 1734 and 1765.

In retrospect, Gronov (1763) would have felt more confident about fish electricity by now. Among other things, another very important letter from the Guianas pointing to eel electricity had just appeared in the same Haarlem journal that had published the letter from Storm van 's Gravesande to Allamand.

FRANS VAN DER LOTT'S LETTER

Frans van der Lott (d. 1804), who is referred to in some texts as a Dutch surgeon and in others as a lawyer/jurist, lived in Essequibo. He probably pursued science in his leisure, as was common among educated and affluent gentlemen in the eighteenth century. He evidently knew Storm van 's Gravesande, and the two Dutchmen seem to have interacted personally.

The title of Van der Lott's (1762a) publication translates as "Short Report on the Conger-Eel, or Tremble Fish; Drawn from a Letter from Frans van der Lott, Dated Rio Essequebo June 7th, 1761, Reported to the Society by One of its Members" (Fig. 13.11). The member who received and edited the letter with its accepted alternative spelling of Essequibo is not named, but this man could well have been Allamand.

Going to press in 1762, this fascinating publication is translated as follows:

> The shock [or "drill"] fish, which is here called the Conger-eel, is very similar in form to an eel, except that the former has ears, resembling those of bats. He has to breathe above the water; and therefore forms bubbles on the water that quickly disappear.
>
> Two varieties may be found, identical in effects but not in power, the black one giving off stronger shocks than the red.

Figure 13.7. Hieronymus David Gaubius (1705–1780), a pupil of Hermann Boerhaave at Leiden and his successor as professor of medicine and chemistry. Gaubius was one of the Dutchman involved with collecting and disseminating written information and good pictures of the Surinam eel.

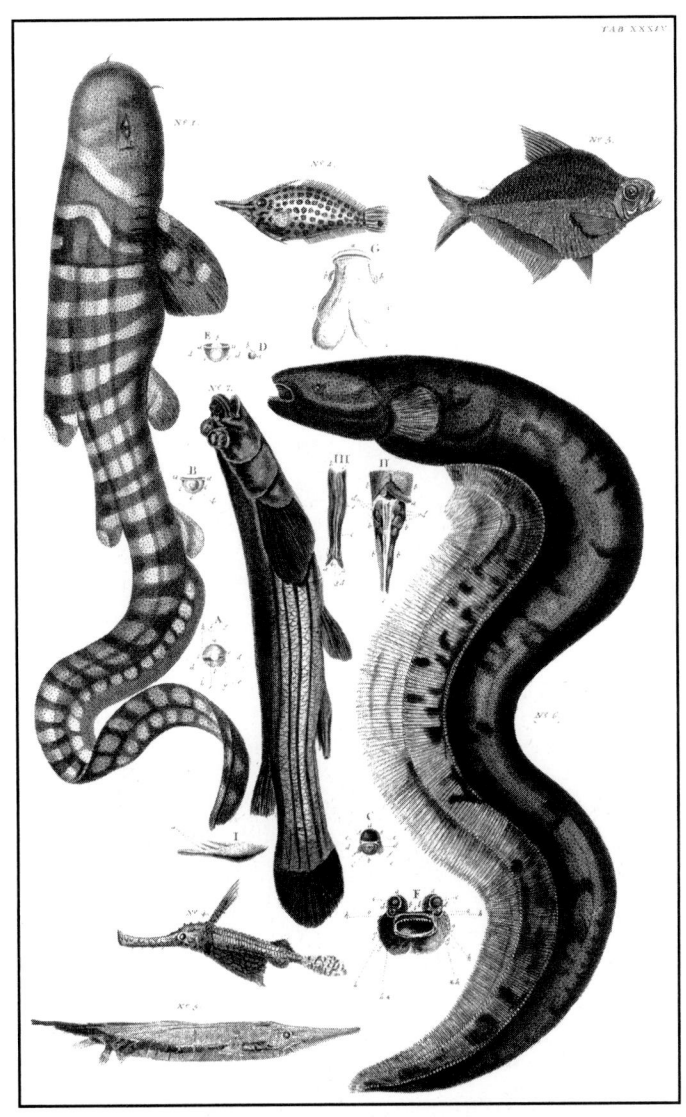

Figure 13.8. The electric eel as illustrated to the right of center in Volume IV of Seba's *Locupletissimi rerum naturalium*... of 1758. Seba had died in 1736, long before this volume was completed, and had been helped by Peter Artedi, who had also died. Gaubius finally completed the volume.

Figure 13.9. A family portrait of Laurens Theodor Gronov (1730–1777) with his two sons. A glass bottle with the electric eel is on the shelf; this was one of the most precious items in his collection (after a painting made in 1775 by Dutch artist Isaac Lodewijk de La Fargue van Nieuwland).

Their sizes vary considerably; one may find them 1 to 5 feet long or longer. Their thickness is proportional to their length. But the longest exert the most power.

They are eaten by many, Whites, Indians, and Negroes. Additionally, they have very bony backs and a fairly solid body, but are quite soft and slimy on the belly.

They mainly stay near rocks and on stony bottoms, although they may also appear in creeks or river branches, but when this is the case few or no other fish are found there because, if a fish approaches too close, it will be immediately shocked to death. And if the Conger-eel itself is in a fyke [a kind of a hoop net or trap] for a long period, it will die, because, as said before, it has to breathe above the water.

The author of this letter had done electrical experiments while in Middelburg [a Dutch city] in 1750. Hence, he reports on experiments that he has undertaken with this fish, and concludes that the power of the fish corresponds with that of electricity, except that the fish does not produce sparks, such as may be observed in electric experiments, and in the case of electricity in very dry and clear weather. In experimenting with the power of this fish, however, water is necessary. In order to prove this he refers to the following experiments:

I touched (he says) such a fish, lying in a tub with water, with a long iron rod, which shocked me enormously. But if I held the said iron rod in a dry cloth and touched the fish again, I did not feel the slightest shock. After wetting or even moistening a cloth, I again felt the stroke like before.

Although the shock also occurs on touching the fish with a copper thread, he supposes that not all copper will serve in this way, because he noticed different effects of different copper pieces (like a skimmer, a candlestick, etc.), in a way such that he did not feel anything at all in one case, and a violent shock in another. Using good English tin, it also shocks very powerfully, but if there is much lead underneath, the blow will be less. Using gold, one will feel it through the whole body; the same is true for silver, but less powerfully. Using lead, tinplate, earthenware, bone, twig, and wax, no effect is observed. The same is observed with dry wood, but it will have the power if the pores are permeated with water.

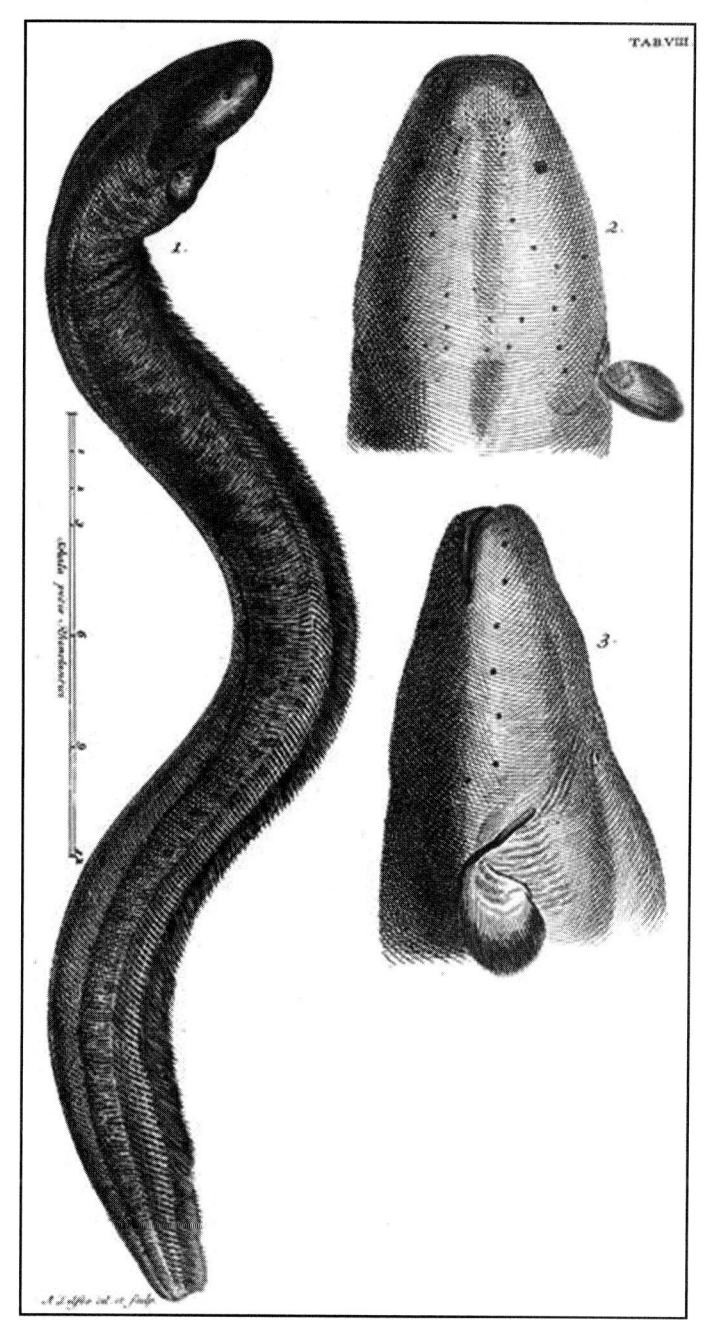

Figure 13.10. Laurens Theodor Gronov's illustration of the South American eel (from Gronov, 1763–64).

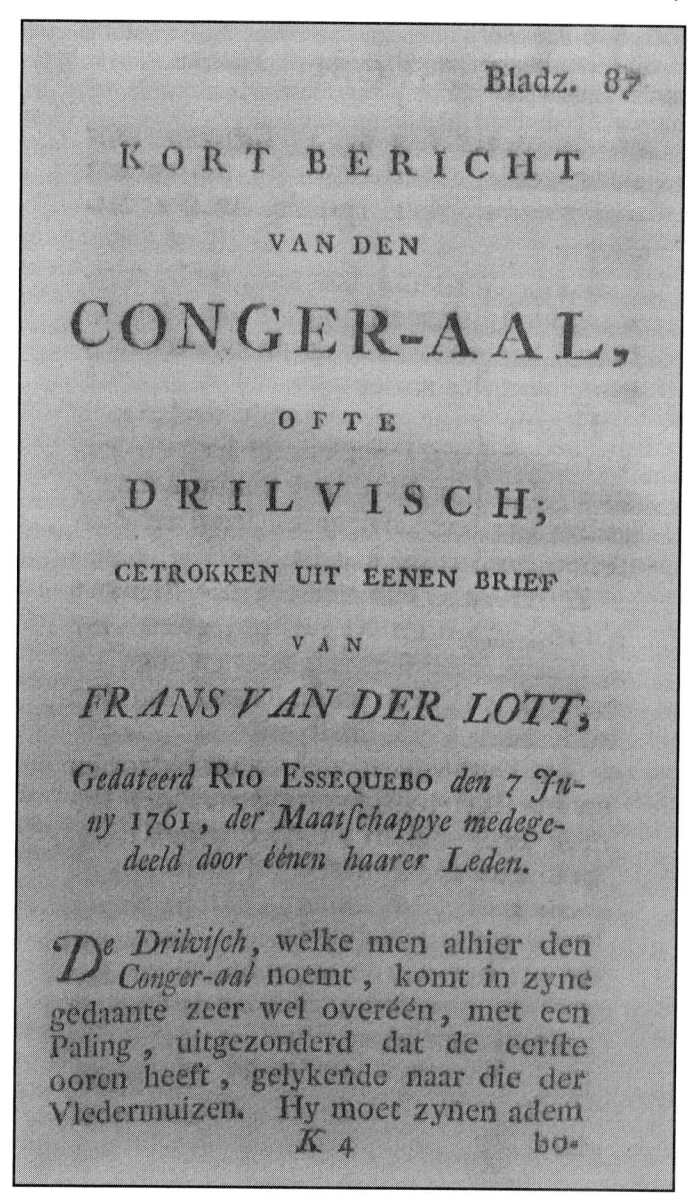

Figure 13.11. Frans Van der Lott's (1762) publication on the South American eel, in which he described these fish as electrical and capable of wonderful cures.

Further, the location where the fish is touched is of importance: The closer to the head, the more powerful the blow. Under the throat, the blow is enormous.

The author reports the following experiments, of which the first was carried out at the house and in the presence of Mr. LAURENS STORM VAN 'S GRAVESANDE, Governor of the Colony. When five persons held each other hand by hand, and the first touched the fish with the point of a sword that had a silver hilt, the blow was felt up to the fifth person, although less powerfully. It should be noted that the shocks in these experiments are felt not further than to the shoulder; that nothing is felt in the other arm or hand even though the shock is felt by another person, who holds the latter's hand.

The second was performed at the plantation in the person of Mr. ADRIAAN SPOOR, counselor and secretary there (who the author calls his dear patron). This gentleman, desirous of experiencing the power of this fish and being diffident to touch it himself, took the author (as he had done several times) to a certain Indian boat constructed from a hollow tree more than 26 feet long and 2.5 feet wide, put water in it until it was almost full, and then a Conger-eel. While the fish was in the front of the enclosure, the gentleman [Spoor] put his hand in the water at the back, roughly 20 feet from the fish, thinking that the blow would not reach him there; then the author touched the fish, and the gentleman said that he felt a very considerable shock.

If a hand is placed three or four inches above the water surface where, at that moment the fish forms a bubble while breathing, it will feel a considerable shock, which most certainly will be caused by the exhaled air.

And, the author tries to prove that this fish, with respect to its effects, is similar to Electricity by the following experiments:

I held by their wings (he said) some chickens suffering from cramps of the tendons in their legs, leading to contraction of the paws in such a way that they could not walk and would die. I put the legs of one on the back of a Conger-eel that was lying in a tub of water. The chicken was so shocked that it started screeching in a most dreadful way. After this was repeated a second time, the chicken seemed to be fully recovered. And continuing [the same experiment] with the others, no more chickens died.

An Indian (he continues) suffered from a Paralysis of the lower body. After he had taken several ineffective external and internal medications, I pressed a Conger-eel that had just been caught in the River, and hence was full of power, against his knees, as witnessed by several of my friends. Its shock was so powerful that two persons, who supported the patient on both sides by holding his arms, almost fell down with him. After I repeated this three times, the person, who had been carried over from a plantation, went back to his plantation completely recovered, without a stick or crutches, and without ever using anything else for it.

Mr. ABRAHAM VAN DOORN, ex-counselor of the same colony, had a Negro boy among his slaves with an estimated age 8 or 9 years. He suffered from obstruction of the nerves in such a way that his arms and legs were crooked. Each day, this gentleman threw the boy in a tub of water with a large Conger-eel of the black variety, which shocked the boy so powerfully that he crept out on all fours. And sometimes, when he was unable to do this, he had to be helped out, although the helper then received some of the shock. The result was that the boy recovered completely from his nervous disorder, but his accompanying malformation of the shinbones remained.

Furthermore, the said gentleman (the author says) carried out the following two experiments:

He likewise threw a slave boy, who suffered from a bad Fever into a tub with a Conger-eel. The boy was shocked in such a way that the gentleman was compelled to help him out of the water. But a few minutes afterwards, the Fever disappeared and did not recur.

The second was on an Indian boy, who had suffered badly from several episodes of a Fever, and had himself shocked with a Conger-eel. The result was the same as in the previous case.

The gentleman also says (the author continues) that when a slave complains of a bad headache, he has them put one of their hands on their head and the other on the fish, and that they will be helped immediately, without exception. (Van der Lott, 1762a; Modified from Koehler, Finger, & Piccolino, 2009)

Frans van der Lott was unequivocal when he referred to the eel's discharges as electrical shocks. He showed that they could pass through a variety of known conductors (e.g., metals, people) but not through non-conductors of electricity. He even found that five individuals holding hands would feel the shock when they completed a circuit with the fish. Further, although a dry stick might not convey a shock, a wet one could certainly do so. But he too never detected a spark, which if he did would have made his electrical conclusions even more noteworthy, even if some of his fantastic statements about fish therapeutics might have seemed unbelievable to his learned cousins back home.

FISH THERAPEUTICS

It is of more than passing interest that Van der Lott devoted so many paragraphs to the medicinal attributes of the shocks. Earlier in this book, we examined how the discharges of torpedoes had been used in the ancient world, starting with Scribonius Largus in the first century (Chapter 4). Although subsequent writers sometimes cited the ancient Romans on these procedures, little new seemed to be added to this medical literature after the fall of Rome, although in 1682 Hiob Ludolf wrote that the Ethiopians sometimes use the discharges of a fish to treat various forms of [malarian] fever, or at least the aches and pains of the vaguely described disease, without providing any information on whether this development had to do with the Europeans now in this part of Africa (Chapter 7). Indeed, prior to the letter by Laurens Storm van 's Gravesande to Jean Allamand, the subject of live fish therapeutics seemed to lie dormant from ancient times until now.

There is nothing in Allamand's 1756 publication, or in Gronov's or Van der Lott's articles, to indicate that the practice of fish therapeutics in the Guianas preceded the Europeans coming to those areas. Further, the earlier reports by missionaries, explorers, and visitors to South America do not mention the use of live eels in medicine. Rather, the historical record suggests that this was a recent, European-inspired development.

Just why fish therapeutics would arise again at this moment in time is an intriguing question. One possibility is that the new practitioners of the art knew what the ancients had written about some fishes with seemingly similar properties, and wanted to see if the eel would show similar attributes. This way of thinking would have fit nicely with the widespread theological notion that God has provided a local cure for every disorder, even in distant exotic lands. Nevertheless, given the secular nature of the letters coming out of the Guianas in the 1750s and 1760s, the physicists for whom they were intended, and the more earthly considerations guiding most scientists during the Age of Enlightenment, other possibilities must be considered.

This realization brings us back to the Leyden jar. As noted, there was a deep conviction that things that feel alike and produce similar effects probably have similar causes. Laurens Storm van's Gravesande had this thought in mind when he informed Allamand that the discharge of the eel "produces the same effect" as a properly charged Leyden jar. This type of subjective experiment was seen as good evidence for eel electricity, even if the fish did not "give off fire or sparks, similar to the apparatus for electricity," to use the words of the Dutch governor.

In the same way, the Leyden jar was now achieving fad status in European medicine (see Chapter 10). Although Benjamin Franklin and Abbé Nollet thought some of the claims for it were suspect, many practitioners and their patients were convinced that electrical therapy could cure palsies, seizures, and a variety of other disorders. If parallel effects could be obtained with eels, such findings would constitute further evidence for these fish being electrical. Van der Lott even tell us that he tried "to prove that this fish, with respect to its effects, is similar to Electricity by the following experiments," before turning to examples of its wondrous cures.

From this perspective, it is worth remembering that Allamand also did experiments on bedside therapeutic electricity. In an article than can be translated as "Cure of a Girl, Suffering from a Kind of Stroke," which appeared in 1754, he found that clinical trials with electricity usually did not produce great medical benefits. He was successful, however, in treating in a young girl with *affectum paralytico-spasmodicum*, who, after being terrified, had become paralyzed on one side, could not speak, and suffered from fits.

Thus, it would appear that fish therapeutics re-emerged in the mid-eighteenth century largely as another test for electricity. But who were the subjects in these clinical trials or experiments?

Occasionally experiments were performed on animals, as evidenced by Van der Lott's comments on how the eels were used to cure chickens of a disorder that had rendered them unable to use their feet for locomotion. With animals, expectations, gullibility, hysteria, and related factors would not enter the picture. Also, the operator could determine the relative safety of a procedure and perhaps fine-tune his methods.

Nevertheless, the real focus for eighteenth-century physicians was on suffering humanity—the greater demand being for clinical trials on sick and injured people. Hence, the Europeans in the New World did not hesitate to use slaves and local Indians as experimental subjects for new procedures (Delbourgo, 2006, pp. 186–187). Van der Lott repeatedly mentions the use of slaves and to a lesser extent Indians. Others subsequently visiting the Dutch colonies would also mention trials involving African slaves and the eels (e.g., Bancroft, 1769a; Fermin, 1769; Schilling, 1772; Flagg, 1786).

Keeping their slaves healthy was economically important, but the Europeans in South America might have also been hoping that the eels could be used to treat themselves. This idea must have been on Laurens Storm van 's Gravesande's mind when he mentioned to Allamand that the eels seemed to provide a safe treatment for gout. This is because gout has always been an affliction closely associated with the well-to-do—people able to afford copious amounts of food and cups filled with wine—not the poor. Well known to ancient Roman aristocrats, gout became virtually endemic among royalty and most successful men during the eighteenth century (Appelboom & Bennett, 1986; Copeman, 1964; Porter, 1994; Rodnan & Benedek, 1963; Schnitker, 1936). Even Franklin, who wrote about the causal roles of too much rich food and fortified wines, was tormented by it (Finger, 2006a, pp. 276–293; Finger & Hagemann, 2008). Simply put, slaves and Indians, with their more modest diets and very different lifestyles, would not have been the subjects for the experiments involving this disorder. But at the same time, the trials on the Europeans would not have occurred if related observations that showed the treatment is safe and had a chance of working had not first been obtained with slaves and Indians.

PIETER VAN MUSSCHENBROEK'S ENDORSEMENT

Pieter van Musschenbroek (Fig. 13.12), who had succeeded Willem Jacob Storm van's Gravesande (Laurens Storm's uncle) as Professor of Mathematics in Leiden (since 1739), helped to promote the case for fish electricity. In 1762, be brought up the information transmitted to Holland on the South American eels in his newest physics book, *Introductio ad philosophiam naturalem*. He cited the work of Allamand but focused more on Gronov's report, including his detailed list of the features of shocks (Van der Lott's paper was not published when the book went to press, although he might have known about it, since it was dated June 1761).

The Dutch physicist interpreted the observations and experiments conducted by the Dutch in Guiana as good evidence for eel electricity. He even asked why the same might not hold for torpedoes. He referred only briefly to Réaumur's (1717) mechanical theory, commenting almost apologetically that "as a matter of fact electricity was at that time unknown" (Van Musschenbroek, 1762, p. 291). Réaumur had died in 1757, and Van Musschenbroek was obviously attempting to justify his break with his now-departed good friend's mechanical theory in the kindest of ways.

Pieter van Musschenbroek was an acknowledged expert on physics and instrumentation, and even Franklin corresponded with and visited him, seeking his advice on matters electrical (Finger, 2006a, p. 133). As for his *Introductio ad philosophiam naturalem*, this book underwent both French and English editions and was widely read by physicists throughout the Western world. Hence, Van Musschenbroek's strong and early endorsement of fish electricity was highly significant in the unfolding drama.

Nevertheless, not everyone agreed with his assessment. The records show, for example, that resistance surfaced when he sent a letter on fish electricity to the *Académie Royale de Sciences* through his physicist-correspondent in Paris, Abbé Jean-Antoine Nollet. The reactions aroused by Van Musschenbroek's letter can be gleaned from the *Histoire de l'Académie* for the year 1760 (*Académie des Sciences*, 1761, p. 22). Monsieur Grand-Jean de Fouchy, *Secrétaire Perpétuel* and editor of the *Histoire*, argued that the sealing wax used to intercept the fish shock might not have been a proper insulating body, because it was humid. Further, he questioned the experiments using metals, because metals could also transmit mechanic vibrations (*ebranlements*) or movements communicated by the fish. Fouchy even maintained that "since there are two thousands leagues from here [i.e., Paris] to Surinam, the facts can be strongly altered during the journey"! Grand-Jean de Fouchy thus chose to remain loyal to Réaumur, whose ghost had not yet departed from the *Académie*.

FURTHER SPREADING THE IDEA

Allamand's and Van der Lott's published letters, the writings of the collectors Gronov and Gaubius, and the physics book by Van Musschenbroek combined in a powerful way to suggest that the shocks of the South American eels are electrical and, most likely, so are those of other torporific fishes. This conclusion was now disseminated in the journals and other publications of the epoch, along with translations and

Figure 13.12. Pieter (Petrus) van Musschenbroek (1692–1761), Dutch inventor of the Leyden jar and one of the leading physicists of his day. His acceptance of fish electricity was extremely important in drawing attention to this possibility. A page from his 1762 treatise of natural philosophy dealing with the electrical nature of these fish is shown on the right.

abstracts of some of the original reports, causing physicists, naturalists, and physicians to think about these fishes in a new way.

The diffusion of the Dutch observations was particularly intense in Germany, where Allamand's 1756 publication was translated and republished multiple times. It appeared in the *Nützliche Samlungen* in 1756 and in the *Neues Hamburgisches Magazin* in 1768 (Allamand, 1756b, 1768). The medical journal *Abhandlungen aus der Naturgeschichte, Praktischen Arzneykunst und Chirurgie* presented it, as well as Van der Lott's paper, in 1775 (Allamand, 1775; Van der Lott, 1775). *Der Artz*, another German medical journal, also covered it, along with an ample discussion of Ingram's paper (St. B. S., 1760). There was also a stream of references to the Dutch eel publications in German dictionaries of science, textbooks of physiology and natural science, and other publications (e.g., Martini, 1774, pp. 1–48; Garn, 1778; Treviranus, 1818, pp. 141, 184; Jéhan, 1852, p. 815).

Van der Lott's (1762a) letter appeared in French in 1762, in the *Bibliothèque des Sciences et des Beaux Arts* (Van der Lott, 1762b). This publication mentioned the therapeutic virtues of the eels. The presenting editor even happily remarked: "We hope that these facts will be more and more ascertained, and, for the welfare of humankind, doctors and, pharmacist excepted, this beneficial eel would come to live in our seas" (Van der Lott, 1762b, pp. 387–388).

The books and articles written by Dutch collectors also spread well beyond the boundaries of the Netherlands. Gronov's monumental *Museum piscium* and *Zoophylacium* and Seba's *Thesaurus* were shelved in many important libraries, where natural philosophers consulted them with particular interest. Further, Gronov's article was translated into Latin, a language more widely known in the "Republic of the Letters" than Dutch or German, for publication in a Swiss scientific journal in 1760. What Gronov wrote about the eel was also cited in other treatises and compilations on fishes.

Linnaeus, the father of taxonomy, was one of the scientists who avidly read Gronov's writings (for more on Linnaeus, see Chapter 18 and Hagberg, 1952; Uggla, 1957; Larson, 1971; Frangsmyr, 1983; Egerton, 2007). He cited what the Dutchman had to say about the eel in his highly influential *Systema naturae*, beginning in 1766 (Linnaeus, 1766, Tome I, p. 427–428). The great Swedish naturalist also cited Allamand's article in this work. But most significantly, and indicative of how perceptions of this fish were now changing, he felt there was sufficient information from reliable sources to designate it *Gymnotus electricus* at this time.

The well-traveled Linnaeus was friendly with Gronov and Gaubius, knew Seba, and had spent considerable time in Leiden and Amsterdam. He held a Dutch medical degree and had even published the first edition of his *Systema naturae* (albeit only 10 pages long) while he was living in the Netherlands (Linnaeus, 1735).[3]

A letter written in French by Jean Allamand in Leiden to Joseph Banks, a Fellow of the Royal Society of London, at the end of 1774 shows that Dutch scientists also exchanged personal communications with others of like mind, communicating the latest findings and offering to help esteemed

---

[3] In 1738, Herman Boerhaave would offer Linnaeus a 2-year appointment as a Dutch government physician in Surinam. This would have allowed the young Swede, who had been in the Netherlands to study the tropical flora and fauna, to become familiar with the dreaded *Gymnotus*. Linnaeus declined the offer, preferring to return to Sweden where his future wife waited for him.

foreigners in worthy new research endeavors (Allamand, 1774). In this letter, Allamand wrote that he was familiar with these animals and would do his best to help Joseph Bank's envoy obtain passage to Surinam so he could study the powerful electric eels for himself and contribute to the advancement of science. He specifically cited the need to do good dissections on their electric organs, like those Hunter performed on torpedoes (see Chapter 16). Allamand even promised to recommend Bank's man to his Dutch friends in South America.

At the time, Allamand was still very pessimistic about live specimens surviving the voyage to Europe from Surinam, and he expressed his doubts. He was also more than a little confused about the eel's domain. He thought it might be easier to get them alive in Senegal, and he cited Adanson, whom he thought had claimed that the eel (rather than a catfish) could be found in the rivers of this part of Africa (see Chapter 12).

CONVERSIONS

Albrecht von Haller, long a professor in Gottingen, was probably the most influential opponent of the notion of animal electricity in the middle of the eighteenth century. He reasoned that precise and localized movements would be impossible if a body were electrical, since electricity could be not contained and would therefore spread across the moist body parts (Piccolino, 2003a; Piccolino & Bresadola, 2003). As he initially viewed it, even specialized fish electricity is a bad idea (Haller, 1762, pp. 484–486).

Nevertheless, even Haller seemed to have a change of heart just a few years later. In the last tome of his *Elementa physiologiae*, published in 1766, he quoted the writings of Allamand and Gronov, and concluded:

> I admit, after the recent experiments made on the numbing eel [*anguilla stuporifera*], that without doubt it rather seems that an electric vapour comes out from the animal. Indeed the fish also permeates the water with her poison, so that she numbs from distance and kills the fish contained in the same barrel together with her. (trans. from Haller, 1766, Tome VIII, p. 176)

The writings of Charles Bonnet (Fig. 13.13), the well-known Swiss naturalist and philosopher, also reveal how the landscape was changing. Bonnet was the author of *Contemplation de la Nature*, one of the bestsellers of mid-eighteenth-century science. This work, dedicated to Haller and Allamand, first appeared in 1764, after which it went though numerous French editions and was translated into many other languages.

In the first edition of this book, Bonnet discussed the possible electrical nature of various physiological processes. He relates electricity to elemental fire, one of the four elements of Greco-Roman antiquity, and shows an awareness of its growing use in medicine, writing in his Preface: "Could one suppose that a piece of Amber, which attracts a straw, might have led to the healing of a paralytic?" (Bonnet, 1764, xi).

In the 1769 edition, we find a footnote added:

> Raja, Torpedo. Linn. Mr De Reaumur has given in the memoirs of the Academy of Sci. (Year 1714) an interesting

Figure 13.13. Charles Bonnet (1720–1793), the great Swiss naturalist and philosopher, whose *Contemplation de la Nature* underwent many editions, providing progressively updated comments on the nature of the torporific fish shocks.

> description of this fish. The numbing effect that it causes seemed to be the effects of a very lively percussion, operated by a curious apparatus of muscles placed under the back of the animal.
>
> A fish of another genus present in the river of Surinam, and which has been described by Gronovius in the Memoirs of the Society of Haarlem, deserves some word here: It produces the effects of violent electricity on those that touch it; that is, it gives them a commotion similar to the Leyden experiment. This effect occurs either if one touches it directly or if the contact is established through a rod of wood or metal; sometimes also by simply dipping the hand in the water in which the fish swims. No similar effect is felt if the fish is touched with a rod of wax. (Bonnet, 1769, Tome II, p. 241)

Thus, Bonnet was opening up to the idea of eel electricity (also see Bonnet, 1779–83). Moreover, even though he brought up Réaumur, he concluded his footnote with a question, one suggesting that he believed that torpedoes are probably also electrical. In his words: "Could one not also suppose a principle of Electricity in the torpedo?" This was a question put firmly on the table by the Dutch, and as shall be seen in the next few chapters, it would be discussed repeatedly around the world, including by a group of experimental natural philosophers at the famed Royal Society of London.

EPILOGUE

Circumstances peculiar to the Netherlands allowed the Dutch to bring the concept of animal electricity to life in the mid-eighteenth century. The invention of the Leyden jar, the tremendous interest in electricity shown by Dutch physicists, their emphasis on experimental methods, and the fact that they possessed colonies with inquisitive men trained in medicine, natural history, and physics all seemed to converge at this moment in time.

Above and beyond what occurred in their colonies, three Dutch cities figured prominently in what transpired. One was Leiden, home to one of the most important universities of the epoch and a center for science and medicine. Allamand, Gaubius, and van Musschenbroek were professors at the university, and Gronov was a part of the town's elite scientific community. All were also members of important foreign scientific societies, including the Royal Society of London and the French *Académie des Sciences*.

Haarlem was a second city repeatedly mentioned. The *Hollandsche Maatschappye der Weetenschappen* (Dutch Society of Sciences and Humanities) was founded here in 1752 (for more on Dutch scientific societies, see Buursma, 1978; Visser, 1975 Van Berkel, 1985). Its initial membership list listed Gaubius and Van Musschenbroek, with Allamand joining 1 year and Gronov 9 years later (Anon., 1902; Bierens de Haan, 1952). The young organization's foreign associates included Albrecht von Haller and Charles Bonnet, as well as John Pringle of the Royal Society of London. Storm van 's Gravesande's and Frans Van der Lott's letters first appeared in this society''s journal.

No to be overlooked, Amsterdam also figured prominently into what happened. It was here that the *Uitgezogte Verhandelingen* began publication in 1757. In line with the rising attention of the *Grand Siècle* for scientific culture, this periodical promoted the circulation of scientific information at an international level, going beyond the narrower limits of academic culture. It was an ideal periodical for collectors, which is why Gronov chose it in 1758.

Several factors made Amsterdam an exceptionally attractive place for Dutch authors and foreigners, such as Carl Linnaeus, who, as already noted, published the first edition of his *Systema naturae* here, and Charles Bonnet, who published his *Contemplation de la Nature* in this city. One was that the city had an advanced typographic industry, which could make books very appealing. Of special interest to collectors, Amsterdam was also home to skilled illustrators capable of making the highest-quality images.

Amsterdam had also become the leading port for obtaining exotic specimens at the end of the seventeenth century. The Latin neologism *Surinamensia* reflects the somewhat overlooked role played by the Dutch, who gladly sent, and were particularly eager to receive, exotic plant and animal specimens from Surinam (as well as from other parts of the world) through Amsterdam. Albertus Seba, to cite but one person, took advantage of what Amsterdam could offer and assembled one of the most important collections of rare and exotic *naturalia* of the time in this port city. He asked individuals in distant ports to ship him specimens, occasionally bought items from ship captains and sailors, and was willing to trade medicines with travelers arriving from foreign lands.

Thus, although the seventeenth century was the Golden Age in the Netherlands, the Dutch Republic remained a very special place for science in the middle of the eighteenth century. For very good reasons, it was in the Netherlands and her colonies that a myriad of factors interwove and provided the fertile conditions that helped electric fishes really begin the transition from mechanical to electrical.

PART V
The Royal Society and the Coveted Spark

Chapter 14
Edward Bancroft's Guiana Eels and London Connections

> From these particulars it is apparent, the shock is produced by an emission of torporific, or electric particles.
>
> Edward Bancroft (1769a, p. 198)

In the previous chapter, we examined how the Dutch began to view the "eels" found in the rivers of their South American colonies as electrical. We also pointed out that the early reports of Allamand (Storm van 's Gravesande), Gronov, and Van der Lott were written in Dutch, not a language that many people understood, and published in rather obscure periodicals, at least outside of the Netherlands. Nevertheless, some of these articles were translated into German and French and noted in books written during the 1760s. As a result, the initial evidence from South America favoring the new theory of fish or torpedinal electricity soon began to spread and, as already mentioned, Linnaeus, who had been in close contact with the Dutch, christened the species *Gymnotus electricus* in 1766.

By no means, however, was everyone convinced that the eels must be electrical. Among other things, no one thus far had been able to draw a spark from an eel, or a torpedo or an electric catfish for that matter. For Franklin and the Franklinists, including Joseph Priestley, attraction and repulsion constituted only weak evidence for electricity (Nollet, in contrast, emphasized these electrical motions), in contrast to fiery sparks and glows, which carried far greater weight (Riskin, 1998). Thus, although these fishes had been unable to attract or repel light objects, such as pith balls, greater emphasis was placed on the fact that they did not emit lightning-like sparks, make crackling sounds like minute thunder, or even emit a phosphorus smell upon discharging.

In addition, the very idea of fish electricity still seemed to defy common sense, not to mention scientific dogma, given all that was known about electricity diffusing through moist bodies. Thus, although the Dutch had broken fertile ground, important scientific and philosophical issues still had to be addressed, and new findings had to be disseminated to an even wider audience.

The next significant steps along the path to fish electricity would involve subjects of the British Empire. The other commonality among these men is that all were, or would soon become, members of the Royal Society of London. Prior to 1776, the British Empire included her North American colonies. This meant that talented North Americans could be elected national Fellows of the Royal Society, along with their brethren in the British Isles, although they would not become "official" until they signed the register in London. As noted in the previous chapter, the Royal Society also granted foreign membership to some of the Dutch natural philosophers involved with understanding the eels.

This chapter centers on what Edward Bancroft, a New Englander who became close to Franklin and was elected a member of the Royal Society with his help, wrote about the South American eels. It extends what one of the authors has recently published about him (see Finger, 2009). During the 1760s, Bancroft had practiced medicine and surgery in Guiana, as had Ingram before him, where he too had the opportunity to study the eels feared by the natives. Franklin had no way of knowing at the time that Bancroft would serve as a double agent during the American War of Independence. But he was quick to appreciate the case made by Bancroft for animal electricity, and fully understood why scientists of the famed Royal Society should mobilize to pursue additional groundbreaking research in this field.

EDWARD BANCROFT

Biographical information about Edward Bancroft (Fig. 14.1), especially in his formative years, when he was studying and beginning to practice medicine, is fuzzy and in some places non-existent, in part because he was secretive about his personal life. What is certain is that he was born in 1744 into a respectable, well-established family in the town of Westfield, Massachusetts. His father died 2 years later and his mother remarried and settled in Hartford, Connecticut, with her children. David Bull, the boy's stepfather, operated the "Bunch of Grapes" tavern, which would be a meeting place for George Washington and other American revolutionaries. It was here that the strategy to trap the British forces at Yorktown, Virginia, was planned, effectively ending the War of Independence.

Bancroft went to school in Groton, Connecticut, where Silas Deane was one of his teachers. He and Deane, who would be one of the American Commissioners during the War of Independence, would be reacquainted in Europe,

Figure 14.1. A portrait of Edward Bancroft (1744–1821), physician, natural philosopher, and later double agent. A true adventurer in the style of the *Grand Siécle*, Bancroft studied electric eels in Guiana and, after moving to London, helped energize Franklin and Walsh to study torporific fishes, starting with torpedoes off the coast of France. (© The Royal Society)

with Franklin's help. Whether Bancroft would actual kill the man who was once his teacher with some poison he had brought back from Guiana is uncertain, although the allegation has been made (Boyd, 1959).

With an interest in the healing arts, he was apprenticed to Dr. Thomas Williams, a "practitioner of physick" in a small Connecticut town, when he was 16 (Anderson & Anderson, 1973). There was no medical school in North America at the time, and there would not be one prior to the establishment of a school in Philadelphia in 1765 (ultimately merged into what would become the University of Pennsylvania; Finger, 2006a). Hence, Bancroft, like many aspiring American medical students, turned to the apprenticeship system to learn his craft (Shryock, 1960; Bell, 1977).

Only a small minority of aspiring physicians in the British North American Colonies had the tenacity, money, and connections to venture across the Atlantic to be trained in the theory of medicine at Edinburgh or Leyden, which attracted some Americans with their leading medical schools. Most were forced to stay close to home, where they took a more hands-on or practical approach to learning medicine. Nevertheless, the apprenticeship system was unregulated in the colonies at this time; not all physicians required their students to have a liberal college education or proficiency in Latin, and there were no set rules for completion (Packard, 1931; Shryock, 1967). Thus, although the ideal was to have liberally educated students, who would work with established physician for about 6 years, observing, diagnosing, bandaging, mixing drugs, and performing simple operations, this was not always the case in the colonies.

Bancroft might have trained for only 3 years with Williams, who was committed to teaching and strengthening the profession, as evidenced by the fact that he tried to organize a medical school in New England (Anderson & Anderson, 1973). But during the summer of 1763, at age 19, the young apprentice abruptly left his master for the Caribbean island of Barbados, and soon afterward for Dutch Guiana.

Bancroft's departure was triggered by some emotional disagreements with Williams. Nevertheless, he felt a need to write to Williams soon after arriving in Guiana, and in his letter of December 11, 1763, he explained that he left because of "Insults receiv'd, a Haughty Disposition, and a Roving fancy." He repeatedly referred to Williams as a friend, and to what transpired as "the departure of friendly unhappy youth" (Bancroft, 1763).

Bancroft might have served as a surgeon's mate to cover his costs at sea. As for the allure of Guiana, wealthy businessmen were building plantations there, because of special tax exemptions granted by the Dutch. The latter, in turn, wanted their enclaves developed for economic, security, and social reasons. Hence, entrepreneurs were building sugar, coffee, and cotton plantations, slaves were being imported from western Africa to work the land, and fortune hunters from many countries were heading to the region, where there was a need for more physicians and surgeons. Hearing these things, and not finding work or the welcome he expected in Barbados, it must have seemed very sensible to Bancroft to sail further south to Guiana (from the Arawak Indian words *wai ana,* meaning "[land of] many waters"), which was then growing rapidly (Fig. 14.2).

After disembarking at the colony of Essequibo, located on the banks of one of Guiana's many waters, the River Demerara, Bancroft met some plantation owners and a physician from Edinburgh. They helped him secure employment as a physician overseeing two large plantations. He was well compensated and provided with a nurse and a servant, and his duties included tending to several hundred slaves.

The gentleman who employed him, he told Williams, treated him "as a companion. I sit at his table, share his diversions, and in effect lead a very agreeable life." Moreover, having become friends with another physician in the area, "I have a good library at command." And with this library and the diversions of the country, he also related to Williams, "my time passes very agreeably" (Bancroft, 1763).

Bancroft used his ample spare time to collect material for a 400-page book he would write about the region. Such a work, he knew, would be welcome in Europe, where there was a growing demand to learn more about distant places, and exotic animals, plants, and people. Knowledge about the region, including its opportunities and its dangers, would be particularly welcome by individuals pondering trade, visits, or settlement in this part of South America. During the Enlightenment, writing a book of this sort would also be a way of securing one's reputation in science and letters and advancing socially. It could also help him obtain more medical training in a leading center, such as London or Edinburgh, should he want to secure a medical degree in the future.

On the second page of his opus, which he titled *An Essay on the Natural History of Guiana in South America,* Bancroft

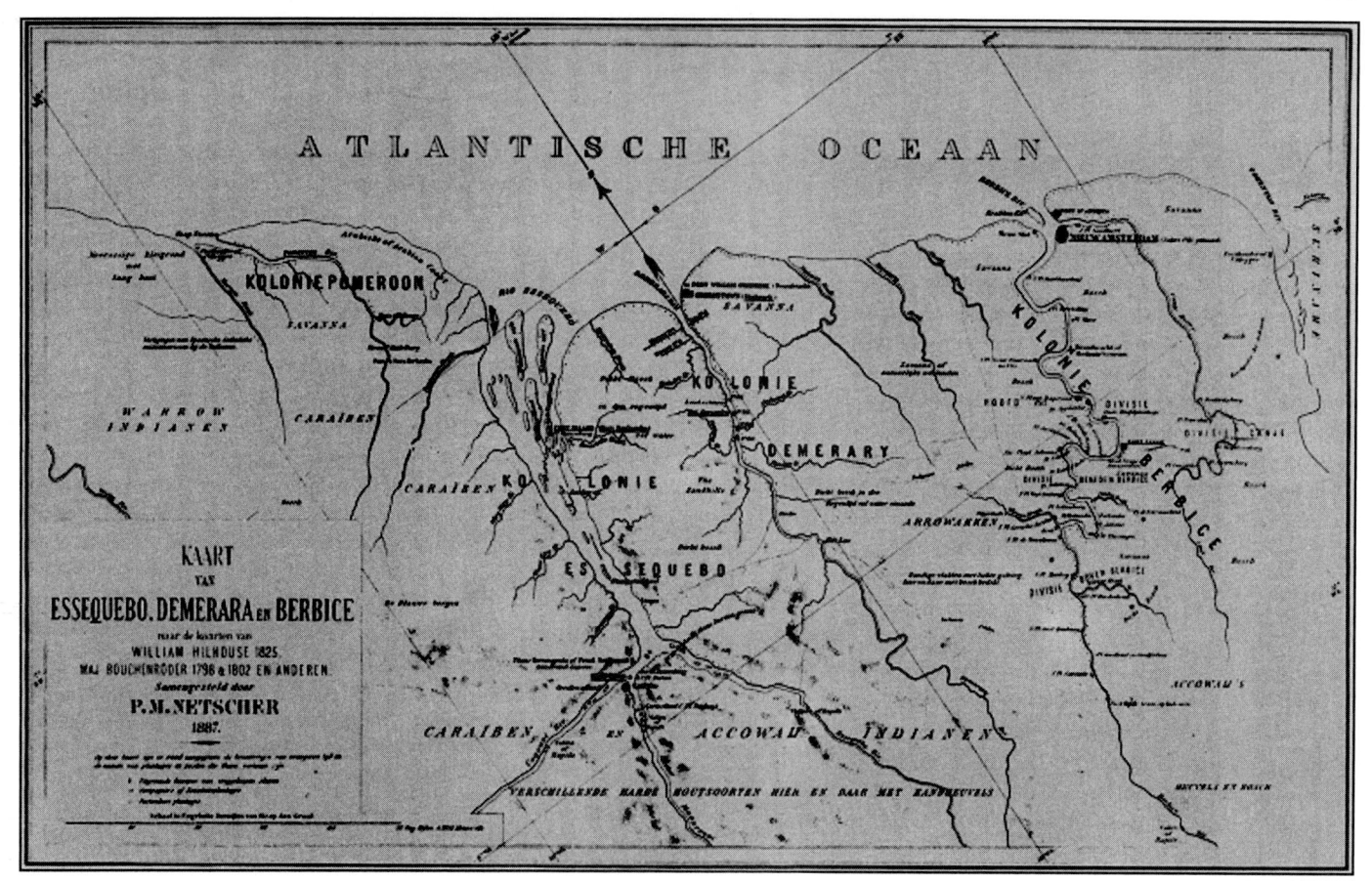

Figure 14.2. A map of the Dutch colonies of Guiana referring to episodes occurring in 1763, the year of Edward Bancroft's arrival. Notice the River Essequebo, which gives its name to the largest of the colonies of the region (from Netscher, 1888).

wrote that this was "a country, which, except its sea coast, and the lands adjacent to its rivers, remains hitherto unexplored by the subjects of any *European* State, and unknown to all, but its aboriginal Natives." This was intended to draw in readers, but it was an exaggeration. There had been earlier books on the region, such as the 1667 English text by George Warren (1667), which was translated into Dutch in 1669 and then heavily plagiarized by Adriaan van Berkel (Warren, 1667, 1669; Van Berkel, 1695) (Chapter 8). But earlier works would pale in comparison to Bancroft's better-written and far more detailed account.

THE ESSAY

Bancroft's lengthy *Essay*, "Containing A Description of many Curious Productions in the Animal and Vegetable Systems of that Country," as well as information about the natives and some medical observations, was printed in London in 1769 (Fig. 14.3). This was 6 years after he had arrived in Essequibo (settled by the Dutch in 1616) and 3 after he left the region, which also included the Dutch colonies of Berbice (founded 1627) and Demerara (founded 1752).

His dedication ("with Respect and Gratitude") was to one of the leading British physicians of the day, "William Pitcairn, M.D., Fellow of the Royal College of Physicians in London, and Physician of St. Bartholomew's Hospital." It is easy to understand why Bancroft dedicated his book to William Pitcairn. After leaving Guiana in 1766, he returned for a while to New England and then sailed off to London. He arrived during the spring of 1767, and within a year undertook additional medical training as a paying apprentice at St. Bartholomew's Hospital. This large institution, founded in 1145, had a new building and was then serving some 450 outpatients each week. Pitcairn had worked at St. Bart's since 1750, seeing patients, training young physicians, making policy, and promoting his medical theories, which were largely based on mechanics (particles in motion). Partly because he helped him with this additional training, but also because Pitcairn was so well known, he was an ideal choice for Bancroft's dedication.

Bancroft informs his readers (on the cover page) that his essay is a work "Containing A Description of many Curious Productions in the Animal and Vegetable Systems of that Country." This tantalizing bit of information is followed by: "Together with an Account of The Religion, Manners, and Customs of several Tribes of its *Indian* Inhabitants." And then, to cap it off: "Interpreted with a Variety of Literary and Medical Observations. In Several LETTERS from A Gentleman of the Medical Faculty, During his Residence in that Country."

Begun as four lengthy letters to his brother—to which "many things have been since added"—Bancroft does everything he promises on his cover page. He gives descriptions of the climate and geography of the region, "bounded on the north and east by the *Atlantic Ocean*, on the west by the great river *Oronoque*, on the south by the river of the *Amazons*, and on the south west by the river *Negro*" (p. 6). He then

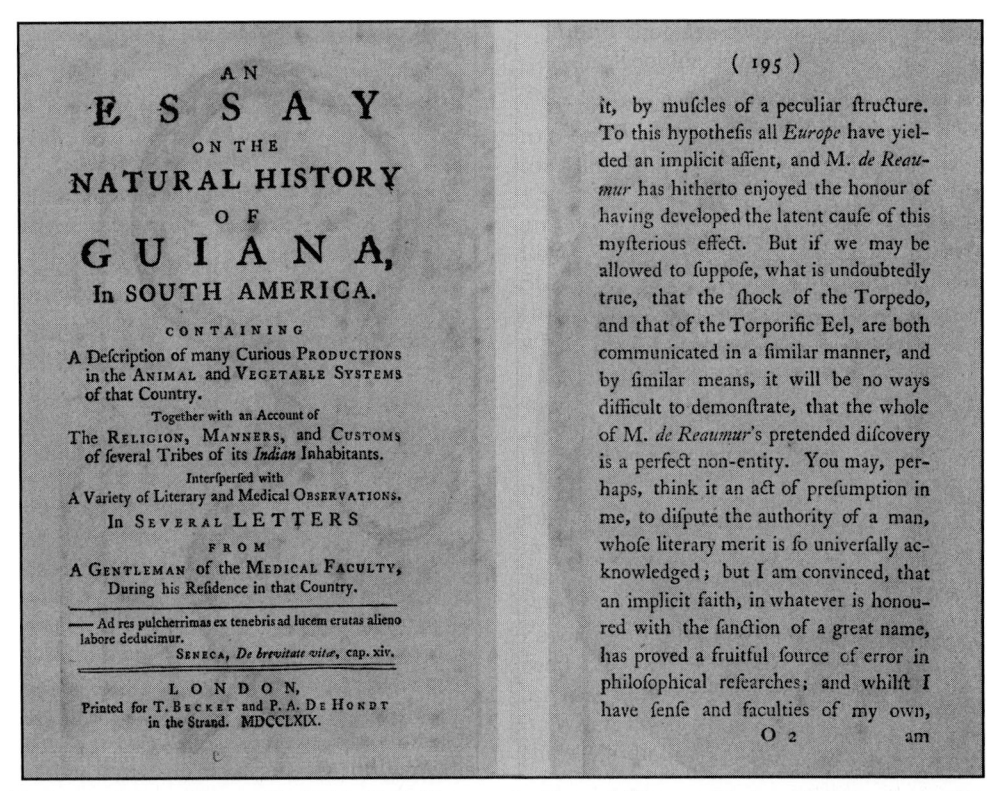

Figure 14.3. The title page of Edward Bancroft's (1744–1820) *Essay on the Natural History of Guiana* (1769) and one of the pages dealing with the "torporific eel."

describes hundreds of plants (Letter 1) and animals, including birds, fishes, and insects (Letter 2), as well as the diverse members of the "Human Species" then occupying the land, from the friendly natives who helped him to more remote tribes that were still practicing cannibalism (Letter 3). His final letter deals with the changes associated with the European settlers, from their plantations dependent upon the importation of African slaves, to the regional diseases and bites, and how they could be treated.

As put by one historian who had written about Bancroft, his eels, and electricity in Colonial America: "Like other colonial American natural histories by Sloane, Catesby, and William Bartram, Bancroft's teemed with natural dangers, from the armies of ticks or 'chiggers' that ravaged the feet of slaves, to the twenty-foot alligators whose skin was impervious to musket shot, to the fierce *Peri* fish who mutilated everything that went into the water, from ducks' feet to human genetalia" (Delbourgo, 2006, p. 175). Bancroft was clearly taken by the poisonous snakes, one of which bit a slave, who "fell to the ground and expired in less than five minutes from receiving the wound" (pp. 217–218). It is telling that Bancroft included an image of a rare two-headed snake on the frontispiece for his book (Fig. 14.4), although that snake was not from Guiana. It would be symbolic of his spying exploits after leaving South America (see below).

A careful reading of Bancroft's essay reveals many things that would not have been known in Europe and North America. With an eye for the unusual, he writes with praiseworthy clarity about how the Indians make and use poisons that quickly cause fatal hemorrhaging. Also, in the domain of diseases, he describes the "Yaws" as a non-fatal disorder marked by pussy eruptions over the whole body, which might be transmitted by the flies that seem attracted to the sores (see Bancroft, 1769a, pp. 385–387; also Gudger, 1910, 1911). Recent evidence suggests that insects can, in fact, transmit this disease, although it is usually contracted by direct skin contact. The *Treponema* bacteria associated with the yaws and syphilis are closely related, and it has recently been suggested that syphilis evolved from the yaws late in the fifteenth century, after Christopher Columbus and his crew brought the genetically older yaws back to Europe (Harper et al., 2008; Zimmer, 2008).

Also intriguing is what he wrote about a fish, "which deserves particular attention, and which I shall beg leave to call the Torporific Eel, till it is distinguished by a more proper name" (Bancroft, 1769a, p. 190). Bancroft was not, however, the first British North American to see these torporific eels or to write something about them. That honor goes to an American seaman.

JOHN GREENWOOD

John Greenwood encountered the eels a decade before Bancroft, and he then wrote about them, but in a less-than-public way. Born in Boston in 1727, Greenwood was destined to become an important American portrait painter and engraver. But prior to achieving lasting fame as an artist, he went to South America and his *Memorandum Book, Dec. 1752–Apr. 1758,* contains various notes and sketches of his experiences.

This diary includes an entry dated "Jan? 1757" about how a sailor was shocked by a large "numbing fish," with Greenwood using the generic term "torpedo" for this river fish:

Going up Para Creek with Capt. S. Pierce of Boston—it rained very hard—we were going on shore for shelter.

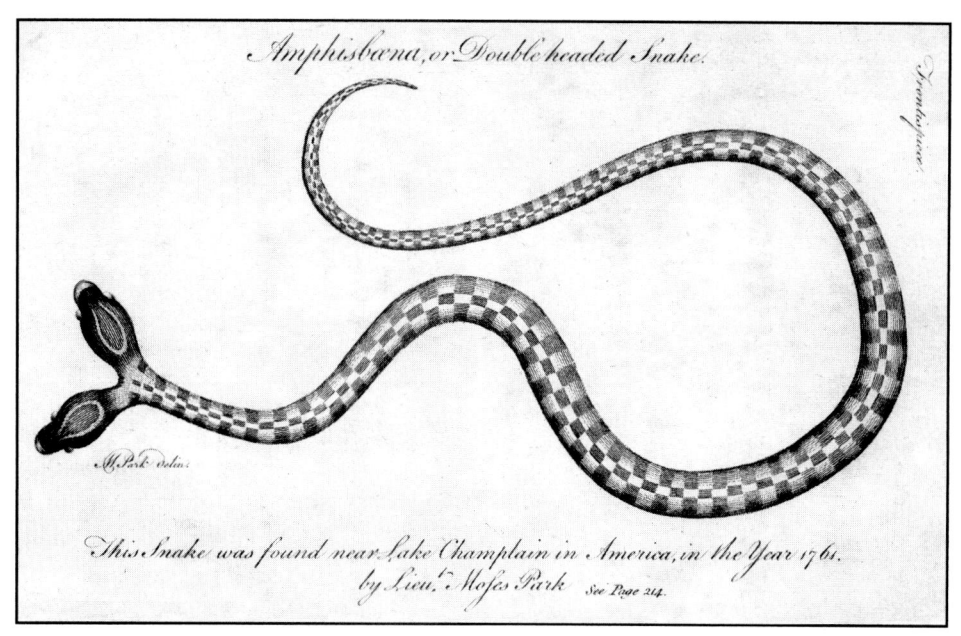

Figure 14.4. The frontispiece of Bancroft's (1769) *Essay*, showing a two-headed snake (albeit not from Guiana). Bancroft became a double agent during the American War of Independence, making this image symbolic of what was yet to come.

Saw a large numbing fish, which we thought was an alligator. The man in the bow of the boat struck it with a boat hook & was knocked down with the shock, when I perceived it was a Torpedo. The man recovered soon. The fish was killed; I measured it eleven feet Dutch, which is [scratched out word] feet less than the English; in the middle as thick as my thigh. (Greenwood, 1752-58, p. 168)

Greenwood used the word "shock" in his description, but nowhere does he brings up the term "electrical" or "electricity." In the terminology of the epoch, to receive a "shock" meant experiencing something that can cause violent shaking movements. It is only with time that the word "shock" would become more firmly and specifically associated with an electrical discharge. With Bancroft, the situation would be different. Electricity would be the central thesis of what he would write about the eels for a very public audience, which would be amazed at what he had to say about the discharges.

THE TORPORIFIC EELS

Bancroft provides numerous details about the eels, some new, in the 12 pages allocated to them in his second letter. He tells his readers that he owed a debt of gratitude to the Akawois Indians, who lived at the source of the Berbice, Demerara, and Essequibo Rivers. The Akawois, he explains, are experts on poisons, including one from the root of a plant that is particularly useful for drugging hard-to-catch fishes.

He writes that the roots come from what the Indians in Guiana call the "Hairree" plant, which he distinguishes from the poisonous Hearee Tree. He does not provide an illustration of this plant, but states that, when they are bruised and tossed into stagnant or sluggish water, they can "inebriate all the fish within a considerable distance, so that, within a few minutes, they float motionless on the surface of the water and are taken with ease" (Bancroft, 1769a, p. 107).

Virtually all of the fishes captured in the rivers of Guiana, he states, were obtained in this way. This almost certainly would have included the feared eels that nobody wanted to touch, if not dead or adequately stunned.

As noted in Chapter 1, in his narrative of his journey to the New World, which would take place more than three decades after Bancroft left Guiana, Alexander von Humboldt would write about the native use of roots to capture lake and river fishes. Humboldt remarked that this method of fishing is called *embarbascar* in what is now Venezuela. This term is derived from the barbasco plant (*Lonchocarpus urucu*), a leguminous species with roots that contain powerful fish and insect killers that are still used today. Another mention of fishing with barbasco roots is given by the Jesuit Joseph Gumilla (1745) in his treatise on the Orinoco region that was published in 1745 (Chapter 8).

Bancroft (1769a, pp. 191–192) now sets forth to describe the eels in more detail than his predecessors. He states that they are "usually about three feet in length, and twelve inches in circumference near the middle," with the head being "equal in size to the largest part of his body" and the mouth devoid of teeth. "On the back part of the head are two small fins, one on each side, which, like the ears of a horse, are either elevated or depressed, as the Fish is pleased or displeased." The round body, which has a long ventral fin, diminishes in size as it extends to the pointed, finless tail. As for the skin, it has a bluish lead colour and it lacks scales. Moreover, it is covered with a "slimey [sic] substance."

Bancroft further informs his readers that this eel "frequently respires"—it elevates its head above the surface every 4 or 5 minutes to gasp air. This gasping behavior had been noted in some Dutch accounts of the eels (Chapter 13) and it would later be described by others, including Alexander von Humboldt (Schilling, 1770; Bajon, 1774; Fahlberg, 1801; Humboldt & Bonpland, 1811). It is an adaptation that is common to freshwater fishes living in habitats where the oxygen content of the water is low, because of mud and rotting vegetation, or with oxygen levels that vary seasonally, due to periods of heavy flooding followed by prolonged dryness. These adaptable fishes can absorb the outside air through various specialized mechanisms, including the skin. In the case of the electric eel, which needs the air to survive in its oxygen-poor habitat, the air is "gulped" into the mouth and absorbed through specialized structures present in the buccal cavity (Farber & Rahan, 1970; Crampton, 1998).

He states that small fishes, earthworms, and roaches are among the foods readily sucked into the eel's mouth. But he admits uncertainty about other things, such as how they generate.

He also mentions that the same eels seem to have been observed by "Mons. de La Condamine," the French explorer who went down the Amazon in 1743–44 and referred to them as a species of lamprey (Chapter 9). La Condamine was still living when Bancroft published his book, and Bancroft even included the great explorer's description, albeit brief, in the original French.

FISH ELECTRICITY

In perhaps his most important sentence of all, Bancroft (1769a, p. 192) states:

> But the most curious property of the Torporific Eel is, that when it is touched either by the naked hand, or by a rod of iron, gold, silver, copper, &c. held in the hand, or by a stick of some particular kinds of heavy American wood, *it Communicates a shock perfectly resembling that of Electricity*, which is commonly so violent, that but few are willing to suffer it a second time. (Italics ours)

In a broader context, he postulates that saltwater torpedoes probably function in a similar way, meaning electrically:

> What affinity there may be between the shock of the Torporific Eel, and that of the Torpedo, I am unable to determine with certainty, having never felt the latter; but from all the particulars which I have been able to collect relative thereto, I think it is pretty evident, that both are communicated in the same manner, and by the same instruments. (1769a, p. 192)

Bancroft, as noted, had access to another physician's library while he was in Guiana, and he was also able to consult books and authorities while he was finishing his *Essay* in London. These resources could have allowed him to familiarize himself with the theory of Réaumur, the French naturalist who had published his important paper on torpedoes in 1717. This honored Frenchman's widely accepted theory, Bancroft writes, is that "the shock of the Torpedo was the effect of a stroke given with great quickness to the limb that touched it, by muscles of a particular structure" (pp. 194–195).

He continues: "To this hypothesis all *Europe* have yielded an implicit assent, and M. *de Reaumur* [sic] has hitherto enjoyed the honour of having developed the latent cause of this mysterious effect" [there is no accent on Réaumur in Bancroft's *Essay* or in its reviews, which are presented below]. "But," he continues, "if we may be allowed to suppose, what is undoubtedly true, that the shock of the Torpedo, and that of the Torporific Eel, are both communicated in a similar manner, and by similar means, it will be no ways difficult to demonstrate, that the whole of M. de Reaumur's pretended discovery is a perfect non-entity . . . Humanity is ever exposed to deception, and the charms of novelty may perhaps have precipitated M. de Reaumur into an error" (pp. 195–196).

Bancroft now argues that "the shock of the Torporific Eel is not the *immediate* effect of *muscular motion*," and therefore, neither is that of a torpedo (p. 196). Rather, "the shock is produced by an emission of torporific, or electric particles" in both cases (p. 198). And to support these strong statements, he lists seven observations and experimental findings, confirming and adding to what the Dutch had done before him:

1. The Torporific Eel, caught by a hook, violently shocks the person holding the line.
2. The same Eel, touched with an iron rod, held in the hand of a person, whose other hand is joined to that of another, &c. communicates a violent shock to ten or a dozen persons thus joining hands, in a manner exactly similar to that of an electric machine.
3. A person holding his finger in the water, at the distance of eight or ten feet from the fish, receives a violent shock, at the instant the fish is touched by another person.
4. This Eel, when enraged, upon elevating its head just above the surface of the water, if the hand of a person is within five or six inches therefrom, frequently communicates an unexpected shock, without being touched.
5. No shock is perceived by holding the hand in the water, near the fish, when it is neither displeased nor touched.
6. This Eel is eat[en] by the *Indians* when dead.
7. The shock is more violent when the fish is highly enraged. (pp. 196–197)

In his mind, Bancroft has now provided very strong scientific evidence in favor of the theory that a living creature can produce electricity, or at least a close cousin of machine-made electricity. Indeed, if the fish really torpifies mechanically, how could it transmit its effects through a line (Point 1), a chain of hand-holding people (Point 2), the water (Point 3), or an air gap (Point 4)? And if it houses and releases some sort of poison, eating it should make people sick, and it simply does not do this (Point 6). In contrast, the violent discharge can easily be communicated via known conductors of electricity, including water and chains of humans holding hands. And as noted by his predecessors, the sensation is "similar to that of an electric machine" (Point 2).

Nevertheless, he is puzzling when he states that the eel's emissions can be conveyed through 5 or 6 inches (roughly 14 cm) of air (Point 4). The discharge of a large electric eel would be unlikely to exceed 700 volts, yet this amount of artificial electricity cannot jump an air gap of even 1 inch. The distance would, of course, be less for an eel in the water, because the water would shunt much of the emission. One is left thinking that what Bancroft reported might have had something to do with water droplets blown into the air, a wet hand moving swiftly up from the water, or someone's vivid imagination. But with no more information provided, this is only speculation.

Right after these seven observations, Bancroft gives four short explanatory statements, all relating to the remarkable ability of this fish to release an electrical charge:

> From these particulars it is apparent, that the shock is produced by an emission of torporific, or electrical particles.

That their emission is voluntary, depending on the will of the animal, who emits them for his defense, either when touched or enraged.

That the existence of these particles depends on that of the Eel, and terminates with its life. And

That they are equally emitted from every part of the body. (p. 198)

Bancroft's first three conclusions would withstand the daunting test of time. But Bancroft was in error if he really meant that the shocks "are equally emitted from every part of the body." They are generated from special organs that extend from behind the head to the tail. In the previous century, as shown, Redi (1671) had written that a sea torpedo's discharge derives from special organs, although these muscle-like structures are arranged top to bottom in columns, not in rows, in these rays. Lorenzini (1678), Réaumur (1717), and other dissectors agreed with Redi, and this discovery, which also has antecedents in some earlier reports (e.g., Oppian and Cardano, see Chapters 3 and 6), was all but universally accepted in the first half of the eighteenth century (Chapter 9).

Yet Bancroft, who had just hypothesized that the eel and torpedoes probably function in the same way, never mentioned specialized organs in his *Essay*. His omission may stem from any of a number of possibilities. One is that he might have regarded the specialized organs described by the Italians and their followers (e.g., Réaumur) as just muscular and mechanical in nature, rather than being electrical. Another possibility is that he might not have dissected an eel, even though he had been trained to use the knife, and was therefore not prepared to discuss these structures. A third possibility is that he might have felt that the idea of special organs was common knowledge. And lastly, he might only have been trying to relate that one did not have to touch specific parts of the eel to experience its shocks.

More importantly, Bancroft now took aim at Réaumur and his mechanical ideas, writing at the end of his brief summaries: "From whence it is self-evident, that either the mechanism and properties of the Torpedo and those of the Torporific Eel are widely different, or that *Mons. de Reaumur* has amused the world with an imaginary hypothesis: and from my own observations, as well as the information which I have been able to obtain on the subject, I am disposed to embrace the latter inference" (pp. 198–199).

THE EEL AND MEDICINE

Bancroft, with his training in bedside medicine, was also interested in the possible therapeutic virtues of the eel's discharges. He did not mention the earlier literature on the sea torpedo or the paper published by Allamand (1756a), which included a statement about how a living eel was used to treat painful gout. But he did bring up the work of "one Vander [sic] Lott, a Surgeon, then in *Essequibo*, but now in *Demerara*, and published in *Holland*" (p. 199). As noted in the last chapter, Frans Van der Lott (1762a) had conducted a series of experiments with the *Gymnotus*, which he published in a Dutch journal. Chronologically, it appeared just months before Bancroft arrived in Guiana.

Van der Lott had claimed that the eel could cure chickens of "the cramps in the tendons of their feet" that had prevented them from walking. And, more significantly, he had maintained that the eel had been used to treat a man unable to walk "without cane or crutches," and a young slave boy "who was afflicted by an obstruction in his nerves in such a way that his arm and legs were drawn crooked." In his published letter, the Essequibo surgeon had also described how slaves and natives were treated for fevers by throwing them "into a tub with a *Conger-aal*," and how the fish were applied to the head to treat severe headaches (Chapter 13).

Bancroft had a difficult time with these claims, and he did not keep his thoughts to himself. He brashly stated that Van der Lott "endows it with many medical properties, which no other person was ever able to discover, particularly for curing nervous fevers, head-achs [sic], &c" (p. 199). To Bancroft, Van der Lott's accounts were simply too fantastic, too magical, and "the marvelous is so abundant, that the Writer, whom I have the honour of knowing, acquired no increase of reputation there from in this Colony."

At this time, as we have already noted in Chapters 10 and 13, much was being made of medical electricity in Europe. Bancroft, who believed "the particles of the Torporific Eel probably produce similar effects to those of Electricity," would have viewed the eel experiments in Guiana as analogous to the clinical trials taking place with electrical generators and charged Leyden jars in Europe and North America—much as did Van der Lott. Similar effects, both men knew, would bolster the argument for similar forces as their causes.

Bancroft probably became even more skeptical about the European claims after he made his way to London, where he completed his book. If he did not know it already, he would have found that, although medical electricity was initially thought to be a panacea for all sorts of disorders, the initial enthusiasm for it was now being met with many negative findings and well-deserved criticism (Bertucci & Pancaldi, 2001). Importantly, Franklin, who was by this time the most admired electrical scientist of all, had found that machine-generated electricity did not cure palsies of long duration, deafness caused by smallpox, and several other disorders (Chapter 10; also see Franklin, 1758; Finger, 2006a, b, c).

Franklin was not alone in his skepticism about electricity being a general cure-all (he thought it worked with hysteria). Although they disagreed when it came to the physics of electricity, Abbé Nollet had also become less than enthusiastic about the cure. After what appeared to be some initial successes with paralytic cases (Nollet, 1746), the leading French experimental natural philosopher failed to support his earlier impressions at a soldier's hospital in Paris. Thus, Bancroft was hardly alone in doubting some of the claims being made with medical electricity, although in his case it involved electricity from the river eels of Guiana, not machines.

Bancroft's choice of words when dealing with these claims is particularly interesting. This is because his language seems to have been taken from Franklin (1758), and could even have reflected Franklin's direct influence on him after both men settled in London. "I have known the Eel frequently touched by paralytic patients," Bancroft wrote in his *Essay*, "though I cannot say with much apparent advantage" (1769a, p. 200). Switch "Leyden jar" for "Eel," and one has virtually the same wording that Franklin used in his letter to the Royal Society on using Leyden jars for curing the palsies. As put by the American electrician: "I never knew any Advantage from Electricity in Palsies that was permanent" (Franklin, 1758, p. 482).

Eel electricity, Bancroft believed, is not negated by negative clinical findings. Neither the living eel nor a charged Leyden jar is a panacea. If anything, the eel very much like the electricity released from a Leyden jar, as suggested by clinical trials showing that the effects of these two treatments have been exaggerated, and that they may not cure many disorders.

A MOST DIFFICULT CARGO

Bancroft wished he could transport some live eels to Europe, where they could be studied by the best experimental natural philosophers, and perhaps be used in clinical tests by leading physicians before reliable witnesses. Shipping live specimens would also allow for direct comparisons to be made between the eel and saltwater torpedoes, which he was sure also shocked, albeit in a less intense way.

That Bancroft did pack some things from Guiana for shipment abroad is certain. We know that he assembled a collection of over 50 different kinds of preserved snakes, largely collected by Indians and slaves, who received some rum for each specimen delivered. He also packed *Woorara*, better known as curare, which the Indians used extremely effectively as an arrow poison. In 1812, Benjamin Brodie would show that animals treated with this poison could survive, provided they are given artificial respiration. Bancroft provided the drug and assisted Brodie in these landmark experiments, which clarified the physiological and toxicological actions of curare (Brodie, 1812; Raghavendra, 2002).

The seemingly attractive idea of shipping living eels to distant localities posed unique problems (see Delbourgo, 2006). One was that the slimy material they excreted made it necessary to change the fresh water in their troughs every day or two. Further, the fish seemed extremely susceptible to bruising. In Bancroft's (1769a, p. 201) own words: "Several attempts have been made to convey these Fish to Europe; but the quantity of fresh water requisite to shift them as often as is necessary, together with the bruises which they must inevitably sustain from the motion of the ship, have hitherto rendered them unsuccessful."

Live eels would not make it to Philadelphia until 1773 (Chapter 17; Williamson, 1775a; Rittenhouse, 1805), to Charleston, South Carolina until 1774 (Chapter 18), and to London until 1776 (Chapter 19; Sonntag, 1999, pp. 345–346). Five years after Bancroft's book was published, one eel of five survived a trans-Atlantic crossing, only to die shortly after its arrival in Falmouth (England), before it could make it to London (Smith, 1821, p. 21).

It was decidedly easier to transport preserved animals for natural history collections. Nevertheless, even dead eels were considered precious collector's items in Europe at this time, and only a few collectors could boast of having one. This is why Laurens Theodor Gronov, the renowned collector of exotica, was not anxious to dissect his prize specimen, even though he wanted to learn more about the source of the eel's shocking powers (see Chapter 13).

In this regard, it is interesting to note that when the great Italian naturalist Lazzaro Spallanzani visited Geneva in the 1770s, he saw a collection of various specimens in the city's public library that had been preserved in alcohol and imported from Surinam. In addition to snakes, there were scorpions, toads, and other exotic curiosities, including one he called a *torpedine*—very likely an electric eel being referred to in this generic way (Spallanzani, 1779).

Somewhat surprisingly, given their allure to collectors and how much natural philosophers might pay for specimens, Bancroft does not mention in his *Essay* anything about taking any live or even dead eels with him when he left Guiana. Further, we could not find his name associated with the transportation of eels in other notes, letters, and publications that we examined.

THE ESSAY'S SUCCESS

Bancroft's *Essay*, more than any of the earlier scientific reports of the eels of South America, attracted broad attention and, as might be expected, it had its greatest impact in the English-speaking world. It appealed to businessmen and was also purchased by the inquisitive public, with its appetite to read about distant lands with exotic flora, fauna, and peoples. But most importantly for us, numerous natural philosophers, physicians, and physicists learned about the strange eel and its electrical virtues from it or the reviews that borrowed liberally from it.

Ten pages were devoted to the *Essay* in a 1769 issue of the *Monthly Review*. The anonymous author of this book review starts off by praising Bancroft as "the ingenious author of the work before us," and calls the *Essay* "a valuable addition to one of the most useful and entertaining branches of philosophy" (*Monthly Review*, 1769, p. 199). He then presents bits and pieces from each of the letters making up Bancroft's book. One of his three selections from "The Animal Kingdom" is "The Torporific Eel." The brief editorial introduction to the eel section brings up electricity and reads:

> The Torpedo of Reaumur; the Gymnotus of the river of Surinam; and the Torporific Eel of our Author; have been considered as possessed of very singular electrical powers. The reality of this particular species of animal electricity has been called in question: but Mr. Bancroft's account of the torpedo Eel seems to determine the point. (*Monthly Review*, 1769, p. 206)

Three things follow this introduction. The first is Bancroft's statement that both the eel and sea torpedoes deliver electrical shocks. The second is his list of reasons for concluding that the eel's discharges are electrical. And the third is his statement that it seems quite likely that "Mons. de Reaumur has amused the world with an imaginary hypothesis."

Bancroft's work was also covered in other popular London periodicals, one being the *Critical Review*. This magazine's anonymous reviewer begins with harsh words for the typical traveler-author, who is "unable to convey a true or satisfactory account of the places he visits." But Bancroft, he proclaims, is not one of these writers, being "well qualified by his medical profession and his residence in the country to correct the mistakes or errors of preceding travelers." He goes on to state that Bancroft "writes with a candour and distrust of himself, which engages at once our attention and confidence" (*Critical Review*, 1769, p. 53). He then devotes considerable space to what

Bancroft had written about the torporific eels, providing a word-for-word transcription of the seven characteristics of the eel's shocks that led Bancroft to conclude that it is electrical.

Bancroft's account of the eels was also reprinted in the section on "Natural History" in the *Annual Register* for the year 1769 (*Annual Register*, 1769). This important London magazine was founded and edited by Edmund Burke, and it too had a broad readership.

In addition, the *London Chronicle* and the *Whitehall Evening Post or London Intelligencer* covered what he had written about the electric eel, again by publishing paragraphs verbatim from the *Essay*. In both periodicals, one can find the list of seven reasons why the eel must be electrical, as well as other important points made by Bancroft as he dismantled Réaumur's mechanical theory (*London Chronicle*, 1769; *Whitehall Evening Post*, 1769).

But perhaps most importantly, it was given special treatment in *The Gentleman's Magazine*, which had covered some of Franklin's electrical discoveries and was still playing an important role in disseminating new scientific and medical findings around the world (*Gentleman's Magazine*, 1769; for more on this periodical, see Chapter 17). Here too the reviewer chose to concentrate on the case Bancroft had made for animal electricity:

> The fishes described in this book have also, most of them, been described before, but in the account given of the Torpedo eel, the author refutes the opinion of Reaumur, that the shock of the Torpedo is the effect of a stroke given with great quickness to the limb that touches it, by muscles of a peculiar structure; he takes for granted that the shock of the Torpedo, and the terperfic [sic] eel are produced in the same manner, and, with respect to the eel, he relates the following facts. (*Gentleman's Magazine*, 1769, p. 146)

Following these words, the reader is presented with four of the seven items on Bancroft's list of experiments and observations that favor an electrical force over Réaumur's theory of rapid mechanical actions.

With the help of these reviews, Bancroft and his eels became even better known, and the sales of his *Essay* rose, along with his royalties. His book was soon translated into German, and two Dutch editions were also published (Bancroft, 1769b, 1782, 1794a).

LONDON ACTIVITIES AFTER THE ESSAY

In 1766, three years before his *Essay* was published, Bancroft went to London where he studied medicine at St. Bartholomew's Hospital. Details of this important period in his life are again scanty. He then went back to Guiana on a business trip and visited New England before settling back in London in 1771. There he strove to enjoy life to its fullest as an enlightened physician, a creative scientist, and a cultured gentleman. Intent on living out his dream, he married, started a family, and went out of his way to associate with the most important scientific, medical, and political figures of the era.

Benjamin Franklin was one of the notables Bancroft began to spend time with, and he, in turn, introduced him to many of his own friends. They included physician John Pringle, who had welcomed Franklin to the Royal Society, and chemist Joseph Priestley, whose monumental *History and Present State of Electricity*, which had been written with Franklin's guidance, had just been published (Chapter 10; Priestley, 1767).

Franklin, as we have seen, loved the thrill of discovery and enjoyed discussing new findings, whether observational or experimental, especially those related to electricity. In addition, he was drawn to bright, creative, and energetic people, and to those who could write clearly and logically. In Bancroft, he found both a fellow colonist with worthy ideas and an exceptionally good writer. The two men born in Massachusetts began meeting more often and we know that they exchanged well over 50 letters.

At the same time, Bancroft began to make money in the textile industry, allowing him to live in a manner befitting a proper gentleman. Drawing on some of what he had learned in Guiana, he busied himself in London studying colors and trying to perfect dyes from plants, bark, and insects. He even introduced some new dyes with commercial appeal.

In 1773, well recognized for his book and chemistry, he was elected into the Royal Society as "A Gentleman versed in natural history and Chymistry, and the author of The Natural History of Guiana" (Fig. 14.5). Among his many sponsors were Nevil Maskelyne, Astronomer Royal, and his cousin John Walsh, who had begun his research on torpedoes with Franklin's guidance (Chapter 15). A third sponsor was William Watson, the highly respected English electrical scientist. And a fourth, as might be expected, was Franklin, who nominated his first candidate to the Royal Society in 1759, and would go on to bring at least 37 other people into the organization, most of whom were involved with electricity. (Franklin and Bancroft were also joint members of the Society of Arts or "Premium Society," an organization founded in 1754 to support the arts, meaning craftsmanship and manufacturing; see Allan, 2000.)

In 1774, Bancroft completed his long-sought M.D. degree, which he received from Marischal College (University of Aberdeen). Well-known physician and chemist George Fordyce, who had obtained a degree there before going to Edinburgh, wrote one of the letters supporting him. The second came from another Franklin friend and the founder of the new Medical Society of London, John Coakley Lettsom, a Quaker who had studied medicine in Britain before receiving his M.D. degree at Leiden. With records lost, whether Bancroft was in residence in Aberdeen and for how long are open questions. The belief is that he remained in bustling London, meeting the requirements for his degree *in absentia*, which was not rare in this era, when some Scottish universities had fallen on hard times and were in need of money (Comrie, 1932).

During the 1770s, Bancroft remained active in Lettsom's Medical Society of London, which included 30 physicians, 30 surgeons, and 30 apothecaries intent on helping the sick poor (Hunt, 1972; Hunting, 2004). He served as secretary of this organization while it met at Crane Court, where the Royal Society also held its meetings. At the same time, he continued to build his private medical practice, while residing on Downing Street, "a doctor's street." He prescribed for Londoners and Americans abroad, including Franklin and Silas Deane. Deane would later write that he had more faith in Bancroft, whom he viewed as a friend, than in almost any other London physician.

Urged on by Franklin, Bancroft wrote pieces for one of the London periodicals that had reviewed his book, the

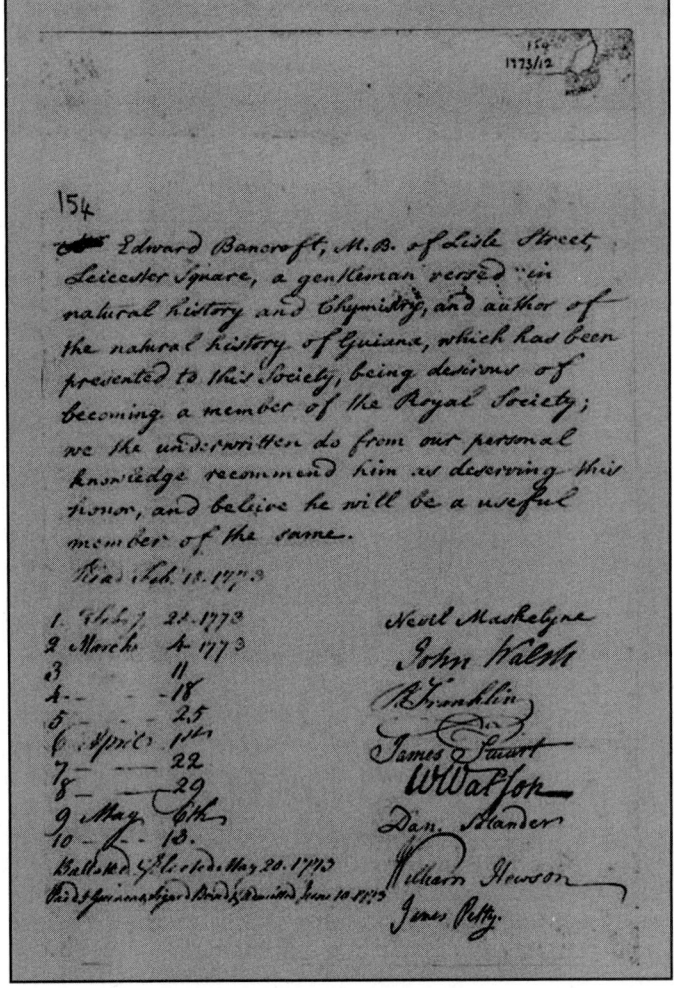

Figure 14.5. Bancroft's certificate of election into the Royal Society of London. Benjamin Franklin and John Walsh were among the Fellows endorsing his candidature (courtesy of the Royal Society of London).

Monthly Review. Some of his pieces, written between 1774 and 1777, were about the British North American Colonies and the rights of its settlers, as suggested and overseen by Franklin. Others had to do with new books and developments in science and medicine. Bancroft had been trying to establish himself as a man of letters prior to this time, and a year after his *Essay* on Guiana went to press he even published a novel (Bancroft, 1770).

THE AUTHORITY ON ELECTRIC FISHES

Bancroft was now a well-known authority on many things, including electric fishes. His eel experiments were mentioned in various books, one of which was Captain John Stedman's *Narrative of a Five Years Expedition against the Revolted Negroes of Surinam, in Guiana on the Wild Coast of South America, from the Year 1772 to 1777*, first printed in 1796 and afterwards published in various editions and multiple languages (Fig. 14.6).

Stedman wrote:

> The other animal I saw at the house of my friend Kennedy: this is what Dr. Bancroft calls the torporific, and others the electric eel, and which Dr. Firmyn [Philippe Fermin] supposes to possess the same qualities with the torpedo. . . . When this animal is touched by the hand, or any rod of metal or hard wood, it communicates a shock, the impulse of which produces the same effect as electricity. . . .
>
> For my own part, all that I can say concerning this animal is, that I saw it in a tub full of water, where it appeared to be about two feet long; that I threw off my coat, and having turned up my shirt-sleeves, tried about twenty different times to grasp it with my hand, but all without effect, receiving just as many electrical shocks, which I felt even to the top of my shoulder, to the great entertainment of Mr. Kennedy, to whom I lost a small wager on the occasion. The electrical eel swims forward or backward at pleasure: it may be eaten with the greatest safety, and is even by many people thought delicious. (Stedman, 1796, vol. 1, pp. 124–125)

Nevertheless, there is no hard evidence to suggest that Bancroft continued to experiment with eels or other electric fishes after he left Guiana. Still, he had to be interested in the experiments that were starting to be conducted by Walsh and others in Franklin's circle at the Royal Society of London. After all, Franklin and Walsh designed these experiments to determine whether sea torpedoes do, in fact, manufacture and release electricity, as Bancroft had postulated.

That some Fellows of the Royal Society now solicited Bancroft's opinions about electric fishes is evident from a note penned by John Walsh. Written in the spring of 1772 and addressed to Franklin at his residence on Craven Street in London, it reads: "Mr. Walsh presents his Compliments to Dr. Franklyn [*sic*] and requests the favour, if Mr. Bancroft is now in England, to know where he resides, as Mr. Walsh is very desirous of making some enquiries concerning his Torporifick Eel" (Willcox, 1975, p. 162). Walsh, about whom more will be written in the next chapter, was well acquainted with what Bancroft had written. In addition, the two men must have had some direct interactions. Both were close to Franklin, and Walsh had co-sponsored Bancroft for membership in the Royal Society, which is something he probably would not have done had they not met previously and been close to Franklin.

OF SPIES AND SPYING

It is now known that Bancroft was paid by the British Secret Service to spy on the Americans prior to, during, and even after the War of Independence (Bemis, 1924; Einstein, 1933; Boyd, 1959; MacNalty, 1944). The American commissioners in France, other than "bilious" Arthur Lee, who suspected almost everyone of spying, accepted Bancroft as one of their own, and Franklin had made him Secretary to the American Commission. He dined with Franklin, John Adams, and Thomas Jefferson, and conversed with naval hero John Paul Jones and another daring military figure, Gilbert du Motier, better known as the Marquis de Lafayette, when they were together in France.

Sometimes using invisible ink between the lines of fictitious love letters that would be deposited in a hollow tree in the garden terraces of the Tuileries, the man who posed as an American patriot wrote hundreds of communications

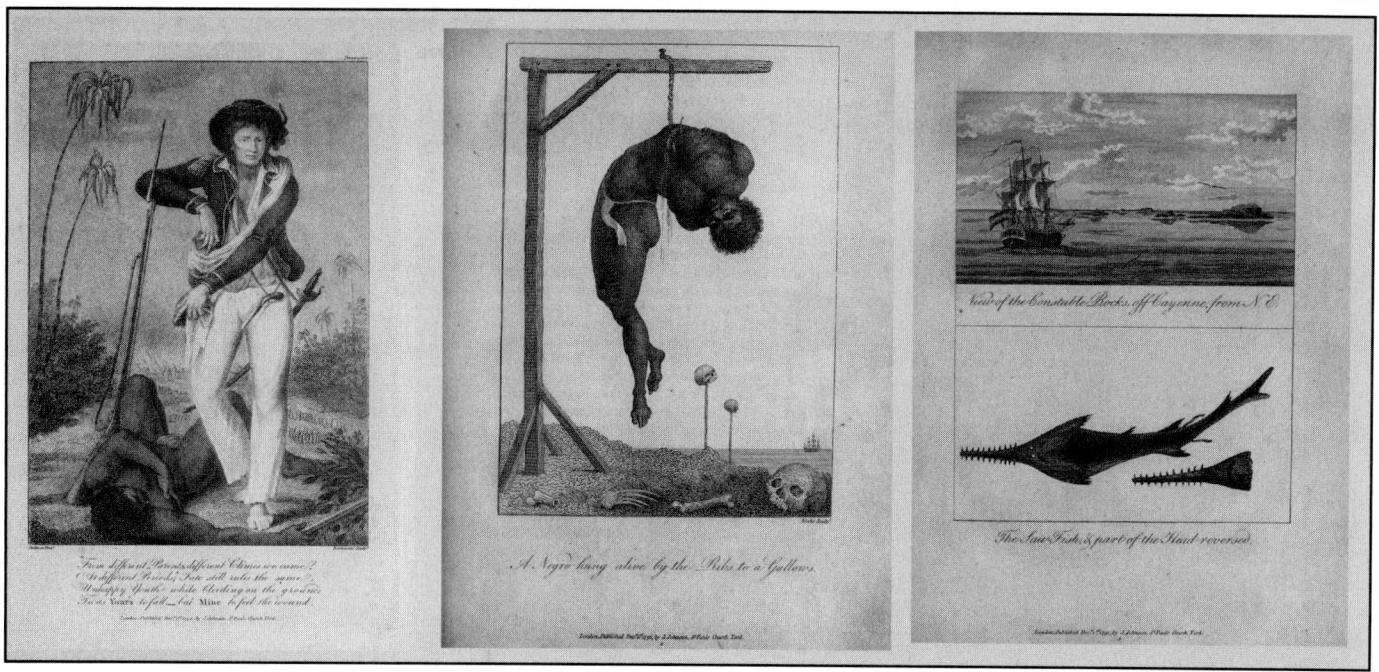

Figure 14.6. The frontispiece and two images from Stedman's 1796 narrative on Surinam, which contains a short report on his electric eel's experiments. The book, with its richness of naturalistic and social observations about the Dutch colony, was especially important for showing the cruelty of colonists towards the slaves and served to promote abolition. Although it contained images on the flora and fauna of the region, it did not have a picture of the electric eel.

about American and French conditions and intentions. These communications traveled so rapidly to England that a copy of the Franco-American alliance of 1778, which was opposed and feared by the British, was in the hands of King George III within 42 hours of its being signed.

George III, however, had a deep distrust of Bancroft. He considered him an opportunistic "stock-jobber" and most likely a double agent, whose only allegiance was to his pocketbook. More than anyone else, the king had it right. Bancroft did play the London stock exchange masterfully, withholding and releasing important political information to his advantage. Most notably, when he learned early on about British General John Burgoyne's disastrous defeat at the Battle of Saratoga in upper New York, he delayed conveying this information to London until after he had set himself up to make a large profit on its stock exchange.

The unknowing Americans also asked Bancroft to spy for them and to run dangerous missions for their new allies, the French. Yet Bancroft did not seem to provide the American commissioners with information of much value. In contrast to what he conveyed to the British, much of what the "double agent" gave to the patriots and their allies seemed to be common knowledge or gossip in London.

Bancroft was clearly "a master of deception and intrigue," possessing a "gambler's instincts and inclinations" (Anderson & Anderson, 1973, p. 356). Unlike most spies, he lived to a ripe old age (77), taking his secrets with him to the grave in 1820. Seventy years would pass before the truth would come out.

It is worth adding that Bancroft was not the first person mentioned in our book to have engaged in spying. Aphra Behn, the English author of *Oroonoko*, visited the Dutch colony of Surinam in the 1660s (Chapter 8). This was just before the English were about to go to war with the Dutch, who dominated trade from the region. King Charles II for whom she later spied, might have sent her there, reasoning that a female novelist would be able to gather valuable information and return home in perfect safety, although this is uncertain.

Scientists were more actively recruited to be spies in the eighteenth century. A major reason for this was that their work often involved travel, which provided them with a perfect alibi. Another is that the scientific brotherhood believed that the gathering and disseminating useful knowledge should know no borders. And a third reason was that they often had the competence to understand technologies that might have little meaning to others.

Jean Antoine Nollet, the electrical scientist introduced in Chapter 10, also did his share of spying. During the 1740s, he journeyed to Italy ostensibly to examine some claims about using electricity to transmit medicines in glass tubes to the human body—claims that seemed too good to be true. Although he reported that the Italian claims stemmed from poor science and vivid imaginations, there was also a clandestine purpose for his trip. The French government had asked him to collect information on silk manufacturing, which had become an important industry in Italy (see Bertucci, 2007).

There are many other instances of scientist-spies in Bancroft's era, but only one more will be mentioned, in part because he corresponded with Bancroft but also because he was so famous. Benjamin Thompson (Fig. 14.7), who was ennobled as Count Rumford in 1792, also came from Massachusetts, spied for the British, and sent at least some of his messages with invisible ink (Brown, 1999). Moving to England in 1776, and also spending considerable time in Bavaria, the Count became well known for his studies on heat production and his inventions, including Rumford fireplaces, Rumford roasters, and Rumford stoves. In 1799, he helped to create the Royal Institution, whose members would include Humphry Davy and Michael Faraday, both of whom would contribute to a better understanding of electricity, including fish electricity (Chapter 24).

This oversight is unfortunate, because Bancroft's *Essay* drew needed scientific attention to these "shocking" South American river fish, especially in English-speaking countries. Even though much of what he wrote merely confirmed earlier Dutch thoughts and findings, he provided more details about the eels and their electrical properties, and he wrote from personal experience in such a compelling and convincing way that he could not be ignored, especially since he was now active in London.

Importantly, Bancroft challenged his contemporaries to determine if the sea torpedo's discharge would also follow the laws of electricity. From what he had observed and done, he was certain that Réaumur's mechanical explanation for a torpedo's actions had to be wrong, and he did not hesitate to predict with confidence that future experiments would show that torpedoes function in precisely the same way as the eels.

Interestingly, and demanding clarification, there is a *Torpedo bancroftii* (*Narcine bancroftii*) in the zoological literature (Fig. 14.8). Although it might seem likely that this

Figure 14.7. Benjamin Thompson (Count Rumford) (1753–1814). Distinguished as an inventor and in many facets of science, as well as for his philanthropy, his help in forming the Royal Institution, and his other British and international affiliations, Thompson (like Bancroft, Aphra Behn, Nollet, and Franklin) was also a spy.

Rumford was so distinguished in so many fields and for his philanthropy that he became an ambassador and a major general, and married chemist Antoine Lavoisier's widow. He was elected a Fellow of the Royal Society of London, honored with the Copley Medal, and made a Foreign Member of the First Class of the French Institute, where Humboldt and Volta were among his distinguished colleagues. And throughout his incredible life, he continued to spy for the British. Today, he is listed as one of the three known Fellows of the Royal Society who spied in the eighteenth century (Jones, 1977). The others were Franklin and Bancroft, key players in our fish story.

POSTSCRIPT

Modern historians tend to remember Bancroft for his closeness to Franklin and the American delegation in France during the War of Independence, and for his spying for the British. A novel has even been written about his clandestine actives called *Spy: America's First Double Agent* (Mullin, 1987). His research, industry, and book on textile colorings (Bancroft, 1794b), which brought him considerable fame and fortune in his day, are virtually forgotten. And few people remember his *Essay* on Guiana, even though it contained so much valuable information about the region and especially its torporific eels.

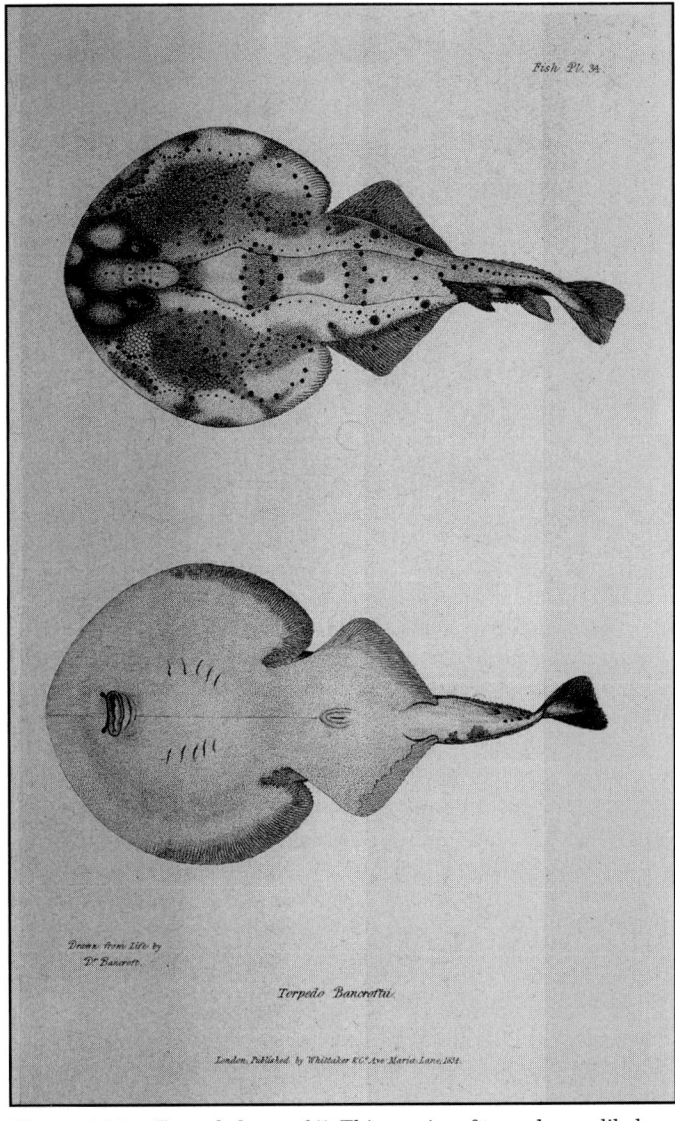

Figure 14.8. *Torpedo bancroftii*. This species of torpedo was likely named after Edward Nathaniel Bancroft, Edward Bancroft's son, also a physician who visited South America (he lived for a number of years in the Caribbean).

name had been given in order to acknowledge Edward Bancroft's important research on electric fishes, this is not the case. The species was most likely named after his surviving son, Edward Nathaniel Bancroft. The younger Bancroft was a physician who assisted the British forces in (among other places) the West Indies and nearby South America, and then practiced medicine in Jamaica, where he died in 1842. He had been affiliated with the College of Physicians in London, was an expert on yellow fever, and was a naturalist particularly interested in parasitology and botany (several plant species have been named after him; see Mabberley, 1981; Desmond, 1994).

The designation *Torpedo bancroftii* first appeared in the 1834 English edition (but not in the original French edition) of Georges Cuvier's monumental "Animal Kingdom," where "Drawn from life by D^r. Bancroft" appears under a dorsal and ventral illustration of the fish. The accompanying text is brief: "We are enabled, through the kindness of Dr. Bancroft, to insert a figure of a torpedo from his drawing, which is new" (Cuvier, 1834, p. 649). Although no more information is given, the figure is that of a torpedo living in the warm, shallow coastal waters off South America, and particularly off Brazil, which has led to its more modern name, *Narcine brasiliensis*.

In the next chapter, we shall look in considerably more detail at how the senior Edward Bancroft's thoughts and experiments on these eels captured Franklin's attention and led to a flurry of activity at the Royal Society. We shall concentrate on John Walsh, who with Franklin's help would first make the idea of torpedo electricity seem even more likely, winning over many scientists and philosophers who still had their doubts.

Chapter 15
John Walsh's Scientific Journey

> ... the effect of the Torpedo appears to be absolutely Electrical ...
> (John Walsh to Benjamin Franklin, July 12, 1772;
> in Willcox, 1975, pp. 204–205)

JOHN WALSH

Colonel John Walsh, who more than anyone else would make electric fishes electrical, was born in Fort St. George, India, in 1726 (he would die in England in 1795; Piccolino, 2003a; Piccolino & Bresadola, 2002) (Fig. 15.1). He was related to Major-General Robert Clive and served as his private secretary in Bengal from 1757 to 1759. He was also a cousin of Neville Maskelyne, the Royal Astronomer, who along with Franklin and four other men sponsored Edward Bancroft for membership in the Royal Society.

Walsh had amassed a fortune, mostly as a result of his share in battle conquests, before he left India for England. As gentleman of means, he began to serve in the House of Commons in 1761, and 5 years later purchased a magnificent country estate, Warfield Park in Berkshire, where he was greeted by his lady-friends when he wanted to retreat from the pressures and partying in London. Although Walsh would continue to hold a chair in Parliament until 1780, his interests went beyond politics, women, and managing his estate, and included natural philosophy. He was elected into the Royal Society in 1770, being "a Gentleman well acquainted with philosophical & polite literature, & particularly versed in the natural history and antiquities of India."

FRANKLIN AND WALSH

After his election into the Royal Society, Walsh interacted with the organization's esteemed natural philosophers and electrical scientists. Of these people, Franklin was singularly important. Not only did Franklin introduce Walsh to other people of merit in the organization, he played the leading role in encouraging Walsh to devote his scientific energies to sea torpedoes, and specifically to test the hypothesis that their discharges are electrical.

Exactly how much Walsh knew about torpedoes before 1772 is difficult to discern. The same can be said for the eel of Surinam, although it is clear that Walsh was familiar with Bancroft's observations, published in 1769, and wanted to meet with him early in 1772. Whether he had previously talked with Bancroft at the Royal Society or elsewhere is uncertain, but we do know that Walsh told Franklin if "Mr. Bancroft is now in England," he would be "very desirous of making some enquiries concerning his Torporific Eel" (Willcox, 1975, p. 162).

By this time in 1772, perhaps at Franklin's urging, Walsh had made up his mind to follow up Bancroft's eel research with experiments on torpedoes, known to be abundant off the coast of France. Bancroft, as noted, had provided good evidence to support the idea that the shock of the eel is electrical, and had postulated that the torpedo's discharge would be found to be no different (except in quality) from that of the eel; that is, that both creatures follow the same basic physiological principles.

The records show that Walsh obtained a copy of Stefano Lorenzini's work on sea torpedoes prior to his trip to France in pursuit of torpedoes. Originally published in 1678 as *Obssevazioni Intorno alle Torpedini . . .*, this work had by this time been published in English (1705) as *The Curious and Accurate Observations of Mr. Stephen Lorenzini . . . on the Dissections of the Cramp-Fish* (see Chapter 9). Most likely, Walsh owned the Italian version, however, since he cited it in one of his papers (Walsh, 1774b, p. 465).

In a note written before he crossed the English Channel, we discover that Walsh had lent this book to Franklin but was eager to retrieve it so as it to check on what Lorenzini had written:

> Mr. Walsh presents his Compliments to Dr. Franklin and shall be glad, if Dr. Franklin has no further occasion for Laurenzini [sic] to have it return'd, as he wishes to look into it for some particulers [sic]. Mr. Walsh hopes for the favour of seeing Dr. Franklin at Coffee on Friday. (Willcox, 1975, pp. 162–163)

Coffeehouses were *en vogue* at this time, and London boasted over 500 such establishments. In general, these establishments were quieter and more genteel than the taverns, particularly the pubs frequented by the local gentry. Hence the men of the Royal Society tended to congregate and socialize at them, while some of London's most sought-after physicians (e.g., Richard Mead) dispensed advice from their

Figure 15.1. A portrait of Colonel John Walsh (1726–1795), the main player of the electric fish story in the second half of the eighteenth century. (Image made by Nicholas Wade from a drawing kindly provided by David Nashford on the basis of a photographic reproduction of an original portrait in Shorland, 1967)

tables at these establishments. The Grecian Coffeehouse, close to Craven Street on the Strand, was one of the coffeehouses frequented by Franklin, and another was the St. Paul's Coffeehouse, which moved to become the London Coffeehouse early in 1772 (Finger, 2006a, p. 131–132). Walsh did not specify the coffeehouse in his letter, and it is also possible that he was inviting Franklin to his residence.

Most likely through Franklin, Walsh had also been in contact with Joseph Priestley, whose huge *History and Current State of Electricity, with Original Experiments* first appeared in 1767. Some years later, when Priestley was planning to publish another monumental work, *The History and Present State of Discoveries Relating to Vision, Light, and Colours* (to be printed in 1772), Walsh would subscribe to the new book, as would Franklin (who would sign up for 20 copies to share with colleagues at home and around the world).

Priestley's (1767) *History of Electricity*, a valuable reference work with an extensive commentary, was written with the assistance of Franklin and some of the electrical scientists he mingled with in London. On June 13, 1772, Priestley wrote a note to Franklin, in which he said that he had borrowed a French translation of his *History* from Walsh and would soon purchase a copy for himself (Willcox, 1975, p. 174). Since Franklin would soon be on his way to Leeds, where Priestley was located, and as he had previously asked to borrow the *Histoire de l'Électricité Traduite de l'Anglois de Joseph Priestley*, Priestley would give this copy to him (Willcox, 1975, p. 126).

Priestley must have been certain that Walsh, who by this time had made his way to France, would have no objections.

Interestingly, Mathurin-Jacques Brisson, a nephew of Réaumur, was the translator behind the three-volume French edition of Priestley's *History*, dated 1771. As for passing books around like this, it was second nature to Franklin. While in Philadelphia, he had founded the first subscription library in the colonies for sharing books that were hard to find or too expensive for most people to own.

Walsh was clearly doing his homework. References in his hardwritten journals, letters, and published articles show that, in addition to Bancroft and Lorenzini, he was familiar with what many previous authors had written about torpedoes. Among the ancients, he knew about Hippocrates, Plato, Socrates, Aristotle, Scribonius Largus, Galen, Oppian, and Claudian. And from the seventeenth and eighteenth centuries, Redi, Jacobaeus, Steno, Kaempfer, and Réaumur stand out.

There is also good evidence showing that Franklin worked with Walsh to plan his experiments on torpedoes. This is not to say that Walsh did not get some ideas from Bancroft or from other sources, or that he was just an obedient foot soldier under Franklin's command. Most likely, he had some ideas of his own, based on his readings and interactions in London, and he seemed to have discussed the literature with others. Moreover, and very importantly, during the intense phase of his torpedo experiments in France, Walsh showed the ability and intelligence to use the results of one experiment to guide what to do next, as well as to choose the best set-ups for gathering needed information about the shocks.

Still, the fact remains that there is a document written by Franklin bearing the title "Instructions for Testing the Torpedo Fish." These instructions were not addressed and Franklin put them on paper while Walsh was traveling in France, *after* having just completed his experiments on the coastal torpedoes. As shall be seen, many of the experiments that Walsh conducted closely followed this plan. This concordance suggests that Franklin had discussed at least some of these needed experiments with Walsh prior to his scientific excursion.

Supporting this possibility, there is an undated and incomplete draft of a piece Franklin wrote called "Directions to Discover whether the Power that gives the Shock, in touching the Torpedo . . . is Electrical or Not" (Willcox, 1975, p. 233). This manuscript contains an earlier version of paragraphs 2 and 3 of the Instructions, which Walsh *could have seen* in England and might have copied for his trip to France. This possibility would explain why Jacques Barbeu-Dubourg at first included only the shorter, earlier document in his French translations, and then separately translated the fourth and fifth paragraphs, giving them the August 12th date—and stating that they were added after Franklin had heard from Walsh.

Thus, various facts would strongly suggest that Franklin worked with Walsh in planning at least some of the electrical experiments. The complete document of August 12, with paragraphs 2 and 3 now refined, which goes beyond what Bancroft had done, reads as follows:

> It has long been supposed that the Stroke given by the Torpedo was the Effect of sudden violent muscular Motion. It is now suspected to be an Effect of the electric or some similar subtil Fluid, which that Fish has a Power of acting upon and agitating at Pleasure.
>
> To discover whether it be the Effect of a subtil Fluid, or of muscular Motion, let the fish be touch'd with the usual

Conductors of Electricity, viz. Iron or other Metals; and with the known Non-conductors of Electricity, dry Wood, Glass, Wax, &c. If the Stroke be communicated thro' the first and not thro' the latter, there is so far a Similarity with the electric Fluid; and at the same time a Proof that the Stroke is not an Effect of mere muscular Motion.

Let it be observed whether the Stroke is sometimes given on the *near Approach* of a conducting Body without actual Contact; if so, that is another Circumstance. Then observe whether in that Case any Snap is heard; and in the Dark whether any Light or Spark is seen between the Fish and the approaching Body. If not, there the Fluids differ.

Let a Number of Persons standing on the Ground, join Hands, and let one touch the Fish so as to receive the Stroke. If all feel it, then let him be laid with his Belly on a Plate of Metal; let one of the Persons so joining Hands touch that Plate, while the farthest from the Plate with a Rod of Metal touches the Back of the Fish; and then observe whether the Force of the Stroke seems to be the same to all in the Circuit as it was before, or stronger.

Repeat the last Experiment with this Variation. Let two of the Persons in the Circuit hold each an uncharg'd electric Phial, the Knobs at the Ends of their Wires touching. After the Stroke, let it be observ'd whether those Wires and attract and repel light Bodies; and whether a Cork Ball suspended by a long silk String so as to hang between the Wires at a small Distance from the Knob of each, will be attracted and repell'd alternately to and from each Knob; if so, the Back and Belly of the Fish are at the Time of the Stroke in different States of Electricity. (Willcox, 1975, pp. 234–235)

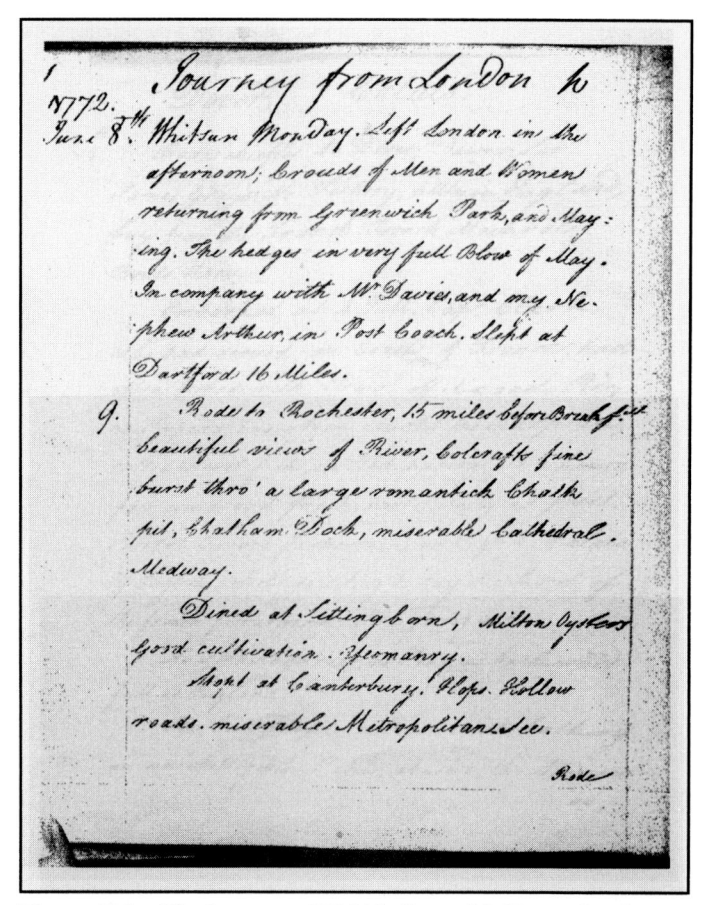

Figure 15.2. The first page of Walsh's *Journal de Voyage* dated 1772, June 8th (© The John Rylands Library, University of Manchester).

THE TRIP TO LA ROCHELLE

Walsh was in his mid-40s when he set forth on his trip to France (under the date July 1, 1772, in his journal of torpedo experiments, he writes from La Rochelle: "my birthday, 46 years old"). Everything of significance and much that was not scientifically important about his trip to La Rochelle was recorded in the first of his manuscripts, *Journey from London to Paris, begun 8th June 1772* (Walsh, 1772a) (Fig. 15.2). He tells us that his ship to France landed in Calais, where he encountered a fishing boat unloading 600 or 700 torpedoes. He took advantage of the situation by conducting some preliminary dissections, but did not provide details.

He also spoke with a few boat captains as he continued his journey. His intentions were to find out what they knew about torpedoes and if he should expect to find them in abundance at La Rochelle. They provided him with some information and assured him that there were plenty in the region.

In addition to these exchanges, and of far greater importance, Walsh wisely spent some time in Paris before heading back to the coast. There he met with leading scientists and naturalists, some of whom were close to Franklin and therefore were ready to receive him. Although England and France had fought each other many times, and were repeatedly locked in mortal combat throughout the eighteenth century, the *First Charter* of the Royal Society, dated 1662, had given its scientists license to have "Correspondence, on Philosophical, Mathematical, or Mechanical subjects, with all sorts of Foreigners" (Hartley, 1962, p. 114). Indeed, internationalism and free circulation of ideas and people was one of the distinctive features of the "Republic of Letters" of the Enlightenment. A "dark" byproduct of this free circulation of *literati* across the frontiers was the fact, mentioned in Chapter 14, that scholars were often charged with spying activity while abroad.

Jean-Baptiste Le Roy, an enlightened physician and an important member of the *Académie Royale des Sciences*, was one of the outstanding foreigners Walsh met in Paris. Le Roy had been among the first French members of the *Académie* to adopt Franklin's electric theory (see Chapter 10). He was particularly interested in hospitals and electric medicine, and was one of the first to apply electricity to disorders of vision and hearing (among his patients was La Condamine, who suffered from deafness; Le Roy, 1761).

In a letter written on June 21, Walsh informed Franklin: "Your friend Le Roy, who presents you his Compliments, has shewn us many Civilities" (Willcox, 1975, pp. 189–190). Among other things, Le Roy discussed electrical instrumentation with Walsh and encouraged him to take every advantage of the opportunity to conduct systematic experiments on live torpedoes. Le Roy would continue to help Walsh in the future, and Walsh would return the favor after returning to England.

Walsh further informed Franklin that, while in Paris, he attended two meetings of the *Académie Royale des Sciences*,

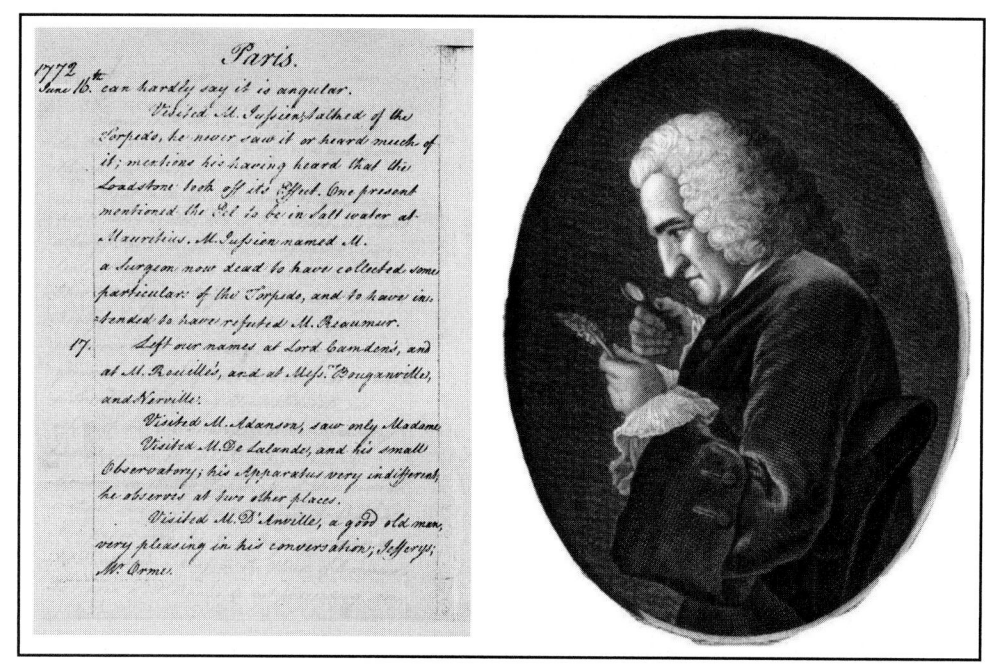

Figure 15.3. The page of Walsh's *Journal de Voyage* dated 1772, June 16th and 17th describing his visit to the *Jardin du Roy* and his conversation with Bernard de Jussieu (1699–1777) (on the right), as well as his attempt to visit Michel Adanson, who had studied the torporific African catfish (© The John Rylands Library, University of Manchester).

where he was politely received. He was surprised that only a few of the academicians knew about torpedoes, "to which animal by the bye, they are almost Strangers here." He also spent time at the *Cabinet du Roy*—that is, at the natural museum of the *Jardin du Roy*, the Royal Botanical Garden, where he was able to view the king's collection "of natural Curiosities." In his travel journal he wrote that the collection included a "Male Torpedo, as called, dryed small about 9 Inches," and "two very small ones in Spirits mention'd from the Isle de Bourbon, two inches in Diameter," but "No trembling eel."

Before leaving Paris, Walsh also met with Monsieur (probably Bernard) Jussieu, a naturalist at the *Jardin du Roy* and another member of the *Académie*. Jussieu's specialty was botany, and he was not personally familiar with torpedoes. But he told Walsh that he had "heard that the Loadstone took off its Effects" (Fig. 15.3). Gottfried Wilhelm Schilling (1770, 1772), a Dutch physician who lived in Paramaribo, the capital of Surinam, was spreading this myth on the basis of his experiments on the electric eel (Fig. 15.4). Schilling's assertion would be contradicted by the experiments of Jan Ingenhousz (1784) on an eel and by those of Spallanzani (1783) on sea torpedoes (see Chapter 19; also Piccolino, 2000a).

Walsh was less successful in personally meeting Michel Adanson (1757), who had written about the African catfish in the 1750s. He must have knocked on his door, since he wrote in his journal that he saw "only Madame" (see Fig. 15.3). He did note, however, that Adanson was present at the June 12 meeting of the *Académie Royale des Sciences* that he attended, as was Jean Baptiste Le Rond d'Alembert, the founder of the monumental French *Encyclopédie* with Denis Diderot, and one of main protagonists of the Enlightenment. Monsieur d'Alembert, we should note, had been instrumental in translating some of Franklin's papers on electricity into French.

Finally, mention must be made of the veritable Condamine, the famous physicist and astronomer, who, as mentioned in Chapter 9, had provided a short description of the electric eel during his scientific journey in South America. Walsh met him and he sold Walsh an electrical machine for his experiments. In his *Journal of Experiments*, Walsh described it as "a Platteau, with Lane's Electrometer fitted to the Prime Conductor" (p. 32). The Platteau ("as they call it, and we a Ramsden") was a circular glass plate that had cushions against each side, and rotated in a vertical plane. It was more efficient than a glass globe or cylinder, and it too generated electricity that could be collected with a Leyden jar.

As for the electrometer that came with the machine, London instrument maker Timothy Lane invented it and described it to Benjamin Franklin in a letter dated 1766, which was published a year later (Lane, 1767) (Fig. 15.5). This instrument allowed Walsh to regulate the intensity of the discharges from the jar. Using a micrometer with a screw, he could alter the size of a gap between the electrometer and the Leyden jar. With the tiniest of separations, the jar would not have to be charged very much for a spark to jump the gap, whereas stronger charges would be needed with larger gaps.

Walsh hoped to mimic the sensations from a torpedo with a properly charged Leyden jar. After all, they had been said to feel alike, suggesting a similar force was involved. He also hoped to be able to observe a spark generated from a torpedo, which he thought would be able to jump a small gap—at least if a healthy ray really discharged electricity. This was one of the key experiments on Franklin's checklist.

Not all of these details were mentioned in the letter Walsh sent Franklin from Paris. But, with regard to torpedoes, he did tell Franklin:

Whether their Effect be Electrical or not, I persuade myself will be soon ascertained. I am at this instant setting out for La Rochelle.... My next [letter] will convey to you the

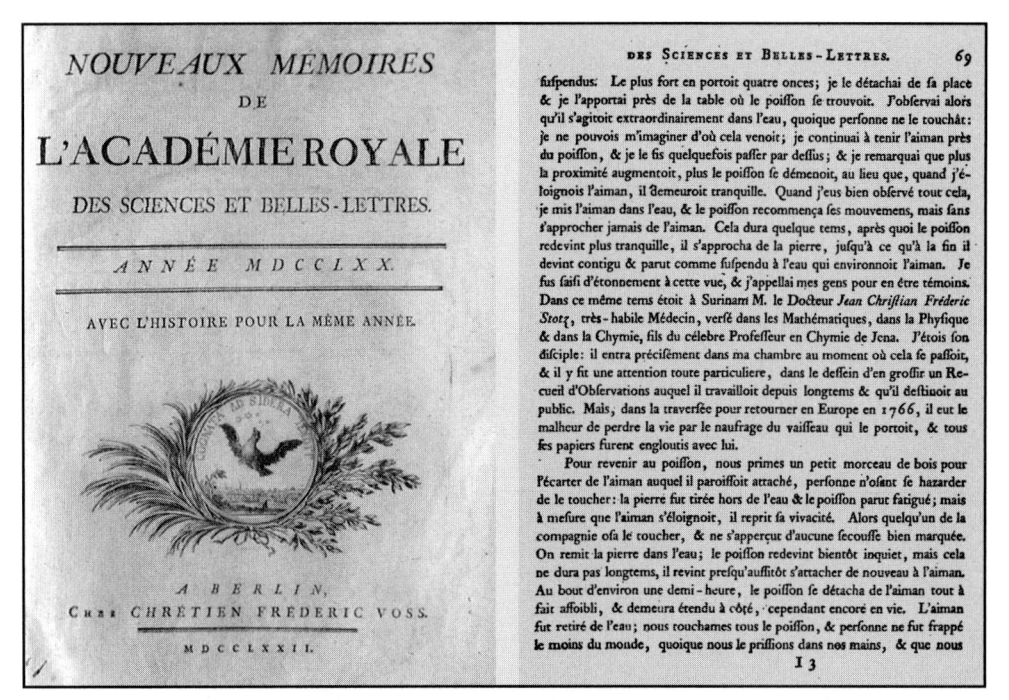

Figure 15.4. The 1772 Proceedings of the Berlin *Académie Royale des Sciences et Belles-Lettres* for the year 1770, with a page of the memoir in which Dutch physician Gottfried Wilhelm Schilling describes his experiments on how the action of a lodestone can dissipate the shocks of the torpifying eel.

Event of our Experiments on the Torpedo. (Wilcox, 1975, p. 190)

The body of Walsh's letter ends here. But the words "our Experiments" are significant. This is because they are consistent with the contention that Franklin gave Walsh scientific guidance. Stated somewhat differently, they provide more evidence that Franklin, working behind the scenes as he often did, was instrumental in setting up the program of experiments that Walsh would soon be conducting.

Walsh's diary of his trip continues to have some entries just before his arrival at the coastal town, including a three-page "Memorandum concerning the Torpedo, made in the Journey from Paris to La Rochelle" (Fig. 15.6). Here Walsh lists his aspirations. He will do experiments "with every Conductor and non-conductor" (Franklin's language) he can think of, and he will try to convey the shock as far as possible in his experiments with people holding hands or wires. Interestingly, this memorandum is written in French (although with several errors), perhaps reflecting Walsh's desire to become more acquainted with the local language as he was approaching his final destination.

SETTING UP AT LA ROCHELLE

On June 26, 1772, Walsh arrived in La Rochelle (Fig. 15.7), the town where Réaumur was born in 1683, and close to where

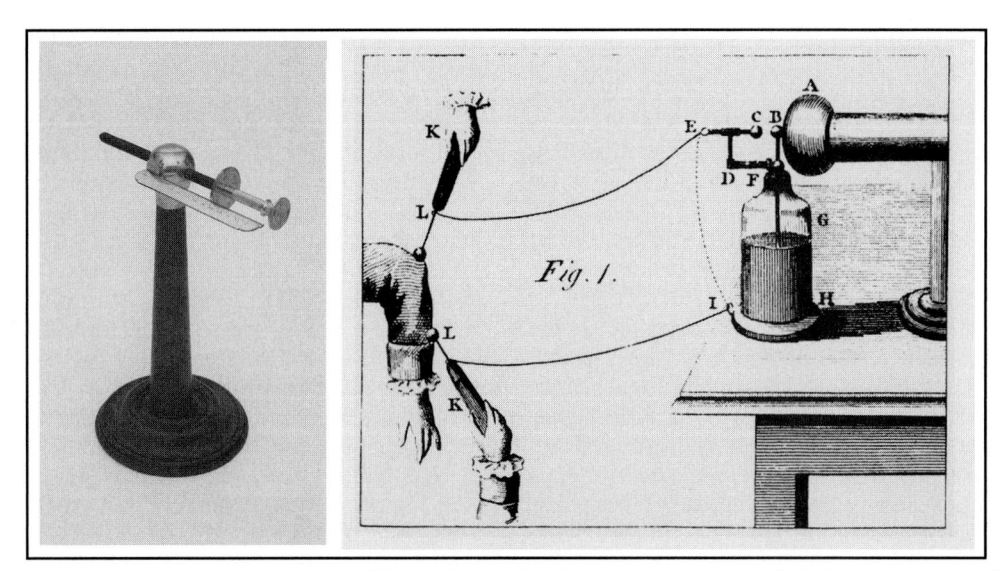

Figure 15.5. Timothy Lane's electrometer and how it could be used with a Leyden jar for applying medical electricity. Lane's electrometer consisted simply of a micrometric system, which allowed for the accurate measurement of the sparking distance from a charged electric device. It was popular in electrical medicine, because it offered practitioners a way to control the strength of the shocks. Walsh used Lane's instrument in some of his torpedo experiments at La Rochelle. (The image on the left is courtesy of the Galileo Museum in Florence; that on the right is from Cavallo, 1795.)

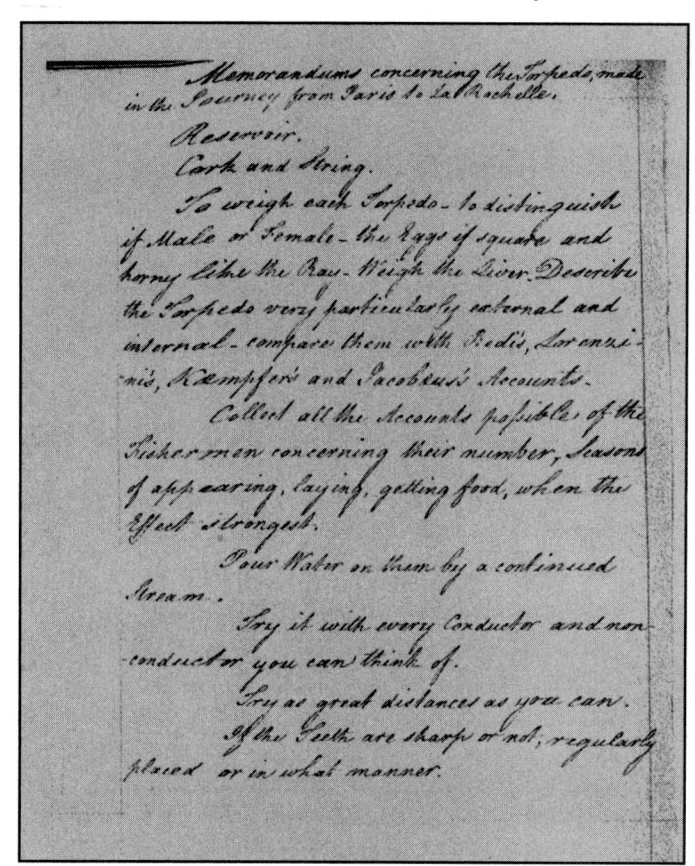

Figure 15.6. The first page of Walsh's *Memorandum* on the experiments to be done on the torpedoes at La Rochelle. This page was inserted just after the page of the *Journal de Voyage* dated 25th June 1772, the day before he arrived at his destination (© The John Rylands Library, University of Manchester).

his French predecessor carried out experiments on torpedoes, which he then explained with his hypothesis of rapidly contracting muscles (Réaumur, 1717). One of his first tasks, after securing lodging, was to meet some of the fishermen of La Rochelle, since he had to count on them to bring him a supply of fresh *la marote ou tremble*, the local name for the *torpille*, the usual French word for a torpedo. There would be times, Walsh anticipated, when he would even want to go out with these fishermen on their boats, the objective being to do experiments on the liveliest and healthiest of torpedoes.

He more than befriended the provincial fishermen by offering them a hefty 24 French *sous* (equal to 13 English pence) per torpedo, or about six times more than they would have gotten in the local markets for the rays. Walsh examined more than 70 torpedoes, and the biggest was not even 16 inches (about 40 cm) across and weighed only approximately 10 pounds (less than 5 kg). Most of his specimens only weighed 3 to 5 pounds, and a large number did not survive a day in captivity.

Walsh knew it was important for him to have witnesses for his experiments. Witnesses would verify his activities and findings, now an important element even for those with academic credentials. In addition, he knew he would need more than one set of hands for some of his experiments; sometimes even an assemblage of volunteers. Thus, as was true of how many experimental natural philosophers worked at the time, his experimental sessions sometimes looked like social gatherings.

Figure 15.7. The old fortress in the harbor of La Rochelle on France's western (Atlantic) coast (By courtesy of Wendy Finger).

One of Walsh's key assistants and witnesses was his nephew, Arthur Fowke, a teenager born in 1756. Although a bachelor, Walsh had been educating Arthur after his sister's death. In his choice of a family member, we are reminded of Franklin, who first flew his famous electrical kite in Philadelphia before a single witness, his son William Franklin. Arthur would be the first member of the group to experience the shock of a torpedo. Mr. David Davis, who also helped out, was Walsh's secretary and a tutor to Arthur.

Respected Frenchmen in the area also assisted. One such person was Monsieur Saunier, a gentleman of La Rochelle, who would receive Walsh's equipment and the flat-bottomed boat he had used for transporting and maintaining live torpedoes as a gift at the end of the experiments. Two other Frenchmen, Messieurs Beauregard and Ranger, were surgeons, and they helped with his dissections and vivisections.

There were also members of the local *Académie de La Rochelle*, including Monsieur Charles-René de Villars, a physician and a corresponding member of the *Académie des Sciences* in Paris. In his recently composed *Éloge de M. Réaumur*, this gentleman, who was also the town librarian, questioned the explanation for the shock given by the town's most famous son. Pierre Henri Seignette, a physicist and pharmacist, also experienced his share of shocks in the experiments. As we shall see, Seignette, who was the town's mayor and perpetual secretary of the local *Académie*, would play a significant role in spreading the news of the successes of Walsh's experiments (Chapters 16 and 20).

As he experienced his successes and failures at La Rochelle, Walsh recorded his research and thoughts in a new journal, one devoted to experiments and some amusing tangential facts, such as that he occasionally dined on boiled torpedoes. This journal was titled *Experiments on the Torpedo or Electric Ray at La Rochelle and l'Isle de Ré—in June–July 1772* (Walsh, 1772b) (Fig. 15.8). It shows day by day and even hour by hour what Walsh was doing with the torpedoes and his thinking process as he went along (see Piccolino, 2003a).

Before turning to the experiments themselves, more must be said about l'Isle de Ré, a small island only 3 kilometers from La Rochelle. Walsh went here with the hope of finding larger or at least more vigorous torpedoes, after not succeeding in seeing a spark with the rays at La Rochelle. He did perform some experiments here, although his research was interrupted when the *Lieutenant du Roy*, Monsieur de Tailler, accused him of being a British spy and asked him to leave. Fortunately, the king's lieutenant changed his mind (the affair turned to be just *un mal-entendu*) and allowed him to do more experiments before returning to nearby La Rochelle, where Walsh gave some grand public demonstrations and then bid his French friends *adieu* on July 28.

THE EXPERIMENTS

On June 28, 2 days after arriving in La Rochelle, exactly 1 month before departing, and upon not finding any *marote*, Walsh wrote in his journal of experiments:

> Two fishermen of La Rochelle came to the Inn, who said that they frequently caught *La Marote ou Tremble*; that the shock was instantaneous; if received on the feet it overset them; if on the hand it left a trembling effect after it, that it could be received through a sword or stick, but not through their Nets in the water. (p. 3)

The visiting English scientist now convinced one of the local fishermen to do a little experiment with him. "Gave one of them a small Shock with the Leyden Phial and repeated it," he wrote, and "the Effect was precisely the same as that of the Torpedo" (pp. 3–4). Cuchon, another of the local fishermen, told Walsh that both young and old fish, and both males and females, could deliver such jolts. He further stated that the shocks were sometimes felt through nets. Walsh knew this was what some earlier writers had reported, although others, including Réaumur, had questioned this assertion.

Gabriel Coyau, still another fisherman, confirmed what Cuchon had said. Walsh would write: "Coyau, on drawing his Nets felt a Shock when the Torpedo was about 12 feet distant, and two or three Shocks more when it got into his Boat" (p. 67). Coyau said the shocks from this unusually large fish (about 14 inches wide, possibly a *Torpedo nobiliana*, a species that can give shocks up to 300 Volts) affected him to his shoulders. Not finished with Coyau, Walsh also did a Leyden jar experiment with him. He also told Walsh that the ensuing sensations from a lightly charged jar were precisely the same as those he had experienced with the torpedo.

Walsh's experiments at La Rochelle had begun about 2 weeks before Coyau told him about his experiences with this large *marote*. Walsh, his nephew, and Saunier first had experienced the shocks of a small female torpedo June 28, all present finding them perceptible but not especially strong or painful. Additional experiments had followed. For example, Walsh tried to determine if one person holding hands with another in contact with the fish would feel the shock, which was also on Franklin's wish list. The results were inconclusive, as was a variation on the theme, connecting two people with a brass rod. The fish, Walsh wrote, "seemed exhausted, with not much life," so the test was of little value.

Still, Walsh wrote on page 7 of his journal, under June 30: "On this, my first experiment on the effects of the

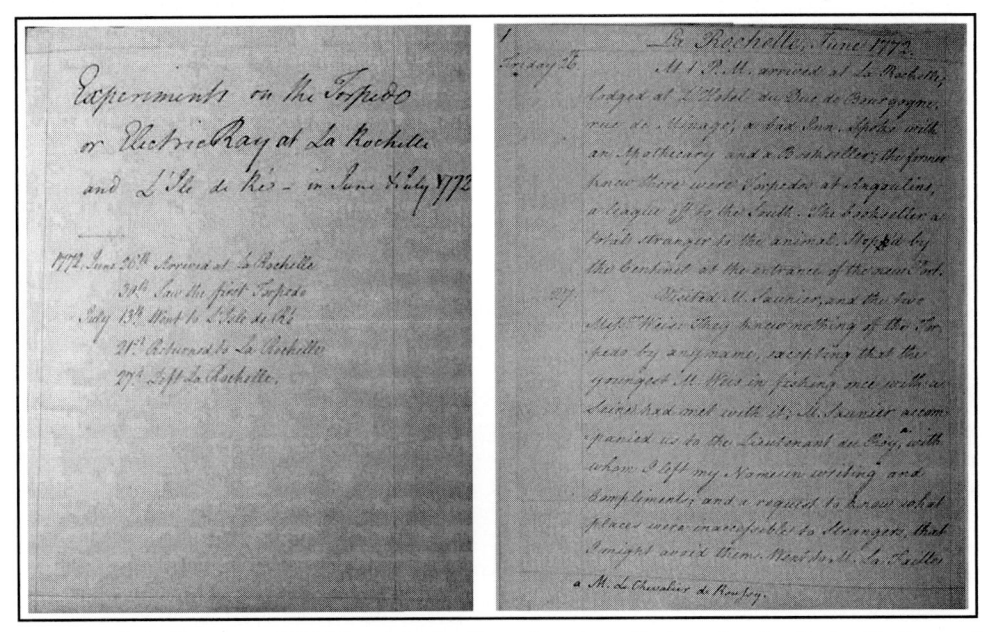

Figure 15.8. The cover page and the first page of Walsh's notebooks of experiments on the torpedoes he obtained in France (© The Royal Society).

Torpedo, I explained this is certainly electricity." But he was perplexed. Hence, he ended his seemingly bold statement with a dash and two words, "but how?"

The ensuing days brought forth more torpedoes for dissection and physiological experiments, which took place primarily at his hotel. For many of the experiments, the torpedo was lifted out of the water, in part because the shocks felt stronger with the fish held in the air or placed on a table than when in a tank. The difference between air and water was calibrated by determining how far up the arm the effect could be felt. Early on, Arthur compared the two, telling his uncle that the sensation only reached the first joint of his thumb when the fish was in water, whereas it was felt above his elbow when held in the air. Walsh would later estimate the shock in water to be about one-fourth the intensity of that in air.

Experiments that followed focused on whether the shocks could be transmitted from person to person in a long chain, as would be true of a shock from a Leyden jar. The connections were again made in various ways. For some tests, the participants held hands; in others, metal rods or brass wires were used as intermediaries; and in still others the participants immersed their hands in "Delphi ware Basins of water" without directly touching one another. Working up from two people, Walsh found that eight people in series could feel the shocks from a large, healthy fish. But there were individual differences: "In weak shocks . . . Persons in the Circuit are not all equally affected" (p. 66).

Also, he found that the charge could be conveyed over long distances with wires, another key feature of electricity that Franklin wanted to be tested. Here, too, he increased the length of the wires as he went along, especially when he obtained large or vigorous specimens. He writes that he was successful with metal wires 10, 13, 29, and even 40 feet (about 12 meters) in length.

There was also a series of experiments in which he set forth to determine, again as Franklin had asked, whether the two sides of the torpedo's body act like oppositely charged surfaces, much like the inside and outside of a charged Leyden jar. These experiments involved touching torpedoes in different places with fingers or metal spoons. When both were placed on just the upper flanks, there were no shocks, and the same was the case when both lower flanks were touched. In contrast, shocks were reliably felt when one upper and one lower flank were touched at the same time. And most convincing of all, a person touching an upper flank and another touching the lower flank felt the shock only when they then joined hands to complete the circuit. Again, the findings—repeated many times and with many variations as demanded of high-quality science—were strongly suggestive of electricity (Fig. 15.9).

Among the experiments done by Walsh, seemingly to follow Franklin's instructions, one carried out on July 7 by Walsh with Arthur and Seignette stands out. In it, "a plate of Tinfoil" was placed under a torpedo and a metallic spoon was used to make contact between the members of the human chain experiencing the fish discharge. Contrary to expectations, this arrangement did not bring about any increase of the severity of the shock.

Walsh and his team also looked for muscle movements from the torpedoes before and during the shocks. Réaumur stressed such movements as fundamental to the shocks being felt, but Walsh failed to detect them. At most, there was a slight movement "of the Cartilage enclosing the Electrical parts," and minor movements "while the Electrical Parts themselves appear totally at rest" (p. 107). Although irrelevant to the shocking action of the fish, the presence of these small movements in the phase immediately preceding the shock became a useful tool for knowing that the fish was producing its discharge, when the experimental design was such that no shock could be perceived (e.g., when some insulating body was placed in chain between the fish and the human subject).

Knowing something about the history of torpedoes, Walsh even examined whether the shocks had a refrigerating quality,

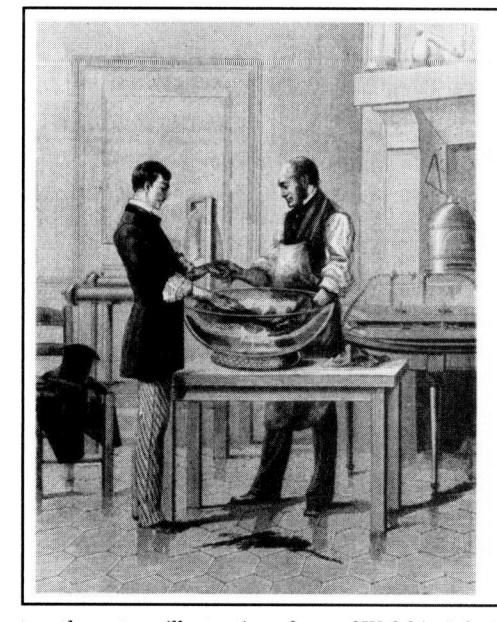

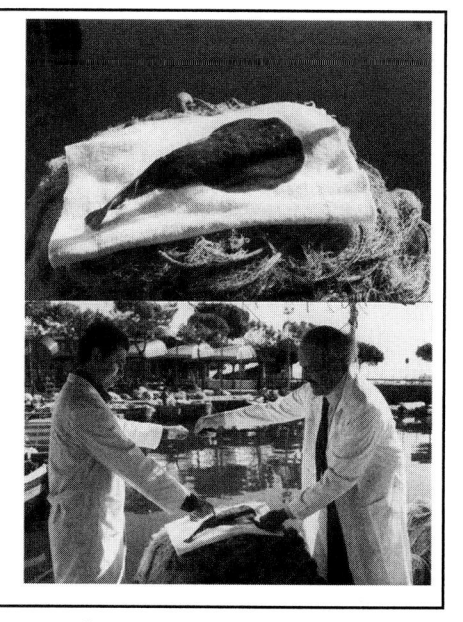

Figure 15.9. A nineteenth-century illustration of one of Walsh's July 9, 1772, torpedo experiments, and a modern attempt to re-enact this experiment (in which the circuit from the torpedo is completed when two participants join hands) in Formia, Italy. The top right shows the detail of the *Torpedo ocellata* (or *Torpedo torpedo*) used in the re-enactment.

as was thought by Galen, Claudian, and others, who had been thinking in terms of a cold "venom." Again, the findings were negative: "All the Experiments this morning strongly marked the effect to be Electrical, and not arising from a muscular stroke, nor from a refrigidating [sic] quality in the animal, of which it has nothing" (p. 29).

Nevertheless, something unusual seemed to be happening, albeit far from the electrical organs. Specifically, the fish's eyes seemed to be signaling an impending shock. "His prominent Eyes always sink down, whenever he gives the Shock," wrote Walsh (p. 35) on July 6. The eye movements, he believed, occur while the animal is preparing to release a shock. This observation led Walsh to the conclusion that the shock is a controlled, voluntary action. In the language of the day, the shocks are "under the will of the animal."

Walsh found utility in this casual observation. He could now track when shocks were released, even if he might not be able to feel them. This would be helpful when working with a feeble specimen, but it also related to his use of "non-conductors" with healthy rays. He could now state with more confidence that the shocks could be not be conveyed by non-conductors (e.g., glass, sealing wax), because he now had a way of knowing that shocks were really being given.

But how many shocks could a healthy ray produce in a minute or more? Walsh attempted to answer this question on the Isle de Ré, and he found that he could count at least 50 shocks in 90 seconds with a healthy torpedo under good experimental conditions. He was amazed by the fact that the torpedo seemed to have some means of replenishing its electricity. That this could occur with a fish out of water implied that it did not draw the electricity out of its natural medium, but rather that it somehow possesses a way to produce it intrinsically.

SOME DISAPPOINTMENTS

Not all of Walsh's experiments at La Rochelle and Isle de Ré started successfully or ended successfully. Clearly discouraging was his inability to provide evidence for electricity using his physical instruments. His attempts to use physical instruments to detect the shocks were probably done at Franklin's suggestion.

The simplest electrometers of his day used foil or very light pith balls that bent or "danced" when hung close to charged electrical bodies. But when Walsh held pith balls over his torpedoes, they did nothing, not even when there was a strong discharge. The Lane electrometer he purchased from Condamine also failed to provide any evidence for electricity (see Fig. 15.5).

More importantly, Walsh was unable to witness a miniaturized thunder and lightning show. Here he used an experimental arrangement based on the use of tinfoil pasted on sealing wax, then cut so as to obtain a very thin separation in the circuit. As he wrote in his journal at l'Isle de Ré:

> Many endeavours to get the Spark, by means of the minutest separation possible, made in Tinfoil pasted on Sealing Wax, and the smallest removal of the Electrometer from the Conductor; but without Effect. No Spring of the Electrical fluid would ensue; no Spark, no ticking noise. (p. 86)

As we shall see in Chapter 19, Walsh would use exactly the same arrangement in 1776 when experimenting with an eel from Surinam. This time, however, he would obtain his coveted sparks, because the eel's shocks are many times stronger than those associated with torpedoes.

Failing to get a spark with one fish, Walsh even tried putting two or even three fish together, with one above the other in a battery-like arrangement. This experiment, which also involved a strip of foil with a thin knife cut through it, was again negative, even though "the separation in the Tinfoil was so very minute, as to be barely discernible to the naked eye" (p. 100). Yet the torpedoes were sensitive to each other's discharges, jumping when another torpedo touching it released a shock.

Even though neither Walsh nor his collaborators were able to see a spark or hear a sudden noise with the cut tinfoil arrangement, Arthur reported that on two occasions he felt a shock by contacting the fish with this experimental arrangement. Walsh, however, did not attribute much importance to this result. He assumed that the humidity might have enhanced communication across the minute tinfoil gap. Throughout his work, he displayed justifiable skepticism and admirable prudence, attributes of a good scientist.

WALSH'S CONCLUSIONS

Walsh concluded that the ray's discharge has to be electrical, even if it does not move pith balls, make a sound, or cause a spark to fly. On July 9, after conducting many experiments with human and metal circuits, and with non-conductors, he penned: *Je l'ai donté* [sic for *dompté*]—meaning "I have tamed it." His wording was based on Claudian, who more than 1,300 years earlier had written: "Who has not heard of the *untamed* art of the marvelous torpedo and of the powers that win it its name?" (Chapter 4) (Figs. 15.10 and 15.11).

Walsh called the natural force "torpedinal electricity," and he associated it with the structures under the skin of the back. He called these structures "honeycombs" because of their morphological appearance. Somehow, they made the torpedo into an "animated Phial" of sorts. He noticed the rich nerve innervations of these structures, but did not go deep when investigating their minute anatomy.

Interestingly, after July 20, corresponding to the last period of his investigations at the Isle de Rè, Walsh began to refer to these structures as "electric organs," probably being the first person to use this phrase, which is still employed today. The day before, he introduced another new phrase in order to denote the electricity responsible for the shock—one also destined to have a long and exciting history. His second new term was "animal electricity." Later in the century, this expression would be used to denote the electricity thought to be involved in various physiological processes of more ordinary animals, especially in nerve and muscle functions.

That the honeycombs or electric organs could be used defensively against more than humans handling these rays was well known to the local fishermen. Bertrand even told Walsh about frequently staged "fights" between torpedoes and lobsters, which were arranged by the fisherman for diversion. The lobster, Bertrand explained, "seizes the Tremble with his Claw, but receiving the Shock, he immediately quit his hold and retires back" (p. 122).

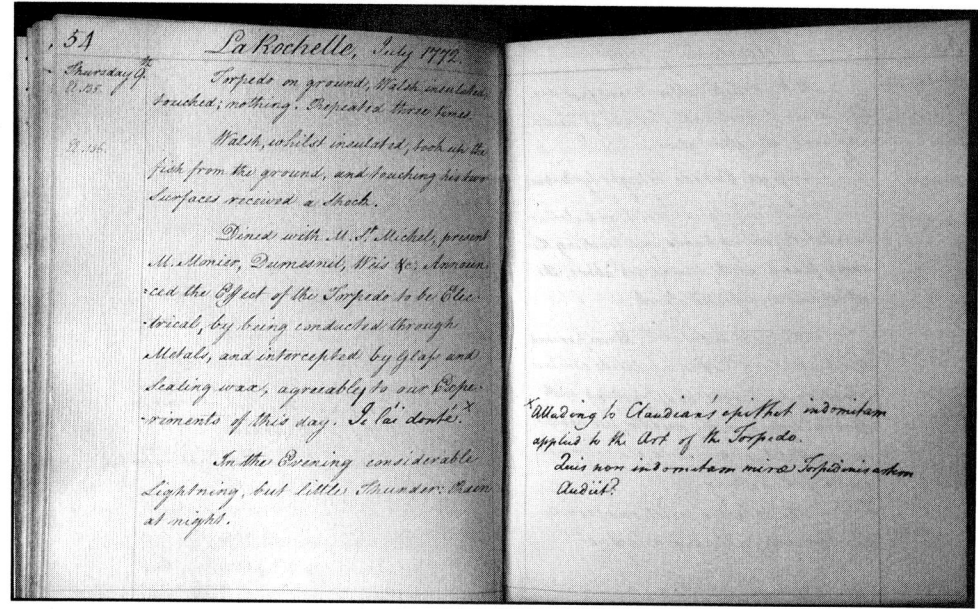

Figure 15.10. The page of Walsh's notebook (dated July 9, 1772) announcing the successes of his torpedo experiments with the marginal notation (written personally by Walsh), *Je l'ai donté*... Walsh is here alluding to Claudian's *epithet indomitam* as applied to the torpedo's art: *Quis non indomitam mirae Torpedinis artem audit?* (© The Royal Society).

Coyau, in the same context, told Walsh that the *marote* kill the *taires* (firefliers or batfish) that are caught with them. And Walsh would include a passage in his journal describing how a vigorous *Serpent de mer* (sea-snake or perhaps pipefish) was put in a trough of water with a torpedo, only to be quickly killed. "The Fishermen all agree that the Torpedo kills all fish that touch it... but is itself killed by the Lobster," he would add to his journal of experiments (pp. 74, 76).

That lobsters could kill torpedoes brings forth images of what Gronov (1758, 1760) reported about electric eels in South America in some of his publications. The terms used by the Dutch scholar to indicate the crustacean capable of killing the fish (*Garnaal* in the Dutch version; *Squilla* in the Latin) make the precise identification of the killer crustacean more ambiguous. It is, however, possible that Gronov was not alluding to some established fact concerning the local electric eel, but was referring to what he had read in some literary source concerning the sea torpedo. This would better correspond to the account of La Rochelle fishermen.

In Walsh's notes, we can also find a statement challenging the notion that, if a person holds his breath, the shock would not be felt. The experimental subject for Walsh's test of this hypothesis, which might have seemed absurd to him, was his nephew. To quote: "Arthur holding his Breath, received Shocks several times, contrary to Kampfer's Observation" (p. 114; see Chapter 12 for more on this observation). Walsh also wrote about the ability of the fish to confine its electricity to certain parts of its body, and in effect not to electrocute itself or lose the charge in the closely surrounding watery medium, which is, after all, also a conductor. He concluded "that Electricity takes many circuits, and does not confine itself to the shortest nor even the best Conductors" (p. 161).

PUBLIC DEMONSTRATIONS

As already mentioned, Walsh interacted with many distinguished people of the local learned community, and particularly

XLIX

The Electric Ray.

Who has not heard of the invincible skill of the dread torpedo and of the powers that win it its name?

Its body is soft and its motion slow. Scarcely does it mark the sand o'er which it crawls so sluggishly. But nature has armed its flanks with a numbing poison and mingled with its marrow chill to freeze all living creatures, hiding as it were its own winter in its heart. The fish seconds nature's efforts with its own guilefulness; knowing its own capabilities, it employs cunning, and trusting to its power of touch lies stretched full length among the seaweed and so attacks its prey. It stays motionless; all that have touched it lie benumbed. Then, when success has crowned its efforts, it springs up and greedily devours without fear the living limbs of its victim.

Should it carelessly swallow a piece of bait that hides a hook of bronze and feel the pull of the jagged barbs, it does not swim away nor seek to free itself by vainly biting at the line; but artfully approaches the dark line and, though a prisoner, forgets not its skill, emitting from its poisonous veins an effluence which spreads far and wide through the water. The poison's bane leaves the sea and creeps up the line; it will soon prove too much for the distant fisherman.

The dread paralysing force rises above the water's level and climbing up the drooping line, passes down the jointed rod, and congeals, e'er he is even aware of it, the blood of the fisherman's victorious hand. He casts away his dangerous burden and lets go his rebel prey, returning home disarmed without his rod.

Figure 15.11. An English translation of Claudian's poem on the torpedo based on a revision of the Latin text in which the torpedo is considered to be terrible (*dirae*) instead of wonderful (*mirae*) (from Claudian, 1985).

Figure 15.12. The *Hôtel de Ville* (town hall) in La Rochelle. This was the usual place for meetings of the local *Académie*, members of which helped Walsh with his studies and served as expert witnesses.

with several members of the local *Académie*, while at La Rochelle (Fig. 15.12). Some of these men, and notably Seignette, played an active role in the torpedo experiments. Seignette would also contribute to disseminating the news of Walsh's successful experiments (see next chapter).

In the last days of his stay, Walsh took the occasion to stage a grand public session before the *Académie* to demonstrate the "power of the Torpedo." His first demonstration was so successful that another one was arranged for the next day at the *Académie*, followed by a third one for the officers of the local military garrison. The academicians formed human chains so they could personally experience the shocks, and they confirmed that the shock could be transmitted through metals and blocked by insulating materials, such as glass and sealing wax.

The academicians failed to see sparks or hear sounds at the time of a torpedo's discharge, and they did not perceive a shock when they put a hand in a large tub of water containing the fish (even though the shock was transmitted through moist bodies or smaller basins full of water). For the Academicians, "a knowledge of the operation of the Torpedo when in Water, was a capital Desideratum." According to them, the electric shock is "a power which this Animal must necessarily be supposed to employ for the great purposes of Life Subsistence and Self Defense," and thus they must be able to employ it in their natural habitat, water.

The Academicians thus raised a serious problem. Water was then considered a good conductor of electricity, almost as good as metals, and as a result it was assumed that, if a torpedo's shock is indeed electrical, it would be shunted by the water immediately surrounding its body. That is, the torpedo's power should not extend for significant distances from its body. As we shall see in the next chapter, this difficulty, as well as other controversial aspects of the shock mechanism, would be explained by Henry Cavendish, the English physicist whom Walsh involved in the torpedo research program along with anatomist John Hunter.

For Walsh, the public demonstrations given at La Rochelle were not simply social courtesies or means of self-promotion. They reflected an important way for communicating new replicable results, and for obtaining public verification of those results. The need for such shows correlated in a positive way with how surprising the results might seem. Walsh would explicitly recognize this fact of scientific life 1 year later, in his article in the *Philosophical Transactions of the Royal Society*:

> I certainly wished to give all possible notoriety to facts, which might otherwise be deemed improbable, perhaps by some of the first rank in science. Great authorities had given a sanction to other solutions of the phaenomena of the Torpedo; and even the Electrician might not readily listen to assertion, which seemed, in some respects, to combat the general principles of electricity. (Walsh, 1773, p. 449)

Walsh, in fact, purposely left a series of documents supporting the result of his experiments, and he drew upon a "virtual witnesses" strategy for his 1773 paper, as we shall see in the next chapter. This strategy of using witnesses had been strongly recommended by Robert Boyle, a leading figure in chemistry and physics (e.g., laws of gasses) during the previous century. Boyle lived in England at a time when checking the validity of experimental results was not easy, making this sort of verification all the more important (see Shapin & Schaffer, 1985).

Walsh, who throughout his stay in La Rochelle had done experiments with both living animals and parallel experiments with Leyden jars, made an important remark at the end of his first demonstration at the *Académie*. This remark,

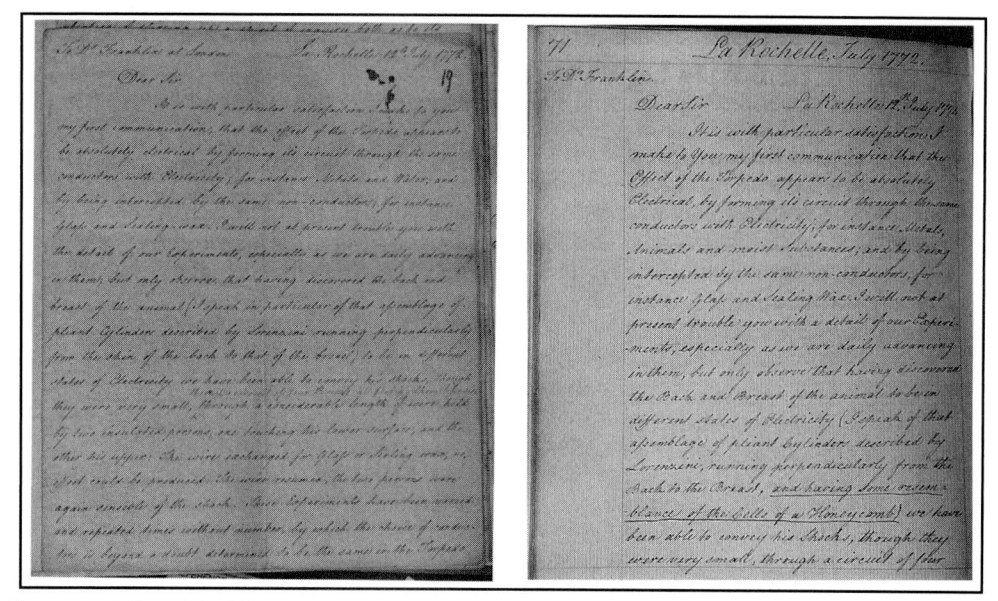

Figure 15.13. The first page of the original letter sent by Walsh to Franklin from La Rochelle (dated 12th July 1772) and its transcription in Walsh's notebook (© The Royal Society).

found in both the journal of his experiments and in his 1773 article (although in slightly different form), is worth quoting here because it highlights the historical importance of electric fish research for the progress of electrical science:

> As artificial Electricity has led to a discovery of some of the operations of the torpedo, the Animal if well considered would lead to a discovery of some truths in artificial Electricity which were at present unknown and perhaps unsuspected. (Walsh, 1772b, p. 145)

FIRST LETTER TO FRANKLIN

Walsh kept his word about keeping Franklin fully informed about what he found and what he believed his data signified. He wrote two notable letters to Franklin. These were more than polite letters of thanks for his help. He provided critical information and was, in effect, corresponding with scientists in the larger Republic of Letters—men throughout the world who communicated in person or in writing with Franklin and other natural philosophers.

The first letter was sent July 12, about 2 weeks after his arrival in La Rochelle, but before he went to Isle de Ré (pp. 71–72 in Walsh's journal of experiments; Willcox, 1975, pp. 204–205) (Fig. 15.13). It would later be published, with minor changes, as a part of his 1773 article in the *Philosophical Transactions* (Walsh, 1772b, 1773).

Jean-Baptiste Le Roy, whom he had met in Paris, also did a French translation (Walsh, 1774a). Le Roy had been elected a foreign associate of the Royal Society of London on June 10, 1773, and two of his supporting letters came from Franklin and Walsh, who were truly appreciative of everything he had done. Walsh handed him the letter informing him of his nomination when he went to Paris (Willcox, 1975, p. 295; 1976, pp. 240–242). For Le Roy, Walsh had material certainly worth conveying—and he now had a perfect opportunity to pay back the favor.

In this letter, Walsh informed Franklin that "the effect of the Torpedo appears to be absolutely Electrical." He based these words on the fact that the discharge could be transmitted through electrical conductors but not various non-conductors. Walsh also let Franklin know that the top and bottom of the ray are in different states of electricity.

In addition, the shocks were like those produced by a Leyden jar or his flat capacitor. "There is not an *Engourdissement* or *Fourmillement* of the Torpedo, that we do not exactly imitate with the Phial by means of Lane's Electrometer" (Walsh, 1772b, p. 71).

After regretting that "We have not yet perceived any Spark or Noise to accompany the Shock, nor that Canton's [pith] Balls were affected by it," Walsh (1772b, p. 71) emphasized that the shocks tended to be weak and "extended seldom farther than the touching finger." Hopefully, he wrote, he would obtain better results with more powerful torpedoes at l'Isle de Ré.

In the remainder of his letter, Walsh told Franklin how the ray seemed to close its eyes before a discharge, how the shocks were easier to perceive with the fish out of water, and how many shocks a fish could deliver one after another.

He then asked Franklin a favor. Specifically, he wrote, "please acquaint Dr. Bancroft of our having thus verified his Prediction concerning the Torpedo; and make any other communications on this Matter you may judge proper" (p. 72).

SECOND LETTER TO FRANKLIN

The second letter to Franklin was written on August 27, after all of the experiments had been completed and Walsh had some time to ponder his results. It was sent from Paris, where he stopped to converse with some of the people he had seen before returning to London. This letter also appeared in Walsh's 1773 publication (Walsh, 1773; Willcox, 1975, pp. 285–289).

This letter differed from his first communication in that it included the results of his experiments at Isle de Ré.

Although he was able to extend some of his findings, he again lamented that he did not succeed in obtaining a spark from the shock of the fish (see next chapter).

Nevertheless Walsh still maintained that torpedoes must be electrical, with the charge coming from "compressed elastic fluid." But now he added an important new thought. The charge has to do with more than just the amount of electricity discharged; it also has to do with how "compressed" or distributed the electric fluid may be—its "spring."

In the next chapter, we shall look at Walsh's communication to the Royal Society, which Franklin helped him write, and how these men managed to gather considerably more support for the electrical hypothesis by enlisting the help of other scientists affiliated with the Royal Society of London.

Chapter 16
The Royal Society and Interdisciplinary Science

> On the whole, I think, there seems nothing in the phenomena of the torpedo at all incompatible with electricity.
>
> Henry Cavendish (1776a, p. 222)

After his intense and largely successful season of experiments with the torpedoes on the Atlantic Coast of France, Walsh did not sail directly back to London. Instead, he took a detour through Paris, arriving there on August 21, 1772. His extant *Journal de Voyage* (the diary of his travels) does not contain any reference to the torpedo experiments in the parts describing his return trip, and suddenly ends without explanation on August 17, while he was at La Ferté, still about 100 miles from Paris (Walsh, 1772b).

Walsh had instructed his correspondents to send mail for him to the French capital, and he found a letter from Franklin waiting for him. It was dated July 28 and it contained the news that his esteemed American colleague had just been elected to the French *Académie des Sciences*. Walsh responded to it on August 27, telling Franklin, "I will not let this Post slip without making to You my hearty congratulations on your being elected by the French Academy of Sciences to be one of it's [sic] eight foreign Members; and I will freely say . . . that Your Election does honour to the Judgement [sic] of the Academy" (Wilcox, 1975, pp. 285–286). He then provided Franklin with more information about his week of experiments at Isle de Ré and his final days at La Rochelle.

Although, as already mentioned, Walsh maintained that he was satisfied with his newer findings, he harped on his negative findings in this letter, lamenting that he was

> not able to force the Torpedinal Fluid across the most Minute Tract of Air; not from one Link of a small Chain to another; not thro' a Separation made by the Edge of a Knife, in a Slip of Tinfoil paste on Sealing-Wax. The Spark therefore and the snapping noise attending it were denied to all our Attempts, either in the Light or compleat Darkness. (Willcox, 1975, p. 287)

He also told Franklin that he "communicated to the Gentlemen of the Rochelle Academy and to many of the principal Inhabitants all that I had observed concerning the Torpedo." This was done "in the Intention of stirring up a Spirit of Enquiry both as to it's [sic] Electricity and general Oeconomy." He continued:

> I did hope that three days of very general Exhibition in this manner would have left me nothing more to do in France, but finding by an absurd Paragraph in the french [sic] Papers that those present are not very capable of relating of what they heard saw and felt; and being a good deal pressed by your Confreres of the Academy to communicate a Fact so interesting and singular, I shall endeavour to conquer my Indolence and Aversion to stepping forth on the publick Stage because indeed I feel that it will be ridiculous in me if I do not. (Willcox, 1975, pp. 286–287)

Walsh was referencing a short report that first appeared in the *Gazette de France*, the official journal of French Government, when he alluded to an "absurd Paragraph in the french Papers." This report, which had come out 2 weeks earlier and had been reprinted in other French and Swiss journals, provided readers with misleading information about his torpedo experiments. In translation, it reads:

> Mr. Walsh, Member of the Parliament of England for the County of Glocester [sic], brought himself to La Rochelle in order to examine the fish denoted as *Torpedo*, which has the property of benumbing those who touch it. He has demonstrated that this fish is endowed with an extraordinary electric force, which he has measured with the electrometer and compared with the electricity of all known bodies. He has placed nine persons situated frontally on a tin wire placed under their feet, everyone dipping his hand in a water basin; with the tip of this wire he has touched the fish while it was swimming in a basin of water, and immediately everybody felt a shock as strong as that with the Leyden jar. He made several other experiments on this fish worthy of the attention of physicists" (*Gazette de France*, August 14, 1772, p. 298) (Fig. 16.1).

This account had to begin with one of the gentlemen of La Rochelle, although the inaccuracies could also have stemmed from the editor's desk. In any case, Walsh must have been particularly astonished to read that he had been successful in picking up the electrical charge with an electrometer, since he viewed this as one of his main experimental failures. Small wonder that he now wanted to set the record straight by "stepping forth on the publick Stage."

Given that his experiments were conducted in France, and knowing that he and Franklin had many friends at the

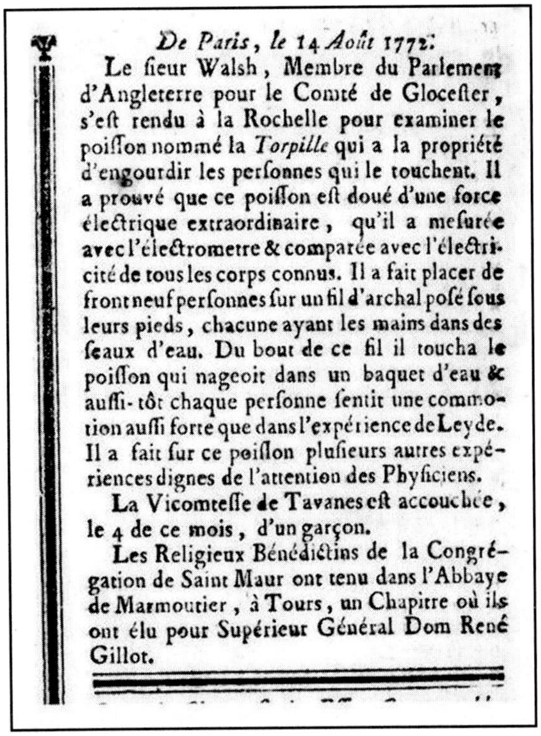

De Paris, le 14 Août 1772.

Le ſieur Walsh, Membre du Parlement d'Angleterre pour le Comté de Glocester, s'eſt rendu à la Rochelle pour examiner le poiſſon nommé la *Torpille* qui a la propriété d'engourdir les perſonnes qui le touchent. Il a prouvé que ce poiſſon eſt doué d'une force électrique extraordinaire, qu'il a meſurée avec l'électrometre & comparée avec l'électricité de tous les corps connus. Il a fait placer de front neuf perſonnes ſur un fil d'archal poſé ſous leurs pieds, chacune ayant les mains dans des ſeaux d'eau. Du bout de ce fil il toucha le poiſſon qui nageoit dans un baquet d'eau & auſſi-tôt chaque perſonne ſentit une commotion auſſi forte que dans l'expérience de Leyde. Il a fait ſur ce poiſſon pluſieurs autres expériences dignes de l'attention des Phyſiciens.

La Vicomteſſe de Tavanes eſt accouchée, le 4 de ce mois, d'un garçon.

Les Religieux Bénédictins de la Congrégation de Saint Maur ont tenu dans l'Abbaye de Marmoutier, à Tours, un Chapitre où ils ont élu pour Supérieur Général Dom René Gillot.

Figure 16.1. The page of the *Gazette de France* of August 14, 1772, with the first printed report of Walsh's experiments at La Rochelle. The report was based on a first communication sent by Henry Seignette, the mayor of the town and also the perpetual secretary of the local *Académie*. Like other members of the learned society, Seignette played an active role in Walsh's experiments.

Académie des Sciences who would welcome a paper on "a Fact so interesting and singular," Walsh thought about publishing in a French journal. But this is not what transpired, and ultimately he published a full account of his torpedo experiments a full year later in the *Philosophical Transactions of the Royal Society of London* (Walsh, 1773). His friend Le Roy, however, translated this paper into French, and it was also published in the prestigious French journal, *Observations sur la Physique*, in 1774 (Walsh, 1774a).

Here we should note that Franklin was happy to comply with Walsh's request for his report to be written in the form of a letter addressed to him. Responding to Walsh on September 6, 1772, he wrote:

> I am glad to find by your Favour of the 27th past, that you are return'd safe and well to Paris after your Expedition to the Sea Coast, and that you intend to publish an Account of your Experiments. Your doing it as you propose in a Letter to me I shall esteem a very great Honour. (Wilcox, 1975, p. 295)

THE FRENCH CONNECTION

Just why Walsh decided to publish his report in English and not in French is not entirely clear and can generate speculation. But several factors could have led to this decision. The most obvious, of course, was that he and Franklin were members of the British Empire and affiliated with its Royal Society. Another is that he seemed to need the help of the men of the Royal Society to understand his findings, especially the negative ones. And a third is that Monsieur Seignette, the Mayor of La Rochelle and Secretary of La Rochelle's *Académie*, quickly attempted to correct the inaccuracies in the *Gazette de France* article with a more accurate report in the same journal.

Seignette's letter in the *Gazette* was dated October 10, 1772 (Fig. 16.2). It provided a more accurate description of the experiments, did not mention anything about detecting torpedinal electricity with an electrometer, and focused on positive conclusions (Seignette, 1772a). Reports based on Seignette's letter also appeared in other French journals in 1772 and 1773, in Walsh's mind correcting the "absurd Paragraph in the french Papers."

Walsh was obviously pleased with Seignette's rendition, because he translated it and included it in his 1773 *Philosophical Transactions* article. Using Walsh's translation, this is what Seignette had written when addressing himself to the *Gazette* editor:

> In the *Gazette* of 14th August, you mentioned the discovery made by Mr. Walsh, Member of the parliament of England, and of the Royal Society of London. The experiment, of which I am going to give you an account, was made in the presence of the Academy of this city [on July 22nd]. A live Torpedo was placed on a table. Round another table stood

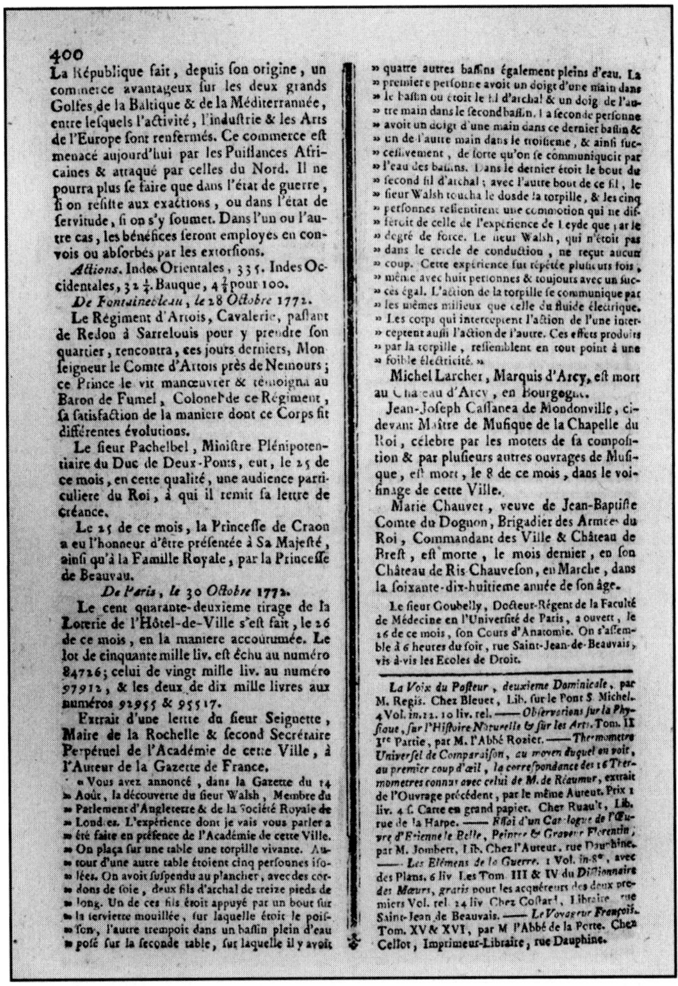

400

La République fait, depuis ſon origine, un commerce avantageux ſur les deux grands Golfes de la Baltique & de la Méditerrannée, entre leſquels l'activité, l'induſtrie & les Arts de l'Europe ſont renfermés. Ce commerce eſt menacé aujourd'hui par les Puiſſances Africaines & attaqué par celles du Nord. Il ne pourra plus ſe faire que dans l'état de guerre, ſi on réſiſte aux exactions, ou dans l'état de ſervitude, ſi on s'y ſoumet. Dans l'un ou l'autre cas, les bénéfices ſeront employés en convois ou abſorbés par les extorſions.

Actions. Indes Orientales, 335. Indes Occidentales, 32 ¼. Bauque, 4 ⅞ pour 100.

De Fontainebleau, le 28 Octobre 1772.

Le Régiment d'Artois, Cavalerie, paſſant de Redon à Sarrelouis pour y prendre ſon quartier, rencontra, ces jours derniers, Monſeigneur le Comte d'Artois près de Nemours; ce Prince le vit manœuvrer & témoigna au Baron de Fumel, Colonel de ce Régiment, ſa ſatisfaction de la maniere dont ce Corps fit différentes évolutions.

Le ſieur Pachelbel, Miniſtre Plénipotentiaire du Duc de Deux-Ponts, eut, le 25 de ce mois, en cette qualité, une audience particuliere du Roi, à qui il remit ſa lettre de créance.

Le 25 de ce mois, la Princeſſe de Craon a eu l'honneur d'être préſentée à Sa Majeſté, ainſi qu'à la Famille Royale, par la Princeſſe de Beauvau.

De Paris, le 30 Octobre 1772.

Le cent quarante-deuxieme tirage de la Loterie de l'Hôtel-de-Ville s'eſt fait, le 26 de ce mois, en la maniere accoutumée. Le lot de cinquante mille liv. eſt échu au numéro 84726; celui de vingt mille liv. au numéro 97912, & les deux de dix mille livres aux numéros 92955 & 95517.

Extrait d'une lettre du ſieur Seignette, Maire de la Rochelle & ſecond Secrétaire Perpétuel de l'Académie de cette Ville, à l'Auteur de la Gazette de France.

« Vous avez annoncé, dans la Gazette du 14 » Août, la découverte du ſieur Walsh, Membre du » Parlement d'Angleterre & de la Société Royale de » Londres. L'expérience dont je vais vous parler a » été faite en préſence de l'Académie de cette Ville. » On plaça ſur une table une torpille vivante. Au- » tour d'une autre table étoient cinq perſonnes iſo- » lées. On avoit ſuſpendu au plancher, avec des cor- » dons de ſoie, deux fils d'archal de treize pieds de » long. Un de ces fils étoit appuyé par un bout ſur » la ſerviette mouillée, ſur laquelle étoit le poiſ- » ſon, l'autre trempoit dans un baſſin plein d'eau » poſé ſur la ſeconde table, ſur laquelle il y avoit » quatre autres baſſins également pleins d'eau. La » premiere perſonne avoit un doigt d'une main dans » le baſſin ou étoit le fil d'archal & un doig de l'au- » tre main dans le ſecond baſſin. La ſeconde perſonne » avoit un doigt d'une main dans ce dernier baſſin & » un de l'autre main dans le troiſieme, & ainſi ſuc- » ceſſivement, de ſorte qu'on ſe communiquoit par » l'eau des baſſins. Dans le dernier étoit le bout du » ſecond fil d'archal; avec l'autre bout de ce fil, le » ſieur Walsh toucha le dos de la torpille, & les cinq » perſonnes reſſentirent une commotion qui ne dif- » féroit de celle de l'expérience de Leyde que par le » degré de force. Le ſieur Walsh, qui n'étoit pas » dans le cercle de conduction, ne reçut aucun » coup. Cette expérience fut répétée pluſieurs fois, » même avec huit perſonnes & toujours avec un ſuc- » cès égal. L'action de la torpille ſe communique par » les mêmes milieux que celle du fluide électrique. » Les corps qui interceptent l'action de l'une inter- » ceptent auſſi l'action de l'autre. Ces effets produits » par la torpille, reſſemblent en tout point à une » foible électricité. »

Michel Larcher, Marquis d'Arcy, eſt mort au Château d'Arcy, en Bourgogne.

Jean-Joſeph Caſſanea de Mondonville, ci-devant Maître de Muſique de la Chapelle du Roi, célebre par les motets de ſa compoſition & par pluſieurs autres ouvrages de Muſique, eſt mort, le 8 de ce mois, dans le voiſinage de cette Ville.

Marie Chauvet, veuve de Jean-Baptiſte Comte du Dognon, Brigadier des Armées du Roi, Commandant des Ville & Château de Breſt, eſt morte, le mois dernier, en ſon Château de Ris Chauveſon, en Marche, dans la ſoixante-dix-huitieme année de ſon âge.

Le ſieur Goubelly, Docteur-Régent de la Faculté de Médecine en l'Univerſité de Paris, a ouvert, le 16 de ce mois, ſon Cours d'Anatomie. On s'aſſemble à 6 heures du ſoir, rue Saint-Jean-de-Beauvais, vis-à-vis les Ecoles de Droit.

La Voix du Paſteur, deuxieme Dominicale, par M. Regis. Chez Bleuet, Lib. ſur le Pont S. Michel. 4 Vol. in-12. 10 liv. rel. —— *Obſervations ſur la Phyſique, ſur l'Hiſtoire Naturelle & ſur les Arts.* Tom. II 1re Partie, par M. l'Abbé Rozier. —— *Thermometre Univerſel de Comparaiſon, au moyen duquel on voit, au premier coup d'œil, la correſpondance des 16 Thermometres connus avec celui de M. de Réaumur*, extrait de l'Ouvrage précédent, par le même Auteur. Prix 1 liv. 4 ſ. Carte en grand papier. Chez Ruault, Lib. rue de la Harpe. —— *Eſſai d'un Catalogue de l'Œuvre d'Etienne le Belle, Peintre & Graveur Florentin*; par M. Jombert, Lib. Chez l'Auteur, rue Dauphine. —— *Les Elémens de la Guerre.* 1 Vol. in-8°, avec des Plans. 6 liv. Les Tom. III & IV du *Dictionnaire des Mœurs*, gratis pour les acquéreurs des deux premiers Vol. rel. 24 liv. Chez Coſtard, Libraire, rue Saint-Jean de Beauvais. —— *Le Voyageur François.* Tom. XV & XVI, par M. l'Abbé de la Porte. Chez Cellot, Imprimeur-Libraire, rue Dauphine.

Figure 16.2. The page of the *Gazette de France* of October 10, 1772, with the announcement of Walsh's experiments at La Rochelle based on the second letter sent by Seignette in order to rectify the previous report published by the same journal.

five persons insulated. Two brass wires, each thirteen feet long, were suspended to the ceiling by silken strings. One of these wires rested by one end on the wet napkin on which the fish lay; the other end was immersed in a basin full of water placed on the second table, on which stood four other basins, likewise full of water. The first person put a finger of one hand in the basin in which the wire was immersed, and a finger of the other hand in the second basin. The second person put a finger of one hand in this last basin, and a finger of the other hand in the third; and so on successively, till the five persons communicated with one another by the water in the basins. In the last basin one end of the second wire was immersed; and with the other end Mr. Walsh touched the back of the Torpedo, when the five persons felt a commotion which differed in nothing from that of the Leyden experiment, except in the degree of force. Mr. Walsh, who was not in the circle of conduction, received no shock. This experiment was repeated several times, even with eight persons; and always with the same success. The action of the torpedo is communicated by the same medium as that of the electrical fluid. The bodies which intercept the action of the one, intercept likewise the action of the other. The effects produced by the Torpedo resemble in every respect a weak electricity. (Walsh, 1773, pp. 467–468)

Walsh might have encouraged Seignette to write this letter, although this should be regarded as only a possibility. Similarly, Walsh might also have personally intervened to see it published or mentioned in some English newspapers. He probably was happy when it appeared in English in the *Middlesex Journal or Universal Evening Post,* a London newspaper, on November 24, 1772 (Seignette, 1772b) and also in general culture magazines of the times (as for instance *The Gentleman's Magazine*). After an inexplicable delay of some 18 years, it also appeared (in abridged form) in the *Public Advertiser,* another London paper, on January 15, 1790 (Seignette, 1790). The wording in these two newspapers was not identical, and both differed somewhat from that found in Walsh's article, suggesting different translators and/or editors intervening.

Thus, Seignette's letter might have lessened the need Walsh could have felt for an immediate publication written by himself in French (and perhaps in English), although it could in no way be considered a full account of his experiments and elaborations on the electricity of the torpedo. In this regard, it fails to justify the lengthy delay that occurred before his more complete report was submitted to the Royal Society on July 1, 1773. Knowing what happened when Walsh returned to the Royal Society, however, can help to make this delay more comprehensible, and it allows us to see the scientific issues faced by Walsh and his colleagues in better perspective.

DILEMMAS

Walsh returned to London on September 17, 1772, as announced by the *Public Advertiser* of the following day. Although he had some negative findings, he was satisfied with the results of the experiments he had conducted on the French coast, which seemed to justify his assertion in his earlier letter to Franklin "that the effect of the Torpedo appears to be absolutely Electrical." He was probably enthusiastic about conveying his findings to friends, supporters, and other interested parties, and perhaps even used the memorable phrase he had originally penned in French in his journal, "I have tamed it."

Nevertheless, in attempting to tidy his experimental notes and prepare the text he wished to present to the Royal Society, Walsh came across a series of difficulties that drew from the clear picture of the "torpedinal electricity" that he wished to provide. These problems undoubtedly contributed in a significant way to the lengthy delay prior to the publication of his final report to the Royal Society.

He understood that the case he wanted to make for torpedinal electricity suffered in two important ways: He failed to obtain a spark and he was unable to detect any electricity with his electrometers. He knew, of course, that sparks would not be obtained in experiments with frictional electricity, if a Leyden jar were only weakly charged. Indeed, in La Rochelle he had conducted experiments with a Leyden jar to verify this point. The last "Reflection" he annotated in his *Journal of the Experiments* for July 16, 1772, was "that the whole of the Torpedo is a weak Electricity." Soon after this annotation, however, he made a "Remark of future date" (that is, one written sometimes after the experiments, probably while he was revising his notes in London). Here he penned, "The Torpedo often gives severe Shocks, his Electricity therefore cannot be deemed weak" (Walsh, 1772b, p. 88).

His last remark obviously contradicts the previous statement. It reveals the conceptual difficulties Walsh was having when pondering torpedinal electricity. After all, the torpedo's electricity was strong enough to numb his hand, and sometimes also his shoulder. But at the same time, it was weak, since it could not jump across a tiny air gap, and it could not produce sparks or sounds. In an attempt to account for this contradiction, Walsh penned a question in his journal, which he also invoked in his letter of August 27 to Franklin: "May he [i.e., the torpedo] not have a better Conductor in his Skin, than in any Space of Air, however small?"

By trying to explain the difficulty in this way, Walsh brought forth yet another dilemma. If one assumes that the fish's skin is so highly conductive as to shunt all of its electricity, how could the shocks affect its prey or a predator some distance away, while it is in its natural habitat, seawater? With reflection and during conversations with his London colleagues, Walsh became strongly convinced of the need for further investigations on the subject, and this required enlisting others for their expertise.

This part of Walsh's intellectual journey would involve three Fellows of the Royal Society, each a giant in his field. One would be Franklin, who had advised Walsh before he left for France, corresponded with him there, and would go through Walsh's original *Journal of the Experiments* on his return to London. In his 1773 paper, Walsh even acknowledged the "crude and bulky state" of the notes Franklin had examined, records that would be edited and transcribed at a later date.

Franklin, to whom Walsh's communication was addressed, also provided Walsh with information and helped organize his material. He was also the person who submitted it to Matthew Maty, then Secretary of the Royal Society. The cover letter Franklin wrote reveals that Walsh had been very busy with more than scientific matters at this time.

These unspecified matters related to his being an active Member of Parliament, the management of his estate, and familial responsibilities (e.g., the well-being of his nephew Arthur and the troubles he was having with the apparent misbehavior of another nephew, Francis Fowke). Dated July 1, 1773, this short document reads:

> Our ingenious and worthy Brother Mr. Walsh, having long had the intention of drawing up from his Minutes a full Account of the numerous Experiments he made on the Torpedo, which Intention his other Avocations have not permitted him to execute, it is but lately that I have obtain'd his Permission to lay before the Society what he had in the meantime been pleased to communicate to me on that very curious and interesting Subject, or I should sooner have put it into your Hands for that purpose. (Wilcox, 1976, p. 257)

The second important person behind the finished product was John Hunter, the London surgeon and anatomist. He would dissect some of Walsh's French torpedoes and some larger ones just caught off the English coast. His dissections would reveal important details about their electric organs and nerve supplies.

The third important source of help at the Royal Society was Henry Cavendish, the eminent but eccentric physicist. His physical measurements and calculations would help Walsh and everyone else understand how a torpedo's electricity could at the same time be "weak" and "not weak." In brief, Cavendish would show that electrical events depend on two different parameters: one being "quantitative" (the "quantity" of electric fluid or electricity released; i.e., the "charge" in today's terminology) and the other "intensive" ("degree of electrification"; "electrical potential" or "voltage" today).

With the assistance of these exceptional individuals in mind, let us now look at Walsh's 1773 publication and then at a second publication on torpedoes by Jan Ingenhousz, another Member of the Royal Society, who had been asked to verify Walsh's findings. After doing these things, we shall examine what Hunter and Cavendish actually did in more detail, because their work affected what was printed in both of these publications, and markedly changed how people would view electric fishes in the future.

THE 1773 PHILOSOPHICAL TRANSACTIONS ARTICLE

After Franklin handed Walsh's materials to the Secretary of the Royal Society, the torpedo paper was read and subsequently published in the *Philosophical Transactions* for 1773 (Walsh, 1773) (Fig. 16.3). It bore the title "On the Electric Property of the Torpedo," appropriately followed by, "In a letter from John Walsh Esq; F.R.S. to Benjamin Franklin, Esq; LL.D., F.R.S., Ac. R. Soc. Ext, &c."

At the beginning of the paper, Walsh justified the study of torpedoes as "a subject not only curious in itself but opening a large field for interesting inquiry, both to the electricians in his walk of physics, and to all who consider, particularly or generally, the animal oeconomy," meaning physiology. History would more than prove Walsh right.

The substantive part of Walsh's publication was an assemblage of documents from 1772. "Loose and imperfect as these informations are, for they were never intended for

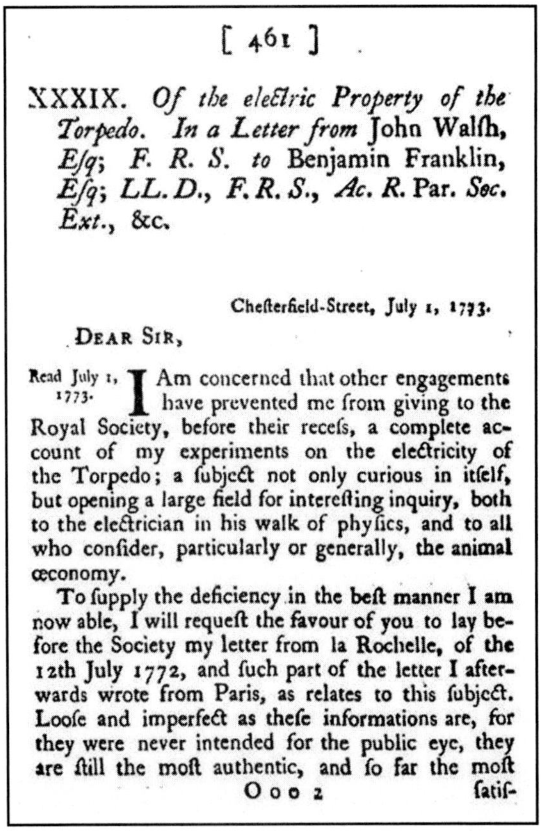
[461]

XXXIX. *Of the electric Property of the Torpedo. In a Letter from* John Walsh, *Esq; F. R. S. to* Benjamin Franklin, *Esq; LL. D., F. R. S., Ac. R.* Par. *Soc. Ext.,* &c.

Chesterfield-Street, July 1, 1773.

DEAR SIR,

Read July 1, 1773. I Am concerned that other engagements have prevented me from giving to the Royal Society, before their recess, a complete account of my experiments on the electricity of the Torpedo; a subject not only curious in itself, but opening a large field for interesting inquiry, both to the electrician in his walk of physics, and to all who consider, particularly or generally, the animal œconomy.

To supply the deficiency in the best manner I am now able, I will request the favour of you to lay before the Society my letter from la Rochelle, of the 12th July 1772, and such part of the letter I afterwards wrote from Paris, as relates to this subject. Loose and imperfect as these informations are, for they were never intended for the public eye, they are still the most authentic, and so far the most

Ooo 2 satis-

Figure 16.3. The title page of John Walsh's (1725–1795) torpedo publication of 1773 based on the letter he sent to Franklin on July 12, 1772 (see Chapter 15).

the public eye," he explained, "they are still the most authentic, and so far as the most satisfactory I can offer" (p. 258). Indeed, the first two documents were his letters to Franklin of July 12 and August 27, 1772, with small but significant changes, reflecting the greater confidence Walsh had now gained about torpedinal electricity. That he would present his results in this form is understandable, since at the time this was a way of showing the authenticity and reproducible nature of the experiments, which were performed before reliable witnesses. To make sure this message got across, Walsh made repeated references to certifiable public demonstrations and included documents other than his letters to Franklin.

The third document was, in fact, a complete English translation of Seignette's letter to the *Gazette de France* (see above). Walsh commented that the "exhibition of the electric powers of the Torpedo, before the Academy of La Rochelle," as described by Seignette, was held "on the 22d July 1772, and stands registered in the journals of the Academy"—thus again certifying the factuality of the events.

In the remainder of his paper, Walsh repeats some of the material in the letters, with the hope that he can convince skeptics that electricity could account for a torpedo's shocks. He particularly insists on experiments indicating that torpedo shock can be transmitted through water. As already noted, this was one of the most controversial features of electricity within the framework of eighteenth-century science. In this respect, it is interesting to note that Walsh acknowledges Franklin for his assistance "in forming hypotheses, how the Torpedo, supposed to be endued with electric properties, might use them in so conducting an element as water." Franklin's support to Walsh after his return to London clearly extended beyond editorial changes in his manuscript.

What stands out as really new in his published paper, as contrasted with his original letters to Franklin, is how Walsh now accounts for the apparent differences between torpedinal and artificial electricity. He lays out the problem:

> But it may be objected, that the effects of the Torpedo, and of the charged Phial [Leyden jar], are not similar in all their circumstances; that the charged Phial occasions attractive or repulsive dispositions in neighbouring bodies; and that its discharge is obtained through a portion of air, and is accompanied with light and sound; nothing of which occurs with respect to the Torpedo. (pp. 474–475)

To overcome these objections Walsh makes reference to electricity as an elastic matter that can exist in a more or less condensed state, clearly in line with the pneumatic theory of electric phenomena, which, as we shall see, was at the center of Cavendish's electrical research. Walsh continues:

> The same quantity of electric matter, according as it is used in a dense or rare state, will produce the different consequences. For example, a small Phial, whose coated surface measures only six square inches, will, on being highly charged, contain a dense electricity capable of forcing a passage through an inch of air, and afford the phaenomena of light, sound, attraction, and repulsion. But if the quantity condensed in this Phial, be made rare by communicating it to three large connected jars, whose coated surfaces shall form together an area 400 times larger than that of the Phial (I instance these jars because they are such as I use); it will, thus dilated, yield all the negative phaenomena, if I may so call them, of the Torpedo; it will not now pass the hundredth part of that inch of air, which in its condensed state it sprung through with ease; it will now refuse the minute intersection in the strip of tinfoil; the spark and its attendant sound, even the attraction or repulsion of light bodies, will now be wanting; nor will a point brought however near, if not in contact, be able to draw off the charge: and yet, with this diminished elasticity, the electric matter will, to effect its equilibrium, instantly run through a considerable circuit of different conductors, perfectly continuous, and make us sensible of an impulse in its passage. (pp. 475–476)

From the way he presents this hypothesis, it is clear that Walsh was personally involved in experiments to verify the circumstances under which a large quantity of electricity in a state of rarefaction (at a low potential, in a modern terminology) might have robust physiological effects, even if it cannot produce sparks or sounds. Soon after, he recognizes the role played by Cavendish in the formulation of this hypothesis: "Let me here remark, that the sagacity of Mr. Cavendish in devising and his address in executing electrical experiments, led him the first to experience with artificial electricity, that a shock could be received from a charge which was unable to force a passage through the least space of air" (p. 476).

Walsh now brings in the anatomy of the electric organ, which also introduces the reader to John Hunter's companion paper:

> But, after the discovery that a large area of rare electricity would imitate the effect of the Torpedo, it may be inquired, where is this large area to be found in the animal? We here approach to that veil of nature, which man cannot remove. This, however, we know, that from infinite division of parts infinite surface may arise, and even our gross optics tell us, that those singular organs, so often mentioned, consist like our electric batteries of many vessels, call them cylinders or hexagonal prisms, whose superficies taken together furnish a considerable area. (p. 476)

Thus, Walsh had acquired the facts he needed after he left France to inform his colleagues at the Royal Society that the absence of light and sound from a torpedo "is no ways repugnant to the laws of electricity" (p. 475).

Walsh concluded his paper with a tribute to Franklin, thanking him for all he had done, displaying wit and a wry sense of humor:

> He who predicted and shewed that electricity winds the formidable bolt of the atmosphere, will hear with attention, that in the deep [seas] it speeds a humbler bolt, silent and invisible. He who analysed the electrical Phial, will hear with pleasure that its laws prevail in animate Phials. He, who by Reason became an electrician, will hear with reverence of an instinctive electrician, gifted in his birth with a wonderful apparatus, and with the skill to use it. (p. 477)

INGENHOUSZ'S TORPEDOES

Franklin took it upon himself to distribute Walsh's 1773 paper to learned colleagues around the world. In 1774, he mailed copies to some friends he knew would have an interest in it, including Giambattista Beccaria in Italy and Jan Ingenhousz, then physician to the Austrian Court and Franklin's closest scientific and medical correspondent (Willcox, 1978, pp. 147–150; Conley & Brewer-Anderson, 1997; Finger, 2006a; also see Chapter 19).

Previously, on September 30, 1773, Franklin had informed Ingenhousz that at least one experimental natural philosopher "has not been able to perceive any certain Signs of Electricity in the Torpedo" (Willcox, 1976, p. 433). That individual was Sir William Hamilton, the intellectual British envoy to Naples. But rather than bothering to defend the electrical hypothesis in his letter to Ingenhousz, Franklin had remained open-minded, as was his custom. Hence, he added: "It is best that there should be two Opinions on this Subject: for all that may occasion a more thorough Examination of it, and finally make us better acquainted with it" (Willcox, 1976, p. 433).

Interestingly, Ingenhousz was able to make experiments on torpedoes in Leghorn, while on a trip to Italy several months before he received this letter, and more than a year before he received the copy of Walsh's publication, sent by Franklin. He had already heard what Walsh had done, and John Pringle, then President of the Royal Society, encouraged

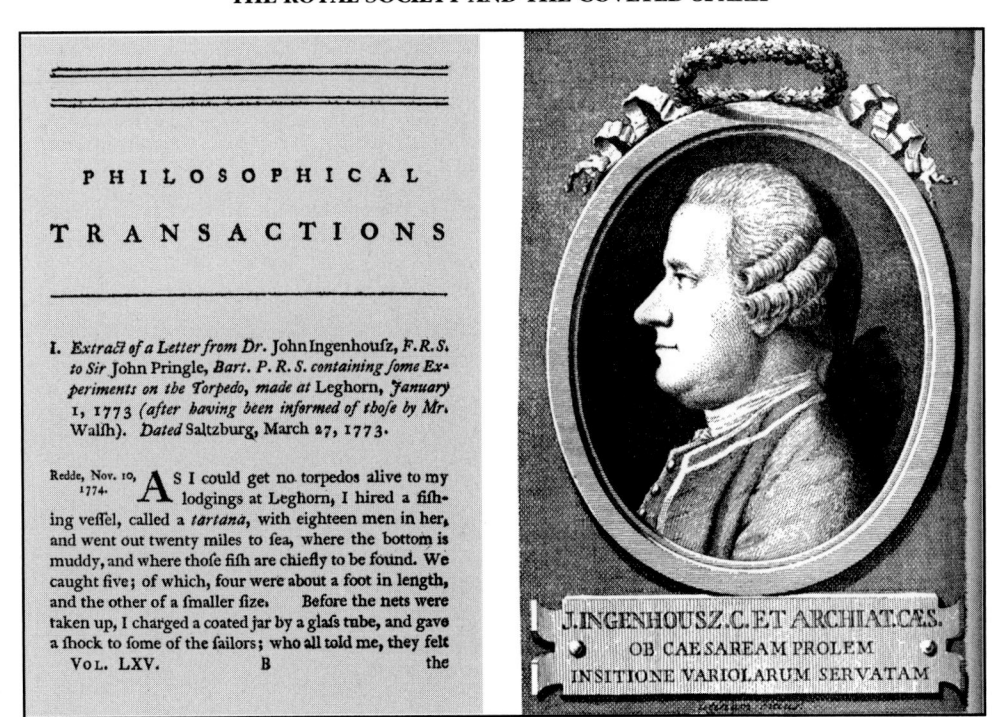

Figure 16.4. Jan Ingenhousz (1730–1799), physician, natural philosopher, and close friend of Franklin, with the first page of his 1775 article dealing with experiments on torpedoes captured in Italy after learning about Walsh's experiments.

him to try to verify Walsh's results. Pringle would later write that Ingenhousz, "being in Italy, when he received a general account of Mr. WALSH's success, at my request repaired to Leghorn, to make some experiments himself upon the torpedo" (Pringle, 1775a, p. 26).

Ingenhousz conducted his experiments on January 1, 1773, 6 months before the Walsh paper was formally submitted to the Royal Society. His informative letter was composed on March 27 and sent to Pringle from Salzburg, Austria. It was not read before the Royal Society until November 1774, and not published in the *Philosophical Transactions* until 1775 (Fig. 16.4).

In the lengthy title to this publication ("Extract of a Letter from Dr. John Ingenhousz . . ."), the reader is informed that these torpedo experiments were made "after having been informed of those by Mr. Walsh." Ingenhousz (or Pringle) wanted to make sure that readers understood that the author was not vying for priority, but rather was following up on what a Fellow of the Royal Society had already done. Ingenhousz also raises the possibility that his results could reflect the cold weather when he was in Leghorn, the fishermen telling him that cold diminishes a torpedo's powers and ability to survive out of water.

Ingenhousz (1775a) found only one similarity between a torpedo's shocks and artificial electricity; namely, both occasioned comparable sensations. Here Ingenhousz writes: "Before the nets were taken up I charged a coated jar by a glass tube, and gave a shock to some of the sailors; who all told me, they felt the same sensation as when they touched the torpedo" (p. 2). Moreover:

I took one of the torpedoes in my hand, so that my thumbs pressed gently the upper side of those two soft bodies at the side of the head, called (perhaps very improperly) *musculi falcati* by REDI and LORENZINI, whilst my forefingers pressed the opposite side. About one or two minutes after, I felt a sudden trembling in my thumbs, which extended no further than my hands: this lasted about two or three seconds. After some seconds more, the same trembling was felt again. . . . These tremors gave me the same sensation, as if a great number of very small electrical bottles were discharged through my hand very quickly once after the other. (Ingenhousz, 1775a, p. 2)

The rest of Ingenhousz's paper cast doubt, although only in an implicit way, on the electric nature of the torpedo's shocks. He writes that the fish does not attract light bodies (such as pith balls), cannot charge a Leyden jar, and does not produce a spark or noise. He also reports that the shock is not transmitted through "a brass chain to the back of the fish" (this finding is also present in Walsh's August 27, 1772, letter to Franklin).

Ingenhousz's possible disbelief in torpedinal electricity also seems to surface where he mentions his dissections of its electrical organs (indicated as "two soft bodies, called *musculi falcati*"). After pointing out (correctly) the four very large bundles of nerves with dense ramifications to these organs, he adds a remark that could suggest an implicit adhesion to the mechanical theory: "I did not observe whether these soft bodies changed in size when the torpedo gives shock; but I suspect they do" (p. 4).

Ingenhousz's letter produces mixed feelings, particularly if one considers the timing of the events leading to it and its relation to Walsh's research and publication. He made these experiments in order to test a hypothesis that Pringle asked him to verify (possibly because of the skepticism circulating among some members of the Royal Society), and his results seemed to oppose it. But when his letter arrived in London, Walsh and his colleagues at the Royal Society were already further substantiating the electrical nature of torpedo's shock! This chronology might explain a footnote Pringle now added to downplay Ingenhousz's negative finding with the brass chain:

> Dr. Ingenhousz means, that he felt no shock, though he saw the animal, by the contortion of its body, give one to the chain. At that time he did not seem to know, that though the shock would be communicated by a rod of any metal, it could not be so by a chain, or where was the least interruption of continuity. (Ingenhousz, 1775a, p. 3)

Cavendish, in collaboration with Walsh, was studying the transmissibility of electricity across metal chains at this time. His findings would account for the negative findings in this experiment.

A decade later, Ingenhousz viewed his experiments in Tuscany as totally supportive "of the truth that Mr Walsh had already demonstrated in a decisive way." This statement appeared in a memoir included in the first volume of his collected works (Ingenhousz, 1784, p. 29). At the time he published his collected works, Ingenhousz, who had by no means been alone in his initial skepticism, had become enthusiastic about fish electricity. This was largely due to Walsh, who would turn to more powerful eels to see a spark in darkness—a spark that Ingenhousz would confirm in experiments of his own with Walsh's eel (Chapter 19).

JOHN HUNTER'S DISSECTIONS

Unlike his brother Dr. William Hunter, who had studied medicine in Scotland before moving to London, John Hunter (Fig. 16.5) was a surgeon who never received a medical degree (Paget, 1897; Kobler, 1960; Dobson, 1969; Qvist, 1981 Moore, 2005; Stone & Goodrich, 2007). Yet "Mister" Hunter was gifted with the knife, garnished a reputation as a skilled dissector, and worked alongside his more erudite brother at their private Hunterian School of Anatomy. There, students paid to attend lectures and dissect fresh human corpses—the latter being an activity frowned upon by the directors of Britain's charity hospitals, who shunned controversy and relied on contributions and good relationships with the public for the survival of their institutions.

Hunter was known by just about everyone in London and well beyond the boundaries of the city by the 1770s. For some, it was because of his skills and attempts to make the practice of surgery modern, by basing it on dissections, experiments, and hard science. For others, it was because he was a "resurrectionist," reputed to obtain body parts and even steal whole bodies if they had special interest. And for those lucky people who managed to see them, it was because of his fantastic collections, comprising an anatomical museum second to none, where men of science and medicine could learn about cruel fates of nature and see exotica that others could only read about in books.

John Bancroft was familiar with John Hunter's writings and ideas when he was composing his essay on Guiana. In the third letter of the essay, in which he described the effects of Indian poisons, he mentioned how these poisons could be absorbed by the lymphatic system and cited Hunter's *Anatomical Lectures* (Bancroft, 1769a, pp. 304–305).

Franklin knew Hunter not only because of his science but also because both men had ties to Dr. William Hewson,

Figure 16.5. John Hunter (1728–1793), one of the most skilled anatomists of his day, who dissected some of Walsh's torpedoes.

a rising star in scientific medicine (Gulliver, 1846). Hewson had married Mary (Polly) Stevenson, Franklin's landlady's daughter, in 1770, 2 years after having won the Copley Medal. Notably, Franklin, John Hunter, and Pringle nominated Hewson into the Royal Society.

Hewson had been teaching anatomy at Great Windmill Street with the Hunters (who had helped train him), but he could not submit to William's demands and left to establish a private museum-school in the very house where his mother-in-law and Franklin were then residing, No. 7 Craven Street (Knapman, 1999). Interestingly, in a letter from Polly Hewson to Franklin dated October 22, 1772, we read that both Walsh and Bancroft were planning to attend lectures at Hewson's school. "Lectures go on briskly," Polly informed Franklin, "a fresh Pupil to day who makes up the half hundred whose names are enter'd, besides some others who have promis'd, among whom are your Friends Mr. Walsh and Mr. Bancroft" (Willcox, 1975, p. 341).

With regard to John Hunter and his involvement with torpedoes, Walsh had preserved some of his French torpedoes in brandy, the most common preservative in the era before formaldehyde. Then, when he brought them back to London, he handed them to Hunter, who was clearly the best man anywhere to study their anatomy. For his part, Hunter felt that these rays were clearly deserving of study, and he may have asked Walsh for them even before Walsh

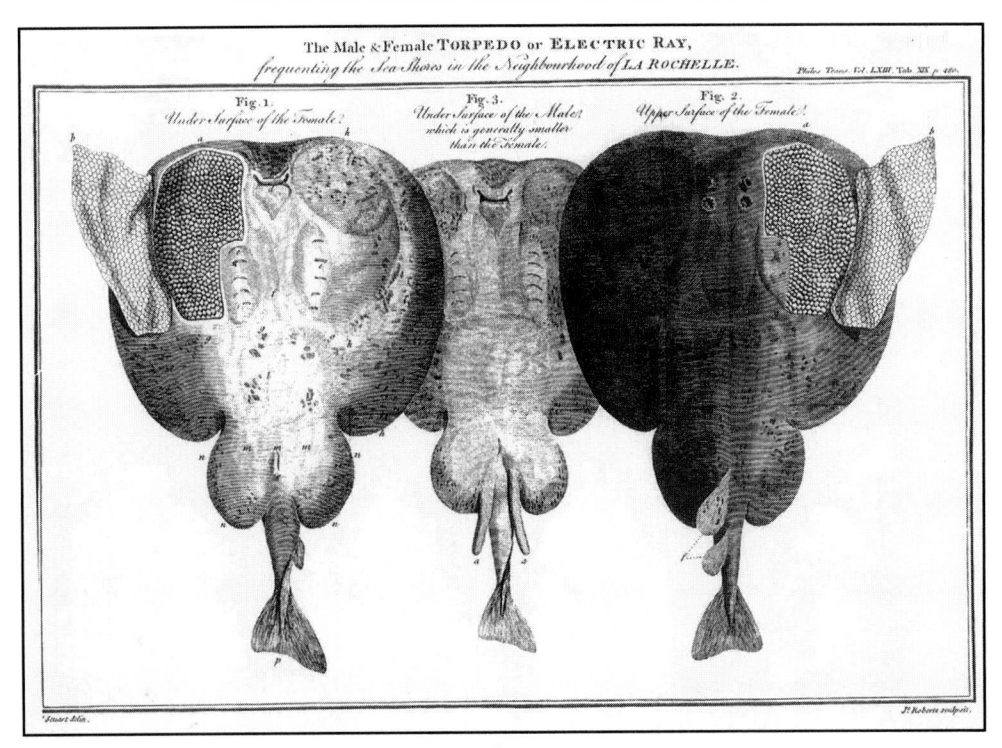

Figure 16.6. Hunter's (1773) torpedo dissections showing the electrical organs.

departed for France. In his 1773 report, which accompanied Walsh's own paper, Hunter writes:

> I was desired for some time since, by Mr. Walsh, whose experiments at La Rochelle had determined the effect of the Torpedo to be electrical, to dissect and examine the particular organs by which that animal produces so extraordinary an effect. This I have done in several subjects furnished to me by that Gentleman. (Hunter, 1773, p. 481)

Hunter verbally described and beautifully illustrated the physical structure of the ray's electric organ (Figs. 16.6 and 16.7). He showed that it was made up of hundreds of perpendicular columns, reaching from the upper to the under surface of the body, and varying in their lengths, according to the thickness of the parts of the body where they are placed. Each organ seemed to have about 470 columns, but this varied with the size of the fish. Most of the columns looked like "either Hexagons or irregular Pentagons." Further, "Each column is divided in horizontal partitions, placed over each other, at very small distances, and forming numerous interstices, which appear to contain a fluid." The partitions "consist of a very thin membrane," and each inch of column is made up of about 150 separated, horizontal disks.

The nerve supply to the electric organs was described as extensive. The nerves, "having entered the organs, ramify in every direction, between the columns, and send in small branches upon each partition" (p. 486). Seeing that it is unlikely that this great abundance of nerves could be accounted for by the sensory or motor needs of these organs, he made an important statement connecting electricity to nerve function:

> If it be then probable, that those nerves are not necessary for the purpose of sensation, or action, may we not conclude that they are subservient to the formation, collection, and management of the electric fluid; especially as it appears evident, from Mr. Walsh's experiments, that the will of the animal does absolutely control the electric powers of it's [sic] body which must depend on the energy of the nerves. How far this may be connected with the power of the nerves in general, or how far it may lead to an explanation of their operations, time and future discoveries alone can fully determine. (p. 487)

Hunter's torpedo dissections extended beyond the electric organs of the French torpedoes, some of which can still be seen at the Science Museum in London (items 2168 to 2179) (Fig. 16.8). In August and then in November 1773, two large torpedoes were caught off England's Torbay Coast. Their capture must have come as a surprise to Walsh, who on July 22, 1772, had told members of the *Académie de La Rochelle*, "Nature had denied the animal to our Country." Thinking there were very few if any torpedoes to be had in England, he even asked Seignette to conduct some additional experiments in his absence, wanting to learn more about the relative transmission of a torpedo's shocks through water, woods, cords made of hemp, and metals (see Piccolino, 2003a).

Interestingly, the opportunity to study the huge English rays came about because Walsh had taken the initiative to make inquiries about the possibility of torpedoes "in some of our southern fishing ports." In other words, he wanted to be informed if torpedoes happened to be caught by English fishermen. This initiative allowed him to acquire two torpedoes from Torbay, on the Derbyshire coast, one in August and the other in November 1773.

These two torpedoes were much larger that the largest one he had in La Rochelle, which had weighed 10 pounds and was less than 2 feet in length. A London apothecary

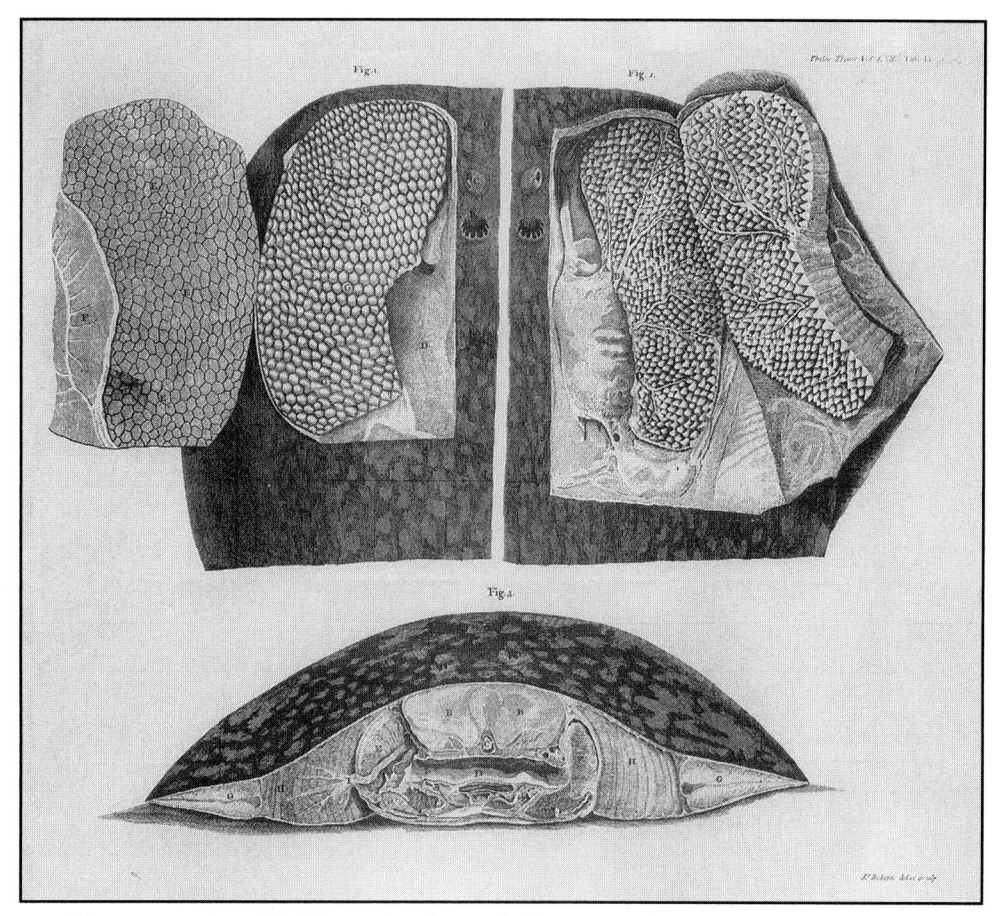

Figure 16.7. A close-up of the cross-section of the Hunter (1773) torpedo illustration showing the columnar organization of the electrical organ.

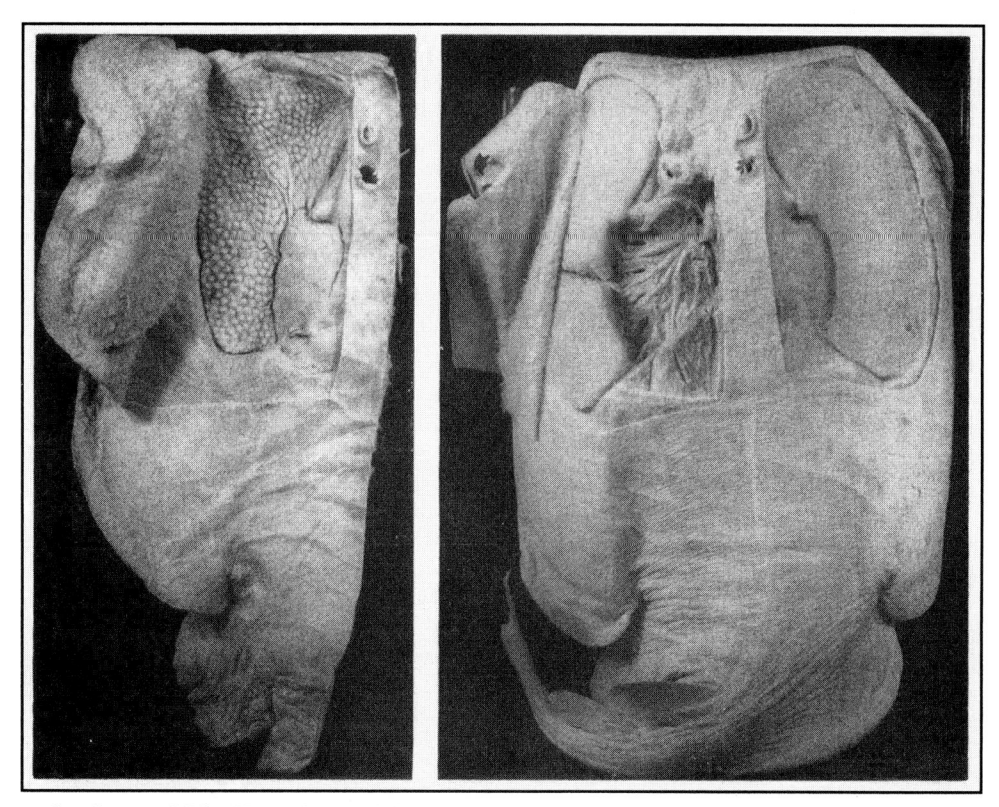

Figure 16.8. Photographs of some of John Hunter's actual dissections using torpedoes he obtained from John Walsh. These were the torpedoes that Walsh had brought to London from La Rochelle, preserved "in brandy". (© The Hunterian Museum at the Royal College of Surgeons)

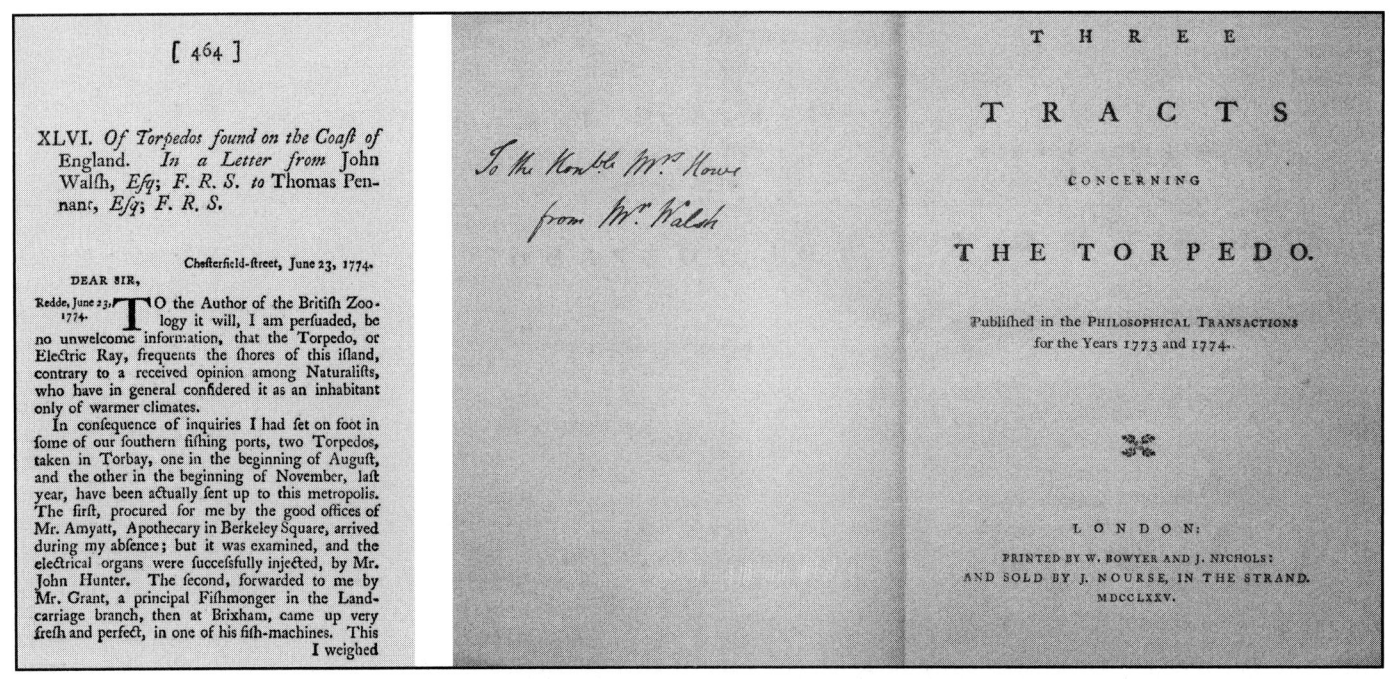

Figure 16.9. The 1774 article by Walsh on the large English torpedoes, with a booklet he published separately in 1775. This booklet, titled *Three Tracts on the Torpedo*, includes Walsh's 1773 and 1774 articles on the fish together with the anatomical paper published by Hunter in 1774. This particularly copy shows an autograph dedication by John Walsh. (© "The Bakken," Minneapolis, for the *Three Tracts*)

procured the first fish for Walsh while he was away. Although it was not weighed, it was thought to be similar to the second torpedo, which was provided by "a principal fishmonger" at Brixham and "came up fresh and perfect in one of his fish-machines." This fish was found "to weigh fifty-three pounds avoirdupois [about 70 pounds or roughly 32 kilograms], and to measure four feet [over a meter] in length, two feet and a half in breadth, and four inches and a half in its extreme thickness" (Walsh, 1774b, pp. 464–465) (Fig. 16.9).

Since the two English rays were delivered dead, Walsh was unable to do physiological experiments on them. But the fishmonger related that "their numbing quality is pretty strong through the net, though much weaker than when they are taken out" (Walsh, 1774b, p. 468). Walsh did, however, examine the gross anatomy of one of them, and he noted that their backs were darker in coloring than the French torpedoes he had studied.

The fine anatomy of both fish was left to Hunter, who included information about these torpedoes in his 1773 publication. Hunter's work on them was also mentioned several times by Walsh in his separate paper on the English torpedoes, where he wrote:

> The electric organs of this torpedo [the larger one] were likewise injected by Mr. Hunter, though not with his first success, from the bursting of the artery in the operation; he determined, however, the number of columns in one organ to amount to 1182, and fully confirmed the observation he formerly made, that their numerous horizontal partitions were very vascular. (Walsh, 1774b, p. 466)

These torpedoes were probably *Torpedo nobiliana*. Carlo Luciano (Charles-Lucien) Bonaparte, the "Emperor of Nature," would describe this variety in an article completed in 1835 for his three-volume *Fauna Italica*, which had 30 parts, the last volume appearing in 1841 (Bonaparte, 1832–41; for more on the Emperor Napoleon's naturalist nephew, see Stroud, 2000). Hoping to become a member of the Royal Society of London, Bonaparte wrote to Richard Owen, "perhaps this favourite subject of theirs may induce them to admit me as a candidate for the first available place among their foreign members" (Stroud, 2000, p. 184). Long after Walsh and Hunter did their studies on torpedoes, they continued to fascinate the British, and just about everyone else, as these words show.

Walsh simply identified his domestic torpedoes as the black variety alluded by Scribonius Largus and others. He does this at the close of the article, after first stating that he had made a trip to Ireland, where he encountered some fishermen also acquainted with it, but who remarked they could go for years without catching a single *Aunghelláw*, the local name for it. Walsh's concluding note again shows his sense of humor, as he equates its medical powers with the more easily obtained Leyden jar (while also showing his opposition to vitalistic notions about torpedo electricity):

> I have thus shewn that Great Britain too claims the Torpedo, or Electric Ray; that ours is the *broad marine sort*, which Socrates, as Meno thought, resembled; and that it is the black Torpedo, whose influence subdues obstinate Head-achs, and the Gout itself [here he inserts a footnote to Scribonius]. In announcing to our Naturalists and Electricians the presence of this wonderful guest, I should certainly felicitate our Invalids on their acquisition, but that the *Leyden phial contains all his magic power*. (Walsh, 1774b, pp. 472–473)

HENRY CAVENDISH'S PHYSICS

In addition to Franklin and Hunter, Walsh, as noted, had been interacting directly with Henry Cavendish, who had been devoting his life to physics and the Royal Society. Cavendish (Fig. 16.10) was an eccentric genius: a fairly reclusive, socially awkward, and shoddy-looking man, whose family wealth allowed him to pursue natural philosophy, and in particular an understanding of electricity (Jungnickel & McCormmach, 1999). His father, Lord Charles, who had received the 1757 Copley Medal and had been Vice-President of the Royal Society, had a special interest in electrical experiments. He extended Franklin's research on the Leyden jar and particularly on the role of glass in electric capacitors, which was also central to his son's interest. In 1752, one of his instruments was used in a famous experiment of electric conduction across the Thames River, which was sponsored by the Royal Society. Lord Cavendish, however, did not have a penchant for publishing, prompting Franklin to write: "It were to be wished, that this noble philosopher would communicate more of his experiments to the world, as he makes many, and with great accuracy" (Labaree, 1966, p. 42).

Franklin's statement is also applicable to Henry Cavendish, who published little and whose articles were brilliant but tortuous. Henry learned mathematics at Cambridge and became a Fellow of the Royal Society in 1760. The greatest English physicist of his epoch, he became particularly famous for being the first to measure the gravitational constant implied in Newton's theory of the universe (using a Coulomb's torsion balance). Yet he left much of his work unpublished and many of his electrical experiments came to light only about a century later, thanks to James Clerk Maxwell, another great English physicist.

Figure 16.10. English physicist Henry Cavendish (1731–1810), whose mathematics and artificial torpedoes helped Walsh, Franklin, and other natural philosophers to understand why real torpedoes may not produce sparks, how their discharges travel through the water, and more.

Among his few published articles, there is one on torpedoes, which solved many of the conceptual problems with fish electricity. As mentioned, the main difficulties with the torpedo concerned the transmission of the shock through water, the inability of the shock to cross even the tiniest air gap, the absence of a spark, and its inability to attract or repel light bodies (e.g., move pith balls), or to register on any other kind of electrometer. In addition, there was the possible lack of transmissibility across a metal chain, alluded to by both Walsh and Ingenhousz. Walsh was aware of these phenomena, and even though he believed that they were not reasons for rejecting the electric hypothesis, he also knew they made it that much harder to accept. What he therefore asked Cavendish was whether these issues could be explained by physics—in effect, whether he would be willing to study these issues with the tools of his trade.

Cavendish (1771) had recently published an important and lengthy paper in which he attempted to account mathematically for some electric phenomena. In doing this, he closely followed Newton's approach for explaining universal gravitation, as laid out in his *Principia*. The similarity with Newton's masterpiece did not end here. As Maxwell later acutely remarked, almost no physicist seems to have fully understood what Cavendish was saying! This was unfortunate for the progress of science, because Cavendish's presentation was the most important mathematical approach to electricity before Siméon-Denis Poisson performed his own mathematical studies on the electrical charge distributions in 1812 (see Maxwell, 1879 Jungnickel & McMormmach, 1999).

The hypothesis Cavendish elaborated was based on Franklin's conception of a single fluid. It assumed that electrical interactions decrease with the square of the distance between charged bodies (another similarity with Newton's thesis of gravitational forces). It could account for attraction and repulsion phenomena, and for the importance of points, in addition to which it could explain the functioning of the Leyden jar and flat capacitors, such as the Franklin square capacitor. Although hardly evident in his published paper, but clear in his laboratory notes, Cavendish's thinking was based on a pneumatic model of the electric charge, in which it could be in more or less compressed states, modifiable by external pressures, as thought to be true of an elastic fluid (Maxwell, 1879 pp. 95–103).

On the basis of the electrical acumen displayed in the 1771 paper, the Royal Society appointed Cavendish to a commission to protect the arsenal at Purfleet from lightning strikes. Franklin and other distinguished electricians joined him on this task. For these and related reasons, Cavendish was viewed by Walsh, and by Franklin as well, as the physicist with the skills needed for dealing with torpedinal electricity.

The discussions between Walsh and Cavendish might have begun late in 1772, as Cavendish's published logs show that in November 1772 he began to measure the resistances of various types of water, including seawater, which was of obvious importance for understanding the electrical operations of the torpedo in its natural habitat (see Maxwell, 1879 p. 262). He extended his measurements in 1773, after Walsh's material was presented to the Royal Society, and he intensified his research again in 1775, as attested by his laboratory notes.

As was the custom among electricians of the epoch, he used his own body in his work, comparing sensations from electric discharges under various physical conditions (see Chapter 22 for more on self-experimentation). As put by Maxwell (1878 p. LVII), who edited Cavendish's notes and lavished praise on him, he "was his own galvanometer... by comparing the intensity of the sensations he felt in his wrist and elbows, he estimated which of the two shocks was the more powerful."

Cavendish also performed detailed mathematical calculations, using Walsh's descriptions of the fish shocks and Hunter's morphological findings, "to examine whether these circumstances [e.g., no sparks] are really incompatible with such an opinion" [torpedoes being electrical]. And he conducted many experiments in the stables of his London residence on Marlborough Street, where he and father had a well-equipped laboratory filled in part with instruments they invented or improved.

None of Cavendish's experiments were, however, on real fish! Rather, he experimented with artificial instruments and developed physical models, which he designated as "artificial torpedoes." As we shall see, this approach would not be unique to Cavendish, and it would culminate in Volta's "artificial electric organ"—that is, his electric battery (see Chapter 22).

Only rarely was somebody admitted to this laboratory. But Henry Cavendish made an important exception on May 27, 1775, because he wanted somebody to witness what he had been able to achieve with his artificial torpedoes. On this occasion, he demonstrated his model to John Hunter, Joseph Priestley, Thomas Ronayne, Timothy Lane, and Edward Nairne—a group of scientists that included some men skeptical about torpedinal electricity.

Word of a recent letter sent by William Henly (d. 1779), an instrument maker (e.g., an electrometer), to William Canton might have contributed to Cavendish opening the doors of his protected scientific refuge. William was the son of John Canton, one of London's experts of electricity and Franklin's friend (see Chapter 10). Henly's letter was received on March 14, 1775, and is worth quoting *in extenso* because its sarcastic style shows that he, for one, was not willing to accept either torpedo electricity or an "artificial torpedo" made with electrical components:

> I beg your acceptance of an artificial torpedo; which I believe you will acknowledge to be a good imitation of the natural one, whom I assure you is composed entirely of conductors: consequently you will have no attraction, no repulsion of light bodies, no snap, no light, nor indeed any sensation, not even the engourdissement [numbness], or fourmillement [swarming], in the discharge. But, shocks in the artificial torpedo may well be accused: for as he is not quarrelsome, nor liable to be destroyed by voracious adversaries; the power of giving a shock, would to him, be useless. Besides as the *electrical shock* of the natural torpedo, can be dispensed at the will of the animal: so in the artificial one, you can only expect him at his pleasure. It may therefore be advisable, not to promise a shock from him, to your acquaintance; but to recommend it to them to be content with his exhibition of the negative phenomena of the torpedo; if any of them should question his power (as being formed of conductors) of giving an electrical shock: all that I can say in his behalf is, that the natural torpedo, in this particular respect, lies under the very same predicament, is that after all the trumpeting we have heard of him, he may be as utterly unable to give a *true electric shock* as the artificial one. (Henly, 1775)

Cavendish's paper was read at Royal Society on January 1776 (as shown in the Royal Society's journal book, although the date indicated near the title is 1775). It was published in 1776 in the *Philosophical Transactions* (Cavendish, 1776a). The first point addressed in this paper concerns the possibility that a torpedo's shock could be transmitted at distance, while the fish is in seawater (the "capital Desideratum" of Walsh's demonstration at La Rochelle). As noted, the difficulty was that water was generally viewed as a pretty good conductor of electricity. This being the case, the expectation might be that the fish's electricity would be shunted through the water immediately adjacent to the fish's body, never reaching distant objects in a perceptible way.

To solve this dilemma, Cavendish referred to studies in progress on the conduction of electricity in various types of water. He explains that rain or distilled water is a far worse conductor than metals (he writes "400 million times" worse, although the difference with very pure rain water is even bigger, more like a billion), and that seawater conducts only about 100 times better than rain water.

Being interested in electric conduction even before his involvement in torpedo research, Cavendish extended his studies and reached conclusions similar to those fully elaborated by Johann Wolfgang Ohm in 1827. He devoted particular attention to the many circuits electricity could take when a charged body (e.g., a torpedo) is connected to various conductors in parallel, or to a spatially distributed conductor (as with a torpedo in water).

Again Cavendish seemed to anticipate future developments. Emil du Bois-Reymond (1848–50, Band I, p. 564) would remark that there would not be any real progress in the theory of current division across many conductors in various arrangements for another 70 years—not until Gustav Robert Kirchhoff would come forth with his famous laws of electrical circuits in the mid-nineteenth century. With regard to Cavendish's anticipation of Ohm's law, two of Cavendish's more recent biographers write that "he arrived at a good many other results in electric conductions that others after him would rediscover." These historians continue: "We make this observation about so many Cavendish's researches that it becomes tiresome, but it is the truth all the same" (Jungnickel & McCormmach, 1999, p. 187).

The sentence in which Cavendish outlines the diffusion of torpedo electricity in water is worth quoting because he refers to an illustration that shows, probably for the first time in the sciences, the spatial diffusion of an electric current—an image that would have a great influence on future developments involving the electrical field (Fig. 16.11). He writes that "if torpedo is immersed in water, the fluid will pass through the water in all directions, and that even to great distances from its body, as represented in Fig. 1, where the full lines represent the section of its body, and the dotted lines the direction of the electric fluid; but it must be observed, that the nearer any part of the water is to the fishes [sic] body, the greater quantity of fluid will pass through it" (Cavendish, 1776a, p. 199).

Having shown that the diffusion of the shock in water poses no theoretical dilemma, Cavendish proceeds to address

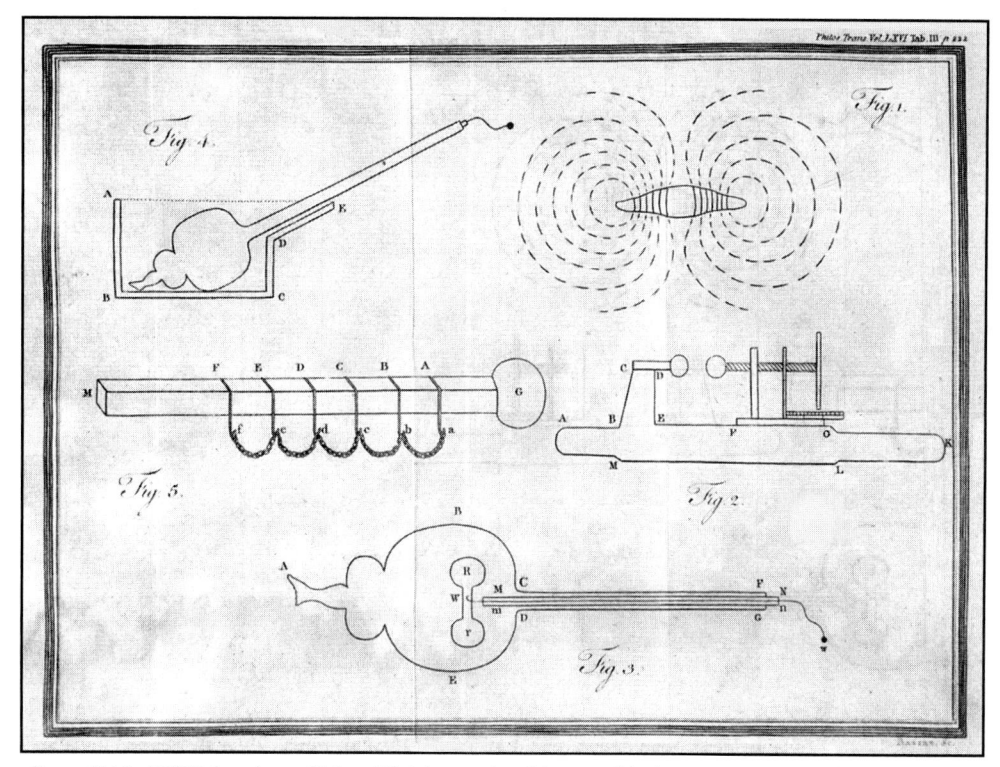

Figure 16.11. Henry Cavendish's (1776) drawings of his artificial torpedo with a graphical representation of the diffusion of electricity in the water.

what some philosophers viewed as even more perplexing: why a torpedo's shocks do not produce sparks and sounds, or jump across the minutest gap in a circuit. Related to this problem is the observation, made by both Walsh and Ingenhousz, that the shock might not even be transmissible via a small metal chain (both men tried brass chains).

The way Cavendish overcame this difficult problem might be the highlight of his torpedo paper, and it is connected to some of the research he had previously done. Several of those studies were presented in his 1771 paper, but others were left unpublished, including a small manuscript from about 1767 titled "Thoughts Concerning Electricity" (see Maxwell, 1878, pp. 94–103; Jungnickel & McCormmach, 1999). Here we find Cavendish clearly recognizing from his careful experiments that the effects of an electric discharge depend on two different parameters that can work independently and be measured separately. He calls one of these parameters "degree of electrification" and refers to the other as the quantity of the charge or of the electric fluid.

Here he compares the sparking distance of a large assembly of Leyden jars connected in parallel (to use today's terminology) with that of a single jar charged to the same degree (as measured with a sensible straw electrometer), and experienced no sensible differences. Cavendish (1776a, p. 201) concludes, "the force with which the fluid endeavours to escape from the single jar is the same as from all the jars together." This force corresponds to the "degree of electrification," and in modern physics it would be its electric potential. With the pneumatic terminology of Cavendish's previous studies, this parameter was called the "pressure" of the electric fluid.

Even though the sparking distance of an ensemble of Leyden jars is the same of that of a single jar charged to the same degree, the shock it produces is stronger, and it increases in proportion to the number of jars connected together.

Cavendish therefore concludes that shock intensity also depends on the quantity of charge present. Indeed, he shows that, under the conditions of his experiments, it depends more on this parameter than on the degree of electrification. Moreover, by increasing the number of Leyden jars and reducing the degree of electrification in an appropriate way, he arrives at a condition in which the shock is strong and comparable to that of the real torpedo.

Since he did not experience a live torpedo himself, Cavendish relied on Walsh's subjective estimates of what a real torpedo feels like, and how this compares to shocks from his collection of Leyden jars, for this part of his work. Oddly, though, he did not include Walsh's name in his published notes, although he did provide the names of some expert witnesses.

In perhaps his most famous demonstration, he used an assembly of 49 Leyden jars connected to a device made of wood that had a shape similar to a torpedo. He called it an "artificial torpedo" and he immersed it in water through an insulated handle. With this setup, and by using a Lane's electrometer (see Chapter 15) and other measurement tools, he demonstrated that an artificial torpedo powered by a great number of Leyden jars that are weakly charged can give a sizeable shock, even if there is no spark and the discharge does not pass though a small metal chain. Thus, Cavendish made the electrical nature of the torpedo's shock more acceptable with his artificial torpedo.

Cavendish knew what Hunter's research had revealed about the torpedo's electric organs, and he used this information in his calculations. The total area of his large assembly of Leyden jars was comparable to the total area of the discs forming the fish's electrical columns. This equivalence was viewed as supportive of his physical model of the fish's electric organ, an organ that has the appearance of stacks of small

Franklin square capacitors (see Chapter 10). This association would reappear at the end of the century (see Chapter 22).

Cavendish modified his artificial torpedo in various ways. For example, he made one with leather soaked with salt water to take the conductive nature of a torpedo's body into account. And with his models, he succeeded in replicating the sensory characteristics of a real torpedo's shocks, as described by Walsh. Among other things, he reproduced the differences in intensity Walsh had felt when he compared the shocks from his fish in water and in air.

He even replicated what "Anteros, a freedman of Tiberius," might have felt when he stepped on a live torpedo at the seashore and was cured of his gouty pain—the story circulated in the first century by Scribonius Largus (Chapter 4). To do this, he buried his artificial torpedo under wet sand, but rather than step on it with his bare feet, "which would have been troublesome," he transmitted the shocks through wet leather straps to his hands. He concluded that a person stepping on a live torpedo in the sand might even fall down, as had been noted in some earlier reports, "considering how much the effect of the shock would be aided by the surprise" (p. 215).

But why do electrometers failed to pick up the discharge? Here Cavendish reasoned that the shock of a living torpedo is such a rapid event that no electrometer known could record the action. "A pair of pith balls . . . will not have time to separate, nor will a fine thread hung near its body have time to move towards it, before the electricity is dissipated" (pp. 103–104). He substantiated his explanation by pointing to Priestley, who did experiments involving an assembly of discharging Leyden jars, yet who also "never could find a pair of pith ball suspended from the discharging rod to separate" (p. 204).

With an ingenious physical method and the use of Franklin's square or planar capacitor (a "plate of coated crown glass," p. 207), Cavendish was able to compare the relative importance of the two electric parameters influencing the intensity of the shock: the degree of electrification and the quantity of charge. He writes in a footnote: "I find, by experiment, that the quantity of electricity which coated glass of different shapes and sizes will receive with the same degree of electrification, is directly as the area of the coating, and inversely as the thickness of the glass; whence the proportion which the quantity of electricity in this battery bears to that in a glass jar of any other size may easily be computed" (p. 206). If we substitute "electric potential" in place of his "degree of electrification," we discover once again that Cavendish has "anticipated" an important scientific milestone, because this is nothing less than the law of the capacitor, a fundamental law of the physics of electricity.

Thus, with his "artificial torpedo" and gift for mathematics, Cavendish showed that "there it seems nothing in the phenomena of the torpedo at all incompatible with electricity" (p. 222). But this is not all. He also provided a clearer, more modern understanding of the laws of electricity.

This brings us back to Walsh, who seemed to anticipate precisely this sort of exchange—one that would benefit both the natural sciences and physics. In his discourse at the La Rochelle *Académie* (later modified in his *Philosophical Transactions* paper), Walsh opined that "as artificial Electricity has led to a discovery of some of the operations of the torpedo, the Animal if well considered would lead to a discovery of some truths in artificial Electricity which were at present unknown and perhaps unsuspected" (Walsh, 1772b, p. 145).

WALSH'S COPLEY MEDAL AND PRINGLE'S SPEECH

In 1774, the Royal Society gave its highest honor, the prestigious (Godfrey) Copley Medal, to Walsh for his research on torpedoes. John Pringle, its President, presented the award,

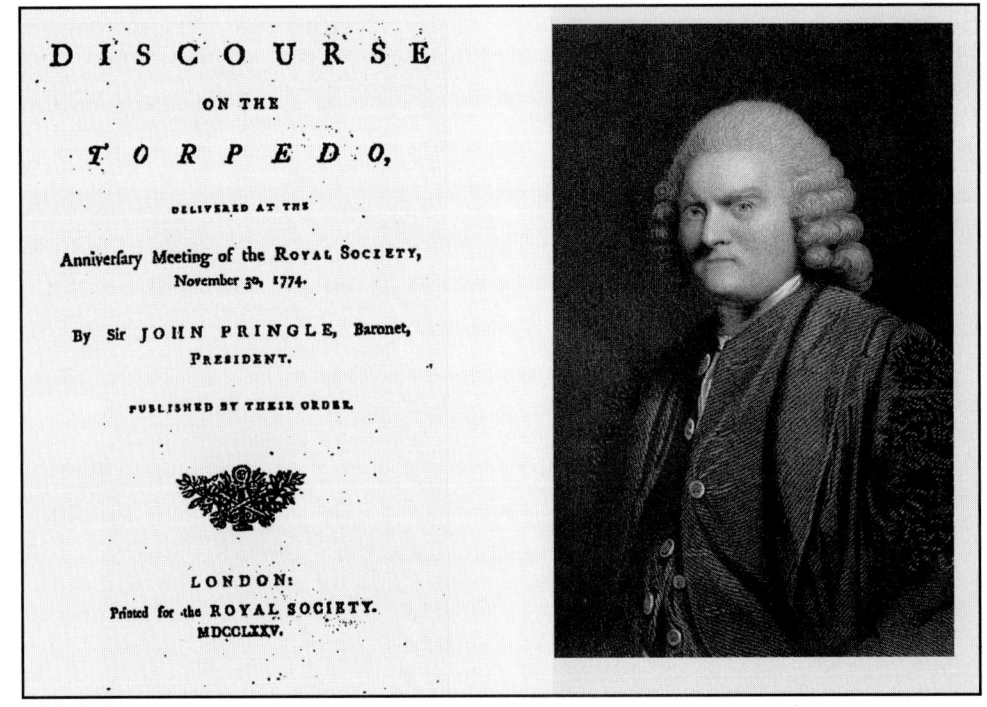

Figure 16.12. John Pringle, who handed Walsh the Royal Society's Copley Medal in 1774, after he gave a discourse on the history of electric fishes, published in 1775 and shown on the left.

which had been established in 1736 and had been bestowed upon Franklin in 1753 (and would be bestowed to Rumford in 1792 and to Alessandro Volta in 1794) (Fig. 16.12). Pringle's speech, delivered at the "Anniversary Meeting of the Royal Society, November 30, 1774," was published the next year (Pringle, 1775a) and was quickly translated into French and Italian (Pringle, 1775b, 1776). It would also be the subject of parodies in libertine pamphlets, as we shall see in Chapter 20.

He starts by mentioning some of the ancients (e.g., Hippocrates, Plato, Aristotle, Hero of Alexandria, Scribonius Largus), who were captivated by electric fishes. He stresses the need for less speculative thinking than some of the authorities of the past had exhibited, and the importance of experimentation. In this context, he points out that Pliny greatly exaggerated things and was "too great a lover of the marvellous" (p. 7); that Aelian provided some of "the lamest and most fabulous accounts of this subject" (p. 8); and that "it does not appear, there has been one, Galen excepted, of all the above mentioned ancient sages, who has ever seen a living torpedo, much less who had made experiments on it; and least of all who had dissected it" (p. 11). It was not until the sixteenth century that Belon, Gessner, and others began to look more closely at torpedoes. This change, Pringle relates, set the stage for the dissections of Redi and Lorenzini. But these men speculated that rapidly contracting special muscles could account for the torporific effects, and in this they were wrong.

Pringle points out that a better understanding of electricity had led to new ways of looking at nature, and specifically at those fishes that now seemed to be comparable to Leyden jars. He writes that Walsh deserves the credit for providing scientists with "not only the first, but a numerous set of the best chosen experiments on the torpedo, for ascertaining its electrical nature" (p. 22).

Elaborating on Walsh's groundbreaking experiments, Pringle cites specific examples, lavishing praise on the experimenter for going to France and doing what he did before reliable witnesses "for the sake of truth and science." He also praises the French academicians, who accepted his findings without envy, as befitting "true lovers of science." And he thanks John Hunter for his wonderful dissections and illustrations of Walsh's torpedoes, revealing their electrical organs and nerve supplies.

"If these reflections be just," Pringle (pp. 29–30) adds, "we may with some probability foretell, that no discovery of consequence will ever be made by future physiologists, concerning the nature of the nervous fluid, without acknowledging the lights they have borrowed from the experiments of MR. WALSH upon the living torpedo, and the dissection of the dead animal by MR HUNTER."

Placing the Copley Medal in Walsh's hand, Pringle concludes by wishing him future successes. He notes, "With pleasure they [the members of the Royal Society] understand that you have already turned your views to the electric gymnotus, that other wonder of the waters" (p. 31). Confident that Walsh will soon uncover other mysteries of nature, he advises him, "Her veil, fear not, SIR, to approach," playing on the phrase, "We here approach to that veil of Nature, which man cannot remove," from Walsh's landmark paper on torpedoes, which in turn referred to the mystery hidden under the veil of Isis, the Egyptian goddess.

Walsh would make some great new discoveries with the electric eel, including obtaining a spark from it. The continuing saga, however, has some unusual twists and turns to it. In the next two chapters, we shall see how some North Americans, about whom almost nothing has been written in this context, helped lay the groundwork for Walsh's ultimate triumph.

Chapter 17
Out of the Guianas: The American Philosophical Society and the Eel

> ... the shocks, such as them as were severe, would pass through us both;
> in which case they doubtless leaped from the point of one wire to the other,
> though we were not so fortunate as to render the spark generally visible.
> (Hugh Williamson, 1773)
> (Note published in Williamson, 1775a, p. 100)

The path of the electric eel as a pivotal organism in understanding how electric fishes might function can be traced to the observations and scientific experiments that came out of South America at the start of the second half of the eighteenth century. In this context, Jean Allamand (1756a) and Laurens Gronov (1758, 1760, 1763–64) related information about electrical experiments that had been performed for them by Storm van 's Gravesande in the Guianas, while Frans van der Lott (1762a), a surgeon in one of the Dutch colonies, presented his own findings. A third person, whose work became better known in London circles, was American-born Edward Bancroft (1769a), a man of many hats who was now closely associated with Franklin and the Royal Society, after leaving Guiana and settling in London.

Some eels would eventually make their way to London, first as dead specimens for dissection and finally alive, but the scientific voyage would not be direct or for that matter easy. This creature would initially debut in the British North American Colonies, where experiments on it would first be carried out in Philadelphia in 1773, and then in Charles Towne (also Charles Town, sometimes hyphenated, and Charlestown after the revolution), the major port in South Carolina and the southern colonies, 1 year later.

Although the American contributions to this story have attracted scant attention in the past, they beg to be presented, because they provided late-eighteenth-century natural philosophers in Europe with important new information about the eels, their electrical discharges, and their electric organs. Even the route the eels would first take from South America to London would be via the Carolinas, where, as we shall see, the sea captain was given very specific instructions about their preservation, should they die, from an American. Thus, before returning to Walsh and his talented associates at the Royal Society, we must examine what was happening in North America, where several colonists would help set the stage for Walsh's ultimate triumph.

To put the North American contributions in perspective, we shall start this chapter with what Joseph Priestley had written about these eels just before news of the North American eel experiments reached England, and before Franklin's forced departure from England in 1775, which effectively marked the end of his direct involvement in the fish experiments. To complete this background picture, mention will again be made of the unique problems associated with shipping live eels out of South America.

After providing this background information, the North American contributions will be able to be appreciated in better perspective. We shall cover the Philadelphia experiments of 1773 in this chapter, and shall turn to what transpired in South Carolina in 1774 in the next chapter, which will culminate with the voyage of five eels across the Atlantic to London, how they were unable survive the trip, and how John Hunter obtained some of them for dissection.

PRIESTLEY'S SYNOPSIS

The third edition of Joseph Priestley's *The History and Present State of Electricity*, which came out in 1775, reflects what was understood about the eel by Franklin's protégé in England in 1773, prior to the North American experiments. Indeed, Franklin helped guide Priestley when he wrote his *History* and, like a good teacher, he generally looked over Priestley's writings before they went to press.

The last section (XV) of Volume 1 bears the title "Miscellaneous Experiments and Discoveries Made within this Period." Here Priestley writes that he "must not omit to mention . . . what the Dutch writers have reported concerning the gymnotus, a fish peculiar to Surinam which very much resembles [in its properties] what naturalists relate concerning the torpedo" (p. 496). He further tells us that "Mr. Muschenbroek says, the gymnotus is possessed of a kind of natural electricity, but different from the common electricity, in that persons who touch it in water are shocked, and stunned by it, so as to be in danger of drowning" (pp. 496–497).

Priestley continues:

> The fish has been taken, and put into a vessel; when experiments were made upon it at leisure; and it was

found, that it might be touched with all safety with a stick of sealing-wax; but if it was touched with the naked finger, or with a piece of metal, and especially a gold ring, held in the fingers, the arm was shocked as high as the elbow. If it was touched with the foot, the sensation reached as high as the knee, and the pain was as great as if the part had been struck with something hard. This kind of electricity is the same by night or by day, when the wind is in every direction, when the fish was put in vessels of any materials, and whether it was in water or out of water. Every part of the body is capable of giving this shock, but more especially the tail. The sensation is the strongest when the fish is in motion, and is transmitted to a great distance; so that if persons in a ship happen to dip their fingers or feet in the sea, when the fish is swimming at the distance of fifteen feet from them they are affected by it. (p. 497)

We have to assume that Priestley meant the river when he referred to "the sea" in this passage, as the eel is never found in salt water. He does, after all, note in another sentence that "The gymnotus is found in the upper part of the river of Surinam." But then he tells his readers that "Other fishes put into the same vessel with it, presently died; but it is itself killed by the lobster" (p. 497). Here he seems to be accepting what Gronov (1758, 1760) had written (see Chapter 15), although perhaps confusing the eel with what Walsh was telling fellow members of the Royal Society about the sea torpedo. And adding to these problems, Priestley writes: "This gymnotus, I suppose, is a different fish from the *Anguille tremblante*, the trembling eel."

On a more positive note, Priestley calls for more research to answer some nagging questions. First, just before a paragraph on the *Anguille tremblante*, he asks, as did Bancroft and others, whether the shocks of the sea torpedo might be qualitatively identical to those of the eel. In his words: "This author proposes a query whether the sensation communicated by the torpedo does not depend upon a similar electricity; since Monsieur Reaumur [no accent] says, that when it is touched, the hand, arm, shoulder are seized with a sudden stupor, which lasts for some time; and is unlike any other sensation" (p. 498).

Then, in his last paragraph (large sentence) on the eels, he calls for experiments to determine whether the eel can attract and repulse pith balls and certain other objects, and whether a spark can be drawn from the fish. "It is to be regretted, that none of the persons who have made experiments on these fishes should have endeavoured to ascertain whether they were capable of exhibiting the phenomena of attraction or repulsion, or the appearance of electric light, as experiments of this kind are of principal consequence, and must have been very easy to make" (pp. 498–499).

These lines in particular would suggest that Priestley was drawing largely from Gronov, Allamand, Van der Lott, and Bancroft when he was writing and revising *The History and Present State of Electricity*. As for Franklin, he probably knew little more than this when overseeing what Priestley was writing. Nevertheless, some very interesting things were happening in Philadelphia while Priestley was writing his volume in 1773, even though there is no mention of the Philadelphia eel experiments in Priestley's volume or in any of the known letters to or from Franklin, including those from later dates.

FRANKLIN'S FAREWELL

The political situation between England and what would soon become the United Colonies of North America was now deteriorating so rapidly that by 1774 Franklin knew that it was just a matter of time before he would be compelled to return home. Why he even stayed in London until 1775 is an open question, and one that has generated considerable debate. But what is clear is that early in 1775 Franklin was still doing electrical experiments and discussing fish electricity (see below), although his mind was more focused on what seemed to be the inevitable war with the mother country. Indeed, he was well aware of the rumors now swirling through London that he might be arrested, tried for treason, and hanged as a traitor.

The changing political climate also affected the people he worked with, forcing some of his more conservative and aristocratic colleagues at the Royal Society to distance themselves from him. Joseph Priestley, however, was not one of these people. He understood and openly sided with the Americans, putting his own life in peril. He not only supported Franklin but even stayed with him during his final days in England, including on his last day. "He dreaded the war," Priestley writes in his autobiography about that somber day in 1775,

> and often said that, if the difference should come to an open rupture, it would be a war of *ten years,* and he should not live to see the end of it. . . . That the issue would be favourable to America, he never doubted. The English, he used to say, may take all our great towns, but that will not give them possession of the country. The last day that he spent in England, having given out that he should leave London the day before, we passed together, without any other company; and much of the time was employed in reading American newspapers . . . the tears trickled down his cheek. (Priestley, 1806/1970, p. 117)

The fighting broke out as Franklin sailed home. And, as a result of being forced to leave London, he would miss Walsh's 1776 spark demonstrations, showing in the most convincing way to date that electric eels are, in fact, electrical.

Franklin was not, however, the only American to play an important role in the fish electricity story. Although Franklin was certainly the best known of the American natural philosophers at this moment in time, several other members of his American Philosophical Society also helped to till the ground for Walsh's greatest achievement.

These men, who interacted with Franklin when they were in the same places, would do this with clever experiments on the South American eels, additional observations, and good knowledge of how best to preserve dead eels. But unlike what Edward Bancroft did on site in the Guianas before joining Franklin in London, these Americans conducted their experiments in Philadelphia and in Charles Towne, South Carolina. Their work stemmed from live eels making the perilous journey to British North America, far from the tepid waters in which they were caught. In effect, what these men did was dependent on improved methods of transporting eels, and on sea captains willing to take significant risks for significant financial rewards.

THE PERILS AND REWARDS OF THE EEL TRADE

As noted in Chapter 14, Edward Bancroft commented on how difficult it was to send live eels from South America to North America or Europe in the 1760s. In his *Essay on the Natural History of Guiana*, he had even written: "Several attempts have been made to convey these Fish to Europe; but the quantity of fresh water requisite to shift them as often as is necessary, together with the bruises which they must inevitably sustain from the motion of the ship, have hitherto rendered them unsuccessful" (Bancroft, 1769a, p. 201).

At least one eel shipped from Guiana survived the trip to Philadelphia in 1773, 7 years after Bancroft left South America and 4 years after his book was published. Alexander Garden, about whom more will be said in the next chapter, provided some information on how the eels managed to make it to South Carolina a year later, and the nature of the eel trade (Garden, 1775a; also see Delbourgo, 2006, for more on animal commerce from South America).

In a paper published in 1775, Garden mentioned George Baker, a British sea captain who was well known for sailing between Surinam, British North America, and London. Baker stands out as one of the first captains to do commerce in live eels, if not the first, while shipping other cargoes as well. Garden (1775a, p. 109) called him "The keeper of this fish," and wrote that he told him that he "catched them in the Surinam river, a great way up, beyond where the salt water reaches; and that they are a fresh water fish only."

Baker was not a naturalist or a gentleman scientist. He was a businessman and wanted to collect and transport delicate eels to more civilized places in North America and Europe, much as he would any other precious cargo. It was all about the monetary rewards such commerce would bring him. Indeed, his prices for a living specimen were viewed as very high, and he also negotiated big prices for dead specimens. Garden (1775a, p. 103), a very frugal Scot now making his home in the Carolinas, complained about this, writing: "The person who owns [the eels] rates them at too high a price (not less than fifty guineas for the smallest) for me to get a dead specimen."

The stop in Charles Towne, where Garden met Captain Baker, was a lucrative port of call on the North Atlantic seaboard. So was more prosperous Philadelphia to the north, the home of Franklin's American Philosophical Society. And with his knowledge of various markets, Baker knew that if he could somehow manage to get even one eel to survive the voyage from South America to bustling and more sophisticated London, he would be able to reap the biggest profits of all.

The matter was a simple one of supply-and-demand economics. It was common knowledge that few if any eels would be able to survive such voyages, so the supply of live eels was then expected to be terribly limited. In contrast, driving up the bidding on the demand side of the ledger, scientists in North America and Europe wanted them for electrical experiments that could be performed only inadequately in the hot, sticky, miasmatic climates of equatorial South America (Delbourgo, 2006). In fact, there seemed to be a race to see who in which country would be able to get them first.

To put this situation in context, Philippe Fermin (1769), who wanted to make a proper dissection to identify the shock organs, found the oppressive climate in Guiana just too overwhelming to carry out his plans (Fig. 17.1). And an American, Henry Collins Flagg, would later complain that his physical and mental powers were too compromised by the climate to do what he wanted to do on location. In Flagg's words, "that relaxation of the mental powers generally consequent upon the lassitude of the body incident to the inhabitants of warm climates, indisposed me to the farther prosecution of the experiments I am now mortified at not having made" (Flagg, 1786, p. 173; see Chapter 18 for more on Flagg, and Parrish, 1996, for more on climate and eighteenth-century American science). Adding to the demand to bring eels out of South America, Baker knew that sickly individuals not trained in medicine or particularly interested in natural history would also pay good money to see and feel these living wonders, which were being associated with miraculous cures.

As a businessman, Baker understood the economics, including both the risks and rewards, of the dicey eel trade. These thoughts must have gone through his mind when laborers, most likely slaves, loaded some captured eels onto his ship in the sweltering Surinam sun. He did everything to maximize his success, putting them in large wood troughs filled with fresh water, and loading an adequate supply of small fishes in separate holding tanks—fishes they could stun and eat on long voyages. The chances of success, he realized, were not very high, and Baker might even have kneeled down and said a prayer for his precious cargo before departing South America, hoping that some eels would survive the perilous journey north and perhaps east, and make him wealthy.

THE ARRIVAL OF AN EEL IN PHILADELPHIA

Of all the people in North America, the members of the American Philosophical Society had to be the most eager to acquire and do experiments on living eels. This was hardly a secret: Baker and other captains involved with South American shipping knew this. So when a ship carrying a live eel arrived in Philadelphia in 1773, two parties had to be elated. One was a group of natural philosophers of the American Philosophical Society. The other was that ship's captain, an unidentified man (quite possibly Baker) who would prove to be a very tough negotiator, knowing that the organization's natural philosophers might think of this as a once-in-a-lifetime chance to study and experience such an exotic and coveted creature.

In the records of the American Philosophical Society, there is an entry dated August 1773 advising Ebenezer Kinnersley, Isaac Bartram, David Rittenhouse, Levi Hollingsworth, and Owen Biddle "to agree with the Owner of the Torpedo, on terms to make a Set of Experiments, with a view to determine the nature of the Shocks which it communicates." The men must have been informed that the unnamed owner would want a very high price, for they are told to make a deal "not exceeding 3 pounds for that privilege." Additionally, "they are requested to call in to their assistance any gentlemen they think proper" (see *American Philosophical Society*, 1885, p. 82; the early notes of the society were not published in the society's journal for over a century).

Just how much was paid for a session with the eel, or perhaps in total for a series of sessions, is unknown. But a deal was struck, because many experiments were performed. Moreover, what transpired in Philadelphia was publicly

Figure 17.1. Philippe Fermin's (1769) map of Surinam. Fermin lived in Surinam and complained in his book that the climate was too oppressive for him to do proper experiments and dissections on the eels.

described in several places, although there are various inadequacies in these records and, in one case, as we shall see, a very lengthy delay prior to publication.

A SHORT 1773 NEWSPAPER CLIPPING

The first bits of public information that we could find about the eels studied in Philadelphia appeared in a brief article that was repeated in several newspapers during the summer of 1773. These newspapers included the *Boston Post* (August 30), the *Massachusetts Gazette* (August 23), and the *Massachusetts Spy* (August 26). The piece was absolutely the same in each case and was reprinted from a New York newspaper on August 23, which obtained the information from someone in Philadelphia.

In its entirety, the clipping reads:

> We hear from Philadelphia that there is now in that City, a live Fish supposed to be a *Torpedo*, which on being touched, gives a very strong electrical Shock, to any number of Persons whose Hands are in Contact with each other; several Persons affected with various Disorders, have found Relief from it, when the common Mode of electerizing [sic] has proved ineffectual. It has been observed that, after receiving a Shock from the Fish, it takes some Time to recover the like Quantity of the electric Fluid; and that if touched sooner, the Shocks are weak or strong in proportion to the Time since the preceeding [sic] Shock. (Reprinted in Lemay, 1964, pp. 108–109)

Two basic problems characterize this piece, not to mention many smaller ones. The most obvious is that it provides next to no details about the experiments. The other is that there is no indication that the "Torpedo" or "Fish" studied was really an electric eel. In fact, the idea that this was an exotic South American eel, as opposed to a sea torpedo ray, might have been far from the average newspaper reader's mind.

PUBLISHED NOTES OF THE EXPERIMENTS

Details of the experiments, however, did make it to press in some notes published in the *Philadelphia Medical and*

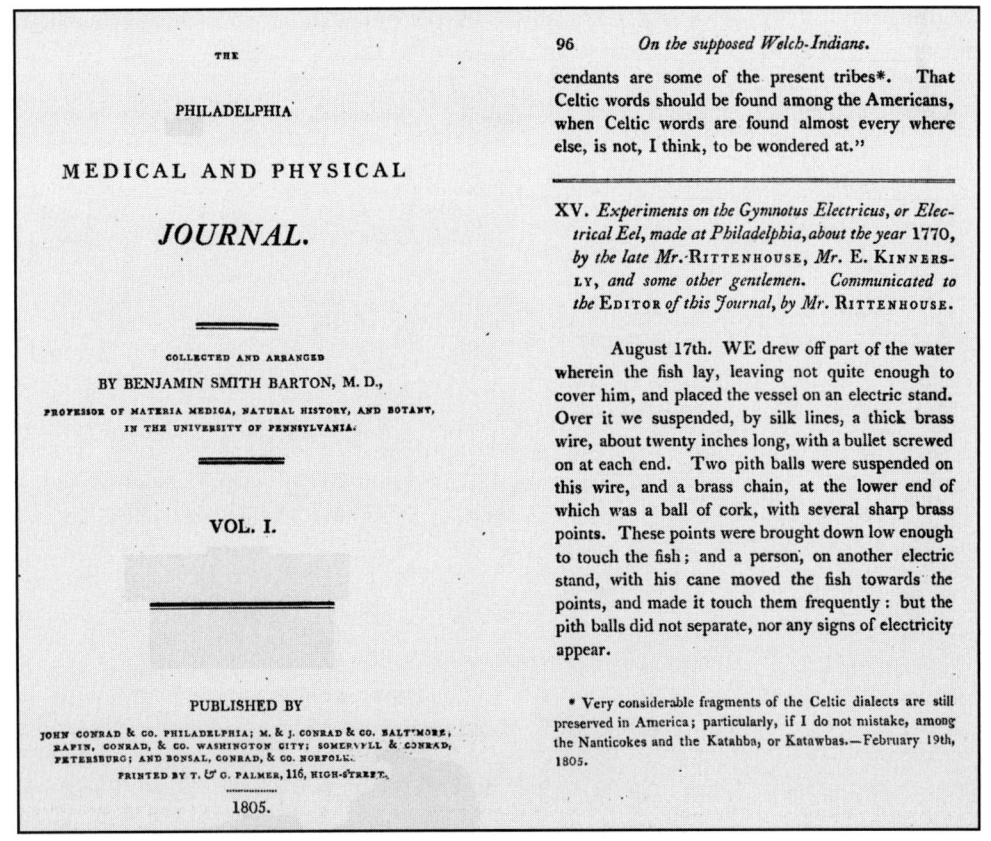

Figure 17.2. The 1805 article dealing with the experiments on the electric eel made in 1773 by the members of the American Philosophical Society.

Physical Journal. Strangely, the American Philosophical Society, which oversaw the experiments, never published these notes. And even more sadly, these original notes, which still failed to mention the sea captain's name, gathered dust for more than 30 years. They did not see the light of day until 1805, by which time fish electricity was well established, Alexander Humboldt had returned to Europe from the Americas, John Walsh was 10 years dead, Luigi Galvani had come forth with his neurophysiology based on electricity, and Alessandro Volta had his new pile or battery modeled on fish electric organs!

The title given to the article is "Experiments on the Gymnotus Electricus, or Electrical Eel, made at Philadelphia, about the year 1770, by the late Mr. RITTENHOUSE, Mr. E. KINNERSLY [sic], and some other gentlemen. Communicated to the EDITOR of this Journal by Mr. Rittenhouse" (Fig. 17.2). In a footnote at the end of the article, we find that Mr. Rittenhouse discussed this subject with the editor on November 22, 1794. This date and the material in the footnote would suggest that this was famed mathematician-astronomer David Rittenhouse (Fig. 17.3), who died in 1796. There was, however, another Rittenhouse who might have inherited these notes and then handed them to the editor, namely David's son, Benjamin Rittenhouse, an instrument and clock maker.

By 1794, David Rittenhouse had become well known for his American land surveys, for working out the transit of the planet Venus, for heading the new United States Mint, and for other things. Although he had been asked to help make a deal for the eel and then did research on it, his role in the electric fish story has been overlooked.

Ebenezer Kinnersley (here spelled Kinnersly), who was also on the committee to get the eel for these experiments, was the only other participant identified in the belated publication. He became well known in the 1740s for conducting landmark experiments on the nature of electricity with Benjamin Franklin, and was now giving lectures and demonstrations on electricity, which he advertised as "Entertainment for the Curious" (see LeMay, 1964, esp. pp. 109–110). As for the "some other gentlemen," they might have been the other people who negotiated for the eel. Whether the group included Hugh Williamson, who was not listed on that committee, and who would publish a somewhat differing report on the eel under his own name, is not clear.

The notes of 14 experiments, followed by a passage alluding to "several other experiments," take the reader through many of the experiments step by step. They provide a wonderful glimpse of how Rittenhouse, Kinnersley, and their unnamed colleagues thought and worked as they conducted their studies on a single eel. In effect, we see them replicating earlier experiments and trying to cover more ground with some new ones.

They begin with a description of how the eel was barely covered with water when placed on an electrical stand, above which was a thick brass wire. Hanging from the wire were some pith balls and some cork balls with sharp brass points, the kind of set-up Priestley had called for to detect an electrical charge. The two goals of these experiments were (1) to determine whether the pith balls would move when brought close to the eel, and (2) to see whether the eel might more readily discharge electricity to points made of metal. "But the pith balls did not separate, nor any signs of electricity appear" (Rittenhouse, 1805, p. 96).

Figure 17.3. David Rittenhouse (1732–1796), an important member of the American Philosophical Society's committee set up to conduct experiments on an electric eel that survived the trip from Surinam to Philadelphia in 1773.

This experiment was followed by one in which one of the researchers touched the fish while holding the pith balls. The unidentified man received repeated shocks, but again the pith balls failed to show signs of electricity.

Forming various circuits with the fish came next. When the participants held hands and the eel was touched to complete a circuit, everyone felt the shocks. Various objects were subsequently included in the circuit, including brass chains. In Experiment 9, when the chain was wet, "we got a shock to pass through it."

This experiment with good results preceded Experiment 10 with a dry brass chain:

> Mr. Kinnersly held a dry brass chain, eighteen inches long, with one hand, and, with the other, took hold of another person, who touched the fish. A third person held the opposite end of the chain, and put his other hand into the water. After many trials, a violent shock passed through the circuit, and each person felt it, in both arms. We were now satisfied, that a chain of wire would conduct a shock from the fish, as well as from the electric phial, provided the fish exerted this wonderful power as strongly as he sometimes did . . . (p. 99)

There were also observations of how the eel stunned some small fishes put in its tank. When it released its discharge, a small fish "turned up its belly, and lay motionless; but on taking it out, in a few minutes it revived" (p. 98). The same "stunning" effect was produced with a discharge from a Leyden jar, further linking the eel's discharge with electricity.

Experiments 13 and 14 are particularly notable, because they were designed to obtain a spark from the fish. In the first experiment,

> a circuit being formed by two persons, one of them held his hand in the water, and with the other touched the fish, each person holding, at the same time, in the other hand, a brass chain, fastened to a thick piece of wire; the ends of the two pieces of wire being rounded off, were made to approach within less than the hundredth part of an inch, but without touching, and secured at that distance. This experiment did not succeed now, though we attended to it, with great patience. But though we had not the pleasure of seeing a spark between the two wires, we had the satisfaction of knowing that none ought to have been seen, because no shock passed through the wires; for the persons who held them always declared, they felt the shock in the hand only which touched the fish or the water. (pp. 159–160)

In Experiment 14:

> Instead of the wires, placed at a small distance from each other, in the last experiment, we made about a quarter of an inch of a line of gilding on a book serve for part of the circuit; but could not make the shock pass through it, and, consequently, saw no appearance of electricity. (p. 160)

The paper ends with these short additions:

> If a single person takes hold of the fish, with both hands, placing one near either extremity, he receives a shock though both arms, and sometimes feels it across the breast likewise. If he touch it with one hand only, he never feels it further than in that hand and arm; generally in the hand only. (p. 160)

> If several persons join hands, and only one of them touch the fish, none of the others touching the water, the shock is never felt by any but the person who touches the fish, and by him in that hand and arm only. (p. 161)

> No shock from the fish can be communicated through dry wood, glass, or any other substance, so far as we have tried, that will not conduct electricity.* (p. 161)

The footnote after the asterisk has the 1794 date in front of it and "Editor" at its end. It states: "Mr. Rittenhouse, in a conversation I had with him, on the subject, seemed very confident, that, in making the preceding experiments upon the Gymnotus, he upon one occasion, saw the ELECTRIC spark" (p. 161).

This statement raises all sorts of questions. For example, if David Rittenhouse had observed the spark, why were the results of the spark experiments presented as being totally negative? And why was this extremely important piece of information not mentioned by Hugh Williamson, Rittenhouse's colleague at the American Philosophical Society, who published his report on this eel in 1775 (see below), or by anyone else in Philadelphia or with connections to this group of natural philosophers for that matter? Even Franklin, who founded the American Philosophical Society, communicated regularly with its members while abroad, and even met with Williamson when he visited England right after these experiments took place, seemed to know nothing about a spark being observed in Philadelphia.

Hence, it is difficult to put much faith in what Rittenhouse supposedly told the Editor 21 years after these experiments took place, and more than 18 years after John Walsh demonstrated the coveted spark, seemingly for the first time anywhere, to his associates at the Royal Society of London (Chapter 19). In brief, whether there had been a real spark, or a possible single instance of what might have been a spark that the experimenters could not replicate, is impossible to tell from this document. As we shall see, this might have been a "false memory" based on subsequent events, to use current terminology, because Hugh Williamson (1775a) would report that there was a sensation across a gap, although no discernable spark at the time. Interestingly, Walsh's nephew Arthur also seemed to feel an occasional sensation across a gap, although Walsh dismissed what Arthur told him because the shock was not sufficiently replicable with their torpedoes (see Chapter 15).

The second part of this footnote is not as interesting for historians, but it bears on how the eel was fed to keep it alive. The editor writes that he was informed by another gentleman who assisted in these experiments that the eel "killed three Cat-fish" with a single discharge, "at least, the fish were instantly stunned, and thrown upon their backs," and that the eel "immediately proceeded to eat the fish." He added that the eel "destroys" many other species of fishes with less difficulty when feeding.

KINNERSLEY'S ARTIFICIAL EEL

Ebenezer Kinnersley subsequently assembled a working model of the eel. In the *Pennsylvania Gazette*, the newspaper once owned by Franklin, there was a notice of his artificial eel (also see Delbourgo, 2006). It appeared in an advertisement on December 29, 1773, promoting his latest lecture-demonstrations:

> Mr. Kinnersley having lately made some considerable additions to his Electrical Apparatus, particularly having added a case of seventy bottles [Leyden jars], each lined and coated with tinfoil, he proposes to exhibit a course of ELECTRICAL EXPERIMENTS, at the College of this city.... The following particulars will, it is presumed, be no considerable part of this Electrical Experiment, viz. Flashes of real lightning visible under water. Iron heated red hot, and even melted by lightning, whilst under, and in contact with, common cold water.... A curious representation of the astonishing electric eel, lately seen in this city, on touching of which, while in the water, an electric shock may be sensibly felt, as from a live one. (see Lemay, 1964, p. 109)

Physicist Henry Cavendish (1776a) of the Royal Society of London, who was then beginning to construct an artificial sea torpedo based on his exceptional grasp of physics, was not alone in his quest to produce an artificial electric fish at this time (see Chapter 16). The talented Cavendish, whose model became far better known than Kinnersley's "representation," clearly had an overlooked counterpart in America, although we know very little about Kinnersley's artificial eel, the physics behind it, and whether he began his project after learning what Cavendish was doing in London.

HUGH WILLIAMSON

Hugh Williamson's article in the *Philosophical Transactions of the Royal Society* of 1775 was, in fact, the only important published source information about the Philadelphia experiments that was widely known by scientists in Europe prior to Walsh's spark experiment (Fig. 17.4). A short note on this paper was published in the September issue of *The Gentleman's Magazine*. It was followed by an equally short note dealing with Garden's experiments in Charles Towne (Fig. 17.5). Similar accounts of these experiments were published afterwards in other magazines, for instance *The Monthly Review* for 1776 (Fig. 17.6). These periodicals, as we have seen, also covered Edward Bancroft's earlier findings on the electric eel (Chapter 14), and they would cover those of Alexander Garden, whose eel research will be described in the next chapter (Chapter 18).

In his *Philosophical Transactions* paper Williamson does not always go into as much detail as can be found in the aforementioned notes that would be published three decades later. But in his presentation of the numerous experiments, Williamson mentions experiencing a sensation across an air gap, a finding that does not appear in the Rittenhouse notes and which would heat up the call for eels in London and affect the race for the elusive spark.

Williamson was one of six pupils to receive the B.A. degree in the first graduating class of the College of Philadelphia (in 1757)—the school Franklin founded in 1753, and which would become the University of Pennsylvania (Hosack, 1820; Jenkins, 1950; Kagarise & Sheldon, 2005; Neal, 1918). He was skilled in languages, philosophy, mathematics, science, and history. Although intending to become a Presbyterian minister after graduation, he chose to further his liberal education, obtaining a master's degree in 1760, after which he was awarded a professorship in mathematics at his *alma mater*.

Williamson began his formal study of medicine in 1764, first at Edinburgh and then in London and the Dutch town of Utrecht. When he returned to Philadelphia a few years later, he had his M.D. degree. Nevertheless, he found the practice of medicine too emotionally exhausting and instead began to spend more time studying natural philosophy, which invigorated him. He was elected into the American Philosophical Society early in 1768 and worked with David Rittenhouse on the transit of Venus. His work on comets and his *Observations on the Climate* furthered whetted his appetite for important new scientific ventures.

In 1772, Williamson went to the West Indies to raise money for the new Academy of Newark (later to become the University of Delaware). While there, he made plans for a tour of England, Scotland, and Ireland for more fundraising. Between trips, he attended meetings of the American Philosophical Society, the last being on August 20, 1773—while the eel was in Philadelphia.

How much Williamson was involved with Rittenhouse and Kinnersley in their experiments is uncertain, because he does not mention these men in his published letters on the subject (below), and because his name does not appear

Figure 17.4. Hugh Williamson (1735–1819), American statesman and physician and amateur scientist, who made experiments on electric eels in Philadelphia in 1773, then met with Franklin, Walsh, Ingenhousz, and Pringle in London, and the first page of his letter in the 1775 issue of the *Philosophical Transactions*.

Figure 17.5. The September 1775 issue of the *Gentleman's Magazine* with the page containing the short notes on the papers of both Williamson's and Garden's experiments on the electric eels.

265

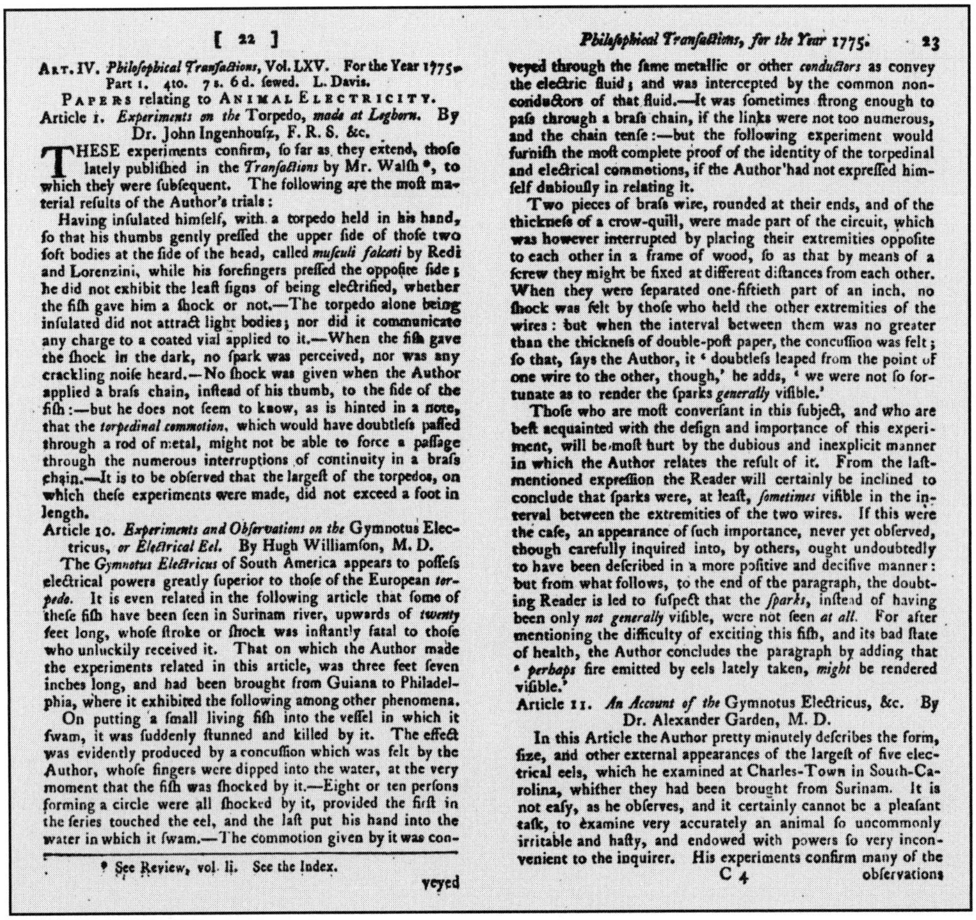

[22]

Art. IV. *Philoſophical Tranſactions*, Vol. LXV. For the Year 1775. Part 1. 4to. 7s. 6d. ſewed. L. Davis.

Papers relating to Animal Electricity.

Article 1. *Experiments on the* Torpedo, *made at Leghorn.* By Dr. John Ingenhouſz, F. R. S. &c.

THESE experiments confirm, ſo far as they extend, thoſe lately publiſhed in the *Tranſactions* by Mr. Walſh *, to which they were ſubſequent. The following are the moſt material reſults of the Author's trials:

Having inſulated himſelf, with a torpedo held in his hand, ſo that his thumbs gently preſſed the upper ſide of thoſe two ſoft bodies at the ſide of the head, called *muſculi falcati* by Redi and Lorenzini, while his forefingers preſſed the oppoſite ſide; he did not exhibit the leaſt ſigns of being electrified, whether the fiſh gave him a ſhock or not.—The torpedo alone being inſulated did not attract light bodies; nor did it communicate any charge to a coated vial applied to it.—When the fiſh gave the ſhock in the dark, no ſpark was perceived, nor was any crackling noiſe heard.—No ſhock was given when the Author applied a braſs chain, inſtead of his thumb, to the ſide of the fiſh:—but he does not ſeem to know, as is hinted in a note, that the *torpedinal commotion*, which would have doubtleſs paſſed through a rod of metal, might not be able to force a paſſage through the numerous interruptions of continuity in a braſs chain.—It is to be obſerved that the largeſt of the torpedos, on which theſe experiments were made, did not exceed a foot in length.

Article 10. *Experiments and Obſervations on the* Gymnotus Electricus, *or Electrical Eel.* By Hugh Williamſon, M. D.

The *Gymnotus Electricus* of South America appears to poſſeſs electrical powers greatly ſuperior to thoſe of the European *torpedo*. It is even related in the following article that ſome of theſe fiſh have been ſeen in Surinam river, upwards of *twenty* feet long, whoſe ſtroke or ſhock was inſtantly fatal to thoſe who unluckily received it. That on which the Author made the experiments related in this article, was three feet ſeven inches long, and had been brought from Guiana to Philadelphia, where it exhibited the following among other phenomena.

On putting a ſmall living fiſh into the veſſel in which it ſwam, it was ſuddenly ſtunned and killed by it. The effect was evidently produced by a concuſſion which was felt by the Author, whoſe fingers were dipped into the water, at the very moment that the fiſh was ſhocked by it.—Eight or ten perſons forming a circle were all ſhocked by it, provided the firſt in the ſeries touched the eel, and the laſt put his hand into the water in which it ſwam.—The commotion given by it was con-

* See Review, vol. li. See the Index.

veyed

Philoſophical Tranſactions, for the Year 1775. 23

veyed through the ſame metallic or other *conductors* as convey the electric fluid; and was intercepted by the common non-conductors of that fluid.—It was ſometimes ſtrong enough to paſs through a braſs chain, if the links were not too numerous, and the chain tenſe:—but the following experiment would furniſh the moſt complete proof of the identity of the torpedinal and electrical commotions, if the Author had not expreſſed himſelf dubiouſly in relating it.

Two pieces of braſs wire, rounded at their ends, and of the thickneſs of a crow-quill, were made part of the circuit, which was however interrupted by placing their extremities oppoſite to each other in a frame of wood, ſo as that by means of a ſcrew they might be fixed at different diſtances from each other. When they were ſeparated one-fiftieth part of an inch, no ſhock was felt by thoſe who held the other extremities of the wires: but when the interval between them was no greater than the thickneſs of double-poſt paper, the concuſſion was felt; ſo that, ſays the Author, it 'doubtleſs leaped from the point of one wire to the other, though,' he adds, 'we were not ſo fortunate as to render the ſparks *generally* viſible.'

Thoſe who are moſt converſant in this ſubject, and who are beſt acquainted with the deſign and importance of this experiment, will be moſt hurt by the dubious and inexplicit manner in which the Author relates the reſult of it. From the laſt-mentioned expreſſion the Reader will certainly be inclined to conclude that ſparks were, at leaſt, *ſometimes* viſible in the interval between the extremities of the two wires. If this were the caſe, an appearance of ſuch importance, never yet obſerved, though carefully inquired into, by others, ought undoubtedly to have been deſcribed in a more poſitive and deciſive manner: but from what follows, to the end of the paragraph, the doubting Reader is led to ſuſpect that the *ſparks*, inſtead of having been only *not generally* viſible, were not ſeen *at all*. For after mentioning the difficulty of exciting this fiſh, and its bad ſtate of health, the Author concludes the paragraph by adding that '*perhaps* fire emitted by eels lately taken, *might* be rendered viſible.'

Article 11. *An Account of the* Gymnotus Electricus, &c. By Dr. Alexander Garden, M. D.

In this Article the Author pretty minutely deſcribes the form, ſize, and other external appearances of the largeſt of five electrical eels, which he examined at Charles-Town in South-Carolina, whither they had been brought from Surinam. It is not eaſy, as he obſerves, and it certainly cannot be a pleaſant taſk, to examine very accurately an animal ſo uncommonly irritable and haſty, and endowed with powers ſo very inconvenient to the inquirer. His experiments confirm many of the

C 4 obſervations

Figure 17.6. As was true for the experiments on the electric eel carried out by the Dutch about 10 years earlier, magazines and general culture journals greatly contributed to spreading the news about the fish experiments made in North America. In these two pages of the January 1776 issue of *The Monthly Review*, Williamson's (1775) and Garden's (1775) reports on the eel are presented, together with those of Ingenhousz (1775), who conducted torpedo experiments while in Italy.

in the American Philosophical Society notes published in 1805. Very likely, these men interacted with each other or at least drew from each other, perhaps with Williamson following up on the Rittenhouse–Kinnersley experiments with some of his own.

Williamson did not submit his rendition for publication right away. This inaction stemmed from the fact that he knew he would soon be sailing for Great Britain. Specifically, his fundraising trip would take him to London, giving him the opportunity to discuss his experiments with a co-founder of American Philosophical Society, namely Franklin, and with Bancroft and Walsh, who would advise him about a possible publication. In effect, why send a letter when one could be delivered by hand a short time later?

The records show that soon after completing his eel experiments, Williamson left for New England *en route* to England. While in Boston that December, he met with the Sons of Liberty (Adams, Warren, Otis, etc.) and witnessed the famous "Boston Tea Party," in which patriots dressed as Indians destroyed a cargo of imported tea in protest over new taxes levied by Parliament in London. He began his trans-Atlantic voyage later that month, being the first person to carry the news of the rebellious event to the irate British—along with a stern warning that there would be a real war if Parliament continued to treat her North American colonies so unjustly.

Soon after setting foot in England early in 1774, Williamson cemented his relationship with Franklin, who he had known from his days in Philadelphia. David Hosack, Williamson's personal friend and first biographer, states that his fundraising efforts did not prevent him "from bestowing a portion of his attention upon scientific pursuits." He continues: "In conjunction with Dr. Ingenhousz, Mr. Walsh, Mr. John Hunter, and Dr. Franklin, he frequently instituted electrical experiments, to which I have often heard him refer with juvenile feelings, at the same time professing his ardent attachment to this branch of knowledge" (Hosack, 1820, pp. 51–52).

Hosack provides neither specific information about the experiments these four men did together nor details about their discussions. Nevertheless, two things are clear. The first is that all of these men were interested in studying the possibility of fish electricity. Thus, all would have been very interested in hearing about the Philadelphia experiments. And the second is that they must have encouraged Williamson to publish his findings, for this is exactly what he did.

The document describing Williamson's North American experiments bears the inscription "Philadelphia" and the date September 3, 1773, revealing it was composed before he left home. Williamson added a cover letter dated February 7, 1775, and it was written while he was in London. Both the cover letter and the text were "Communicated by John Walsh" to the Royal Society right after he received them, and they were published later that year (Williamson, 1775a) (see Fig. 17.3). There were no accompanying figures.

The opening line of Williamson's cover letter reads: "As the electrical eel has lately engaged the public attention, and yours in particular, I have taken the liberty of sending you some experiments which I made on that fish: they are the same that I had the pleasure of shewing you last winter, on my arrival from Pensylvania" [sic] (Williamson, 1775a, p. 94). The wording here is ambiguous. On the one hand, these words might convey the impression that Williamson had an eel in London in February 1775, allowing him to replicate his experiments before an audience of Walsh, Hunter, and others. On the other hand, the sentence could be read as stating nothing more than that he showed Walsh his notes from 1773 after he arrived in 1774. The latter interpretation is without question the correct one; that is, that there were no living eels in London at this time.

Williamson then explained that his experiments in Philadelphia had not been conducted under ideal circumstances: "The eel being sickened by the change in climate, its owner refused to let us take it out of the water, for the purpose of making experiments, on reasonable terms; and there were many experiments which I could not make on it in water, to my own satisfaction" (pp. 94–95). His cover letter did, however, mention an accidental but still memorable experience, and Williamson's important conclusion:

> Perhaps it may deserve notice, that a small hole being bored in the vessel in which the eel was swimming, one person provoked the eel so as to receive a shock; another person at the same time, not in contact with him, but holding his finger in the stream that spouted from the vessel, received a shock also in that finger. From this and other sundry experiments, I am induced to believe, that the *gymnotus* has powers greatly superior to, or rather different from, those of the *torpedo,* which you have examined with so much attention. (p. 95)

The earlier communication with the "sundry other experiments" performed in Philadelphia in 1773 follows, beginning with the lines:

> Some weeks ago, a sea-faring man brought to this city a large eel, that had been caught in the province of Guiana, a little to the westward of Surinam. It had the extraordinary power of communicating a painful sensation, like that of an electrical shock, to the people who touched it, and of killing its prey at a distance.... The eel was three feet seven inches long and about two inches thick near the middle. (Williamson, 1775a, pp. 95–96)

Williamson did not identify the mysterious mariner. But he stated that he thought this was the first time such a creature had made it to the colonies, and further, that he was quite certain none had yet made it alive to Europe. He then mentioned 20 experiments, beginning with "On touching the eel with one of my hands, I perceived such a sensation in the joints of my fingers as I received on touching a prime conductor or charged phial." He also related how he had used conductors and non-conductors, finding, as had others before him, that only known conductors of electricity would convey the shocks. Water, he noted, could carry it several feet. He reported that, with his finger in the water, he had no trouble feeling the shocks that the eel used to stun some small fishes that it then devoured.

Most of Williamson's experiments and conclusions are comparable to those described in the Rittenhouse notes. This parallel extends to Experiment 18, where he explains that, as much as the first 17 experiments suggested that the eel is electrical, the creature does not exhibit every property that an electrified body should show. Specifically, "he exhibited no marks of a *plus* state of electricity, nor would cork balls, suspended by silken threads, give any marks of it, either when they were suspended over the eel's back, or when touched by the insulated person at the instant he received the shock" (p. 99). Further, in Experiment 19, he mentioned that he also failed to charge a Leyden jar with the shocks, which were felt by an assistant in the circuit between the eel and the bottle.

This set up Experiment 20, on trying to get the charge to jump a gap in a wire. This experiment corresponded to Experiments 13 and 14 in the notes that Rittenhouse had published in 1805, and Williamson also starts off negatively:

> Two pieces of brass wire, about the thickness of a crow's quill, were screwed, in opposite directions, into a frame of wood, some as to some within less than the hundredth part of an inch of contact; they were rounded at the point. I held the removed end of one of those wires, while an assistant held the other; in the mean while, one of us putting his hand into the water near the eel, touched it so as to receive a shock. We repeated this experiment fifteen or twenty times with different success: when the points of the wires were even screwed asunder, to the fiftieth part of an inch, the shock never passed in the circle ... (p. 100)

In the continuation of this gap experiment, however, Williamson states that the shock could, in fact, jump a tiny gap. Using his words, when the points of the wires "were screwed up within the thickness of double-post paper, the shocks, such as them as were severe, would pass through us both; in which case they doubtless leaped from the point of one wire to the other, though we were not so fortunate as to render the spark generally visible" (p. 100).

Williamson emphasized that his eel "was not easily provoked, and appeared to be in bad health." In other words, had a stronger, healthier eel been used, there would have been a better chance of seeing a spark in the gap. Still, by getting the charge to jump in a reliable way from one brass wire to another, another basic and extremely important criterion for the force being electrical could be checked off the list.

The final part of Williamson's Philadelphia communication gives his conclusions in list form. The first four are: (1) the eel can communicate painful shocks; (2) the shocks are dependent on the will of the eel; (3) they are not the direct result of muscular actions, because they can be communicated at a distance and only by conductors of electricity; and (4) "the shock must therefore depend on some fluid, which the eel discharges from its body" (p. 101). Although he did not specifically single out his partial success with the gap experiment (feeling the shock, but no spark), his fifth conclusion pretty much summarizes things and would have been warmly welcomed by Walsh and Bancroft, who was also cited in his paper, and by others in Franklin's circle of experimental natural philosophers:

> 5. That as the fluid discharged by the eel affects the same parts of the human body that are affected by the electrical fluid; as it excites sensations perfectly similar; as it kills or stuns animals in the same manner; as it is conveyed by the same bodies that convey the electric fluid, and refuses to be conveyed by other bodies that refuse to convey the electric fluid, it must also be the true electrical fluid; and the shock given by this eel must be the true electrical shock. (p. 101)

A STOP IN NEW YORK?

The eel studied in Philadelphia might have been displayed in New York later in 1773. Among the (Joseph) *Banks Letters* housed in the British Library, there is a manuscript titled, "The Electrical Fish shewn at Philadelphia & New York. 1773 Communicated by Capt. Spehen [*sic*?] Payn Adye."

Adye was a major in the Royal Artillery (Dawson, 1958, p. 6). With only this limited information to draw on, it is not clear whether he played a direct role in transporting, exhibiting, or experimenting with the eel or was just a witness who took notes.

The contents of this communication do not reveal any more information about a New York stopover, such as how long it was there or the names of the natural philosophers who might have been present to see it or experiment with it. In its entirety, this overlooked document reads as follows:

> This fish is of the cartilaginous kind, breathing through two apertures or Nostrils, & often rises to the top of the water for that purpose. It has a slender eel shap'd body like the Lamprey, but has two pectoral fins like the Conger Eel, which the Lamprey wants, & has also a ventral Fin, or rather a Webb [*sic*] running almost from the head to the Tail, & there unites with the Caudal Fin, which is perpendicular to the body; but it has no dorsal fin, as all other Eels have. It is three feet eight inches long & about fourteen in circumference in the thickest part, which is near the head.
>
> It has a surprising power of giving a shock to any person touching it, & communicating it through a Circle of any Number of persons, resembling the strokes of Electricity.
>
> The Shock is more or less severe according to the pressure or inconvenience given to the fish by the person taking hold of it.
>
> So great a shock was given to a man in moving it from one Tub of water to another, as to knock him down & render him insensible for some time. By this power it kills small fishes before it devours them.
>
> It is a Native of South America, & was found in a fresh water River, at some distance from the Ocean. It's [*sic*] colour inclines more to an Ash colour than that to the common Eel. (Adye, 1773)

As we shall see in the next chapter, much more can be written about how some eels made it to South Carolina in 1774, what transpired there, and their subsequent voyage to London.

Chapter 18
Alexander Garden: A Linnaean in South Carolina and Captain Baker's Eels

> The person who owned these fish carried them from hence [Charles Towne] to England . . . [But] lest they should die by the way, I desired him to put them in a small keg of rum . . . for the purpose of preserving them. . . .
> (Alexander Garden to John Ellis, 1775)
> (In Smith, 1821, Vol. 1, p. 604)

In the previous chapter we saw the first of the important North American contributions to the electric fish story, specifically a series of experiments conducted by members of the American Philosophical Society, including one showing that the discharge from an eel could jump a minute gap in a wire, even though no spark was perceived at that time. We now turn to what transpired in South Carolina, and specifically the role played by Alexander Garden in the continuing saga of how these eels increasingly became to be perceived as electrical in America and in Europe. Garden, as we shall see, was an ardent follower of Carl Linnaeus (Fig. 18.1), and although not an experimentalist of the first rank, he was a superb naturalist with exceptionally sharp eyes who was very interested in classification and preservation, especially when it came to nature's oddities from the New World.[1]

ELECTRIC EELS IN CHARLES TOWNE

During the summer of 1774, five *Gymnoti* survived the voyage from Surinam to Charles Towne (also Town), South Carolina. Since there were five fish, whereas only one was mentioned in the Philadelphia experiments the previous summer, this was undoubtedly a new shipment of eels from South America, rather than one that went from Surinam up past the Carolinas to Philadelphia, perhaps on to New York, and then back down the coast to Charles Towne. Indeed, it would have been almost impossible for these fish to survive the cold of winter in these northern cities.

In Garden's case, we know definitively that George Baker was the captain of the ship and the owner of the eels, because Garden says so (Garden, 1775a; also see Delbourgo, 2006). As noted, no mariner's name was provided in any of the publications on the Philadelphia eels that we discussed in the previous chapter. Nevertheless, when it came to protecting his eel from overly zealous philosophers, Baker laid down protective rules in Charles Towne similar to those that were enforced in Philadelphia. This similarity, the profit motive, and to some extent the experience factor (e.g., knowing the eels would have the best chance of surviving if they arrived in the heat of the summer) would indirectly suggest that Baker, a seaman who carried commodities and other cargoes between South and North America, might have been the captain and owner of this precious cargo in both instances.

In contrast to what transpired at the American Philosophical Society in Philadelphia, the eels were a public happening in South Carolina. For those willing to pay, an amalgam that included physicians, gentlemen scientists, and many inquisitive laymen and laywomen, there was ample opportunity to observe them. An anonymous promotional piece in the June 20, 1774, edition of the *South Carolina Gazette* informed readers that they can be observed "At the House of Thomas Adamson, In Meeting-Street, at the Sign of the Horse Mask, opposite Edward Rutledge, Esq'rs, at any Hour of the Day" (this piece is cited in Sanders, 1997, and Delbourgo, 2006) (Fig. 18.2).

Bearing the title "The wonderful Electrical FISHES" in large letters right after mentioning the location, interested parties were even promised bewildering demonstrations:

> They are Natives of the Southernmost Part of North America, and have never been seen before that we know of. These Fish have the surprising Power of darting the Shock thro' a Circle formed by any Number of Persons, the same as the Electrical Phial doth in the Leyden Experiment: This Fish by its Power likewise kills small Fish when put in the Water to it, before he devours them—Many Ingenious Experiments have been made by a Committee appointed for that Purpose, by a Philosophical Society.
>
> Gentlemen and Ladies who wish to gratify their Curiosity by viewing this extraordinary Production of Nature, at the small expense of ONE DOLLAR each, are

---

[1] A preliminary report on Garden and his eels can be found in Finger (2010).

Figure 18.1. Carl Linné (Linnaeus; 1707–1778), the famous Swedish taxonomist, who had people sending him both common and unusual specimens from around the world, including Alexander Garden from South Carolina.

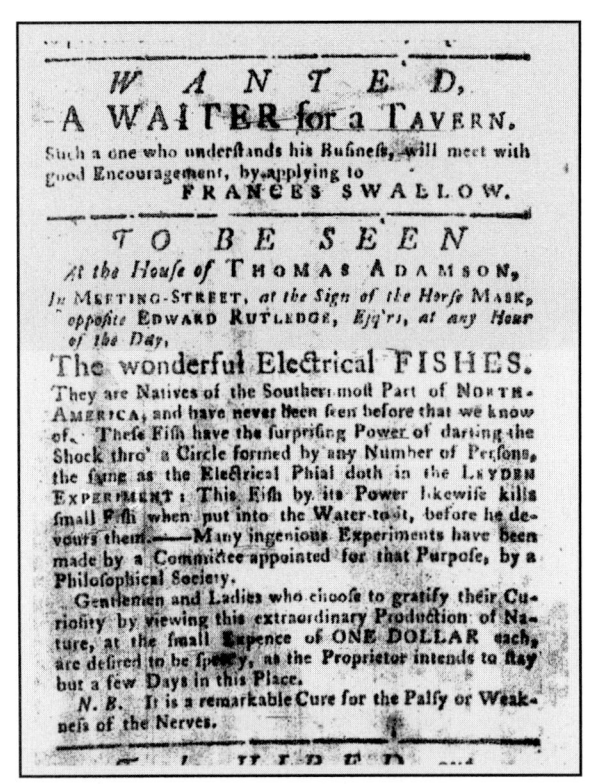

Figure 18.2. The notice in the *South Carolina Gazette* of June 20, 1774, announcing the arrival of the South American eels that could shock like "the LEYDEN EXPERIMENT," and where they could be observed for a price.

desired to be speedy, as the Proprietor intends to stay but a few Days in this Place.

(*South Carolina Gazette*, 1774)

This not-to-be-missed, truly wonderful opportunity was made even more appealing and sensational by the final line of the notice: "It is a remarkable Cure for the Palsy or Weakness of the Nerves." Could any curious or enlightened mind, or a person with an illness, possibly ask for more?

ALEXANDER GARDEN

These were indeed the "very curious fish" that Alexander Garden, who was mentioned in connection with the eel trade in Chapter 17, saw and then wrote about. In his words: "I was both surprised and delighted to observe their strange shape, and experience their wonderful properties" (Garden, 1775a, p. 102).

Garden was born in January 1730 in an austere farming region of northeastern Scotland near Aberdeen.[2] His father was a minister in the parish of Birse, where times were hard and money scarce. His education began in a small parish school, where Hugh Rose, a graduate of the University of Aberdeen, was the more-than-competent schoolmaster.[3] Garden's education then continued at Marischal College of the University of Aberdeen, where he followed the philosophy curriculum (liberal arts) and also studied medicine. Edward Bancroft would later receive his medical degree from Marischal College (Chapter 14). In Aberdeen, Garden served as an apprentice to Dr. James Gordon, Professor of Medicine, who stimulated his interest in botany and more generally in natural history, for which he was forever grateful.

After 3 years in Aberdeen, Garden felt ready to start his own career. His plan was to begin as a surgeon's mate on a naval ship and, after acquiring sufficient experience, to start a practice on shore. In 1746, he passed a surgical examination in London but did not immediately secure a position in the British Navy. Rather, he returned to Gordon for more surgical training. Following a second examination 2 years later, he became a surgeon's first mate in the British Navy. Purchasing his own instruments and medicines from Apothecaries' Hall, he now boarded the first of several ships on which he would serve and joined the senior surgeon in the infirmary.

Garden left the Navy in 1750 to pursue his medical studies at the University of Edinburgh, even though he had enough money for only 1 year of education. His highly regarded instructors included Dr. Alexander Monro, Professor of Anatomy; Dr. John Rutherford, Professor of the Practice of

---

[2] The best biography on Garden is the one written by Berkeley and Berkeley (1969). Other biographical sources consulted are Jenkins (1928), Denny (1948), Waring (1964), Stearns (1970), and Sanders (1997).

[3] Rose would later practice surgery in Norwich, England, and join the literary circle of Samuel Johnson, David Hume, and other notables.

Physic; and Dr. Robert Whytt, Professor of the Theory of Medicine. But it was Dr. Charles Alston, the inspiring Professor of Botany and Materia Medica, and the King's Botanist, who had an even greater influence on his life.

Garden qualified for his medical degree in 1751, although he could not or would not pay the fee for his diploma. Hearing about greater opportunities for physicians in the New World, and worried about consumption (tuberculosis) in his family, he left cold and wet Scotland to practice medicine in warmer South Carolina. He sailed to Charles Towne, located on the peninsula between the Ashley and Cooper rivers, in April 1752, 2 years before the arrival of his diploma.

CHARLES TOWNE

Founded in 1670 by English and Barbadian colonists, Charles Towne had about 8,000 people in the 1750s, half being slaves. Nevertheless, it had a small but vibrant scientific community, made mostly of physicians (Waring, 1964).[4] But even though South Carolina had some enlightened physicians, most were not well trained, theoretical medicine and treatments left much to be desired, and there was no regulation or licensing of the profession there. Thus, we find an anonymous freeholder from the mid-1750s quipping: "I shall only observe, that it has often been a Question, whether Lawyers or Physicians did the greatest Mischief to Men: one impairs their Estates, the other their Health and Life" (Waring, 1964, p. 65).

Garden, who was not a bashful man, complained that there were significant pockets of ignorance in natural philosophy in South Carolina, particularly in botany:

> Ever since I have been in Carolina I have never been able to set my eye upon one who had barely a regard for Botany. Indeed, I have often wondered how there should be a country abiding with almost every sort of plant, and almost every species of the animal kind, and yet that it should not have pleased God to raise up one botanist. Strange indeed that this creature should be so rare! (Smith, 1821, vol. 1, p. 477)

Garden was arrogant, and he diminished the importance of some of the other naturalists who studied South Carolina's botany and living nature during the Colonial Era. Indeed, from its start, the new South Carolinians were collecting and sending specimens back to Britain (Sanders & Anderson, Jr., 1999). Although artist-naturalist Mark Catesby (1731, 1743), author of *The Natural History of Carolina, Florida, and the Bahama Islands,* and a correspondent of Hans Sloan in London, had lived in Charles Towne in the 1720s and was long gone (yet still despised by Garden), Garden's physician friend William Bull shared this scientific interest with him. Bull corresponded with famed botanist Peter Collinson in England, helped Garden familiarize himself with the local plants, and even lent him Linnaeus' *Classes plantarum* and *Fundamenta botanica,* which he studied intently (Berkeley & Berkeley, 1969).[5] Thomas Walter (1788) was another South Carolina botanist overlapping Garden, although his extensive *Flora Caroliniana,* with approximately 1,000 species, many new, was not published prior to the War of Independence.

Nevertheless, there was some truth to what Garden wrote. Because South Carolina's hot and humid climate was deemed so dangerous, learned and gifted men, who would have loved to study its flora and fauna, did not want to settle there or even spend extended time in the region, risking exposure to its thick and dangerous miasmatic airs. In fact, South Carolina had long been reputed to be the unhealthiest of the British North American Colonies, with pools of stagnant water, bugs everywhere, and miserable endemic febrile diseases (e.g., chronic malaria) that could quickly bring the fittest men down.[6] Garden, who worried about his fragile health before leaving Europe, would constantly be ill in Charles Towne—a hardship he attributed to the effects the challenging climate had on both his body and mind.

Tall, thin, animated, and with a sense of purpose, Garden first assisted physician William Rose. After 2 years, in 1754, he went into partnership with Scottish native Dr. David Olyphant, who had worked with John Lining and experimented with electricity, and was now retiring. When Olyphant moved soon after Garden joined him, Garden took over his large practice.

THE COLLECTOR

Garden was happiest as a botanist and, more generally, a naturalist. Early on, Garden (1771) began sending botanical specimens and plant descriptions back to Edinburgh, much as did some other Edinburgh-trained physicians now living in South Carolina, including John Moultrie and John Lining. Scholarly Europeans were eager for these reports and specimens, wanting to know more about the plants and other natural wonders of North America, where family and acquaintances were settling, and where there seemed to be excellent opportunities for profitable commerce. Indeed, collecting had become the rage in Europe, with people anxious to fill their cabinets with exotica from around the world.[7]

Garden worked hard to expand his network of correspondents beyond the few professors he first communicated with in Edinburgh. In 1754, he began to write to Steven Hales (1733), the multifaceted clergyman-scientist who had written *Vegetable Staticks* (Garden named a tree after him). And starting with his second year in the colonies, he began his traveling with a trip to Florida to collect a wider variety of plants and animal specimens for himself and his correspondents.

In 1754, he traveled to New York and Philadelphia, eager to get away from the sweltering summer heat. In New York, he met with Cadwallader Colden, also a former student of

---

[4] William Bull, for example, was the first native-born American to receive the M.D. degree (at Leiden). John Lining, with his Edinburgh degree, arrived in 1728 and began correlating weather changes with diseases and metabolic functions about a decade later, also publishing an important treatise on *American Yellow Fever* in 1756 (Mendelsohn, 1960). Lionel Chambers, another physician of merit, was also an Edinburgh graduate, and he studied meteorology and epidemics, as well as tetanus (Waring, 1964, pp. 188–197).

[5] In 1773, Bull, then President of the Library Society, even appointed a special committee for collecting specimens, establishing a museum (the first of its type in America), and promoting natural history in the province. Garden was not, however, a member of the Library Society.

[6] In 1779, Alexander Hewatt would write that the summer "air [was] so poisoned by marshy swamps, that no European without hazard, can endure the fatigues of labouring in the air" (Parrish, 1996).

[7] For more on collecting, see Olmi (1992), Bredekamp (1995), Pietsch & Anderson, Jr. (1997), Impey & MacGregor (2001), and Koehler, Finger, & Piccolino (2009).

Charles Alston in Edinburgh, and a noted botanist with a large library. It was at Colden's estate overlooking the Hudson River that Garden, only 24 years old at the time, first opened Linnaeus' *Genera plantarum, Critica botanica,* and *Systema naturae.* He also read the letters that the great Swedish classifier and systematizer of nature exchanged with his correspondent in America. The letters of Peter Collinson in England, Johannes Frederick Gronovius in the Netherlands, and others with a passion for collecting also fascinated Garden, who became even more determined to expand his own network of correspondents.

Garden met botanist John Bartram, who paid a surprise visit, while he was with Colden, and he visited him in Philadelphia on the way home. He also went out of his way to meet Benjamin Franklin when he was in Philadelphia, discussing electricity and other subjects with him. Bartram and Franklin had co-founded the American Philosophical Society in 1743 (it began meeting in 1744), and Garden became a corresponding member of this organization in 1768. In that same year, he was elected into the rival American Society for Promoting and Propagating Useful Knowledge. The two groups would merge in 1769.

While awaiting the departure of his ship for London, Franklin penned a note to Garden, telling him, "I wish I had brought some of your ingenious Letters with me, that I might have consider'd them more fully." He added that he hoped he could stop to see him and other friends in South Carolina on the way home (Labaree, 1967, p. 183).

Garden would become well known in Britain and Sweden for the specimens he would send, his detailed descriptions, his adept classifications, and his personal and institutional affiliations. He became corresponding member of the Royal Society in 1755 and would become a Fellow in 1773, after being recommended by Franklin and others. And he was the first colonial correspondent of the Royal Society of Arts, which aspired to improve society through new inventions, discoveries, and commerce. William Shipley had founded this society in 1754, and Franklin also became active in it after he arrived in London (Allan, 1968). In addition, Garden could boast of being a member of the Royal Society of Uppsala (1763), which was where Carl Linnaeus, the greatest systematizer and classifier of all, held court.

Linnaeus, who was a foreign member of the Royal Society, became a foreign member of the American Philosophical Society in 1770. He actively corresponded with Garden after being advised to do so by John Ellis, another member of the Royal Society and a scientist who served Linnaeus from London (Groner & Cornelius, 1996). It was Ellis who also convinced Linnaeus to name the Cape jasmine plant, with its spectacular flowers, the "gardenia," in 1760. It was a fitting way to honor the South Carolinian who had been so willing to send him specimens, and had adopted and promoted his classificatory system (Smith, 1821, Vol. 1, pp. 129–130, 501).

Indeed, Garden had become a tenacious defender of Linnaeus' *Systema naturae,* which first appeared 25 years before this honor was granted and had gone through an amazing 10 editions by this time (Linnaeus, 1735, 1758; Garden ordered Linnaeus' books as they came out) (Fig. 18.3). Although he had not been the discoverer of the gardenia, he rejoiced when he received the news that his name would now be attached to this magnificent flowering plant, and in his elation he promised to send even more specimens to Linnaeus.

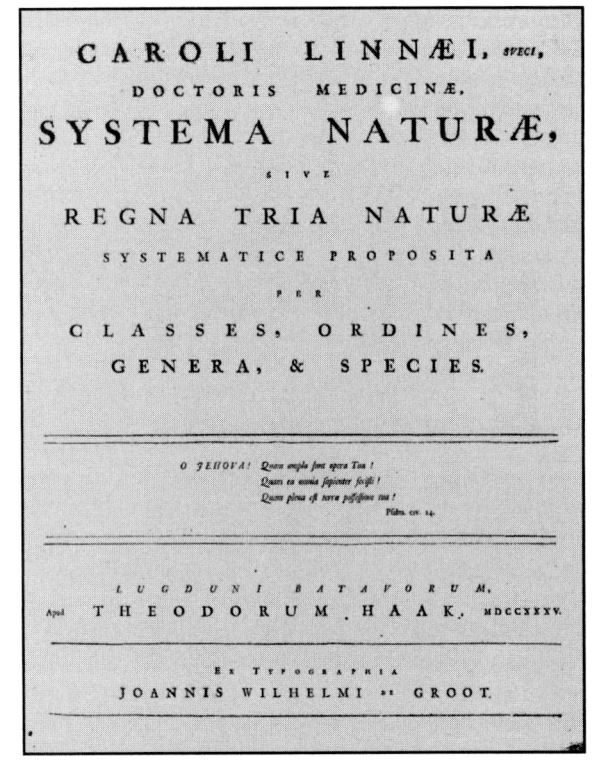

Figure 18.3. Cover of the first (1735) edition of Linnaeus' *Systema naturae.*

Three decades later, the President of the Linnean Society said that Garden's name appeared more frequently than any other name in the last edition (the 12th, 1766) of Linnaeus' *Systema naturae* (Smith, 1791). Among other things, he had 33 new species of fishes associated with his name (Denny, 1948, p. 172).

Aspiring scientists strove to affiliate with both masters and revered institutions at this time. They offered to collect material, to preserve, package, and send them to distant ports, and even to perform needed experiments. In addition to participating in the scientific Enlightenment, which meant acquiring useful knowledge (e.g., botanicals with medical promise), these activities nourished minds and allowed people in remote places to make internationally recognized names for themselves as practicing naturalists (Parrish, 1996).

The natural philosophers receiving these "gifts" were expected to send books, instruments (e.g., microscopes), and supplies back to their correspondents, and perhaps even sponsor them in their elite organizations, in return for the specimens they received (Denny, 1948). As put by historian Susan Scott Parrish (1996, p. 106) when discussing trans-Atlantic correspondence and the advent of scientific networks involving North Americans:

> Colonials were hungry for institutional connections, for print matter and scientific equipment, and in certain ways for publicity, whereas Londoners [among others] needed the stuff of nature and connections with people reliable enough to send steady shipments and accurate descriptions of that matter.... Though printed volumes like Mark Catesby's *Natural History,* collections like those of Sir Hans Sloan (which formed the basis of the British Museum), journals like the *Philosophical Transactions,* or definitive catalogs like Linnaeus's

Figure 18.4. Johann Frederik Gronovius (1686–1762), the father of Laurens Theodor, and his 1742 article in the *Philosophical Transactions* on how to preserve fish specimens. Garden followed these guidelines and many of the specimens he sent to Linnaeus are still in good condition today.

Systema Naturae were the public and final products, the continual, busy, informal, and sometimes familiar world of letter writing and specimen exchange made up the day-to-day practice of natural history upon which more institutional or public achievements were based. (Parrish, 1996, pp. 106–107)

LINNAEUS, FISHES, AND UNUSUAL SPECIMENS

Once Linnaeus began to communicate with Garden, he asked him to send him more unusual fishes and animal specimens, not just plants. In a return letter to Linnaeus dated January 2, 1760, Garden wrote:

Before the receipt of your letter, I had scarcely paid any attention to our fishes; but all your wishes are commands to me, so that I wish you never to have occasion to repeat them. Nothing would grieve me more than to disappoint you in any respect, and therefore I immediately set about procuring all the kinds I can. I have caused their skins to be dried, by which I think you will be able to see the true situation of the fins. This will be more satisfactory to you than a bare, and perhaps inaccurate, description of mine. Nevertheless, I will subjoin their characters, however imperfectly described. (Smith, 1821, vol. 1, p. 300)[8]

Thus, Alexander Garden, "probably the ablest of the Charles Town community of scientists before the close of the Old Colonial Era" (Stearns, 1970, p. 599), became even more interested in collecting and preserving fishes. He was "one of that eighteenth-century international scientific circle whose members exchanged ideas, information, and specimens, and encouraged each other in the continuing pursuit of scientific enlightenment" (Berkeley & Berkeley, 1969, p. 325).

Garden's preservation method was taken from Johannes Gronovius (1744), who had described the full-day preparation in detail in the previous decade in a two-page paper in the *Philosophical Transactions of the Royal Society of London* (Fig. 18.4). The technique involved cutting the fish dorsally from head to tail, pinning and drying (by sun or fire) half the fish with vertical fins attached, removing the remaining flesh, and pressing the skin.[9] When shipped, they were packed with colocynth and aloe to ward off insects.

Linnaeus received four fish collections from Garden, via Ellis in London, the last being in 1771 (Wheeler, 1985; see Sanders, 1997, for Garden's name, the current scientific and popular names for each item sent, and some pictures). There also were some lost and confiscated shipments, as well as shipments of individual items (Smith, 1821, vol. 1, p. 421; Günther, 1899; Sanders, 1997). Garden's dried specimens were exceptionally well prepared, as evidenced by the fact that long after many (but not all) of Linnaeus' other dried fish specimens had deteriorated, those he prepared with the Gronovius (1744) method still exist in reasonably good shape.[10]

---

[8] James Edward Smith was the first President of the Linnean Society of London, and he published two volumes of letters to and from Linnaeus and other naturalists associated with him, including Garden. Smith would purchase Linnaeus' library and specimens, which would go to the Linnaean Society after his death in 1828 (Wheeler, 1985). In 1785, Garden would help nominate Smith for membership in the Royal Society of London.

[9] Garden's surviving specimens also show the skins varnished and placed on herbarium paper or cards, although good reasons have been given (e.g., the uniformity of the papers, no notes on them by Garden) to suspect that the papers they are now on might have been added or changed after the specimens reached Europe (Wheeler, 1985).

[10] Eighty-five specimens can still be found in the Linnaean Society's fish collection of 168 dried specimens (Wheeler, 1985; Sanders, 1997), and 16 others are remain in the Laurens Gronovius collection acquired by the British Museum (Natural History) (Wheeler, 1958, 1989). For more on Garden's fishes in the

Garden did more, however, than just send dried skins. He also sent fishes preserved in rum and in spirits of wine. Further, he went out of his way to supply Linnaeus with critical written information about each specimen, serving as one of his most important sources about fishes, and certainly his most important source from North America.

As for Linnaeus, he was convinced that God had a well-designed plan for the preservation of each of his creations, no matter how strange (Hagberg, 1952; Larson, 1971; Egerton, 2007). And he was thrilled with the dried fishes' skins Garden skillfully prepared, with his whole fishes shipped in rum or wine, and with his notes conveying important information about his specimens.

THE RIGHT MAN FOR THE EELS

In 1774, when the live eels arrived in Charles Towne, Garden had to be the perfect person to describe them in or near this southern port city. First he had an inquisitive mind. Second, he wanted to publish a report with the Royal Society, which had elected him a Fellow in 1773.[11] Third, he wanted Linnaeus to know more about these fish. And fourth, he was just starting to recover from scarlet fever and might have wanted to be shocked by the eels, because they were said and advertised to be therapeutic.

To these observations, knowledge of fish electricity and electricity in general must be added. Regarding the former, Garden was familiar with some of the experiments that had been conducted on electric fishes. This is suggested by the fact that he consulted Laurens Gronov's books and was communicating with him prior to 1774. Before this time, Gronov had written an article in Dutch followed by a Latin translation in which he presented evidence for the eels being electrical (Ch. 13; Gronov, 1758, 1760). He had a preserved *Gymnotus* in his cabinet and had included a picture of it in his magnificent *Zoophylacium* (Gronov, 1763–64). He was also familiar with the writings of Allamand, who published on the subject in 1756, and eagerly wanted to serve the Dutch physicist as one of his foreign correspondents. Further, some of the earlier experiments on fishes were mentioned in readily available texts, including those by Swiss naturalist Charles Bonnet (1769), whose *Contemplation de la Nature* underwent many editions and circulated widely among naturalists.

With regard to electricity in general, Garden kept in contact with John Lining, whose practice he had taken over and who discussed electrical issues with Franklin, including how to capture lightning from the heavens (Labaree, 1962, pp. 521–526; Finger, 2006a, pp. 109–110). One of Franklin's letters to Lining was even included in a revised edition of his *Experiments and Observations on Electricity*, which appeared in the very year the eels came to South Carolina (Franklin, 1774, pp. 319–328). Garden discussed electricity when he met with Franklin in Philadelphia, and he read his publications. He also received articles on electricity from Ellis (see Smith, 1821, vol. 1, p. 407). Further, as mentioned, he was reading the works of Allamand, one of the leading "electricians" of the day, and those of Peter van Musschenbroek, with whom Allamand first experienced the Leyden jar (Chapter 10). Hence, in addition to having some knowledge about electric fishes, Garden had some familiarity with the progress being made in this important branch of natural philosophy.

Thus, many things might have converged on Garden's mind when he read about the eels in the *South Carolina Gazette*. In an early biographical sketch, we read: "Botany, and some of the more obscure departments of Zoology, especially fishes and reptiles, were his constant resources for amusement and health" (Smith, 1821, vol. 1, p. 282). Thus, Garden mustered the energy to leave his sickbed to go to see the fish and take detailed notes. He was fearful that if he did not do so quickly he would miss a golden opportunity, as the eels were not going to stay in Charles Towne very long.

OBSERVATIONS, EXPERIMENTS, AND GARDEN'S 1774 LETTER

Garden penned a letter from "Charles-Town, South Carolina, Aug. 14, 1774," and sent it to John Ellis in London. This was precisely 1 year after the Philadelphia experiments described in the previous chapter were performed. Ellis, as stated, had helped Garden to become a member of the Royal Society and was an active correspondent for Linnaeus in London. Moreover, like Linnaeus and Garden himself, Ellis was especially focused on flora and fauna that might have some "service" or utility. The eel, with its strong shocks and possible use in medicine, was a creature most worthy of research, and Garden knew that both Ellis and Linnaeus would have great interest in it.

The letter to Ellis was written before Garden had received a copy of the *Philosophical Transactions* that contained Walsh's (1773) paper on the French torpedoes. He would later state that he had wished he had seen Walsh's report (note to Ellis of March 12, 1775, in Smith, 1821, vol. 1, p. 605). As for Bancroft (1769a), his book came into his possession while he was writing this letter, and he was relieved to find that his own physical descriptions of the fish and its behaviors far exceeded what Bancroft wrote and, for that matter, all earlier reports.

Garden was not just being egotistical about his descriptions. He had an eye for nature and he did not miss the fine details. This can be appreciated by reading what he wrote about the physical features of the eels and how they swam around the tank. He begins by stating that they ranged in size from about 2 feet long to 3 feet 8 inches long (Williamson's eel was "three feet seven inches long") and writes:

> The whole body, from about four inches below the head, seems to be clearly distinguished into four different longitudinal parts or divisions. The upper part or back is roundish, of a dark colour, and separated from the other parts on each side by the *lateral lines;* which, taking their rise at the base of the head, jut above the pectoral fins, run down the sides, gradually converging, as the fish

---

Linnaean collections, see Goode & Bean (1885), Günther (1899), Wheeler (1958, 1978, 1985, 1989, 1991), Sanders (1997), and Sanders & Anderson (1999).

[11] The letter of recommendation on his behalf was dated March 4, 1773, and reads: "Alexander Garden M.D. of Charles Town South Carolina, A Gentleman very well known for his many new and curious discoveries in Natural History, being very desirous of becoming a Member of the Royal Society, We from his general good Character and his publications, do recommend him as likely to prove a very useful one." The signers included John Ellis, Daniel Solanderr, Benjamin Franklin, Thomas Pennant and London physician William Pitcairn.

grows smaller, to the tail, and make so visible a depression or furrow in their course, as to distinguish this from the second part or division, which may be properly called the body, or at least, appears to be the strong muscular part of the fish. This second division is of a lighter and more clear bluish colour than the upper or back part, and seems to swell out somewhat on each side, from the depression of the *lateral lines;* but, towards the lower or upper part, is again contracted, or sharpened into the third part, or *carina.* This *carina,* or heel, is very distinguishable from the other two divisions, by its thinness, or its apparent laxness, and by the reticulated skin of a more grey and light colour, with which it is covered. When the animal swims gently in deep water, the rhomboidal reticulations of the skin of this *carina* are very discernible; but when the water is shallow, or the depth of the *carina* is contracted, these reticulations appear like many irregular longitudinal *plicae.* (Garden, 1775a, pp. 104–105)

There is no reason to give Garden's complete description of the eel here, as it goes on for five full pages (without a new paragraph) and deviates from the real focus of this book, its shocks. Yet it is interesting to note that Garden expressed dismay, because he was not allowed to handle the fish for a more thorough examination (the same complaint voiced by Williamson), and because he did not have access to a dead one for dissection.

The rest of his paper deals with various observations and experiments on the shocking capability of the eels; they are reminiscent of some of the experiments performed the previous summer by Rittenhouse, Kinnersley, Williamson, and other members of the American Philosophical Society in Philadelphia (Williamson, 1775a). Here he points out that the shocks can pass from person to person when they hold hands to form a chain. He also notes that a circuit can be completed without actually touching the fish with both hands, writing that with the hand of one person touching the fish and that of the other person at the end of the human chain immersed in the water near the fish, everyone will feel its shock. Further, the "stroke" is communicated by known conductors of electricity and blocked by non-conductors.

"The shock which our Surinam fish gives," concludes Garden, "seems to be wholly electrical; and all the phaenomena or properties of it exactly resemble those of the electric aura of our atmosphere when collected, as far as they are discoverable from the several trials made on this fish" (p. 109).

Ellis was so impressed when he received Garden's report that he sent it to Daniel Solander, a Swede who studied with Linnaeus and was sent on a voyage that took him to London in 1760, with the hope that he would transmit new information and worthy specimens back to Sweden (Duyker & Tingbrand, 1995; Duyker, 1998).[12] Solander was warmly welcomed in London, because he understood and explained Linnaeus' methods and systems, and because the Enlightenment goal of sharing and spreading useful knowledge was universal. In fact, he so enjoyed London that he made it his new home, apparently offending Linnaeus, who hoped he would return to Sweden to become his successor at Uppsala rather than staying an independent researcher in Britain.

When Garden's report arrived, Solander had already circled the globe on the *Endeavour* with Captain James Cook and naturalist Joseph Banks, who covered Solander's expenses. On this voyage, Solander showed himself to be more than a knowledgeable botanist; he collected some 222 previously unknown fishes in the Pacific Basin alone. On his return, he continued to work for Banks, was an active member of the Royal Society, and held positions at the British Museum, including Keeper of the Natural History Department.

Solander showed Garden's report to John Pringle, the President of the Royal Society who was so close to Franklin that the king would soon have him replaced by the more Loyalist and aristocratically sensitive Banks. In a letter from Solander to Ellis dated October 13, 1774, we read: "I have read through Dr. Garden's account of the Electric Eel, and think it a paper well worth publishing. I have also shewn it to Sir John Pringle, who is of the same opinion" (in Duyker & Tingbrand, 1995, p. 338). Hence, as Garden had hoped, his paper was found to be novel enough to be read aloud before the Royal Society, and then published in the 1775 *Philosophical Transactions* as "An Account of the Gymnotus Electricus, or Electric Eel," where several thousand people would encounter his worthy contribution (Fig. 18.5).

This report, however, did help to sustain an erroneous myth. Gardner wrote at the end of his letter: "I am told, that some of them have been seen in Surinam river upwards of twenty feet long, whose stroke or shock proved instant death to any person that unluckily received it" (p. 110). Clearly, fantastic rumors and misunderstandings were still a part of

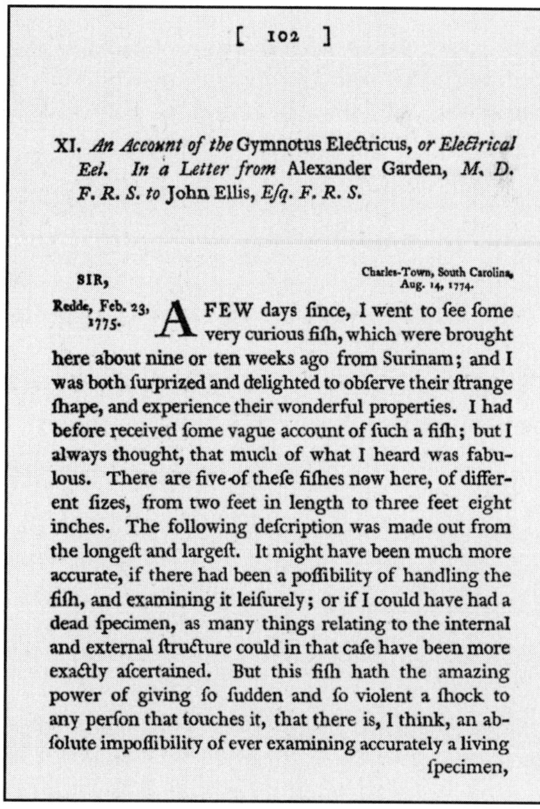
[102]

XI. *An Account of the* Gymnotus Electricus, *or Electrical Eel. In a Letter from* Alexander Garden, *M. D. F. R. S. to* John Ellis, *Esq. F. R. S.*

SIR, Charles-Town, South Carolina, Aug. 14, 1774.

Redde, Feb. 23, 1775.

A FEW days ſince, I went to ſee ſome very curious fiſh, which were brought here about nine or ten weeks ago from Surinam; and I was both ſurprized and delighted to obſerve their ſtrange ſhape, and experience their wonderful properties. I had before received ſome vague account of ſuch a fiſh; but I always thought, that much of what I heard was fabulous. There are five of theſe fiſhes now here, of different ſizes, from two feet in length to three feet eight inches. The following deſcription was made out from the longeſt and largeſt. It might have been much more accurate, if there had been a poſſibility of handling the fiſh, and examining it leiſurely; or if I could have had a dead ſpecimen, as many things relating to the internal and external ſtructure could in that caſe have been more exactly aſcertained. But this fiſh hath the amazing power of giving ſo ſudden and ſo violent a ſhock to any perſon that touches it, that there is, I think, an abſolute impoſſibility of ever examining accurately a living ſpecimen,

Figure 18.5. The cover page of Alexander Garden's letter on the eels that arrived in South Carolina in 1774.

[12] Solander, who was fellow of the Royal Society from 1764, had supported Garden's election in 1773.

the landscape, since these fish never reach even half this length. More likely than not, the creature observed was a large South American river snake, which kills by constriction, although not instantly. Less likely, but also a possibility, it could have been a school of eels. To be fair to Garden, however, he might not have believed the myth he was passing along. In his concluding paragraph, he writes: "I shall be on the watch to procure a more accurate knowledge of, and acquaintance with, this animal: and if I can learn any thing farther about it, you may depend on my communicating it" (p. 110). His use of the words "more accurate knowledge" is revealing of his skepticism about many things he, as an enlightened man of science, personally could not confirm.

THE FATE OF THE FIVE EELS

It had taken several weeks for the eels to make the trip from Guiana to South Carolina, and Garden thought Captain George Baker had little chance of getting them alive to his ultimate destination, more lucrative London. In fact, Baker had told him that, "when they were first caught, they could give a much stronger shock by a metalline conductor than they can at present" (Garden, 1775a, p. 108). Their loss of vigor was obviously a concern to Baker, as this translated into smaller profits and raised questions about their longer-term survivability. It was partly for this reason that Baker had instituted such strict rules for handling these fish (of course, he also knew they bruised easily, etc.).

Aware of the situation and what was at stake, Garden told Baker to take the dead eels to Ellis, should they fail to make it alive to London. "But as it is very uncertain whether they will arrive in health and all alive, I have recommended to him to get a small cask of rum, with a large bung, into which he may put any of them that may die, and so preserve them for the inspection and examination of the curious when he arrives" (p. 103).

In a follow-up letter to Ellis dated March 12, 1775, Garden wrote about his severe fever and how he feared that his earlier letter on the eels from Surinam was "very confused." He also mentions receiving some volumes of the *Philosophical Transactions*, "with many curious papers in them, particularly one on the electric power of the torpedo" (Walsh's 1773 paper). More importantly, he wrote about an agreement he had made with Captain Baker:

> The person who owned these fish carried them from hence to England, but whether they arrived alive, I really know not. I desired him, if they arrived safe, immediately to go to you, and I gave him your address; but lest they should die by the way, I desired him to put them in a small keg of rum, which I advised him to prepare, for the purpose of preserving them till he got home. I wish to hear both of the fate of my letter and of these fish, whether they arrived and reached you; and, if they did, what satisfaction they gave. (Smith, 1821, vol. 1, p. 604)

Garden was a realist who knew a lot about living nature, and he was correct in his assessment. When Baker left North America for England, he still had five live eels. Four never made it alive to Falmouth on the English coast. The fifth was barely alive when the ship landed there in November 1774. Although several inquisitive people experienced some weak shocks from the frail survivor, it too failed to make it alive to London.

But these eels still had stimulating effects in another, very different way. Thankfully, Baker followed Garden's advice about preservation. As each of the four eels died during the trans-Atlantic crossing, they were put into a vat of rum. Daniel Solander wrote the following to John Ellis on November 7, 1774:

> The man who shewed the Electric Eels in Carolina, as described by Dr. Garden, is now arrived; but, unluckily, all the five eels died during the voyage, or at least before he came up to London. One was alive when he landed in Falmouth, where several persons felt the electrical shocks; but that one died on the passage from thence to London. However, he has benefited by Dr. Garden's advice, to put them in spirits in case they should die. He has brought us four complete specimens, well preserved. . . . Mr. John Hunter danced a jig when he saw them, they are so complete and well preserved.
>
> We shall have a drawing made of one of the Electric Eels; and John Hunter has promised an anatomical description, to accompany Dr. Garden's account, when presented to the Royal Society. (Smith, 1821, vol. 2, p. 21)

When John Walsh learned about the arrival of the preserved eels, he bought three of them. Solander mentioned what Walsh did, and Walsh's motivation, in a letter to Banks in broken English. It was dated November 10, 1774:

> Walsh is come to town, and I have told him our plan about Electrical Eels; he entered into the Scheme with Spirit, and when he this night at the Royal Society, heard an account from W. Hope in Amsterdam, that his friends had sent him several but that they all died in the passage; Walsh though that they may perhaps be brought over to Holland preserv'd in Spirit, he came up to me and said: England must have the honour of first displaying the Electrical Eel to the world, as well as the Torpedo.

Solander continues:

> He gave me leave to write him down the purchaser of 3 of the Eels at 20 guineas a piece—so they may soon be examined and dissected by John Hunter and that account delivered as soon as possible to the Royal Society and of course we be before hand with the Dutch. This I thought generous and proper and therefore promise to procure him the largest and 2 of the lesser. I don't doubt of easily receiving the rest to make it up 50 pounds and if I am not mistaken you thought that to be a proper sum. Mr Cavendish promised to subscribe and 2 or 3 more that heard of it. I hope you approve of the above. If you don't come to town soon, Walsh & Hunter seem to be bent upon beginning with opening one at least the beginning of next week; therefore let me know if you don't think as I do. . . . Walsh intends 1 of his eels for the Br[itish] Museum, 1 for the Royal Society and one for himself. (Duyker & Tingbrand, 1995, pp. 342–343)

HUNTER'S EEL DISSECTION

John Hunter (1775, p. 395) confirmed that the specimen he dissected "was procured by that Gentleman [Walsh], and at

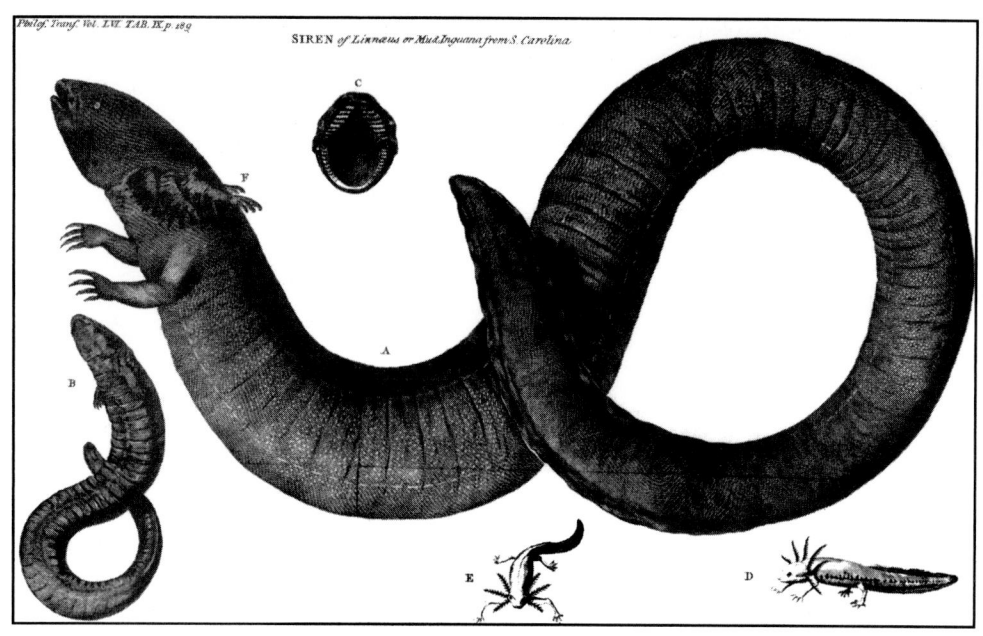

Figure 18.6. Alexander Garden's amphibious bipes (*Siren lacertina*) as it appeared in an article by Ellis in the 1766 *Philosophical Transactions*.

his request this dissection was performed." And as Solander had warned Banks, London's famous "Knife Man" wasted little time before beginning his dissection (Moore, 2005). He began on November 15, 1774, within days of obtaining his specimens.

This was not the first dissection linking Hunter to Garden. In 1765, Garden had sent a local siren, or mud iguana, to Europe (Fig. 18.6). Because it had only front legs, Linnaeus wanted proof that this strange animal was not just a larva, despite Garden's assertions that he had seen this swamp creature in various sizes, but always in the same form (Smith, 1821, vol. 1, pp. 327). Hunter was asked to do a proper dissection, which revealed reproductive organs supportive of Garden's non-larval, non-juvenile interpretation, and everything was published in the 1766 *Philosophical Transactions* (Ellis, 1766; Hunter, 1766). The mud iguana was, in fact, Hunter's first communication to the Royal Society, and it led to his election into that organization.

Returning to the eel dissection, Solander was a witness. He wrote the following to Banks, providing a rare behind-the-scenes look at how rapidly Hunter worked, as well as some often-overlooked problems of scientific illustration at this time:

> This morning John Hunter began the dissection of the Electric Eels; we found a much greater quantity of Electric organs than was expected—suppose the annexed figure a cross section—all what is marked with lines is electrical repositories and the white part is the muscular. It has no lungs only gills—therefore a fish. Walsh is now busy to get the drawings made. I believe Roberts [probably English engraver James Roberts] is to draw the minute and Anatomical part and he wants much to get Miss [Gertrude] Metz to draw the external appearance or the figure proper. I told him I could not promise her to make a finished drawing without first asking your leave, but that I would ask her to give an outline as Mr. Stuart of Leicester . . . named her as the best shaper of fish, of any draughts person now in London. He thinks the fish piece she painted some time ago had great merit. If you will spare her 2 or 3 days to make a compleat [*sic*] drawing that is to say to shade the drawing let me know, I think it would be right as I really didn't now know a good person to do it . . . and old [John] Miller has quarreled some time ago with Walsh and Roberts can give no life to his drawings. (Duyker & Tingbrand, 1995, p. 344)

An entry from Solander's diary, dated 20 Jan. 1775, reads: "An Electrical Eel (*Gymnotus Electricus Linn.*) from Mr Walsh - The electrical organs laid open by Mr John Hunter" (Solander, 1775).

Hunter presented his findings to the Royal Society on February 23, 1775—not coincidently on the very same day that Garden's paper from South Carolina was read—and they were published later that year (Fig. 18.7). The famed anatomist wrote that the eel possesses two sets of paired organs, a larger one that has an anterior and lateral location, and a smaller one ventral to it. These organs, he noted, account for about half of the flesh in which they are placed and more than a third of the whole eel. They are made of parallel rows of flat disks separated by thin membranes, much like the vertical columns in torpedoes.

The nerves, he noted, originate in the medulla, come out of the spinal cord, and branch when they enter the organs. Torpedoes have a denser nerve supply than the eel: "Perhaps when experiments have been made upon this fish, equally accurate with those made upon the torpedo, the reason for this difference may be assigned" (Hunter, 1775, p. 404). Five figures accompanied Hunter's verbal descriptions (Figs. 18.8 and 18.9).

There is also an entry in Daniel Solander's dairy dated June 16, 1775, which reads: "Mr John Hunter of Jermyn Street presents a Transverse section of the Electrical Eel (*Gymnotus Electricus*) accompanied with an explanation of the different organs" (Solander, 1775).

Hunter's eel specimens have survived, even though large parts of Hunter's collections were destroyed in World War II. They exist as items 2185 and 2186 in the Hunterian Museum

[395]

XXXIX. *An Account of the Gymnotus Electricus.*
By John Hunter, F. R. S.

Redde, May 11, 1775.

TO Mr. WALSH, the first discoverer of animal electricity, the learned will be indebted for whatever the following pages may contain, either curious or useful. The specimen of the animal which they describe was procured by that Gentleman, and at his request this dissection was performed, and this account of it is communicated.

This fish, on the first view, appears very much like an eel, from which resemblance it has most probably got its name; but it has none of the specific properties of that fish. This animal may be considered, both anatomically and physiologically, as divided into two parts; viz. the common animal part; and a part which is superadded, viz. the *peculiar organ*. I shall at present consider it only with respect to the last; as the first explains nothing relating to the other, nor any thing relating to the animal œconomy of fish in general. The first, or common animal part, is so contrived as to exceed what was necessary for itself, in order to give situation, nourishment, and most probably the peculiar property to the second. The last part, or peculiar organ, has an immediate connexion with the first; the body affording it a situation; the heart, nourishment; and the brain, nerves and probably its peculiar powers. For the first of these purposes, the body is extended out in length, being much longer

Ggg 3   than

Figure 18.7. The first page of John Hunter's anatomical study of the electric eel.

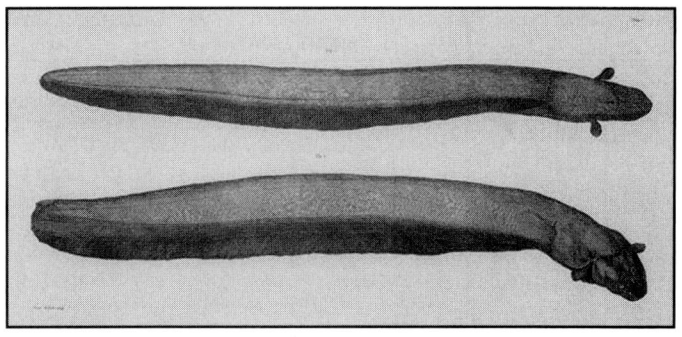

Figure 18.8. John Hunter's (1775) illustration of intact electric eels.

of the Royal College of Surgeons in London. The respective labels are "A Large Electric Eel" and "A Segment of the Anterior Part of the Trunk of an Electric Eel."

Walsh had used the catchy term "animal electricity" in his notebooks (see Piccolino, 2003a), but not in his published 1773 and 1774 papers on the torpedo. Hunter (1775, p. 395) now introduced this term in his published paper on the eels, writing: "To Mr. WALSH, the first discoverer of animal electricity, the learned will be indebted for whatever the following pages may contain, either curious or useful."

Actually, Hunter had not been convinced that the eel's discharges are electrical prior to their arrival in London (Piccolino, 2003a; Walker, 1937). But his stance was now changing. Walsh's spark experiment would be the event that would dramatically change many other agnostic and skeptical minds. And here too the arrival of Garden's dead eels played a role.

In the November 7, 1774, letter written by Solander to Ellis, he basically told his friend, colleague, and fellow Linnaean that having dead eels was a first step, but hardly the golden grail he and other members of the Royal Society so badly coveted. Hence, live ones must somehow be obtained. With a focus on this ultimate goal, "we propose to raise by

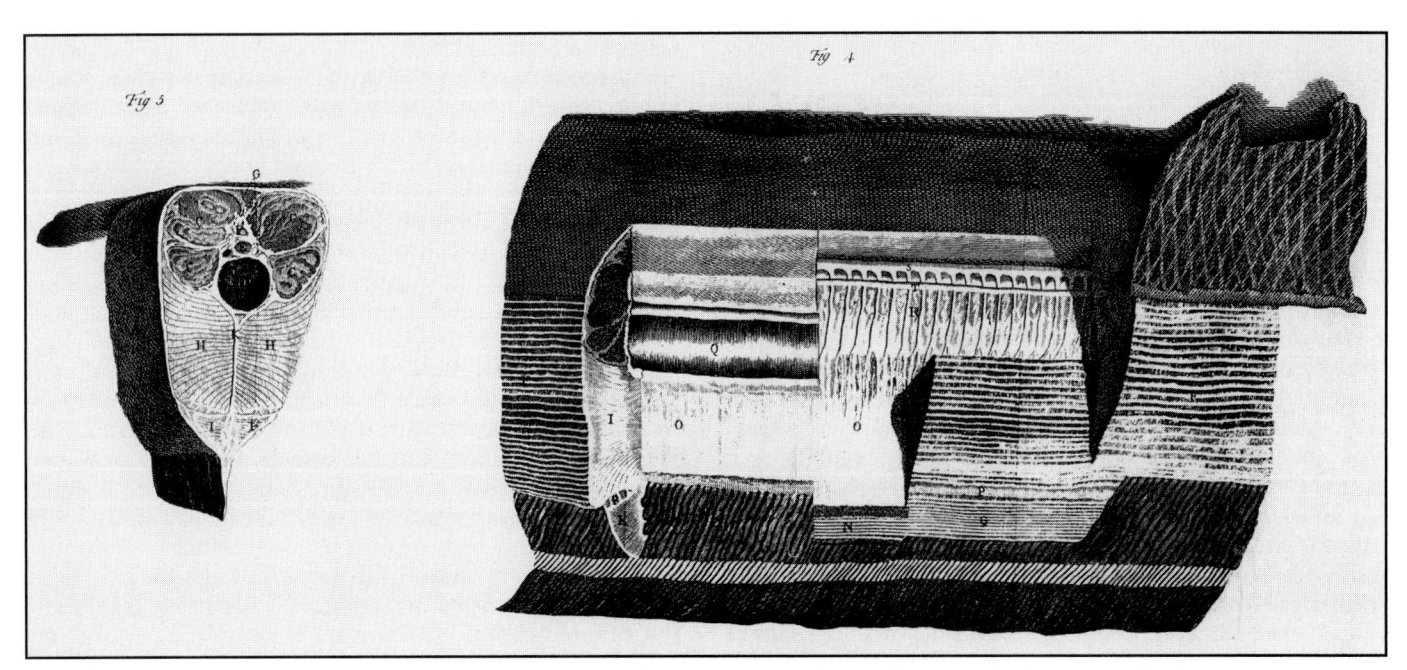

Figure 18.9. Left: A cross-section of the electric eel showing its organs. The large electric organs are labeled H and the smaller ones below it are labeled I. Right: A longitudinal section through the eel's body showing its electric organs. The large electric organ is labeled F and L, and the small organ is G and N. R represents nerves going to the electric organs (from Hunter, 1775).

subscription, or some other method, a sum of money, to enable the man [Baker] to go out again" (Smith, 1821, vol. 2, p. 21).

Baker would indeed go back to Surinam. And he would return to London in 1776 with live eels, although this fact is not mentioned in the electric fish literature, leading one historian who recently wrote on Garden to conclude that nothing came of the idea to encourage or underwrite Baker to try again (Delbourgo, 2006). In the next chapter, we shall examine the excitement Baker's return trip created in London, and how it enabled Walsh to show Hunter, Cavendish, and other members of the Royal Society of London that the eel really can produce a tiny spark when it unleashes its weapon—a spark that can be seen across a minute gap in darkness, even in cooler London, far from its equatorial climate.

AN AMERICAN POSTSCRIPT

Although Garden backed the Americans in some of their grievances against the Crown (e.g., against the Stamp Act), he remained a Loyalist during the American War of Independence, even sending an ill-timed letter congratulating Lord Charles Cornwallis after his victory at Camden in 1780 (Berkeley & Berkeley, 1969; Sanders, 1997). A year later, Cornwallis' surrender at Yorktown effectively ended the war, but not in the way Garden had wanted. The South Carolina Legislature now banished the Loyalists and confiscated their properties, sending Garden, his wife Elizabeth, and their two daughters sailing to London with few worldly possessions.

Garden arrived in London in 1783 and spent considerable time trying to receive compensation (ultimately successfully) from the Crown for his lost properties. He also signed the *Charter Book of the Royal Society*, now presided over by Joseph Banks, making his membership official in the organization that had published his eel paper. But although he now attended its meetings and "was a most welcome addition to the scientific circles in London," he was no longer capable of pursuing hands-on natural history (Smith, 1821, vol. 1, p. 283).

In part, this was due to his declining health. But a second factor was that his intellectual circle had grown smaller. Ellis and Collinson died in 1776, Linnaeus and Bartram passed away in 1777, and other friends had also been buried. One survivor was naturalist Thomas Pennant, who had also received specimens from him (but not electric eels), and seemed delighted to find him now living in London (Pennant would name a heron after him).

In 1788, Garden visited the European continent, and a year later he made his way back to his childhood home near Aberdeen. He returned to London debilitated and died in 1791, possibly from a tubercular disorder, much as he feared when he first set sail for South Carolina (Jenkins, 1928).

Obituaries appeared in many newspapers and periodicals, some being *Scots Magazine, European Magazine and London Review*, and *The Gentleman's Magazine*. The notice on page 3 of the April 18, 1791, issue of *The London Times* was just one sentence long: "Early on Friday morning last departed this life, after a lingering illness, Dr. Alexander Garden, of Cecil Street, in the Strand, late of Charlestown, South Carolina." There was no mention of Garden's illustrious contributions to natural history, his association with Linnaeus, his long-distance exchanges with the Linnaeans and other natural philosophers in London, or his electric eels, which by this time (as we shall see) had radically changed how people were thinking about nerve and muscle physiology.

In retrospect, Garden followed several other North Americans who played notable roles in the electric fish story. They include Benjamin Franklin, John Bancroft, David Rittenhouse, and Hugh Williamson. They would not be the only North Americans doing electric fish research in the second half of the eighteenth century. Henry Collins Flagg, who was briefly mentioned in the last chapter, and William Bryant, also dealt with the electric eel, and what they reported in 1786 also merits some attention, in part because their reports make it clear that electric eels were just as exciting to some Americans immediately after the Revolution as they were to their cousins still living at home and in England during this turbulent time.

Henry Collins Flagg, who was born in Rhode Island in 1742, probably practiced medicine for several years in South America before he moved to Charles Towne, South Carolina, where he served as a surgeon during the War of Independence (Waring, 1964). In 1782, 1 year before the Treaty of Paris was signed, he sailed to Guiana. His account of the electric eel was published in the *Transactions of the American Philosophical Society* in 1786.

Flagg made the interesting point that some people seemed immune to its shocks; something he had witnessed first hand. His statement was that, "if a number of persons join hands, and one touch[es] the eel, they are all equally shocked, unless there should happen to be one of the number incapable of being affected by the eel" (Flagg, 1786, p. 171). The insensible party was "a very worthy lady" who could "handle this fish at will." Surprised that she could "treat the fish with so much familiarity while in a perfect state of health," he wondered if she were really healthy. In a footnote, he relates that, when he met her, she "was far gone in an hectic fever . . . I did not think to enquire if she could treat the fish with so much familiarity while in a perfect state of health" (Flagg, 1786, p. 171).

This lady, however, was not alone in her capacity to tolerate the shocks. Although not in her social set, Flagg relates that some Indians and slaves could do the same when in perfect health. In his words: "I have seen negroes take hold of it, at first very cautiously, receiving many light shocks, but presently grasp it very hard and take it out of water." Flagg probably had the mindset that Indians and African slaves might be less sensitive to pain, a notion widespread on North and South American plantations. The basic idea was that black Africans and Indians are inferior racial groups that share more traits with wild animals than do Caucasians.

This, of course, is not to say that all members of these "inferior" groups are insensitive to pain, or that they did not boast of their abilities to tolerate pain. Flagg tells the story of how:

> a negro fellow formerly being bantered by his companions for his fear of this eel, determined to give proof of his resolution, and attempted to grasp it with both hands. The unhappy consequence was, a confirmed paralysis of both arms. (Flagg, 1786, p. 171)

Flagg, who was clearly skeptical, at first questioned this man's honesty. But in a footnote (on p. 171), he tells us: "This account was afterwards confirmed by me, with further information, that after several years the negro recovered the use of his arms by slow degrees, and I think without assistance from medicine."

As for the nature of the discharge, Flagg states that he, for one, is convinced that "the electric and torporific particles are the same" (p. 172). Nevertheless, when he tried to see if the eel would affect a compass needle, he was unsuccessful.

William Bryant, also a physician, was born in 1730 and had lived in Trenton, New Jersey, just across the Delaware River from Franklin's Philadelphia. In 1786, the year in which he died, he wrote that Surinam

> abounds with as many natural curiosities as any country in the world. But that which I look upon to be as surprising as any in it, and which I believe has not yet been accurately described, is a fish of the species of eel, and is caught there in nets among other fish; generally in muddy rivers, and I believe is found in most of the neighboring provinces. It is called by the Dutch *Beave Aal*, and by the English, the Numbing Eel. (Bryant, 1786, p. 166)

He then described how touching the eel in a tub of water led to "a sudden and violent shock, in all respects like that which is felt on touching the prime conductor, when charged with electrical fluid" (Bryant, 1786, p. 167). The shock more severely affected his fingers and elbow, but even his breast and shoulders felt it. It was also felt by seven people when they formed a chain with the eel, "in the same manner as I remember to have been in the electrical [Leyden jar] experiment, when several persons take hold of the wire and the equilibrium is restored by the fluids passing through their bodies" (Bryant, 1786, p. 167). Also repeating previous work with conductors and non-conductors, Bryant confirmed that the shock could pass through a metal sword blade but not sealing wax or glass. He did, however, relate what he believed was one new observation worthy of mention (although it too has some history):

> For one morning, while I was standing by, as a servant was emptying the tub, which he had lifted intirely [*sic*] from the ground, and was pouring off the water to renew it, and the fish left almost dry, the negro received so violent a shock as occasioned him to let the tub fall, and calling another to his assistance, I caused them both to lift the tub free from the ground, when pouring off the remains of the water they both received smart shocks and were obliged to desist from emptying the tub in that manner. This I afterwards tried myself and received the like shock. (Bryant, 1786, p. 168)

Bryant was perplexed when it came to the source of the eel's electricity. He lamented "that whether it has an unaccountable faculty of collecting a quantity of the fluid from the surrounding waters, or through the body of the person touching it, or has in its own body a large fund which it can discharge at pleasure, I am greatly at a loss to think or imagine" (Bryant, 1786, p. 168). These thoughts were reported in the *Transactions of the American Philosophical Society*, and his wonder was shared by natural philosophers around the world, although 9 years before this time the coveted spark had been drawn from an eel in London, leaving no doubt about this feared creature being electrical, as we shall now see.

Chapter 19
Sparks in Darkness and the Eel's Electrical Sense

> I have forgot whether I mentioned Mr Walsh's having drawn sparks of fire from those animals [eels]. He has, & I am one of the numerous witnesses of that curious circumstance.
> John Pringle to Albrecht von Haller, December 13, 1776
> (Sonntag, 1999, pp. 348–349)

As we have seen, Walsh (1773) never produced a spark indicative of electricity with his French torpedoes. No sparks seemed to occur naturally; none jumped even the smallest gaps he could make in tinfoil; and there was not even the expected crackling sound when the fish emitted a shock. Additionally, the living torpedo did not attract pith balls or other light objects, and his Lane electrometer provided no evidence for electricity. Hence, although Walsh felt sure he was dealing with an electrical phenomenon, not all of the proofs he hoped to present to Franklin and the Royal Society on the French torpedoes fell into place.

The absence of a spark or an electrometer measurement had two effects. One was on his colleagues, some of whom demanded more evidence before accepting animal electricity, even after he and others, most notably Henry Cavendish, did their best to explain the absence of a spark or why the pith balls would not move. The other effect was on Walsh himself. He still wanted to complete the job he had started. He knew that lingering doubts about the electrical nature of the fish discharge would dissipate if he could just show an accompanying spark. It was with this mindset that Walsh now focused on the more powerful eel.

A PLACE FOR THE EEL

The eel was better choice for the next phase of his research, because a healthy eel's shocks are far more powerful than those of common torpedoes. Thus, even though both the eel and torpedoes discharge electricity, only the eel's might show certain effects, because ordinary torpedoes produce a discharge too weak to be detected with the technologies then available.

The difference seemed obvious. When a torpedo is held, as Walsh knew firsthand, its shock might be confined to the fingers or a hand, with only a really vigorous ray producing shocks that would be felt up the arm to the shoulder. But in the case of the eel, such a discharge might characterize only a very weak and sickly specimen. A healthy eel could affect the whole body, causing a strong adult to buckle over in great pain.

The difference between the two species could also be quantified in a different way, again without any tools or fancy machines: All one had to do was see how far down a hand-holding chain of people the shocks would be felt. Walsh was not able to exceed seven or eight people in his experiments with the French rays, and in all but one of his experiments the number was considerably less than this—often just one or two people. In contrast, an eel's shock could be felt by three times this number of people lined up to form a "human electrometer."

Today, with more sophisticated instrumentation, we know that the voltage difference between the shocks of these two creatures, at least when healthy, really is about tenfold, as suggested by the human chain experiments. A typical torpedo, such as most of those caught in the region of La Rochelle, will emit a shock of less than 50 Volts. In contrast, a strong South American electric eel can emit shocks that can be 600 to 700 Volts.

Hence, when the eighteenth-century scientists set forth to try to obtain a spark from a living creature, the eel was clearly the far better choice. But, as Bancroft had pointed out in his 1769 publication, the South American eels, which require almost daily freshwater changes and extreme care, had been unable to survive long trips.

Early in the 1770s, the eel remained a dream for experimental natural philosophers in Europe, whose appetites were whetted even more by reports from South America. But by 1774, the situation had become more hopeful, when George Baker was able to get one of his five eels to survive the voyage from Charles Towne, South Carolina, to England, although it died before it could reach London (Chapter 18). At that time, as noted in the previous chapter, some optimistic members of the Royal Society started a subscription to have Baker try again.

This time Baker was more successful. The local papers even announced the arrival of the eels in London during the summer of 1776:

> The curious in this City have now an opportunity of examining the wonderful powers of the Gymnotus Electricus,

commonly called the Electrical Eel, as Mr. George Baker, mariner, who last year made an unsuccessful attempt to bring some alive from the equinoctial [equatorial] parts of America to London, indefatigably to gratify the public in this particular, has now succeeded in landing here five living Gymnoti, all taken by himself 150 miles up the river of Surinam: he is probably the first who has landed any of the kind alive in Europe; and has thereby afforded a striking instance of the industry of this country in matters of science, particularly as the Gymnotus, though common in the American Colonies of other European states, is not to be met in those of Great Britain.

This announcement appeared in the *Gazetteer and New Daily Advertiser* and in the *Public Advertiser* on Monday, August 5, 1776. The identical announcement appeared a day later in the *Morning Chronicle and London Advertiser*.

THE SPARK

The protocol for the critical experiment was similar to the one Walsh had employed with torpedoes in France in 1772 and that used by Williamson with eels in Philadelphia in 1773 (see Williamson, 1775a). It involved trying to get a discharge from the eel to jump a tiny gap in a piece of metal foil. To make the detection of the spark easier, the test would have to take place in the dark, as recommended by Franklin.

The first successful spark experiment took place on Thursday, August 1, 1776. The same newspapers that announced that Baker had succeeded in bringing five eels to London covered Walsh's successful experiment in the same story. Hence, readers were informed about it on August 5 and 6 in the *Gazetteer and New Daily Advertiser*, the *Public Advertiser*, and the *Morning Chronicle and London Advertiser*, and more than likely in some other newspapers as well. The continuation and conclusion of the newspaper article presented above reads:

> On Thursday, the President and many Gentleman of the Royal Society were present at an exhibition of the effects of these extraordinary fish; and on Friday Mr. Walsh, whose Observations on the Electricity of the Torpedo have been published in the Philosophical Transactions, made some experiments on them, and obtained from the *Gymnotus* the electrical spark, which he never could procure from the torpedo; by which event an entire agreement in the natural effect of these animals, and the artificial effect of the Leyden phial, is established.

This was not the only demonstration before credible and important witnesses in London. There was another one on Friday, November 15, 1776. This time the *St. James's Chronicle or the British Evening Post* (November 16) and *Public Advertiser* (November 19) conveyed this information:

> On Friday last, his Grace the Duke of Devonshire, the Marquis of Rockingham, Lord Charles Cavendish, the President [John Pringle], and about thirty Members of the Royal Society, many Members of Parliament, and other Gentlemen of Note, Lovers and Encouragers of Science, met at Mr. Baker's Exhibition in the Haymarket to see the astonishing Phenomena of the Gymnotus Electricus, similar to those of Lightning; when all present expressed Pleasure and Surprize [sic] at the vivid Flashes produced by that Animal; and about seventy of the Company joining Hands, all felt its Electrical Stroke in the same Instant of Time. Mr. Baker, who brought over these Wonderful Fishes from Surinam, a Dutch Settlement in America, intends to make an Exhibition of their luminous Effect at his new Apartments in Piccadilly, nearly opposite to St. James's street, every Day exactly at One o'clock, when it is supposed that such Gentlemen and Ladies as have a true Relish for new Discoveries in Nature will take the Opportunity of Satisfying their Curiosity, while these Animals, the Inhabitants of a very warm Climate, and never before introduced alive into Europe, continue in Life and Vigour. Admittance for each Person at the stated hour will be 2s. 6d. A select Party may be gratified at any other Time for 2 Guineas.

The preceding note did not mention Walsh's name, but Walsh was there, as is made clear in a note dated November 16, 1776, from Daniel Solander to Joseph Banks. "Yesterday Mr Walsh made his second Public Experiment with the *Gymnotus* shewing the Electrical Sparks," wrote Solander, adding: "The shock was felt by a circle of 71 people" (Solander, 1776; also cited in Duyker & Tigbrand, 1995, p. 373). Various experiments were performed that day (see below). The experiment with 71 people did not involve an air gap in the circuit. A spark seemed, however, to jump an air gap in a different experiment in which a chain of 27 people felt the shock.

NO WALSH PUBLICATION

Unfortunately, Walsh did not leave a journal of his experiments like the one he kept in France, and he never published an account of what he had done. For this reason, even some of the leading historians of animal electricity have overlooked his eel experiments. To give but one example, I. Bernard Cohen (1953, p. 22), an historian with a penchant for coming forth with new facts and fascinating details about electricity, wrote that John Walsh's "only contribution to science was his study of the torpedo."

Why Walsh, who had won the Copley Medal and was well regarded, did not go public is a mystery, although there have been conjectures (Piccolino, 2003a). One theory is that first he wanted to collect more data on the eels. This hope, however, was increasingly impeded by his various business and political activities, which were taking more and more time and eating away at his once-sizeable fortune.

Another is that he might still have been perplexed about how these fishes actually produce electricity, or stymied by why they do not electrocute themselves. He did, after all, state in his 1773 paper that the results of his experiments seemed "to combat the general principles of electricity" (Walsh, 1773, p. 469). This might have continued to be the case, and even with further proof that electricity is involved, he might still have been hoping for a good explanation of how such things could be accomplished by a living creature.

A third factor that could well have played a role in his decision not to pick up his pen was that he suffered from some sort of a writer's or mental block. As we have seen, Franklin had to go to unusual lengths to bring his initial work on torpedoes to press. And for the eel experiments, Walsh might have felt that the news would be spread with or without his pen, namely by those members of the Royal Society and the broader "Republic of Letters" who attended his demonstrations. As noted, Franklin, who had earlier taken him under his wing and had exceptional writing skills, was no longer present in London to help him.

Walsh would have been right about word spreading. As noted, it was carried in several London newspapers. Some of those invited to see the spark provided written accounts of what had transpired, while others seemed content to spread the word orally. And, as might be expected, a few people did both.

PRINGLE'S LETTERS TO HALLER

John Pringle spread the word orally and in a series of unpublished letters. One of his many correspondents was Albrecht von Haller, whom he had first met in the 1720s, before Haller even began practicing medicine in Switzerland (their letters appear in Sonntag, 1999). Haller went on to secure an appointment in the newly founded University of Göttingen, where he became world famous for his approach to medicine, and he was elected a Fellow of the Royal Society of London in 1749.

The correspondence of particular interest occurred between 1774 and 1777, while Haller was suffering from stones and on laudanum (opium) to control his pain (he died in December 1777). These letters are especially interesting because they reveal some pertinent facts that do not appear in formal publications. In this regard, they fill in some gaps about Walsh, how he obtained his the eels, his spark experiments, and how the eels rose to celebrity status in London.

In a letter dated April 9, 1774, Pringle told Haller that Walsh's paper on torpedoes had just come out in the *Philosophical Transactions* and that he expected Walsh to receive the next Copley Prize. He then added, "A friend in Holland has undertaken to procure us from Surinam a living *Gymnotus* or electric eel," ending his letter at this point (Sonntag, 1999, p. 300).

Later that year, Pringle told Haller about his *Discourse* honoring Walsh at the Copley Medal ceremony, also informing him: "One of our Sailors lately brought a *gymnotus electricus* from Cayenne, or Surinam, to Falmouth . . . but it died there in a few days, but not before it had shewn several feats of superior energy to the to[r]pedo" (p. 317).

On August 27, 1776, a year after Pringle's (1775a) *Discourse on the Torpedo* was published, Haller was told that some eels were now alive in London—and that Walsh had finally obtained his coveted spark:

A common English Sailor has brought no less than 5 electric eels (*gymnoti electrici*) alive from Surinam. One died some days after coming to London, but the rest are still living, though in a weak condition. I saw them all, & had the satisfaction of receiving a sensible shock, in the circle made by 5 or 6 of us, who taking one another by the hand, one at one end put his finger only into the water, whilst he at the other extremity dipt his hand far enough into the tub, as slightly to touch one of the animals. A tolerably charged Leyden bottle could not have behaved better.

Next day Mr Walsh went alone & with proper apparatus was sensible of a spark crossing the inter-section of a strip of tin foil . . . an electrical feat which he had never been able to obtain from the torpedo. But he has not been able to repeat the experiment; on account of the weakness of his eel, which he purchased from the importer. (Sonntag, 1999, pp. 345–346)

On December 13, 1776, Pringle told Haller that the four eels were now in good health, the change for the better occurring after they realized that the eels required warmer water. The most significant news, however, was that Walsh had been able to draw sparks again, now before witnesses, including himself:

I have forgot whether I mentioned Mr Walsh's having drawn sparks of fire from those animals. He has, & I am one of the numerous witnesses of that curious circumstance. In your new edition of the Physiology you may safely assert it. The sparks were vivid & repeated, & were seen, the last time I attended, by about 70 people at once; and in another experiment the shock was conveyed through the bodies of the same number of people, when they joined hands to form a circle, & the last two completed the circle by each of them, with the disengaged hand, touching the animal at different parts of his body, at a time. (pp. 348–349)

Pringle added that one must touch a flat torpedo on the back and underside to receive the strongest shock, but that it was different for the elongated eel. One hand must touch the eel near the head, and the other near the tail, to "receive a violent concussion."

A few weeks later, Pringle shared his thoughts on the subject of the "animal oeconomy" with Haller. With his empirical orientation, he was not ready to conclude that the new research on fishes proved that nerves must work by electricity. Instead, on January 28, 1777, he advised caution before making the leap, writing that he and Franklin were in agreement that more had to be learned.

You did wisely not to enter into the affair of electricity, in so far as the animal oeconomy is concerned; for Franklin himself, all successful as he had been in finding the facts turn out in favor of his theory, at large, in the earth & heavens; yet never durst apply it to an animal body, further than thinking that in general the electrical fluid might have a share in the mechanism, but in what manner he professed a total ignorance. (p. 351)

A postscript on the status of the eels was added to the margin of the first page of this letter, and it reveals a lot about what was happening in London. It tells us that *anyone* with the means to pay for the experience could feel *and see* the spark. The sailor (Captain Baker), who owned three of the eels had learned how to do the critical experiment and had the business acumen to demand high fees from those inquisitive, well-placed individuals, who called for public demonstrations.

Of the 5 electrical Eels (*gymnoti electrici*) 4 are still alive & yielding when properly excited sparks of fire. The water is kept at about 75 deg. of Fahrenheit. Mr Walsh purchased one which is a great favourite with that gentleman: the rest are shewn for money by the sailor who brought them home; & he gets half a crown from every person who is allowed to get a shock from them, & a whole crown if they desire to see the fire[.] (p. 352)

That March, Pringle told Haller more about the eel's electrical apparatus, guiding him to Hunter's (1775) article in the *Philosophical Transactions*. He then repeated his warning about jumping to broad conclusions and making physiological conjectures that might not withstand the test of time. He again warns Haller that:

> since Dr Franklin could never satisfy himself about the manner of action of those electrical organs, . . . but was only convinced (even before the discovery of the spark of fire in the gymnotus) that the shock, given by them, was by the same fluid as that in the Leyden bottle, . . . I should judge it safest for the physiol[og]ists, not to push their conjecture much further at present, lest they should run into an hypothesis, which subsequent discoveries will probably contradict. (p. 354)

Pringle's last letter to Haller on the eels was dated June 28, 1777, and it provides a rare glimpse of what he expected to happen next. Specifically, he tells Haller that he is anticipating that Walsh will have an article on these sparking eels in the *Philosophical Transactions*. In Pringle's words, "The most curious memoire of the second part will be, I imagine, that other of Mr Walsh, upon the Gymnotus electricus or Surinam eel, from which I told you, he drew sparks of fire; & of which he has other interesting circumstances to tell" (p. 361). Moreover, "if there are any unbelievers with regard to the igneous emissions, they need only come to my neighborhood, where they will see this little Jove, for a crown piece, darting his lightning to a large company, once in the week" (p. 361).

As pointed out earlier, the expectation that Walsh would publish the spark experiment—an expectation probably shared by many members of the Royal Society of London—was never met.

JAN INGENHOUSZ

In addition to Pringle, Jan Ingenhousz was a second witness who helped to spread the word. He sent Franklin a letter dated November 15, 1776. Not knowing his American friend was already in France, it was sent to Philadelphia, where his mail was collected and forwarded to him.

Most of Ingenhousz's letter dealt with the fact that America, which had been "the seat of tranquility," had now become "the seat of horror and bloodshed" (Willcox, 1983, p. 7). After expressing his concerns and fears for his close friend, he wrote: "Mr. Walsh has at last found out the method of making an Electric spark visible from a *Gymnotus* of Surinam." This is consistent with Pringle's dating of the event to 1776. Ingenhousz (1775a) had previously been somewhat skeptical about fish electricity, because his own experiments on torpedoes had shown that they do not attract pith balls, cannot charge a Leyden jar, and do not give a crackle or a spark when releasing a shock.

Ingenhousz did more than act as a conduit for this important news—he was afterwards invited to replicate Walsh's experiment for himself. In 1778, Walsh gave Ingenhousz access to his eel, which was still at his house on Chesterfield Street. Ingenhousz showed up accompanied by Arnould-Barthélemi Beerenbroeck, a Flemish surgeon then living in England. Other commitments precluded Walsh from being present for the replication and extension of Ingenhousz's experiments, which were published as a part of a larger treatise 4 years later (Ingenhousz, 1784).

One of the new experiments Ingenhousz performed was with a lodestone. As noted in an earlier chapter, the Dutch physician and critic of animal electricity G. W. Schilling (1770, 1772) had stated that a lodestone would attract these eels and also block their natural ability to deliver shocks. Ingenhousz's experiments in 1778 showed that such notions were nonsense.

LE ROY'S PUBLICATION

One scientist who saw the need for a more public communication was Walsh's friend, the Frenchman Le Roy. Unlike Pringle and Ingenhousz, he was not present for the critical spark experiment, but he was told about it in a since-lost letter from Walsh. Le Roy was also in communication with other English scientists, men who verified what had transpired.

Thanks to Le Roy, some of Walsh's own wording appeared in 1776 (Le Roy, 1776). His words were also translated and published in an article in an important French scientific journal, *Observations sur la Physique*, and the article appeared in Italian a year later (Le Roy, 1777) (Figs. 19.1 and 19.2) and in many other journals and magazines in different countries. Translated from the French of 1776, we read:

> It is with pleasure that I inform you that they [i.e., the electric eels] have given me an electric spark, perceptible in its passage through a small gap or separation made in a tin lamina pasted on a glass. These fish were in the air, since this experience has not succeeded in water; their electricity is very much stronger than that of the Torpedo, and there are some considerable differences in their electrical effects. (Le Roy, 1776, p. 331)

CAVALLO'S BOOK

Yet another disseminator of the spark experiment, albeit a later one, was Tiberius Cavallo. Born in 1749, he was an Italian "amateur" who went to London for business reasons and stayed there, dying in 1809. Cavallo enjoyed experimental natural philosophy, especially electricity, and carefully followed developments in the field, synthesizing and publishing them. His publications were particularly influential in bringing British and Italian lines of research together (Fig. 19.3).

A Complete Treatise of Electricity was Cavallo's most widely circulated book, undergoing many revisions and being translated into several languages. The fourth English edition of 1795 stands out. Cavallo writes in the Preface that "The rapid advances made in the Science of ELECTRICITY since the first publication of this Treatise, have furnished considerable additions for every subsequent edition." He adds: "Besides a great many other articles, the Reader will find in this volume an account of the new subject of ANIMAL ELECTRICITY, which may be justly considered as one of the greatest discoveries made in the present century" (Cavallo, 1795, pp. iii–iv).

Appendix VII in Volume 2 covers this subject and is titled "Of the Electric Properties of the Torpedo, *Gymnotus*

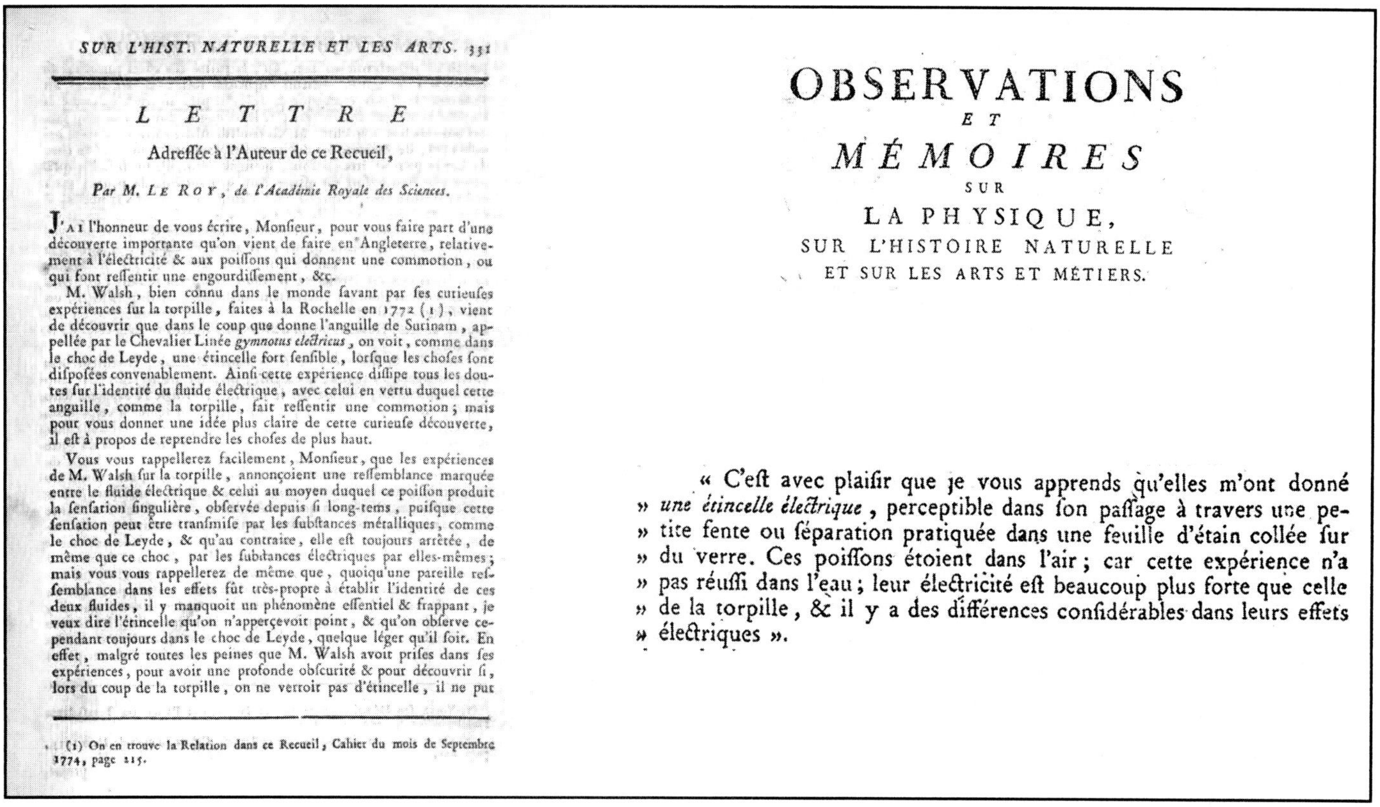

SUR L'HIST. NATURELLE ET LES ARTS. 331

L E T T R E

Adreſſée à l'Auteur de ce Recueil,

Par M. LE ROY, de l'Académie Royale des Sciences.

J'AI l'honneur de vous écrire, Monſieur, pour vous faire part d'une découverte importante qu'on vient de faire en Angleterre, relativement à l'électricité & aux poiſſons qui donnent une commotion, ou qui font reſſentir une engourdiſſement, &c.

M. Walsh, bien connu dans le monde ſavant par ſes curieuſes expériences ſur la torpille, faites à la Rochelle en 1772 (1), vient de découvrir que dans le coup que donne l'anguille de Surinam, appellée par le Chevalier Linée *gymnotus electricus*, on voit, comme dans le choc de Leyde, une étincelle fort ſenſible, lorſque les choſes ſont diſpoſées convenablement. Ainſi cette expérience diſſipe tous les doutes ſur l'identité du fluide électrique, avec celui en vertu duquel cette anguille, comme la torpille, fait reſſentir une commotion; mais pour vous donner une idée plus claire de cette curieuſe découverte, il eſt à propos de reprendre les choſes de plus haut.

Vous vous rappellerez facilement, Monſieur, que les expériences de M. Walsh ſur la torpille, annonçoient une reſſemblance marquée entre le fluide électrique & celui au moyen duquel ce poiſſon produit la ſenſation ſingulière, obſervée depuis ſi long-tems, puiſque cette ſenſation peut être tranſmiſe par les ſubſtances métalliques, comme le choc de Leyde, & qu'au contraire, elle eſt toujours arrêtée, de même que ce choc, par les ſubſtances électriques par elles-mêmes; mais vous vous rappellerez de même que, quoiqu'une pareille reſſemblance dans les effets fût très-propre à établir l'identité de ces deux fluides, il y manquoit un phénomène eſſentiel & frappant, je veux dire l'étincelle qu'on n'apperçevoit point, & qu'on obſerve cependant toujours dans le choc de Leyde, quelque léger qu'il ſoit. En effet, malgré toutes les peines que M. Walsh avoit priſes dans ſes expériences, pour avoir une profonde obſcurité & pour découvrir ſi, lors du coup de la torpille, on ne verroit pas d'étincelle, il ne put

(1) On en trouve la Relation dans ce Recueil, Cahier du mois de Septembre 1774, page 215.

OBSERVATIONS
ET
MÉMOIRES
SUR
LA PHYSIQUE,
SUR L'HISTOIRE NATURELLE
ET SUR LES ARTS ET MÉTIERS.

« C'eſt avec plaiſir que je vous apprends qu'elles m'ont donné » *une étincelle électrique*, perceptible dans ſon paſſage à travers une petite fente ou ſéparation pratiquée dans une feuille d'étain collée ſur » du verre. Ces poiſſons étoient dans l'air; car cette expérience n'a » pas réuſſi dans l'eau; leur électricité eſt beaucoup plus forte que celle » de la torpille, & il y a des différences conſidérables dans leurs effets » électriques ».

Figure 19.1. Left: The title page of Jean-Baptiste Le Roy's 1776 announcement of John Walsh's landmark spark experiment with the eel (in French). Right (under the title of the journal): The passage of Walsh's letter addressed to Le Roy as quoted in the article.

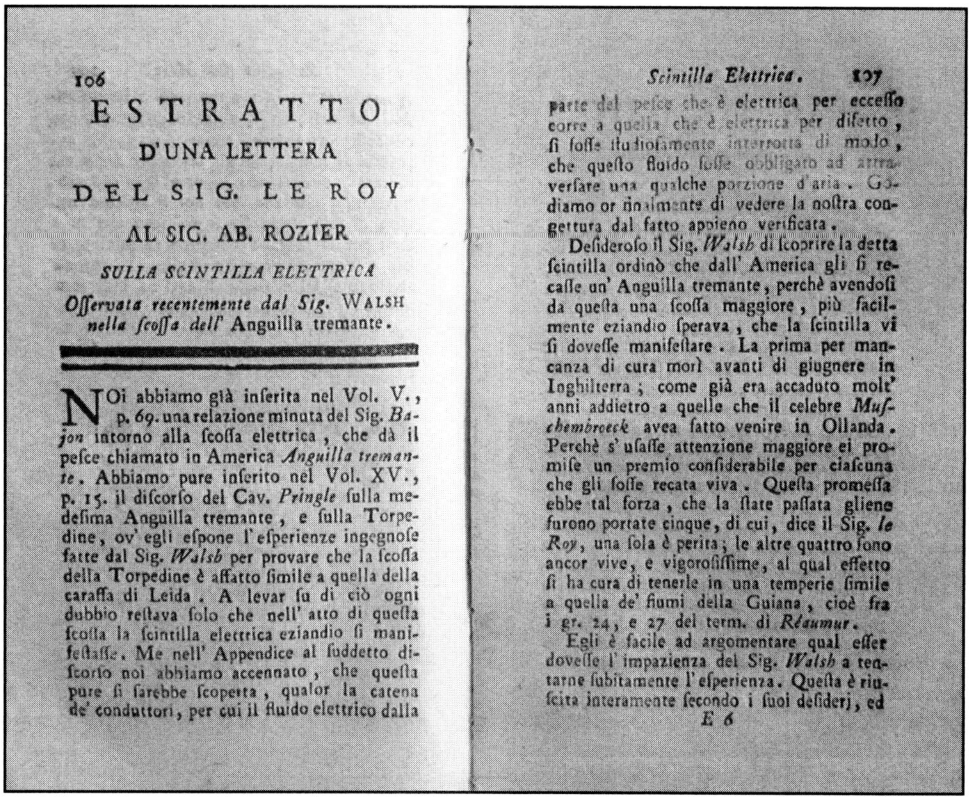

106

ESTRATTO
D'UNA LETTERA
DEL SIG. LE ROY
AL SIG. AB. ROZIER
SULLA SCINTILLA ELETTRICA
Oſſervata recentemente dal Sig. WALSH *nella ſcoſſa dell'* Anguilla tremante.

NOi abbiamo già inſerita nel Vol. V., p. 69. una relazione minuta del Sig. *Bajon* intorno alla ſcoſſa elettrica, che dà il peſce chiamato in America *Anguilla tremante*. Abbiamo pure inſerito nel Vol. XV., p. 15. il diſcorſo del Cav. *Pringle* ſulla medeſima Anguilla tremante, e ſulla Torpedine, ov' egli eſpone l'eſperienze ingegnoſe fatte dal Sig. *Walsh* per provare che la ſcoſſa della Torpedine è affatto ſimile a quella della caraffa di Leida. A levar ſu di ciò ogni dubbio reſtava ſolo che nell' atto di queſta ſcoſſa la ſcintilla elettrica eziandio ſi manifeſtaſſe. Me nell' Appendice al ſuddetto diſcorſo noi abbiamo accennato, che queſta pure ſi ſarebbe ſcoperta, qualor la catena de' conduttori, per cui il fluido elettrico dalla

Scintilla Elettrica. 107

parte del peſce che è elettrica per eccesſo corre a quella che è elettrica per difetto, ſi foſſe ſtudioſamente interrotta di modo, che queſto fluido foſſe obbligato ad attraverſare una qualche porzione d'aria. Godiamo or finalmente di vedere la noſtra congettura dal fatto appieno verificata.

Deſideroſo il Sig. *Walsh* di ſcoprire la detta ſcintilla ordinò che dall' America gli ſi recaſſe un' Anguilla tremante, perchè avendoſi da queſta una ſcoſſa maggiore, più facilmente eziandio ſperava, che la ſcintilla vi ſi doveſſe manifeſtare. La prima per mancanza di cura morì avanti di giugnere in Inghilterra; come già era accaduto molt' anni addietro a quelle che il celebre *Muſchembroeck* avea fatto venire in Ollanda. Perchè s' uſaſſe attenzione maggiore ei promiſe un premio conſiderabile per ciaſcuna che gli foſſe recata viva. Queſta promeſſa ebbe tal forza, che la ſtate paſſata gliene furono portate cinque, di cui, dice il Sig. *le Roy*, una ſola è perita; le altre quattro ſono ancor vive, e vigoroſiſſime, al qual effetto ſi ha cura di tenerle in una temperie ſimile a quella de' fiumi della Guiana, cioè fra i gr. 24, e 27 del term. di *Réaumur*.

Egli è facile ad argomentare qual eſſer doveſſe l' impazienza del Sig. *Walsh* a tentarne ſubitamente l' eſperienza. Queſta è riuſcita interamente ſecondo i ſuoi deſiderj, ed

E 6

Figure 19.2. An Italian abridged version of Le Roy's announcement of Walsh's success in obtaining a spark with an eel. This piece appeared in 1777 in a widely spread cultural and scientific magazine, *Scelta degli opuscoli interessanti*, published in Milan.

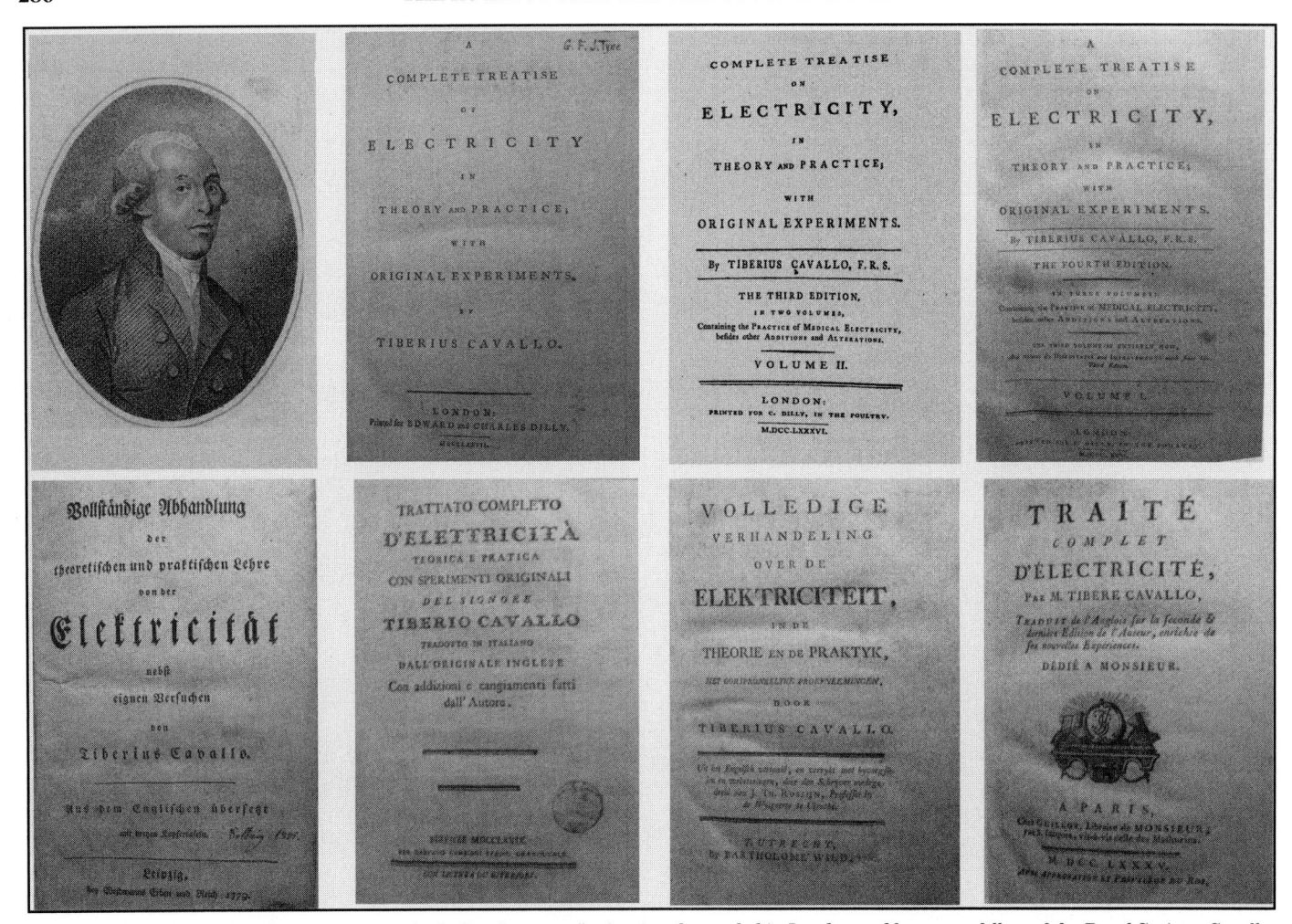

Figure 19.3. Tiberius Cavallo (1749–1809), the Italian "amateur" scientist who settled in London and became a fellow of the Royal Society. Cavallo was a prolific writer of treatises on electricity and its medical applications. In this image, he is depicted as if he were looking at three English editions of his *Complete Treatise of Electricity,* with German, Italian, Dutch, and French translations below them. It is only in the fourth edition, published in 1795, that we find Walsh's spark experiments with the eel. (Title pages photographed with permission at "The Bakken" in Minneapolis.)

Electricus, and *Silurus Electricus.*" In the first paragraph, Cavallo writes that three types of fishes are known to give shocks: torpedoes, the electric eel of South America, and the electric catfish from North Africa. He then states that the ancients were "utterly ignorant of the cause of it," and that "Two distinguished writers of the last century endeavoured to explain this property upon mechanical laws, but their ingenuity was insufficient to account for the phenomenon" (pp. 287–288).

He continues:

> The principal discoveries, relating to the identity of the above-mentioned property of these fishes, and the electric shock, were made by JOHN WALSH, Esq. F.R.S. to whose ingenuity we are indebted for the demonstration of the power of the torpedo agreeing with the electric shock in all the points of comparison in which it could be examined; and also for almost all the other discoveries which were since made relating to those animals. (p. 288)

After discussing Walsh's physiological experiments and Hunter's anatomical studies on torpedoes, Cavallo turns to the *Gymnotus electricus,* correctly noting that it "has been frequently called *electric eel*, on account of its superficial resemblance to the common eel; though, when accurately examined, it is found to have none of the properties of that animal" (p. 299). He goes on: "A few of these animals were brought back to England about eighteen years ago, which, as far as I know, were the first of the kind brought to Europe."

Cavallo adds that, "I have often received shocks, which I felt not only up in my arms, but even very forcibly in my breast" (p. 308). As for the spark of this creature:

> The strongest shocks of the gymnotus will pass a very short interruption of continuity in the circuit. Thus they will be conducted by a chain, especially when it is not very long, and is stretched, so as to bring its links into better contact. When the interruption is formed by the incision made by a pen-knife on a slip of tin-foil that is pasted on glass, and that slip is put into the circuit, the shock in passing through that interruption, will shew a small but vivid spark, plainly distinguishable in a dark room. (p. 309)

The importance of Walsh's spark experiment is made very clear, before Cavallo concludes that "The gymnotus possesses all the electrical properties of the torpedo, but in superior degree" (p. 308):

> The subject of Animal Electricity was considerably advanced by the discovery of the spark, with which the

shock of the gymnotus was attended; for, notwithstanding the previous discoveries relating to the torpedo, and the actual possibility of imitating the effects of that animal's extraordinary power by means of a large battery weakly charged with artificial Electricity, yet the scrupulous philosophers still suspected the power of the torpedo might be something different from Electricity, since the two principal characteristics of Electricity, namely the spark and the attraction, had never been discovered in the torpedo; and at the same time it was difficult to conceive the manner in which the electric fluid might be generated, accumulated, and discharged in an animal, which at least in its usual state of existence is a Conductor of Electricity, and is surrounded by a fluid which is likewise a Conductor of that power. This indeed still remains a profound secret; and it is difficult to say, whether any future experiments will ever disclose it. But the spark having been discovered with the gymnotus, the analogy between its power and Electricity is rendered considerably more evident . . . (pp. 300–301)

AN ELECTRICAL SENSE?

The scientists who spoke with Walsh and those who witnessed his experiments also learned about another unusual characteristic of the eel, and it merits comment. This eel, they were surprised to discover, seems to possess an unusual sense, one unlike anything the humans experimenting with them might have.

The unexpected sensory capacity of the eel came to the fore in several ways. In one experiment, each of two people put one hand in the water with the eel, which did not elicit a barrage of shocks. But when they used their free hands to form a human chain, the eel began to shock vigorously. In a related experiment, two wires were put into the tank with the eel, each of which ended in a separate bowl filled with water. The eel seemed to pay no attention to the wires under this condition. But when a person completed the circuit by putting a hand into each of the two bowls, the eel swam to the wires and immediately began discharging its electrical shocks. Remarkably, the eel did not have to be able to see the circuit being completed for this "shocking" effect to take place.

Ingenhousz included what Walsh had told him about this surprising new sense in his 1784 publication (Fig. 19.4). He wrote:

For example, a group of 10 persons, hand by hand, situate themselves in such a way so the fish can see only the two extreme ones. Of these, one puts the finger in the water near the head of the fish, while, at the same time, the other does the same near the tail. When the persons connect one to the other by joining hands, or, said with the words of the electrician, close the electric circuit, the fish will absolutely notice it and will discharge his electric shock. At the moment that the persons separate from each other, the fish will immediately understand that and will no longer shoot his dart, aware of its ineffectiveness. This is the peculiar characteristic that Mr. Walsh has observed. (Ingenhousz, 1784, pp. 275–276; trans. in Piccolino, 2003a, p. 202)

The sense was also mentioned in Cavallo's book on electricity, where we read:

Mr. Walsh made another remarkable discovery with the gymnotus, which he shewed at his house to various ingenious persons: it was a new sort of sense in the animal, by which he knew when the bodies which came near him, were such as could receive the shock (*viz.* Conductors) and when they were of the contrary nature; in the former of which cases the animal gave the shock, but not in the latter.

In order to shew this wonderful property, divers experiments were made, but the most convincing one was the following:—the extremities of two wires were dipped into the water of the vessel wherein the animal was kept, then they were bent, and extended a great way, and lastly terminated in two separate glasses full of water. These wires being supported by non-conductors at a considerable distance from each other, it is plain that the circuit was not complete: but if a person put the fingers of both his hands into the glasses wherein the wires terminated, *viz.* those of one hand into one, and those of the other hand into the other glass, then the circuit became complete.

Now it was constantly observed, that whilst the above-described circuit remained interrupted, the animal never went purposely near the extremities of the wires, as he used to do when willing to give the shock: but the moment that the circuit was completed, either by a person or any other Conductor, the animal immediately went towards the wires, and gave the shock; though the completion of the circuit was performed quite out of his sight. (Cavallo, 1795, vol. 2, pp. 309–311)

We also find a good description of the eel's newly discovered sense in a letter from Volta dating from 1782 (trans. in Piccolino & Bresadola, 2002, p. 57). But unlike Ingenhousz and Cavallo, Volta used the phrase *senso elettrico*, meaning "electrical sense," for this phenomenon, this being perhaps the first appearance of this remarkable expression in the history of science.

Sadly, the experiments performed on the eel's unusual sense seem to have been forgotten. One reason for the forgetting could well be that researchers seemed far more interested in the strong shocks discharged by these specialized eels, rays, and catfishes than in the newly discovered strange sense. This at least seemed to be the case prior to the 1950s, when "electroreception" was warmly welcomed as a great new discovery (see Chapter 25; Lissman, 1963; Bennett, 1971a, b; Moller, 1995).

Today, electroreception is an object of intense study. We now know that there are many types of electric fishes that send out weak electrical pulses (just a few volts) to obtain sensory information about their environments, including detecting nearby fishes of the same species. The electric eel is just one fish that emits and responds to weak electrical discharges for orientation; other knife-fishes differ from it in not being able to give strong shocks, but they too can discharge pulses of about 1 Volt for orienting to the environment. This form of electroreception is considered "active" in that these fishes emit electrical signals and then pick up and respond to distortions in the electric fields.

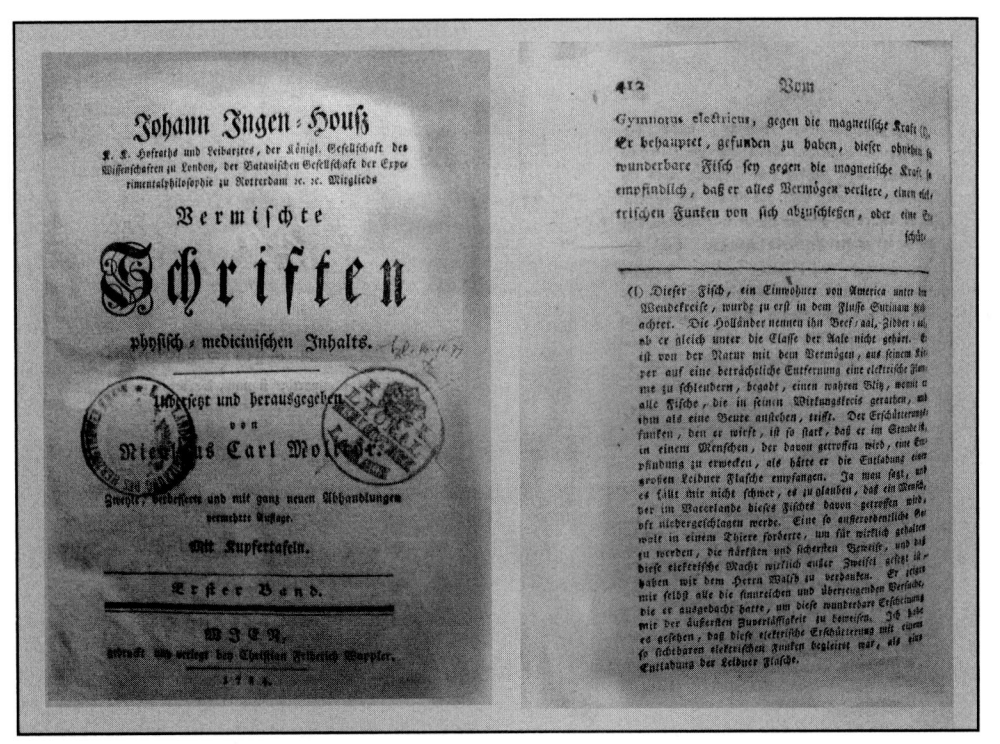

Figure 19.4. The cover of Jan Ingenhousz's 1784 book of collected works and the passage describing spark experiments with an eel in London. Ingenhousz confirmed the spark on his own with Walsh's eel, and he showed that a lodestone has no effect on the shock, dispelling a myth.

We also know that many fishes with an electrical sense are not capable of active electroreception. These fishes use what is called passive electroreception, meaning detecting signals of external origin, not signals of their making, in their activities. In addition, some electric fishes have both active and passive electroreceptors, and this dual reception possibility can be found among both strongly and weakly electric species.

We shall discuss these newer discoveries in Chapter 25. For now, the important point is that both active and passive electroreception systems allow fishes to function where vision may fail. Indeed, many of the fishes that can detect weak electrical signals are found in murky water. The electric eel, which lives in muddy river water, can emit weak electrical pulses for gathering sensory information about obstacles, prey, and predators in its environment, in addition to possessing the ability to give "torporific" shocks that are 50 times stronger. This is because it has two sets of discharging organs. Torpedoes, in contrast, do not emit low-intensity pulses, although they do have "ampullae of Lorenzini" that can be used for passive electroreception. Electric catfishes also lack the ability to generate weak pulses but have electroreception capabilities.

All of these fishes, we must add, also have lateral line organs for mechanically detecting objects that might not be seen. Thus, they are not solely dependent on electroreception in dark or murky water.

IMMEDIATE IMPACT

The news of Walsh's critical experiment had a great impact on other scientists. To illustrate this, let us examine the reaction of a previously very skeptical member of the Royal Society, and then two relatively unknown naturalists who had conducted experiments of their own in South America.

William Henly, who invented the quadrant electrometer, had scoffed at the notion of animal electricity prior to the Walsh's demonstration. We already mentioned in Chapter 16 that Henly wrote a letter to Canton dated May 21, 1775, in which he opined: "Indeed, I must say, that when a Gentleman can so give up his reason as to believe in the possibility of an accumulation of electricity among conductors sufficient to produce the effects ascribed to the Torpedo, he need not hesitate to embrace as truths the greatest contradictions that can be laid before him" (Henly, 1775, p. 105; underlining his) (Fig. 19.5, left).

After seeing the spark, however, Henly had a quite different frame of mind. Although mistakenly writing "torpedo" instead of "eel," in an undated letter that is also in the *Canton Papers of the Royal Society* (probably written in 1776), we find:

> I suppose you have heard of Walsh's exhibition of the torpedo [really eel]. I was there, & indeed saw the light very brilliant many times. I have contrived a little apparatus to determine the direction of it in a simple way: it is in hand and will be done very soon; I shall then wait for your summon to try it. (Henly, 1776?, p. 103) (Fig. 9.5, right)

Among the naturalists who had previously experienced these eels in South America, Ramón María Termeyer stands out. He was a Spanish Jesuit missionary of Dutch ancestry who lived in the Santa Fe province of Argentina from 1764 to 1767, the year in which the Jesuits were expelled, forcing his return to Italy. Termeyer tells us that he encountered the *anguilla tremante* in the Saladillo River, close to his mission (today, the eel is not found this far south). Familiar with the published letters of Allamand (Van's Gravesande)

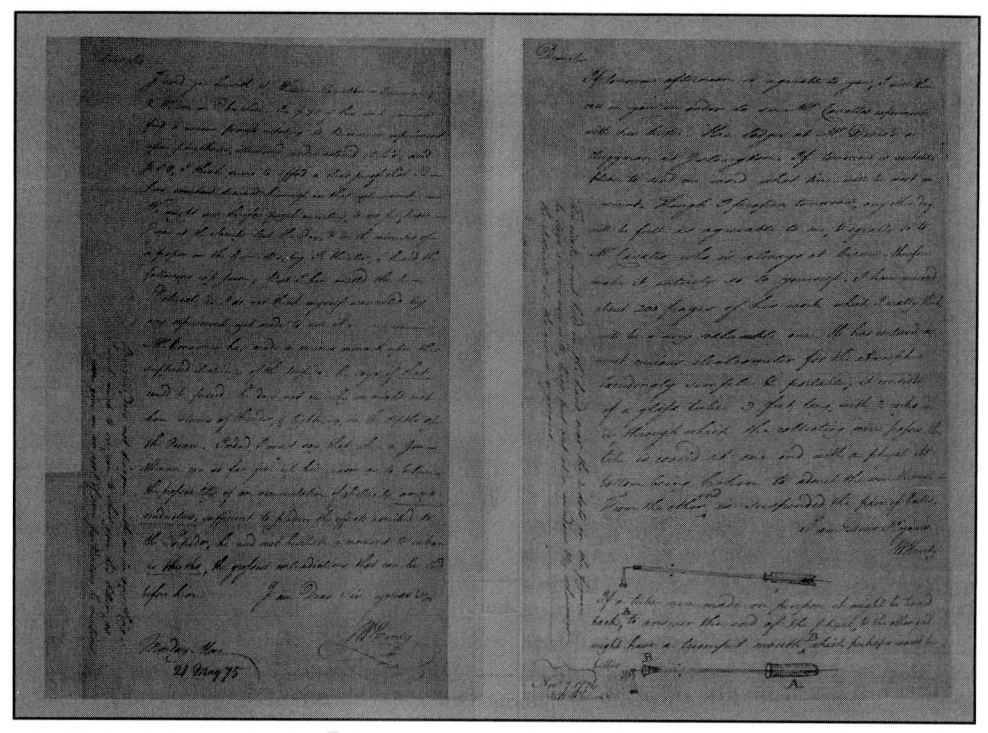

Figure 19.5. William Henly's two letters on the electric fish experiments. Left: The letter first expressing his incredulity. Right: His later acceptance of the electric nature of the shock. The drawings at the bottom on the right side depict an instrument Henly contrived to ascertain the polarity of the fish shocks. (© The Royal Society, London)

and Van der Lott, he had conducted some experiments but did not publish his findings at the time. Only later, after Le Roy's letter describing what Walsh had found was translated into Italian, did he publish his own report (Termeyer, 1781; see Asúa, 2008).

Only now does Termeyer tells us that he had conducted some 16 experiments with reasonably good electrical equipment on these eels, although he never obtained a spark or movements of pith balls or gold leaf. His paper is by no means a significant contribution to the literature: his experiments were not novel, his anatomy was flawed, and he did not equate the *fluido ginnotico* ("gymnotic fluid") with machine-produced "electrical fluid" (also see Termeyer, 1810). Nevertheless, the fact that Termeyer published after such a lengthy delay reflects the stir that had been created by Walsh's success.

Bertrand Bajon, a surgeon who lived in Cayenne from 1763 to 1776, was also influenced by Le Roy's report. Bajon had been a correspondent of the French *Académie des Sciences*, and he provided that organization with a report on *l'anguille tremblante* in 1773 that was thereafter published (Fig. 19.6). In his communication, he did many things with his eel that Walsh had published with torpedoes. For example, he listed materials that would and would not transmit the eel's discharge, and he mentioned how people who formed a circuit with the eel would feel its powerful shock. He also did some experiments to see a spark, only to fail in this domain.

After he returned to Paris, Bajon read Le Roy's article and felt it necessary to explain why he had been unable to draw a spark in his own experiments (Asúa, 2008). In his publication of 1778, he alluded to the humid air in Cayenne, hypothesizing that it could have attracted the electrical fluid away from the conductors (Bajon, 1778, p. 307).

REWRITING THE PAST

Walsh's work also led people to present some earlier descriptions in a new light, bringing up electricity as if the author had recognized it, when there was no mention of electricity in the original description. A good example of this rewriting of history takes us back to Francis Moore, who sailed from England to Africa in 1730, as "a writer in the service of the Royal African company." In 1738, he described his 4 years abroad, mentioning a fish caught 5 years earlier in Gambia (Fig. 19.7). Chronologically, this was 110 years after Captain Jobson and his crew had encountered the electric silurus on the River Gambia and tricked the unsuspecting cook into touching it while it was still alive (Jobson, 1623; see Chapter 7).

The location of Moore's fish was a lake near a factory belonging to the Royal African Company. Like those before him, Moore did not provide a good anatomical description of the specimen, but he wrote that his "torpedo" or "numbfish" looked "something like a Gudgeon" but larger, thereby clearly distinguishing it from the ray. In his own words:

On the 1st of December [1733], some of the Natives having got a Net, came and desired I would go along with them to fish in the Lake overagainst *Yamyamacunda* [by the factory]. We caught a great Number of Fish, and amongst the rest one something like a Gudgeon, but much larger. None of them cared to touch it, neither would they suffer me to come near it, telling me that it would kill me. Some of them got long Sticks, and touched the Fish with it; but as they found the Effect was not so very strong as they imagined, they cut the Sticks shorter and shorter, and even at six Inches Length the Fish had no Effect; but at last, when they touched it with their Fingers they could not bear it the twentieth Part of a Minute. By this time

SUR L'HIST. NATURELLE ET LES ARTS. 47

temps, si l'on n'a pas la précaution de la garder dans des bouteilles à col très-étroit, & de la préserver du contact de l'air, en la couvrant d'un peu d'huile, à moins qu'au lieu d'eau pure on ait employé de l'eau distilée, telle que seroit celle de plantain.

M É M O I R E

Sur un Poisson à commotion électrique, connu à Cayenne sous le nom d'*Anguille tremblante;*

Par M. BAJON, *Médecin à Cayenne* (1).

LA nature toujours variée dans ses opérations, semble prendre un plaisir singulier à les couvrir des voiles du mystère; & souvent l'esprit humain peut à peine soulever les coins de ce voile. Occupons-nous donc des effets; considérons-les attentivement; rassemblons-les, accumulons-les; ils serviront peut-être un jour de matériaux pour établir un sistême général de l'Univers.

La Physique, cette science immense dans son objet, embrasse la nature entière, s'enrichit chaque jour d'une foule de découvertes dues plutôt à un hasard heureux qu'à une sage réflexion; mais elle sait les perfectionner & les expliquer à nos besoins.

Telles ont été les découvertes des lunettes, de l'éguille aimantée, de l'électricité; & aidé de ses lumières, le Physicien regarde aujourd'hui avec étonnement la distance immense qu'il y a d'un morceau d'ambre au fluide électrique du tonnerre. Voilà cependant le point presque imperceptible d'où il est parti pour parvenir à cette sublime découverte. Maître, pour ainsi dire, du plus terrible & du plus effrayant de tous les météores, il peut à volonté diriger ses effets & se soustraire à sa fureur par les pointes & les conducteurs placés sur la maison qu'il habite. Célèbre Franklin, votre nom passera à la postérité, porté sur les aîles de l'admiration & de la reconnoissance: moins hardi que vous, je me contente en ce moment d'examiner ce qui s'exécute sous mes yeux, & de constater par des expériences les effets singuliers de l'anguille tremblante.

(1) Voyez dans le volume, *in*-12 1771, c'est-à-dire tome I, part. II, page 259; dans celui de Septembre, tome II, part. I, page 176, les faits rapportés sur l'Electricité de quelques poissons; ils forment un Précis de nos connoissances sur ce point important: le Mémoire de M. Bajon le complète.

1774. *JANVIER.*

69

DESCRIZIONE D'UN PESCE

Che dà la scossa elettrica, conosciuto a Cayenne sotto nome d' Anguilla tremante

DEL SIG. BAJON

MEDICO A CAYENNE. (*)

LA natura ognora diversa nelle sue operazioni, sembra prendersi un singolar diletto di coprirle co' veli del mistero; e sovente l'accorgimento umano può giugnere a pena a rialzare gli angoli di questo velo. Volgiamo dunque le nostre mire agli effetti; consideriamoli attentamente, e aduniamoli; serviranno forse quandoche sia di materiali per fabbricare un sistema dell' universo.

La

(*) Isola lontana 5. gradi all' incirca dall' Equatore.

Figure 19.6. The 1774 article of Bernard de Bajon dealing with his experiments on the *anguille tremblante* (electric eel) and an Italian abridged version published in 1775. Both the original paper and the Italian version were published in the same journals that would carry Le Roy's announcement of Walsh's 1776 spark experiment (see Fig. 19.2).

TRAVELS

INTO THE

Inland Parts of *AFRICA*:

CONTAINING A

DESCRIPTION

OF THE

Several Nations for the space of Six Hundred Miles up the River GAMBIA; their Trade, Habits, Customs, Language, Manners, Religion and Government; the Power, Disposition and Characters of some NEGRO Princes; with a particular Account of JOB BEN SOLOMON, a *Pholey*, who was in *England* in the Year 1733, and known by the Name of the *African*.

To which is added,

Capt. STIBBS's Voyage up the GAMBIA in the Year 1723, to make Discoveries;

WITH

An Accurate MAP of that River taken on the Spot:

And many other Copper Plates.

ALSO

EXTRACTS from the *Nubian*'s Geography, LEO the *African*, and other Authors antient and modern, concerning the NIGER, NILE, or GAMBIA, and Observations thereon.

By FRANCIS MOORE, *Factor several Years to the Royal* African *Company of* England.

LONDON:

Printed by EDWARD CAVE, at *St John's Gate*, for the AUTHOR, and sold by J. STAGG, in *Westminster Hall*; and at *St John's Gate* aforesaid. M,DCC,XXXVIII.

176 TRAVELS *into the*

1733. cold, Mornings and Evenings, but in the middle of the Day still hot.

December. ON the 1st of *December*, some of the Natives having got a Net, came and desired I would go along with them to fish in the Lake overagainst *Yamyamacunda*. We caught a great Number of Fish, and amongst the rest one something like a Gudgeon, but much larger. None of them cared to touch it, neither would they suffer me to come near it, telling me that it would kill me. Some of them got long Sticks, and touched the Fish with it; but as they found the Effect was not so very strong as they imagined, they cut the Sticks shorter and shorter, and even at six Inches Length the Fish had no Effect; but at last, when they touched it with their Fingers, they could not bear it the twentieth Part of a Minute. By this time I understood it was a Torpedo, or Numb-Fish, and had the Curiosity of touching it with one of my Fingers, but in a Moment's time my Arm was dead quite up to my Elbow; as soon as I withdrew my Hand, my Arm came to itself again. I touched it a great many times, and found it have the same Effect, even after the Fish was dead. Then I order'd one of the People to skin it, and found that the Quality lay in the Skin only; which, when dried, had no Effect at all.

Author catches a Torpedo, or Numb-Fish.

Its Quality.

ON the 20th arrived the *Fame* Sloop on her way to *Fatatenda*, having brought me Forkillas and other things to rebuild this Factory-House.

THE

(453)

voulût toucher, ni même m'en laisser approcher, me disant qu'il me feroit mourir. Quelques-uns s'armèrent de longs bâtons avec lesquels ils touchèrent ce poisson; mais comme ils n'éprouvèrent aucun effet, ils raccourcirent peu à peu les bâtons. Voyant qu'à six pouces même de longueur, le poisson ne leur faisoit rien ressentir, ils hazardèrent de le toucher avec les doigts; mais la douleur les leur fit retirer aussi-tôt. — J'appris alors que c'étoit une *torpille*, et j'eus la curiosité d'y porter un de mes doigts. En un instant mon bras fut engourdi jusqu'au coude; il revint dans sa situation naturelle aussi-tôt que je retirai ma main. Je renouvellai l'expérience plusieurs fois, et j'éprouvai toujours le même effet, même quand le poisson fut mort. Après l'avoir fait écorcher par un de mes gens, je trouvai que sa vertu *électrique* résidoit dans sa peau seulement, et que

Ff 3.

Figure 19.7. Left and Middle: The title page of Francis Moore's 1738 book on Africa, which has a passage on a river torpedo, which can only be an electric catfish. Right: An 1803 French translation of Moore's book with an allusion to a *vertu électrique*, which does not appear in the original.

I understood it was a Torpedo, or Numb-Fish, and had the Curiosity of touching it with one of my Fingers, but in a Moment's time my Arm was dead quite up to my Elbow; as soon as I withdrew my Hand, my Arm came to itself again. I touched it a great many times, and found it have the same Effect, even after the Fish was dead. Then I ordered one of the People to skin it, and found that the Quality lay in the Skin only; which, when dried, had no Effect at all. (p. 176)

Moore's statement that the fish could produce its effects even after death contradicts Jobson, who emphasized that only a living fish will have this power. This raises the possibility that his catfish was not quite dead or that its nerves and electric organs still maintained some viability immediately after it expired. His related observation, that the fish "Quality lay in the Skin only," is more intriguing because, unlike the sea torpedo and electric eel, the electric organs of African catfishes are derived from subcutaneous muscle and, as a consequence, their electrical powers disappear when they are skinned (Fessard, 1958; Moller, 1995.

Much like Jobson's *Golden Trade*, Moore's *Travels in the Inland Parts of Africa* was frequently included in some form in edited collections about voyages. Nevertheless, as so often happened with such texts, Moore's account was sometimes edited, and not always in an accurate way. In this case, we need only look at a French collection of travels in Africa from 1804, after Walsh's case for specialized fish electricity had become widely accepted.

Moore's notation about the "quality" of the fish residing exclusively in the skin is now given as "*sa vertu électrique résidoit dans sa peau seulement,*" meaning "its electric virtue resided only in the skin" (Lallemant, 1804, vol. 1, p. 453). The added word *électrique* clearly does not belong here. Moore never mentioned electricity in his aforementioned paragraph, which was written more than a decade before it was suggested that torporific fishes, which were then thought to convey their effects mechanically, might be electrical (Chapter 11).

But with the acceptance of torpedinal electricity, editors began to see things that were not there before and to interpret matters accordingly. And while what happened to Moore's description in 1804 provides a "shocking" example of this sort of a change, changes of this sort were by no means unique to later writings.

Chapter 20
Public Knowledge: Newspapers, Magazines, and "Shocking" Poetry

> What mighty difference should there be
> Between this wanton Eel and me?
> From *The Torpedo* (An Anonymous poem by James Perry, 1777b, p. 3)

One of the more interesting questions that can be asked about the developments that occurred with torpedoes and electric eels during the 1770s is: Just how much of this information trickled down to the laity? To scientists and physicians, of course, the new evidence for animal electricity was a monumental achievement, one destined to cause a paradigm shift in physiology, while also providing more justification for "therapeutic" electrical shocks to be applied to the sick and maimed. Further, and on a more fundamental level, it also revealed the pervasiveness of a natural force that had captured their imagination. Electricity could be generated from static machines and it abounds in the atmosphere, especially during storms—it could even be observed in the mineral tourmaline. And now, as we have seen, a good case could be made for it being the agent used by a few specialized living creatures to benumb their prey and repel potential enemies. But what did the public know about the electric fish discoveries of the 1770s and their significance?

Examining every possible source covering developments in science and medicine in Europe and the North America, and each piece of literature using the terms "torpedo," "electric eel," or their synonyms, would, of course, be the ideal way of judging what filtered down to the literate public. The downside of this approach, however, is the overwhelming amount of material that would have to be gathered for examination. A more pragmatic alternative is to limit such a survey to several sources of information, and this is the strategy we shall adopt in this chapter. We shall begin with the London newspapers, since many of the key developments took place in that city and involved members of the Royal Society of London. Next we shall look at a very important magazine, one known for its coverage of developments in science and medicine and with a readership extending well beyond the British Isles, namely *The Gentleman's Magazine*, (also referred to more informally as *Gentleman's Magazine*). Then we shall examine some poetry and literature from this era. And in closing, we shall turn to how the word "torpedo" began to be used in everyday language with negative connotations.

THE LONDON NEWSPAPERS

Scanning the contents of the London newspapers shows that they readily passed on information about torpedoes to the public—the stranger the better. Moreover, and perhaps somewhat surprisingly, they often did so in ways suggesting that their readers had a fairly good grasp of what they were writing about, which is significant in itself.

For example, as noted in Chapter 16, they covered some of Walsh's most important experiments on the French torpedoes. Specifically, the *Middlesex Journal* provided readers with a complete translation of Seignette's 1772 letter, which mentioned how a torpedo's shock was felt through a circuit of hand-holding people—even before that letter appeared in English in Walsh's 1773 paper (Seignette, 1772a; Walsh, 1773). This is the letter that ends with the provocative statement, "The effects produced by the Torpedo resemble in every respect a weak electricity." After an inexplicable delay of 14 years, it also appeared in abridged form in the *Public Advertiser* (Seignette, 1790). It would also appear in *The Gentleman's Magazine* (see below).

There was also a curious note in the *Morning Chronicle and London Advertiser* on May 15, 1773, dealing with the ability of torpedoes to cure a high fever. The source was not revealed, and it has the flavor of the ancient idea of treating with opposites. Specifically, readers are told: "We are informed that an able physician, indefatigable in medical researches, has discovered that the Torpedo is a fish of so cold a nature, that when applied to the body of a person who is in a burning fever, it will assuage and remove the same" (*Morning Chronicle,* 1773). There is nothing about electricity in this sentence or in the Latin sentence that follows, and no comparisons are made to popular claims of bedside cures using Leyden jars. Further, nothing is written about how torpedoes had been used in the past to treat fevers and other disorders, or about how electric eels were now being used to treat sick people in tropical America (Chapters 4, 5, and 13). Nevertheless, the idea conveyed is that torpedoes are deserving of a place in medicine.

When a torpedo was netted in the Thames River in 1774, it received much more coverage; in fact, it created such a sensation that it seemed as if every London newspaper was compelled to cover the story. The note in the *St. James's Chronicle or the British Evening Post* on June 23 is identical to that in the *General Evening Post* of the same date, and is the same as the article that appeared in the *London Evening Post* 2 days later. It reads:

> Wednesday morning, as Richard and John Piper, fishermen at Eton, were drawing the river for salmon, they caught a most extraordinary fish, by name torpedo, between Maidenhead and Windsor bridges, which kind of fish was never known to be taken in the river Thames since the memory of man. John Piper's hand was void of sensation for more than an hour, in extracting the fish from the net.

Late in 1779, two boys caught another torpedo not far from London with similar effects, and the story was covered in several places, including the *Public Advertiser* on December 28 of that year, again without specifically mentioning electricity. In this case, people read: "Friday last two Lads fishing for Flounders between Battersea and Putney, caught in their Net one of the Fish called Torpedo: The Lad that took it out of the Net had his Hand and Arm swelled to a great Size, and became useless for several Hours."

Not surprisingly, when some huge torpedoes were now caught off the British Coast, they also made the papers, with some fanfare. These were the English torpedoes that Walsh wrote about in the *Philosophical Transactions* of 1774. As we pointed out in Chapter 16, Walsh (1774b) was surprised by the sizes and weights of the British torpedoes (one was 4 feet long and 2.5 feet wide and weighed 53 pounds avoirdupois, and he was told some could reach 80 pounds), since the largest of the small torpedoes he had experimented with off the coast of France had been only a fraction as big. Yet the newspapers did not bring up their sizes, although they were described as electrical in the identical article that appeared in the *General Evening Post* of December 27, the *Public Advertiser* of December 28, and the *Middlesex Journal and Evening Advertiser* of December 29, 1774.

> Newcastle, Dec. 24. It has lately been discovered, that the extraordinary fish called torpedo, or electric ray, frequents the shores of this Island, contrary to a received opinion among naturalists, who have in general considered it as an inhabitant only of warmer climates. Three of these fish were taken the beginning of this month near Torbay, and several have for some weeks past been caught near Brixham.

John Pringle's discourse on electric fishes, which was given in 1774 in celebration of John Walsh's Copley Medal, received the greatest coverage of all the electric fish stories that ran in the London newspapers of the 1770s. It must have been evident to the people running the papers, or at least to their editors and advisors, that this was a noteworthy event. As succinctly put in the *General Evening Post* on November 29, the *Daily Advertiser* on December 1, and the *Public Advertiser* on December 1, 1774:

> Yesterday being St. Andrew's day, the Royal Society held their anniversary meeting at their house in Crane-court, Fleet-Street, when the President, John Pringle, Bart. in the name of the Society, presented the gold medal (called Sir Godfrey Copley's) to John Walsh, Esq; for his most curious experiments and observations on the fish torpedo: the president on this occasion made a learned discourse on the matters contained in Mr. Walsh's paper. Afterwards the Society proceeded to the choice of Council and Officers for the ensuing year . . . And the Members of the Society dined together as usual on their anniversary day.

Longer articles, which included paragraphs from Pringle's speech and lavish praise for Pringle and Walsh, appeared in the *Morning Chronicle and London Advertiser* on January 17 and 20, 1775, and in the *London Chronicle or Universal Evening Post* on February 7, 1775. But this was not all. There was also a parody based on the article signed J.P. (playing on Pringle's name), which appeared on Christmas Eve in the *London Evening Post* (J.P., 1774).

The theme of this parody is that John Pringle resembles a torpedo, because he also put people to sleep! J.P. begins by stating that he will "give you a cool speech upon that cold fish, which fish is called a Torpedo." He describes it as "an odd fish," much as "there are odd fishes in Crane Court" (home of the Royal Society). Halfway through his talk, he asks his audience if his speech "has not produced a kind of torpor in your nerves already?" He then confuses torpedoes with the electric eel (Pringle actually confused an electric catfish with the eel!) and says: "Lords and Gentlemen, if you want to see that fish, you must even go and take your passage in the East India ships; not that I wish to ship you off from Crane Court, but that you might go and prove, in person, what I have said upon this important subject." J.P. then thanks his "ingenious electricity friend, Dr. Benjamin Franklin, who has frequently, with his electrical fire, recovered me from that numbness which I naturally have myself, and create in others, when even I do not describe this same Torpedo." And he concludes with some words about how even this discourse must now be putting people to sleep, much like too much port—or a torpedo.

The year 1776 proved to be a banner year for torpedoes in the London newspapers. As noted in Chapter 18, the *Gazetteer and New Daily Advertiser* (August 5), *the Public Advertiser* (August 5), and the *Morning Chronicle and London Advertiser* (August 6) announced the arrival of five live torpedoes in London, along with Walsh's spark experiment before expert witnesses. Four months later, the *St. James's Chronicle or the British Evening Post* (November 16) and the *Public Advertiser* (November 19) covered the second major demonstration of this wondrous phenomenon. These were not the only newspapers covering these events, and this broad coverage shows that the London newspapermen understood or at least sensed from their contacts that these demonstrations were highly significant in the history of science.

In the above-mentioned notices of the second spark experiment, readers were further informed that Captain Baker, who brought the fish from Surinam, was going to do the same sort of thing in London that he had done so successfully in South Carolina. That is, he was going to give public and private demonstrations with his eels—for fees, of course. Baker did precisely this, and his shows were advertised repeatedly in the London papers. The following appeared in the *Gazetteer and New Daily Advertiser* (August 12 and then in other issues) under an uncharacteristically sophisticated heading: "THE ELECTRICAL EEL, or NUMBING FISH of SOUTH AMERICA. GYMNOTUS ELECTRICUS of LINNAEUS."

> The fish, which has [n]ever been exposed to public view. This natural exhibition of Electrical Phaenomena is possessed of all the properties of the Leyden Phial. It is hoped that the curious will not lose the present opportunity of seeing this miracle of nature, as the cold season is advancing, which will necessarily put a period to its existence. The profits arising from the exhibition are intended to reward Mr. George Baker, who was sent at the expense of several Gentlemen of the Royal Society, for the sole purpose of bringing it over alive. To be seen in the Haymarket, late Ford's Auction Room, from nine in the morning till eight in the evening. Admittance Five Shillings each person.

A notice with slightly different wording can be found in the *Daily Advertiser* (August 14). And very different one, with a new address, appeared late in 1776 in the *Gazetteer and New Daily Advertiser* (November 29 and December 14) and into the spring of the next year in the *Public Advertiser* (February 19), and the *Morning Chronicle and London Advertiser* (March 10). This time it bore an even lengthier heading: "THE GYMNOTUS ELECTRICUS, or the ELECTRICAL EEL, at Mr. Baker's New Apartment in Piccadilly, nearly opposite St. James's Street."

It reads:

> Mr. Baker presents his respectful compliments to the learned, and public in general who have honoured him with their presence and approbation at his exhibition of the above natural electrical phaenomenon, and begs leave to acquaint them he is obliged to draw the spark or vivid flashes from his fish three times a week, that is Mondays, Wednesdays, and Fridays, on account of the danger of their being exhausted by the too repeated experiments of every day; for which reason he hopes the indulgent public will not think such an imposition. To be exactly at one o'clock. The fish may be seen every day, and the electrical shock felt from ten till four, at 2s. 6d. per person.

Thus, the London papers carried some of the highlights of the British contributions to the electric fish story in the 1770s, and more. As might be expected, they emphasized the sensational, ranging from torpedoes being caught in the Thames to Walsh's landmark spark experiment with Captain Baker's eels. They did not venture into Hunter's dissections, which would have bored readers, or into the importance of Cavendish's physics, which would have been over the heads of even their best-educated editors. Yet from this coverage, it can be safely said that the laity was kept well informed about the latest British developments with the wondrous fishes that really seemed to generate electricity.

The Gentleman's Magazine

The Gentleman's Magazine or Trader's Monthly Intelligencer was established in 1731 by Edward ("Ned") Cave, an astute and at times even ruthless businessman who would later publish Franklin's famous pamphlet on electricity (Chapter 10; Franklin, 1751; Carlson, 1938; Locke & Finger, 2007; Finger

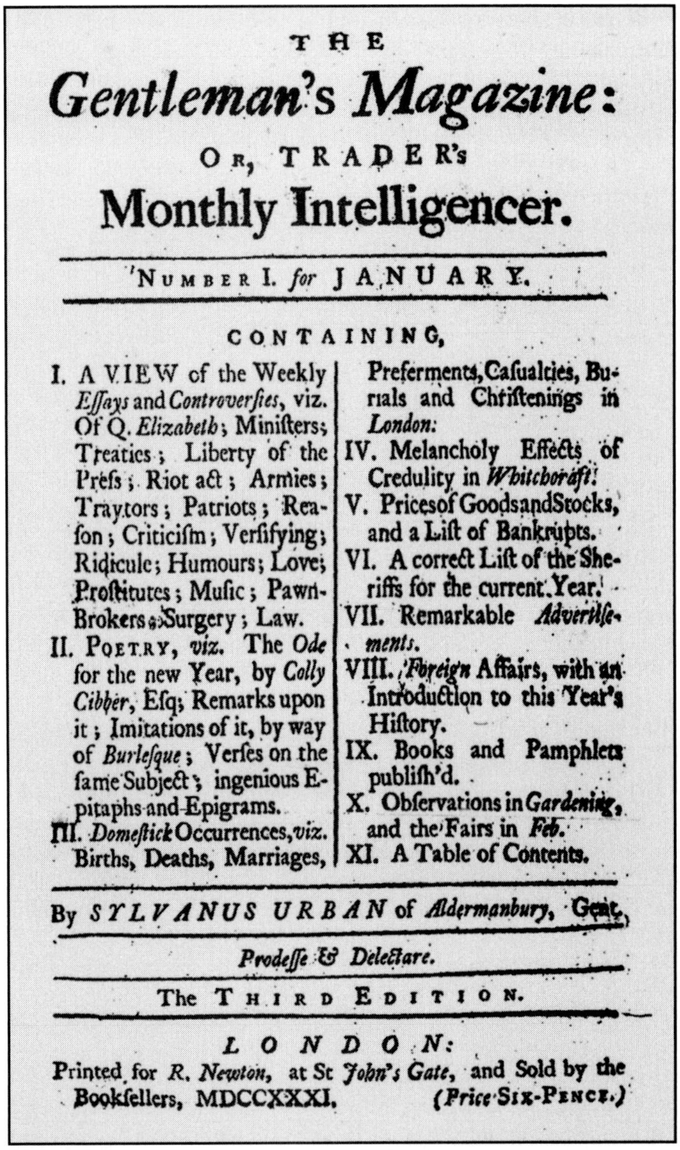

THE
Gentleman's Magazine:
OR, TRADER's
Monthly Intelligencer.
NUMBER I. *for* JANUARY.
CONTAINING,

I. A VIEW of the Weekly *Essays* and *Controversies*, viz. Of Q. *Elizabeth*; Ministers; Treaties; Liberty of the Press; Riot act; Armies; Traytors; Patriots; Reason; Criticism; Versifying; Ridicule; Humours; Love; Prostitutes; Music; Pawn-Brokers; Surgery; Law.	Preferments, Casualties, Burials and Christenings in *London:*
II. POETRY, *viz.* The *Ode* for the new Year, by *Colly Cibber*, Esq; Remarks upon it; Imitations of it, by way of *Burlesque*; Verses on the same Subject; ingenious Epitaphs and Epigrams.	IV. Melancholy Effects of Credulity in *Whitchcraft!*
III. *Domestick* Occurrences, viz. Births, Deaths, Marriages,	V. Prices of Goods and Stocks, and a List of Bankrupts.
	VI. A correct List of the Sheriffs for the current Year.
	VII. Remarkable *Advertisements.*
	VIII. *Foreign* Affairs, with an Introduction to this Year's History.
	IX. Books and Pamphlets publish'd.
	X. Observations in *Gardening*, and the Fairs in *Feb.*
	XI. A Table of Contents.

By *SYLVANUS URBAN* of *Aldermanbury*, Gent.
Prodesse & Delectare.
The THIRD EDITION.
LONDON:
Printed for *R. Newton*, at St *John's Gate*, and Sold by the Booksellers, MDCCXXXI. (*Price* SIX-PENCE.)

Figure 20.1. Cover page of the first issue of Edward Cave's *Gentleman's Magazine.*

& Ferguson, 2009) (Fig. 20.1). Cave's initial goal was to provide city information to people living in rural areas, and information from more rural places to people living in the city. His pseudonym, Sylvius Urban, meaning forest and city, reflected this objective of bringing these two worlds together.

Published in London, *The Gentleman's Magazine* has been heralded as the first modern magazine and one of the most important periodicals of the Enlightenment. It covered everything from politics to medicine, and from voyages of discovery to the latest developments in the arts. Its editors scanned and summarized reports from scientific societies around the world, although concentrating mostly on the Royal Society of London, and also encouraged eyewitness accounts and testimonials from others, including "amateurs" not affiliated with national academies, leading medical schools, and the like.

Cave's concept proved extremely successful, and even though his initial focus was on readers in the British Isles, he increasingly attracted a large number of readers from distant places with his affordable monthly coverage of such a variety of material. With some remarkable people on his staff, including literary great Dr. Samuel Johnson, who wrote some medical biographies for this publication (McHenry, 1959), and poet Alexander Pope, Cave's magazine would boast over 10,000 subscribers. This, of course, must be regarded as a low estimate of the periodical's real readership, because many subscribers (Franklin and his associates in the New World being good examples) shared copies of this magazine, readily passing it around to others.

Experiments on the nature or physics of electricity were a common feature in *The Gentleman's Magazine* at mid-century. In addition, Cave, his replacement after his death in 1754, and their staffs covered some of what was being claimed with therapeutic electricity, starting shortly after its birth in the 1740s (Locke & Finger, 2007). For Cave, of course, these were fascinating developments that helped sell his periodical, and in this context letters from readers praising what the magazine covered were sometimes published. One from 1747 reads: "Amongst the agreeable variety which your Magazine contains, I shall always esteem that part the most useful, which can contribute towards relieving those whom it has pleased providence to afflict with sickness" (Elles, 1747, p. 77). Cave or one of his associates replied: "To find that we have in any degree contributed to the benefit of the public, or of any individual, gives us great pleasure" (Anon., 1747, p. 362).

Most of the science and medical news articles were no more than a paragraph or two in length, but others were considerably longer. More importantly, the editors displayed a clear preference for empirical findings and a bias against loose theorizing, reflecting a key feature of the scientific Enlightenment. Further, even when the contributors did not agree with one another, they were always non-argumentative and inviting of further discussion. As put by one historian of Cave's periodical:

> In these pages where dull prose and duller verse alternate with accounts of brilliant discovery and extracts from poetical masterpieces, where politics jostle with theology and the sciences with art, here we have a kaleidoscopic view of the culture of a century. We sense the development of popular taste, we feel the surge of popular opinion. (Carlson, 1938, p. 58)

Cave and his staff recognized early on that articles on fishes that could numb people would fascinate readers, even people with little knowledge of science. Hence, they began to cover these wonders of nature in 1750, although not at first associating their discharges with electricity (Finger & Ferguson, 2009).

The first mention of one of these torporific fishes appeared in an abstract of "Mr. Anson's Voyage," a reference to George Anson, who sailed around the world from 1740 to 1744 and whose exploits were published in book form in 1748 by Richard Walter (see Chapter 12). Here the reader was informed about some powerful torpedoes caught off the Mexican coast:

> At *Chequetan*, and no where else, they found the Torpedo, which causes a numbness in those who touch it. If this climate be different from those of *Italy* and *France*, where the same fish has been taken notice of by Messrs *Lorenzini* and *Reaumur*, its effects in *America* are much stronger than in *Europe*. Our author avers, that the numbness seizes the whole body; his arm was violently affected by one which he leaned on with his cane for some time, and believes it would have been more so, had not the fish been near expiring. Neither is this numbness so transient as naturalists have supposed; Mr *Walter* had some feeling of it the next day; so that this *American* torpedo is in its effects like that which M. *Reaumur* observed at *Cayenna* [French Guiana] but of a very different figure, being like a thornback [a type of ray; *Raja clavata*], and the other was like a conger, or an eel. (Anon., 1750, p. 67)

The anonymous author of the aforementioned piece was incorrect in stating that Réaumur studied eels in Guiana, revealing one of the shortcomings of some of the reports in *The Gentleman's Magazine*. Réaumur worked through many correspondents, one being Jacques François Artur, who was in French Guiana and sent him a wealth of information about the flora and fauna in the region (Chaïa, 1968). As for his own work, it was on French torpedoes, as we discussed in Chapter 9. Interestingly, although Réaumur is named in the piece, the underlying nature of the torpedo's shock, which Réaumur thought is mechanical, was not brought up.

In 1763, 6 years after he died, Réaumur's name came up again in rather lengthy article celebrating his contributions to science (Anon., 1763). He is praised as "a man of great ingenuity and learning, of the strictest integrity and honour, the warmest benevolence, and the most extensive liberality" (Anon., 1763, p. 116). In contrast to the earlier article, this one includes a brief statement about his rapid muscular contraction theory, which he had maintained could account for the torpedo's numbing effects. It reads: "M. *Reaumur*, after many experiments made with the Torpedo, or Numb-fish, discovered that its effect was not produced by an emission of torporific particles, as some have supposed, but by the great quickness of a stroke given by this fish to the limb that touches it, by muscles of a most admirable structure, which are adapted to that purpose" (Anon., 1763, p. 112). In the writer's opinion, "These discoveries, however, are chiefly matters of curiosity," and he rightfully questions whether direct contact with the fish is needed to feel its "stroke," as had been claimed by Réaumur.

The next article on electric fishes follows a hiatus of 6 years and raises the possibility that its discharges are electrical. It is dated 1769 and it is a review of Edward Bancroft's (1769a) *Essay* on Guiana (*Gentleman's Magazine*, 1769). Here readers are informed, in Bancroft's own words and without commentary, of some of the observations and experiments that he conducted on the eels, and his firm opinion that they are definitely electrical (see Chapter 15).

But what about Walsh's experiments to determine whether torpedoes are also electrical? Like some of the London newspapers, *The Gentleman's Magazine* promptly provided a full translation of Seignette's account of Walsh's endeavor at La Rochelle. This was presented in November 1772 as an "Extract of a Letter from the Sieur Seignette, Secretary to the Academy at Rochelle" (Seignette, 1772b; Chapter 15; also see Fig. 20,2). It appeared in the "Historical Chronicle" section of the periodical.

Readers were subsequently informed about how Walsh (1774b) went on to inform fellow members of the Royal Society that some huge torpedoes had been caught off the English coast (Walsh, 1775). The awarding of the 1774 Copley

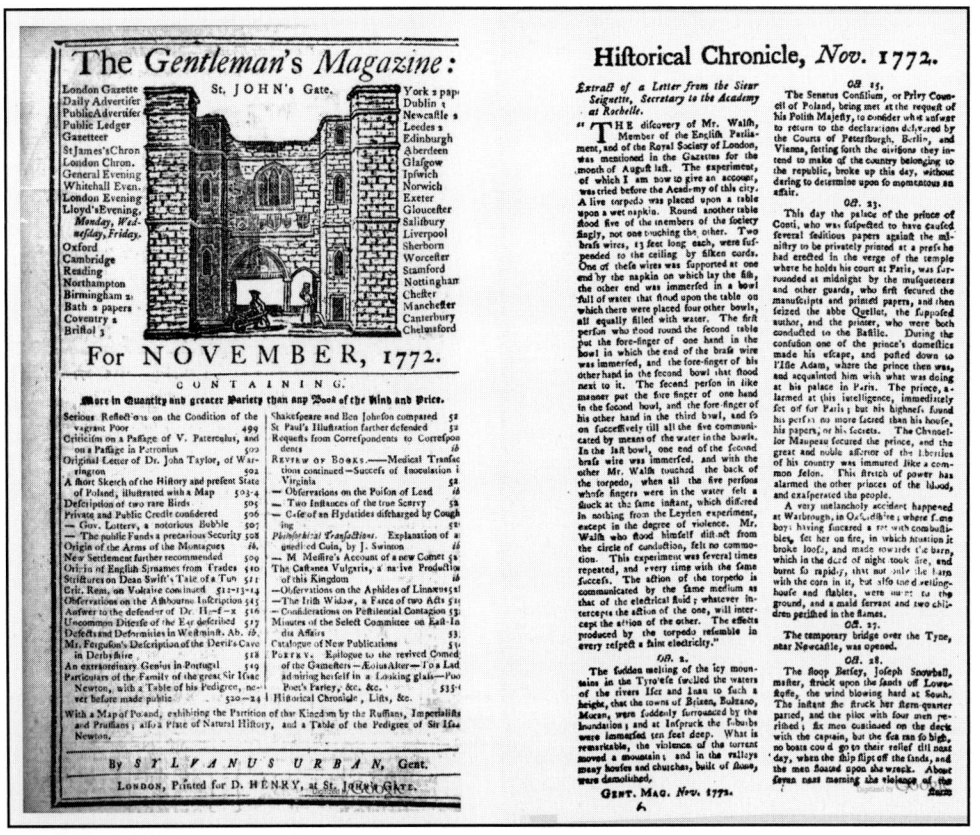

Figure 20.2. Cover page of the November 1772 issue of the *Gentleman's Magazine* with the page providing a translation of Seignette's communication to the *Gazette de France* about Walsh's torpedo experiments at La Rochelle.

Medal to Walsh, and the fact that the fish was dissected by John Hunter, were also noted in this piece, which is brief and reads in its entirety:

> Naturalists have generally considered the torpedo, or electric ray, as an inhabitant only of warmer climates; but, contrary to this received opinion, two of these fish, taken in Torbay, were sent up, in 1773, to London, one of them weighing 53 pounds avoirdupois, which is much larger than any that this writer [Walsh] ever saw or read of in the bay of Biscay, the Mediterranean, &c. Their electrical organs were injected by Mr. John Hunter. Accounts of several others caught on the coast of Cornwall, &c. and some curious particulars relating to them, are subjoined, for which the society have, this year, adjudged to Mr. Walsh their prize medal. (Walsh, 1775, p. 83)

The editors of *The Gentleman's Magazine* were not concerned with esoteric details, which most people could care less about and even fewer might understand. Their aim was to provide the latest news in a readable and interesting way—the understanding being that those who wanted to know more could always go elsewhere. This aim to be selective, brief, informative, and entertaining is best exemplified by the fate of Henry Cavendish's (1776a) contributions about the physics of the discharge, based in part on his artificial torpedoes (Chapter 16). Faced with what to report without putting readers to sleep, the editors provided Cavendish's original title ("An Account of some Attempts to Imitate the Effects of the Torpedo by Electricity") followed by the author's name and affiliation (By the Hon. Henry Cavendish, F. R. S.), and then directed their interested readers to the primary source, with the words: "We must refer our electrical readers to the article, as we find in it none of the *utile* [usefulness], and little of the *dulce* [sweetness]" (Cavendish, 1776b, p. 561).

The same strategy of sending interested readers to the source was used if there were questions about the originality, intrinsic value, or newsworthiness of a scientific report. Consequently, although the *Philosophical Transactions* published Jan Ingenhousz's (1775a) four-page letter about his torpedo experiments in Italy, this was now reduced to just two sentences, with readers again being directed to the original report (Ingenhousz, 1775b, p. 436).

Eels from Guiana, with their decidedly stronger shocks, now made their way back into the magazine, reflecting the fact that some were finally being studied outside the wilds of South America (Chapters 17 and 18). Hugh Williamson (1775a,b), who in 1773 conducted electrical experiments with a live eel in Philadelphia, had his 1775 letter extracted, although the piece failed to mention his most important finding, namely that he was able to feel its shocks across an air gap (Fig. 20.3). Notice of Alexander Garden's (1775a) submission from South Carolina to the Royal Society followed on the very same page (Garden, 1775b). Although Garden's statement that "The largest was three feet eight inches long, and some, it is said, have been seen upwards of twenty feet, whose stroke, or shock, was instant death," appeared in this brief price (p. 437), the editors did not mention that Garden had also expressed serious doubts about what "some" have said about its length and the need for more information.

The readers of *The Gentleman's Magazine* would be misled 11 years later, when Lieutenant William Paterson reported to Joseph Banks at the Royal Society that he had

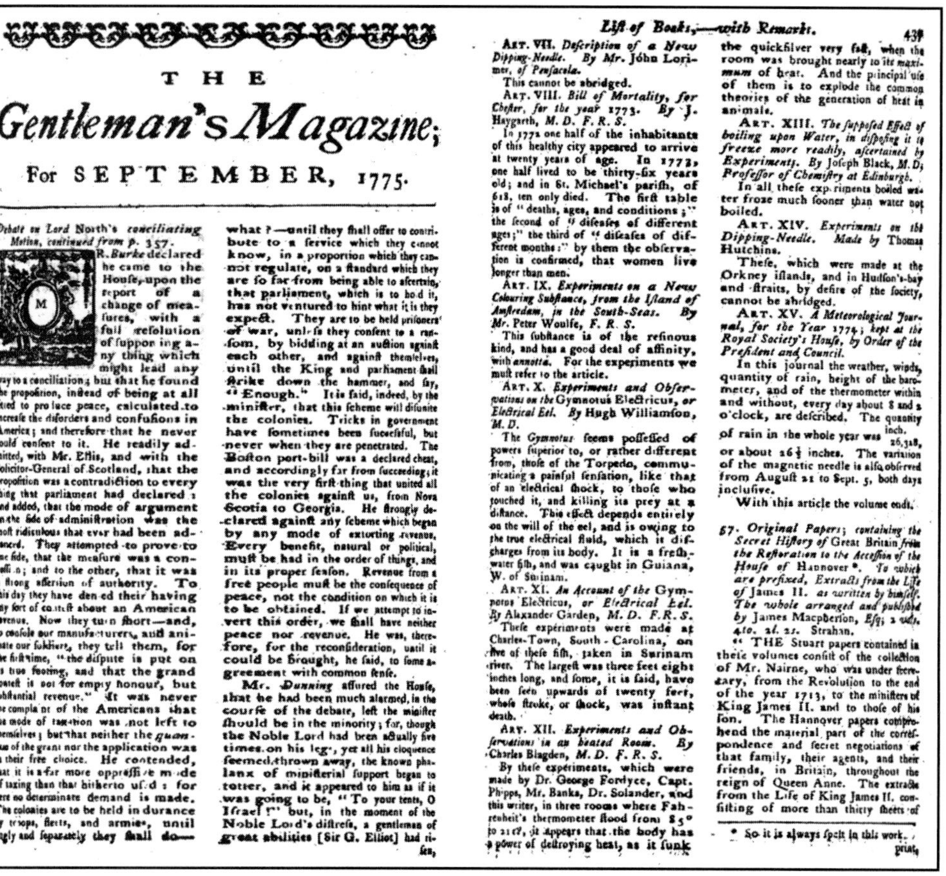
THE

Gentleman's Magazine;

For SEPTEMBER, 1775.

Debate on Lord North's *conciliating Motion, continued from p.* 357.

MR. *Burke* declared he came to the Houſe, upon the report of a change of meaſures, with a full reſolution of ſupporting any thing which might lead any way to a conciliation; but that he found the propoſition, inſtead of being at all fitted to produce peace, calculated to increaſe the diſorders and confuſions in America; and therefore that he never could conſent to it. He readily admitted, with Mr. Ellis, and with the Solicitor-General of Scotland, that the propoſition was a contradiction to every thing that parliament had declared: and added, that the mode of argument on the ſide of adminiſtration was the moſt ridiculous that ever had been advanced. They attempted to prove to one ſide, that the meaſure was a conceſſion; and to the other, that it was a ſtrong aſſertion of authority. To this day they have denied their having any ſort of conteſt about an American revenue. Now they turn ſhort—and, to conſole our manufacturers, and animate our ſoldiers, they tell them, for the firſt time, "the diſpute is put on its true footing, and that the grand conteſt is not for empty honour, but ſubſtantial revenue." It was never the complaint of the Americans that the mode of taxation was not left to themſelves; but that neither the *quantum* of the grant nor the application was in their free choice. He contended, that it is a far more oppreſſive mode of taxing than that hitherto uſed: for here no determinate demand is made. The colonies are to be held in durance by troops, fleets, and armies, until ſingly and ſeparately they ſhall do—what?—until they ſhall offer to contribute to a ſervice which they cannot know, in a proportion which they cannot regulate, on a ſtandard which they are ſo far from being able to aſcertain, that parliament, which is to hold it, has not ventured to hint what it is they expect. They are to be held priſoners of war, unleſs they conſent to a ranſom, by bidding at an auction againſt each other, and againſt themſelves, until the King and parliament ſhall ſtrike down the hammer, and ſay, "Enough." It is ſaid, indeed, by the miniſter, that this ſcheme will diſunite the colonies. Tricks in government have ſometimes been ſucceſsful, but never when they are penetrated. The Boſton port-bill was a declared cheat, and accordingly far from ſucceeding; it was the very firſt thing that united all the colonies againſt us, from Nova Scotia to Georgia. He ſtrongly declared againſt any ſcheme which began by any mode of extorting revenue. Every benefit, natural or political, muſt be had in the order of things, and in its proper ſeaſon. Revenue from a free people muſt be the conſequence of peace, not the condition on which it is to be obtained. If we attempt to invert this order, we ſhall have neither peace nor revenue. He was, therefore, for the reconſideration, until it could be brought, he ſaid, to ſome agreement with common ſenſe.

Mr. *Dunning* aſſured the Houſe, that he had been much alarmed, in the courſe of the debate, leſt the miniſter ſhould be in the minority; for, though the Noble Lord had been actually five times on his legs, yet all his eloquence ſeemed thrown away, the known phalanx of miniſterial ſupport began to totter, and it appeared to him as if it was going to be, "To your tents, O Iſrael!" but, in the moment of the Noble Lord's diſtreſs, a gentleman of great abilities [Sir G. Elliot] had riſen,

Liſt of Books,—with Remarks. 437

ART. VII. *Deſcription of a New Dipping-Needle. By Mr.* John Lorimer, *of Penſacola.*

This cannot be abridged.

ART. VIII. *Bill of Mortality, for Cheſter, for the year* 1773. *By* J. Haygarth, *M. D. F. R. S.*

In 1772 one half of the inhabitants of this healthy city appeared to arrive at twenty years of age. In 1773, one half lived to be thirty-ſix years old; and in St. Michael's pariſh, of 618, ten only died. The firſt table is of "deaths, ages, and conditions;" the ſecond of "diſeaſes of different ages;" the third of "diſeaſes of different months:" by them the obſervation is confirmed, that women live longer than men.

ART. IX. *Experiments on a New Colouring Subſtance, from the Iſland of Amſterdam, in the South-Seas. By Mr.* Peter Woulfe, *F. R. S.*

This ſubſtance is of the reſinous kind, and has a good deal of affinity, with *annotta*. For the experiments we muſt refer to the article.

ART. X. *Experiments and Obſervations on the* Gymnotus Electricus, *or Electrical Eel. By* Hugh Williamſon, *M. D.*

The *Gymnotus* ſeems poſſeſſed of powers ſuperior to, or rather different from, thoſe of the Torpedo, communicating a painful ſenſation, like that of an electrical ſhock, to thoſe who touched it, and killing its prey at a diſtance. This effect depends entirely on the will of the eel, and is owing to the true electrical fluid, which it diſcharges from its body. It is a freſh-water fiſh, and was caught in Guiana, W. of Surinam.

ART. XI. *An Account of the* Gymnotus Electricus, *or Electrical Eel. By* Alexander Garden, *M. D. F. R. S.*

Theſe experiments were made at Charles-Town, South-Carolina, on five of theſe fiſh, taken in Surinam river. The largeſt was three feet eight inches long, and ſome, it is ſaid, have been ſeen upwards of twenty feet, whoſe ſtroke, or ſhock, was inſtant death.

ART. XII. *Experiments and Obſervations in an heated Room. By* Charles Blagden, *M. D. F. R. S.*

By theſe experiments, which were made by Dr. George Fordyce, Capt. Phipps, Mr. Banks, Dr. Solander, and this writer, in three rooms where Fahrenheit's thermometer ſtood from 85° to 21[illegible]°, it appears that the body has a power of deſtroying heat, as it ſunk the quickſilver very faſt, when the room was brought nearly to its *maximum* of heat. And the principal uſe of them is to explode the common theories of the generation of heat in animals.

ART. XIII. *The ſuppoſed Effect of boiling upon Water, in diſpoſing it to freeze more readily, aſcertained by Experiments. By* Joſeph Black, *M. D. Profeſſor of Chemiſtry at Edinburgh.*

In all theſe experiments boiled water froze much ſooner than water not boiled.

ART. XIV. *Experiments on the Dipping-Needle. Made by* Thomas Hutchins.

Theſe, which were made at the Orkney iſlands, and in Hudſon's-bay and ſtraits, by deſire of the ſociety, cannot be abridged.

ART. XV. *A Meteorological Journal, for the Year* 1774; *kept at the Royal Society's Houſe, by Order of the Preſident and Council.*

In this journal the weather, winds, quantity of rain, height of the barometer, and of the thermometer within and without, every day about 8 and 2 o'clock, are deſcribed. The quantity of rain in the whole year was inch. 26,318, or about 26⅓ inches. The variation of the magnetic needle is alſo obſerved from Auguſt 21 to Sept. 5, both days incluſive.

With this article the volume ends.

57. *Original Papers; containing the Secret Hiſtory of* Great Britain *from the Reſtoration to the Acceſſion of the Houſe of* Hannover*. *To which are prefixed, Extracts from the Life of* James II. *as written by himſelf. The whole arranged and publiſhed by* James Macpherſon, *Eſq;* 2 *vols.* 4*to.* 2*l.* 2*s.* Strahan.

"THE Stuart papers contained in theſe volumes conſiſt of the collection of Mr. Nairne, who was under ſecretary, from the Revolution to the end of the year 1713, to the miniſters of King James II. and to thoſe of his ſon. The Hannover papers comprehend the material part of the correſpondence and ſecret negotiations of that family, their agents, and their friends, in Britain, throughout the reign of Queen Anne. The extracts from the Life of King James II. conſiſting of more than thirty ſheets of print,

* So it is always ſpelt in this work.

Figure 20.3. The September 1775 issue of *The Gentleman's Magazine* with brief notes of both Hugh Williamson's and Alexander Garden's experiments on the electric eel, which had just been published in the *Philosophical Transactions.*

discovered a new electric fish in the East Indies (Paterson, 1786a) (Fig. 20.4). The magazine editors, much like the men of the Royal Society (Fig. 20.5), did not realize that Paterson's toby was only opening and closing its mouth rapidly, while vibrating its fins in bursts, so as to produce a high-frequency whirring sound with vibrations (Ritterbush, 1964, p. 42). So this time the editors were in good company. The editors even reproduced a picture of the mighty 7-inch fish "that gave him a severe electrical shock, which obliged him to quit his hold," this being the only picture of an "electric fish" to appear in this era in this popular magazine (Paterson, 1786b, p. 1007).

Thus, the editors of *The Gentleman's Magazine* made their share of errors of commission, which made for more exciting reading, but which helped to promulgate some unsubstantiated myths. Conversely, they also made some errors of omission, leaving out important sentences from some of the articles they summarized, as well as notable reports, including those from the Dutch on fish electricity prior to Bancroft. The most notable omission of all had to be Walsh's demonstration of the spark (Chapter 19), perhaps because it failed to appear in the *Philosophical Transactions*, even though it was witnessed by prominent members of the Royal Society and was even covered in many London newspapers.

In summary, enlightened "gentleman" around the world could keep abreast of some of the developments taking place with electric fishes via the first modern magazine, *The Gentleman's Magazine.* Nevertheless, as was true for the electrical fish papers published in the *Philosophical Transactions* in this era, nothing was written about what the new discoveries meant in the bigger scheme of things. In this regard, the material that the editors of *The Gentleman's Magazine* chose to unveil was less than a dazzling canvas with thick coats of colored paint, although the idea that some fishes seem to be electrical was sketched out with some notable lines in black and white.

"SHOCKING" POETRY: SOME PIECES EVEN LEWD

As revealed in Chapter 4, poems had been written about torpedoes well before they began to be viewed as electrical. For example, during the second century, Oppian published a lengthy poem in which he wrote that a torpedo:

> Quick thro' the whole it shoots the rushing Pain,
> Freezes the Blood, and thrills in ev'ry Vein;
> Strikes all that dare approach with strange Surprise,
> Stiffens the Fin, and dims the mazed Eyes.

Similarly, Claudian asked: "Who has not heard of the untamed art of the marvelous torpedo, and of the powers that win it its name?" Torpedoes thereafter appeared in poetry that drew or borrowed from these early writers. Manuel Philes wrote one such poem in about 1300 (see Philes, 1730, p. 21; Kellaway, 1946, p. 121).

During the 1780s, when Aphra Behn was writing her novel *Oroonoko*, about a royal slave in Guiana, with its wonderful passages about what happened to him while swimming in a river infested with torporific eels (Chapter 8), some people were

Figure 20.4. The toby, as illustrated in *The Gentleman's Magazine* in 1786. Described as an electric fish that can delivers strong shocks in this periodical and in a more complete report by Lieutenant William Paterson that appeared in the *Philosophical Transactions* earlier that year, this small saltwater fish was clearly misrepresented.

reading a new poem by Edward Taylor. Called *Upon a Wasp Chilled with Cold*, it was composed after Taylor had emigrated from England to North America. Here we find torpedoes still associated with the quality cold, an association that goes back to Plutarch and Galen in the Roman period, if not earlier:

> The bear that breathes the northern blast
> Did numb, torpedo-like, a wasp
> Whose stiffened limbs encramped, lay bathing
> In Sol's warm breath and shine as saving,
> Which with her hands she chafes and stands
> Rubbing her legs, shanks, thighs, and hands.
> Her pretty toes, and fingers' ends
> Nipped with this breath, she out extends
> Unto the sun, in great desire

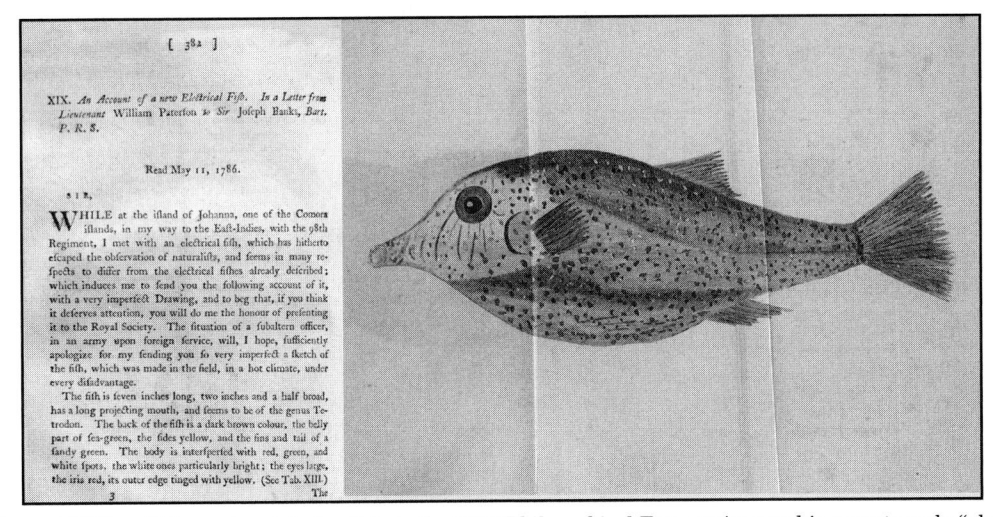

Figure 20.5. Lieutenant William Paterson's original article in the 1786 *Philosophical Transactions* on his new strongly "electric" fish.

To warm her digits at that fire.
(In Whittaker, 1992, p. 487)

Taylor is not regarded as one of the great British or American poets, but his lines are beautiful. They stand in stark contrast with the poetry being published about these fishes in John Walsh and John Hunter's day. But although the newer poetry was by no means beautiful, it is "revealing" in several ways. One is that it shows more of what was filtering down to non-scientists about electric fishes right after the successful spark experiment. A second way in which they are revealing will soon become evident and was intended to bring a wry smile to the reader knowing the characters involved.

James Perry became the Poet Laureate of the electric fish world at this crucial point in time. Born in 1756, Perry was a radical Scottish journalist and political writer, who died in 1821. He published his poetry about electric fishes either anonymously or under one of two pseudonyms, Adam Strong and Lucretia Lovejoy—and for good reason. Perry's poems were sarcastic, revealing, libertine, and rather pornographic in nature. That is, unlike the poets of the Roman Empire who tried to convey information about these marvelous creatures that could benumb the unwary, Perry's intention was, quite literally, to shock his readers and get what he described as a rise out of the men, and perhaps some actions in the women too, with his social innuendoes.

Three scientific themes triggered erotic writings in England at this time (Peakman, 2003). One derived from advances in botany; a second was a better understanding of the biology of reproduction; and a third was the rapidly growing natural world of electricity. With scientists abuzz with the possibility that some fishes might be electrical, some poets jumped on the thought that humans might also be capable of "electrical" actions—especially how female electricity might stir an ordinarily flaccid male organ into action. That the electric eel bears a certain likeness to a large penis was central to some of this poetry, and by no means lost on its readers.

Julie Peakman (2003, p. 89), who has examined the pornographic poetry of this era, writes, "Part of the intention of the 'electrically' themed poems was not merely to satirize the recent experiments (although this was obviously a main part of the joke) but to gossip openly about notorious libertines and recent scandals." James Perry, who clearly exemplifies this orientation, repeatedly uses electricity as a metaphor for sexual attraction and does not hesitate to mention adulterous aristocrats and socialites with loose morals in his poems. Allusions to their names can be found alongside those of the leading natural philosophers of the Royal Society, including John Pringle, in reality a rather austere Scot, whose mind was focused more on helping suffering humanity than on clandestine sexual affairs.

Perry's (1777a) first poem in the series was *The Electrical Eel, or, Gymnotus Electricus*, and it appeared in March 1777 (Fig. 20.6, left). This poem was, in fact, dedicated to the "learned members of the R***l S******" [Royal Society]. Perry's thesis was that the electric eel "is the original SERPENT OF SIN"! Modeled after John Milton's seventeenth-century masterpiece *Paradise Lost*, and using the entertaining yet appropriate pseudonym Adam Strong, Perry writes of Eve, that

None her attention drew,
'Till from the pond the lengthen'd Eel,
Bewitch'd, began her pow'r to feel,
And from the mud he grew.
He grew erect—he won her eyes,
Both by his sleekness and his size,
And thus the Dame address'd:
"Fair, beauteous lady—give your hand
"For you alone can make me stand,
"As you have made me bledd'd." (pp. 6–7)

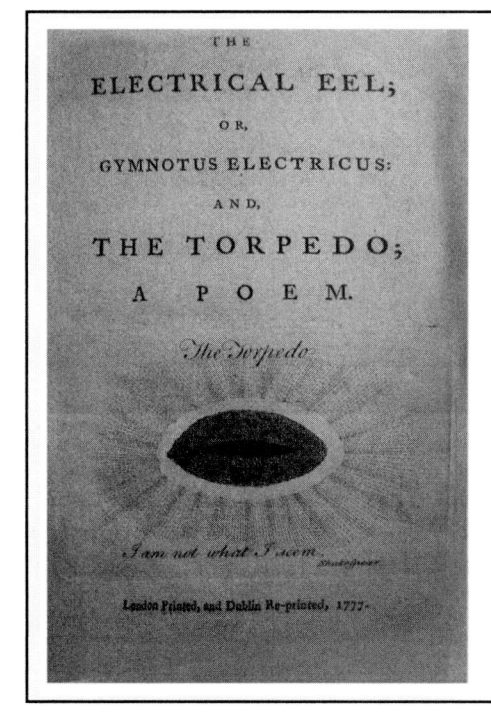

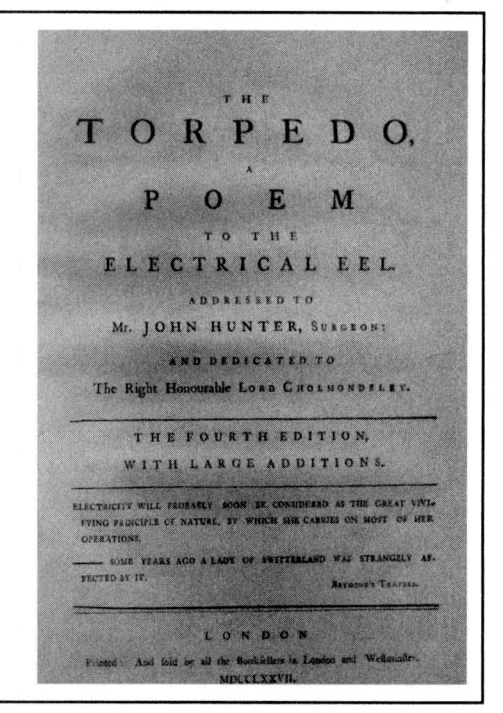

Figure 20.6. Covers of James Perry's first two poems, *The Electric Eel* and *The Torpedo*. Both debuted in 1777 and were shocking in multiple ways.

... Th' electrick fire soon warmed her heart,
The fire to all she did impart,
And nature own'd the feel:
From that gay period up to this,
The Wife, the Matron, and the Miss,
Hallow'd—th' Electrick Eel. (p. 8)
Therefore to end this long dispute,
And make all classick blockheads mute,
Who now abuse each other:
Let them this orthodoxy feel,
This arbor vitae was an Eel,
And long'd for by our mother. (p. 10)

Even the Advertisement at the end of this poem addressed the eel's electricity, which when viewed as the male sexual organ, is presented as quite superior to the electricity that a woman might experience from a Leyden jar:

Ye maids pray make the trial:
It hath the properties of wine,
Of fire, of love,—the true divine,
It beats the *Leyden Phial*. (p. 33)

Perry's (1777b) next poem, *The Torpedo*, followed that April (Fig. 20.6, right). It should be particularly interesting to historians of science and medicine because it is "Addressed to Mr. John Hunter, Surgeon." Hunter, the best-known anatomist in London, had by this time dissected both torpedoes (French and British) and South American eels (Chapters 16, 18). He was also feared by the laity as a "resurrectionist," a body snatcher intent on gathering unusual specimens from the grave for his massive cabinets.

Perry (again anonymously) begins this poem with some lines clearly describing Hunter:

O THOU! whose microscopic eye
Can every living thing decry,
And search Dame Nature's womb!
Whose power can raise the lifeless clay,
Drag the pale spectres into day,
And starve the hungry Tomb! (p. 1)

He then goes on to ask, with less-than-subtle reference to the male reproductive organ,

What mighty difference should there be
Between this wanton Eel and me?
What greater feats can he do? (p. 3)

And he answers:

Yet all confess it, when they feel,
That tho' in form it looks an Eel.
It, Serpent-like, can sting.
Spite of the pain which they endure,
'Tis his, I grant, to work a cure
For palsies and for age:
Those master-strokes, I know not how,
Active as C[ar]LT[o]n, or as H[o]WE,
Amid the battle's rage. (pp. 3–4)

As for other lines of note:

These are his Arts, we all must own,
Which, spread thro' Country and thro' Town,
Immortalize his name:
But Sir JOHN PR[in]GLE must agree,
That fish's Electricity
From the Torpedo came. (p. 9)
What tho' Lord CH[o]LM[o]D[e]LY may conceal
A most enormous length of Eel,
Admired for size and bone:
This mighty thing when lank, depress'd,
A mere noun adjective at best,
Is useless when alone. (p. 16)

The Semi-Globes or Electrical Orbs followed in July, and was dedicated to the lecherous Earl of H[arrington] (Perry, 1777c) (Fig. 20.7, left). It deals with "the vivifying powers" of a woman's semi-globular breasts. In this poem, Perry (pp. iii–iv) anonymously contends that animal electricity can be witnessed in the "Swelling beauties of the Female breast" and in what he calls "the touch electric" ("For *Nature's Electricity* is there!"). Neither the eel nor the torpedo appears in the poem itself, but they are mentioned in the dedication, where we read, "That after all the obscene allusions made in some late publications, to the *Gymnotus Electricus*, on the one hand, and the so famous *Torpedo*, on the other, I cannot help determining that the *true Electricity of Nature*, actually resides in the ELECTRIC ORBS," (pp. iii–iv).

Perry next felt compelled to add *An Elegy on the Lamented Death of the Electrical Eel, or Gymnotus Electricus* to his list of poetry (Perry, 1777d) (Fig. 20.7, right). This time his pseudonym is Lucretia Lovejoy, a woman identified as Adam Strong's sister. The elegy and the epitaph lament a oncegreat eel that can no longer rise to the occasion!

Perry's poetry went through many editions, and it was heavily advertised in the London newspapers. *The Electrical Eel*, for instance, was promoted in the *Morning Post and Daily Advertiser* on March 21, 1777 (just before it was published), March 25 ("This Day is published"), and March 31. Advertisements for it can also be found in the *St. James Chronicle or the British Evening Post* (April 1, July 3), the *Gazetteer and New Daily Advertiser* (April 12 and June 28), and the *London Evening Post* (June 24 and July 17). *The Torpedo* was advertised in the *Morning Post and Daily Advertiser* (April 21). There was a notice announcing "another curious production ... the *Semi Globes* or *Electrical Orbs*" in the *Morning Post and Daily Advertiser* (July 1), and notices promoting a new edition of *An Elegy on the Lamented Death of the Electrical Eel*, including two in the *General Advertiser and Morning Intelligencer* (March 6 and 23, 1779).

Some of the London papers also carried take-offs and editorials about Perry's poetry. For example, in the *Morning Post and Daily Advertiser* of May 20, 1777, we encounter this tongue-in-cheek piece of "news":

We are credibly informed, that Lord H—r—ng—ton has not *felt as usual*, since the publication of the second edition of the celebrated Poem of the TORPEDO; and so delirious is he of suppressing those *touching* instances of his sensibility, that he has made some very lucrative offers for the omission of those stanzas, in the next edition; but to the honors of this age be it spoken, we are informed that they have been refuted.

In addition, Perry's poetry clearly rubbed some people the wrong way:

The Semi-Globes may be a proper companion for the Electrical Eel and Torpedo, says a correspondent, but will either of these celebrated productions be of any service to the cause of literature? Surely not; because Wit, exercised

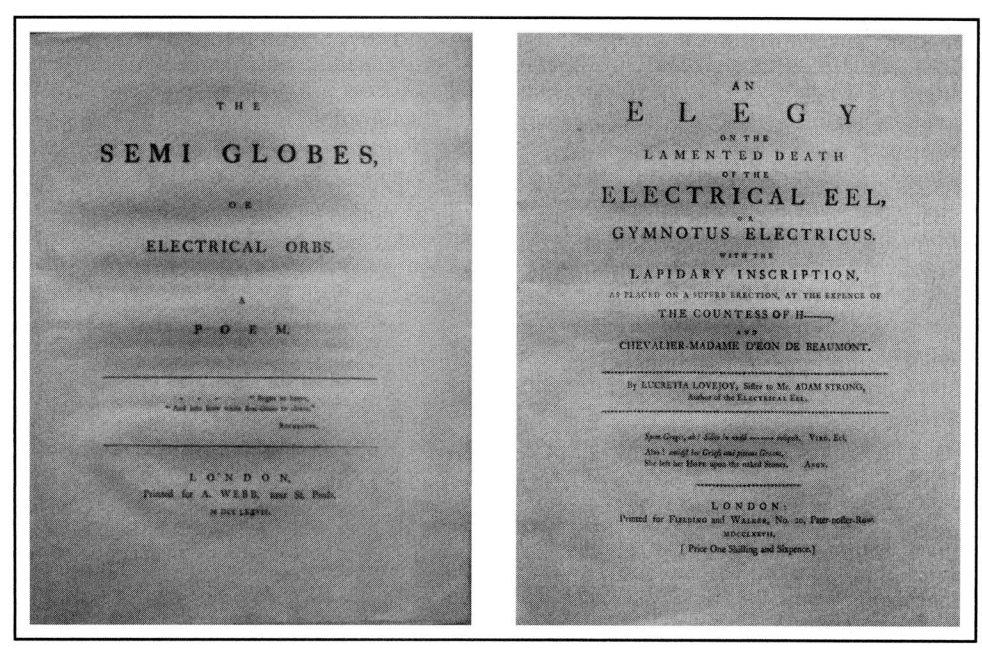

Figure 20.7. Perry's follow-up "electrico-libertine" poetry from 1777—with clear sexual allusions in their titles.

at the expense of Decency, can never be countenanced by as the effect of sterling genius. (*Morning Post and Daily Advertiser*, July 4, 1777)

It is interesting to note that the editors of *The Gentleman's Magazine* also responded negatively to the publication of *The Electric Eel* and another libertine poem in 1777 (Anon., 1777, p. 389) (Fig. 20.8). The following sentence appeared after the titles and prices for the two pieces: "We mention these vermin merely to caution our readers against them—*latet anguis in herba*—and if they should inadvertently touch either of them, they will most certainly receive a shock

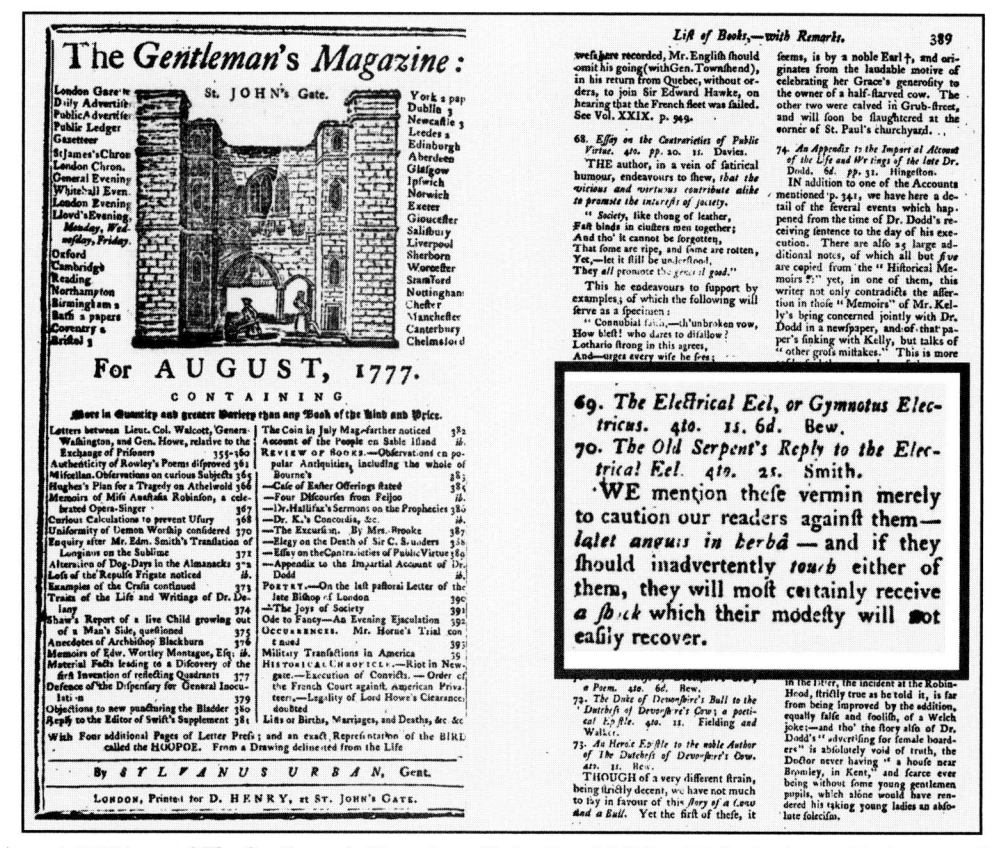

Figure 20.8. The August 1777 issue of *The Gentleman's Magazine* with (on the right) the detail of a short critical note on Perry's *Electric Eel*. The note also addresses a text that appeared in the same year in reply to Perry's *Eel* bearing the title *The Old Serpent's Reply to the Electric Eel*. The author of this second text might be Perry himself. He might also be William Combe, the author of comic plays or imaginary letters often written under various pseudonyms.

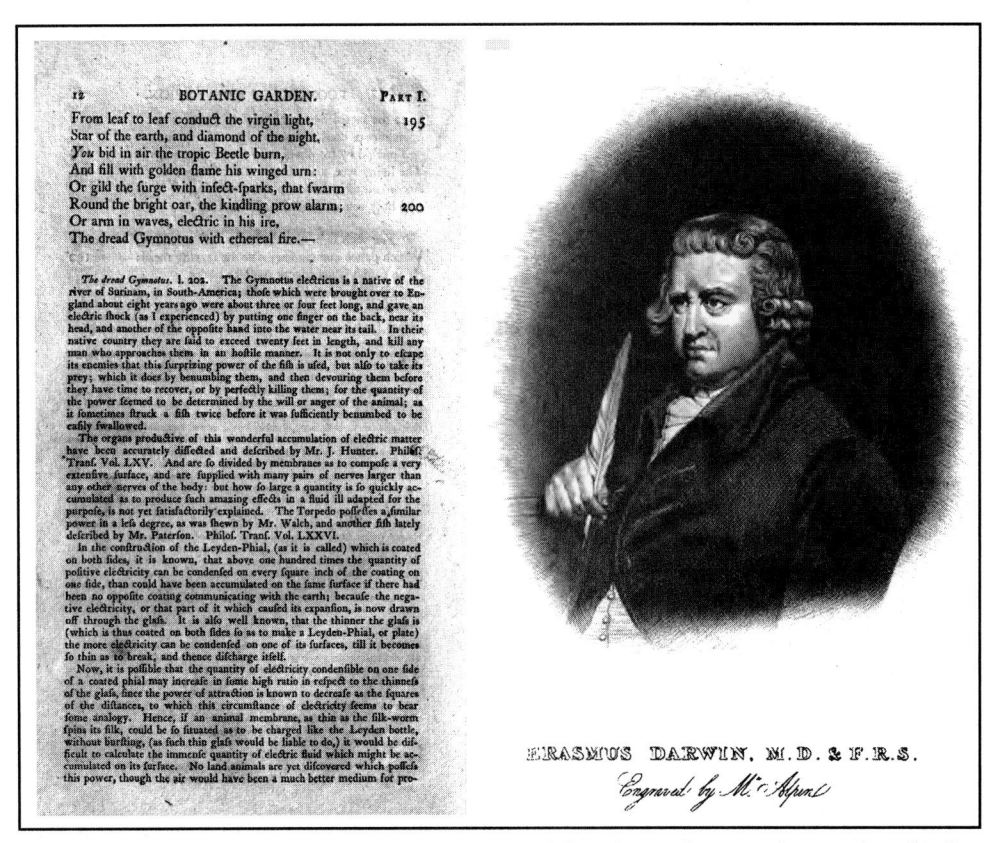

Figure 20.9. Erasmus Darwin (1731–1802), celebrated physician, scientist, philosopher, and poet, and a page from his *Botanical Garden* about "The dread *Gymnotus* with ethereal fire."

which their modesty will not easily recover." For readers not familiar with Latin, the four offset and italicized words in this sentence translate as "a snake in the grass."

A very different and notably higher form of poetry with sections about electric fishes followed Perry's, to the delight of the *literati* and other well-educated people at the end of the eighteenth century. This higher poetry was written by Erasmus Darwin, a physician and natural philosopher from England's Midlands, and one of the greatest minds of the Enlightenment. Darwin was born in 1731 and lived until 1802, and he was a friend and correspondent of Benjamin Franklin, and notably, the grandfather of Charles Darwin.

In 1791 Darwin completed *The Botanic Garden*, two massive poems with philosophical notes (Fig. 20.9). The first, called *Economy of Vegetation*, deals with the operations of the four elements of antiquity and contemporary scientific questions, especially matters relating to botany. The second, *The Loves of the Plants*, deals with the sexual classification system used by Linnaeus, on which Darwin had worked for years.

In Canto 1 of Darwin's *Economy of Vegetation* we encounter much about the element fire, with some lines and notes about animal luminescence, followed by these four lines (201–204) on the electric eel:

Or arm in waves, electric in his ire,
The dread Gymnotus with ethereal fire.—
Onward his course with waving tail he helms,
And mimic lightnings scare the watery realms. . . .

Darwin tells his readers that he had personally experienced the eel's shocks. "The Gymnotus electricus," he writes in a footnote to Line 202, "is a native of the river of Surinam in South America; those which were brought over to England about eight years ago were about three or four feet long, and gave an electric shock (as I experienced) by putting one finger on the back near its head, and another of the opposite hand into the water near the tail."

After making the mistake of repeating the myth that they can "exceed twenty feet in length," he further informs his readers:

> The organs productive of this wonderful accumulation of electric matter have been accurately dissected and described by Mr. J. Hunter. Philos. Trans. Vol. LXV. And are so divided by membranes as to compose a very extensive surface, and are supplied by many pairs of nerves larger than any other nerves of the body; but how so large a quantity is so quickly accumulated as to produce such amazing effects in a fluid ill adapted for the purpose is not yet satisfactorily explained. The Torpedo possesses a similar power in a less degree, as was shewn by Mr. Walch [Walsh]. . . .

Darwin also discusses the nature of electricity and its possible therapeutic applications. One part reads:

> You crowd in coated jars the denser fire,
> Pierce the thin glass, and fuse the blazing wire;
> Or dart the red flash though the circling band
> Of youths and timorous damsels, hand in hand.—
> Starts the quick Ether through the fibre-trains
> Of dancing arteries, and of tingling veins,
> Goads each fine nerve, with new sensation thrill'd,
> Bends the reluctant limbs with power unwill'd;
> Palsy's cold hands the fierce concussion own,

And Life clings trembling on her tottering throne.—
So from dark clouds the playful lightning springs,
Rives the firm oak, or prints the Fairy-rings.
(Lines 359–370)

The footnote to Palsy's cold hands is particularly informative. Darwin rightfully distinguishes between voluntary and involuntary movements, telling his readers that "Paralytic limbs are in general only incapable of being stimulated by the power of the will . . . and it commonly happens, when paralytic people yawn and stretch themselves (which is not a voluntary motion,) that the affected limb moves at the same time." He adds that "the temporary motion of a paralytic limb is likewise caused by passing the electric shock through it, which would seem to indicate some analogy between the electric fluid, and the nervous fluid . . . diffused along the nerves for the purposes of motion and sensation." He does not specify that the therapeutic shock must come from an electrical machine.

Darwin returns to electric fishes in a later work, *The Temple of Nature*, published in 1803. In Canto III, "Progress of the Mind," there are lines (22–26) about electric streams from "fierce Gymnotus" lurking beneath the waves, and torpedoes with "benumbing charm," and later on (111–112) on how "The tropic eel, electric in his ire, Alarms the waves with unextinguish'd fire." In the corresponding footnote to the first citation, we are even told that "The electric shocks given by the torpedo and by the gymnotus, are supposed to be similar to those of the Galvanic pile." This is a reference to Volta's newly invented battery, which we shall discuss in Chapter 22.

Despite its high cost, *The Botanic Garden* was a best-seller in its day, and *The Temple of Nature* also did very well for a book of its type. In his own way, Darwin dealt with the emerging concept of animal electricity, bringing new scientific discoveries and philosophical thoughts to highly literate and inquisitive readers, including other physicians and natural philosophers, not in Perry's pornographic way, but with many fact-filled verses and extremely informative footnotes.

LANGUAGE AND CULTURE

Originally, the Latin word "torpedo" was used specifically to describe certain round and flat rays, and then more generically to describe any electric fishes, before it reverted back to the rays again. The idea that some humans have a torpedo-like ability over others, however, can be traced back to the Greeks, who used the word *nárkē*. In translation, Plato has his character Meno state (he says in jest) that Socrates possesses a "power over others to be like the flat torpedo fish, who torpifies those who come near him and touch him, as you have now torpified me" (Chapter 3).

Torpedoes swam further into mainstream literature during the Elizabethan period, but now with more negative connotations (Whittaker, 1992, pp. 486–487). The term "torpedo" stands out particularly negatively in Christopher Marlowe's *Edward the Second*, where the queen is warned, "Fair queen forbear to angle for the fish . . . I mean that vile torpedo, Gaveston." This play was completed in 1590. The word also comes up in a less-than-flattering way in John Donne's *Eligie on Prince Henry*, finished in 1613, where he writes:

For Whom what Princes angles (when the tryed)
Mett a Torpedo, and were stupefied.

With the attention given to electric fishes in the 1700s, the word "torpedo" began to be used more often to describe people so incredibly boring that they numb others to sleep. "An Essay on Conversation," which appeared in the *Public Advertiser* on July 22, 1765, reflects this trend (Anon., 1765). The anonymous author writes: "Some, to know the Sublimity of their Knowledge, speak what neither others nor themselves can apprehend; on the Wings of furious Bombast they soar above the Clouds, and scale the blazing Battlements of Heaven: and some creep in Insignificance, and are so dull, that, by Infection, they doze the Faculties of their Hearers, and numb us like the Torpedo." Similarly, in response to a dull speech given at the Robinhood Society, there was an anonymous piece in the *General Evening Post* on January 31, 1771, which begins: "There are certain speakers who have the qualities of the Torpedo, who, by their drowsy periods and droning accents benumb the senses, and produce all the apparent effects of a lethargy" (Anon., 1771).

In response to a "generous" review of a comedy called the *Cholerick Man*, we find a comparable sentence under "Postscript. The Theatre," in the *St. James's Chronicle or British Evening Post* of December 27, 1774 (Anon., 1774a). It reads: "After the first Act, which really promises something, and is spirited and pleasing, the Author seems to have been struck with a Torpedo, for each succeeding Scene grows duller than another, and Improbability rises upon Improbability." A year later, as noted above, John Pringle was likened to a torpedo, possessing an uncanny ability to put audiences to sleep (J.P., 1774). And jumping ahead to 1796, we find, "*Helen* is another character which acted as a torpedo on the audience" (Anon., 1796).

As we noted earlier in this chapter, a torpedo was caught in the Thames River in 1774. Not then mentioned by us was that some of the London newspapers used the event in pithy editorials lampooning their leaders. One such example can be found in the *London Evening Post* of June 25, 1774: "The torpedo taken near Windsor on Wednesday morning last, was observed by some bargemen floating on the water early that morning, just under the wall of Richmond-gardens, as a Great Personage was taking his morning walk, which, in a great measure, *naturally* explains for that *stupor* with which he was seized about noon, in being *insensible* to the solicitations of so large a body of his people" (Anon., 1774b).

Clearly, the overall usage of the word "torpedo," with its various negative connotations in the common vocabulary, accelerated in the eighteenth century, paralleling what was happening in the scientific literature. In 1775, for instance, we find one unnamed person calling a certain lawmaker a "blockhead" in the *Middlesex Journal and Evening Advertiser*—to which he adds, he is "an absolute torpedo to an audience, and a better opiate than a poppy" (Anon., 1775).

The now-common analogy was also applied to political assemblies as a whole. To quote: "When I come into this House, it appears to me as if I came into an Assembly where a number of persons had been benumbed by a torpedo, or struck by a paralytick affection, which palsies all their limbs, cripples their energies and powers of action, and leaves nothing free and vigorous about them but their tongues" (Anon, 1797a). And at a later time: "The Resolutions of the House of Commons upon this subject [new taxes], were to that parish, like the touch of the Torpedo" (Anon., 1797b).

Criticizing the leadership had long attained the status of an art form in Great Britain, which brings us back to poetry, in this case a piece about the mental "sloth" and "corruption" common to Batavia, a part of London where men assemble to cramp the mind, resembling torpedoes with their icy venom. This short but pithy poem was published in the *Universal Spectator and Weekly Journal* on June 12, 1731 (and in the *Echo or Edinburgh Weekly Journal* on June 23):

As *Story* tells us, in the *Indian Seas*,
Of finny Race, the cold *Torpedo* plays;
Who to the Hook, by greedy Hunger brought,
The Fish and Fisher both at once are caught:
The Icy Venom from his Rod assails,
And soon thro' all his curdling Blood prevails:
Nature's remotest Springs now feel its Charms,
And sympathetic Stupefaction reigns.
Hands, Feet, and Tongue, forget their former Use,
He stand—a Subject fit for *Ovid*'s Muse
Tho' Fable this—Yet ah! too oft we find,
Sloth, like a true *Torpedo*, cramps the Mind,
Numbers submit to its inactive Sway,
A lazy Lethargy benumbs their Souls,
And all its Faculties, like *Rust* controuls.
So in *Batavia*, muddy Lakes are seen,
Whose sluggish Waters wear a loathsome Green
While in their Bosom dire Corruption's bred,
And from their *Streams* the soul infection's spread.
(Anon., 1731a,b)

Before the century was over, the word "torpedo" would appear even in obituaries. One can only wonder what John Cleland (d. 1789), the subject of one obituary, would have thought about this assessment: "As a writer he shewed himself best in novels, song-writing, and the lighter species of authorship; but when he touched politics, he touched it like a torpedo, he was cold, benumbing, and soporific" (Anon., 1789).

FISHES AND THE COMMON MAN

One did not have to be an experimental natural philosopher, a traveler to some distant land, or a learned physician to learn the latest about torpedoes or South American eels and the mounting case for fish electricity. All that was needed in the 1760s and 1770s was a London newspaper, a subscription to one of the most informative and popular magazines of the day, *The Gentleman's Magazine*, or a willingness to buy some rather lewd poetry by a man who was following what was happening at the Royal Society but was not particularly anxious to reveal his name on his otherwise revealing poems. Moreover, for those readers who wanted a more scholarly treatise for their shelves, Erasmus Darwin's books of poetry provided a wonderful education, with insightful footnote commentaries about the presented information.

It is impossible to guess how many people might actually have read these newspapers, *The Gentleman's Magazine*, James Perry's poetry, or Darwin's masterpieces in this era. But it cannot be denied that the concept of fish electricity was spreading from the halls of science to more public domains, a contention supported by how the word "torpedo" in particular had now taken on various meanings and a life of its own in the common vocabulary.

In the next chapters, we shall look at what Erasmus Darwin called the "analogy between the electric fluid, and the nervous fluid" in the second of his 1791 poems. We shall see that this was an emotionally charged subject late in the eighteenth century, when the case was beginning to be made for electricity being more than a property of a few specialized and rather bizarre fishes.

PART VI
From Fish to Nerve Physiology and Back

Chapter 21
Galvani's Animal Electricity

> ... and still we could never suppose that fortune were to be so friendly to us, such as to allow us to be perhaps the first in handling, as it were, the electricity concealed in nerves, in extracting it from nerves, and, in some way, in putting it under everyone's eyes.
>
> (Translated from Luigi Galvani, 1791)

The evidence that some fish can produce electricity, or at least a fluid that closely resembles electricity, helped lead to a new way of thinking about how the nerves might work in more than a few specialized creatures. It acted mainly by undermining through a *de facto* argument the existing objections against the electric theories of nerve and muscle function, and particularly those raised by Albrecht von Haller and by his followers. As we saw in Chapter 11, Haller pointed to what he considered the impossibility that an electrical fluid could play a specific role in animal physiology, because of the electrically conductive nature of bodily tissues. To some natural philosophers, the difficulty seemed even stronger with fish, since, in addition to being internally conductive, they live in a habitat that would rapidly disperse any electrical disequilibrium that could be generated in their bodies. For a man of the Enlightenment to conceive of the possibility of electric fish was just as absurd as to conceive that a Leyden jar could maintain its charge underwater.

Despite this logic, electric fish research, and particularly Walsh's spark experiment with the eel, showed that some fish might indeed work electrically. In the Newtonian *Zeitgeist* of the eighteenth century, experimental demonstrations were of greater value than any *a priori* objections unsupported by sound experimental evidence. After Walsh, therefore, it became logical to think that more ordinary animals might use electricity for physiological functions. As discussed in Chapter 11, nerve conduction was undoubtedly one of the physiological processes in which electricity seemed to be an appropriate agent. But even though electric fish research showed clearly that some animals could function electrically, it did not provide any clear-cut indications of just how electricity might be involved in such processes.

The idea present in many of the works dealing with animal electricity around the 1780s, and particularly in the influential texts of Italian physician Francesco Giuseppe Gardini and French physicist Abbé Pierre Bertholon, was that the electricity present in organisms (both animal and plants) derived largely from a universal electric fluid present throughout the physical world (somewhat like the element fire, one of the four basic elements of classical science: see Gardini, 1780; Bertholon, 1780, 1786). In Bertholon's words, "this fluid, so active and penetrating, cannot exist in the atmosphere which surrounds us, and in which we are immersed like fish in water, without it acting on us, and thus without having some influence on our bodies" (Bertholon, 1786, vol. I, p. XIV). Bodily processes, and particularly blood flow, these men believed, might even increase the internal electricity, mainly by friction, much like the operation of an electric machine. In both Gardini's and Bertholon's conceptions, it was because of its intrinsic power that the universal electrical fluid, largely absorbed through the skin pores, plays a role in the animal economy.

Both Gardini and Bertholon attempted to construct physiological and medical schemas with diseases dependent on either excesses or deficiencies of electricity. Bertholon's "evidence" for the involvement of electricity in nervous function was vague and inconsistent, especially when examined through modern eyes. He related it to how aristocratic women became particularly nervous on very dry days and during storms, when there is intense electricity in the atmosphere. He also pointed to how some men endowed with particularly strong electricity could generate enough light to illuminate their way in the night and could produce sparks during sexual intercourse, like sparking electrical eels. Indeed, most of the evidence invoked by Bertholon for the existence of "animal electricity" (a phrase attributed to him, albeit incorrectly) was simply an expression of a physical electricity, which could be excited in animal bodies as well as in inanimate physical objects.

Alessandro Volta recognized this flaw in thinking and pointed it out in a particularly clear way in an already mentioned private letter written in 1782 to a French lady, Mme. Lenoir de Nanteuil (Volta, 1918, pp. 14–25). According to Volta, the expression "animal electricity" was unsuited to those forms of electricity that could be produced "by rubbing the back of a cat, by currying a horse," or to the electricity "that has been observed arising spontaneously from the feathers of a living parrot." Volta would stress similar considerations 10 years later in his "First Memoir" on animal

Figure 21.1. Luigi Galvani (1737–1798), who supported the case for animal electricity with experiments reported in his *De viribus electricitatis in motu musculari, Commentarius* of 1791 (Ferrara, private collection).

electricity, which would be stimulated by the publication of Galvani's research (see Chapter 22; Volta, 1918, pp. 3–7).

ELECTRIC FISH AS GALVANI'S STARTING POINT

Luigi Galvani (Fig. 21.1) merits the lion's share of the credit for expanding the doctrine of animal electricity beyond the singular powers of a few fish, and for doing this in a way that would reshape the life sciences and markedly influence medicine. With Galvani, animal electricity would become the agent of one of the fundamental process of animal physiology, namely nervous conduction.

Galvani's work stands in deep contrast to the neuroelectric speculations put forward during the pre-Walsh era, and also to subsequent loose conceptions, such as those of Gardini and Bertholon, who cited electric fish, in addition to unsubstantiated anecdotal reports, as evidence for their electrical systems of medicine. Moreover, these earlier theories did not lead to any coherent or acceptable models of how electricity could perform its role in nerve conduction. Equally notably, they did not deal in effective ways with Haller's important criticisms of the neuro-electric theory (Chapter 11).

In contrast, Galvani's views were based on an extensive series of experiments conducted over the course of several years with great intelligence and advanced methodologies. His research led to a coherent and effective model of how electricity could appropriately function as the agent of nervous transmission, despite the conductive nature of bodily tissues, which was of such great concern to Haller.

As we shall see, Galvani's endeavor in Bologna was the outcome of a multiplicity of historical and personal factors. But without question it was stimulated by electric fish studies, and notably by Walsh's achievements with his torpedoes and electric eels. There is indeed a clear temporal relationship between the start of Galvani's investigations on neuromuscular function (around 1780) and the spread of information about Walsh's successful spark experiment. An extract of Le Roy's article dealing with this experiment appeared in an influential Italian scientific magazine in 1777, and we know that books mentioning Walsh's experiments were housed in Galvani's private library (see Bresadola, 1998). Further, the final hypothesis developed by Galvani to account for the involvement of electricity in neuromuscular function was based on a model (the "minute animated Leyden jar") that echoed the "animated phial" image presented by Walsh when explaining the power of the torpedo. Galvani's model, however, would be far more fully developed than the one presented by Walsh.

With this as prelude, let us now examine how Galvani's investigations led him to his revolutionary ideas about animal electricity, as expressed in his landmark publication of 1791. Let us see how animal spirits, the elusive messengers of the soul that carried sensory and motor signals in classical and even later physiology, would only now start to be banished from science and medicine by newer electrical ideas.

MISREPRESENTATIONS OF GALVANI

Galvani has been one of the most misrepresented scientists in both popular culture and in the history of science. Following a tradition started soon after his death, he has been considered as no more than an uncertain pioneer who by chance stumbled upon some unexpected contractions with dead frogs, which led him to suppose that a particular form of electricity must exist inside animal tissues. Continuing this line of thought, Galvani's theory would eventually be proven wrong, but it had, nevertheless, the merit of stimulating the interest of Alessandro Volta, the great physicist of Pavia. With his own research, Volta would be able to claim that the electricity responsible for the contractions is not intrinsic to nerves and muscles, but derived instead from the metals used by Galvani to connect nerves to muscles, in opposition with Galvani's idea of the neuromuscular preparation as a biological Leyden jar.

Such misrepresentations can be traced back to the first *éloge* of Galvani, which was written by Jean Louis Alibert (1801) at the start of the new century. Here we find that Galvani started his research because he observed some chance contractions in the frog legs he was preparing as a nourishing food for the fragile health of his beloved wife, Lucia Galeazzi! Emil du Bois-Reymond, about whom more will be written in Chapter 24, was one of the subsequent scientists who, despite some admiration for Galvani, most contributed to misrepresent his achievements. He was probably driven by the desire to present himself as the true discoverer of animal electricity (Du Bois-Reymond, 1848; see Piccolino & Bresadola, 2003).

Very few scholars, even in recent times, have critically addressed these time-honored views, and the stereotypes persist in compilations, including virtually all biographical "dictionaries" of scientists. In some cases, contemporary historians have even added new distortions to the story about the doctor of Bologna (e.g., Pera, 1986). Many historians, in fact, still trust in an uncritical way the stereotyped image of the purely physical origin of the electricity involved in Galvani's experiments with metals.

Regrettably, many of Galvani's texts are poorly known, even to historians of science. Most writers rely almost exclusively on his 1791 memoir, written in Latin and published in the *Commentarii* of the *Accademia delle Scienze* of Bologna. The most frequently cited English edition of this text was published in 1953 and contains many inaccuracies, starting with the title itself (Galvani, 1953a; Cohen, 1953). With his title *De viribus electricitatis in motu musculari, Commentarius* (Fig. 21.2), Galvani was not simply announcing a "Commentary on the Effects of Electricity on Muscular Motion," as the above mentioned English translation reads, but rather was implying from the outset that electricity (or "electric forces") might be involved in muscle motion. Interestingly, a less frequently cited English edition published in the same year bears the identical misleading title (Galvani, 1953b).

To revise this traditional but mistaken image of Galvani, it is necessary to go beyond *De viribus* and to consider all of his writings, including many manuscripts preserved in the *Istituto delle Scienze* of Bologna, most of which are still unpublished (some appeared in 1937, on the occasion of the bicentennial of Galvani's birth). These manuscripts show far more clearly the day-by-day progress of his experimental research, and they contain his physiological reflections and mention various significant attempts to prepare a text for publication prior to his *De viribus* of 1791.

It is also crucial to situate Galvani's endeavor within the particularly intense period of scientific development then taking place in Bologna, after a lengthy decline of its old and glorious university (Heilbron, 1991). Further, it is necessary to interpret some of Galvani's experiments in the light of modern electrophysiological notions, because only by doing so can we fully understand the very real scientific problems he encountered and the basis of his polemics with his countryman Volta (see Chapter 22).

GALVANI, THE *ISTITUTO*, AND THE *ZEITGEIST*

Although information on Galvani's life is rather sparse, it recently has been examined thoroughly and presented in a more accurate way (Bresadola, 1997, 1998; also see Piccolino & Bresadola, 2003). Galvani was born in Bologna in 1737 and spent almost his entire life in his native town, where he died in 1798. The historical tradition tends to depict him as a mild-tempered, modest, and pious individual who shied away from publicity and did not particularly strive to play an important public role in the Republic of Letters. But whereas there are elements of truth to this image, it certainly does not provide an exhaustive and faithful portrait of Galvani's human and scientific personality.

Despite his high religious and moral standards, Galvani was also an ambitious man in many ways. Although his social status was rather low for the standards of *ancien régime* society (he was the son of a jeweler), he managed to climb many steps in the social and cultural ladders of a town still dominated by aristocracy and high clergy. He became a professor at the University of Bologna, the oldest continuing university in the Western world, and a professor and full member of the *Istituto delle Scienze* (Fig. 21.3), a second prestigious cultural institution of the city. He also became a full member of the *Collegio Medico*, serving for some time as the *protomedico* (chief physician) of the college.

In addition to his undoubted intelligence and hard work habits, his advancement was helped when he married Lucia Galeazzi (Fig. 21.4). She was the daughter of Domenico Gusmano Galeazzi, one of the most authoritative members of Bologna's scientific milieu, being professor at both the university and the *Istituto,* and the leader of the most prestigious private school of anatomy in the city (Fig. 21.5). Nevertheless, as we shall see, it is in his scientific work that the strong and unyielding nature of Galvani's character emerges most clearly, challenging the idea of a mild and submissive man soon to be confronted by an aggressive adversary, Alessandro Volta (Piccolino & Bresadola, 2003; Bresadola, 2008a). Further, despite the modesty repeatedly expressed in his publications, Galvani clearly mastered effective rhetorical and communicative techniques to promote his experiments and insights, and, after 1791, he did not hesitate to make recourse to anonymous publications in order to counteract the objections of his adversaries (e.g., Galvani, 1794a,b).

Among the activities that helped shape Galvani's skills in experimental research after his graduation from the University of Bologna in 1759 is his long training period as

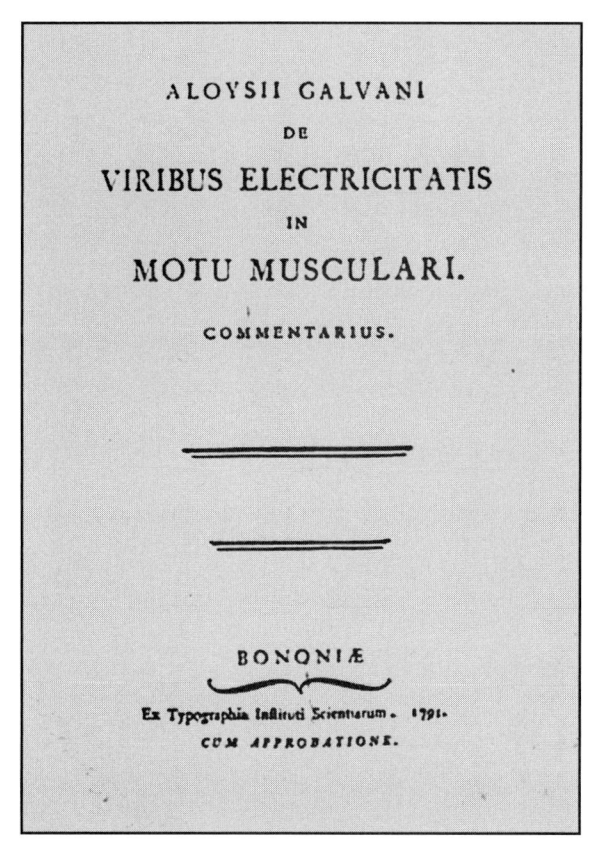

Figure 21.2. The title page of Galvani's *De viribus electricitatis in motu musculari*, published in 1791 in the eighth and last volume of the *Commentarii* of the *Istituto delle Science* of Bologna. This was the most important eighteenth-century work to promote the notion that nerve and muscle physiology involves an electric mechanism. The work was separately reprinted in 1792 in Modena, with a long introduction due in part to Galvani's nephew, Giovanni Aldini.

Figure 21.3. The emblem of the Bologna *Istituto*.

Figure 21.4. Lucia Galeazzi (1743–1790), Galvani's wife, who along with other members of the family participated actively in various experiments (Ferrara, private collection).

Figure 21.5. Domenico Gusmano Galeazzi (1686–1775), Galvani's father-in-law, was an important member of Bologna's medical establishment. A professor at both the university and the *Istituto*, he ran the most important private school of anatomy of the city.

assistant surgeon in the hospitals of the town. As noted by historian Marco Bresadola (1998), surgery in Bologna was then changing from a low-level practice, mainly assigned to barbers and blood-letters, to a more sophisticated, scientific discipline requiring a higher level of knowledge and more training. By assisting an important surgeon of the time, Giovanni Antonio Galli (one of the first to introduce forceps in obstetrics in Italy), Galvani, already a trained anatomist, acquired the skills needed for dealing with live bodies.

Studies of the urinary and auditory systems of birds stand among Galvani's research endeavors prior to the publication of *De viribus*. Although his chosen approach was basically anatomical, in both cases he showed a particular interest in the physiological implications of his results. This way of approaching his subjects was akin to the more dynamic approach to the study of living organisms that was prevailing in eighteenth-century science. Lazzaro Spallanzani, Galvani's correspondent and supporter in his research on animal electricity, was the main champion of this orientation in Italy (see Chapter 22). There is also evidence that Galvani conducted some physiological investigations before starting his research program on neuromuscular physiology (Bresadola, 1998; Piccolino & Bresadola, 2003).

Galvani's professional and scientific education was markedly influenced by his enrollment in the *Istituto delle Scienze*, a research and teaching facility founded in 1714 by Count Luigi Ferdinando Marsili. Marsili's objective was to bring modern science to Bologna, hoping to counteract the decline of the old university (Cavazza, 1990, 1996). As a geographer, diplomat, soldier, and author of scientific texts (e.g., Marsili, 1725), Marsili had wide-ranging international experiences, and as of a man of the Enlightenment, he was guided by truth and public utility. He also had energy and the resources of a rich and capable aristocrat and politician to promote his agenda.

The *Istituto delle Scienze* included a scientific and a literary academy, and was largely inspired by Baconian scientific ideals (Chapter 10). It was a suitable place where culture, art, and science could be performed in a public and organized way. It became one of the main scientific and research institutes of the eighteenth century (the word "institute" to designate an institution dedicated to experimental science was, in fact, introduced by Marsili for his creation). It was organized in "chambers" (*camere*), each devoted to a specific discipline, where the professors could take advantage of instruments and materials suitable for the practical and experimental teaching of their disciplines. In Galvani's time, some of the chambers were devoted to geography, navigation, astronomy, optics, physics, anatomy, natural history, and obstetrics.

Two fundamental features of the *Istituto* were its interdisciplinary attitude and emphasis on experimentation, with members periodically exchanging information on their experiments and on the new knowledge rapidly emerging in the lively culture of the Enlightenment. As we shall see, Galvani's investigative approach would be interdisciplinary. In particular, during one phase of his research on neuromuscular physiology, he temporarily abandoned his investigations on the effects of electricity and turned instead to chemical studies of nerves to obtain further information on the nature of the agent involved in nerve function.

Marcello Malpighi, Marsili's teacher and one of the greatest anatomists of his era, was the reference personality for members of the *Istituto*. Malpighi had been the main exponent in Italy of the renovation of life sciences along the principles championed by Galileo Galilei. Against the raw empiricism that was dominating important sectors of seventeenth-century medicine, Malpighi promoted *medicina rationalis* (rational medicine). By this he signified the need for medicine to be based not simply on empirical observations, but also on modern scientific investigations of the underlying physiological mechanisms, acquired by using advanced experimental logic and modern tools (see Malpighi, 1967; Cavazza, 1997a; Bresadola, 1998; Piccolino, 1999, 2005a). This attitude would play an important role in Galvani's decision to understand the role of electricity in animal physiology.

During the period of Galvani's scientific training, the *Istituto* was involved in an important debate on Haller's physiology, particularly his irritability theory (Haller, 1756–60; Cavazza, 1997b). In 1757, Giacinto Bartolomeo Fabri, a member of the *Istituto*, translated Haller's dissertation on irritability from Latin into Italian. He published it together with various memoirs *pro* and *contra* Haller's views, some of which were written by other members of Bologna's scientific community (Fabri, 1757).

One of them, Tomaso Laghi, strongly criticized Haller's experiments and conclusions and advocated a revised version of the classical theory of animal spirits, one in which electricity would play a central role and would be the effective cause of muscle contraction (Laghi, 1757a,b). Haller's main supporters in Bologna were Marc'Antonio Caldani (1757) and Felice Fontana (1757), who argued instead that contractions depend on muscle irritability. In their views, any external or internal stimulus (for instance, nerve action) causes contractions simply because it is able to stimulate irritable muscles, not because it is the effective cause of muscle shortening.

On the basis of experiments performed in the private laboratory of Laura Bassi and Giuseppe Veratti, two members of the *Istituto*, both Caldani and Fontana recognized that electricity is the strongest stimulant of contraction (see Cavazza, 1995, for more on Bassi's physical cabinet). Fontana pointed in a clear way to the disproportion that could exist between a weak stimulus and a strong contraction, making the analogy to a small spark and the explosion of a store of gunpowder. But at the same time, both men denied that electricity of internal origin could be the effective cause of muscle contraction.

The debates on Haller's theory that took place in Bologna starting in the late 1750s set the stage for Galvani's research, which actually began several decades later. Thus, even though the final hypothesis elaborated by the physician of Bologna put electricity at the center of neuromuscular physiology, the framework of Haller's views was of paramount importance in the development of crucial phases of his investigations.

In 1772, while Galvani held the prestigious chair of anatomy at the *Istituto*, he presented a memoir on *Irritabilità Halleriana*, which unfortunately has been lost. We know that he had started experimenting on irritability 2 years before writing this memoir, using mechanical and chemical stimuli similar to those used by Haller in many of his studies.

An intervening event that led Galvani to start his neuromuscular experiments in about 1780 was his involvement with the "public anatomical function" of that year (see Galvani, 1937, pp. 134–137). According a well-established tradition, the human dissection (public anatomy) took several days and was performed in front of members of the university and ladies and gentlemen of the town. At the conclusion of the dissection, Galvani alluded to the corpse in front of the audience and asked in a rhetorical way: "Where is it then, that very noble electrical fluid upon which motion, sensation, blood circulation, and life itself seems to depend?" He continued by saying that the answer to such a question could not be provided by simple anatomical investigations. In a way, he was implicitly and perhaps unconsciously alluding to what would become the scientific research program that he would pursue to the end of his life.

As mentioned, the other time-related event that contributed to Galvani's decision to start his research at this time was Walsh's successes with electric fish. In addition to the information that he could derive from Italian periodicals and academic correspondence, Galvani could have known about contemporary electric fish investigations from a treatise on the nervous system published in 1778 by Swiss physician Daniel de la Roche, which he also possessed in his personal library (Roche, 1778; see Bresadola, 1997). In the second volume of this work, there is a rather detailed account of these investigations, with references not only to Walsh, but also to Bancroft, Hunter, and Cavendish.

La Roche pointed to the particular importance of the electric eel spark experiment. To deal with Haller's objections against the neuro-electric theory, he assumed that the electric fluid must have a particular attraction for nervous tissue, so it does not spread from the nerves to surrounding organs. La Roche's opinion was that the electric fluid involved in the nervous system might have a general and pervasive role in human physiology. He thought that it might be empowered by blood movement through a friction-type process. These speculations were in line with similar arguments expressed by other scholars, including Gardini and Bertholon, and they also influenced Galvani in his subsequent investigations.

Finally, the waxing and waning of electrical medicine in the second half of the century also contributed to Galvani's decision to follow this new line of physiological research. Considered as a kind of panacea at its outset, the practice was still amply praised by some authors (for instance, Bertholon, Gardini, Mauduyt, and others), who presented impressive lists of diseases that could be treated with electrical therapies. Still, as we have seen, criticisms of some of these findings soon emerged from highly respected members of the scientific community, including Nollet and Franklin (Chapter 10).

The *Istituto* of Bologna had been involved in the discussion about electrical medicine from its start in the 1740s. In 1747, Giuseppe Veratti carried out, upon the request of his colleagues at the *Istituto*, a study on the efficacy of electric treatments aimed particularly at verifying the pretended extraordinary effectiveness of the electric methods of Giovanni Francesco Pivati, a Venetian lawyer (Pivati, 1747; see Chapter 10). A year later, Veratti (1748) published his findings. He had applied electricity to patients suffering from various types of disorders and largely supported Pivati's treatments, particularly for diseases of nerves and muscles, thinking these disorders stemmed from "a hardening and accumulation of internal humours." In his mind, the electrical treatments were effective because of the "extreme thinness and energy of the electric fluid" (Veratti, 1748, foreword).

Afterwards, this issue generated intense debate, particularly after Pivati had been obliged to recognize the uncertain efficacy of his electric therapies, following the stringent criticisms raised by Abbé Nollet, who visited Italy in 1749 and had been unable to confirm the reported effects (see Bertucci, 2007).

Thus, by 1780 there were many good reasons for Galvani to turn to physiological studies of electricity, one being that such studies fit well with the idea of "rational medicine," which was inspiring the medical work of other members of the *Istituto*. Galvani's project was, however, far more ambitious than that of Veratti. More in line with Malpighi's original elaboration of *medicina rationalis*, he sought not simply to examine the efficacy of electric treatments—he aimed at verifying and understanding the possible involvement of electricity in animal physiology. For these reasons, he chose to perform physiological experiments on animal preparations rather than doing electrical treatments with patients, although human disorders remained on his mind.

GALVANI'S EXPERIMENTS

Let us now reconstruct the intense research that eventually led Galvani to conceive of electricity in a condition of disequilibrium between the interior and the exterior parts of excitable fibers, as if the fibers were miniature Leyden jars. Here the study of Galvani's texts, and particularly his unpublished manuscripts, demolishes the legend of him being a bumbling pioneer, a representative of an old scientific approach at his leisure with anatomy and medicine, but unable to understand the physical laws underlying the phenomena he was studying. To the contrary, it shows the endeavors of an informed, intelligent investigator, capable of mastering both physical and anatomical approaches, a scientist extremely attentive to his animal preparations and the various arrangements of his instruments and experiments.

Galvani's experimental attitude blended the long-established Bologna tradition of anatomy (with Malpighi as its main reference) with a new, more dynamic approach to the study of living organisms. This approach involved the newest instruments in physics and reflected the theoretical interests that emerged with post-Newtonian science. From Galvani's writings and from the exceptional plates illustrating his 1791 *De viribus*, we can envision Galvani's room being much more like a *cabinet de physique* of a leading natural philosopher of the late-eighteenth century than the outdated dissection room of some Renaissance anatomist (Fig. 21.6).

His first extant experimental logs are dated November 6, 1780, but Galvani probably started his research on the effects of electricity on muscle contractions a bit earlier. The initial experiments concerned the effects of artificial electricity; that is, the frictional electricity produced by electrical machines and stored in capacitors, including the Leyden jar and the Franklin square capacitor. The square capacitor (used in his first experiment) was more commonly used in this period, probably because its shape also allowed it to serve as a convenient support for the frog preparation (his first reference to a Leyden jar is in a protocol dated December 6, 1780).

The frog was Galvani's most frequently used laboratory animal, and it was prepared so that nerve and muscle tissues

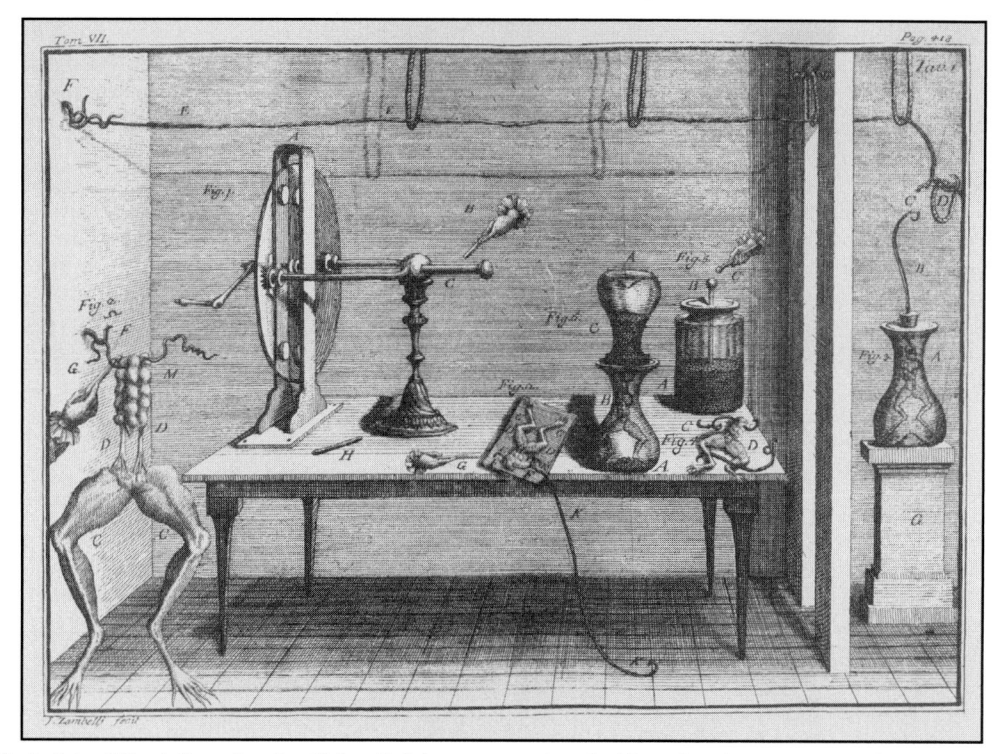

Figure 21.6. The first plate of *De viribus*, showing Galvani's laboratory equipped with various instruments, including a frictional machine for generating electricity and Leyden jars. Note also the prepared frog on the left. This image alludes to the observation made in January 1781, when the frog preparation jumped when a spark was generated by an electric machine situated some distance away. This was the crucial observation reported in the opening of *De viribus* (from Galvani, 1791).

could be easily accessed. Cutting the spinal cord precluded sensibility and voluntary muscular contractions. Galvani's "prepared frog," an experimental animal preparation partially derived from those extensively used by Swammerdam and Malpighi in their anatomical and physiological investigations, would remain popular in electrophysiological experiments into modern times.

The electrophore (the atypical electric generator invented by Volta in 1775 and mentioned for the first time by Galvani on February 7, 1781) was among the devices he used for producing and maintaining electric power to stimulate the frog preparations. Besides electric machines, capacitors, and electrophores, and in addition to his surgical instruments and especially contrived tools (*macchinette*, i.e., small machines), Galvani relied heavily on metal arcs (i.e., the tools normally employed to discharge electrical machines or capacitors) and metal wires to connect various parts of his animals to electrical sources.

Galvani's interests initially seemed limited to ascertaining the impairments induced by strong electric discharges on the neuromuscular system (i.e., how electricity can extinguish the muscle or nerve "force"). The experimental questions he was addressing, however, were more physiological and reflected the debate on the neuro-electric theory and the objections of the Hallerians, as will now become apparent.

On November 22, he compared the effects of the electrical stimulus on a frog preparation in which one crural nerve was ligated and the other set free. The procedure was clearly aimed at ascertaining the validity of Haller's findings with ligatures. As noted in Chapter 11, Haller had commented that "a ligature on the nerve takes away sense and motion, but cannot stop the motion of a torrent of electrical matter" (Haller, 1786/1966, p. 221). Three days later, Galvani started

verifying another important nerve property of the neuro-electric theory, the ability of nervous tissue to conduct electricity more or less freely. The results of these experiments convinced Galvani that electricity could flow across nervous tissue, but its passage might not be as easy and free as it is across metals or with other highly conductive bodies. This conclusion was in accord with assumptions made by some of the supporters of the neuro-electric theory (e.g., La Roche), who used it to account for why the electric fluid does not spread from the nerves to surrounding tissues.

Another significant result of Galvani's initial experiments was the demonstration that contractions could be evoked by extremely weak electric stimuli, such as those provided by a flat capacitor or a Leyden jar almost completely discharged (so as not to give any clear-cut electric signs, such as sparks, sounds, etc.). Galvani's main aim in starting these experiments was to verify the neuro-electric theory and to assess its implications (this appears from some annotations in his journal of experiments and is explicitly declared in the introduction to his then-unpublished memoir, *Sulla forza nervea* ["On the Nervous Force"], dated 1782; see Galvani, 1937). Nevertheless, he still seemed to consider artificial electricity simply as a way to excite nerves and muscles and to produce contractions, and thus only as an external agent of the phenomenon, in his initial experiments.

The possible involvement of an electric fluid *internal* to the animal body emerges more clearly with the research carried out at the beginning of 1781. It becomes dominant after a fundamental chance observation of January 26 of that year. As declared at the beginning of *De viribus*, this observation became the starting point of all further investigations, and Galvani emphasizes its fortuitous character, somewhat distorting the representation of the events that stem from his laboratory notes. A frog preparation contracts when somebody ("my wife or other," he writes in the experimental protocols) extracts the spark from an electrical machine that is not connected to the frog by any type of conductor (Galvani, 1937, p. 254).

In subsequent experiments, Galvani tried to determine the circumstances under which this phenomenon occurs, so as to realize the essential conditions for it. He noted that somebody must touch the nervous tissue with a conductive body (e.g., a metal, fingers, etc.) at the moment when a spark flies from a distant electric machine. No contraction occurs if nervous tissue is touched with an insulating body, such as glass or old bone, he annotated. Importantly, contractions are less easily produced if the conductive body is put in contact with the muscles rather than with the nerves. This observation was at odds with Haller's irritability doctrine, which stipulated that the underlying force responsible for a contraction is intrinsic to the muscles.

Over the following months, Galvani varied the conditions of his experiments in a remarkable number of ways. Most of the experiments made up at the beginning of 1783 appeared to be variations of his spark experiment, and were evidently carried out to ascertain the underlying physiological mechanism. Within this period, he showed his ability to develop new and more complex experimental arrangements, some based on unique tools that he designed (Bresadola, 2003; Piccolino & Bresadola, 2003).

In the course of Galvani's experiments, the prepared frog progressively became a key component of many complex arrangements, which involved electrical sources, metal wires, and sometimes the experimenter himself. These arrangements helped Galvani to identify the circuits followed by the electric fluid in producing the contractions.

Thus, the contractions produced by distant sparks directed Galvani's attention toward the frog as the source of a most subtle fluid, which can be "excited by pushing, by vibration, by the impulse of the spark" (p. 18). The electric nature of the fluid responsible for muscle contractions seemed to be contradicted, however, by the difficulties encountered when electrical stimulation was applied directly to the muscles. One of the predictions of neuro-electric theory was that the muscles should readily contract in response to direct electric stimulation, because the physiological agent of contraction is the electricity brought to muscles by nerves.

At the end of 1782, Galvani wrote a memoir on the nervous force in order to summarize the results of 2 years of experiments, with the aim of deriving a coherent picture from them. His term "nervous force" was the noncommittal expression he used to designate the nervous agent, as there was still uncertainty in his mind about its nature and role in muscle contraction. Galvani's choice of words reflected his difficulty in elaborating a comprehensive theory capable of accounting for the involvement of electricity in neuromuscular function at this time. It indicates, nonetheless, his belief that electricity acts mainly on the nerves rather than on the muscle. Thus, it shows how Galvani was already distancing himself in some ways from Haller's doctrine of irritability. Most of the memoir is devoted to describing the conditions in which artificial electricity is effective in producing muscle contractions, with Galvani not attempting to propose a mechanistic explanation of neuromuscular physiology.

Two points are stressed in a particularly clear way. The first is the necessity that the electrical stimulus is of sudden onset and offset, since contractions were not usually observed when the frog preparation was connected to a continuously operated machine, which would produce a constant and uniform flow of electricity. Sparking electricity is particularly effective, because its time characteristics suit the temporal requirements for nerve excitability.

The other important point brought up by Galvani in his 1782 memoir concerns the relationship between stimulus intensity and the strength of the contractile response. Although contractions grow stronger with more intense electrical stimuli, there is no simple proportionality. Contractions appear only when the stimulus intensity exceeds a certain minimal value. A further increase in its strength results in stronger contractions, but only within a given range; more intense stimuli do not necessarily result in stronger effects.

These properties further pointed to the animal preparation as the site of the force responsible for the contractile responses. In other words, the electrical stimulus seemed not to be the effective agent of contractions, but only the stimulus capable of putting an internal force responsible for them into motion. As we have seen, Haller's irritability doctrine focused on the relative lack of a direct relationship between the intensity of the stimulus and the strength of muscular response. But in contrast to Haller, Galvani was now inclined to situate the internal force aroused by the external electrical agency in the nerves rather than in the muscles, showing he could think independently of prevailing intellectual dogma.

Lastly, Galvani stressed the possible recovery of excitability in preparations fatigued by repetitive electrical stimulation. This was seen if the preparation were left at rest for a while. This observation further suggested that the contraction is mainly the expression of an internal agent—a force that might become exhausted after prolonged stimulation.

EXPERIMENTS WITH METALS

In Galvani's *De viribus*, the results are organized in three parts devoted respectively to experiments on artificial, atmospheric, and animal electricity. The impression one gets is of a logical and temporal sequence of experiments carried out at rather defined and regular paces. The experimental logs, however, suggest a different view. Galvani carried out the experiments with artificial electricity between November 1780 and February 1783, and he moved on to his investigations of the effects of atmospheric electricity only in April 1786, some 3 years later. In September of that year, he also began the last phase of his investigations, largely based on the use of metals. The passage from the second to the third phase of his program is relatively poorly defined in the experimental protocols, as compared to what appeared his 1791 memoir. This is only one of the occasions in which the picture of the events recorded in the laboratory experimental protocols contrasts with that in his published writings (see Piccolino & Bresadola, 2003).

In the period from 1783 to 1786, Galvani conducted a series of physicochemical investigations on animal bodies, along the lines of works on the "airs," which were attracting the attention of many eminent scientists (e.g., Priestley and

Lavoisier), and which would eventually culminate in a revolution in chemistry (see Seligardi, 1997, 2002). These experiments were probably aimed at determining whether a principle other than electricity might underlie neuromuscular function (see Bresadola, 1998; Piccolino & Bresadola, 2003).

When Galvani returned to his electrophysiological studies in 1786, he profited from some of the results obtained during this physicochemical period. In particular, he found that nerve tissue could produce a great quantity of "inflammable air" (i.e., hydrogen). He deduced from this finding that the nerves probably contain an abundant amount of oily matter. This revelation would eventually justify his model of the nerve, which he viewed as made up of a central conductive core surrounded by electrically insulating matter (not just a basic assumption of the minute Leyden jar hypothesis of neuromuscular physiology, but a fundamental tenet of modern neuroscience).

The experiments on the actions of atmospheric electricity described in the second part of *De viribus* were important because they proved that effects similar to those of artificial electricity could be produced from a natural source (i.e., electricity associated with thunder and lightning). The illustration of these experiments (Fig. 21.7), with the frog preparation on a table in Galvani's home terrace with long wires pointing toward the sky, is famous and has even been an inspiration for various cinematographic versions of *Frankenstein*. With these experiments, Galvani proved that the natural electricity from storms could produce muscle contractions, and that these contractions follow the same laws as those induced by artificial electricity.

The experiments described in the third part of the *De viribus* begin with a chance observation made in September 1786. A frog preparation with a metal hook inserted in its spinal cord was suspended on the iron fence of the balcony on a day that he described in *De viribus* as clear and calm (though much less so in his unpublished manuscripts) (Fig. 21.8). The purpose was to ascertain if the weak atmospheric electricity on a non-stormy day could also stimulate contractions. This was in line with the contemporary interest in small degrees of electricity.

What followed was frequently cited in physics textbooks from the nineteenth century. The suspended frog legs failed to move for a long while. Eventually Galvani (or possibly his nephew Camillo, according to the protocols) started manipulating the preparations, which is when something unexpected occurred. Contractions appeared when the metal hooks touched the iron bars of the railing, with no relation whatsoever to any atmospheric events.

To exclude any intervention of atmospheric electricity, the experiments were repeated within in a closed room, with the same outcome. What was needed for obtaining contractions was simply to establish contact between muscle and nerve tissue via a metallic conductor (normally a metallic arc) (Fig. 21.9). Nothing happened when using an insulating body or if the metallic contact were intercepted by the interposition of a non-conductive material.

The different efficacies of various metals seemed to correlate with their conductive power. Water and electrically conductive liquids could also be used, although they were less efficient than metals. An effective circuit could be formed by a chain of people connected together, with the two at the end touching the nerve and muscle of the animal with a metallic body. Some of these experiments bear an obvious visual resemblance to those Walsh performed in La Rochelle (Fig. 21.10). As Galvani wrote in *De viribus*, these experiments led him to suspect the presence, between

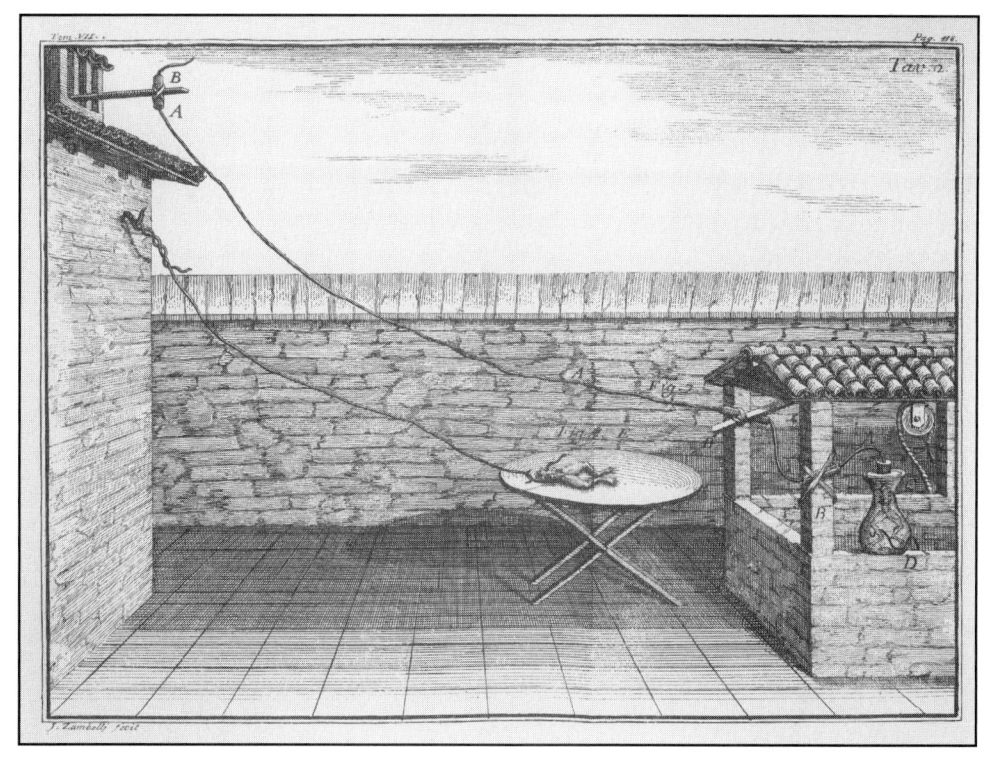

Figure 21.7. The second plate of *De viribus*. Galvani's experiments with frogs, especially those with long wires to electrical machinery (in this case, in the house), provided some of the *Zeitgeist* that inspired Mary Shelley to compose her saga *Frankenstein*, which she completed in Switzerland in 1817 (from Galvani, 1791).

nerves and muscles, "of a flow of an extremely tenuous nervous fluid... similar to the electric circuit which develops in a Leyden jar" (Galvani, 1791, p. 378) (Fig. 21.11).

As in the first phase of his investigations on artificial electricity, Galvani performed a great number of experiments with metals, varying their designs with great imagination. Some of these experiments were described in great detail, as if to bring his readers into his "room of experiments." This is the case with a "jumping" frog preparation—the contractions of the frog's leg break the circuit with the metal contact, which is re-established when the leg relaxes, resulting in the repetitive movements.

This way of presenting results corresponds with a communication strategy aimed at accentuating the factual and veridical character of the narrated events. Within the framework of this strategy (corresponding to the so-called virtual testimony already noted in Walsh's publication), we must also consider Galvani's frequent recourse into long sentences. His writing style, with its lively expressions, tends to accentuate the immediateness and unexpectedness of what was about to happen. The lack of historical criticism and also poor familiarity with Latin (the language of *De viribus*) had led even highly respected historians to misjudge Galvani's language and style (e.g., Cohen, 1953). Linguists familiar with the scientific literature of the era have since

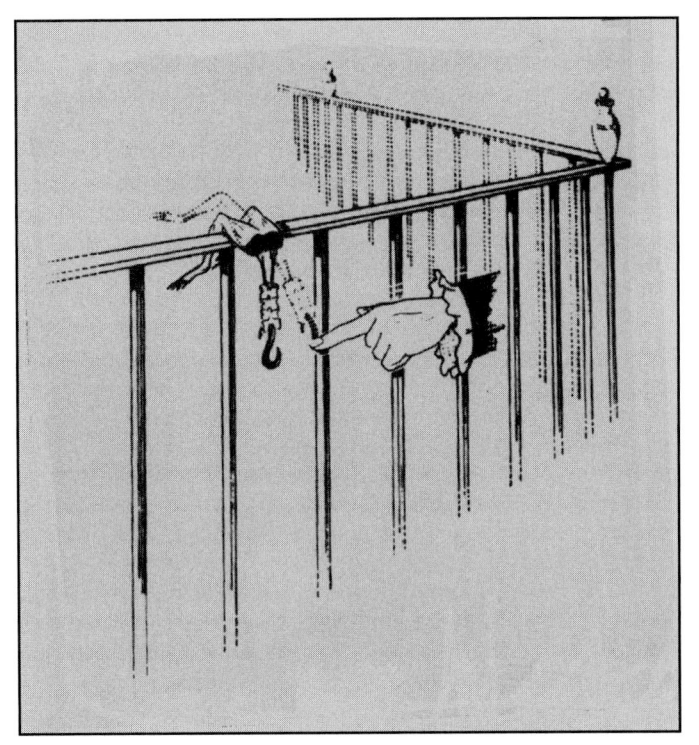

Figure 21.8. Galvani's frog legs on hooks on a metal railing, another crucial observation in his path to animal electricity (from Sirol, 1939).

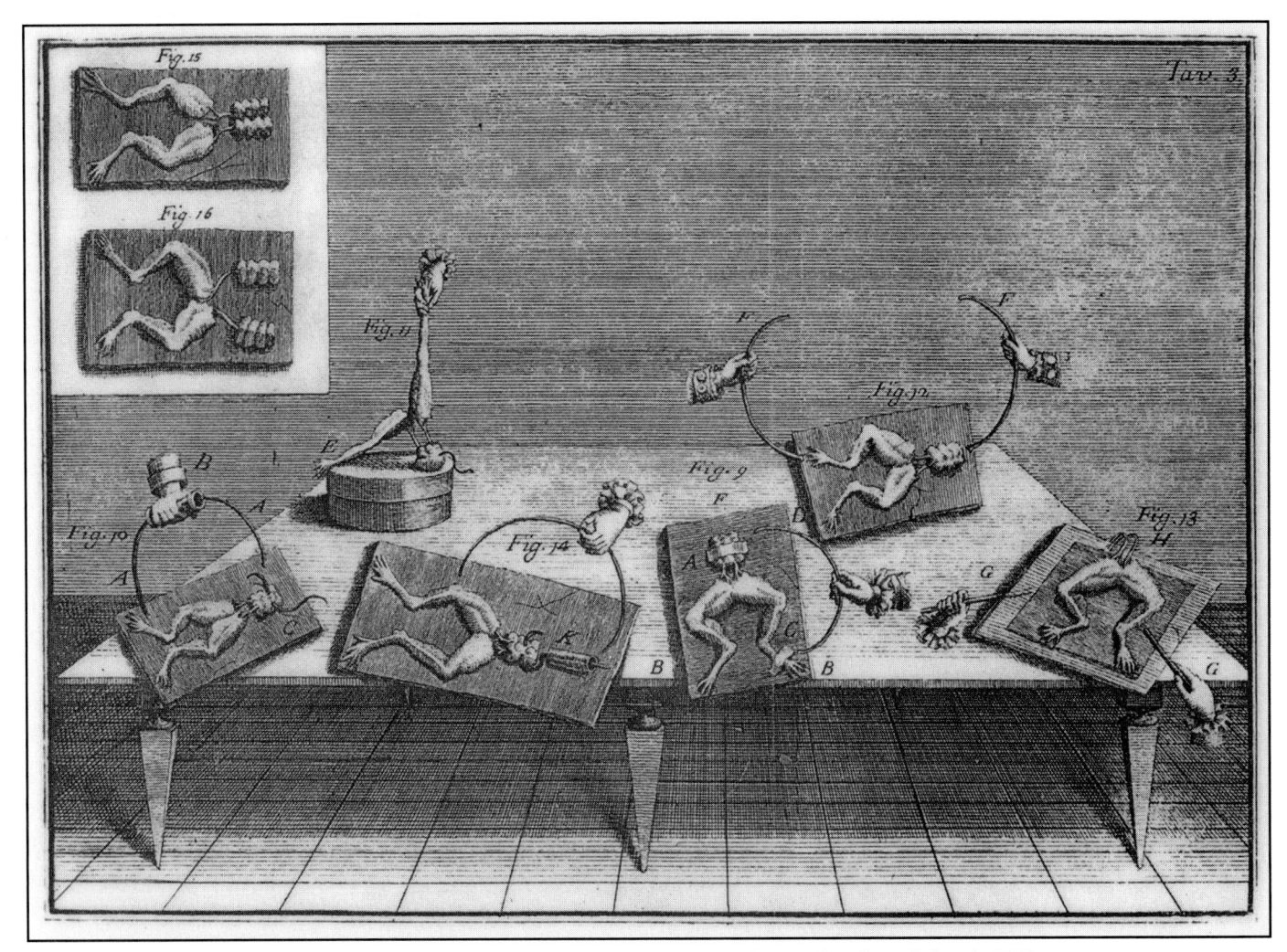

Figure 21.9. Galvani's experiments with metal arcs, as depicted in the third plate of *De viribus* (from Galvani, 1791).

Galvani's Animal Electricity

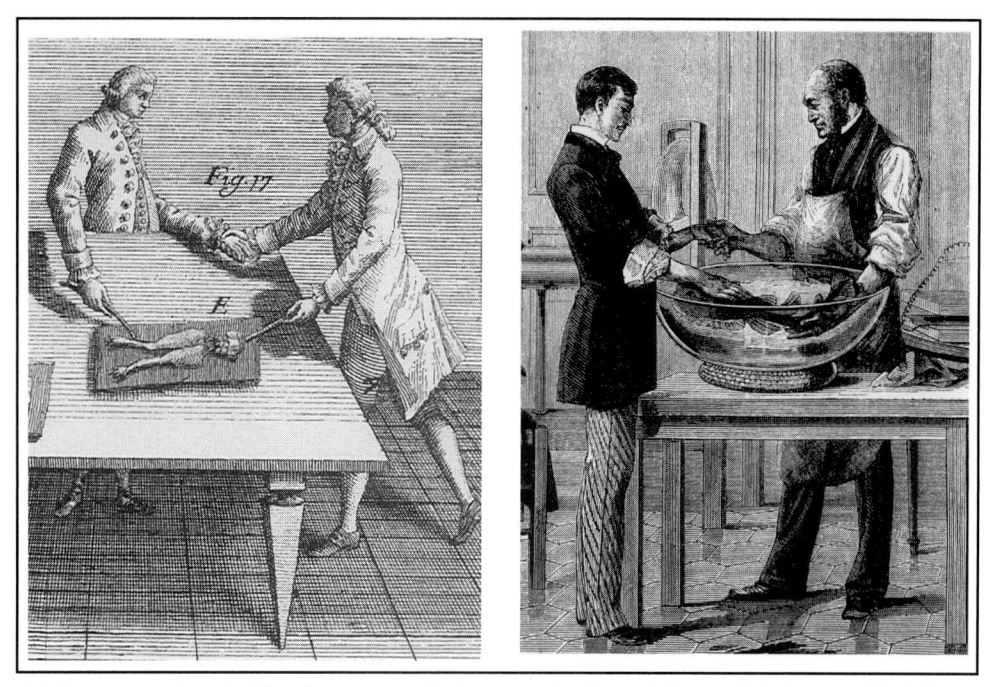

Figure 21.10. Galvani doing an experiment with a frog preparation in a way that is reminiscent of the image of Walsh's experiment, in which he had two people holding hands to complete a circuit with the torpedo (see the image in Chapter 15; from Galvani, 1791).

corrected these misguided negative opinions (e.g., Altieri-Biagi, 1998).

Compared to the period of his investigations with artificial electricity, Galvani started from safer grounds in his attempts to demonstrate the electric nature of the neuromuscular fluid in his experiments with metals. There appeared to be no evident sources of electricity external to the preparation. The principle responsible for the contraction was thus very likely internal to the animal. Moreover, since this principle is capable of circulating through various material bodies following the same laws as electricity, it was logical for him to consider it to be electrical in nature.

Before conceiving that the electricity is indeed internal to the animal, Galvani seriously considered the possibility that it might instead originate from the metals used to connect the nerve and muscle tissues. Yet, on the basis of a series of

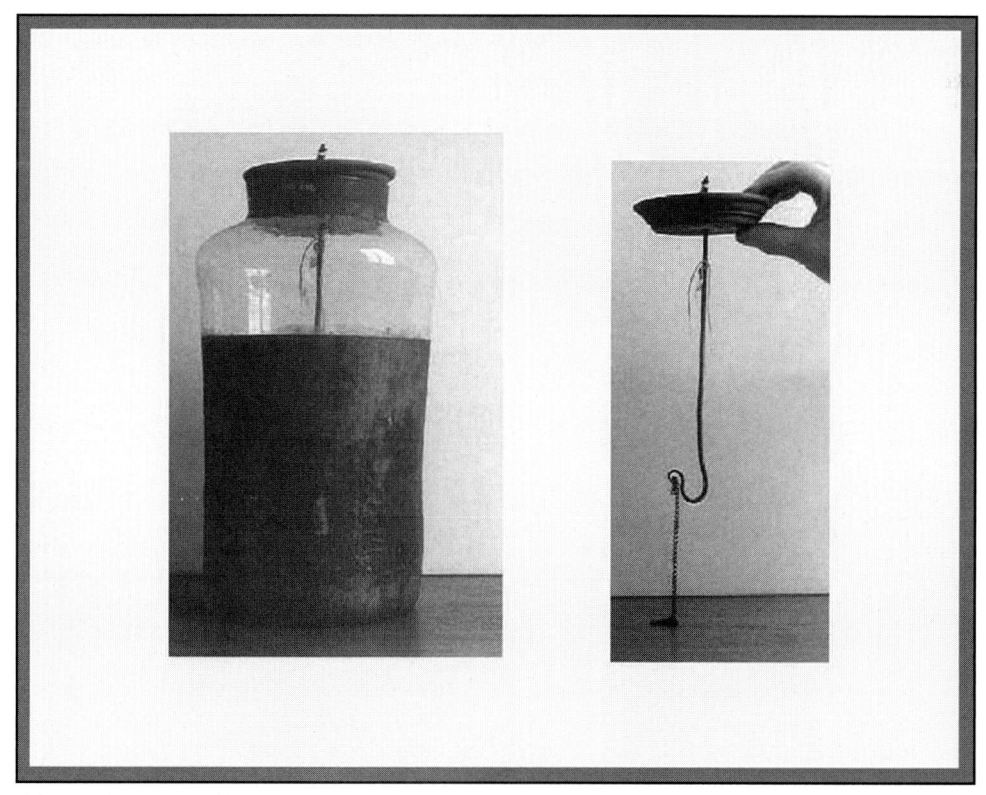

Figure 21.11. One of the Leyden jars used by Galvani in his experiments.

experiments, and drawing on the laws of physics, he excluded such a possibility.

We shall return to Galvani's experiments with metals and to this important point in the next chapter. For now, we shall concentrate on the logical and experimental itinerary that led Galvani to his conclusive model, starting from the moment the electric nature of the neuromuscular fluid seemed to be safely established in his mind. As we have been doing, we shall do this primarily by analyzing the various unpublished texts that he wrote during this period.

THE ELECTROPHORE AND TOURMALINE

From the moment he became convinced that the electricity responsible for muscle contraction is intrinsic to the animal organism, Galvani entered into an extremely exciting phase in his research program. The contractions obtained through a metal contact between the nerve and muscle led him to suppose that there must be a flow of electricity from nerve to muscle, in a way similar to how electricity circulates between the external and internal surfaces of a Leyden jar via a metallic conductor (Galvani, 1937, p. 166).

The Leyden jar as a mental image for the hypothetical electric circuit between nerve and muscle, alluded to in the *De viribus*, first appeared in an unpublished memoir that Galvani wrote at the end of October 1786 (i.e., a few months after his first experiments with metals; Galvani, 1937, pp. 162–193). In the Leyden jar, as Galvani noted, the electricity flows because of the presence of positive and negative electricity, situated in the two metallic plates (or armatures) of the jar (the inner and outer ones). The problem was to identify the site of "this double and opposite electricity, i.e., positive, as it is said, and negative," within the animal tissue.

After a series of experiments, Galvani concluded that "no doubt can exist that, out of the said two forms of electricity, one is situated in the muscle and the other in the nerve" (p. 176). But despite the important evidence that accumulated for this conclusion, Galvani decided not to publish his memoir at this time. He also decided not to publish another memoir, dated August 16, 1787. It was only in 1791, 11 years after beginning his studies, that he publicly announced his discovery of animal electricity.

So why did this happen? A possible answer to this question can be found by following the itinerary that led Galvani to his model of the neuromuscular system as a minute Leyden jar, beginning with his initial assumption about positive and negative electricity being situated in muscle and nerve.

There is an important difference between the model presented in *De viribus* and the ideas presented in his earlier, inconclusive attempts to publish. In the final memoir, and only in it, Galvani provides a model seemingly capable of accounting in a rational way for the problem that he was eagerly investigating for so many years: the mechanism whereby electricity is involved in neuromuscular function. It appears evident that the identification of the localization of the two forms of electricity in nerve and muscle, despite the experimental evidence supporting it, did not provide a satisfactory explanation for neuromuscular physiology. That is, it did not lead to a mechanism by which electricity would flow to produce contractions under physiological conditions. It was difficult, moreover, to conceive of how an electrical disequilibrium could exist between nerve and muscle, given the conductive nature of body fluids. Indeed, it appeared physically impossible for an electric difference to exist between two different parts of a conductive body.

Galvani invoked this argument (central to the objections the Hallerians raised against the neuro-electric theory) in his 1786 memoir. With it, he dismissed the possibility that positive and negative forms of electricity could be located inside the metal of the arc used to connect nerve and muscle. He was, however, aware of a possible exception to this rule. As he stated, the presence inside of a conductive body "of a double polarity, one positive and the other negative, this is a fact that physicists admit for tourmaline." Yet, he noticed, this does not happen for metals, and thus he concluded that double electricity could not be situated inside the matter of the metallic arc.

With the localization of electricity inside the animal body having been being firmly established in his mind, Galvani now considered other possibilities (as he also narrates in *De viribus*) as to the specific localization of the positive and negative electricity. In particular, he alludes to the possibility that both forms of electricity might exist within muscle tissue. This appeared likely, since "there is in muscles a big quantity of substance, which for its nature may be apt to develop and hold electricity, despite the presence inside it of a conductive matter." And he continues by saying, "this is not unlike what we saw happening in electrophores that are made of analogous substances. If that were to happen, it would perhaps be justified to call muscles *animal electrophores*" (Galvani, 1937, p. 169; italics ours).

This passage is interesting because it alludes to his first physical model of the underlying neuromuscular physiology. An electrophore is made of disks of different substances, some insulating and others conductive. Thus, it could provide a visual suggestion for a possible biological model of electricity, in view of the striated (and thus apparently heterogeneous) nature of the muscles. Nevertheless, Galvani did not elaborate on this possibility in his 1786 memoir. Instead, he concluded that the two forms of electricity (i.e., positive and negative) should still be localized in the muscles and nerves, respectively. (The generation of electricity in an electrophore actually depends on a complex and coordinated series of mechanical manipulations, the occurrence of which can be difficult to envision in muscle tissue.)

In his other unpublished memoir dated 1787, Galvani reflected further on the problem of the possible localization of the intrinsic electricity (Galvani, 1937, pp. 190–212). This is of particular interest because it offers an important cue about the itinerary culminating in his 1791 model. The argument now centers on the analysis of an electrical tool already considered *en passant* in the text of 1786, namely tourmaline. This stone interested Galvani for several reasons. It is able to produce unequivocal signs of double electricity upon heating; however, unlike most other electric devices (and similar to the prepared frogs), it does not produce sparks when touched. For Galvani there could be other important analogies between the neuromuscular complex and tourmaline, as shown in this passage:

> Our electricity has much in common with that of tourmaline, for what concerns its localization, distribution, and characteristics of parts. In this stone we observe indeed a double matter, transparent and reddish being the first

one, opaque and colorless being the other; this second one is arranged in stripes; the other is situated laterally with respect to these stripes. Nobody ignores that nerves are interspersed between layers of muscle fibers, and they appear transparent when devoid of blood, while nerves are opaque. In tourmaline, the poles of the double electricity appear to be situated along the same opaque line; similarly in the muscles they appear to be situated along in the same direction as in the nerves. The double electricity of tourmaline is not just situated in the entire stone, it is in every fragment. Similarly, in muscles, the admitted double electricity does not belong only to the entire muscle body, but to every part of it. (Galvani, 1937, p. 194)

Galvani now considered tourmaline as a physical analogy to the neuromuscular system. It could serve as a model, both operatively and structurally, one capable of accounting for the generation of electricity inside living organisms. Franz Aepinus had recently made a study of this stone and made important analogies between its power and magnetism. As in the case of a magnetic body, the attracting properties and the capability of generating a double pole do not reside in its external aspect, nor in the way of cutting it, but in its internal structure and in the very constitution of the stone itself (Aepinus, 1756, 1762). In accord with this observation, Galvani had noticed that animal electricity shows its effects both in the entire muscle and in every part of it just separated from the animal (Galvani, 1937, p. 195).

As in the case of the muscle as an animal electrophore, Galvani was particularly sensitive to visual suggestions, and he now saw a visual similarity between the muscles, with their striated and heterogeneous parts, and tourmaline. He suggested that electricity might arise from the contact between a single muscle fiber and a single nerve fiber within the muscle mass. In this way, he retained his previous idea about muscle and nerve as the respective sites of the double electricity, while shifting his attention from the macroscopic to the microscopic level.

GALVANI'S LEYDEN JAR MODEL

Notwithstanding its attractiveness, Galvani eventually discarded the tourmaline analogy. Even though it could provide insights about a possible mechanism of animal electricity generation, it did not allow him to conceive how electricity could be involved in the processes of nerve conduction and muscle contraction in a physically sound way. Further, he had noticed an important property of animal electricity that pointed toward a different physical instrument as a model of neuromuscular function. This was the Leyden jar.

Specifically, he had discovered that contractions are more vigorous (and could be excited more easily) if muscle and nerve tissues are wrapped with a thin metallic sheet (silver, brass, gold, orichalc, and particularly tinfoil). Galvani described this power of metallic sheets in both his 1786 and 1787 memoirs; he mentioned a series of experiments in which the sheets were wrapped in various ways around muscles, spinal cord, isolated nerves, and even exposed brain.

There is an important linguistic difference between these two memoirs. In the first one, the metallic sheets are presented exclusively as laminas or foils (*lamine* or *fogli*), whereas in the second memoir a different phrase appears from the outset in relation to these experiments—"metallic armatures" (*armature metalliche*) (Piccolino, 2008). In the electric terminology of the epoch, "armature" was the term commonly used to designate the thin laminas coating the internal and external glass surfaces of a Leyden jar. They were conceived as the sites in which positive and negative charges accumulated, due to the effect of the glass insulator.

Galvani's frequent use of the term "armature," together with the verb "to arm" (*armare*) in his 1787 memoir, strongly suggest that his attention was moving to the Leyden jar as a plausible electrical model of neuromuscular function in the 1786 to 1787 period. The word "armature" was also used to designate the metallic laminas coating the surfaces of the Franklin square capacitor (Chapter 10). Nevertheless, Galvani rarely mentioned the square capacitor in his published and unpublished memoirs, even though it was frequently used in his experiments. Very likely, this was because its simple shape did not evoke the same strong visual images of a possible model for the involvement of electricity in neuromuscular function.

The Leyden jar represented a fundamental step in Galvani's path toward his final model of the neuromuscular system (Fig. 21.12). In addition to its operative characteristics, it had a strong visual attractiveness, as Galvani recognized in an explicit way: the frog leg, with the nerves emerging from the muscle tissue, bore a strong visual resemblance to the Leyden jar, with its metallic conductor protruding from the jar's mouth. In the Leyden jar, the discharge is normally obtained by establishing a contact between its outer armature and its conductor (i.e., the metallic wire connected to the inner armature); however, the double electricity does not accumulate between the outer armature and the conductor, but between the external and internal armatures.

If the neuromuscular complex resembles a Leyden jar from an operative point of view (as suggested by the effect of armatures), then electricity should accumulate in its entirety (i.e., both in its positive and negative forms) in the muscle, rather than between the muscle and the nerve (as Galvani had initially assumed). This elaboration is explicitly expressed in the fourth part of the *De viribus* (Galvani, 1791, p. 395). The problem he now faced was how the two forms of electricity might be situated inside the mass of the muscle without violating the laws of physics. Could insulating matter be present inside the muscles?

Galvani had already considered these issues in his 1786 memoir, when he wrote about the presence of "a big quantity of substance," which because of its "rubbery" nature might be able to generate and hold electricity (despite the prevalence of conductive matter inside the muscle mass). He had at that time suggested the electrophore as a possible model of the neuromuscular system. As we have seen, however, after 1786 Galvani had found tourmaline to be a better model of how muscle tissue, with its striated and fibrous aspects, might store electricity in its interior. And now he focused on the Leyden jar.

There were three important logical steps for Galvani to move from the tourmaline idea to the final minute animal Leyden jar model. One was to envision where an insulating substance could exist within a muscle at a microscopic level. As Galvani speculated in *De viribus*, a likely possibility is

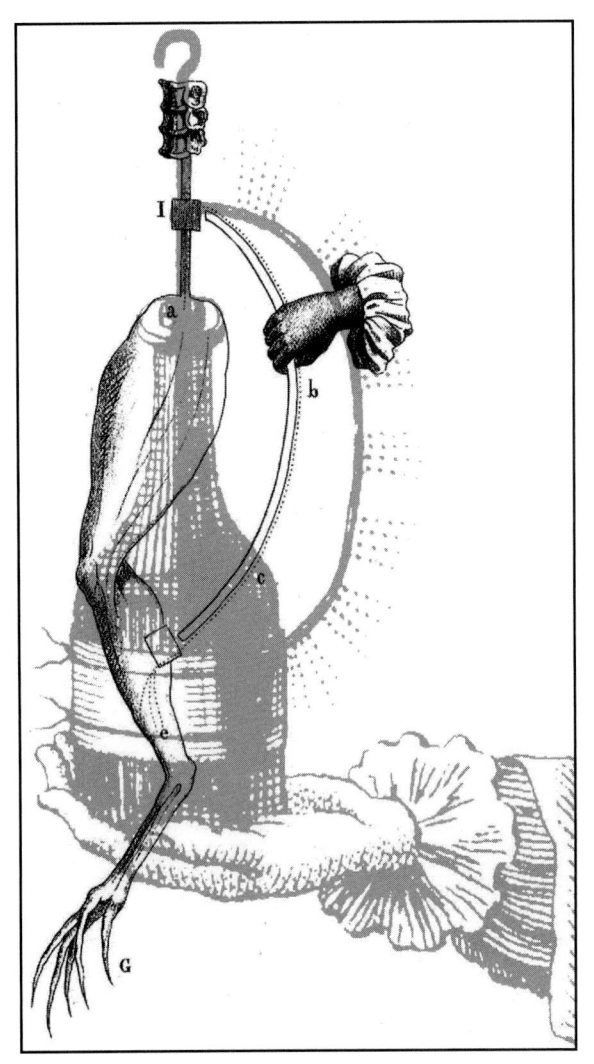

Figure 21.12. Galvani's view of the visual and functional similarity of the frog preparation with a Leyden jar (image kindly provided by Nicholas J. Wade).

that this substance might be situated at the surface of separation of the interior and the exterior of every muscle fiber:

> It is not far from the truth to admit that the fiber itself has two surfaces, opposite one another; and this stems from the consideration of the cavity that not a few people admit in it; or it can be because of the diversity of the substances, which the fiber is composed of, as we have said; a diversity which necessarily implies the presence of various small cavities, and thus surfaces. (Galvani, 1791, p. 196)

With this bold conjecture, Galvani placed the electric disequilibrium between the interior and the exterior surface of an excitable fiber, adhering to the laws of physics. This manipulation allowed him to face the recurrent objection raised by the Hallerians against the neuro-electric theory.

Galvani put forward his conjecture in an epoch in which there was no cell theory and the concept of fiber was the only available microscopic approximation to the elementary constitution of living tissues. It would recur afterwards, although in more modern ways—for example, with Bernstein's membrane theory of bioelectric potential at the beginning of the twentieth century (Bernstein, 1902; see Chapters 24 and 25).

Having situated the two forms of electricity at the two faces of the surface of the muscle fiber (and having attributed an insulating character to this surface), Galvani was now able to take his second important step. He assumed that the nerve fiber penetrates inside the muscle fiber, much as the conductor of a Leyden jar penetrates inside the bottle, in order to allow for a possible outflow of the internal electricity. Compared to the tourmaline stone model, this seemed to be just a small rearrangement of the relationship between the nerve and muscle fiber.

Lastly, he addresses how electricity might be delimited in its flow from the interior of the muscle to the nerve fiber, notwithstanding the conductive nature of the matter surrounding the nerves. He refers to his previous physicochemical studies, which showed a particular richness of oily matter present in the nervous tissue. He assumes that this oily matter forms an insulating sheet around the central conductive core of the nerve fiber. With this final conjecture, Galvani is able to circumvent another fundamental objection of the adversaries of the neuro-electric theory. He enunciates this explicitly in *De viribus* in the form of an unsolvable dilemma:

> As a matter of fact, either nerves are of an idioelectric [i.e., insulating] nature, as many admit, and they could not then behave as conductors; or they are conductors, and were this the case, how could an electric fluid be contained inside, one which would not spread and diffuse to nearby parts, with a negative consequence on muscle contractions? (Galvani, 1791, pp. 398–399)

The dilemma is solved in a clear way in the next passage:

> But this difficulty can be easily faced by supposing that nerves are hollow in their internal part, or at least made up of matter fit for the passage of electric fluid, and exteriorly [made up] of an oily substance or of another matter capable of hindering the passage and the dispersion of the electric fluid, which flows inside them. (Galvani, 1791, p. 399)

A muscle fiber delimited by an insulating substance that separates the two forms of electricity at its two faces, and a nerve fiber that penetrates inside it with its inner conductive core and its insulating surface! This is the final and conclusive model of the minute animal Leyden jar, with which Galvani laid down the foundations of modern electrophysiology more than two centuries ago.

EVER THE PHYSICIAN

It is worth noting that Galvani, as a physician, did not hesitate to apply his new theory to the healing arts. He did this in the conclusions of *De viribus*. In fact, he did this in two ways. One was to try to explain certain disorders in the context of his knowledge of animal electricity, and the other was to promote therapies based on what he now believed he knew about the brain, the nerves, and the muscles.

Yet despite the freshness of his thinking about animal electricity as the no-longer-mysterious fluid of the nerves, Galvani's medicine still had roots that reflected iatromechanical (mechanical medicine) notions, which in turn were based on some far older ideas that can be traced back to Greco-Roman times. In particular, he still believed that many disorders stemmed from sluggish humors that could obstruct or block the nerve fluid, thereby affecting various functions. In Galvani's case, however, the supposedly blocked nerve fluid was viewed as electrical, not as some spirituous matter or *pneuma*, and not as an ill-defined liquid.

This mixing of new ideas (electricity) with older conceptions (blockages of hollow conduits or tubes) is most apparent when he discusses paralyses and related disorders. In the final pages of his 1791 opus, he writes: "These must be caused by a stoppage in the circuit of nerveo-electric fluid flowing from either the muscle to the nerve or from the nerve to the muscle" (Galvani, 1791/1953, p. 83).

Galvani also tried to explain epilepsy, the "sacred disease" of old, opining: "Animal electricity so contaminated, which floods either from the muscles or other parts through the nerves to the cerebrum, can sometimes produce epilepsy and apoplexy according to the degree of its force and impact against the substance of the brain or nerves and the extent of its contamination" (Galvani, 1791/1953, p. 84).

With regard to therapeutics, he explained that properly charged Leyden jars and animal electricity should produce the same effects. This parallel, as we have noted, can be found in the clinical trials that Van der Lott and others had conducted with the eels of South America (Chapters 12 and 13). Galvani proceeds more cautiously in this domain, and he warns therapists to exercise extreme care with both manmade and biological electrical machines. Still, he is convinced that successful therapies must somehow alter the flow of the electrical force within the nerve circuits:

> From our experiments it seems to be extremely clear that whatever remedies are employed to alleviate these diseases (including even externally applied electricity) it is fitting that all of these, if they are to do any good, should particularly exert their force on animal electricity and should either augment, diminish, or change it and its circuit in some way. For this reason a physician must of necessity hold this electricity and its state before his eyes, especially in treatment. (Galvani, 1791/1953, p. 85)

The Bolognese physician specifically cites the treatment of rheumatic contractions as a part of a broader discussion on when to use medical electricity for movement disorders. He explains that electrical treatments would not be effective if the brain and nerves are physically damaged, as might occur with a war wound or an accident. But he speculates that external electricity has the potential for dealing with problems caused by blockages, as well as for "increasing the strength of [inordinately weak] animal electricity" (Galvani, 1791/1953, p. 87).

SUNSHINE AND IMPENDING STORMS

In retrospect, the Leyden jar model introduced by Galvani in 1791 represented a formidable step forward in the progress of eighteenth-century science toward an understanding of neuromuscular physiology, and it also had medical implications. Galvani's achievement was indeed a major landmark in the history of physiology, even though much work would still be needed to arrive at a full elucidation of nerve conduction—an elucidation that would not correspond in all ways with his views (see next chapters and particularly Chapter 24; also Piccolino & Bresadola, 2003).

The doctor of Bologna was conscious of the momentous importance of his discovery and what it meant for the ancient doctrine of animal spirits. In the conclusion of *De viribus* he overcame his modesty and expressed his justified pride with these words:

> If it will be so, then the electric nature of animal spirits, until now unknown and for long time uselessly investigated, perhaps will appear in a clear way. Thus being these things, after our experiments, certainly nobody would, in my opinion, cast doubt on the electric nature of such spirits . . . and still we could never suppose that fortune were to be so friendly to us, such as to allow us to be perhaps the first in handling, as it were, the electricity concealed in nerves, in extracting it from nerves, and, in some way, in putting it under everyone's eyes. (Galvani, 1791, p. 402)

Galvani's physiological message was hard to ignore: It was that scientists could now discard the ancient theory of animal spirits, along with its newer modifications. Explosive viscous fluids and the idea of nerve vibrations (the speculations of Newton and Hartley), he thought, could also be abandoned. Moreover, there was no longer any reason to bring vitalistic metaphysical notions into the equation. Rather, the physical laws of nature, which could be approached scientifically, govern the intrinsic nerve force, which is quite obviously electrical.

Copies of Galvani's 1791 publication were sent to research centers in Italy and around the world. Extracts of it circulated, as did descriptions of it from colleagues to correspondents. In brief, leading scientists around the world soon knew what he had done and what he was proposing, and many set forth to replicate his various experiments with frogs of their own.

Emil du Bois-Reymond, the leading neurophysiologist in the mid-nineteenth century, and a tireless researcher who also did important research on electric fishes, stated that Galvani created a scientific storm that was equaled only by the political revolutions in the late-eighteenth century. Tongue in cheek, he expressed dread for the future of European "city" frogs, the old "martyrs of science" (Du Bois-Reymond, 1848, 1851, p. 37; Dierig, 2000). With thousands of zealous scientists out to catch, dismember, and electrify frogs in order to understand the properties of the nervous system, he feared the worst for these creatures. The common frog, with its unique sounds that can fill the air on a lovely summer's night, might, he worried, be headed for extinction!

European frogs survived the physiological onslaught and this feared threat of extinction. In contrast, Du Bois-Reymond was hardly overstating the case when he wrote that Galvani's work created a great scientific storm. Within a few short years, natural philosophers were divided by what Galvani had written. Although some did hail his achievement as a landmark in the history of physiology and medicine, others now harshly criticized his experiments and conclusions.

Giovanni Aldini (Fig. 21.13) figured prominently among those who supported Galvani. Born in 1762, Aldini was Galvani's nephew and he collaborated with his uncle on the second edition of *De viribus*, published in 1792. He also conducted many experiments himself, including some described in his widely dispersed *Essai* published in Paris in 1804 (Aldini, 1794, 1804) (Figs. 21.14 and 21.15). The fame of this work stemmed in part from some of the macabre electrical

Figure 21.13. Giovanni Aldini (1762–1834), Galvani's nephew, who took an active part in his uncle's experiments and was one of the main supporters of his theory of animal electricity.

Figure 21.14. One of Aldini's texts on animal electricity, published in Bologna in 1794.

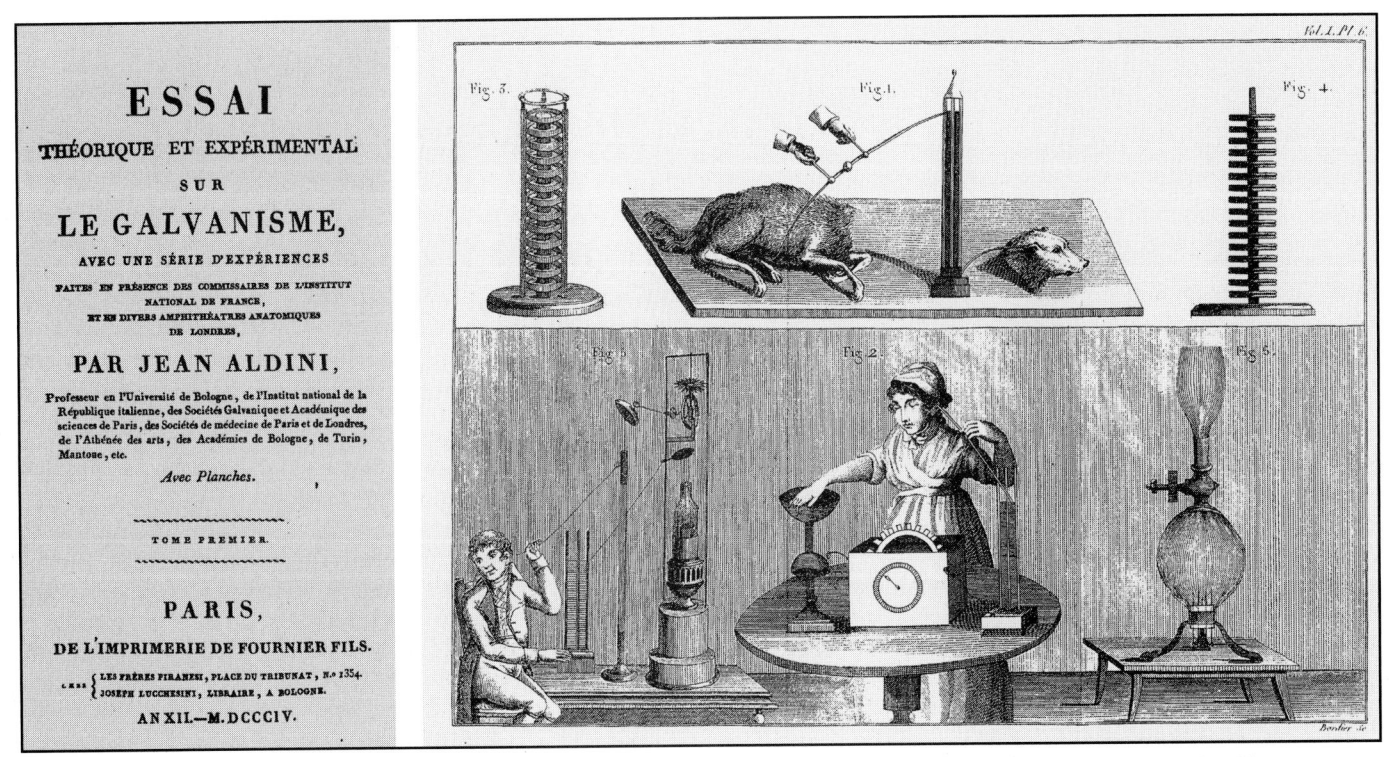

Figure 21.15. The title page of Aldini's 1804 *Essai*, with a plate showing some of the equipment in his laboratory and animal and human experiments (from Aldini, 1803).

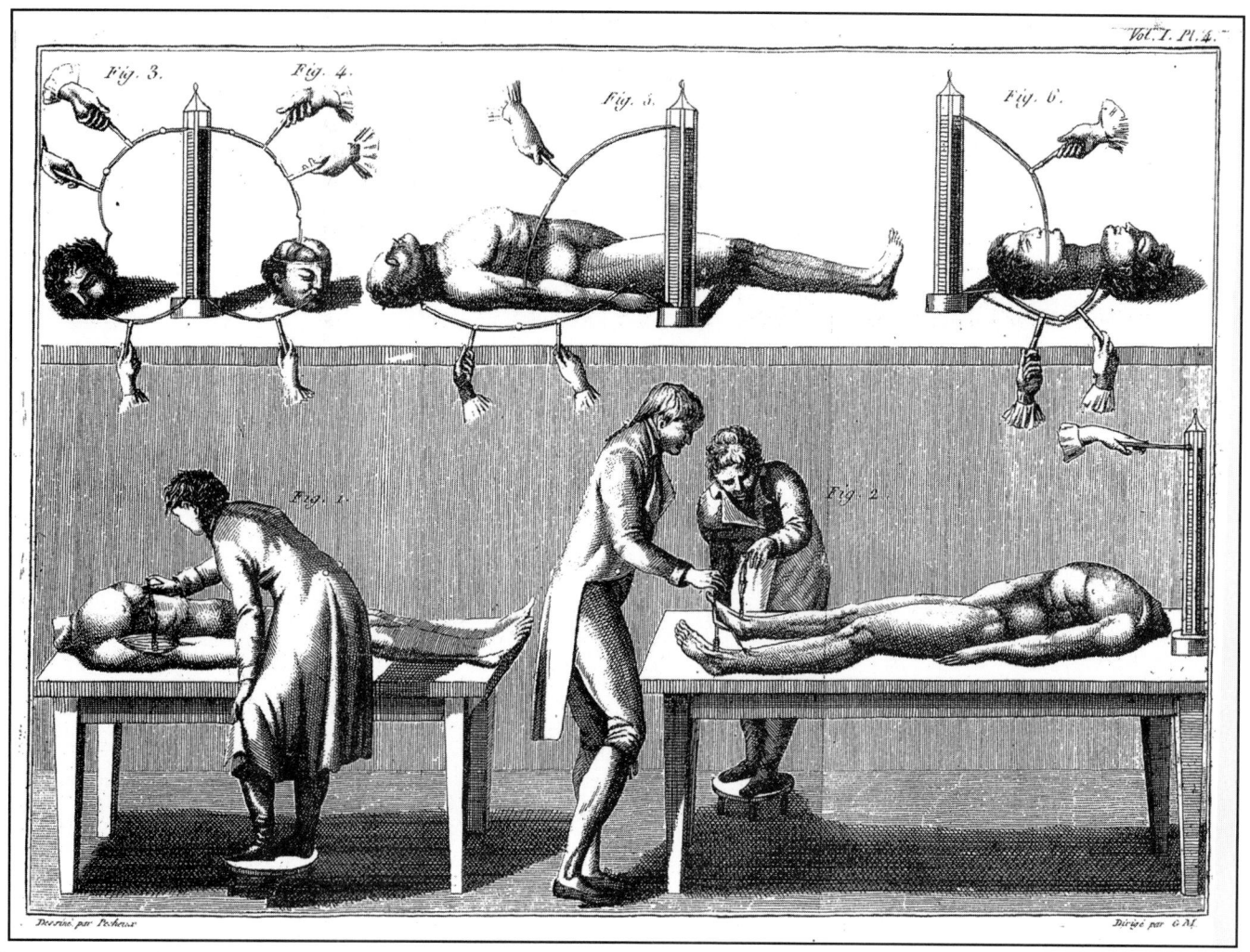

Figure 21.16. A plate from Aldini's 1804 *Essai* illustrating some of his gruesome experiments on humans after execution. Aldini did such experiments on people recently guillotined or hanged, common occurrences in France and England, two of the countries he visited in his travels.

experiments he conducted and described in detail—research involving the body parts of criminals executed only minutes earlier, which, like his uncle's equipment-filled laboratory, also created the atmosphere for *Frankenstein* (Fig. 21.16). Aldini would also be involved in electrical medicine (Fig. 21.17) and would outlive his uncle by 36 years, dying in 1834.

Eusebio Valli, a Tuscan physician who conducted important experiments on animal electricity and spent his life traveling around the world (he worked for years in Paris and in London), was another important soldier in Galvani's army (Fig. 21.18). In his main opus, published in London in 1793, Valli asked whether fish electricity might be due to some mechanism specific to these fish, or whether it might be the expression of a more general property of living tissues. He concluded by stating: "We cannot avoid being convinced that the shock of the torpedo, and the shock and spark of the Gymnotus, are effects of the same cause, which produce the movements in the frogs, fowls, cats, dogs, and horses made the subject of experiment" (Valli, 1793, p. 113). For Valli, the power responsible for an electric fish's shocks "is common to

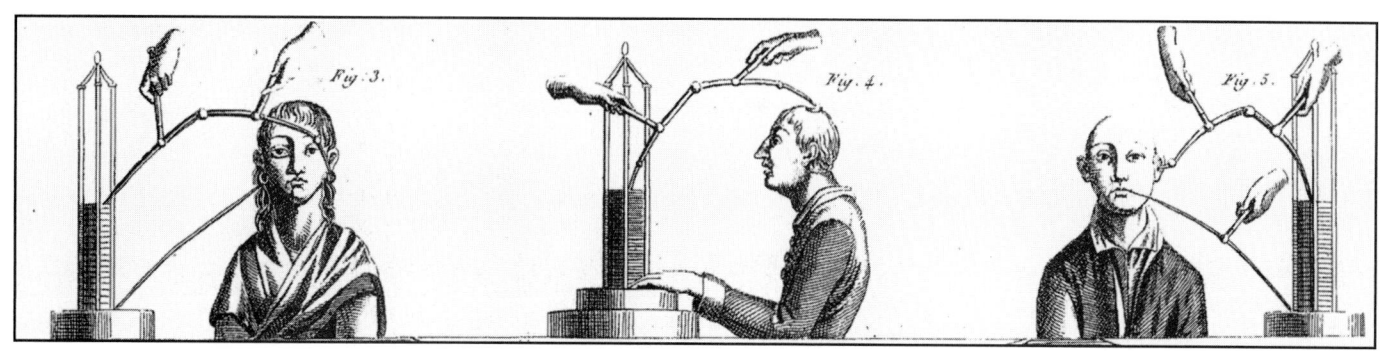

Figure 21.17. Diagram of some of Aldini's (1804) attempts to apply electricity to the head when treating "mad" patients (see Beaudreau & Finger, 2006).

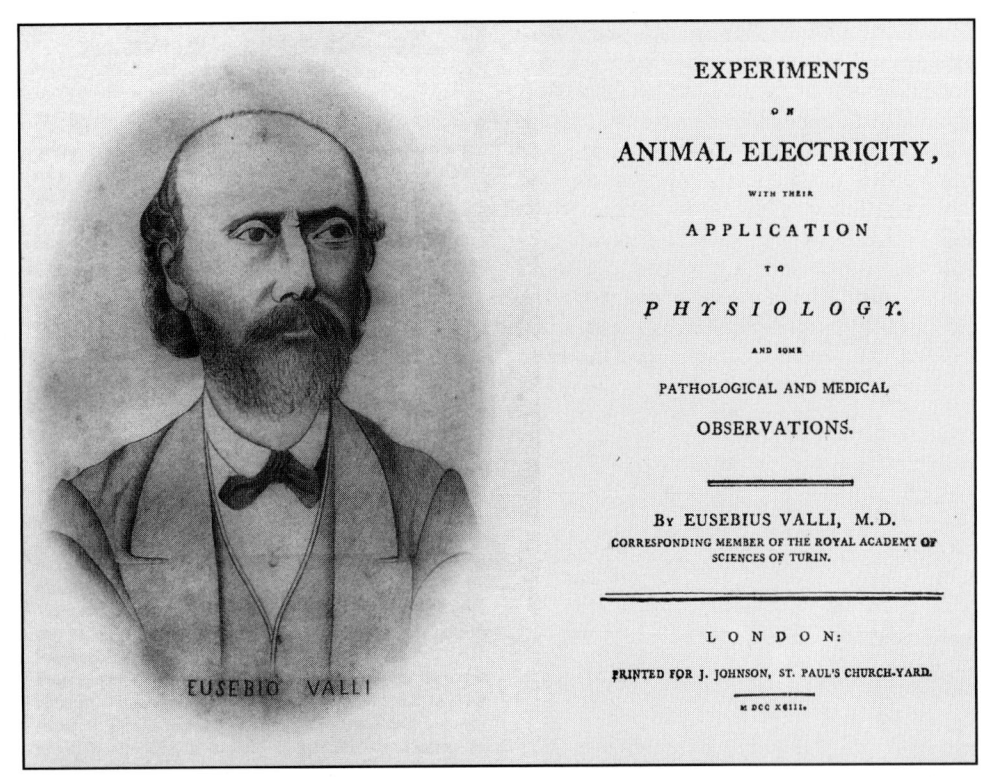

Figure 21.18. Eusebio Valli (1755–1816), the Italian physician who was one of the strongest supporters of Galvani's theory of animal electricity, along with the title page of his *Experiments on Animal Electricity*, which was published in London in 1793.

all animals in which the phenomena of electricity are apparent," such that muscles should be considered to be "electrical machines."

Another of Galvani's supporters was German physician, physiologist, and anthropologist Johann Friedrich Blumenbach (Fig. 21.19), who in 1795 glowingly wrote:

> By the combined labours of experimental physiologists in different parts of the world, this branch of science was at length matured for giving birth to another discovery, which will probably be found of equal importance, in explaining the phenomena, and in removing the diseases of the animal system, with that which consigned to immortality the name of the illustrious Harvey. The discovery to which I wish at present to direct the attention of the reader is that of, what is usually called "animal electricity," or, the existence and operation of a fluid extremely similar to electricity in the living animal system. For the fortunate Galvani, professor of anatomy at Bologna, was referred the honour of lighting by accident on this beautiful and divine discovery—a discovery which entitles its author to be ranked with the great promoters (of) science and the essential benefactors of man. (Blumenbach, 1795, pp. 217–218)

In contrast, there were those who sided against Galvani. The individual soon to become the acknowledged leader of this opposing army was Alessandro Volta, professor of physics in Pavia. Volta agreed with Galvani about some fish being electrical, but now challenged his more general formulations about animal electricity, including the idea that we too are electrical—and he had a considerable following.

The clash between Volta and Galvani would be one of the most important in the history of science. As we shall see in our next chapters, the debate over Galvani's theory of animal electricity would lead Volta to his new electric battery, which was modeled on what he knew about the electric organs of a few strange fish. Volta's ideas and opinions would also stimulate Galvani and his supporters to conduct additional clever experiments, which they would construe as

Figure 21.19. Johann Friedrich Blumenbach (1752–1840), professor of medicine in Göttingen, who wrote in 1795 that Galvani's work on animal electricity was of equal importance to William Harvey's discovery of the circulation of the blood.

supportive of animal electricity's genuine physiological origin.

As we shall also show, some of Galvani's new experiments would be done on live torpedoes—electric fishes being the only animals thought capable of producing electricity or something like electricity by Galvani's opponents at the end of the eighteenth century. Whether the fish (or animal) fluid was really genuine electricity, or only something superficially resembling it, would remain an important issue. Alexander von Humboldt, among others, was far from sure about the identity of the animal force, which he called galvanism, when he conducted his electrical experiments in Europe, and then while in the swamps of South America, where he finally obtained some live eels for study in 1800 (Chapter 1).

Ultimately, Galvani's research and ideas would give rise to scientific investigations that would allow researchers to clarify the basic mechanisms underlying nervous conduction. This story has a long time course, and it will take us well into the twentieth century. We shall summarize some of the highlights of this phase of more modern physiology. But first we must begin with the especially harsh debate on animal electricity that pitted Galvani and his followers against Alessandro Volta—a pivotal event in the history of electrophysiology and that of physics, and one with further ties to electric fishes.

Chapter 22
Electric Fishes in Volta's Path to the Battery

Alexander Volta, in re electrica princeps, vim raiae torpedinis meditatus, naturae interpres et aemulus.
[Alexander Volta, the prince of electrical science, having meditated the force of the torpedo ray, interpreter and emulator of Nature.]
Inscription by Pietro Configliachi under Volta's portrait
(See Volta, 1918, Frontispiece)

Alessandro Volta (Fig. 22.1), the Italian physicist who would be celebrated for the invention of the electric battery ("Voltaic pile"), was Galvani's most influential and visible antagonist in the debate over whether electricity might exist in more than a few strange fishes. In fact, the invention of the electric battery was an unexpected but fortuitous consequence of Volta's research on the debated subject of animal electricity.

There are few circumstances in which the intricacies and complexities of the routes of scientific discovery emerge as clearly as in the path leading Volta to the invention of his epoch-making device, which was based on the actions of dissimilar metals. As we now know, particularly after chemist Humphry Davy of the Royal Institution of London carried out his research at the beginning of the nineteenth century, Volta's instrument is capable of converting the chemical energy of saline solutions into electric energy. For these reasons, the electrical battery obviously belongs to the provinces of physics and technology. Volta's device, however, also stemmed from physiological studies of electric phenomena in living organisms. In this regard, the paths leading to its invention can be directly related to the history of physiology and medicine, in addition to physics.

Its ties to animal physiology are, in fact, twofold. First, Volta's battery was invented toward the end of 1799, at the conclusion of an intense period of research started 7 years earlier. It was in 1792, upon being informed of Galvani's (1791) experiments and publication on animal electricity, that he decided to try to verify the findings and conclusions of his countryman by replicating his experiments and carrying out some new ones. Most of these experiments were performed on frogs, following and varying the protocols Galvani used in Bologna. But Volta was also guided in a significant way by his knowledge of the anatomy and physiology of the electric organs in torpedoes and the eel, especially in the final phase of his research program. The inspiration he drew from electric fishes was so important that, in March 1800, he referred to his device as an "artificial electric organ", this being in his letter announcing the invention of his battery to the President of the Royal Society of London (Volta, 1800).

Indeed, in its common columnar version, Volta's battery visibly resembled the electrical organs of torpedo fishes. Both were constructed of flat disks arranged in vertical columns (the disks are assembled in horizontal rows in the eels, though still following the same principle). Moreover, Volta believed that the electricity depended on similar phenomena in these fishes and with his "artificial" instrument, which imitated nature. Both, in brief, required contact between dissimilar substances.

It was not only by chance that Volta discovered the electrical property of metallic contacts, a fundamental step leading him to the battery. There was another important consideration that merits attention, and it has to do with the complex nature of the electrical mechanisms underlying nerve conduction. This factor has become fully understood only with twentieth-century research on nerve conduction. As we shall see, it is with this modern understanding of nerve function that we can begin to appreciate the difficulties that Galvani and Volta encountered in their attempts to arrive at a shared opinion about the origins of the electricity involved in their animal experiments.

Anatomy, physiology, physics, and chemistry will be connected historically and logically in the story we shall narrate in this and the remaining chapters of our book. To find an Arianna's thread capable of guiding us through this complex and fascinating labyrinth, we shall start with the research that immediately followed the publication of Galvani's *De viribus*. Specifically, we shall begin this chapter with Volta's initial forays into animal electricity, which will lead him to metallic electricity and his repeated assaults on Galvani's experiments and theories. We shall then show how Volta's obsession with weak electricity arising from different metals motivated him to find ways of multiplying the metallic force, culminating in the invention of the electric battery, an acknowledged wonder of the scientific Enlightenment and an invention with important past and future associations with electric fishes.

Figure 22.1. Alessandro Volta (1745–1827), the Italian physicist who accepted electricity in some specialized fishes but questioned Galvani's broader concept of animal electricity.

ALESSANDRO VOLTA: FROM LOVE OBSESSION TO SCIENCE LUST

Early in the spring of 1792, Galvani's *De viribus* reached Pavia, then the site of one of the most important universities of the Austrian Empire. The secretary of the Bologna *Istituto* had sent a copy of Galvani's *opus* to Bassiano Carminati, professor of medicine and pharmacology at the University of Pavia. Encouraged by Carminati, Alessandro Volta, professor of physics at the university, soon started to repeat Galvani's experiments.

Recognized as a brilliant autodidact, lacking a regular university education, Volta was already one of Europe's most famous electricians (for biographical material, see Polvani, 1942; Pancaldi, 2003). He was also a member of many prestigious scientific societies, including the elite Royal Society of London, to which he had just been elected in 1791. He was particularly well known for the invention of atypical electric instruments, one being the electrophore mentioned in the previous chapter. Called the *elettroforo perpetuo* by Volta, this device (with its insulating "cake" of resin and wax in a metal dish, and with a metal plate with an insulated handle on top of it) could seemingly produce perpetual electricity (Fig. 22.2). It functioned through a variable capacitance effect, unlike the frictional mechanisms found in more typical electric machines. The condenser-electroscope for revealing relatively weak electricity (again through the action of a variable capacitance) was another of his notable electrical inventions (Fig. 22.3).

Volta was interested in many aspects of science, including both inanimate and living nature, and it is easy to envision why and how he was drawn to the new and surprising features of animal electricity, as revealed by Galvani with his many experiments. But besides his obvious scientific interests, there were also reasons of a more personal nature that made him want to throw himself into a program of intense experimental research at this moment in time.

Volta was then in a state of deep emotional distress, caused by the termination of a torrid love affair with an opera singer. As a handsome, elegant, and accomplished man from a noble family, and with his great culture and easy refined conversation, Volta had all the characteristics needed for catching the attention of beautiful, sophisticated women. Further, he was certainly very sensitive to the feminine mystique, regardless of social level. But like a true Casanova, he had been decidedly unwilling to give up his free lifestyle in exchange for matrimonial bonds.

Suddenly in 1789, however, everything changed. The precipitating event was the arrival of Marianna Paris, the *prima donna* in an opera at a local theater (Paisiello's *Il Barbiere di Siviglia*). In an unexpected blaze of passion, the respectable physics professor suddenly decided to marry the singer, despite the very low status of singers and actors in *Ancien Régime* society.

The infatuated lover's decision provoked intense reactions from members of his respected family, which included many priests and nuns. The torrid affair even involved the Austrian governor of Lombardy and the Emperor himself. But this marriage was not to be, and Volta eventually recanted early in 1792. With his heart broken, he now desperately hoped to overcome his feelings for his beloved Marianna.

It was precisely at this time that Galvani's memoir reached Pavia. As recently and colorfully put by Paolo Mazzarello (2009), an historian who has written in depth about the "scandalous" affair between *Il Professore e la Cantante*, Galvani's frogs now entered into Volta's life to occupy his rapt attention and fill the vacuum left by Marianna's departure.

After reading Galvani's *De viribus*, Volta now dutifully replicated many of the experiments of his Bologna colleague. His interest became particularly lively when he realized that the prepared frog preparations were capable of detecting even weaker electricity than what could be revealed by his most sensitive electrometers. It is important to realize at this juncture that there was a great interest in weak electricity during the closing decades of the eighteenth century. This was because many physical and chemical processes were now thought capable of generating electricity, although the electricity they produced seemed very feeble compared to lightning or the discharges of a frictional machine that could be collected in a Leyden jar. Because of this interest, physicists, Volta among them, were displaying great efforts to invent more sensitive electrometers.

On April 3, after just a few days of experimenting, Volta published a short account of his initial experiments on animal electricity. Appearing in a scientific journal published in Pavia, this account represented the first of a series of scientific writings on the subject. One month later, he wrote a more extensive and systematic memoir, in which he praised Galvani and compared the importance of his discovery of animal electricity to Franklin's discovery of the electrical

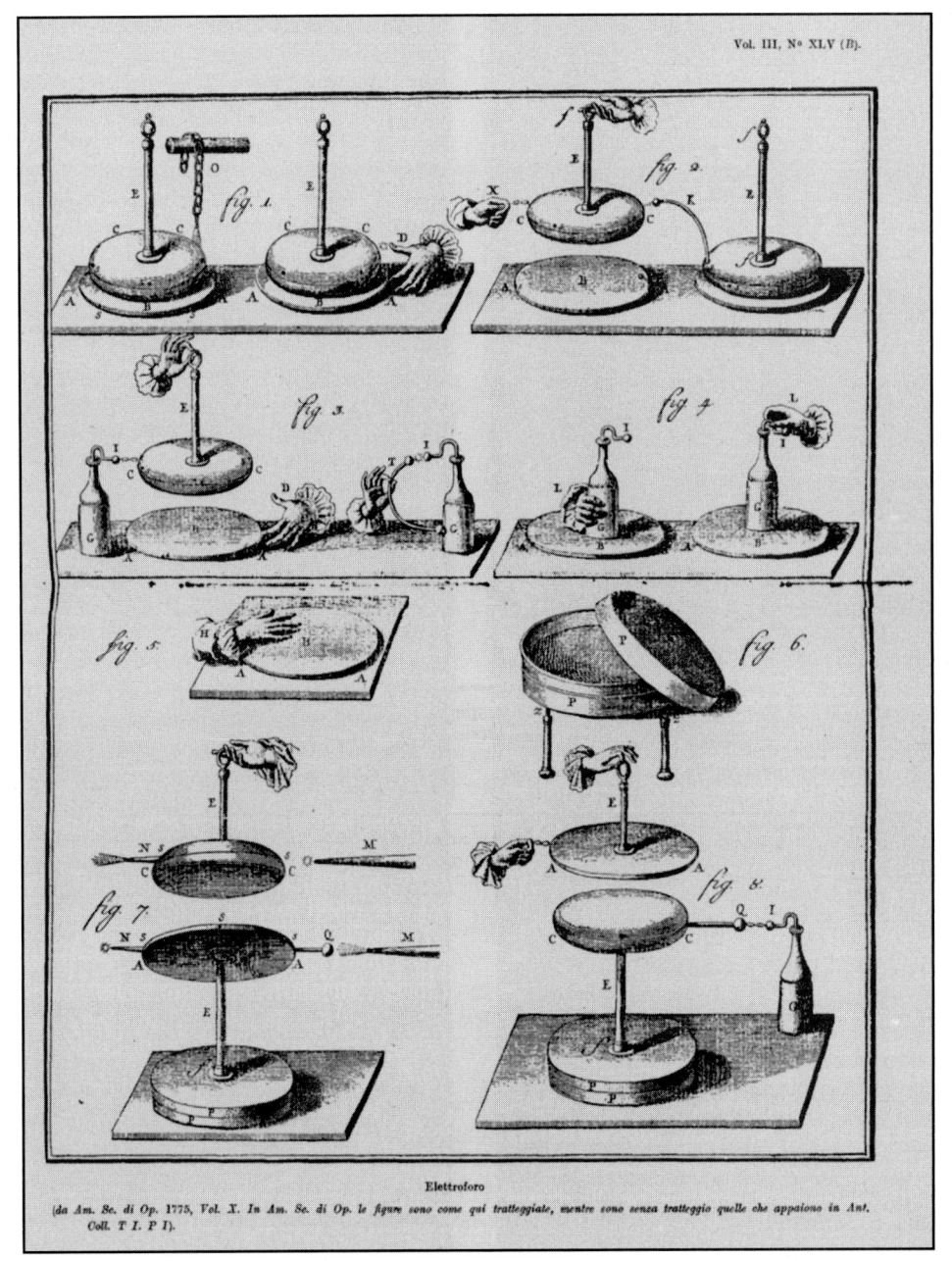

Figure 22.2. Volta's *electrophores* and the way they were used to produce and accumulate electricity (from Volta, 1926).

nature of the lightning (Volta, 1918). After examining the theories of nerve function elaborated prior to Galvani, Volta focused on the possible medical applications of Galvani's results. As a man of the Enlightenment and a member of elite scientific societies that stressed utility and helping suffering humanity, that his mindset would include practical applications for both Galvani's and Franklin's fundamental discoveries is eminently understandable.

By expanding some of the arguments developed in his private letter of 1782 to Mme. Le Noir de Nanteuil (see Chapter 21), Volta pointed to a clear distinction between genuine animal electricity (i.e., that produced inside animal bodies as a consequence of a physiological process) and electricity that could be produced by physical means with living or even dead bodies (e.g., "by rubbing the back of a cat, by currying a horse").

To his list of electric fishes, the only animals he had previously thought capable of producing genuine electricity, Volta could now add the subtle electricity that Galvani had recently discovered in the nerves and muscles of frogs and other animals. Many authors, who provided only vague and uncertain speculations, had suspected this was the case. But now, he continued, it had clearly been detected, thanks to Galvani's efforts. As a consequence, Volta continued, it is exclusively to Galvani that we should attribute "the merit and originality of this great and stupendous discovery" (Volta, 1918, p. 23).

Many historians have examined Volta's shift from his initial unconditioned praise for Galvani and admiration for his discovery, as revealed at this time, to his biting critiques of Galvani's interpretations and conclusions, which are expressed in a continuous barrage and crescendo in

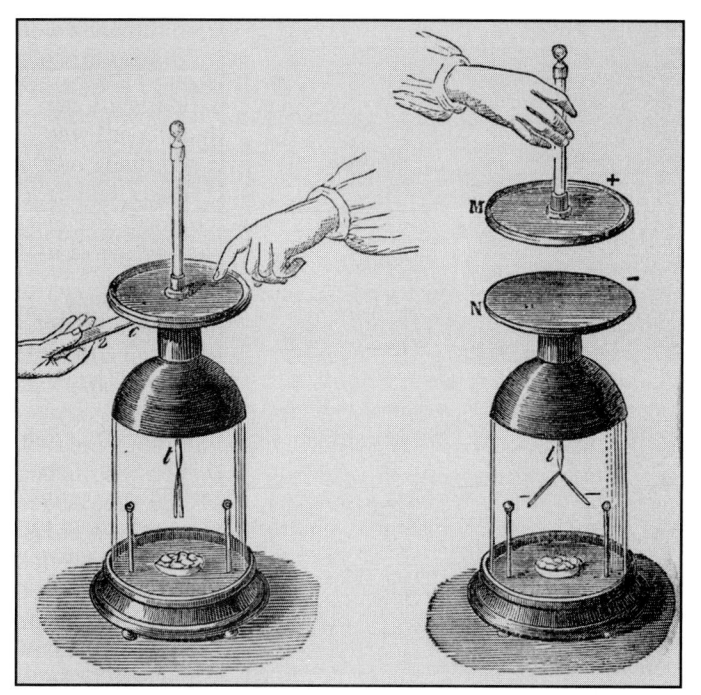

Figure 22.3. Volta's *condensatore*. This instrument is capable of revealing weak electricity by the action of a variable electric capacity (from Ganot, 1882).

subsequent publications (e.g., Polvani, 1942; Pera 1986; Piccolino & Bresadola, 2003; Bresadola, 2008a, b). We shall limit ourselves to the analysis of those phases of his research that are particularly relevant to the main theme of our book—that is, to those that relate to electric fishes and the emergence of a new nerve physiology.

EMERGING DOUBTS

Volta gradually started to conceive of the frog as an extraordinarily sensitive electroscope, rather than as a source of electricity, over the course of his investigations. He was astonished to see that an almost completely discharged Leyden jar, one to which his most sensitive electrometer was insensitive, could still effectively stimulate a prepared frog's nerves, so as to produce observable contractions.

In previous experiments, some of which were carried out with Lavoisier and other *académiciens* in Paris 10 years earlier, Volta had been surprised to discover that weak electricity could be developed by a variety of physical processes, including the evaporation of a liquid. Thus, it was possible, he speculated, that a tiny amount of electricity might originate from an accidental external source in the frog experiments.

Volta gained insight into the nature of this possible external electricity when he realized that the contractions were produced much more effectively when the arcs used to establish the connection between the nerve and muscle were made of two dissimilar metals (Fig. 22.4). This observation further led him to suspect that the electricity responsible for stimulating the contractions might not be intrinsic to the animal preparation at all—that it might instead originate from the contact of two or more different metals. During these investigations, he studied a variety of different metals, and he derived a scale on which the different metals were arranged according to their efficacy in producing the presumptive electricity when put in contact with a metal standard.

Another finding that led Volta to question the validity of Galvani's interpretations involved an experiment in which he produced a contraction by applying both tips of a bimetallic arc to a nerve (instead of to a nerve and a muscle, as was

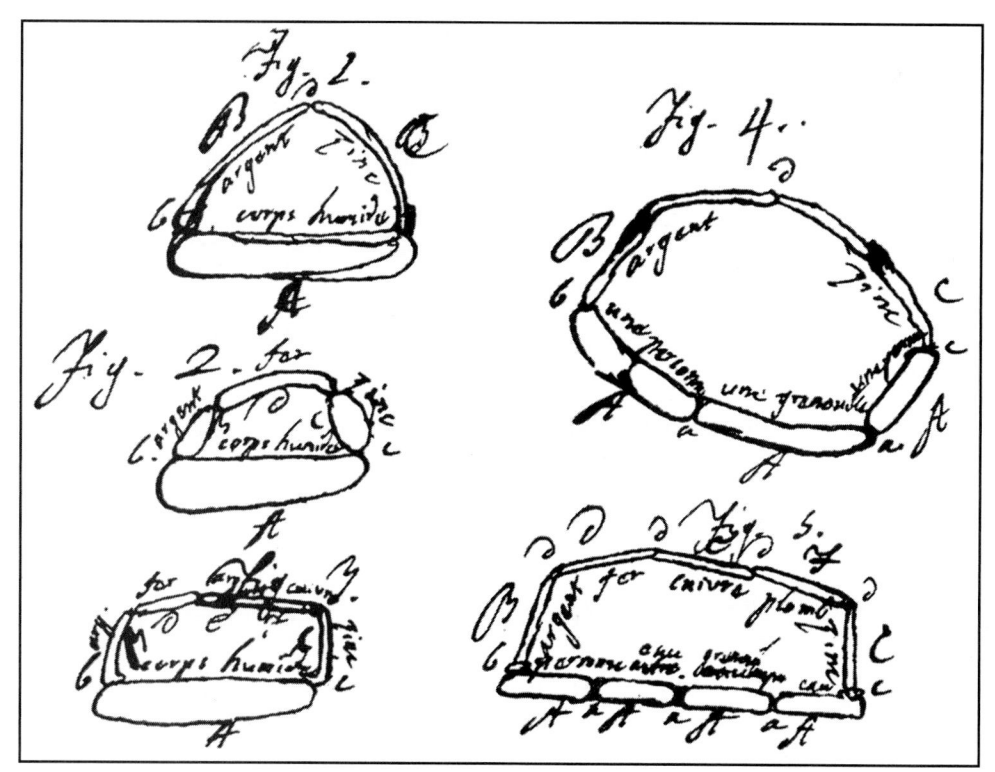

Figure 22.4. Volta's experiments with electric circuits, consisting of different metallic arcs, moist bodies, frog preparations, and the body of the experimenter (from Volta, 1918).

typical in Galvani's experiments). This effect could not be reconciled with Galvani's Leyden jar model, since the nerves in his model exclusively represented the conductors of the electricity stored in the interior of muscle fibers. Accordingly, no flux of electricity should be produced in the frog in the absence of some communication between nerve and muscle.

Volta had a great proclivity to vary his preparations in diverse ways. In contrast to Galvani, he often used intact animals and even his own body as experimental tools. In one of his experiments (made with the aim of producing contractions of intact human muscles), he applied a bimetallic lamina to his own tongue. Instead of a muscle contraction, he experienced a taste sensation. The gustatory sensation changed from acid "to rather alkaline, pungent, with a tendency to bitter" when he inverted the polarity of the metallic arc (Volta, 1918, p. 63). Volta correctly interpreted these results as due to stimulation of gustatory nerves in the tongue, but this conclusion again did not fit with the Leyden jar model of the neuromuscular system that had been presented by Galvani.

Volta soon understood that the physiological effects of nerve stimulation did not necessarily depend on the nerve's relationship with muscle, as suggested by Galvani. Rather, they were related to the type of nerve stimulated: contraction with a motor nerve, and a specific sensation with a given sensory nerve. He supported this conclusion with additional experiments, in some of which he employed bimetallic laminas to stimulate the visual and auditory systems, as well as the skin senses. On the basis of his results, he formulated a law of the physiological effects of electrical stimulation, anticipating the conclusions drawn by Charles Bell in 1811, which would be fully elaborated as the "law of the nervous energies" (or "law of specific nerve energies") by Johannes Müller shortly thereafter (Bell, 1811; Müller, 1826, 1840; for English, see Müller 1843, pp. 711–713; Piccolino, 2000b; Piccolino & Bresadola, 2003; Finger & Wade, 2002a, b).

Having a great tendency to design experiments in which his results could be demonstrated in particularly expressive and lively ways, Volta combined the different effects of the electricity derived by metallic contacts in a myriad of ways. For example, he put one of two different metals on the tip of his tongue and the other on his conjunctiva. When the two metals were connected, he experienced both a light and a taste sensation.

By forming a chain that included a frog preparation in addition to the experimenter's eye and tongue, this experiment was made even more complex. Closing the circuit now elicited the double sensation (light and taste) and contractions of the muscles of the frog's leg. By turning to yet other arrangements, Volta simultaneously produced a taste sensation in one subject, a visual sensation in another, and the contraction of the frog's leg. Displays of this sort became particularly rich and expressive after the invention of the battery, because the stronger electrical force generated by this instrument allowed him to combine even more elements into a single experiment—one that could demonstrate virtually the entire range of physiological effects (Piccolino, 2000b).

VOLTA *CONTRA* GALVANI: PHYSICS OPPOSING PHYSIOLOGY?

An older tradition, revived by some recent historical studies, presents Volta as exclusively a physicist only interested in the physical implications of the phenomena he studied, not in the life sciences. He is placed in stark contrast to Galvani, a physician and anatomist uniquely interested in physiology and medicine but not well acquainted with the laws of physics (see Arago, 1854; Polvani, 1942; Pera, 1986). Accordingly, Volta would have been inclined to see a physical agent behind Galvani's experiments, because he was a physicist. In contrast, Galvani would have been more prepared to see intrinsic animal electricity, because of his training in physiology and medicine and his supposed lack of interest in physics.

What the historical record really shows, however, does not support this simplistic view. It reveals that there was little disciplinary separation in eighteenth-century science, with leading natural philosophers typically showing interests in a wide variety of natural phenomena, both inanimate and animate, and this included both Galvani and Volta. In Volta's case, this historiographic narrowness is particularly contradicted by the continuous interest he showed in the life sciences since his early years, and by his repeated references to the possible physiological and medical implications of electrical research in his writings (see Piccolino, 2000b; Piccolino & Bresadola, 2003).

Some of Volta's many physiological achievements have already been mentioned, and they include his anticipation of the law of specific nerve energies and his demonstrations of the polar (positive or negative) effects associated with electrical stimulation of the sensory nerves. Volta was able to show, moreover, the different excitabilities of various types of sensory and motor nerves, the transient nature of the response of some of them to a steady electrical stimulus (e.g., motor nerves, the optic nerve), and the maintained (tonic) actions of others (e.g., those involved with taste and pain sensations). Using electrical stimuli, he was also able to differentiate between two different types of pain sensations, which are today associated with fast-conducting myelinated fibers and slow-conducting non-myelinated fibers.

After the invention of his electric battery, Volta was even able to determine with astonishing precision the time in which a prolonged electrical stimulus could summate in the production of motor responses in human beings ("one minute third" or 1.6666 ms *vs.* 17 ms for his transcutaneous stimulus, using modern estimates). If we consider that he succeeded in doing this during a period in which there were no clocks capable of effectively measuring times smaller than one second, we have to admire how far-reaching Volta really was in what we would now call his electrophysiological investigations.

As mentioned above and noted in the context of his first memoir on animal electricity, Volta also showed a genuine interest in electrical medicine. One year after his first memoir on animal electricity, when commenting on some experiments in which he used a bimetallic lamina to stimulate luminous sensations in the eye, he pointed to a possible diagnostic application of his results, writing:

> On the other hand, I am persuaded that the experiment would succeed even in persons blinded because of cataracts, or any other fault, except for insensibility or paralysis of optic nerves. Therefore, these trials could be of some utility, allowing one to discover if such fault exists. Moreover, who knows if, being well administered, they could be of some help in this same paralysis, both initially or more or less advanced. (Volta, 1918, p. 222)

The idea of separating the world of the physicist from that of the physiologist is therefore far from a constructive and helpful key to understanding Volta's electrical endeavors. In fact, some of his experiments show so much interplay between physics and physiology that it can be difficult, if not impossible, to classify them into one or the other discipline.

This interplay between physics and life sciences is especially notable in Volta's path to the battery. Here, as already hinted, Volta's reflections on electric fishes were extremely important, especially in the final phase of his work. To understand why and how this happened, we must examine Volta's experiments and elaborations after he started criticizing Galvani's "flawed" experiments and theories. Far from being the source of an internal electrical disequilibrium, as the doctor of Bologna assumed, the prepared frog would now be viewed as no more than a detector of the weak physical electricity generated from an external source, and here he would point to contacts between different metals.

GALVANI'S "CONTRACTIONS WITHOUT METALS" AND VOLTA'S EXTENSION OF THE THEORY OF METALLIC ELECTRICITY

Volta critically addressed Galvani's views in the numerous memoirs on animal electricity that he wrote starting in the spring of 1792, following his praise for his countryman's discovery. Although he still commended Galvani for the beauty of his experiments and the interest they generated, he severely attacked his conclusions, arguing that they stemmed from the use of metals.

The news of Volta's experiments apparently demonstrating the metallic origin of Galvani's "animal electricity" spread rapidly through Europe, and in 1794 he was awarded the Copley Medal for his research on metallic electricity. This was the same prize that had been given exactly 20 years earlier to John Walsh for his research on torpedoes, and to Franklin 21 years before Walsh for his studies on the nature of electricity (Chapters 10 and 16).

The year 1794 was an important one in the development of the polemics between Galvani and Volta. Convinced that the electricity in the experiments using metals originated exclusively from the contact between dissimilar metals, Volta tried to undermine the relevance of some of the experiments in which the adherents of Galvani's theory had demonstrated that contractions could also be produced with an arc made of only a single metal. He did so by arguing about the possible non-homogeneous nature of the metal used to connect nerve and muscle. This argument seemed, however, to be ill founded, particularly in relation to some experiments in which Galvani's supporters succeeded in eliciting contractions with highly purified quicksilver (Valli, 1793; Aldini, 1794).

In 1794, Galvani published, albeit anonymously, the *Trattato dell'Arco Conduttore* (actually with a longer title see Fig. 22.5). In this text, which was written in Italian, he responded to Volta's criticisms on both experimental and theoretical grounds. The *Trattato* is particularly important because it contains a very clever experiment that he and his followers regarded as a *de facto* refutation of Volta's arguments and theories. Using a prepared frog, Galvani (1794b) succeeded in eliciting a contraction by establishing a direct contact between a nerve and a muscle without a metal or any other extraneous matter in the circuit (Fig. 22.6).

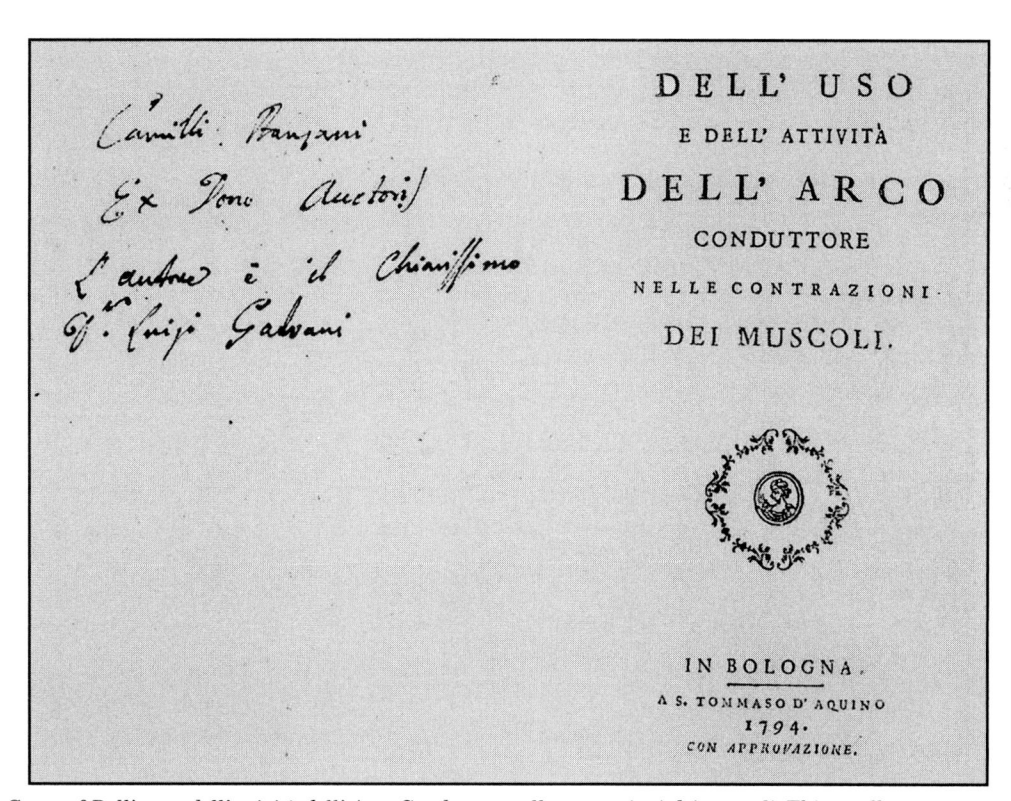

Figure 22.5. The Cover of *Dell'uso e dell'attività dell' Arco Conduttore nelle contrazioni dei muscoli*. This small treatise was published anonymously in 1794, but everyone in Bologna knew Galvani was the author. This is made evident from the inscription on the back cover of this copy (on left), which says that the book was a gift of the author and "The author is the celebrated Prof. Luigi Galvani."

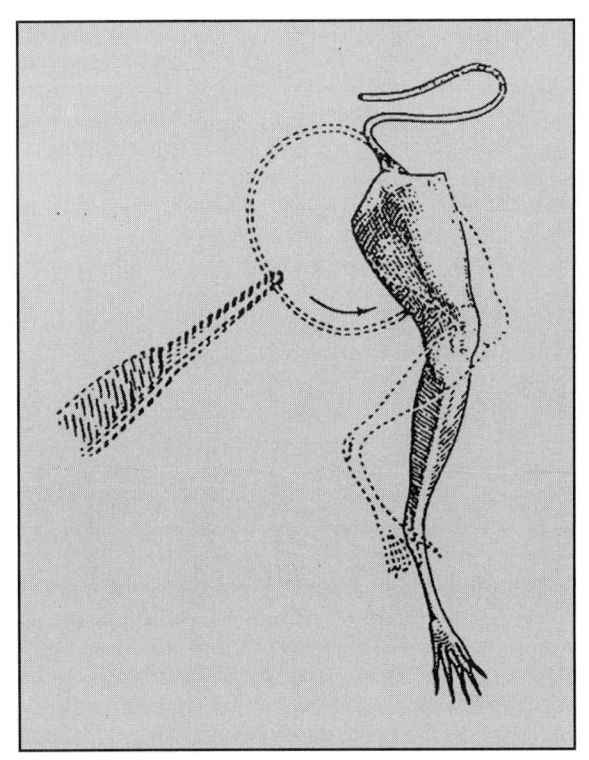

Figure 22.6. Galvani's experiment on *contrazioni senza metallo*, meaning contractions without metal (from Sirol, 1939).

Volta recognized the significance of this experiment and even lamented that, because of it, many adherents of his own ideas crossed over to Galvani's side. Soon afterwards, however, with a kind of *ad hoc* hypothesis, he attempted to counteract the logical implications of the experiment by stating that, even in the case of the "contractions without metal," the electricity involved must be of a purely physical origin.

As Volta now saw it, the electricity could originate from contact between two dissimilar matters of any kind, in this case the substance of the nerve and the substance of the muscle. It really does not matter if the substances are animal in nature. With this thought, Volta now extended his electric hypothesis of metallic contacts to come forth with a more general theory—electricity generated by the contacts of different substances of any kind.

By taking this step, Volta felt convinced that he had put Galvani and his supporters in their places, because, as he remarked in 1795, "they will never succeed in showing that convulsions can be excited with the making of the [electrical] circle of conductors, all of the same species" (Volta, 1918, p. 325). Indeed, this difficulty seemed insurmountable, particularly because a key requirement of Galvani's Leyden jar model was that contact must be made between nerve and muscle tissues. This demand necessitated heterogeneous matters, something fundamental to Volta's new stance.

Fully understanding the call for experimental facts in scientific controversies, Galvani succeeded in doing something that initially seemed logically impossible. It appeared in 1797, in the second of a series of memoirs on animal electricity addressed to Lazzaro Spallanzani, one of the most famous biologists of the era and Volta's colleague in Pavia. To put this important experiment in chronological context, it is important to remember that it appeared in the same year in which Humboldt published his two-volume treatise on galvanism (Chapter 1). Humboldt (1797) included several experiments pointing to the possibility of obtaining contractions in frog preparations even in the absence of metals—for instance, using moist animal tissue to connect nerve and muscle. Nevertheless, Humboldt's experiments still involved contacts between heterogeneous substances.

Responding to Volta's challenge to come forth with what might be regarded as an "impossible experiment," meaning one not involving heterogeneous matters, Galvani separated and prepared the two legs of a frog with their respective sciatic nerves sectioned near where they emerged from the vertebral canal (Fig. 22.7). He then placed them apart, and with a glass rod moved the nerve corresponding to one of the legs so that it bent to form a small arc that touched the other nerve in two different places. With great care taken during the manipulation, so that one of the parts of the first nerve used to establish the contact was its cut mouth, he found that a contraction could be elicited in the first leg and frequently also in the second one (Galvani, 1797, p. 16)!

MULTIPLYING WEAK ELECTRICITY

Despite its importance, the aforementioned experiment did not particularly attract Volta's attention. This was mainly because the physicist of Pavia had just succeeded in obtaining a definite *physical* demonstration of the electromotive power of metals (i.e., of the capability of two metals to generate electricity at their contacts).

As first communicated to German chemist Friedrich Albrecht Carl Gren, Volta measured this electricity "with the help of my *condensatore* of electricity, and better with the spinning duplicator of Nicholson, based on the same principles of *condensatore*" (Volta, 1918, pp. 419–420). These two

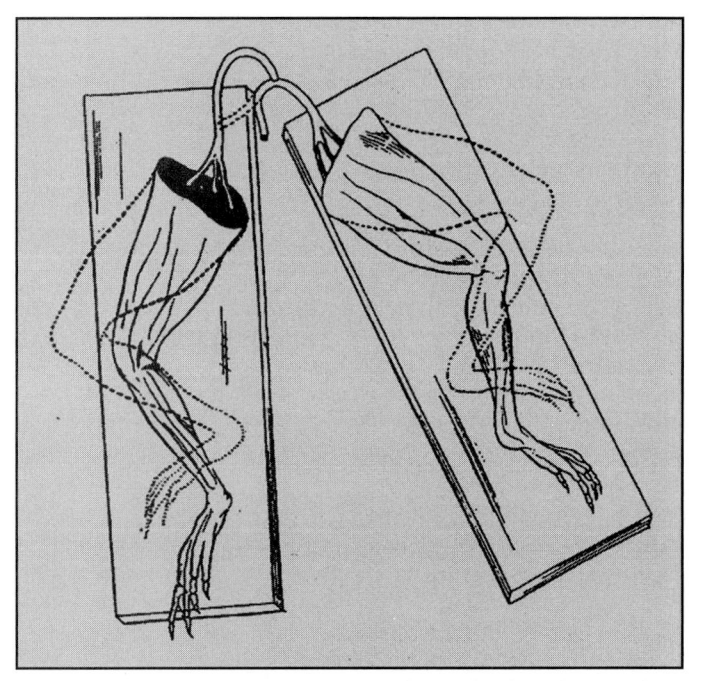

Figure 22.7. Galvani's experiments with two frog legs showing that connecting two nerves together could cause contractions. This clever experiment was particularly important because it showed that it was not necessary to put dissimilar conductors in contact to obtain contractions, as Volta had been contending (from Sirol, 1939).

instruments were the most sensitive physical detectors of electricity at the time, both being based on a variable capacitance effect, like the aforementioned electrophore. Unlike Volta's *condensatore,* which called for certain manual actions to change the capacitance, Nicholson's duplicator did this automatically through a rotating mechanism. With these more sensitive instruments, Volta was able to measure "electricity undoubtedly tiny, and below that degree needed to give sign in the common electrometers."

Nicholson's instrument was also important for Volta because it required the sources of the electricity to have round shapes. In Volta's laboratory, the metals, previously in the forms of arcs and wires, now started to become disks, a further step towards the invention of his battery (Pancaldi, 1990).

By now being able to measure subtle metallic electricity with these physical instruments, Volta no longer needed a biological preparation to detect the electricity in his experiments with metals. He also felt he was on safer ground when he asserted that the power of metallic arcs to generate electricity could alone explain the contractions elicited with frogs. From this premise, he concluded (although wrongly, as we now know) that there is no animal electricity—absolutely no electricity intrinsic to the nervous system in Galvani's frog experiments.

Starting with when he became sure of the physical origin of "his" metallic electricity, Volta began an extraordinary and even hectic phase of new research studies. He was, however, not totally satisfied with his results, even though he was well convinced of the importance of what he had achieved: having introduced a new and unexpected principle in physics, namely that metals, far from being uniquely conductors of electricity, could act as "electromotors" (i.e., that they could generate electricity upon contact). The problem that continued to bother him was that the metallic electricity was very feeble. Detectable only in a well-equipped laboratory, it was a wonder of nature that seemed to have no apparent practical importance.

Volta tried in various ways to add or multiply the electricity produced by a single bimetallic contact. He arranged different types of metals in various chains, but even with his complex arrangements the electricity remained weak and could be detected only with his ultrasensitive electrometers or via the physiological effects it produced. The metallic electricity did not generate sounds or sparks, did not attract pith balls or other light bodies, did not produce any clear chemical effects, and could not be used to heat a thin wire. It seemed to be, as people started to say, not true or genuine electricity, like that produced from frictional machines or electricity drawn from the heavens. Rather, it seemed to be something a bit different.

In fact, to many natural philosophers the metallic force seemed to have more in common with the torpedo's powers and with what others were calling animal electricity in their experiments with frogs and even people. These forces, real or envisioned, also did not display the typical signs of electricity. Rather ironically for Volta, the words often used by many people for this quasi-electrical force generated by metals and animals were "galvanism" and "galvanic." These were rather noncommittal terms that pointed to Galvani's supposed importance and priority in the discovery (Chapter 1)!

Among the many attempts Volta made to multiply his metallic electricity, there were some that involved connecting several bimetallic pieces to one another. Such arrangements were particularly easy to contrive after Volta started using metallic disks in his experiments, as demanded by Nicholson's duplicator. It was indeed simple enough to stack several disks in a column. These arrangements, however, were unsuccessful; the electricity remained as weak as that produced by a single bimetallic pair. Volta soon understood the reason for this. For instance, in a column made up of alternating disks of zinc and copper, the copper disks would become positive with respect to both the zinc below and the zinc above it. This effectively precluded any summation of the electricity generated at the different contacts.

Despite this setback, Volta continued his trials with great determination and even stubbornness. "If nature has succeeded in contriving something," he would write in 1802 (albeit in a somewhat different context), then "art should also be able to do it" (Volta, 1923, p. 62). Art in this case referred to human endeavors, a category that included technology (for more on wonders of nature inspiring "art" or technology, see Daston & Park, 1998, pp. 255–301).

Happily, nature stepped forth to assist Volta at the end of 1799. The solution to his problem came forth when his attention was drawn to electric fishes. As noted, Volta had long accepted that torpedoes and electric eels are electrical, although he rejected the notions of frog or human electricity. Notably, the organs of these fishes were also made up of many disk-like elements stacked in parallel columns (torpedoes) or rows (eels).

We have already discussed Volta's longstanding interest in electric fishes and have pointed out that references to electric fishes can found in his letters and other writings on animal electricity. Here we must interject that references to electric fishes, which had been rather frequent in Volta's first writings on animal electricity, had diminished in his published texts and manuscript after he began to attack Galvani. Suddenly, however, the fish notations escalated again, starting in 1798, when Volta was trying to multiply his metallic electricity. Let us now examine why this happened.

GALVANI'S TORPEDOES

Volta's newfound attention to electric fishes stemmed in good measure from one of the few documented trips Galvani made outside his native Bologna. As already mentioned, in 1797 Galvani had published a series of memoirs on animal electricity, the second of which described his "crucial" experiment in which two nerves touching each other produced contractions. Interestingly, it was not this memoir but the fifth one in the collection that had the greater impact on Volta, and that memoir concerned Galvani's experiments on live torpedoes.

Electric fishes had played a significant role in Galvani's decision to start his frog experiments in 1780. They were also invoked in his *De viribus* to substantiate his animal electricity hypothesis, involving nerve conduction and muscle contraction (Galvani, 1791). According to Galvani, this was because the electricity revealed in his experiments on frogs (and other ordinary animals) had many similarities with that of these fishes. At the time, however, Galvani was referring to these fishes only on the basis of what he knew from the literature, because he did not have the opportunity to work on them.

Eventually, in 1795, while embroiled in the controversy with Volta, he set forth "for leisure together with some honest friends to the shores of the Adriatic Sea." His destination was less than 100 miles from Bologna, and it provided him with the opportunity to study the anatomy and physiology of the local torpedoes, which were easily caught in the warm, shallow waters of the Adriatic at certain times of the year. Over the few days of his brief sojourn, Galvani (1797, 1937b) was able to conduct a series of experiments. And in his mind they confirmed the essential similarity of fish and frog electricity, which he viewed as basically similar to the electricity produced with frictional machines.

Galvani's experimental results largely replicated findings obtained by previous authors (including Walsh and Spallanzani, the addressee of the memoirs), particularly regarding how the nerves to the electric organs are essential for the torpedo's shocks. In his initial anatomical dissections, Galvani confirmed John Hunter, who wrote about a particular abundance of nerves to the electric organs (Galvani would write about "a boundless abundance of nerves"). He also confirmed the findings obtained by one of Spallanzani's correspondents, anatomist Michele Girardi, who noted that some branches of the nerves directed to the electric organ also supply adjacent muscle tissues. Convinced as he (along with Hunter) was that "the brain is the laboratory of production and the main reservoir of electricity" and "nerves are the conductors" of the electricity directed to the organs, Galvani accepted these findings as a confirmation of his views on the similarity of the electrical fluid directed to the fish organs and the nervous fluid directed to more ordinary animal tissues. In other words, it was highly unlikely that the same nerve could contain two fluids of two different natures that would go to two different destinations. If the one is electrical, the law of parsimony dictated that the other should also be electrical. As stated by Newton (1729/1971, vol. 2, p. 398) in the "Rules of Reasoning" found in his *Principia*: "We are to admit no more causes of natural things than such as are both true and sufficient to explain their appearances . . . for Nature is pleased with simplicity. . . ."

Galvani next investigated the effects of abolishing the nerve supplies to the electric organs. This was done by cutting nerves or by ablating the brain, the supposed source of the electricity. In both cases the shocks stopped instantaneously, despite the fact that the torpedo could maintain its vitality for quite a while. The absence of shocks did not depend on a more systemic weakening of vital forces. In accordance with this observation, removing a torpedo's heart reduced the fish's vitality but did not have an immediate effect on its shock production.

Galvani also conducted several experiments aimed at comparing the characteristics of the electricity produced by the fish's electric organs with the electricity seemingly intrinsic to frogs. He confirmed many of the similarities that he had already discussed in *De viribus*, and insisted that, in order to be effective, both must circulate through complete conductive circuits. This property seemed to differentiate this form of electricity from the frictional type, in which simply bringing one's finger close to an electric machine could produce effects without the need of any apparent circuit.

Among the differences noted between torpedoes and frogs, in addition to the obvious one, which was the stronger intensity of a torpedo's electricity, was that the strength of the shocks did not increase when a torpedo was covered with metallic laminas. These laminas, however, seemed to strengthened the frog's electricity in a significant way. Galvani accounted for this difference (in an anti-Volta way) by assuming that the metallic "armatures" do not produce any electricity and succeed only in producing a small intensification of the existing animal electricity. For Galvani, this effect is detectable only when the existing current is weak (as it is with nerves and muscles), not when "electricity is by its nature strong and robust, as is the case with torpedo" (Galvani, 1797, pp. 68–69).

In some of his experiments, Galvani put frog preparations on the body of a live torpedo. He was surprised to see the frog's legs contract, even when the fish did not give any overt signs of releasing a shock. This finding was interpreted by Galvani as evidence for a continuous flux of weak electricity from the fish organs. It indirectly supported his opinion, previously expressed in the *Trattato*, that the animal electricity of frogs circulates continuously inside living tissues, although it does not ordinarily produce visible contractions because of its weakness (Galvani, 1794b).

Profiting from the great sensitivity obtained with prepared frogs, Galvani was able to show that a torpedo's shocks could diffuse at considerable distances through the water. They could even be transmitted through the humidity present on the experimental table. This was something that Walsh (1773) was unable to observe at La Rochelle (Chapter 15).

As already indicated, Galvani was not only interested in showing that the electricity of frogs is similar in nature to that of torpedoes, but he also wished to demonstrate that both are identical to artificial electricity. In this regard, he established an interesting parallel between fish and artificial electricity in some experiments in which he investigated the effects of both types of electricity on the legs and heart of a frog. The finding was the same with both forms of electricity: leg muscles contracted more readily than heart muscles.

In commenting on these experiments to Spallanzani, he wrote: "If I am not wrong, from all these facts you can see very clearly and without any doubt that the stimulating quality of the electric fluid of the torpedo is entirely similar to that of common electricity." He added that he could "also further confirm the analogy of the torpedo with the magic square," meaning Franklin's planar or square capacitor (Galvani, 1797, pp. 75–76).

Hence, Galvani concluded from a variety of experiments that the three types of electricity he was considering are fundamentally similar. He opined that apparent differences depended only on specific properties or conditions that modified the strengths or other characteristics of the electric phenomena. In particular, any differences between fish, frog, and artificial electricity could be ascribed to the "particular machine" present in living organisms. This machine would impart the specific properties necessary to suit the physiological needs to the electric fluid present in living tissues.

This was a concept that he had already elaborated in the *Trattato*, where, when discussing the source of the "disequilibrium" responsible for electric fluid movement along the nerves, he reasoned:

Such disequilibrium in the animal either must be there naturally or should result from some kind of artifice. If it is there naturally, we should admit that in the animal there is a particular machine capable of generating such a disequilibrium, and it will be convenient to refer to this form of electricity as an animal electricity in order to denote, not an electricity whatsoever, but a particular one referred to a particular machine. (Galvani, 1794b, pp. 70–71)

Well conscious of the difficulty of ascertaining the operating mechanisms of this machine with the practical and theoretical tools of his epoch, Galvani added a notation, which reflects Newton's continuing influence:

But what will it be, this animal machine? We cannot establish it with certitude; it remains totally occult to the most acute sight; we can do nothing else than figure out its properties, and, from these, somewhat envision its nature. (Galvani, 1794b, p. 76)

Galvani regarded the assembly of the disks as basic to the operations of the fish's electric organs. He likened the disks to a series "of small magic squares," meaning Franklin's square capacitors. He thus endorsed the idea that had emerged particularly with Walsh (1773), and which was then elaborated upon by Cavendish (1776a), of a columnar assembly of minute planar capacitors.

In addition to the specific reflections it stimulated in Volta, the importance of Galvani's memoir on torpedoes was that it brought the image of electric fishes back to Volta's attention. As noted, this happened during a period in which his gaze seemed to be concentrated exclusively on the physics of metals and not on living nature. This reawakening is well documented in Volta's writings, and particularly in a 1798 letter addressed to Johann Peter Frank, an illustrious physician of the epoch and Volta's colleague in Pavia, who had since moved to Vienna.

NICHOLSON'S MODEL AND VOLTA'S REACTION

In the same year in which Galvani published his collection of memoirs on animal electricity, including the one on torpedoes, William Nicholson published a largely theoretical study. In it he revived the hypothesis that the electric organs of fishes work like an assembly of minute planar capacitors. In a communication titled "Observations on the Electrophore, Tending to Explain the Means by which the Torpedo and Other Fish Communicate the Electric Shock," which appeared in the *Journal of Natural Philosophy, Chemistry and the Arts* in 1797, Nicholson proposed a physical model based on stacking of minute disks made of mica or resin. This model was inspired by the structure of the electrophore, the instrument invented by Volta.

As with the previous models of Walsh (1773) and Cavendish (1776a), Nicholson's (1797) model necessarily implied the presence of some sort of insulating material inside the animal tissue—an insulating layer being an essential characteristic of all capacitor devices. This assumption, however, stood in direct contrast with another line of thinking at the time: that all animal tissues, especially moist tissues, are good conductors of electricity. Thus, while helping to redirect Volta's attention to electric fish organs and even to other animal tissues as possible sources of electricity, Nicholson's model also stimulated justifiable skepticism in Volta.

As Volta saw it, there is nothing insulating in animal tissues, and therefore electric fish organs could not work like an assembly of capacitors. Rather, it was more likely that they worked following the new physical principle that he had discovered, the principle underlying the electromotive action of metals, because metals are, as is true of bodily tissues, electrically conductive. Yet some electric fishes, such as torpedoes and South American eels, are able to produce strong electrical effects, while bimetallic pairs or assemblies of metallic disks can generate only feeble electricity. And this, in short, was clearly problematic.

VOLTA'S ARTIFICIAL ELECTRICAL ORGAN

As mentioned above, Volta had confidence in the principle that art could imitate nature. This confidence, especially when Galvani's and Nicholson's memoirs had brought electric fishes back into his mental elaborations, seemed to play a critical role in the "great step" that now led him along the final pathway of his journey to his battery. The main difference between the assembly of the disks making up the electric organs, so powerful that they could produce electricity that could torpify at a distance, and the minimally effective assembly of metallic disks had to lie in the nature and assembly of the disks—the metals in the device contrived by Volta and the humid disks in the fish organs.

Eventually, around the fall of 1799, Volta decided (nobody knows precisely how and why, perhaps not even Volta himself), to interpolate humid disks in the form of circles of paper or cloth soaked in salt or acid solutions or even simply in water in between his bimetallic disks. The result was the culmination of a long-sought dream, one dating back to when Volta had become convinced of the electromotive actions of metal contacts. He now discovered that assembling numerous (50 or more) bimetallic pairs (e.g., tin and copper) together could indeed produce strong electrical effects—provided that humid material separated the coupled disks from each other.

The apparatus that Volta contrived in various forms could not only produce powerful physiological effects, but could also generate heat, visible sparks, and chemical effects. That is, it could produce all the actions expected from true electricity. The difference between the frictional electric machines (and other traditional devices capable of generating strong discharges) was that Volta's new instrument produced its electrical effects in a continuous, seemingly unending stream, whereas these earlier devices exhausted their charges in an explosive moment's time.

With his new instrument, Volta was now able to provide a particularly effective demonstration of the physiological effects of metallic electricity, as he made clear in 1800, in the letter in which he announced his invention to Joseph Banks, then President of the Royal Society (Fig. 22.8):

But the most curious of all these experiments is, to hold the metallic plate between the lips and in contact with the tip of tongue; since, when you afterwards complete the circle in the proper manner, you excite at once, if the apparatus is sufficiently large and in good order, and the electric current sufficiently strong and in good course, a sensation of light in the eyes, a convulsion in the lips, and even in the

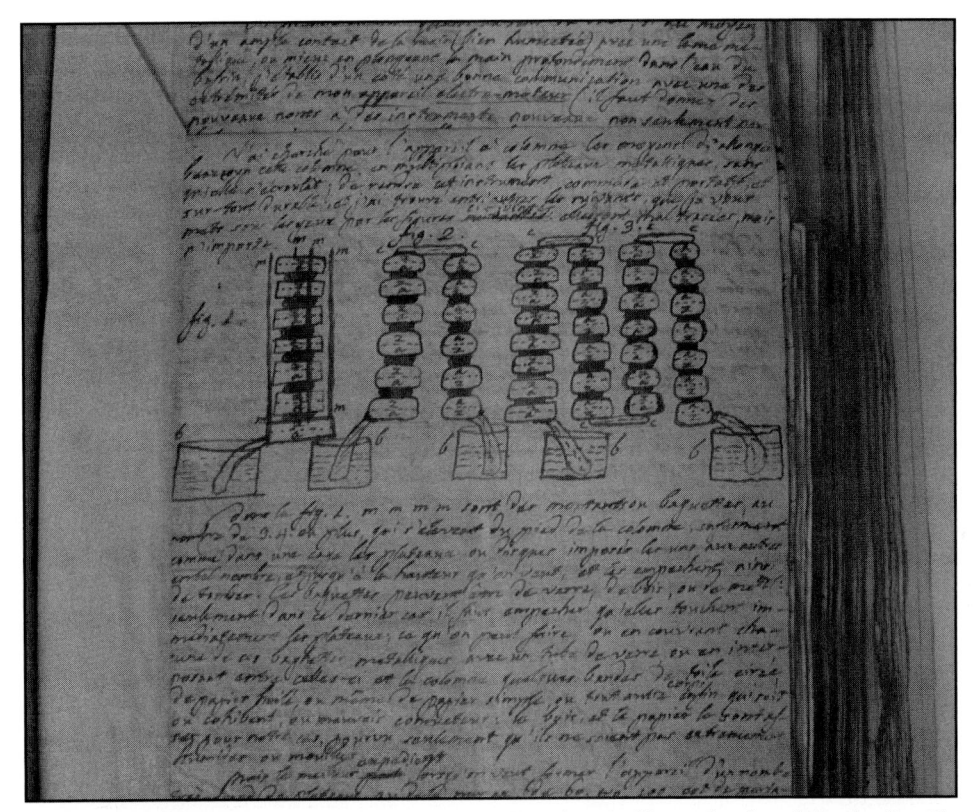

Figure 22.8. A detail of a page of Volta's 1800 communication written in French and addressed to Joseph Banks, President of the Royal Society. The image shows a schema of Volta's new pile or battery, which was based on the organs of electric fish (see Chapter 1 for another illustration of the pile's components from this same letter; © The Royal Society).

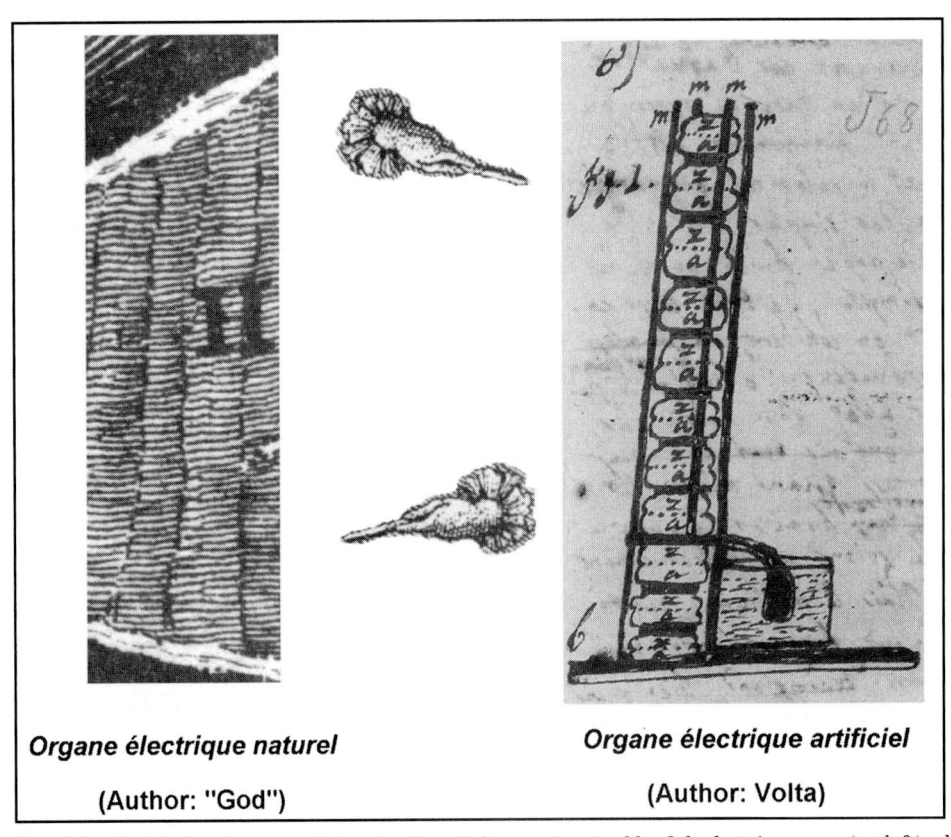

Organe électrique naturel

(Author: "God")

Organe électrique artificiel

(Author: Volta)

Figure 22.9. "Natural" and "artificial electric organs," showing how Volta was inspired by fish electric organs (on left) when he invented his battery (on right) and how he drew from his invention the principles to account for the electric organ functioning (from Piccolino, 2000b).

Figure 22.10. Left: The first page of Volta's (1800) article in the *Philosophical Transactions*, based on the communication sent to Banks. Notice the English title and the text in French. William Nicholson provided a full English translation of the text in the *Philosophical Magazine*, which he edited (image on right).

tongue, and a painful prick at the tip of it, followed by a sensation of taste (Volta, 1800, pp. 426–427).

The stronger power of the new device, as opposed to his previous constructions using metals, also had the potential to be useful in medicine. In 1802, Volta applied the electricity from one of his batteries to a 15-year-old girl completely deaf from the moment of birth. The girl was given electrical shocks in both ears over several days, one every second in 10-minute sessions. Although Volta was not completely satisfied with the results, he remarked that his young patient "has acquired the hearing to distinguish various sounds, even of rather low intensity, and from a distance of some feet" (Volta, 1923, p. 181). Among Volta's opinions on medical electricity, was his belief that, in order to produce needed physiological and medical effects, the stream of electricity that could now be applied with his battery should be presented in an intermittent, discontinuous way.

To Volta, it was his new apparatus, and not the model contrived by Cavendish in 1776 or Nicholson in 1797, that provided the best model for the electrical organs of the torpedo and electric eel (Fig. 22.9). In part, this was because the battery was made exclusively from conductive materials, just like the animal organ. This similarity was not, however, all to the picture he was painting. There was also morphological similarity in the design of his device, and functional similarity in the fact that it, much like a real electric fish, could produce a train of shocks in succession, unlike a Leyden jar that would produce a single shock and would have to be refilled after each discharge. For these reasons, and with homage to the electric fishes that had provided him with his insights, he called his new device *organe électrique artificiel* in the dissertation letter in which, on March 20, 1800, he described his invention (in French) to the Royal Society of London (Fig. 22.10).

With his battery, Volta had contrived one of the most important tools in the history of the physical sciences

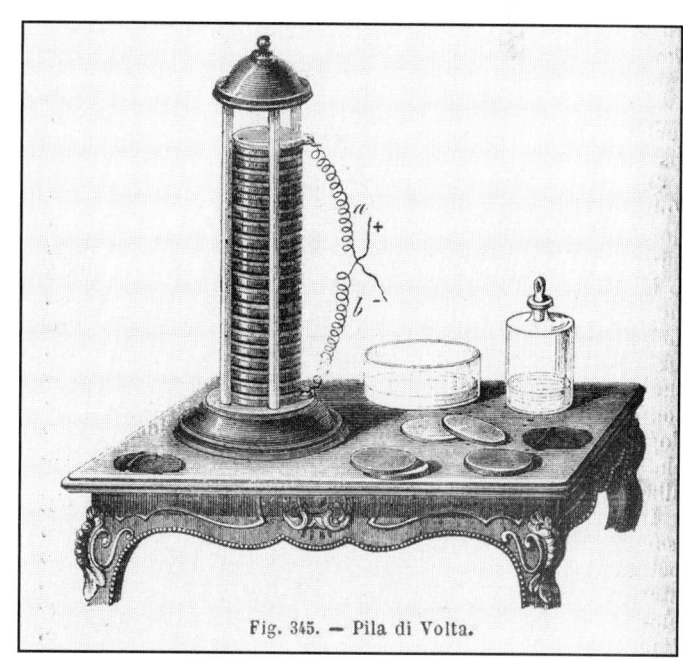

Figure 22.11. A Voltaic pile (from Ganot, 1882).

(Fig. 22.11), and one that would find a myriad of applications in science, medicine, and everyday life. The physical principles underlying the functioning of the battery would be of fundamental importance in providing insights into the electrical forces in the physical constitution of matter. But more than this, the main reason why Volta seemed to be so proud of his invention was that he thought that he would now be able to solve one of the most challenging mysteries of living nature: disentangling the mechanisms underlying the power of electric fishes. This appears particularly clearly in the passage concluding his long letter to Joseph Banks, where electric fishes are again presented prominently:

To what electricity then, or to what instrument ought the organ of the torpedo or electric eel, &c. to be compared? To that which I have constructed according to the new principle of electricity, discovered by me some years ago, and which my successive experiments, particularly those with which I am at present engaged, have so well confirmed, viz. that conductors are also, in certain cases, exciters of electricity in the case of the mutual contact of those of different kinds, &c. in that apparatus which I have named the *artificial electric organ*, and which being at bottom the same as the natural organ of the torpedo, resembles it also in its form, as I have advanced. (Volta, 1800, p. 311)

Chapter 23
Galvanism *contra* "Voltaism": Electric Fishes and the "Unsolvable" Dilemma

> [This] immortal work, which inaugurated a new epoch in the whole field of science, is a most brilliant illustration of the extreme fruitfulness of an intimate combination of the exploration of the laws of inanimate nature with the study of the properties of living organisms.
> Niels Bohr, 1937 (Celebrating Galvani's and Volta's achievements, on the occasion of the 200th anniversary of Galvani's birth)

With the invention of his battery—that is, his artificial electrical organ—Volta (Fig. 23.1) obtained the kind of celebrity and fame that is granted only to the greatest contributors to the sciences. He became a scientific hero, a giant celebrated alongside other such giants, including Nicolaus Copernicus, Johannes Kepler, Galileo Galilei, and Isaac Newton. In 1831, Dominique François Jean Arago, speaking before the *Académie de France* on the occasion of Volta's death, declared the battery to be of even greater historical importance than the telescope or the steam engine! Eighteen years later, Volta was listed among the scientist "saints" in Auguste Comte's *Calendrier Positiviste* (Pancaldi, 2003).

The ascent of Volta's celebrity well outside the circles of the natural philosophers of the epoch had begun well before his death in 1827. One of the crucial events in Volta's apotheosis occurred in 1801, when Emperor Napoleon Bonaparte attended the meeting of the newly established *Institut de France*, which had invited Volta to demonstrate his new device (Fig. 23.2). At this celebration, Franklin was lauded as the landmark figure behind the first main phase of electrical science, which was characterized by frictional electricity and explosive discharges, like those associated with the heavens and Leyden jars. The second phase belonged to Volta, and it was associated with the continuous flow of electricity, a development stemming from his discovery that electricity could be generated with metallic contacts, such as those now making up his revolutionary battery.

After his successful and impressive demonstration, the *Institut* awarded Volta its new, special prize for discoveries in the field of electricity. Napoleon also showered Volta with numerous honors and gifts. He made him a *Chevalier de la Legion d'honneur* and gave him the title of Count of the Kingdom of Italy (Fig. 23.3). So impressed was Napoleon I with Volta's device, genius, and loyalty to the new republic that he even gave him a lifetime pension, in addition to other cash and gold awards.

HUMBOLDT'S CONVERSION AND VOLTA'S IMPACT

Among natural philosophers, Alexander von Humboldt was one of the greatest admirers of Volta's achievements. In the introduction to his memoir on the electric eels of South America, he praised the scholar of Pavia, who "by the force of his genius and an unrivalled sagacity, unveiled the mystery under which the Galvanic phenomena remained enveloped during a long series of years."(Humboldt, 1811, p. 49)

Humboldt learned about Volta's battery before he returned to Europe from the Americas in 1804. Humboldt even went to meet with Volta in Milan and Como in 1805. With Volta's help, Humboldt now recognized that several factors could affect whether electricity could be transmitted across a piece of bone, some charcoal, or a flame. This also held true for electricity from frictional machines, the atmosphere, different metals, and some fishes. The limited amount of current generated, he now realized, had a lot to do with his own negative findings, both as reported in his book of 1797, which did not include experiments on electric fishes, and thereafter with his experiments on electric fishes. In brief, whereas Humboldt had earlier believed that the animal force and metals produced something like electricity, a force that did not have all of the features of true electricity (see Chapter 1), he left Volta convinced that there is but a single electricity in nature. In Humboldt, and by no means in him alone, Volta had a convert.

A year after the two men had met, Volta responded warmly to a letter sent by the German baron, telling him that "In comparison to your own researches, such a little thing is my one." Volta politely pointed out that his own major contribution was contained "in a small sphere," whereas Humboldt had been successful in learning about "the three kingdoms of nature, embracing the sky, the earth, and the atmosphere." If Volta had ever felt coldly about Humboldt with his *Naturphilosophie* and different way of viewing Nature, and

Figure 23.1. Volta (1745–1827) later in life with his battery and electrophore, two of the instruments that won him great fame as the eighteenth century ended and the nineteenth century began.

with his "galvanism" in particular, there was certainly no evidence of this now.

When Humboldt wrote up his eel encounter for his narrative and other publications, as well as when he described the torpedo experiments he conducted off Naples soon after he returned to Europe, he showed this new understanding of electricity. He also repeatedly brought up Volta's name as the leading authority on the physics of electricity. For example, right after providing information on the sizes and weights of his eels, he speculated on the possible functional important of the abundant mucous matter covering the fish's skin.

"Might it be that this mucous matter contributes to make the fish's skin more apt to propagate the influence of the electric organs?" he wrote, "since, as Volta has proved, all the fluids produced inside the living bodies conduct electricity twenty or thirty times better than pure water"? (Humboldt 1811, p. 61, translated from the French). This sentence is notable in a second respect as well, because Humboldt did not use the term "animal electricity" or "galvanic fluid," or any such synonym in it. Instead, the force he had previously wrestled with is now designated as "electric," without any adjectives or other modifier attached to this word. This new language can also be found where he discusses the eel's "*electric organs,* which occupy more than two-thirds of the animal's body" (italics ours). Here too his choice of words shows that his earlier thinking about a quasi-electrical force was no longer guiding him.

Humboldt even mentions the battery several times in his passages about the electric eel, as for instance when he writes: "To perceive the difference that exists between the sensation produced by the Voltaic battery and electric fishes, the latter should be touched when they are in a state of extreme weakness" (p. 67). For Humboldt, the perceived differences that can occur between a fish and an electrical machine are now readily explainable. Alluding to some of

Figure 23.2. Volta showing his battery to Napoleon Bonaparte (1761–1821) in Paris during a *séance* of the *Institut de France*.

Figure 23.3. Left: Volta made a *Chevalier de la Legion d'honneur* by Napoleon. Right: The document by which Napoleon donned Volta "Count of our Kingdom of Italy" during the epoch when the French controlled the Lombardy region.

the painful self-experimentation he had performed on his blistered shoulders in the 1790s, he tells his readers:

> ... the sensation caused by the feeble shocks of an electric eel appeared to me analogous to that painful twitching produced by silver and zinc to sores on the back and the hand. These sores, which I had made on myself, the first by means of cantharides, the second by means of a superficial cutting, have provided some very convincing evidence on the relations that exist between the effects of electric fishes and those of a galvanic current arising from the application of heterogeneous metals on human organs. (p. 67)

For Humboldt, the action of the fish on human organs is transmitted and intercepted by the same bodies that transmit and intercept the electrical current of the charged conductor of a Leyden jar, or of a Voltaic battery. Some of the anomalies observed in his electric eels experiments, are easily explained, he writes, when one recollects that even metals (as is proved from their incandescence when exposed to the action of a battery) present a slight obstacle to the passage of the electricity; and that a bad conductor annihilates the effects of a feeble electricity on our organs, whereas it transmits to us the effect of a very strong one.

After discussing some circumstances in which even a relatively good conductor can interfere with the physiological effects produced by a weak current, Humboldt concludes that these effects are dependent on three variables: the energy of the electromotive apparatus, the conductibility of the medium, and the irritability of the organs that receive the impressions. In his opinion, it is because experiments have not been sufficiently multiplied with a view to these three variable elements, that, in the action of electric eels and torpedoes, accidental circumstances have been taken for absolute conditions, without which the electric shocks are not felt.

Humboldt explains that very dry wood, horn, and even bones could prevent the shocks of the eel from being felt, and that it takes a very strong discharge to be felt through imperfect conductors. Further, he recognized that the extension and strength and other conditions of the contact could influence the transmission of the shock, citing, for example, some experiments in which the *Gymnotus* was put in a wet pot made from brown clay. As for the differences between the shocks of a torpedo and the eel, in his opinion the cause of these anomalies is most likely due to the inequalities of the electric powers in these fishes, rather than to substantial differences between the mechanisms responsible for their shocks. With regard to the bigger picture: "The difference observed between the effects of the Voltaic battery or a feebly-charged Leyden and electric fishes do not indicate any heterogeneity in the cause [responsible for them]" (p. 84).

AN INSTRUMENT FOR THE AGES

Above and beyond the honors bestowed upon Volta by the French and by others, and even with the concrete example of Humboldt's rather dramatic conversion from an ill-defined animal force to a single electricity, the importance of Volta's artificial electric organ in the progress of both science and technology cannot be overestimated. One need only examine the scientific literature after 1800, when Volta's description of his new device that structurally and functionally resembled the electric organs of some specialized fish was published in the *Philosophical Transactions of the Royal Society of London* (Volta, 1800).

In the same year as Volta's letter was published, William Nicholson and Anthony Carlisle and other fellows of the Royal Society employed the new instrument for the electrochemical decomposition of water (Fig. 23.4). After this, there was a significant thrust to study the chemical effects of the battery and the chemical phenomena basic to its functioning (phenomena that Volta never properly acknowledged, convinced as he was of the purely conductive role of moist disks in his device). Humphry Davy and Michael Faraday, Davy's pupil and successor at the Royal Institution of London (where an enormous battery of more than 2,000 bimetallic couples was constructed), performed some of the most important of these studies (Figs. 23.5 and 23.6). Davy isolated a number of new chemical elements (sodium, potassium, calcium, magnesium, barium, strontium) with the battery. Not one to be

Figure 23.4. The article from 1800 by Nicholson, Carlisle, Cruickshank, and other Fellows of the Royal Society dealing with the dissociation of the water using Volta's battery. Notice that the term "Galvanic" is here applied to the battery's electricity. This article was published in the same issue of *The Philosophical Magazine* in which Volta's letter describing his new battery in French was translated into English (see Fig. 22.10).

Figure 23.5. Humphry Davy (1778–1829), who isolated many new elements with a massive battery housed at the Royal Institution of London, showed that the electricity generated by Volta's battery depends on chemical mechanisms. He also had an interest in electric fishes (see Chapter 24).

outdone, Faraday eventually developed his fundamental laws of electrochemical actions based on it.

Danish physicist Hans Christian Oersted's discovery of magnetic phenomena associated with electric currents, together with the demonstration of the reciprocity of the magnetic and electrical forces (put in the proper light by Faraday), would open new and previously unexpected horizons for the technological development of electricity (Fig. 23.7). With the construction of powerful electrical alternators and dynamos, these seminal discoveries would lead to the industrial production of electricity on a grand scale, and to the development of ever more powerful electrical engines. Eventually they would prompt the discovery of electromagnetic waves, thus marking the birth of the modern technology of telecommunications.

Also, electrochemical studies, particularly those carried out first by Faraday in England and afterwards by physical chemist Walther Nernst in Germany, provided insights into the close association of electricity with the basic structure of matter. Their insights would eventually pave the way for the development of modern atomic physics.

Even if Volta had some awareness of the importance of his discovery, which he did, he could never have imagined the great extent to which scientific progress would benefit from his artificial electrical organ, subsequently referred to by him as his electromotor or electric battery (and by the French as the Voltaic *pilière* or *pile Voltaïque*). This certainly would have added further gratification to his self-satisfaction, which started after he looked carefully at Hunter's drawings of the electric organs of some European torpedoes and South American electric eels!

THE NAGGING ISSUE OF GENUINE ELECTRICITY

There is, however, no glory without some pain in the lives of human beings. Despite his overwhelming successes, Volta soon started to be dissatisfied by the way the electricity from his battery was viewed by some of the experts in the field. He was particularly discontented by the fact that, unlike Humboldt and certain others, there were still some scientists who did not view his newly discovered metallic electricity as a genuine electricity—one identical to the artificial electricity of friction machines and the natural electricity witnessed during lightning storms. Rather, these scientists

Figure 23.6. Michael Faraday (1791–1867), Humphry Davy's pupil and successor at the Royal Institution of London. He also isolated new elements with the huge battery and developed his fundamental laws of electrochemical actions with it. Like Davy, Faraday's interests extended to electric fishes (see Chapter 24).

Figure 23.7. Hans Christian Ørsted (Oersted; 1777–1851), the Danish physicist who associated magnetic phenomena with electrical currents, a development that had major effects on industry, communications, and various aspects of everyday life.

continued to see it as a somewhat different force or fluid, similar to genuine (frictional or atmospheric) electricity only in some of its characteristics, yet different in others.

Despite the strong physiological action of the battery on animals and their organs, and even on the bodies of human subjects, both living and recently executed (Aldini, 1804), it was, in fact, relatively difficult to generate sparks or sounds with Voltaic piles. Further, there were other differences, as with the ability to move pith balls and other light bodies. These differences could be explained only in part by differences in "tension" (i.e., potential)—that is, by taking the higher tensions of these more established forms of electricity into account. And yet another of the characteristics of the battery that seemed to suggest a different force was that it often seemed to produce stronger physiological effects than a Leyden jar charged to exactly the same tension.

Some of these difficulties were similar to those that had prevented the acceptance of the electric nature of the shock of torpedoes and electric eel for decades. As a matter of fact, it was widely remarked that the pile's actions had many features in common with animal electricity—that is, the electricity generated by electric fishes and, as Galvani asserted, the intrinsic electricity underlying nerve conduction and muscle contractions in frogs, cattle, and even humans. After all, living bodies did not produce sparks or attract pith balls either. This, as we have already noted, led to the expressions "galvanic fluid" or "galvanism" to designate the fluid stemming from metals and consequently the battery (Fig. 23.8). Less commonly, but in the same way, expressions such as "metallic electricity" or "metallic irritation" could be found in the literature, following the wording that Humboldt had originally used in his two volumes on animal electricity dating from 1797 (*galvanisme* was, however, the term used in the French translation from the German; see Chapter 1).

To face this problem, Volta was obliged to reconsider the subject thoroughly, and in this context he again made frequent allusions to electric fishes. Moreover, he did not limit himself to rhetorical arguments in support of his conviction of the perfect "identity of the electric fluid with the so-called galvanic fluid." He also conducted a series of experiments aimed at demonstrating that his new device, and not the "artificial torpedoes" contrived before him (particularly by Cavendish and Nicholson, see Chapters 15 and 22), best imitated the natural organs of these fishes.

This was particularly well expressed in a long memoir that Volta composed between 1801 and 1805—the contents of which he would have discussed with Humboldt when the younger man visited him in 1805. This document, however, was not published at that time. It finally appeared in 1814, and when it finally came out it bore not his name but that of his student Pietro Configliachi (Fig. 23.9).

In this document, Volta frequently referred to the "celebrated English physicist . . . Lord Kavendish" [*sic*], and to "his small machine, that he liked to denote as *artificial Torpedo*." In spite of being capable of imitating some of

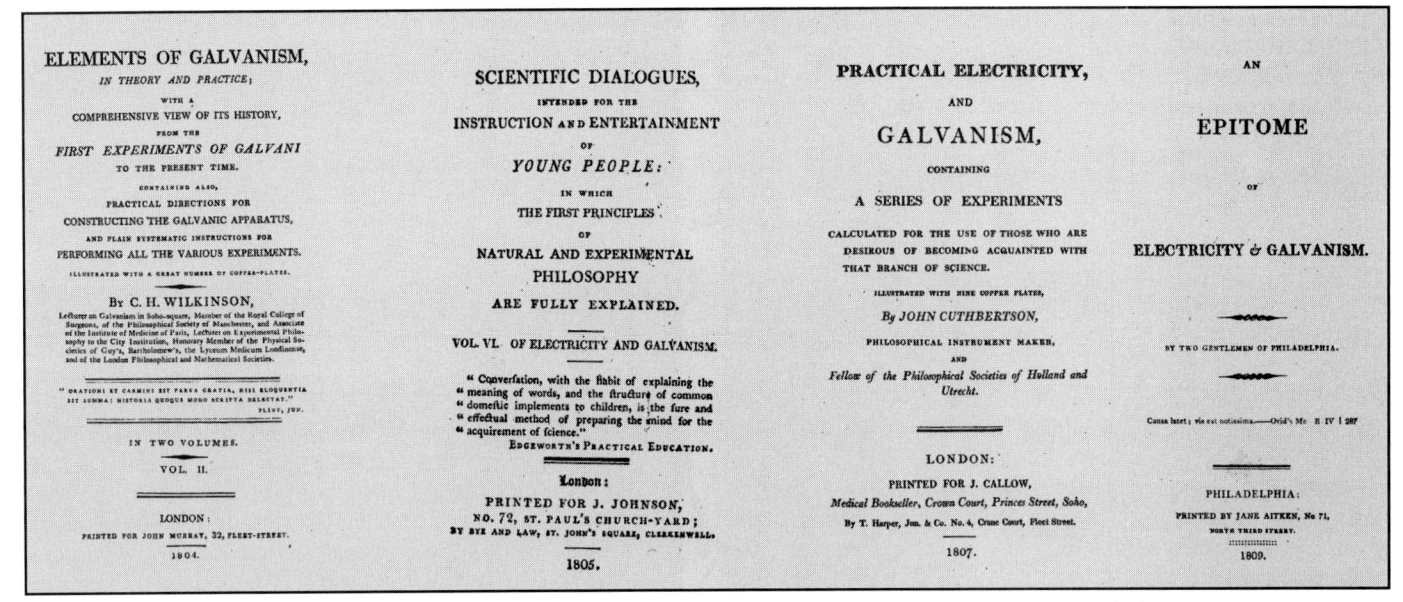

Figure 23.8. Four of the numerous volumes published after Volta invented his battery in which the term "galvanism" was used to refer to the electricity produced by the new instrument. Volta was upset by this term, as it conveyed the mistaken impression that Luigi Galvani had discovered metallic electricity, and also because it led his metallic electricity to be associated with what others were calling animal electricity in more than a few strange fishes.

the effects of the natural torpedo, Cavendish's contrivance "did not imitate the natural one intrinsically, not for any proper virtue or action with which the last one is endowed." Thus, although

> the experiments and observations of Kavendish [are] very beautiful and instructing, they lack much of what, to their completion, was added by our Italian [read Volta], with the discovery of the Electromotor [Voltaic battery], and with the application of this extraordinary instrument to the true, or at least most probable, explanation of Torpedo's phenomena, and, moreover, to the most perfect and complete imitation of the same effects (Volta, 1923, p, 268).

For Volta, the reason for the superiority of his device in imitation of the natural torpedo (and thus making it a more appropriate *artificial torpedo*) had much to do with its construction. Unlike the older models, his battery did not use capacitors made with two conductive bodies separated by an insulating substance. In brief, Volta was convinced that the natural electric organ is based on the electromotive actions between conductive substances of different types, and he believed that insulating substances did not have a place either in this picture or in any model worthy of the designation "artificial torpedo" or "artificial electrical organ."

IMITATING TORPEDOES

In an impressive series of experiments that would terminate only at the end of his scientific life, Volta worked hard to show that he could imitate all the phenomena of a torpedo with his electric battery. His device could produce shocks even if immersed in water, and it could produce its electrical effects for considerable distances under the water. Along with torpedoes and the electric eel, the battery also gave powerful shocks that affected animals and humans, even though it was rather ineffective in producing the classical

Figure 23.9. The memoir in which Volta asserted that the fluid produced by the battery is, like that of the electric fishes, genuine electricity. This work was published in 1814 under the name of his pupil, Pietro Configliachi.

signs of frictional electricity. These characteristics in common, and differences from charged Leyden jars, were explained by assuming that both the battery and the fishes could produce intense electric currents but at relatively low tensions (i.e., small electrical potentials). Following in Cavendish's footsteps, and using a sophisticated tool especially contrived by him (e.g., the spincterometer), Volta made an accurate estimate of the dependence of the sparking distance on this tension.

If we convert the tension Volta measured with his primitive (although sensitive) electrometers in these experiments to modern units of electrical potential (i.e., Volts [V]), we can better understand why Walsh could obtain a spark from the eel but not from his torpedoes (Chapters 15 and 19). At a tension corresponding to about 330 V (a value within the range of the discharge of the electric eel), Volta estimated that the discharge might cross 1/16 of a line (i.e., approximately 0.1325 mm), or roughly "the thickness of a paper sheet." In contrast, the discharge from a small torpedo, like those that he had in La Rochelle, is only about 40 to 50 V. For a potential in this range, Volta calculated an explosive distance of only about "1/160 of a line" (approximately 1/10th that of an eel and something like 13 µm in modern units). When Walsh made a cut in a metal circuit with his penknife while at La Rochelle, the gap was too great for him to obtain a spark with his torpedoes.

By reinforcing the argument already developed in his 1800 letter to Banks, Volta, under Configliachi's name, concluded his comparison between the electric fish organs and his pile by writing:

> We are induced further to believe that fundamentally, i.e., as to the essential construction, they are the same, namely that their virtue and activity come from the general principle established by Volta. This principle states that different conductors, when put in mutual contact, are also motors of electricity. Keeping this in mind, it is easy to suppose that the small laminas, or pellicles, stacked one above the other in great number in many small columns or tubes (of which these organs are made), differ one from the other. This might happen in such a guise that two and three different substances follow one after the other, in an alternate fashion, together with some type of humour, by which they would be interpolated two by two. In sum, they would be in a convenient order, as indeed are the double metallic laminas, and interpolated by a third conductor in the [Voltaic] batteries: to which the name of artificial electric organs proposed by Volta would be appropriate. (trans. from Volta, 1923, p. 268)

Nevertheless, Volta had some difficulties in conceiving the mechanism of an electric fish's shocks in the light of his hypothesis. The main problem was to figure out how electric fishes could produce their electrical actions voluntarily, whereas a powerful shock could be produced in the battery by simply closing the circuit between the poles, the device being constantly charged and active.

To account for this difference, Volta suggested that the discs or membranes of the natural electric organs in the resting condition might not be in a situation of effective contact, a prerequisite for a powerful electromotive action. This might be due to an inappropriate spatial arrangement or perhaps because of some other reason or reasons. As he wrote in 1805, in a letter addressed to Configliachi, who was then planning experiments on torpedoes in the Italian Riviera, the shock would be produced every time that "either as consequence of a voluntary effort of the animal, or in some other manner, a congruous contact would be established or made, or a complete communication would be completed, between those parts of the organs that, in the natural state, with our fish free and quiet, happen to be disjointed or ill-communicating" (Volta, 1923, p. 195).

Volta attached considerable importance to the mechanical movements sometimes observed in torpedoes at the moment of the shock. These were the movements on which Lorenzini and Réaumur had based their mechanical hypotheses for the rays' shocks (Chapter 9). In Volta's opinion, these movements were an expression of the animal's effort to bring the discs of the electric organ into close contact, in order to produce a strong electric effect. If this were the case, Volta argued, there would be only a small flux of electricity, insufficient to produce a shock, yet detectable with some sensitive measuring device, with the animal at rest.

In this context, Volta recalled the "beautiful experiments of Galvani," the research he conducted on torpedoes caught off Italy's Adriatic coast. In these experiments, Galvani had been able to show that there was "some continuous passage of electric fluid from the back to the belly," even when the torpedoes did not produce any overt shocks (see Chapter 22). The doctor of Bologna discovered this using prepared frogs as highly sensitive electrometers.

In keeping with his own hypothesis of the torpedo's shock, Volta suggested a possible way to force a fish to produce a continuous electricity, so as to make it measurable with a sensitive instrument like his *condensatore*. In his letter to Configliachi, he wrote that "by placing, on the back of the Torpedo laid down with its belly on the basin, or directly on the wet cloth, a weight that might compress it enough, one might, I think, oblige those organs to act in a continuous way" (Volta, 1923, p. 195). He was convinced that the difficulties that people were experiencing in the measurement domain depended mainly on the transient character of the fish's electricity. This contrasts with the continuous nature of the current generated by the battery.

Configliachi performed his planned research on torpedoes, but his subsequent publication was stopped during the editorial work.[1] Nevertheless, we know its title and eventual publication year, which was 1811, as well as that it contained three engraved plates and was dedicated to Prince Eugène Napoleon (de Beauharnais), then viceroy of Italy (Configliachi, 1811; see Zantedeschi, 1845). More importantly, Configliachi did not succeed in measuring torpedinal electricity, even though he probably followed Volta's advice to squeeze the ray. Since the mechanisms whereby electric fishes produce their shocks differ substantially from Volta's hypothesis, it is in any case extremely unlikely that physical compression of the torpedo's body would have intensified or

---

[1] We have an echo of Configliachi's research on torpedoes, in a short mention made by Humboldt in his narrative of the experiments on the electric eels at Calabozo. Humboldt, who met him in Pavia on the occasion of his visit to Volta, writes that Configliachi spoke to him of his results proving that the torpedo' shock can be transmitted at distance through a moist cloth (Humboldt, 1811, pp. 89–90). Similar results had been already obtained by Galvani in his experiments on the torpedoes caught in the Adriatic sea (see Chapter 22).

prolonged the shock's duration. It might perhaps only force a fish to produce more than an isolated shock, because of the irritation and the discomfort it would be experiencing when being tested.

ONE ELECTRICITY

During the period in which Volta was trying to show that his new device could account for how fish electric organs function, he obtained a series of results that was relevant for both the physics of electricity and for physiology. For reasons of space and given the objectives of this book, we cannot examine all of these findings in detail. It is important to state, nonetheless, that Volta's concern at this time was more than just to demonstrate that fish electricity is one and the same with the electricity produced by his battery. He also strove to provide evidence to show that his battery's output is electrical, and thus comparable to the artificial electricity of friction machines and the natural electricity characterizing atmospheric events. Stated somewhat differently, he believed that all had to be expressions of the same force, which could acquire different characteristics depending on specific physical conditions. (As we shall see in the next chapter, this would remain a particularly "stormy" issue until the 1830s, when Faraday conducted some revealing experiments on an electric eel in London.)

As we have noted, one of the main issues Volta had to face in this context related to the different physiological effects produced by a battery and a Leyden jar charged to the same tension. He succeeded in explaining the stronger efficacy commonly exhibited by the battery on the basis of the continuous currents it generates, as contrasted with the transient nature of the Leyden jar discharge.

To support this interpretation, Volta accurately studied the time courses of Leyden jar discharges of different capacities and of assemblies of Leyden jars in parallel, as well as the laws underlying the temporal summation of the physiological effects of short-duration events. In these investigations, he put forward the idea that the intensity of a physiological effect, and particularly a sensation, is the product of tension and the quantity of the charge. By so doing, he anticipated a fundamental tenet of sensory physiology, namely the dependence of the sensation experienced on the total energy of the stimulus. This notion underlies a psychophysical law first formulated for visual perception ("Bloch's Law"), according to which stimulus time and intensity are interchangeable parameters in evoking a sensation (Bloch, 1885; see Piccolino & Bresadola, 2003).

The fruitful exchange between physics and physiology that guided Volta in his investigations of animal electricity is reflected in the conclusion of his 1805 letter to Configliachi. Here we can find an expression of pride for his discovery:

> I have no doubt that when you examine and fathom the electric organs of the Torpedo more, you cannot but discover a greater resemblance with my batteries, or even an essential conformity with those that I call of the third type [i.e., composed exclusively of humid conductors]. The altogether singular construction of these organs was long a mystery for both physicists and physiologists, and it is still so for many. But it ended being so for me from the moment that I succeeded in building my motor apparatuses, namely the above-mentioned batteries of the third type, which are, I endeavor to say, fundamentally the same as those organs. The experiments and researches that I proposed to you are aimed at verifying and confirming this in every way, so as eventually to convince those who might still have doubts or objections. If such experiments succeed well, as I hope they will, they will show how one could obtain from Torpedoes out of water, or even from their electric organs alone, besides the shocks already known, all the other phenomena that my batteries present.

He continues:

> I have already shown how, in a reciprocal way, the batteries also perfectly imitate torpedoes in water by shocking and benumbing the hand plunged into the water itself. They shock it even before it could touch the fish's body, even at a considerable distance; something that was not previously understood, and which I can explain and confirm with other experiments in a way so as not to leave any uncertainty. In imitation of the torpedoes, which discharge their shocks in their native element, I activated these batteries of mine and similarly made them give shocks, even under water and to a plunged hand, which does not, however, actually touch them. In sum, I reduce them [i.e., my batteries] to true artificial Torpedoes. (Volta, 1923, pp. 202–203)

GALVANI, VOLTA, AND THE "UNSOLVABLE" DILEMMA

An analysis of the electrophysiological studies that have led to a full elucidation of the mechanisms underlying nerve conduction is of particular importance for what remains to be covered in this book. This is not only because such an analysis would provide a better understanding of the electrophysiological revolution that began in the second half of the eighteenth century, but because it also led to a better understanding of the mechanisms underlying the production of the fish shocks (as we shall see in the next chapters). The newer work is also particularly noteworthy because, as we shall now show, it throws more light on the controversy over animal electricity that took place between Galvani and Volta, and their respective supporters.

Although not well recognized, both Galvani and Volta attempted to arrive to a compromise in their debate—that is, both protagonists made efforts to recognize the importance of their opponent's discoveries on several occasions. In fact, both men always nourished reciprocal esteem, despite the somewhat harsh tones that sometimes characterized the scientific debate. What impeded such reconciliation, however, was the very nature of the controversy itself, which seemed to lead toward an unavoidable and unsolvable dilemma, one that demands our attention at this juncture.

From Galvani's perspective, the question could be put in the following terms: Since experiments, particularly those revealing "contractions without metal" (i.e., those based on direct nerve–muscle contact or on contact between two nerves), clearly demonstrate the existence of an electrical disequilibrium inside animal bodies, why should one evoke, as Volta is doing, the agency of an external electricity excited by metallic contacts?

As Volta saw it, the question was pretty much the same, although inverse in nature. Numerous experiments, particularly those made without recourse to the animal preparation, have proved beyond any doubt that different metals can generate electricity at their contacts. It is, moreover, well established that frog preparations are easily stimulated by external electricity. Why, then, should one invoke an electric disequilibrium intrinsic to the animal preparation in order to account for the contractions produced by bimetallic arcs?

Both Galvani and Volta recognized the problem they faced as a true dilemma. Indeed, each accused the other of violating the first and most fundamental law of the scientific method, the law of parsimony. This was a law firmly reasserted by Newton himself: the interdiction of multiplying the causes of a phenomenon without sufficient reason, once a true and sufficient cause had been found (see Chapter 21).

In classical logic, a dilemma stems from two mutually exclusive propositions, pitting an idea that is true against one that is flawed or erroneous. But given this definition, was the Galvani and Volta dispute a true dilemma? We shall argue that it was not, because there was a third logical possibility (*tertium datur*). In most of Galvani's and Volta's experiments, two sources of electricity were needed to produce muscle contractions: one intrinsic to the animal, as postulated by Galvani, and the other generated by the metals put in contact with the excitable tissues of the animals, the force postulated by Volta. A good understanding of this double involvement would emerge only with the electrophysiological research carried in the nineteenth and twentieth century, culminating in 1952 with the landmark biophysical studies of Alan Lloyd Hodgkin and Andrew Fielding Huxley, which we shall examine in Chapter 26.

The matter is not trivial, and it reflects the huge physical difficulties that animals had overcome in the course of evolution in order to exploit the advantages conferred by electrical signal conduction in the nervous system. Fast conduction of the nerve signals underlying sensation, motion, and the cognitive elaborations of the brain is, of course, basic to organisms capable of complex adaptive behaviors. And electricity would be a suitable agent for these functional demands. Still, the path toward an electrical conduction system using nerve fibers would involve many steps, and one can only imagine how many of nature's failures fell by the wayside as nervous systems ever so slowly advanced.

The main difficulty in this natural history saga is that bodily liquids and tissues are far less conductive than metals, even though this was not thought to be the case when electrical science was still in its formative years. Only a few scholars during the Enlightenment, among whom we can count Cavendish and Volta, were aware of the enormous differences in conducting properties favoring metals over moist tissues, bodily humors, and other liquids. With modern measurements, we now know that a cylindrical column of 1 centimeter in length, made with matter of the same composition as the intracellular fluid of nerve cells would have the same resistance as a metallic cable of the same diameter with a length of about 1000 kilometers! (about 600 miles). Moreover, we must consider that, in order to allow for a great overall capacity of a nerve trunk to transmit signals, the fibers need to be quite small in diameter (in the range of microns, i.e., thousandths of a millimeter). This makes the electrical conductive task of a nerve system even more difficult.

Alan Hodgkin alluded to this problem in a particularly expressive way in a lecture he gave at Liverpool University in 1961, two year before being awarded the Nobel Prize with Andrew Huxley for their studies on nerve conduction. Based on the physical characteristics of biological tissues, a long and thin nerve fiber would have – according Hodgkin's calculation – an electrical resistance comparable to that of a large metallic cable stretching ten times the distance from the Earth to Saturn (the most distant of the seven planets of the classical cosmology). To transmit an electric signal in a fiber having so much electric resistance would at first glance seem to be impossible (Fig. 23.10).

For a long time, engineers working in the field of communications have been confronted with the problem of transmitting signals over larger and larger distances with available forms of energy. This, of course, has demanded

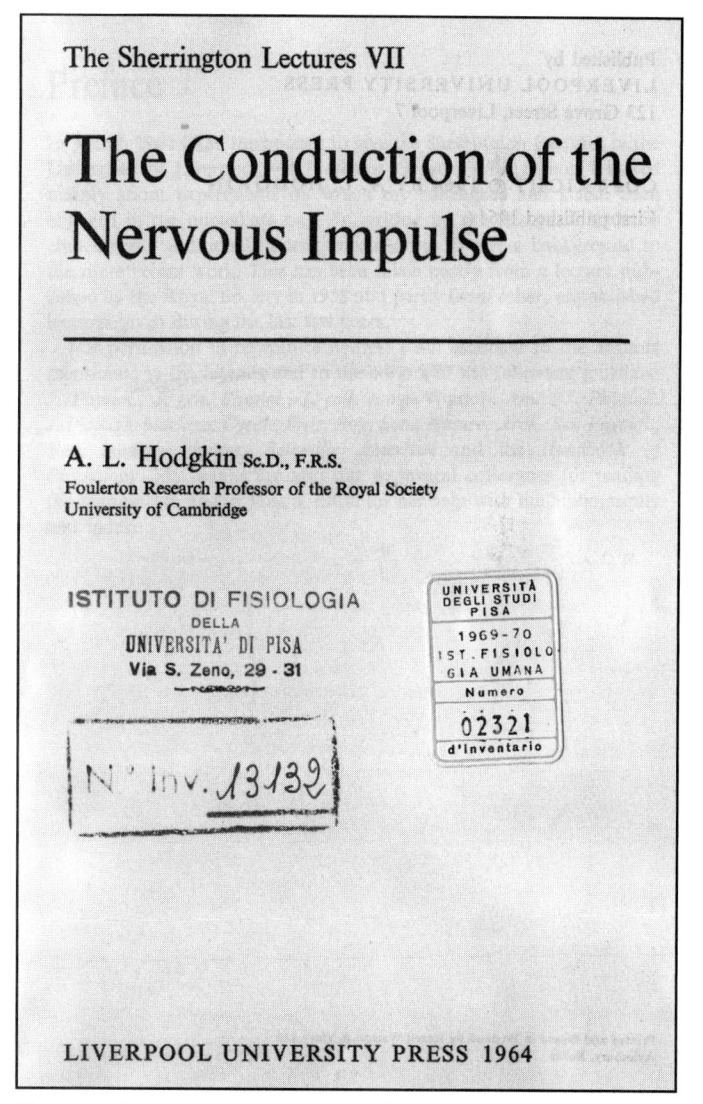

Figure 23.10. Alan Hodgkin's small volume in which he summarized the experimental evidence showing why nerve signaling should not be regarded as simply the passive conduction of electricity. The basic research conducted by Hodgkin with Andrew Huxley shed new light on the dynamic biophysical events underlying nerve conduction, the subject of his book.

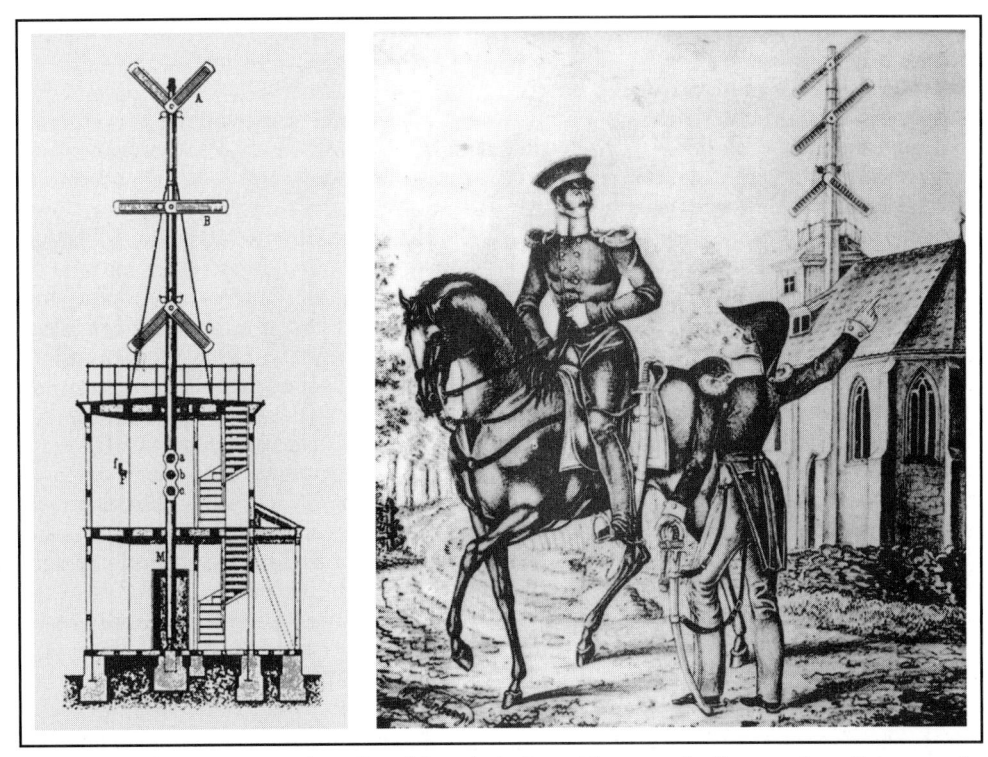

Figure 23.11. The optical telegraph was a system that allowed for relatively rapid communication over long distances prior to the development of modern telecommunication systems. It was based on a series of posts separated at appropriate distances from each other, which acted as reception-transmission relays. The Prussian system, one of a number of optical telegraph systems used in Europe, is shown here.

technological innovations and the development of sophisticated strategies. This has happened, for instance, with optical-type telegraphs and then with telegraphs based on electrical cable transmission (Fig. 23.11).

Let us consider how an engineer might proceed with the distance issue in electric cable transmission, one calling for the transmission of an electric signal over a space of 10,000 miles (about 16,000 kilometers) using metal cables capable of transmitting only up to about 1,000 miles. Our engineer could proceed in a way somewhat similar to that used in the construction of old optical telegraph systems (and other traditional low-technology communication systems). Specifically, he (or she) will interpose a series of electric stations along the transmission line, at a distance of 1,000 miles from each other (in the optical equivalent, this could correspond to series of appropriately distanced towers). When a large electric signal generated from the first station arrives, attenuated but still detectable, the receiving station will enhance the signal and send it on to the next station, and so on to the last station. For this solution to be successful, each station must have the means to boost the signal before sending it on to the next station, and must have a local source of electrical energy.

This system could be automated using an appropriate device (an electric or electronic *relay*) capable of generating large electric signals upon the arrival of smaller incoming ones, through the use of local energy. With such arrangements, however, what really circulates from the first to the last station is not the energy of the initial signal; it is only the information associated with it—that is, the command leading to its generation. Using eighteenth-century wording, one could say that nothing of the electric "fluid" present at the first station would reach the last one, although the arrival of the fluid (signal) at the end of the transmission line is dependent upon the electric actions occurring at the initial station.

The mechanism allowing for the transmission of electric signals along nerve fibers, particularly in the large fibers of mammals, is, in fact, similar to that in the model outlined above.

Like the electrical cables used in underwater transmissions, these fibers need to be enveloped by a highly insulating material, because of the conductive and shunting nature of the fluids surrounding them. The biological coating serving this role is the myelin sheath, which is rich in lipids (the oily, potentially insulating substance envisioned by Galvani in his experiments; Chapter 21). The insulating coating, however, is not continuous in these nerve fibers. There are interruptions that are much shorter in length than the coated segments, where the fiber membrane is exposed to the extracellular space. In 1878, French histologist Louis-Antoine Ranvier gave the first good description of the bead-like appearance of covered nerve fibers (Ranvier, 1878). Today, we know how glial cells that wrap around the axon form the myelin sheath, and we use the eponym "nodes of Ranvier" to denote the gaps in the axon's coating where there is no myelin.

On a functional level, the nodes of Ranvier are the reception-transmission stations in the process of conveying an electric signal along a long axon. The myelinated segments, in contrast, correspond to the insulated cables for the transmission of the signal from station to station, until the end is reached. Every node is endowed with local energy, due to the asymmetric distribution of ions at the membrane sites (see next chapters), along with the complex machinery needed for the production of relatively large-amplitude electric impulses (about 100 millivolts [mV], or 0.1 Volt).

An impulse is automatically activated once a small electric signal (less than about 10 mV) reaches the node.

Under physiological conditions, the activating signal comes from the adjacent, previously activated node via the transmission through the internodal segment of the fiber. The impulse is transmitted to the following node in the sequence in a partly attenuated way and regenerated there. The process then continues until the fiber terminates, the repeated "leaping" of the event from node to node inspiring the phrase "saltatory" (meaning leaping, hopping, or jumping) conduction (Fig. 23.12).

Under experimental conditions, the activation of a node (with the resulting local discharge of an impulse) can be brought about by an external electrical stimulus capable of producing a small but adequate modification of the nodal membrane potential. And this bit of information brings us back to Galvani and Volta, and to their "dilemma."

In most of the experiments that the two Italian scholars conducted, the external electrical stimulus was provided by the application of a metallic arc. The electric potential generated at the contact of two metals is generally of the order of 1 V, and although the current flow is largely dispersed in the passage across the muscle and nerve tissues, this can still be sufficient to trigger activation at the nodes of Ranvier. Also, the contact of biological liquids with a metal or metals (and under some conditions direct contact between living tissues without any intermediate body), can produce a sufficient electric stimulus, as we shall show in the next chapters.

In this context we need to note that muscle fibers are also capable of generating electrical impulses in a way that is fundamentally similar to nerves, even though muscles are generally much less sensitive to external stimuli. But muscle fibers do not have an insulating myelin coating with nodes of Ranvier, so the propagation and transmission of the electric impulse takes place in a continuous way. These are not, however, just characteristics of muscle fibers. The smallest nerve fibers of vertebrates, and all nerve fibers in invertebrates, lack myelin and conduct the electric impulse in a continuous, non-saltatory way.

Importantly, the relationships between nerves and muscles in both vertebrates and invertebrates is normally nothing like the tight electric relationship between muscle and nerve that Galvani envisioned with his Leyden jar model. Nerve and muscle fibers communicate only via chemical signals, and both have a local reserve of energy necessary for the conduction of their own electric signals.

So now we can appreciate why the Galvani–Volta dilemma was not a true dilemma, and why both Galvani's animal electricity and Volta's metallic electricity were required to bring about nerve stimulation and consequent muscle contractions. Under their experimental conditions, the metallic electricity would have activated only the local processes in the nodes, which might eventually result in a full impulse. The energy of the full electric signal, however, could not be derived from this external source, but depended on the local "animal" energy accumulated in the membrane of the node. In the absence of the intrinsic electricity at the node, the metallic electricity would have been transmitted for only a short distance (of the order of microns or perhaps a few millimeters, depending on the fiber type). Thus, it would not have reached the end of the fiber, so as to bring about even the smallest muscle contraction, because of the extremely high electric resistance of the nerve fibers involved.

AN EXTREMELY SENSITIVE ELECTRIC AMPLIFIER

We have already remarked that both Galvani and Volta had been struck by the great sensitivity of the prepared frogs to weak electric current. This had been of particular concern to Volta, who was engaged in research on weak electricity, a topic of great interest to scientists in the second half of the eighteenth century. Until the development of his *condensatore* and Nicholson's duplicator, the electrometers of the epoch were barely capable of revealing electric potentials of less than 100 V. Since the electricity generated at the contact of two different metals was of the order of a single volt or less, it would have escaped detection with the more primitive devices.

Because both Volta's *condensatore* and Nicholson's duplicator could reveal electric potentials of less than a volt, it might be asked whether the use of Galvani's frog preparations was all that important for the discovery of the electric power of metallic contacts and subsequently the battery. After all, it would seem like the discovery of metallic electricity was simply waiting for either of these two sensitive physical devices to be used in this situation.

This use of the latest tools, however, did not play out in this more direct way. As a matter of fact, Volta made use of the two devices only after he was convinced of metallic electricity from his experiments using frog preparations as described by Galvani. There are important reasons for this being the case, which make it unlikely that metallic electricity would have been discovered as quickly as it was, were it not for Galvani's cherished frogs.

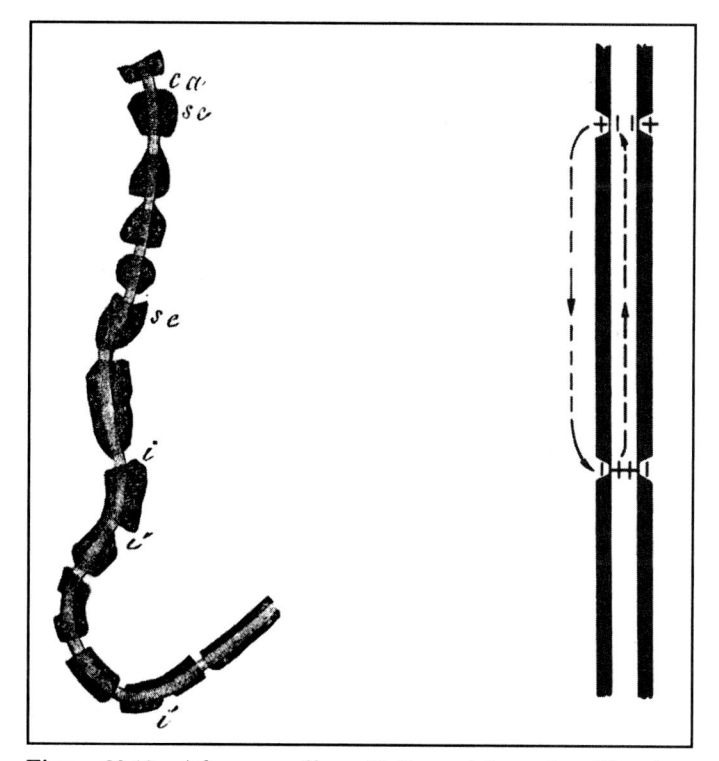

Figure 23.12. A frog nerve fiber with its overlying nodes of Ranvier, illustrating how the current involved in nervous conduction jumps from one node to another by saltatory conduction (left, from Ranvier, 1878; right, from Hodgkin, 1964).

A major difficulty with the sensitive detectors designed by Volta and particularly Nicholson was the tendency to give spurious readings indicative of electricity from an electrically devoid source. Along with the instructions (and warnings) given in the electrical treatises of the epoch for the correct use of Nicholson's duplicator, there are ways of trying to minimize the "residual electricity" that the instrument itself produces. As Volta himself had noted, in 1789 Nicholson had supposed, on the basis of some experiments made with his duplicator, that an electrical effect could arise from the contact of different metals. But the invention of the electric battery did not follow from Nicholson's supposition.

The lack of any major scientific consequences stemming from Nicholson's device can be attributed only in part to the fact that scientists recognized that his instrument could be unreliable, tending to produce many "false positives," using today's terminology. That is, there was a second important reason: Neither his duplicator nor Volta's *condensatore* was capable of revealing the most striking characteristic of the metallic electricity—the continuous nature of its current flow. Hence, Galvani's frogs were needed at this time, because the mechanisms present in the neuromuscular systems of frogs and other animals were much more sensitive to a stream of weak current than even the most sophisticated physical electrometers of the era.

The mechanism making the "galvanoscopic" frog so sensitive can be thought of as an extremely effective molecular machine. Its evolution permitted the organism to overcome the aforementioned physical difficulties associated with electric conduction along thin nerve fibers. The inherent mechanism is capable of detecting potential changes of the order of few millivolts, and it is an essential component of the system necessary for regenerating the weakened electric impulse in the course of its propagation down the nerve fiber.

Hence, it was the extremely sensitive electric amplifier present in the nerves and muscles of common animals that allowed Volta to establish the continuous character of the current produced by metallic contacts. In this context, it is interesting to reflect on the fact that even though some nerves can be stimulated in a constant way by a continuous current, most respond far better to electric stimuli of very short duration. This is because ordinary nerve fibers conduct electric impulses of short duration (normally less than a millisecond). Therefore, to allow for the effective propagation of these signals, they have to be especially sensitive to short-duration electric signals.

The temporal character of this sensitivity explains why Galvani found that a frog muscle contracted when a spark was thrown from a distant electric machine, this being in 1781. Because of capacitive and possibly inductive electric effects, the spark triggered the flow of a short-duration current across the tissues of his frog preparation. As Galvani remarked at the beginning of *De viribus*, this experiment played a crucial role in stimulating his interest in the possible involvement of electricity in animal physiology. The special electric amplifier present in the nerves and muscles of prepared frogs also played an important role at the beginning of the discovery path that culminated in the invention of the battery (Chapter 22).

The amplifier present in the excitable tissues of many animals is in many respects even more refined than even the latest technological devices. In this regard, it would be inaccurate to say that there was no device capable of revealing extremely weak electricity at the end of the eighteenth century. There were indeed many, and the one that emerged in an outstanding way was Galvani's frog preparation!

In the chapter to come we shall look at some of the complex and fascinating research conducted after Galvani and Volta that has contributed to our current understanding of neuromuscular electrophysiology. As we shall see, electric fishes, which have been so important thus far in this story, will continue to play important roles in various phases of this continuing and exciting path of discovery, with its broad ramifications for the life sciences and medicine.

Chapter 24
Electric Fishes in the Nineteenth Century

> I have seen a GLOWORM!!! water-spouts torpedo in the museum at the Academy del Cimento as well as St. Peter's and some of the antiquities here and a vast variety of things far too numerous to enumerate.
> Michael Faraday's April 14, 1814, letter to his mother during his trip to Italy with Humphry Davy (From Bowers & Symons, 1991, p. 83)

> I was extremely curious to see what the result would be of my experiments on Torpedo.
> Emil du Bois-Reymond (1885; 1887 trans., p. 436)

A growing understanding of the electrical nature of the nervous signal would remain strongly dependent on research on electric fishes long after Volta's (1800) invention of the battery. Moreover, the interactions that would take place between the two investigative fields—namely studies involving electricity in these unusual fishes and neuromuscular events in animals in general—would become even more tightly associated in certain ways than they had been during the second half of the eighteenth century.

In the nineteenth century, the interest in the involvement of electricity in animal economy was an important aspect of the general interest in all the dimensions of electricity, which grew even more intense than in the Enlightenment. Electricity was extensively present in the agendas and intellectual elaborations of the great scientists of the epoch. Even the most superficial looks at the biographies of these scientists will confirm this view, regardless of their primary fields of interest. Almost everyone was doing electric experiments, studying the interactions—and possible reciprocal transformations—of electricity with other forms of energy (magnetic, chemical, thermal, luminous, mechanical). Many were attempting to invent new electric tools (e.g., electroscopes, galvanometers, various types of electric batteries). The new instruments (and particularly the new apparatuses capable of measuring electric currents with high sensitivity) would be of paramount importance not only for the study of artificial electricity, but also for the study of animal electricity in electric fishes and the neuromuscular systems of more ordinary animals.

Many scientists through the century showed a genuine interest in torpedoes and other electric fishes. Even though not all were able to perform experiments on these singular creatures, most of the savants of the epoch could not help discussing the extraordinary powers of these fishes and trying to provide satisfactory interpretations of the origins of these powers. Another important and continuing issue was the relationship of fish electricity to the other, more usual forms of electricity.

By showing that electricity can be derived by the appropriate contact of liquids and metals, Volta's discovery of the battery had provided a needed basic insight into the relation of electricity with the deep constitution of inanimate matter. In a somewhat analogous way, the electricity of some fishes seemed to open an important window on what would become a tight association of this form of energy with the phenomena of life.

The latter idea appeared in a clear way in a part of a treatise on electricity and magnetism published in 1836 by Antoine Caesar Becquerel, a French physicist who would make important contributions to electric fish studies. To quote: "I have paid particular attention to everything that concerns electric fishes, since if some day one would happen to discover that electric fluid intervenes in the phenomena of life, this would likely be after having studied the singular property that these fishes have" (Becquerel, 1836, vol. IV, p. XIV).

More publications would now be devoted to these fishes and to nerve physiology. These overlapping fields would continue to follow the trend started late in the eighteenth century, when such research began to become increasingly dominated by highly trained professionals at research institutions, as opposed to well-meaning "amateurs" wishing to contribute to natural philosophy or natural history in the best spirit of the Enlightenment. This shift can be related to the changing demand characteristics of the research endeavors.

Over time, the needed studies required more technological sophistication, and the problems and findings were becoming more mentally challenging. New universities and other research and teaching institutions would be created, as for instance the Royal Institution of London, one of the leading centers for electrical research in Europe. Particularly during the second half of the century, research structures would be created with facilities and organizations typical of modern laboratories. This would be especially the case in

Figure 24.1. Left: A portrait of Alexander von Humboldt (1769–1859) made in 1806, the year in which he published his account of the torpedo experiments he made with Joseph Louis Gay-Lussac. Right: The initial page of the German article dealing with these experiments, which shows only Humboldt's name.

Germany, which would become one of the main centers of scientific development around the middle of the century, due in part to the wise and energetic policies of Alexander von Humboldt, the doyen of German science during the first half of the century (Fig. 24.1), and then others who followed in his path.

Another aspect of this transformation is that some of the most important scientists of the nineteenth century, whose work we shall examine in this chapter, came from middle class and even humble origins. This was true of Michael Faraday, one of the greatest scientific personalities of the epoch and also, as we shall see, an important player in the electric fish story. Faraday was the son of a blacksmith.

The warm shores of the Mediterranean (and other southern European seas) would continue to serve as a kind of Holy Land or Mecca for these scientific pilgrims, especially during the first half of the nineteenth century, much as they did in the late-seventeenth century and throughout the eighteenth century. Later in the nineteenth century, live electric eels and electric catfishes would be imported from Africa and South America into Europe, making pilgrimages to Western Europe's scientific holy sites and voyages to sweltering and perilous parts of the globe less critical.

These changes will become apparent in this chapter, in which we shall focus our attention on electric fish research during the nineteenth century. These studies would involve attempting to get sparks from torpedoes, determining the polarity of the electric organs, trying to understand why these fishes do not electrocute themselves, and much more. As will be shown here and in the final chapters of this book, fish studies and research on nerve and muscle physiology would be viewed as having much in common, and they would continue to draw on each other.

GREAT MEN, NOTABLE FAILURES, AND TORPEDINAL ELECTRICITY

A number of outstanding scholars studied torpedoes in the opening decades of the 1800s, but not always with the sort of success that one would ordinarily associate with their revered names. One of these men was, in fact, Alexander von Humboldt, who carried out a series of experiments on torpedoes caught in the Gulf of Naples in 1805 and first published his results in 1806 (see Fig. 24.1).

Humboldt's experiments were conducted a year after he returned to Europe from the Americas (see Chapter 1). Once he was back, and having experimented with electric eels in Spanish Guiana (Venezuela), he decided to study the torpedoes that had eluded him in the 1790s. His torpedo experiments were made while on a trip to Italy with multiple objectives. With his broad worldview, Humboldt also wanted to visit archaeological museums, meet with numerous scientists (including Volta), and study the culture and the features of land. One highlight of his trip occurred when he was able to witness an eruption of Mount Vesuvius on August 12, 1805, which unlike Pliny in 79 A.D., he survived.

Humboldt carried out his experiments with another great scientist of this era, Joseph Louis Gay-Lussac, a professor of physics at the Sorbonne and of chemistry in the *École Polytechnique* in Paris, and a second article (in French) dealing with their achievements bears the names of both scholars (Fig. 24.2) (Humboldt & Gay-Lussac, 1805: Humboldt, 1806, Humboldt & Bonpland, 1811, 1852; see Crosland, 1978). In 1804, Gay-Lussac had made some balloon ascents to study the earth's magnetism and atmospheric changes, and a year later, with Humboldt joining him, he reported that the relative composition of the atmosphere does not

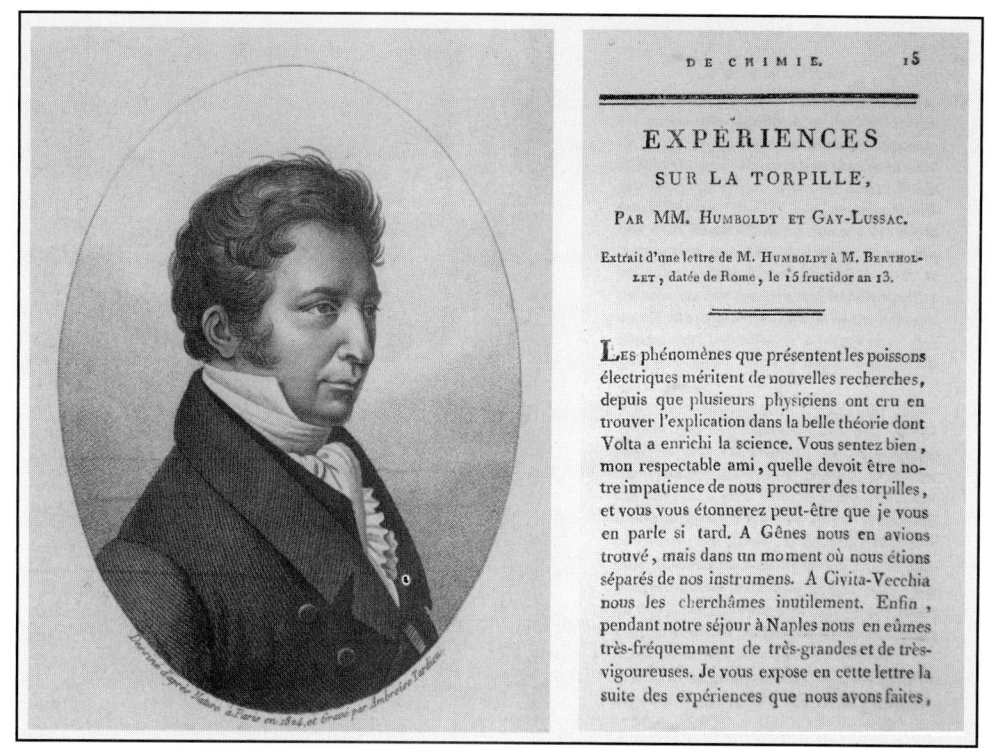

Figure 24.2. The great French physicist and chemist Joseph Louis Gay-Lussac (1778–1850) with the title page of the 1806 French article dealing with the torpedo experiments he and Humboldt made in Naples.

change with increasing altitude, and that water is composed of two parts oxygen and one part hydrogen. Gay-Lussac would go on make other notable discoveries, including the elements boron and *iode* (iodine) and two chemico-physical laws, one concerning the chemistry of gases (First Law of Gay-Lussac), and the other the variation of the pressure of a gas with variations in temperature when volume is kept constant (Second Law of Gay-Lussac).

Humboldt and Gay-Lussac found that even a very sensitive electroscope (Volta's *condensatore*) could not pick up their torpedoes' discharges. They were also unable to see a spark and could not detect the transmission of a torpedo's shock through the water (a regression with respect to Galvani's 1797 memoir; see Chapter 23). Moreover, they remarked, "it is indispensable that the contact be direct: no shock is felt when a conducting body, for instance a metal, is interposed between the finger and organ of the fish . . . the animal may be touched with impunity with a key or any other metallic instrument" (Humboldt & Gay-Lussac, 1805, p. 18). A flame in the circuit, they further reported, also blocked transmission of the shock.

Hence, even though two remarkable scientists had joined forces to study torpedinal electricity, their results were essentially negative, contributing little beyond what John Walsh (1773) had found in La Rochelle back in 1772. Their various failures can be partially explained with recourse to the feebleness of the torpedoes studied. They could also reflect the still significant barriers (e.g., vitalistic beliefs) that made some people (particularly Humboldt) hesitant to accept the reality of animal electricity.

In the conclusion of his torpedo publication, Humboldt seemed to refute Volta's idea that a torpedo's electricity originates from the contact of different conductive substances (i.e., through a mechanism similar to that of the battery).[1] As noted in Chapter 23, he would, however, go on to change his mind about the torpedo's discharge probably not being true electricity. This important change from his original thinking comes forth in his New World narrative, in the section in which he describes his adventures with the electric eels of Spanish Guiana. He also wrote about his torpedo experiments in this part of his narrative, which was published in an extended form in 1811, some 6 years after he and Gay-Lussac had conducted their torpedo experiments in Naples, and also after he had met with Volta in Como.

Humphry Davy was another great scientist of the era (for biographical information, see Treneer, 1963; Hartley, 1966; Knight, 1992; Lamont-Brown, 2004). After showing that the electricity generated by the Voltaic battery depends on chemical mechanisms, he used Volta's new device as an instrument of discovery, particularly in the field of chemistry. Davy too was a discoverer of many chemical elements, mostly through the electrochemical actions of the extremely powerful electric battery housed at the Royal Institution of London.

Davy, who made an impressive series of experiments on various aspects of electricity, had a particularly strong

---

[1] Humboldt sent his memoir on the electric eels for publication in the Proceeding of the French *Académie* on September 27 1806, as documented by a letter he addressed to the president of the scientific institution. This letter is conserved in the *pochette* (file) of the *Académie* archives. In the same *pochette* there is another letter sent by Humboldt to his friend, *académicien* Jacques Thenard from Naples, on August 14 1805. This long letter contains a rather detailed account of the events in Naples, and notably of the earthquake, and also of the various experiments made by Humboldt and Gay-Lussac. A long description, including a schematic drawing, is devoted to the experiments on the torpedoes. Despite the fact that Humboldt writes that the torpedo "has presented the phenomena of the galvanic chain," he remarks that in some respects fish electricity is more like that of the Leyden jar than the galvanic battery.

Figure 24.3. A young Michael Faraday (1791–1867), who accompanied Davy to Italy, where they studied torpedoes together in 1814. Faraday, one of the greatest physicists of the century, would go on to study electrical induction and other electromagnetic phenomena, and in the 1830s would come forth with important new findings about the electric eel using techniques he had developed.

interest in electric fishes. While making some of his many voyages to the Continent, he studied torpedoes, largely to satisfy his own scientific curiosity. The first occasion was in the period 1813–15, when Napoleon's admiration for his studies on "galvanism" allowed him to become one of the first Englishman to be admitted to France during a period characterized by wars and political instability. Davy traveled with his new wife and a most remarkable assistant, Michael Faraday, who was only 23 when the party left London in October 1813 (Fig. 24.3). Happily, Faraday recorded the most remarkable events of the trip (Bowers & Symons, 1991; James, 200: Hamilton, 2002; . His notes, together with other historical documents, most of which were collected and edited by Humphry Davy's brother, John Davy, reveal their great interest in electric fishes (Davy, 1839–40).

From these records we learn that Davy met several important members of the scientific establishment in Paris, including Humboldt and Gay-Lussac (both Davy and Faraday attended Gay-Lussac's chemical lectures at the *École Polytechnique*). As would occur with Davy's two subsequent "grand tours," most of the time, however, was spent in Italy—seeing monuments, fishing and shooting, observing nature in its various forms, conducting various experiments in physics and chemistry (particularly in Florence, Rome, and Naples), and notably studying torpedoes.

Davy conducted his first experiments on these torpedoes in Genoa on March 4, 1814. They were made in collaboration with a naturalist and botanist, Domenico Viviani, who was a professor at the local university. The aim of their first trials was to show that the torpedo's electricity, like that of the Voltaic battery, could have chemical effects, one being the decomposition of water. The experiments were, however, entirely unsuccessful, a result that Faraday thought could have been due to the weakness of the fish. After dissecting a large but dead torpedo in Genoa, Davy conducted more experiments with the same purpose in Naples and in Mola di Gaeta (now Formia; Fig. 24.4), and that June in Rimini, where he had access to a large live torpedo. These experiments were also unsuccessful. In Rimini, Davy also failed to obtain a visible spark at the time of discharge.

Despite his failures, Davy did not conclude in his article, which was published many years later, in 1829, that a torpedo's shock is not electrical, or that its electricity is different from artificial electricity. This was in part due to a discussion that he had with Volta in Milan in June 1814. The Italian

Figure 24.4. Left: The initial page of Humphry Davy's article on the torpedo published in 1829 but dealing with experiments generally done more than 10 years before. Right: An 1820 view of Mola di Gaeta (or *Gaieta*, now Formia), the small town midway between Naples and Rome where he made experiments on the fish in 1815 (the image of Mola di Gaeta is from Treglia, 2008).

physicist suggested that Davy's negative findings might not be incompatible with the organic and the inorganic forces still being the same. Notably, Volta explained that a battery composed of only moist conductive substances might likewise fail to produce these telltale signs of electricity, even though it could still generate a feeble shock.

Davy's interest in torpedoes became almost obsessive during his third voyage to the Continent, which concluded with his premature death in Geneva in May 1829. His main purpose this time had been to show that a torpedo's electricity could produce magnetic effects, and that the discharge could be measured using a "multiplier" or "galvanometer," an instrument invented in 1820 by German chemist Johann Solomo Schweigger (and also by others almost at the same time) for measuring electrical current (Hackmann, 1985).

Davy made experiments in Ravenna in 1827, and afterwards in Trieste and in Rome, but again without success. In Ravenna, he also conducted experiments on frogs, using his galvanometer to try to detect the electricity that might be associated with neuromuscular actions and attempting to magnetize metals with that electricity. Again, his experiments were unsuccessful, as were some related experiments in which he tried to determine whether sunlight could magnetize metals.

Humphry Davy's involvement with torpedoes persisted until the very end of his life. In February 1829, while his health was failing, he wrote to his brother John, a surgeon in the English Navy, asking him to continue the research he had started on the torpedo. His letter contained directions on how to conduct the needed experiments and even where to find torpedoes ("Civita Vecchia" and "Fumicina"; i.e., Fiumicino near Rome, a name that became "Tormicina" in the paper published in 1829, soon after Humphry's death; J. Davy, 1839–40, vol. I, p. 405).

John dutifully honored Humphry's request. Soon after reaching his brother in Rome, he started to perform dissections of torpedoes, while Humphry was slowly recovering. He showed his results to Humphry, "all which," he remarked, "not only amused, but interested him deeply" (J. Davy, 1839–40, vol. IX, p. 409). But with his health improving, Humphry seemed less interested in John's dissections. In his recollections, John wrote: "Now that he was intent on recovery, he no longer took the same deep interest in *my* examination of the torpedo, as if he looked forward to the time *he* should be able to enter in the investigation actively again" (J. Davy, 1839–40, vol. I, p. 411; Italics Davy's).

JOHN DAVY: TORPEDOES BECOME MORE ELECTRICAL

John Davy did many experiments on torpedoes after Humphry's death. He had graduated from Edinburgh with a medical degree in 1814 and entered English Army Medical Department as a surgeon. Having been awarded the post of Inspector General of Hospitals, he traveled throughout the British Empire, from Ceylon and India to Barbados in the West Indies. His stops included a lengthy stay in Malta, where many of his experiments were conducted. In 1814, the year of his graduation in medicine, he was elected a Fellow of the Royal Society, his brother Humphry being one of his sponsors.

The younger Davy's torpedo studies were published in 1832 and 1834 in the *Philosophical Transactions of the Royal Society* (Davy, 1832, 1834) (Fig. 24.5). He would go on to write over 100 papers, many in the fields of comparative anatomy and physiology (e.g., Davy, 1839). In contrast to his more celebrated brother, John Davy succeeded in obtaining many indicators of torpedinal electricity, among which were

Figure 24.5. John Davy (1790–1868) and his first article on the torpedo (Davy, 1832). In contrast to his older brother Humphry, John Davy obtained many new findings with his torpedoes, some of which he studied while on Malta. They included chemical transformations and heating and magnetizing effects.

a variety of chemical transformations, the heating of metal wires, and even the magnetization of a needle.[2]

John Davy was also able to detect fish electricity with a galvanometer. He tried to establish the polarity of the shock by comparing the shift of the galvanometer needle produced by the torpedo's shock with that from a Voltaic battery. He was, however, incorrect in his conclusions, attributing the negative polarity to the back of the fish and the positive polarity to its belly (see later).

Prior to John Davy, another report of a galvanometric measurement of torpedo's electricity had already appeared in print, although it is problematic and was not written by the men who conducted the experiment. One of the experimenters was French zoologist and anatomist Henri Marie Ducrotay de Blainville, professor at the Natural History Museum of Paris, and the other was Louis Benjamin Fleuriau de Bellevue, a geologist and naturalist of La Rochelle, and the founder of an important natural science museum (which still exists in this small town on the French Atlantic coast). The report was published in 1832, appearing in the second edition of one of the most influential textbooks of physics of the nineteenth century, the *Elémens de Physique Expérimentale et de Météorologie,* of Claude Pouillet, an important member of French *Académie*. After mentioning Walsh's previous experiments on torpedoes, Pouillet referred to the new experiments that were also made at La Rochelle:

> It was interesting (*curieux*) to assay the torpedo's effects on Schweiger's [*sic*] multiplier [meaning galvanometer], in order to ascertain once more, in a more definite way, the relationship of this phenomenon with the electric discharges. In the month of August 1828, M. de Blainville has carried out in La Rochelle, together with M. Fleuriau de Bellevue, some experiments of this kind, and the needle [*scil* of the multiplier] pirouetted for more than a half-circumference when the two needles connected to the multiplier wires were inserted inside the organ of the torpedo. (Pouillet, 1832, vol. I, p. 32)

The text is vague and does not give further details. It does not even state if the effects on the multiplier appeared only during the shocks. Further, no subsequent report by the authors seems to have appeared, not even by Blainville, who was a prolific writer, even though he referred to the results of his dissections made on torpedoes "fresh and almost alive" in his *Cours de Physiologie Générale et Comparée* (Blainville, 1833, vol. II, p. 447). Because the supposed achievement of these two Frenchmen is secondhand, vague, and questionable at best, John Davy merits the distinction of being the first to obtain a recognized galvanometric measurement of the torpedo's discharge.

Notwithstanding his many successes in revealing torpedo electricity, Davy still fell short when it came to obtaining sparks. Also, he could not detect torpedinal electricity with a classical electroscope, a device that could reveal the presence of electricity by the divergence of two metallic laminas (due to the repulsive actions of electrical charges).

John Davy (1834) also investigated the chemical composition of this fish's electrical organs, and other aspects of its physiology and anatomy. His anatomical studies confirmed John Hunter's findings of a very rich nerve supply and, as previously shown by Girardi and Galvani, he also found that some branches of the electrical organ's nerves ramify toward more ordinary tissues, such as muscles. He further confirmed the findings of Galvani and Spallanzani, who had previously noted that a torpedo cannot produce a shock when the nerves to the electric organs are cut or after the brain has been "entirely extracted."

Davy (1834, p. 548) wrote that electric organs do not exhibit movements when there is a discharge: "I have not witnessed in the Torpedos of the Mediterranean; nor, indeed, have I been able to associate any visible sign, any apparent movement of the fish, with the electrical discharge." Also citing his morphological studies, he concluded that there must be a fundamental dissimilarity between muscle fibers and the components of the electrical organ. Later research would show that the muscle fibers and the specialized cells making up the electric organs are, in fact, related, despite their morphological differences and the lack of contractility in the cells of the electrical organ (see Chapter 26). Based on some of the differences from manmade electricity that he had found, John Davy (1834, p. 548) would also conclude that "the electricity of the Torpedo is specific and peculiar." In other words, he thought it represented just one type within a larger family of electricities.

THE POLARITY OF A TORPEDO'S DISCHARGE

Understanding who should be given the priority for the discovery of the polarity of a torpedo's discharge is not an easy matter. This is because there were discoveries that were not made public, information that was only vaguely described, and egos involved during the 1830s. What is clear, however, is that the polarity was correctly established during this decade, with several investigators recognizing that a torpedo's back is positive and its belly is negative at the time of discharge.

The first person to get the polarity correct seems to have been Swiss physicist Jean Daniel Colladon. He used a galvanometer to investigate the discharge in 1831 (he also used an electroscope but was unsuccessful with it). Interestingly, Colladon conducted his experiments at La Rochelle on the French coast, the same place where Réaumur and Walsh had carried out their important research on the torpedoes that abound in these waters. He worked with a *Monsieur Lebrun*, professor of physics at the local college.

The problem that we encounter with Colladon (1836) is that he did not publish his discovery until 5 years later. This delay left the door open for others whose experiments followed his own to go public before he did. In fact, most researchers learned of his experiments only because Emil du Bois-Reymond, the German physiologist who dominated electrophysiological investigations in the second half of the century (see below), made repeated references to them at later dates.

The situation with Leopoldo Nobili, the talented Italian physicist about whom much more will be said later in this chapter, is similar to that of Colladon in some ways. Nobili conducted his torpedo experiments in November 1834 in

---

[2] Before John Davy, Pietro Configliachi, Volta's student (see Chapter 23), had obtained the latter effect. Although he had announced it in a letter dated September 3, 1827, Configliachi never published this letter, in which he also stated that he had been unable to detect any signs of fish electricity with a galvanometer (see Zantedeschi, 1845, v. II, p. 320).

Figure 24.6. Leopoldo Nobili (1784–1835) with a page of one of the four letters he sent to Charles Lucien Bonaparte dealing with the electromagnetic experiments he made on a torpedo in Leghorn, Italy, in 1834. Bonaparte (1841) would later name the large *Torpedo nobiliana* after Nobili. In addition to making significant discoveries with the torpedo, Nobili would also make important contributions to nerve and muscle physiology. (© Bibliothèque Centrale M.N.H.N. Paris, 2005)

Leghorn, the same town in Italy where Jan Ingenhousz had investigated torpedoes in the previous century (Chapter 16). Because torpedoes are seasonal and his timing was unfavorable, Nobili could get only a single live fish. Nevertheless, he was able to conclude that the top becomes positive and the belly negative at the time of a discharge from his torpedo. Thus, Nobili was correct, even though, strangely enough, he wrote that his result "matches that of Gio. Davy" (who had the polarity wrong!).

Nobili's observations and elaborations on the torpedo are contained in four letters that he sent to Charles Lucien Bonaparte, the dedicated naturalist related to Emperor Napoleon Bonaparte (see Chapter 16) (Fig. 24.6). They included his work on the polarity of the discharge, as well as his successes in detecting the discharge with his galvanometer, and magnetizing small pieces of iron and steel with it. Because of Nobili's interest in electric fish, Charles Lucien Bonaparte (1841) would later dedicate the *Torpedo nobiliana*, the large torpedo he described in the third tome of his *Iconografia della Fauna Italica*, to him. Unfortunately, even though Bonaparte realized their importance and wanted to see them published, Nobili's (1835a,b) letters remained in manuscript form through the 1830s and they have yet to be published (see *Atti della prima Riunione . . .*, 1840, p. XVIII).

This brings us to Gilbert Breschet and Antoine Caesar Becquerel, who studied torpedoes caught near Venice in 1835. These Frenchmen extended Davy's results on the effects of torpedo electricity on the decomposition of various chemical compounds, and were able to measure the electricity associated with a torpedo's shock with a sensitive galvanometer (Becquerel & Breschet, 1835; Breschet & Becquerel, 1836) (Fig. 24.7). They also correctly established the polarity, attributing John Davy's mistake about the polarity to artifacts stemming from electrical actions independent of the fish's shocks (Becquerel, 1836). It is possible, however, that Davy's incorrect interpretation could involve other factors (see Pianciani, 1838, pp. 17–18).

Becquerel and Breschet briefly announced what they had done to the French *Académie* on October 19, 1835. Nevertheless, their first published note is brief and devoid of details. There was not even a statement in this communication about the part of the animal that became positive and the part that became negative at the moment of the shock.

On July 11, 1836, Carlo Matteucci presented his own report on the torpedo to the same organization (the report also included the work of Linari; see below). His report to the *Académie* included the correct polarity and other important achievements, including the production of a spark (Matteucci, 1836a).

After learning of Matteucci's claim, Becquerel wanted to set the record straight, at least with regard to the polarity. On August 8, 1836, at a session of the *Académie*, he claimed priority for himself and Breschet for the polarity discovery. Although his 1835 report did not have the details, he pointed to a book he had recently sent to the organization that contained more detailed information.[3]

Colladon's communication, it should be noted, was presented at the *Académie* a few months later in 1836. Becquerel was one of the *commissaires* charged to examine it (as had been the case for Matteucci's communication dealing with the production of the spark). Notably, the production of a spark from a torpedo had been one of the goals that Becquerel and Breschet had set out for themselves but failed to achieve.

---

[3] Matteucci complains in some of his writings about Becquerel's attitude toward him on this and other occasions. In particular, he remarks that Becquerel's book describing the experiments on torpedo made in Venice by Breschet and Becquerel had actually appeared *after* the communication on the torpedo he (Matteucci) had sent to the *Académie*. The book alluded to by Becquerel is his *Traité Expérimental de l'Électricité et du Magnétisme*, published in 1836.

Figure 24.7. Antoine Caesar Becquerel (1788–1878) and Gilbert Breschet (1799–1850) studied torpedoes caught near Venice and were able to detect their electricity and establish the correct polarity of the discharge.

THE TORPEDO'S SPARK

The spark, first seen by Walsh with an electric eel in London during 1776 (Chapter 19), still had never been witnessed with a torpedo (or even an electric catfish). This milestone, which had long been sought by Walsh, Humboldt, and others, was finally obtained in Italy in the 1830s.[4]

This distinction goes to a (now) rather obscure Italian physicist, Santi Linari, a professor at the University of Siena and a Piarist (the first Catholic order specializing in educational activities). Studying torpedoes caught off the coast of Tuscany in March 1836, Linari succeeded in obtaining visible sparks (Linari, 1836; see Matteucci, 1836a). He did this with an apparatus inspired by physicist Michael Faraday's studies on electromagnetic phenomena, one that depended on the inductive effects of currents passing through long metal wires coiled around a magnetic body (Fig. 24.8). Linari was then associated with Carlo Matteucci, the Italian physicist mentioned above, to whom we shall soon return. It was Matteucci who first communicated the long-desired achievement to the *Académie* in 1836, based on Linari's and his own experiments (Matteucci, 1836a; see Becquerel, 1837, p. 797).

Matteucci, who had corresponded with Faraday since 1833, knew about his experiments on electric induction. Further, he had worked in the scientific cabinet of the Grand Duke's Museum in Florence, where Leopoldo Nobili first obtained a spark from the current induced by the variable magnetic field that was generated by the movement of a lodestone. Nobili, who used Faraday's apparatus, had preceded Faraday himself in this achievement (Nobili & Antinori, 1831; Bianchi, 1874, pp. 45–46; Sbrighi, 1985: see below).

The reasons why a spark could be produced from a torpedo with Faraday's inductive circuit, despite the relatively low voltage of the fish's shock, needs to be considered here. The physical mechanism of spark production upon suddenly interrupting an electric circuit involves an "auto-induction" phenomenon, a particular expression of the law of electromagnetic induction (one of Faraday's many laws in the field of electricity). This law stipulates that a conductive circuit develops a current when immersed in a temporally varying magnetic field (the intensity of the current being a function of the rapidity of change of the magnetic field, and of some characteristics of the medium and the circuit). A varying magnetic field can be produced by either moving a magnet rapidly toward or away from the circuit, or by passing a rapidly varying current through another circuit situated in close proximity to the first one. This is because, as first shown in the spring of 1820 by Danish natural philosopher Hans Christian Ørsted, a current flowing along a circuit generates a magnetic field around it and, if the current changes, the magnetic field changes in accordance with it.

The point is that when a timely variable current flows along a circuit, the resulting variable magnetic field can also produce a current in the same circuit in which the current flows, this auto-induced current being of opposite sign to the initial current. If the variation of the current is very rapid (as it happens, for instance, upon suddenly interrupting the circuit), the voltage produced by this auto-inductive effect can be much higher than the original voltage responsible for

---

[4] In this context it is worth remarking that before Linari and Matteucci a spark from the torpedo's shock had been already obtained by Francesco Giuseppe Gardini, an Italian doctor of Piedmont and a pupil of Giovan Battista Beccaria. Gardini's achievement concerned, however, a single experiment in which he happened to observe the spark on *only one* occasion. This is the way he reported on his scientific *unicum* obtained in a fish investigated in Genoa: "Only once during the shock I saw the spark, and I heard the noise, and even the attending persons also observed it; the Torpedo indeed was not in water, but placed on an electric stool" (Gardini, 1792, pp. 100–101). Despite the exceptional character of the event, it is possible that Gardini was not deceiving himself, because the experiment was done on a very big torpedo (possibly a *Torpedo nobiliana*, which can produce shocks up to 300 Volts). Moreover, the fish was investigated outside the water, which results in stronger shocks because of the absence of the shunting effect of the liquid medium. Gardini found that the shocks produced by his big torpedo were stronger than all those he had experienced with a Leyden jar and could overthrow him.

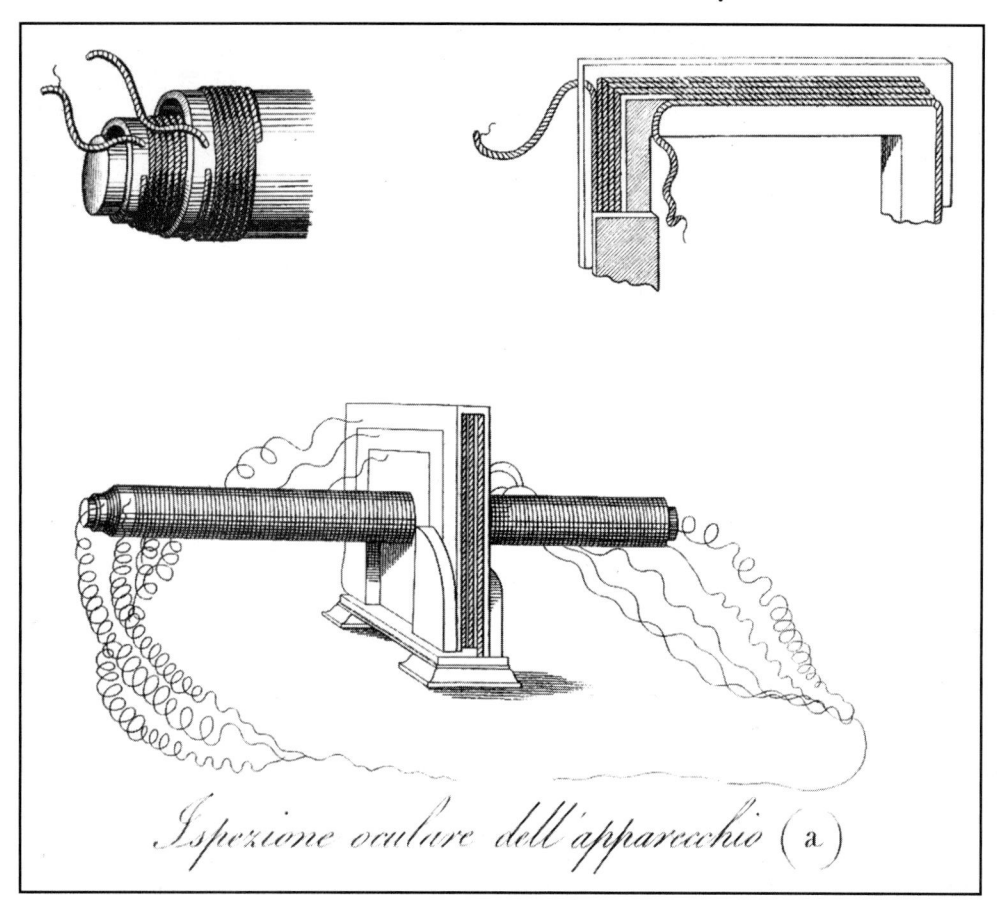

Figure 24.8. The induction coil used by Santi Linari in March 1836, when he first began his experiments on spark production with torpedoes caught off the coast of Tuscany.

the current flow. This situation can lead to the appearance of a spark at the moment of circuit interruption. This is why we can obtain sparks from a car battery that has low voltage.

The electric induction effect is particularly prominent with long coiled circuits, such as those contrived by Faraday to reveal induction currents, and then by Linari and Matteucci in their experiments with torpedoes. Moreover, since the voltage produced through induction depends on the speed of change of current, it is particularly strong when a large current is suddenly interrupted.

One arrangement used to derive a large current from a low-voltage device is to establish a low-resistance, large-surface contact between the device and the circuit. This is exactly what Linari and Matteucci did (as did others after them; e.g., du Bois-Reymond)—compressing a torpedo between two large metallic laminas. In his experiments aimed at producing a spark from a torpedo, Linari used an arrangement that allowed him to open the circuit with great rapidity by manipulating the tip of a wire near the top of a vertical glass tube filled with mercury. With the circuit interrupted right after the discharge, he was often able to see a spark.

In subsequent experiments made that September, Linari witnessed sparks by using a mercury tube to interrupt the discharge circuit without recourse to the inductive coil. In these experiments, the time relation between the shock and the spark was more evident, because there was no delay typical of inductive phenomena (Fig. 24.9). The spark was sometimes so strong that "it could be seen distinctly even in the clear daylight" (Linari, 1836, p. 2; also see 1837, p. 68).[5] This particular brightness (and even the production of the spark itself in the absence of the inductive coil) might in

---

[5] There are considerable historical problems in referring to Linari's publications on the torpedo. The initial announcement of the spark and other experiments on the fish were made in a letter by Matteucci to the *Académie*, reported in the *Comptes Rendus* of the French *Académie* on July 11, 1836. In this report Matteucci puts Linari's and his own experiments together (Matteucci, 1836a). Linari, who considered Matteucci's behavior improper, hurried to have a publication announcing his experiments independently of Matteucci (and somewhat polemical with him). What is probably the first publication of this type appeared as a *Supplemento* to the issue of December 13, 1836, of the *Indicatore sanese*, a local magazine of Siena of general character. Before the title, this article bears the indication that it is a Note to the publication of the *Académie*. Although this publication is generally referred to as "Linari, 1836," it must be noted that it was written in a third-person style. This occurs in most of the numerous articles dealing with Linari's achievements in a great number of Italian and foreign scientific journals and magazines through 1839 (and also later). Furthermore, it is uncertain that the *Supplemento* really appeared in 1836, because it was not inserted in the main body of the *Indicatore*. This also occurred for the second paper, dealing with Linari's experiments of 1838, also written in third person and also bearing the indication that it was a second *Supplemento* to the December 13, 1836, article in the *Indicatore sanese*. The second paper bears, before the title, the indication that it was published in 1838 (no corresponding indication is present in the first paper). Both articles were reprinted in the *Giornale scientifico-letterario* of Perugia in 1837 and 1838, respectively. Importantly, in Issue 59 (January to March 1839), this journal published the first original memoir of Linari (the only one written in first person) with an editor's footnote saying that this was the unpublished memoir of Linari used by Matteucci for writing his 1836 communication to the *Académie* (Linari, 1839). This memoir was also printed as a separate booklet with a somewhat different title and different pagination (Linari, 1839), adding to the confusion.

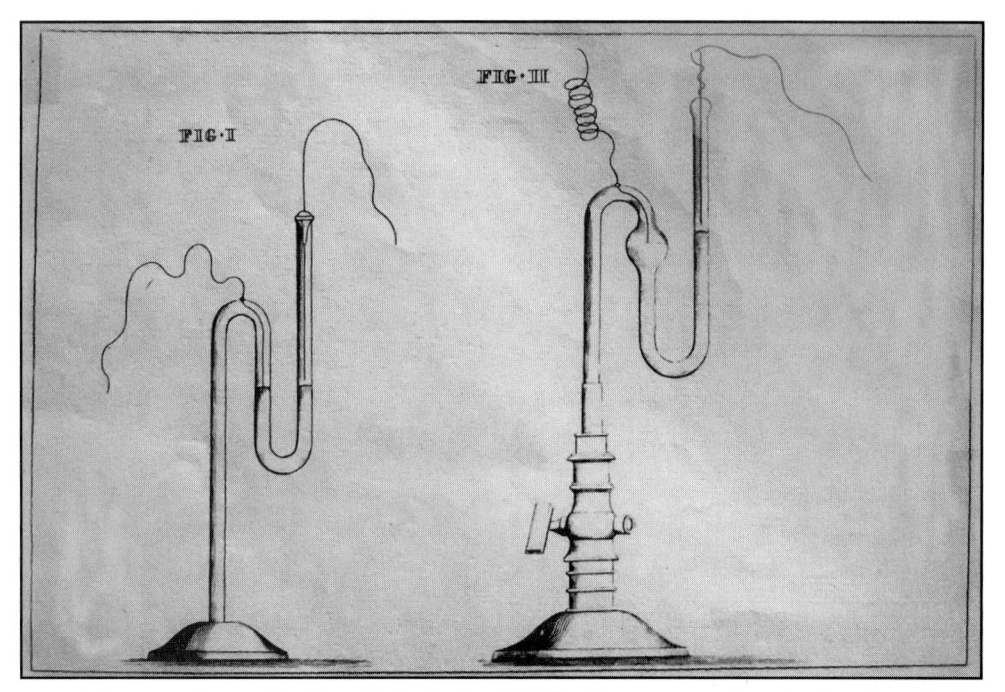

Figure 24.9. The mercury tube apparatus used by Linari in September 1836 for his newer experiments of spark production with torpedoes. Using this apparatus, Linari (1836) would write that the spark "could be seen distinctly even in the clear daylight."

part be the physical consequence of the passage of electricity through the mercury vapor present in the tube. With his powerful sparks, Linari was even able to ignite an explosive gas mixture.

A few months after Linari, Carlo Matteucci, who made his torpedo experiments on the Adriatic coast of Italy, also saw sparks (Matteucci, 1836a). A harsh debate sprung up between Linari and Matteucci over who had first produced the spark (Fig. 24.10) And after Linari and Matteucci, many others succeeded in this achievement by modifying the methods used by both of them (see Zantedeschi, 1845). One of the first was Giovan Battista Pianciani, a Jesuit professor in Rome, who studied two live torpedoes in 1837 in Fiumicino, on the Tyrrhenian Sea near Rome (the same place where Humphry Davy was planning his experiments in 1829). Pianciani (1838) also obtained other notable effects, particularly various electromagnetic phenomena, including magnetizing metals, induced currents, etc.

Pianciani's experiments were performed in the same place where his colleague and friend, Saverio Barlocci, had investigated torpedoes 2 years earlier. Barlocci, a physicist, did his experiments in collaboration with Pietro Peretti, a pharmacologist, and Pietro Carpi, a mineralogist, also professors in Rome. In addition to electromagnetic actions, Barlocci had obtained chemical effects (not investigated by Pianciani), but he had been unable to produce a visible spark (see Barlocci, 1841, pp. 217–222).[6]

---

[6] At the time Barlocci was famous because of experiments he first performed in 1830, with the aim of confirming the idea that electricity might be associated with sunlight. These experiments did not involve torpedoes but were based on the use of a galvanoscopic frog to detect the possible electricity. Barlocci split apart the violet and red rays of the sunlight with a glass prism and separately directed them onto two metallic laminas. The galvanoscopic frog started jumping as soon as its nerves and muscles were connected to the two laminas, which he viewed as supportive of his thesis (Barlocci, 1831, 1841, pp. 295–315; also see Zantedeschi, 1838, 1845). An account of these experiments, which represent a particular instance of the great interplay between physics and physiology in the science of the epoch, circulated in the journals and magazines of the times, including the *Quarterly Journal Of Science, Literature and Art* of London and the *New York Medical and Physical Journal*.

LINARI'S MORE COMPLETE PICTURE

Between 1836 and 1838, Linari was able to complete the picture he wished to paint of the electrical nature of a torpedo's shock (Linari, 1836, 1838). In addition to obtaining various chemical and galvanometric measurements of the shock—all converging to establish the correct polarity of the discharge—Linari also demonstrated the heating action of a single shock using a thermoelectric device based on the Peltier effect (i.e., on temperature changes occurring at the contact of two metals with the passage of electric current). Moreover, he used torpedinal electricity to make colored rings appear on his metal laminas, the result of an electrochemical oxidation process discovered by Leopoldo Nobili in 1826 and referred to as the "metallochromic effect" (see Nobili, 1830a; Pedeferri, 1989).

Linari also succeeded in demonstrating that the electric fluid involved in the torpedo's shocks could have electrostatic attractive and repulsive effects, just as is true of artificial electricity. These phenomena had long eluded the endeavors of more than a few talented scientists. Linari did this using Volta's *condensatore*. As noted, this device is based on a sensitive electroscope connected to a capacitor, with a movable handle allowing it to function as a variable capacitor. As Volta had remarked in his letter to Configliachi, the difficulty in trying to use an electroscope to detect the electric charge generated by a torpedo is mainly due to the extreme short duration of the shock. It was necessary to counteract the possibility that the torpedo's electricity, transferred onto the electroscope's laminas at the moment of discharge, could rapidly be dissipated by flowing back

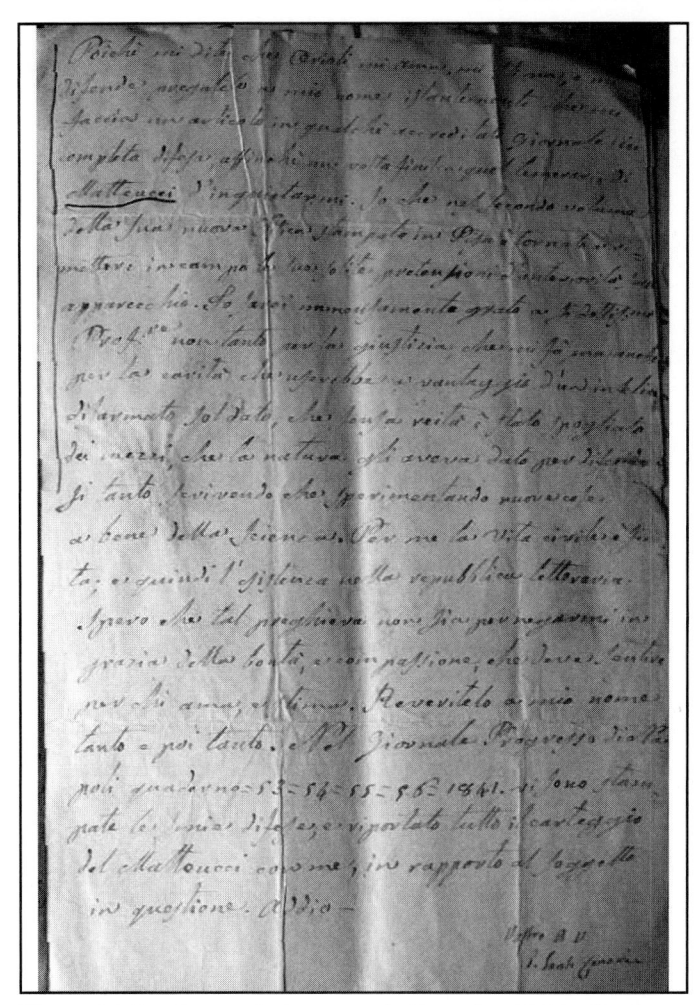

Figure 24.10. A page of a letter sent in December 1841 by Santi Linari in Naples to physicist Eusebio Giorgi, a member of his religious order, in Florence. Linari complains about Matteucci's "customary pretenses of priority" in the construction of the instrument used to obtain sparks from the torpedo. Matteucci's "pretenses" had been reiterated in 1841, in the second volume of one of his treatises (*Lezioni di fisica—II. Fenomeni elettrici*). Linari asked whether a colleague, physicist Francesco Orioli, could write an article in his defense against Matteucci (from the "Archivio degli Scolopi," Firenze).

through the circuit and the fish's body as soon as the shock was over. As mentioned in the previous chapter, Volta had suggested that one should try to get the fish to produce its electricity in a more continuous way, perhaps by squeezing its electric organs, in order to overcome this difficulty. This view, as we have seen, was consistent with Volta's interpretation of the battery mechanism. The strategy now used by Linari was, however, quite different. It was based on interrupting the discharging circuit soon after the shock by using a specially contrived device that avoided the back-flowing of the electricity that was just discharged.[7]

The results of Linari's experiments were initially published as a supplement in an obscure magazine from Siena (Linari, 1836, 1838). They were afterwards translated into various languages and published or reported in both scientific journals and general culture publications (e.g., *Annalen der Physiks, Gazette de France, Bibliothèque Universelle de Genève, Penny Cyclopaedia, Nederlandsch Magazijn, Årsberättelser om Nyare Zoologiska*). Some publications, however, erroneously referred to his "electric eel experiments," missing the important point that he had worked on less powerful sea torpedoes. Linari's fame would grow even more in subsequent years, when he and Luigi Palmieri would become the first individuals to produce electricity through a magneto-telluric action (i.e., by simply rotating circular conductors in the earth's magnetic field) (Palmieri, 1842, 1845). They detected the weak electricity produced by this action with an apparatus similar to the one Linari used when studying his torpedoes, and they were also able to produce sparks.

With Linari's achievements, there now appeared to be nothing left wanting for the complete "electrification" of the torpedo rays. That is, all of the basic demands that had been listed by eighteenth-century experimental natural philosophers had finally been met.

COMPARISONS WITH THE EEL

In the period 1838–39, Faraday was able to obtain most of these electrical effects from a more exotic electric fish, this being part of the research program he had been pursuing since 1833 with the aim of demonstrating the basic similarity of all known forms of electricity (Faraday, 1833, 1838, 1839a,b). As already mentioned, Faraday had been with Humphry Davy during Davy's first journey to Italy. In a letter sent to his mother in 1814, he had listed torpedoes as being among the wonders he had admired during his trip to Europe, together with the glowworm; the Louvre, the Luxembourg Palace, and the *Jardin des Plantes* in Paris; the Academy of Cimento in Florence; and Saint Peter's Church in Rome (see opening quotation to this chapter; Bowers & Symons, 1991, p. 88).

Faraday's decision to investigate the electric eel many years later was undoubtedly stimulated by Alexander von Humboldt, who in 1834 had sent him a letter dealing with electric fishes. Humboldt acquainted Faraday with some literature on electric fishes that the Englishman had apparently ignored (see Faraday, 1833), and he encouraged him to apply his experimental talents to this field, because, as Humboldt wrote, "I am convinced that with the knowledge of electro-magnetism and physiology that we possess today, the study of the phenomena of the gymnoti should shed strong light on the functions of nerves and the muscular movements of man" (see Faraday, 1993, p. 189).

Initially Faraday attempted to import eels from Guyana through the Colonial Office, and Humboldt gave him useful indications on how to increase their probability of survival during the long trans-Atlantic trip (Faraday, 1993, pp. 214, 220). Eventually, however, Faraday decided to investigate an eel imported from Guyana for commercial purposes and kept in an aquarium in the Gallery of Adelaide Street in London. When finally performing his study, he was aware of Matteucci's successful research on spark production with a torpedo, because Matteucci had personally told him what he had done soon after obtaining the spark in May 1836 (Faraday, 1993, p. 151).

---

[7] Many years after Linari, French doctor and physiologist Armand Moreau, a pupil and collaborator of Claude Bernard, conducted a similar experiment with an improved technique. Moreau (1862) used an automated system in which an electrical stimulus delivered to the electric nerve (in order to induce the shock) was followed, after a short delay, by an interruption of the circuit connecting a torpedo to an electric capacitor.

Faraday had to exercise special precautions so as not to threaten the life of the eel, once he had access to it. In particular, since he did not wish to take the eel out of the water, he developed a complex system to measure its electricity while it remained in the water. With his arrangement, which was afterwards also used by du Bois-Reymond in his studies on the *Malapterurus*, Faraday obtained galvanometric measurements, chemical effects, and thermoelectric actions, and was able to magnetize metals. He also established the polarity of the eel's shock (the head region is positive with respect to the tail at the moment of the shock; Faraday, 1839a,b). Moreover, he was able to obtain sparks, first by using an inductive coil and afterwards with short wires. He achieved this by rubbing two metallic files used to complete the discharge circle together, thus causing rapid interruptions and closures in the discharge circuit.

Faraday's interest in and fascination with the electric eel are also attested to by the fact that he gave a lecture, *Discourse on the Gymnotus*, as one of his famous Friday evening public lectures at the Royal Institution. Very large audiences typically attended these lectures. During his discourse on January 18, 1839, he displayed the organs of one of the eels from the Adelaide Gallery, which had been prepared for him by Richard Owen, the famous British anatomist and paleontologist (see Faraday, 1993, p. 547).

In 1839, German-Swiss chemist Christian Friedrich Schönbein, the discoverer of ozone (and another of Faraday's correspondents), also investigated a live eel from the Adelaide Gallery. Schönbein (who was in England in the second half of the year to attend a scientific conference and had met Faraday) published his results in 1841. He confirmed most of Faraday's findings but was unable to obtain a spark corresponding to the shock (Schönbein, 1841).

In addition to a series of chemical and physical effects, a particular prominent spark was obtained in 1844–45 with another electric eel. This eel was imported from Brazil and was housed in the aquarium of the King of Naples. Two Italian physicists, Domenico de Miranda and Giacomo Maria Paci, who worked at the Royal Physical Cabinet, studied it (Miranda & Paci, 1845a,b). They successfully used an apparatus that was based on an inductive coil, one similar to that used by Linari on torpedoes and by Faraday and Schönbein on their eels (also see Matteucci, below). Interestingly, the eel's shock was used (apparently with great success) for the medical treatment of two servants of the King's House.

Hence, there was firm evidence by 1840 that the shocks from torpedoes and eels are in all respects electrical, and that they are identical other than in the strengths of their electric powers. Indeed, Faraday's experiment on the *Gymnotus* was a part of his program aimed at showing that the same effects could be obtained with artificial electricity (in its various forms), atmospheric electricity, and animal electricity. But how firmly could these findings be tied to frog and perhaps even human nerve function?

MATTEUCCI AND ANIMAL ELECTRICITY

There is a deep conviction in the texts of many scientists who studied or wrote about electric fishes in the first half of the nineteenth century (e.g., Humboldt, Becquerel, Faraday) that the investigation of fish electricity, as well as that of more common animals, would provide important clues for understanding nerve function. Nevertheless, attempting to derive information from studies of electric fishes that could shed direct light on nerve function and on other important aspects of animal physiology would take additional time.

In the domain of nerve physiology, a common theme to emerge from the fish research mentioned above, and from other early-nineteenth-century studies (such as the findings published by John Tweedie Todd, an English surgeon and naturalist who investigated torpedoes in both South Africa and at La Rochelle; Todd, 1816, 1817), was the disappearance of the shock upon cutting the nerves innervating the organs or destroying the brain itself. Following the idea expounded by John Hunter in his anatomical investigations of torpedoes and electric eels, the general assumption was that nerves transmit electricity fabricated in the brain to the fish's electric organs. The prevailing thought was that this electricity would accumulate and thereby intensify within the prisms making up the electric organs, so as to produce the shocks typical of the species.

A major step in the research path linking electric fishes to nerve physiology (and to the muscles and other excitable tissues) came from the investigations of Carlo Matteucci, who was born in Forlì, a town near the Adriatic coast of Italy, and was appointed Professor of Physics at the University of Pisa in 1840 (Fig. 24.11). A recommendation letter sent by Alexander von Humboldt to the Grand Duke of Tuscany, who had asked for his advice on the matter, helped Matteucci obtain the professorship in Pisa (Moruzzi, 1964, 1996 p. 79; Bianchi, 1874, pp. 72–73). In his letter, cited by François Arago during a *séance* of the French Academy in 1837, Humboldt had declared that the scientific findings that "had most impressed him in recent times" were the results of Matteucci's experiments on the nervous control of the torpedo's shocks (see Becquerel, 1837, p. 797).[8]

Matteucci's life was relatively short (he died in 1868 at age 57) but very intensive. In addition to studying electrical phenomena in living animals, he made important discoveries in physics and chemistry and took part in the *Risorgimento*, the political and military movement that led to the unification of Italy and its liberation from foreign powers (Bianchi, 1874). As noted by Giuseppe Moruzzi, the twentieth-century neurophysiologist who studied Matteucci and his electrophysiological work, Matteucci's commitment

---

[8] The correspondence between Matteucci and Arago that is still extant and the archives of the French *Académie des Sciences* (and particularly a letter written by Matteucci from Ravenna on August 1, 1839) also makes it likely that Arago intervened directly with the Grand Duke to support Matteucci's professorship at Pisa. One of the reasons why the Grand Duke was at that time particularly sensitive to the influence of great European scientists was the occasion of the first meeting of Italian scientists that he was organizing in Pisa in 1839. One of the political aims of this meeting was to put Tuscany in a prominent position in the process of the reunification of Italy. Italy was at the time divided into a number of small states, some of which were under foreign control. The complete reunification would take place in various step during the second half of the nineteenth century. Matteucci and many other scientists played a direct role in the cultural, political, and military processes leading to the reunification (the *Risorgimento*). In addition to Arago and Humboldt, Jean-Baptiste Dumas was another strong supporter of Matteucci among the members of the *Académie*. He was a chemist interested in living processes and a professor in various Paris institutions, and he eventually became a minister in the French government and a senator. Matteucci expresses his devotion and respect for Dumas in letters to him conserved in the archives of the *Académie*, and sometimes call him "*Mon cher Maitre*" (My dear Master).

Figure 24.11. Left: German-Swiss chemist Christian Friedrich Schönbein (1799–1868), the discoverer of ozone and nitrocellulose, studied the electric eel in London in 1839 and published his results in 1841. Right: Although he observed most of the fish phenomena, he was unable to obtain a spark at the moment of the shock.

to science began early. He published his first scientific work at age 16, 1 year before graduating in physics from the University of Bologna; he succeeded in earning the esteem of scientists, including Arago and Becquerel, before he was 20; and he thereafter had the close attention of both Faraday and Humboldt.

Matteucci's interest in torpedoes was by no means a fleeting endeavor. Having published his first work on the subject in the Proceedings of the Paris *Académie des Sciences* in 1836 (at age 25), he published the last article of a long series on the same subject in the same journal in 1865, i.e., three years before his death. Moreover, he was waiting for live torpedoes from a fishmonger in La Spezia on the very last day of his life, while at his vacation house near Leghorn (Bianchi, 1874).

Matteucci's 1836 publication on this electric fish, written in form of a letter to *Académie* secretary Arago, dealt with the torpedo experiments performed by himself and by Linari in the very same year. Both Matteucci, working on the shores of the Adriatic Sea, and Linari, working on the Tuscan coast, had obtained sparks from their catches. A second reason why Matteucci dealt with Linari's experiments, which were made a few months before his own, was that he had suggested the methods and equipment first used by Linari, and he was now following the same protocols. Because of the importance of the achievement, and possibly for other reasons as well, Matteucci's communication led to a rather harsh dispute over the priority of these experiments, which had a particularly strong echo within the French *Académie* (see Bianchi, 1874; Moruzzi, 1964, 1996; Becquerel, 1837; Blanch, 1841).

Matteucci (1836a) wrote about having succeeded in obtaining chemical, thermal, and magnetic effects, as well as being able to measure the torpedo's electricity with a galvanometer and establishing its polarity (Fig. 24.12). But he did more than use the latest instruments to study fish electricity; he also employed "prepared frogs," following in Galvani's footsteps (Fig. 24.13). As mentioned in Chapter 22, Galvani had used frog preparations for detecting weak torpedinal electricity. With his "galvanoscopic" frogs, Matteucci was able to show that the shock of a torpedo kept in a basin of salt water could diffuse over more than 3 feet (1 meter). His 1836 communication ended with a brief note on the deleterious effects of cutting the nerves on the discharge.

The nerves associated with the shocks remained the subject of Matteucci's (1836b) second communication, which was sent to the same scientific society and also published in 1836. The nerves would also be addressed in later communications (Matteucci, 1837, 1844, 1865). Besides confirming that destroying the brain and severing the nerves annihilates a torpedo's shocking powers, Matteucci identified a specific part of the brain controlling the shocks. Here he pointed to the "fourth lobe," a term meaning the medulla oblongata.

He further showed that an electric organ isolated from the brain could still discharge a shock if he stimulated the nerves reaching it, even though the torpedo's organs are themselves unresponsive to direct applications of electricity. This is an important characteristic of torpedoes and a notable difference between these rays and other strongly electric fish, which would later be clarified with electrophysiological research (see Chapter 26). Also, when he stimulated specific branches of the nerve to the electric organs, the effects were restricted to the corresponding parts of the organ.

In one of his experiments, a variation on the latter theme, Matteucci cut many branches of the nerves to an electric organ, leaving only a few intact. Inspired by Galvani's work, he then covered the fish's body with some galvanoscopic frogs and stimulated the main nerve trunk. The activation of only the remaining innervated part of the fish's electrical

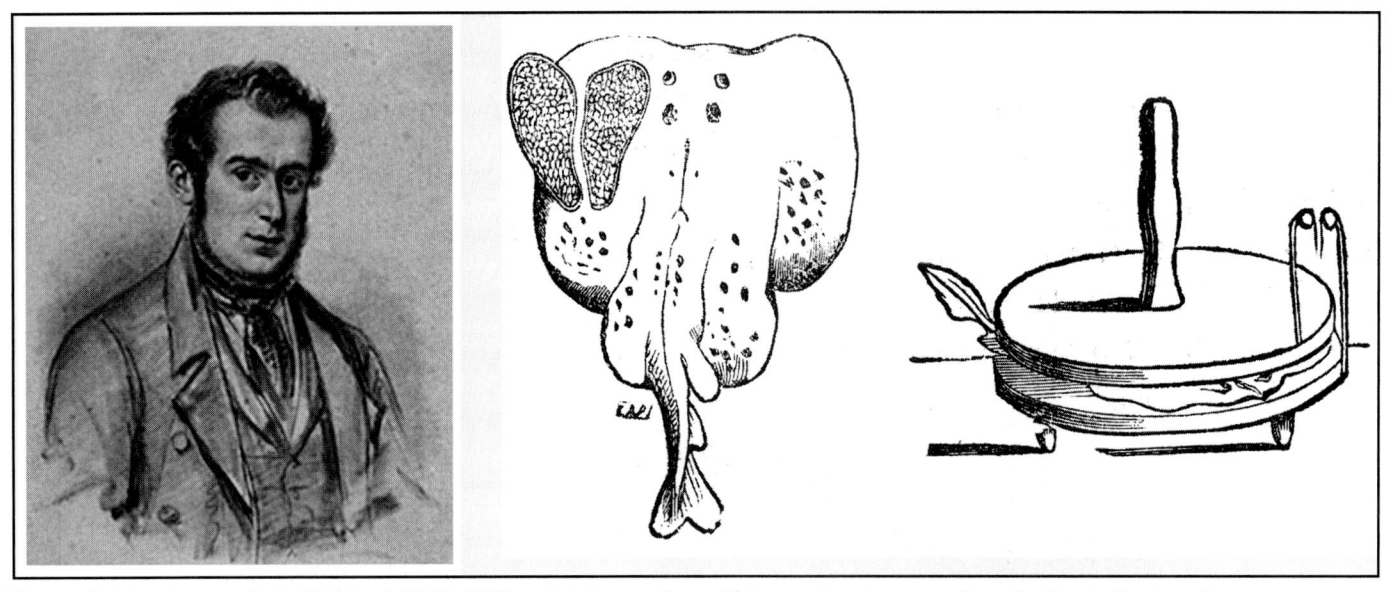

Figure 24.12. A young Carlo Matteucci (1811–1868) and an image of one of his experiments on spark production in the torpedo.

organ made for a striking demonstration, with only one of his galvanoscopic frogs responding in a reliable way to stimulation of the nerve trunk.

Matteucci discussed his earlier torpedo studies in his *Traité des Phénomènes Electro-Physiologiques des Animaux*, published in 1844. His treatise was supplemented with a careful anatomical study by Paolo Savi, an Italian zoologist and botanist, and Matteucci's colleague at Pisa (Fig. 24.14). Matteucci's *Traité* was the most important work on animal electricity after Galvani's *Commentarius* of 1791, and it remained so until du Bois-Reymond's two-volume monumental work on the same subject appeared between 1848 and 1884 (see below).

According to Matteucci, the production of fish's shock is essentially a reflex phenomenon controlled by specific parts of the central nervous system. Moreover, an electric organ is not simply an accumulator of electric fluid originating in the brain and transported by the nerves. On the contrary, it is

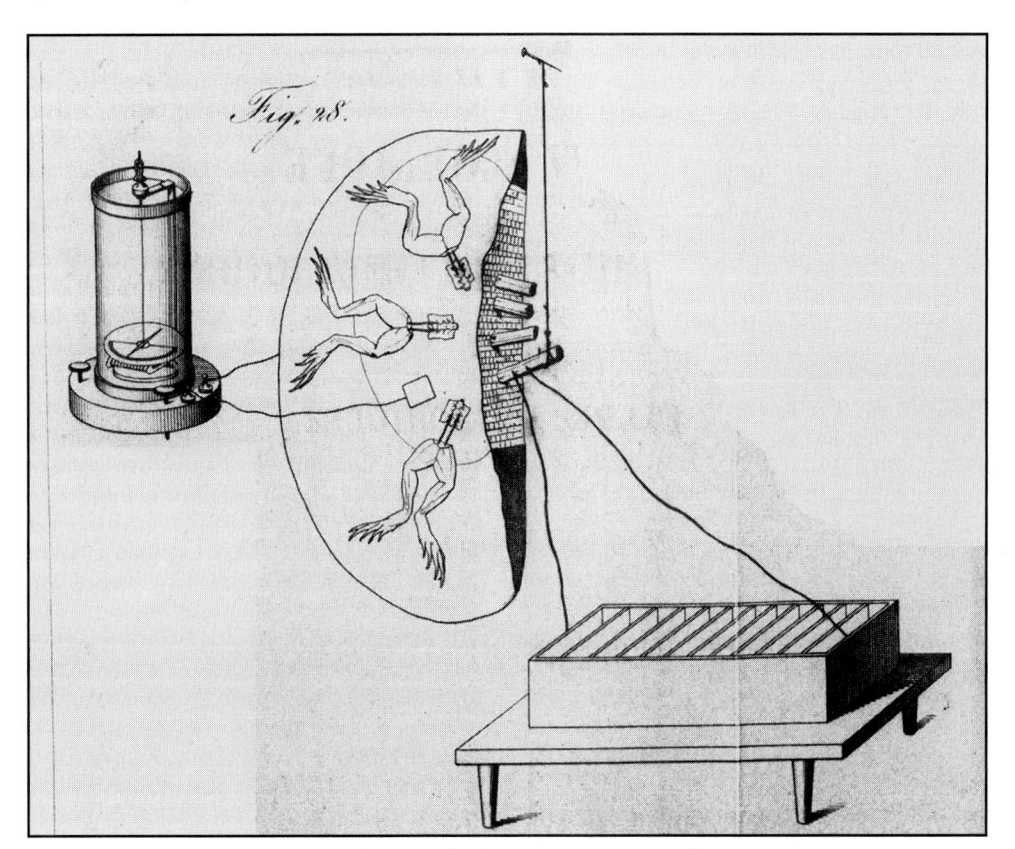

Figure 24.13. An image showing how Matteucci obtained physical, physiological, and chemical effect from the torpedo's shock. When the fish produced its bolt, the galvanometer needle moved, the frog jumped, and electrochemical effects were produced in the solution contained in the receptacle on the right.

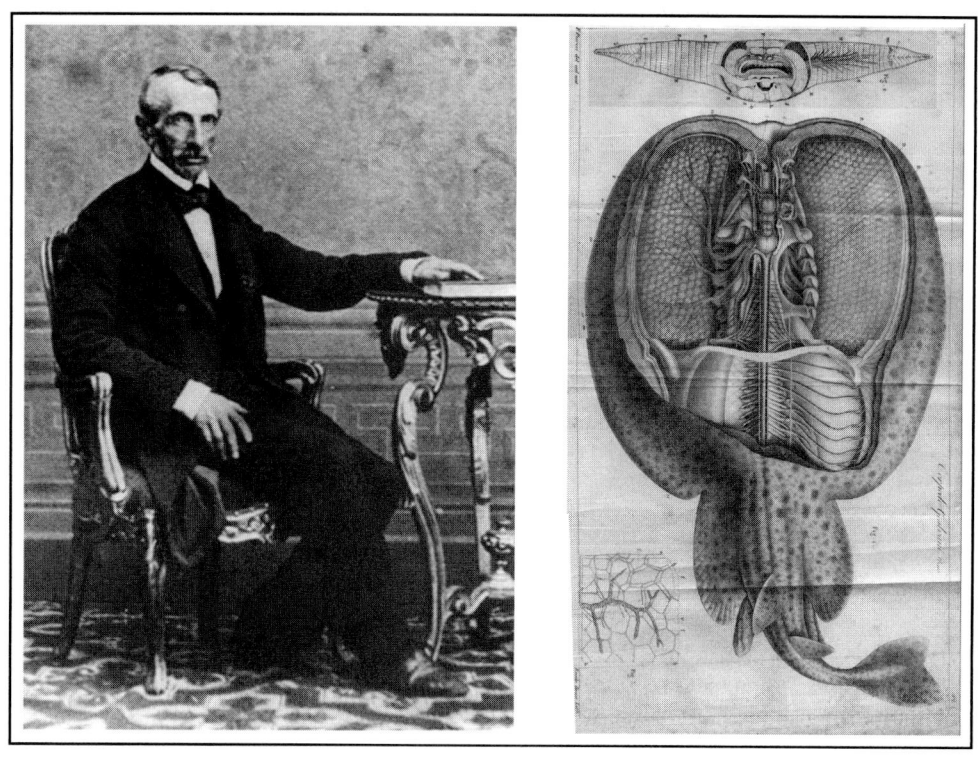

Figure 24.14. Paolo Savi (1798–1871), the naturalist of Pisa who collaborated with Matteucci in his electric fish studies, and the image of a torpedo dissection made by Savi and inserted in Matteucci's *Traité des Phénomènes Électro-Physiologiques des Animaux*, first published in 1844.

the place where the electricity involved in the discharge is produced.[9]

Among the experiments Matteucci did in order to show that the discharge is not accompanied by mechanical or other such phenomena was one in which he demonstrated that the volume of a torpedo's body does not change during production of a shock. To this purpose, he enclosed the fish in a glass bottle with a wide mouth and used a galvanoscopic frog to track its shocks. The experiment was somewhat reminiscent of Jan Swammerdam's experiments from the seventeenth century, but for the fact that a torpedo and not a frog muscle was the subject of the volumetric analysis (see Chapter 11).

In a series of experiments involving cutting the electric organ, Matteucci was able to prove that the separated columns (or "prisms") of the organ could give the shock when stimulated through their corresponding nerve filaments. The shock was also observed, although smaller in intensity, when the organ was sectioned horizontally, thus leading to hemi-prisms of decreased height in the two separated parts. From these and other experiments, Matteucci clearly recognized the flat cellular elements making up the organ as the units producing the electricity. As he wrote in his 1845 communication to the French *Académie*, "The cell which composes the prisms of the organ of the electric fish is the very elemental electric organ, which, in order to function, needs only the excitation of the corresponding nerve filament, and the integrity of the albuminous substance it contains" (Matteucci, 1844, p. 494).

Matteucci's interest in electric fishes was directly connected to the study of electricity in the nerves and muscles of more ordinary animals. In fact, his most important physiological achievement is the first indisputable instrumental measurement of electricity from tissues of ordinary animals. The way that he did this is worth considering in some detail, not only because it is a landmark in physiology, but also because it helps us to follow his thinking and appreciate features of the scientific climate at this moment in time. For it was only then that investigations of ordinary animal electricity started to be conducted again, following a lull stemming from Volta's negative thinking on the subject, a temporary calm that had spanned the opening decades of the new century.

Historically, galvanometric measurements of frog electricity had actually been obtained several years before Matteucci's work. Leopoldo Nobili, mentioned above, who was born near Bologna and was long-time Professor of Physics at the Museum of Physics and Natural History of Florence, had done it. He was interested in various aspects of electric science, particularly the magnetic and thermal effects of electricity, and he had a special talent for inventing or improving scientific instruments. To his credit, Nobili succeeded in markedly increasing the sensitivity of a galvanometer by diminishing the influence of terrestrial magnetism on its needle (Nobili, 1825) (Fig. 24.15). In 1827, Nobili recorded weak electricity by connecting the poles of his new "astatic" galvanometer to the liquid solutions of two glass containers containing frog parts. In one he had immersed

---

[9] This last view would receive strong support by a nice experiment performed in 1861 by Moreau, who worked initially in Naples and afterwards near Montpellier. Moreau cut the electric organ's nerve of one side and stimulated repetitively the distal stump so as to exhaust the torpedo's capacity to produce shocks. After a sufficiently long rest, the fish was found to recover the capacity to produce shocks in response to electric stimulation of the distal stump of the nerve, despite the absence of any communication with the central nervous system. The shocks produced after the recovery in the organ separated from the central nervous system were of the same intensity of those that could be elicited by the stimulation of the intact nerve on the other site (Moreau, 1861, 1862).

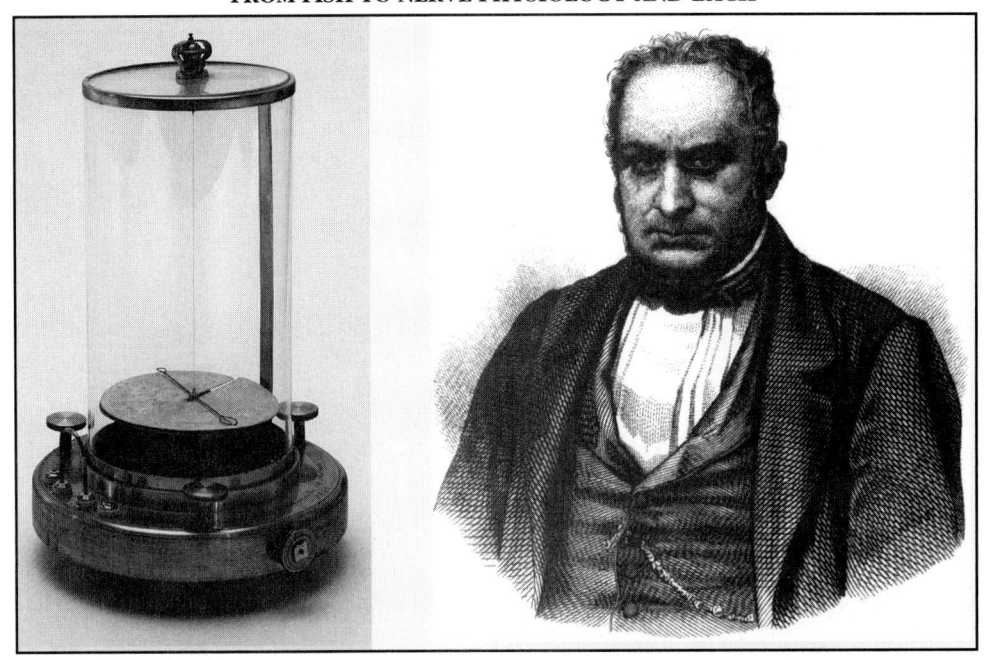

Figure 24.15. Matteucci later in life and the Nobili-type astatic galvanometer that belonged to him and that he used in his experiments. (© Museo Galileo, Firenze)

the legs, and in the other the spinal cord, of a prepared frog.[10]

When publishing the results of his experiments, Nobili (1828, 1830b) did not, however, attribute the origin of this electricity (which he denoted as "frog current" or "proper current") to a genuine physiological phenomenon. He assumed instead that the measured current was due to a purely physical phenomenon of thermoelectric origin (see Moruzzi, 1964, 1996; Mazzolini, 1985). Because of the greater surface–volume relationship of the nerves to the muscles in his frog preparations, he reasoned that the nerves would cool more intensely and rapidly than the muscles, due to the evaporation of the liquids impregnating them. He opined that this non-homogeneous cooling would lead to thermoelectric currents, which his galvanometric needle now picked up as "frog currents."

Although Nobili's reasoning can be difficult for us to understand today, the thermal effects of electricity was a topic of great interest in the first half of the eighteenth century. This was particularly true after Estonian physicist Thomas Johann Seebeck (1822) published his findings on the electrical currents produced by differentially heating two metals in contact with each other. Also, and not to be overlooked, the science of this epoch was still dominated by Volta's idea that there is no genuine electricity in the bodies of ordinary animals.

Matteucci repeated Nobili's experiments and varied them in diverse ways while he was in Ravenna, before moving to Pisa. He showed that the main source of the recorded current stemmed from the muscle mass. Importantly, he demonstrated that the current would appear only when one of the electrodes of his galvanometer was placed on an intact surface of the muscle, while the other was put on the cut (or otherwise injured) surface. No current could be recorded if the electrodes were both on the intact, or both on the injured, surface.

Further, Matteucci also established the polarity of the "muscle current," showing that the injured part was normally negative with respect to the intact surface.

As mentioned in Chapter 23 (and to be discussed in more detail later on), the electricity involved in nerve conduction and muscle excitation is due to the different charges on the two sides of the fiber's membrane, the interior being negative with respect to the exterior. As we now know, the cuts in Matteucci's experiments produced an easy path for the outward flow of current from the inside of the muscle fiber to the recording electrode (see Moruzzi, 1964, 1996; Piccolino & Bresadola, 2003). This was why the injured exterior surface was now negative relative to an intact exterior surface.

Matteucci discounted the possibility that the injury current he recorded (afterwards also denoted as the "demarcation current," see later) might be due to experimental artifacts. Besides thermal effects (like those invoked by Nobili), the main source of artifacts was the electricity stemming from the contacts between the recording electrodes and the biological liquids bathing and impregnating the animal tissues. This type of artifact posed major problems before scientists, and particularly du Bois-Reymond, contrived special tools, such as "unpolarisable" electrodes, to reduce its occurrence.

Having used many experimental arrangements to reduce these artifacts, Matteucci completely circumvented the problem by constructing a "pile" of frog half-thighs (Fig. 24.16). The deflection of the galvanometer needle increased in progressive steps with every new half-thigh that was added to the pile, while the number and type of liquid–metal junctions was held constant. There was no increase in the needle if the intact elements of the biological pile were directly connected to one another, or if just the cut parts were so connected. The galvanometric needle rose only when direct contacts were made between cut and intact surfaces.[11]

---

[10] Nobili also used his sensitive galvanometer in his experiments on the torpedo in Leghorn mentioned above.

[11] From two letters written by Matteucci in 1857 and 1860 (conserved respectively at the library of the Paris *Institut* and at the archives of the *Académie*)

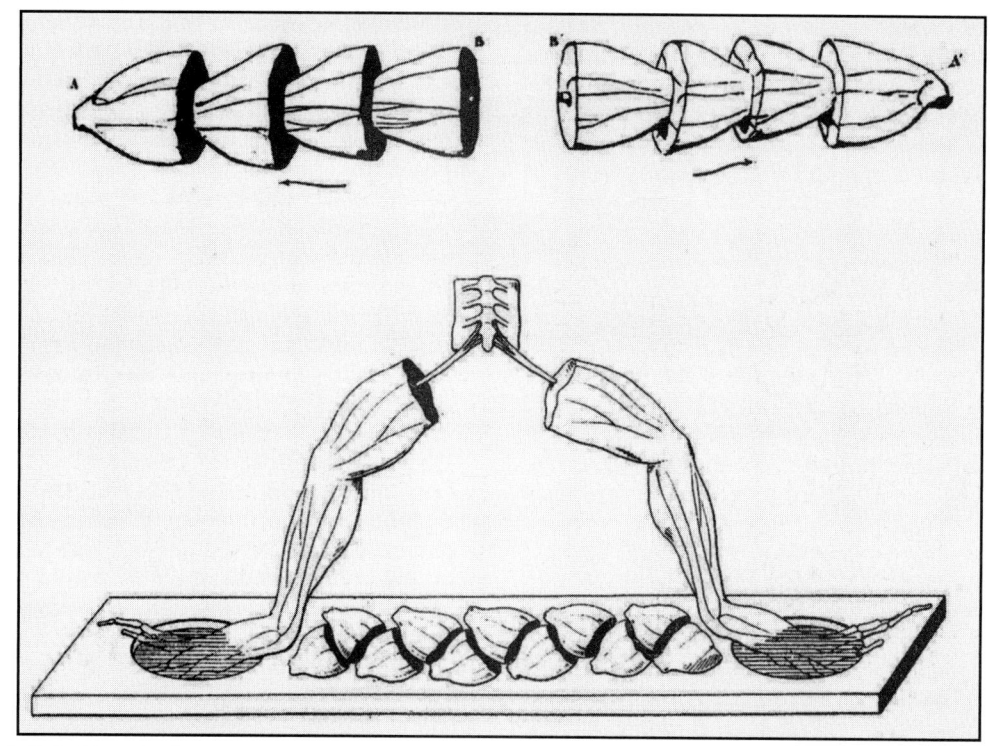

Figure 24.16. The "pile" of frog thighs Matteucci used in some of his galvanometric measurements of animal electricity. With this experimental arrangement, Matteucci provided the first unequivocal evidence of the biological origin of the current recorded with a galvanometer between the intact and the cut surface of a muscle. A demonstration of this experiment was repeated by Matteucci in Paris in the presence of Humboldt and other members of the *Académie*. In the lower part of the figure the electricity produced by the frog pile is used to stimulate the frog leg preparation (from Matteucci, 1844).

Matteucci confirmed the muscle pile findings with different warm-blooded and cold-blooded animals, producing electrical currents that could be detected with his physical instruments. In a sense, we could say that Matteucci even produced an "electric eel." He did this by using the body parts from real eels; that is, from the snake-like aquatic creatures found locally, as contrasted with the South American fish that can deliver very strong shocks, but which only superficially resembles an eel.

Seen from these perspectives, is easy to understand why Matteucci and not Nobili is given the lion's share of the credit for being the first person to provide an acceptable instrumental measurement of electricity using ordinary animals. And not to be overlooked, besides affecting the galvanometer needle and somewhat imitating Volta's battery and the organs of strongly electric fish, Matteucci's muscle piles were associated with electrochemical decomposition, the generation of heat, and other electric effects.

As he had done in his electric fish studies, Matteucci went on to study the actions of physical and biological agents on the muscle currents of frogs and his other animals. Importantly, he found that the muscle injury current disappears when the muscles enter a state of maintained contraction (i.e., in a condition of "tetany," as, for instance, following the application of strychnine). On the basis of modern electrophysiological studies with excitable fibers, we now know that this happens because the negativity of the electric potential of the intracellular compartments disappears during the discharge. But Matteucci did not fully grasp the meaning of this important observation, which would provide a key insight into the mechanisms underlying the electrical phenomena of the muscles and nerves.

Besides using preparations made exclusively with muscle tissues, Matteucci was also able to measure electric currents in frogs in which the muscles and nerves maintained relatively intact relations. The nerve current (corresponding to Nobili's "proper current" or "frog current") was typically the same or of a smaller amplitude than the muscle current. This observation led Matteucci to hypothesize wrongly that, at least with a typical Galvanic frog preparation, the "frog current" originates exclusively (or almost so) from muscle tissues, whereas the nerves act simply as conductors of the muscle-derived current.

Matteucci deliberately designed experiments aimed at investigating whether the nerves could produce electricity, but he failed to obtain any clear evidence of "nerve electricity." All told, he became increasingly inclined to believe that the nervous agent (or "force") is not truly electrical in nature. The effects of nerve ligatures were among the experimental observations he invoked to support this contention.

---

we know that Matteucci had made a convincing demonstration of this experiment in Paris. It took place around 1840 in the laboratory of Jean-Baptiste Dumas, an important member of the *Académie*. In addition to Dumas (the addressee of the second letter), Humboldt and Henri Milne-Edwards, a zoologist and also member of the *Académie*, were among those in attendance. In the 1857 letter, addressed to astronomer and *académicien* Jean Joseph Le Verrier, Matteucci wrote that Dumas praised his frog muscle pile experiments, saying their importance was comparable to Faraday's experiments on electromagnetic induction and to the discovery of the planet Neptune made by Leverrier (*sic*) (Le Verrier had assumed the existence of the new planet on the basis of the perturbation it produced on the orbit of Uranus). In another letter addressed to Dumas on March 8, 1856, Matteucci speaks of the "benevolences and encouragements" given to him by his correspondent when "in the presence of M. de Humboldt and other scholars I repeated in your laboratory my experiments of electrophysiology". This letter too in the archives of the *Académie*.

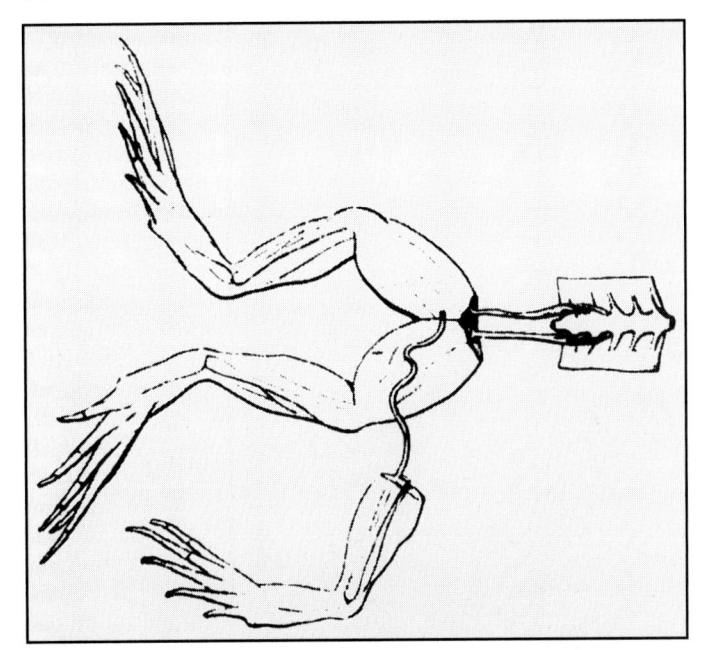

Figure 24.17. The arrangement for Matteucci's induced-twitch experiments with prepared frogs. These experiments provided the first demonstration that the electric signal propagating along nerve and muscle fibers is capable of exciting the zone ahead in the direction of the propagation. Matteucci, however, did not fully grasp the importance of his crucial observation (from Matteucci, 1847).

Expanding on what Haller had found in the previous century (Chapter 11), and using a sensitive galvanometer, he confirmed that electricity could pass down a ligated nerve in a neuromuscular preparation, but noted that no muscle response was triggered in this condition (this suggesting that electricity could not be the nervous agent). These ligature experiments were conducted with both frogs and torpedoes with similar results.

Another landmark observation made by Matteucci in his research on neuromuscular physiology was the so-called induced twitch (Fig. 24.17). He placed the nerve of one galvanoscopic-type frog preparation above the thigh of a typical frog preparation. When a contraction was induced in the first frog (by any kind of stimulation: electrical, chemical, mechanical), the galvanoscopic leg also moved. Matteucci discounted the possibility that the contraction of the second galvanoscopic frog depended on mechanical irritation of its nerves, induced by the movement of the first frog. Initially, following the indication of Becquerel, the reporter of his communication to the French *Académie*, he assumed that the effect was the consequence of an electric current propagated along the excited muscle (of the first preparation), which acted as a stimulant of the nerve in the second preparation.

Matteucci varied the induced-twitch experiment in various ways. A rabbit neuromuscular preparation could be used in place of the first frog. Moreover, when a series of frog preparations was arranged in a pile-like way, the induced twitch of the second frog might stimulate the nerve of the third preparation, and thus cause contraction in the corresponding leg, and the process could be repeated sometimes up to the fourth frog. The arrangements Matteucci used were again inspired by Galvani's research using frog preparations to monitor torpedinal electricity. The difference was that, in Matteucci's case, the source of electricity was the frog preparation and not a torpedo.

As noted by Moruzzi in his study of Matteucci, with his induced-twitch experiments Matteucci had what he needed for understanding the electrical mechanisms underlying nerve and muscle functions. The experiments showed that an advancing electric wave is capable of acting as an electric stimulus for the resting fiber ahead. Moreover, the dispositions with several nerve–muscle preparations arranged in a pile suggested that the traveling electric signal does not attenuate over the course of its propagation. But as with the disappearance of the muscle current during intense muscle stimulation, Matteucci again missed the real meaning of his findings. After his initial assumption about the electric nature of the signaling passing from the muscle of the first frog to the nerve of the second frog, he changed his mind—and eventually denied that this signal could be electrical.

Various reasons can be given for this failure, and they too merit some attention. One, as we have already stated, was that Matteucci had repeatedly been unable to detect any electrical current in the nerves with his galvanometer. He was also influenced by the outcome of his ligature experiments, which also worked against an electrical conception of the nervous signal. But more generally, what led Matteucci to his seemingly stubborn attitude against a neuroelectric theory was the difficulty he had in conceiving how electricity could be confined to nerve fiber without leakage (again in a way somewhat reminiscent of Haller).

This mitigating factor, seemingly overcome by Galvani (1791) with his evidence of a fatty insulating material (see Chapter 21), resurfaces in a particularly clear way in the following passage from Matteucci's *Traité*:

> I will add, moreover, that because of the known properties of electricity, and because of the laws of its propagation, it will be impossible for us to conceive of the existence of an electric current confined to nerves. This current should be strong enough to go from one side to the other of the nervous system. It should be of an intensity comparable to that of a true electric current, capable of exciting contractions in muscles similar to those produced by an act of the will. A current like this would not be confined in a nerve unless we admit that the nerve is similar to a metallic wire covered with an insulating coating, which is very far from the truth. (Matteucci, 1844, p. 255)

EMIL DU BOIS-REYMOND'S ELECTRIC FISHES

Emil du Bois-Reymond studied physiology in Berlin under Johannes Müller, who made him his assistant in 1840 (Boruttau, 1922; (Estelle) du Bois-Reymond, 1927; Rothschuh, 1964; Finkelstein, 2000, 2003). For his starting point, Müller gave him Matteucci's writings on the nerves and muscles, and he rose to the occasion and never looked back (Du Bois-Reymond, 1843) (Fig. 24.18). His massive volumes on animal electricity, titled *Untersuchungen über thierische Elektricität* ("Research on Animal Electricity"), are widely recognized as classics in the field, although, because of their style and prolixity, they are now very difficult to read (Du Bois-Reymond, 1848–84).

Du Bois-Reymond initially confirmed Matteucci's findings. More importantly, he succeeded in measuring the electrical events associated with nerve excitation using an extremely

Figure 24.18. Emil du Bois-Reymond (1818–1896), the Berlin physiologist who studied all three kinds of strongly electric fishes. Although a towering figure in neuromuscular electrophysiology in the nineteenth century, many of his results in this field were flawed. He is shown here with the second part of his large text on electrophysiology, dated 1849.

sensitive galvanometer, one that he designed and assembled. He referred to the electrical event accompanying muscle and nerve excitation as the *negative Schwankung*, meaning "negative variation" or "negative oscillation," and he provided the first instrumental recordings of what were to be called action currents and action potentials (Du Bois-Reymond, 1843b, 1848–84; see du Bois-Reymond & Bence Jones, 1852 for a short English *résumé* of his work).

The term "negative" described two related aspects of the electrophysiological responses. Negative, meaning "diminution," was used to indicate that the difference in the potential between an intact and an injured surface decreases (or vanishes) when electricity flows through the intact tissue. It also meant that the outer membrane surfaces of nerves and muscles become negative with respect to (distant) inactive regions during excitation.

In addition to his experiments with laboratory animal preparations, du Bois-Reymond used sophisticated experimental techniques to prove that electricity is also produced during muscle contractions, even in living human subjects (Fig. 24.19). These last experiments were, however, coldly received by French *académiciens* when he went to Paris to demonstrate them (Du Bois-Reymond, 1849; Lenoir, 1986; Finkelstein, 2003). This cold reception illustrates the mental set that existed at mid-century, when many scientists still did not believe that a human nervous or mental act could produce physically measurable electricity. It also reflects on the enormous importance of research on electric fishes at this time, since only they appeared capable of providing seemingly indisputable evidence of the role of electricity in animal physiology.

Du Bois-Reymond maintained a life-long interest in electric fishes. He provided a scholarly history of electric fishes in his doctoral dissertation, which was filled with Greek and Latin citations and quotations from ancient writings

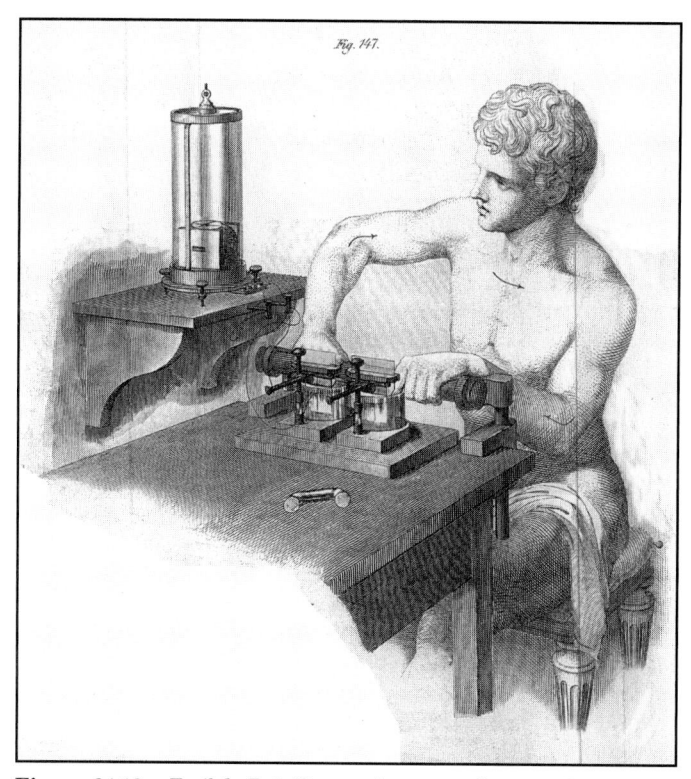

Figure 24.19. Emil du Bois-Reymond attempted to record electrical currents in healthy people as well as in frogs and other animals. This illustration shows one of his experimental setups with a man serving as the subject. The results obtained in these experiments encountered strong criticisms in some scientific circles of this epoch, particularly in France. In fact, du Bois-Reymond's contention that there is a stable electric difference between the proximal and the distal part of a man's arm or forearm is largely artifactual (see Chapter 25). (From du Bois-Reymond, 1848–1884, zweiten Bandes zweite Abtheilung.)

Figure 24.20. Andrea Ranzi (1810–1859), an Italian surgeon who was in Egypt from 1853 to 1855 (as a personal physician of the viceroy), where he studied the electric catfish. Matteucci guided him in his work, helping Ranzi to establish the correct polarity of the discharge. He also published a note based on Ranzi's letters, the beginning of which is illustrated in the right (from Matteucci & Ranzi, 1855b).

(Du Bois-Reymond, 1843a; see Debru, 2006). He also discussed electric fish phenomena in his first experimental publication on animal electricity (Du Bois-Reymond, 1843b). In subsequent years, he would conduct and direct experiments on all three kinds of fishes capable of giving strong shocks: electric catfishes, torpedoes, and the electric eel.

In 1857, after having worked on frogs at his home and at the University of Berlin, he was able to obtain his first live specimens of *Malapterurus* (Dierig, 2000). The wife of a Scottish missionary to Africa had brought three catfish back to Edinburgh and a friend took one live specimen and then two more with him on visits to Germany. With more specimens from other sources following, du Bois-Reymond delved into his research on them, "convinced that comparative studies of both physiological systems [frogs and fishes] were necessary to gain deeper insights into the physical laws of animal electricity" and that "the general physics of muscles and nerves could only be fathomed if both the electric organs and the frog nerve-muscle preparation were investigated and results collated" (Dierig, 2000, p. 7).

His catfish experiments were described in numerous books and papers over many years (e.g., du Bois-Reymond, 1857, 1858b, 1859, 1877, 1884, 1885, 1887). To study the characteristics of its shocks, he built a *Froschwecker* or "frog alarm" and a *Froschunterbrecher* or "frog circuit breaker," comprising biological "switches" (frog neuromuscular preparations). These switches could be activated by a fish's discharge, much as they could be triggered by more traditional electric stimuli.

Using his physiological rheoscopes, as well as the physical instruments perfected by him, he made a series of important observations on this fish's shocks. Among other things, he confirmed the polarity of the electric catfish's discharge, which had been discovered in 1855 by Andrea Ranzi, an Italian surgeon. Ranzi did these experiments at the request and upon the advice of Carlo Matteucci, while he served as physician to Viceroy Abbas I in Cairo (see Matteucci & Ranzi, 1855a,b). He demonstrated that the catfish's tail becomes positive with respect to the head region during a shock (Fig. 24.20).

At almost the same time, German physician Theodor Bilharz, also working in Cairo, carried out a careful anatomical investigation of the fine structure of the shocking Nile catfish, including the nerves innervating its electric organs (Bilharz, 1857) (Fig. 24.21). With regard to the polarity of the catfish's electrical organs, Bilharz, unlike Ranzi, concluded that its head is positive relative to its tail. This error stemmed from what he knew about the nerve fibers and the flat cells of the electric organ in torpedoes and the electric eel. As we shall see in Chapter 26, electric catfishes do not obey the so-called "Pacini law," which stipulates that the side of the electric cell that is innervated will be negative, and that the non-innervated side will be positive (Fig. 24.22). Among Bilharz's important achievements with *Malapterurus* is his demonstration that the discharge of each hemi-organ is made by a single, huge nerve fiber. Although du Bois-Reymond praised his work, Bilharz is far better remembered today for his discovery of *schistosoma*, the parasite responsible for the eponymous disease bilharziosis, an endemic condition associated with severe urinary bleeding, than he is for his studies on the electric catfish.

Du Bois-Reymond was also able to obtain a myriad of chemical and physical effects (sparks, electromagnetic effects, decomposition of potassium iodide, decomposition of water, fusion of metals, attraction and repulsion, etc.) in his experiments on the electric catfish. He was particularly interested in ascertaining how any electric fish could endure the potentially harmful effects of its own shocks. Humboldt and Colladon, as well as other scholars, had been intrigued by this immunity. Having proved that a part of the fish's electricity passes through its internal organs, du Bois-Reymond was able to show that the *Malapterurus* is much less sensitive than other fishes to the artificial electric shocks he sent coursing through the water in the aquarium. For instance, it was minimally disturbed by an alternating current capable

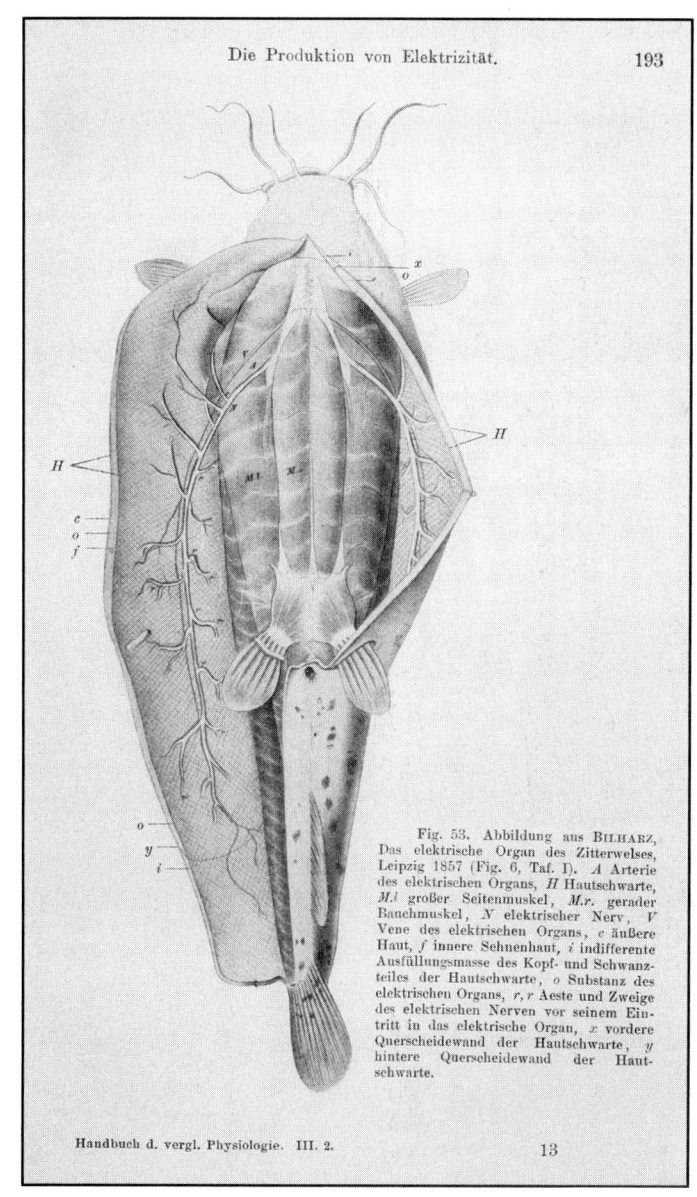

Figure 24.21. A *Malapterurus* dissection made by Theodor Bilharz (1857), a German physician who had been in Cairo. Bilharz was wrong about the polarity of the discharge, but he was correct about a single large nerve fiber branching and controlling each electric organ.

of killing other nearby fish. Similarly, it was apparently insensitive to strong constant currents. On the basis of these studies and his research on other electric fishes (see below), du Bois-Reymond would attribute their apparent immunity to all electric shocks to the insulating property of the fatty tissues abundant in the bodies of these fishes, and to the thick sheathing of their nerves.

By examining the effects of the catfish discharge on his galvanometer and on his galvanoscopic frogs, du Bois-Reymond was able to show that an apparently prolonged shock from the fish is actually composed of a train of many rapid discharges. Among his other interesting observations, we find some suggesting that *Malapterurus* can sense an electric field in its environment (see Chapter 19 for the discovery of the eel's electrical sense).

Stressing the need for large numbers of electric fishes ("nearly every new experiment on the electric organ needs a new preparation"), du Bois-Reymond organized an expedition to Venezuela in 1876 with funding provided by the Humboldt Foundation. His plan was to study electric eels in the same region of the *Llanos* where Humboldt had studied them at the beginning of the century (Chapter 1). Nevertheless, he did not personally go to South America. Instead, he sent Carl Sachs, his young assistant, with the physical tools of the trade and a finely detailed list of experiments to perform.

Despite the logistical difficulties encountered in the wilds of the *Llanos*, Sachs attempted to replicate the very investigations that du Bois-Reymond had been pursuing on *Malapterurus*, albeit now on the *Gymnotus* (Figs. 24.23 and 24.24). He made recourse to similar techniques, including the use of the *Froschwecker* and *Froschunterbrecher*, to monitor the eel's discharges (although, in the absence of suitable frogs, he was obliged to use readily available giant toads, *Bufo marinus*). He found he could obtain a variety of reactions using the shocks from his eels, including magnetization of metals, chemical and thermal effects, and galvanometric measurements. Among the matters that Sachs attempted to clarify, but without success, was the envisioned origin of the electric organ elements from muscle tissue.

Sachs also investigated the apparent immunity of electric fishes to their own discharges and also to artificial electricity. He found that isolated muscles seemed to be normally excitable by electrical stimuli. He tended therefore to attribute the electric immunity to particular characteristics of the fish nerves, specifically "a more solid molecular constitution" and a "more stable equilibrium" than the nerves of other animals. When bringing up the "molecular constitution" of the eel's nerves, Sachs was referring to the "molecular theory" put forward by du Bois-Reymond to account for nerve and muscle excitability (see later). This theory served as a frame of reference for all of the electric fish studies conducted by du Bois-Reymond and his students, including Carl Fritsch, who was one of the better-known researchers involved with the fish studies.

Sachs obtained preliminary results suggesting that the electric discharge induced by stimulation of the electric nerve could be blocked by curare, as also happens with muscle contractions (Sachs, 1877; Sachs & du Bois-Reymond, 1881; see Dierig, 2000). Curare did not interfere, however, with the shock induced by direct stimulation of the electric organ. In other experiments, Sach was unsuccessful in observing a spark when an eel discharged, and he also failed to get its electricity to pass across a Geissler tube (i.e., a glow tube filled with low-pressure gas, invented 20 years earlier by German physicist Johann Heinrich Geissler).

Sachs died in an accident in 1878, soon after his return to Europe. This was before he could publish his *Gymnotus* findings in a complete form. A preliminary account of his results had, however, been published in 1877, based on the letters he sent to du Bois-Reymond from Venezuela (Sachs, 1877). One year after Sachs' death, a book was published that dealt mainly with the naturalistic and geographic observations he had made during his journey in pursuit of the eels (Sachs, 1879). Finally, in 1881, du Bois-Reymond, who organized and edited Sachs' scientific notes and enriched them with additional materials, published a more complete account of Sachs' research (Sachs & du Bois-Reymond, 1881).

Gustav Fritsch, who had already made a name for himself with his co-discovery of the dog's motor cortex (Fritsch & Hitzig, 1870), and was now investigating the comparative

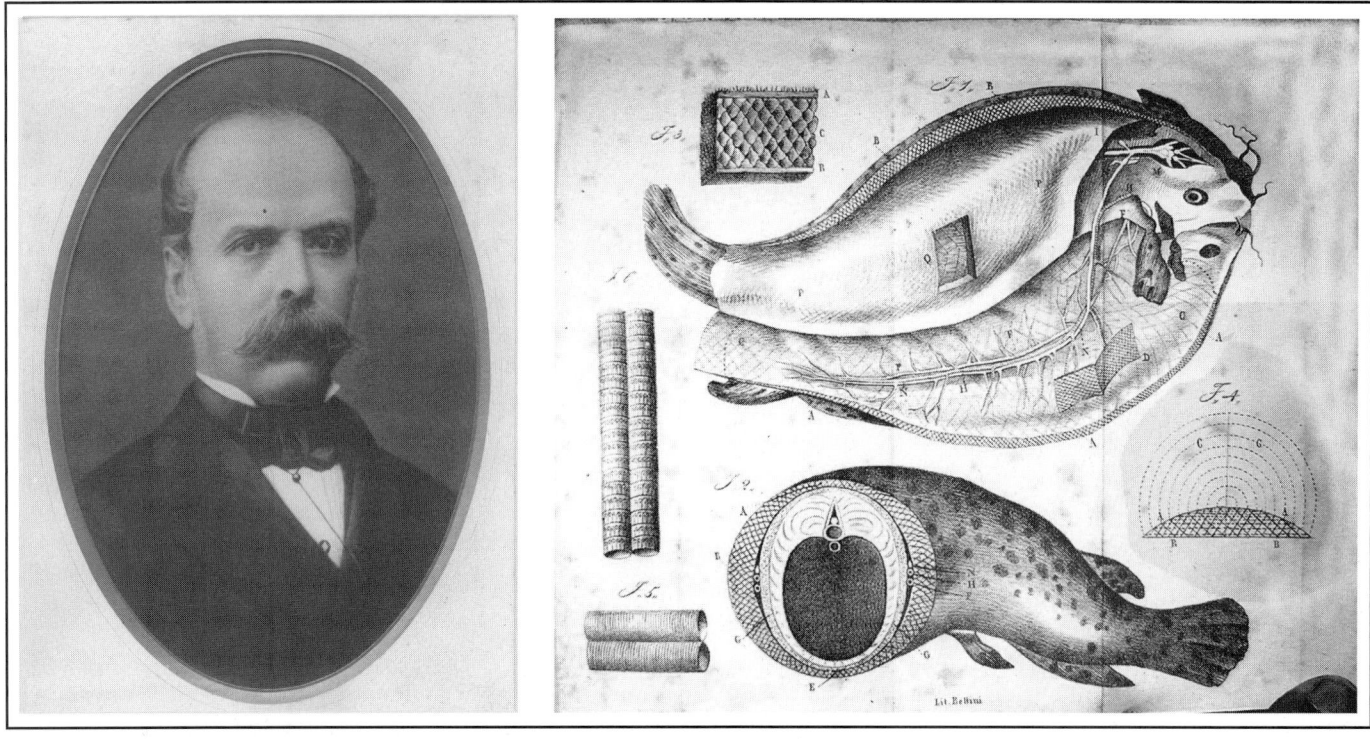

Figure 24.22. Filippo Pacini (1812–1883), the Italian anatomist known for his discovery of the mechanical receptors named after him ("Pacini corpuscles"), and an image of his 1852 dissection of the electric catfish, *Malapterurus*. Pacini studied all three types of strong electric fish and elaborated a theory on the relationship between the polarity of the innervation of the electric cells and the electrical polarity of the shock. *Malapterurus*, however, was found not to obey "Pacini's law."

anatomy of the electric organs of various fishes (Fritsch, 1881), assisted du Bois-Reymond with this publication. In the appendix, we also find the results of Fritsch's research in progress, which included work on some of the eels Sachs brought to Europe. In subsequent years, Fritsch (1883, 1887, 1890 would continue this line of research at an intense pace, focusing on the embryological origins of the electric organs and their innervations in the different species of

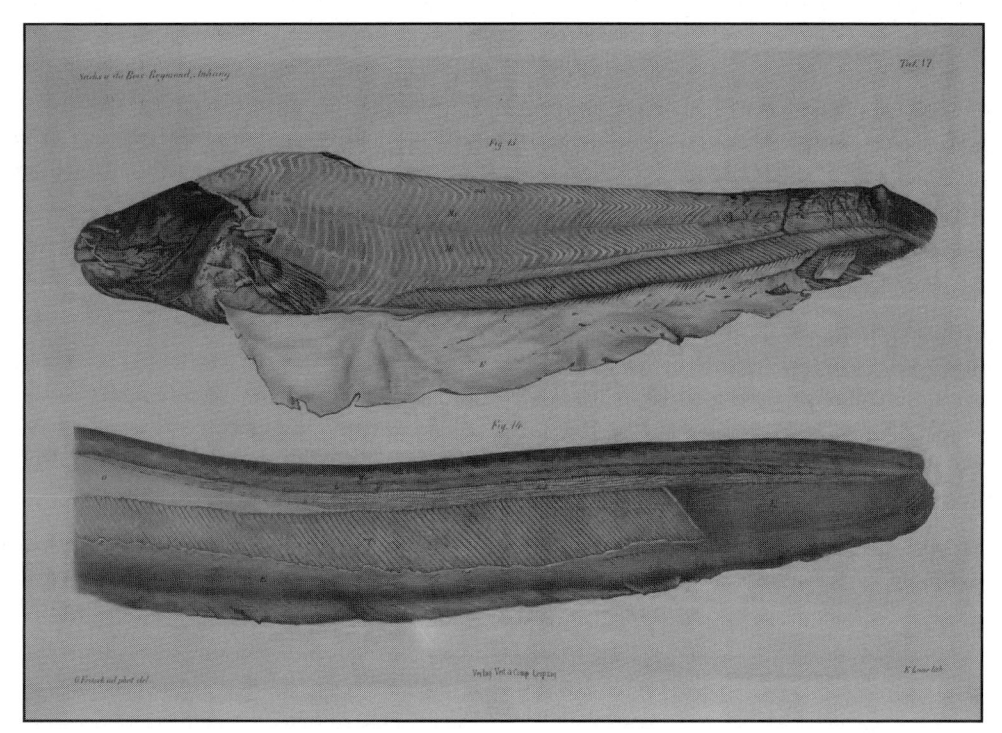

Figure 24.23. Following Emil du Bois-Reymond's instructions, Carl Sachs conducted experiments on electric eels in the same part of Venezuela where Humboldt did his research in 1800. Sachs died in 1878, soon after his return from South America, but his findings were enriched with new material (he brought some eels back to Berlin) and published in 1881 in a book with Sachs and du Bois-Reymond presented as co-authors. This illustration of a dissected *Gymnotus* comes from that book.

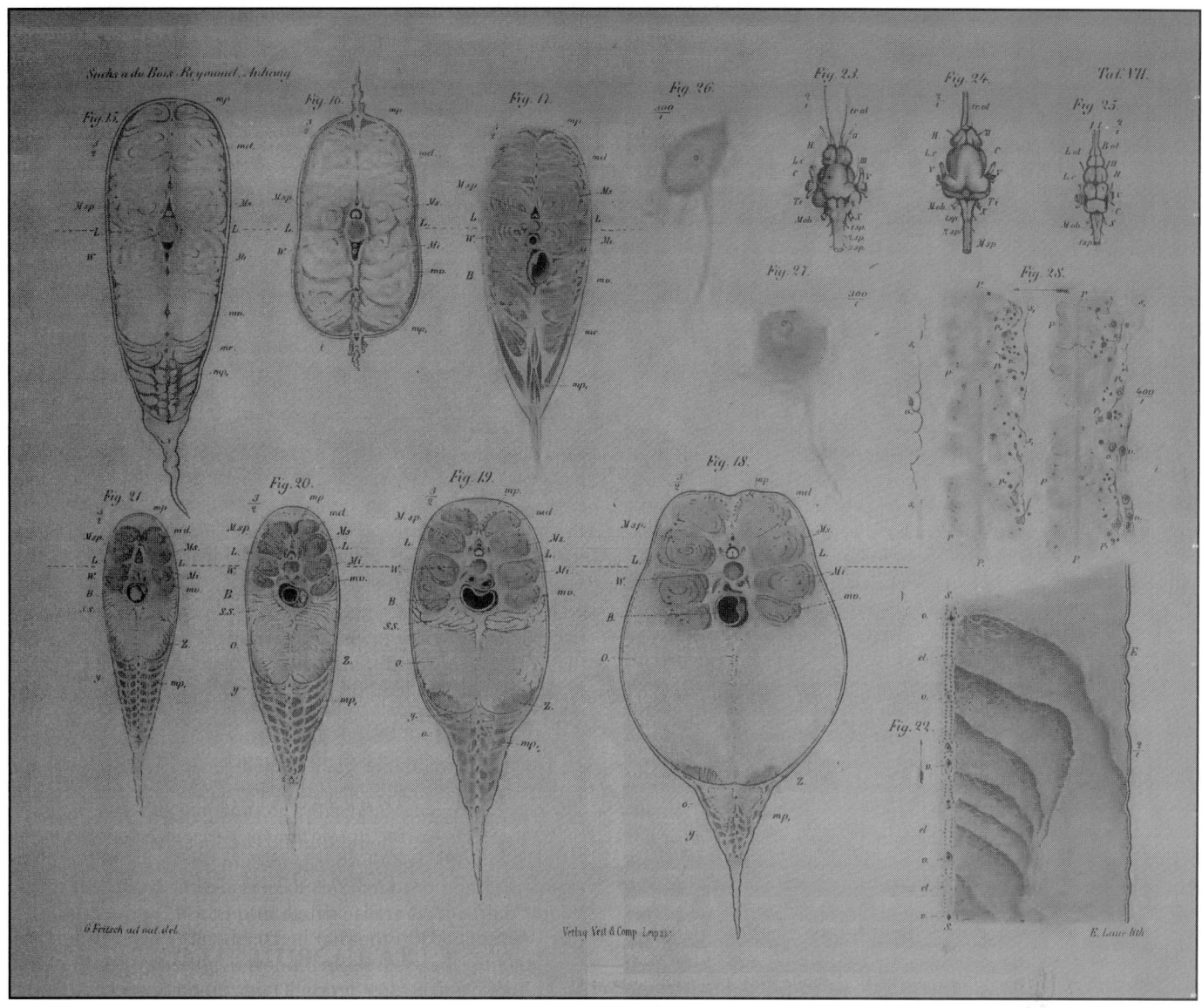

Figure 24.24. Another plate from Sachs and du Bois-Reymond's (1881) book, showing cross-sections of the electric eel. In addition to studying the eel's large electric organs (one on each side), Sachs also identified a smaller, accessory electric organ on each side, which is now called "Sachs' organ."

electric fishes (for an extensive bibliography of Fritsch's studies on electric fish, see Ballowitz, 1899).

Among the important results of Sachs' investigations, one stands out as being tied to his name today because of its originality. It is his accurate anatomical description of an accessory electric organ, now known by the eponym "Sachs' organ." Because of the great thickness of its electric cells, this organ would prove valuable for electrophysiological studies of the discharge mechanisms at the cellular level (Keynes & Martins-Ferreira, 1953). As we shall see in Chapter 26, Sachs' organ does not contribute in a major way to the strong shocks produced by electric fish, which are due primarily to the action of the main organ.

Sachs had attempted to bring six live electric eels back to Germany in 1877. One of the fish died during the trans-Atlantic voyage. The remaining ones were seriously injured during the train trip from Bremen to Berlin and proved to be unfit for physiological experiments, although they were used for anatomical studies. The situation is somewhat reminiscent of what happened to Captain Baker when he first tried to ship some live electric eels to London in 1775, after first showing them in South Carolina (Chapter 18).

Returning to du Bois-Reymond, he had also been able to make a few experiments on an electric eel prior to this time, although not in Germany or South America. They were, in fact, his first experimental investigations with electric fish, and they took place during a sojourn to England in 1852. Working with Michael Faraday, he observed that a *Gymnotus* maintained in the bustling city appeared to be rather resistant to artificial electric shocks (see du Bois-Reymond, 1887). But since he considered his results less than conclusive, he wanted to repeat his experiments, as well as to expand them to *Malapterurus* and torpedoes, when the opportunity arose.

With regard to sea torpedoes (the third of the three electric fishes he studied), after an unsuccessful attempt to use the facilities of Anton Dohrn's newly founded *Stazione Zoologica* in Naples (the prototype of modern marine biological laboratories, see Partsch, 1980), du Bois-Reymond was able to maintain and study live torpedoes in Berlin

(Du Bois-Reymond, 1885, 1887, 1889). These fish were caught on the Adriatic coast of the Mediterranean Sea and were initially kept in the aquarium of the Zoological Station of Trieste. This institution was frequented by German scientists, including Sigmund Freud, who had done his first laboratory study there, an investigation of the male gonads of (true) eels (an early commitment to sex?) (see Ricci, 1995). After they arrived in Berlin, du Bois-Reymond's torpedoes were housed in the aquarium in his new institute, which received oxygenated water from the Spree River via pipes and a pump (Dierig, 2000).

Prior to du Bois-Reymond, the great French histologist Louis Ranvier had obtained live torpedoes in Paris from the French Atlantic coast, and he carried out an accurate microscopic study of how the nerves terminate in the electric organs (see Ranvier, 1878, vol. II, pp. 85–213). Ranvier used his live torpedoes mainly to investigate how the nerves associated with the electric organs regenerated.

The problems connected to the difficulties of maintaining live torpedoes far from the seas and oceans where they normally live found a creative solution in the experimental fantasy of another great French scholar, Etienne-Jules Marey. Throughout his life, he was particularly interested in building instruments capable of revealing the time courses of physiological phenomena. Marey, who was first to obtain a graphic recording of a torpedo's discharge, initially studied live torpedoes in Naples in 1871, where he attempted to determine the conduction times of the electric nerves and the duration of the shocks (Marey, 1871, 1872). In 1879, after an attempt to study a live but weak electric eel in Paris, he used a telephone system in Paris to record the electric discharges produced by torpedoes living in the Aquarium of Concarneau, located on the French Atlantic coast (Marey, 1879).

After Marey, other researchers also employed telephone technologies. This was not done with the aim of transmitting the shocks from fish at a distance, but as a way to study the temporal patterns of the discharges (e.g., Schönlein, 1896). Late in the nineteenth century, telephones were increasingly used by researchers to detect fast electrical events that seemed undetectable with conventional electrophysiological tools, because of the slow time responses of the galvanometers.

Du Bois-Reymond confirmed and extended the observations of previous authors on torpedoes. He discussed his findings in relation to both nerve and muscle physiology, and noted the adaptive specializations that these fishes possess, which enable them to thrive in their habitats. In particular, he remarked that the flat shape of the sea torpedo, as compared to the elongated catfish and electric eel, is a beneficial adaptation for the different conductivity of sea water as opposed to fresh water.

Du Bois-Reymond was intrigued by a new discovery that he called *irreciprok Widerstand* (literally "irreciprocal resistance"). It refers to the fact that, when electricity is applied to the fish, it is conducted better in one direction than another. Specifically, he noted that externally applied electricity is conducted better from the top to bottom in a torpedo, and from the tail to the head in an electric catfish. Although he tried to understand this phenomenon, its underlying mechanism was not within his reach, particularly because of the lack of knowledge of the complex electric properties of nerve membranes. Similarly, it was not within the reach of other scientists of the time, including English physiologist Francis Gotch, who conducted experiments at the Zoological Station of Arcachon on the French Atlantic coast (Gotch, 1887). Only with the development of more modern electrophysiological tools would it become possible to understand irreciprocal resistance as an expression of the complex electric properties of the membranes of an electric organ's cells (Chapter 26).

The studies conducted by du Bois-Reymond, his students, and others in the second half of the nineteenth century clearly led to a better understanding of the anatomy and some aspects of the physiology of electric fishes. But they did not produce the needed breakthrough when it came to understanding the mechanisms whereby a few strange fishes could produce a shock so much stronger than the ordinarily imperceptible electricity generated by laboratory animal and human muscle and nerve fibers. Moreover, despite the great interest that these fishes long had for many scientists and naturalists, after the middle of the century the study of the neuromuscular system was increasingly taking center stage. The most important experiments and lively discussions were now about how nerve and muscle fibers might generate their electricity, and how this electricity might be involved in their excitation.

The theories developed in the field of neuromuscular physiology would now be applied to the mechanisms underlying the fish discharges, making them more understandable, despite some missteps along the way. Du Bois-Reymond, for one, tried to interpret electric fish phenomena by making recourse to an "electric molecules" model he developed for nerve and muscle fibers. This was consistent with his philosophy, and with the philosophy of many others at this time, that the challenge of understanding "nerves, muscles and electric organs is not divisible but has to be regarded as a whole" (Dierig, 2000, p. 6).

In the next chapter, we shall look at developments in nerve and muscle physiology dating from the second half of the nineteenth century. Many of these developments took place in Germany, where du Bois-Reymond was a key player, although a controversial one, in what was transpiring. Not to be viewed as a digression, this information is necessary to present an accurate picture of the dynamic interplay that was taking place between developments in nerve and muscle physiology on the one hand and fish studies on the other. Indeed, it is only by now turning to nerve and muscle physiology that we can appreciate the bumpy historical route that led physiologists to an understanding of how electric fishes can do what they do to those who dare to touch them, and to other animals in their sea, river, and lake environments that inadvertently venture to close to them.

Chapter 25
The Changing Neurophysiological Setting

> An unreal and incomprehensible nervous agent, more like a mythological Deity than a force of Nature, no longer dominates in the nerve fibers in an arbitrary way. On the contrary, Helmholtz has pushed the excitation of the nerve fiber back into the ranks of measurable processes, and du Bois-Reymond has made it recognizable through the magnetic needle. We now know that it propagates in the nerve fiber as a wave of a given length and that it can be directly measured in an electrical way.
> Julius Bernstein (1871, p. 2; Translated from German)

The studies conducted by du Bois-Reymond, his students, and many important scientists in the second half of the nineteenth century led to a better understanding of the anatomy and some aspects of the physiology of fish electric organs. But they did not produce the needed breakthrough when it came to understanding the mechanisms whereby a few strange fishes could produce shocks much stronger than the minute electricity generated by muscle and nerve fibers. Moreover, despite the great interest that these fishes had for physiologists, studies of the neuromuscular system increasingly took center stage after the middle of the century. The most important experiments and lively discussions now concerned how nerve and muscle fibers might generate electricity, and how this electricity might be involved in nerve events.

From the era of Galvani and Volta, and through the time of Matteucci, Italians had played leading roles when it came to studying the nature of nerve and muscle electricity. By the middle of the nineteenth century, talented Germans with well-equipped laboratories were dominating the physical and life sciences, including physiology. To a great extent, Johannes Müller's students were now shaping the course physiology would follow (for more on Müller, his science, and his influence, see du Bois-Reymond, 1858a; Haberling, 1924; Ebbecke, 1951; Koller, 1958; Lohff, 1995, 2001; Finger & Wade, 2002a,b; Otis, 2007).

Müller was Professor of Anatomy and Physiology and Curator of the Anatomical Museum in Berlin, a beloved teacher, and the author of a fundamental treatise on physiology, his *Handbuch der Physiologie des Menschen für Vorlesungen* (Müller, 1834-40; English edition, 1840-43). This two-volume book was translated into many languages, and it was by far the most influential physiology textbook at mid-century.

But in contrast to Müller, who incorporated metaphysical views associated with German *Naturphilosophie* into his science, his most influential students found no place whatsoever for his vitalistic views (notions of soul), or any metaphysics for that matter, in the "new" physiology they hoped to promote. Considering themselves "organic physicists," they strove for a more physically and chemically grounded nerve physiology (Cranefield, 1957; Otis, 2007).

Emil du Bois-Reymond was one of these students, and he studied nerve and muscle physiology in addition to conducting research on various electric fishes (Chapter 24). For du Bois-Reymond, the two fields were tightly coupled, and this was consistent with his philosophy that the challenge of understanding "nerves, muscles and electric organs is not divisible but has to be regarded as a whole" (Dierig, 2000, p. 6).

Du Bois-Reymond strove to come forth with an "electric molecules" model for nerve and muscle fibers, one that would also be applicable to electric fish organs. But although his conception forced other scientists to think about the issues, while generating additional research, the particular model he developed in order to account for electric phenomena in nerves and muscle was flawed, and in this regard one could make the case that it hindered progress in electrophysiology.

In this chapter, we shall examine du Bois-Reymond's model and its impact. We shall also look at Ludimar Hermann's criticisms of this model and the path that led Julius Bernstein, who also had a strong interest in electric fish, to his insightful membrane theory. We begin, however, with Hermann Helmholtz (later von Helmholtz), who was both Müller's student and du Bois-Reymond's friend, and his landmark estimates of the speed of nerve conduction and what this work signified. In the next chapter, we shall return to main subject of this book, electric fishes, with the physiological background needed to appreciate how scientists were finally able to explain how they could produce their torporific discharges.

HELMHOLTZ AND THE SPEED OF NERVE CONDUCTION

Although du Bois-Reymond was clearly the dominant figure in nerve physiology in the second half of the nineteenth century, he was not the first scientist to provide an accurate

Figure 25.1. Hermann Helmholtz (1821–1894), one of the greatest scientists of the nineteenth century, whose research helped shape the fields of physiology, physics, and psychology. He was the first to measure the speed of nerve conduction in frogs and humans in a accurate way.

estimate of the speed of nerve conduction. Hermann Helmholtz (Fig. 25.1), who had been close to du Bois-Reymond since his student days in Berlin, was the researcher who did this, and he accomplished this feat in various ways (for more on Helmholtz, see McKendrick, 1899; Koenigsberger, 1902, 1906; Hall, 1912; Kahn, 1971; Cahan, 1993a; Meulders, 2001; Finger & Wade, 2002a,b).

The thought that the nerve impulse could be measured accurately seemed contrary to almost everyone's expectations during the 1840s. Nerve signal propagation now appeared to be instantaneous, and it seemed well beyond the scope of the measuring instruments then available. Helmholtz's teacher, Johannes Müller, was among those convinced that it could never be measured. He called it an "imponderable nervous principle," writing, "We shall probably never attain the power of measuring the velocity of nervous action; for we have not the opportunity of comparing its propagation through immense space, as we have in the case of light" (Müller, 1840, p. 729).

Helmholtz (1843, 1848) had already published important papers on fermentation, digestion, and muscle contraction, challenging Müller's "outdated" vitalistic physiology and agreeing with du Bois-Reymond that "the phenomena of life are not essentially different from physical phenomena" (Holmes, 1994, p. 12). To account for some of his physiological findings, he formulated the law of the conservation of energy (i.e., the first principle of thermodynamics) (Helmholtz, 1847; Meulders, 2001). Then, after having briefly served in the army, as an assistant in Müller's anatomical museum, and as a lecturer in anatomy at Berlin's *Kunstschule* (Academy of Arts), he took on the "imponderable nervous principle" in his new position as "Extraordinary Professor of Physiology" at Königsberg (then Germany, now Kalinigrad, Russia).

Helmholtz's initial solution to the question of how to estimate nerve conduction speed was based on a method suggested by du Bois-Reymond in 1845, which in turn depended on the technique contrived by Claude Pouillet to measure the speed of a cannonball. The critical timing feature in Helmholtz's experiment was a sensitive galvanometer that turned on at the start of an electrical stimulus to a nerve and was shut off by a muscle contraction (Olesko & Holmes, 1993) (Fig. 25.2).

These experiments began near the end of 1849. From his time recordings and measurements of the distance from where the nerve was stimulated to the muscle, Helmholtz's simple mathematics revealed that the rate of motor nerve conduction was much slower than light and even slower than sound. Nevertheless, his early results were variable (from approximately 25 to 38 meters per second), probably because some of his frogs had been weakened by months of cold and captivity. After he improved his procedures, his data revealed that the mean speed of motor nerve conduction was approximately 27 meters per second, both with Pouillet's "precision method" and his own *Myographion* (an adaptation of Carl Ludwig's *Kimographion*, an apparatus originally used for measuring blood pressure; see Ludwig, 1852) (Fig. 25.3).

Helmholtz's (1850a,b) first reports on nerve conduction appeared in 1850. Müller referred to his student's surprising accomplishment as a "great stride," and he encouraged him to expand his program to measure conduction in other kinds of nerves and reflexes (Olesko & Holmes, 1993, p. 90).

Helmholtz began to work on the sensory nerves in the spring of 1850. Human subjects were asked to make movements with a hand or with their teeth as quickly as possible when a weak shock was applied to a body part. He recorded significant increases in response time as he stimulated body parts increasingly distant from the brain, writing that "a message from the big toe arrives about one thirteenth of a second later than from the ear or face" (see de Jaager, 1865; trans. in Brozek & Sibinga, 1970, p. 42). He first thought that the human sensory nerves conduct at about 60 meters per second, but additional studies soon convinced him that human sensory nerves conduct at only about half this rate, more like his frog motor nerve findings (Hirsch, 1862; Jaager, 1865). "Happily," Helmholtz later declared, "the distances our sense-perceptions have to traverse before they reach the brain are short, otherwise our consciousness would always lag far behind the present" (Koenigsberger, 1906, p. 71; Piccolino, 2003b).

Lastly, he turned to human motor nerve conduction. He and Russian scientist Nikolai Baxt immobilized an arm and stimulated the nerve to the thumb either at either the wrist or the elbow. When the automatic twitching of the thumb was measured with a myograph, human motor nerve conduction was found to be about 33 meters per second, in line with his other conduction estimates (Helmholtz, 1867).

Helmholtz's achievements were milestones in the history of science. For the first time, a manifestation of a nervous phenomenon that had previously been associated with an immaterial entity, namely a vital spirit or soul, had been measured with precision. And also importantly for our story, his finding that the speed of nerve conduction slowed in a substantial way with cooler room temperatures was indicative of a process involving a chemical mechanism or mechanisms.

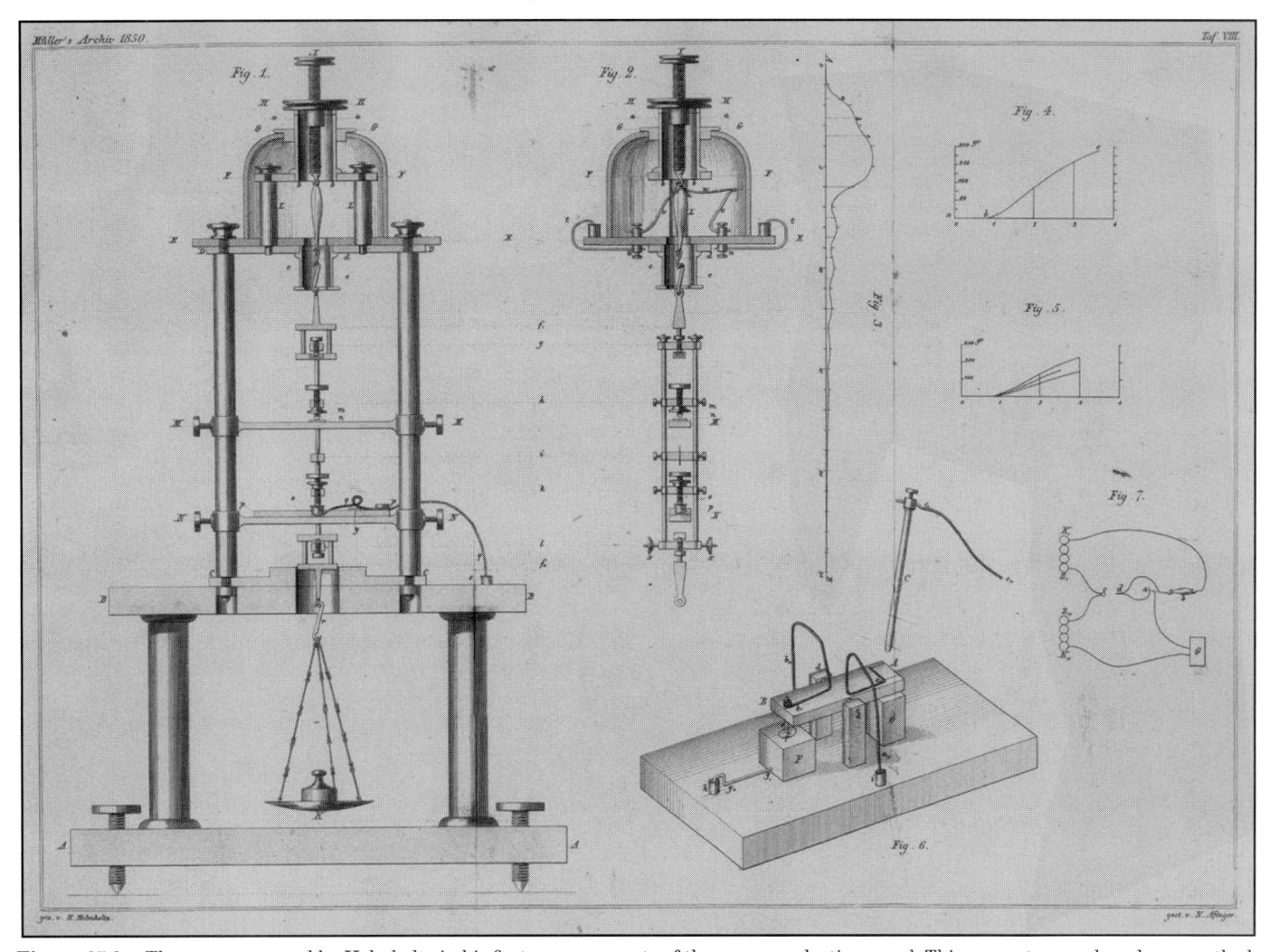

Figure 25.2. The apparatus used by Helmholtz in his first measurements of the nerve conduction speed. This apparatus was based on a method developed by French physicist Claude Pouillet to measure extremely fast phenomena, such as the speed of cannonballs. Basically, the electrical stimulus applied to the nerve innervating a frog muscle starts a precise time-measurement device. This is switched off mechanically by muscle contraction. Some of Helmholtz's machinery, drawings, and calculations can be seen in this figure (from Helmholtz, 1852a).

BERNSTEIN'S MEASUREMENTS OF THE NERVE IMPULSE

Although Helmholtz's experiments on nerve conduction had been prompted by du Bois-Reymond's electrical thinking, his results cast some doubts on the electrical nature of nervous fluid, because his values were far slower than the speed of electricity along physical conductors. To re-establish confidence in the nerve signal being electrical, it was critical to measure the speed of the negative variation that du Bois-Reymond had been studying in his Berlin laboratory to see if it corresponded with Helmholtz's aforementioned estimates. Du Bois-Reymond understood this need, but hampered by the slow responsiveness of his electrometers he was unable to measure the speed of the negative variation (Lenoir, 1986; Piccolino, 1998, 2003b).

The task was passed to Julius Bernstein, who first studied medicine and physiology at the University of Breslau (1858–60) and then returned to Berlin to study under both du Bois-Reymond and Helmholtz (Lenoir, 1986; Seyfarth, 2006).

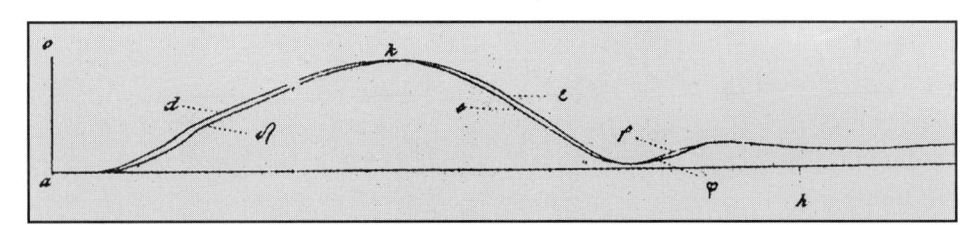

Figure 25.3. Tracings of Helmholtz's measurement of nerve conduction speed based on the graphic recording of muscle contraction obtained with his *Myographion*. This device uses a rotating smoked drum and a pen connected mechanically to the contracting muscle, inscribing the time course of the muscle movement. Compared to the previous method, it shows in a more immediate and demonstrative way the increase in the latency of muscle contraction when the nerve is stimulated at a farther distance from the muscle. It helped make Helmholtz's finding more easily accepted (from Helmholtz, 1852a).

Although he is far better known for his work leading to the formulation of an ionic theory of the membrane potential, Bernstein would be the first scientist to propose an acceptable model to account for how the fish's electrical organ might produce its strong shocks, as we shall see in our next chapter (Bernstein, 1912, pp. 121–122). This model represents just one of his many great achievements in the field of electrophysiology.

When Helmholtz accepted a professorship in Heidelberg, Bernstein followed him there as an assistant (in 1864) and quickly rose through the ranks to become *Dozent* (in 1865), then *Ausserordentlicher* [ie. extraordinary] *Professor* (in 1869), and finally interim head of the Physiology Department upon Helmholtz's departure for Berlin in 1871. He then accepted a professorship in Halle (from 1873 to 1911), where he became the first Jewish rector of a German university. It was there that he concluded his long and remarkable scientific life.

Bernstein's measurement of the negative variation took place in Heidelberg in 1868. This difficult technical achievement was made possible with an instrument he had invented, his "differential rheotome," the word "rheotome" meaning "current slicer" (Bernstein, 1868; also see Hoff & Geddes, 1957, 1960; Schuetze, 1983; Zett, 1983; Lenoir, 1986; Piccolino, 1998; Seyfarth, 2006) (Figs. 25.4 and 25.5). This apparatus was probably inspired by the rapidly rotating, toothed-wheel device used by French physicist Hippolyte Louis Fizeau in 1849 to measure the speed of light. Through the action of a rotating mechanical device, it made electrical recordings based on a "timing, sampling and holding procedure"—recordings that would otherwise be impossible because of the inertia of the galvanometric needle.

Bernstein measured the negative event at a given point on a nerve by superimposing the repeated effects of nerve stimulation a known distance away. Conduction speed was calculated by altering the distance of the stimulus from the recording electrode (in a way following what Helmholtz had done using muscle twitches). The changes in the latencies of the event provided a conduction rate of 29 meters per second (see Fig. 25.5).

Referring to Helmholtz's studies on the speed of the excitation wave, Bernstein wrote, "the excitation process and the process of the negative oscillation have one and the same speed." His conclusion was that it is justified to assume that the electrical excitation wave "is nothing other than the shape [*Bild* in the original German] of the excitation process that flows through the nerve" (Bernstein, 1868, pp. 188 and 199; see also Bernstein, 1871).

Besides demonstrating the similarity of conduction speeds between the nerve signal and the propagated electrical wave, Bernstein came forth with the first faithful instrumental recordings of the time course of the excitation wave ("action current" or "action potential" to use today's terminology). Again using frogs, he showed that it is a transient phenomenon lasting only about a millisecond. From his recordings it was, moreover, clear that the nerve event is much faster than the mechanical muscle contraction in a neuromuscular preparation.

One of Bernstein's most significant findings was that the amplitude of the current during the excitation phase exceeds the level of the resting current that he measured between

Figure 25.4. Julius Bernstein (1839–1917), the major contributor to the physiology of nervous conduction late in the nineteenth and into the twentieth century. His first major achievements in this field were the recording of the time course of the electrical signal in nerves ("negative variations") and the measurement of their conduction speed. Bernstein's figure for the negative variation speed corresponded closely to Helmholtz's estimates of the nerve signal responsible for muscle contraction. Bernstein's achievements were published in 1871 in an extensive monograph, the title page of which is illustrated on the right. This text came out in the same year of Bernstein's move to Halle from Heidelberg (where the material for his text was collected).

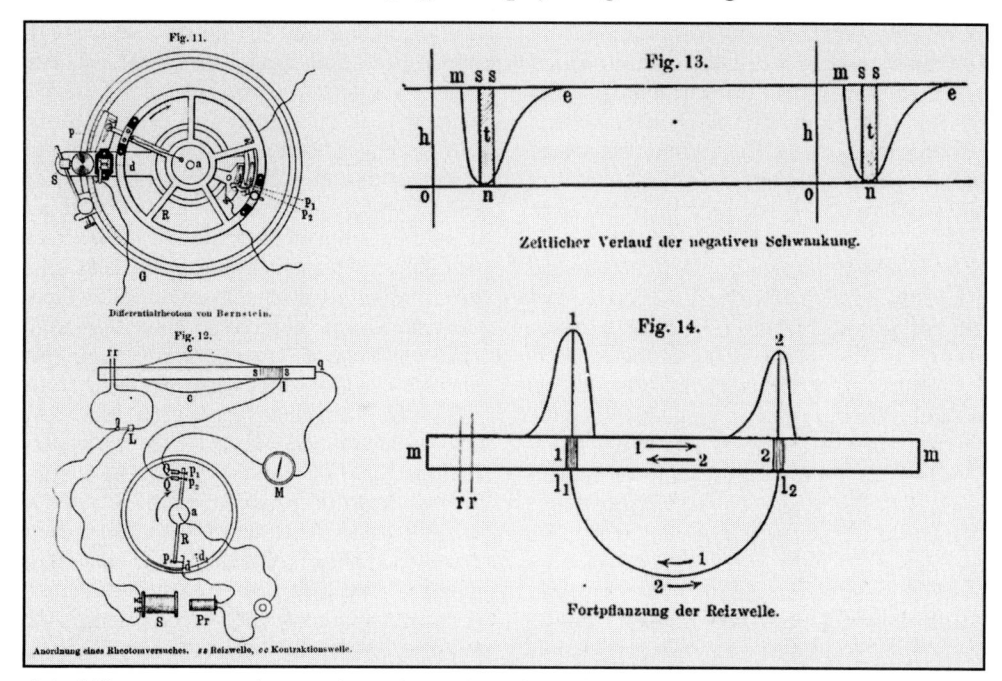

Figure 25.5. Bernstein's ability to measure the negative variation depended on better instrumentation, and specifically his "differential rheotome," or "current slicer," shown on the left in this schematic. This instrument made possible what at first seemed impossible: to measure an event, like the electrical impulse in nerve fibers, that lasts less than a millisecond by using a galvanometric system, which *per se* cannot measure signals lasting less than about a second.

an intact and an injured segment of the nerve in some of his recordings. In translation from Bernstein's original article:

> These experiments demonstrate sufficiently enough that the growth of the negative variation does not reach its limit at the point at which the recorded current becomes zero. *Rather, with stronger stimuli the strength of the negative variation can become larger than that of the nerve current, and even exceed it by many times.* (Bernstein, 1868; p. 194; emphasis Bernstein's)

This unexpected overshoot indicated that the excitatory event was more like an inversion of the resting electric difference between the intact and injured nerve surface, rather than a simple neutralization or "destruction" of the resting condition. Later on, Bernstein would play down the occurrence of the overshoot, despite the fact that it was also observed by other scientists of the time, one being his colleague Ludimar Hermann and another being English physiologist Francis Gotch (see Gotch, 1900).

One of the reasons Bernstein (1874, 1912) gave in his later works for his personal neglect of the nerve current "overshoot" was that he observed the phenomenon exclusively in nerves, whereas in muscle fibers the negative variation was never seen to exceed the resting current of injured fibers. As we shall discuss later on, there probably were also other thoughts behind Bernstein's failure to appreciate what he had discovered. Very likely, he was influenced by du Bois-Reymond's electrophysiological theories, which could have given him the mindset to see nothing beyond the elimination of the resting condition during nerve excitation.

Bernstein's electric "overshoot" or reversal of the injury current would be slowly forgotten over the ensuing decades. It would reappear again in mainstream science only after a lengthy delay. We shall encounter it again in the studies of Hodgkin and Huxley starting from 1939, which are vital to understanding electric fishes and will be dealt with in the next chapter.

DU BOIS-REYMOND'S "ELECTRIC MOLECULES"

The experiments of Helmholtz and Bernstein suggested that although the nerve signal is always accompanied by an electrical event, the propagation of this event does not follow simple laws for passive propagation of an electrical current—that is, they hinted that a more complex and specific mechanism must be involved. In this context, we have already mentioned the reasons behind Carlo Matteucci's reluctance to accept the electric hypothesis of nervous conduction (Chapter 24), and Matteucci was not alone. In Germany, Carl Ludwig (1861), another Müller student and "organic physicist," also questioned the idea that the nerve signal is electrical in nature. He considered the longitudinal resistance to be too high to permit effective propagation of a weak electrical signal (i.e., passive electrical conduction) down a cable-like nerve.

Helmholtz's demonstration that the speed of nerve conduction is strongly affected by changes of temperature is particularly significant here. This finding served as a strong indication that nerve conduction and the passive propagation of an electric current down a wire cable must be fundamentally different processes (Helmholtz, 1850b, 1852b). On the basis of thermodynamics, a field to which Helmholtz had made important contributions dating back to 1847, a phenomenon's strong temperature dependency indicated that chemical mechanisms are basic to the event.

Since his initial electrophysiological studies, which were initiated in 1841 and published in 1843, du Bois-Reymond had adopted a view of electrical signal conduction in nerves and muscles basically different from the simplistic idea that this occurs in a way similar to electric conduction along purely physically cables. He even developed a complex model, which he presented in many of his publications, including his monumental *Untersuchungen*. He was so committed to this model that he would adhere tenaciously to it, even after one of his former students, Ludimar Hermann, showed that the basic observations on which it was based were largely due to experimental artifacts.[1]

As he makes clear at the beginning of his 1843 article, du Bois-Reymond started his experiments on the advice of Johannes Müller. He did this to pursue the electrophysiological investigations of Nobili and Matteucci on the current measured in frog preparations with a galvanometer. As we have mentioned, Matteucci had provided a clear demonstration that a current of genuine biological origin could be measured between the intact and the injured surfaces of muscle fibers, and that this current decreases or disappears during excitation. Matteucci, could not, however, detect a similar current in nerves.

By using more sensitive galvanometers, du Bois-Reymond was able to extend Matteucci's findings to nerves, this representing his most important contribution to electrophysiology. Because of the extreme sensitivity of his galvanometers, and also for other, more elusive, reasons, du Bois-Reymond also detected, however, a current (albeit of a small amplitude) in apparently *uninjured* muscle preparations. This observation, mostly of artifactual origin (and in any case based on phenomena completely unrelated to nerve and muscle physiology), became the starting point for the elaboration of his model of bioelectric phenomena.

Du Bois-Reymond attributed particular relevance to the fact that, in order to record this current (that we now know to be artifactual), he had to situate one of the two recording electrodes on the tendinous extremity of the intact muscle, while the other electrode could be positioned on different locations on the lateral surface of the muscle (Fig. 25.6). The tendinous extremity was found to be negative in these experiments, and thus similar in polarity to that of the injured surface. The observation that this resting current seemed to increase in amplitude when he increased the distance between the electrode on the tendinous extremity and the one placed on the muscle's lateral surface was important for du Bois-Reymond in his further elaborations. He usually found a maximum amplitude when the second recording electrode was on the center of the muscle belly.

Du Bois-Reymond further observed a small current between the mouth and tail of the intact animal, this further suggesting that the presence of a bioelectric current does not necessarily require a lesion in excitable fibers. On the basis of these and other results, he concluded that the tendinous extremity of an intact muscle is similar to a cut muscle, and he suggested that it should be considered as a kind of "natural transverse section," corresponding to the artificial section created by an injury.[2]

Another observation of great relevance to him must also be mentioned. He found that an injury current could be measured wherever the fiber was cut or otherwise injured. It was usually possible, moreover, to renew this current after it had spontaneously declined, by making a fresh cut nearer to the intact (recording) side. This finding led him to assume that the power of generating this current in the resting fiber is distributed along the fiber, somewhat like the attractive and repulsive properties of a magnetic body. Even when broken into pieces, the magnetized parts exhibit the same properties, including two opposing poles.

Inspired by the work of André-Marie Ampère and Michael Faraday on electromagnetic phenomena, du Bois-Reymond constructed a model that he thought would be capable of accounting for the supposed existence of a steady current in the resting state, and for the processes of electrical excitation and conduction in both muscles and nerves (Fig. 25.7). In his schema, the excitable fiber is made up of a multitude of parallel cylinders (or prisms), each one containing electric molecules in a longitudinal alignment (similar to the magnetic molecules in Ampère's model for magnetized bodies). Each molecule, now designated as a "peripolar molecule," was assumed to have a spherical shape. Each was also assumed be negative at its two poles (along the longitudinal direction of the cylinder) and positive in its equatorial belt.[3]

In the resting condition, du Bois-Reymond theorized, this molecular arrangement would produce an electrical field that would have a positive charge on the lateral surface of the

---

[1] The historian who looks retrospectively at Emil du Bois-Reymond's work on animal electricity cannot but arrive at a somewhat sad conclusion. Without any doubts, he was a great scientist, a profound intellect, and a man of considerable culture. He was capable of intense experimental work, he possessed an extraordinary capability to contrive new scientific instruments and methods, and he prevailed in the scientific panorama of his time with his strong personality, his university chair in Berlin, and his prolific writings. Yet relatively little still stands of his many "laws" and "principles" in mainstream science. This is in part because he was unable to abandon his faulty theoretical framework throughout his long scientific career. In contrast to the praise of many contemporary philosophers of science for the importance of rhetoric in science, his case also illustrates the potential *deleterious* effects of rhetoric in scientific discourse. His unfair and exceedingly harsh remarks about the experimental results and interpretations of his scientific adversaries—Matteucci and Hermann, in particular—seem more like the language used by a trial lawyer than that of a scholar devoted to discovering the truth. Du Bois-Reymond seemed to be entangled in his own polemics, and it is likely that they blinded him to other possibilities. The great Galileo, who unfortunately in the last years of his live lost his vision in a different, non-metaphorical, way, had warned scientists of the potential dangers of "rhetorical illations," meaning inferences or conclusions. He wrote "in natural sciences . . . we must be careful not to defend falsity, because one thousand Demosthenes and one thousands Aristotles could not prevail against any lesser intelligence, who would happen to adhere to the truth" (Galilei, 1632, pp. 45–46).

[2] In his subsequent work, du Bois-Reymond would repeatedly measure intact animal currents in a variety of species and also in men (including himself). He would replicate these experiments and elaborate on them, particularly during his polemics with Hermann, hoping to show that the presence of a resting current does not require any injury to animal tissues. He interpreted these whole animal currents largely as the spatial sum of the individual currents produced by the various muscles of the body. To explain how the ensemble of muscles could produce a total current with a definite spatial polarity, he had to assume (and try to demonstrate) that there is a definite patter of resting current between the two tendinous extremities of each muscle. He concluded that in most animals in which the head region is positive to the tail, the upper tendinous extremity of a muscle is positive to the lower extremity. In humans (where he found that the shoulder is negative with respect to the hand) he had to assume (and thought to have proved) that the reverse is true (i.e., the upper tendinous extremity is negative). All of the experimental measurements on which he based these conclusions were flawed by artifacts or reflect marginal physiological phenomena that are of little relevance for understanding muscle and nerve physiology.

[3] Du Bois-Reymond wrote that he assumed the spherical shape of the electric molecules only for reasons of simplicity. Other forms could be conceived, provided that electric charges are arranged in such a way as to produce an electric field with a positivity on the lateral surface of the fiber and a negativity on the transverse section (see du Bois–Reymond, 1848-1884, vol. I, p. 671).

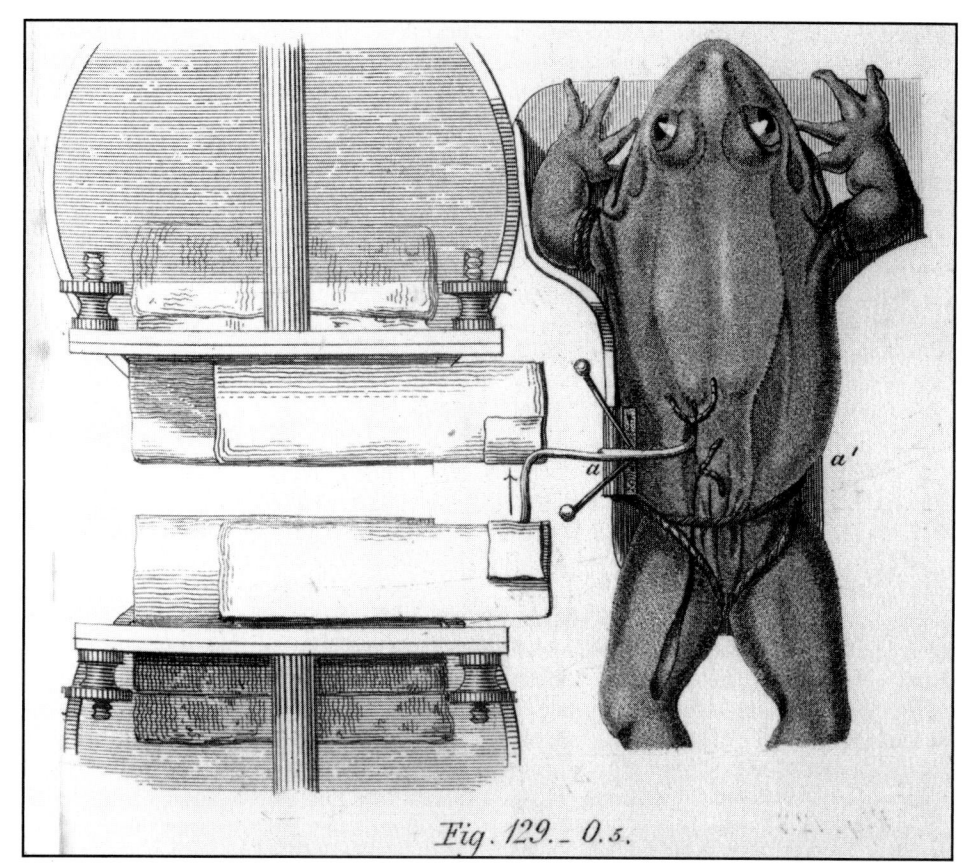

Figure 25.6. Emil du Bois-Reymond was able to record current changes in nerves, and he also claimed that he could record small electrical changes in uninjured muscles. To accomplish the latter, he situated one electrode on the tendinous extremity of an intact muscle and another on the lateral surface of the muscle, as seen here. Although these "muscle currents" were actually artifacts, they stimulated him to develop his molecular theory of bioelectric currents (from du Bois-Reymond, 1848-84, zweiten Bandes erste Abtheilung).

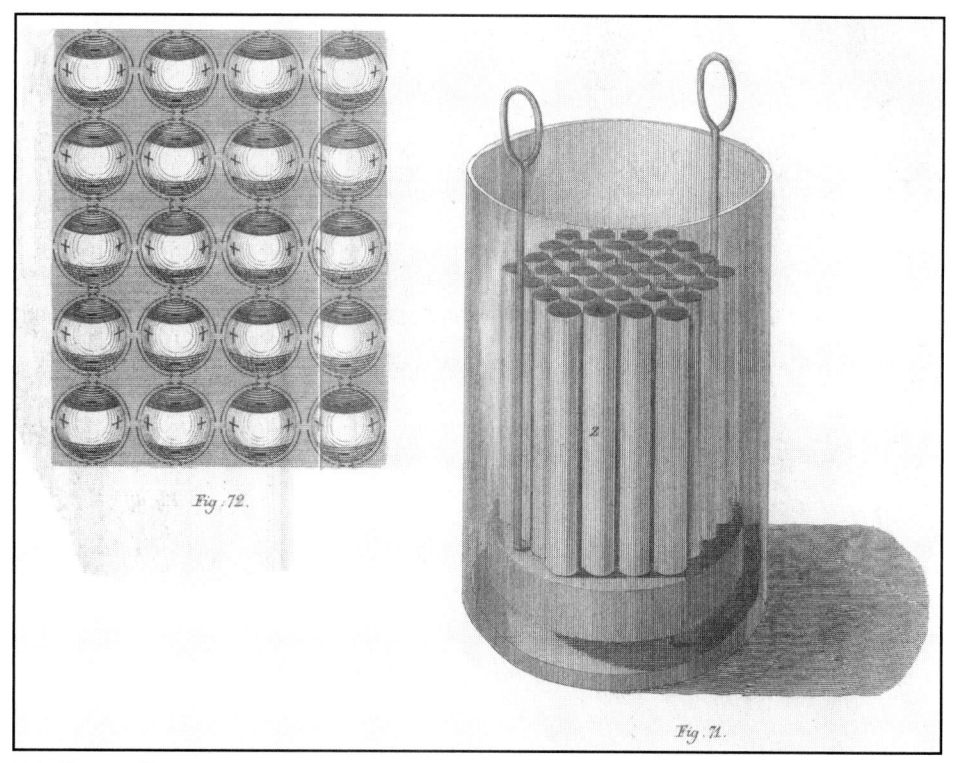

Figure 25.7. Emil du Bois-Reymond's molecular model for nerves and muscles. He theorized that each excitable fiber is made up of parallel cylinders (on right), each containing electric molecules that would be negative at the poles and positive across, and which would be lined up longitudinally (on left). The disruption of this orderly arrangement, he argued, would result in the disappearance of the resting current and result in the *negative Schwankung*, the propagated excitation wave (from du Bois-Reymond, 1848-84, erster Bande).

fiber (where, for spatial reasons, the influence of a positive charge would be stronger), and it would have a negative one at the transverse cut surface of the cylinders. In muscle fibers, this negativity could also be detected in the distal ends near the tendons (i.e., in his "natural transverse section").

In a further elaboration of this model, du Bois-Reymond assumed that each peripolar molecule could assume the form of two dipolar molecules. They would be united under ordinary conditions at the internal positive sites (and thus similar to the peripolar molecules). They would, however, be capable of moving independently under the influence of a stimulus, particularly an electrical one.

An electric stimulus, he maintained, would initiate a process of an electrolytic nature, causing a transformation of the molecules from the peripolar to the dipolar conformation. As a result, there would also be a disruption of the ordered ("polarized") condition, and thus the disappearance of the resting current. This, he assumed, is the basis of the *negative Schwankung*, the propagated excitation wave.

For du Bois-Reymond, the *negative Schwankung* was basically the abolition of the pre-existing electric condition. It would be conducted along the fiber at a rate much slower than the flow of current along a metallic cable. The reason for this slow speed is that movements of the charged molecules would require more time than that involved in the flow of electricity along an ordinary conductor. Helmholtz was inspired to try to measure the speed of propagation of the nervous signal, despite Müller's initial skepticism about this ever being done successfully, with this model in mind.

The assumption that the fiber is in a definite electrical state in the absence of an adequate stimulus (of external or physiological origin) is basic to du Bois-Reymond's conception. This electrical state is potentially capable of producing continuous current flow. Within this theoretical framework, excitation was viewed as the disruption of the resting condition.

These assumptions would form the theoretical background for Bernstein's later (1902, 1912) "membrane theory." As we shall see, Bernstein's theory would, however, differ profoundly from du Bois-Reymond's conception in that it would make the cell membrane the centerpiece in the generation of bioelectric phenomena. This emphasis accounts for why Bernstein used the phrase "membrane theory" to designate his new conception. In du Bois-Reymond's model, the membranes of muscle and nerve fibers do not play any relevant role whatsoever in the resting polarization of the fiber or in the mechanisms of excitation.[4]

Bernstein's model would also differ from du Bois-Reymond's conception in the importance attributed to movements of the charged particles in the cellular and extracellular liquid milieux. In fact, du Bois-Reymond's terminology reveals his idea that the basic structure leading to the generation of bioelectric phenomena has a "solid" architecture. In this context, it is interesting to recall that "cylinders" and "prisms" had also been used to designate components of the electric organs of fishes and the Voltaic battery. This may reflect Volta's influence on du Bois-Reymond's thinking.

The virtual absence of the cell membrane in du Bois-Reymond's model merits comment. In his elaborations, the German scientist seems to have had no place for cell theory, despite the fact that this theory was already established when he was developing his mature views on muscle and nerve phenomena. In contrast, well before the emergence of the scientific idea of the cell as the anatomical and functional unit of organization, Galvani invoked the existence of a membrane, viewed as a barrier separating the interior from the exterior of the muscle fiber, to account of his ideas about animal electricity. By comparing the different attitudes of these men, it is difficult to escape the conclusion that the way in which scientists view and interpret their findings is strongly influenced by their personal theoretical conceptions.

HERMANN'S "ALTERATION THEORY"

The observation on which du Bois-Reymond had built much of his theory—i.e., the existence of a "resting current" (*Ruhestrom*) in the intact muscle fiber—was strongly challenged on an experimental basis by Ludimar Hermann. Hermann had been a student of Müller and of du Bois-Reymond (and initially a friend of Bernstein), and had worked in Zurich from 1868 to 1884, where he did his most important work before moving on to Königsberg.

Hermann had been unable to replicate du Bois-Reymond's observation on the resting current in intact fibers, a finding that Helmholtz had already contested in the 1850s (Helmholtz, 1852b). Convinced that the detection of the resting current required a lesion of the fiber surface, Hermann attributed the existence of the small resting currents in du Bois-Reymond's "intact" preparations to small accidental injuries in the tendinous extremes of the fibers.

This interpretation marked the beginning of a harsh dispute, one characterized by strong personal attacks, that would continue to the end of du Bois-Reymond's scientific life. The dispute, however, went well beyond these protagonists. On Hermann's side, it also involved Eduard Pflüger and Ewald Hering, among others. And on du Bois-Reymond's side, it drew Isidor Rosenthal, Ernst von Fleischl-Marxow, and importantly Bernstein, into the fray (Lenoir, 1986; Finkelstein, 2006; De Palma & Pareti, 2010).

On the basis of numerous experiments, Hermann proposed a theory almost diametrically opposed to that of his former mentor. His theory was based on the assumptions that there is no pre-existing electric condition in the resting fiber, and that electricity is produced as a consequence of chemical processes consequent to the injury of the fiber's surface or to the process of excitation (Fig. 25.8). A stimulus, he maintained, would produce a phenomenon on the intact fiber surface similar to one that would occur with the death of the tissue. It would, he maintained, break the chemical bonds of an "energy-producing" substance (*krafterzeugende* or *inogenic*). The electricity generated as a consequence of these processes would flow from the excited part of the fiber to the nearby zones through a current-propagation process indicated as *Kernleitertheorie*, or "core conductor theory." Here Hermann drew on a physical model formulated by Lord Kelvin, William Thompson, for conduction along underwater electric cables (Fig. 25.9).

---

[4] One of the moments in which the inessentiality of cell membrane in du Bois-Reymond appears in a very clear way is when he says that while the artificial transverse section of the muscle produces a much stronger current than the natural one, the longitudinal artificial section behaves like the natural one. Surprisingly, the German scientist seems to be unaware that, because of the longitudinal course in the muscle fiber bundles of connective fascicles and cell membranes, it is somewhat unlikely that one would cause a lesion to the fiber membrane when trying to create a "natural longitudinal section."

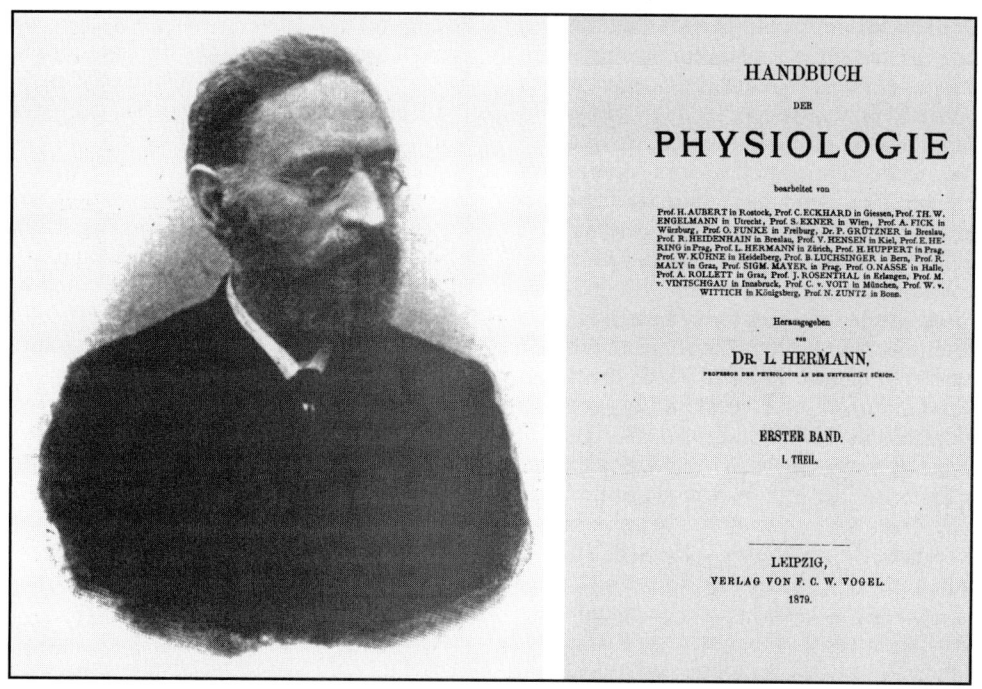

Figure 25.8. Ludimar Hermann (1838–1914) and the first volume of his *Handbuch der Physiologie*. Hermann correctly criticized du Bois-Reymond's contention of the existence of a measurable current at the surface of intact nerves or muscles at rest, but he was wrong when he argued against the existence of an electric state in resting fibers. Importantly, he drew some attention to the role that might be played by the fiber's membrane and to local changes that could spread along the membrane (his *Kernleitertheorie* or core conduction theory).

Hermann's conception was based on the idea that a nerve fiber consists of a conductive core separated from the external fluid by a partially insulating material. A second key element was the thought that any electrical disturbance originating on the nerve could influence nearby regions through local current loops involving three elements: the internal core, the insulating sheet, and the external fluid. This theory, which in Hermann's view could explain various electrical phenomena in muscle and nerve fibers, is now called Hermann's "local circuit theory."

Although Hermann made reference to state-of-the-art biochemistry, the precise nature of the chemical mechanisms involved in the excitation and propagation of the nerve and muscle events remained somewhat undetermined in his model. Since the model did not imply any electric state in a resting intact fiber, Hermann criticized du Bois-Reymond's use of expressions such as *Nervenstrom*, *Muskelstrom*, or *Ruhestrom* which indicated the existence of a current in unexcited resting fibers. Instead, he used phrases such as *Demarcationsstrom* ("demarcation current") to indicate that the electricity develops as a consequence of processes occurring at the surface of separation between the injured and the intact zones of the fiber (later indicated as "injury current" in the English scientific literature). Further, he proposed *Actionsstrom*, meaning "action current," as a better term than du Bois-Reymond's *negative Schwankung*. In all such instances, he wished to make it clear that the current produced in the excited fiber is the expression of a previously nonexistent process, a transformational activity triggered by the stimulus.

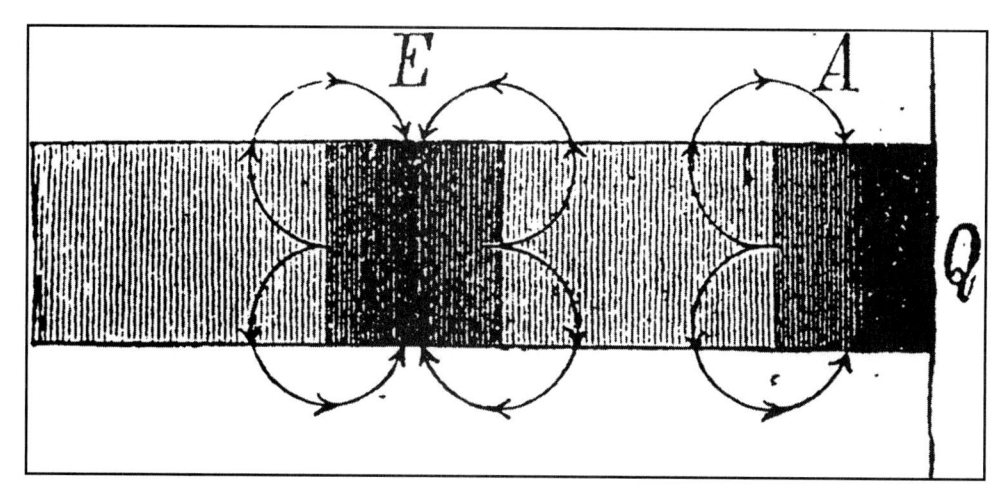

Figure 25.9. Ludimar Hermann's illustration of how local events could affect nearby areas, his local circuit theory. Basically the same type of local electric circulation is established between a resting and an active zone (E) as between an intact and an injured region (Q). In Hermann's elaboration, the metabolic processes responsible for nerve (or muscle) excitation are similar to those involved in injury processes (from Hermann, 1879).

Urged by Hermann, du Bois-Reymond conducted a series of experiments to support his contention about the presence of a resting current in intact preparations and also in entire animals. Moreover, he modified his theory in order to account for the scarcity or non-visibility of these currents. With an *ad hoc* hypothesis, he assumed the existence of a "parelectronomic" (from a Greek expression meaning "contrary to law") layer of electric molecules oriented in an opposite way at both extremities of muscle fiber, which would mask the electrical effects of ordinary molecules in a partial or complete way.

He then embarked on a series of (apparently) well-controlled experiments to substantiate his new hypothesis. In the course of his trials, he became convinced that the parelectronomic layer could change its characteristics with the physiological conditions of the preparation. He maintained that it could weaken or disappear after some physical or chemical treatments. Alternatively, it could become intensified after other treatments, notably after keeping frogs at low temperatures, which he thought would cause an overall positivity of the tendinous extremity compared to the muscle belly.[5]

The polemics between the backers of the "pre-existence theory" (du Bois-Reymond's) and those favoring the "alteration theory" (Hermann's) raged for many years, and it involved a plethora of publications from both sides. When addressing his old student, du Bois-Reymond showed even more acrimony and causticity than he had in his polemics with Matteucci (Chapter 24). On the other side of the fence, Hermann was not particularly diplomatic either.

In 1884, the polemics took on new life with the publication of the last volume of du Bois-Reymond's *Untersuchungen*. Here he criticized Hermann's explanation of electrotonus and tried to show that it could be accounted for with his electric molecules hypothesis. In his reply, published in the same year, Hermann spoke of the *bankruptcy of the molecular hypothesis* and, with bitter humor, considered its formulations as belonging to "the science of the future." Further, he stigmatized the efforts of his antagonist to come forth with *ad hoc* hypotheses to defend his theory, and he did this in print with biting sarcasm. To quote:

> The molecular theory owes its production to the error of fact that uninjured muscles show as strong a muscle-current at their natural cross-section as injured muscles do at their artificial cross-sections. When this was found to be an error, the theory was not given up, but the mischievous compromise of `parelectronomy' was invented. (Hermann, 1884, pp. 159–160)

The fighting seemed to terminate toward the end of the century, with the almost complete dominance of Hermann's views (see Biedermann, 1895, and Boruttau, 1904, for contemporary accounts). The ground was now being tilled by parallel studies in various fields of science, from thermodynamics to chemistry, from biophysics to physiology, and even botany.

The results of these parallel studies, some of which can be considered among the greatest achievements of nineteenth-century science, would ultimately converge at the beginning of the new century with the development of Julius Bernstein's "membrane theory."

Before considering Bernstein's elaboration and the way he arrived at it, we should note that, despite their obvious differences, some similarities exist between the pre-existence and the alteration theories. One is that both du Bois-Reymond and Hermann viewed signal conduction along nerve and muscle fibers as different from passive electric conduction along a conductive wire or cable. Another is that in both cases (and particularly in Hermann's) a fundamental link in the signal propagation is a localized negativity on the fiber surface. This negativity, generated by the excitation process (albeit in different ways in the two theories), is similarly capable of acting as an exciting stimulus for the activation of a resting segment in the direction of the propagation. In other words, the electric negativity of the external surface brought about by the excitation process (*negative Schwankung* or *Actionsstrom*) would play the key role of the element closing the loop between the (electrical) triggering stimulus and the (electrical) outcome of the elementary excitatory mechanism. For du Bois-Reymond, this would appear particularly clearly in the correct way he interpreted Matteucci's induced-twitch experiments, and the corresponding results he personally was able to obtain in nerves.[6] Despite his indeterminacy about the chemical processes underlying the production of electricity during signal propagation, there is also a clear awareness of the need for energetic mechanisms for signal progression along the fiber in Hermann's theory, and here it is even better expressed.

BERNSTEIN'S ELECTROCHEMICAL THEORY OF BIOELECTRIC PHENOMENA

While at Halle, Julius Bernstein, who had first succeeded in measuring the speed of propagation of the negative variation in Heidelberg, now made his second monumental contribution to the field, opening the path to the main developments of electrophysiology of the twentieth century. This achievement stemmed from his long interest in theories of bioelectric phenomena, which became particularly intense and compelling with the polemics between du Bois-Reymond and Hermann.

As already mentioned, Bernstein camped on the side of his former teacher, although he kept a more open mind and tried to modify his views as new scientific facts started emerging. Most importantly, he was able to keep his sights on the development of various fields of physics and chemistry, which were laying the groundwork for epochal advancements in the sciences in Germany and other European countries. Of particular relevance here is Bernstein's great versatility in

---

[5] An experiment conducted by Hermann, and considered by him as crucial evidence in favor of the alteration theory, consisted in making a rapid cut of the fiber and calculating the time needed for the development of the demarcation. Since this current took about 300 milliseconds to appear, Hermann considered the mechanism responsible for its generation not pre-existent to the lesion. Bernstein, who repeated this experiment in collaboration with Tschermak, found a latency of only about 1 millisecond and considered Hermann's inference less than conclusive (Bernstein & Tschermak, 1902).

[6] Du Bois-Reymond performed the induced-twitch (or secondary twitch) experiments in a nerve by placing the nerve of a neuromuscular preparation in contact with another isolated nerve, which was stimulated in a repetitive way. When the contact was close and wide enough, the muscle contracted when the isolated nerve was excited. He interpreted his results as evidence that the wave of excitation traveling along the nerve (i.e., the *negative Schwankung*) is capable of exciting the unexcited nerve region ahead of it along the path of propagation.

mathematical physics, and his familiarity with the most advanced and sophisticated physical conceptions of the time. He was especially well versed in thermodynamics, Helmholtz being his other teacher.

Bernstein had expressed a deep commitment to du Bois-Reymond's molecular theory dating back to the publication of his own 1868 article, which dealt with his rheotome measurements of the speed of the *negative Schwankung*. Six years later, in the middle of the controversy between du Bois-Reymond and Hermann, he nevertheless wrote that he did not see the molecules as the reality, but as "a representation of the forces that we imagine under their image" (Bernstein, 1874, p. 58).

Years later, when Hermann's views were dominating German electrophysiology, Bernstein (1888) published a long memoir in which he tried to incorporate Edward Pflüger's views on the chemical processes of living matter into the molecular theory. Pflüger had been another of du Bois-Reymond's students, and he was a talented experimentalist and the founder of the famous *Archiv für die gesammte Physiolologie* (a physiology journal that is still is published, although under a different name). He had become famous for his studies of *Elektrotonus* (a term introduced into physiology by du Bois-Reymond).[7] This phenomenon, which reflects a change of the excitability of a muscle or nerve fiber under the action of a maintained electrical stimulus, had been difficult to account for with the molecular theory in its initial version (Pflüger, 1859).

Unlike Bernstein, Pflüger was one of the supporters of Hermann's theory. Nevertheless, he promoted his own views on bioelectric phenomena. Specifically, he connected them in an essential way to the fundamental chemical processes of life, based on the concept of a "living protein" (Pflüger, 1859, 1875).

Bernstein derived the idea of the importance of energetic processes based on oxygen and oxidation from Pflüger. He too saw these mechanisms as key steps in the process by which the electric molecules of du Bois-Reymond's conception could change conformation and thereby produce the *negative Schwankung*. He also used an idea developed by Pflüger, namely that an excitable fiber could stay in only two thermodynamic states, both stable but characterized by different energy levels. According to Pflüger, the excitation process would consist of the sudden passage from one to the other of these states, and metabolic energy would be required to return to the resting state after the end of the excitation (see Lenoir, 1986, for a clear account of Pflüger's conception). Within this framework, Bernstein could provide a better explanation for the propagation of electrical excitation and also for the process by which the resting condition is re-established at the end of the excitation phase. He could also account for the changes in fiber excitability accompanying Pflüger's state of *Elektrotonus*.

With these elaborations, Bernstein thought he was able to integrate the chemical processes typical of the alteration theory (in its various formulations) within the electrical molecules theory promoted by du Bois-Reymond. This is why he referred to his version of du Bois-Reymond's conception as his "molecular electrochemical theory." Thus, he preserved du Bois-Reymond's fundamental assumptions about an electrical state already being present in the resting condition, adding that the excitation process releases, via oxidative processes, a form of electricity, which although constrained in some way, is nevertheless ready to manifest itself. No wonder, then, that the new theory was at one and the same time a "pre-existence theory" and an "alteration theory" in Bernstein's view.

The year 1888, the year Bernstein introduced his "new theory," was a fatal year for the electrochemical theory. This was because Walter Nernst (1888, 1889) started publishing his articles on electric batteries, presenting a new theory based on ion diffusion in this context (Fig. 25.10). Nernst's thinking would lead physiologists to the concept of an electrochemical potential. As we shall see, this concept would prove to be of great importance for Bernstein in his subsequent explanations of the genesis of electric potential across a cell membrane. Nernst's studies were, however, only one of the elements in the chain of events underlying Bernstein's "electrochemical" membrane theory, which was elaborated between 1902 and 1912. There were many other elements derived from the parallel developments in various and sometimes disparate research fields, which converged in his landmark achievement.

IONS, CELLS, MEMBRANES, THEORIES, AND NOBEL PRIZES

We cannot delve in detail into the various physical and life science contributions that led to Bernstein's membrane theory, and instead refer the interested reader to two important articles on this subject (Lenoir, 1986; De Palma & Pareti, 2010). We shall, however, provide a synopsis of the main aspects of this fascinating history, since it is almost totally ignored in books and other publications dealing with the foundations of modern physiology and the neurosciences.

With a mechanistic orientation, with great interest in the physical phenomena that could underlie life processes, and hoping to replicate artificially the processes of cell growth, in 1867 German chemist Moritz Traube succeeded in synthesizing membranes permeable to water but impermeable to various solutes (see Fig. 25.10). These semipermeable membranes were developed as models for plant cell membranes, and they were used in the osmotic studies of plant physiology that Wilhelm Pfeffer (1877) carried out a decade later (see Fig. 25.10).

Thanks to Dutch botanist Hugo De Vries (1884) (see Fig. 25.10), Pfeiffer's research came to the attention of another Dutch scientist, Jacobus Henricus van't Hoff, who would be the first winner of the Nobel Prize for chemistry (in 1901). In 1887, 1 year before Nernst's first article on electrochemical diffusion, Van't Hoff published an important article on osmosis in which the properties of solutions were considered similar to those of compressed gases (Van't Hoff, 1887).

In the same first issue of the *Zeitschrift für physikalische Chemie*, where Van't Hoff's paper appeared, there was also another important contribution in the story of bioelectric phenomena. Svante Arrhenius, who would be awarded the Nobel Prize in chemistry 2 years after Van't Hoff, wrote this

---

[7] The issue of electrotonus, which led Pflüger to propose a law of electrical excitability for nerve and muscle, was one of the most debated problems of late-nineteenth-century physiology. Hermann's theory could explain it better, particularly on the basis of the "core conductor model" elaborated by him.

Figure 25.10. Some of the scientists who, with their achievements in their respective disciplines, contributed to lay the grounds on which Bernstein build his membrane theory of electric phenomena in excitable cells. From left to right and top to bottom: chemist Moritz Traube (1826–1894), the inventor of artificial membranes; plant physiologist Wilhelm Friedrich Pfeffer (1845–1920); botanist Hugo De Vries (1848–1935); physico-chemists Svante Arrhenius (1859–1927) and Walther Nernst (1864–1941).

article (see Fig. 25.10). It dealt with electrolytic dissociation, a phenomenon by which a solute can separate into two ion types bearing opposite charges—that is, positively charged cations (such as Na$^+$ or K$^+$) and negatively charged anions (such as Cl$^-$ or HCO3$^-$), each capable of independently diffusing and contributing to the osmotic effects.

The *Zeitschrift für physikalische Chemie* would continue to be important in our history, particularly because of an article that appeared in it in 1890, although this paper was written more as a commentary or discussion than a laboratory report. Wilhelm Ostwald, chief editor of the journal with Van't Hoff, and yet another German chemist who would receive a Nobel Prize (in 1908), wrote this paper (Fig. 25.11). To his credit, Ostwald had grasped the importance of bringing various research fields together for interpreting bioelectric phenomena in plant and animal cells. He was especially drawn to Arrhenius' work on the dissociation of solutes in charged particles, Nernst's theory of ion diffusion, and the newest studies on artificial semipermeable membranes and osmotic phenomena in plant cells.

Drawing on this work, Ostwald tried to figure out how an electric potential could arise between two different electrolytic solutions separated by a semipermeable membrane with ions-sieve properties. He also considered how these new concepts could be relevant to understanding various biological events and "first of all the electrophysiological phenomena that have been studied so amply particularly by du Bois-Reymond" (Ostwald, 1890, p. 76).

Ostwald followed these words with an important statement that proved to be a somewhat prophetical anticipation of what would be the outcome of biophysical research in subsequent years. In fact, Bernstein (1902) would quote these words from Ostwald (1890, p. 80) in his first formulation of his membrane theory. In translation from the German: "It is perhaps not too daring to express here the hypothesis that, through the properties of semipermeable membranes here discussed, it would be possible to find an explanation not only for the currents in muscles and nerves, but also—and particularly—for the enigmatic effects of electric fish."

Soon after doing this, Ostwald turned to how new ideas on the permeability of membranes could help to solve the old problem of accounting for the presence of different electric potentials in animal tissues, humid tissues long being known to be electrically conductive. "By knowing that humid tissues, being wrapped by semipermeable membranes, can act as perfect insulators," he wrote with great insight, "this difficulty disappears" (p. 76).

Various authors attempted with limited successes to pursue the research path Ostwald had outlined. They included Napoleon Cybulski (the "father" of Polish physiology), Ukrainian physiologist Vasilij Jur'evich Chagovec, Finnish scientist Maximilian Oker-Blom, and English physiologist

Figure 25.11. Jacobus van't Hoff (1852–1911) and Wilhelm Ostwald (1853–1932) in Ostwald's laboratory. In addition to the importance of their achievements in the field of physico-chemistry the two scientists also contributed to the scientific discussion in the field through the *Zeitschrift für physikalische Chemie* they co-edited.

John Smith MacDonald (Lenoir, 1986; De Palma & Pareti, 2010). Nernst (1899) also wrote an important article illustrating how his theory of electrochemical diffusion could account for the genesis of bioelectric potentials.

The door, however, was left open for Bernstein to explore the physiological path outlined by Ostwald, and to come forth with the first physically coherent version of an electrochemical theory of membrane phenomena. Before examining Bernstein's 1902 elaboration, however, it is important to say that 1902 also stands out in this history for another reason.

CHARLES OVERTON AND THE "SCANDAL" OF THE MUSCLE THAT DID NOT LOVE SUGAR

In 1902, Charles Overton, an English physiologist then working in Germany, published two important papers on the effects of bathing excitable fibers in external solutions (Overton, 1902a,b) (Fig. 25.12). Overton had previously been able to show with clarity that the membranes of a variety of biological cells are permeable to lipophilic solutes. This work had led him to conceive of lipids, and particularly of lecityne-cholesterol, as fundamental to the membrane structure.

In the second of his 1902 papers, Overton reported the results of a study stimulated by an observation so surprising that it had the flavor of a scientific scandal. Specifically, a frog muscle preparation was found to lose its contractility completely when the bathing solution was changed from salty solution (0.6% NaCl) to a sugar solution of similar osmotic pressure. Almost by chance, Overton decided to test the effects of adding a small quantity of NaCl to this same preparation, which had been abandoned in the laboratory after losing its excitability. Following this maneuver, and to his "great satisfaction," the muscle rapidly recovered and displayed almost normal excitability and contractility.

Repeating this basic experiment in a variety of ways, and being able to dismiss the possibility that sugar might have toxic effects, Overton next decided to determine the essential chemical requirements of the extracellular medium for sustaining normal muscle excitability. Muscles became unexcitable when perfused with "iso-osmotic solutions of glucose, fructose, lactose, mannite, alanine, etc.," but recovered their excitability when appropriate amounts of NaCl were added to these solutions. It now seemed evident that the presence of NaCl in the extracellular medium (or at least one of its ions) must be fundamental for normal muscle contractility.

In further experiments, Overton showed that, in the presence of NaCl-free solutions, muscles lost not only their contractility but also the ability to produce and conduct an electrical excitation wave. By using different sodium salts, he further ascertained that Na^+ is the necessary factor. More specifically, only Li^+ in adequate concentrations was found to be a good substitute for Na^+, whereas high solutions of K^+ had a paralyzing action. In contrast, various anions could substitute for the Cl^-.

On the basis on these and other experiments (and the results of previous studies showing that Na^+ is the prevailing extracellular cation, whereas K^+ dominates inside the muscle fiber), Overton (1902b, p. 383) suggested that with the onset of excitation "during a certain time lapse (likely very short) muscle fibers might become permeable to sodium and potassium ions." Further, he correctly envisioned that, from an exchange between the Na^+ (or Li^+) ions of the extracellular medium "and the potassium ions of muscle fiber, an electric tension could be produced, which perhaps represents one of the sources of the *action currents*" (Overton, 1902b, p. 383). Overton considered this conclusion possible despite the fact that a mechanism of this type might involve the transfer of an enormous number of ions, particularly

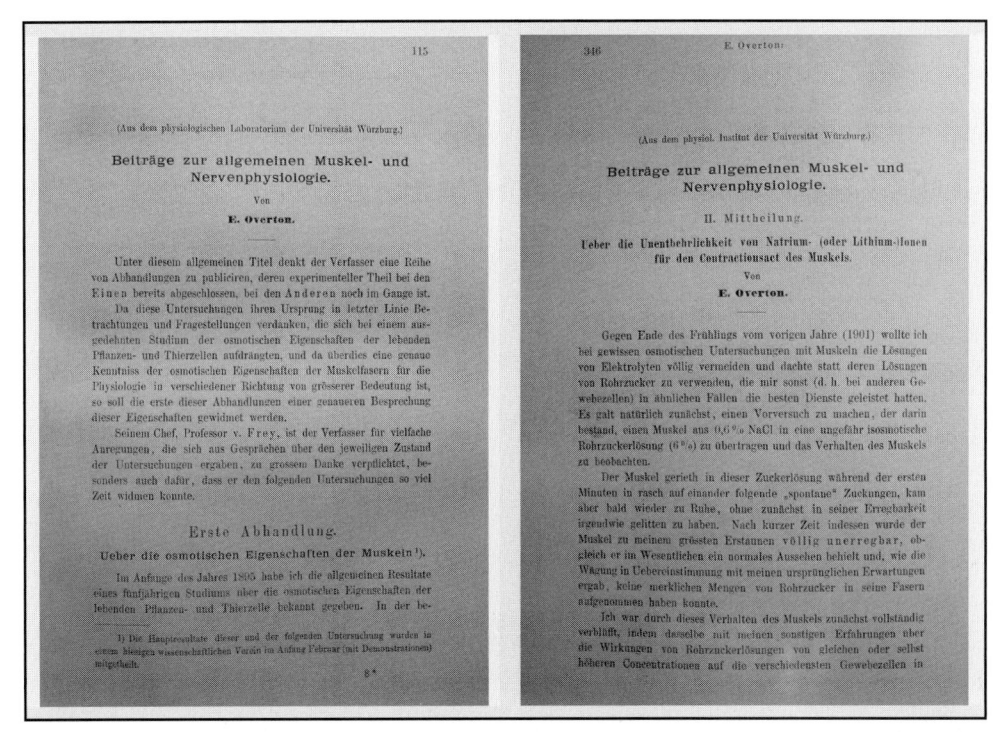

Figure 25.12. The initial pages of Charles Overton's two 1902 articles dealing with the effects of different ionic solutions on muscle excitability.

for a muscle like the heart, which contracts many times a minute all life long.[8]

BERNSTEIN, THERMODYNAMICS, AND "MEMBRANE THEORY": A KEY TO THE CHEMISTRY OF VITAL PROCESSES

Given these advances in the sciences, the *Zeitgeist* at the beginning of the new century could not have been more favorable for a better theory of bioelectric phenomena, one that would combine what was now known about ion dissociation and diffusion with what had been learned about osmosis through artificial and natural membranes. Particularly after the publication of Overton's 1902 studies, it was relatively easy to integrate the movements of specific ions into a diffusion theory of membrane potentials. With his long-term involvement in the study of nerve and muscle excitability, and with his great familiarity with physico-chemical studies, Bernstein was the right person, at the right time, and with the right tools, to initiate this new phase of electrophysiology. He did not miss the chance!

The way in which he developed his new conception in 1902, with his knowledge of the thermodynamics of different types of electric batteries, sheds some light on the importance he placed on this goal (Fig. 25.13). Bernstein did not simply aim at explaining the origin of bioelectric potentials and concluding, 7 years after du Bois-Reymond's death, the long controversy between the pre-existence and the alteration theories. With his membrane theory, he wished to make it clear that

"*the thermodynamic treatment of bioelectric currents can be of great importance to get an insight to the internal chemism [Chemismus] of the internal organs,* in a way that we could not get by any chemical investigation of them" (Bernstein, 1902, pp. 530–531; emphasis Bernstein's). In other words, the new theory was Bernstein's way of pursuing his program aimed at giving life phenomena a firmer and more advanced physico-mathematical background.

This was indeed the same program that had been inaugurated by Emil du Bois-Reymond and Ernst Brücke in the 1840s, when they pledged an oath to show that "no other forces than the common physical chemical ones are active within the organism" (Du Bois-Reymond, 1927, p. 19). Helmholtz was the third member of this group, and it was du Bois-Reymond who "introduced young Helmholtz to the Society, where he was warmly welcomed as its greatest ornament" (Koenigsberger, 1906, p. 31). Thus, both of Bernstein's teachers were early members of this small but dedicated club, which emphasized "organic physics," sophisticated instrumentation, and mathematical analyses—a group that became the *Berliner Physikalische Gesellschaft* in 1845 (Bernfield, 1944).

Helmholtz's *Über die Erhaltung der Kraft* ("On the Conservation of the Force"), published in 1847, helped define the course the "new" German science would now take. Described by du Bois-Reymond as "a fellow who has devoured chemistry, physics, and mathematics with a spoon, who completely holds our worldview, and who is rich in thoughts and new ways of looking at things" (Cahan, 1993b, p. 27), Helmholtz was only 26 years old at the time he formulated the first principle of thermodynamics.[9] He took on the subject in order to account physically and mathematically for the energy transformations underlying muscle contractions, refusing to accept the vitalistic and metaphysical views of

---

[8] Curiously, in the course of these experiments Overton noticed that, unlike muscles, nerves remain excitable even if kept for many hours in sodium-free solutions. As we now know, this surprising result can be explained by the presence of thick layers of connective tissue around the nerves, which prevent ion diffusion.

[9] Besides, he was convalescent following a typhoidal infection.

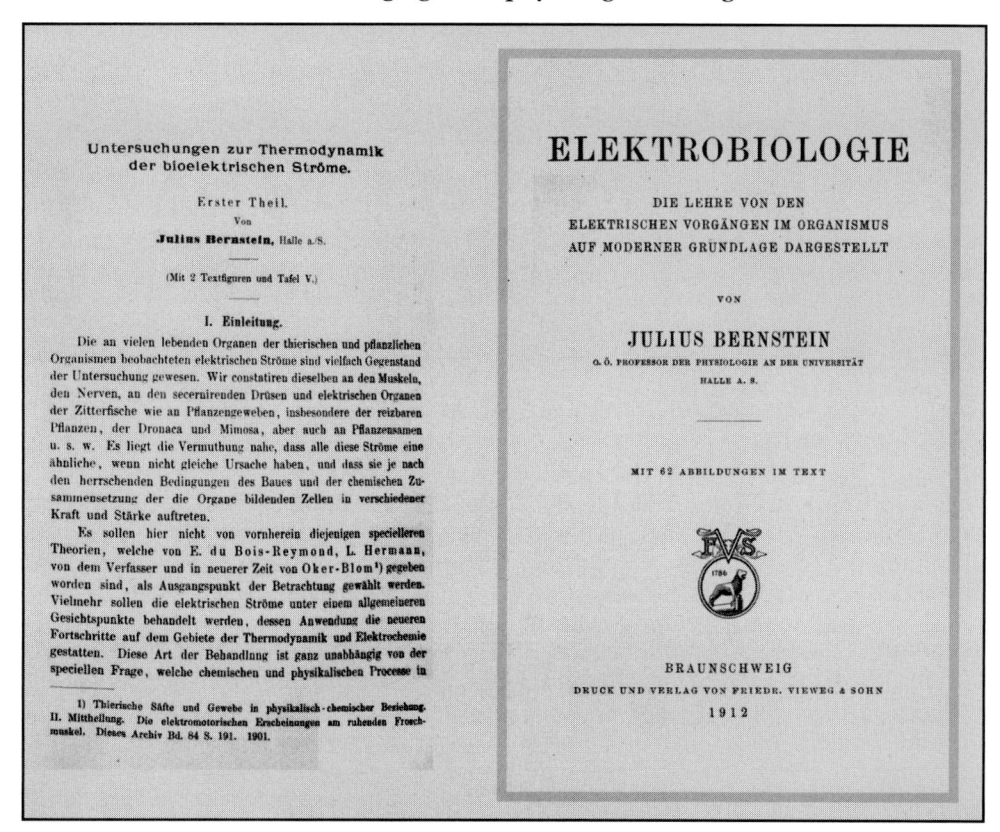

Figure 25.13. Julius Bernstein drew more attention to the importance of the cell membrane in an article published in 1902 and a book that appeared in 1912, both shown here. Bernstein accepted some of what du Bois-Reymond was stating about resting currents and molecules, and some of Hermann's ideas about the membrane and local events. His formulation, based on the movement of ions across a semipermeable membrane, is normally referred to as "membrane theory."

Naturphilosophie, including those held by his own mentor, Johannes Müller.

In this context it is interesting to note that in 1902, just before publishing his landmark paper on membrane theory, Bernstein had published an article dealing with the energy underlying muscle excitation. In this paper, written in collaboration with his Austrian student Armin Tschermak, he tried to develop his previous "electrochemical theory" and incorporate in it some aspects of the chemistry deemed important by the supporters of the opposing alteration theory, particularly by Pflüger (Bernstein & Tschermak, 1902). Bernstein had already been firmly committed to a physico-mechanical conception of life phenomena, with particular reference to thermodynamic principles. One need only examine his *Lehrbuch der Physiologie*, which went through three editions between 1894 and 1910, to see this strong and lasting commitment.

The way in which Bernstein developed the foundations of his membrane theory shows that he was not stressing the importance of chemical processes in life phenomena. As a matter of fact, among the three different thermodynamically different batteries that he considered as possible models for bioelectric phenomena, he chose the one in which an electric energy is produced solely on the basis of ion movements, with no chemical reactions being involved whatsoever (a model indicated as *Concentrationkette*, meaning "concentration cell"). On the basis of the thermodynamic principles elaborated by Helmholtz, and further developed by American chemist and physicist Josiah Willard Gibbs, this type of battery could be distinguished experimentally from the two other kinds, both of which depend on chemical processes, on the basis of purely thermodynamic criteria. The electric potential generated by a concentration cell should increase in a linear way by increasing the temperature and, moreover, the temperature of the battery should decrease when it is discharged through an external circuit.

Bernstein presented the preliminary results of some experiments he made on both muscles and nerves in the experimental part of his article. Nevertheless, they corresponded in only a limited way to his theoretical predictions for the actions of a concentration cell. In muscles, the electromotive force (measured with a galvanometer as an injury potential) was found to increase with the temperature in the range studied (0° to +31°C), but this dependence did not accurately follow the slope predicted by the theory. In nerves, there was even less agreement with the theory, because the electromotive power increased between 0° and +18°C but decreased with further increases of the temperature up to +32°C. Maintaining that these divergences between theory and experimental findings could be explained by taking the complexity of the biological preparations into account, Bernstein concluded that the mechanism responsible for the bioelectric potential corresponds to that involved in a concentration cell.

Referring afterwards to Ostwald and Pfeiffer, Bernstein then put forward the idea that the resting potential recorded in the injured muscle cell could be a consequence of an unequal distribution of electrolytes on the two sides of the muscle membrane. In his own words, it could be due to *"the electrolytes already contained inside the uninjured muscle fiber, and thus particularly the inorganic salts,* for example KH_2PO_4 [potassium diphosphate], *are the causes of the potential differences in the injured muscle*" (Bernstein, 1902, p. 542; Bernstein's emphasis). As a consequence of the outward diffusion of K^+ ions (not accompanied by phosphate anions, to which the membrane is assumed to be impermeable), "a double electric layer [*elektrische Doppelschicht*] would be formed on the surface of the fibril, which would have a negative tension toward the interior and a positive one toward the exterior."

Bernstein wrote that "for sake of brevity" he would call his conception *Membrantheorie*, meaning "theory of the membrane" or "membrane theory," after remarking with reference to the du Bois-Reymond and Hermann dispute that he viewed it as a version of the pre-existence theory. This was because the electrical states responsible for both the resting and excitation potentials exist in the cells prior to excitation (and independently of any previous alteration of the fiber).

In the concluding section of his article, however, Bernstein softened his wording, which had reflected his bias toward his teacher's position, and also wrote favorably about Hermann's views. He now explained that membrane theory could also be interpreted in the framework of alteration theory by assuming the ions involved in bioelectric phenomena are produced at the moment of the excitation (or injury): "It is evident that the negative oscillation (or action currents) induced by the stimulation could be explained equally well on the basis of the two theories: on the basis of the alteration theory, with the formation and disappearance of the corresponding organic electrolyte; on the basis of the membrane theory through an increase of the permeability for the blocked ion" (Bernstein, 1902, p. 560). In the framework of the pre-existence version of the membrane theory (undoubtedly the most likely one for Bernstein), the "chemism" of the process is thus only reduced to the possible chemical modification at the surface of the fiber underlying the change of the ionic permeability (ultimately responsible for the electric excitation).

Bernstein's 1902 article is usually referred to as his "membrane theory" paper. Yet this can be somewhat deceiving, because of how poorly the membrane really materializes along the pages of his text. Although he referred to Ostwald's and Peiffer's work, he did not attempt to provide any information about the membrane's structure and composition. There was nothing even about lipids, which were at the center of Ostwald's work.

Another somewhat surprising absence concerns sodium ions, which Overton had put on the center stage when discussing bioelectric phenomena in the same year. One thought is that Bernstein might not have been acquainted with Overton's work when he wrote his own paper. This cannot be the whole story, however, because sodium ions would remain in the shadows in the mature formulation of Bernstein's theory, which would be expounded 10 years later. In retrospect, the great German scientist seems to have had the intellectual power and material chances to formulate a complete membrane theory of bioelectric phenomena (with potassium permeability prevailing in the resting state, and sodium during the excitation phase) at the beginning of the new century. But he decided not to take the larger step at this time, which is unfortunate, because it would have given his innovative theory an even more modern flavor.

ELECTRIC FISHES, MEMBRANES, AND A COOLING BATTERY

Before trying to understand why Bernstein stopped where he did, it is important to state that, soon after publishing his 1902 paper, he tried to obtain additional experimental evidence to support his new position on the origin of bioelectric potentials. Since the energy processes responsible for the mechanical phenomena of muscle contraction largely depend, from a thermodynamic point of view, on electric mechanisms, Bernstein needed to find a biological preparation in which the production of electricity overwhelms mechanical processes (and possibly other types of energy-involving mechanisms).

What could be better than electric fishes, in effect nature's living electric machines? To make matters more attractive, these unusual fishes could now be obtained rather easily, even in Germany. Still in collaboration with Tschermak, Bernstein started a new series of experiments on torpedoes. He presented some of his findings at the Berlin Academy in 1904, published them in full in 1906, and incorporated and discussed them in his *Elektrobiologie* of 1912.

In our next chapter, we shall examine Bernstein's endeavors with electric fishes in some detail, particularly because he would formulate a theory capable of explaining the unique capacity of these fishes in his *Elektrobiologie*—the ability to sum the electricity produced by a multitude of electrically excitable cells and to convey it outside its body—in his *Elektrobiologie*. Here we shall only briefly mention these experiments, given their importance for the formulation of the mature form of his membrane theory.

In one set of studies, he stimulated the nerve to an electric organ isolated from a fish, while varying the temperature and other experimental conditions. He recorded the heat associated with the production of the shock and the electrical energy discharged with specially designed devices (Fig. 25.14), and measured the shock intensity with a galvanometer. With these experiments, Bernstein hoped to decide between two possible alternatives for production of electricity, in accord with the asymmetrical ion distribution mentioned in the conclusion of his 1902 article.

On the one hand, if ion asymmetry were produced at the moment of the shock by the production of an excess of ions on one side of the membrane (through a chemical process of scission and oxidation, and as implied by the alteration hypothesis), then for thermodynamic reasons there should be a substantial production of heat at the moment of the shock. This would be similar to what happens with muscle during the contraction process. On the other hand, if the concentration difference exists prior to the excitation phase (and excitation is only due to changes in the sieve property of the membrane), then there should be minimal heat production. This is because the production of electricity in a diffusion cell does not produce this sort of extra heat, apart from that due to the passive flow of current in the discharge circuit.

The experimental results favored the second alternative, leading Bernstein and Tschermak (1906, p. 503) to conclude that "the electric organ represents a concentration cell of a special type." That is, heat production was minimal in most preparations, and importantly, the discharge was accompanied by a cooling of the organ in some cases. This last result was especially relevant, because it parallels the behavior of a

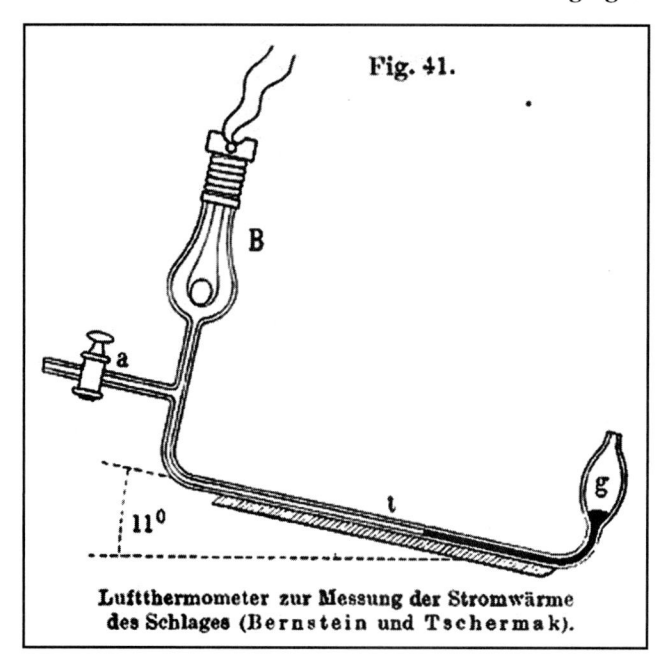

Figure 25.14. Bernstein's approach to membrane phenomena in excitable cells (nerve, muscles, electrocyte) was mainly of a thermodynamic type. This was largely the result of the influence of one of his two main teachers, Helmholtz (the other being du Bois-Reymond). This approach required examining heat production (or absorption) during excitation, the direction and the degree of the change being viewed as clues about the processes involved in the activation of excitable cells. The *Luftthermomether*, shown here, was developed by Bernstein especially for these types of studies and was used in a particularly effective way in the study of the torpedo's shock (from Bernstein, 1912).

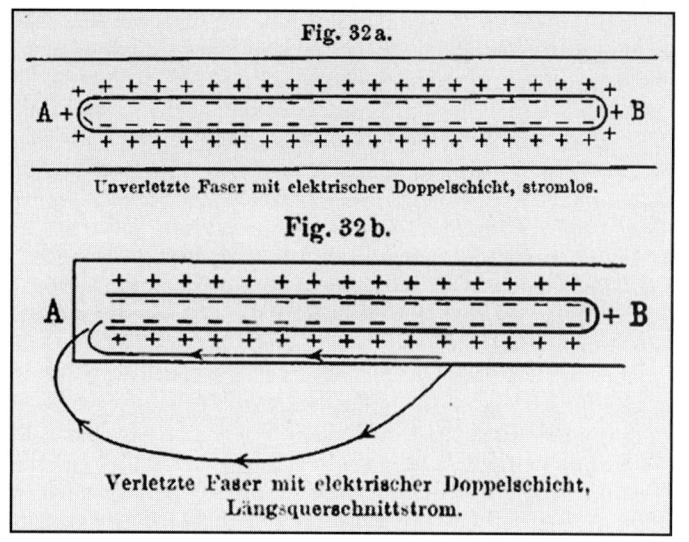

Figure 25.15. Bernstein's original figure illustrating his *Membrantheorie*, in which the inside of the cell would normally be negative relative to the outside. This figure appeared in Bernstein's later (1912) text on the subject. He stressed the importance of positively charged potassium ion movements from inside the cell across the membrane, but failed to recognize the role played by an influx of also positively charged sodium ions across the membrane during the excitatory nerve event (from Bernstein, 1912).

concentration cell when the quantity of electrical energy generated is larger than the energy expended. So as not to violate the principles of thermodynamics, the system must absorb the excess energy needed for the transformation in the form of thermal energy, this leading to a temperature decrease.

These and other results from his electric fish studies supported Bernstein's idea that the production of electricity in biological tissues—from muscles and nerves to the specialized organs of fishes—depends on transmembrane ion gradients in excitable tissues at rest (Fig. 25.15). They also suggested that chemical processes do not intervene in an important way at the moment of excitation. In other words, they favored the pre-existence theory. In Bernstein's new view, however, the idea of pre-existence was inserted into a framework characterized by ions and their movements in liquid media. In this regard, Bernstein's conception was diametrically opposed to the "solid" molecular model proposed by du Bois-Reymond many years earlier.

In 1912 Bernstein discussed another of his experiments, which in his view also supported pre-existence against the alteration theory on the basis of thermodynamic considerations (already considered in the 1910 edition of his *Lehrbuch*). A muscle was cut at one end and it was differentially heated, either in the injured region or in the intact zone. Heating the injured region did not appreciably modify the longitudinal current flowing from the intact to the cut region. In contrast, the current increased appreciably when the temperature of the intact side was increased, a result that corresponded with the thermodynamic predictions of the membrane theory.

As Bernstein saw it, if the current flow were due mainly to chemically-active processes in the injured zone, the current should show a substantial change when this zone is heated, which was not the case. The experiment was interpreted as evidence that the mechanisms responsible for the production of bioelectric currents are better represented by the electric condition of the intact zone. In Bernstein's view the lesion allows only the mechanisms responsible for electrical condition existing in the intact fiber to manifest themselves, albeit in an imperfect way. This is because the lesion provides an access to the internal electric condition of the fiber.

This was just the opposite of what Hermann had assumed when formulating his alteration theory. According to Hermann, the excitation of the intact zone involves chemical reactions, which are similar to those that occur in the injured zone (and also during the death or degeneration of the fiber). For him, what occurs in the intact fiber during excitation (or *Actionsstrom* in his terminology) is only a partial expression of what happens in the injured region.

A "CONSERVATIVE REVOLUTION"

Because of Bernstein's strong insistence on the elements of pre-existence in his membrane theory, it seems correct to say that if its formulation represented a scientific revolution, it was a kind of "conservative revolution." At its foundation, it tried to preserve in a modern and physically plausible way some of the basic assumptions of du Bois-Reymond's pre-existence theory (De Palma & Pareti, 2010).

The "conservative" aspects of Bernstein's theoretical elaboration gives us hints about the motives underlying his choice not to elaborate fully on all of the implications of the ionic theory, given the evidence at hand. As mentioned, membrane theory, even its most mature, 1912 form, does not take into account the role played by sodium ions when dealing with cell excitability. Consequently, it fails to provide a full picture of the excitation phase of bioelectric phenomena. As we now know, the membrane potential goes from a condition with the internal potential negative with respect to the external medium, to one in which the cell interior becomes positive to the exterior during excitation (see Chapter 25).

On the basis of Nernst's equation, the resting state is largely accounted for by the selective membrane permeability to potassium ions, which (as Bernstein knew) are more concentrated in the intracellular medium. The positivity of the cell's interior during the active excitatory phase is now known to depend on a change in the membrane's characteristics, whereby it becomes specifically permeable to sodium ions. The peak of the positive wave during the action potential corresponds rather closely to Nernst's electrochemical potential for sodium ions.

Bernstein, of course, could only speculate about the changes in the membrane permeability responsible for the passage from the resting to the active phase. Particularly by 1912, he knew well, however, that sodium is indispensable for electrical excitability, at least in muscle fibers, and possibly also in nerve cells. Further, he knew that this ion predominates in the extracellular compartment compared to the cell's interior, and thus that its influx during excitation could potentially lead to an internally positive membrane potential.

In fairness, we cannot accuse Bernstein of not having recognized sodium as the ion of the excitatory phase. Nor can we blame him for not having elaborated in his writings how the fiber potential would change if the membrane became specifically permeable to sodium ions during the active phase. As he pointed out in his initial rheotome recordings of the *negative Schwankung* in muscle fibers (see above), the electrical event of the excitation phase does not correspond to a simple destruction of the resting current. Rather, as we would express it today, the current could reverse polarity during the action potential.

That a reversal could actually happen had become more evident around the turn of the century, particularly with the introduction of more sensitive and faster electrical recording devices (e.g., the "capillary electrometer" and the "spring galvanometer"). In 1891, John Burdon-Sanderson and Francis Gotch reported some experiments conducted with a capillary electrometer that showed the action current was often of a larger amplitude than the resting current. Nevertheless, Bernstein considered the results of the two English physiologists to be inconclusive. And when it came to his earlier investigations with the rheotome, he thought that the findings could have been due to some kind of experimental artifact.

By looking retrospectively at the setting and full history behind Bernstein's membrane theory, it is difficult to escape the conclusion that what prevented him from making a more dramatic, revolutionary jump toward an even more modern electrophysiology was probably his firm adhesion to the pre-existence theory. As had happened to his former teacher, du Bois-Reymond, and despite his own theoretical and experimental ingenuity, he was unable to conceive of the electrical phenomena in the excited fiber as anything more than the destruction of a *pre-existing* electrical state—a state already present in the resting condition. An inversion of the polarity of membrane potential (consequent to a selective increase of sodium permeability) might have appeared to Bernstein to be somewhat of an active process, but this conflicted with what psychologists would call his "mental set" about pre-existence, a preconceived notion that he was unwilling or unable to renounce.

In defense of Bernstein, we must remember that he found the "overshoot" of the action potential (again using modern terminology) only in nerves and not in muscles. Since membrane theory appeared to be applicable to both types of fibers, at least in terms of its thermodynamics, this could have influenced his later rejection of the "overshoot" as an inconstant phenomenon, a potential artifact, or a possibility of little significance.

It was relatively easy to conceive how a particular conformation of the membrane in the resting state (i.e., one allowing for selective ion permeability) could be lost due to some molecular disturbance. But it was more difficult to figure out how the configuration of the membrane could pass from a particularly orderly arrangement to a different but similarly ordered conformation during the excitation process. In other words, it was easier to conceive that a membrane specifically permeable to potassium ions could become permeable to all ions, because it would lose its selective permeability as a result of some structural derangement. Overton, for one, had theorized that the membrane becomes permeable to both sodium and potassium ions at the time of the *Actionsstromes*.

In this context, it is interesting to note that Bernstein (1912) wrote that the membrane modifications occurring during excitation are similar to the *Absterben* or necrosis of the fiber. After all, prolonged and intense excitation can lead to the death of nerves and muscles. The change of the membrane properties occurring in the active zone, which is responsible for the undifferentiated permeability to all ions, must therefore have something to do with the destruction or dissolution of the membrane. The difference is that the changes taking place during excitation are of a transient and reversible character, and also occur to a lesser degree.

With these things in mind, and perhaps even unconsciously, Bernstein integrated some of the postulates of Hermann's alteration theory into the pre-existence theory that he had favored. In the first half of the twentieth century, particularly in the English scientific literature, the events occurring in the membrane during the excitation phase would be viewed as a kind of instantaneous and reversible "breakdown" of the membrane. It would take time and considerable new experimental evidence for this view to be abandoned.

Bernstein's adherence to some postulates of the pre-existence theory might not be the only explanation for this state of affairs. At least with regard to the lack of recognition of the role of sodium, it is possible that, as recognized by Andrew Huxley (1999), this depended on the apparent immunity of the nerve to sodium ion depletion, as shown in Overton's experiments. In any case, the story unfolded as stated, and it was not until half a century after Bernstein's first formulation of his membrane theory that both the presence of the overshoot and the significance of sodium in nerve physiology came to be fully and properly recognized.

In our next chapter, we shall see how scientists finally achieved what we can rightly call a modern understanding of the processes underlying nerve and muscle physiology. The breakthroughs that occurred near the middle of the twentieth century will in turn allow us to explain the actions of our remarkable electric fishes, although as will also be shown, the nerve and electric organ arrangements of these fishes differ from one another.

Our focus shall be on discoveries made in England, the country that took the lead from Germany, which had by the middle of the nineteenth century taken the lead from Italy, in this field of science. It is with some exciting and sophisticated experiments of modern electrophysiology that we shall complete our lengthy journey, which began in Africa along the Nile and took many twists and turns before taking us to where we are now with our more detailed understanding of electric fishes—those oddities of nature that are still as wondrous today as they were in ancient times.

Chapter 26
Understanding the Shock Mechanisms: A Twentieth-Century Odyssey

> It is perhaps not too bold to express here the conjecture that, through the properties of the semi-permeable membrane discussed until now, one could find the explanation, not only of the currents in muscles and nerves, but also particularly the enigmatic effects of electric fishes.
> (Wilhelm Ostwald, 1890, p. 80)

> The large voltages and currents generated by electric organs turn out not to be a result of membrane properties very different from those of nerve and muscle; rather they are the result of modification of the single cells to maximize external current and of arrangements of many cells in series and in parallel.
> (Michael Bennett, 1970, p. 475)

As shown in previous chapters, evidence continued to accumulate throughout the nineteenth century to support the view that the shocks produced by three kinds of strongly electric fishes are, in fact, electrical. Specifically, the shocks from torpedoes, electric catfishes, and electric eels were each found capable of producing the myriad of effects associated with electricity generated by batteries and other devices, be they physiological, magnetic, chemical, thermal, mechanical, luminous, or even chromatic.

Interestingly, electric catfishes, which had long remained in the shadows of experimental science, had become the most studied of all of the electric fishes during the second half of the century, when du Bois-Reymond was directing his integrated research program on the electric organs of three different fishes and nerve and muscle physiology. But although he and other German physiologists broke new ground when it came to understanding frog and even human nerves and muscles, for a rather long while they did not come forth with a monumental breakthrough when it came to understanding the cellular mechanisms underlying the shocks of their catfishes, torpedoes, and eels. In this domain, even the greatest physiologists of the day could only express wonder.

Over time, an inversion had taken place in the fruitful interchange between electric fish research and electrical investigations of nerve and muscle fiber physiology. Late in the eighteenth century and into the middle of the nineteenth century, research on the South American eel and European sea torpedoes had stimulated and served as reference points for important neuromuscular studies. But in subsequent decades, the great strides that were being made in nerve and muscle physiology programs were guiding thinking about the special organs found in electric fish.

This trend would continue into the twentieth century. As we shall see in some detail in this chapter, the elucidation of the cellular mechanisms underlying physiological electricity would first be reached with nerve cells, and the principles formulated from them would then be applied to the cellular elements of the electric organs in our fishes, which would be called "electrocytes." This cross-fertilization would prove successful because the cellular mechanisms that combine to produce the shocks in such fishes are modifications of those that generate electric potentials in nerve and muscle fibers. This relationship would become clearer in the early 1950s, when the first electrophysiological studies of the electric eel's electrocytes would be conducted; this occurring right after Alan Hodgkin and Andrew Huxley published their landmark studies on the biophysical foundations of nerve excitability (Keynes & Martins-Ferreira, 1953).

This relational view of the nature of things is in accord with a thought that emerged in the nineteenth century, only to be fully confirmed by later studies. It is that the cellular elements of the large electric organs in all three kinds of strongly electric fishes are derived from muscle cells. There are, however, notable differences between muscles and a fish's electric organs, one being the lack of contractility in the latter. As noted in Chapter 9, seventeenth- and early-eighteenth-century researchers, who promoted mechanical explanations for the fishes' shocks, had thought that they stemmed from specialized muscles that moved or vibrated with great rapidity. We now know they were right about the origins of the electric organs but wrong when it came to this sort of overt moving.

Additional differences have to do with the shape and characteristics of the elements in the electric organs. Together with their specific arrangements (e.g., their stacking in columns or rows), these functional adaptations make them especially well suited for summing up the minute electricity generated by the individual cells, each of which might account for only about a tenth of a volt. This summation process is clearly necessary for these creatures to produce formidable shocks, their most impressive feature and the one that has allowed them to thrive in their unique environments. For it is with these shocks that they gained an adaptive advantage in the great Darwinian battle for survival.

We shall focus on several related twentieth-century developments in this chapter. We shall examine how researchers achieved a modern understanding of nerve and muscle electrophysiology. Then we shall discuss how this development affected thinking about the electric organs of strongly electric fishes. Further, and in the same context, we shall examine the specific anatomical and functional changes that make it possible for strongly electric fishes to stun prey, ward off predators, and do other things that have allowed them to thrive in their different watery habitats.

But before we do these things, we must start off by showing that more than some African catfishes, one species of so-called eel, and the family of sea torpedoes possess electric organs, since this discovery will add needed color to the bigger evolutionary and physiological picture that we intend to paint. Hence, we must begin with some words about a number of fishes that are currently designated as "weakly electric fishes."

ELECTRIC ORGANS IN WEAKLY ELECTRIC FISHES

In 1844, James Stark, a Scottish surgeon, discovered that electric organs are present in more than the three types of strongly electric fishes that we have been discussing. In a communication to the Royal College of Edinburgh, Stark (1844) related that no one except Étienne Geoffroy Saint-Hilaire (1802) had previously endeavored to show that skates have electric organs, adding that the great French naturalist had mistakenly alluded to an organ "quite distinct from the electrical organs."[1] Stark said that his attention was drawn from the body of the skate to its tail by another "circumstance," about which no information was given in his article. Most likely, he was referring to remarks made by fishermen, who claimed that they detected weak shocks when they handled common rays by the tail (Burdon-Sanderson & Gotch, 1888).[2]

When Stark removed the skin from the tail of a flapper skate, he discovered an organ on each side very similar in structure to that possessed by torpedoes and the electric eel, although it was considerably smaller. His report was limited to its morphology, particularly its columnar arrangement and extensive nerve supply. He did not conduct physiological studies on living specimens. Nevertheless, he felt confident that he had discovered "an electrical apparatus," with the caveat that it was not equally developed in all rays. "In most of them it was merely rudimentary, not being thicker than a crow-quill or common pen, and consisting of only four or five columns, similarly divided into cells by cross membranous septa" (Stark, 1844, p. 2).

Two years later, Charles-Philippe Robin, a French physician and microscopist, studied the same tail organs (Robin, 1846, 1847). Then, after a lengthy delay of 19 years, Robin (1865) wrote that the organs of common rays are capable of producing electricity, although it is weaker in intensity than that in the other electric fishes. He conducted this more significant physiological phase of his research with a galvanometer and galvanoscopic frogs.

The presence of electric organs was afterwards demonstrated in many other fishes, including some odd-looking African fishes belonging to the genus *Mormyridae*. In 1877, the Russian scientist Aleksandr Ivanovich Babukhin (or Babuchin) discovered the electric capabilities of the latter genus. Using metal wires to connect a prepared frog's nerve and muscle to an aquarium containing a small mormyrid fish, he was taken aback by the appearance of sudden and unexpected muscle contractions (Babuchin, 1877).

The presence of electric organs in a rapidly expanding number of fishes incapable of giving strong shocks or even perceptible shocks, at least to the human hand (and therefore referred to by du Bois-Reymond as pseudo-electric or "imperfectly-electric" fish), posed a biological mystery. If their discharges are too weak to stun other fishes for food or to serve as weapons of defense, why might they have these organs?

This mystery drew more attention when Darwin's theory of evolution and natural selection was gaining wide attention. In 1859, in the first edition of his monumental *On the Origin of Species*, Darwin presented electric fishes as one of the great challenges facing his own thinking about evolution and change. "The electric organs of fishes offer another case of special difficulty," he wrote, explaining that "it is impossible to conceive by what steps these wondrous organs have been produced" (Darwin, 1859, p. 192).

Darwin was perplexed in part because he did not realize, at least when his book went to press, that common rays could give mild shocks. As a result, he could not point to any missing links or intermediate forms between fishes that produce strong electric shocks and fishes that do not shock at all. Here Darwin was misled by Matteucci, writing in his best-selling book that "it has lately been shown that Rays have an organ closely analogous to the electric apparatus, and yet do not, as Matteuchi [sic] asserts, discharge any electricity" (Darwin, 1859, p. 193).

Darwin also mentioned another difficulty: that the electric organs could be found in a number of fishes that differ considerably from each other. This seemed to go against his concept of a common ancestor or progenitor, and hence the idea that all electric fishes are fairly closely related to each other. In this context, Darwin had no reason to believe that almost all fishes once had electric organs, which, as he expressed it, "most of their modified descendents have lost."

The mystery posed by the electric organs of weakly electric fishes remained unsolved for more than a century. Its solution represented one of the most important discoveries of modern zoology, and it opened up a fascinating and fruitful field of inquiry that is at the crossroads of several fields, including the neurosciences, ethology, evolutionary science, and biocybernetics. It also gave investigators the clues they needed to answer another nagging question: Why are most

---

[1] Attempts to identify structures corresponding to the electric organs of ordinary rays had actually been made well before Geoffroy Saint-Hilaire (e.g, Monro, 1785). Nevertheless, Geoffroy Saint-Hilaire seems to have been the first researcher to consider the possibility that they might be akin to the electric organs of torpedoes and other electric fishes.

[2] Remarks of this sort, based on what fisherman were saying, circulated fairly widely during this era. Although Burdon-Sanderson and Gotch (1888) and others cited Stark's article in the *Proceedings of the Royal Society of Edinburgh*, this information did not appear in Stark's 1844 article or the account of his work that appeared a year later (Stark, 1844, 1845). Stark might have included it in his oral communication to the Royal Society of Edinburgh or from some other venue that we have been unable to find.

fishes that are capable of producing electricity also able to detect changes in the electrical fields in their environments?

During the 1770s, John Walsh had discovered the eel's remarkable ability to detect the presence of a circuit of conductors in the water, as opposed to a non-conducting circuit (see Chapter 19). In one of his studies, he extended two wires from a holding tank into two separate bowls of water, observing that this had no effect on his eel's behavior. But when a person put one hand into each of the bowls, thereby completing the circuit, the eel swam to the submerged wires and immediately began its incessant discharging. As for the phrase *sense électrique* (meaning "electrical sense)," this term seems, as we have seen, to date back to 1782, when Alessandro Volta commented on what Walsh had discovered (Chapter 19).

A century later, Emil du Bois-Reymond noticed the existence of an electric sense in a species of electric catfish. Subsequent studies revealed the widespread presence of electric sensibility in a great number of fishes, especially cartilaginous species (e.g., rays and sharks). We now know that it is present even in some amphibians, such as salamanders.

Modern morphological and physiological studies have clarified the mechanisms responsible for this sense. Detection depends on the presence of specialized receptors, today denoted as "electroreceptors." These receptors come in various sizes and forms, and they can be extremely sensitive (Bennett, 1971a; Fessard & Bullock, 1974; Bullock & Heiligenberg, 1986; Kalmijn, 1988; Moller, 1995; Kapoor & Khanna, 2004; Bullock, 2005).

Interestingly, one class of electroreceptors corresponds to the ampullary formations that Stefano Lorenzini first noticed late in the eighteenth century with torpedoes (Chapter 9). Lorenzini had observed pits on the skin that corresponded to the "mouths" of the long *canaliculi* typical of these receptors. These structures now go by a fitting eponym, "ampullae of Lorenzini" (Murray, 1962, 1974). In addition to torpedoes, ampullae of Lorenzini are present in rays and sharks, and somewhat similar formations can be found in siluroid fishes, including the electric catfishes.

There is also another type of electroreceptor, commonly called "tuberous" because of its morphology. Tuberous receptors are found in gymnotiform fishes—the class to which the electric eel belongs. These receptors can assume many forms and come in several physiological types.

Returning to our mystery, the solution to why some fishes have only weak electric organs started to emerge during the middle of the twentieth century. It began with some ethological observations on the behavior of some African fishes endowed with these organs (notably *Gymnarchus niloticus*). Hans Werner Lissmann and his collaborators in Cambridge, England, observed that such fishes often swim backwards, avoiding obstacles even in a turbid environment, and that they can locate prey at surprisingly long distances under conditions in which visual cues are of little utility (Lissmann, 1951, 1958; Lissmann & Machin, 1958). Their fishes were, moreover, capable of detecting metals and other conductors in the water, and also of reacting to the opening and closing of an electrical circuit. Further, they could distinguish between objects of constant shape and size, but with differences in electrical conductivity (e.g., porous earthenware pots filled with different concentrations of salt water). Remarkably, these weakly electric fishes could even do such things in total darkness.

Electrical recording with tools that were not available to early-twentieth-century researchers showed that the electric organs of these fishes produce a pattern of high-frequency discharges that varies with the experimental conditions. Notably, the electrical activity increases when they are put into the dark.[3] Lissmann further found that, with the recordings properly amplified and redirected to electrodes immersed in the water, these fishes could react to their own discharges. He was historically inaccurate, however, when he dated the "the first discussion of the possibility of the 'electrical sense' " to German anatomist Franz Boll in the 1870s (Lissmann, 1958, p. 182).

These mid-twentieth-century findings led to the inescapable conclusion that fishes with weak electric organs and very sensitive electroreceptors use electricity for two main purposes. One is to provide them with a means to scan their environments in a non-visual, non-tactile way, through a mechanism now denoted as "electrolocation." The other is for fish-to-fish communication. As put by Lissmann (1958, p. 188), "the electric pulses have a social significance." This includes schooling, attracting mates and courtship, and determining whether to stay, advance, or retreat in the presence of other fishes, based on size, shape, and other features. Both electrolocation and electrocommunication have since been studied in a wide variety of fishes under a myriad of conditions, and the number of studies continues to grow (see Bennett, 1970, 1971a; Kapoor & Khanna, 2004).

Two forms of electroreception have now been distinguished. The type described above, in which a fish emits electrical discharges that are used for scanning and detecting objects in the environment, is called "active electroreception." Some fishes, however, lack the organs for these discharges but can still echolocate by "passive electrolocation." They are able to detect only objects capable of producing their own electrical potentials (e.g., other fish contracting their muscles). Sharks, for example, use passive electrolocation. They have electroreceptors, but lacking special organs for generating electricity they cannot actively create electrical maps of the objects in their environments (Kalmijn, 1974, 1978, 1988).

The electrical potentials emitted by weak electric organs differ with the species, and some are of very high frequency.[4] They can also vary with gender, with other features of the individual, and with the presence of other fish generating similar pulses in the environment.[5] Fishes that emit long-duration waves of constant frequency with relatively simple quasi-sinusoidal forms are sometimes called "wave fishes" or "hummers." Those that emit intermittent pulses with short durations and variable frequencies (with more or less complex shapes) are often referred to as "pulse fishes" or "buzzers."

Many combinations and permutations of electric organs and electroreceptors exist in nature. Rather surprisingly, common torpedoes of the Mediterranean Sea (e.g., *Torpedo*

---

[3] The presence of these potentials, easily recordable with an amplifier connected to a loudspeaker, would prove useful in catching *Gymnarchus* and other weakly electric fishes in their murky natural environments.

[4] Some weakly electric fishes are capable of discharges of about 1,800 impulses per second (e.g., sternarchids, weakly electric fish with organs derived from nerves instead of muscles; see Pappas, Waxman, & Bennett, 1975).

[5] The "jamming avoidance reaction" is among the interesting phenomena that can be observed among fishes that emit weak electric discharges. When such a fish comes near another emitting signals with the same frequencies, it can modify its own discharge to avoid interference or jamming.

torpedo, *Torpedo marmorata*) lack active electrolocation mechanisms, although they possess powerful electric organs for stunning prey and warding off predators. In contrast, weak electric organs for electrocommunication and electrolocation can be found in some torpediniform fishes (e.g., *Narcine brasiliensis*) from the waters off Central and South America, as well as in electric catfishes found in African rivers. In the electric eel, there are two smaller electric organs in addition to the very large shocking organ. These "accessory organs" are called Hunter's organ and Sachs' organ, and they produce much weaker electrical discharges that could also be useful for actively perceiving the environment (Kapoor & Khanna, 2004; Bullock, 2005).

Astroscopus, better known as the "stargazer," merits special attention. It has electric organs but apparently lacks receptors for sensing electrical currents. Its organs are situated near its eyes and originate from its eye muscles. They produce shocks of few volts, which make them strong enough to be used for stunning small marine animals. On the basis of the intensity of their discharges, *Astroscopus* occupies an intermediate position between strongly and weakly electric fishes, like a missing link in the great Darwinian world of nature.

Reviewing all of these fascinating newer studies is well beyond the purview of this book. Because of the morphological and functional variations present in nature, not to mention the finer physiological and biophysical details, even a rather superficial survey would take a great many pages. There are many good reviews of this literature that cite the primary sources, and the interested reader is directed to these books and articles, (e.g., Fessard, 1958; Bennett, 1971a,b; Fessard & Bullock, 1974; Bullock & Heiligenberg, 1986; Moller, 1995; Kapoor & Khanna, 2004; Bullock, 2005).

For us, the important point is that there really are an astounding number of fishes with electric organs that differ morphologically and physiologically. As Darwin recognized well before there was an extensive literature on weakly electric fish, the fact that electric organs exist in even a few unrelated fishes shows how a successful adaptation might emerge along parallel lines. In his own words:

> I am inclined to believe that in nearly the same way as two men have sometimes independently hit on the very same invention, so natural selection, working for the good of each being and taking advantage of analogous variations, has sometimes modified in very nearly the same manner two parts in two organic beings, which owe little of their structure in common to inheritance from the same ancestor. (Darwin, 1859, pp. 193–194)

Darwin's thinking about electric fish lacking tight evolutionary connections was later embraced by Lissmann (1958, p. 188), who provided more details about how these organs could have evolved in stages. In his words: "The easiest explanation for the evolution of strong electric organs would appear to start from such muscular action potentials, and proceed via weak electric organs used for orientation, to the powerful offensive and defensive electric organs." These ideas about the powerful electric organs starting from muscles and becoming weak electric organs before evolving into more powerful ones are widely accepted today. But how and why did these changes come about in several distinct lines, including the torpedoes, which are cartilaginous fishes, unlike more recently evolved electric catfishes and the electric eel?

The modern explanation is based on the idea that, because of their great adaptive value in electrolocation and electrocommunication, weak electric organs appeared in a great many fishes. Afterwards, mutations capable of transforming these weak organs into stronger organs that could be used as offensive or defensive weapons occurred independently in the lines leading to the three types of fishes that have been on center stage in the drama we have been constructing. On the one hand, the appearance of the powerful organs led to the disappearance of the weaker ones (and consequently active electrolocation and electrocommunication) in common Mediterranean torpedoes. On the other, the weaker organs remained in the torporific catfishes and the eel, and also in some *Narcine*.

The emergence of powerful electric organs from weak ones that were likely used for communication and location illustrates how an organ adapted for one purpose could end up serving a totally different purpose. But it is also worth noting that the appearance of these organs probably occurred well after the emergence of electroreception. Not only is electroreception more common than active electrolocation, but this idea is also supported by the presence of electroreceptors in lampreys, a primitive form of aquatic animal considered an intermediate form or "link" between invertebrates and vertebrates.

Mentioning the existence of many fishes with weak electric organs and providing these evolutionary perspectives helps us to understand strongly electric fishes more fully. With this knowledge, we can now turn to the physiological mechanisms underlying their shocks and better appreciate the differences that exist among these strongly electric fishes, which are not closely related on the evolutionary tree. As we do this, we should keep in mind that the properties of the electrocytes as electrogenic systems in these strongly electric fishes are, nevertheless, less complex than they are in many weakly electric fishes (Kapoor & Khanna, 2004).

LUCAS AND ADRIAN

As mentioned previously, unequivocal evidence that the nervous signal is fundamentally an electrical event, as Galvani had supposed at the end of the eighteenth century, had to wait for the modern epoch of electrophysiology. It was finally achieved with the studies of Hodgkin and his collaborators in Cambridge, who elucidated the mechanisms underlying the generation and propagation of the nerve signal. The ground for this mid-twentieth-century development was, however, tilled by the work of other English physiologists, particularly Keith Lucas and his student Edgar Adrian, who was Hodgkin's professor at Cambridge. The newer physiological studies would, in turn, finally allow researchers to understand the shocking mechanisms of electric fishes.

The recognition of the all-or-nothing (or all-or-none) character or "law" of electrical excitation in nerve and skeletal muscle fibers was one of Lucas and Adrian's greatest achievements. As is true of other developments in the history of physiology, this accomplishment had a great deal to do with clever preparations and exceptional instrumentation skills, in addition to receptive minds (Lenoir, 1986; Frank, 1988, 1994).

The all-or-nothing phenomenon had already started to emerge in 1871 with Henry Pickering Bowditch's studies on heart muscle. Bowditch was an American who performed his experiments while working in Carl Ludwig's laboratory in Leipzig, Germany (Bowditch, 1871; see Cannon, 1922). He noticed that if an electrical stimulus applied directly to the isolated apex of frog heart were strong enough to produce a contraction, further increasing the intensity of the stimulus would not increase the strength of the contraction.

Bowditch's findings on heart were extended by French scientist Étienne-Jules Marey (1876), who recognized the refractory period of insensitivity to further stimulation that followed the full contraction of heart muscle. But for several years, the all-or-nothing phenomenon appeared to belong exclusively to heart muscles. This was because the strength of the contraction elicited in skeletal muscles seemed to increase in a relatively smooth way as experimenters increased the intensity of a stimulus applied either to the skeletal muscles or to their motor nerves.

Perceptions started to change in the opening years of the new century, when Francis Gotch (1902) suggested that this pattern of excitability might be an artifact reflecting an increase in the number of fibers excited, rather than an increase of the response amplitude in the excited fibers. Gotch's idea registered on Keith Lucas, and he set forth to discover whether skeletal muscles follow the same all-or-nothing law as do heart muscles (for more on Lucas, see Bayless, 1917–19) (Fig. 26.1).

Lucas (1905), who was at Cambridge, cut the dorsocutaneous muscle of the frog into small strips, each having 12 to 30 fibers. He then excited the muscle directly with a weak electrical stimulus and increased the strength of the stimulus. A staircase function emerged as the individual fibers in the muscle joined in the response, with the number of steps in the staircase never exceeding the number of muscle fibers in a strip. His results showed that a partial muscle shortening is due to some muscle fibers contracting fully while the remaining fibers do not respond at all.

Four years later, Lucas elicited muscle contractions with a stimulus applied to a motor nerve. Again the contractile response increased in discrete steps (Fig. 26.2). "In each muscle-fibre the contraction is always maximal regardless of the strength of the stimulus which excites the nerve-fibre," he wrote (Lucas, 1909, p. 133).

Lucas now hoped to determine whether nerve fibers also respond maximally or not at all, and he asked his student Edgar Douglas Adrian to work on the problem (Hodgkin, 1979; Finger, 2000; Piccolino, 2003c) (Fig. 26.3). Adrian's initial strategy was to isolate a long nerve from a frog and to place a segment of it in a chamber with "narcotizing" alcohol vapors. He then exposed the nerve to just enough alcohol to weaken the impulse, but not enough to stop it from making it through the block. Adrian reasoned that the weakened impulse should jump back to full strength once it comes out of the alcohol block, if the all-or-nothing conduction principle holds. To test this hypothesis, he measured how strong a second alcohol block has to be to stop the impulse further down the axon.

Adrian's (1912) findings showed that the nerve impulse immediately returns to full strength after it passed through the first block: the second alcohol block has to be just as strong to stop the previously "narcotized" impulse as it has to be to stop a nerve impulse that was never blocked. He also found that it did not matter if the nerve impulse had been triggered by a barely adequate stimulus or a very strong stimulus. So long as a stimulus could trigger an impulse, the action potential shot down the axon at full strength (Piccolino, 2003c; Piccolino & Bresadola, 2003). Although Japanese physiologist Gen-Ichi Kato criticized these studies on experimental grounds, Adrian's conclusions held and were of great importance for the future of electrophysiology (Piccolino, 2003c).

Figure 26.1. Keith Lucas (1879–1916), the Cambridge physiologist who asked whether the all-or-nothing principle, which he helped establish in skeletal muscles in the opening decade of the twentieth century, also occurs in nerves. Having a great talent for technology, he served as an expert in aerial navigation during World War I and died in 1916 in a military airplane accident. His seminal work on nerve conduction was published by his pupil Edgar Adrian, who further developed his research interests.

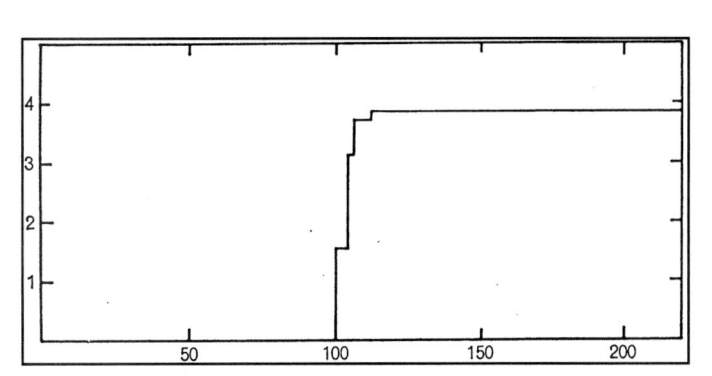

Figure 26.2. Lucas obtained a step-like function in muscles as he increased the strength of his stimulus, which in effect caused more individual muscle fibers to contract in an all-or-nothing manner. This figure, from Lucas' 1909 publication, plots the strength of the contraction in a thin frog muscle (ordinate) as a function of the intensity of the current applied to the motor nerve (abscissa). There is no muscle response if the current intensity does not reach a given "threshold" value. With stronger stimuli the contractions appear suddenly and increase in a step-like fashion when the current is augmented.

Figure 26.3. Edgar Adrian (1889–1977), Lucas' student, who took on the task of trying to determine whether nerves behave in the same all-or-nothing fashion as muscles. Adrian, who would share the 1932 Nobel Prize in Physiology or Medicine with Charles Sherrington, would go on to make monumental contributions in neurophysiology, from studying the all-or-nothing features of nerves themselves to exploring the nature of the electroencephalogram and identifying new cortical sensory areas. (© Trinity College, Cambridge)

Notably, Adrian (1912, 1914) showed that the amplitude and time course of the nerve signal depend upon the local conditions of the nerve where the recordings are made. Lucas (1917) also pointed to the local energy distributed along the nerve fiber as support for the progression of the nervous signal, denoted by him only as "propagated disturbance" when he resurrected the gunpowder track analogy invoked in Galvani's times (see Chapter 21) and also considered by Helmholtz in his electrophysiological studies (Meulders, 2001). Like Lucas, Adrian had initially avoided referring to the nerve signal as an electrical event. To put the reticence of these Cambridge physiologists in perspective, we have to remember that theories of nerve conduction emphasizing chemical mechanisms were still dominant early in the new century.

Adrian and his co-workers now extended the all-or-nothing phenomenon to sensory nerves by progressively dividing the sternocutaneous muscle of the frog until they had only one sense organ, which was then stimulated by stretching the muscle while recordings were made from a single sensory nerve fiber (Adrian & Zotterman, 1926) These studies, and those carried out in the following years (Adrian, 1932; Adrian & Moruzzi, 1939), helped to make it clear that an impulse of constant amplitude is the basic signal by which information is encoded and transmitted in the nervous system of vertebrates (today's frequency coding: (Figs. 26.4 and 26.5), even though there might be exceptions, particularly for some classes of small-size cells of the central nervous system, the so-called non-spiking neurons.

The research conducted by Adrian and others at this time also made it increasingly evident that electrical stimuli that might *not* generate a propagated disturbance ("sub-threshold" stimuli in modern terminology) are nonetheless capable of increasing the electrical excitability of a narrow segment of nerve near the point of application, at least for a limited time. Thus, a stimulus that is sub-threshold could produce a fully developed nerve signal if it occurs in temporal or spatial combination with other sub-threshold stimuli. These findings revealed something very important: that local responses might not follow the all-or-nothing law like the propagated responses.

Notwithstanding the obvious differences between local and propagated responses, the two classes of phenomena seemed to be related. First, the local response could develop into a fully propagated response if the stimulus is increased beyond the sub-threshold value. Moreover, if a nerve event is partially blocked by some treatment to a short segment of its axon, an increase in excitability could be detected in the region beyond the block upon the arrival of the propagated disturbance. These findings served to indicate that some event, similar to that induced by artificial local stimulation, accompanies the propagated signal—a finding in line with Ludimar Hermann's local circuit theory (Chapter 25). Under normal conditions, the leading wave of the propagated disturbance seemed to be able to increase the excitability of the nerve region ahead of it to such a degree that a new propagated disturbance could be generated there.

ALAN HODGKIN AND THE NERVE IMPULSE

The "modern" phase of research on the electrical phenomena of excitable cells began in Cambridge around 1934. As with Adrian, who had conducted his breakthrough study of the all-or-nothing event using nerve blocks when he was an undergraduate at Trinity College, Cambridge, Alan Hodgkin, his undergraduate assistant there, began by studying the effects of nerve blocks in frogs (see Hodgkin, 1992)[6] (Fig. 26.6). Hodgkin induced a local block by cooling a short segment of the sciatic nerve, and investigated the effects induced by electrically stimulating the proximal region of the nerve and using muscle contraction as a measure of nerve excitation. He positioned electrodes in different locations beyond the blocked region to test the consequences of a "conditioning" stimulus applied before the block that could produce a full-blown impulse. By stimulating the region down the axon soon after applying a conditioning stimulus, he found that a weak test electrical stimulus, incapable by itself of exciting a propagated impulse under normal conditions, could trigger a full-blown electric impulse, if applied right after an action potential had been produced proximal to the block.

Hodgkin correctly interpreted the phenomenon as an expression of a local increase of excitability, a finding

---

[6] As with his mentor, Hodgkin was just 23-years old when he began to publish the results of his first studies in the *Journal of Physiology*. This was in 1937, exactly two centuries after Galvani's birth (Hodgkin, 1937a,b).

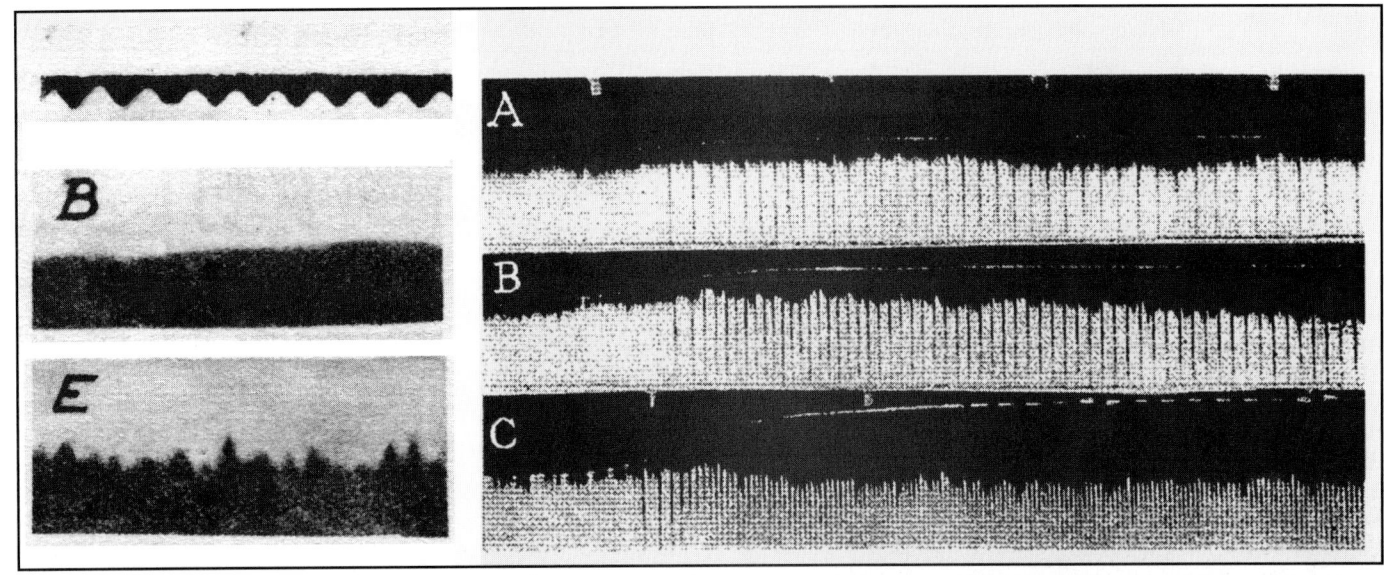

Figure 26.4. Left: The first recording of the electric activity of single nerve fibers as obtained in 1926 by Adrian in a sensory nerve with a sensitive amplifier (capillary electrometer). The fibers innervate mechanical receptors situated in the muscle, which measure the stretching of the muscle. B illustrates the control condition with the muscle in a relaxed state without any weight stretching it. In E a weight of 100 grams has been applied to the muscle. The resulting stretch is sufficient to evoke electric responses in the nerve fibers, visible as small bumps in the voltage recording (ordinate). Right: Unitary action potentials obtained with improved techniques (valve-tube amplifier) by Adrian in 1933 in a study of the sensory fibers coming from the insufflation receptors of the lung. Increasing lung volume (from A to C) resulted in an increase in the firing frequency with no substantial change of the impulse amplitude. The nerve signals appear as downward-going deflections of the black trace.

consistent with Hermann's local circuit theory. He also found that the increased excitability observed beyond the blocked region was accompanied by a measurable local electrical response, which had the characteristics of current spreading passively along the nerve fiber, similar to the spread accompanying the increase in excitability.

Besides further supporting Ludimar Hermann's local circuit theory, Hodgkin's findings helped to strengthen the case for a close relationship between signal propagation and subsequent electrical events in the nerve. The results of further research aimed at investigating how changes in the conductivity of the extracellular medium could affect conduction speed in crab and squid nerves provided more evidence for this relationship (Hodgkin, 1937a,b). More research was needed, however, to relate the sub-threshold excitatory responses to the full-blown nerve response, and to elucidate how the localized potential changes that accompany the leading edge of this response figure in the conduction process.

Figure 26.5. Left: The first single-unit recording of electric impulses produced by a neuron of the brain's cortex. The tracings were obtained by Adrian and Moruzzi in 1939 when recording from the fibers of the pyramidal tract of the cat. These fibers are the axons of the cells of the motor cortex. Their discharges were evoked in a reflex way by various forms of sensory stimulation. What also changes in this case is the frequency of the discharge, not the amplitude of the impulses. Right: Adrian with Giuseppe Moruzzi in Italy in 1954.

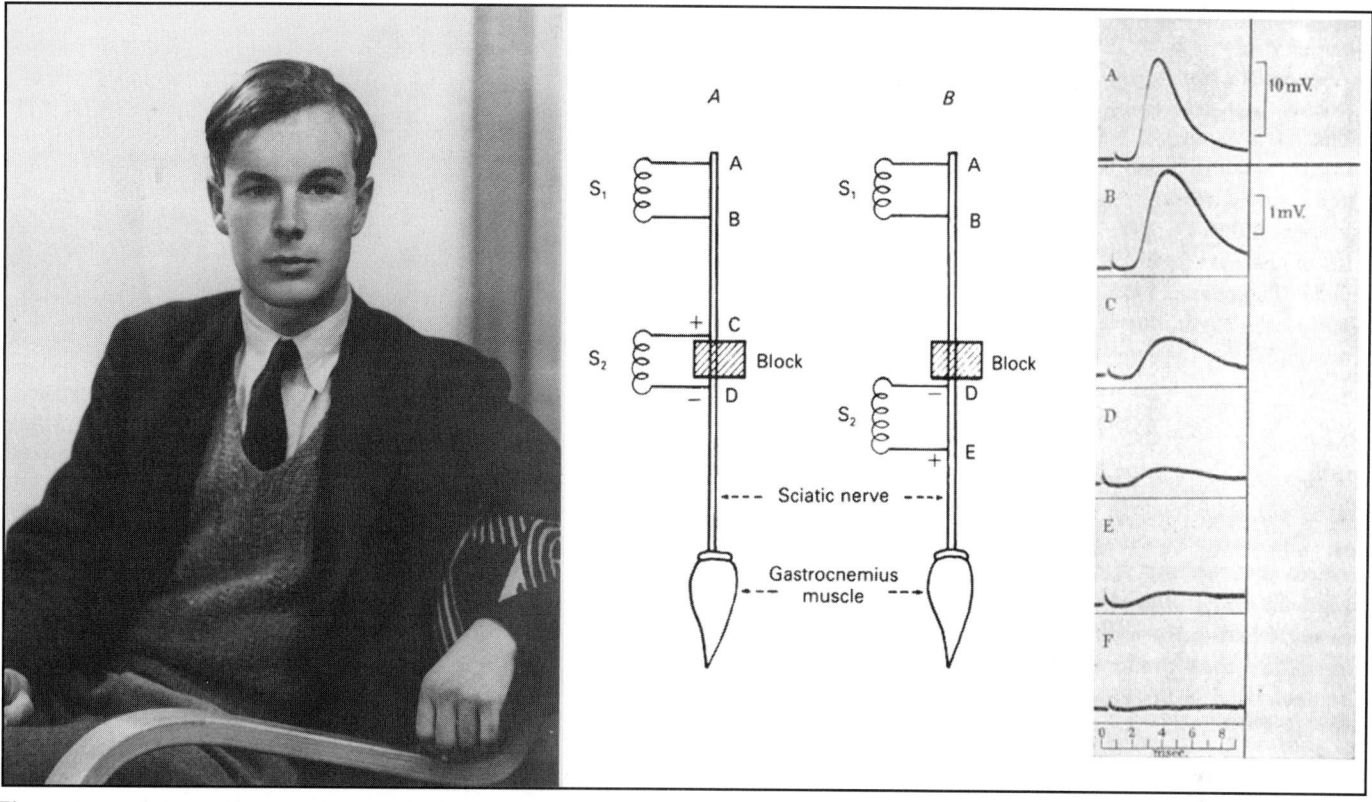

Figure 26.6. A young Alan Hodgkin (1914–1998) and his first experiments on nerve conduction with cold blocks. His findings on the right (A before the block and B-F at different distances beyond the block) showed sub-threshold increases in electrical excitability, consistent with Hermann's local circuit theory (from Hodgkin, 1937b).

This next step followed English zoologist John Zachary Young's rediscovery of the squid's giant axon in the 1930s (Young, 1936a-c). Because of its size, the giant axon is ideal for electrophysiological recordings. In 1939, two Americans, Kenneth Cole and Howard Curtis, using electrodes positioned near the outer face of the cell membrane, provided clear evidence that the squid's action potential is accompanied by a decrease in membrane resistance (Cole & Curtis, 1939). Afterwards Hodgkin and Huxley on one side of the Atlantic, and Cole and Curtis on the other, succeeded in inserting a recording electrode inside the axon. This allowed them to record the transmembrane potential directly, both in the resting state and during excitation (Hodgkin & Huxley, 1939; Cole & Curtis, 1940) (Fig. 26.7).

The existence of the resting membrane polarization, which had been postulated by Bernstein, was confirmed by these studies (the interior of the cell being measured as about 50 mV negative with respect to the exterior). It appeared, however, that the potential overshoots the zero level by several tens of millivolts during excitation. This finding was in line with some observations initially made by Bernstein, namely that the amplitude of the action current might sometimes exceed that of the injury current, although Bernstein thought this might be an artifact (see Chapter 25).

The interpretation of this unexpected discovery was postponed because of World War II. Afterwards, in 1949, Hodgkin and Bernard Katz succeeded in showing that the action potential of the squid nerve decreases in amplitude when the extracellular concentration of sodium ions is reduced (Hodgkin & Katz, 1949) (Figs. 26.8 and 26.9). These experiments strongly suggested that a selective increase in membrane permeability for positively-charged sodium ions from the extracellular fluid accompanies the action potential. The change between the resting state, in which the membrane is semipermeable for potassium, and that of the action potential, with its accompanying influx of sodium, now appeared to be a consequence of

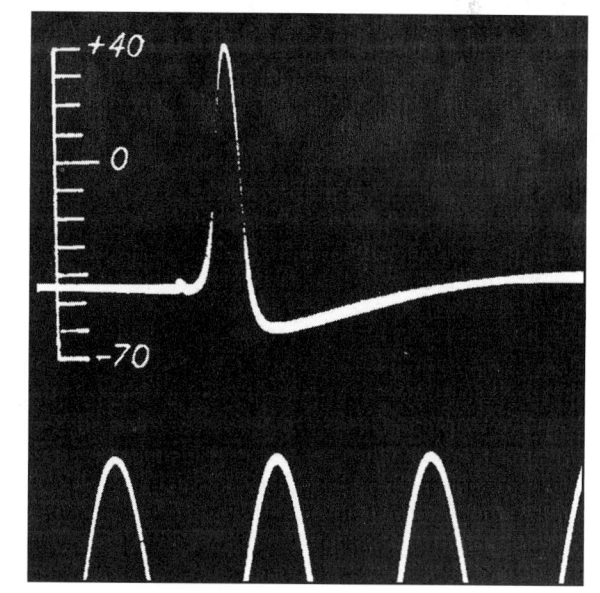

Figure 26.7. Hodgkin and Huxley's (1939) recording of an action potential with an electrode in a squid's giant axon. These researchers were able to record a negative resting membrane potential, and further observed that the potential overshoots the zero level and goes well into the positive range during excitation.

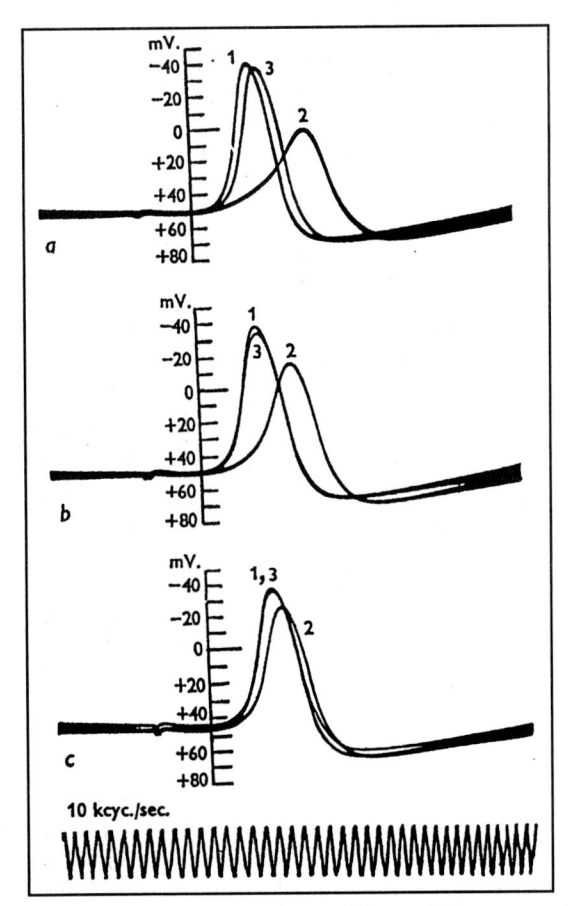

Figure 26.8. In 1949 Alan Hodgkin and Bernard Katz reported that the squid's action potential decreases in amplitude if the extracellular concentration of sodium is reduced. This finding, backed by recordings like this one, suggested that an influx of sodium is fundamental to the nerve event.

Figure 26.9. Bernard Katz (1911–2003), the German refugee who worked in England and Australia on physico-chemical mechanisms underlying neuromuscular events. Katz, a biophysicist by training, conducted experiments with Alan Hodgkin showing sodium's importance in nerve electrophysiology, and he shared the 1970 Nobel Prize in Physiology or Medicine for his own work on neurotransmitter release at the synapse.

specific changes in the cell membrane induced by the electrical stimulus.

The theory that the membrane becomes selectively permeable to sodium at the peak of the action potential called for a revision of Bernstein's theory. This is because Bernstein assumed that there would be a nonselective passage of all ions during excitation that would result in an almost complete disappearance of the negative potential. To find out how well the newly revised theory could account for nerve excitation and conduction, it was now necessary to study those changes in membrane currents and permeability that accompany the appearance of action potentials in more precise ways.

Research of this kind was difficult with conventional techniques, because the regenerative and explosive properties of the action potential precluded an analysis of the relationship between the membrane current and the membrane voltage during the excitation process. It was necessary to circumvent the explosive character of this event by forcing the membrane potential to assume and hold certain levels, despite its proclivity to change when its threshold value is exceeded.

The sought-after procedure that made the study of membrane events underlying the generation of action potentials possible is called the voltage-clamp technique. Kenneth Cole and George Heinemann Marmont came forth with it independently in 1949 (Cole, 1949; Marmont, 1949). In the epoch of Lucas and Adrian, the explosive character of the propagated response had been somewhat subdued by using procedures based on partial blocks of small segments of nerve, and their procedures could give little information about what would happen once the threshold was exceeded. With the voltage clamp, the block of the electrical events underlying the explosive character of the action potential is not based on the application of narcotic agents or temperature changes, but instead relies on the use of a feedback circuit capable of setting the potential at any desired value.

Hodgkin and Huxley, in collaboration with Bernard Katz, used the voltage clamp in a masterful way in a series of studies on the squid's giant axon that they published in 1952 (Hodgkin & Huxley, 1952a-d) (Figs. 26.10 and 26.11). They provided the final, unequivocal evidence that nerve conduction is fundamentally an electrical event, as Galvani had maintained a century and a half earlier, albeit with indirect evidence and a theory weak in its finer details (Chapters 22 and 23). As Hodgkin formulated it, the action potential is not just an electrical sign of the impulse—it is the causal agent in its propagation.

Following the landmark studies of Hodgkin and Huxley, it was no longer possible to view the electric flow across the membrane after the application of an external electric stimulus as a purely passive physical consequence of the stimulus. On the contrary, it now appeared to be an active process that depends on a particular form of electrical energy. This energy accumulates between the interior and the exterior of the nerve fiber as a consequence of physiological processes clearly

Understanding the Shock Mechanisms 403

Figure 26.10. Left: Alan Hodgkin, one of the towering figures of mid-twentieth-century neurophysiology and biophysics. Hodgkin's research with Andrew Huxley led to a much better appreciation of the dynamics of the resting potential and the nerve impulse, and ideally concluded the research path initiated by Galvani with his frog experiments in the second half of the eighteenth century. Right: Some pieces of the "voltage-clamp" apparatus used by Hodgkin and Huxley in their studies.

belonging to the domain of life—true animal electricity with features confirming Galvani's fundamental intuition.

Moreover, nervous signaling now appeared to be a genuine electrical phenomenon for another important reason. Electricity not only represented the source of energy used in producing the nervous signal, but it also appeared to be the causal link needed to release this energy during signal propagation. The electricity stored between the interior and the exterior of the membrane cannot flow under normal conditions, because of the relative impermeability of membrane to certain ion movements. To change this condition, and thus to make the membrane electrically conductive, the membrane potential must undergo a modification (the interior potential moving in a positive direction) under the influence of an *electrical stimulus*. This stimulus represents electricity's second type of causal involvement in nervous signaling, working as an integrated mechanism that is like an electrically operated relay for controlling the electrical energy discharge.

Figure 26.11. A picture taken in Ferrara, Italy, in 2002 at a celebration of the 50th anniversary of the Hodgkin–Huxley experiments. Andrew Huxley (b. 1917) is at the center of the figure with, at his right, his wife, Lady Jocelyn Richenda Pease. Huxley shared the 1963 Nobel Prize in Physiology or Medicine with Alan Hodgkin and John Eccles. He collaborated with Hodgkin to study the roles played by sodium and other ions in excitation and signal conduction in nerve fibers. Besides contributing to nerve signaling investigations, Huxley conducted important research elucidating the molecular mechanisms underlying muscle contractions.

The membrane modification involved in this mechanism consists of an increase in the permeability to sodium ions, which allows for the entrance of these ions in accord with the electrochemical gradient, through a process that would be later denoted as gating. Hodgkin and Huxley assumed the increase in permeability induced by depolarizing stimuli results from an electrical influence affecting the charged particles present in the membrane. This regulates the membrane capability and allows for the selective passage of sodium ions via a mechanism denoted as activation.

The voltage-dependent, ion-permeability changes involved in the nervous impulse link electricity to this event in a fundamental way. An apparently paradoxical consequence of this mechanism is that positive charges (brought about by sodium ions) tend to enter across the membrane when the internal potential is displaced toward positivity, a maneuver that on a purely physical basis should favor the *exit* of positive charges. With this process linking electricity to the nerve signal, hypotheses that consider electrical changes as epiphenomena or secondary phenomena of nerve excitation become highly unlikely.

A basic consequence of the gating properties of the membrane permeability for sodium is that the entry of sodium ions, brought about through the influence of depolarization on the membrane, results in a net increase of the positive charge influx, and therefore in a further depolarization, which in turn causes an additional increase in sodium permeability. This positive feedback cycle controlling the entry of sodium ions activated by membrane depolarization, later denoted as the "Hodgkin cycle," is at the heart of the mechanism of nervous signal generation and accounts for the twofold electrical nature of this signal: a signal generated by the release of an intrinsic electrical energy, yet one that requires an electrical trigger to be released (Fig. 26.12).

The model of the nerve signal generation proposed by Hodgkin and Huxley on the basis of the activation characteristics and the kinetic properties of the membrane currents identified in their studies accounts for most of the properties of the excitation and conduction processes along the nerve fiber. In this model, the equilibrium potentials of the various membrane currents correspond to the values of the electrical membrane potential, in which the tendency of the ions to move, as a consequence of their transmembrane concentration gradients, is counteracted by the electrical field. These equilibrium potentials for the ionic mechanisms of the membrane can be modeled by three electrochemical batteries (i.e., by three electric batteries, depending on an asymmetric distribution of ionic charges). Being distributed along the axon fiber, these batteries provide the local energy for the propagation of the nervous impulse—an energy that corresponds to the gunpowder in the analogy which had been proposed by Fontana in the eighteenth century (Chapters 21 and was revived different forms in many texts of later epochs (Helmholtz, 1850b; Bernstein, 1894, p. 17; see Lucas above).

The electrical behavior of the axon membrane during the generation and conduction of the action potential is accurately described by the Hodgkin–Huxley model in all its basic aspects. Among other properties, the model accounts for the all-or-nothing character of the nervous impulse, with a threshold and a refractory period. It describes the time course of both the local and propagated electrical signals, and the accompanying conductance changes, in a satisfactory way. Moreover, the temperature dependence of the processes of excitation and conduction can be explained as a result of the strong influences that temperature exerts on the gating mechanisms controlling the ion currents involved in action potential generation.

Most of the basic predictions of the Hodgkin–Huxley model have been confirmed in subsequent studies. It was shown that the early inward and the delayed outward current, carried by sodium and potassium ions, respectively, depends on two specific, independent membrane mechanisms, because they can be blocked by different pharmacological agents (e.g., the sodium current by the puffer fish poison tetrodotoxin, and the delayed potassium current by tetraethylammonium; see Hille, 2001). Moreover, as assumed by Hodgkin and Huxley, the ion currents involved in the electrical excitation of nerve and muscle membranes are due to fluxes of ions along their electrochemical gradients, and active metabolic phenomena ("ion pumps") do not intervene directly in producing the currents. Metabolic processes intervene only indirectly in electrical membrane phenomena, because they serve to establish and maintain the ion gradients, from which the electrical potentials at rest and during excitation originate (Fig. 26.13).

The questions left largely unresolved by the Hodgkin–Huxley studies concerned the mechanisms of ion permeation through the membrane. Researchers oscillated between a diffusion process through a pore-type device and a permeation mechanism involving a carrier molecule. Later studies have clarified the mechanisms of ion permeation involved in membrane electrical excitability with a molecular resolution, and this is probably the most important achievement of contemporary membrane electrophysiology.

The success of these studies was made possible largely by the invention of a new electrophysiological technique, the "patch clamp," in 1976. Erwin Neher and Bert Sakmann developed the patch-clamp technique, which allows researchers to record the fundamental electrical event of the membrane (Fig. 26.14); that is, current passing through a single molecule of the permeation mechanism present in the membrane structure (Neher & Sakmann, 1976). Patch-clamp studies have provided definite evidence that membrane currents involved in the electrical behavior of cells are carried by ions passing through protein structures embedded in the membrane. These structures, indicated as ion channels, provide paths that ions that are hydrophilic (i.e., that attract

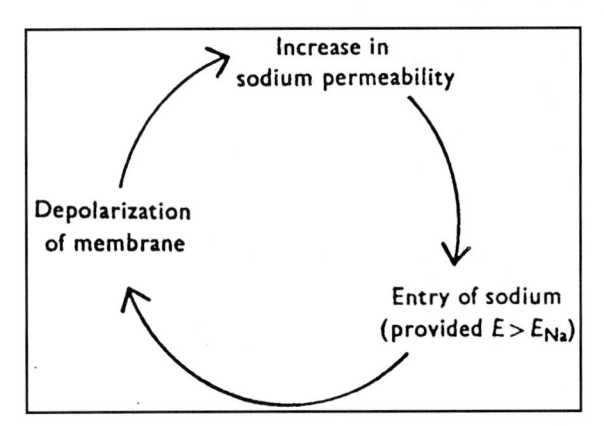

Figure 26.12. A scheme of the positive feedback cycle controlling the entry of sodium ions activated by membrane depolarization. These interrelated events are sometimes referred to as the "Hodgkin cycle."

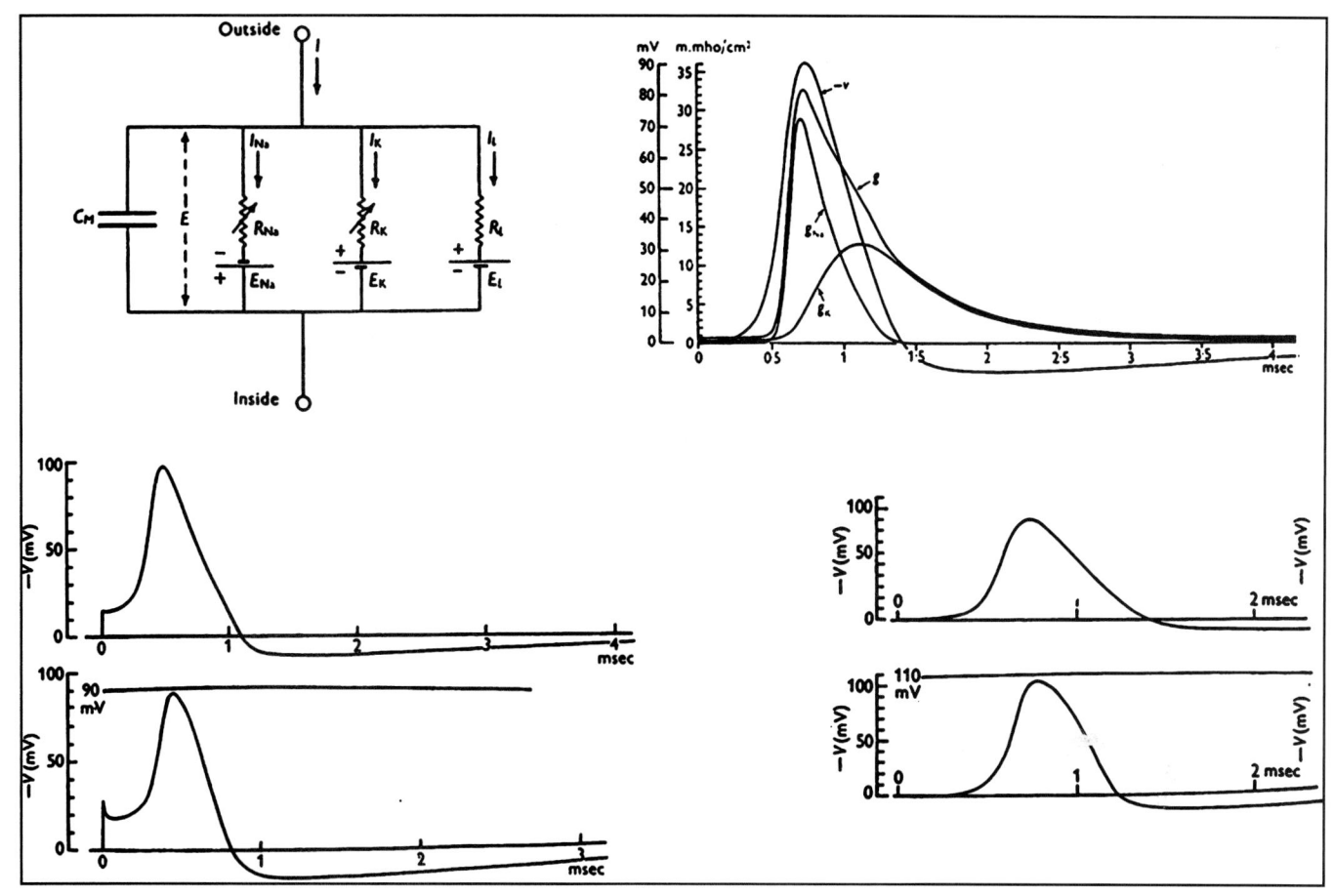

Figure 26.13. A diagram and computed curves aimed at illustrating the behavior of the Hodgkin–Huxley model of nerve conduction, in comparison with the actual recordings from the squid axon, as presented by these researchers in 1952. With these studies, "animal electricity" took on a new, more dynamic meaning. The comparison of the actual recordings with the predictions of the model showed a more-than-satisfactory correspondence.

water) can pass through, despite the hydrophobic properties of the lipid constituents of the membrane (Sakmann & Neher, 1995; Hille, 2001).

The research path initiated by Neher and Sakmann with their extraordinary technique and their first recording of single channel current represents in an ideal way the most modern phase of the studies initiated by Galvani in 1780. No wonder, therefore, that Sakmann was one of the scientific personalities commemorating Galvani in 1998, on the occasion of the bicentennial of his death (Fig. 26.15).

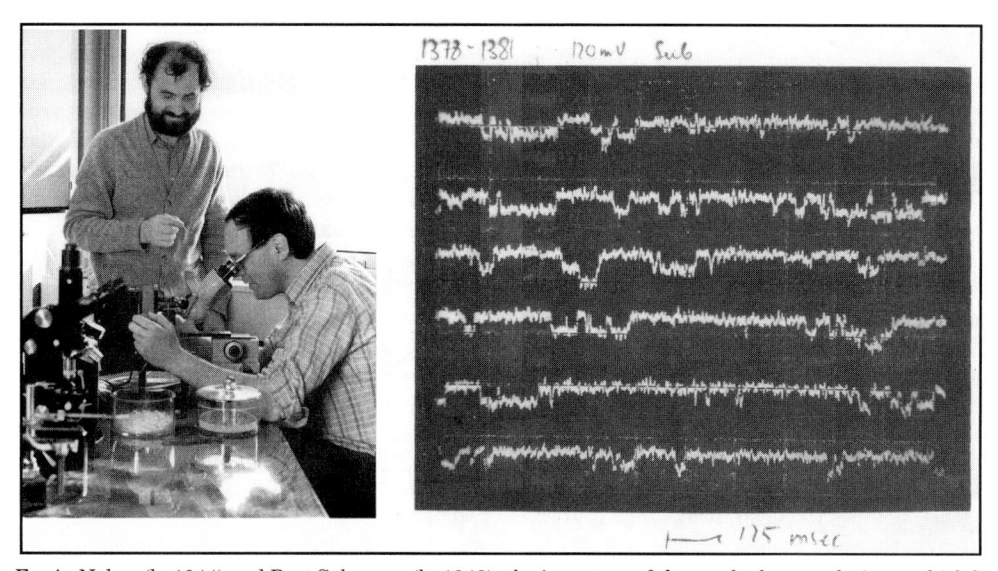

Figure 26.14. Left: Erwin Neher (b. 1944) and Bert Sakmann (b. 1942), the inventors of the patch-clamp technique, which has allowed for the measurement of the current passing through a single ionic channel of the plasma membrane. Right: This originally unpublished figure shows the tracings obtained by the two German biophysicists on November 8, 1975, and was considered by them as their first clear evidence of single-channel currents (by courtesy of Erwin Neher; left picture, © MPG-Pressebild/Lüthje).

Figure 26.15. Bert Sakmann (at the center) visiting the *Istituto delle Scienze* of Bologna in 1998, during the celebrations of the bicentennial of Galvani's death. This was one of the places where Galvani reported his experiments on animal electricity. The inscription on the left celebrates Luigi Ferdinando Marsili, the founder of the *Istituto* (courtesy of Masaki Sakai of the Okayama University, Japan).

ELECTRIFYING THE CELLS OF THE ELECTRIC ORGANS

As mentioned in our discussion of Galvani's work and that of subsequent nineteenth-century scientists, including du Bois-Reymond, there had long been a firm belief that the mechanisms underlying the production of electricity in fish electric organs would prove to be the same as those for more ordinary tissues. In this unifying sense, people, frogs, barnyard animals and humans that do not shock must have nerves and muscles that share some features with torpedoes, the electric eel, and electric catfishes, all of which can deliver perceptible and sometimes very strong shocks.

Most nerve and muscle cells are capable of discharging impulses of about a tenth of volt. If we consider that there are about 10 trillion nerve cells in the human body, this would mean that, if we could sum their individual signals, we could produce a shock that would measure billions of volts, without even including the potentials from our muscle fibers and other electrically excitable cells that are found in our bodies!

Of course, we are not electric animals in the same sense as electric fishes. Even though compound electrical potentials can be recorded from the surfaces of our bodies with electroencephalographs, electrocardiographs and other sensitive devices, these potentials are of very small intensity (less than a millivolt in most cases). The main reason why we and most other animals are not capable of producing large electric potentials has much to do with symmetry.

Let us consider the problem in more detail by trying to imagine the difficulties that we would encounter if we were trying to assemble an electric organ using ordinary electrically excitable cells. Since electric organs are generally derived from muscle, we can assume that this would amount to trying to force an ensemble of muscle cells to sum their tiny individual potentials so as to produce a formidable bolt like one of our fishes can do. The first step might be to change the shape of the cells—making them flat and wide, instead of long and thin like normal nerve and muscle cells. This would allow the cells to be stacked one above the other, so they could add their electric signals. Let's call the cells we modify in this way our electrocytes.

Suppose too that we want to limit the lateral dispersion of the current generated by each electrocyte so there will be more effective summation. Again, let us consider a number of our flattened cells stacked in a vertical column and focus, say, on the first one from the bottom. At rest, its internal potential is about –90 mV, as is typical for muscle cells with respect to the extracellular space below and above the cell. When the organ is excited, the internal potential of this particular cell and all the other cells in the column becomes positive with respect to the exterior by about 50 mV. This means that the overall change or signal generated will be about 140 mV.

But potential changes occur in both halves of the membrane, the bottom and the top half. Consequently, the interior of this particular cell will become positive with respect both to the extracellular spaces above and below it, which will remain at the zero level. Even in the absence of any lateral dispersion of the current, there would be no effective summation of the potential generated by a single cell. For an electrode that penetrates the membrane upward from below, any step towards positivity that might be encountered when it enters the inner compartment of our electrocyte would be offset by an opposite-sign change when it exits from the cell's interior by crossing the upper membrane leaflet. Therefore, there will not be a significant advance in the production of an electric shock if we cross an entire electrocyte thus constituted.

There are other ways to illustrate why the expected summation of the potentials generated by a stacked column of excited electrocytes is bound to fail. Let's suppose for a moment that the contrary is true, and that with sufficient excitation the entire organ becomes polarized—its innervated side developing a large negative potential with respect to the opposite side. This condition could be brought about by an electrical configuration in which the intracellular compartment of every electrocyte is positive relative to the two extracellular compartments, namely the one below and the one above it. This being the case, one should also expect that the opposite configuration (i.e., the intracellular compartment being negative to the extracellular one) should lead to a shock of opposite sign in the entire organ. Some reflection will reveal that this is the configuration present in the electric organ in the resting situation. This means that the electric organ would be continuously in a condition to produce an opposite-polarity shock—that is, an endless electrical discharge in the resting condition. This would be an absurd situation, because such a mechanism would soon exhaust the animal.

Another way to illustrate the ineffectiveness of the electric summation in stacked columns of flat excitable cells is to consider how such cells would resemble two small Voltaic batteries connected in series, but with the positive pole of one connected to the positive pole of the other. With such an arrangement, no current would pass through an external circuit, if we connect wires or even our hands to the free poles of the batteries, since both would have similar negative potentials. Moreover, if we connect an ensemble of these double inverted batteries in series, there would be no summation.

In short, the impossibility of summing the potentials of a stacked grouping of flat electrocytes is due to the symmetry of the potential changes generated at the two membranes of every electrocyte. It is analogous to what Volta encountered while he was trying to sum the potentials of zinc and copper couples by simply stacking many similar couples one above the other (Chapter 22). In Volta's case, the electric effect occurring at bottom face of a zinc disk in contact with a copper disk was annulled by the similar but spatially symmetrical contact of the upper face of the zinc disk with the copper just above it. At least, this is what happened before Volta decided to interpose humid disks between his metallic couples.

Let us consider a comparable step in our efforts to summate the discharges of the electrocytes in the column. Since we have to overcome the symmetry, we might, for instance, limit the excitation process to a single side of the electrocyte membrane. To this end, we can keep the lower leaflet excitable while removing or somehow inactivating the machinery (ionic channels) underlying the electrical excitability of the upper leaflet. We should, moreover, make the upper leaflet strongly electrically conductive, so the internal potential change induced by the excitation of our electrocyte could be transmitted with little attenuation from the internal compartment to the extracellular compartment just above it. Ideally, the upper membrane would provide only a minimal barrier to the current flow, thus making the extracellular potential just above our given electrocyte practically equal to the intracellular potential of that electrocyte.

To do this, we can endow the upper membrane with a great density of channels permeable to all ions. Here we can choose between two possibilities: either we can keep these channels permanently open or they can open (or increase their openings) when the internal potential turns positive, as happens during the excitation process. To make for more effective electric conduction from the intracellular compartment of every electrocyte to the extracellular space above it, we can also increase the upper membrane surface.

Following the plan outlined above, there would be no need to innervate the upper leaflet of the electrocyte membrane, because this leaflet does not have to be activated by the incoming nerve signals controlling the organs discharge. This will amount to a greater asymmetry between the two membranes of the electrocyte—the lower one being innervated and capable of undergoing electrical excitation, and thus producing a positive potential in the intracellular compartment, and the upper membrane being non-innervated and endowed only with ionic channels capable of passively conducting electric currents as a consequence of the intracellular potential changes brought about by the activity of the inner membrane.

PACINI'S OBSERVATIONS

The morphological and functional schema presented above follows the basic principles to which the real organs of electric fishes conform, even though the individual organs show considerable diversity, reflecting their evolutionary roots. This morphological asymmetry in the constitutive elements of the electrical organs had, in fact, been clearly recognized near the middle of the nineteenth century.

Italian histologist Filippo Pacini, who is better known to pathologists for his discovery of *Vibrio cholerae*, the bacterium responsible for cholera, and to histologists and neuroscientists for his description of the onion-shaped sensory end-organ bearing his name (Pacinian corpuscle), investigated the organs of the three strongly electric fishes with his microscopes (Pacini, 1846, 1852). He found that nerves innervate only one of the two surfaces of the cells making up their organs. This innervation is on the bottom surface of the cells in a torpedo's columnar organ and on the tail side of the cells aligned in rows in both the electric eel and the catfish organs.

Pacini also showed that the membrane opposite the innervated side, while devoid of nerve terminations, is characterized by numerous infoldings. As a consequence of these infoldings, the surface of the upper membrane is greatly augmented. As noted in the construction of our model, increasing the surface of the upper membrane creates a condition that favors the passive spread of electric current, and thus reduces the voltage drop across the membrane.

At the time of Pacini's studies, the polarity of the shock had been established only for torpedoes and the electric eel (Chapter 24). It was thought that the organs become negative at the innervated side and positive at the opposite side in these fishes. On the basis of these observations, Pacini assumed that the innervated side becomes negative at the moment of the shock compared to the other side, a reasonable hypothesis that was soon called "Pacini's law." When, following the studies of Ranzi and du Bois-Reymond, it was discovered that the electric organs of *Malapterurus* become negative on the tail side, the electric catfish (only one species was then recognized) emerged as an exception to Pacini's so-called law.

BERNSTEIN'S MODEL OF THE DISCHARGE

Although Pacini speculated about the mechanisms involved in the functioning of the electric organs, he did not propose a real model for the production of the shock. In 1912, Julius Bernstein realized the importance of the asymmetry in the genesis of the electric organ discharge and provided a model that also drew on his membrane theory of muscle and nerve excitability (Chapter 25). In so doing, he recognized the impossibility of summation of the individual potentials of the electric cells, if these cells strictly followed the behavior of ordinary excitable cells, and particularly if both leaflets of their plasmatic membranes underwent the same excitation process.

In developing his conception, Bernstein assumed that the elements of the electric organ are derived from excitable cells, similar in particular to muscle cells. This assumption fit well with previous studies on the development of torpedo and eel electric organs (Babuchin, 1876; Fritsch, 1883). Another important source of support for this assumption came from electrical measurements performed on torpedoes by German physiologist Karl (or Carl) Schönlein (1896), who worked at the Zoological Station of Naples late in the nineteenth century.

By using a variation of the rheotome method, initially contrived by Bernstein, Schönlein was able to measure the intensity of a fish's shocks with considerable accuracy. By dividing the average value found (about 34 Volts in modern electrical units) by the average number of the elements in the columns of the organs, Schönlein provided an accurate estimate of the potential discharged by a single element of the electric organ. This turned out to correspond closely with the discharge of a single muscle fiber (initially measured by Bernstein, see Chapter 25). In introducing his own theory of the electric organ discharge, Bernstein praised Schönlein's measurements, saying they represented "a remarkable fact, which is important for the theory" (Bernstein, 1912, p. 115). Three years after Schönlein's torpedo studies, Gotch and Burch reached similar conclusions for *Malapterurus* using a capillary electrometer (Gotch & Burch, 1896; Gotch, 1899). They found that the discharge of a single "disk" of the fish's organ was in the range of 40 to 50 mV, and thus similar to that of nerves.

Bernstein proposed his asymmetry theory of the discharges during a period in which the great interest in studying these fishes had been fading away, and when electrophysiological researchers were mainly concerned with nerve and muscle mechanisms. As acutely pointed out by Harry Grundfest (1957), who conducted many important studies on electric fish in the middle of the twentieth century, this relative neglect might have been due to the invention of more sensitive electrometers (e.g., capillary electrometers, spring galvanometers). With these instruments, it was relatively easy to detect weak electrical phenomena associated with the functioning of ordinary excitable cells, making the need for research on strongly electric fish less important for physiologists. As for Bernstein, however, a few years before putting forward his asymmetry theory, he had been studying the thermodynamics of the torpedo's shock, hoping to verify that the biological battery responsible for the shocks is similar to that he assumed to be responsible for muscle and nerve excitation (i.e., a "concentration battery," in which the energy for the production of electricity comes exclusively from ion diffusion processes; Bernstein & Tschermak, 1906; see Chapter 25).

Bernstein (1912) hypothesized that a symmetrical condition is present only in the resting state, with both membrane leaflets of the flat cells making up electric organs being electrically polarized. The inner compartment would be negative relative to the external compartment because of selective membrane permeability to a specific class of ions. He assumed that upon the arrival of the nervous influx exclusively to the innervated site, only the innervated leaflet would become excited (the membrane becoming permeable to all ions). This would cause the negative intracellular potential to move in a positive direction. Because of the persisting resting condition of the opposite site, the change in the internal potential would produce a current that would not be negated by a current of opposite sign in the other side. This would allow the potentials generated by the individual cells to summate, with the entire electric organ becoming negative on the innervated side compared to the opposite side.

In Bernstein's view, the functional asymmetry had to be a direct consequence of the morphological asymmetry. An implicit assumption needed to account for the effectiveness of his model is that, unlike nerve and muscle cells, the excitation wave would remain circumscribed at its origin, without propagating like the negative *Schwankung*. To meet this requirement, Bernstein (1912) considered two possibilities: either the rim separating the two faces must be incapable of conducting the excitation from the innervated to the opposite face, or the non-innervated face must be intrinsically unexcitable.

When considering the historical path that led Bernstein to formulate his theory for the shock, an important point must be made. There is little doubt that beside his great theoretical acumen, this crucial achievement depended heavily on his thinking about the various types of electric batteries that had been developed by nineteenth-century physicists (Bernstein, 1902, 1912; Bernstein & Tschermak, 1906). One of the main principles guiding the assemblage of a battery from a combination of metals, solutions, and other types of electrical conductors had been that the chain of the connected materials must not be symmetrical. Any symmetrical combination would prove to be ineffective in producing a current capable of circulating in the external circuit.

The principle of asymmetrical functioning assumed by Bernstein applies to all electrical organs, both in strongly and weakly electric fishes. Nevertheless, the actual mechanisms are different in the various electric fishes, and they only partially conform to his model, which was very innovative and extremely advanced for the time. Among the strongly electric fishes, the main problem concerned *Malapterurus*, which, as noted, does not obey Pacini's law and thus seemed to violate one of Bernstein's main predictions—namely that the organ will become negative on the innervated side compared to the opposite side. Modern studies have now allowed researchers to account for the apparent paradox of *Malapterurus'* organ, as we shall see later in this chapter. But first, let us look at how Bernstein's principle of electrocyte asymmetry was first confirmed with experiments on electric eels in the 1950s.

DEMONSTRATING THE FUNCTIONAL ASYMMETRY

The discovery of the cellular mechanisms involved in the electric eel's discharge, and specifically the asymmetrical properties of the electrocytes, brought two research teams together. One was based in Rio de Janeiro, where Carlos Chagas, who had a long-term commitment to studying electric eels, had established his physiological laboratory. The other was in Cambridge, where Hodgkin, Huxley, and their collaborators

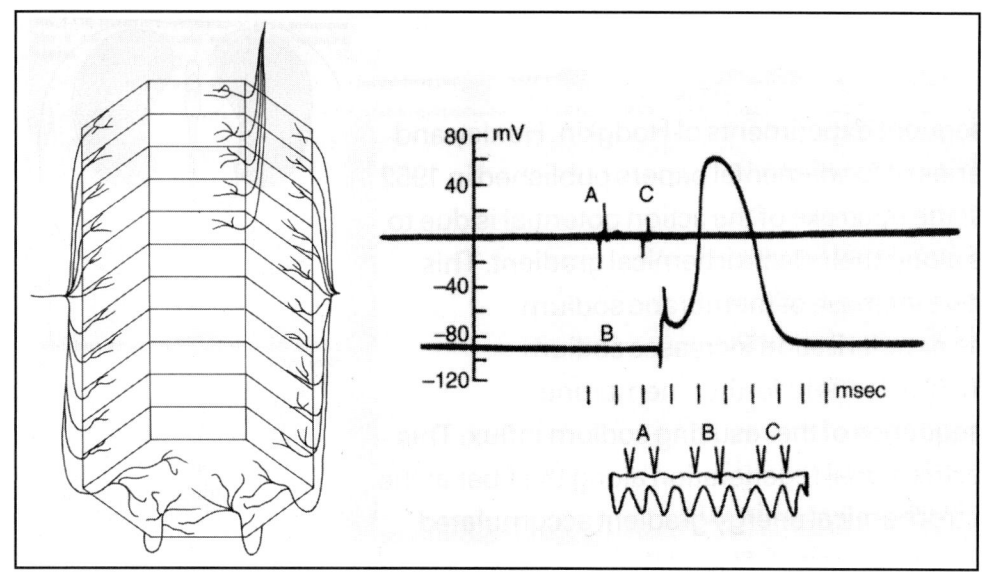

Figure 26.16. A diagram showing the asymmetrical innervation of the electrocytes of fish electric organs, and the results of some experiments by Richard Keynes and His Martins-Ferreira (1953) on the eel's electrocytes. These researchers showed that the opposite sides of the electrocyte behave differently when stimulated, in accord with Bernstein's idea that they must be functionally asymmetrical. They also showed that the internal electrical potential overshoots zero and goes into positive territory when the electrocyte is activated, supporting the idea that the electrocyte membrane, like nerve and muscle cells, is selectively permeable to sodium during excitation.

had just succeeded in elucidating the mechanisms of nerve excitability.[7] The electrophysiological studies of the electric eel organs were carried out in Rio de Janeiro by one of Hodgkin's collaborators, Richard Darwin Keynes (Charles Darwin's great-grandson), and by His Martins-Ferreira, a new member of Chagas' laboratory (Keynes & Martins-Ferreira, 1953).[8]

Keynes and Martins-Ferreira aspired to test the basic predictions of Bernstein's electric fish model with electrophysiological techniques. In particular, they set forth to determine the behavior of the two opposite membranes of an electrocyte during the excitation process. Small sections of the electric organs containing a limited number of electrocytes were therefore isolated and placed vertically on a recording stage with the innervated (tail) side at the top. By carefully inserting two glass microelectrodes, they found it possible to record the potentials on the two sides of an electrocyte's membrane. Those electrocytes came from the Sachs' organ, because these electrocytes are separated by wide cleft (about 2 mm), whereas they are more densely packed in the large electrical organ. Excitation was achieved by passing a current through extracellular microelectrodes across the entire preparation.

During the resting condition, they found an intracellular potential of about –85 mV. Following excitation, when one electrode was in the intracellular compartment and the other was outside the electrocyte on the innervated side, the intracellular potential became positive (approximately +50 mV). In contrast, no such change occurred when one electrode was in the intracellular compartment while the other was situated outside the electrocyte on the non-innervated side (Fig. 26.16). Attempts to measure the electric resistance of the non-innervated side gave low values, consistent with the idea that this side acts as only a minimal barrier to the diffusion of ions. Thus, the experiments in Brazil fully confirmed the asymmetrical behaviors of the two membrane leaflets during the excitation process, the most important feature of Bernstein's model.

Experiments carried out with the main organ confirmed the general findings obtained in the Sachs' organ, although the full action potential was of smaller amplitude and had a slightly shorter duration. Hunter's organ was investigated on only one occasion, because of the particularly tight arrangement of the electrocytes, but the results corresponded to those from the two other organs. Further, the excitation was accomplished by nerve stimulation in a few experiments, with the results being basically the same, when the delay in the transmission from nerve to electrocytes was taken into account.

Nevertheless, Keynes and Martins-Ferreira had some results that did not conform to Bernstein's model, and they concerned the characteristics of the action potential generated at the innervated face. These differences were not, however, specific to electric organs. Rather, they reflected the limits of Bernstein's membrane theory in predicting the actions of excitable cells in general, and of nerves and muscles in particular. As discussed above, Bernstein's theory was based on the idea that the membrane would become permeable to

---

[7] Carlos Chagas was the son of the discoverer of the tropical illness known as Chagas disease (Carlos Chagas Sr.). At the time, the younger Chagas was known for his studies on the effects of curare on the eel's electric organs. He had established a long-term collaboration in electric fish research with Alfred Fessard, a pioneer of modern electrophysiology in France. Fessard and his collaborators at the marine laboratory of Arcachon had discovered that, similar to what occurred in the neuromuscular system, the transmission from the nerve to electric organs in torpedoes involves the neurotransmitter acetylcholine. After confirming that acetylcholine plays a similar role in the electric eel, Chagas tried to isolate the molecular structures of the eel electrocyte membrane responsible for its action (afterwards denoted as acetylcholine receptors or acetylcholine channels). Since it was known that curare acts on these structures, Chagas used radioactive curare to detect and isolate them in his research. This was an important step in the path leading to the identification of the molecular configurations of the ionic channels (Chagas, 1959; Hasson & Chagas, 1959; see Mello, 2002; Fontoura de Almeida, 2003).

[8] As remarked by Hodgkin in his autobiography, a remarkable byproduct of Keynes' sojourn in South America was his excellent writing on the voyage of the *Beagle*, performed by Keynes' ancestor, Charles Darwin, and the discovery of the drawings made by Conrad Martens, who was Darwin's artist during his expedition (Keynes, 1979; Hodgkin, 1992).

all ions during the excitation process, with the potential going to zero at the peak of the action potential. Contrary to this prediction, the electrocyte usually became positive by about 50 mV. This value corresponds to that of nerve (and muscle) cells, supporting their idea that the electrocyte membrane becomes selectively permeable to sodium ions during excitation. In confirmation of the Hodgkin-Huxley model, Keynes and Martins-Ferreira also reduced extracellular sodium concentrations in some of their eel experiments and observed a drastic reduction in the amplitude of the action potentials.

Thus, although there were some differences in their findings, Keynes (Fig. 26.17) and Martins-Ferreira provided the first really good evidence in support of Bernstein's basic conceptions. Rather surprisingly, the German physiologist's views had not been given much attention and had achieved surprisingly little favor from investigators of electrical fishes prior to this time (see Cox, Coates, & Brown, 1945: Fessard, 1946).

SYNAPTIC TRANSMISSION AND LIGAND-GATED CHANNELS

The physiological mechanisms underlying the torpedo's shocks are somewhat different and functionally less complex than those of the electric eel. To understand this difference,

Figure 26.17. Richard Darwin Keynes (1919–2010) in a picture taken during a conference he organized in March 2001 at the Royal Institution of London. This conference, initially promoted by Alan Hodgkin to commemorate in 1998 the bicentennial of Galvani's death, turned out to be a celebration of both Galvani and Hodgkin because of Hodgkin's death in December 1998. The Royal Institution was an important center for research on electricity in the first half of the nineteenth century, particularly for the endeavors of Humphry Davy and Faraday (whose statue is partially visible at right of the picture).

we must lead off with a few words about chemical transmission and the diversity of ionic channels and mechanisms responsible for electrical signals.

After a long and interesting period marked by speculations, early in the twentieth century it started to become increasingly evident that communication between cells is not brought about by the direct passage of electric impulses from one cell to another. The experiments of Henry Dale and Otto Loewi on an adrenalin-like transmitter (nor-adrenalin) and acetylcholine served as important early steps in this understanding. So did Santiago Ramón y Cajal's neuron doctrine and Charles Scott Sherrington's conception of the synapse. Most of the early work was on the neuromuscular junction. Proofs that chemical messengers (neurotransmitters) are also involved in neuron-to-neuron communication came forth only near the middle of the century (for newer historical reviews, see Shepherd, 1991, 2010; Finger, 2000, Bennett, 2001; Ochs, 2004; Valenstein, 2005).

The transmission process is normally unidirectional, and, when there is an electrical impulse, neurotransmitter is released from vesicles inside the nerve terminals into the small gaps that Sherrington had called *synapsis* in 1897. Once released, neurotransmitter molecules diffuse across the cleft and within a fraction of a millisecond they reach receptors on the post-synaptic side, triggering a series of events that could excite or inhibit the receiving cell.

There are many different types of transmitters and synaptic receptors, and the same neurotransmitter can interact with a variety of different synaptic receptor types, with different and sometimes opposite physiological effects. Adding to the complexity, there are also differences in how post-synaptic potentials are brought about after transmitters bind to receptors.

Acetylcholine was one of the first chemical agents studied, and by the 1930s it appeared to be the transmitter involved in neuron-to-skeletal muscle synapses. In the previous century, French physiologist Claude Bernard had found that curare (an Indian arrow poison) and its derivatives could cause paralysis without interfering with the electric properties of nerve or muscle cells. After the discovery that acetylcholine is involved in neuromuscular transmission, it became increasingly clear that curare works by blocking the acetylcholine receptors on the muscle fibers. In all strongly electric fish, and for almost all weakly electric fish, acetylcholine is the neurotransmitter from the nerves to the electric organs. This fact reflects the muscular origins of their electric organs.

Neurotransmitters modify the openings and closings of the ionic channels on the postsynaptic membrane, thereby affecting the post-synaptic potentials. The simplest way to achieve this action occurs in the case of acetylcholine transmission to the muscle fibers. Here the sites capable of interacting with the acetylcholine molecules and the ionic channels responsible for the post-synaptic potential are part of the same molecular complex. This specific molecular complex is called the "nicotinic acetylcholine receptor," because it is specifically activated by nicotine, the alkaloid normally present in coffee, tea, and cigarette smoke (there are also "muscarinic" acetylcholine receptors, but not on skeletal muscles). The changes in the post-synaptic potential that take place are in the positive direction, with amplitudes that can vary from less than one millivolt to several millivolts, leading toward the disappearance of the intracellular negativity.[9,10]

The nicotinic acetylcholine receptor/channel is the prototype of a class of ionic channels that open or close when binding with a specific chemical substance. It belongs to a wider ensemble of channels called "ligand-gated channels," because specific ligand molecules control them.[11] Ligand-gated channels are distinct from the voltage-gated channels, which are controlled by changes in the membrane potential.

We have already discussed voltage-gated channels when outlining the mechanisms of membrane electric excitability, and particularly the sodium and potassium channels that open sequentially during the discharge of electric impulses. The membranes of the electric eel's organ are extremely rich sources of voltage-gated sodium channels, and this characteristic was exploited in investigations leading to the isolation and molecular characterization of these channels. The extraordinary density of sodium channels in the electric eel's organ has much to do with the need to produce a strong current density during the discharges. Rather surprisingly, sodium channels are not strictly needed to produce a powerful electric discharge, as will now become clear with torpedoes.

THE TORPEDO PARADOX: SHOCKS FROM ELECTRICALLY NON-EXCITABLE CELLS

Nothing can illustrate better the differences between the mechanisms responsible for the shocks of torpedoes and those of the electric eel than the complete absence of voltage-gated sodium channels in the membranes of a torpedo's electrocytes. This surprising molecular difference corresponds to the specific and somewhat paradoxical characteristics of the torpedo's electric organs—that is, the electrical inexcitability of their electrocytes (by non-excitability, we mean only that torpedo electrocytes do not respond to electric stimuli, since these cells do produce, of course, electric potentials).

Since the time of Matteucci, it had appeared evident that experimenters had to apply electrical stimuli to the associated nerves in order to elicit shocks from a torpedo's electric organs (see Chapter 24). That is, no effect was produced if current were passed directly through an isolated electrical organ. Under normal conditions with intact nerves, as can be imagined, the electric non-excitability to direct stimulation would have been more difficult to ascertain, because electricity applied to the organ might also stimulate the nerve to it, creating a secondary effect.

Siegfried Garten (1899, 1910), a collaborator of Carl Ludwig at the Physiological Institute of Leipzig, demonstrated the non-excitability of the torpedo's electrical organs in a convincing way. He showed that the organs become totally unresponsive to electrical stimuli after the nerves to them degenerate. About 50 years later, Alfred Fessard (1946) confirmed this result with more sophisticated experiments in France.

Torpedo and electric eel physiology contrast in some ways, as du Bois-Reymond pointed out when commenting on the electric eel experiments conducted by Sachs during his trip to Venezuela (Sachs & du Bois-Reymond, 1881; see Chapter 24). Sachs had found that an eel's electric organ could indeed be excited both by electrical stimulation of its nerves and by application of currents directly to the organ. Among the points emphasized by du Bois-Reymond was that the excitability of the eel organ to direct stimulation persists for a much longer time in exhausted (or otherwise weakened) eels, compared to the effects induced by nerve stimulation. Also, the shocks induced by nerve stimulation in the eel can be blocked by curare, although responses to direct organ stimulation with currents of appropriate polarity will persist (this also came from Sachs' preliminary investigations in Venezuela; see Sachs & du Bois-Reymond, 1881). In fact, even before the important study of Keynes and Martins-Ferreira (1953), Martins-Ferreira and Antonio Couceiro (1951) had conclusively demonstrated the direct excitability of the eel's electrocytes. They showed that electrical stimuli could elicit the eel's shocks even after the nerves to its electric organs degenerated.

Although the electrical non-excitability of the torpedo's organ to direct stimulation was well established during the nineteenth century, doubts persisted about another possible difference between torpedoes and the eel. As mentioned in the previous chapter, Moreau (1861, 1862) had reported that the shocks induced by nerve stimulation seemed to persist in torpedoes treated with curare. Franz Boll, the discoverer of "visual purple" (a retinal pigment), and Francis Gotch also noted the resistance of the torpedo's shock to curare (Boll, 1873; Gotch, 1888). But contrary to these earlier findings, the blockage of the transmission from the nerve to the electric organ by curare was thereafter firmly established by Garten (1899, 1910) and amply confirmed many years later by several research teams (Auger & Fessard, 1941; Chagas & Albe-Fessard, 1954; Altamirano et al., 1955).

Modern investigations have now established beyond a doubt that the transmission from the nerves to the electric organs is blocked by curare in all three strongly electric fishes (and in almost all weakly electric fishes thus far investigated). This is because the nervous control of the electric organs involves acetylcholine acting on nicotinic-type receptors that are sensitive to curare blocks. Moreover, as first shown by Wilhelm Feldberg in conjunction with Alfred Fessard, acetylcholine mimics the effect of nerve stimulation in eliciting a torpedo's shocks (Feldberg & Fessard, 1942).

TORPEDO ELECTROPHYSIOLOGY

Today we can account for many properties and specificities of a torpedo's electric organ when compared to other

---

[9] This is brought about by the passage of positively charged ions, particularly sodium and potassium, across the open channels. In contrast to voltage-gated sodium and potassium channels, this type of channel does not discriminate between different cations. The tendency of sodium to enter and thus to make the internal potential strongly positive is partially counteracted by the tendency of potassium to exit from the cell, the equilibrium between the two process eventually being reached with about a zero transmembrane potential.

[10] Not all ligand-gated channels behave in as simple a way as the acetylcholine receptor complex, which contains both the transmitter receptor and the ionic channel in a single molecular complex. In other ligand-gated channels, the two structures are physically separated and interact in a complex way, often through intermediary chemical molecules. Normally the latter molecules diffuse through the cell's interior and are therefore generally denoted as internal or secondary transmitters.

[11] In general, the opening state of a ligand channel is not appreciably modified by changes in the membrane potential, this representing a fundamental difference with respect to the voltage-gated channels mainly responsible for the discharges of action potentials in nerve and muscle cells.

electric fishes. The view that has emerged is that the electric discharge of a single torpedo electrocyte is nothing more than a large-amplitude synaptic potential. This potential is brought about by the actions of acetylcholine on the synaptic receptors situated on the innervated side of the electrocyte membrane.

Synaptic potentials like those produced by torpedo electrocytes are also produced at the synapses between the nerve and the electric organ of the eel, as well as at synapses between nerves and muscle fibers. In the electric eel's electrocytes (and likewise in muscle fibers), however, these potentials are just the initial phase of the electric response. Once they have exceeded a threshold value, they trigger the discharge of action potentials by activating voltage-gated sodium and potassium channels located throughout the innervated membrane of the electrocyte. It is through this mechanism that the full-blown discharge of the eel's electrocytes, and consequently of the entire electric organ, is achieved. In the case of the muscles, the full-blown discharge is due to the activation of the entire muscle fiber, since the appropriate voltage-gated ion channels are widespread over the entire plasma membrane.

In torpedoes, there are no such electrically excitable channels. Nevertheless, their electrocytes are still capable of developing large-amplitude potentials. This is because of the extremely rich innervation on the innervated (ventral) surface. It is almost as if this surface is transformed into an extremely wide synaptic area.

The torpedo's mechanism contrasts with the much more reduced surface of contact between nerve terminals and the electrocytes in the eel's electric organ. This difference can account for the bigger size of the torpedo's electric nerves, which are much more fully developed and extensively branched than the eel's nerves, as John Hunter (1773, 1775) first reported to the Royal Society in the 1770s (Chapter 16). At a microscopic level, the ventral face of every single torpedo electrocyte is covered by an extensive plexus of arborizations coming from five to seven nerve terminals, which penetrate the spaces between the electrocytes from the surround of every column constituting the organ.

Harry Grundfest and his collaborators at the Marine Biological Laboratory at Woods Hole, Massachusetts, studied the discharges of the torpedo's electrocytes. They found that the discharges correspond to a large-amplitude postsynaptic potential in the absence of any action potential. Their first studies were carried out on the giant *Torpedo nobiliana,* which can give extremely powerful shocks of several hundred volts in amplitude and tenths of amperes in intensity (Bennett, Wurzel, & Grundfest, 1961). The electrocytes of this torpedo are particularly thick and wide, which facilitated these electrophysiological investigations. The electric recordings were achieved with either one or two independent glass microelectrodes, while nerve terminals innervating single electrocytes were stimulated with fine metal electrodes. The preparation itself consisted of small fractions of the organ column.

In the resting condition, the intracellular potential was −50 to −70 mV with respect to the extracellular medium. Nerve stimulation brought about a large drop in the intracellular potential, which could in some cases exceed the zero level. This response had a latency of about 3 ms, reflecting the synaptic transmission process.[12] With the innervated side upward, the resting (negative) internal potential suddenly disappeared as the electrode was lowered, while the depolarizing response to nerve stimulation persisted at practically full amplitude. This happened when the electrode tip passed across the non-innervated membrane. The persistence of a large-amplitude response to nerve stimulation was an indication of the much lower resistance of this membrane.

On further advancing the microelectrode, the negative intracellular potential reappeared, but the response evoked by nerve stimulation was approximately halved, suggesting that the microelectrode tip had entered a new cell, which was only indirectly affected by the response generated in the adjacent electrocyte. The drastic reduction in the response amplitude suggested that in contrast to the passive, non-innervated face, the innervated membrane had a high electric resistance, which limited the propagation of potential drop generated by the excitation of the adjacent electrocyte. With the preparation turned in the opposite way (non-innervated side upwards), the sequence was reversed, as was the polarity of the response induced by electric stimulation.

The summation of the responses produced by the individual elements of the torpedo's organ was demonstrated in experiments in which microelectrodes were separately inserted in two adjacent electrocytes. With appropriate nerve stimulation, it was possible to excite either one or both electrocytes, the two responses combining in the latter case. The summation process did not occur in a simple way, however, because the change in the resistance of the innervated membrane precluded a simple linear addition of the separate response components.

In contrast to the responses evoked by neural stimulation (and also with what had been observed by Keynes and Martins-Ferreira in the eel), torpedo electrocytes were not excited by direct stimulation of their membranes, irrespective of the polarity and of the mode of current application, provided the stimulus did not excite the nerves. Importantly, the amplitudes of the responses evoked by nerve stimulation were significantly altered by a concomitant injection of current across the innervated membrane. Specifically, currents that increased the negativity of the intracellular side also increased the amplitude of the evoked responses, while the reverse occurred with currents of opposite sign. By progressively increasing the intensity of the latter currents, the responses induced by nerve stimulation were nullified and eventually reversed in sign. This behavior of the electrocyte responses corresponds closely to what had been observed with synaptic transmission involving other nicotinic acetylcholine receptors, and particularly transmission from nerve to muscle fibers, provided the preparations were treated

---

[12] A poorly known contribution of electric fish studies to the general physiology of excitable cells concerns the measure of the latency of synaptic transmission. When electrophysiological investigations depended heavily on the study of the time course of muscle contractions induced by stimulation of the nerve in the neuromuscular preparation, it was difficult to measure the latency of synaptic transmission. This was because this latency is very short compared to the slow time course of the mechanical response of muscle. Electric fish discharges provided a more favorable preparation. The latency of transmission from the electric nerve to the organ was estimated to be about 0.5 ms on the basis of the studies carried out by Gotch on the torpedo and by Gotch and Burch on *Malapterurus* (Gotch, 1887; Gotch & Burch, 1896).

with toxins capable of blocking the discharges of action potentials.

The involvement of acetylcholine in the electrocyte responses induced by nerve stimuli has been supported by experiments in which potentials similar to those induced by nervous excitation are brought about by application of acetylcholine and other chemicals capable of mimicking its actions. It was also found that these potentials are potentiated by the application of substances capable of counteracting the physiological deactivation of acetylcholine. Importantly, this also occurred with electrocyte responses evoked by nerve stimulation. Two plant-based alkaloids, *d*-tubocurarine and dihydro-β-erythroidine, were among the agents used to block acetylcholine action. Both also eliminated the electrocyte response to nerve stimulation. In the case of either activators or blockers of acetylcholine receptors, the effects are observed only when the products are applied near to the innervated side.

One conclusion that can be reached from these experiments is that a torpedo's electrocyte discharge is an acetylcholine-induced, large-amplitude synaptic potential that is generated exclusively across the innervated side. Another is that the non-innervated side acts as a low- resistance leaky barrier, allowing for the spreading of the potentials generated across the innervated side. These two mechanisms account for the main distinguishing characteristic of a torpedo's electric organ, compared to the organs of the eel and also, as we shall see, of *Malapterurus*—namely the non-excitability of a torpedo's organ to directly applied currents.

As repeatedly mentioned, the synaptic potentials produced by acetylcholine-receptor activation are followed by the discharge of action potentials in the eel's electrocytes. In the eel organs, these action potentials sum because of the asymmetric membrane activation (and also because of other contrivances that we cannot deal with in detail), leading to the powerful bolt used for defensive and offensive purposes. The overall current density produced by the synaptic action is not in itself sufficient to produce a strong shock. This is because of the limited extent of the synaptic areas compared to the total surface of each electrocyte. In the case of a torpedo, synaptic currents can and do produce strong shocks. This happens because of the great surface areas occupied by the synapses and, moreover, because of the great amount of acetylcholine released by the nerve excitation. The current density produced in any single electrocyte is therefore very large, thus making the development of action potentials functionally unnecessary.

The absence of action potentials in the discharge of the torpedo's electrocytes is a secondary embryological and evolutionary adaptation, since the electrocytes are modified muscle cells, and muscle cells are normally capable of discharging action potentials. During their transformation into electrocytes, the primitive muscle cells of a torpedo's organ change shape, lose their striation and the contractility typical of muscle fibers, become spatially asymmetrical in their electric behaviors, and lose the capability of discharging spikes. The loss of spikes is effectively offset by the enormous increase in the nerve supply and the extension and efficacy of the synaptic transmission.

We have previously alluded to du Bois-Reymond's supposition that the difference in the shape and structure of the electric organs (and the overall body configuration) between the "flat" torpedo and the elongated eel (and also *Malapterurus*) are adaptations to differences in habitats, and particularly to differences in the electrical conductivity of sea water as compared to fresh water. Because of the stronger conductivity of its saltwater habitat, a torpedo's shocks cannot propagate for long distances from the fish body. With its unique electric organ, which unlike the other two strongly electric fishes comprises a large number of short columns, it emits a shock of relatively low voltage and very high current density.

Despite the short range of its propagation, the current density caused by this shock is very great in proximity to a torpedo's body. As a matter of fact, physical calculations indicate that the overall electric power emitted by big torpedoes, particularly huge exemplars of *Torpedo nobiliana*, may be stronger than in electric eels.

In this context, it is important to note that, as with other electrical effects, the physiological effects of an electric shock will become more intense as more electric power is transferred from the electric source to the target body. The flat organs of a torpedo, composed of many columns containing a relatively small number of electrocytes, are very well adapted to have a greater electric influence on organisms situated in its proximity in a strongly conductive medium. The long organs of the eel, made of a smaller number of columns containing many more electrocytes, will perform their defensive and offensive functions better in a relatively less conductive medium, such as non-salty river or lake water.

Results similar to those described in studies of *Torpedo nobiliana* have been obtained in *Narcine brasiliensis*, a much smaller and less powerful ray (Bennett & Grundfest, 1961a). This fish has a maximum potential of approximately 30 volts, as compared to 300 volts and even more in some of the bigger torpedoes. Since *Narcine* fish possess an accessory organ besides the main one, it is possible to compare the behavior of the electrocytes of the two different organs. This has been done, and the only significant physiological difference to emerge is the longer duration of the discharges of the electrocytes from the accessory organ when compared to those of the main one.

The mechanisms described in *Torpedo nobiliana* and *Narcine brasiliensis* are also at work in the other torpedoes and in other marine electric fishes, such as the stargazer *Astroscopus* (Bennett & Grundfest, 1961b). All are characterized by the absence of action potentials in the electrocyte discharge.

A remarkable difference found in a group of weakly electric rays is that the resistance of the non-innervated surface of the electrocytes seems to be high in the resting state. It decreases appreciably, however, when the electrocyte is excited (Brock & Eccles, 1958; Bennett, 1961). A decrease in membrane resistance induced by membrane depolarization is a rather common property of muscle and nerve cells, and it is referred to as "rectification," a term derived from electric circuit science (Hille, 2001). This mechanism seems to be useful for saving energy.[13]

Regarding the energy features of the discharge, comparative studies indicate that the organs of weakly electric fish are

---

[13] This is because a low-resistance membrane is leaky to ions and metabolic energy is eventually needed to restore the ionic composition of the intracellular compartment after significant ionic losses.

better optimized for the rectification adaptation. This is because weak organs tend to be more continuously active, and thus they have to use energy more efficiently so as not to be exhausted. In contrast, this energy-sparing mechanism is less important in strongly electric fishes, because they shock only occasionally and require a mechanism that would not diminish the intensity of the discharge (see Kapoor & Khanna, 2004).[14]

THE SURPRISING ELECTRIC CATFISHES

We now come to the surprising mechanisms of *Malapterurus*, the third of our strongly electric fishes, of which many species are now recognized. Some of the scientists involved in studies of the electric eel and torpedoes investigated some of these African river fishes at almost the same time as they studied the other two fishes. In 1959, Carlos Chagas organized a symposium on "bioloelectrogenesis" in Rio de Janeiro, and the results of some of the critical experiments performed by Keynes, Bennett, and Grundfest on a species of electric catfish were presented there orally and in print 2 years later (Chagas & Paes de Carvalho, 1961; Keynes, Bennett, & Grundfest, 1961).

One of the most intriguing problems for Keynes, Bennett, and Grundfest was this catfish's apparent violation of Pacini's law. That is, in contrast to the other types of strongly electric fishes, the catfish's electric organ seemed to become positive on the innervated (tail) side and negative on the non-innervated side at the time of discharge. Also intriguing was the fact that, in contrast to plentiful nerve innervations of the electric organs of the other two types of strongly electric fishes (and particularly torpedoes), electric catfishes have just a single axon innervating the organ on each side of its body. As with the differences between the electric eel and the torpedo's mechanisms, these peculiarities suggested considerable evolutionary independence.

To understand the results of their electrophysiological experiments, it is important to outline the morphology of the catfish's organ and its nerve supply in more detail. In contrast to torpedoes and the eel, the electric organ of the catfish occupies a relatively peripheral position with respect to the main mass of the animal, having the appearance of a thick sheath surrounding its body. This "peripheral" arrangement seems to interfere minimally with the ordinary behavior of this river fish with its distinctive movement patterns.

For a long time, the prevailing view was that its electric organ might have originated from cutaneous glandular tissue rather than from muscles, adding to the mystique of this fish (Fritsch, 1887; Schnakenbeck, 1955). This thinking, however, was found to be erroneous. It now appears that the *Malapterurus* electrocytes also have a muscular origin, most likely being derived from a small part of a muscle in the pectoral girdle (Johnels, 1956). Its possible origins from a very limited number of ancestral muscle cells (likely innervated by a single axon), which could change shape and greatly multiply in number, provides an evolutionary explanation for the surprising fact that each electric organ, which contains of about 1 million cells in an adult fish, is innervated by just a single axon.

The axon itself is very big relative to other fishes (its diameter is about 40 microns), and a thick connective sheath surrounds it, so that the single fiber in its electric nerve has an overall thickness of several hundred microns. This nerve originates from the spinal cord and has an initial course similar to that of the motor fibers directed to ordinary muscle fibers (i.e., it exits the spinal cord from the ventral side). Upon reaching the electric organ, the axon enters the caudal end and then branches repeatedly, so as to provide terminal arborizations to the entire electrocyte population.[15]

The catfish contacts differ considerably from those of torpedoes and the eel, being somewhat analogous to those present in some weakly electric fish (all the mormyrids and some of the gymnotids). They are established at the tip of a long stalk formation protruding from the caudal face of the electrocyte, giving them a flower-like appearance. In contrast to torpedoes in particular, but also the eel, the surfaces of the contacts are rather restricted, and this imposes some constraints on the mechanisms of electrocyte excitation.

At the outset of their experiments, Keynes, Bennett, and Grundfest (1961) examined the direct excitability of the catfish electrocytes. Contrary to the older findings of Gotch and Burch (1896), which pointed to the need for nerve stimulation in order to produce shocks in this species, they found that currents applied directly to the electric organ could excite the electrocytes. In this regard, its electrocytes were found to be functionally similar to those of the eel and different from those of torpedoes. Indeed, the responses to current passed across the entire organ persist even after degeneration of the innervating axon. The electrical excitability of catfish electrocytes also made it possible to stimulate them by direct current applied through an electrode situated inside the cell or in the immediate proximity of its membrane.

Keynes, Bennett, and Grundfest also confronted the old idea that the excitation of the catfish organ might involve a mechanism opposite to that of the other electric fish, and hence to nerves and muscles. This hypothesis had been used to account for the "reverse polarity" of the catfish organ by assuming that the intracellular potential is positive to the exterior in the resting state, and that the active side of the electrocyte inverts its polarity upon stimulation (see Garten, 1910). These researchers found that this was an erroneous explanation for the anomalous polarity of electrocyte discharge. They discovered that the electrocytes, as with almost all excitable cells, have a negative intracellular potential at rest. The paradoxical behavior of the catfish electrocytes could be accounted for by the general principles of membrane theory, supplemented with Bernstein's asymmetry hypothesis (although the actual mechanisms are significantly different in some ways from those of the eel and torpedoes).

They also found that excitation of an electrocyte (either by applying direct current or via nerve stimulation) leads to

---

[14] Here we might think of the differences between a standard automobile and a racing car. The racing car has maximal speeds that far exceed those of an ordinary car, but it also has a far greater energy burden when racing, burning fuel much faster than an ordinary car—with obvious consequences.

[15] Among other things, the singular characteristics of electric fish nerves offered the possibility of verifying whether transmission of impulses in nerves is necessarily unidirectional or could be bidirectional (at least in experimental conditions). Babuchin showed that a discharge of the entire organ could be obtained by stimulating the ramifications of the electric nerve at various positions. The effect necessarily implied that the impulse could be transmitted in both centrifugal and centripetal directions (Babuchin, 1877).

the production of an action potential of conventional polarity (i.e., the cell's interior becomes positive relative to the extracellular space) on both the innervated (stalk) face of the electrocyte and the opposite side. The crucial experiment was performed using three different microelectrodes placed in various sites of an isolated preparation. One was inserted inside an electrocyte, while the other two were placed in the extracellular compartments, one close to the stalk membrane and the other near the external face of the opposite membrane.

Measurements of the differences across the two membranes at the time of discharge revealed impulses of similar (positive) polarity in both membranes. Despite the common excitability of the two membranes, the overall behavior of the electrocyte appeared, however, to be largely asymmetric. This was because the stalk membrane produces a potential that is smaller in amplitude and much shorter in duration (0.3 versus 2.0 milliseconds) than the one that the non-innervated membrane generates. Consequently, only the non-stalk membrane is polarized (with the intracellular face positive with respect to the extracellular compartment) for most of the excitation cycle, whereas the stalk membrane is inactive after the initial, short-duration impulse. Thus, the non-stalk face of a catfish electrocyte, although it is not innervated, actually behaves very much like the *innervated* face of the eel's electrocyte. This observation explains why a catfish electrocyte (and really the entire organ) develops a positive polarity on the extracellular side of the non-innervated membrane during excitation, thus contradicting Pacini's so-called law.

As to the excitability of both sides of the electrocyte membrane that distinguishes *Malapterurus* from other strongly electric fish (but is generally present in the weakly electric fish having a stalk-type innervation), there are various possible explanations. The most likely one is that the short-duration potential of the innervated side is a local action potential capable of triggering (by passive propagation) the longer-lasting, full action potential on the non-innervated side. In this way, a highly localized nerve input could excite a large surface membrane. This would represent an evolutionary strategy to deal with the problem of how to innervate a multitude of electrocytes via a single nerve cell.

There is another interesting explanation for the possible advantage of the functional asymmetry brought about by differential excitation of both electrocyte surfaces, rather than by the excitation of a single surface. It has to do with the fact that the discharge of an action potential is followed by a relatively long phase in which the electric resistance of the membrane decreases in many excitable cells (this is normally due to the activation of a particular type of potassium channel and represents the mechanism of membrane rectification; see above). It could be that the short-lasting potential discharged at the innervated face is a device allowing this surface to have a low resistance only during the excitation of the electrocyte and not in the resting condition.

As we discussed when describing the weak organs of some rays, the low resistance of a membrane at rest is an energy-consuming mechanism, because this normally implies a continuous transmembrane flow of ions, which must eventually be counterbalanced by an active transporting mechanism (a process requiring metabolic energy). Hence, this might be one of the mechanisms whereby *Malapterurus*, generally considered the most perfect of the strongly electric fishes, produces powerful electric shocks without being rapidly exhausted, as can happen with torpedoes.

SYNOPSIS

In conclusion, this discussion of the mid-twentieth-century discoveries concerning the mechanism of the shocks of our three strongly electric fishes shows how nature is capable of producing fairly similar outcomes with different mechanisms. At one extreme we find the torpedoes, which produce shocks with a mechanism that does not involve discharging action potentials. A torpedo's shock is dependent upon an organ that receives an extremely rich nerve innervation with an abundant release of acetylcholine, and the presence of a multitude of synaptic channels and other chemical devices common to synaptic transmission. At the other extreme we find the catfish *Malapterurus*. In its case, both electrocyte membranes produce action potentials, and there is only a limited nervous and synaptic investment in the peripheral control of the organ (a single nerve fiber for each organ, a very low release of acetylcholine, and sparse chemical machinery for acetylcholine-based synaptic transmission).

These differences, and those characterizing the electric eel, reflect the diverse evolutionary histories of our three strongly electric fishes. They also shed light on the weak electric organs from which the powerful electric machines of the strongly fishes seem to have evolved. In this regard, these adaptations relate to the diverse habitats of these fishes, and they have had obvious survival value in the past while still allowing these fishes to thrive in their challenging environments.

Epilogue

> But what will be this animal machine?
> (Luigi Galvani, 1794b, p. 76)

> Primary among the rewards of a scientific explorer is the discovery of a new phenomenon. Only he who has had the experience knows the thrill of it.
> (Walter Cannon, 1945, p. 205)

At the start of this volume, we described Alexander von Humboldt's pursuit of electric eels in the swamps and pools of the *Llanos* in Spanish Guiana, a part of present-day Venezuela. By portraying in such an extremely colorful and vivid way the extraordinary scene of how these frightening creatures were captured with horses and mules as "bait," Humboldt did more than produce a fascinating piece of scientific literature; he wrote a piece that virtually every nineteenth-century European student would encounter in his or her studies. Humboldt would have his place with Dante, Cervantes, Shakespeare, Goethe, Byron, Poe, Balzac, Dickens, Flaubert, Tolstoy, and Mark Twain. His book, or at least this one part of it, would also become a favorite part of the school curriculum throughout the Americas and even elsewhere.

By representing the incredible combat between the powerful pack animals and the daunting eels with the rhetorical richness of an epic battle, Humboldt also communicated much about himself. Here, perhaps even more than anywhere else in his volume, he conveyed his love of nature, his excitement about his research and his dedication to it, and his sense of history, to a readership that, like the great explorer himself, spoke many languages, spanned much of the globe, and was captivated by what they were learning, in this case about some very powerful electric fish.

Humboldt's involvement with electric eels represented just one small part of his many interests in science and explorations of nature, but it touched upon two of the greatest scientific debates that were raging at the end of the eighteenth century and into the nineteenth century. These debates figured prominently into our book.

One was whether fish electricity is true electricity or only a close cousin to "real" electricity; that is, the electricity that Benjamin Franklin captured from the skies above Philadelphia with his wired kite, and the electricity that Dutchman Pieter van Musschenbroek collected from his frictional machine with his accidentally-discovered Leyden jar. The other was whether these fishes are unique, or whether other fishes, frogs, cattle, and humans also possess some sort of an electrical fluid, or what was then called animal electricity, in their nerves and muscles. This was an important physiological question, since, as we have seen, it would lead to a plethora of investigations that would provide the foundations for modern electrophysiology, a development that would help shape the neurosciences, cardiology, physiotherapy, and at its core a whole new way of looking at and treating animate bodies.

By choosing Humboldt for our opening, we also wished to signify, even if only indirectly, that his era would pretty much mark the end of the detailed and increasingly narrative part of our electric fish history. In this context, he was put on center stage to remind us of a time when science was dominated by the individuality of the researcher, both at a personal level and as a representative of a colorful but decreasingly aristocratic society. Could any person or any endeavor be chosen as more representative on both fronts than Humboldt with his scientific journey to the New World? Not only was he an exceptionally well-educated German baron and a learned scientist communicating with others around the world about nature, but he literally invested everything he had in the enterprise: his reputation, his inherited fortune, his well-being, and even parts of his own body, which he repeatedly used as a scientific instrument. Although human commitments would continue to be an important feature of scientific endeavors well after his time, it would be futile to try to find a narrative or autobiographical account as important, as thrilling, as absorbing, and as influential as Humboldt's had been in the first half of the nineteenth century.

In the end, we opted to work our way into twentieth-century science more than we initially thought we would. We did so because we realized that the story we were telling needed better closure, even if some of the newer material would be more difficult to follow and could not be covered with the same style or in the same depth. Specifically, we felt a need to satisfy some readers' curiosity for understanding how, in contrast to all known animals, three fishes with moist insides could produce and discharge electricity in a watery habitat—observations that seemed to defy common sense and a reality that left physiologists, physicists, and pretty much everyone else scratching their heads into Humboldt's era. Moreover, we also wanted to show how research on electric fishes had changed since Humboldt's voyage, and to illustrate in a concrete way that these fishes have also played roles in some newer, particularly noteworthy developments (about which more will be stated below). As so well put by Galileo, a work "needs not to be restricted so strongly to that unit, which leaves no field open for the introduction of new episodes" (Galilei, 1632, pp. 155–156).

Science became progressively more professionalized after Humboldt's time, with an exponential increase in the number of the institutions where science was taught and practiced, in the number of scientists, in the standardization of instruments and measures, and, of course, in the number of publications. Along with these developments, articles and other scientific texts started to be written in a more impersonal, technical style, which clearly veered away from divulging the deep human feelings associated with scientific investigations and discoveries. These changes, so evident in today's research reports and reviews, are in some ways regrettable. They tend to make it more difficult for historians to reconstruct the paths of scientific discovery with the *humano aroma* that Nobel Prize-winning Spanish neuroanatomist Santiago Ramón y Cajal beautifully displayed in his colorful autobiography and fully appreciated as a feature of every scientific endeavor.

Hence, by choosing Humboldt, we began our book with a more open sort of science than today's, one dominated at all levels by humanity and culture, one wishing to be expressed in all its dimensions. The German baron's life and writings, and his research on the South American eels and later on torpedoes, reflected this *Zeitgeist*, and little is left hidden about his personal goals, ideas, sense of adventure, and values at this pivotal moment in history.

The choice of Humboldt thus anchored us for a trip back in time and then a somewhat different one forward—an odyssey in which we first looked at the history of strongly electric fishes from ancient Egypt into the opening decades of the nineteenth century, and then sailed on into the mid-twentieth century, when the elusive mechanisms behind their frightening shocks were finally understood. In a sense, we might say that our own voyage into this facet of the history of science and medicine, not to mention fields as varied as magic and physics, had the growing knowledge about electric fishes as its Arianna's thread, allowing us to deal with the meanders in this labyrinth of history. For us, Alexander von Humboldt played the role of a companion and a guide, somewhat like Virgil did for Dante in his journey in the *Divina Commedia*.

NOW-ANSWERED QUESTIONS

As with any voyage of discovery, and especially those still in progress, only the port of departure is ever really certain at the time of leaving. In this case, our thirst for history took us to many ports of call and involved some very famous people, as well as a number of individuals whose names have long been forgotten, although many are deserving of some resurrection. The route we followed was anything but straight, with one stop or discovery not always leading directly to the next, and not always with crystal-clear statements about what each of our players might really have been thinking when encountering these fishes or philosophizing from a distance about them. It was indeed difficult to foresee from the outset, for example, that by following the path to the (re)discovery of the electric catfish by the Portuguese in the seventeenth century, we would have been transported back in time to the biblical world of King Solomon and the Queen of Sheba, the fascinating myths about Prester John, some heated disputes over the "heresy" of Ethiopian Christianity, and the wrath of the Catholic Counter-Reformation.

Even though we have sometimes been blown off the course we thought we had plotted, these unexpected stops provided new insights, allowing us to answer many of the questions that we raised in our Introduction, and many more that came to us along the way. Hence, we can state that the earliest known pictorial records of an electrical fish date from about 2,400 B.C., when the great pyramids were being constructed, although catfish symbols (perhaps reflecting *Malapterurus*?) predate the Old Kingdom and have been traced to predynastic times in Egypt.

We can also state that the ancient Greeks seem to have been the first people to describe the shocking powers of torpedoes in their texts, with Plato alluding to them in his *Meno* and then the Aristotelians writing more scientifically about these rays and their torporific actions. We can also point to live torpedoes being used in the healing arts (e.g., for treating gout and headache) in the first century A.D., although future research may show that this history, like so many other parts of this story, might be even deeper and have more ancient roots than the documents we currently have would suggest.

We further asked how people first measured the powers of these seemingly inexplicable fishes: basically what instruments they might have used. Here we discovered that human bodies served as the first "electrometers." The intensities of the shocks were directly compared by how much of a hand, an arm, or even a torso was numbed by the discharge. Equally interestingly, although strings, pith balls, and other electrometers were used to detect electricity during the eighteenth century, these physical instruments still lacked the sensitivity to do the job when Galvani was conducting his experiments, in contrast to "prepared" frogs, meaning exposed frog nerve and muscle preparations. It was only during the nineteenth century that physical instruments became sensitive enough to provide accurate quantitative data about their discharges.

We also wanted to know when scientists began to think that the powers of these strongly discharging fishes are electrical, what cultural changes led to this new way of thinking, and what kinds of findings were presented as proofs. As we have shown, the first hints that any of these fishes might be electrical came forth in the middle of the eighteenth century, which we loosely referred to as the "electrical century," since many things in nature (from terrifying earthquakes and storms to beautiful auroras and even plants and minerals) seemed to become electrical at this time. More concretely, fish electricity was born at mid-century when some Europeans, who had sailed to the Dutch colonies in South America, began to do some simple experiments on the feared river eels, and when their discharges began to be compared to the newly invented Leyden jar.

From what we can gather, English surgeon Dale Ingram, who had lived in Surinam, might have been the first to make this association and back it up with at least some experimental evidence, publishing his brief note in an English periodical in 1750. His and other experiments of the 1750s and 1760s showed two things: that the discharges really feel like those from Leyden jars, and that they can be transmitted by known conductors of electricity (e.g., metal wires) and blocked by nonconductors (e.g., wax). Whether the eel's painful discharges might cure deformities and various disorders (a claim being made by some medical practitioners using Leyden jars) was, however, a much more contentious issue.

Yet for many natural philosophers at this time, the growing evidence for fish electricity remained rather superficial, and the thought that a moist living body, particularly one in water, could contain electricity without harming itself seemed absurd. In this context, a landmark event took place in 1776, when John Walsh showed that the eel does, in fact, give off a tiny spark when it discharges—this being perhaps the most important piece of evidence to date, and the one that natural philosophers everywhere had been waiting for. Walsh's spark experiment, which was replicable and witnessed by members of the Royal Society of London, served as the death knell for the ancient theory that torporific fishes might be emitting a cold poison, ideas based on magic and occult properties, and the mechanical theories that had begun to appear in the second half of the seventeenth century. The latter had emphasized extremely rapid (vibratory) muscle actions, and in some cases the accompanying release of noxious particles or corpuscles that could somehow travel through water and still penetrate the skin.

In effect, give or take the finer details, this was what was known about electric fishes when Humboldt disembarked in South America in 1800. Still left unanswered, and very much on his mind, was whether there is only one electricity or a family of electricities, and the matter of whether other animals also possess electricity, or at least something in their nerves having many of the features of true electricity. In this context, we took a hard and fresh look at Luigi Galvani, who had in 1791 first argued for a more general role of electricity in animal physiology, and at his equally famous adversary, Alessandro Volta, who believed that only a few fishes are electrical (these fishes being the only creatures with identifiable electrical organs). We also looked more carefully at Humboldt, who initially agreed with Galvani that frogs and humans also possess something like electricity, but with the caveat that it is probably not true electricity. As we have seen, Humboldt appeared to change his mind after returning from the New World and meeting with Volta, who had just modeled his pile (battery) after fish electric organs, a development that revolutionized physics and one that led the German baron to think in a more positive way about a universal electrical fluid.

SOME NEW DEVELOPMENTS

In the final chapters of this book, we examined how electric fishes shed much-needed light on nerve and muscle physiology, first in Italy and Germany during the nineteenth century, and then to a greater extent in England and America during the twentieth century. We also emphasized how the insights gained from studies on frogs and then the squid's giant axon helped reveal the basic mechanisms by which these fishes produce their shocks. In this context, we discussed evolution and the evidence strongly suggesting that the electric organs of the eel, electric catfishes, and torpedoes evolved independently, although in all cases the organs originated from nerve-innervated muscles.

Today there is more interest in weakly electric fishes than the strongly electric fishes that we have focused on in this volume, but the two are very much related. As we have noted, the powerful electrical organs of our strongly electric fishes seem to have emerged from weaker electrical organs. But although the idea that there might be weakly electric fishes began to appear in print before 1850, these fishes did not become the center of attention until more than a century later, with recognition of their various mechanisms of electroreception, electrocommunication, and electrolocation.

We now know that there are many more kinds of weakly electric fishes than strongly electric fishes, each with distinctive behavioral and morphological features: adaptations sculpted by natural selection for survival in their unique habitats. This literature has grown exponentially and is already enormous. Yet unlike the early literature on strongly electric fishes, it is readily accessible and has been the subject of many informative reviews (see Chapter 25 for references).

It would be wrong to think, however, that the story of strongly electric fishes ended during the 1950s with the well-publicized discovery of many weakly electric fishes. Not only has more been learned about the more powerful fishes since that time, but these truly torporific fishes have continued to play significant roles in understanding nerve and muscle actions, and in developments in the field of medicine.

In a way, these newer developments might be thought of as yet another shift in the interplay between studies of electric fishes and those on nerve and muscle research. As we have seen, electric fishes basically guided nerve and muscle studies late in the eighteenth century and pretty much through the first half of the nineteenth century. Then, in the closing decades of the nineteenth century and into the middle of the twentieth century, the table turned and nerve and muscle studies opened the needed doors for understanding how fish electric organs actually function. In the second half of the twentieth century, the give-and-take between these closely related fields seemed to shift again, with new studies of electric fishes guiding and confirming thinking about nerve and muscle ion channels, and more.

To appreciate these newer developments, we must imagine ourselves in the post-Hodgkin–Huxley era, when pharmacologists, biochemists, and biophysicists were attempting to determine how ions could permeate cell membranes. A major difficulty facing researchers had to do with the fact that these membranes are protected by double-lipid barriers, which are impermeable to water-soluble molecules (i.e., ions). The idea that started to emerge was that the process must involve discrete molecular formations, each having specific permeation characteristics and gating properties.

In particular, there was now tremendous interest in acetylcholine-mediated synaptic transmission, because acetylcholine was by this time the established transmitter between nerves and skeletal muscle fibers (Dale & Feldberg, 1934; Brown, Dale, & Feldberg, 1936; Dale, Feldberg, & Vogt, 1936; also see Feldberg, 1977). Not only was this an important fact in itself, but it was also a discovery that was guiding synaptic research. This is because it was clear that synaptic transmission could be more easily studied at the neuromuscular synapse, with its single transmitter, than within the central nervous system, with its cells more densely packed, multiple transmitters, and even nerve-to-nerve synapses.

Electric fishes entered the scene of ionic channel research at this time for two main reasons: first, because their electric organs are extremely rich sources of ionic channel molecules; and second, because their electric organs are purely cholinergic (for reviews, see Whittaker, 1992, 1998; Keesey, 2005). The discovery of cholinergic transmission in torpedo electrical organs dates back to 1939, although the publication by Wilhelm Feldberg and Alfred Fessard (1942) on this subject was delayed for 3 years because of World War II. In 1963, Richard Keynes expressed the important belief that fish electric organs are richer sources of acetylcholine than brain or ordinary muscle—far richer than anyone had imagined (Whittaker, 1998).

Keynes was correct. Containing as much as 1,000 times the cholinergic content of ordinary muscle, electric fish organs proved to be ideal sources for isolating these channels, much as the squid had been found to be an ideal subject for intracellular electrical recordings in the 1940s because of its special attribute, its giant axon. In effect, torpedoes and electric eels were now sought to study cholinergic transmission (synthesis, transmitter storage, release, post-synaptic actions, inactivation, and reuptake), with torpedoes providing the best chance of determining the underlying molecular events, due to their more extensive synaptic formations and greater number of ionic channels.

The acetylcholine channels of torpedoes were, in fact, the first ionic channels to be isolated chemically and to be "sequenced" with the newly developed techniques of molecular biology and molecular engineering. The departure points were the isolation of this channel from the torpedo's electrical organ and the subsequent identification of the sequence of amino acids with conventional biochemical techniques.

The torpedo work was aided by the use of a toxin called alpha-bungarotoxin, which is derived from snake venom. Unlike many other natural poisons, it is capable of binding with high affinity and with near irreversibility to the ionic channel. Chuan-Chiung Chang and Chen-Yuan Lee, pharmacologists at the National Taiwan University, isolated this useful toxin in the 1960s from snakes common to their island, under very difficult and trying conditions (Chang, 1999; Chu, 2005).

From the identified sequence of the amino acids, it became possible to isolate and sequence the DNA coding for the cholinergic channel molecule with molecular biology. And from the DNA sequence it became possible to establish the sequence of amino acids in the various subcomponents of the channel protein, a landmark achievement first obtained by Shozaku Numa and his colleagues in Kyoto, Japan (Noda et al., 1982; Changeux, Devillers-Thiery, & Chemouilli, 1984; see Hille, 2001). Subsequent studies on the channel with more advanced techniques have led to an even better understanding of its structure and manner of functioning.

Electrophorus electricus, the electric eel, while less generous than torpedoes in providing acetylcholine channels, has also figured into this newer and exciting picture. In 1970, French neuroscientist Jean-Pierre Changeux and his associates isolated the nicotinic acetylcholinergic receptor in the eel, again using the Taiwanese snake toxin (Changeux, Kasai, & Lee, 1970).

The eel has also been a valuable source of voltage-gated sodium channel molecules and a rich source of other molecules involved in basic membrane mechanisms, including the so-called sodium-potassium pump, which is responsible for differences in the concentrations of these two ions across the cell membrane. The amino acid sequence of eel-derived sodium channels was established with molecular biological techniques similar to those used in the case of the acetylcholine channel in torpedoes, again in the laboratory of Shosaku Numa (Noda et al., 1984). The initial isolation and identification of the channel molecule was made possible by the use of a powerful ligand of the sodium channel, tetrodotoxin, a toxin extracted from puffer fish (Agnew et al., 1978). So here, too, scientists continued to profit from nature's resources, using them, as John Walsh and John Pringle in the 1770s liked to say, to unveil her deepest secrets.

Following this initial work on acetylcholine and sodium channels, a number of other ionic channels were isolated, sequenced, and investigated. The results obtained are among the most important modern scientific achievements, and they have had significant medical applications. Even a succinct account of these developments would be beyond the scope of our book, but one stands out and must be mentioned.

FROM SHOCKING FISH TO MODERN MEDICINE

Myasthenia gravis is a chronic, progressive condition characterized by weakness and fatigue of the skeletal (voluntary) muscles (Keesey, 2002). The muscles of the face, arms, and legs, and those involved in chewing, swallowing, and talking, are among the most affected. There are several forms of this disorder, including a genetic one, but the most common type is autoimmune myasthenia gravis, in which the immune system produces antibodies that block or destroy the acetylcholine receptors on the muscle membrane.

This basis of myasthenia gravis began to be understood in the 1970s, when radioactive alpha-bungarotoxin, the same toxin that had been used in electric fish studies to isolate nicotinic acetylcholine receptors, was used to study the muscle fibers of myasthenia gravis sufferers. In 1973, researchers at Johns Hopkins University reported a reduction in the number of acetylcholine receptors in their patients (Fambrough, Drachman, & Saryamurti, 1973).

Simultaneously, another group at the Salk Institute in La Jolla, California, took some purified acetylcholine receptor protein from an electric eel and injected it into some rabbits, hoping to generate antibodies to acetylcholine receptors. They found that their rabbits started to fatigue and display muscle paralyses with repeated injections of the eel receptor protein, and that an anticholinesterase agent could counteract these effects (Patrick & Lindstrom, 1973).

Further studies by the California team showed elevated amounts of antibodies in the sera of myasthenia gravis patients, thus confirming the now-suspected relationship between "experimental autoimmune myasthenia gravis" in laboratory animals and the progressive disease that can so devastate humans (Lindstrom et al., 1976; Keesey, 2002; Chu, 2005). This recognition, which was so dependent on electric fish findings throughout its history, has led to a clinical test for this disorder and to new research on treatment options.

On the subject of myasthenia gravis, one of the authors of this book (M.P.) cannot forget a clinical case that has remained the most significant in his memory. It occurred when he was a trainee for his medical degree at a local hospital in Italy. One night, a foreign tourist was brought into the emergency room in a condition of extreme weakness, with his life in immediate danger. After a rapid medical examination and a reconstruction of his medical history (made difficult because of the patient's severe condition and the fact that he was a Swede trying to communicate with an Italian physician in yet a third language, English), it became clear that he had myasthenia gravis and was undergoing acute deterioration. The patient was now given a pharmaceutical preparation based on an acetylcholinesterase inhibitor. Because acetylcholinesterase is the enzyme that degrades acetylcholine in the synapse, the inhibitor increased the concentration of acetylcholine in his neuromuscular synapses, and thereby prolonged its action. What happened in the next few minutes was like the biblical story of Lazarus' death and resurrection. Progressively, the patient's body started to reanimate, and in less than an hour he could stand upright and felt himself ready to resume his life!

For a young medical student, this episode was an impressive demonstration of what could be accomplished when the knowledge gained from experimental science can be combined with what is observable in a hospital setting (for more on how myasthenia gravis was first treated pharmacologically, see Walker, 1934, 1935, 1973). In retrospect, all of us who now know something about the history of research on synaptic transmission can fully appreciate the importance of animal experimentation in this domain—a history that takes us back to Galvani's prepared frogs and to Hodgkin and Huxley's squid axons, and then into the use of electric fishes to isolate acetylcholine and understand other ionic channels.

A NEVER-ENDING STORY

In the opening paragraph of his 1773 report to the Royal Society of London on the torpedoes he had studied in La Rochelle, John Walsh wrote that torpedo electricity is "a subject not only curious in itself, but [one] opening a large field for interesting inquiry" (Walsh, 1773, p. 461). Walsh died in 1795 at age 69. Yet we cannot but wonder how he would have responded had he lived long enough to see developments such as those just mentioned, which are based on discoveries involving the electric fishes that had captured his imagination and the attention of his colleagues at the Royal Society of London.

The story of electric fishes is, of course, a never-ending story. Indeed, given how frequently the authors of this book found "surprises" about these fishes in the books, articles, letters, and other communications they consulted, we must emphasize that even the lengthy story we have told here cannot by any stretch of the imagination be regarded as complete. What we have tried to do is to show the complex and fascinating routes of scientific progress that have led to our present understanding of these fishes and something about the people and the cultures involved. Of course, we also have tried to present this story chronologically and in an accurate and, we hope, interesting way.

Without question, documents we have overlooked will surface in the future, and more detailed analyses of some of the material we have covered will provide new perspectives and fresher insights about who did what and why. Clearly, we did not even attempt to get into the newer literatures on the social, courtship, and hunting behaviors of these fishes, their developmental changes, and much more that we know is worthy of attention. Instead, we chose to focus on these strongly electrical fishes as wonders, which they are, exploring what people believed about their discharges from ancient times through the middle of the twentieth century in a way that made sense to us, given our interests in the history of the neurosciences and especially neurophysiology and medicine.

Looking forward, as is true with almost all scientific, medical, and even technological developments, it is impossible to predict in what ways future research with electric fishes will further affect the sciences, medicine, or any of the other fields alluded to in this book. If the past is prologue, as the Bard once contended, it seems safe to say only that some new developments will come from currently unexpected sources and explorations, quite possibly by chance or accident. It is this indisputable fact that makes science and medicine, as well as historical studies of specific discoveries, phenomena, and people, such as Walsh, Galvani, Volta and Humboldt, so exciting.

With these thoughts in mind, we might think back to what Galvani wrote in 1794, when pondering the possible role of electricity in the "animal machine." He asked: "But what will be this animal machine?" He then continued: "We cannot ascertain it with certitude: it remains totally occult to the most acute sight; we cannot but recognize its properties, and, in some way, conjecture its nature" (Galvani, 1794b, p. 76).

We have now ventured far beyond Galvani's own bold conjectures about this "animal machine" being electrical. Moreover, as we have seen, our current understanding stems in good measure from three types of strongly electric fishes, those wonders of nature that have fascinated people from ancient times, through the Enlightenment, and into the present. For us, they too should be regarded as heroes in this epic tale with its many important ramifications.

In closing, we hope that you, the reader, found this history of electric fishes as captivating as we did while conducting our historical research and putting the story together. Much as we hope that this book will fill in many gaps, correct some errors, and lead to a fuller appreciation of electric fishes and the roles they have played in physiology, medicine, and even physics, we would be especially pleased if what we have written will serve as a worthy stimulus for further historical research that could extend, make clearer, and possibly correct, the picture of the electric fish research that we have portrayed in our book. There are treasures still awaiting scientific explorers in libraries and archives, as well as in the rivers, lakes, and oceans of the world, some of which may indeed prove shocking, and they are begging to be brought from obscurity to light.

Appendix
Names with Birth and Death Dates

'Abbās, 'Ali ibn al-Majusi (10th cent.)
Abbas I (Viceroy of Egypt) (1813–1854)
Abbas, Haly (see 'Abbās, Ali ibn al-Majūsi)
Abbeville, Claude de (d'Abbeville) (d. 1632)
'Abd Allatif (see al-Baghdadi)
Abelard, Peter (1079–1142)
Abu-l-Walid Muhammad ibn Ahmad ibn Rushd (1126–1198)
Achard, Franz Karl (1753–1821)
ad-Damiri, Muhammad Ibn Musa Kamal ad-din (1341–1405)
Adams, John (1735–1826)
Adanson, Michel (1727–1806)
Adelard of Bath (fl. 1116–1142)
Adrian, Edgar (1889–1977)
Adye, Spehen Payn (fl. 1770s)
Aelian (Claudius Aelianus) of Praeneste (c. 170–c. 235)
Aelianus, Claudius (see Aelian)
Aepinus, Franz (1724–1802)
Aesop (620–560 B.C.)
Aëtius of Amida (c. 530–560)
Agrippa, Henricus Cornelius (1486–1535)
Agrippa, Marcus Vipsanius (c. 63 B.C.–12 B.C.)
al-Baghdadi, 'Abd al-Latif ('Abd Allatif) (1162–1231)
al-Baitar (see Ibn al-Bītār)
al-Bītār (see Ibn al-Bītār)
al-Ghazi, Ahmad ibn Ibrihim ("Gragn") (1506–1543)
al-Idrisi, Muhammad (1100–c. 1165)
al-Kindi, Abū Yūsuf Ya'qūb ibn Ishāq (c. 801–873)
al-Latif (see al-Baghdadi)
al-Qazwini, Hamdullah al-Mustafa (c. 1281–after 1339)
al-Qazwini, Yahya Zakariya' ibn Muhammad (1203–1283)
al-Razi, Abu Bakr Muhammad ibn Zekariya (865–925)
Al-Zahrawi, Abu al-Qasim (936–1013)
Alberti, Leon Battista (1404–1472)
Albertus Magnus (Albert the Great) (c. 1193–1280)
Albucasis (see Al-Zahrawi)
Alcmaeon of Croton (c. 500 B.C.)
Aldini, Giovanni (1762–1834)
Aldrovandi, Ulisse (1522–c. 1605)
Alexander III (Pope) (c. 1100–1181)
Alexander of Aphrodisias (c. 2nd century B.C.)
Alexander of Tralles (525–605)
Alexander the Great (356–323 B.C.)
Alexander VI (Pope) (1431–1503)
Alibard, Thomas-François (1703–1799)
Alibert, Jean Louis (1768–1837)
Alighieri, Dante (1265–1321)
Alkindus (see al-Kindi)
Allamand, Jean Nicolas Sébastien (1731–1787)
Allatif (see al-Baghdadi)
Almeida, Manoel de (1580–1646)
Alston, Charles (1683–1760)
Alvares (Alvarez), Francisco (1465–1540)
Alvarez, Francisco (see Alvares)
Amenemhat I (1991–1962 B.C.)

Ampère, André-Marie (1775–1836)
Anglicus, Bartholomeus (before 1203–1278)
Anson, George (1697–1762)
Anteros (c. 1st cent.)
Antigonos of Carystos (241–197 B.C.)
Antony, Marc (Marcus Antonius; 83–30 B.C.)
Apollodorus of Athens (c. 180–after 120 B.C.)
Apollonius of Perga (c. 262–c. 190 B.C.)
Apuleius of Madaura (c. 125–180)
Aquinas, Thomas (1225–1274)
Arago, Dominique François Jean (1786–1853)
Archimedes of Syracuse (c. 287–c. 212 B.C.)
Aristotle (374–322 B.C.)
Arrhenius, Svante August (1859–1927)
Artedi, Peter (1705–1735)
Artimedorus of Daldianus or Ephesius (fl. 140 A.D.)
Artur, Jacques François (1708–1779)
Atahaulpa (Inca king) (d. 1533)
Athenaeus of Naucratis (ca. 200)
Augustine of Hippo (354–430)
Augustus (63 B.C.–14 A.D.)
Aurelius, Marcus (121–180)
Averroës (see Abu-l-Walid Muhammad ibn Ahmad ibn Rushd)
Avicenna (see Ibn Sina)

Babuchin, Aleksandr Ivanovich (1835–1891)
Babukhin, Aleksandr Ivanovich (see Babuchin)
Bacci, Andrea (1524–1600)
Bachiacca, Francesco (see Ubertini)
Bacon, Francis (Baron Verulam) (1561–1626)
Bacon, Roger (c. 1214–1294)
Baerle, Caspar van (see Van Baerle)
Bajon, Bertrand (fl. 1760s–80s)
Baker, George (fl. 1770s)
Baltasar (Baltazar, Balthasar), Juan de (1601–1658)
Baltazar, Juan de (see Baltasar)
Balthasar, Juan de (see Baltasar)
Balzac, Honoré de (1799–1850)
Bancroft, Edward (1744–1820)
Bancroft, Edward Nathaniel (1772–1842)
Banks, Joseph (1743–1820)
Barbaro, Ermolao (1453/54–1493)
Barberini, Francesco (1597–1679)
Barbeu-Dubourg, Jacques (1709–1799)
Barlaeus, Caspar (see Van Baerle)
Baron Verulam (see Bacon, F.)
Barreto, João Nuñes (c. 1510–1562)
Bartholin, Caspar (1655–1738)
Bartholin, Christopher (1657–1714)
Bartholin, Thomas (1616–1680)
Bartram, Isaac (1725–1801)
Bartram, John (1699–1777)
Bassi, Laura (1711–1778)
Baxt, Nikolai (1843–1904)

Beauharnais, Eugène Napoleon de (see Napoleon, E.)
Beccaria, Giambattista (1716–1781)
Becquerel, Antoine Caesar (1788–1878)
Bede (672/73–735)
Beerenbroeck, Arnould-Barthélemi (1751–1825)
Behn, Aphra (1640–1689)
Belcher, Jonathan (1682–1757)
Bell, Charles (1774–1842)
Bellini, Lorenzo (1643–1704)
Belon, Pierre (1517–1564)
Benedetti, Giovanni Battista (1530–1590)
Bennett, Michael V.D. L. (b. 1931)
Berkel, Adriaan van (see Van Berkel)
Bermudes (Bermudez), João (d. 1570)
Bermudez, João (see Bermudes)
Bernard, Claude (1813–1868)
Bernouilli, Jean (1667–1748)
Bernstein, Julius (1839–1917)
Bertholon, Pierre (1741–1800)
Bertruccio, Niccolò (d. 1347)
Biddle, Owen (1737–1799)
Bilharz, Theodor (1825–1862)
Bilroth, Theodor (1829–1894)
Blainville, Henri Marie Ducrotay de (1777–1850)
Bloch, A. M. (fl. 1880s)
Bloch, Marcus Elieser (1723–1799)
Blumenbach, Johann Friedrich (1752–1840)
Boaz, George (1891–1980)
Boerhaave, Herman (1668–1738)
Boëthius, Anicius Manlius Severinus (c. 480–c. 524)
Bohr, Niels (1885–1962)
Bois-Reymond, Emil du (1818–1896)
Boll, Franz (1849–1879)
Bolos Democritus (see Bolos of Mendes)
Bolos of Mendes, the Democritic (3rd cent. B.C.)
Bonaparte, Carlo Luciano (Charles-Lucien) (1803–1857)
Bonaparte, Napoleon (1761–1821)
Bonnet, Charles (1720–1793)
Bonpland, Aimé Alexandre (Goujoud) (1733–1858)
Borelli, Giovanni Alfonso (1608–1679)
Botticelli, Sandro (c. 1445–1510)
Bowditch, Henry Pickering (1840–1911)
Boyle, Robert (1621–1691)
Bracciolini, (Gian Francesco) Poggio (1380–1459)
Braulio, Bishop of Saragossa (d. 650)
Breasted, James (1865–1935)
Breschet, Gilbert (1784–1845)
Brisson, Mathurin-Jacques (1723–1806)
Brodie, Benjamin (1783–1862)
Broussonet, Pierre Marie Auguste (1761–1807)
Browne, Thomas (1605–1682)
Brunelleschi, Filippo (1377–1446)
Bruno, Giordano (1548–1600)
Bry, Theodoris de (see de Bry)
Bryant, William (1730–1786)
Brydone, Patrick (1741–1819)
Buonaparte, Napoleone di (see Bonaparte, Napoleon)
Bull, William (1683–1755)
Buonarroti, Michelangelo di Lodovico (1475–1564)
Burdon-Sanderson, John Scott (1828–1905)
Burgoyne, John (1722–1792)
Buridan, John (before 1300–c. 1361)
Burke, Edmund (1729–1797)
Byron, George Gordon (1749–1832)

Cabral, Pedro Álvares (c. 1467–1520)
Cajal, Santiago Ramón y (see Ramón y Cajal)
Calcidius (fl. 321)

Caldani, Marc'Antonio (1758–1794)
Caligula (Gaius Julius Caesar Augustus Germanicus (12–41 A.D.)
Callistus, Gaius Julius (1st cent.)
Campanella, Tommaso (1568–1639)
Cannon, Walter (1871–1945)
Cantharus (fl. 422 B.C.)
Canton, John (1718–1772)
Capella, Martianus (fl. 5th cent.)
Cardano, Girolamo (Hieronymus Cardanus) (1501–1576)
Cardim, Fernão (ca. 1548–1625)
Carlisle, Anthony (1768–1842)
Carminati, Bassiano (1750–1830)
Carpi, Pietro (d. 1861)
Carus, Titus Lucretius (see Lucretius)
Casseri, Giulio Cesare (1552–1616)
Cassini, Gian Domenico (Jean-Dominique)(1625–1712)
Cassiodorus, Flavius Magnus Aurelius (c. 485–c. 585)
Castanhoso, Miguel de (fl. 1540s)
Catsby, Mark (1682–1749)
Cavallo, Tiberius (Tiberio) (1749–1809)
Cave, Edward (1691–1754)
Cavendish, Charles (c. 1700–1783)
Cavendish, Henry (1731–1810)
Celsus (c. 25 B.C.–50 A.D.)
Cervantes, Miguel de (1547–1616)
Cesalpino, Andrea (1519–1603)
Cesi, Federico (1585–1630)
Chagas, Carlos (Jr.) (1910–2000)
Chagas, Carlos (Sr.) (1879–1934)
Chagovec, Vasilij Jur'evich (1873–1941)
Chanbers, Lionel (1715–1777)
Chang, Chuan-Chiung (b. 1928)
Changeau, Jean-Pierre (b. 1936)
Charcot, Jean-Martin (1825–1893)
Charles II (of England) (1630–1685)
Charles XI (of Sweden) (1655–1697)
Chaucer, Geoffrey (c. 1343–1400)
Chauvin, Étienne (1640–1725)
Cheyne, George (1671–1743)
Chircher, Athanasius (see Kircher)
Choffin, David Etienne (1703–1773)
Cicero, Marcus Tullius (106–43 B.C.)
Cigoli, Ludovico (1559–1613)
Claudian (Claudius Claudianus) (c. 370–410)
Claudino, Giulio Cesare (1550–1618)
Claudius (of Ethiopia) (see Gelawdewos)
Claudius (of Rome) (10 B.C.–54 A.D.)
Clearchus of Soli (fl. 4th cent. B.C.)
Cleland, John (d. 1789)
Cleopatra VII (69–30 B.C.)
Clive, Robert (1725–1774)
Cobo, Bernabé (1580–1657)
Colden, Cadwallader (1688–1776)
Cole, Kenneth (1900–1984)
Colladon, Jean Daniel (1802–1893)
Collinson, Peter (1694–1768)
Columbus, Christopher (Cristoforo Colombo) (1451–1506)
Combe, William (1741–1823)
Commodus (161–192)
Comte, Auguste (1798–1857)
Condamine, Charles-Marie de la (1701–1774)
Configliachi, Pietro (1779–1844)
Constantine I (c. 272–377)
Constantinus Africanus (c. 110–1087)
Cook, James (1728–1779)
Copernicus, Nicolaus (1473–1543)
Cornwallis, Charles (1738–1805)
Count Rumford (see Thompson)

Coyau, Gabriel (fl. 1770s)
Crescenzi, Pietro (de) (c. 1233–c. 1321)
Cruickshank, William (d. 1810 or 1811)
Croone, William (1633–1684)
Cuchon,? (fl. 1770s)
Cunaeus, Andreas (1712–1788)
Curtis, Howard (1906–1972)
Cuvier, Georges Leopold Chretien Frédéric Dagobert (1769–1832)
Cybulski, Napoleon Nikodem (1854–1919)

d'Abbeville, Claude d' (see Abbeville)
d'Alibard, Thomas-François (see Alibard)
d'Abano, Pietro (c. 1210–c. 1316)
d'Ascoli, Cecco (1257–1327)
d'Aurillac, Gerbert (946–1003)
da Gama, Cristóvão (c. 1516–1542)
da Gama, Vasco (c. 1460–1524)
da Vinci, Leonardo (1452–1519)
Dale, Henry (1875–1968)
Dante (see Alighieri)
Dapper, Olfert (c. 1635–1689)
Darwin, Charles (1809–1882)
Darwin, Erasmus (1731–1802)
Davy, Humphry (1778–1829)
Davy, John (1790–1868)
de Beauvais, Vincent (1190–c. 1254)
de Vries, Hugo (1848–1935)
de' Medici, Lorenzo di Pierfrancesco (1463–1503)
Deane, Silas (1737–1789)
de Bry, Theodoris (1528–1598)
della Francesca, Piero (1415–1492)
della Porta, Giambattista (see Porta)
Demetrius of Ixion (2nd cent. B.C.)
Democritus of Abdera (460–370 B.C.)
Dengel, Lebna (Emperor) (1508–1540)
Descartes, René (1596–1650)
Dias, Bartolomeu (c. 1451–1590)
Dickens, Charles (1812–1870)
Dioscorides of Anazarbos (fl. 60–70 A.D.)
Diphilus of Laodicea (1st cent.)
Djoser (see Nesu Netjerikhet)
Dohrn, Anton (1840–1909)
Donne, John (1572–1631)
Dos Santos, João (c. 1564–1622)
Dryden, John (1631–1700)
du Bois-Reymond, Emil (1818–1896)
Dubourg, Jacques Barbeu (see Barbeu-Dubourg)
DuFay, Charles François de Cisternay (1698–1739)
Dumas, Jean-Baptiste (1800–1884)
Duns Scotus, John (1265/66–1308)
Dürer, Albrecht (1471–1528)

Ebers, Georg (1837–1898)
Eccles, John Carew (1903–1997)
Elizabeth I (of England) (1533–1603)
Ellis, John (1710–1776)
Empedocles (490–430 B.C.)
Epicharmus of Megara (c. 540–c. 450 B.C.)
Epicurus (341–270 B.C.)
Erastus, Thomas (1524–1583)
Euclid (fl. 300 B.C.)
Eugene of Palermo (1130–1202)
Evans, Cadwalader (1716–1773)

Faber, Johannes (see Schmidt)
Fabiano, Alexander (17th cent.)
Fabri, Giacinto Bartolomeo (c. 1726–c. 1783/94)

Fabricius (ab Acquapendente), Hieronymus (Girolamo Fabrici d'Acquapendente) (1537–1619)
Faraday, Michael (1791–1867)
Feldberg, Wilhelm (1900–1993)
Felici, Costanzo (1625–1585)
Ferdinando I (de' Medici) (1549–1609)
Ferdinando II (de' Medici) (1610–1670)
Fermin, Philippe (1720–1790)
Fernandes (Fernández), Antonio (1569–1642)
Fernández, Antonio (see Fernandes)
Fernel, Jean (1497–1558)
Fessard, Alfred (1900–1982)
Fibonacci, Leonardo (c. 1130–c. 1250)
Fichte, Johann Gottlieb (1762–1814)
Ficino, Marsilio (1433–1499)
Ficinus, Marsilius (see Ficino)
Finch, John (1626–1682)
Fizeau, Hippolyte Louis (1819–1896)
Flagg, Henry Collins (1742–1801)
Flaubert, Gustave (1821–1880)
Fleischl-Marxow, Ernst von (1846–1891)
Fontana, Felice (1730–1805)
Fordyce, George (1706–1802)
Forlì, Jacopo da (d. c. 1414)
Forsskål, Petrus (1732–1763)
Fothergill, John (1712–1780)
Fouchy, Grand-Jean de (1707–1788)
Fowke, Arthur (1756–1775)
Fracastoro, Girolamo (1478–1553)
Francesco I (de' Medici) (1541–1587)
François de Tournon (Cardinal) (1489–1652)
Frank, Johann Peter (1745–1821)
Franklin, Benjamin (1706–1790)
Franklin, William (1731–1813)
Frederick I of Barbarossa (1122–1190)
Frederick II of Hohenstaufen (1194–1250)
Frederick II, the Holy Roman Emperor and King of Sicily (1197–1250)
Freising, Otto von (c. 1112–1158)
Freud, Sigmund (1856–1939)
Fritsch, Gustav (1837–1927)
Frumentius (Saint) (d. 383)
Fuchs, Leonhart (1501–1556)
Fumagalli, Pietro (fl. 1550s)

Gabriel, João (Giovanni Gabrielli, fl. 1554–1626)
Galeazzi, Domenico Gusmano (1686–1775)
Galeazzi, Lucia (1743–1790)
Galen (c. 129– c. 200)
Galilei, Galileo (1564–1642)
Galileo (see Galilei)
Galli, Giovanni Antonio (1708–1782)
Galvani, Luigi (1737–1798)
Gama, Cristóvão da (see da Gama, C.)
Gama, Vasco da (see da Gama, V.)
García, Matías (d. 1691)
Garden, Alexander (1730–1791)
Gardini, Francesco Giuseppe (1740–1816)
Garten, Siegfried (1871–1923)
Gaubius, Hieronymus David (1705–1780)
Gay-Lussac, Joseph Louis (1788–1850)
Gaza, Theodorus (Theodore Gazis) (c. 1400–1475)
Geissler, Johann Heinrich (1814–1879)
Gelawdewos (Claudius; of Ethiopia) (1540–1559)
Geoffroy Saint-Hilaire, Étienne (1772–1844)
George III (of England) (1738–1820)
Gerard of Cremona (c. 1114–1187)
Gerardus Cremonensis (see Gerard of Cremona)

Gervase of Tilbury (c. 1150–c. 1228)
Gessner, Conrad (1516–1565)
Gibbs, Josiah Willard (1839–1903)
Gilbert, William (1544–1603)
Gilles, Pierre (1490–1555)
Giorgi, Eusebio (d. 1845)
Giovio, Paolo (1483–1552)
Girardi, Michele (1731–1797)
Glisson, Francis (1597–1677)
Glycas, Michael (fl. 12th cent.)
Gmelin, Johann Friedrich (1748–1804)
Gockel, Eberhard (1636–1703)
Godinho, Nicolao (1556–1616)
Goethe, Johann Wolfgang von (1749–1832)
Góis, Damião de (1502–1574)
Golpaia, Camillo della (15th–16th cents.)
Gordon, James (fl. 1740s)
Gorgoreyos, Abba (fl. 1680)
Gotch, Francis (1853–1913)
Goujaud, Aimé Alexandre (see Bonpland)
Gragn (see al-Ghazi)
Gratian (Gratianus), Flavius (359–383)
Gravesande, Jonathan Samuel Storm (see Van 's Gravesande, J.)
Gravesande, Laurens Storm van 's (see Van 's Gravesande, L.)
Gravesande, Willem Jacob (see Van 's Gravesande, W.)
Gray, Stephen (1666–1736)
Greenwood, John (1727–1792)
Gregory XIII (Pope) (1502–1585)
Gregory, Richard (b. 1923)
Gren, Friedrich Albrecht Carl (1760–1798)
Grévin, Jacques (1538–1570)
Gronov (Gronow; Gronovius), Laurens Theodor (1730–1777)
Gronovius, Laurentius Theodorus (see Gronov)
Gronovius, Jan (Johann) Fredrick (1686–1762)
Gronow, Laurens Theodor (see Gronov)
Grosseteste, Robert (c. 1175–1253)
Grundfest, Harry (1904–1983)
Guericke, Otto von (1602–1686)
Guerreiro, Fernão (1567–1617)
Gumilla, Joseph (1686–1750)
Gutenberg, Johannes (1400–1468)

Hakluyt, Richard (ca. 1562–1616)
Hales, Stephen (1677–1761)
Hall, David (1714–1772)
Haller, Albrecht von (1708–1777)
Hals, Frans (1580–1666)
Hamilton, William (1734–1803)
Harpocration of Alexandria (c. 2nd cent. A.D.)
Hartley, David (1705–1757)
Harvey, William (1578–1657)
Hawksbee, Francis (1666–1713)
Helmholtz, Hermann (1821–1894)
Henly, William (d. 1779)
Henrique o Navegador (see Henry the Navigator)
Henry the Navigator (Henrique o Navegador) (1394–1460)
Her-watet-khet (Watet Hathor) (c. 2340 B.C.)
Heraclides, Ponticus (390–310 B.C.)
Hering, Karl Ewald Konstantin (1834–1918)
Herman of Carinthia (1114–1187)
Hermann, Ludimar (1838–1914)
Hero of Alexandria (c. 10–70)
Herodotus of Halicarnassus (c. 484–c. 425 B.C.)
Hewatt, Alexander (fl. 1770s)
Hewson, William (1739–1774)
Hieronymus, Sophronius Eusebius (see Saint Jerome)
Hippocrates (c. 460–c. 370 B.C.)
Hispanus, Petrus (1215–1277)

Hodgkin, Alan Lloyd (1914–1998)
Hodierna, Giovanni Battista (1597–1660)
Hohenheim, Theophrastus Bombastus von (Paracelsus) (1493–1541)
Hollingsworth, Levi (1739–1824)
Hondius, Jodocus (1563–1612)
Honorius (384–423)
Hooke, Robert (1635–1703)
Hopkinson, Thomas (1709–1751)
Horapollo Niloticus (c. 600)
Hosack, David (1769–1835)
Humboldt, Alexander von (1769–1859)
Humboldt, Wilhelm von (1767–1835)
Hume, David (1711–1776)
Hunter, John (1728–1793)
Hunter, William (1718–1783)
Huxley, Andrew Fielding (b. 1917)
Huxley, Thomas (1825–1895)
Huygens, Christian (1620–1695)

Ibn al-Bītār (d. 1248)
Ibn Rushd, Abdul Walid Muhammad ibn Ahmad (see Averroës)
Ibn Sina (Avicenna) (980–1037)
Imhotep (c. 2650–2575 B.C.)
Ingenhousz, Jan (1730–1799)
Ingram, Dale (1710–1793)
Innocent IV (Pope) (c. 1195–1254)
Isaac, Antoine (see Silvestre de Sacy, Baron)
Isidore of Seville (ca. 560–636)
Iulianus, Flavius Claudius (see Julian the Apostate)

Jacobaeus, Oligerus (see Jocobsen)
Jallabert, Jean (1712–1768)
James I (of Britain) (1566–1625)
Jaucourt, Louis de (1704–1779)
Jefferson, Thomas (1743–1826)
Jobson, Richard (fl. early 1600s)
Jocobsen (Jacobaeus), Holger (1650–1701)
John XXI, Pope (see Hispanus)
Johnson, Samuel (1709–1784)
Jones, John Paul (1747–1792)
Jovian (Iovianus), Flavius (331–364)
Juba II (King of Numidia) (29–27 B.C.)
Julian the Apostate (331–363)
Julius III (Pope) (1487–1555)
Justinian, Flavius Anicius (Justinian the Great) (c. 482–565)
Jussieu, Bernard de (1699–1777)

Kaempfer (Kämpfer), Englebert (1651–1716)
Kant, Immanuel (1724–1804)
Katz, Bernard (1911–2003)
Kepler, Johannes (1571–1630)
Keymis, Lawrence (c. 1564–1618)
Keynes, Richard Darwin (1919–2010)
Khnum-Hotep (c. 2400 B.C.)
Kinnersley, Ebenezer (1711–1778)
Kircher, Athanasius (Atanasio) (1602–1680)
Kirchhoff, Gustav Robert (1824–1887)
Kleist, Ewald Georg von (1715–1759)
Kolben, Peter (1675–1726)
Kratzenstein, Christian Gottlieb (1723–1795)
Krüger, Johann Gottlob (1715–1759)
Kuhn, Thomas (1922–1996)
Kühn, Karl Gottlob (1754–1840)

La Roche, Daniel de (see Roche)
Lacépède (Bernard-Germain-Etienne de la Ville-sur-Illon, Compte de La Cépède) (1756–1825)
Lafayette, Marquis de (Gilbert du Motier) (1757–1834)

Laghi, Tommaso (1709–1764)
Lane, Timothy (1743–1807)
Langrish, Browne (d. 1759)
Laurentius (late 2nd–early 3rd cents.)
Le Roy, Jean-Baptiste (1726–1779)
Le Verrier, Jean Joseph (1811–1877)
Lee, Arthur (1740–1792)
Lee, Chen-Yuan (1915–2001)
Leeuwenhoek, Anton van (see Van Leeuwenhoek)
Leo III, Pope (d. 816)
Lettsom, John Coakley (1744–1815)
Leucippus (ca. 430 B.C.)
Ligozzi, Jacopo (1547–1627)
Linari, Santi (1777–1858)
Lindestolpe, Johan (1676–1724)
Lining, John (1708–1760)
Linnaeus, Carolus (see Linné)
Linné (Linnaeus), Carl (1707–1778)
Lissmann, Hans Werner (1909–1995)
Lobo, Jerónimo (1593–1678)
Locke, John (1602–1734)
Loewi, Otto (1873–1961)
Logan, James (1674–1751)
Lorenzini, Stefano (1645–1725)
Lott, Frans van der (see Van der Lott)
Louis XV (of France) (1710–1774)
Lovejoy, Lucretia (see Perry)
Lovell, Robert (c. 1630–1690)
Lower, Richard (1631–1691)
Lucas, Keith (1879–1916)
Lucretius, (Titus Lucretius Carus) (ca. 99 BC–ca. 55 BC).)
Ludolf, Hiob (1624–1704)
Ludwig, Carl (1816–1895)
Lusitanus, Amatus (João Rodrigues de Castelo Branco) (1511–1568)

MacDonald, John Smith (1867–1941)
Macrobius, Ambrosius Theodosius (395–423)
Magalhães, Fernão (see Magellan)
Magellan, Ferdinand (Fernão Magalhães) (1480–1521)
Magnus, Albertus (see Albertus Magnus)
Maimonides, Moses (Rabbi Moshe ben Maimon) (c. 1137–1204)
Makeda (Queen of Sheba) (10th cent. B.C.)
Malpighi, Marcello (1628–1694)
Manasse, Constantine (c. 1130–c. 1187)
Mandeville, John (14th cent.)
Manuel I (King of Portugal) (1469–1521)
Manuzio, Aldo (1449–1515)
Marcellus (Empiricus) of Bordeaux (c. 400 A.D.)
Marcellus of Side (2nd cent. A.D.)
Marcgraf, Georg (1610–1643/4)
Marey, Etienne-Jules (1830–1904)
Mariette, Auguste (1821–1881)
Marlowe, Christopher (1564–1593)
Marmont, George Heinemann (1914–1983)
Marsili, Luigi Ferdinando (1638–1730)
Martens, Conrad (1801–1878)
Martins-Ferreira, His (b. 1920)
Maskelyne, Nevil (1732–1811)
Massari, Francesco (Francesco Maser) (fl. 1542)
Matteucci, Carlo (1811–1868)
Mattioli, Pier'Andrea (1501–1577)
Maty, Matthew (1718–1776)
Mauricius, Jan Jacob (1692–1768)
Maurus, Rabanus (Hrabanus) (c. 780–856)
Maximilian (Holy Roman Emperor) (1459–1519)
Maxwell, James Clerk (1831–1879)
Mayow, John (1640–1679)
Megenberg, Konrad von (1309–1374)

Mendes, Afonso (1579–1656)
Menes (c. 3100–3000 B.C.)
Mercuriale, Girolamo (Geronimo) (1530–1606)
Mereruka (Meri) (c. 2340 B.C.)
Meri (see Mereruka)
Meriteti (c. 2340 B.C.)
Messalina (wife of Claudius) (c. 17–48 A.D.)
Meyssonnier, Lazare (1602–1672)
Michelangelo (see Buonarroti)
Milne-Edwards, Henri (1880–1885)
Milton, John (1608–1664)
Miranda, Domenico de (ca. 1845)
Monro, Alexander (1698–1767)
Montaigne, Michel Eyquem de (1533–1592)
Moore, Francis (fl. 1730s)
Moreau, Armand (1823–1881)
Moruzzi, Giuseppe (1910–1986)
Motier, Gilbert du (see Lafayette)
Müller, Johannes (1801–1858)
Musschenbroek, Pieter van (see Van Musschenbroek)

Nairne, Edward (1726–1806)
Napoleon (Emperor) (see Bonaparte, Napoleon)
Napoleon, Eugène (1810–1835)
Narmer (c. 3150 B.C.)
Neferikara (c. 2400 B.C.)
Neher, Erwin (b. 1944)
Nemesius (c. 390)
Nernst, Walther (1864–1941)
Nero (37–68 A.D.)
Nesu Netjerikhet (Djoser) (c. 2650–2575 B.C.)
Nettesheim, Cornelius Agrippa von (1486–1535)
Newton, Isaac (1643–1727)
Niankh-Khnum (c. 2400 B.C.)
Nicholson, William (1753–1815)
Niebuhr, Carsten (1733–1815)
Nobili, Leopoldo (1784–1835)
Nollet, Jean-Antoine (Abbé) (1700–1770)
Nonius Marcellus (fl. 300 A.D.)
Numa, Shozaku (1929–1992)
Nyuserra (Niuserre) (c. 2400 B.C.)

Ockham, William of (c. 1280– c. 1349)
Octavian (Octavianus, Augustus) (63 B.C.–14 A.D.)
Odoric of Pordenone (Odorico di Pordenone) (c. 1286–1331)
Oersted, Hans Christian (see Ørsted)
Ohm, Johann Wolfgang (1789–1854)
Ojeda, Alonso de (1465–1515)
Oker-Blom, Maximilian Ernst Gustaf (1863–1917)
Olyphant, David (1720–1804)
Oppian (Oppianus) of Corycus (2nd cent.)
Oppianus (see Oppian)
Oresme, Nicole of (c. 1320–1382)
Oribasius (c. 320–400)
Orioli, Francesco (1778–1852)
Origen(es) (c. 185–254)
Ørsted, Hans Christian (1777–1851)
Ostwald, Wilhelm (1853–1932)
Otto IV (Holy Roman Emperor) (c. 1175–1218)
Overton, Charles Ernest (1865–1933)
Oviedo y Valdez, Gonzalo Fernández de (1478–1557)
Oviedo, Andrés de (1518–1577)
Owen, Richard (1804–1892)

Paci, Giacomo Maria (ca. 1845)
Pacini, Filippo (1812–1883)
Paez (Paéz, Pêro Pais), Pedro (Pêro) (1564–1622)
Pais, Pedro (see Paez)

Pais, Pêro (see Paez)
Palmieri, Luigi (1807–1896)
Panofsky, Erwin (1892–1968)
Paracelsus (see Hohenheim)
Pasteur, Louis (1822–1895)
Paterson, William (1755–1810)
Paul of Aegina (Paulus Aegineta) (c. 625–690)
Paulus Aegineta (see Paul of Aegina)
Pease, Jocelyn Richenda (1925–2003)
Pecham, John (c. 1230–1292)
Penn, William (1644–1714)
Peretti, Pietro (1781–1864)
Perry, James (1756–1821)
Peter the Great (1672–1725)
Pfeffer, Wilhelm Friedrich Philipp (1845–1920)
Pflüger, Eduard Friedrich Wilhelm (1829–1910)
Philip of Macedonia (382–336 B.C.)
Philoponus of Alexandria, John (490–570)
Philoxenos of Citera (c. 436–c. 380 B.C.)
Pian del Carpine, Giovanni da (c. 1180–1252)
Pianciani, Giovan Battista (1784–1862)
Piccolos, Nikolaos Savva (see Pikkolos)
Pikkolos, Nikolaos Savva (or Piccolos) (1792–1865)
Piso Willelm (or Wilhelm) (1611–1678)
Pitcairn, William (1712–1791)
Pivati, Giovanni Francesco (1689–1764)
Pizarro, Francisco (1474–1541)
Pizzetta, Jules (1820–1900)
Plato (c. 428–c. 348 B.C.)
Plato Comicus (fl. 400 B.C.)
Pletho, Gemistus (c. 1355–1452)
Pliny the Elder (23–79)
Pliny the Younger (c. 61–c. 112 A.D.)
Plotinus (c. 205–270)
Plutarch(us) (of Chaeronea) (c. 46–127)
Pococke, Edward (1648–1727)
Poe, Edgar Allan (1809–1849)
Poisson, Siméon-Denis (1781–1840)
Polo, Marco (1254–1324)
Pomponazzi, Pietro (1462–1525)
Ponticus, Evagrius (345–399)
Ponzetti, Ferdinando (1444–1527)
Pope, Alexander (1688–1744)
Porphyrius (235–305)
Porta, Giambattista della (c. 1535–1615)
Pouillet, Claude (1791–1868)
Praeconimus, Aelius Stilo (154–74 B.C.)
Prévost, Antoine François (1697–1763)
Priestley, Joseph (1733–1804)
Pringle, John (1707–1782)
Prior, Matthew (1664–1721)
Ptolemy (c. 90–c. 168)
Pulci, Luigi (1432–1484)
Purchas, Samuel (1577–1626)
Pythagoras (c. 580–500 B.C.)

Queen of Sheba (see Makeda)
Quibell, James E. (1867–1935)

Raleigh, Walter (c. 1552–1618)
Ramée, Pierre de la (1515–1572)
Ramón y Cajal, Santiago (1852–1934)
Ranvier, Louis-Antoine (1835–1922)
Ranzi, Andrea (1810–1859)
Réaumur, René-Antoine Ferchault de (1683–1757)
Redi, Francesco (1626–1698)
Regius, Henricus (1598–1679)
Rhazes (see al-Razi)

Richer, Jean (1630–1696)
Riolan, Jean (1577–1657)
Rittenhouse, Benjamin (1740–1825)
Rittenhouse, David (1732–1796)
Ritter, Johann Wilhelm (1776–1810)
Robin, Charles-Philippe (1821–1885)
Roche, Daniel de La (fl. 1770s)
Rodrigues de Castelo Branco, João (see Lusitanus)
Roger II, King (1095–1154)
Ronayne, Thomas (d. ca 1800)
Rondelet, Guillaume (1507–1566)
Rose, William (fl. 1740s)
Rosenthal, Isidor (1836–1915)
Rowlandson, Thomas (1756–1827)
Rubruck, William of (1220–1293)
Rutherford, John (1695–1779)
Ruysch, Frederik (1638–1731)

Sachs, Carl (1853–1878)
Saint Augustine of Hippo (354–430)
Saint Jerome (c. 347–420)
Saint Louis IX (of France) (1214–1270)
Saint-Hilaire, Étienne Geoffroy (see Geoffroy Saint-Hilaire)
Sakmann, Bert (b. 1942)
Saladin (Salah ad-Din Yusuf ibn Ayyub) (c. 1138–1193)
Salviani, Ippolito (1514–1572)
Sano, Buckor (fl. early 1600s)
Santorio, Santorio (1561–1636)
Santos, João dos (see Dos Santos)
Savi, Paolo (1798–1871)
Scaligero, Giulio Cesare (1484–1558)
Schelling, Friedrich Wilhelm Joseph (1775–1854)
Schiller, Friedrich (1759–1805)
Schilling, Gottfried Wilhelm (1725–1799)
Schmidt (Faber), Johannes (1574–1629)
Schönbein, Christian Friedrich (1799–1868)
Schönlein, Karl (1858–1899)
Schweigger, Johann Solomo (1779–1857)
Scribonius Largus (c. 3 B.C.–c. 54 A.D.)
Seba, Albert(us) (1665–1736)
Sebond, Raymond (d. 1436)
Seebeck, Thomas Johann (1770–1831)
Seignette, Pierre Henri (1734–1808)
Senbi (1991–1962 B.C.)
Seneca, Lucius Annaeus (c. 4 B.C.–65 A.D.)
Suetonius (Gaius Suetonius Tranquillus) (c. 69–after 130)
Severino, Marco Aurelio (1580–1656)
Shakespeare, William (1547–1616)
Sherrington, Charles Scott (1857–1952)
Shipley, William (1714–1803)
Silvester II, Pope (see d'Aurillac)
Silvestre de Sacy, Antoine Isaac, Baron (1758–1838)
Simocatta, Theophylact (d. 640)
Simplicus of Cilicia (c. 490–c. 560)
Sittardus, Cornelius (1515–1544)
Sloan, Hans (1660–1753)
Smith, Edwin (1822–1906)
Smith, James Edward (1759–1828)
Socrates (470–399 B.C.)
Solander, Daniel (1733–1782)
Solinus, Julius Gaius (fl. 350 A.D.)
Solomon (of Israel) (c. 1011–c. 932 B.C.)
Spallanzani, Lazzaro (1729–1799)
Spencer, Adam (fl. 1740s)
Stark, James (1811–1890)
Stedman, John (1744–1797)
Steno, Nicolaus (1638–1687)
Stensen, Niels (see Steno)

Stephen of Pisa or Antioch (early 12th cent.)
Stertinius (see Xenophon Stertinius)
Stevenson, Mary (Polly) (1734–1795)
Stilicho (359–408)
Strabo (c. 64 B.C.–c. 24 A.D.)
Strato of Lampsacus (ca. 340–268 B.C.)
Strong, Adam (see Perry)
Suetonius (Gaius Suetonius Tranquillus) (c. 69–c. 122)
Swammerdam, Jan (1637–1680)
Sydenham, Thomas (1624–1689)

Taylor, Edward (1642–1729)
Telesio, Bernardino (1509–1588)
Telles, Balthasar (1596–1675)
Termeyer, Ramón María (1737–c. 1814)
Teti (c. 2340 B.C.)
Thales of Miletos (624–546 B.C.)
Thénard, Jacques Louis (1777–1857)
Theodore of Antioch (c. 350–428)
Theodosius, Flavius (Theodosius the Great) (347–395)
Theophrastos of Eresos (c. 371–c. 287 B.C.)
Thévenot, Melchisédech (1620–1692)
Thomas de Cantimpré (1201–1272)
Thompson, Benjamin (Count Rumford) (1753–1814)
Thompson, D'Arcy (1860–1948)
Ti (c. 2400 B.C.)
Tiberius, Emperor (42 B.C.–37 A.D.)
Titian (Tiziano Vecellio) (c. 1488–1576)
Tiziano Vecellio (see Titian)
Todd, John Tweedie (1789–1840)
Tolstoy, Leo (1828–1910)
Torreblanca, Villalpando Francisco (d. 1645)
Torricelli, Evangelista (1608–1647)
Traube, Moritz (1826–1894)
Tschermak von Seysenegg, Armin (1870–1952)
Turner, Robert (fl. 1746)
Twain, Mark (Samuel Clemens) (1835–1910)

Ubertini, Francesco (1497–1557)
Unzer, Johann August (1727–1799)
Urban VIII, Pope (1568–1644)
Urreta, Luis de (ca. 1570–1636)

Valdez, Gonzalo Fernández de Oviedo y (see Oviedo y Valdez)
Valens, Flavius Julius (328–378)
Valeriano (Giovan Pietro dalle Fosse) (1477–1558)
Valeriano, Piero (1477–1558)
Valli, Eusebio (1762–1816)
Van Baerle, Caspar (1584–1648)
Van Berkel, Adriaan (ca. late 17th cent.)
Van der Horst, Gijsbert (16th cent.)
Van der Lott, Frans (fl. 1750s, d. 1804)

Van Helmont, Jean Baptiste (1579–1644)
Van Leeuwenhoek, Anton (1632–1723)
Van Musschenbroek, Pieter (Petrus) (1692–1761)
Van 's Gravesande, Jonathan Samuel Storm (1723–1750)
Van 's Gravesande, Laurens Storm (1704–1775)
Van 's Gravesande, Willem Jacob (1688–1742)
Van't Hoff, Jacobus Henricus (1852–1911)
Vanini, Giulio Cesare (1585–1619)
Varro, Marco Terentius (116–27 B.C.)
Veratti, Giuseppe (1707–1793)
Verdi, Guiseppe (1813–1901)
Verulam (Baron) (see Bacon, F.)
Vesalius, Andreas (1514–1564)
Vespucci, Americo (1454–1512)
Villars, Charles-René de (b. 1698)
Ville-sur-Illon, Bernard-Germain-Etienne de la (see Lacépède)
Vincent, Levinus (1658–1727)
Vincent of Beauvais (see de Beauvais)
Vinci, Leonardo da (see da Vinci)
Vitelo (see Witelo)
Viviani, Domenico (1772–1840)
Volta, Alessandro (1745–1827)

Walker, Mary Broadfoot (1888–1974)
Walsh, John (1726–1795)
Walter of (or de) Henley (fl. 1280)
Walter, Richard (ca. 1716–1785)
Walter, Thomas (c. 1740–1789)
Wansleben, Johann Michael (1635–1679)
Warburg, Aby Moritz (1866–1929)
Warren, George (17th cent.)
Washington, George (1732–1799)
Watet Hathor (see Her-watet-khet)
Watson, William (1715–1787)
Wesley, John (1703–1791)
Whytt, Robert (1714–1766)
William I of Sicily (1120–1166)
William II of Sicily (1155–1189)
William of Rubruck (see Rubruck)
Williams, Thomas (fl. mid-18th cent.)
Williamson, Hugh (1735–1819)
Willis, Thomas (1621–1675)
Witelo, Erazmus (c. 1230–before 1314)
Woolf, Virginia (1882–1941)

Xenophon of Athens (c. 430–355 B.C.)
Xenophon Gaius Stertinius (c. 10 B.C – 54 A.C)

Young, John Zachary (1907–1997)

Zabarella, Jacopo (1532–1589)

References

[Notice that the indication *Anon.* for anonymous works is used, at the beginning of the References list, to indicate articles, pieces of poetry, letters, and responses to pieces without the author's name being provided. For advertisements, news and notices, and for various other works for which we cannot provide and author's name, the pieces will be listed alphabetically under the title of the periodical or the volume in which they appeared].

Anon. (1731a). [No Title: Poem about torpedo-like corruption.] *Echo or Edinburgh Weekly Journal,* June 23, Issue CXXVIII.

Anon. (1731b). [No Title: Poem about torpedo-like corruption.] *Universal Spectator and Weekly Journal,* June 12, Issue CXL.

(Anon.) (1745). An historical account of the wonderful discoveries, made in Germany, &c concerning Electricity. *Gentleman's Magazine,* 15, 192–197. & Hall.

Anon. (1747). [Response to a letter by a G. M. editor.] *Gentleman's Magazine,* 17, 77.

Anon. (1750). Abstract of Mr. Anson's voyage, Part III, containing the account of the difference of countries and people; with remarks. *Gentleman's Magazine,* 20, 64–68.

Anon. (1755). *The Gentleman and Lady's Palladium and Chronologer for the Year of Our Lord 1755, containing, besides what is usual, a Poetical View of Commodore Anson's Voyage round the World Inscribed to the Lords of the Admiralty.* London: Printed for W. Owen, pp. 42–45.

Anon. (1763). Some account of the life of the celebrated French academist Monsieur de Reaumur [sic]. *Gentleman's Magazine,* 33, 112–116.

Anon. (1765). An Essay on Conversation. *Public Advertiser,* July 22, Issue 9585.

Anon. (1771). [No Title: Article on a speaker with torpedo-like effects, under "The Literary, Political, and Miscellaneous Repository."] *General Evening Post,* January 31, Issue 5821.

Anon. (1774a). Review of Cholerick Man under "Postscript. The Theatre." *St. James's Chronicle or the British Evening Post,* December 27, Issue 2165.

Anon. (1774b). [No title: The stupor of a great personage.] *London Evening Post,* December 24, Issue 8230.

Anon. (1775). [Piece lambasting a lawmaker, calling him a torpedo.] *Middlesex Journal and Evening Advertiser,* June 13, Issue 970.

Anon. (1777). [An announcement of the publication of two poems by Perry]. *Gentleman's Magazine,* 47, 389.

Anon. (1789). Memoirs of the late John Cleland, Esq. *General Evening Post,* January 27, Issue 8614.

Anon. (1796). Remarks on Mr. Coleman's Preface to the Iron Chest. *Morning Post and Fashionable World,* September 6, Issue 7640.

Anon. (1797a). [Article calling an assembly torpedo-like or palsied.] *St. James's Chronicle or the British Evening Post,* March 23, Issue 6115.

Anon. (1797b). [Article about a parish torpified by new taxes.] *Craftsman or Say's Weekly Journal,* December 9, 1797; Issue 3109.

Anon. (1902). *Herdenking van het Honderdvijftigjarig Bestaan van de Hollandsche Maatschappij der Wetenschappen op 7 Juni 1902.* Martinus Nijhoff, 's Gravenhage.

Abbās, Ali Ibn (1877). *Kāmil as-Sinā'a at-Tibbiyya.* Cairo: Bulaq Government Press.

Abbeville, C. de (1614). *Histoire de la Mission des Pères Capucins en l'isle de Maragnan et Terres Circonvoisines.* Paris: Impr. de F. Huby.

Académie des Sciences. (1761). *Histoire de l'Académie Royale des Sciences. Année MDCCLX.* Paris: Imprimerie Royale.

ad-Damīrī, ibn M. (1906–08). *A Zoological Lexicon* (2 vols). Trans. by A. S. G. Jayakar. London: Luzac.

ad-Damīrī, M. (c. 1975). *Hayāt-al-Hayawān . . .* Cairo: al-Bābī al-Halabī Co.; 5th printing.

Adams, G. (1785). *An Essay on Electricity: Explaining the Theory and Practice of that Useful Science; and the Mode of Applying it to Medical Purposes. With an Essay on Magnetism. The Second Edition. Corrected and considerably enlarged by George Adams, . . .* London: Printed at the Logographic Press for the author, and sold by him.

Adams, G. (1792). *An Essay on Electricity: Explaining the Principles of the Useful Science and Describing the Instruments . . .: To which is now added, a Letter to the Author, from Mr. John Birch, Surgeon, on the Subject of Medical Electricity.* London: Printed for the author by R. Hindmarsh.

Adams. F. (1844). *The Seven Books of Paulus Aegineta, tr. from the Greek, with a Commentary Embracing a Complete View of the Knowledge Possessed by the Greeks, Romans, and Arabians on all Subjects Connected with Medicine and Surgery.* London: The Sydenham Society.

Adanson, M. (1757). *Histoire Naturelle du Sénégal.* Paris: Chez Claude-Jean-Baptiste Bauche.

Adanson, M. (1759). *A Voyage to Senegal, The Isle of Goreé, and the River Gambia.* London: Printed for J. Nourse and W. Jonhston [sic].

Adelard of Bath. (1998). *Adelard of Bath, Conversations with his Nephew: on the Same and the Different, Questions on Natural Sciences and on Birds.* (C. Burnett, ed., Trans.). Cambridge: Cambridge University Press.

Adelmann, H. (1966). *Marcello Malpighi and the Evolution of Embryology* (5 vols.). Ithaca, NY: Cornell University Press.

Adelmann, H. (1975). *The Correspondence of Marcello Malpighi.* Ithaca, NY: Cornell University Press.

Adrian, E. D. (1912). On the conduction of subnormal disturbances in normal nerve. *Journal of Physiology,* 45, 389–412.

Adrian, E. D. (1914). The all-or-none principle in nerve. *Journal of Physiology,* 47, 460–474.

Adrian, E. D. (1932/1965). The activity of the nerve fibres. In *Nobel Lectures: Physiology or Medicine, 1922–1941.* Amsterdam: Elsevier Publishing Co., pp. 293–300.

Adrian, E. D., and Zotterman, Y. (1926). The impulses produced by sensory nerve endings. Part 2: The response of a single end-organ. *Journal of Physiology,* 61, 151–171.

Adrian E. D., and Moruzzi, G. (1939). Impulses in pyramidal tract. *Journal of Physiology,* 97, 53–99.

Adye, S. P. (1773). The Electrical Fish shewn at Philadelphia & New York. 1773. Communicated by Capt. Spehen Payn Adye. The Banks Letters of the British Library, Additional Manuscripts No. 9084, pp. 9–10.

Aelian. (1562). *De historia animalium libri XVII*. Lugduni: Apud Guillielmum Rouillium.

Aelian. (1611). *Klaudiou Ailianou Peri zoon idiote[to]s biblia [iota zeta i.e. heptadeka] = Claudii Aeliani, De animalium natura libri XVII*. (Gilles, P., and Gessner, C., Eds., Trans.). Geneva: Apud Ioann Tornaesium.

Aelian. (1959). *On the Characteristics of Animals. With an English Translation by A. F. Scholfield*. Cambridge, MA: Harvard University Press.

Aepinus, F. U. T. (1756). Mémoire concernant quelques nouvelles expériences électriques remarquables. *Histoire de l'Académie Royale des Sciences et Belles-Lettres de Berlin*, 12, 101–121.

Aepinus, F. U. T. (1762). *Recueil de Différents Mémoires sur la Tourmaline*. Saint-Petersbourg: Imprimerie de l'Académie des Sciences.

Aetius. (1542). *Aetii medici Graeci contractae ex veteribus medicinae tetrabiblos* . . . Basileae: Impensis Hier. Frobenii, et Nic. Episcopii.

Agnew, W. S., et al. (1978). Purification of the tetrodotoxin-binding component associated with the voltage-sensitive sodium channel from *Electrophorus electricus* electroplax membranes. *Proceedings of the National Academy of Sciences (USA)*, 75, 2606–2610.

Agrippa, C. von Nettesheim (1533). *Henrici Cornelii Agrippae de Nettesheym: De occulta philosophia libri tres.* . . . [*Coloniae?*].

al-Qazwini (1983). *Ajā'ib al-Makhlūqāt wa-Gharā'ib al-Mawjūdāt* (5th printing). Beirut: Dar al-Afaq al Jadida.

Albertus Magnus. (1891). *Alberti Magni opera omnia/11*. Animalium Lib. XXVI (pars prior, I-XII). Parisiis: Vivès.

Albertus Magnus (1999). *Albertus Magnus on Animals: A Medieval Summa Zoologica (Vol. 2)*. Translated and annotated by K. Kitchell, Jr. and I. M. Resnick. Baltimore: Johns Hopkins University Press.

Aldini, G. (1794). *De animali electricitate dissertationes duae*. Bononiae: ex typographia Instituti Scientiarum.

Aldini, G. (1803). *An Account of Late Improvements in Galvanism*. London: Printed for Cuthell and Martin . . . by Wilks and Taylor. . . .

Aldini, G. (1804). *Essai Théorique et Expérimental sur le Galvanisme*. Paris: Imprimerie de Fournier Fils.

Aldrovandi, U. (—). Aldrovandi manuscript collection. Bologna: Biblioteca Universitaria.

Aldrovandi, U. (1599). *Ulyssis Aldrovandi Ornithologiae, hoc est, De avibus historiae libri XII*. Bononiae: Apud Franciscum de Franciscis Senensem.

Aldrovandi, U. (1613). *Ulyssis Aldrovandi De piscibus libri V et de cetis liber unus*. Bononiae: Apud Bellagambam.

Aldrovandi, U. (1640). *Ulyssis Aldrovandi patricii Bononiensis serpentum, et draconu[m] historiae libri duo*. Bononiae: Sumptibus Marci Antonii Bernie.

Alexandre de Tralles. (1556). *Alexandri Tralliani medici, libri XII, graeci et latini*. . . . Basileae: Henricus Petrus.

Alfonso Mola, M., and Martínez Shaw, C. (2004). Pedro Páez y la misión jesuítica en Etiopía en el contexto de la unión de las Coronas de España y Portugal. *Espacio, Tiempo y Forma, Serie IV, Historia Moderna*, 17, 59–75.

Alibert, J.-L. (1801). Éloge historique de Louis Galvani. *Mémoires de la Société Médicale d'Émulation, Séante a l'École de Médicine de Paris pour l'an VIII*, I-CLXVI. Paris: Chez Richard, Caille et Ravier.

Allamand, J. N. S. (1754). Bericht der Geneezinge van een Meisken, met een zeker soort van beroerdheid bezet, dewelke door hulp van electriciteit hersteld is. *Verhandelingen der Hollandsche Maatshappye der Weetenschappen, Haarlem*, 1, 485–497.

Allamand, J. N. S. (1756a). Kort verhaal van de uitwerkzelen, welke een Americaanse vis veroorzaakt op de geenen die hem aanraaken. *Verhandelingen Hollandsche Maatshappye der Weetenschappen, Haarlem*, 2, 372–379.

Allamand, J. N. S. (1756b). Kurze Erzählung der Würkungen, welche ein gewisser amerikanischer Fisch denen verursachet, welche ihn berühren. *Nützliche Samlungen*, 2.T: 429–432.

Allamand, J. N. S. (1768). Kurzer Bericht von den Wirkungen, welche ein amerikanischer Fisch bey denenjenigen verursacht, die denselben anrühren. *Neues Hamburgisches Magazin*, 1767–81, Band 4, 178–183.

Allamand, J. N. S (1774). [Letter written December 27, 1774, to Joseph Banks of the Royal Society of London.] British Library Additional Manuscripts, 8094, pp. 45–46.

Allamand, J. N. S (1775). Von den sonderbaren Wirkungen welche die Berührung eines gewissen amerikanischen Fisches verursachet. *Abhandlungen aus der Naturgeschichte, praktischen Arzneykunst und Chirurgie . . .*, Band 1: 99–104.

Allan, D. G. C. (1968). *William Shipley: Founder of the Royal Society of Arts*. London: Hutchinson.

Allan (2000). "Dear and serviceable to each other": Benjamin Franklin and the Royal Society of Arts. *Proceedings of the American Philosophical Society*, 144, 245–266.

Allan, D. G. C., and Schofield, R. E. (1970). *Stephen Hales: Scientist and Philanthropist*. London: Scolar Press.

Altamirano, M., et al. (1955). Electrical activity in electric tissue. III. Modifications of electrical activity by acetylcholine and related compounds. *Biochemical and Biophysical Acta*, 16, 449–463.

Altenmüller, H. (1973). Bemerkungen zum Hirtenlied des Alten Reiches. *Chronique d'Égypte*, 48, 211–231.

Altieri-Biagi, M. L. (1998). La lingua della scienza nel secolo di Luigi Galvani. In M. Poli (Ed.), *Luigi Galvani (1737–1798)*. Bologna: Fondazione del Monte di Bologna e Ravenna, pp. 39–60.

Alvares, F. (1540). *Ho Preste Ioam das Indias: verdadera informaçam das terras do Preste Ioam*. Lisboa: em casa de Luis Rodriguez liureiro de sua alteza.

Amari, M. (1868). *Storia dei musulmani di Sicilia 3,1*. Firenze: Le Monnier.

American Philosophical Society. (1885). August 1773. [No title; note advising members to strike a deal with the owner of an electric eel]. *Proceedings of the American Philosophical Society*, 22, 82.

Anderson, G. T., and Anderson, D. K. (1973). Edward Bancroft, M.D., F.R.S., aberrant "practitioner of physick." *Medical History*, 4, 356–367.

Annual Register (1769). ["The following account of an extraordinary fish of the eel tribe, which the author calls the Torporific Eel, is taken from Mr. Bancroft's ingenious Essay on the Natural History of Guiana, lately published. . . ."] *Annual Register* 12, 88–91. (Book Review)

Appelboom, T., and Bennett, C. (1986). Gout of the rich and famous. *Journal of Rheumatology*, 13, 618–622.

Arago, F.-J. (1854). Éloge historique d'Alexandre Volta lu à la séance publique du 26 Juillet 1831. In J.-A. Barral (Ed.), *Œuvres Complètes, Vol 1*. Paris: Gide et J. Baudry, pp. 187–240.

Aristotle. (1476). *De animalibus, Theodori Gazae interprete*. Venetiis: Per Iohannem de Colonia sociūq; eius Iohannē Māthen de Gherretzē.

Aristotle. (1863). *Aristotelous Peri zōōn historias biblia VIII*, Nicolas Sava Piccolos (Ed.). Paris: Firmin Didot Frères, Fils, et Cie.

Aristotle. (1969). *Histoire des Animaux, Tome III*. P. Louis (Ed.). Paris: Belles Lettres.

Aristotle. (1965). *Historia animalium, vol. 1*. A. L. Peck (Ed.). Cambridge, MA: Harvard University Press.

Aristotle. (1910). *The Works of Aristotle, Vol. IV, Historia animalium*. D. W. Thompson (Trans.). Oxford: Oxford University Press.

Artemidorus. (1644). *The Interpretation of Dreams (4th ed.)*. London: Printed by Bernard Alsop.

Asúa, M. de (2008). The experiments of Ramón M. Termeyer SJ on the electric eel in the River Plate Region (c. 1760) and other early accounts of Electrophorus electricus. *Journal of the History of the Neurosciences*, 17, 160–174.

Athenaeus (1556). *Dipnosophistarum, siue Coenae sapientum libri XV*. Venetiis: Apud Andream Arriuabenum, ad signum putei.

Athenaeus. (1929). *The Deipnosophists, Vol. 3*. C. B. Gulick (Ed., Trans.). Cambridge, MA: Harvard University Press.

Athenaeus (2001). *I deipnosofisti, i dotti a banchetto* (5 vols.). L. Canfora and C. Jacob (Eds.) Roma: Salerno Editrice.

Ashworth, W. B. Jr. (1990). Natural history and emblematic world view. In D. Lindberg and R. Westman (Eds.), *Reappraisals of the Scientific Revolution*. Cambridge: Cambridge University Press, pp. 303–332.

Atti della prima Riunione degli Scienziati italiani tenuta in Pisa nell'Ottobre del 1839. (1840) (2nd ed). Pisa: Tipografia Nistri.

Aubert, D., and Whitaker, H. (1996). David Hartley's model of vibratiuncles as a contribution to localization theory of brain functions, with a side note on short-term memory. *History and Philosophy of Psychology Bulletin*, 8, 14–16.

Auger, D., and Fessard, A. (1941). Actions du curare, de l'atropine, et de l'ésérine sur le pouvoir électrogène de la Torpille. *Comptes Rendus de la Société de Biologie*, 135, 76–78.

Avicenna. (1564). *Avicennae: principis, et philosophi....* Trans. by Gerardo Cremonensi. Venetiis: Apud Vincentium Valgrisium.

Babuchin, A. I. (1876). Uebersicht der neueren Untersuchungen über Entwicklung, Bau und physiologische Verhältnisse der elektrischen und pseudoelektrischen Organe. *Archiv für Anatomie, Physiologie und wissenschaftliche Medicin*, 501–542.

Babuchin, A. I. (1877). Beobachtungen und Versuche am Zitterwelse und *Mormyrus* des Niles. *Archiv für Anatomie und Physiologie*, 250–274.

Bacon, F. (1857–74). *The Works of Francis Bacon* (14 vols.). J. Spedding, R. L. Ellis, and D. D. Heath, Eds. London: Longman, Green, Longman, & Roberts.

Bacon, F. (2000). *The New Organon*. L. Jardine and M. Silverthorne, Eds. Cambridge: Cambridge University Press. (Originally published in 1620.)

Bailey, D. R. S. (1971). *Cicero*. London: Duckworth.

Bajon, B. (1774). Mémoire sur un poisson à commotion électrique, connu en Cayenne sous le nom d'Anguille tremblante. *Observations sur la Physique*, 3, 47–58.

Bajon, B. (1775). Descrizione di un pesce che dà la scossa elettrica, conosciuto a Cayenne sotto il nome di Anguilla tremante. *Scelta di opuscoli interessanti tradotti da varie lingue*, 1, 158–165.

Bajon, B. (1778). Sur un poisson à commotion électrique connu à Cayenne sous le nom d'anguille tremblante. In *Mémoires por Servir à l'Histoire de Cayenne et de la Guiane Française, Vol 2*. Paris: Chez Grangé.

Baldwin, B. (1992). The career and works of Scribonius Largus. *Rheinisches Museum*, 135, 54–82.

Ballowitz, E. (1899). *Das elektrische Organ des afrikanischen Zitterwelses*. Jena: Gustav Fischer.

Bancroft, E. (1763). Letter of December 11, 1763 to Dr. Thomas Williams, Lebanon, Connecticut. Historical Society of Pennsylvania (Simon Gratz Collection), Philadelphia.

Bancroft, E. (1769a). *An Essay on the Natural History of Guiana in South America*. London: Becket and De Hondt. (Reprinted by Arno Press & the New York Times, New York, 1971.)

Bancroft, E. (1769b). *Naturgeschichte von Guiana in Süd-amerika*. Frankfurt und Leipzig: J. Dodsley und Compagnie.

Bancroft, E. (1770). *The History of Charles Wentworth, Esq*. London: Printed for T. Becket.

Bancroft, E. (1782). *Proeve over de Natuurlyke Geschiedenis van Guiana*. Utrecht: Abraham Van Paddenburg.

Bancroft, E. (1794a). *Beschryving van Guiana: en een Bericht van de Rivieren en Plantagien Berbice, Essequebo en Demerary*. Amsterdam: G. Roos.

Bancroft, E. (1794b). *Experimental Researches Concerning the Philosophy of Permanent Colours and the Best Means of Producing Them by Dying, Calico Printing, &tc*. London: Printed for T. Cadell, Jun., and W. Davies in the Strand. Band. Leipzig:

Baratti, G. (pseudonym). (1670). *The Late Travels of S*. Giacomo Baratti, an Italian Gentleman *into the Remote Countries of the Abissins, or of Ethiopia Interior: ...* London: Printed for Benjamin Billingsley.

Barbaro, E., Soter, J., and Egnazio, G. B. (1529). *Hermolai Barbari Patritii Veneti et Aquileiensis Patriarchae, in Dioscoridem Colollariorum libri qvinqve: Adiectus est index eorum quae hisce libris explicantur, quem post Dioscoridis indices consulto locauimus*. Coloniae: Apud Ioan. Soterem.

Barlocci, S. (1831). Congetture sulla origine dell' elettricità atmosferica. *Poligrafo, Giornale di Scienze*, 4, 35–52.

Barlocci, S. (1841). *Lezioni di Fisica Sperimentale, Tomo II*. Roma: Tipografia dell'Ospizio Apostolico.

Barocchi, P. (1962). *Trattati d'arte del Cinquecento fra Manierismo e Controriforma. ... 3, C. Borromeo, Ammannati, Bocchi, R. Alberti, Comanini*. Bari: Laterza.

Barrère, P. (1741). *Essai su l'Histoire Naturelle de la France Équinoxiale....* Paris: Chez Piget; de l'imprimerie de la Veuve Delatour.

Bartholin, T. (1961). *On the Burning of his Library and On Medical Travel*. (C. D. O'Malley, trans.). Lawrence: University of Kansas Libraries. (First published in Latin in 1674.)

Bausi, A. (1989). I manoscritti Etiopici di J. M. Wansleben nella Biblioteca Nazionale Centrale di Firenze. *Rassegna di Studi Etiopici*, 33, 5–33.

Bayliss, W. M. (1917–19). Keith Lucas, 1879–1916. *Proceedings of the Royal Society of London*, **90**, xxxi–xlii

Beagon, M. (1992). *Roman Nature: The Thoughts of Pliny the Elder*. New York: Oxford University Press.

Beaudreau, S. A., and Finger, S. (2006). Medical electricity and madness in the 18th century: the legacies of Benjamin Franklin and Jan Ingenhousz. *Perspectives in Biology and Medicine*, 49, 330–345.

Beauvais, V. de (1624). *Bibliotheca mundi. Vincentii Burgundi,... Speculum quadruplex, naturale, doctrinale, morale, historiale, in quo totius naturae historia, omnium scientiarum encyclopedia, moralis philosophiae thesaurus, temporum et actionum humanarum theatrum ... exhibetur*. Duaci: ex officina B. Belleri.

Beazley, C. R. (1894). *Prince Henry the Navigator, the Hero of Portugal and of Modern Discovery....* London: G. P. Putnam's Sons.

Beccari, C. (1903–17). *Rerum aethiopicarum scriptores occidentales inediti a saeculo XVI ad XIX* (15 vols.). Roma: C. de Luigi.

Beccaria, G. (1753). *Dell'Elettricismo Artificiale e Naturale: Libri Due*. Torino: Nella Stampa di Filippo Antomio Campana.

Beck, H. (1959–61). *Alexander von Humboldt* (2 vols.). Wiesbaden: Steiner.

Beckingham, C. F., and Huntingford, G. W. B. (1954). *Some Records of Ethiopia, 1593–1646 ...* London: The Hakluyt Society.

Becquerel, A. C. (1836). *Traité Expérimental de l'Électricité et du Magnétisme et de leurs Rapports. Tome IV*. Paris: Firmin Didot.

Becquerel, A. C. (1837). Rapport sur un mémoire de M. Ch. Matteucci ayant pour titre: Recherches physiques, chimiques et physiologiques sur la Torpille; et sur diverses Notes, relatives aux contractions de la Grenouille. *Comptes Rendus Hebdomadaires de l'Académie des Sciences de Paris*, 5, 788–797.

Becquerel, A. C., and Breschet, G. (1835). Expériences sur la commotion électrique de la torpille. *Comptes Rendus Hebdomadaires de l'Académie des Sciences de Paris*, 1, 242–244.

Bedini, S. (1965). The evolution of science museums. *Technology and Culture*, 6, 1–29.

Behn, A. (1688/2000). *Oroonoko; or, The Royal Slave*. Boston, New York: Reprinted by Bedford/St. Martin's Press.

Bell, C. (1811). *Idea of a New Anatomy of the Brain; Submitted for the Observations of His Friends*, London: Strahan and Preston. (Reprinted in Wade, N. J. [Ed.], The Emergence of Neuroscience in the Nineteenth Century, Vol. 1. Containing the Nervous System. Idea of a New Anatomy of the Brain. London: Routledge/Thoemmes.)

Bell, W. J. Jr. (1977). *The Colonial Physician and Other Essays*. New York: Science History Publications.

Belloni, L. (1967). *Opere Scelte di Marcello Malpighi*. Torino: UTET.

Belon, P. (1551). *L'histoire Naturelle des Estranges Poissons Marins auec la vraie peincture & description du Daulphin, & de plusieurs autres de son espece* (sic) . . . . Paris: R. Chaudière.

Belon, P. (1553). *Petri Bellonii cenomani de aquatilibus libri duo cum inconibus ad vivam ipsorum effigem quoad eius fieri potuit expressis*. Parisiis: Apud Carolum Stephanum.

Bemis, S. F. (1924). British Secret Service and the French-American Alliance. *The American Historical Review*, 29, 474–495.

Bennett, M. R. (2001). *History of the Synapse*. Amsterdam: Overseas Publishers Association (Harwood Academic Publishers).

Bennett, M. V. L. (1961). Modes of operations of electric organs. *Annals of the New York Academy of Sciences*, 94, 458–509.

Bennett, M. V. L. (1970). Comparative physiology: electric organs. *Annual Review of Physiology*, 32, 471–528.

Bennett, M. V. L. (1971a). Electroreception. In W. S. Hoar and D. J. Randall (Eds.), *Fish Physiology, Vol. 5*. New York: Academic Press, pp. 493–574.

Bennett, M. V. L. (1971b). Electric organs. Sensory systems and electric organs. In W. S. Hoar and D. J. Randall (Eds.), *Fish Physiology, Vol. 5*. New York: Academic Press, pp. 347–491.

Bennett, M. V. L., and Grundfest, H. (1961a). The electrophysiology of electric organs of marine electric fishes. II. The electroplaques of main and acccessory organs of *Narcine brasiliensis*. *Journal of General Physiology*, 44, 805–818.

Bennett, M. V. L., and Grundfest, H. (1961b). The electrophysiology of electric organs of marine electric fishes. III. The electroplaques of the stargazer, *Astroscopus y-graecum*. *Journal of General Physiology*, 44, 819–843.

Bennett, M. V. L., Wurzel, M., and Grundfest, H. (1961). The electrophysiology of electric organs of marine electric fishes. I. Properties of electroplaques of *Torpedo nobiliana*. *Journal of General Physiology*, 44, 757–804.

Beretta M., and Citti, F. (2008). *Lucrezio, la Natura e la Scienza*. Firenze: Olschki.

Berkeley, E., and Berkeley, D. S. (1969). *Dr. Alexander Garden of Charles Town*. Chapel Hill: University of North Carolina Press.

Bermudes, J. (1565). *Esta he hu[m]a breue relação da embaixada q[ue] o Patriarcha dõ Ioão Bermudez trouxe do Emperador da Ethiopia chamado vulgarmente Preste Ioão* . . . . Lisboa: F. Correia.

Bernardi, W. (2008). *Il paggio e l'anatomista: scienza, sangue e sesso alla corte del granduca di Toscana*. Firenze: Le Lettere.

Bernardi, W., and Guerrini, L. (Eds.). (1999). *Francesco Redi, un protagonista della scienza moderna*. Firenze: Olschki.

Bernfield, S. (1944). Freud's earliest theories and the school of Helmholtz. *Psychoanalatic Quarterly*, 13, 341–362.

Bernstein, J. (1868). Über den zeitlichen Verlauf der negativen Schwankung des Nervenstromes. *Pflüger's Archiv für der gesamte Physiologie des Menschen under der Tiere*, 1, 173–207.

Bernstein, J. (1871). *Untersuchungen über den Erregungsvorgang im Nerven und Muskelsystem*. Heidelberg: C. Winter.

Bernstein, J. (1874). Ueber Electrotonus und die innere Mechanik der Nerven. *Archiv für die gesammte Physiologie des Menschen und der Thiere*, 8, 40–59.

Bernstein, J. (1888). Neue Theorie der Erregungsvorgänge und elecktrischen Erscheinungen an der Nerven- und Muskelfaser. *Untersuchungen aus dem physiologischen Institut der Universität Halle*, 1, 27–103.

Bernstein, J. (1894). *Lehrbuch der Physiologie des thierischen Organismus, im speciellen des Menschen* (First edition). Stuttgart: F. Enke. (Second edition 1900, third edition 1910)

Bernstein, J. (1902). Untersuchungen zur Thermodynamik der bioelektrischen Ströme. *Pflüger's Archiv für der gesamte Physiologie des Menschen under der Tiere*, 92, 521–562. (English translation in J. W. Boylan [Ed.], *Founders of Experimental Psychology*. München: J. F. Lehmanns Verlag, 1971, pp. 258–299.)

Bernstein, J. (1912). *Elektrobiologie. Die Lehre von den elektrischen Vorgängen im Organismus auf moderner Grundlage dargestellt*. Braunschweig: Vieweg & Sohn.

Bernstein, J., and Tschermak, A. (1902). Ueber die Beziehung der negativen Schwankung des Muskelstromes zur Arbeitsleistung des Muskels. *Archiv für die gesammte Physiologie des Menschen und der Thiere*, 89, 289–331.

Bernstein, J., and Tschermak, A. (1904). Ueber das thermische Verhalten des elektrischen Organs von Torpedo. *Sitzungsberichte der Königlich Preussischen Akademie der Wissenschaften*, 113, 301–313.

Bernstein, J., and Tschermak, A. (1906). Untersuchungen zur Thermodynamik der bioelektrischen Ströme. Uber die Nature der Kette des elektrischen Organs bei Torpedo. *Archiv für die gesammte Physiologie des Menschen und der Thiere*, 112, 439–521.

Bertelsen, L. (1992). Journalism, Carnival, and *Jubilate Agno*. *English Literary History*, 59, 357–384.

Bertholon, P. (1780). *De l'Électricité du Corps Humain Domain dans l'Etat de Santé et de Maladie*. Paris: Chez Didot.

Bertholon, P. (1786). *De l'Électricité du Corps Humain Domain dans l'Etat de Santé et de Maladie* (2nd ed., 2 vols.). Paris: Chez Croulbois . . .; Lyon: Chez Bernuset. . . .

Bertholon, P., and Kühn, K. G. (1788). *Anwendung und Wircksamkeit der Electrizitaet zur Erhaltung und Wiederheilung der Gesundheit des menschlichen Koerpers*. Weissenfels: Severin.

Bertucci, P. (2001a). The electrical body of knowledge: Medical electricity and experimental philosophy in the mid-eighteenth century. In P. Bertucci and G. Pancaldi (Eds.), *Electric Bodies: Episodes in the History of Medical Electricity*. Bologna: Università di Bologna, pp. 43–68.

Bertucci, P. (2001b). A philosophical business: Edward Nairne and the patent medical electrical machine (1782). *History of Technology*, 23, 41–58.

Bertucci, P. (2006). Back from wonderland: Jean Antoine Nollet's Italian tour (1749). In R. J. W. Evans and A. Marr (Eds.), *Curiosity and Wonder from the Renaissance to the Enlightenment*. Aldershot: Ashgate Publishing Ltd., pp. 193–211.

Bertucci, P. (2007). *Viaggio nel paese delle meraviglie: Scienza e curiosità nell'Italia del Settecento*. Torino: Bollati Boringhieri.

Bertucci, P., and Pancaldi G. (2001). *Electric Bodies: Episodes in the History of Medical Electricity*. Bologna: Università di Bologna.

Beullens, P., and Gotthelf, A. (2007). Theodore Gaza's translation of Aristotle's *De animalibus*: content, influence, and date. *Greek, Roman, and Byzantine Studies*, 47, 469–513.

Bianchi, N. (1874). *Carlo Matteucci e l'Italia del suo Tempo*. Firenze: Fratelli Bocca.

Biedermann, W. (1895). *Elektro-physiologie* (2 vols.). Jena: G. Fischer. (English translation by F. A. Welby, 1896–98, 2 vols, London: Macmillan & Co., 1896–98.)

Bierens de Haan, J. A. (1952). *De Hollandsche Maatschappij der Wetenschappen. 1752–1952*. Haarlem: Gedrukt door Joh. Enschedé en Zonen.

Bilharz, T. (1857). *Das electrische Organ des Zitterwelses*. Leipzig: W. Engelmann.

Bini, C. (1967). *Atlante dei Pesci delle Coste Italiane, vol. 1*. Roma: Mondo Sommerso Editrice, pp. 127–134.

Birch, J. (1792). A letter to Mr. George Adams. In *An Essay on Electricity . . . by George Adams* (4th ed.). London: Printed for the Author by R. Hindmarsh and sold by the author, pp. 519–573.

Birembaut, A. (et al.). (1962). *La Vie et l'œuvre de Réaumur*. Paris: Presse Universitaire de France, pp. 1683–1757.

Bishop, G. (1998). *A Lion to Judah: The Travels and Adventures of Pedro Paez. . . .* Anand: Gujarat Sahitya Prakash.

Bitterling, R. (1959). *Alexander von Humboldt*. München: Deutscher Kunstverlag.

Blackman, A. M. (1914). *The Rock Tombs of Meir. Vol. 1: The Tomb-Chapel of Ukh-Hotep's Son Senbi*. London: William Clowes and Sons.

Blainville, H. M. D. de (1833). *Cours de Physiologie Générale et Comparée* (3 vols). Paris: G. Baillière.

Blanch, L. (1841). Sulla scoperta della scintilla tratta dalla scarica della torpedine. *Il Progresso delle Scienze, Lettere ed Arti (Nuova Serie)*, 28, 127–138. (The text of this article is probably due to a large extent to Linari.)

Bloch, A. M. (1885). Expériences sur la vision. *Comptes Rendus de Séances de la Société de Biologie (Paris)*, 37, 493–495.

Bloch, M. E. (1785–97). *Ichthyologie, ou Histoire Naturelle Générale et Particulière des Poissons* (6 vols). (Trans. by J.-C. Thiébault de Laveaux). Berlin: Chez l'Auteur et Chez François de La Garde.

Bloch, M. E. (1786–87). *Naturgeschichte der auslaendischen Fische*. Berlin: Aus Kosten der Verfassers, und in Commission in der Buchhandlung der Realschule.

Bloch, M. E. (1801). *Histoire Naturelle des Poissons, Avec les Figures Dessinées d'Après Nature. Ouvrage Classé par Ordres, Genres et Espéces, d'Aprés le Systeme de Linné avec les Caractères Génériques* (René Richard Castel, Ed.). Paris: Chez Déterville.

Blumenbach, J. F. (1795). *Elements of Physiology*. Philadelphia: Thomas Dobson at the Stone-House.

Blumenbach, J. F. (1810). *Abbildungen naturhistorischen Gegenstand*. Göttingen: H. Dieterich.

Boaz, G. (1993). *The Hieroglyphics of Horapollo*. Princeton: Princeton University Press.

Bodson, L. (1981). L'incubation bucco-pharyngienne de *Sarotherodon niloticus* (Pisces: Ciclidae) dans la tradition Grecque ancienne. *Archives Internationales d'Histoire des Sciences*, 31, 5–25.

Boerhaave, H. (1743). *Dr. Boerhaave's Academical Lectures on the Theory of Physic. Being a Genuine Translation of his Institutes and Explanatory Comment, Collated and Adjusted to Each Other, as they were Dictated to his Students at the University of Leyden. . . .* London: Printed for Innys.

Bohr, N. (1937). Biology and atomic physics. In *Celebrazione del Secondo Centenario della Nascita di Luigi Galvani*. Bologna: Luigi Parma, pp. 68–78.

Boll F. (1873). Beiträge zur Physiologie von Torpedo. *Archiv für Anatomie und Physiologie und wissenschaftliche Medicin*, 76–102.

Boll, F. (1874). Ein historischer Beitrag zur Kenntniss von Torpedo. *Archiv für Anatomie und Physiologie*, 16, 152–158.

Bonaparte, C. -L. (1832– 41). *Iconografia della Fauna Italica per le Quattro Classi degli Animali Vertebrati* (3 vols). Roma: Dalla Tipographia Salviucci. Napoli: Stamperia Reale.

Bonnet, C. (1764). *Contemplation de la Nature* (1st ed., 2 tomes). Amsterdam: Chez Marc-Michel Rey.

Bonnet, C. (1769). *Contemplation de la Nature* (2nd ed., 2 tomes). Amsterdam: Chez Marc-Michel Rey.

Bonnet, C. (1779-83). *Collection Complète des Œuvres d'Histoire Naturelle et de Philosophie de Ch. Bonnet* (8 tomes). Neuchatel: Samuel Fauche Libr. Du Roi.

Borelli, G. (1680–81). *De motu animalium* (2 vols). Roma: ex typographia Angeli Bernabò.

Borelli, G. (1685). *De motu animalium. . . .* Lugduni in Batavis: Apud Cornelium Boutesteyn, Danielem à Gaesbeeck, Johannem de Vivie & Petrum vander Aa (Trans. by P. Maquet as *On the Movement of Animals*. Berlin: Springer Verlag, 1989.)

Borgnet, A. (1891). *Alberti Magni Opera omnia 11, Animalium Lib. XXVI* (pars prior, I–XII). Parisiis: Vivès.

Boruttau, H. J. (1904). Zur Geschichte und Kritik der neueren bioelektrischen Theorien, nebst einigen Bemerkungen über die Polemik in der Elektrophysiologie. *Archiv für die gesammte Physiologie des Menschen und der Thiere*, 427–443.

Boruttau, H. J. (1922). *Emil du Bois-Reymond*. Vienna: Springer Verlag.

Botting, D. (1973). *Humboldt and the Cosmos*. New York: Harper & Row.

Boulenger, G. A. (1907). *The Fishes of the Nile*. London: H. Rees Ltd.

Bowditch, H. P. (1871). Eigenthümlichkeiten der Reizbarkeit, welche die Muskelfasern der Herzens zeigen. *Bericht der Sächsische Gesellschaft (Akademie) der Wissenschaften*, 23, 2–39.

Bowers, B., and Symons, L. (Eds.). (1991). *Curiosity Perfectly Satisfied: Faraday's Travels in Europe, 1813–1815*. London: Peter Peregrinus.

Bowers, J. Z. (1966). Engelbert Kaempfer: Physician, explorer, scholar and author. *Journal of the History of Medicine and Allied Sciences*, 21, 237–259.

Bowers, J. Z., and Carrubba, R. W. (1970). The doctoral thesis of Engelbert Kaempfer. *Journal of the History of Medicine and Allied Sciences*, 25, 270–310.

Boxer, C. (1991). *The Portuguese Seaborne Empire, 1415–1825*. Manchester, UK: Carcanet Press.

Boyd, J. P. (1959). Silas Deane: Death by a kindly teacher of treason. *William and Mary Quarterly*, 16, 515–550.

Boyle, R. (1738). *The Philosophical Works of the Hounorable Robert Boyle Esq., Vol. 1, 2nd ed.* (P. Shaw, Ed.). London: Printed for W. Innys, and R. Manby . . .; and T. Longman.

Braund, D., and Wilkins, J. (Eds.) (2000). *Athenaeus and his World: Reading Greek Culture in the Roman Empire*. Exeter: University of Exeter Press. Braunschweig: Vieweg & Sohn.

Brazier, M. (1984). *A History of Neurophysiology in the 17th and 18th Centuries*. New York: Raven Press.

Brazier, M. A. B. (1988). *A History of Neurophysiology in the 19th Century*. New York: Raven Press.

Breasted, J. (1906–07). *Records of Ancient Egypt . . .*, (Vol. IV). Chicago: University of Chicago Press.

Breasted, J. (1912). *The Development of Religion and Thought in Ancient Egypt*. New York: Charles Scribner's Sons.

Breasted, J. (1930). *The Edwin Smith Surgical Papyrus*. Chicago: University of Chicago Press.

Bredekamp, H. (1995). *The Lure of Antiquity and the Cult of the Machine: The Kunsthammer and the Evolution of Nature, Art and Technology*. (Trans. by A. Brown). Princeton: Marcus Weiner.

Bresadola, M. (1997). La biblioteca di Luigi Galvani. *Annali di Storia delle Università Italiane*, 1, 167–197.

Bresadola, M. (1998). Medicine and science in the life of Luigi Galvani (1737-1798). *Brain Research Bulletin*, 46, 367–380.

Bresadola, M. (2003). At play with nature: Luigi Galvani's experimental approach to muscular physiology. *Archimedes*, 7, 67–92.

Bresadola, M. (2008a). Animal electricity at the end of the eighteenth century: the many facets of a great scientific controversy. *Journal of the History of Neurosciences*, 17, 8–32.

Bresadola, M. (2008b). L'elettricità contesa: Luigi Galvani, Alessandro Volta e la nascita dell'elettrofisiologia moderna. In M. Piccolino (Ed.), *Neuroscienze controverse*. Torino: Bollati Boringhieri, pp. 75–106.

Breschet, G., and Becquerel, A. C. (1836). Expériences sur la torpille. *Annales des Sciences Naturelles, Série II, (Zoologie)*, 6, 123.

Brett-Smith, S. C. (1994). *The Making of Bamana Sculpture: Creativity and Gender.* Cambridge: Cambridge University Press.

Brewer, D. J. (1986). Cultural and Environmental Change in the Faiyum Egypt. Doctoral dissertation, University of Tennessee.

Brewer, D. J., and Friedman, R. F. (1989). *Fish and Fishing in Ancient Egypt.* Warminster: Aris and Phillips.

British Apollo (1711). [Question and answer about "Nummfish"] *British Apollo*, Wednesday, January 11, Issue 123.

Brock, L. G., and Eccles, R. M. (1958). The membrane potentials during rest and activity of the ray electroplate. *Journal of Physiology*, 42, 251–274.

Brodie, B. C. (1812). Further experiments and observations on the action of poisons on the animal system. *Philosophical Transactions of the Royal Society of London*, 102, 205–227.

Broussonet, P. M. A. (1774). Sur le trembleur, espèce peu connue de poisson électrique. *Histoire de l'Académie Royale des Sciences, pour l'Année 1782*, 692–698.

Brown, G. (1999). *Count Rumford - Scientist, Soldier, Statesman, Spy - The Extraordinary Life of a Scientific Genius.* Sutton Publishing: Stroud in Gloucestershire.

Brown, G. L., Dale, H. H., and Feldberg, W. (1936). Reaction of the normal mammalian muscle to acetylcholine and to eserine. *Journal of Physiology*, 87, 394–424.

Browne, T. (1646). *Pseudodoxia Epidemica: Or, Enquiries into Very Many Received Tenents, and Commonly Presumed Truths.* London: Printed by T.H. for E. Dod. (Reprinted in 1972: Menston UK: Scolar Press).

Brozek, J., and Sibinga, M. S. (1970). *Origins of Psychometry: Johan Jacob de Jaager on Reaction Time and Mental Processes.* Nieuwkoop, Holland: de Graff.

Bruce, J. (1790). *Travels to Discover the Source of the Nile, in the Years 1768, 1769, 1770, 1771, 1772 & 1773.* (5 vols.). Edinburgh: Robinson.

Bruhns, K. (Ed.). (1873). *Life of Alexander von Humboldt.* J. Lassell and C. Lassell (trans.) (2 vols.). London: Longmans, Green & Co.

Brunet, F. (1933). Biographie d'Alexandre de Tralles: Sa personne. Ses écrits. In *Oeuvres Médicales d'Alexandre de Tralles, le Dernier Auteur Classique des Grands Médecins Grecs de l'Antiquité.* Tome I: Les Douze Lives de Médecine. F. Paris: Librairie Orientaliste Paul Geuthner, pp. 1–90.

Brunet, F. (1937). *Oeuvres Médicales d'Alexandre de Tralles, le Dernier Auteur Classique des Grands Médecins Grecs de l'Antiquité* (Tome IV). F. Paris: Librairie Orientaliste Paul Geuthner.

Bry, J. T. de. (1602). *Americae nona et postrema pars, qua De Ratione elementorum . . .* Francofurti: De Bry.

Bryan, C. P. (1974). *Ancient Egyptian Medicine: The Papyrus Ebers.* Chicago: Ares Publishers Inc. (Reprint from 1930.)

Bryant, W, (1786). An account of an electric eel, or the torpedo of Surinam. *Transactions of the American Philosophical Society*, 2, 166–169.

Brydone P (1757). An instance of the electrical virtue in the cure of a palsy. *Philosophical Transactions of the Royal Society of London*, 50(1), 392–395.

Buckingham, H. W., and Finger, S. (1996). David Hartley's psychobiological associationism and the legacy of Aristotle. *Journal of the History of the Neurosciences*, 6, 21–37.

Buffon, G. L. L. (1788). *Histoire Naturelle des Mineraux, Tome V, Traité de l'Aimant et de ses Usages.* Paris: Imprimerie des Batimens du Roi.

Bullock, T. H. (Ed.). (2005). *Electroreception.* New York: Springer.

Bullock, T. H., and Heiligenberg, W. (1986). *Electroreception.* New York: Wiley Interscience.

Burdon-Sanderson, J., and Gotch, F. (1888). On the electric organ of the skate. *Journal of Physiology*, 9, 137–166.

Burdon-Sanderson, J., and Gotch, F. (1891). Excitatory change in muscle. *Proceedings of the Physiological Society*, 5, xliii–liii.

Bury, J. B. (1958). *History of the Later Roman Empire, From the Death of Thoedosius I to the Death of Justinian.* New York: Dover Publications.

Buursma, J. H. (1978). *Nederlandse Geleerde Genootschappen Opgericht in de 18e Eeuw. (Discom. 7.)* Nijmegen: Bibliografische Bijdragen.

Cahan, D. (Ed.). (1993a). *Hermann von Helmholtz and the Foundations of Nineteenth-Century Science.* Berkeley: University of California Press.

Cahan, D. (Ed.) (1993b). *Letters of Hermann von Helmholtz to his Parents: 1837–1846.* Stuttgart: Franz Steiner Verlag.

Caldani, L. M. A. (1757). Sur l'insensibilité et l'irritabilité de Mr. Haller. Seconde lettre de Mr. Marc Antoine Caldani. In A. von Haller, *Mémoires sur la Nature Sensible et Irritable des Parties du Corps Animal, Vol III.* Lausanne: M.-M. Bousquet, pp. 343–490.

Cameron, A. (1970). *Claudian: Poetry and Propaganda at the Court of Honorius.* New York: Oxford University Press.

Camerota, M. (2004). *Galileo Galilei e la cultura scientifica nell'età della Controriforma.* Roma: Salerno.

Camerota M. (2008) Galileo, Lucrezio e l'atomismo, in M. Beretta and F. Citti edrs, *Lucrezio, la natura e la scienza*, Firenze: Olschki, pp. 141–75.

Campanella, T. (1629). *Astrologicorum libri VI, in quibus astrologia, omni superstitione Arabum & Iudæorum eliminata, physiologicè tractatur. . . .* Lugduni: Sumptibus Iacobi, Andreæ, & Matthæi Prost.

Canfora, L. (1990). *The Vanished Library.* Berkeley: University of California Press.

Cannon, W. B. (1922). Henry Pickering Bowditch (1840–1911). *National Academy of Sciences*, 17, 181–196.

Cannon, W. B. (1945). *The Way of an Investigator.* New York: W. W. Norton.

Cantimpré, T. de (1973). *Liber de natura rerum: editio princeps secundum codices manuscriptos.* H. Boese (Ed.). Berlin: Walter de Gruyter.

Canton, J. (1753). Electrical experiments with an attempt to account for their several phaenomena. *Philosophical Transactions of the Royal Society of London*, 48, 350–358.

Canton, J. (1754). Concerning some new electrical experiments. *Philosophical Transactions of the Royal Society of London*, 48, 780–785.

Caraman, P. (1985). *The Lost Empire: The Story of the Jesuits in Ethiopia.* Oxford: Oxford University Press.

Cardano, G. (1557). *De rerum varietate.* Basel: Per Henrichum Petri.

Cardano, G. (1560). *De subtilitate rerum . . ., Book XXI.* Parisiis: ex officina M. Fezandat and R. Granion.

Cardano, G. (1962). *The Book of My Life* (J. Stoner, trans.). New York: Dover Publications.

Cardim, F. (1925). *Tratados da Terra e Gente do Brasil.* Rio de Janeiro: Leite & Cia.

Cardim, F. (1965). *A Treatise of Brazil, Written by a Portugal who Had Long Lived There.* In S. Purchas, *Hakluytus Posthumus, or Purchas his Pilgrimes, Vol. 16.* New York: AMS Press, pp. 417–502.

Carlson, C. L. (1938). *The First Magazine: A History of The Gentleman's Magazine.* Providence: Brown University Press.

Carruba, R. W., and Bowers, J. Z. (1982). Engelbert Kaempfer's first report of the torpedo fish of the Persian Gulf in the late seventeenth century. *Journal of the History of Biology*, 15, 263–274.

Carvalho, M. R. de, and Randall, J. E. (2003). Numbfishes from the Arabian Sea and surrounding gulfs, with the description of a new species from Oman (*Chondrichthyes: Torpediniformes: Narcinidae*). *Ichthyological Research*, 50, 59–66.

Casseri, G. C. (1600). *Iulii Casserii Placentini De vocis auditvs[que] organis historia anatomica singvlari fide methodo ac indvstria concinnata: tractatibvs dvobvs explicata, ac variis iconibvs ære excvsis illvstrata.* Ferrariæ: Ex. Victorius Baldinus, typographus Cameralis, sumptibus Unitorum Patavii.

Casseri, G. C. (1609). *Ivlii Casserii . . . Pentæstheseion, hoc est de qvinqve sensibvs liber, organorvm fabricam variis iconibvs fideliter ære incisis illustratam, nec non actionem et vsum, discursu anatomico & philosophico accuratè explicata continens.* Venetiis: Apud Nicolavm Misserinvm.

Castanhoso, M. de (1564). *Historia das cousas que o muy esforcado capitão Dom Christouão da Gama fez nos Reynos do Preste Ioão com quatroce[n]tos portugueses que consigo leuou.* Lisboa: Ioã de Barreyra.

Castellani, C. (1975). Ippolito Salviani. In: *Dictionary of Scientific Biography (Vol. 12).* New York: Scribner, pp. 89–90.

Catesby, M. (1731, 1743). *The Natural History of Carolina, Florida, and the Bahama Islands. . . .* London: Printed at the Expence of the Author, and Sold by W. Innys and R. Manby . . . by Mr. Hauksbee, at the Royal Society House, and by the Author, at Mr. Bacon's in Hoxton.

Caton-Thompson, G., and Gardner, E. W. (1934). *The Desert Fayum.* London: The Royal Anthropological Institute of Great Britain and Ireland.

Cavallo, T. (1795). *A Complete Treatise of Electricity in Theory and Practice with Original Experiments, Vol. 2 (4th ed.).* London: Printed for C. Dilly.

Cavazza, M. (1990). *Settecento Inquieto. Alle Origini dell'Istituto delle Scienze di Bologna.* Bologna: il Mulino.

Cavazza, M. (1995). Laura Bassi e il suo gabinetto di fisica sperimentale. Realtà e mito. *Nuncius,* 10, 715–753.

Cavazza, M. (1996). L'Istituto delle Scienze di Bologna negli ultimi decenni del Settecento. In G. Barsanti, V. Becagli e R. Pasta (Eds.), *La Politica della Scienza. Toscana e Stati Italiani nel Tardo Settecento.* Firenze: Olschki, pp. 435–450.

Cavazza, M. (1997a). The uselessness of anatomy. Mini and Sbaraglia versus Malpighi. In D. Bertoloni Meli (Ed.), *Marcello Malpighi, Anatomist and Physician.* Firenzi: Olschki, pp. 129–145.

Cavazza, M. (1997b). La recezione della teoria halleriana dell'irritabilità nell'Accademia delle Scienze di Bologna. *Nuncius,* 12, 359–377.

Cavendish, H. (1771). An attempt to explain some of the principal phaenomena of electricity, by means of an elastic fluid. *Philosophical Transactions of the Royal Society,* 61, 584–677.

Cavendish, H. (1776a). An account of some attempts to imitate the effects of the torpedo by electricity. *Philosophical Transactions of the Royal Society,* 66, 196–225.

Cavendish, H. (1776b). Account of some attempts to imitate the effects of the torpedo by electricity. *Gentleman's Magazine,* 46, 561.

Celsus, A. C., Hippocrates, Marcellus Empiricus, Melanchthon, P., Ruel, J., Scribonius Largus, Vindicianus, Wechel, C. (1529). *De re medica libri octo . . . ad veterum & recentium exemplarium fidem, necnon doctorum hominum judicium, summa diligentia excusi.* Parisiis: Apud Christianum Wechel.

Cerulli, E. (1943–47). *Etiopi in Palestina; Storia della Comunità Etiopica di Gerusalemme* (2 vols). Roma: Libreria dello Stato.

Cerulli, E. (1968). *La Letteratura Etiopica: l'Oriente Cristiano nell'Unità delle sue Tradizioni.* Firenze: Sansone.

Chagas, C. (1959). Studies on the mechanism of curarization. *Annals of the New York Academy of Sciences,* 81, 345–357.

Chagas, C., and Albe-Fessard, D. (1954). Action de divers curarisants sur l'organe électrique de l'*Electrophorus electricus* (Linnaeus). *Acta Physiologica Latino-Americana,* 4, 49–60.

Chagas, C., and Paes de Carvalho, A. (1961). *Bioelectrogenesis; A Comparative Survey of its Mechanisms with Particular Emphasis on Electric Fishes; Proceedings of the Symposium on Comparative Bioelectrogenesis,* held in Rio de Janeiro, August 1959. Amsterdam and New York: Elsevier Publishing Co.

Chaïa, J. (1968). Sur une correspondence inedite de Réaumur avec Artur, premier Médecin du Roy a Cayenne. *Episteme,* 2, 36–57.

Chambers, E. (1743). *Cyclopaedia: Or, An Universal Dictionary of Arts and Sciences.* London: Printed for W. Innys and J. Richardson, *et al.*

Chang, C. C. (1999). Looking back on the discovery of alpha-bungarotoxin. *Journal of Biomedical Science,* 6, 368–375.

Chang, C. C., and Lee, C.-Y. (1963). Isolation of neurotoxins from the venom of *Bungarus multicinctus* and their modes of neuromuscular blocking action. *Archives Internationales de Pharmacodynamie et de Thérapie,* 144, 241–257.

Changeux, J. P., Devillers-Thiéry, A., and Chemouilli, P. (1984). Acetylcholine receptor: an allosteric protein. *Science,* 225, 1335–1345.

Changeux, J. P., Kasai, M., and Lee, C. Y. (1970). Use of a snake venom toxin to characterize the cholinergic receptor protein. *Proceedings of the National Academy of Sciences (USA),* 67, 1241–1247.

Cheyne, G. (1733). *The English Malady; Or a Treatise of Nervous Diseases of all Kinds, . . .* London: Printed for G. Strahan, and J. Leake at Bath.

Choffin, D. E. (1749–50). *Amusemens Philogiques.* Halle: La Maison des Orphelins.

Chu, N-S. (2005). Contributions of a snake venom toxin to myasthenia gravis: the discovery of alpha-bungarotoxin in Taiwan. *Journal of the History of the Neurosciences,* 14, 138–148.

Cicero, M. T. (1958). *De natura deorum.* Edited and notes by A. S. Pease. Cambridge, MA: Harvard University Press.

Cicero, M. T. (1997). *The Nature of the Gods.* Translated and notes by P. G. Walsh. Oxford: Clarendon Press.

Clark-Kennedy, A. E. (1929). *Stephen Hales, D.D. F.R.S.: An Eighteenth Century Biography.* Cambridge: Cambridge University Press.

Clarke, E. (1968): The doctrine of the hollow nerve in the seventeenth and eighteenth centuries. In L. Stevenson and R. Multhaui (Eds.), *Medicine, Science and Culture: Historical Essays in Honor of Owsei Temkin.* Baltimore: Johns Hopkins University Press, pp 123–141.

Claudian, C. (1598). *Cl. Claudiani poetae celeberrimi opera . . .* Lugduni: Apud Ant. Gryphium.

Claudian, C. (1922). *Claudian, with an English Translation by Maurice Platnauer.* London: W. Heinemann.

Claudian, C. (1985). *Claudii Claudiani carmina.* (Ed. by J. Barrie Hall.) Leipzig: B. G. Teubner.

Clower, W. T. (1998). The transition from animal spirits to animal electricity. *Journal of the History of the Neural Sciences,* 7, 201–218.

Cobo, B. (1956). *Historia del Nuevo Mundo, Vol. 1.* Madrid: Ediciones Atlas.

Codellas, P. S. (1932). Alcmaeon of Croton: His life, work, and fragments. *Proceedings of the Royal Society of Medicine,* 25, 1041–1046.

Cohen, I. B. (1943). Benjamin Franklin and the mysterious "Dr. Spence." *Journal of the Franklin Institute,* 235, 1–25.

Cohen, I. B. (1953). Introduction. In *Luigi Galvani: Commentary on the Effects of Electricity on Muscular Motion.* Norwalk: Burndy Library, pp. 9–41.

Cohen, I. B. (1990). *Benjamin Franklin's Science.* Cambridge, MA: Harvard University Press.

Cole, F. J. (1944). *A History of Comparative Anatomy from Aristotle to the Eighteenth Century.* London: Macmillan & Co.

Cole, K. S. (1949). Dynamic electrical characteristics of the squid axon membrane. *Archives des Sciences Physiologiques,* 3, 253–258.

Cole, K. S., and Curtis, H. (1939). Electric impedance of the squid giant axon during activity. *Journal of General Physiology*, 22, 649–170.

Cole, K. S., and Curtis, H. (1940). Membrane action potentials from the squid giant axon. *Journal of Cellular and Comparative Physiology*, 15, 147–157.

Colladon, J. D. (1836). Expériences sur la Torpille. *Annales des Sciences Naturelles, Sér. II (Zoologie)*, 6, 255–256.

Comrie, J. D. (1932). *History of Scottish Medicine (2nd ed.)*. London: Welcome Historical Medical Museum.

Condamine, C.-M. La (1745). *Relation Abrégée d'un Voyage fait dans l'Interieur de l'Amérique Méridionale depuis la Côte de la Mer du Sud, Jusqu'aux Côtes du Brésil et de la Guiane, en Descendant la Rivière des Amazone*. Paris: Chez la Veuve Pissot.

Condamine, C-M. La (1749). Relation Abrégée d'un Voyage fait dans l'intérieur de l'Amérique méridionale depuis la Côte de la Mer du Sud, jusqu'aux Côtes du Brésil et de la Guiane, en descendant la Rivière des Amazones. *Histoire de l'Académie des Sciences pour l'Année 1745 avec le Mémoires de Mathématique et de Physique pour la même Année*, 391–492.

Condamine, C-M. La (1751). *Journal de Voyage fait par Ordre du Roi à l'Equateur Servant d'Introduction Historique à la Mesure des Trois Premiers Degrés du Méridien*. Paris: Imprimerie Royale.

Condamine, C.-M. La (1754). *Mémoire sur l'Inoculation de la Petite Vérole. Lu a l'Assemblée Publique de l'Académie Royale des Sciences. . . .* Paris: Durand.

Condamine, C.-M. La (1759). *Second Mémorie sur l'Inoculation de la Petite Vérole, contenant son Histoire depuis l'Année 1754 lu a l'Assemblée Publique de l'Académie des Sciences. . . .* Genève: Chez Emanuel Du Villard.

Configliachi, P. (1811). *Memorie sui Fenomeni Elettrici della Razza Torpedine*. Pavia: Giovanni Capelli.

Configliachi, P., and Volta, A. (1814). *L'identita' del fluido elettrico col cosi detto fluido galvanico vittoriosamente di mostrata con nuove esperienze ed osservazioni*. Pavia: G. G. Capelli.

Conley, T. K., and Brewer-Anderson, M. (1997). Franklin and Ingenhousz: a correspondence of interests. *Proceedings of the American Philosophical Society*, 141, 276–296.

Conte, G. B. (1984). L'inventario del Mondo: ordine e linguaggio della natura nell'opera di Plinio il Vecchio. In: *Plinio Gaio Secondo, Storia Naturale, con un'introduzione di Italo Calvino, Vol. 1*. Torino: Einaudi.

Conti Rossini, C. (1894). *Storia di Lebna Dengel, re d'Etiopia, sino alle prime lotte contro Ahmad Ben Ibrahim*. Roma: Reale Accademia del Lincei.

Conti Rossini, C. (1928). *Storia d'Etiopia: Parte Prima, dalle Origini all'Avvento della Dinastia Salmonide*. Bergamo: Istituto Italiano D'Arti Grafiche.

Conti Rossini, C. (1940). Portogallo ed Etiopia. In *Relazioni Storiche fra Italia e Portogallo: Memorie e Documenti*. Roma: Reale Accademia d'Italia.

Conti Rossini, C. (1941). Le sorgenti del Nilo Azzurro e Giovanni Gabriel. *Bollettino della Società Geografica Italiana*, Serie VII, VI, Fasc. 1.

Copeman, W. (1964). *A Short History of the Gout and the Rheumatic Diseases*. Berkeley: University of California Press.

Copenhaver, B. P. (1990). Natural magic, hermetism, and occultism in early modern science. In D. Lindberg and R. Westman (Eds.), *Reappraisals of the Scientific Revolution*. Cambridge: Cambridge University Press, pp. 261–302.

Copenhaver, B. P. (1991). A tale of two fishes: Magical objects in natural history from antiquity through the scientific revolution. *Journal of the History of Ideas*, 52, 373–398.

Copenhaver, B. P. (1992). Did science have a Renaissance? *Isis*, 83, 387–407.

Cox, P. (1983). The Physiologus: A poiēsis of nature. *Church History*, 52, 433–443.

Cox, R. T., Coates, C. W., and Brown, M. V. (1945). Relations between the structure, electrical characteristics and chemical processes of electric tissue. *Journal of General Physiology*, 28, 187–212.

Crampton, W. G. R. (1998). Effects of anoxia on the distribution, respiratory strategies and electric signal diversity of gymnotiform fishes. *Journal of Fish Biology*, 53 (Suppl. A), 307–330.

Cranefield, P. F. (1957). The organic physics of 1847 and the biophysicists of today. *Journal of the History of Medicine and Allied Sciences*, 12, 407–423.

Critical Review. (1769). *An Essay on the Natural History of Guiana. Critical Review*, 27, 52–59. (Book Review)

Crombie, A. C. (1952). *Augustine to Galileo; the History of Science, A.D. 400–1650*. London: Falcon.

Croone, W. (1664). *De ratione motus musculorum*. London: Excudebat J. Hayes, prostant venales S. Thomson. . . .

Croone, W. (2000). *William Croone, On the Reason of the Movement of the Muscles*. P. Maquet (Trans.), M. Nayler (Intro.). Philadelphia: American Philosophical Society.

Crosland, M. (1978). *Gay-Lussac, Scientist and Bourgeois*. Cambridge: Cambridge University Press.

Crowther, J. G. (1950). *Francis Bacon: The First Statesman of Science*. London: Cresset Press.

Crummey, D. (2000). *Land and Society in the Christian Kingdom of Ethiopia from the Thirteenth to the Twentieth Century*. Oxford: Oxford University Press.

Cuvier, G. L. C. F. D. (1823). Pline. In *Biographie Universelle, Tome XXXV*. Paris: Michaud, p. 70.

Cuvier, G. (1834). *The Animal Kingdom Arranged after its Conformity with its Organization. 10: The Class Pisces. With Supplementary Additions by E. Griffith and H. C. Smith*. London: Whittaker and Co.

Cuvier, J. L. (1995). *Historical Portrait of the Progress of Ichthyology from its Origins to Our Time*. (Ed. by T. W. Pietsch; Trans. by A. J. Simpson). Baltimore: Johns Hopkins University Press.

Daily Advertiser. (1774). [No title: Pringle's speech and Walsh's medal.] *Daily Advertiser*, December 1, Issue 13712.

Daily Advertiser. (1776). THE ELECTRICAL EEL, or numbing FISH of South America, Gymnotus Electricus of Linnaeus. *Daily Advertiser*, August 14, Issue 14245.

Dale, H. H., and Feldberg, W. (1934). Chemical transmission at motor endings in voluntary muscle? *Journal of Physiology*, 18, 39P–40P.

Dale, H. H., Feldberg, W., and Vogt, M. (1936). Release of acetylcholine at voluntary motor nerve endings. *Journal of Physiology*, 86, 353–380.

Dapper, O. (1670). *Gedenwaerdig bedryf der Nederlandsche Oost-Indische Maetschappye, op de kuste en in het keizerrijk van Taising of Sina . . . beschrevens door Dr. O. Dapper*. Amsterdam: Jacob van Meurs. . . .

Darby, W. J., Ghaliounghui, P., and Grivetti, L. (1977). *Food: The Gift of Osiris* (2 vols.). London: Academic Press.

Darnton, R. (2003). *George Washington's False Teeth: An Unconventional Guide to the Eighteenth Century*. New York: W. W. Norton.

Darwin, C. (1859). *On the Origin of Species by means of Natural Selection: or, Preservation of favoured Races in the Struggle for Life*. London: J. Murray.

Darwin, E. (1791). *The Botanic Garden*. London: Printed for J. Johnson.

Darwin, E. (1803). *The Temple of Nature*. London: Printed for J. Johnson.

Daston, L. (1991). The ideal and reality of the Republic of Letters in the Enlightenment. *Science in Context*, 4, 367–386.

Daston, L., and Park, K. (1998). *Wonders and the Order of Nature 1150–1750*. New York: Zone Books.

Davy, H. (1829). An Account of some Experiments on the Torpedo. *Philosophical Transactions of the Royal Society of London*, 119, 15–18.

References

Davy, H. (1839–1840). *The Collected Works of Sir Humphry Davy, Bart, edited by his Brother, John Davy*. London: Smith, Elder, and Company.

Davy, J. (1832). An Account of Some Experiments and Observations on the Torpedo (Raia Torpedo, Linn.). *Philosophical Transactions of the Royal Society of London*, 122, 259–278.

Davy, J. (1834). Observations on the Torpedo, with an account of some additional experiments on its electricity. *Philosophical Transactions of the Royal Society of London*, 124, 531–550.

Davy, J. (1839–40). *The Collected Works of Sir Humphry Davy* (9 vols.). London: Smith, Elder & Co.

Davy, J. (1839). *Researches, Physiological and Anatomical*. London: Smith, Elder, and Co.

Dawson, W. R. (1958). *The Banks Letters*. London: The British Museum.

Debru, A. (2006). The power of torpedo fish as a pathological model to the understanding of nervous transmission in Antiquity. *Comptes-Rendus Biologies,* 329, 298–302.

Debru, A. (2008). Les enseignements de la torpille dans la médecine antique. In I. Boehm and P. Luccioni (Eds.), *Le Médecin initié par l'Animal*. Lyon: Maison de l'Orient et de la Méditerranée - Jean Pouilloux.

Debus, A. G. (1966). *The English Paracelsians*. New York: Franklin Watts.

Delbourgo, J. (2006). *A Most Amazing Scene of Wonders: Electricity and Enlightenment in Early America*. Cambridge, MA: Harvard University Press.

Delatte, L. (1942). *Textes Latins et Vieux Français Relatifs aux Cyranides*. Liège: Faculté de Philosophie et Lettres.

Den Heijer, H. (2002). *De Geschiedenis van de WIC (3rd ed.)*. Zutphen: Walburg.

Denny, M. (1948). Linnaeus and his disciple in Carolina: Alexander Garden. *Isis*, 38, 161–174.

De Palma, A., and Pareti, G. (2011). Bernstein's long path to membrane theory. Radical change and conservation in nineteenth-century German electrophysiology. *Journal of the History of Neurosciences (In the press)*

Descartes, R. (1644). *Principia philosophiae*. Amstelodami: Apud Ludovicum Elzevirium.

Descartes, R. (1649). *Les Passions de l'Ame*. Amstelodami: Apud Ludovicum Elzevirium. (Trans. by S. Voss. as *The Passions of the Soul*. Indianapolis: Hackett Publishing Co., 1989.)

Descartes, R. (1662). *De homine figuris et latinitate donatus a Florentio Schuyl*. Leyden: Franciscum Moyardum and Petrum Leffen. (Trans. by T. S. Hall as *Treatise of Man by René Descartes*. Cambridge, MA: Harvard University Press, 1972.)

Desmond, R. (1994). *Dictionary of British & Irish Botanists & Horticulturists, Including Plant Collectors, Flower Painters and Garden Designers*. London: Taylor & Francis.

De Vries, H. (1884). Eine Methode zur Analyse der Turgorkraft. *Jahrbücher für wissenschaftliche Botanik*, 14, 427–601.

Dewhurst, K. (1982). Thomas Willis and the foundations of British Neurology. In F. C. Rose and W. F. Bynum (Eds.), *Historical Aspects of the Neural Sciences*. New York: Raven Press, pp. 327–346.

Dierig, S. (2000). Urbanization, place of experiment and how the electric fish was caught by Emil du Bois-Reymond. *Journal of the History of the Neurosciences*, 9, 5–13.

Dioscorides, P. (1527). *De medicinali materia libri quinque. De virulentis animalibus, & venenis cane rabioso, eorum notis, ac remediis libri quatuor*. Venetiis: Per Johannes Antonium et fratres de Sabio.

Dioscorides, P. (1559). *Les six livres de Pedacion Dioscoride d'Anazarbe de la matiere medicinale*. Lyon: Chez Thibault Payan.

Dioscorides, P., and Ruel, J. (1516). *Pedacii Dioscoridis Anazarbei de medicinali materia: libri quinq[ue]. De virule[n]tis animalibus, et venenis cane rabioso, et eorum notis, ac remediis libri quattuor*. Parisiis: In officina Henrici Stephani.

Dioscorides, P., and Vergilius, M. (1529). *Opera Graece et Latine; interpr. M. Vergilio cum comm*. Coloniae: Opera et Impensa Joannis Soteris.

Di Pasquale, G. (2004). *Tecnologia e meccanica: trasmissione dei saperi tecnici dall'età ellenistica al mondo romano*. Firenze: Olschki.

Dobell, C. (1960). *Anthony van Leeuwenhoek and his Little Animals*. New York: Dover Publications.

Dobson, J. (1969). *John Hunter*. Edinburgh: E & S Livingstone.

Donati, A. and Pasini, P. (1997). *Pesca e pescatori nell'antichità*. Milano: Leonardo arte.

Doresse, J. (1957). *L' Empire du Prêtre-Jean*. Paris: Plon.

Dorsman, C., and Grommelin, C. A. (1957). The invention of the Leyden jar. *Janus*, 46, 275–280.

Dos Santos, J. (1609). *Ethiopia Oriental e Varia Historia de Cousas Notaveis do Oriente*. Evora: Manoel de Lyra.

Douglas, A. E. (1968). *Cicero*. Oxford: Clarendon Press.

Du Bois-Reymond, E. (1843a). *Quae apud veteres de piscibus electricis existant argumenta*. Berlin: Nietackianis.

Du Bois-Reymond, E. (1843b). Vorläufiger Abriss einer Untersuchung über den sogenannten Froschstrom und über die elektromotorischen Fische. *Annalen der Physik und der Chemie*, 58, 1–30.

Du Bois-Reymond, E. (1848–84). *Untersuchungen über thierische Elektricität* (2 vols). Berlin: G. Reimer.

Du Bois-Reymond, E. (1849). Nouveaux détails sur les expériences de M. E. du Bois Reymond concernant l'électricité développée par le fait de la contraction musculaire. *Comptes Rendus Hebdomadaires de l'Académie des Sciences de Paris*, 28, 641–643.

Du Bois-Reymond, E. (1851). Über thierische Bewegung. In: Du Bois-Reymond (1912), *Reden von Emil du Bois-Reymond, Vol. 2*. Leipzig: Veit & Co., p. 29–55.

Du Bois-Reymond, E. (1857). Nachricht von einem nach Berlin gelangten lebenden Zitterwels. *Monatsberichte der Königlichen preussische Akademie des Wissenschaften zu Berlin*, 424–429.

Du Bois-Reymond, E. (1858a). Gedächtnisrede auf Johannes Müller gehalten in der Leibniz-Sitzung der Akademie der Wissenschaften am 8. Juli 1858. In E. Du Bois-Reymond (1912), *Reden von Emil Du Bois-Reymond, Vol. 2*, Leipzig: Veit, pp. 143–334.

Du Bois-Reymond, E. (1858b). Note sur le Malaptérure électrique. *Annales de Chimie et de Physique*, 52, 124–125.

Du Bois-Reymond, E. (1859). Bemerkungen über die Reaction der elektrischen Organe und der Muskeln. *Archiv für Anatomie, Physiologie und wissenschaftliche Medicin*, 846–853.

Du Bois-Reymond, E. (1877). Beobachtungen und Versuche an lebend nach Berlin gelangten Zitterwelsen (*Malopterurus* [sic] *electricus*) (Neue Abhandlung). In *Gesammelte Abhandlungen zur allgemeinen Muskel-und Nervenphysik, Zweiter Band*. Leipzig: Veit & Co., pp. 601–647.

Du Bois-Reymond, E. (1884). Lebende Zitterrochen in Berlin. *Sitzungsberichte der Königlich Preussischen Akademie der Wissenschaften*, 1, 181–242.

Du Bois-Reymond, E. (1885). Lebende Zitterrochen in Berlin. *Sitzungsberichte der Königlich Preussischen Akademie der Wissenschaften*, 2, 691–750.

Du Bois-Reymond, E. (1887). Researches relating to the electric organ of the *Malapterurus* and of the *Torpedo*. In J. Burdon-Sanderson (Ed.) *Translations of Foreign Memoirs: Memoirs on the Physiology of the Nerve, of Muscle and of the Electric Organ*. Oxford: Clarendon, pp. 370–547.

Du Bois-Reymond, E. (1889). Bemerkungen über einige neuere Versuche an Torpedo. *Archiv für Physiologie*, 316–344.

Du Bois-Reymond, E., and Bence Jones, H. (1852). *On animal electricity: being an abstract of the discoveries of Emil du Bois-Reymond*. London: Churchill.

Du Bois-Reymond, E(stelle) (Ed.) (1927). *Zwei Grosse Naturforscher des 19 Jahrhunderts. Ein Briefweschel zwischen Emil du Bois Reymon und Karl Ludwig*. Leipzig: Johann Ambrosius Barth.

Duchesneau, F. (1982). *La Physiologie des Lumières. Empirisme, Modèles et Théories*. The Hague: Nijhoff.
Duell, P. (1938). *The Mastaba of Mereruka, by the Sakkarah Expedition; Field Director, Prentice Duell*. Chicago: University of Chicago Press.
DuFay, C. F. de C. (1735). Quatrième mémoire sur l'électricité. *Mémoires de l'Académie Royale des Sciences*, 457–476.
Duffy, M. (1977). *The Passionate Shepherdess: Aphra Behn, 1640–89*. London: Cape.
Durant, W. (1953). *The Renaissance: A History of Civilization in Italy from 1304–1576 A.D*. New York: Simon and Schuster.
Duyker, E. (1998). *Nature's Argonaut: Daniel Solander 1733–1782: Naturalist and Voyager with Cook and Banks*. Melbourne: Melbourne University Press.
Duyker, E., and Tingbrand, P. (1995). *Daniel Solander: Collected Correspondence 1753–1782*. Melbourne: Melbourne University Press.
Ebbecke, U. (1951). *Johannes Müller, der große rheinische Physiologe, mit einem Neudruck von Johannes Müllers Schrift: Über die phantastischen Gesichtserscheinungen*. Hannover: Schmorl & von Seefeld.
Eckermann, J. P. (1836). *Gespräche mit Goethe in den letzten Jahren seines Lebens.: 1823–32*. Lepzig: Brockhaus.
Egerton, F. N. (1973). Changing concepts of the balance of nature. *Quarterly Review of Biology*, 48, 322–350.
Egerton, F. N. (1983). Latin Europe. In *Landmarks of Botanical History* (2 vols.). Stanford: Stanford University Press, pp. 444–452.
Egerton, F. N. (2001). A history of the ecological sciences, Part 2: Aristotle and Theophrastos. *Bulletin of the Ecological Society of America*, 82, 149–152.
Egerton, F. N. (2002). A History of the Ecological Sciences. Part 7: Arabic Languages Science: Botany, Geography, and Decline. *Bulletin of the Ecological Society of America*, 83, 261–266.
Egerton, F. N. (2003). A history of the ecological sciences. Part 11: Emergence of vertebrate zoology during the 1500s. *Bulletin of the Ecological Society of America*, 84, 206–212.
Egerton, F. N. (2005). A history of the ecological sciences. Part 17: Invertebrate zoology and parasitology during the 1600s. *Bulletin of the Ecological Society of America*, 86, 133–144.
Egerton, F. N. (2006). A history of the ecological sciences. Part 21: Réaumur and his history of insects. *Bulletin of the Ecological Society of America*, 87, 212–224.
Egerton, F. N. (2007). A history of the ecological sciences. Part 23: Linnaeus and the economy of nature. *Bulletin of the Ecological Society of America*, 88, 72–88.
Einstein, L. (1933). *Divided Loyalties: Americans in England During the War of Independence*. London: Cobden-Sanderson.
Eiseley, L. (1973). *The Man who Saw through Time*. New York: Charles Scribner's Sons.
Elles, N. (1747). Of methods of exercise within doors. *Gentleman's Magazine*, 17, 77.
Ellis, J. (1766). An account of an amphibious bipes. *Philosophical Transactions of the Royal Society of London*, 56, 191–192.
Estienne, H. (1567). *Medicae artis principes, post Hippocratem et Galenum. . . .* Parisiis: Henricus Stephanus illustris, Huldrichi Fuggeri typographus.
Evans, C. (1757). A relation of a cure performed by electricity. *Medical Observations and Inquiries*, 1, 83–86.
Evans, James Allan. (2005). *The Emperor Justinian and the Byzantine Empire*. Westport, CT: Greenwood Press.
Everitt, A. (2001). *Cicero: A Turbulent Life*. London: John Murray.
Fabri, G. B. (1757). *Sulla insensitività ed irritabilità Halleriana. Opuscoli di Vari Autori* (2 vols.). Bologna: Corciolani ed Eredi Colli.
Fabrici, G. d'Acquapendente (1600). *Hieronymi Fabricii ab Aquapendente, de Visione, voce, auditu*. Venetiis: Apud Franciscum Bolzettam.

Fahlberg, S. (1801). Beskrifning öfver elektriska Ålen, *Gymnotus electricus*. Linn. *Kongliga Vetenskaps Academiens Nya Handlingar*, 22, 122–156.
Fakhry, M. (2001). *Averroes (Ibn Rushd): His Life, Works and Influence*. Oxford: Oneworld.
Fambrough, D. M., Drachman, D. B., and Saryamurti, S. (1973). Neuromuscular junction in myasthenia gravis: decreased acetylcholine receptors. *Science*, 176, 189–190.
Faraday, M. (1833). Experimental researches in electricity. Third series. Identity of electricities derived from different sources. Relation by measure of common and voltaic electricity. *Philosophical Transactions of the Royal Society of London*, 123, 23–54.
Faraday, M. (1838). Experimental Researches in Electricity. Fifteenth Series. Note of the character and direction of the electric force of the *Gymnotus*. *Abstracts of the Papers Printed in the Philosophical Transactions of the Royal Society of London*, 4, 111–112.
Faraday, M. (1839a). Experimental researches in electricity. Fifteenth series. Notice on the character and direction of the electric force of the *Gymnotus*. *Philosophical Transactions of the Royal Society of London*, 129, 1–12.
Faraday, M. (1839b). *Experimental Researches in Electricity*. London: R. and J. E. Taylor.
Faraday, M. (1993). *The Correspondence of Michael Faraday (Vol. 2)*. F. A. J. L. James (Ed.). London: The Institution of Electrical Engineers.
Farber, J., and Rahn, H. (1970). Gas exchange between air and water and the ventilation pattern in the electric eel. *Respiratory Physiology*, 9, 151–161.
Feder, T. H. (1978). *Great Treasures of Pompeii and Herculaneum*. New York: Abbeville Press.
Feldberg, W. (1977). The early history of synaptic and neuromuscular transmission by acetylcholine: Reminiscences of an eye witness. In A. L. Hodgkin et al. (Eds.), *The Pursuit of Nature: Informal Essays on the History of Physiology*. Cambridge: Cambridge University Press, pp. 65–83.
Feldberg, W., and Fessard, A. (1942). Cholinergic nature of the nerves to the electric organ of the *Torpedo*. *Journal of Physiology*, 101, 200–216.
Felici, C. (1982). *Lettere a Ulisse Aldrovandi* (G. Nonni, Ed.). Urbino: Quattro Venti.
Fermin, P. (1769). *Déscription Générale, Historique, Géographique et Physique de la Colonie de Surinam* (2 vols). Amsterdam: E. van Harrevelt.
Fermin, P. (1781). *An Historical and Political View of the Present and the Ancient State of the Colony of Surinam in South America*. London: W. Nicoll.
Fessard, A. (1958) Les organes électriques. In P. P. Grassé (Ed.), *Traité de Zoologie, Vol. 13, Agnates et Poissons. Anatomie, Éthologie, Systématique*. Paris: Masson, pp. 1143–1238.
Fessard, A., and Bullock, T. H. (Eds.). (1974). *Electroreceptors and Other Specialized Receptors in Lower Vertebrates. Handbook of Sensory Physiology, Vol. III*. Berlin and New York: Springer Verlag.
Fessard, A. (1946). Some basic aspects of the activity of electric plates. *Annals of the New York Academy of Sciences*, 47, 501–516.
Festugière, A. J. (2006). *La Révélation d'Hermès Trismégiste*. Paris: Les Belles Lettres.
Figuier, L. G. (1869). *Les Poissons, les Reptiles et les Oiseaux, par Louis Figuier*. Paris: Hachette.
Findlen, P. (1994). *Processing Nature: Museums, Collecting, and Scientific Culture in Early Modern Italy*. Berkeley: University of California Press.
Findlen, P. (1997). Francis Bacon and the reform of natural history in the Seventeenth Century. In D. R. Kelley (Ed.), *History and the Disciples: The Reclassification of Knowledge in Early Modern Europe*. Rochester: University of Rochester Press.

Finger, S. (1995). Descartes and the pineal gland in animals: A frequent misinterpretation. *Journal of the History of the Neurosciences,* 4, 166–182.

Finger, S. (2000). *Minds behind the Brain: A History of the Pioneers and their Discoveries.* New York: Oxford University Press.

Finger, S. (2006a). *Doctor Franklin's Medicine.* Philadelphia: University of Pennsylvania Press.

Finger, S. (2006b). Benjamin Franklin, electricity, and the palsies: On the 300th anniversary of his birth. *Neurology,* 66, 1559–1563.

Finger, S. (2006c). Benjamin Franklin and the neurosciences. *Functional Neurology,* 21, 67–75.

Finger, S. (2007). Benjamin Franklin and the electrical cure for disorders of the nervous system. In H. Whitaker, C. U. M. Smith, and S. Finger (Eds.), *Brain, Mind and Medicine: Essays in Eighteenth Century Neuroscience.* Boston: Springer, pp. 245–256.

Finger, S. (2009). Edward Bancroft's (1769) "Torporific Eels." *Perspectives in Biology and Medicine,* 52, 61–79.

Finger, S. (2010). Dr. Alexander Garden, a Linnaean in Colonial America, and the travels of some "electric eels." *Perspectives in Biology and Medicine.,* 53, 388–406.

Finger, S., and Ferguson, I. (2009). The Role of *The Gentleman's Magazine* in the Dissemination of Knowledge about Electric Fish in the Eighteenth Century. *Journal of the History of the Neurosciences,* 18, 347–365.

Finger, S., and Hagemann, I. S. (2008). Benjamin Franklin's risk factors for gout and stones: From genes and diet to possible lead poisoning. *American Philosophical Society Proceedings,* 152, 189–206.

Finger, S., and Wade, N. (2002a). The neuroscience of Helmholtz and the theories of Johannes Müller. Part 1: Nerve cell structure, vitalism, and the nerve impulse. *Journal of the History of the Neurosciences,* 11, 136–155.

Finger, S., and Wade, N. (2002b). The neuroscience of Helmholtz and The theories of Johannes Müller. Part 2: Sensation and perception. *Journal of the History of the Neurosciences,* 11, 234–254.

Finger, S., and Zaromb, F. (2006). Benjamin Franklin and shock-induced amnesia. *American Psychologist,* 61, 240–248.

Finkelstein, G. (2000). The ascent of man? Emil du Bois-Reymond's reflections on scientific progress. *Endeavour,* 24, 129–132.

Finkelstein, G. (2003). M. du Bois-Reymond goes to Paris. *British Journal of the History of Science,* 36, 261–300.

Finkelstein, G. (2006). Emil du Bois-Reymond vs. Ludimar Hermann. *Comptes Rendus de l'Académie de Sciences (Biologies),* 329, 340–347.

Fischer, H. (1966). *Conrad Gessner (26 März 1516 - December 13, 1565): Leben und Werk.* Zurich: Kommissionsverlag Leeman.

Fitzgerald, R. P. (1968). The Form of Christopher Smart's *Jubilate Agno. Studies in English Literature, 1500–1900,* 8, 487–499.

Flagg, H. C. (1786). Observations on the numb fish, or torporific eel. Letter from South Carolina, Oct. 8, 1782. *Transactions of the American Philosophical Society,* 2, 170–173.

Folena, G. (1963-64). Per la storia dell'ittionimia volgare: tra cucina e scienza naturale. *Bollettino dell'atlante linguistico mediterraneo,* 5–6, 61–137.

Fontana, F. (1757). *Dissertation Épistolaire [. . .] adréssée au r.p. Urbain Tosetti.* In A. von Haller, *Mémoires sur la nature sensible et irritable des parties du corps animal, Vol III.* Lausanne: Bousquet, pp. 157–243.

Fontoura de Almeida, D. (2003). Carlos Chagas Filho, September 12, 1910 - February 16, 2000. *Proceedings of the American Philosophical Society,* 147, 77–82.

Forsskål, P (edited by Niebuhr, C.) (1775). *Descriptiones animalium: avium, amphibiorum, piscium, insectorum, vermium; quæ in itinere orientali observavit Petrus Forskål.* Hauniae: Mölleri.

Fortenbaugh, W. W., and Gutas, D. (Eds.) (1992). *Theophrastus: His Psychological, Doxographical, and Scientific Writings.* New Brunswick: Transactive Books.

Fortenbaugh, W. W., and Sharpless, R. W. (Eds.) (1988). *Theophrastean Studies: On Natural Science, Physics, and Metaphysics, Ethics, Religion, and Rhetoric.* New Brunswick: Transactive Books.

Fortenbaugh, W. W., Huby, P. M., and Long, A. A. (Eds.) (1985). *Theophrastus of Eresus: On his Life and Work.* New Brunswick: Transactive Books.

Fortenbaugh, W. W., Huby, P. M., Sharpless, R. W., and Gutas, D. (Eds.) (1992). *Theophrastus of Eresus: Sources for his Life, Writings, Thought, and Influence* (2 vols.). Leiden: E. J. Brill.

Forty, J. (1998). *Ancient Egyptian Pharaohs.* London: PRC Publishing, Ltd.

Foucault, M. (1965). *Madness and Civilization.* New York: Vintage Books.

Fracastoro, G. (1530). *Hieronymi Fracastorii syphilis sive morbvs gallicvs.* Veronae: [Stefano dei Nicolini da Sabbio].

Fracastoro, G. (1539). *Syphilis sive morbus gallicus. Alfonsi Ferri, ... de Ligni sancti multiplici medicina et vini exhibitio[n]e, libri quatuor. Quibus nunc primu[m] additus est Hieronymi Fracastorii Syphilis, sive morbus gallicus.* Parisiis: Veneunt in vico Iacobaeo a Ioanne Foucher sub scuto Florentio.

Fracastoro, G. (1546) *De Contagione. . . .* In: *De sympathia et antipathia rerum liber unus, De contagione et contagiosis morbis et curatione libri III.* Venetiis: Apud heredes Lucaeantonii Juntae Fiorentini.

Frame, D. M. (1958). *The Complete Essays of Montaigne.* Stanford: Stanford University Press.

Frangsmyr, T. (Ed.). (1983). *Linnaeus: The Man and His Work.* Berkeley: University of California Press.

Frank, R. G. Jr. (1988). The telltale heart: Physiological instruments, graphic methods, and clinical hopes, 1854-1914. In W. Coleman and F. L. Holmes (Eds.), *The Investigative Enterprise: Experimental Physiology in Nineteenth-Century Medicine.* Berkeley: University of California Press, pp. 211–290.

Frank, R. G. Jr. (1994). Instruments, nerve action, and the all-or-none principle. *Osiris,* 9, 208–235.

Franklin, B. (1751). *Experiments and Observations on Electricity. . . .* London: Ed. Cave.

Franklin, B. (1758). An account of the effects of electricity in paralytic cases. In a letter to John Pringle M.D. F.R.S. from Benjamin Franklin, Esq; F.R.S. *Philosophical Transactions of the Royal Society of London, 50*(2), 481–483.

Franklin, B. (1774). *Experiments and Observations on Electricity.* London: Newberry.

Franklin, B. (1996). *Autobiography of Benjamin Franklin.* Mineola: Dover Publications.

Frederick II (Emperor). (1596). *Reliqua librorum de arte venandi cum avibus: cum Manfredi regis additionibus.* Augvstae Vindelicorum: Apud Ioannem Prætorium.

Frederick II (Emperor). (2000). *De arte venandi cum avibus: edizione e traduzione italiana del ms. lat. 717 della Biblioteca Universitaria di Bologna, collazionato con il ms. Pal. lat. 1071 della Biblioteca Apostolica Vaticana.* A. L. Trombetti Budrieri (Ed.). Roma-Bari: Laterza.

Freedberg, D. (2002). *The Eye of the Lynx: Galileo, His Friends, and the Beginnings of Modern Natural History.* Chicago: University of Chicago Press.

French, R., and Greenaway, F. (Eds.) (1986). *Science in the Early Roman Empire: Pliny the Elder, his Sources and Influence.* Totowa: Barnes and Noble Books.

Fritsch, G. (1881). Vergleichend-anatomische Betrachtung der elektrischen Organe von Gymnotus electricus. In: Sachs and DuBois-Reymond, *Dr. Carl Sachs, Untersuchungen am Zitteraal Gymnotus electricus,* Anhang II. Leipzig: Veit and Company, pp. 347–400.

Fritsch, G. (1883). Bericht über die Fortsetzung der Untersuchungen an elektrischen Fischen. Beiträge zur Embryologie von Torpedo. *Sitzungsberichte der Königlich Preußischen Akademie der Wissenschaften zu Berlin,* 1, 205–209.

Fritsch, G. (1887). *Die elektrischen Fische nach neuen Untersuchungen anatomisch-zoologisch dargestellt. Vol. I. Malopterurus [sic] electricus.* Leipzig: Veit & Comp.

Fritsch, G. (1890). *Die elektrischen Fische nach neuen Untersuchungen anatomisch-zoologisch dargestellt. Vol. 2. Die Torpedineen.* Leipzig: Veit & Comp.

Fritsch, G., and Hitzig, E. (1870). Über die elektrische Erregbarkeit des Grosshirns. *Archiv für Anatomie und Physiologie,* 300–332.

Frixione, E. (2007). Irritable glue: The Haller-Whytt controversy on the mechanism of muscle contraction. In H. Whitaker, C. U. M. Smith, and S. Finger (Eds.), *Brain, Mind and Medicine: Essays in Eighteenth Century Neuroscience.* Boston: Springer, pp. 115–124.

Fuchs, L. (1542). *De historia stirpium commentarii insignes.* . . . Basileae: Ex Officina Isingriniana.

Fullmer, J. Z. (1969). *Sir Humphry Davy's Published Works.* Cambridge, MA: Harvard University Press.

Gaillard, M. C. (1923). *Faune Égyptienne Antique. Recherches sur les poissons représentés dans quelques tombeaux Égyptiens de l'Ancien Empire. Vol. 51: Mémoires de l'Institut Français d'Archéologie Orientale.* Cairo: Imprimerie de l'Institut Français d'Archéologie Orientale, pp. 75–78.

Galen. (1576). *Galeni librorum septima classis: curativam methodum tum diffuse tum breviter descriptam.* . . . Venetiis: Apud Juntas.

Galen. (1969). *Galen on the Parts of Medicine, On Cohesive Causes, On Regimen in Acute Diseases in Accordance with the Theories of Hippocrates.* M. Lyons (Trans.). Berlin: Akademie-Verlag.

Galen. (1821–33). *Claudii Galeni Opera omnia. Medicorum graecorum opera quae exstant* (20 vols). Karl Gottlob Kühn, editor and translator. Lipsiae: Officina Libraria Cnoblochii.

Galen. (1586). *Librorum quarta classis.* Venetijs: Apud Iuntas.

Galilei, G. (1632). *Dialogo di Galileo Galilei sopra i due Massimi Sistemi del Mondo.* Fiorenza: Per Gio. Batista Landini.

Galilei, G. (1638). *Discorsi e dimostrazioni matematiche, intorno à due nuove scienze attenenti alla mecanica e i movimenti local.* . . . Leida: Appresso gli Elseviri.

Gallagher, C., and Stern, S. (2000). Introduction: Culture and historical background. In *Aphra Behn: Oroonoko; or, They Royal Slave.* New York: Bedford/St. Martin's Press.

Galvani, L. (1791). De viribus electricitatis in motu musculari commentarius. *De Bononiensi Scientiarum et Artium Instituto atque Academia Commentarii,* 7, 363–418.

Galvani, L. (1794a). *Dell'uso e dell'attività dell'arco conduttore nelle contrazioni dei muscoli.* Bologna: A San Tommaso d'Aquino.

Galvani, L. (1794b). *Supplemento al trattato dell'uso e dell'attività dell'arco conduttore nelle contrazioni dei muscoli.* Bologna: A San Tommaso d'Aquino.

Galvani, L. (1797). *Memorie sulla elettricità animale di Luigi Galvani P. Professore di Notomia nella Università di Bologna al celebre Abate Lazzaro Spallanzani Pubblico professore nella Università di Pavia.* Bologna: Per le stampe del Sassi.

Galvani, L. (1937a). *Memorie ed Esperimenti Inediti: Con la iconografia di lui e un saggio di bibliografia degli scritti.* Bologna: Licinio Cappelli.

Galvani, L. (1937b) *Il "Taccuino" di Luigi Galvani.* Zanichelli: Bologna.

Galvani, L. (1953a). *Commentary on the Effects of Electricity on Muscular Motion.* Translated by M. G. Foley with a foreword by I. B. Cohen. Norwalk, CT: Burndy Library.

Galvani, L. (1953b). *A Translation of Luigi Galvani's De viribus electricitatis in motu muscularis commentarius: Commentary on the Effect Of Electricity on Muscular Motion.* By Robert Montraville Green, with a foreword by G. Pupilli. Cambridge, MA: Licht.

Gamble, D. P., and Hair, P. E. H. (1999). *The Discovery of the River Gambia (1623) by Richard Jobson.* London: The Hakluyt Society.

Gammer-Wallert, I. (1970). *Fische und Fischkulte im Alten Ägypten.* Wiesbaden: O. Harrassowitz.

Ganot, A. (1882). *Corso di fisica* (3rd Italian edition). Milano: Pagnoni.

Garden, A. (1771). An account of the India pink. *Essays and Observations, Physical & Literary,* 3, 145–153.

Garden, A. (1775a). An account of the *Gymnotus electricus,* or electrical eel. *Philosophical Transactions of the Royal Society,* 65, 102–110.

Garden, A. (1775b). An account of the *Gymnotus electricus,* or electrical eel. *Gentleman's Magazine,* 45, 437.

Gardiner, A. (1978). *Egyptian Grammar, Being an Introduction to the Study of Hieroglyphs* (3rd ed., rev). Oxford: Ashmolean Museum.

Gardini, F. G. (1780). *De effectis electricitatis in homine dissertatio.* Genova: Haeredes Adae Scionici.

Gardini, F. G. (1792). *De electrici ignis natura dissertatio.* . . . Mantuae: Typis Haeredis Alberti.

Garn, J. A. (1778). *Dissertatio inauguralis medica de torpedine recentiorum genere anguilla. Præside G. A. Langguth.* Wittenberg: Litteris Caroli Christ. Dürrii.

Garten, S. (1899). Beiträge zur Physiologie des elektrischen Organes des Zitterrochen. *Abhandlungen der Mathematisch-Physischen Classe der Königlich Sächsischen Gesellschaft der Wissenschaften,* 25, 251–366.

Garofalo, I. and Vegetti, M. (1978). *Opere scelte di Galeno.* Torino: Unione Tipografico-Editrice Torinese.

Garten, S. (1910). Die Produktion von Elektrizität. In H. Winterstein (Ed), *Handbuch der vergleichenden Physiologie,* IV Band, II Hälfte. Jena: Fisher, pp. 105–224.

Gazette de France. (1772). [Article about Walsh's torpedo experiments]. *Gazette de France,* August 14, 1772, p. 298.

Gazetteer and New Daily Advertiser. (1776). [No title: Baker's eels and Walsh's spark.] *Gazetteer and New Daily Advertiser,* August 5, Issue 14805.

Gazetteer and New Daily Advertiser. (1776). THE ELECTRICAL EEL, or NUMBING FISH of SOUTH AMERICA. GYMNOTUS ELECTRICUS of LINNAEUS. *Gazetteer and New Daily Advertiser,* August 12, Issue 14811. (Notice repeated August 14, Issue 14814.)

Gazetteer and New Daily Advertiser. (1776). THE GYMNOTUS ELECTRICUS, or the ELECTRICAL EEL, at Mr. Baker's New Apartments in Piccadilly, nearly opposite St. James's Street. *Gazetteer and New Daily Advertiser,* November 29, Issue 14906. (Notice repeated December 14, Issue 14919.)

Gazetteer and New Daily Advertiser. (1777). Advertisement for Adam Strong's *The Electrical Eel. Gazetteer and New Daily Advertiser,* April 12, Issue 15108; June 28, Issue 15084.

General Advertiser and Morning Intelligencer. (1779). Advertisement for *An Elegy on the Lamented Death of the Electrical Eel. General Advertiser and Morning Intelligencer,* March 6, Issue 628; March 23, Issue 642.

General Evening Post. (1774). [No title: A torpedo caught in the Thames.] *General Evening Post,* June 25, Issue 6348.

General Evening Post. (1774). [No title: Pringle's speech and Walsh's medal.] *General Evening Post,* November 29, Issue 6368.

General Evening Post. (1774). [No title: Torpedoes caught of the English coast.] *General Evening Post,* December 27, Issue 6380.

Gentleman's Magazine. (1769). *An Essay on the Natural History of Guiana in South America,* by Edward Bancroft. 39, 145–149. (Book Review)

Geoffroy Saint-Hilaire, E. (1802). Mémoire sur l'anatomie comparée des organes électriques de la *Raie torpille,* du *Gymnote engourdissant*, et du *Silure trembleur*. *Annales du Muséum National d'Histoire Naturelle*, 1, 392–407.

Gessner, C. (1545). *Bibliotheca vniuersalis, siue Catalogus omnium scriptorum locupletissimus: in tribus linguis, Latina, Græca, & Hebraica: extantium et non extantiu[m]* . . . Tiguri: Apud Christophorum Froschouerum.

Gessner, C. (1551). *Historia animalium. 1, De quadrupedibus viviparis*. Tiguri: Apud Christophorum Froschouerum.

Gessner, C. (1558). *Conradi Gesneri medici Tigurini historiae animalium liber IIII qui est de piscium et aquatilium animantium natura*. . . . Tiguri: Apud Christophorum Froschouerum.

Gessner, C. (1587). *Conradi Gesneri Tigurini medicinae et philosophiae professoris . . . Historiae animalium lib. V. qui est de serpentium natura.: ex variis schedis et collectaneis eiusdem compositus per Iacobum Carronum*. . . . Tiguri: In Officina Froschoviana.

Gessner, K., and Simmler, J. (1574). *Bibliotheca instituta et collecta primum a Conrado Gesnero*. Tiguri: Apud Christophorum Froschouerum.

Geyer, B. (1955). *Alberti Magni Opera omnia. T. 12, Liber de natura et origine animae; Liber de principiis motus processivi; Quaestiones super de animalibus*. Monasterii Westfalorum: Aschendorff.

Gilbert, W. (1600). *De magnete*. Londini: Excudebat Petrus Short.

Gilfillan, S. C. (1965). Lead poisoning and the fall of Rome. *Journal of Occupational Medicine*, 7, 53–60.

Gilfillan, S. C. (1990). *Rome's Ruin by Lead Poison*. Long Beach: Wenzel Press.

Gill, T. N. (1864). [No formal title: "Several points in ichthyology and conchology"]. Proceedings of the Academy of Natural Sciences of Philadelphia, 16, 151–152.

Gilles, P. (1533). *Ex Aeliani historia*. Leyden: Apud Seb. Gryphium.

Giovio, P. (1524). *De romanis piscibus*. Romae: F. M. Calui.

Glassman, R. B., and Buckingham, H. W. (2007). David Hartley's neural vibrations and psychological associations. In H. Whitaker, C. U. M. Smith, and S. Finger (Eds.), *Brain, Mind and Medicine: Essays in Eighteenth Century Neuroscience*. Boston: Springer, pp. 177–190.

Glisson, F. (1677). *Tractatus de ventriculo et intestinis. Cui præmittitur alius, de partibus continentibus in genere, et in specie, de iis abdominis*. Londini: Typis E.F., prostat venalis apud Henricum Brome. . . .

Glycas, M. (1836). *Michaelis Glycae Annales*. Editio Emendatior Et Copiosior, Consilio B. G. Niebuhrii C. F. Instituta. Bonnae: impensis Ed. Weberi.

Gmelin, J. F. (1789). *Caroli a Linné . . . Systema naturae per regna tria naturae, secundum classes, ordines, genera, species; cum characteribus, differentiis, synonymis, locis. Editio decimo tertia, aucta, reformata* (13th Ed., Vol. 2). Lugduni: Apud J.-B. Delamolliere.

Godinho, N. (1615). *De Abassinorum rebus*. . . . Lugduni: Horatij Cardon.

Godwin, J. (1979). *Athanasius Kircher: A Renaissance Man and the Quest for Lost Knowledge*. London: Thames and Hudson.

Goethe, J. W. von. (1791–92). *Beiträge zur Optik* (2 vols.). Berlin: A. Weimar.

Goethe, J. W. von. (1810). *Zur Farbenlehre*. Weimar. (Translated by C. L. Eastlake as *Goethe's Theory of Colours*. Frank Cass & Company, 1967.)

Gombrich, E. H. (1945). Botticelli's mythologies: a study in the Neoplatonic symbolism of his circle. *Journal of the Warburg and Courtauld Institutes*, 8, 7–60.

Gombrich, E. H. (1960). *Art and Illusion: A Study in the Psychology of Pictorial Representation*. New York: Pantheon Books.

Góis, D. de. (1541). *Fides, religio, moresque Aethiopum sub imperio Preciosi Ioannis (quem vulgoÌ Presbyterum Ioannem vocant)*. . . . Parisiis: Apud Christianum Wechelum.

Goode, B. B., and Bean, T. H. (1885). On the American fish in the Linnaean Collection. *Proceedings of the United States National Museum*, 8, 193–208.

Goodman, L. E. (1992). *Avicenna*. London: Routledge.

Goreau, A. (1980). *Reconstructing Aphra: A Social Biography of Aphra Behn*. New York: Dial Press.

Gotch, F. (1887). The Electromotive properties of the electrical organ of *Torpedo marmorata*. *Philosophical Transactions of the Royal Society of London, B*, 178, 487–537.

Gotch, F. (1888). Further observations on the electromotive properties of the electrical organ of *Torpedo marmorata*. *Philosophical Transactions of the Royal Society of London, ser. B*, 179, 329–363.

Gotch, F. (1900). Nerve. In E. A. Schäfer (Ed.), *Text-Book of Physiology* (Vol. 2). Edinburgh & London: Young J. Pentland, pp. 451–560.

Gotch, F. (1902). The submaximal electrical response of nerve to a single stimulus. *Journal of Physiology*, 28, 395–416.

Gotch, F., and Burch, G. J. (1896). The electromotive properties of *Malapterurus electricus*. *Philosophical Transactions of the Royal Society of London, ser. B*, 187, 347–407.

Gotthelf, A. (Ed.). (1985). *Aristotle on Nature and Living Things*. Pittsburgh: Mathesis Publications Inc.

Gow, A. S. F. (1968). On the Halieutica of Oppian. *Classical Quarterly*, 18, 60–68.

Grafton, A. T. (1993). Foreward. In G. Boaz (Ed., Trans.), *The Hieroglyphics of Horapollo*. Princeton: Princeton University Press, pp. xi–xxi.

Gray, S. (1731). A letter to Cromwell Mortimer. *Philosophical Transactions of the Royal Society*, 37, 18–44.

Grayeff, F. (1956). The problem of the genesis of Aristotle's text. *Phronesis*, 1, 105–122.

Greenwood, J. (1752–58). *Memorandum Book, Dec. 1752 - Apr. 1758*. New York Historical Society Manuscripts Collection, p. 168.

Greenwood, P. H. (1976). Fish fauna of the Nile. In J. Rzóska (Ed.), *The Nile, Biology of an Ancient River*. The Hague: W. Junk B.V.

Gregory, R. L. (2005). *Eye and Brain. The Psychology of Seeing* (5th ed.). New York: Oxford University Press.

Grévin, J. (1567–68). *Deux Livres des Venins*. . . . Anvers: Christofle Plantin.

Groner, J., and Cornelius, P. F. S. (1996). *John Ellis: Merchant, Microscopist, Naturalist, and King's Agent*. Pacific Grove: Boxwood Press.

Gronov, J. F. (1742). A method of preparing specimens of fish by drying their skins, as practiced by John. Frid. Gronovius, M.D., at Leyden. *Philosophical Transactions of the Royal Society of London*, 42, 57–58.

Gronov, L. T. (1754–56). *Museum ichthyologicum, sistens piscium indigenorum & quorundam exoticorum . . .* (2 vols). Lugduni Batavorum: Apud Theodorum Haak.

Gronov, L. T. (1758). Brief van D. Laurens Theodorus Gronovius aan C. N. over niew-ontdekte Zeediertjes en Byzonderheden van den Siddervis of Beef-Aal (uit het Latyn vertaald) (Leide October 1758). *Uitgezogte Verhandelingen uit de Nieuwste Werken van de Societeiten der Setenschappen in Europa en van Andere Geleerde Mannen*, 3, 464–478.

Gronov, L. T. (1760). Laur. Theodori Gronovii . . . Gymnoti tremuli descriptio, atque experimenta cum eo instituta. *Acta Helvetica Physico-Mathematico-Anatomico-Botanico-Medica . . . in Usus Publicos Exarata*, 4, 27–35.

Gronov, L. T. (1763–64). *Zoophylacii Gronoviani fasciculi*. . . . Lugduni Batavorum: Sumptibus Auctoris.

Gronov, L. T. (1854). *Catalogue of Fish Collected and Described by Laurence Theodore Gronow now in the British Museum*. (J. E. Gray, ed.) London: Published by Order of the Trustees.

Grundfest, H. (1957). The mechanisms of discharge of the electric organs in relation to general and comparative electrophysiology. *Progress in Biophysics and Biophysical Chemistry*, 7, 1–85.

Gudger, E. W. (1910). An early note on flies as transmitters of disease. *Science*, 31, 31–32.

Gudger, E. W. (1911). Further early notes on the transmission by flies of a disease called yaws. *Science*, 33, 427–428.

Gudger, E. W. (1924). Pliny's *Historia naturalis*. The most popular natural history ever published. *Isis*, 6, 269–281.

Gudger, E. W. (1934). The five great naturalists of the sixteenth century: Belon, Rondolet, Salviani, Gessner and Aldrovandi: A chapter in the history of ichthyology. *Isis*, 22, 21–40.

Guericke, O. von (1672). *Experimenta nova (ut vocantur) Magdeburgica. . . .* Amstelodami: Apud Joannem Janssonium à Waesberge.

Guerrini, L. (1999). Contributo critico alla biografia rediana. Con uno studio su Stefano Lorenzini e le sue Osservazioni intorno alle torpedini. In Bernardi, Walter, and Luigi Guerrini. *Francesco Redi: un protagonista della scienza moderna : documenti, esperimenti, immagini*. Firenze: L.S. Olschki, pp. 48–69.

Guerreiro, F. (1611). *Relaçam annal das cousas que fizeram os padres da Companhia de Iesus, nas partes da India Oriental . . . com mais hua addiçam aì relaçam de Ethiopia*. Lisboa: Impresso por Pedro Crasbeeck.

Gulliver, G. (1846). *The Collected Works of William Hewson*. London: The Sydenham Society.

Gumilla, J. (1745). *El Orinoco Ilustrado y Defendido*. Madrid: Manuel Fernandez. (Reprinted, Caracas: Biblioteca de la Academia Nacional de la Historia, 1963.)

Gunderson, L. (1980). *Alexander's Letter to Aristotle about India*. Meisenheim am Glan: Hain.

Gunther, R. T. (1959). *The Greek Herbal of Dioscorides, Illustrated by a Byzantine, A.D. 512; Englished by John Goodyer, A.D. 1655*. New York: Hafner Publishing Co.

Günther, A. C. L. G. (1899). The President's Anniversary Address. *Proceedings of the Linnean Society of London, 111th Session*, 15–39.

Haberling, W. (1924). *Johannes Müller, das Leben des rheinischen Naturforschers*. Leipzig: Akademische Verlagsgesellschaft. (Grosse Männer, Studien zur Biologie des Genies, 9.)

Hackmann, W. D. (1978). *Electricity from Glass*. Alphen aan den Rijn: Sijthoff & Noordhoff.

Hackmann, W. D. (1985). Leopoldo Nobili and the beginnings of galvanometry. In G. Tarozzi (Ed.), *Leopoldo Nobili e la Cultura Scientifica del suo Tempo* Bologna: Istituto per i beni Artistici, Culturali, Naturali della Regione Emilia-Romagna, pp. 183–199.

Hackmann, W.D. (1972). The researches of Dr. Martinus van Marum (1750-1837) on the influence of electricity on animals and plants. *Medical History*, 16, 11–26.

Hagberg, K. (1952). *Carl Linnaeus*. (A. Blair, trans.). London: Jonathan Cape.

Hales, S. (1733). *Statical Essays: Containing Haemasticks* (2 vols). London: W. Innys, and others.

Hall, G. S. (1912). Hermann L. F. von Helmholtz. In: *Founders of Modern Psychology*. New York: Appleton, pp. 247–308.

Haller, A. von (1753). De partibus corporis humani sensibilibus et irritabilibus. *Commentarii Societatis Regiae Scientium Gottingensis* 2: 114–58 (Translated as, *A Dissertation on the Sensible and Irritable Parts of Animals with an Introduction by Oswei Temkin*). Baltimore: Johns Hopkins University Press, (1936).

Haller, A. von (1756-60). *Mémoires sur la Nature Sensible et Irritable des Parties du Corps Animal* (4 vols.). Lausanne: M.-M. Bousquet.

Haller, A. von (1762). *Elementa physiologiae corporis humani. Tomus IV: Cerebrum. Nervi. Musculi*. Lausanne: Sumptibus Francisci Grasset et Sociorum.

Haller, A. von (1766). *Elementa physiologiae corporis humani*. Bernae: Sumptibus Societatis Topograficae.

Haller, A. von (1769). *Elementa physiologiae corporis humani. Tomus V: Sensus externii internii*. Lausanne: Sumptibus Francisci Grasset et Sociorum.

Haller, A. von (1771). *Primae lineae physiologiae: in usum praelectionum academicarum (IV ed)*. Lausanne: Apud Franciscum Grasset et Socios.

Haller, A. von (1786) *First Lines of Physiology*. Edinburgh: Printed for Charles Elliot. (New York: Johnson Reprint Company, 1966.)

Hamilton, J. S. (1986). Scribonius Largus on the medical profession. *Bulletin of the History of Medicine*, 60, 209–216.

Hamilton, J. (2002). *A Life of Discovery: Michael Faraday, Giant of the Scientific Revolution*. New York: Random House.

Harper, C., et al. (2008). On the origin of the treponematoses: A phylogenetic approach. *PLoS Neglected Tropical Diseases*, 2, 1–13.

Hartley, D. (1746). *Various Conjectures on the Perception, Motion, and Generation of Ideas*. London. (Edited and trans. by R. E. A. Palmer and M. Kallich. Los Angeles: Augustan Reprint Society, 1959.)

Hartley, D. (1749). *Observations on Man, His Frame, His Duty, and his Expectations* (2 vols). London: Printed by S. Richardson for James Leake and Wm. Frederick.

Hartley, H. (1962). The tercentenary of the Royal Society's Charter. *Notes and Records of the Royal Society of London*, 17, 111–116.

Hartley, H. (1966). *Humphry Davy*. London: Thomas Nelson and Sons.

Hartung, E. F. (1954). History of the use of colchicum and related medicaments in gout. *Annals of Rheumatoid Diseases*, 13, 190–220.

Harvey, W. (1628). *Exercitatio anatomica de motu cordis et sanguinis in animalibus*. Francofurti: Sumptibus Guilielmi Fitzeri.

Haskins, C. H. (1928). Latin literature under Frederick II. *Speculum*, 3, 129–151.

Haskins, C. H. (1939). *The Renaissance of the Twelfth Century*. Cambridge: Cambridge University Press.

Hassan, F. (1986). Holocene lakes and prehistoric settlements of the Western Faiyum. *Journal of Archeological Science*, 13, 483–501.

Hasson, A., and Chagas, C. (1959). Selective capacity of components of the aqueous extract of the electric organ to bind curarizing quaternary ammonium derivatives. *Biochemical and Biophysical Acta*, 36, 301–308.

Hawass, Z. (2008). *Tutankhamun: The Golden King and the Great Pharaohs*. Washington, DC: National Geographic Society.

Hawksbee, F. (1709). *Physico-Mechanical Experiments on Various Subjects, Containing an Account of Several Surprising Phenomena Thouching Light and Electricity. . . .* London: Printed for the author.

Hayduck, M. (1888). *Asclepii in Aristotelis metaphysicorum libros A-Z commentaria. Commentaria in Aristotelem Graeca*, Vol. 6, Pt. 2. Berolini: Typis et impensis G. Reimeri.

Healy, J. F. (1999). *Pliny the Elder on Science and Technology*. New York: Oxford University Press.

Heathcote, N. H. de V. (1955). Franklin's introduction to electricity. *Isis*, 46, 29–35.

Heiberg, I. L. (1894). *Simplicii in Aristotelis De caelo commentaria*. Berlin: G. Reimeri.

Heilbron, J. L. (1977) Franklin, Haller, and Franklinist History. *Isis*, 68: 539–549.

Heilbron, J. L. (1979). *Electricity in the Seventeenth and Eighteenth Centuries*. Berkeley: University of California Press.

Heilbron, J. L. (1991). The contributions of Bologna to Galvanism. *Historical Studies in the Physical Sciences*, 22, 57–85.

Helck, H., and Westendorf, W. (1982). Fische, profan. *Lexicon der Äegyptologie*, 4, 348–349.

Helferich, G. (2004). *Humboldt's Cosmos*. New York: Gotham Books.

Helmholtz, H. (1843). Ueber das Wesen der Fäulniss und Gährung. *Archiv für Anatomie, Physiologie und wissenschaftliche Medizin*, 453–463.

Helmholtz, H. (1847). *Ueber die Erhaltung der Kraft, Eine Physicalische Abhandlung.* Berlin: G. Reimer.

Helmholtz, H. (1848). Ueber die Wärmeentwickelung bei der Muskelaction. *Archiv für Anatomie, Physiologie und wissenschaftliche Medizin,* 147–164.

Helmholtz, H. (1850a). Ueber die Fortpflanzungsgeschwindigkeit der Nervenreizung. *Monatsberichte de Königlich Preussischen Akademie der Wissenschaften zu Berlin,* January 21, 1850.

Helmholtz, H. (1850b). Messungen über den zeitlichen Verlauf der Zuckung animalischer Muskeln und die Fortpfanzunggeshwindigkeit der Reizung in den Nerven. *Archiv für Anatomie, Physiologie und wissenschaftliche Medizin,* 276–364.

Helmholtz, H. (1852a). Messungen über Fortpflanzungsgeschwindigkeit der Reizung in den Nerven. Zweite Reihe. *Archiv für Anatomie, Physiologie und wissenschaftliche Medizin,* 199–216.

Helmholtz, H. (1852b). Die Resultate der neueren Forschungen über thierische Elektricität. *Allgemeine Monatsschrift fur Wissenschaft and Literatur.* Reprinted in Helmholtz, H. (1883), *Wissenschaftliche Abhandlungen,* Vol. 2, Leipzig: J. A. Barth, pp. 888–923.

Helmholtz, H. (1867). Mittheilung, betreffend Versuche über die Fortpflanzungsgeschwindigkeit der Reizung in den motorischen Nerven des Menschen, welche Herr N. Baxt aus Petersburg im Physiologischen Laboratorium zu Heidelberg ausgeführt hat. *Monatsberichte de Königlich Preussischen Akademie der Wissenschaften zu Berlin,* 228–234.

Helmont, J. B. van, and Helmont, F. M. van. (1652). *Joannis Baptistae van Helmont ortus medicinae . . .* Amsterodami: Elzevir.

Helmreich, G. (1889). *Marcelli, de medicamentis.* Lipsiae: B. G. Teubneri.

Henly, W. (1775). Letter to William Canton, received March 14, 1775. *Canton Papers of the Royal Society,* 2, 105.

Henly, W. (c. 1776). Letter to William Canton. *Canton Papers of the Royal Society,* 2, 103.

Hercher, R. (1858). *Erotici Scriptores Graeci, recognovit Rudolph Hercher.* Leipsiae: In aedibus B.G. Teubneri.

Hermes Trismegistus (1471). *Pimander de potestate et sapientia Dei.* M. Ficino (Ed., Trans.). Treviso: Gerardus de Lisa.

Hermes Trismegistus [Mercurio Trismegisto]. (1549). *Il Pimandro.* Firenze: [s.n.]

Hille, B. (2001). *Ion Channels of Excitable Membranes* (3rd ed.). Sunderland: Sinauer Associates.

Hippocrates (1525). *Hippocratis Coi octoginta volumina: quibus maxima ex parte latina lingua caruit, Graeci vero, Arabes scripta sua illustrarunt, nunc tandem per M. Fabium Calvum latinate donata.* Romae: Ex aedibus Francisci Minitii Calvi.

Hirsch, A. (1862). Expériences chronoscopiques sur la vitesse des différentes sensations et de la transmission nerveuse. *Bulletin de la Société des Sciences Naturelles de Neuchâtel,* 6, 100–114.

Hochadel, O. (2001). "My patient told me how to do it": The practice of medical electricity in the German Enlightenment. In P. Bertucci and G. Pancaldi (Eds.), *Electric Bodies: Episodes in the History of Medical Electricity.* Bologna: Università di Bologna, pp. 69–90.

Hodgkin, A. L. (1937a). Evidence for electrical transmission in nerve. I. *Journal of Physiology,* 90, 183–210.

Hodgkin, A. L. (1937b). Evidence for electrical transmission in nerve. II. *Journal of Physiology,* 90, 211–232.

Hodgkin, A. L. (1979). Edgar Douglas Adrian, Baron Adrian of Cambridge. *Biographical Memoirs of Fellows of the Royal Society,* 25, 1–73.

Hodgkin, A. L. (1992). *Chance and Design: Reminiscences of Science in Peace and War.* Cambridge: Cambridge University Press.

Hodgkin, A. L., and Huxley, A. F. (1939). Action potentials recorded from inside a nerve fibre. *Nature,* 144, 710–711.

Hodgkin, A. L., and Huxley, A. F. (1952a). Currents carried by sodium and potassium ions through the membrane of the giant axon of *Loligo. Journal of Physiology,* 116, 449–472.

Hodgkin, A. L., and Huxley, A. F. (1952b). The components of membrane conductance in the giant axon of *Loligo. Journal of Physiology,* 116, 473–496.

Hodgkin, A. L., and Huxley, A. F. (1952c). The dual effect of membrane poential on sodium conductance in the giant axon of *Loligo. Journal of Physiology,* 116, 497–506.

Hodgkin, A. L., and Huxley, A. F. (1952d). A quantitative description of membrane current and its application to conduction and excitation in nerve. *Journal of Physiology,* 117, 500–544.

Hodgkin, A. L., and Katz, B. (1949). The effect of sodium ions on the electrical activity of the giant axon of the squid. *Journal of Physiology,* 108, 37–77.

Hoff, H. E., and Geddes, L. A. (1957). The rheotome and its prehistory: A study in the historical interrelation of electrophysiology and electromechanics. *Bulletin of the History of Medicine,* 32, 212–234, 327–347.

Hoff, H. E., and Geddes, L. A. (1960). Ballistics and the instrumentation of physiology. The velocity of the projectile and the nerve impulse. *Journal of the History of Medicine,* 15, 133–146.

Hofmeister, A. [Otto Bishop of Freising] (1912). *Ottonis episcopi fringensis chronica; sive, historia de duabus civitatibus.* Hannoverae, Lipsiae: Impensis Bibliopolii Hahniani.

Home, R. W. (1970). Electricity and the nervous fluid. *Journal of the History of Biology,* 3, 235–251.

Holmes, F. L. (1994). The role of Johannes Müller in the formation of Helmholtz's physiological career. In: Krüger L, ed., *Universalgenie Helmholtz: Rückblick nach 100 Jahren.* Berlin: Akademie Verlag, pp. 3–21.

Holthuis, L. B. (1996). Original watercolours donated by Cornelius Sittardus to Conrad Gesner, and published by Gesner in his (1558–1670) works on aquatic animals. *Zoologische Mededelingen,* 70, 169–196.

Hondius, J. (1598). *Caert Thresoor; inhoudende de tafelen des gantsche Werelts Landen, met beschryvingen verlicht . . .* By B. Langenes. [Maps engraved by J. Hondius & P. Kaerius]. Middleburgh: Cornelis Claess.

Hooke, R. (1665). *Micrographia: or some Physiological Descriptions of Minute Bodies Made by Magnifying Glasses.* London: Printed by J. Martyn and J. Allestry.

Hopkinson, B. R. (1972). Electrical treatment of incontinence using an external stimulator with intra-anal electrodes. *Annals of the Royal College of Surgeons of England,* 50, 92–111.

Hopkinson, B. R. (1975). Electrical activation of the spincters in the treatment of rectal prolapse. *Proceedings of the Royal Society of Medicine,* 68, 21–22.

Horace Q. F. (1825). *Quinti Horatii Flacci opera omnia,* H. J. C. Zeune (Ed.). Londini: Curante et Imprimente A. J. Valpy.

Horapollo (1543). *Orus Apollo de Aegypte, de la signification des notes hieroglyphiques des AEgyptiens. . .* Paris: Par Jacques Kerver.

Horapollo. (1599). *Hōrou Apollōnos Neilōou Hieroglyphika eklekta = Hori Apollinis selecta Hieroglyphica: imagines vero cum priuilegio.* Romae: Sumtibus Iulij Franceschini.

Horapollo, Du Pré, G., Ruelle, J., and Cousin, J. (1574). *Ori Apollinis Niliaci, De sacris Aegyptiorum notis, Aegyptiacè expressis libri duo. . . .* Paris: Apud Galeotum à Prato, & Ioannem Ruellium.

Hornung, E. (1982). *Conceptions of God in Ancient Egypt.* Ithaca: Cornell University Press.

Hosack, D. (1820). *Biographical Memoir of Hugh Williamson, M.D., LL.D.* New York: C. S. Van Winkle.

Humboldt, F. W. H. A. von (1797). *Versuche über die gereizte Muskel- und Nervenfaser nebst Vermutungen über den chemischen Prozess des Lebens in der Tier- und Pflanzenweldt* (2 vols.). Posen: Decker und Compagnie; Berlin: H. A. Rottmann.

Humboldt, F. W. H. A. von (1806). Versuche über die elektrischen Fische. *Gilberts Annalen der Physik*, 22, 1–13.

Humboldt, F. W. H. A. von (1807). Jagd und Kampf der electrischen Aale mit Pferden. *Gilberts Annalen der Physik*, 25, 34–43.

Humboldt, F. W. H. A. von (1819). Sur les gymnotes et autres poissons électriques. *Annals de Chimie et de Physiques*, 11, 408–437.

Humboldt, F. W. H. A. von (1820). Account of the electrical eels and of the method of catching them in South America by means of wild horses (Abridged from Humboldt's Personal Narrative). *Edinburgh Philosophical Journal*, 2, 242–249.

Humboldt, F. W. H. A. von, and Bonpland, A. J. A. (1811). *Recueil d'Observations de Zoologie et d'Anatomie Comparée faites dans l'Océan Atlantique, dans l'Intérieur du Nouveau Continent et dans la Mer du Sud, Pendant les Années 1799–1803* (vol 1). Paris: F. Schoell, G. Dufour, J. Smith et Gide.

Humboldt, F. W. H. A. von, and Bonpland, A. (1852). *Personal Narrative of Travels to the Equinoctial Regions of America, during the Years 1799–1804* (3 vols). Trans. from the French by T. Ross. Reprinted in 1971 by Benjamin Blom, Inc., New York. (The eel encounter is in Vol. 2).

Humboldt, A. von, and Gay-Lussac, J. L. (1805). Expériences sur la Torpille. *Annales de Chimie*, 66, 15–23.

Hunt, T. (1972). *The Medical Society of London*. London: William Heinemann.

Hunter, J. (1766). Supplement to the account of the amphibious bipes; by John Ellis . . . being the animal description of the said animal. *Philosophical Transactions of the Royal Society of London*, 56, 307–310.

Hunter, J. (1773). Anatomical observations on the torpedo. *Philosophical Transactions of the Royal Society*, 63, 481–488.

Hunter, J. (1775). An account of the *Gymnotus electricus*. *Philosophical Transactions of the Royal Society*, 65, 395–407.

Hunting, P. (2004). *The Medical Society of London 1773–2003*. London: The Medical Society of London.

Huxley, L. (1900). *Life and Letters of Thomas Henry Huxley* (2 vols.). London: Macmillan.

Huxley, A. F. (1999). Overton on the indispensability of sodium ions. *Brain Research Bulletin*, 50, 307–308.

Ibn l-Bītār (1874). *Al-Jāmi 'li mufradāt al-adwiya wa-l-aghdhiya* (Vol. 2). Cairo: Bulaq Government Press.

Ibn Rushd. (1994). *Jawāmi' al-samā' wa-l-'ālam*. . . . Beirut: Markaz al-Nashr al-Jami'i. Dar al-Fikr al-Lebnani.

Ibn Sina (1294 AH). *al-Qanun* (Vol. I). Cairo: Matbaat Bulaq.

Impey, O., and MacGregor, A. (2001). *The Origin of Museums: The Cabinets of Curiosities in Sixteenth- and Seventeenth-Century Europe* (2nd ed.). London: House of Stratus.

Ingenhousz, J. (1775a). Extract of a letter from Dr. John Ingenhousz, F.R.S. to Sir John Pringle, Bart. P.R.S. containing some experiments on the torpedo, made at Leghorn, January 1, 1773 (after having been informed of those by Mr. Walsh). Dated Saltzburg, March 27, 1773. *Philosophical Transactions of the Royal Society*, 65, 1–4.

Ingenhousz, J. (1775b). Extract of a letter from Dr. John Ingenhousz F.R.S. to Sir John Pringle. *Gentleman's Magazine*, 45, 436.

Ingenhousz, J. (1784). *Vermischte Schriften phisisch-medizinische Inhalts*. Wien: J. P. Krauss.

Ingram, D. (1750a). New experiments concerning the torpedo. *The Student, Or the Oxford Monthly Miscellany*, 2, 49–52.

Ingram, D. (1750b). Von dem Torpedo, oder Krampf-Fisch, aus dem Englischen, des Oxford. Student. N. 2. *Hannoverische Gelehrte Anzeigen*, 1, 83–84.

Ingram, D. (1770). Von dem Torpedo oder Krampfisch, aus dem Englischen. *Neue physikalische Belustigungen, Sechste Abhandlung*, 288–293.

Isidorus Hispalensis (Isidore of Seville). (1483). *Libri etymologiarum XX et de summo bono libri tres*. Venetiis: Per Petrum Loslein de Langencen.

Isler, H. (1968). *Thomas Willis (1621–1675): Doctor and Scientist*. New York: Hafner.

Izadpanah, A., and Hosseini, S. (2005). Comparison of electrotherapy of hemorrhoids and Ferguson hemorrhoidectomy in a randomized prospective study. *International Journal of Surgery*, 3, 258–262.

J. P. (1774). [No title: A Parody of Pringle's Discourse.] *London Evening Post*, December 24, Issue 8230.

Jaager, J. C. de (1865). *De Physiologische Tijd bij Psychische Processen*. Dissertation: Medical Faculty of the University of Utrecht, July 1, 1865.

Jacob, C. (2001). Ateneo o il dedalo delle parole. In L. Canfora and C. Jacob (Eds.), *Athenaeus, I deipnosofisti, i dotti a banchetto*. Roma: Salerno Editrice, pp. 81–83.

Jacobaeus, H. (1675). Anatome piscis torpedinis motusque tremuli examen. De scorpione observationes, serpentum et viperarum anatome. In Bartholin, T.: *Acta Medica et Philosophica Hafniensia Anno 1673*. Hafniae: G. Godiani, pp. 253–272.

Jacobaeus, H. (1686). *De ranis et lacertis observationes*. Hafniae: Impensis Johannis M. Lieben.

Jaeger, W. W. (1948). *Aristotle; Fundamentals of the History of his Development*. (Engl. Transl. by Richard Robinson). Oxford: Clarendon Press.

Jallabert, J. L. (1748). *Experiences sur l'electricité, avec quelques conjectures sur la cause de ses effets*. Geneve: Chez Barrillot & Fils.

James, F A. J. L. (2000). *The Manuscripts of Michael Faraday (1791–1867)*. East Ardsley, Wakefield, West Yorkshire, England: Microform Academic Publishers.

Jardine, N., Secord, J. A and Spary, E. C. (1996). *Cultures of Natural History*. Cambridge: Cambridge University Press.

Jaucourt, L. de. (1765). Torpille. In: *Encyclopédie*, vol. XVI. Paris: Briasson, David, Le Breton, Durand, pp. 428–431.

Jéhan, L. F. (1852). *Dictionnaire ou Histoire Naturelle de Quatre Grands Embranchement du Regne Animal. . . . Dictionnaire de Théologie Chrétienne* (Tome 2). Paris: Mignes aux Ateliers Catholiques.

Jenkins, H. (1950). *The Versatile Dr. Hugh Williamson: 1735–1789*. Master's Thesis. Chapel Hill: University of North Carolina.

Jenkins, P. G. (1928). Alexander Garden, M.D., F.R.S. (1728–1791), Colonial Physician and Naturalist. *Annals of Medical History*, 10, 149–158.

Jobson, R. (1623). *The Golden Trade: Or, A Discovery of River Gambia*. London: Printed by Nicholas Okes.

Johnels, A. G. (1956). On the origin of the electric organ in *Malapterurus electricus*. *Quarterly Journal of Microscopical Science*, 97, 455–463.

Johnson, R. B. (Ed.). (1892). *The Poetical Works of Matthew Prior* (Vol. 2). London: Bell and Sons.

Jones, R. V. (1977). Benjamin Franklin. *Notes and Records of the Royal Society of London*, 31, 201–225.

Jones, W. H. S. (1979). Regimen II. In *Hippocrates* (vol. IV). Cambridge, MA: Harvard University Press.

Jouanna, J. (1999). *Hippocrates. Medicine & Culture*. Baltimore: Johns Hopkins University Press.

Jungnickel, C., and McCormmach, R. (1999). *Cavendish: The Experimental Life*. Lewisburg: Bucknell University Press.

Kaempfer, E. (1712). *Amoenitatum exoticarum politico-physicomedicarum fasciculi V, quibus continentur variae relationes, observationes et descriptiones rerum Persicarum . . .* Lemgoviae: typis & impensis H. W. Meyeri.

Kagarise, M. J., and Sheldon, G. F. (2005). Hugh Williamson, M.D., LL.D. (1735-1819): Soldier, surgeon, and founding father. *World Journal of Surgery*, 29, S80–84.

Kahn, R. (Ed.) (1971). *Selected Writings of Hermann von Helmholtz*. Middletown: Wesleyan University Press.

Kaimakis, D. V. (1976). *Die Kyraniden. Beiträge zur klassischen Philologie, Heft 76*. Meisenheim am Glan: Hain.

Kalmijn, A. J. (1974). The detection of electric fields from inanimate and animate sources other than electric organs. In A. Fessard (Ed.), *Handbook of Sensory Physiology*, Vol. III. *Electroreceptors and Other Specialized Receptors in Lower Vertebrates*. Berlin, New York: Springer, pp. 147–200.

Kalmijn, A. J. (1978). Electric and magnetic sensory world of sharks, skates and rays. In E. S. Hodgson and R. F. Mathewson (Eds.), *Sensory Biology of Sharks, Skates, and Rays*. Arlington: Office of Naval Research, Department of the Navy, pp. 507–528.

Kalmijn, A. J. (1988). Detection of weak electric fields. In J. Atema, R. R. Fay, A. N. Popper, and W. N. Tavolga (Eds.). *Sensory Biology of Aquatic Animals*. New York: Springer, pp. 151–186.

Kaplan, E. L., et al. (2009). History of the recurrent laryneal nerve: From Galen to Lahey. *World Journal of Surgery*, 33, 386–393.

Kapoor, B. G., and Khanna, B. (2004). *Ichthyology Handbook*. New York: Springer.

Keesey, J. C. (2002). *Myasthenia Gravis: An Illustrated History*. Roseville: Publishers Design Group.

Keesey, J. C. (2005). How electric fish became sources of acetylcholine receptor. *Journal of the History of the Neurosciences*, 14, 149–164.

Kehlmann, D. (2005). *Die Vermessung der Welt: Roman*. Reinbek: Rowohlt. English translation by Carol Brown Janeway. *Measuring the World*. New York: Pantheon Books, 2006.

Keighley, M. R., and Matheson, D. M. (1981). Results of treatment for rectal prolapse and fecal incontinence. *Diseases of the Colon and Rectum*, 24, 449–453.

Kellaway, P. (1946). The part played by electric fish in the early history of bioelectricity and electrotherapy. *Bulletin of the History of Medicine*, 20, 112–137.

Kellner, L. (1963). *Alexander von Humboldt*. London: Oxford Univesity Press.

Kettenmann, H. (1997). Alexander von Humboldt and the concept of animal electricity. *Trends in Neurosciences*, 20, 239–241.

Keynes, R. D. (1956). The generation of electricity by fishes. *Endeavour*, 15, 215–222.

Keynes, R. D. (1979). *The Beagle Record*. Cambridge: Cambridge University Press.

Keynes, R. D., and Martins-Ferreira, H. (1953). Membrane potentials in the electroplates of the electric eel. *Journal of Physiology*, 119, 315–351.

Keynes, R. D., Bennett, M., and Grundfest, H. (1961). Studies on the morphology and electrophysiology of electric organs. II. Electrophysiology of the electric organ of Malapterurus electricus. In C. Chagas (Jr) and A. Paes de Carvalho (Eds.), *Bioelectrogenesis; a Comparative Survey of its Mechanisms*. Amsterdam and New York: Elsevier Publishing Co, pp. 103–112.

King, L. S. (1966). Introduction. In *First Lines of Physiology by the Celebrated Baron Albertus Haller, MD* . . . . New York: Johnson Reprint Company, pp. ix–lxxii.

Kircher, A. (1652–54). *Oedipus Aegyptiacus, hoc est universalis hieroglyphicae veterum doctrinae*. . . . (4 vols.). Roma: ex typographia Vitalis Mascardi.

Kircher, A. (1667). *Magneticum naturae regnum*. . . . Romae: Typis Ignatij de Lazaris.

Kitchell, K. F., and Resnick, I. M. (1999). *Albertus Magnus on Animals: A Medieval Summa Zoologica*. Baltimore: Johns Hopkins University Press.

Knapman, P. (1999). Benjamin Franklin and the Craven Street bones. *Transactions of the Medical Society of London*, 116, 9–17.

Knight, D. (1992). *Humphry Davy: Science and Power*. Cambridge: Cambridge University Press.

Kobler, J. (1960). *The Reluctant Surgeon: A Biography of John Hunter*. Garden City: Doubleday & Co.

Koehler, P. J., Finger, S., and Piccolino, M. (2009). The eels of South America: Mid-eighteenth-century Dutch contributions to the theory of animal electricity. *Journal of the History of Biology*, 42, 235–251.

Koenigsberger, L. (1902). *Hermann von Helmholtz* (3 vols.). Braunschweig: F. Vieweg und Sohn.

Koenigsberger, L. (1906). *Hermann von Helmholtz*. F. A. Welby, trans. Oxford: Clarendon Press.

Kolben, P. (1741). *Description du Cap de Bonne-Espérance, où l'on Trouve tout ce qui Concerne l'Histoire Naturelle du Pays* . . . Amsterdam: Chez Jean Catuffe.

Koller G (1958). *Das Leben des Biologen Johannes Müllers, 1801–1858*. Stuttgart: Wissenschaftliche Verlagsgesellschaft. (Grosse Naturforscher, 23).

Kratzenstein, C. G. (1745). *Abhandlung von dem Nutzen der Electricität in der Arzneywissenschaft*. Halle: Hemmerde.

Kudlien, F. (1970). Aëtius of Amida. In: *Dictionary of Scientific Biography* (Vol. 1), C. C. Gillispie (Ed.). New York: Charles Scribner's Sons, pp. 68–69.

Kühn, C. G. (Ed.). (1821–1833). *Claudii Galeni opera omnia* (20 vols). Lipsiae: Prostat in officina libraria Car. Cnobloch.

Kuhn, T. S. (1962). *The Structure of Scientific Revolutions*. Chicago: University of Chicago Press.

Krüger, J. G. (1744). *Zuschrift an seine Zuhörer, worinnen er Gedancken von der Electricität mittheilt und Ihnen zugleich seine künftigen Lectionen bekannt macht*. Halle: C. Hemmerde.

Labaree, L. W. (1960). *The Papers of Benjamin Franklin* (Vol. 2). New Haven: Yale University Press.

Labaree, L.W. (1961a). *The Papers of Benjamin Franklin* (Vol. 3). New Haven: Yale University Press.

Labaree, L. W. (1961b). *The Papers of Benjamin Franklin* (Vol. 4). New Haven: Yale University Press.

Labaree, L. W. (1962). *The Papers of Benjamin Franklin* (Vol. 5). New Haven: Yale University Press.

Labaree, L. W. (1965). *The Papers of Benjamin Franklin* (Vol. 8). New Haven: Yale University Press.

Labaree, L. W. (1966). *The Papers of Benjamin Franklin* (Vol. 10). New Haven: Yale University Press.

Labaree, L. W. (1967). *The Papers of Benjamin Franklin* (Vol. 7). New Haven: Yale University Press.

Lacépède, B-G. de (1781). *Essai sur l'Électricité Naturelle et Artificielle*. Paris: l'imprimerie de Monsieur.

Lacépède, B-G. de (1803). *Histoire Naturelle des Poissons* (Tome V). Paris: Plassan.

Laet, J. de (1625). *Nieuwe wereldt, ofte Beschrijvinghe van West-Indien . . . door Joannes de Laet*. Tot Leyden: in de druckerye van J. Elzevier.

Laet, J. de (1633). *Novus orbis, seu, Descriptionis Indiae Occidentalis, libri XVIII*. Lugduni Batavorum: Apud Elzevirios.

Laghi, T. (1757a). Cl. Viro D. Cesareo Pozzi [Epistola]. In G. B. Fabri (Ed.), *Sulla Insensitività ed Irritabilità Halleriana: Opuscoli di Vari Autori* (Vol. 2). Bologna: Corciolani ed Eredi Colli, pp. 110–116.

Laghi, T. (1757b). De sensitivitate, atque irritabilitate halleriana. Sermo alter. In G. B. Fabri (Ed.), *Sulla Insensitività ed Irritabilità Halleriana: Opuscoli di Vari Autori* (Vol. 2). Bologna: Corciolani ed Eredi Colli, pp. 326–345.

Lallemant, A. J. N. (1804). *Voyages de MM. Ledyard et Lucas en Afrique . . . Traduits de l'Anglais par A.J.N. Lallemant* (2 vols.). Paris: Chez Xhrouet . . . Déterville.

Lamont-Brown, R. (2004). *Humphry Davy, Life Beyond the Lamp*. Sutton Publishing Co.

Lane, T. (1767). Description of an electrometer invented by Mr. Lane with an account of some experiments made by him with it. *Philosophical Transactions of the Royal Society*, 57, 461–470.

Langrish, B. (1747). The Croonean Lectures on Muscular Motion, Read before the Royal Society in the year 1747. *Philosophical Transactions of the Royal Society of London*, 44, Suppl. to Part ii, 1–66.

Lanza, C. (1881). Electric fish. *Science,* 2, 26–31.
Lanza, D., and Vegetti, M. (1971). *Opere biologiche di Aristotle.* Torino: Unione Tipografico Editrice Torinese.
Larson, J. L. (1971). *Reason and Experience: The Representation of Natural Order in the Work of Carl von Linné.* Berkeley: University of California Press.
Lauer, J.-P. (1976). *Saqqara: The Royal Cemetery of Memphis: Excavations and Discoveries since 1850.* London: Thames and Hudson.
Leaman, O. (1998). *Averroes and his Philosophy.* Richmond: Curzon.
Lefevre, R. (1944). Riflessi Etiopici nella Cultura Europea del Medioevo e del Rinascimento. *Atti Lateranensi,* 8, 9–89.
Lefevre, R. (1945). Riflessi Etiopici nella Cultura Europea del Medioevo e del Rinascimento. *Atti Lateranensi,* 9, 332–444.
Lefevre, R. (1947). Riflessi Etiopici nella Cultura Europea del Medioevo e del Rinascimento. *Atti Lateranensi,* 11, 255–342.
Leibowitz, J. O. (1954). Avicenne en Hébreu et le manuscrit de Bologne. Comptes Rendus du XIII Congrès International d'Histoire de la Médecine, Nice, 1952. Bruxelles.
Leibowitz, J. O. (1957). Electroshock therapy in Ibn-Sina's Canon. *Journal of the History of Medicine and Allied Sciences,* 12, 71–72.
Lemay, J. A. L. (1964). *Ebenezer Kinnersley, Franklin's Friend.* Philadelphia: University of Pennsylvania Press.
Lenoir, T. (1986). Models and instruments in the development of electrophysiology. *Historical Studies in the Physical and Biological Sciences,* 17, 1–54.
Le Roy, J. B. (1761). Mémoire ou l'on rend compte de quelques tentatives que l'on a faites pour guérit plusieurs maladies par l'Electricité. *Histoire de l'Académie Royale des Sciences pour l'Année 1755 avec les Mémoires de Mathématique & de Physique pour la meme Année. Memoires,* pp. 60–98.
Le Roy, J-B. (1776). Lettre adressée a l'auteur de ce recueil par M. Le Roy. *Observations sur la Physique,* 8, 331–335.
Le Roy, J-B. (1777). Estratto di lettera del Signor Le Roy al Sig. abate Rozier sulla scintilla elettrica ... *Scelta di opuscoli interessanti tradotti da vane lingue,* 25, 106–108.
Levy, T. E., et al. (1995). New light on King Narmer and the protodynastic Egyptian presence in Canaan. *Biblical Archaeologist,* 58, 26–36.
Licht, S. (1967). History of electrotherapy. In S. Licht (Ed.), *Therapeutic Electricity and Ultraviolet Radiation* (2nd ed.). Baltimore: Waverly Press, pp. 1–70.
Linari, S. (1836). Vera scintilla elettrica: effetti d'elettrica tensione, di proprietà chimica e calorifica ottenuti dalla scossa della torpedine ..., Suppl. a *L'Indicatore Sanese,* Anno 4, No. 50.
Linari, S. (1837). Vera scintilla elettrica: effetti d'elettrica tensione, di proprietà chimica e calorifica ottenuti dalla scossa della torpedine ..., *Giornale Arcadico di Scienze Lettere ed Arti,* 70, 50–71.
Linari, S. (1838). *Sperimenti sopra le proprietà elettriche della torpedine,* Suppl. 2 a *L'Indicatore Sanese,* Anno 4, n. 50.
Linari, S. (1839). *Scintilla ed altri fenomeni d'elettriche correnti, ottenuti per azione induttiva dell'elettriche scariche del pesce Torpedine (raja torpedo) dal Padre Santi Linari delle Scuole Pie....* Fascicolo di Gennajo, Febbrajo, Marzo, 35–46.
Lindberg, D. C. (1976). *Theories of Vision from al-Kindi to Kepler.* Chicago: University of Chicago Press.
Lindstrom, J. M., et al. (1976). Antibody to acetylcholine receptor in myasthenia gravis. Prevalence, clinical correlates, and diagnostic value. *Neurology,* 26, 1054–1059.
Linnaeus, C. (1735). *Systema naturae....* Lugduni Batavorum: Apud Theodorum Haak, ex typographia Joannis Wilhelmi de Groot.
Linnaeus, C. (1758). *Systema naturae....* Holmiae: Impensis Direct. Laurentii Salvii.
Linnaeus, C. (1766). *Systema naturae..., Editio XII Reformata.* Holmiae: Impensis Direct. Laurentii Salvii.

Lissmann, H. W. (1951). Continuous electric signals from the tail of a fish, *Gymnarchus niloticus* Cuv. *Nature,* 167, 201.
Lissmann, H. W. (1958). On the function and evolution of electric organs in fish. *Journal of Experimental Biology,* 35, 156–191.
Lissman, H. W. (1963). Electric location by fishes. *Scientific American,* 208, 50–59.
Lissmann, H. W., and Machin, K. E. (1958). The mechanism of object location in *Gymnarchus niloticus* and similar fish. *Journal of Experimental Biology,* 35, 451–486.
Littré, É. (1839-61). *Oeuvres Complètes d'Hippocrate.* Paris, London: Chez J.B. Baillière.
Lobo, J. (1728). *Relation Historique d'Abissinie ... Traduite du Portuguais, Continuée et Augmentée de Plusieurs Dissertations, Lettres et Mémoires par M. Le Grand. ...* Paris: Veuve de A. V. Coustelier et J. Guérin.
Locke, H. S., and Finger, S. (2007). Gentleman's Magazine, the advent of medical electricity and disorders of the nervous system. In H. Whitaker, C. U. M. Smith, and S. Finger (Eds.), *Brain, Mind and Medicine: Essays in Eighteenth Century Neuroscience.* Boston: Springer, pp. 257–270.
Lohff, B. (1995). "... in Berlin eine würdige Stätte schaffen": Anatomie und Physiologie in der Zeit von Johannes Müller. In: P. Schneck and H. E. Lammel (Eds.), *Die Medizin an der Berliner Universität und an der Charité zwischen 1810 und 1850.* Husum: Mathiessen, pp. 31–51.
Lohff, B. (2001). Facts and philosophy in neurophysiology. The 200th anniversary of Johannes Müller (1801–1858). *Journal of the History of the Neurosciences,* 10, 277–292.
London Chronicle. (1769). An Essay on the Natural History of Guiana in SOUTH AMERICA.... *London Chronicle,* February 4, Issue 1895. (Book Review)
London Chronicle and Universal Evening Post. (1775). [No title: Pringle's Speech and Walsh's Medal.] *London Chronicle and Universal Evening Post,* February 7, Issue 2835.
London Evening Post. (1774). [No title: A torpedo caught in the Thames.] *London Evening Post,* June 25, Issue 8152.
London Evening Post. (1777). Advertisement for Adam Strong's The Electrical Eel. *London Evening Post,* June 24, Issue 8623; July 17, Issue 8633.
Lorenzini, S. (1678). *Osservazioni intorno alle Torpedini....* Firenze: Per l'Onofri.
Lorenzini, S. (1705). *The Curious and Accurate Observations of Mr. Stephen Lorenzini ... on the Dissections of the Cramp-Fish.* Trans. J. Davis. London: Printed for, and sold by, Jeffery Wale.
Lortet, L., and Gaillard, C. (1908). *La Faune Momifée de l'Ancienne Égypte,* 4ᵉ Sér. Lyon: Henri Georg.
Lovejoy, A. O. (1936). *The Great Chain of Being; A Study of the History of an Idea.* Cambridge, MA: Harvard University Press.
Lovell, R. (1661). *Panzooryktologia sive, Panzoologicomineralogia. Or A Compleat History of Animals and Minerals....* Oxford: Printed by Hen: Hall, for Jos. Godwin.
Löwenberg, J., Avé-Lallemant, R., and Dove, A. (1873). *Life of Alexander von Humboldt.* London: Longmans, Green, and Co.
Lower, R. (1669). *Tractatus de corde: item de motu colore sanguinis et chyli in eum transitu.* London: Jo. Redmayne for Jacob Allestry.
Lucas, K. (1905). On the gradation of activity in a skeletal muscle-fibre. *Journal of Physiology,* 33, 124–137.
Lucas, K. (1909). The "all-or-none" contraction of the amphibian skeletal muscle-fibre. *Journal of Physiology,* 38, 113–133.
Lucas, K. (1917). *The Conduction of the Nervous Impulse.* (E. D. Adrian, Ed.). London: Longmans Green.
Ludolf, H. W. (1681). *Historia Aethiopica, sive brevis et succincta descriptio regni Habessinorum....* Francofurti ad Moenum: Zunner et Wustii.
Ludolf, H. W. (1682). *A New History of Ethiopia: Being a Full and Accurate Description of the Kingdom of Abessinia, Vulgarly, though Erroneously Called the Empire of Prester John.* London: Printed for Samuel Smith.

References

Ludolf, H. W. (1691). *J. Ludolphi . . . ad suam Historiam Aethiopicam antehac editam Commentarius. . . .* Francofurti ad Moenum: Sumptibus J. D. Zunneri.

Ludwig, C. (1852). *Lehrbuch der Physiologie des Menschen* (Vol. 1). Heidelberg, Leipzig: Winter.

Ludwig, C. (1861). Über die Krafte der Nervenprimitivenrohr. *Wiener medizinische Wochenschrift*, 729, 129.

Luff, R. M., and Bailey, G. (2000). The aquatic basis of ancient civilizations: The case of *Synodontis schall* and the Nile Valley. In G. Bailey, R. Charles, and N, Winder (Eds.), *Human Ecodynamics*. Oxford: Oxbow Books, pp. 100–113.

Lusitanus, A. (João Rodrigues de Castelo Branco) (1558). *Amati Lusitani, Enarrationes in Dioscoridis de Medica Materia Libros V*. Lugduni Batavorum: B. Arnolleti Witwe.

Mabberley, D. J. (1981). Edward Nathaniel Bancroft's obscure botanical publications and his father's plant names. *Taxon*, 10, 7–17.

Maccagni, C. (1985). *Le due edizioni della "Demonstratio proportionum motuum localium contra Aristotelem et omnes philosophos" di Giovanni Battista Benedetti: Convegno Internazionale di Studio su "Giovan Battista Benedetti e il Suo Tempo."* Venezia: Istituto Veneto di Scienze, Lettere ed Arti.

MacNalty, A. S. (1944). Edward Bancroft, M.D., F.R.S., and the War of American Independence. *Proceedings of the Royal Society of Medicine*, 38, 7–15.

Malek, J. (1986). *In the Shadow of the Pyramids: Egypt During the Old Kingdom*. Norman: University of Oklahoma Press.

Malphighi, M. (1666). De cerebri cortice. In *De Viscerum Structura Exercitatio Anatomica*. Bononiæ: ex typographia Jacobi Montii.

Malpighi, M. (1698). *Marcelli Malpighi . . . Opera posthuma quibus praefationes & animadversiones addidit . . . Editio novissima*. Venetiis: ex typographia A. Poleti.

Malpighi, M. (1967). *Opere scelte*. L. Belloni (Ed). Torino: Unione Tipografico-Editrice Torinese.

Manzoni, T. (1998). The Cerebral ventricles, the animal spirits and the dawn of brain localization of function. *Archives Italiennes de Biologie*, 136, 103–152.

Manzoni, T. (2001). *Il Cervello Secondo Galeno*. Ancona: Il Lavoro Editoriale.

Marcgraf, G. (1648). *Historia naturalis Brasiliae: auspicio et beneficio illustriss. I. Mauriti Com. Nassau illius provinciae et maris summi praefecti adornata. . . .* Lugduni Batavorum: Apud Franciscum Hackium, et Amstelodami: Apud Lud. Elzevirium.

Marcocci, G. (2005). Gli umanisti Italiani e l'impero Portoghese: una interpretazione della "Fides, Religio, Moresque Aethiopum" di Damião de Góis. *Rinascimento*, 45, 307–366.

Marey, E.-J. (1871). Du temps qui s'écoule entre l'excitation du nerf électrique de la torpille et la décharge de son appareil. *Comptes Rendus Hebdomadaires des Séances de l'Académie des Sciences*, 73, 918–921.

Marey, E.-J. (1872). Mémoire sur la torpille. *Annales Scientifiques de l'Ecole Normale Supérieure*, II Sér. 1, 85–114.

Marey, E.-J. (1876). Des excitations électriques du coeur. In *Travaux du Laboratoire de M. Marey* (vol. 2). Paris: G. Masson, pp. 63–86.

Marey, E.-J. (1879). Nouvelles recherches sur les poissons électriques: Caractères de la décharge du gymnote; effets d'une décharge de torpille lancée dans un téléphone. *Comptes Rendus Hebdomadaires de l'Académie des Sciences*, 88, 318–321.

Markus, F. (Ed.). (1983). *Girolamo Cardano (1501–1576): Physician, Natural Philosopher, Mathematician, Interpreter of Dreams*. Boston: Birkhäuser.

Marmont, G. H. (1949). Studies on the axon membrane. *Journal of Cellular and Comparative Physiology*, 35, 351–382.

Marsili, L. F. (1725). *Histoire Physique de la Mer*. Amsterdam: Aux Dépens de la Compagnie.

Martini, F. H. W. (1774). *Allgemeine Geschichte der Natur in alphabetischer Ordnung, mit vielen Kupfern nach Bomarischer Einrichtung* (Vol. 1). Berlin: Schlesinger.

Martins-Ferreira, H., and Couceiro, A. (1951). Comportement du tissue électrique de l'*Electrophorus electricus* en consequence de la dénervation. *Anais da Academia Brasileira de Ciências*, 23, 377–385.

Massari, F. (1542). *Francisci Massarii, . . . In nonum Plinii de naturali historia librum castigationes et annotationes*. Parisiis: ex officina M. Vascosani.

Mattern, S. P. (1999). Physicians and the Roman imperial aristocracy: The patronage of therapeutics. *Bulletin of the History of Medicine*, 73, 1–18.

Matteucci, C. (1836a). Poissons électriques: Extrait d'une lettre de M. Matteucci à M. Arago, contenant les résultats des expériences faites par ce physicien et par M. Linari, professeur à Siène. *Comptes Rendus Hebdomadaires de l'Académie des Sciences*, 13, 46–49.

Matteucci, C. (1836b). Électricité animale. Expériences de M. Matteucci sur la Torpille. *Comptes Rendus Hebdomadaires de l'Académie des Sciences*, 13, 430–431.

Matteucci, C. (1837). Recherches physiques, chimiques et physiologiques sur la Torpille. *Annales de Chimie et de Physique*, 66, 396–437.

Matteucci, C. (1844). *Traité des Phénomènes Électro-Physiologiques des Animaux Suivi d'Études Anatomiques sur le Système Nerveux et sur l'Organe Électrique de la Torpille par Paul Savi*. Paris: Fortin, Masson et Cie.

Matteucci, C. (1847). *Lezioni di fisica, di Carlo Matteucci (3a edizione)*. Pisa: R. Vannucchi.

Matteucci, C. (1865). Électricité animale. Expériences de M. Matteucci sur la Torpille. *Comptes Rendus Hebdomadaires de l'Académie des Sciences*, 61, 627–629.

Matteucci, C., and Ranzi, A. (1855a). Sul Siluro del Nilo. *Il Nuovo Cimento*, 1, 297.

Matteucci, C., and Ranzi, A. (1855b). Esperienze sulla scarica elettrica del Siluro del Nilo. *Il Nuovo Cimento*, 2, 447–448.

Mattioli, P.-A. (1555). *I discorsi di M. Pietro Andrea Matthioli . . . ne i sei libri della materia medicinale di Pedacio Dioscoride Anazarbeo: con i veri ritratti delle piante & de gli animali, nuovamente aggiuntivi dal medesimo*. Vinegia: Nella bottega d'Erasmo, appresso Vincenzo Valgrisi.

Mattioli, P.-A. (1565). *Petri Andreæ Matthioli . . . Commentarii in sex libros Pedacii Dioscoridis Anazarbei De medica materia . . . Cum locupletissimis indicibus tum ad rem herbariam, tum medicamentariam pertinentibus*. Vinegia: Nella bottega d'Erasmo, appresso Vincenzo Valgrisi.

Maxwell, J. C. (1879). *The Electrical Researches of the Honourable Henry Cavendish F.R.S., written between 1771 and 1781*. Cambridge: Cambridge University Press.

Mayow, J. (1674). *Tractatus quinque medico-physici*. Oxonii: e Teatro Sheldoniano.

Mazzarello, P. (2009). *Il Professore e la Cantante: La Grande Storia d'Amore di Alessandro Volta*. Torino: Bollati Boringhieri.

Mazzolini, R. G. (1985) Il contributo di Leopoldo Nobili all'elettrofisiologia. In *Leopoldo Nobili e la cultura scientifica del suo tempo*. Tarozzi, G. ed. Bologna: Istituto per i Beni Artistici, Culturali, Naturali della Regione Emilia-Romagna, pp. 183–199.

McConnell, A. (1991). La Condamine's scientific journey down the River Amazon, 1743–1744. *Annals of Science*, 48, 1–19.

McHenry, L. C. (1959). Dr. Samuel Johnson's medical biographies. *Journal of the History of Medicine and Allied Sciences*, 14, 298–310.

McKendrick, J. G. (1899). *Hermann Ludwig Ferdinand von Helmholtz*. New York: Longmans, Green.

McPhee, I., and Trendall, A. D. (1987). *Greek Red-Figured Fish-plates.* Basel: Vierzehntes Beiheft zur Halbjahresschrift Antike Kunst Herausgegeben von der Vereinigung der Freunde antiker Kunst.

Megenberg. K. von (1475). *Buch der Natur.* Augsburg: Johann Bämler.

Meier-Lemgo, K. (1968). *Die Reisetagebücher Engelbert Kaempfers.* Weisbaden: Steiner.

Mello, F. de (2002). Neurosciences in Brazil: Rio de Janeiro. *Cellular and Molecular Neurobiology,* 22, 475–478.

Melzack, R., and Wall, P. D. (1965). Pain mechanisms: a new theory. *Science,* 150, 171–179.

Melzack, R., and Wall, P. D. (1996). *The Challenge of Pain* (2nd ed.). London: Penguin Books.

Mendelsohn, E. (1960). John Lining and his contribution to Early American science. *Isis,* 51, 278–292.

Mercante, A. S. (1978). *Who's Who in Egyptian Mythology.* New York: Clarkson N. Potter.

Mercuriale. (1571). *Hieronymi Mercurialis Variarum lectionum libri quatuor.* . . . Venetiis: Gratiosus Perchacinus excudebat, sumptibus Pauli & Antonii Meieti frat.

Merlan, Philip. (1970). Alexander of Aphrodisias. *Dictionary of Scientific Biography* (Vol. 1). New York: Charles Scribner's Sons, pp. 117–120.

Meulders, M. (2001). *Helmholtz. Des Lumières aux Neurosciences.* Paris: O. Jacob.

Meyboom, P. G. P. (1977). I *mosaici* pompeiani con figure di pesci. *Mededelingen van het Nederlands Historisch Instituut te Rome,* 39, 49–93.

Meyboom, P. G. P. (1978). A Roman fish mosaic from Populonia. *Bulletin Antieke Beschaving,* 52-53, 209–221.

Meyssonnier, L. (1639). *Pentagonum philosophico-medicum.* . . . Lugduni: Iacobi et Petri Prost.

Middlesex Journal and Evening Advertiser. (1774). [No title: Torpedoes caught of the English coast.] *Middlesex Journal and Evening Advertiser,* December 29, Issue 899.

Middleton, W. E. K. (1971). *The Experimenters: A Study of the Accademia del Cimento.* Baltimore: Johns Hopkins University Press.

Miller, G. (1939). *Albrecht von Haller's Controversy with Robert Whytt.* Baltimore: Johns Hopkins University Press.

Miranda, D. de, and Paci, G. M. (1845a). *Esperimenti Istituiti sul Ginnoto Elettrico.* Napoli: Stamperia reale.

Miranda, D. de, and Paci, G. M. (1845b). Expériences sur le Gymnote électrique. *Archives de l'Électricité,* 5, 496–515.

Moller, P. (1995). *Electric Fishes, History and Behavior.* London: Chapman & Hall.

Monceaux, F. (1683). *Disquistio de Magia Divinatrice et Operatrice.* . . . Francofurti et Lipsiae: Joh. Christiani.

Monro, A. (1785). *The Structure and Physiology of Fishes Explained and Compared with those of Man and other Animals.* Edinburgh: Printed for C. Elliot and G. G. J. and J. Robinson.

Montaigne, M. E. de (1652). *Les Essais de Michel Seigneur de Montaigne: Nouvelle Édition.* . . . Paris: Chez Augustin Courbé.

Monthly Review. (1769a). An Essay on the Natural History of Guiana. *Monthly Review,* 40, 198–207. (Book Review)

Monthly Review. (1776). Papers relating to animal electricity. *Monthly Review,* 54, 22–23.

Moore, F. (1738). *Travels into the Inland parts of Africa, Containing a Description of the Several Nations for the Space of Six Hundred Miles up the River Gambia.* . . . London: Printed by E. Cave for the author, and sold by J. Stagg.

Moore, W. (2005). *The Knife Man: The Extraordinary Life and Times of John Hunter, Father of Modern Surgery.* New York: Broadway Books.

Moreau, F. A. (1861). L'électricité de la decharge de la torpille peut être recueillie et conservée dans un appareil de physique. *Comptes Rendus Hebdomadaires de l'Académie des Sciences,* 53, 512–515.

Moreau, F. A. (1862). Recherches sur la nature de la source électrique de la torpille et manière de recueillir l'électricité produite par l'animal. *Annales des Sciences Naturelles,* 18, 6–30.

Morel, F. (1591). *Marcelli Sidetae medicis, De Remediis ex piscibus.* F. Morel (trans). Lutetiae: Apud Fed. Morellum.

Morning Chronicle and London Advertiser. (1773). [No title: torpedo used to cure high fevers.]. *Morning Chronicle and London Advertiser,* May 15, Issue 1242.

Morning Chronicle and London Advertiser. (1775). [No title: Pringle's Speech and Walsh's Medal.] *Morning Chronicle and London Advertiser,* January 17, Issue 1764.

Morning Chronicle and London Advertiser. (1775). [No title: Pringle's Speech and Walsh's Medal.] *Morning Chronicle and London Advertiser,* January 20, Issue 1767.

Morning Chronicle and London Advertiser. (1776). [No title: Baker's eels and Walsh's spark.] *Morning Chronicle and London Advertiser,* August 6, Issue 2250.

Morning Chronicle and London Advertiser. (1777). To be seen at Mr. BAKER'S New Apartments in Piccadilly, nearly opposite St. James's Street, THE GYMNOTUS ELECTRICUS, or the ELECTRICAL EEL. *Morning Chronicle and London Advertiser,* March 10, Issue 2435.

Morning Chronicle and London Advertiser. (1777). "News": Lord H after publication of *The Torpedo. Morning Post and Daily Advertiser,* May 20, Issue 1429.

Morning Chronicle and London Advertiser. (1777). Advertisement for Adam Strong's *The Electrical Eel. Morning Post and Daily Advertiser,* March 21, Issue 1378; March 25, Issue 1381; March 31, Issue 1386.

Morning Chronicle and London Advertiser. (1777). Advertisement for *Semi Globes. Morning Post and Daily Advertiser,* July 1, Issue 1464.

Morning Chronicle and London Advertiser. (1777). Advertisement for *The Torpedo. Morning Post and Daily Advertiser,* April 21, Issue 1404.

Morning Chronicle and London Advertiser. (1777). Editorial on *Semi Globes. Morning Post and Daily Advertiser,* July 4, Issue 1467.

Moruzzi, G. (1964). L'opera elettrofisiologica di Carlo Matteucci. *Physis,* 4, 101–140.

Moruzzi, G. (1996). The electrophysiological work of Carlo Matteucci. *Brain Research Bulletin,* 40, 69–91.

Müller, J. (1826). *Zur vergleichenden Physiologie des Gesichtssinnes des Menschen und der Thiere, nebst einem Versuch über die Bewegung der Augen und über den menschlichen Blick.* Leipzig: C. Knobloch.

Müller, J. (1834–40). *Handbuch der Physiologie des Menschen für Vorlesungen* (2 vols). Bonn: Verlag von J. Hölscher.

Müller, J. (1840–43). *Elements of Physiology* (2 vols.). (W. Baly, trans.) London: Taylor and Walton.

Mullin, A. (1987). *Spy: America's First Double Agent.* Santa Barbara: Capra Press.

Murray, R. W. (1962). The response of the ampullae of Lorenzini of elasmobranchs to electrical stimulation. *Journal of Experimental Biology,* 39, 119–128.

Murray, R. W. (1974). The Ampullae of Lorenzini. In: *Handbook of Sensory Physiology.* A. Fessard (Ed.). New York: Springer, pp. 125–145.

Müsch, I. (2005). *Introduction to Seba, A. Cabinet of Natural Curiosities.* . . . Cologne: Taschen.

Musitelli, S. (2002). *L'Elettricità Animale dalle Origini alla Polemica Galvani-Volta.* Pavia: La Goliardica Pavese.

Must, G. (1960). A Gaulish incantation in Marcellus of Bordeaux. *Language,* 36, 193–197.

Neal, J. W. (1919). *Life and Public Service of Hugh Williamson.* Historical Papers, Ser. III. Durham: Trinity College Historical Society.

Neher, E., and Sakmann, B. (1976). Single-channel currents recorded from membrane of denervated frog muscle cells. *Nature*, 260, 799–802.

Nelson, J. S. (2006). *Fishes of the World* (4th ed.). Hoboken: John Wiley & Sons, Inc.

Nernst, W. (1888). I: Theorie der Diffusion. *Zeitschrift für Physikalische Chemie*, 2, 613–637.

Nernst, W. (1889). Die elecktromotorische Wirksamkeit der Ionen. *Zeitschrift für Physikalische Chemie*, 4, 129–181.

Nernst, W. (1899). Zur Theorie der elektrischen Reizung. *Nachrichten der Königlichen Gesellschaft der Wissenschaften zu Göttingen, Mathematisch-physikalische Klasse*, 104–108.

Netscher, P. M. (1888). *Geschiedenis van de koloniëun Essequebo, Demerary, en Berbice, van de vestiging der Nederlanders aldaar tot op onzen tijd.* 's-Gravenhage: M. Nijhoff.

Newton, I. (1704). *Opticks*. London: Printed for Samuel Smith, and Benjamin Walford.

Newton, I. (1713). *Philosophiae naturalis principia mathematica*. 2 vols. Cantabrigiae: [s.n.]

Newton, I. (1729). *Mathematical Principles of Natural Philosophy*. 2 vols. translated by A. Motte from the 1687 Latin edition. London: for Benjamin Motte.

Newton, I. (1730). *Opticks: Or a Treatise of the Reflections, Refractions, Inflextions, and Colours of Light*. London: Printed for William Innys. (Reprint: New York: Dover Publications, 1952.)

Newton, I. (1803). *The Mathematical Principles of Natural Philosophy* (vol. 2). *A New Edition translated by A. Motte*. London: Printed for H. D. Symonds.

Newton, I. (1972). *Philosophiae naturalis principia mathematica*. Edited by I. Bernard Cohen, and Alexandre Koyré. Cambridge, Mass.: Harvard University Press.

Nicholson, W. (1797). Observations on the electrophore, tending to explain the means by which the Torpedo and other fish communicate the electric shock. *Journal of Natural Philosophy, Chemistry and the Arts*, 1, 355–359.

Nicholson, W., Carlisle, J., and Cruickshank (1800). *Experiments in galvanic electricity. The Philosophical Magazine*, September, 337–350.

Nicolson, M. (1987) Alexander von Humboldt, Humboldtian science and the origins of the study of vegetation. *History of Science*, 25: 167–194.

Nix, L., and Schmidt, W. (1899). *Heronis Alexandrini Opera quae supersunt omnia. Mit einer Einleitung ueber die Heronische Frage und Anmerkungen. ...Vol. I: Pneumatica et Automata*. Leipzig: In aedibus B.G. Teubneri.

Nobili, L. (1825). Descrizione di un nuovo galvanometro. *Giornale de Fisica, Chimica, e Storia Naturale*, 8, 278–282.

Nobili, L. (1828). Comparaison entre les deux galvanomètres les plus sensibles, la grenouille et le multiplicateur à deux aiguilles, suivie de quelques résultats nouveaux. *Annales de Chimie et de Physique*, 38, 225–245.

Nobili, L. (1830a). Sui colori in generale ed in particolare sopra una nuova scala cromatica dedotta dalla metallocromia ad uso delle scienze e delle arti. *Antologia*, 39, 1–39.

Nobili, L. (1830b). Analyse expérimentale et théorique des phénomènes physiologique s produits par l'électricité sur la grenouille; avec une appendice sur la nature du tétanos et de la paralysie, et sur les moyens de traiter ces deux maladies par l'électricité. *Annales de Chimie et de Physique*, 44, 60–94.

Nobili, L. (1835a). *Lettre à Charles Lucien Bonaparte*. Muséum National d'Histoire Naturelle (Paris) Mss. 2608, 2435.

Nobili, L. (1835b). *Lettre à Charles Lucien Bonaparte*. Muséum National d'Histoire Naturelle (Paris) Mss. 2608, 2436.

Nobili, L., and Antinori, V. (1831). Sopra la forza elettromotrice del magnetismo. *Antologia*, 131, 149–151.

Noda, M., et al. (1982). Primary structure of alpha-subunit precursor of *Torpedo californica* acetylcholine receptor deduced from cDNA sequence. *Nature*, 299, 793–797.

Nollet, J.-A. (1746). *Essai sur l'Electricité des Corps*. Paris: Les Frères Guérin.

Nollet, J.-A. (1749a). Extract of a letter from the Abbé Nollet, FRS &c to Charles Duke of Richmond, FRS accompanying an examination of certain phaenomenona in electricity, published in Italy. *Philosophical Transactions of the Royal Society*, 46, 368–397.

Nollet, J.-A. (1749b). Expériences et observations en différens endroits d'Italie. *Mémoires de l'Académie des Sciences de Paris*, 444–488.

Norris, S. M. (2002). *A revision of the electric catfishes, Family Malapteruridae (Teleostei Siluriformes) with erection of a new genus and descriptions of fourteen new species and an annotated bibliography*. Tervuren, Belgique: Musée Royal de l'Afrique Centrale. (Series Zoologische Wetenschappen, No. 289.)

North, D. D. (1989). *Stars, Minds, and Fate: Essays in Ancient and Medieval Cosmology*. London: Hambledon Press.

Nriagu, J. O. (1983a). *Lead Poisoning in Antiquity*. New York: John Wiley & Sons.

Nriagu, J. O. (1983b). Saturnine gout among Roman aristocrats. *New England Journal of Medicine*, 38, 660–663.

Nunn, J. F. (1996). *Ancient Egyptian Medicine*. Norman: University of Oklahoma Press.

Nutton, V. (1983). The seeds of disease. *Medical History*, 27, 1–34.

Nutton, V. (1995). Scribonius Largus: The unknown pharmacologist. *Pharmaceutical Historian*, 25, 5–8.

O'Malley, C. D. (1964). *Andreas Vesalius of Brussels, 1514–1564*. Berkeley: University of California Press.

O'Malley, C. D., and Saunders, J. B. de C. M. (1952). *Leonardo da Vinci on The Human Body*. New York: Henry Schuman.

Ochs, S. (2004). *A History of Nerve Functions: From Animal Spirits to Molecular Mechanisms*. Cambridge: Cambridge University Press.

Olesko, K. M., and Holmes, F. L. (1993). Experimentation, quantification, and discovery. In D. Cahan (Ed.), *Hermann von Helmholtz and the Foundations of Nineteenth-Century Science*. Berkeley: University of California Press, pp. 50–108.

Olivieri. A. (1935). *Aetius Amideni Libri medicinales 1 - 4 = 8,1*. Leipzig: B. G. Teubner Verlag.

Olmi, G. (1992). *L'Inventario del Mondo: Catalogazione della Natura e Luoghi del Sopare nella Prima età Moderna*. Bologna: il Mulino.

Olmi, G., and Tongiorgi-Tomasi, L. (1993). *De piscibus: la bottega artistica di Ulisse Aldrovandi e l'immagine naturalistica*. Roma: Edizioni dell'Elefante.

Olmsted, J. W. (1942). The scientific expedition of Jean Richer to Cayenne. *Isis*, 34, 117–128.

Olmsted, J. W. (1960) The voyage of Jean Richer to Acadia in 1670: a study in the relations of science and navigation under Colbert. *Proceedings of the American Philosophical Society*, 104, 612–634.

Olschki, L. S. (1937). *Storia letteraria delle scoperte geografiche; Studi e ricerche*. Firenze: L. S. Olschki.

Oppenheimer, J. M. (1936). Guillaume Rondelet. *Bulletin of the Institute of the History of Medicine*, 4, 817–834.

Oppian (1722). *Oppian's Halieuticks, of the Nature of Fishes and Fishng of the Ancients in V. Books*. (Jones and Diaper, trans.) Oxford: Printed at the Theatre.

Ostwald, W. (1890). Elektrische Eigenschaften halbdurchlässiger Scheidewände. *Zeitschrift für physikalische Chemie*, 6, 71–82.

Ostwald, W. (1896). *Elektrochemie: Ihre Geschichte und Lehre*. Leipzig: Veit & Comp.

Otis, L. (2007). *Müller's Lab*. New York: Oxford University Press.

Overton, C. E. (1899). Über die allgemeinen osmotischen Eigenschaften der Zelle, ihre vermuthlichen Ursachen und ihre

Bedeutung für die Physiologie. *Vierteljahrschrift der Naturforschung Gesellschaft in Zürich*, 44, 88–135.

Overton, C. E. (1902a). Beiträge zur allgemeinen Muskel- und Nervenphysiologie. *Archiv für die gesammte Physiolologie des Menschen und der Tiere*, 92, 115–280.

Overton, C. E. (1902b). Beiträge zur allgemeinen Muskel- und Nervenphysiologie, II. Mittheilung: Ueber die Unentbehrlichkeit von Natrium- (oder Lithium-) Ionen für den Contractionsact des Muskels. *Archiv für die gesammte Physiolologie des Menschen und der Tiere*, 92, 346–386.

Oviedo y Valdez, G. F. de (1959). *Historia General y Natural de las Indias, Vol. 1. Biblioteca de Autores Españoles, vol. 117*. Madrid: Ediciones Atlas.

Pacini, F. (1846). Sopra l'organo elettrico del Siluro del Nilo. comparato a quello della Torpedine e del Gimnoto e sull'apparecchio del Weber del siluro comparato a quello dei Ciprini. *Nuovi Annali delle Scienze Naturali*, Ser. II, 6, 41–61.

Pacini, F. (1852). *Sulla Struttura intima dell'Organo elettrico del Gimnoto e di altri Pesci elettrici . . . Memoria del dott. Filippo Pacini . . . letta alla R. Accademia dei Georgofili nella Seduta del 19 Settembre 1852*. Firenze: Mariano Cecchi.

Packard, F. R. (1931). *History of Medicine in the United States*. New York: Hoeber.

Páez, P. (1905–06). *Historia Aethiopiae. . . .* (2 vols.). In *Rerum Aethiopicarum scriptores Occidentales inediti a saeculo XVI ad XIX*. C. Beccari. ed., Vols. II and III. Romae: C. De Luigi.

Paget, S. (1897). *John Hunter: Man of Science and Surgeon. 1728–1793*. London: T. Fisher Unwin.

Palmieri, L. (1842). Nuove esperienze sull'induzione del magnetismo terrestre. *Rendiconto dell' Accademia delle Scienze Fisiche e Matematiche*, 1, 337–341.

Palmieri, L. (1845). Nuovo apparecchio di induzione tellurica. *Rendiconto dell' Accademia delle Scienze Fisiche e Matematiche di Napoli*, 5, 173–178.

Pancaldi G. (1990). Electricity and life: Volta's path to the battery. *Historical Studies in the Physical and Biological Sciences*, 21, 123–160.

Pancaldi, G. (2003). *Volta: Science and Culture in the Age of Enlightenment*. Princeton: Princeton University Press.

Panofsky, E. (1954). *Galileo as a Critic of the Arts*. The Hague: Nijhoff.

Pappas, G. D., Waxman, S. G., and Bennett, M. V. L. (1975). Morphology of spinal electromotor neurons and presynaptic coupling in the gymnotid *Sternarchus albifrons*. *Journal of Neurocytology*, 4, 469–478.

Parrish, S. S. (1996). *American Curiosity: Cultures of Natural History in the Colonial British Atlantic World*. Chapel Hill: University of North Carolina Press.

Partsch, K.-J. (1980). *Die Geschichte der Zoologischen Station Neapel von der Gründung durch Anton Dohrn bis zum ersten Weltkreig*. Göttingen: Vandenhoeck & Ruprecht.

Paszewski, A. (1968). Les problèms physiologiques dans *De vegetabilus et plantis libri VII* d'Albert von Lauingen. *Actes du XIe Congrès International d'Histoire des Sciences*, 5, 323–330.

Paterson, W. (1786a). An account of a new electric fish. *Philosophical Transactions of the Royal Society*, 76, 382–383.

Paterson, W. (1786b). An account of a new electric fish. *Gentleman's Magazine*, 60, 2, 1007.

Patrick, J., and Lindstrom, J. (1973). Autoimmune response to acetylcholine receptor. *Science*, 180, 871–872.

Paulus Aegineta. (1921-1924). *Paulus Aegineta*. (I. L. Heiberg, Ed.) Lipsiae: Berolini.

Peakman, J. (2003). *Mighty Lewd Books: The Development of Pornography in Eighteenth-Century England*. New York: Palgrave Macmillan.

Pedeferri, P. (1989). Le apparenze elettrochimiche di Leopoldo Nobili e la luce del titanio. *Politecnico*, 2, 3–12.

Pellegrino, E. D., and Pellegrino, A. A. (1988). Humanism and ethics in Roman medicine: Translation and commentary on a text of Scribonius Largus. *Literature and Medicine*, 7, 22–38.

Pennec, H. (2003). *Des Jésuites au royaume du prêtre Jean (Ethiopie): stratégies, rencontres et tentatives d'implantation, 1495–1633*. Paris, Lisboa: Centre Culturel Calouste Gulbenkian et Fondação Calouste Gulbenkian.

Pera, M. (1986). *La rana ambigua. La controversia sull'elettricità animale tra Galvani e Volta*. Torino: Einaudi.

Pera, M. (1992). *The Ambiguous Frog: The Galvani-Volta Controversy on Animal Electricity*. Princeton: Princeton University Press.

Pérez-Ramos, A. (1988). *Francis Bacon's Idea of Science and the Maker's Knowledge Tradition*. Oxford: Clarendon Press.

Perry, J. (1777a). *The Electrical Eel, or, Gymnotus Electricus. Inscribed to the Honourable Members of the R***l S****** by Adam Strong, Naturalist* (3rd ed.). London: Printed for J. Bew.

Perry, J. (1777b). *The Torpedo: A Poem to the Electrical Eel. Addressed to Mr. John Hunter, Surgeon, and Dedicated to the Right Honourable Lord Cholmondely*. London: Printed [anonymously] and sold by all the booksellers in London and Westminster.

Perry, J. (1777c). *The Semi-Globes or Electrical Orbs*. London: Printed [anonymously] for A. Webb.

Perry, J. (1777d). *An Elegy on the Lamented Death of the Electrical Eel, or Gymnotus Electricus: With the Lapidary Inscription, as placed on a Superb Erection, at the Expence of the Countess of H_____, and Chevalier-Madame d'Eon de Beaumont. By Lucretia Lovejoy, sister to Mr. Adam Strong, Author of the Electric Eel*. London: Printed for Fielding and Walker.

Peiffer, W. (1877). *Osmotische Untersuchungen: Studien zur Zellmechanik*. Leipzig: Engelmann.

Pflüger, E. (1859). *Untersuchung über die Physiologie des Electrotonus*. Berlin: Hirschwald.

Pflüger, E. (1875). Beiträge zur Lehre von der Respiration, I: Ueber die physiologische Verbrennung in den lebendigen Organismen. *Archiv für die gesammte Physiologie des Menschen und der Thiere*, 10, 251–367.

Petit, G., and Théodoridès, J. (1962). *Histoire de la Zoologie*. Paris: Hermann.

Philes, M. (1730). *De propietate animalium*. Utrecht: Guilielmus Stouw.

Pianciani, G. B. (1838). Saggio d'applicazione del principio dell'induzione elettro-dinamica a' fenomeni elettro-fisiologici e in particolare a quelli della torpedine. *Memorie di Matematica e Fisica della Società Italiana delle Scienze: Memorie della Fisica*, 22, 4–47.

Piccolino, M. (1998). Animal electricity and the birth of electrophysiology: The legacy of Luigi Galvani. *Brain Research Bulletin*, 46, 381–407.

Piccolino, M. (1999). Marcello Malpighi and the difficult birth of modern life sciences. *Endeavour*, 23, 175–179.

Piccolino, M. (2000a). Lazzaro Spallanzani e le ricerche sui pesci elettrici nel Secolo dei Lumi. In P. Manzini & P. Tongiorgi, Eds., *Edizione Nazionale delle Opere di Lazzaro Spallanzani*. Moderna: Mucchi, pp. 359–456.

Piccolino, M. (2000b). The bicentennial of the Voltaic battery (1800-2000): The artificial electric organ. *Trends in Neurosciences*, 23, 47–51.

Piccolino, M. (2003a). *The Taming of the Ray, Electric fish Research in the Enlightenment, from John Walsh to Alessandro Volta*. Firence: Olschki.

Piccolino, M. (2003b). A "lost time" between science and literature: the *"temps perdu"* from Hermann von Helmholtz to Marcel Proust. *Audiological Medicine*, 1, 261–270.

Piccolino, M. (2003c). Nerves, alcohol and drugs, the Adrian-Kato controversy on nervous conduction: deep insights from a "wrong" experiment? *Brain Research Reviews*, 43, 257–265.

Piccolino, M. (2005a). From ambiguous torpedo to animal and physical electricity. *Audiological Medicine*, 3, 124–132.

Piccolino, M. (2005b). *Lo Zufolo e la Cicala: Divagazioni Galileiane tra la Scienza e la sua Storia*. Torino: Bollati Boringhieri.

Piccolino, M. (2008). Visual images in Luigi Galvani's path to animal electricity. *Journal of the History of Neurosciences*, 17, 335–348.

Piccolino, M., and Bresadola, M. (2002). Drawing a spark from darkness: John Walsh and electric fish. *Trends in Neuroscience*, 25, 51–57.

Piccolino M. (2010) "La battipotta, ovvero Boccaccio al mercato del pesce". *Sapere*, 5, 86–94.

Piccolino, M., and Bresadola, M. (2003). *Rane, Torpedini e Scintille: Galvani, Volta e l'Elettricità Animale*. Torino: Bollati Boringhieri.

Piccolino, M., & Wade, N. J. (2008a). Galileo Galilei's vision of the senses. *Trends in Neurosciences*, 31, 585–590.

Piccolino, M., & Wade, N. J. (2008b). Galileo's eye: a new vision of the senses in the work of Galileo Galilei. *Perception*, 37, 1–29.

Pietsch, T. W., and Anderson, W. D., Jr. (Eds.). (1997). *Collecting and Building Ichthyology and Herpetology*. Lawrence: American Society of Ichthyologists and Herpetologists, Special Pub. 3.

Pine, M. (1968). Pomponazzi and the Problem of *Double Truth*. *Journal of the History of Ideas*, 29, 163–176.

Piso, W. (1648). *Guilielmi Pisoni, . . . de medicina Brasiliensi libri quatuor . . .* In W. Piso, G. Marcgraf, J. de Laet, *Historia naturalis Brasiliae*. Leiden: F. Hackius.

Pivati, G. F. (1747). *Della elettricità medica. Lettera del chiarissimo signore Gio. Francesco Pivati . . . al celebre signore Maria Zanotti*. Lucca: [s.n.]

Pizzetta, J. (1893). *Galerie des naturalistes. Histoire des sciences naturelles depuis leur origine jusqu'à nos jours*. Paris: A. Hennuyer.

Platnauer, M. (1922). Introduction. In *Claudian*. London: W. Heinemann.

Plato. (1892). *The Dialogues of Plato. Translated into English, with Analyses and Introductions by B. Jowett*. New York: Macmillan and Co.

Pliny (the Elder). (1543). *Historia naturale di C. Plinio Secondo di Latino in volgare tradotta per Christophoro Landino et nuovamente in molti luoghi . . . emendata, et . . . corretta per Antonio Brucioli*. Venetia: Gabriel Jolito di Ferrarii.

Pliny (the Elder). (1779). *Historia naturalis* (vol. 1). Parisiis: Typis J. Barbou.

Pliny (the Elder). (1855–57). *The Natural History of Pliny*. J. Bostock and H. T. Riley (Trans., Annot.). 6 vols. London: H. G. Bohn.

Pliny (the Elder). (1963). *Natural History*. Trans. by W. H. S. Jones. Cambridge, MA: Harvard University Press.

Pliny (the Elder). (1983). *Natural History* (2nd ed.). Trans. by H. Rackham. Cambridge, MA: Harvard University Press.

Pliny (the Younger). (1832). *Le lettere di Caio Plinio Cecilio Secondo, Tome 1* [With an Italian translation by G. Bandini]. Parma: Rossetti.

Plotinus. (1995). *Ennead IV. With an English Translation by A. H. Armstrong*. Cambridge, MA: Harvard University Press.

Plutarch. (1957). *Plutarch's Moralia. With an English Translation by Harold Cherniss and William C. Helmbold, in Fifteen Volumes* (Vol. XII). Cambridge, MA: Harvard University Press.

Polvani, G. (1942). *Alessandro Volta*. Pisa: Domus Galilaeana.

Pomponazzi, P. (1567) *De naturalium effectuum causis sive de incantationibus*. In Pomponazzi, P., *Opera*. Basle: G. Grataroli.

Ponzetti, F. (1521). *Melfitensis, De venenis libri tres. . . .* Basileae: H. Petri und H. Perna.

Porter. R. (1987). *Mind-Forg'd Manacles*. Cambridge, MA: Harvard University Press.

Porter, R. (1994). Gout: Framing and fantasizing disease. *Bulletin of the History of Medicine*, 68, 1–28.

Pouillet, C. (1832). *Elémens de Physique Expérimentale et de Météorologie* (2nd ed., 2 vols). Paris: Béchet.

Prévost, A. F. et al. (1746–59). *Histoire Générale des Voyages* (15 vols). Paris: Chez Didot.

Priestley, J. (1767). *History and Present State of Electricity, with original Experiments*. London: Printed for J. Dodsley, J. Johnson, B. Davenport, and T. Cadell.

Priestley, J. (1775). *History and Present State of Electricity, with Original Experiments* (3rd ed.). London: Printed for C. Bathurst and T. Lowndes, in Fleet Street.

Priestley, J. (1806). *The Memoirs of Dr. Joseph Priestley to the Year 1795*. London: J. Johnson. (Reprinted by Fairleigh Dickinson University Press, Teaneck, NJ, 1970.)

Pringle, J. (1775a). *Discourse on the Torpedo Delivered at the Anniversary Meeting of the Royal Society, November 30th, 1774*. London: Printed for the Royal Society.

Pringle, J. (1775b). Discours sur la torpille. *Observations sur la Physique*, 5, 241–252.

Pringle, J. (1776). *Discorso sulla Torpedine: Recitato nell'Adunanza Annuale della Società Regale di Londra nel Giorno 30 Novembre 1774 dal Cavaliere Giovanni Pringle*. Napoli: [s.n.].

Public Advertiser. (1774). [No title: Pringle's speech and Walsh's medal.] *Public Advertiser*, December 1, Issue 14094.

Public Advertiser. (1774). [No title: Torpedoes caught of the English coast.] *Public Advertiser*, December 28, Issue 14116.

Public Advertiser. (1776). [No Title: Baker and the spark.] *Public Advertiser*, November 19, Issue 13135.

Public Advertiser. (1776). [No title: Baker's eels and Walsh's spark.] *Public Advertiser*, August 5, Issue 14614.

Public Advertiser. (1777). To be seen at Mr. BAKER'S New Apartments in Piccadilly, nearly opposite St. James's Street, The GYMNOTUS ELECTRICUS, or the ELECTRICAL EEL. *Public Advertiser*, February 19, Issue 13214.

Public Advertiser. (1779). [No title: A torpedo caught between Battersea and Putney.] *Public Advertiser*, December 28, Issue 14109.

Pugliese Carratelli, G. et al. (1998). *Pompei: Pitture e Mosaici. Vol. VIII, Regio VIII - Regio IX, Parte I*. Roma: Istituto della Enciclopedia Italiana.

Pulci, L. (1482). *Morgante Maggiore*. Florence: Francesco di Dino.

Purchas, S. (1613). *Purchas his Pilgrimage, or Relations of the World and the Religions Observed in all Ages and Places Discovered, from the Creation unto this Present. . . .* (4th ed.). London: Stansley.

Purchas, S. (1619). *Purchas his Pilgrim: Microcosmus, or the Historie of Man. . . .* London: Fetherstone.

Purchas, S. (1625). *Hakluytus Posthumus, or Purchas his Pilgrimes: Contayning a History of the World in Sea Voyages and Lande Travells by Englishmen and Others*. (4 vols.). London: Printed by W. Stansby for H. Fetherstone. (Reprinted as a 22-volume series in 1905–07: Glasgow: J. MacLehose and Sons.)

Purchas, S. (1626). *Purchas his Pilgrimage, or Relations of the World and the Religions Observed in all Ages and Places Discovered, from the Creation unto this Present. . . .* (4th ed.). London: Fetherstone.

Puschmann, T. (1963). *Alexander von Tralles: Original-text und Uebersetzung nebst einer einleitenden Abhandlung: ein Beitrag zur Geschichte der Medicin*. Amsterdam: Hackert.

Quegnon, H. (1925). *L'Abbé Nollet Physicien*. Paris.

Qvist, G. (1981). *John Hunter 1728–1793*. Heinemann: London.

Radcliffe, W. (1974). *Fishing from the Earliest Times*. Chicago: Ares Publishers.

Raghavendra, T. (2002). Neuromuscular blocking drugs: discovery and development. *Journal of the Royal Society of Medicine*, 95, 363–367.

Ranvier, L. A. (1878). *Leçons sur l'Histologie du Système Nerveux, par M. L. Ranvier, . . . Recueillies par M. Ed. Weber*. (2 Vols.). Paris: F. Savy.

Read, W. A. (1945). Some fish names of Indian origin. *International Journal of American Linguistics*, 11, 234–238.

Réaumur, R. A. F. de (1714). Des effets que produit le poisson appellé en Latin Torpille, sur ceux qui le touchent, et de la cause dont ils dépendent. *Académie des Sciences, Procès Verbaux*, Tome 33, 311–376.

Réaumur, R. A. F. de (1717). Des effets que produit le poisson appellé en français Torpille, ou Tremble, sur ceux qui le touchent; et de la cause dont ils dépendent. *Mémoires de l'Académie Royale des Sciences pour l'Annee 1714*, 344–360.

Redi, F. (1664). *Observazioni Intorno alle Vipere . . .* Firenze: Stella.

Redi, F. (1668). *Esperienze Intorno alla Generazione degl'Insetti*. Florence: All'insegna della Stella.

Redi, F. (1671). *Esperienze Intorno a Diverse Cose Naturali e Particolarmente Interno a quelle che ci son Portate dall'Indie. . . .* Firenze: All'insegna della Nave.

Redi, F. (1988). *Francesco Redi on Vipers* (P. K. Knoefel, trans. and ed.). Leiden: E. J. Brill.

Rees, G. (2000a). Francis Bacon. In W. Applebaum (Ed.), *Encyclopedia of the Scientific Revolution from Copernicus to Newton*. New York: Garland, pp. 65–69.

Rees, G. (2000b). Baconism. In W. Applebaum (Ed.), *Encyclopedia of the Scientific Revolution from Copernicus to Newton*. New York: Garland, pp. 69–71.

Reiff, J. G. (1805) *Artemidori Oneirocritica 1, Textum, varias lectiones atque ipsum Artemidorum spectantes indices continens*. Lipsiae: Crusius.

Reilly, C. (1974). *Athanasius Kircher, S. J., Master of a Hundred Arts*. Rome: Edizioni del Mondo.

Renna, E. (1995). I pesci elettrici. In O. Longo, F. Ghiretti, and E. Renna (Eds), *Aquatilia: animali di ambiente acquatico nella storia della scienza: da Aristotele ai giorni nostri*. Napoli: Procaccini, pp. 147–171.

Rescigno, A. (2000). Alessandro di Afrodisia e Plotino: il caso della θαλαττία νάρκη. *Koinonia*, 24, 199–230.

Reverte, J. (2001). *Dios, el Diablo y la Aventura: La Historia de Pedro Páez, el Español que Descubrió el Nilo Azul*. Barcelona: Plaza y Janés.

Ricci, G. (1995). *Le Città di Freud: Itinerari, Emblemi, Orizzonti di un Viaggiatore*. Milano: Jaca Book.

Richer, J. (1693). *Recueil d'Observations Faites en Plusieurs Voyages par Ordre de sa Majesté, pour Perfectionner l'Astronomie et la Geographie, avec Divers Traitez Astronomiques*. Paris: De l'Imprimerie Royale.

Richer, J. (1729). Observations astronomiques et physiques faites en l'isle de Caienne. *Mémoires de l'Académie Royale des Sciences 1666–1699*, 7(1), 231–326.

Riddle, J. M. (1985). *Dioscorides on Pharmacy and Medicine*. Austin: University of Texas Press.

Rippa Bonati, M. and Pardo-Tomás, J. (2004). *Il teatro dei corpi: le pitture colorate d'anatomia di Girolamo Fabrici d'Acquapendente*. Milano: Mediamed.

Riskin, J. (1998). Poor Richard's Leyden jar: Electricity and economy in Franklinist France. *Historical Studies in the Physical and Biological Sciences*, 28, 301–336.

Rittenhouse, B. (1805). Experiments on the Gymnotus Electricus, or Electric Eel, made at Philadelphia, about the year 1770, by the late Mr. Rittenhouse, Mr. E. Kinnersley, and some other gentlemen. *Philadelphia Medical and Physical Journal*, 96–100, 159–160.

Ritterbush, P. C. (1964). Electricity: The soul of the universe. In *Overtures to Biology: The Speculations of Eighteenth-Century Naturalists*. New Haven: Yale University Press.

Robin, C. (1846). Recherches sur un organe particulier qui se trouve sur les poissons du genre des raies (Raia, Cuv.). *Comptes Rendus Hebdomadaires des Séances de l'Académie des Sciences*, 22, 821–822.

Robin, C. (1847). *Recherches sur un Appareil qui se Trouve sur les Poissons du Genre des Raies (Raia, Cuv.) et qui Présente les Caractères Anatomiques des Organes Électriques*. Paris: L. Martinet.

Robin, C. (1865). Mémoir sur la démonstration expérimentale de la production d'électricité par un appareil propre aux poissons du genre des raies. *Journal de l'Anatomie et de la Physiologie*, 23, 577–645.

Rocca, J. (2003). *Galen on the Brain: Anatomical Knowledge and Physiological Speculation in the Second Century AD*. London: Brill.

Roche, D. de la (1778). *Analyse des Fonctions du Système Nerveux, pour Servir d'Introduction à un Examen Pratique des Maux de Nerfs* (2 Vols.). Genève: Du Villard Fils & Nouffer.

Rodnan, G. P., and Benedek, T. G. (1963). Ancient therapeutic arts in the gout. *Arthritis and Rheumatism*, 6, 317–340.

Rolland, E. (1919). *Faune Populaire de la France. Vol. XI, (1ère Partie) Reptiles et Poissons*. Paris: Maisonneuve et Cie.

Romm, J. S. (1989). Aristotle's elephant and the myth of Alexander's scientific patronage. *American Journal of Philology*, 110, 566–575.

Romo, E. J. A. (2006). Andrés de Oviedo, Patriarca de Etiopía. *Peninsula*, 3, 215–231.

Rondelet, G. (1554). *Gulielmi Rondeletii doctoris medici . . . de piscibus marinis in quibus verae piscium effigies expressae sunt*. Lyon: Matthiam Bonhomme.

Rondelet, G. (1555). *Universae aquatilium historiae cum veris ipsorum imaginibus*. Lyon: Matthiam Bonhomme.

Rondelet, G. (1558). *L'histoire Entière des Poissons*. Lion: Macé Bonhomme, à la Masse d'Or.

Rosen, E. (1956). The invention of eyeglasses. *Journal of the History of Medicine and Allied Sciences*, 11, 13–46, 183–218.

Rosenfeld, L. C. (1940). *From Beast-Machine to Man-Machine*. New York: Oxford University Press.

Rosenthal, J. W. (1996). *Spectacles and other Vision Aids*. San Francisco: Norman Publishing.

Rossi, P. (1960). *Clavis Universalis: arti della memoria e logica combinatoria da Lullo a Leibniz*. Milano-Napoli: Ricciardi.

Rossi, P. 1998. *La nascita della scienza moderna in Europa*. Fare l'Europa. Roma-Bari: Laterza. (Engl. Transl. By Cynthia DeNardi Ipsen, *The bird of modern science*. Oxford: Blackwell, 2001).

Rossi, P. (2002). *I filosofi e le macchine 1400–1700*. Milano: Feltrinelli.

Rossi, P. (2006). *Il tempo dei maghi. Rinascimento e modernità*. Milano: Cortina.

Roth, W. E. (Ed., Trans.) (1941). *Adriaan van Berkel's Travels in South America between the Berbice and Essequibo Rivers and in Surinam, 1670–1689*. Georgetown, British Guiana: Daily Chonicle, Ltd.

Rothschuh, K. (Ed.). (1964). *Emil Du Bois-Reymond (1818–1896) und die Elektrophysiologie der Nerven*. In Von Borhaave bis Berger. Stuttgart: Fischer.

Royal Society of Arts. *Proceedings of the American Philosophical Society*, 144, 245–266.

Ruelle, C-E. (1898). *Les Lapidaires de l'Antiquité et du Moyen Age. . . . Vol. 3: Les Lapidaires Grecs*. Paris: E. Leroux.

Russell, D. A. (2001). *Plutarch*. London: Duckworth.

Russell, P. E. (2000). *Prince Henry "the Navigator": A Life*. New Haven: Yale University Press.

Sachs, C. (1877). Beobachtungen und Versuche am südamerikanischen Zitteraale (*Gymnotus electricus*). *Archives für Physiologie*, 66–95.

Sachs, C. (1879). *Aus dem Llanos: Schilderung einer naturwissenschaftlichen Reise nach Venezuela*. Lepzig: Veit.

Sachs, C., and Du Bois-Reymond, E. (1881). *Dr. Carl Sachs, Untersuchungen am Zitteraal, Gymnotus electricus*. Leipzig: Veit.

Sakmann, B., and Neher, E. (1995). *Single-Channel Recording* (2nd ed.). New York: Plenum Press.

Salviani, I. (1554). *Aquatilium animalium historiae....* Romae: Apud eundem Hippolytum Salvianum.

Salviani, I. (1557). *Aquatilium animalium formae aere excusae.* Roma: Apud eundem Hippolytum Salvianum

Salviani, I. (1558). *Aquatilium animalium historiae, liber primus,: cum eorumdem formis, aere excusis.* Roma: Apud eundem Hippolytum Salvianum.

Sanders, A. E. (1997). Alexander Garden (1730-1791), pioneer naturalist in Colonial America. In T. W. Pietsch and W. D. Anderson, Jr. (Eds.), *Collecting and Building Ichthyology and Herpetology.* Lawrence: American Society of Ichthyologists and Herpetologists, Spec. Pub. 3, pp. 409–437.

Sanders, A. E., and Anderson, W. D. (1999). *Natural History Investigations in South Carolina from Colonial Times to the Present.* Columbia: University of South Carolina Press.

Sarton, G. (1954). *Galen of Pergamon.* Lawrence: University of Kansas Press.

Sarton, G. (1955). *The Appreciation of Ancient and Medieval Science During the Renaissance.* Philadelphia: University of Pennsylvania Press.

Sbrighi, G. (1985). La controversia Nobili-Faraday sull'induzione elettromagnetica. In *Leopoldo Nobili e la cultura scientifica del suo tempo* Tarozzi, G. ed., pp. 183–199. Bologna: Istituto per i beni artistici, culturali, naturali della Regione Emilia-Romagna.

Scarborough, J. (Ed.). (1985). *Symposium on Byzantine Medicine.* Washington: Dumbarton Oaks Research Library.

Schaffer, S. (1983). Natural philosophy and public spectacle in the eighteenth century. *History of Science,* 21, 1–43.

Schaffer, S. (1993). The consuming flame: electrical showmen and Tory mystics in the world of goods. In: Brewer, J. and Porter, R. (Eds.), *Consumption and the World of Goods.* London: Routledge, pp. 489–526.

Schaffer, S (1994). Self evidence. In Chandler, J, Davidson, A. I., and Harootunian, H. (Eds.), *Questions of Evidence: Proof, Practice, and Persuasion across the Disciplines.* Chicago: University of Chicago Press, pp. 56–104.

Schechter, D. C. (1971). Origins of electricity. *New York State Journal of Medicine,* 997–1008.

Schilling, G. W. (1770). *Diatribe de morbo in Europa pene ignoto, quem Americani vocant jaws, adjecta est hecas casuum rariorum . . . observatorum, nec non observatio physica de torpedine pisce.* Trajecti ad Rhenum: J. C. ten Bosch.

Schilling, G. W. (1772). Sur le phénomènes de l'Anguille tremblante. *Nouvelle Mémoires de l'Académie Royale des Sciences et des Belle- Lettres de Berlin,* 68–74.

Schinz, H. R., and Joseph Brodtmann, R. (1836). *Naturgeschichte und Abbildungen der Fische: Nach den neuesten Systemen zum gemeinnuützigen Gebrauch entworfen und mit Beruücksichtigung fuür den Unterricht der Jugend bearb. [2] Atlas.* Leipzig: Schaffhausen.

Schnakenbeck, W. (1955). Acrania-Cyclostoma-Pisces. In J.-G. Helmcke and H. von Lengerken (eds.), *Handbuch der Zoologie,* Band VI, Helfte 1, Lief 7. Berlin: De Gruyter.

Schnitker, M. A. (1936). A history of the treatment of gout. *Bulletin of the History of Medicine,* 4, 89–120.

Schönbein, C. F. (1841). Observations sur les effets électrique du Gymnote. *Archives de l'Électricité,* 1, 445–467.

Schönlein, K. (1896). Beobachtungen und Untersuchungen über den Schlag von Torpedo. Zweite Mittheilung. *Zeitschrift für Biologie* (neue Folge), 15, 408–461.

Schuetze, S. M. (1983). The discovery of the action potential. *Trends in Neuroscience,* 6, 164–168.

Sconocchia, S. (2001). Alcune note sulle *Compositiones* di Scribonio Largo. In A. Debru, N. Palmieri, and B. Jacquinod (Eds.), *Docente Natura.* Saint-Étienne: Publications de l'Université de Saint-Étienne, pp. 1–18.

Scribonius Largus. (1983). *Scribonii Largi Compositiones.* S. Sconocchia (Ed.). Leipzig: B. G. Teubner.

Seba, A. (1734-65). *Locupletissimi rerum naturalium thesauri accurata descriptio* (4 Vols.). Amstelodami: Apud Janssonio-Waesbergios et Apud Arksteum & Schouter.

Seba, A. (2005). *Cabinet of Natural Curiosities Locupletissimi Rerum Naturalium Thesauri 1734-1765: Based on the Copy in the Koninklijke Bibliotheek, The Hague/Albertus Seba.* (With an Introduction by Irmgard Müsch). Cologne: Taschen.

Seebeck, T. J. (1822). Magnetische Polarisation der Metalle und Erze durch Temperatur-Differenz. *Abhandlungen der Königlich Preussischen Akademie der Wissenschaft,* 289–346.

Seignette, P. H. (1772a). Extract of a letter from Sieur Seignette, Secretary to the Academy of Rochelle, dated Oct. 30. *Middlesex Journal or Universal Evening Post,* November 24, Issue 571.

Seignette, S. (1772b). Extract of a letter from the Sieur Seignette, Secretary to the Academy at Rochelle. *Gentleman's Magazine,* 42, 567.

Seignette, P. H. (1790). *A curious Experiment for evincing the Circuit of the ELECTRICAL MATTER from the Torpedo; in presence of the Academy at Rochelle. Middlesex Journal or Universal Evening Post,* January 15, Issue 17314.

Seligardi, R. (1997). Luigi Galvani e la chimica del settecento. In F. Calascibetta (Ed.), *Atti del VII Convegno Nazionale di Storia e Fondamenti della Chimica.* Roma: Accademia Nazionale delle Scienze detta dei XL, pp. 147–162.

Seligardi, R. (2002). *Lavoisier in Italia: La Comunità Scientifica Italiana e la Rivoluzione Chimica.* Firenze: L.S. Olschki.

Serrai, A. (1990). *Conrad Gesner.* Roma: Bulzoni.

Seyfarth, E.-A. (2006). Julius Bernstein (1839-1917): Pioneer neurobiologist and biophysicist. *Biological Cybernetics,* 94, 2–8.

's Gravesande, L. S. van (1911). *Storm van 's Gravesande; the Rise of British Guiana.* Translated and edited by C. Alexander Harris, and John Abraham Jacob De Villiers. London: Printed for the Hakluyt Society.

Shapin, S. and Schaffer. (1985). *Leviathan and the Air-Pump: Hobbes, Boyle, and the Experimental Life: Including a Translation of Thomas Hobbes, Dialogus physicus de natura aeris by Simon Schaffer.* Princeton, NJ: Princeton University Press.

Shepherd, G. M. (1991). *Foundations of the Neuron Doctrine.* New York: Oxford University Press.

Shepherd, G. M. (2010). *Creating Modern Neuroscience: The Revolutionary 1950s.* New York: Oxford University Press.

Shorland, E. (1967). *The Pish (Parish) of Warfield and Easthampstead, which includes Old Bracknell (Berks.).* Bracknell: The Author.

Shryock, R. H. (1960). *Medicine and Society in America, 1650–1965.* New York: New York University Press.

Shryock, R. H. (1967). *Medical Licensing in America, 1650–1965.* Baltimore: Johns Hopkins University Press.

Sigerist, H. E. (1961). *A History of Medicine; Vol. II: Early Greek, Hindu, and Persian Medicine.* New York: Oxford University Press.

Silvestre de Sacy, A. I. (1810). *Relation de l'Égypte par 'Abd al-Latif, Médecin Arabe de Bagdad.* Paris: Dreuttel et Würtz.

Singer, C. (1952). *Vesalius on the Human Brain.* London: Oxford University Press.

Simocatta, T. (1835). *Theophylacti Simocattae Quaestiones physicas et epistolas.* J. F. Boissonade (Ed.). Paris: J. A. Mercklein.

Sirol, M. (1939). *Galvani et le Galvanisme. L'Électricité Animale....* Paris: Vigot Frères.

Slavin, K. V. (2008). Peripheral nerve stimulation for neuropathic pain. *Journal of the American Society for NeuroTherapeutics,* 5, 100–106.

Slessarev, V. (1959). *Prester John: The Letter and the Legend.* Minneapolis: University of Minnesota Press.

Smith, C. U. M. (1987). David Hartley's Newtonian neuropsychology. *Journal of the History of the Behavioral Sciences,* 23, 123–136.

Smith, J. E. (1791). Introductory Discourse on the Rise and Progress of Natural History (1788 Presidential Address). *Transactions of the Linnean Society of London,* 1, 1–57.

Smith, J. E. (1821). *A Selection of the Correspondence of Linnaeus and other Naturalists from the Original Manuscripts* (2 vols). London: Longman, Hurst, Rees, Orme, and Brown.

Snorrason, E. (1968). The studies of Nicholas Steno 1659 in Copenhagen libraries. In G. Scherz (Ed.), *Steno and Brain Research in the Seventeenth Century Oxford.* Pergamon Press, pp. 69–93.

Solander, D. (1775). Diaries of Daniel Solander, M.D., D.C.L., Assistant Keeper of Natural History. *British Museum Diaries and Registers,* IX, No. 45875 (April 2, 1773 to April 26, 1782).

Solander, D. (1776). [Letter of November 16, 1776 to Joseph Banks.] Department of Botany of the British Museum. *Correspondence of Joseph Banks, Vol. 1,* pp. 133–134.

Somogyi, J. (1957). Medicine in ad-Damiri's Hayat Al-Hayawan. *Journal of Semitic Studies,* 2, 62–91.

Sonntag, O. (Ed.). (1999). *John Pringle's Correspondence with Albrecht von Haller.* Basel: Schwabe & Co.

South Carolina Gazette. (1774). The wonderful Electrical FISHES. (Under New Advertisements). *The South Carolina Gazette,* June 20, 1774.

Spallanzani, L. (1783). Lettera al Sig. Marchese Lucchesini. *Opuscoli scelti sulle Scienze e sulle Arti,* 6, 73–104.

Spallanzani, L. (1779). Breve relazione delle cose più rilevanti de' Musei di Storia Naturale di Ginevra, e della Svizzera, vedute da me Lazzaro Spallanzani Regio. Professore nel viaggio fatto in quelle parti l'estate del 1779. Biblioteca Panizzi, Reggio Emilia. (Unpublished manuscript; Mss. Regg. B51.)

St. B. S. (1760), [Untitled letter to the editor.] *Der Artz - Eine medicinische Wochenschrift, Zweyter Theil, Zwote Auflage* Hamburg: G. C. Grunds Witwe.

St. James's Chronicle or the British Evening Post. (1774). [No title: A torpedo caught in the Thames.] *St. James's Chronicle or the British Evening Post,* June 23, Issue 2085.

St. James's Chronicle or the British Evening Post. (1776). [No Title: Baker and the spark.] J*St. James's Chronicle or the British Evening Post,* November 16, Issue 2450.

St. James's Chronicle or the British Evening Post. (1777). Advertisement for Adam Strong's *The Electrical Eel. St. James's Chronicle or the British Evening Post,* April 1, Issue 2506; July 3, Issue 2546.

Stannard, J. (1973). Marcellus of Bordeaux and the beginnings of Medieval materia medica. *Pharmacy in History,* 15, 47–54.

Stannard, J. (1978). Natural history. In D. C. Lindberg (Ed.), *Science in the Middle Ages.* Chicago: University of Chicago Press, pp. 429–460.

Stark. J. (1844). On the existence of an electrical apparatus in the flapper skate and other rays. *Proceedings of the Royal Society of Edinburgh,* 25 (Vol. 2), 1–3.

Stark. J. (1845). On the existence of an electrical apparatus in the flapper skate and other rays. *Annals and Magazine of Natural History,* 15, 122.

Stearns, R. P. (1970). *Science in the British Colonies of America.* Urbana: University of Illinois Press.

Stedman, J. G. (1796/1988). *Narrative of a Five Years Expedition against the Revolted Negroes of Surinam, in Guiana on the Wild Coast of South America, from the Year 1772 to 1777.* Reprinted in 1988: Baltimore: Johns Hopkins University Press.

Steno, N. (1669). *Discours de Monsieur Stenon sur L'Anatomie du Cerveau.* Paris: Chez Robert de Ninville. (Translated in 1733 as *A Dissertation on the Anatomy of the Brain by Nicolaus Steno.* Reprinted in Copenhagen by Nyt Nordsk Forlag, 1950).

Steno, N. (1675). Ova viviparum spectantes observationes. *Acta Medica Philosophica Hafneiensia,* 2, 219–232.

Stephenson, J. (Ed., Trans., Annot.) (1928a). *The zoological section of the Nuzhatu-l-Qulub (Hamdullah al-Mustafi al-Qazwini).* London: The Royal Asiatic Society.

Stephenson, J. (1928b). The Zoological Section of the *Nuzhatu-l-Qulub. Isis,* 11, 285–315.

Stone, J., and Goodrich, J. T. (2007). John Hunter's contributions to neuroscience. In H. Whitaker, C. U. M. Smith, and S. Finger (Eds.), *Brain, Mind and Medicine: Essays in 18th-Century Neuroscience.* Boston: Springer, pp. 66–86.

Stratton, G. M. (1917). *Theophrastus and the Greek Physiological Psychology before Aristotle.* London: Allen and Unwin.

Strickland, S. W. (1998). The ideology of self-knowledge and the practice of self-experimentation. *Eighteenth-Century Studies,* 31, 453–471.

Stroud, P. T. (2000). *The Emperor of Nature; Charles-Lucien Bonaparte and his World.* Philadelphia: University of Pennsylvania Press.

Stukeley, W. (1750). *The Philosophy of Earthquakes, . . .* London: C. Corbett.

Swammerdam, J. (1758). Experiments in the particular motions in the muscles of the frog. . . . In *The Book of Nature.* London: C. G. Seyfert, pp. 122–125.

Suetonius. (1543). *Suetonij Tranquilli Dvodecim Caesares, ex Erasmi recognitione.* Parisiis: Apud Simonem Colinœum.

Taylor, B. (1859). *The Life Travels and Books of Alexander von Humboldt.* London: Sampson Low, Son & Co.

Tedeschi, S. (1992). Giacomo Baratti e il suo Presunto Viaggio in Abissinia. *Africa,* 47, 1–27.

Telles, B. (1660). *Historia geral da Ethiopia a alta ou Abassia ou Preste Ioam e do que nella Obraram os Padres da Companhia de Iesus.* Coimbra: Manoel Diaz.

Temkin, O. (1964). The classical roots of Glisson's doctrine of irritation. *Bulletin of the History of Medicine,* 38, 297–328.

Temkin, O. (1991). *Hippocrates in a World of Pagans and Christians.* Baltimore: Johns Hopkins University Press.

Termeyer, R. (1781). Esperienze e riflessioni sulla torpedine. *Raccolta Ferrarese di Opuscoli Scientifici et Letterari,* 8, 23–70.

Termeyer, R. (1810). Intorno ad un'Anguilla, Ossia Ginnoto Americano. . . . *Opuscoli Scientifici d'Entomologia, di Fisica e d'Agricoltura, Vol. 5.* Milano: Carlo Dova, pp. 105–173.

Terra, H. de (1955). *Humboldt: The Life and Times of Alexander von Humboldt, 1769–1859.* New York: Knopf.

Thompson, D. W. (1928). On Egyptian fish names used by Greek writers. *Journal of Egyptian Archeology,* 14, 27.

Thomson, D. W. (1947). *A Glossary of Greek Fishes.* London: Oxford University Press.

Thorndike, L. (1923-58). *A History of Magic and Experimental Science.* New York: Columbia University Press.

Tjomsland, A. (1938). Niels Stensen: His tercentenary. *Annals of the History of Medicine,* 10, 491–507.

Todd, J. T. (1816). Some Observations and Experiments Made on the Torpedo of the Cape of Good Hope in the Year 1812. *Philosophical Transactions of the Royal Society of London,* 10, 120–126.

Todd, J. T. (1817). Account of Some Experiments on the *Torpedo Electricus,* at La Rochelle. *Philosophical Transactions of the Royal Society of London,* 107, 32–35.

Tongiorgi-Tomasi, L. (2000). L'immagine naturalistica: tecnica e invenzione. In Olmi, G., Tongiorgi-Tomasi, L., and Zanca, A. *Natura-cultura: l'interpretazione del mondo fisico nei testi e nelle immagini: Atti del convegno internazionale di studi.* Firenze: Olschki, pp. 133–151.

Torlais, J. (1954). *L'Abbé Nollet.* Paris: Sipuco.

Torreblanca, V. F. (1623). *Daemonologia sive de Magia Naturali Daemoniaca et Illicita.* . . . Mainz: Schoenwetter.

Traube, M. (1867). Experimente zur Theorie der Zellenbildung und Endosmose. *Archiv für Anatomie und Physiologie und wissenschaftliche Medizin,* 87–165.

Treneer, A. (1963). *The Mercurial Chemist, A Life of Sir Humphry Davy.* London: Methuen.

Treglia, A. (2008). *Mola e Castellone di Gaeta oggi Formia. Memorie e immagini dei viaggiatori del Grand Tour.* Formia: Graficart.

Treviranus, G. R. (1818). *Biologie, oder Philosophie der lebenden Natur für Naturforscher und Aerzte.* Göttingen: J. F. Röwer.

Trillat, E. (1995). Conversion disorder and hysteria. In G. E. Berrios and R. Porter (Eds.), *A History of Clinical Psychiatry.* New York: New York University Press, pp. 433–450.

Tugnoli Pattaro, S. (1981). *Metodo e Sistema delle Scienze nel Pensiero di Ulisse Aldrovandi.* Bologna: CLUEB.

Turner, R. (1746). *Electricology: Or, a Discourse upon Electricity.* Worcester: Th. Olivers.

Uggla, A. H. (1957). *Linnaeus* (Trans. from the Swedish by A. Blair). Stockholm: Swedish Institute.

Urreta, L. de (1610). *Historia eclesiastica y politica, natural y moral, de los grandes y remotos Reynos de la Etiopia.* . . . València: Mey.

Urreta, L. de (1611). *Historia de la Sagrada Orden de Predicadores, en los remotos Reynos de la Etiopia.* . . . Valencia: Chrysostomo Garriz.

Valdes, D. (1579). *Rhetorica Christiana.* Perusa.

Valenstein, E. S. (2005). *The War of the Soups and the Sparks.* New York: Columbia University Press.

Valeriano, G. P. I. (1556). *Hieroglyphica, sive, De sacris Aegyptiorvm literis commentarii.* Basileae: Palma Michael Isengrin.

Valeriano, G. P. I. (1615). *Les Hieroglyphiques de Ian-Pierre Valerian, vulgairement nommé Pierius* . . . Lyon: Par Paul Frellon.

Valeriano, G. P. I. (1625). *I Ieroglifici ouero commentarii delle occulte significationi de gl'Egittij.* . . . Venice: Presso Giovanni Battista Combi.

Valli, E. (1793). *Experiments on Animal Electricity with their Applications to Physiology and Some Pathological and Medical Observations.* London: J. Johnson.

Van Baerle, C. (1660). *Rerum per octennium in Brasilia et alibi gestarum . . . historia* (2nd ed.). Clivis: Tobiæ Silberling.

Van Berkel, A. (1695). *Amerikaansche voyagien, Behelzende een reis na rio de Berbice, gelegen op het vaste land van Guiana, aande wilde-kust van America.* . . . Amsterdam: J. ten Hoorn.

Van Berkel, K. (1985). *In het voetspoor van Stevin. Geschiedenis van de natuurwetenschap in Nederland 1580–1940.* Amsterdam: Boom.

Van der Lott, F. (1762a). Kort bericht van den conger-aal, afte drilvisch. *Verhandelingen Hollandsche Maatschappye der Weetenschappen, Haarlem,* 6, 87–94.

Van der Lott, F. (1762b). Extrait d'une lettre sur le congre, écrite de Rio Essequebo le 7 Juin 1761 par M. Van der Lott, qui est Chirurgien. *Bibliothèque des Sciences et des Beaux Arts,* 17, 386–388.

Van der Lott, F. (1775). Kurzer Auszug eines Briefs aus Rio Issequebo, den Zitterfisch oder sogenannten Kongeraal betreffend. *Abhandlungen aus der Naturgeschichte, praktischen Arzneykunst und Chirurgie aus den Schriften der Harlemer und anderer holländischen Gesellschaften.* Band 1, 105–110.

Van Doren, C. (1938). *Benjamin Franklin.* Cleveland, OH: World Publishing Company.

Van Helmont, J. B. (1621). *De magnetica vulnerum curatione.* . . . Paris: Victor Le Roy.

Van Helmont, J. B. (1648). *Ortus medicinae* . . . . Amsterdam: Elzevir.

Van Leeuwenhoek, A. (1674). More observations from Mr. Leewenhook, in a letter of Sept. 7. 1674. sent to the publisher. *Philosophical Transactions of the Royal Society of London,* 9, 178–182.

Van Leeuwenhoek, A. (1719). *Epistolæ ad Societatem Regiam Anglicam et alios illustres viro.* . . . Lugduni Batavorum: Apud Joh. Arnold Langerak.

Van Musschenbroek, P. (1746). Observations de Monsieur Musschenbroek lues par Monsieur de Reaumur. *Académie Royale des Sciences, Procès Verbaux,* 65, 4–6.

Van Musschenbroek, P. (1762). *Introductio ad philosophiam naturalem.* Lugduni Batavorum: Apud Sam. et Joh. Luchtmans.

Van Neer, W. (1986). Some notes on the fish remains from Wadi Kubbaniya (Upper Egypt; late Paleozoic). In D. Brinkhuizen and A. Clason (Eds.), *Fish and Archeology. British Archeological Reviews, International Series,* 294, 103–113.

Van Neer, W. (1989). Fishing along the prehistoric Nile. In L. Krzyżaniak and M. Kobusiewicz (Eds.), *Late Prehistory of the Nile Basin and the Sahara (Studies in African Archaeology, Vol. 2).* Poznań: Poznań Archaeological Museum, pp. 49–56.

Vansleb, P. (1677). *Nouvelle Relation en Forme de Journal d'un Voyage fait en Egypt par P. Vansleb, R. D. en 1672 and 1673.* Paris: Michallet.

Van't Hoff, J. H. (1887). Die Rolle des osmotischen Druckes in der Analogie zwischen Lösungen und Gasen. *Zeitschrift für physikalische Chemie,* 1, 481–508.

Varchi, B. (1962-63). Lezzione nella quale si disputa della maggioranza delle arti e quale sia la più nobile, o la pittura o la scoltura. In P. Barocchi (Ed.). *Trattati d'arte del Cinquecento fra Manierismo e Controriforma,* Vol. I. Bari: G. Laterza, pp. 1–58.

Veratti, G. (1748). *Osservazioni Fisico-Mediche Intorno alla Elettricità.* Bologna: Lelio dalla Volpe.

Vesalius, A. (1543). *De humani corporis fabrica.* Basilae: J. Oporini.

Veith, I. (1970). *Hysteria.* Chicago: University of Chicago Press.

Vincent, L. (1719). *Elenchus tabularum, pinacothecarum, atque nonnullorum cimeliorum, in gazophylacio Levini Vincent.* Harlemi Batavorum: Sumptibus auctoris.

Visser, R. P. W. (1975). De Nederlandse Geleerde genootschappen in de achttiende eeuw. In: *Documentatieblad Werkgroep, 18e Eeuw,* 1–10 (Reprint: HES: Utrecht, pp. 175–186).

Volkmann, R. (1884). *Plotini Enneades: praemisso Porphyrii de vita Plotini deque ordine librorum eius libello. II.* Leipzig: B.G. Teubner.

Volta, A. (1800). On the electricity excited by the mere contact of conducting substances of different species. Letter to Sir Joseph Banks, March 20, 1800. *Philosophical Transactions of the Royal Society,* 90, 403–431.

Volta, A. (1918). *Le Opere di Alessandro Volta, Edizione nazionale* (Vol. I). Milano: Hoepli.

Volta, A. (1923). *Le Opere di Alessandro Volta* (Vol. II). Milano: Hoepli.

Volta, A. (1926). *Le Opere di Alessandro Volta* (Vol. III). Milano: Hoepli.

Voltaire (François-Marie Arouet) (1759). *Candide, ou l'Optimisme.* Paris: Cramer; English translation by John Everett Butt. 1947. *Candide: or, Optimism.* Penguin classics. London: Penguin Books.

Wade, N. J. (Ed.) (2003). *Müller's Elements of Physiology* (4 vols). Bristol: Thoemmes.

Wade, N. J. (2005). The persisting vision of David Hartley (1705-1757). *Perception,* 34, 1–6.

Wagner, B. (2000). *Die Epistola Presbiteri Johannis, Lateinisch und Deutsch.* . . . Tübingen: Niemeyer.

Walker, M. B. (1934). Treatment of myasthenia gravis with physostigmine. *Lancet* 1934, i, 1200–1201.

Walker, M. B. (1935). Case showing the effect of prostigmin on myasthenia gravis. *Proceedings of the Royal Society of Medicine,* 28, 759–761.

Walker, M. B. (1973). Some discoveries on myasthenia gravis: The background. *British Medical Journal, 2,* 42–43.

Walker, W. C. (1936). The detection and estimation of electric charges in the eighteenth century. *Annals of Science, 1,* 66–100.

Walker, W. C. (1937). Animal electricity before Galvani. *Annuals of Science, 2,* 84–113.

Wall, P. D., and Sweet, W. H. (1967). Temporary abolition of pain in man. *Science, 155,* 108–109.

Wallace, S. L. (1968). Colchicum. *Bulletin of the New York Academy of Medicine, 49,* 130–135.

Wallace, W. (2003). The vibrating nerve impulse in Newton, Willis, and Gassendi: First steps in a mechanical theory of communication. *Brain and Cognition, 51,* 66–94.

Wallis Budge, E. A. T. (1928). *A History of Ethiopia, Nubia & Abyssinia....* London: Methuen.

Walsh, J. (1772a). *Journey from London to Paris, begun 8th June 1772.* John Rylands Library, Manchester, UK, unpublished manuscript 724.

Walsh, J. (1772b). *Experiments on the Torpedo or Electric Ray at La Rochelle and l'Isle de Ré—in June-July 1772.* Library of Royal Society of London, unpublished manuscript 609.

Walsh, J. (1773). On the electric property of the torpedo. In a letter from John Walsh Esq; F.R.S. to Benjamin Franklin, Esq; LL.D., F.R.S., Ac. R. Soc. Ext. &c. *Philosophical Transactions of the Royal Society, 63,* 461–480.

Walsh, J. (1774a). Lettre de M. Walsh, Ecuyer & de la Société Royale de Londres; à M. Benjamin Franklin, Ecuyer.... *Observations sur la Physique, 4,* 207–219.

Walsh, J. (1774b). Of torpedoes found on the coast of England. In a letter from John Walsh Esq. F.R.S. to Thomas Pennant, Esq., F.R.S. *Philosophical Transactions of the Royal Society, 64,* 464–473.

Walsh, J. (1775). Of torpedos found on the coast of England. *Gentleman's Magazine, 45,* 83.

Walter, R. (1748). *A Voyage Round the World, in the Years MDCCXL, I, II, III, IV Compiled from Papers and other Materials by the Right Honourable George Anson, and Published under his Direction, by Richard Walter, M.A.* London: Printed for the Author by John and Paul Knapton.

Walter, R. (1756). *A Voyage Round the World, in the Years MDCCXL, I, II, III, IV by George Anson, Esq, now Lord Anson ...* London: Printed for D. Browe, J. Osborne, and others.

Walter, T. (1788). *Flora Caroliniana.* London: John Fraser.

Wansleben, J. M. (1678). *The Present State of Egypt, or, A New Relation of a Late Voyage into that Kingdom Performed in the Years 1672 and 1673.* Englished by M.d., B.D. London: Printed by R.E. for John Starkey.

Wansleben, J. M. (1679). *A Brief Account of the Rebellions and Bloudshed Occasioned by the anti-Christian Practices of the Jesuits and other Popish Emissaries in the Empire of Ethiopia/ Collected out of a Manuscript History Written in Latin by Jo. Michael Wansleben, a Learned Papist.* London: Edwin.

Warburg, A. (1912/1922). *Italienische Kunst und internationale Astrologie im Palazzo Schifanoia zu Ferrara.* Paper read at The 10th Annual International Congress of Art History in 1912; printed in 1922 in *L'Italia e l'arte straniera.* Atti del X Congresso Internazionale di Storia dell'arte, 1922. Roma: Unione editrice.

Warburg, A. (1999). *The Renewal of Pagan Antiquity: Contributions to the Cultural History of the European Renaissance.* K. W. Forster, Intro., David Britt, trans. Los Angeles: Getty Research Institute for the History of Art and the Humanities.

Waring, J. I. (1964). *A History of Medicine in South Carolina, 1670-1825.* Spartanburg: The Reprint Company, Publishers.

Warren, G. (1667). *An Impartial Description of Surinam upon the Continent of Guiana in America, with a History of Several Strange Beasts, Birds, Fishes, Serpents, Insects, and Customs of that Colony, &c.* London: Printed by William Godbid for Nathaniel Brooke.

Warren, G. (1669). *Een onpartydige beschrijvinge van Surinam, gelegen op het vaste landt Guiana in Africa [sic for America]....* Amsterdam: Pieter Arentsz.

Watson, W. (1751). An account of Mr. Benjamin Franklin's treatise, lately published, intituled, Experiments and Observations on Electricity, Made at Philadelphia, in America. *Philosophical Transactions of the Royal Society, 47,* 202–211.

Wear, A. (2000). *Knowledge and Practice in English Medicine, 1550–1680.* Cambridge, Cambridge University Press.

Weisheipl, J. A. (Ed.). (1980). *Albertus Magnus and the Sciences: Commerative Essays.* Toronto: Pontifical Institute of Medieval Studies.

Wendorf, F., and Schild, R. (1976). *The Prehistory of the Nile Valley.* New York: Academic Press.

Wellmann, M. (1928). *Die Physika des Bolos Demokritos und der Magier Anaxilaos aus Larissa.* Berlin: Preussische Akademie der Wissenschaften.

Wellmann, M. (1934). *Marcellus von Side als Arzt und die Koiraniden des Hermes Trismegistos.* Leipzig: Dieterich'sche Verlagsbuchhandlung.

Wenke, R., et al. (1983). The Fayum Archeological Project: Preliminary report of the 1981 season. *American Research Center in Egypt Newsletter, 122,* 24–34.

Wenke, R., Long, J., and Buck, P. (1988). Epipaleolithic and neolithic subsistence and settlement in the Fayyum Oasis of Egypt. *Journal of Field Archeology, 15,* 29–51.

Wesley, J. (1760). *The Desideratum: Or Electricity made Plain and Useful.* London: W. Flexney.

Westendorf, W. (1999). *Handbuch der altägyptischen Medizin (Band 2).* Brill: Leiden-Boston-Köln.

Wheeler, A. (1958). The Gronovius fish collection: a catalogue and historical account. *Bulletin of the British Museum (Natural History), Historical Series, 1,* 185–249.

Wheeler, A. (1978). The sources of Linnaeus's knowledge of fishes. *Svenska Linnésällskapet Arsskrift, 1,* 156–211.

Wheeler, A. C. (1985). The Linnaean fish collection in The Linnean Society of London. *Zoological Journal of the Linnaean Society, 84,* 16–76.

Wheeler, A. C. (1989). Further notes on the fishes from the collection of Laurens Theodore Gronovius (1730-1777). *Zoological Journal of the Linnaean Society, 95,* 205–218.

Wheeler, A. C. (1991). The Linnaean fish collection in the Zoological Museum of the University of Uppsala. *Zoological Journal of the Linnaean Society, 103,* 145–195.

Wheeler, W. M. (1926). *R.-A. F. de Réaumur.* New York: Knopf.

White, J. C., and Sweet, W. H. (1969). *Pain and the Neurosurgeon.* Springfield: Charles C. Thomas.

Whitehall Evening Post (1750). [Advertisement for *The Student; or, The Oxford Monthly Miscellany*]. February 27, Issue 632.

Whitehall Evening Post (1769). Natural History and Description of the TORPORIFIC EEL, with a curious relation of its benumbing quality. From Mr. Bancroft's *History of Guiana,* lately published. *Whitehall Evening Post or London Intelligencer,* February 7, Issue 3568.

Whiteway, R. S. (1902). *The Portuguese Expedition to Abyssinia in 1541–1543, as Narrated by Castanhoso, the Short Account of (João) Bermudez, and Certain Extracts from Correa, with some Contemporary letters.* London: Hakluyt Society.

Whittaker, V. P. (1992). *The Cholinergic Neuron and its Target: The Electromotor Innervation of the Electric Ray Torpedo as a Model.* Boston: Birkhauser.

Whittaker, V. P. (1998). Arcachon and cholinergic transmission. *Journal of Physiology (Paris), 92,* 53–57.

Willcox, W. B. (1973). *The Papers of Benjamin Franklin* (Vol. 17). New Haven: Yale University Press.

References

Willcox, W. B. (1975). *The Papers of Benjamin Franklin* (Vol. 19). New Haven: Yale University Press.

Willcox, W. B. (1976). *The Papers of Benjamin Franklin* (Vol. 20). New Haven: Yale University Press.

Willcox, W. B. (1978). *The Papers of Benjamin Franklin* (Vol. 21). New Haven: Yale University Press.

Willcox, W. B. (1983). *The Papers of Benjamin Franklin* (Vol. 23). New Haven: Yale University Press.

Williams, S., and Friell, G. (1994). *Theodosius: The Empire at Bay.* New Haven: Yale University Press.

Williamson, H. (1775a). Experiments and observations on the *Gymnotus Electricus,* or electrical eel. *Philosophical Transactions of the Royal Society,* 65, 94–101.

Williamson, H. (1775b). Experiments and observations on the *Gymnotus Electricus (sic),* or electrical eel. *Gentleman's Magazine,* 45, 437.

Willis, T. (1664). *Cerebri anatome: cui accessit nervorum descriptio et usus.* London: J. Martyn and J. Allestry. (Trans. by S. Portage in 1681 and reprinted as *Thomas Willis, The Anatomy of the Brain and Nerves.* Montreal: McGill University Press, 1965.)

Willis, T. (1672). *De anima brutorum.* Oxonii: R. Davis. (Trans. by S. Pordage as *Two Discourses Concerning the Soul of Brutes in Practice of Physick,* London: Dring, Harper and Leigh, 1683).

Willis, T. (1676–80). *Opera medica & physica.* Genevae: apud Samuelem de Tournes.

Woodcroft, B. (1971). *The Pneumatics of Hero of Alexandria.* London: MacDonald.

Wu, C. H. (1984). Electric fish and the discovery of animal electricity. *American Scientist,* 72, 598–607.

Yates, F. A. (1964). *Giordano Bruno and the Hermetic Tradition.* London: Routledge and Kegan; Chicago: University of Chicago Press.

Yerkes, D. (1993). Franklin's vocabulary. In J. A. L. Lemay (Ed.), *Reappraising Benjamin Franklin.* Newark: University of Delaware Press, pp. 396–411.

Young, J. Z. (1936a). Structure of nerve fibres and synapses in some invertebrates. *Cold Spring Harbor Symposia on Quantitative Biology,* 4, 1–6.

Young, J. Z. (1936b). The structure of nerve fibres in cephalopods and crustacea. *Proceedings of the Royal Society of London, ser. B,* 121, 319–337.

Young, J. Z. (1936c). The giant nerve fibres and epistellar body in cephalopods. *Quarterly Journal of Microscopical Science,* 78, 367–386.

Zaganelli, G. (1992). *La lettera del Prete Gianni.* Parma: Pratiche.

Zand, K. H., Videan, J. A., and Videan, I. E. (1965). *The Eastern Key. Kitab al-ifadah wa'l-i'tibar of 'Abd al-Latif al Baghdadi.* London: George Allen and Unwin Ltd.

Zantedeschi, F. (1838). *Ricerche sul Termo-Elettricismo Dinamico, e Luci-Magnetico ed Elettrico.* Milano: Pirotta.

Zantedeschi, F. (1845). *Trattato dell'Elettricità e del Magnetismo* (Vol. II). Venezia: Tipografia Armena di San Lazzaro.

Zarncke, F. (1879–83). *Der Priester Johannes* (2 vols.) Leipzig: Sächsische Akademie der Wissenschaften.

Zett, L. (1983). J. Bernstein—Leben, Persönlichkeit und wissenschaftliches Werk. In L. Zett and B. Nilius (Eds.), *Bernstein Symposium 1981.* Beiträge zur Universitätsgeschichte der Martin-Luthor-Universität Halle-Wittenberg 32. Halle a. d. Saale, pp. 7–22.

Zimmer, C. (2008). Isolated tribe gives clues to the origins of syphilis. *Science,* 319, 272.

Zimmermann, J. G. (1755). *Das Leben das Herrn von Haller.* Zurich: Heidegger und Compagnie.

Zirkle, C. (1967). The death of Gaius Plinius Secundus (213–79 A.D.). *Isis,* 58, 553–559.

Zophy, J. W. (1997). *A Short History of Renaissance Europe: Dances over Fire and Water.* Upper Saddle River: Prentice Hall.Bonaparte, C-L. (1832– 41). *Iconografia della Fauna Italica per le Quattro Classi degli Animali Vertebrati* (3 vols). Roma: DallaTipographiaSalviucci.Napoli:StamperiaReale.*Gentleman's Magazine.* (1769). *An Essay on the Natural History of Guiana in South America, by Edward Bancroft.* 39, 145–149. (Book Review)Newton, I. (1713). *Philosophiae naturalis principia mathematica.* 2 vols. Cambridge: Cambridge University Press.

Newton, I. (1729). *Mathematical Principles of Natural Philosophy.* 2 vols. translated by A . Motte from the 1687 Latin edition. London: for Benjamin Motte.

Index

Note: Page numbers followed by '*f*' denote figures and footnotes are indicated with '*n*'.

Abassinorum rebus, 116*f*, 119
   Ethiopia's freshwater torpedo (*Malapterurus electricus*), description, 117*f*
'Abbās, Ali ibn al-Majūsi, 79
Abbeville, Claude de, 130
Abd Allatif, 81
Abelard, Peter, 89
Abu-l-Walid Muhammad ibn Ahmad ibn Rushd, 82
Académie (Royale) des Sciences of Paris, 7*n*134, 156–59, 174, 211, 214, 232–35, 244
Académie Royale de La Rochelle, 235, 246, 250, 256
Academies and Scientific Institutions. *See:*
   Académie (Royale) des Sciences of Paris
   Académie Royale de La Rochelle
   Accademia dei Lincei
   Accademia del Cimento
   Accademia delle Scienze di Bologna
   American Philosophical Society
   Istituto delle Scienze di Bologna
   Royal Institution
   Royal Society of London
Accademia dei Lincei, 146
Accademia del Cimento, 147, 148
Accademia delle Scienze di Bologna, 11, 309. See also *Istituto delle Scienze di Bologna*
"Account of Egypt," 81
Acetylcholine, 409–13, 418–20
   muscarinic acetylcholine receptors, 410
   nicotinic acetylcholine receptors, 412, 419
Achard, Franz Karl, 11
Action current, 369, 378, 383, 387, 390, 392, 401
Action potential, 369, 378, 392, 397–99, 400, 401–2, 409–13, 415
Adams, John, 226
Adanson, Michel, 199, 200, 213, 233
   African catfish, 198–200
   shocking freshwater fish, description, 199*f*
Ad-Damiri, Muhammad Ibn Musa Kamal ad-din, 81, 84
Adelard of Bath, 88, 89
Adrian, Edgar Douglas, 397–99, 399*f*, 402
   electric activity of single nerve fibers, 400*f*
Adye, Spehen Payn, 268
Æcumene, 59

Aelian (Claudius Aelianus) of Praeneste, 68*n*, 100*n*, 257
   myths about torpedoes and water, 54–55
   zoological poems, title page of 1562 Latin edition of, 50*f*
Aeneid, 56*n*
Aepinus, Franz, 319
Aesop, 25, 67
   Fables, 25
Aëtius of Amida, 68, 70–71
   medical treatise, 70*f*
African torpedo, western description of, 114, 116, 126
Agriculture/husbandry works, animal description, 91
Agrippa, Henricus Cornelius, 141*f*
Agrippa, Marcus Vipsanius (B.C.), 39
Al-Baghdadi, 'Abd al-Latif (Abd Allatif), 81
Al-Baitar. *See* Ibn al-Bītār
Alberti, Leon Battista, 95
Albertus Magnus (Albert the Great), 78, 92, 93
   criticisms on Aristotle, 90
   promoting nature, 89
   and stupefying torpedo, 91–94
   teaching his pupils, 92
   treatise on animals dealing with torpedo, 93*f*
Albucasis, 73, 149
Alcmaeon of Croton, 179
Aldini, Giovanni, 321, 322*f*, 331, 343
   attempts to apply electricity to the head when treating "mad" patients, 323*f*
   texts on animal electricity, 322*f*
   title page of Aldini's 1804 *Essai*, 322*f*
Aldrovandi, Ulisse, 95, 99–100
   extending Aristotle and Pliny, 107–10
   portrait from *Ornithologiae*, 108*f*
   treatise on fishes, *De piscibus*, 109
Alexander of Aphrodisias, 52, 82, 88, 138
   torpedo's shock transmission, 53
Alexander of Tralles, 68, 72, 74, 75, 77
   and magic of torpedo, 71–73
   medical books, torpedoes in, 72
   therapeutic principles, 72
   torpedo prescriptions, 72
   treatise, 71*f*
   writings, 71, 87
Alexander the Great, 33–34
Alexander VI (Pope), 112
Al-Ghazi, Ahmad ibn Ibrihim (Gragn), 113
Alibard, Thomas-François, 170–71
Alighieri, Dante, 416, 417

De alimentorum facultatibus, 49–50
Al-Kindi, Abu Yusuf Ya'qub ibn Ishaq, 88
Allamand, Jean Nicolas Sébastien, 166, 201–5, 210, 211, 274. *See also* Leyden Jars
   commentary, 204
   letters, published, 211
   and natural history, 201
   publication on eels of Surinam, 203*f*
Allatif. *See* Al-Baghdadi, 'Abd al-Latif (Abd Allatif)
All-or-none principle, 397–99
Almeida, Manoel de, 122, 123, 126
   torporific catfish, rediscovering, 120
Alpha-bungarotoxin, 419
Al-Qazwini, Hamdullah al-Mustafa, 83
Al-Qazwini, Yahya Zakariya' ibn Muhammad, 84
   ambiguities in, 82–83
Alston, Charles, 271, 272
Alvares (Alvarez), Francisco, 113, 128
Al-Zahrawi, Abu al-Qasim, 73
Amenemhat I, 21
American Philosophical Society
   arrival of eel in Philadelphia, 260–61
   and eel, 258
   experiments on electric eel, 262*f*
   Franklin's farewell, 259
   Hugh Williamson, 264–68
   Kinnersley's artificial eel, 264
   perils and rewards of eel trade, 260
   Priestley's synopsis, 258–59
   published notes of experiments, 262–64
   short 1773 newspaper clipping, 261
Ampère, André-Marie, 380
Amphibious bipes, 277*f*
Ancient theories of nerve function, 179–80
Anglicus, Bartholomeus, 90
Animal electricity, 10–13, 195, 238, 278, 281, 284, 340
   Aldini's text on, 322
   Bancroft on, 225
   Galvani's, 307–25
   Galvani's memoir on, 11, 326
   Hodgkin–Huxley model of nerve conduction and, 405
   Humbolt's uncertainty on, 12–13
   Matteucci and, 362–68
   Valli on, 323
   Volta on, 195, 326
   Walsh and, 238, 278, 286

Index

Animal machine (in Galvani's theory of animal electricity), 420
Animal spirits and physiology, 179
　ancient theories of nerve function, 179–80
　Haller's irritability theory, 189–90
　impact of electricity, 188–89
　problems with earlier theories, 185–88
　Renaissance and Early Modern Period, 180–83
　vibrations, 183–85
Animals that Live in Holes, 60
Anson, George, 196, 197–98, 295
　Torpedo revisited, 197–98
Anteros, 46, 47, 88, 256
Antony, Marc, 39, 43
Aphrodisiac (and anti-aphrodisiac) effects of electric fishes, 82, 151
Aphrodite belt, 76
Apollodorus of Athens, 67*n*
Apollonius of Perga, 144
Apology for Raymond Sebond, 98
Apuleius of Madaura, 139*n*
Aquatilium animalium historiae, 110
Aquinas, Thomas, 89, 90, 92, 94, 138, 142, 180
Arabic words in European vocabulary, 65
Arago, Dominique François Jean, 339, 362, 363
Archimedes of Syracuse, 144
Aristotle, 33–38, 92, 179
　De caelo, 53, 82
　inherent sensory attributes of reality, 145
　Meteorologica, 52
　natural philosophy, 33–36
　on Plato and truth, 72, 82
　terrestrial/celestial physics, conceptions, 89
　vacuum existence, 52
　works, translation, 65
Arrhenius, Svante August, 385–86
Artedi, Peter, 202, 205, 208
Artificial electricity, 241, 256, 313, 315–16, 334, 371
Asymmetrical innervation of electrocytes of fish electric organs, 409*f*
Athenaeus of Naucratis, 34*n*, 35, 36, 55, 67, 100, 104, 111
　and bibliomaniac gourmands, 59–61
　title page of 1556 edition of *Deipnosophistae*, 59*f*
Atmospheric electricity, 10, 178, 314, 315
Autobiography of Benjamin Franklin, 170
Averroës. *See* Abu-l-Walid Muhammad ibn Ahmad ibn Rushd
Avicenna, 80*f*. *See* Ibn Sina (Avicenna)

Babuchin, Aleksandr Ivanovich, 395, 408, 414
Bachiacca, Francesco. *See* Ubertini, Francesco
Bacon, Francis (Baron Verulam), 110, 135, 146, 163, 172, 173, 174
Bacon, Roger, 88, 89, 92, 94
Baerle, Caspar van. *See* Van Baerle, Caspar
Bajon, Bertrand, 289
　experiments on electric eel, 290*f*

Baker, George, 260, 281, 294
　eels, 269
Baltasar (Baltazar, Balthasar), Juan de, 118–19, 119*f*, 123
Bancroft, Edward Nathaniel, 259, 274, 295
　authority on electric fishes, 226
　certificate of election to the Royal Society of London, 226*f*
　eel research and medicine, 223–24, 230
　electric theory, 160
　Essay on the Natural History of Guiana, 219–20, 224–25
　fish electricity, 222–23
　frontispiece of Bancroft's (1769) *Essay*, 221*f*
　Guiana eels and London connections, 217
　John Greenwood, 220–21
　live eels, transportation, 224
　London activities after essay, 225–26
　postscript, 228–29
　of spies and spying, 226–28
　torporific eels, 221–22
Banks, Joseph, 10, 11, 212, 268, 275, 279, 335
Barbaro, Ermolao, 100, 107
Barberini, Francesco, 147
Barbeu-Dubourg, Jacques, 231
Barreto, João Nuñes, 113, 119
Bartram, Isaac, 220, 260
Bartram, John, 272
Basilisk, 40*f*, 43, 48, 140*n*, 142
Bassi, Laura, 311
Bastet (sun goddess), 22
Baxt, Nikolai, 376
Beccaria, Giambattista, 169, 188, 189, 247, 358*n*
Becquerel, Antoine Caesar, 351, 358*f*, 363
　torpedo studies, 357
Bede, 88, 111
Behn, Aphra, 133–34, 227, 297
　Oroonoko, 132–33
Belcher, Jonathan, 177
Bell, Charles, 218, 230
Bellini, Lorenzo, 153, 155
Belon, Pierre, 87
　books, torpedoes, 101–3
　De aquatilibus, torpedoes images from, 103*f*
Benedetti, Giovanni Battista, 88
Bennett, Michael V.D. L., 211, 287, 394, 396, 397, 410, 412–14
Berkel, Adriaan van. *See* Van Berkel, Adriaan
Bermudes (Bermudez), João, 113
Bernard, Claude, 233, 361*n*, 410
Bernouilli, Jean, 155
Bernstein, Julius, 320, 378*f*, 382, 388, 389*f*
　ability to measure negative variation, 379*f*
　approach to membrane phenomena, 391*f*
　electrochemical theory of bioelectric phenomena, 384–85
　measurements of nerve impulse, 377–79
　membrane theory, scientists who assisted in, 386*f*
　model of discharge, 407–8
　original figure, 391*f*

　thermodynamics, and membrane theory, 388–90
Bertholon, Pierre, 50, 194, 195
Bertruccio, Niccolò, 149
Bibliomaniac gourmands, 59–61
Bibliotheca universalis, 106
Bident fishing, 20
Bilharz, Theodor, 370
　Malapterurus dissection made by, 371*f*
Bilroth, Theodor, 76
Bishop Braulio, Saint Isidore of Seville with his pupil, 78*f*
Blainville, Henri Marie Ducrotay de, 356
Bloch, Marcus Elieser, 197, 200
Blumenbach, Johann Friedrich, 61–62, 324*f*
Boaz, George, 25
Boerhaave, Herman, 185, 187, 207, 212
Boëthius, Anicius Manlius Severinus, 68, 88
Bohr, Niels, 339
Bois-Reymond, Emil du, 4, 6, 14, 73*n*, 254, 308, 321, 369*f*, 372*f*, 381*f*
　electric fishes, 368–74
　recording electrical currents, 369*f*
Boll, Franz, 396, 411
Bologna *Istituto*, emblem, 310*f*
Bonaparte, Carlo Luciano (Charles-Lucien), 35*n*, 252, 357*f*
Bonaparte, Napoleon, 13, 339, 340, 357
Bonnet, Charles, 213*f*, 214, 274
Bonpland, Aimé Alexandre (Goujoud), 3, 4, 5*f*, 10, 13–14, 221, 252
Book of Healing, The, 79
Borelli, Giovanni Alfonso, 137, 146*n*, 151, 153*n*, 155–57, 186, 198
　De motu animalium, second volume, 187*f*
Botticelli, Sandro, 140
Bowditch, Henry Pickering, 397–98
Boyle, Robert, 143, 146
Bracciolini, (Gian Francesco) Poggio, 143
Brain's cortex, electric impulses, 400*f*
Brazil, 130, 362, 409
　coast, 129*f*
　South American eels, early descriptions from, 127–31
Breasted, James, 22–23
Breschet, Gilbert, 357, 358*f*
Brisson, Mathurin-Jacques, 231
British Apollo, 156
Brodie, Benjamin, 224
Broussonet, Pierre Marie Auguste, 200
Browne, Thomas, 174, 175, 188
Bryant, William, 279, 280
Brydone, Patrick, 177
Bry, Theodoris de, 128*f*
Bull, William, 271
Buonarroti, Michelangelo, 95
Burdon-Sanderson, John Scott, 392, 395
Buridan, John, 89
Burke, Edmund, 225
Butterfly (*Saturnia pyri*) from botanical image depicting *Euphorbia*, 101*f*
Byzantine writings, 64
　Aëtius of Amida, 70–71
　Alexander of Tralles and magic of torpedo, 71–72

Index

Early Middle Eastern writings, 79–82
encyclopedias and cultural transmission to Middle Ages, 67–68
Islamic writers, 83–84
Kyranides, 75–77
Marcellus of Bordeaux, 69–70
Paulus Aegineta, 73–75
Roman Empire's end/dark ages beginning, 64–66
Byzantine Empire, 64–65
cultural exchange with Eastern Empire, 65
Byzantine Era, 73

Cabral, Pedro Álvares, 127, 128*f*
Cajal, Santiago Ramón y, 417
Calcidius, 54
Caldani, Marc'Antonio, 311
Caligula (Gaius Julius Caesar Augustus Germanicus), 40*n*
Callistus, Gaius Julius, 45
Campanella, Tommaso, 140, 142
Canon of Medicine, The, 80
Canton, John, 164, 254, 288
Capella, Martianus, 68*n*
Cardano, Girolamo, 142, 233
Cardim, Fernão, 130
Carlisle, Anthony, 341
Carminati, Bassiano, 327
Carpi, Pietro, 360
Casseri, Giulio Cesare, 96*n*, 97*f*
anatomical plate of organs of speech, 97*f*
Cassini, Gian Domenico (Jean-Dominique), 134
Cassiodorus, Flavius Magnus Aurelius, 68
Castanhoso, Miguel de, 113
Catfishes, electric *See also* Catfish of the Nile, *Malapteruridae, Malapterurus electricus*, Torporific African catfishes, Torporific catfishes, rediscovering
and chisel, 23
depiction, 20*f*
in mythology, 22
stylized images of, 23
surprising, 414–15
thunder, origin, 19
vs. ra'ad, 19
Catfish King, 24
Catfish of Nile, shocking, 19
depictions, 19–22
Ebers Papyrus, 22–23
Hieroglyphica of Horapollo, 24–27
magic, religion, and tomb art, 22
mummies and sacred animals, 28
Narmer's catfish hieroglyph, 23–24
Cavallo, Tiberius (Tiberio), 284, 286
book, 284–87
Cave, Edward, 170, 294, 295
The Gentleman's Magazine, cover page of first issue of Edward Cave's, 294*f*
Cavendish, Charles, 11, 14, 240, 243, 246, 247, 249, 343, 347
Cavendish, Henry, 253*f*, 256, 264, 281, 296
drawings of artificial torpedo, 255*f*
Cayenne, 5, 134–35, 283, 289
Cell membrane, 382, 385, 389, 401
Cell theory, 320, 382

Celsus, Aurelius, 68, 97
sixteenth-century edition medical treatise, 46*f*
Cerebri anatome, 182*f*
Cesi, Federico, 147
Chagas, Carlos (Jr.), 408, 409, 411
Chagas, Carlos (Sr.), 409
Chagovec, Vasilij Jur'evich, 386
Chang, Chuan-Chiung, 419
Charcot, Jean-Martin, 76
Charles II (of England), 132, 172, 227
Charles Towne, 258–59, 260, 264, 271
electric eels in, 269–70
Charles XI (of Sweden), 191
Chaucer, Geoffrey, 79
Cheyne, George, 183
Choffin, David Etienne, 191, 198–99
general culture work of, 199*f*
Christianity, 24, 49, 64, 65, 69, 113–14, 117, 127, 140, 417
Cicero, Marcus Tullius, 36*n*, 45, 67
Cigoli, Ludovico, 95, 111
Cimento, 148
Claudian (Claudius Claudianus), 45, 51, 55–56, 57–58, 133, 196, 238
poems, 59*f*, 239*f*
Claudino, Giulio Cesare, 149
Clearchus of Soli, 35, 38, 60, 70
Clive, Robert, 230
Cobo, Bernabé, 130
Coin (ancient) showing Roman Emperor Justinian I, 65*f*
Colchicum, 75*n*
Colden, Cadwallader, 271, 272
Cold venom theory, 50–51, 57, 58, 193
Cole, Kenneth, 103, 401
Colladon, Jean Daniel, 356, 370
Collegio Romano, 147
Collinson, Peter, 168, 170, 171, 175, 205, 271, 272
Columbus, Christopher (Cristoforo Colombo), 108, 111, 127
Combe, William, 301
Commentaries *vs.* original works, 68
Commentarius (Ludolf's), 123, 308, 364
title page, 124*f*
Commodus, 56
Compendiosa brevitas, 68
Complete Treatise of Electricity, 286*f*
Compositiones medicae, 45–46, 50
Condamine, Charles-Marie de la, 157, 158, 159, 232–33
Relation Abregée d'un Voyage...., 158*f*
Configliachi, Pietro, 326, 343, 345, 346
Conservative priest/Florentine polymath/ torpedo, 146–48
Conservative revolution, 391–93
Constantinus Africanus, 79
Cook, James, 275
Copernicus, Nicolaus, 96, 135, 339
Copley Medal
in 1774 on fish electricity, Pringle to Walsh, 25
in 1753 to Franklin for his studies on electricity, 171
to Rumford, 228
in 1757 to Charles Cavendish, 253
1794 to Volta, 331

Cornwallis, Charles, 279
Corpus Aristotelicum, 35, 37*f*
Corpus Hermeticum, 138, 139*f*
Corpus Hippocraticum, 32, 46, 51
title page, 31*f*
and torpedoes, 31–33
Coyau, Gabriel, 236, 239
Croone, William, 182, 183
Cruickshank, William, 342*f*
Cultural legacies, earlier, 87–89
Culture of Renaissance, 94–98
Cunaeus, Andreas, 166
Curtis, Howard, 401
Cuvier, Georges Leopold Chretien Frédéric Dagobert, 43, 99
Cybulski, Napoleon Nikodem, 386

D'Abano, Pietro, 89
D'Abbeville, Claude d', 130
Da Gama, Cristóvão, 113
Da Gama, Vasco, 113
D'Alibard, Thomas-François. *See* Alibard, Thomas-François
Dante. *See* Alighieri, Dante
Dapper, Olfert, 123
Dark ages, 64–66
Darwin, Charles, 14, 156, 302, 395, 409
Darwin, Erasmus, 302*f*
Da Vinci, Leonardo, 95, 101, 142*n*, 180
Davy, Humphry, 227, 326, 341–43, 342*f*, 353
article on torpedo, initial page, 354*f*
Davy, John, 355–56, 355*f*
first article on torpedo, 355*f*
Deane, Silas, 217, 225
De animalibus, 92
De animalium natura, 54
De aquatilibus
torpedoes images from, 103*f*
De arte venandi cum avibus by Emperor Frederick II, 90*f*
De Beauvais, Vincent, 90, 91, 111
Speculum naturale, thirteenth-century codex of, 91*f*
De Bry, Theodore (Theodorus), 128*f*
De coelo, 53
De fabrica corporis humani, 96*f*
Degree of electricity, 164
De historia stirpium, 97*f*
Deipnosophistae, 55, 59, 60
title page of 1556 edition of Athenaeus', 59*f*
Della Francesca, Piero, 95
Dell'uso e dell'attività dell'Arco Conduttore nelle contrazioni dei muscoli, 331*f*
De locis affectis, 51
De magnete, 163*f*
De materia medica, 47–48, 65
De medicamentis, 69
De' Medici, Lorenzo di Pierfrancesco, 138
De medicina, 68
Demerara, 218, 221
Demetrius of Ixion, 60, 67
Democritus of Abdera, 146
De motu animalium, 155, 187*f*
Dengel, Lebna (Emperor of Ethiopia), 113
De occulta philosophia, 141*f*
De puero epileptico, 50

De rerum natura, 143
De romanis piscibus, 99
    Paolo Giovio with page from, 97*f*
Descartes, René, 116, 143, 151, 156, 180*f*, 181, 186
    cross-section of brain by, 181*f*
Desideratum: Electricity Made Plain and Useful, 177
De symptomatum caussis, 51
De torpedine, 57
De usu respirationis, 51
De viribus, 313*f*, 315*f*
De vita, 140
De Vries, Hugo, 386*f*
Dialogo sopra i due Massimi Sistemi del Mondo, 53
Dialogues of Plato, 33
Dias, Bartolomeu, 112
Dickens, Charles, 416
Dietetic food, torpedo as, 45
Dioscorides of Anazarbos, 48
    Materia medica, 48*f*
    medical works, 47*f*
Diphilus of Laodicea, 55, 60, 104
    and experimentally localizing torpedo's power, 55
Discontinuous vacuum, 52
Discorsi e Dimostrazioni matematiche sopra due nuove Scienze, 53
Dos Santos, João, 114–15, 116, 119, 120, 122, 126
Dryden, John, 132
Du Bois-Reymond, Emil, 4*f*, 6*f*, 14, 73, 254, 356, 359, 366, 368, 369
    electric molecules, 379–82
Dubourg, Jacques Barbeu. *See* Barbeu-Dubourg, Jacques
DuFay, Charles François de Cisternay, 169
Dumas, Jean-Baptiste, 362, 367
Duns Scotus, John, 89
Dürer, Albrecht, 25, 95, 101, 105
Dutch colonies of Guiana, 219*f*
Dutch, eel, electricity, 201
    conversions, 213
    Dutch collectors and zoological cabinets, 205–7
    epilogue, 214
    fish therapeutics, 210–11
    Frans Van Der Lott's letter, 207–10
    Pieter Van Musschenbroek's endorsement, 211
    Storm Van's Gravesande and Jean Allamand, 201–5

Early Middle Eastern writings, 79–82
Eastern Empire, 65
    cultural exchange with Byzantine Empire, 65
    Greek language and culture, 65
Ebers, Georg, 19, 23, 73
Ebers Papyrus, 22–23
Eccles, John Carew, 403, 413
Edwin Smith Papyrus, 23
Eels. *See also* Electric eels
    arrival in Philadelphia, 260–61
    capturing, 5–7
    electrical sense of, sparks in darkness and, 287–88
    electric fishes comparisons with, 361–62
    and medicine, 223–24
    trade, perils and rewards of, 260
    transportation, live, 224
Electrical sparks, creation, 164
Electrical world of Benjamin Franklin, 163
    foremost among electricians, 170–71
    Franklin and electricity, 168–70
    history of electricity, 163–65
    Leyden jar, 165–67
    medical electricity, 175–78
    myths, 174–75
    new *Zeitgeist*, 178
    Royal Society of London, 171–74
    vulgar errors scrutinized, 174
    wonders and amusements, 167–68
Electricam, 163
Electric amplifier, extremely sensitive, 349–50
Electric atmospheres, 197
Electric Eel, The, 299*f*
Electric eels, 5
    in Charles Towne, 269–70
    cross-section of, 278
    experiments on (article), 262*f*
    Hunter's illustration of intact, 278*f*
    as illustrated in Seba's *Locupletissimi rerum naturalium . . .*, 208*f*
    remedies, 23
Electric fishes
    allure of, 3–5, 7–8
    authority on, 226
    capturing eels, 5–7
    Edward Bancroft's authority on, 226
    electric organs in weakly, 395–97
    Humboldt's initial forays with electricity, 11–12
    membranes, and cooling battery, 390–91
    need for, 12–13
    need for electric fish, 12–13
    odyssey into past, 14–15
    philosopher with scientific agenda, 8–10
    South American experiments, 13–14
    Volta's artificial electrical organ, 10–11
Electric fishes in nineteenth century, 351–52
    comparisons with eel, 361–62
    Emil Du Bois-Reymond's electric fishes, 368–74
    great men, notable failures, and torpedinal electricity, 352–55
    John Davy, 355–56
    Linari's more complete picture, 360–61
    Matteucci and animal electricity, 362–68
    polarity of torpedo's discharge, 356–57, 358–60
    torpedo's spark, 358–60
Electric fishes in Volta's path to battery, 326
    Alessandro Volta: from love obsession to science lust, 327–29
    emerging doubts, 329–30
    Galvani's contractions without metals, 331–32
    Galvani's torpedoes, 333–35
    multiplying weak electricity, 332–33
    Nicholson's model, 335
    Volta *contra* Galvani, 330–31
    Volta's artificial electrical organ, 335–38
    Volta's extension of theory of metallic electricity, 331–32
    Volta's reaction, 335
Electric fluid, 232, 261
Electric instruments
    capillary electrometer, 392, 400, 408
    condensatore (Volta's instrument), 329, 332, 333, 345, 349, 350, 353, 360
    electric machines, 313, 327, 335, 390, 415
    electrometers, 3, 4, 9, 238, 245, 256, 327, 333, 345, 349, 377, 408
    galvanometers, 351, 374, 380
    Nicholson's duplicator or multiplier, 333, 349, 350
    sensitive electrometers, 327, 333, 345, 408
Electricity, 54
    application, sick or disabled patients, 176*f*
    etymology. *See Electricam*
    history of, 163–65
    impact of, 188–89
    speculation about, 45
Electric molecule model (of du Bois-Reymond), 374, 375, 379–82, 384, 385
Electrico-libertine poetry from 1777, 301*f*
Electric organs in weakly electric fishes, 395–97
Electric ray, 8*f*
Electric virtue, 291
Electrifying cells of electric organs, 406–7
Electrochemical potential, 385, 392
Electrocytes, 394, 406–7, 409, 411–15
Electrophore, 318–19
Electrophysiology, 14, 320, 325, 350, 369, 378, 384–85, 395, 398, 411–14
Electrotherapy *ante litteram*, 45
Emperors of Late Roman period, 69*f*
Empiricus (Marcellus), 69
Enarrationes, 107n
Encyclopedias and cultural transmission to middle ages, 67–68
Enneades, 53, 54
Epic battle between horses and eels, *Llanos*, 6*f*
Epicharmus of Megara, 60
Epicurus, 52, 145
Epitomes iatrikes biblio hepta, 73
Erastus, Thomas
Essai (Aldini's work on animal electricity), 322*f*
    experiments on humans after execution, 323*f*
Essay on the Natural History of Guiana in South America, 218
Essequibo (or Essequebo), 134, 201, 202, 203, 207, 218, 219, 221, 223
Ethiopia, 111, 121*f*
    in 1500s, 112–13
    chapter on from *Purchas his Pilgrimage*, 123
    Christianity and religious intolerance in, 125–26
    freshwater torpedo's description, 117*f*
    second Jesuit mission to, 114–15
    torporific catfishes in, 8

Index

Ethiopia Oriental
  title page of, João Dos Santos', 114f
Etymologies, 67, 77–78
Euclid, 66n, 144
Eugene of Palermo, 66n
Euphorbia, butterfly from botanical image depicting, 101f
Evans, Cadwalader, 177

Fabiano, Alexander, 147
Fabrica corporis humanis, 96
Fabricius (ab Acquapendente), Hieronymus (Girolamo Fabrici d'Acquapendente), 96, 97
Fabri, Giacinto Bartolomeo, 311
Faraday, Michael, 227, 341–42, 343f, 352, 353f, 354, 358, 359, 361, 373
Feldberg, Wilhelm, 411, 418, 419
Felici, Costanzo, 109–10
Female hysteria, 51n
Fermin, Philippe, 131, 133, 211, 260
Fernandes (Fernández), Antonio, 115–16
Fessard, Alfred, 8, 291, 396–97, 410, 411
Fibonacci, Leonardo, 88
Ficino, Marsilio, 110, 138, 139f, 140–41
Finch, John, 198
Fish books, illustrated, 100–101
Fish electricity, 222–23
Fish electricity, first steps toward, 191
  Adanson's African catfish, 198–200
  flounder electrical, 194–95
  Ingram's 1750 publication, 195–97
  Kaempfer's analogy to lightning, 191–94
  Walter and Anson's torpedo revisited, 197–98
Fishes symbolizing different things in tomb paintings, 22
Flagg, Henry Collins, 211, 260, 279, 280
Flaubert, Gustave, 416
Fleischl-Marxow, Ernst von, 382
Flounder electrical, 192, 194–95
Fontana, Felice, 311, 404
Fordyce, George, 225
Forlì, Jacopo da, 89
Forsskål, Petrus, 82
Fortress in harbor of La Rochelle on France's western coast, 235f
Fothergill, John, 170, 171
Fowke, Arthur, 235
Fracastoro, Girolamo, 142, 143
Francesco I (de' Medici), 101, 149
François de Tournon (Cardinal), 102, 103
Frank, Johann Peter, 335
Franklin, Benjamin, 9, 23, 38, 47, 143, 168f, 194, 223–25, 230–32, 243–44, 246 249, 252, 259, 264, 266, 272
  1774 edition of pamphlet on electricity, 172f
  electrical world of, 163–78
  and electricity, 168–70
  experiments, 236–38
  first page of original letter sent by Walsh to, 241f
  foremost among electricians, 170–71
  Franklin's farewell, 259
  Franklin's kite experiment, 171f
  illustration of his electrical equipment, 169f
  and John Walsh, 230–32
  La Rochelle, setting up at, 234–36
  letters from, 242
  letters (Walsh's) to, 241–42
  magic square, flat version of first electric capacitor, 169f
  sentry box with tall pole for capturing electricity, 170f
  trip to La Rochelle, 232–34
Franklin squares, 9
Frederick I, Barbarossa, 66, 118
Freising, Otto von, 117
French Guiana with Island of Cayenne, 135f
Freud, Sigmund, 76, 374
Frigida podagra, 47n
Fritsch, Gustav, 371, 372, 408
Frog nerve fiber, 349f
Froschunterbrecher (in du Bois-Reymond's experiments), 370, 371
Froschwecker (in du Bois-Reymond's experiments), 370, 371
Frost (*frigus*), 51
Frumentius (Saint), 113
Fuchs, Leonhart, 95, 96, 98f
Fumagalli, Pietro, 110

Gabriel, João (Giovanni Gabrielli), 115, 119, 120
Galeazzi, Domenico Gusmano, 310f
Galeazzi, Lucia, 308, 310f
Galen, 29, 71–73, 79, 80f, 83–84, 143f, 174
  cold venom theory, 50–51
  title page of Venetian edition of works, 50f
  torpedinal medicine, 49–50
Galilei, Galileo, 53n, 110, 116, 138, 339, 417
Galvani, Luigi, 308f
  contractions without metals, 331–32
  contra Volta, 330–31
  De viribus electricitatis in motu musculari, 309f
  experiment on *contrazioni senza metallo*, 332f
  experiments with metal arcs, 316f
  father in law, 310f
  frog legs on hooks experiment, 316f
  frog preparation, 317f
  Leyden jars used in experiments, 317f
  nerve contractions, experiments on, 332f
  similarity of frog preparation/Leyden jar, 320f
  torpedoes, 333–35
  wife, 310f
Galvani and Volta
  controversy between, 11, 324–25, 330–32
  reciprocal dependency in their work, 326, 328
  stereotypes, 308
Galvani's animal electricity, 307–8
  electric fish as Galvani's starting point, 308
  electrophore and tourmaline, 318–19
  experiments with metals, 314–18
  Galvani as physician, 320–21
  Galvani's experiments, 312–14
  Galvani's Leyden jar model, 319–20
  The *Istituto*/The *Zeitgeist*, 309–12
  misrepresentations of Galvani, 308–9
  sunshine and impending storms, 321–25
Galvanism *Contra* Voltaism, 339
  extremely sensitive electric amplifier, 349–50
  Galvani, Volta, and unsolvable dilemma, 346–49
  Humboldt's conversion and Volta's impact, 339–41
  imitating torpedoes, 344–46
  instrument for the ages, 341–42
  nagging issue of genuine electricity, 342–44
  on electricity, 346
Galvanism, term usage, 342f, 344f
Gama, Vasco da. *See* Da Gama, Vasco
Garden, Alexander, 260, 264, 270–71
  American postscript, 279–80
  amphibious bipes, 277f
  collector, 271–73
  electric eels in Charles Towne, 269–70
  experiments on electric eel, 296f
  fate of five eels, 276
  Hunter's eel dissection, 276–79
  letter on eels, cover page, 275f
  Linnaean in South Carolina/Captain Baker's eels, 269
  Linnaeus, fishes, and unusual specimens, 273–74
  observations, experiments, and Garden's 1774 letter, 274–76
  right man for eels, 274
Gardini, Francesco Giuseppe
Garten, Siegfried, 411
Gate control theory, 47
Gaubius, Hieronymus David, 207
Gay-Lussac, Joseph Louis, 353f
Gaza, Theodorus (Theodore Gazis)
  initial page of Corpus Aristotelicum, 37f
  portrait of, 37f
Gazette de France, 244f
Gelawdewos (Claudius) of Ethiopia, 113
Gelido veneno, 58
Gentleman's Magazine. *See The Gentleman's Magazine*
Geoffroy Saint-Hilaire, Étienne, 395
Geographia, 66n
George III (of England), 227
Gerard of Cremona, 65, 73
Gervase of Tilbury, 90
Gessner, Conrad, 100, 101, 105–7, 109, 110
  massive ichthyology work, 104–7
  portrait, 105f
Gilbert, William, 163f
  frontispiece for Gilbert's *De Magnete* of 1600, 164f
Gilles, Pierre, 100, 105
Giorgi, Eusebio, 361
Giovio, Paolo, 30n, 99f, 100, 102n, 113
  book on fishes, 99–100
Girardi, Michele, 356
Glisson, Francis, 185, 186
Glycas, Michael, 37n, 54
Gmelin, Johann Friedrich, 8f
Gnosis, 25

Godigno, Nicolao, 115–16, 119, 123, 126, 151
  Abassinorum rebus, 116*f*, 119
  book, 119
Goethe, Johann Wolfgang von, 13, 54*n*, 416
Góis, Damião de, 113, 125
Golden Trade: Or, A Discovery of River Gambia, The, 121*f*
Goujaud, Aimé Alexandre. *See* Bonpland, Aimé Alexandre (Goujoud)
Gout, 46, 47. *See also* Podagra
  and arthritis, 73
Gout and Arthritis treatise, 73
Gragn. *See* Al-Ghazi, Ahmad ibn Ibrihim (Gragn)
Gratian (Gratianus), Flavius
Gray, Stephen, 164
  suspended boy experiment, 166*f*
Greenwood, John, 220–21
Gregory, Richard, 108
Gregory XIII (Pope), 108
Gren, Friedrich Albrecht Carl, 332
Grévin, Jacques, 143
Gronovius, Jan (Johann) Fredrick, 205, 213, 272, 273*f*
Gronov, Laurens Theodor, 205–7, 208, 208*f*, 213, 217, 224, 258
  illustration of South American eel, 209*f*
Grosseteste, Robert, 88, 89
Grundfest, Harry, 5, 8, 408
Guericke, Otto von, 164*f*
  1672 treatise with page illustrating his electric experiments, 165*f*
Guerreiro, Fernão, 115
Guiana, 127, 129*f*, 131, 134, 135*f*, 211, 223
  Dutch colonies of, 219*f*
  eels, 217
  with Island of Cayenne, French, 135*f*
Gumilla, Joseph, 159
  book on Orinoco region of South America, 159*f*
Gutenberg, Johannes, 36
Guyana, 3, 5, 14, 127, 131, 361
Gymnotus electricus (electric eel), 5, 7*f*, 35, 72*n*, 212, 217, 262, 275, 277, 281–284, 286, 294, 299–300

Hairstylist, The, 60
Hakluyt, Richard, 122
Hales, Stephen, 188*f*, 271
Halieutica, 56
  Florentine codex dated 1478 of Latin edition of Oppian's, 58*f*
  sixteenth-century Greek manuscript of Oppian's, 57*f*
Haller, Albrecht von, 168, 186, 187*f*, 213–214, 283–284, 308, 311–314
  irritability theory, 189–90
Hallerian irritability, 313, 318, 320
Hamilton, William, 45, 247, 354
Harpocration of Alexandria, 75
Harpoon fishing. *See* Bident fishing
Hartley, David, 172, 184, 232, 353
Harvey, William, 131, 173

Hawksbee, Francis, 164, 165
  frictional apparatus for producing electricity, 165*f*
Hearing faculty, 53
Hellenes, 33
Hellenistic period, 33, 67
Helmholtz, Hermann, 376*f*, 378, 379, 382, 388, 391
  apparatus used in first measurements of nerve conduction speed, 377*f*
  and speed of nerve conduction, 375–76
  tracings of measurement of nerve conduction, 377*f*
Hemicrania continua, 23
Hemorrhoids, 48
Henly, William, 254, 288
  two letters on electric fish experiments, 289*f*
Henrique o Navegador, 112
Heraclides, Ponticus, 67
Hering, Karl Ewald Konstantin
Hermann, Ludimar, 383*f*
  alteration theory, 382–84
  local circuit theory, 383*f*
Herman of Carinthia, 185
Hermetic corpus, 139
Hermetic thinking, 138–41
Herodotus of Halicarnassus, 114
Hero of Alexandria, 52
Her-watet-khet (Watet Hathor)
Heterobranchus bidorsalis, 28
Hewatt, Alexander, 271
Hewson, William, 249
Hierakonpolis, 23
Hieroglyphica of Horapollo, 24–25
  title page, 26*f*
Hieroglyphica (Valeriano), 25–26
  frontispiece and initial dedicatory page of, 27*f*
  frontispiece of 1615 French edition of, 27*f*
Hippocrates, 32, 80*f*
  coin from island of Cos depicting head of, 31*f*
  depiction in coin, 31*f*
Histoire de la Mission des Pères Capucins en l'Isle de Maragnan et Terres Circonvoisines, 130*f*
Historia Aethiopica, 123–24
Historia animalium, 26, 34, 35, 36, 37, 43, 105
  unicorn, 106*f*
Historia del Nuevo Mundo, 130
Historia eclesiastica, 117*f*, 118
Historia geral da Ethiópia a Alta, 120*f*
Historia naturalis Brasiliae, 131*f*
Historiae animalium, 105
Hodgkin, Alan Lloyd, 401*f*, 402*f*
  Hodgkin cycle, 404*f*
  and nerve impulse, 399–405
  nerve signaling, passive conduction of electricity, 347*f*
Hodgkin–Huxley experiments, 403
Hodgkin–Huxley model of nerve conduction, 405*f*
Hohenheim, Theophrastus Bombastus von (Paracelsus), 181

Hollingsworth, Levi, 260
Hondius, Jodocus, 129*f*
Honorius, 57
Hooke, Robert, 146
Hopkinson, Thomas, 49
Horapollo Niloticus, 24
Hosack, David, 264
Hôtel de Ville (town hall) in La Rochelle, 240*f*
Humboldt, Alexander von, 3, 5, 5*f*, 6, 8–9, 11, 13, 60, 221, 262, 339, 352*f*, 361
  brother. *See* Humboldt, Wilhelm von
  experiment, 352
  illustration of fishes studied in Venezuela, 7*f*
  initial forays with electricity, 11–12
  use of roots to capture lake and river fishes, 221
Humboldt, Wilhelm von, 13*f*
Hunter, John, 11, 195, 213, 240, 246, 249*f*
  anatomical study of electric eel, first page, 278*f*
  close-up of cross-section, 251*f*
  dissections using torpedoes obtained from Walsh, 251*f*
  eel dissection, 276–79
  illustration of intact electric eels, 278*f*
Hunter, William, 249
Huxley, Andrew Fielding, 347, 392, 394, 401–2, 403*f*
Huygens, Christian, 134

Ibn al-Bītār, 84
Ibn Rushd, Abdul Walid Muhammad ibn Ahmad. *See* Abu-l-Walid Muhammad ibn Ahmad ibn Rushd
Ibn Sina (Avicenna), 79, 81, 83, 84
Ichthyology, 94, 98, 99, 102
  Renaissance interest in, 94
Ichthyology work, Gessner's massive, 104–7
Iliad, 56*n*
Imhotep, 20
Induction coil used by Santi Linari, 359*f*
Ineb-Hedj (The White Wall), 19–20
  Malapterurus -like images, 19
Inflammations, treating, 23
Ingenhousz, Jan, 4, 233, 246– 249, 248*f*, 253, 255, 284
  spark experiments with eel description, 288*f*
Ingram, Dale, 195–98, 217, 417
  Publication on Torpedo [i.e. electric eel], 195–97
Innocent IV (Pope), 118
Interpretation of the Dreams, 56*n*
Ionic channels
  ligand-gated channels, 410–11
  potassium channels, 411
  sodium channels, 411, 419
  voltage-gated channels, 411
Ions, cells, membranes, theories, and Nobel Prizes, 385–87
Isaac, Antoine. *See* Silvestre de Sacy, Antoine Isaac, Baron
Isidore of Seville, 66, 68, 77–79
Islamic writers, 73–84
Islam, rise, 66

Istituto delle Scienze di Bologna, 309–12, 406. *See also* Accademia delle Scienze di Bologna
  Commentarii of the, 11, 309

Jallabert, Jean, 175
James I (of Britain), 172
Jaucourt, Louis de, 160
Jefferson, Thomas, 226
Jesuit mission to Ethiopia, second, 114–15
Jobson, Richard, 121–24
  cover page of *The Golden Trade*..., 121*f*
  Gambia, 121–22
Johnson, Samuel, 270*n*, 295
Joint swelling/pain, avoiding, 73*n*
Jones, John Paul, 32, 39, 228
Jovian (Iovianus), Flavius, 69
Juba II (King of Numidia), 42*n*
Julius III (Pope), 113
Jussieu, Bernard de, 233
Justinian, Flavius Anicius (Justinian the Great), 64
Justinian I (Roman Emperor), ancient coin showing, 65*f*

Kaempfer (Kämpfer), Englebert, 191–94, 192*f*
  analogy to lightning, 191–94
  cover page to 1712 book, 192*f*
  illustration of *Torpedo sinus persici*, 192*f*
Kant, Immanuel, 170
Katz, Bernard, 402*f*
Kepler, Johannes, 339
Keymis, Lawrence, 131
Keynes, Richard Darwin, 5, 373, 394, 409–10, 410*f*, 411, 412, 414
Khnum-Hotep, 21
Kinnersley, Ebenezer, 260, 262, 264
  artificial eel, 264
Kircher, Athanasius (Atanasio), 120, 136, 147, 148, 150, 151, 155
Kirchhoff, Gustav Robert, 254
Kleist, Ewald Georg von, 166
Kolben, Peter, 198
Kratzenstein, Christian Gottlieb, 175–76
  successes of medical application of electricity, 176*f*
Krüger, Johann Gottlob, 175
  book on electricity and its utility in medicine, 176*f*
Kühn, Karl Gottlob, 50, 51
Kuhn, Thomas, 137
Kulliyat, 82
Kyranides, 64, 68, 72, 75–76, 75–77, 82, 87, 102

Lafayette, Marquis de (Gilbert du Motier), 226
Laghi, Tommaso, 188
Lake Tana, 8, 115
Lane, Timothy, 233, 238
  electrometer and usage with Leyden jar, 234*f*
Langrish, Browne, 188
La Roche, Daniel de. *See* Roche, Daniel de La

La Rochelle, 134, 154, 156, 232–33, 234–36, 238, 240, 243–44, 250, 251, 334, 356
Late Travels of S. Giacomo Baratti, The, 125*f*
Laurens Storm van 's Gravesande, 203*f*
Leo III, Pope, 65
Le Roy, Jean-Baptiste, 232, 241
  Walsh's landmark spark experiment with eel, 285*f*
Leucippus, 52, 143
Le Verrier, Jean Joseph, 367
Leyden jars, 165–67, 166*f*
  Galvani's, 319–20
  with metallic arc connects inner and outer plates, 166*f*
  shocks to those of eels of Surinam, 204
  used by Galvani in his experiments, 317*f*
Lezzione, 95
L'histoire Naturelle des Estranges Poissons Marins, 102
Liber de medicamentis, 69
Libri Medicinales, 70, 71
Linari, Santi, 358–63
  induction coil used by, 359*f*
  letter to physicist Eusebio Giorgi, 361*f*
  mercury tube apparatus used by, 360*f*
Lining, John, 163, 170, 271*n*, 274
Linnaean in South Carolina, 269
Linné (Linnaeus), Carl, 270*f*
  cover of first (1735) edition of *Systema naturae*, 272*f*
  fishes, and unusual specimens, 273–74
Lissmann, Hans Werner, 396–97
Literary digression, 98–99
Litteromantic, 75
Llanos, epic battle between horses/eels, 6*f*
Lobo, Jerónimo, 115
Local circuit theory (of Hermann, in nervous conduction), 383, 399, 400, 401
Locke, John, 117, 294, 295
Locupletissimi rerum naturalium thesauri accurata descriptio, 207*f*
Logan, James, 176–77
Longobardia, 65*n*
Lorenzini, Stefano, 87, 145–46, 150–56, 160, 231
  Osservazioni intorno alle Torpedini, anatomical drawings from, 152*f*
Louis XV (of France), 167
Lovejoy, Lucretia. *See* Perry, James
Lovell, Robert, 49, 174
Lower, Richard, 168
Lucas, Keith, 397, 398*f*
  step-like function in muscles, 398*f*
Lucretius, (Titus Lucretius Carus), 68, 143, 144
Ludolf, Hiob, 83, 123, 124*f*, 126
Lunary fishes, 141*f*
Lyceum (School), 34, 35, 52

Magalhães, Fernão, 112
Magellan, Ferdinand (Fernão Magalhães), 112
Magic
  allegations, 141

  and occult qualities, 110
  religion, and tomb art, 22
  Renaissance, 110, 138, 141, 142
Magna curia, 66
Magneticum naturae regnum, 147
Magnus, Albertus. *See* Albertus Magnus (Albert the Great)
Maimonides, Moses (Rabbi Moshe ben Maimon), 81
Makeda (Queen of Sheba), 113
Malabar, 83
Malapteruridae (family: electric catfishes), 19, 25
Malapterurus electricus, 19, 61. *See also* Catfish of the Nile, Catfishes, electric, *Malapteruridae*, Torporific African catfishes, Torporific catfishes, rediscovering
  dissection made by Theodor Bilharz, 371*f*
  nineteenth-century popular treatise, 7*f*
  in Pliny, 42
  to wider audience, bringing, 122–25
Malphigi, Marcello, 183*f*
Mandeville, John, 91
Manuel I (King of Portugal), 112
Manuzio, Aldo, 97
Marcellus (Empiricus) of Bordeaux, 68–69, 69–70, 70, 74, 77
  vs. Scribonius, 69
Marcgraf, Georg, 130–31, 202
Marey, Etienne-Jules, 374
Mariette, Auguste, 20
Marlowe, Christopher, 303
Marmont, George Heinemann, 402
Marsili, Luigi Ferdinando, 311
Martens, Conrad, 409
Massari, Francesco (Francesco Maser), 100*n*
Mastaba (table-like) Tomb of Ti, 20
Matteucci, Carlo, 364*f*
  animal electricity, 362–68
  galvanometric measurements of animal electricity, 367*f*
  induced-twitch experiments with prepared frogs, 368*f*
  Nobili-type astatic galvanometer, 366*f*
  physiological and chemical effect from the torpedo's shock, 364*f*
Mattioli, Pier'Andrea, 30*n*, 107*n*
Maty, Matthew, 245
Mauricius, Jan Jacob, 195
Maurus, Rabanus, 90, 91*f*
Maximilian (Holy Roman Emperor), 25
Maximilian, Holy Roman Emperor, 25
Maxwell, James Clerk, 253–55
Mayow, John, 183
Medical electricity, 175–78
Medical formulas/treatments, 94
Medical treatises (1567 edition), 68*f*
Medicine
  electric, 197, 232
  empiric, 311
  rational, 311, 312
Megenberg, Konrad von, 98*f*
Membrane theory, 389*f*
Membrantheorie, 391*f*
Memphis, 19–20
Mendes, Afonso, 115

Menelik (mythical first Ethiopia Emperor), 113
Menes, 19, 23
Meno, 33
Meno (Socratic dialogue written by Plato), 33, 61
Mercuriale, Girolamo, 142
Mercury tube apparatus used by Linari, 360*f*
Mereruka (Meri), 21
  Malapterurus depictions, 21
  Mastaba burial complex of, 21–22
Meriteti, 21
Messalina (wife of Claudius), 45
Meteorologica, 52
Metallic (or conductive) arc (in electric experiments), 166, 315, 330, 349
Meyssonnier, Lazare, 142*n*
Michelangelo. *See* Buonarroti, Michelangelo
Middle Ages, 64, 66
  revival, 89–90
Milne-Edwards, Henri, 367*n*
Milton, John
Mirabilia, 148
Miranda, Domenico de, 362
Misericordia, 45
Monk Urreta and Prester John's fabulous empire, 116–18
Monro, Alexander, 270
Monthly Review, The, 265*f*
  reports on eel, 266*f*
Moore, Francis, 289–91
  Africa, 1738 book on, 290*f*
  French translation with allusion to *vertu électrique*, 290*f*
Moruzzi, Giuseppe, 362, 363, 366, 368
Mosaics
  with aquatic scenes, 63
  from Pompeii depicting a sea scene, 62*f*
Motier, Gilbert du. *See* Lafayette, Marquis de (Gilbert du Motier)
Mouthbreeding tilapia, depiction in tomb, 22
Mozambique, 112, 114, 122
Mullets, depiction in tomb, 22
Mummies and sacred animals, 28
Mundus Novus, 112
Museum (*gazophylacium*) of Levinus Vincent, 206*f*
Musschenbroek, Pieter van. *See* Van Musschenbroek, Pieter (Petrus)
Myths and reality in seventeenth century, 126

Nairne, Edward, 254
Napoleon (Emperor). *See* Bonaparte, Napoleon
Nárkē, 30. *See also* Torpedoes
  words derived from, 30
Nárkē thalattía/potamía, 31
Narmer, 23–24
  catfish hieroglyph, 23–24
  plate/palette, 23
  serekh, 23, 24
Naturalis historia, 37–38, 41–43, 67, 79, 100*n*
  title page of 1543 Italian edition of, 43*f*
Neápolis — Naples, 61

Neferikara, 20
Negative Schwankung, 369, 381*f*, 382, 383, 384, 385, 392, 408
Neher, Erwin, 405*f*
Nernst, Walther, 386*f*
Nerve function, ancient theories of, 179–80
Nerve ligature experiments, 367
Nervous (or nerve) force, 313, 314
Nervous fluid, 154, 188, 304
Nervous signal (or nerve impulse, or action potential), 351, 369, 377–79, 403
  analogy with telegraphic transmission, 51
  conduction speed (measurement by Helmholtz), 376, 377*f*, 378
  mechanisms of generation and conduction, 404
Nesu Netjerikhet (Djoser), 20
Nettesheim, Cornelius Agrippa von, 141
Neuroelectric hypothesis, 313, 368
Neurophysiological setting, changing, 375
  Bernstein's electrochemical theory of bioelectric phenomena, 384–85
  Bernstein's measurements of nerve impulse, 377–79
  Bernstein, thermodynamics, and membrane theory, 388–90
  Charles Overton and the "scandal" of "muscle that did not love sugar", 387–88
  conservative revolution, 391–93
  Du Bois-Reymond's electric molecules, 379–82
  electric fishes, membranes, and cooling battery, 390–91
  Helmholtz and speed of nerve conduction, 375–76
  Hermann's alteration theory, 382–84
  ions, cells, membranes, theories, and Nobel Prizes, 385–87
Newton, Isaac, 89, 134, 155, 170, 174, 183, 184*f*, 189, 339, 347
Niankh-Khnum, 21
Nicholson, William, 332, 335, 337, 342, 350
  Nicholson's model, 335
Nicole of Oresme, 89
Niebuhr, Carsten, 82
Nightmarish/wonderful torpedo of poets, 56–59
Nile River, 22, 26, 41, 60, 81, 83
  Blue Nile, 8, 83, 115, 119–20, 147
  sources of Nile, 8, 42*n*, 83, 113, 115
Niliacus, Horapollo, 24
Nobili, Leopoldo, 357*f*
Nollet, Abbé Jean-Antoine, 167*f*
Non-conductive bodies, 54
Novum Organum, 173
Numb Eele, 132
Numbness, 23, 30*f*, 51, 54–55, 254
Nuzhatu-l-Qulûb, 83

Occult
  conservative priest, a florentine polymath, and torpedo, 146–48
  Francesco Redi, 148–49
  hermetic thinking, 138–41
  to mechanical theories of discharge, 137–38

qualities during Renaissance, 141–46
Réaumur's sphere of influence, 157–60
Redi's torpedo and mechanical theory, 149–51
René-Antoine Ferchault De Réaumur, 156–57
Stefano Lorenzini, 152–54
torpedo and occult qualities during Renaissance, 141–46
torpedo scientists at Medici Court, 155–56
Ockham, William of, 89
Odoric of Pordenone (Odorico di Pordenone), 91
Oersted, Hans Christian. *See* Ørsted, Hans Christian
Ohm, Johann Wolfgang, 254
Ojeda, Alonso de, 127
Oker-Blom, Maximilian Ernst Gustaf, 386
Olyphant, David, 271
Oneirokritika, 56*n*
On Poisonous Animals and on Animals that Sting, 60
On Regimen in Acute Diseases in Accordance with the Theories of Hippocrates, 32, 49
On Sympathies and Antipathies. *See Phūsika*
"On the Vacuum," 52
Oppian (Oppianus) of Corycus, 55
Optical telegraph, 348*f*
"Organic Remedies," 23
Oribasius, 70–71
Oriente Cristiano, 65
Origen(es), 67*n*
Orioli, Francesco, 361
Ornithologiae
  Aldrovandi's portrait from, 108*f*
Oroonoko, 132–33
Ørsted, Hans Christian, 343*f*, 358
Osiris, myth of, 22
Osservazioni intorno alle Torpedini, anatomical drawings from, 152*f*
Osservazioni intorno alle Vipere, 149
Ostwald, Wilhelm, 386, 387*f*, 389, 390
Overton, Charles Ernest, 387–388, 390, 392
  effects of ionic solutions on muscle excitability, 388*f*
  and scandal of "muscle that did not love sugar", 387–88
Oviedo, Andrés de, 113, 119
Oviedo y Valdez, Gonzalo Fernández de, 101, 127–28, 130, 135
Owen, Richard, 252, 362

Paci, Giacomo Maria, 362
Pacini, Filippo, 370, 372*f*, 407
Paestum, 61
Paez (Paéz, Pêro Pais), Pedro (Pêro), 114, 115, 119, 120, 122
  unpublished reaction, 119–20
Pain, Paulus' recipe for treatment of, 74
Palmieri, Luigi, 361
Panofsky, Erwin, 95*n*, 111, 138
Paracelsus, 181*f*. *See also* Hohenheim, Theophrastus Bombastus von (Paracelsus)

Pasteur, Louis, 149
Patch-clamp, 404
Paterson, William, 296–97, 298
 Philosophical Transactions on electric fish, 298f
Paulus of Aegina (Paulus Aegineta), 73–75, 84
 medical treatises of, title page of, 74
Pease, Jocelyn Richenda, 403n
Pecham, John, 88
Peretti, Pietro, 360
Perry, James, 292, 299, 300, 301
Peter the Great, 205
Pez temblador, 159f
Pfeffer, Wilhelm Friedrich Philipp, 385–86
Pflüger, Eduard Friedrich Wilhelm, 382, 385, 389
Phanus, 60
Philadelphia, 163, 168
 arrival of eel in, 260–61
Philip of Macedonia, 33
Philippe Fermin, 226, 260
 map of Surinam, 261f
Philoponus of Alexandria, John, 88
Philosophical Transactions (*of the* Royal Society of London), 171, 240, 244, 264, 273, 293, 298f, 341
 1773 Article by Walsh, 246–47
 September 1775 issue with brief notes of both Hugh Williamson's/Alexander Garden's experiments on electric eel, 296f
Philosophy/medicine, Jewish contribution, 65
Philoxenos of Citera, 60
Phthísis, 32
Phūsika, 75n
Physiologus, 67
 of Bern, page from, 67f
Physiology, animal spirits and, 179
 ancient theories of nerve function, 179–80
 Haller's irritability theory, 189–90
 impact of electricity, 188–89
 problems with earlier theories, 185–88
 Renaissance and Early Modern Period, 180–83
 vibrations, 183–85
Pianciani, Giovan Battista, 357, 360
Pian del Carpine, Giovanni da, 118
Pikkolos, Nikolaos Savva (or Piccolos), 36
Pimander, 139
Pisa, 107, 144, 150, 153, 366
Pitcairn, William, 219, 274
Pivati, Giovanni Francesco, 177, 312
Pizarro, Francisco, 10, 127
Pizzetta, Jules, 105
Places where electric fishes were found or studied. See also Ethiopia, Guyana, La Rochelle, Nile River, Pisa, Surinam
 Berbice River, 134, 201, 204, 219, 221
 Calabozo, 3, 4, 5, 8, 345
 Cayenne, 5, 134, 135, 204, 283, 289
 Demerara River, 218, 221
 Essequibo (or Essequebo) River, 134, 219f, 221

 Isle de Ré, 236, 238, 241, 243
 Lake Tana, 8, 115
 Leghorn, 30, 149, 150, 153, 247, 248, 357
 Mozambique, 112, 114, 122
 Oronoque River, 219
 Rastro de Abaxo, 3, 5, 6, 10, 14
 Senegal, 198–99, 213
 Tacazee River, 120
Plagiarist, 134
Plato, 29, 31, 34, 38, 53
 school of learning, Academy, 33
 Socrates–torpedo analogy, 33
Plato Comicus, 60
Pletho, Gemistus, 138
Plethoras, 32
Pliny the Elder, 26
 death during eruption of Vesuvius, 39f
Pliny the Younger, 38
 legacy, 42–44
 Malapterurus in, 42
 Naturalis historia, 38–41
 title page of 1543 Italian edition of *Naturalis historia*, 43f
 torpedo prescriptions, 41–42
Plotinus, 53–54, 138–39
Plutarch(us) (of Chaeronea), 37–38, 53, 55, 298
Pneumatica, 52, 66n
Pococke, Edward, 81
Podagra, 46, 69
 paroxysmal phase of bilious gout, 74
 torpedo for treating, 69, 73
Poisson, Siméon-Denis, 253
Polo, Marco, 91, 111, 118
Pomponazzi, Pietro, 142, 143
Ponticus, Evagrius, 67
Ponzetti, Ferdinando, 143
Poor Richard's Almanack, 23
Pope, Alexander, 195, 295
Porphyrius, 54
Pouillet, Claude, 356, 376, 377f
Practica chirurgica, 71
Practice of Medicine, 72
Praeconimus, Aelius Stilo, 67
Prester John, 91, 112
 fabulous Ethiopia Emperor, 116–18
 letters of, 111
 the myth of, 118
Prévost, Antoine François, 197
Priestley, Joseph, 172f, 217, 225, 231, 254, 256, 258, 262, 314
 synopsis, 258–59
Primeval knowledge (*prisca sapientia*), 25
Principia philosophiae, 151, 183
Pringle, Sir John, 256f
 Speech, 256–57
 torpedoes, description, 29
Prior, Matthew, 146, 194, 201, 356, 374
Problemata, 52
Prolapsus (*prolapsus ani*), 48, 49, 50, 70, 71, 72, 76, 77, 84, 104
"Protector of fishes," 25
Ptolemy, 65–66
Public knowledge, newspapers, magazines, and shocking poetry, 292
 fishes and common man, 304
 language and culture, 303–4

 London newspapers, 292–94
 shocking poetry, 297–303
 The Gentleman's Magazine, 294–97
Pulci, Luigi, 100
Purchas his Pilgrimage
 first page of chapter on Ethiopia, 123f
Purchas, Samuel, 82, 112, 121–25, 130
Pythagoras, 31

Qanun, of Avicenna, 80f
Quaestiones naturales, 67, 88–89
Queen of Sheba. *See* Makeda (Queen of Sheba)
Quibell, James E., 23
Quinta essentia, 53

Ra'ad (electric fish name, Arabic), 19
 anatomy, 20
 associated deity, 22
Ra'ash (trembler), 19
Rabanus Maurus, Benedictine monk, 91f
Raleigh, Walter, 3, 131
Ramée, Pierre de la, 172
Ranvier, Louis-Antoine, 348–49, 374
Ranzi, Andrea, 370f, 407
"Ray, On the," 76
Réaumur, René-Antoine Ferchault de, 156–57
 1714 memoir on torpedo, 157f
 sphere of influence, 157–60
 torpedo's *organes électriques* and their construction, 158f
"Receptivity" in body, 53
Re (deity associated with sun), 22
Red-figured fish plate, torpedo, 62f
Redi, Francesco, 148–49
 torpedo and mechanical theory, 149–51
 written in response to Father Kircher, 150f
Regimen, 32
Regius, Henricus, 98, 103, 180, 185
Relation Abregée d'un Voyage...., 158f
Religio, 67
Remora, 98
Remora, mid-seventeenth-century image of, 40f
Renaissances, 66, 144
 torpedo and occult qualities during, 141–46
Rescigno, Andrea, 53, 54
Rete mirabile, 179
Richer, Jean, 127, 134–35
 initial page of report, 135f
 voyage, 134–35
Riolan, Jean, 142
Rittenhouse, Benjamin, 262
Rittenhouse, David, 224, 262, 263f
Ritter, Johann Wilhelm, 12
Robin, Charles-Philippe, 395
Roche, Daniel de La, 312–13
Roger II, King, 66n
Romagna, 65n
Roman Empire
 collapse, 66
 emperors of late Roman period, 69f
Ronayne, Thomas, 254

Rondelet, Guillaume, 99–101, 103–107, 109, 142n
  text, torpedoes, 103–4
  torpedoes various types of depicted by, 103f
Rosenthal, Isidor, 97, 382
Rose, William, 271
Royal Institution, 227, 228, 326, 341, 342, 343, 362, 410
Royal Society of London, 29, 36, 59, 65–66, 97, 171–74, 205, 213–14, 243–44, 294, 326, 341. *See also* Copley Medal
  dilemmas, 245–46
  French Connections, 244–45
  Henry Cavendish's physics, 252–56
  Ingenhousz's torpedoes, 247–49
  Interdisciplinary Science, 243–44
  John Hunter's dissections, 249–52
  1773 Philosophical Transactions article, 246–47
  Walsh's Copley Medal and Pringle's speech, 256–57
Rubruck, William of, 91, 111, 118
Rutherford, John, 270
Ruysch, Frederik, 205

Sachs, Carl, 4, 7, 371–73, 411
Saint Isidore of Seville with his pupil, Bishop Braulio, 78f
Sakmann, Bert, 405f, 406f
Saladin (Salah ad-Din Yusuf ibn Ayyub), 81
Salviani, Ippolito, 104
  Torpedo ocellata (*occhiatella*) as shown by, 105f
Sano, Buckor, 121
Santorio, Santorio, 154
Santos, João dos. *See* Dos Santos, João
Saqqara (Sakkara), 19–20. *See also* *Mastaba* (table-like) Tomb of Ti
  fish images in Tomb of Mereruka at, 21f
  Malapterurus-like images, 19
  ruler Ti, depiction of, 20f
  step pyramid, 20
Saturnia pyri (butterfly) from botanical image depicting *Euphorbia*, 101f
Savi, Paolo, 364, 365f
Scaligero, Giulio Cesare, 142
Scelta degli opuscoli interessanti, 285f
Schelling, Friedrich Wilhelm Joseph, 12
Schilling, Gottfried Wilhelm, 211, 221, 233, 234f
Schmidt (Faber), Johannes, 52
Schönbein, Christian Friedrich, 362, 363f
Schönlein, Karl, 374, 408
Schweigger, Johann Solomo, 355
Scot, Michael, 66
Scribonius Largus, 45
  vs. Marcellus, 69n
Sea torpedo, 48
  medical treatises of Paulus of Aegina, 74f
  ointment, 74f
  ointment from Alexander of Tralles' medical treatise, 71f
  prescriptions for patients with headache and *prolapsus ani*, 71f
Seba, Albert(us) 205, 207, 214

Sebond, Raymond, 98
Second Jesuit mission to Ethiopia, 114–15
Seignette, Pierre Henri, 235, 240, 244–46, 294–95
  Walsh's torpedo experiments at La Rochelle, 296f
Senegal, 198, 199, 200, 213
Serekh, 23
Shocking catfish of Nile, 19
  depictions, 19–22
  Ebers Papyrus, 22–23
  Hieroglyphica of Horapollo, 24–27
  magic, religion, and tomb art, 22
  mummies and sacred animals, 28
  Narmer's catfish hieroglyph, 23–24
Shock mechanisms, understanding, 394–95
  Bernstein's model of discharge, 407–8
  demonstrating functional asymmetry, 408–10
  electric organs in weakly electric fishes, 395–97
  electrifying cells of electric organs, 406–7
  Pacini's observations, 407
  surprising electric catfishes, 414–15
  torpedo electrophysiology, 411–14
  torpedo paradox, 411
Sicily, 45, 63, 90
  role in intercultural transmission, 66
Silurus electricus, 7–8
Silvestre de Sacy, Antoine Isaac, Baron, 81
Simocatta, Theophylact, 37, 54
Sittardus, Cornelius, 106, 107n
Sloan, Hans, 271
Smith, Edwin, 23
Socrates, 33, 61, 231, 252
  teaching style, 33
Solander, Daniel, 275, 276, 277, 278, 282
Solinus, Julius Gaius, 68
Solomon (of Israel), 113, 118
South American eels
  Aphra Behn's *Oroonoko*, 132–33
  early descriptions from Venezuela/Brazil, 127–31
  George Warren's description of Surinam, 131–32
  Jean Richer's voyage, 134–35
  plagiarist, 134
  theoretical void, 135–36
  voyages to South America, 127
South Carolina Gazette
  arrival of South American eels (announcement), 270f
Spain, 3, 13, 77
  centers of Islamic culture, 66
  as intercultural center, 65
  translations(ors), 65
Spallanzani, Lazzaro, 149, 224, 233, 311, 334, 356
Sparks in darkness and eel's electrical sense, 281
  Cavallo's book, 284–87
  electrical sense, 287–88
  immediate impact, 288–89
  Jan Ingenhousz, 284
  Le Roy's publication, 284

  a place for eel, 281–82
  Pringle's letters to Haller, 283–84
  rewriting past, 289–91
  the spark, 282
Spear fishing, 20
Specific nervous energies, law of the
  anticipated by Volta, 330
  of Johannes Müller, 330
Speculum historiale, 111
Speculum naturale
  thirteenth-century codex of Vincent de Beauvais', 91f
Spencer, Adam, 168
Split (*Spalatum*), 73n
Squid's giant axon, 401f
  action potential with, 402f
Stark, James, 395
Stedman, John, 226
  frontispiece and two images from book on Surinam, 227f
Steno, Nicolaus, 150, 153, 156, 186f
Stensen, Niels, 150
Stephen of Pisa or Antioch, 79
Stertinius Xenophon, 45
Stevenson, Mary (Polly), 249
Stilicho, 57
Stinging catfish, 24
Stinging fish, 23
Strato of Lampsacus, 52
Strong, Adam. *See* Perry
The Structure of Scientific Revolutions, 137
Student, images of January 1750 issue of, *The*, 196f
Stupefactor (torpedo, Latin name), 93f
De subtilitate rerum, 142n
Suetonius (Gaius Suetonius Tranquillus)
Surinam, 131, 195, 201, 207f, 230, 233
  Allamand's publication on eels of, 203f
  Behn's novel and, 132–33
  Bryant on, 280
  eels from, 13, 276, 284
  George Warren's description of, 131–32
  Philippe Fermin map of, 261f
  specimens from, 214, 224
  Stedman's 1796 narrative on, 227f
Swammerdam, Jan, 185, 212
Sydenham, Thomas, 143
"Sympathy," 54
Symposium, 59
Synaptic transmission and ligand-gated channels, 410–11
Synodontis, 28
Systema naturae, 272f

Tabula Rogeriana, 66n
Tacazee River, 120
Taylor, Edward, 298–99
Telegraphs
  electric telegraph, 28
  optical telegraphs, 348
Telesio, Bernardino, 142
Telles, Balthasar, 115, 120–21
  Historia geral da Ethiópia a Alta, 120f
  torporific catfish, rediscovering, 120
Termeyer, Ramón María, 127, 288, 289
Teti, 21
Tetrodotoxin, 404, 419

Index

Thales of Miletos, 31
The Gentleman's Magazine, 294–97
   August 1777 issue of, 301*f*
   of both Williamson's/Garden's experiments on electric eels, 265*f*
   brief notes of on Garden's experiments on electric eel, 296*f*
   cover page of November 1772 issue of, 296*f*
   Edward Cave's, cover page of first issue of, 294*f*
Thénard, Jacques Louis, 353
Theodore of Antioch, 66*n*
Theodosius, Flavius (Theodosius the Great), 69
Theophrastos of Eresos, 26
Therapeutica ("Twelve Books on Medicine"), 71, 72
Therapeutic herbs, 48*f*
Therapeutic shocks/theories of discharge, 45
Thesaurus Linguae Grecae, 75
Thévenot, Melchisédech, 186
Thomas of Cantinpré, 90
Thompson, Benjamin (Count Rumford), 19, 22, 228*f*
Thompson, D'Arcy, 19, 22, 63
Tiberius, Emperor, 284, 286*f*
Timaeus, 33–34
Toby, The, 297, 298*f*
Todd, John Tweedie, 362
Tolstoy, Leo, 416
Tomb of Mereruka at Saqqara, fish images in, 21*f*
Tomb of the Two Brothers, 21
Tomb paintings, 22
Torpedinal electricity, 352–55
Torpedinal medicine, 49–50
Torpedinal therapy, birth of, 45–49
Torpedo bancroftii, 228*f*
Torpedoes, 8*f*, 87. *See also* Torpedoes in the Greco-Roman World and specific torpedo types
   Albertus and stupefying, 91–94
   in Alexander's medical books, 72
   Belon's books, 101–3
   Claudian's piece, 58
   Corpus Hippocraticum and, 31–33
   cosmos, and vision, 52–54
   cultural legacies, 87–89
   De aquatilibus, images in, 103*f*
   depiction, inspired by Claudian's poem, 59*f*
   discharge, polarity of, 356–57
   electrophysiology, 411–14
   and fine arts, 61–63
   Gessner's book/ichthyology work, 104–7, 106*f*
   illustrated fish books, 100–101
   Latin term, 73
   less-than-realistic image of, 107*f*
   literary digression, 98–99
   magic and occult qualities, 110
   Middle Ages revival, 89–90
   myths about, 54–55
   names (popular) for, 30*n*
   as *narcos*, 91*f*
   Paolo Giovio's book on fishes, 99–100
   paradox, 411
   and Renaissance, 94–98, 141–46
   Salviani, 104
   as sleepers and tremblers, 30–31
   spark, 358–60
   swimming in Tyrrhenian Sea, 30*f*
   and torporific African catfish, 30–31
   use of, depiction, 71*f*
   various types of, depicted by Rondelet, 103*f*
   zoological texts of late middle ages, 90–91
Torpedoes in Greco-Roman world, 29–30
   Aelian's myths about torpedoes and water, 54–55
   Aristotelian natural philosophy, 33–36
   Athenaeus and bibliomaniac gourmands, 59–61
   birth of torpedinal therapy, 45–49
   Corpus Hippocraticum, 31–33
   dahlias and localizing torpedo's power, 55
   Galen's cold venom theory, 50–51
   Galen's torpedinal medicine, 49–50
   Malapterurus in Pliny, 42
   nightmarish/wonderful (poets'), 56–59
   Plato's Socrates–torpedo analogy, 33
   Pliny and *Naturalis historia*, 38–41
   Pliny's legacy, 42–44
   Pliny's torpedo prescriptions, 41–42
   sleepers/tremblers, 30–31
   therapeutic shocks/theories of discharge, 45
   torpedo, cosmos, and vision, 52–54
   torpedoes and fine arts, 61–63
   transmissible power, 36–38
   "un-nameable qualities," 51–52
Torpedo nobiliana, 93
Torpedo oculata, 102–3
Torpedo, The, 299*f*
Torporific African catfishes, 30–31. *See also* Catfishes, electric, Catfish of the Nile, *Malapteruridae*, *Malapterurus electricus*, Torporific catfishes, rediscovering
Torporific catfishes, rediscovering
   Almeida and Telles, 120
   bringing *Malapterurus* to wider audience, 122–25
   Ethiopia in 1500s, 112
   Ethiopian Christianity/religious intolerance, 125–26
   Fernandes' letter, 115–16
   Godinho's book, 119
   Jobson's Gambia, 121–22
   Monk Urreta/Prester John's fabulous empire, 116–18
   myths and reality in seventeenth century, 126
   Paez's unpublished reaction, 119–20
   published western description of African torpedo, 114
   second Jesuit mission to Ethiopia, 114–15
   voyages to distant lands, 112
Torporific eels, 221–22
Torreblanca, Villalpando Francisco, 142
Torricellian tube, 147
Torricelli, Evangelista, 147
Tourmaline, 292, 318–19
Translations of Classics
   in Middle Ages, 34*n*, 43, 47, 66, 68, 73, 89–91, 147
   in the Renaissance, 25–26, 38, 43, 47, 94–98, 100–101, 147
Transmissible power, 36–38
Traube, Moritz, 386*f*
Tschermak von Seysenegg, Armin, 384, 389, 390
Turner, Robert, 11, 194–95
Twain, Mark (Samuel Clemens), 416
Twelfth-Century Renaissance, 64, 66

Ubertini, Francesco, 95
University of
   Aberdeen, 225, 270
   Berlin, 8, 370, 380*n*
   Bologna, 89, 107, 109, 188, 309, 310*f*, 363
   Breslau, 377
   Cairo, 84
   Delaware, 264
   Edinburgh, 270
   Göttingen, 9, 189, 283
   Gregorian University, 147. *See also Collegio Romano*
   Halle, 198
   John Hopkins University, 419
   Leiden, 165, 191, 214
   Manchester, 232, 233, 235
   Montpellier, 89, 103, 105
   Okayama, 406
   Oxford, 89, 122
   Padua, 92, 107
   Paris, 89, 92, 167
   Pavia, 10, 142, 327
   Pennsylvania, 218, 264
   Pisa, 144, 362
   Taiwan, 419
"Un-nameable qualities," 51–52
Unzer, Johann August, 197
Urban VIII, Pope, 140, 144
Urreta, Luis de, 117, 118, 119
   Historia eclesiastica, 118*f*

Valens, Flavius Julius, 69
Valeriano, Piero, 26, 140
Valli, Eusebio, 324*f*
Van Baerle, Caspar, 124
Van Berkel, Adriaan, 134, 214, 219
Van der Horst, Gijsbert, 106*n*
Van der Lott, Frans, 207–12, 209*f*, 217, 259
Van Helmont, Jean Baptiste, 181*f*
Vanini, Giulio Cesare, 142
Van Musschenbroek, Pieter (Petrus), 166, 167, 211, 214, 212*f*
   1746 Leyden jar communication to Réaumur, 167*f*
   endorsement, 211
Van 's Gravesande, Jonathan Samuel Storm, 201, 202
Van 's Gravesande, Laurens Storm, 201, 202*f*, 203*f*, 204
Van't Hoff, Jacobus Henricus, 385, 386, 387*f*

Vapors, 143
Varro, Marco Terentius, 64, 67
Venezuela, 3, 4, 6f, 127, 128, 221, 371, 411
  map, 4f
  South American eels, descriptions, 127–31
Venom (*gelido veneno*), 51
Veratti, Giuseppe, 312
Vernacular names of torpedo, electric eel, catfish, or electric fish
  Croatian: *trpigna*, 73n
  Dutch: *Siddervis, Sidderrog, Beef-Aal, Beave Aal, Conger-aal*, 5, 30, 202, 205, 223, 280
  English: *cramp-fish, numb-fish*, 56, 57, 122, 125, 230, 291
  French: *dourmilleuse, torpille, turpille, trembleur*, 30, 104, 107, 160, 200, 235
  German: *Zitterrochen, Zitterfish, Krampfish*, 30, 72n, 107n, 123n
  Greek (late vernacular names): *marga, margotirem*, 76, 102
  Italian: *battipotta* (or *battipota*), *crampo, fotterigia, fumicotremula, mostargo, sgranfo, tremola, tremolo, tremoriza, tremula, terpina, treppina, trippina*, 30, 30n 73, 73n, 99, 106, 107n, 150
  Middle Ages and late-Latin vernacular names: *barachi, barkis, barq, tourpaena, tourpaína, turpana, turpena*, 71f, 72, 74f, 79, 93, 93f, 100n
  Natives in Africa: *adenguêz, ouaneiar, thinta*, 114, 119, 123
  Natives in South America: *pouraké, puraquê, pouraquam*, 130, 131, 132, 159
  Portuguese: *peixe tremedor, tremedeira, tremelga, peixe viola*, 30, 114, 131
  Spanish: *hugia, tremielga, temblador* (also used for the electric eel)., 5, 30, 107n, 159f
Verulam, Baron (Francis Bacon), 172
Vesalius, Andreas, 95–96, 180
  De fabrica corporis humani, 96f
Vespucci, Americo, 127
Villars, Charles-René de, 235
Vincent, Levinus, 90, 111
  museum (*gazophylacium*) of, 206f
Vincent of Beauvais. See De Beauvais, Vincent

Virtual witness (or testimony) strategy (as a scientific communication strategy), 240
Visitator Indiarum, 115
Vitelo. See Witelo, Erazmus
Viviani, Domenico, 354
Volta, Alessandro, 327f
  article in Philosophical Transactions, 337f
  artificial electrical organ, 10–11, 335–38
  Chevalier de la Legion d'honneur, 341f
  communication to Joseph Banks, 336f
  condensatore, 329f
  contra Galvani, 330–31
  drafts of letter announcing invention of pile/battery, 10
  electrophores, 328f
  experiments with electric circuits, 329f
  extension of theory of metallic electricity, 331–32
  fluid produced by battery/electric fishes/ genuine electricity, 344f
  with his battery and electrophore, 340f
  from love obsession to science lust, 327–29
  natural and artificial electric organs, 336f
  reaction, 335
  showing his battery to Napoleon Bonaparte, 340f
  Voltaic pile, 337f
Voltage-clamp, 402, 403
Voyages to distant lands, 112

Walker, Mary Broadfoot, 164, 193, 420
Walsh, Colonel John, 230, 231f
  conclusions of, 238–39
  Copley Medal, 29, 256–57
  disappointments, 238
  experiments on torpedoes in France, 236f
  Franklin and, 230–32, 241–42
  Hunter dissections using torpedoes, 251f
  Journal de Voyage, 232f, 233f
  landmark spark experiment with eel, announcement, 285f
  on large English torpedoes (article), 252f
  letter sent to Franklin, 241f
  Memorandum, 235f
  public demonstrations, 239–41

  scientific journey of, 230–42
  Seignette's communication to *Gazette de France*, translation, 296f
  successes of torpedo experiments, 239f
  torpedoes experiments, 237f
  torpedo publications, 246f
Walter of (or de) Henley, 91n, 427
Walter, Richard, 196, 197
  and Anson's torpedo revisited, 197–98
  book, with page dealing with experiments on torpedo, 198f
Wansleben, Johann Michael, 122, 125
Warburg, Aby Moritz, 138, 140
Warren, George, 131–32, 219
  description of Surinam, 131–32
Washington, George, 217
Watet Hathor. See Her-watet-khet (Watet Hathor)
Watson, William, 171, 173f
Wesley, John, 177
Western description of African torpedo, published, 114
Western empire, collapse, 64
Western/Mediterranean civilizations, 64
Whytt, Robert, 177, 271
William I of Sicily, 66n
William II of Sicily, 66n
William of Rubruck. See Rubruck, William of
Williamson, Hugh, 264–68, 265f
  experiments on electric eel, 296f
Willis, Thomas, 182f
  Cerebri anatome, 182f
Witelo, Erazmus, 88
Wonderful/nightmarish torpedo of poets, 56–59
Wonders and amusements, 167–68
Woolf, Virginia, 132

Xenophon of Athens, 56, 59

Young, John Zachary, 401

Zabarella, Jacopo, 142
Zagazabo (Ṣägä Zä´ Äb, Ethiopian monk), 113
Zitterochen (torpedo in German), 72n
Zoological texts of Late Middle Ages, 90–91